STUDENT SOLUTIONS MANUAL

FINITE MATHEMATICS AND CALCULUS WITH APPLICATIONS SERIES
EIGHTH EDITION

Margaret L. Lial
America River College

Raymond N. Greenwell
Hofstra University

Nathan P. Ritchey
Youngstown State University

PEARSON

Addison
Wesley

Boston San Francisco New York
London Toronto Sydney Tokyo Singapore Madrid
Mexico City Munich Paris Cape Town Hong Kong Montreal

Reproduced by Pearson Addison-Wesley from QuarkXPress® files.

Copyright © 2008 Pearson Education, Inc.
Publishing as Pearson Addison-Wesley, 75 Arlington Street, Boston, MA 02116.

ISBN-13: 978-0-321-45598-7
ISBN-10: 0-321-45598-3

2 3 4 5 6 BB 10 09

CONTENTS

CHAPTER 3 LINEAR PROGRAMMING: THE GRAPHICAL METHOD

CHAPTER 4 LINEAR PROGRAMMING: THE SIMPLEX METHOD

CHAPTER 5 MATHEMATICS OF FINANCE

CHAPTER 6 LOGIC

CHAPTER 7 SETS AND PROBABILITY

CHAPTER 8 COUNTING PRINCIPLES; FURTHER PROBABILITY TOPICS

CHAPTER 9 STATISTICS

CHAPTER 10 NONLINEAR FUNCTIONS

CHAPTER 11 THE DERIVATIVE

CHAPTER 12 CALCULATING THE DERIVATIVE

CHAPTER 13 GRAPHS AND THE DERIVATIVE

CHAPTER 14 APPLICATIONS OF THE DERIVATIVE

CHAPTER 15 INTEGRATION

CHAPTER 16 FURTHER TECHNIQUES AND APPLICATIONS OF INTEGRATION

CHAPTER 17 MULTIVARIABLE CALCULUS

CHAPTER 18 PROBABILITY AND CALCULUS

PREFACE

This book provides solutions for many of the exercises in *Finite Mathematics and Calculus with Applications*, Sixth Edition, by Margaret L. Lial, Raymond N. Greenwell, and Nathan P. Ritchy. Solutions are included for odd–numbered exercises. Solutions are not provided for exercises with open–response answers. Sample tests are provided at the end of each chapter to help you determine if you have mastered the concepts in a given chapter.

This book should be used as an aid as you work to master your coursework. Try to solve the exercises that your instructor assigns before you refer to the solutions in this book. Then, if you have difficulty, read these solutions to guide you in solving the exercises. The solutions have been written so that they are consistent with the methods used in the textbook.

You may find that some of the solutions are presented in greater detail than others. Thus, if you cannot find an explanation for a difficulty that you encountered in one exercise, you may find the explanation in the solution for a similar exercise elsewhere in the exercise set.

In addition to solutions, you will find a list of suggestions on how to be successful in mathematics. A careful reading will be helpful for many students.

The following people have made valuable contributions to the production of this *Student's Solutions Manual:* LaurelTech Integrated Publsihing Services, editors; Judy Martinez and Sheri Minkner, typists; and Joe Vetere, Senior Author Support/Technology Specialist..

We also want to thank Tommy Thompson of Cedar Valley Community College for the essay "To the Student: Success in Mathematics."

TO THE STUDENT

TO THE STUDENT: SUCCESS IN MATHEMATICS

The main reason students have difficulty with mathematics is that they don't know how to study it. Studying mathematics *is* different from studying subjects like English or history. The key to success is regular practice.

This should not be surprising. After all, can you learn to play the piano or to ski well without a lot of regular practice? The same thing is true for learning mathematics. Working problems nearly every day is the key to becoming successful. Here is a list of things you can do to help you succeed in studying mathematics.

1. *Attend class regularly.* Pay attention in class to what your instructor says and does, and make careful notes. In particular, note the problems the instructor works on the board and copy the complete solutions. Keep these notes separate from your homework to avoid confusion when you read them over later.

2. Don't hesitate to ask questions in class. It is not a sign of weakness, but of strength. There are always other students with the same question who are too shy to ask.

3. *Read your text carefully.* Many students read only enough to get by, usually only the examples. Reading the complete section will help you to be successful with the homework problems. Most exercises are keyed to specific examples or objectives that will explain the procedures for working them.

4. Before you start on your homework assignment, rework the problems the instructor worked in class. This will reinforce what you have learned. Many students say, "I understand it perfectly when you do it, but I get stuck when I try to work the problem myself."

5. Do your homework assignment only *after* reading the text and reviewing your notes from class. Check your work with the answers in the back of the book. If you get a problem wrong and are unable to see why, mark that problem and ask your instructor about it. Then practice working additional problems of the same type to reinforce what you have learned.

6. Work as neatly as you can. Write your symbols clearly, and make sure the problems are clearly separated from each other. Working neatly will help you to think clearly and also make it easier to review the homework before a test.

7. After you have completed a homework assignment, look over the text again. Try to decide what the main ideas are in the lesson. Often they are clearly highlighted or boxed in the text.

8. Use the chapter test at the end of each chapter as a practice test. Work through the problems under test conditions, without referring to the text or the answers until you are finished. You may want to time yourself to see how long it takes you. When you have finished, check your answers against those in the back of the book and study those problems that you missed. Answers are referenced to the appropriate sections of the text.

9. Keep any quizzes and tests that are returned to you and use them when you study for future tests and the final exam. These quizzes and tests indicate what your instructor considers most important. Be sure to correct any problems on these tests that you missed, so you will have the corrected work to study.

10. Don't worry if you do not understand a new topic right away. As you read more about it and work through the problems, you will gain understanding. Each time you look back at a topic you will understand it a little better. No one understands each topic completely right from the start.

SOLUTIONS
TO
ODD-NUMBERED EXERCISES

ALGEBRA REFERENCE

R.1 Polynomials

1. $(2x^2 - 6x + 11) + (-3x^2 + 7x - 2)$
$= 2x^2 - 6x + 11 - 3x^2 + 7x - 2$
$= (2 - 3)x^2 + (7 - 6)x + (11 - 2)$
$= -x^2 + x + 9$

2. $(-4y^2 - 3y + 8) - (2y^2 - 6y - 2)$
$= (-4y^2 - 3y + 8) + (-2y^2 + 6y + 2)$
$= -4y^2 - 3y + 8 - 2y^2 + 6y + 2$
$= (-4y^2 - 2y^2) + (-3y + 6y)$
$\quad + (8 + 2)$
$= -6y^2 + 3y + 10$

3. $-6(2q^2 + 4q - 3) + 4(-q^2 + 7q - 3)$
$= (-12q^2 - 24q + 18) + (-4q^2 + 28q - 12)$
$= (-12q^2 - 4q^2) + (-24q + 28q) + (18 - 12)$
$= -16q^2 + 4q + 6$

4. $2(3r^2 + 4r + 2) - 3(-r^2 + 4r - 5)$
$= (6r^2 + 8r + 4) + (3r^2 - 12r + 15)$
$= (6r^2 + 3r^2) + (8r - 12r)$
$\quad + (4 + 15)$
$= 9r^2 - 4r + 19$

5. $(0.613x^2 - 4.215x + 0.892)$
$\quad - 0.47(2x^2 - 3x + 5)$
$= 0.613x^2 - 4.215x + 0.892$
$\quad - 0.94x^2 + 1.41x - 2.35$
$= -0.327x^2 - 2.805x - 1.458$

6. $0.5(5r^2 + 3.2r - 6) - (1.7r^2 - 2r - 1.5)$
$= (2.5r^2 + 1.6r - 3) + (-1.7r^2 + 2r + 1.5)$
$= (2.5r^2 - 1.7r^2) + (1.6r + 2r) + (-3 + 1.5)$
$= 0.8r^2 + 3.6r - 1.5$

7. $-9m(2m^2 + 3m - 1)$
$= -9m(2m^2) - 9m(3m) - 9m(-1)$
$= -18m^3 - 27m^2 + 9m$

8. $(6k - 1)(2k - 3)$
$= (6k)(2k) + (6k)(-3) + (-1)(2k)$
$\quad + (-1)(-3)$
$= 12k^2 - 18k - 2k + 3$
$= 12k^2 - 20k + 3$

9. $(3t - 2y)(3t + 5y)$
$= (3t)(3t) + (3t)(5y) + (-2y)(3t) + (-2y)(5y)$
$= 9t^2 + 15ty - 6ty - 10y^2$
$= 9t^2 + 9ty - 10y^2$

10. $(9k + q)(2k - q)$
$= (9k)(2k) + (9k)(-q) + (q)(2k)$
$\quad + (q)(-q)$
$= 18k^2 - 9kq + 2kq - q^2$
$= 18k^2 - 7kq - q^2$

11. $\left(\frac{2}{5}y + \frac{1}{8}z\right)\left(\frac{3}{5}y + \frac{1}{2}z\right)$
$= \left(\frac{2}{5}y\right)\left(\frac{3}{5}y\right) + \left(\frac{2}{5}y\right)\left(\frac{1}{2}z\right) + \left(\frac{1}{8}z\right)\left(\frac{3}{5}y\right)$
$\quad + \left(\frac{1}{8}z\right)\left(\frac{1}{2}z\right)$
$= \frac{6}{25}y^2 + \frac{1}{5}yz + \frac{3}{40}yz + \frac{1}{16}z^2$
$= \frac{6}{25}y^2 + \left(\frac{8}{40} + \frac{3}{40}\right)yz + \frac{1}{16}z^2$
$= \frac{6}{25}y^2 + \frac{11}{40}yz + \frac{1}{16}z^2$

12. $\left(\frac{3}{4}r - \frac{2}{3}s\right)\left(\frac{5}{4}r + \frac{1}{3}s\right)$
$= \left(\frac{3}{4}r\right)\left(\frac{5}{4}r\right) + \left(\frac{3}{4}r\right)\left(\frac{1}{3}s\right) + \left(-\frac{2}{3}s\right)\left(\frac{5}{4}r\right)$
$\quad + \left(-\frac{2}{3}s\right)\left(\frac{1}{3}s\right)$
$= \frac{15}{16}r^2 + \frac{1}{4}rs - \frac{5}{6}rs - \frac{2}{9}s^2$
$= \frac{15}{16}r^2 - \frac{7}{12}rs - \frac{2}{9}s^2$

13. $(2 - 3x)(2 + 3x)$
$= (2)(2) + (2)(3x) + (-3x)(2) + (-3x)(3x)$
$= 4 + 6x - 6x - 9x^2$
$= 4 - 9x^2$

14. $(6m + 5)(6m - 5)$
$= (6m)(6m) + (6m)(-5) + (5)(6m)$
$\quad + (5)(-5)$
$= 36m^2 - 30m + 30m - 25$
$= 36m^2 - 25$

15. $(3p-1)(9p^2+3p+1)$
$$= (3p-1)(9p^2) + (3p-1)(3p)$$
$$+ (3p-1)(1)$$
$$= 3p(9p^2) - 1(9p^2) + 3p(3p)$$
$$- 1(3p) + 3p(1) - 1(1)$$
$$= 27p^3 - 9p^2 + 9p^2 - 3p + 3p - 1$$
$$= 27p^3 - 1$$

16. $(3p+2)(5p^2+p-4)$
$$= (3p)(5p^2) + (3p)(p) + (3p)(-4)$$
$$+ (2)(5p^2) + (2)(p) + (2)(-4)$$
$$= 15p^3 + 3p^2 - 12p + 10p^2 + 2p - 8$$
$$= 15p^3 + 13p^2 - 10p - 8$$

17. $(2m+1)(4m^2-2m+1)$
$$= 2m(4m^2 - 2m + 1) + 1(4m^2 - 2m + 1)$$
$$= 8m^3 - 4m^2 + 2m + 4m^2 - 2m + 1$$
$$= 8m^3 + 1$$

18. $(k+2)(12k^3-3k^2+k+1)$
$$= k(12k^3) + k(-3k^2) + k(k) + k(1)$$
$$+ 2(12k^3) + 2(-3k^2) + 2(k) + 2(1)$$
$$= 12k^4 - 3k^3 + k^2 + k + 24k^3 - 6k^2$$
$$+ 2k + 2$$
$$= 12k^4 + 21k^3 - 5k^2 + 3k + 2$$

19. $(x+y+z)(3x-2y-z)$
$$= x(3x) + x(-2y) + x(-z) + y(3x) + y(-2y)$$
$$+ y(-z) + z(3x) + z(-2y) + z(-z)$$
$$= 3x^2 - 2xy - xz + 3xy - 2y^2 - yz + 3xz - 2yz - z^2$$
$$= 3x^2 + xy + 2xz - 2y^2 - 3yz - z^2$$

20. $(r+2s-3t)(2r-2s+t)$
$$= r(2r) + r(-2s) + r(t) + 2s(2r)$$
$$+ 2s(-2s) + 2s(t) - 3t(2r) - 3t(-2s) - 3t(t)$$
$$= 2r^2 - 2rs + rt + 4rs + 2st - 6rt + 6st - 3t^2$$
$$+ 2rt - st + t^2$$
$$= 2r^2 + 2rs - 5rt - 4s^2 + 8st - 3t^2$$

21. $(x+1)(x+2)(x+3)$
$$= [x(x+2) + 1(x+2)](x+3)$$
$$= \left[x^2 + 2x + x + 2\right](x+3)$$
$$= \left[x^2 + 3x + 2\right](x+3)$$
$$= x^2(x+3) + 3x(x+3) + 2(x+3)$$
$$= x^3 + 3x^2 + 3x^2 + 9x + 2x + 6$$
$$= x^3 + 6x^2 + 11x + 6$$

22. $(x-1)(x+2)(x-3)$
$$= [x(x+2) + (-1)(x+2)](x-3)$$
$$= (x^2 + 2x - x - 2)(x-3)$$
$$= (x^2 + x - 2)(x-3)$$
$$= x^2(x-3) + x(x-3) + (-2)(x-3)$$
$$= x^3 - 3x^2 + x^2 - 3x - 2x + 6$$
$$= x^3 - 2x^2 - 5x + 6$$

23. $(x+2)^2 = (x+2)(x+2)$
$$= x(x+2) + 2(x+2)$$
$$= x^2 + 2x + 2x + 4$$
$$= x^2 + 4x + 4$$

24. $(2a-4b)^2 = (2a-4b)(2a-4b)$
$$= 2a(2a-4b) - 4b(2a-4b)$$
$$= 4a^2 - 8ab - 8ab + 16b^2$$
$$= 4a^2 - 16ab + 16b^2$$

25. $(x-2y)^3$
$$= [(x-2y)(x-2y)](x-2y)$$
$$= (x^2 - 2xy - 2xy + 4y^2)(x-2y)$$
$$= (x^2 - 4xy + 4y^2)(x-2y)$$
$$= (x^2 - 4xy + 4y^2)x + (x^2 - 4xy + 4y^2)(-2y)$$
$$= x^3 - 4x^2y + 4xy^2 - 2x^2y + 8xy^2 - 8y^3$$
$$= x^3 - 6x^2y + 12xy^2 - 8y^3$$

R.2   Factoring

1. $7a^3 + 14a^2 = 7a^2 \cdot a + 7a^2 \cdot 2$
$$= 7a^2(a+2)$$

2. $3y^3 + 24y^2 + 9y = 3y \cdot y^2 + 3y \cdot 8y + 3y \cdot 3$
$$= 3y(y^2 + 8y + 3)$$

3. $13p^4q^2 - 39p^3q + 26p^2q^2$
$$= 13p^2q \cdot p^2q - 13p^2q \cdot 3p + 13p^2q \cdot 2q$$
$$= 13p^2q(p^2q - 3p + 2q)$$

4. $60m^4 - 120m^3n + 50m^2n^2$
$$= 10m^2 \cdot 6m^2 - 10m^2 \cdot 12mn$$
$$+ 10m^2 \cdot 5n^2$$
$$= 10m^2(6m^2 - 12mn + 5n^2)$$

5. $m^2 - 5m - 14 = (m-7)(m+2)$

since $(-7)(2) = -14$ and $-7 + 2 = -5$.

6. $x^2 + 4x - 5 = (x+5)(x-1)$

since $5(-1) = -5$ and $-1 + 5 = 4$.

7. $z^2 + 9z + 20 = (z+4)(z+5)$

since $4 \cdot 5 = 20$ and $4 + 5 = 9$.

8. $b^2 - 8b + 7 = (b - 7)(b - 1)$

since $(-7)(-1) = 7$ and $-7 + (-1) = -8$.

9. $a^2 - 6ab + 5b^2 = (a - b)(a - 5b)$

since $(-b)(-5b) = 5b^2$ and
$-b + (-5b) = -6b$.

10. $s^2 + 2st - 35t^2 = (s - 5t)(s + 7t)$

since $(-5t)(7t) = -35t^2$ and $7t + (-5t) = 2t$.

11. $y^2 - 4yz - 21z^2 = (y + 3z)(y - 7z)$

since $(3z)(-7z) = -21z^2$ and
$3z + (-7z) = -4z$.

12. $6a^2 - 48a - 120 = 6(a^2 - 8a - 20)$
$$= 6(a - 10)(a + 2)$$

13. $3m^3 + 12m^2 + 9m = 3m(m^2 + 4m + 3)$
$$= 3m(m + 1)(m + 3)$$

14. $3x^2 + 4x - 7$

The possible factors of $3x^2$ are $3x$ and x and the possible factors of -7 are -7 and 1, or 7 and -1. Try various combinations until one works.

$$3x^2 + 4x - 7 = (3x + 7)(x - 1)$$

15. $3a^2 + 10a + 7$

The possible factors of $3a^2$ are $3a$ and a and the possible factors of 7 are 7 and 1. Try various combinations until one works.

$$3a^2 + 10a + 7 = (a + 1)(3a + 7)$$

16. $4a^2 + 10a + 6 = 2(2a^2 + 5a + 3)$
$$= 2(2a + 3)(a + 1)$$

17. $15y^2 + y - 2 = (5y + 2)(3y - 1)$

18. $21m^2 + 13mn + 2n^2$
$$= (7m + 2n)(3m + n)$$

19. $24a^4 + 10a^3b - 4a^2b^2$
$$= 2a^2(12a^2 + 5ab - 2b^2)$$
$$= 2a^2(4a - b)(3a + 2b)$$

20. $24x^4 + 36x^3y - 60x^2y^2$
$$= 12x^2(2x^2 + 3xy - 5y^2)$$
$$= 12x^2(x - y)(2x + 5y)$$

21. $x^2 - 64 = x^2 - 8^2$
$$= (x + 8)(x - 8)$$

22. $9m^2 - 25 = (3m)^2 - (5)^2$
$$= (3m + 5)(3m - 5)$$

23. $10x^2 - 160 = 10(x^2 - 16)$
$$= 10(x^2 - 4^2)$$
$$= 10(x + 4)(x - 4)$$

24. $9x^2 + 64$ is the *sum* of two perfect squares. It cannot be factored. It is prime.

25. $z^2 + 14zy + 49y^2$
$$= z^2 + 2 \cdot 7zy + 7^2y^2$$
$$= (z + 7y)^2$$

26. $s^2 - 10st + 25t^2$
$$= s^2 - 2 \cdot 5st + (5t)^2$$
$$= (s - 5t)^2$$

27. $9p^2 - 24p + 16$
$$= (3p)^2 - 2 \cdot 3p \cdot 4 + 4^2$$
$$= (3p - 4)^2$$

28. $a^3 - 216$
$$= a^3 - 6^3$$
$$= (a - 6)[(a)^2 + (a)(6) + (6)^2]$$
$$= (a - 6)(a^2 + 6a + 36)$$

29. $27r^3 - 64s^3$
$$= (3r)^3 - (4s)^3$$
$$= (3r - 4s)(9r^2 + 12rs + 16s^2)$$

30. $3m^3 + 375$
$$= 3(m^3 + 125)$$
$$= 3(m^3 + 5^3)$$
$$= 3(m + 5)(m^2 - 5m + 25)$$

31. $x^4 - y^4 = (x^2)^2 - (y^2)^2$
$$= (x^2 + y^2)(x^2 - y^2)$$
$$= (x^2 + y^2)(x + y)(x - y)$$

32. $16a^4 - 81b^4$
$$= (4a^2)^2 - (9b^2)^2$$
$$= (4a^2 + 9b^2)(4a^2 - 9b^2)$$
$$= (4a^2 + 9b^2)[(2a)^2 - (3b)^2]$$
$$= (4a^2 + 9b^2)(2a + 3b)(2a - 3b)$$

R.3   Rational Expressions

1. $\dfrac{5v^2}{35v} = \dfrac{5 \cdot v \cdot v}{5 \cdot 7 \cdot v} = \dfrac{v}{7}$

2. $\dfrac{25p^3}{10p^2} = \dfrac{5 \cdot 5 \cdot p \cdot p \cdot p}{2 \cdot 5 \cdot p \cdot p} = \dfrac{5p}{2}$

3. $\dfrac{8k + 16}{9k + 18} = \dfrac{8(k + 2)}{9(k + 2)} = \dfrac{8}{9}$

4. $\dfrac{2(t-15)}{(t-15)(t+2)} = \dfrac{2}{(t+2)}$

5. $\dfrac{4x^3 - 8x^2}{4x^2} = \dfrac{4x^2(x-2)}{4x^2}$

$\qquad\qquad\;\; = x - 2$

6. $\dfrac{36y^2 + 72y}{9y} = \dfrac{36y(y+2)}{9y}$

$\qquad\qquad = \dfrac{9 \cdot 4 \cdot y(y+2)}{9 \cdot y}$

$\qquad\qquad = 4(y+2)$

7. $\dfrac{m^2 - 4m + 4}{m^2 + m - 6} = \dfrac{(m-2)(m-2)}{(m-2)(m+3)}$

$\qquad\qquad\quad = \dfrac{m-2}{m+3}$

8. $\dfrac{r^2 - r - 6}{r^2 + r - 12} = \dfrac{(r-3)(r+2)}{(r+4)(r-3)}$

$\qquad\qquad\quad = \dfrac{r+2}{r+4}$

9. $\dfrac{3x^2 + 3x - 6}{x^2 - 4} = \dfrac{3(x+2)(x-1)}{(x+2)(x-2)}$

$\qquad\qquad\quad = \dfrac{3(x-1)}{x-2}$

10. $\dfrac{z^2 - 5z + 6}{z^2 - 4} = \dfrac{(z-3)(z-2)}{(z+2)(z-2)}$

$\qquad\qquad\quad = \dfrac{z-3}{z+2}$

11. $\dfrac{m^4 - 16}{4m^2 - 16} = \dfrac{(m^2+4)(m+2)(m-2)}{4(m+2)(m-2)}$

$\qquad\qquad\quad = \dfrac{m^2 + 4}{4}$

12. $\dfrac{6y^2 + 11y + 4}{3y^2 + 7y + 4} = \dfrac{(3y+4)(2y+1)}{(3y+4)(y+1)}$

$\qquad\qquad\qquad = \dfrac{2y+1}{y+1}$

13. $\dfrac{9k^2}{25} \cdot \dfrac{5}{3k} = \dfrac{3 \cdot 3 \cdot 5k^2}{5 \cdot 5 \cdot 3k} = \dfrac{3k^2}{5k} = \dfrac{3k}{5}$

14. $\dfrac{15p^3}{9p^2} \div \dfrac{6p}{10p^2}$

$\qquad = \dfrac{15p^3}{9p^2} \cdot \dfrac{10p^2}{6p}$

$\qquad = \dfrac{150p^5}{54p^3}$

$\qquad = \dfrac{25 \cdot 6p^5}{9 \cdot 6p^3}$

$\qquad = \dfrac{25p^2}{9}$

15. $\dfrac{3a+3b}{4c} \cdot \dfrac{12}{5(a+b)} = \dfrac{3(a+b)}{4c} \cdot \dfrac{3 \cdot 4}{5(a+b)}$

$\qquad\qquad\qquad\qquad = \dfrac{3 \cdot 3}{c \cdot 5}$

$\qquad\qquad\qquad\qquad = \dfrac{9}{5c}$

16. $\dfrac{a-3}{16} \div \dfrac{a-3}{32} = \dfrac{a-3}{16} \cdot \dfrac{32}{a-3}$

$\qquad\qquad\qquad = \dfrac{a-3}{16} \cdot \dfrac{16 \cdot 2}{a-3}$

$\qquad\qquad\qquad = \dfrac{2}{1} = 2$

17. $\dfrac{2k-16}{6} \div \dfrac{4k-32}{3}$

$\qquad = \dfrac{2k-16}{6} \cdot \dfrac{3}{4k-32}$

$\qquad = \dfrac{2(k-8)}{6} \cdot \dfrac{3}{4(k-8)}$

$\qquad = \dfrac{1}{4}$

18. $\dfrac{9y-18}{6y+12} \cdot \dfrac{3y+6}{15y-30}$

$\qquad = \dfrac{9(y-2)}{6(y+2)} \cdot \dfrac{3(y+2)}{15(y-2)}$

$\qquad = \dfrac{27}{90} = \dfrac{3 \cdot 3}{10 \cdot 3} = \dfrac{3}{10}$

19. $\dfrac{4a+12}{2a-10} \div \dfrac{a^2-9}{a^2-a-20}$

$\qquad = \dfrac{4(a+3)}{2(a-5)} \cdot \dfrac{(a-5)(a+4)}{(a-3)(a+3)}$

$\qquad = \dfrac{2(a+4)}{a-3}$

20. $\dfrac{6r - 18}{9r^2 + 6r - 24} \cdot \dfrac{12r - 16}{4r - 12}$

$$= \dfrac{6(r-3)}{3(3r^2 + 2r - 8)} \cdot \dfrac{4(3r - 4)}{4(r - 3)}$$

$$= \dfrac{6(r-3)}{3(3r - 4)(r + 2)} \cdot \dfrac{4(3r - 4)}{4(r - 3)}$$

$$= \dfrac{6}{3(r + 2)}$$

$$= \dfrac{2}{r + 2}$$

21. $\dfrac{k^2 + 4k - 12}{k^2 + 10k + 24} \cdot \dfrac{k^2 + k - 12}{k^2 - 9}$

$$= \dfrac{(k+6)(k-2)}{(k+6)(k+4)} \cdot \dfrac{(k+4)(k-3)}{(k+3)(k-3)}$$

$$= \dfrac{k - 2}{k + 3}$$

22. $\dfrac{m^2 + 3m + 2}{m^2 + 5m + 4} \div \dfrac{m^2 + 5m + 6}{m^2 + 10m + 24}$

$$= \dfrac{m^2 + 3m + 2}{m^2 + 5m + 4} \cdot \dfrac{m^2 + 10m + 24}{m^2 + 5m + 6}$$

$$= \dfrac{(m+1)(m+2)}{(m+4)(m+1)} \cdot \dfrac{(m+6)(m+4)}{(m+3)(m+2)}$$

$$= \dfrac{m + 6}{m + 3}$$

23. $\dfrac{2m^2 - 5m - 12}{m^2 - 10m + 24} \div \dfrac{4m^2 - 9}{m^2 - 9m + 18}$

$$= \dfrac{2m^2 - 5m - 12}{m^2 - 10m + 24} \cdot \dfrac{m^2 - 9m + 18}{4m^2 - 9}$$

$$= \dfrac{(2m+3)(m-4)(m-6)(m-3)}{(m-6)(m-4)(2m-3)(2m+3)}$$

$$= \dfrac{m - 3}{2m - 3}$$

24. $\dfrac{4n^2 + 4n - 3}{6n^2 - n - 15} \cdot \dfrac{8n^2 + 32n + 30}{4n^2 + 16n + 15}$

$$= \dfrac{(2n+3)(2n-1)}{(2n+3)(3n-5)} \cdot \dfrac{2(2n+3)(2n+5)}{(2n+3)(2n+5)}$$

$$= \dfrac{2(2n-1)}{3n - 5}$$

25. $\dfrac{a+1}{2} - \dfrac{a-1}{2}$

$$= \dfrac{(a+1) - (a-1)}{2}$$

$$= \dfrac{a + 1 - a + 1}{2}$$

$$= \dfrac{2}{2} = 1$$

26. $\dfrac{3}{p} + \dfrac{1}{2}$

Multiply the first term by $\frac{2}{2}$ and the second by $\frac{p}{p}$.

$$\dfrac{2 \cdot 3}{2 \cdot p} + \dfrac{p \cdot 1}{p \cdot 2} = \dfrac{6}{2p} + \dfrac{p}{2p}$$

$$= \dfrac{6 + p}{2p}$$

27. $\dfrac{6}{5y} - \dfrac{3}{2} = \dfrac{6 \cdot 2}{5y \cdot 2} - \dfrac{3 \cdot 5y}{2 \cdot 5y}$

$$= \dfrac{12 - 15y}{10y}$$

28. $\dfrac{1}{6m} + \dfrac{2}{5m} + \dfrac{4}{m}$

$$= \dfrac{5 \cdot 1}{5 \cdot 6m} + \dfrac{6 \cdot 2}{6 \cdot 5m} + \dfrac{30 \cdot 4}{30 \cdot m}$$

$$= \dfrac{5}{30m} + \dfrac{12}{30m} + \dfrac{120}{30m}$$

$$= \dfrac{5 + 12 + 120}{30m}$$

$$= \dfrac{137}{30m}$$

29. $\dfrac{1}{m - 1} + \dfrac{2}{m}$

$$= \dfrac{m}{m}\left(\dfrac{1}{m-1}\right) + \dfrac{m-1}{m-1}\left(\dfrac{2}{m}\right)$$

$$= \dfrac{m + 2m - 2}{m(m - 1)}$$

$$= \dfrac{3m - 2}{m(m - 1)}$$

30. $\dfrac{5}{2r + 3} - \dfrac{2}{r}$

$$= \dfrac{5r}{r(2r + 3)} - \dfrac{2(2r + 3)}{r(2r + 3)}$$

$$= \dfrac{5r - 2(2r + 3)}{r(2r + 3)}$$

$$= \dfrac{5r - 4r - 6}{r(2r + 3)}$$

$$= \dfrac{r - 6}{r(2r + 3)}$$

31. $\dfrac{8}{3(a - 1)} + \dfrac{2}{a - 1}$

$$= \dfrac{8}{3(a - 1)} + \dfrac{3}{3}\left(\dfrac{2}{a - 1}\right)$$

$$= \dfrac{8 + 6}{3(a - 1)}$$

$$= \dfrac{14}{3(a - 1)}$$

32. $\dfrac{2}{5(k-2)} + \dfrac{3}{4(k-2)} = \dfrac{4 \cdot 2}{4 \cdot 5(k-2)} + \dfrac{5 \cdot 3}{5 \cdot 4(k-2)}$

$$= \dfrac{8}{20(k-2)} + \dfrac{15}{20(k-2)}$$

$$= \dfrac{8+15}{20(k-2)}$$

$$= \dfrac{23}{20(k-2)}$$

33. $\dfrac{4}{x^2+4x+3} + \dfrac{3}{x^2-x-2}$

$$= \dfrac{4}{(x+3)(x+1)} + \dfrac{3}{(x-2)(x+1)}$$

$$= \dfrac{4(x-2)}{(x-2)(x+3)(x+1)}$$

$$+ \dfrac{3(x+3)}{(x-2)(x+3)(x+1)}$$

$$= \dfrac{4(x-2)+3(x+3)}{(x-2)(x+3)(x+1)}$$

$$= \dfrac{4x-8+3x+9}{(x-2)(x+3)(x+1)}$$

$$= \dfrac{7x+1}{(x-2)(x+3)(x+1)}$$

34. $\dfrac{y}{y^2+2y-3} - \dfrac{1}{y^2+4y+3}$

$$= \dfrac{y}{(y+3)(y-1)} - \dfrac{1}{(y+3)(y+1)}$$

$$= \dfrac{y(y+1)}{(y+3)(y+1)(y-1)}$$

$$- \dfrac{1(y-1)}{(y+3)(y+1)(y-1)}$$

$$= \dfrac{y(y+1)-(y-1)}{(y+3)(y+1)(y-1)}$$

$$= \dfrac{y^2+y-y+1}{(y+3)(y+1)(y-1)}$$

$$= \dfrac{y^2+1}{(y+3)(y+1)(y-1)}$$

35. $\dfrac{3k}{2k^2+3k-2} - \dfrac{2k}{2k^2-7k+3}$

$$= \dfrac{3k}{(2k-1)(k+2)} - \dfrac{2k}{(2k-1)(k-3)}$$

$$= \left(\dfrac{k-3}{k-3}\right) \dfrac{3k}{(2k-1)(k+2)}$$

$$- \left(\dfrac{k+2}{k+2}\right) \dfrac{2k}{(2k-1)(k-3)}$$

$$= \dfrac{(3k^2-9k)-(2k^2+4k)}{(2k-1)(k+2)(k-3)}$$

$$= \dfrac{k^2-13k}{(2k-1)(k+2)(k-3)}$$

$$= \dfrac{k(k-13)}{(2k-1)(k+2)(k-3)}$$

36. $\dfrac{4m}{3m^2+7m-6} - \dfrac{m}{3m^2-14m+8}$

$$= \dfrac{4m}{(3m-2)(m+3)} - \dfrac{m}{(3m-2)(m-4)}$$

$$= \dfrac{4m(m-4)}{(3m-2)(m+3)(m-4)}$$

$$- \dfrac{m(m+3)}{(3m-2)(m-4)(m+3)}$$

$$= \dfrac{4m(m-4)-m(m+3)}{(3m-2)(m-4)(m+3)}$$

$$= \dfrac{4m^2-16m-m^2-3m}{(3m-2)(m+3)(m-4)}$$

$$= \dfrac{3m^2-19m}{(3m-2)(m+3)(m-4)}$$

$$= \dfrac{m(3m-19)}{(3m-2)(m+3)(m-4)}$$

37. $\dfrac{2}{a+2} + \dfrac{1}{a} + \dfrac{a-1}{a^2+2a}$

$$= \dfrac{2}{a+2} + \dfrac{1}{a} + \dfrac{a-1}{a(a+2)}$$

$$= \left(\dfrac{a}{a}\right)\dfrac{2}{a+2} + \left(\dfrac{a+2}{a+2}\right)\dfrac{1}{a} + \dfrac{a-1}{a(a+2)}$$

$$= \dfrac{2a+a+2+a-1}{a(a+2)}$$

$$= \dfrac{4a+1}{a(a+2)}$$

38. $\dfrac{5x+2}{x^2-1} + \dfrac{3}{x^2+x} - \dfrac{1}{x^2-x}$

$$= \dfrac{5x+2}{(x+1)(x-1)} + \dfrac{3}{x(x+1)} - \dfrac{1}{x(x-1)}$$

$$= \left(\dfrac{x}{x}\right)\left(\dfrac{5x+2}{(x+1)(x-1)}\right) + \left(\dfrac{x-1}{x-1}\right)\left(\dfrac{3}{x(x+1)}\right)$$

$$- \left(\dfrac{x+1}{x+1}\right)\left(\dfrac{1}{x(x-1)}\right)$$

$$= \dfrac{x(5x+2)+(x-1)(3)-(x+1)(1)}{x(x+1)(x-1)}$$

$$= \dfrac{5x^2+2x+3x-3-x-1}{x(x+1)(x-1)}$$

$$= \dfrac{5x^2+4x-4}{x(x+1)(x-1)}$$

R.4   Equations

1. $0.2m - 0.5 = 0.1m + 0.7$
   $10(0.2m - 0.5) = 10(0.1m + 0.7)$
   $2m - 5 = m + 7$
   $m - 5 = 7$
   $m = 12$

   The solution is 12.

2. $\frac{2}{3}k - k + \frac{3}{8} = \frac{1}{2}$

   Multiply both sides of the equation by 24.

   $$24\left(\frac{2}{3}k\right) - 24(k) + 24\left(\frac{3}{8}\right) = 24\left(\frac{1}{2}\right)$$
   $$16k - 24k + 9 = 12$$
   $$-8k + 9 = 12$$
   $$-8k = 3$$
   $$k = -\frac{3}{8}$$

   The solution is $-\frac{3}{8}$.

3. $2x + 8 = x - 4$
   $x + 8 = -4$
   $x = -12$

   The solution is -12.

4. $5x + 2 = 8 - 3x$
   $8x + 2 = 8$
   $8x = 6$
   $x = \frac{3}{4}$

   The solution is $\frac{3}{4}$.

5. $3r + 2 - 5(r + 1) = 6r + 4$
   $3r + 2 - 5r - 5 = 6r + 4$
   $-3 - 2r = 6r + 4$
   $-3 = 8r + 4$
   $-7 = 8r$
   $-\frac{7}{8} = r$

   The solution is $-\frac{7}{8}$.

6. $5(a + 3) + 4a - 5 = -(2a - 4)$
   $5a + 15 + 4a - 5 = -2a + 4$
   $9a + 10 = -2a + 4$
   $11a + 10 = 4$
   $11a = -6$
   $a = -\frac{6}{11}$

   The solution is $-\frac{6}{11}$.

7. $2[3m - 2(3 - m) - 4] = 6m - 4$
   $2[3m - 6 + 2m - 4] = 6m - 4$
   $2[5m - 10] = 6m - 4$
   $10m - 20 = 6m - 4$
   $4m - 20 = -4$
   $4m = 16$
   $m = 4$

   The solution is 4.

8. $4[2p - (3 - p) + 5] = -7p - 2$
   $4[2p - 3 + p + 5] = -7p - 2$
   $4[3p + 2] = -7p - 2$
   $12p + 8 = -7p - 2$
   $19p + 8 = -2$
   $19p = -10$
   $p = -\frac{10}{19}$

   The solution is $-\frac{10}{19}$.

9. $x^2 + 5x + 6 = 0$
   $(x + 3)(x + 2) = 0$
   $x + 3 = 0$ or $x + 2 = 0$
   $x = -3$ or $x = -2$

   The solutions are -3 and -2.

10. $x^2 = 3 + 2x$
    $x^2 - 2x - 3 = 0$
    $(x - 3)(x + 1) = 0$
    $x - 3 = 0$ or $x + 1 = 0$
    $x = 3$ or $x = -1$

    The solutions are 3 and -1.

11. $m^2 = 14m - 49$
    $m^2 - 14m + 49 = 0$
    $(m)^2 - 2(7m) + (7)^2 = 0$
    $(m - 7)^2 = 0$
    $m - 7 = 0$
    $m = 7$

    The solution is 7.

12. $2k^2 - k = 10$
    $2k^2 - k - 10 = 0$
    $(2k - 5)(k + 2) = 0$
    $2k - 5 = 0$ or $k + 2 = 0$
    $k = \frac{5}{2}$ or $k = -2$

    The solutions are $\frac{5}{2}$ and -2.

13. $12x^2 - 5x = 2$

$12x^2 - 5x - 2 = 0$

$(4x + 1)(3x - 2) = 0$

$4x + 1 = 0 \quad \text{or} \quad 3x - 2 = 0$

$4x = -1 \quad \text{or} \quad 3x = 2$

$x = -\dfrac{1}{4} \quad \text{or} \quad x = \dfrac{2}{3}$

The solutions are $-\frac{1}{4}$ and $\frac{2}{3}$.

14. $m(m - 7) = -10$

$m^2 - 7m + 10 = 0$

$(m - 5)(m - 2) = 0$

$m - 5 = 0 \quad \text{or} \quad m - 2 = 0$

$m = 5 \quad \text{or} \quad m = 2$

The solutions are 5 and 2.

15. $4x^2 - 36 = 0$

Divide both sides of the equation by 4.

$x^2 - 9 = 0$

$(x + 3)(x - 3) = 0$

$x + 3 = 0 \quad \text{or} \quad x - 3 = 0$

$x = -3 \quad \text{or} \quad x = 3$

The solutions are -3 and 3.

16. $z(2z + 7) = 4$

$2z^2 + 7z - 4 = 0$

$(2z - 1)(z + 4) = 0$

$2z - 1 = 0 \quad \text{or} \quad z + 4 = 0$

$z = \dfrac{1}{2} \quad \text{or} \quad z = -4$

The solutions are $\frac{1}{2}$ and -4.

17. $12y^2 - 48y = 0$

$12y(y) - 12y(4) = 0$

$12y(y - 4) = 0$

$12y = 0 \quad \text{or} \quad y - 4 = 0$

$y = 0 \quad \text{or} \quad y = 4$

The solutions are 0 and 4.

18. $3x^2 - 5x + 1 = 0$

Use the quadratic formula.

$x = \dfrac{-(-5) \pm \sqrt{(-5)^2 - 4(3)(1)}}{2(3)}$

$= \dfrac{5 \pm \sqrt{25 - 12}}{6}$

$x = \dfrac{5 + \sqrt{13}}{6} \quad \text{or} \quad x = \dfrac{5 - \sqrt{13}}{6}$

$\approx 1.4343 \qquad\qquad \approx 0.2324$

The solutions are $\frac{5+\sqrt{13}}{6} \approx 1.4343$ and

$\frac{5-\sqrt{13}}{6} \approx 0.2324$.

19. $2m^2 - 4m = 3$

$2m^2 - 4m - 3 = 0$

$m = \dfrac{-(-4) \pm \sqrt{(-4)^2 - 4(2)(-3)}}{2(2)}$

$= \dfrac{4 \pm \sqrt{40}}{4} = \dfrac{4 \pm \sqrt{4 \cdot 10}}{4}$

$= \dfrac{4 \pm \sqrt{4}\sqrt{10}}{4}$

$= \dfrac{4 \pm 2\sqrt{10}}{4} = \dfrac{2 \pm \sqrt{10}}{2}$

The solutions are $\frac{2+\sqrt{10}}{2} \approx 2.5811$ and

$\frac{2-\sqrt{10}}{2} \approx -0.5811$.

20. $p^2 + p - 1 = 0$

$p = \dfrac{-1 \pm \sqrt{1^2 - 4(1)(-1)}}{2(1)}$

$= \dfrac{-1 \pm \sqrt{5}}{2}$

The solutions are $\frac{-1+\sqrt{5}}{2} \approx 0.6180$ and

$\frac{-1-\sqrt{5}}{2} \approx -1.6180$.

21. $k^2 - 10k = -20$

$k^2 - 10k + 20 = 0$

$k = \dfrac{-(-10) \pm \sqrt{(-10)^2 - 4(1)(20)}}{2(1)}$

$k = \dfrac{10 \pm \sqrt{100 - 80}}{2}$

$k = \dfrac{10 \pm \sqrt{20}}{2}$

$k = \dfrac{10 \pm \sqrt{4}\sqrt{5}}{2}$

$k = \dfrac{10 \pm 2\sqrt{5}}{2}$

$k = \dfrac{2(5 \pm \sqrt{5})}{2}$

$k = 5 \pm \sqrt{5}$

The solutions are $5 + \sqrt{5} \approx 7.2361$ and

$5 - \sqrt{5} \approx 2.7639$.

22. $5x^2 - 8x + 2 = 0$

$$x = \frac{-(-8) \pm \sqrt{(-8)^2 - 4(5)(2)}}{2(5)}$$

$$= \frac{8 \pm \sqrt{24}}{10} = \frac{8 \pm \sqrt{4 \cdot 6}}{10}$$

$$= \frac{8 \pm \sqrt{4}\sqrt{6}}{10} = \frac{8 \pm 2\sqrt{6}}{10}$$

$$= \frac{4 \pm \sqrt{6}}{5}$$

The solutions are $\frac{4+\sqrt{6}}{5} \approx 1.2899$ and

$\frac{4-\sqrt{6}}{5} \approx 0.3101$.

23. $2r^2 - 7r + 5 = 0$

$(2r - 5)(r - 1) = 0$

$2r - 5 = 0 \quad$ or $\quad r - 1 = 0$

$2r = 5$

$r = \dfrac{5}{2} \quad$ or $\qquad r = 1$

The solutions are $\frac{5}{2}$ and 1.

24. $2x^2 - 7x + 30 = 0$

$$x = \frac{-(-7) \pm \sqrt{(-7)^2 - 4(2)(30)}}{2(2)}$$

$$x = \frac{7 \pm \sqrt{49 - 240}}{4}$$

$$x = \frac{7 \pm \sqrt{-191}}{4}$$

Since there is a negative number under the radical sign, $\sqrt{-191}$ is not a real number. Thus, there are no real number solutions.

25. $3k^2 + k = 6$

$3k^2 + k - 6 = 0$

$$k = \frac{-1 \pm \sqrt{1 - 4(3)(-6)}}{2(3)}$$

$$= \frac{-1 \pm \sqrt{73}}{6}$$

The solutions are $\frac{-1+\sqrt{73}}{6} \approx 1.2573$ and $\frac{-1-\sqrt{73}}{6} \approx -1.5907$.

26. $5m^2 + 5m = 0$

$5m(m + 1) = 0$

$5m = 0 \quad$ or $\quad m + 1 = 0$

$m = 0 \quad$ or $\qquad m = -1$

The solutions are 0 and -1.

27. $\dfrac{3x - 2}{7} = \dfrac{x + 2}{5}$

$$35\left(\frac{3x - 2}{7}\right) = 35\left(\frac{x + 2}{5}\right)$$

$$5(3x - 2) = 7(x + 2)$$

$$15x - 10 = 7x + 14$$

$$8x = 24$$

$$x = 3$$

The solution is $x = 3$.

28. $\dfrac{x}{3} - 7 = 6 - \dfrac{3x}{4}$

Multiply both sides by 12, the least common denominator of 3 and 4.

$$12\left(\frac{x}{3} - 7\right) = 12\left(6 - \frac{3x}{4}\right)$$

$$12\left(\frac{x}{3}\right) - (12)(7) = (12)(6) - (12)\left(\frac{3x}{4}\right)$$

$$4x - 84 = 72 - 9x$$

$$13x - 84 = 72$$

$$13x = 156$$

$$x = 12$$

The solution is 12.

29. $\dfrac{4}{x - 3} - \dfrac{8}{2x + 5} + \dfrac{3}{x - 3} = 0$

$$\frac{4}{x - 3} + \frac{3}{x - 3} - \frac{8}{2x + 5} = 0$$

$$\frac{7}{x - 3} - \frac{8}{2x + 5} = 0$$

Multiply both sides by $(x - 3)(2x + 5)$. Note that $x \neq 3$ and $x \neq -\frac{5}{2}$.

$$(x-3)(2x+5)\left(\frac{7}{x-3} - \frac{8}{2x+5}\right) = (x-3)(2x+5)(0)$$

$$7(2x + 5) - 8(x - 3) = 0$$

$$14x + 35 - 8x + 24 = 0$$

$$6x + 59 = 0$$

$$6x = -59$$

$$x = -\frac{59}{6}$$

Note: It is especially important to check solutions of equations that involve rational expressions. Here, a check shows that $-\frac{59}{6}$ is a solution.

30. $\dfrac{5}{p-2} - \dfrac{7}{p+2} = \dfrac{12}{p^2-4}$

$\dfrac{5}{p-2} - \dfrac{7}{p+2} = \dfrac{12}{(p-2)(p+2)}$

Multiply both sides by $(p-2)(p+2)$. Note that $p \neq 2$ and $p \neq -2$.

$(p-2)(p+2)\left(\dfrac{5}{p-2} - \dfrac{7}{p+2}\right) = (p-2)(p+2)\left(\dfrac{12}{(p-2)(p+2)}\right)$

$(p-2)(p+2)\left(\dfrac{5}{p-2}\right) -$

$(p-2)(p+2)\left(\dfrac{7}{p+2}\right) = (p-2)(p+2)\left(\dfrac{12}{(p-2)(p+2)}\right)$

$(p+2)(5) - (p-2)(7) = 12$

$5p + 10 - 7p + 14 = 12$

$-2p + 24 = 12$

$-2p = -12$

$p = 6$

The solution is 6.

31. $\dfrac{2m}{m-2} - \dfrac{6}{m} = \dfrac{12}{m^2-2m}$

$\dfrac{2m}{m-2} - \dfrac{6}{m} = \dfrac{12}{m(m-2)}$

Multiply both sides by $m(m-2)$. Note that $m \neq 0$ and $m \neq 2$.

$m(m-2)\left(\dfrac{2m}{m-2} - \dfrac{6}{m}\right) = m(m-2)\left(\dfrac{12}{m(m-2)}\right)$

$m(2m) - 6(m-2) = 12$

$2m^2 - 6m + 12 = 12$

$2m^2 - 6m = 0$

$2m(m-3) = 0$

$2m = 0$ or $m - 3 = 0$

$m = 0$ or $m = 3$

Since $m \neq 0$, 0 is not a solution. The solution is 3.

32. $\dfrac{2y}{y-1} = \dfrac{5}{y} + \dfrac{10-8y}{y^2-y}$

$\dfrac{2y}{y-1} = \dfrac{5}{y} + \dfrac{10-8y}{y(y-1)}$

Multiply both sides by $y(y-1)$.
Note that $y \neq 0$ and $y \neq 1$.

$y(y-1)\left(\dfrac{2y}{y-1}\right) = y(y-1)\left[\dfrac{5}{y} + \dfrac{10-8y}{y(y-1)}\right]$

$y(y-1)\left(\dfrac{2y}{y-1}\right) = y(y-1)\left(\dfrac{5}{y}\right)$

$+ y(y-1)\left[\dfrac{10-8y}{y(y-1)}\right]$

$y(2y) = (y-1)(5) + (10-8y)$

$2y^2 = 5y - 5 + 10 - 8y$

$2y^2 = 5 - 3y$

$2y^2 + 3y - 5 = 0$

$(2y+5)(y-1) = 0$

$2y + 5 = 0$ or $y - 1 = 0$

$y = -\dfrac{5}{2}$ or $y = 1$

Since $y \neq 1$, 1 is not a solution.
The solution is $-\frac{5}{2}$.

33. $\dfrac{1}{x-2} - \dfrac{3x}{x-1} = \dfrac{2x+1}{x^2-3x+2}$

$\dfrac{1}{x-2} - \dfrac{3x}{x-1} = \dfrac{2x+1}{(x-2)(x-1)}$

Multiply both sides by $(x-2)(x-1)$.
Note that $x \neq 2$ and $x \neq 1$.

$(x-2)(x-1)\left(\dfrac{1}{x-2} - \dfrac{3x}{x-1}\right) = (x-2)(x-1)$

$\cdot\left[\dfrac{2x+1}{(x-2)(x-1)}\right]$

$(x-2)(x-1)\left(\dfrac{1}{x-2}\right)$

$- (x-2)(x-1)\cdot\left(\dfrac{3x}{x-1}\right) = \dfrac{(x-2)(x-1)(2x+1)}{(x-2)(x-1)}$

$(x-1) - (x-2)(3x) = 2x+1$

$x - 1 - 3x^2 + 6x = 2x + 1$

$-3x^2 + 7x - 1 = 2x + 1$

$-3x^2 + 5x - 2 = 0$

$3x^2 - 5x + 2 = 0$

$(3x-2)(x-1) = 0$

$3x - 2 = 0$ or $x - 1 = 0$

$x = \dfrac{2}{3}$ or $x = 1$

1 is not a solution since $x \neq 1$.
The solution is $\frac{2}{3}$.

34.
$$\frac{5}{a} + \frac{-7}{a+1} = \frac{a^2 - 2a + 4}{a^2 + a}$$

$$a(a+1)\left(\frac{5}{a} + \frac{-7}{a+1}\right) = a(a+1)\left(\frac{a^2 - 2a + 4}{a^2 + a}\right)$$

Note that $a \neq 0$ and $a \neq -1$.

$$5(a+1) + (-7)(a) = a^2 - 2a + 4$$
$$5a + 5 - 7a = a^2 - 2a + 4$$
$$5 - 2a = a^2 - 2a + 4$$
$$5 = a^2 + 4$$
$$0 = a^2 - 1$$
$$0 = (a+1)(a-1)$$
$$a + 1 = 0 \quad \text{or} \quad a - 1 = 0$$
$$a = -1 \quad \text{or} \quad a = 1$$

Since -1 would make two denominators zero, 1 is the only solution.

35.
$$\frac{5}{b+5} - \frac{4}{b^2 + 2b} = \frac{6}{b^2 + 7b + 10}$$

$$\frac{5}{b+5} - \frac{4}{b(b+2)} = \frac{6}{(b+5)(b+2)}$$

Multiply both sides by $b(b+5)(b+2)$.
Note that $b \neq 0$, $b \neq -5$, and $b \neq -2$.

$$b(b+5)(b+2)\left(\frac{5}{b+5} - \frac{4}{b(b+2)}\right)$$

$$= b(b+5)(b+2)\left(\frac{6}{(b+5)(b+2)}\right)$$

$$5b(b+2) - 4(b+5) = 6b$$
$$5b^2 + 10b - 4b - 20 = 6b$$
$$5b^2 - 20 = 0$$
$$b^2 - 4 = 0$$
$$(b+2)(b-2) = 0$$
$$b + 2 = 0 \quad \text{or} \quad b - 2 = 0$$
$$b = -2 \quad \text{or} \quad b = 2$$

Since $b \neq -2$, -2 is not a solution. The solution is 2.

36.
$$\frac{2}{x^2 - 2x - 3} + \frac{5}{x^2 - x - 6} = \frac{1}{x^2 + 3x + 2}$$

$$\frac{2}{(x-3)(x+1)} + \frac{5}{(x-3)(x+2)} = \frac{1}{(x+2)(x+1)}$$

Multiply both sides by $(x-3)(x+1)(x+2)$.
Note that $x \neq 3$, $x \neq -1$, and $x \neq -2$.

$$(x-3)(x+1)(x+2)\left(\frac{2}{(x-3)(x+1)}\right)$$

$$+ (x-3)(x+1)(x+2)\left(\frac{5}{(x-3)(x+2)}\right)$$

$$= (x-3)(x+1)(x+2)\left(\frac{1}{(x+2)(x+1)}\right)$$

$$2(x+2) + 5(x+1) = x - 3$$
$$2x + 4 + 5x + 5 = x - 3$$
$$7x + 9 = x - 3$$
$$6x + 9 = -3$$
$$6x = -12$$
$$x = -2$$

However, $x \neq -2$. Therefore there is no solution.

37.
$$\frac{4}{2x^2 + 3x - 9} + \frac{2}{2x^2 - x - 3} = \frac{3}{x^2 + 4x + 3}$$

$$\frac{4}{(2x-3)(x+3)} + \frac{2}{(2x-3)(x+1)} = \frac{3}{(x+3)(x+1)}$$

Multiply both sides by $(2x - 3)(x + 3)(x + 1)$.
Note that $x \neq \frac{3}{2}$, $x \neq -3$, and $x \neq -1$.

$$(2x - 3)(x+3)(x+1)$$

$$\cdot \left(\frac{4}{(2x-3)(x+3)} + \frac{2}{(2x-3)(x+1)}\right)$$

$$= (2x - 3)(x+3)(x+1)\left(\frac{3}{(x+3)(x+1)}\right)$$

$$4(x+1) + 2(x+3) = 3(2x - 3)$$
$$4x + 4 + 2x + 6 = 6x - 9$$
$$6x + 10 = 6x - 9$$
$$10 = -9$$

This is a false statement. Therefore, there is no solution.

R.5 Inequalities

1. $x < 4$

Because the inequality symbol means "less than," the endpoint at 4 is not included. This inequality is written in interval notation as $(-\infty, 4)$. To graph this interval on a number line, place an open circle at 4 and draw a heavy arrow pointing to the left.

2. $x \geq -3$

Because the inequality sign means "greater than or equal to," the endpoint at -3 is included. This inequality is written in interval notation as $[-3, \infty)$. To graph this interval on a number line, place a closed circle at -3 and draw a heavy arrow pointing to the right.

3. $1 \le x < 2$

The endpoint at 1 is included, but the endpoint at 2 is not. This inequality is written in interval notation as $[1, 2)$. To graph this interval, place a closed circle at 1 and an open circle at 2; then draw a heavy line segment between them.

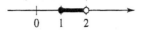

4. $-2 \le x \le 3$

The endpoints at -2 and 3 are both included. This inequality is written in interval notation as $[-2, 3]$. To graph this interval, place an open circle at -2 and another at 3 and draw a heavy line segment between them.

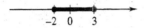

5. $-9 > x$

This inequality may be rewritten as $x < -9$, and is written in interval notation as $(-\infty, -9)$. Note that the endpoint at -9 is not included. To graph this interval, place an open circle at -9 and draw a heavy arrow pointing to the left.

6. $6 \le x$

This inequality may be written as $x \ge 6$, and is written in interval notation as $[6, \infty)$. Note that the endpoint at 6 is included. To graph this interval, place a closed circle at 6 and draw a heavy arrow pointing to the right.

7. $[-7, -3]$

This represents all the numbers between -7 and -3, including both endpoints. This interval can be written as the inequality $-7 \le x \le -3$.

8. $[4, 10)$

This represents all the numbers between 4 and 10, including 4 but not including 10. This interval can be written as the inequality $4 \le x < 10$.

9. $(-\infty, -1]$

This represents all the numbers to the left of -1 on the number line and includes the endpoint. This interval can be written as the inequality $x \le -1$.

10. $(3, \infty)$

This represents all the numbers to the right of 3, and does not include the endpoint. This interval can be written as the inequality $x > 3$.

11. Notice that the endpoint -2 is included, but 6 is not. The interval shows in the graph can be written as the inequality $-2 \le x < 6$.

12. Notice that neither endpoint is included. The interval shown in the graph can be written as $0 < x < 8$.

13. Notice that both endpoints are included. The interval shown in the graph can be written as $x \le -4$ or $x \ge 4$.

14. Notice that the endpoint 0 is not included, but 3 is included. The interval shown in the graph can be written as $x < 0$ or $x \ge 3$.

15. $6p + 7 \le 19$
$$6p \le 12$$
$$\left(\frac{1}{6}\right)(6p) \le \left(\frac{1}{6}\right)(12)$$
$$p \le 2$$

The solution in interval notation is $(-\infty, 2]$.

16. $6k - 4 < 3k - 1$
$$6k < 3k + 3$$
$$3k < 3$$
$$k < 1$$

The solution in interval notation is $(-\infty, 1)$.

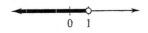

17. $m - (3m - 2) + 6 < 7m - 19$

$m - 3m + 2 + 6 < 7m - 19$

$-2m + 8 < 7m - 19$

$-9m + 8 < -19$

$-9m < -27$

$-\dfrac{1}{9}(-9m) > -\dfrac{1}{9}(-27)$

$m > 3$

The solution is $(3, \infty)$.

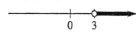

18. $-2(3y - 8) \geq 5(4y - 2)$

$-6y + 16 \geq 20y - 10$

$-6y + 16 + (-16) \geq 20y - 10 + (-16)$

$-6y \geq 20y - 26$

$-6y + (-20y) \geq 20y + (-20y) - 26$

$-26y \geq -26$

$-\dfrac{1}{26}(-26)y \leq -\dfrac{1}{26}(-26)$

$y \leq 1$

The solution is $(-\infty, 1]$.

19. $3p - 1 < 6p + 2(p - 1)$

$3p - 1 < 6p + 2p - 2$

$3p - 1 < 8p - 2$

$-5p - 1 < -2$

$-5p < -1$

$-\dfrac{1}{5}(-5p) > -\dfrac{1}{5}(-1)$

$p > \dfrac{1}{5}$

The solution is $\left(\dfrac{1}{5}, \infty\right)$.

20. $x + 5(x + 1) > 4(2 - x) + x$

$x + 5x + 5 > 8 - 4x + x$

$6x + 5 > 8 - 3x$

$6x > 3 - 3x$

$9x > 3$

$x > \dfrac{1}{3}$

The solution is $\left(\dfrac{1}{3}, \infty\right)$.

21. $-11 < y - 7 < -1$

$-11 + 7 < y - 7 + 7 < -1 + 7$

$-4 < y < 6$

The solution is $(-4, 6)$.

22. $8 \leq 3r + 1 \leq 13$

$8 + (-1) \leq 3r + 1 + (-1) \leq 13 + (-1)$

$7 \leq 3r \leq 12$

$\dfrac{1}{3}(7) \leq \dfrac{1}{3}(3r) \leq \dfrac{1}{3}(12)$

$\dfrac{7}{3} \leq r \leq 4$

The solution is $\left[\dfrac{7}{3}, 4\right]$.

23. $-2 < \dfrac{1 - 3k}{4} \leq 4$

$4(-2) < 4\left(\dfrac{1 - 3k}{4}\right) \leq 4(4)$

$-8 < 1 - 3k \leq 16$

$-9 < -3k \leq 15$

$-\dfrac{1}{3}(-9) > -\dfrac{1}{3}(-3k) \geq -\dfrac{1}{3}(15)$

Rewrite the inequalities in the proper order.

$-5 \leq k < 3$

24. $-1 \leq \dfrac{5y + 2}{3} \leq 4$

$3(-1) \leq 3\left(\dfrac{5y + 2}{3}\right) \leq 3(4)$

$-3 \leq 5y + 2 \leq 12$

$-5 \leq 5y \leq 10$

$-1 \leq y \leq 2$

The solution is $[-1, 2]$.

25. $\dfrac{3}{5}(2p+3) \geq \dfrac{1}{10}(5p+1)$

$10\left(\dfrac{3}{5}\right)(2p+3) \geq 10\left(\dfrac{1}{10}\right)(5p+1)$

$6(2p+3) \geq 5p+1$

$12p+18 \geq 5p+1$

$7p \geq -17$

$p \geq -\dfrac{17}{7}$

The solution is $[-\frac{17}{7}, \infty)$.

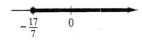

26. $\dfrac{8}{3}(z-4) \leq \dfrac{2}{9}(3z+2)$

$(9)\dfrac{8}{3}(z-4) \leq (9)\dfrac{2}{9}(3z+2)$

$24(z-4) \leq 2(3z+2)$

$24z-96 \leq 6z+4$

$24z \leq 6z+100$

$18z \leq 100$

$z \leq \dfrac{100}{18}$

$z \leq \dfrac{50}{9}$

The solution is $(-\infty, \frac{50}{9}]$.

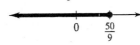

27. $(m-3)(m+5) < 0$

Solve $(m-3)(m+5) = 0$.

$(m-3)(m+5) = 0$

$m = 3 \quad \text{or} \quad m = -5$

Intervals: $(-\infty, -5), \ (-5, 3), (3, \infty)$

For $(-\infty, -5)$, choose -6 to test for m.

$(-6-3)(-6+5) = -9(-1) = 9 \not< 0$

For $(-5, 3)$, choose 0.

$(0-3)(0+5) = -3(5) = -15 < 0$

For $(3, \infty)$, choose 4.

$(4-3)(4+5) = 1(9) = 9 \not< 0$

The solution is $(-5, 3)$.

28. $(t+6)(t-1) \geq 0$

Solve $(t+6)(t-1) = 0$.

$(t+6)(t-1) = 0$

$t = -6 \quad \text{or} \quad t = 1$

Intervals: $(-\infty, -6), \ (-6, 1), \ (1, \infty)$

For $(-\infty, -6)$, choose -7 to test for t.

$(-7+6)(-7-1) = (-1)(-8) = 8 \geq 0$

For $(-6, 1)$, choose 0.

$(0+6)(0-1) = (6)(-1) = -6 \not\geq 0$

For $(1, \infty)$, choose 2.

$(2+6)(2-1) = (8)(1) = 8 \geq 0$

Because the symbol $\geq$ is used, the endpoints -6 and 1 are included in the solution, $(-\infty, -6] \cup [1, \infty)$.

29. $y^2 - 3y + 2 < 0$

$(y-2)(y-1) < 0$

Solve $(y-2)(y-1) = 0$.

$y = 2 \quad \text{or} \quad y = 1$

Intervals: $(-\infty, 1), \ (1, 2), (2, \infty)$

For $(-\infty, 1)$, choose $y = 0$.

$0^2 - 3(0) + 2 = 2 \not< 0$

For $(1, 2)$, choose $y = \frac{3}{2}$.

$\left(\dfrac{3}{2}\right)^2 - 3\left(\dfrac{3}{2}\right) + 2 = \dfrac{9}{4} - \dfrac{9}{2} + 2$

$= \dfrac{9 - 18 + 8}{4}$

$= -\dfrac{1}{4} < 0$

For $(2, \infty)$, choose 3.

$3^2 - 3(3) + 2 = 2 \not< 0$

The solution is $(1, 2)$.

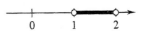

30. $2k^2 + 7k - 4 > 0$

Solve $2k^2 + 7k - 4 = 0$.

$$2k^2 + 7k - 4 = 0$$
$$(2k - 1)(k + 4) = 0$$

$$k = \frac{1}{2} \quad \text{or} \quad k = -4$$

Intervals: $(-\infty, -4), \left(-4, \frac{1}{2}\right), \left(\frac{1}{2}, \infty\right)$

For $(-\infty, -4)$, choose -5.

$$2(-5)^2 + 7(-5) - 4 = 11 > 0$$

For $\left(-4, \frac{1}{2}\right)$, choose 0.

$$2(0)^2 + 7(0) - 4 = -4 \not> 0$$

For $\left(\frac{1}{2}, \infty\right)$, choose 1.

$$2(1)^2 + 7(1) - 4 = 5 > 0$$

The solution is $(-\infty, -4) \cup \left(\frac{1}{2}, \infty\right)$.

31. $x^2 - 16 > 0$

Solve $x^2 - 16 = 0$.

$$x^2 - 16 = 0$$
$$(x + 4)(x - 4) = 0$$

$$x = -4 \quad \text{or} \quad x = 4$$

Intervals: $(-\infty, -4), (-4, 4), (4, \infty)$

For $(-\infty, -4)$, choose -5.

$$(-5)^2 - 16 = 9 > 0$$

For $(-4, 4)$, choose 0.

$$0^2 - 16 = -16 \not> 0$$

For $(4, \infty)$, choose 5.

$$5^2 - 16 = 9 > 0$$

The solution is $(-\infty, -4) \cup (4, \infty)$.

32. $2k^2 - 7k - 15 \le 0$

Solve $2k^2 - 7k - 15 = 0$.

$$2k^2 - 7k - 15 = 0$$
$$(2k + 3)(k - 5) = 0$$

$$k = -\frac{3}{2} \quad \text{or} \quad k = 5$$

Intervals: $\left(-\infty, -\frac{3}{2}\right), \left(-\frac{3}{2}, 5\right), (5, \infty)$

For $\left(-\infty, -\frac{3}{2}\right)$, choose -2.

$$2(-2)^2 - 7(-2) - 15 = 7 \not\le 0$$

For $\left(-\frac{3}{2}, 5\right)$, choose 0.

$$2(0)^2 - 7(0) - 15 = -15 \le 0$$

For $(5, \infty)$, choose 6.

$$2(6)^2 - 7(6) - 15 \not\le 0$$

The solution is $\left[-\frac{3}{2}, 5\right]$.

33. $x^2 - 4x \ge 5$

Solve $x^2 - 4x = 5$.

$$x^2 - 4x = 5$$
$$x^2 - 4x - 5 = 0$$
$$(x + 1)(x - 5) = 0$$

$$x + 1 = 0 \quad \text{or} \quad x - 5 = 0$$
$$x = -1 \quad \text{or} \quad x = 5$$

Intervals: $(-\infty, -1), (-1, 5), (5, \infty)$

For $(-\infty, -1)$, choose -2.

$$(-2)^2 - 4(-2) = 12 \ge 5$$

For $(-1, 5)$, choose 0.

$$0^2 - 4(0) = 0 \not\ge 5$$

For $(5, \infty)$, choose 6.

$$(6)^2 - 4(6) = 12 \ge 5$$

The solution is $(-\infty, -1] \cup [5, \infty)$.

34. $10r^2 + r \le 2$

Solve $10r^2 + r = 2$.

$$10r^2 + r = 2$$
$$10r^2 + r - 2 = 0$$
$$(5r - 2)(2r + 1) = 0$$

$$r = \frac{2}{5} \quad \text{or} \quad r = -\frac{1}{2}$$

Intervals: $\left(-\infty, -\frac{1}{2}\right), \left(-\frac{1}{2}, \frac{2}{5}\right), \left(\frac{2}{5}, \infty\right)$

For $\left(-\infty, -\frac{1}{2}\right)$, choose -1.

$$10(-1)^2 + (-1) = 9 \not\le 2$$

For $\left(-\frac{1}{2}, \frac{2}{5}\right)$, choose 0.

$$10(0)^2 + 0 = 0 \le 2$$

For $\left(\frac{2}{5}, \infty\right)$, choose 1.

$$10(1)^2 + 1 = 11 \not\le 2$$

The solution is $\left[-\frac{1}{2}, \frac{2}{5}\right]$.

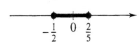

35. $3x^2 + 2x > 1$

Solve $3x^2 + 2x = 1$.

$$3x^2 + 2x = 1$$
$$3x^2 + 2x - 1 = 0$$
$$(3x - 1)(x + 1) = 0$$
$$x = \frac{1}{3} \quad \text{or} \quad x = -1$$

Intervals: $(-\infty, -1), \left(-1, \frac{1}{3}\right), \left(\frac{1}{3}, \infty\right)$

For $(-\infty, -1)$, choose -2.

$$3(-2)^2 + 2(-2) = 8 > 1$$

For $\left(-1, \frac{1}{3}\right)$, choose 0.

$$3(0)^2 + 2(0) = 0 \not> 1$$

For $\left(\frac{1}{3}, \infty\right)$, choose 1.

$$3(1)^2 + 2(1) = 5 > 1$$

The solution is $(-\infty, -1) \cup \left(\frac{1}{3}, \infty\right)$

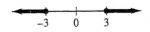

36. $3a^2 + a > 10$

Solve $3a^2 + a = 10$.

$$3a^2 + a = 10$$
$$3a^2 + a - 10 = 0$$
$$(3a - 5)(a + 2) = 0$$

$$a = \frac{5}{3} \quad \text{or} \quad a = -2$$

Intervals: $(-\infty, -2), \left(-2, \frac{5}{3}\right), \left(\frac{5}{3}, \infty\right)$

For $(-\infty, -2)$, choose -3.

$$3(-3)^2 + (-3) = 24 > 10$$

For $\left(-2, \frac{5}{3}\right)$, choose 0.

$$3(0)^2 + 0 = 0 \not> 10$$

For $\left(\frac{5}{3}, \infty\right)$, choose 2.

$$3(2)^2 + 2 = 14 > 10$$

The solution is $(-\infty, -2) \cup \left(\frac{5}{3}, \infty\right)$.

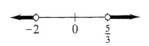

37. $9 - x^2 \le 0$

Solve $9 - x^2 = 0$.

$$9 - x^2 = 0$$
$$(3 + x)(3 - x) = 0$$
$$x = -3 \quad \text{or} \quad x = 3$$

Intervals: $(-\infty, -3), (-3, 3), (3, \infty)$

For $(-\infty, -3)$, choose -4.

$$9 - (-4)^2 = -7 \le 0$$

For $(-3, 3)$, choose 0.

$$9 - (0)^2 = 9 \not\le 0$$

For $(3, \infty)$, choose 4.

$$9 - (4)^2 = -7 \le 0$$

The solution is $(-\infty, -3] \cup [3, \infty)$.

38. $p^2 - 16p > 0$

Solve $p^2 - 16p = 0$.

$$p^2 - 16p = 0$$
$$p(p - 16) = 0$$
$$p = 0 \quad \text{or} \quad p = 16$$

Intervals: $(-\infty, 0), \ (0, 16), \ (16, \infty)$

For $(-\infty, 0)$, choose -1.

$$(-1)^2 - 16(-1) = 17 > 0$$

For $(0, 16)$, choose 1.

$$(1)^2 - 16(1) = -15 \not> 0$$

For $(16, \infty)$, choose 17.

$$(17)^2 - 16(17) = 17 > 0$$

The solution is $(-\infty, 0) \cup (16, \infty)$.

39. $\dfrac{m - 3}{m + 5} \leq 0$

Solve $\dfrac{m - 3}{m + 5} = 0$.

$$(m + 5)\frac{m - 3}{m + 5} = (m + 5)(0)$$
$$m - 3 = 0$$
$$m = 3$$

Set the denominator equal to 0 and solve.

$$m + 5 = 0$$
$$m = -5$$

Intervals: $(-\infty, -5), \ (-5, 3), \ (3, \infty)$

For $(-\infty, -5)$, choose -6.

$$\frac{-6 - 3}{-6 + 5} = 9 \not\leq 0$$

For $(-5, 3)$, choose 0.

$$\frac{0 - 3}{0 + 5} = -\frac{3}{5} \leq 0$$

For $(3, \infty)$, choose 4.

$$\frac{4 - 3}{4 + 5} = \frac{1}{9} \not\leq 0$$

Although the $\leq$ symbol is used, including -5 in the solution would cause the denominator to be zero.
The solution is $(-5, 3]$.

40. $\dfrac{r + 1}{r - 1} > 0$

Solve the equation $\dfrac{r + 1}{r - 1} = 0$.

$$\frac{r + 1}{r - 1} = 0$$
$$(r - 1)\frac{r + 1}{r - 1} = (r - 1)(0)$$
$$r + 1 = 0$$
$$r = -1$$

Find the value for which the denominator equals zero.

$$r - 1 = 0$$
$$r = 1$$

Intervals: $(-\infty, -1), \ (-1, 1), \ (1, \infty)$

For $(-\infty, -1)$, choose -2.

$$\frac{-2 + 1}{-2 - 1} = \frac{-1}{-3} = \frac{1}{3} > 0$$

For $(-1, 1)$, choose 0.

$$\frac{0 + 1}{0 - 1} = \frac{1}{-1} = -1 \not> 0$$

For $(1, \infty)$, choose 2.

$$\frac{2 + 1}{2 - 1} = \frac{3}{1} = 3 > 0$$

The solution is $(-\infty, -1) \cup (1, \infty)$.

41. $\dfrac{k - 1}{k + 2} > 1$

Solve $\dfrac{k - 1}{k + 2} = 1$.

$$k - 1 = k + 2$$
$$-1 \neq 2$$

The equation has no solution.
Solve $k + 2 = 0$.

$$k = -2$$

Intervals: $(-\infty, -2), \ (-2, \infty)$

For $(-\infty, -2)$, choose -3.

$$\frac{-3 - 1}{-3 + 2} = 4 > 1$$

For $(-2, \infty)$, choose 0.

$$\frac{0 - 1}{0 + 2} = -\frac{1}{2} \not> 1$$

The solution is $(-\infty, -2)$.

42. $\dfrac{a-5}{a+2} < -1$

Solve the equation $\dfrac{a-5}{a+2} = -1$.

$$\frac{a-5}{a+2} = -1$$
$$a-5 = -1(a+2)$$
$$a-5 = -a-2$$
$$2a = 3$$
$$a = \frac{3}{2}$$

Set the denominator equal to zero and solve for a.

$$a+2 = 0$$
$$a = -2$$

Intervals: $(-\infty, -2), \left(-2, \frac{3}{2}\right), \left(\frac{3}{2}, \infty\right)$

For $(-\infty, -2)$, choose -3.

$$\frac{-3-5}{-3+2} = \frac{-8}{-1} = 8 \not< -1$$

For $\left(-2, \frac{3}{2}\right)$, choose 0.

$$\frac{0-5}{0+2} = \frac{-5}{2} = -\frac{5}{2} < -1$$

For $\left(\frac{3}{2}, \infty\right)$, choose 2.

$$\frac{2-5}{2+2} = \frac{-3}{4} = -\frac{3}{4} \not< -1$$

The solution is $\left(-2, \frac{3}{2}\right)$.

43. $\dfrac{2y+3}{y-5} \le 1$

Solve $\dfrac{2y+3}{y-5} = 1$.

$$2y+3 = y-5$$
$$y = -8$$

Solve $y-5 = 0$.

$$y = 5$$

Intervals: $(-\infty, -8), (-8, 5), (5, \infty)$

For $(-\infty, -8)$, choose $y = -10$.

$$\frac{2(-10)+3}{-10-5} = \frac{17}{15} \not\le 1$$

For $(-8, 5)$, choose $y = 0$.

$$\frac{2(0)+3}{0-5} = -\frac{3}{5} \le 1$$

For $(5, \infty)$, choose $y = 6$.

$$\frac{2(6)+3}{6-5} = \frac{15}{1} \not\le 1$$

The solution is $[-8, 5)$.

44. $\dfrac{a+2}{3+2a} \le 5$

For the equation $\dfrac{a+2}{3+2a} = 5$.

$$\frac{a+2}{3+2a} = 5$$
$$a+2 = 5(3+2a)$$
$$a+2 = 15+10a$$
$$-9a = 13$$
$$a = -\frac{13}{9}$$

Set the denominator equal to zero and solve for a.

$$3+2a = 0$$
$$2a = -3$$
$$a = -\frac{3}{2}$$

Intervals: $\left(-\infty, -\frac{3}{2}\right), \left(-\frac{3}{2}, -\frac{13}{9}\right), \left(-\frac{13}{9}, \infty\right)$

For $\left(-\infty, -\frac{3}{2}\right)$, choose -2.

$$\frac{-2+2}{3+2(-2)} = \frac{0}{-1} = 0 \le 5$$

For $\left(-\frac{3}{2}, -\frac{13}{9}\right)$, choose -1.46.

$$\frac{-1.46+2}{3+2(-1.46)} = \frac{0.54}{0.08} = 6.75 \not\le 5$$

For $\left(-\frac{13}{9}, \infty\right)$, choose 0.

$$\frac{0+2}{3+2(0)} = \frac{2}{3} \le 5$$

The value $-\frac{3}{2}$ cannot be included in the solution since it would make the denominator zero. The solution is $\left(-\infty, -\frac{3}{2}\right) \cup \left[-\frac{13}{9}, \infty\right)$.

45. $\dfrac{2k}{k-3} \le \dfrac{4}{k-3}$

Solve $\dfrac{2k}{k-3} = \dfrac{4}{k-3}$.

$$\frac{2k}{k-3} = \frac{4}{k-3}$$

$$\frac{2k}{k-3} - \frac{4}{k-3} = 0$$

$$\frac{2k-4}{k-3} = 0$$

$$2k-4 = 0$$

$$k = 2$$

Set the denominator equal to 0 and solve for k.

$$k-3 = 0$$

$$k = 3$$

Intervals: $(-\infty, 2), (2, 3), (3, \infty)$

For $(-\infty, 2)$, choose 0.

$$\frac{2(0)}{0-3} = 0 \text{ and } \frac{4}{0-3} = -\frac{4}{3}, \text{ so}$$

$$\frac{2(0)}{0-3} \not\le \frac{4}{0-3}.$$

For $(2, 3)$, choose $\frac{5}{2}$.

$$\frac{2\left(\frac{5}{2}\right)}{\frac{5}{2}-3} = \frac{5}{-\frac{1}{2}} = -10 \text{ and } \frac{4}{\frac{5}{2}-3} = \frac{4}{-\frac{1}{2}} = -8, \text{ so}$$

$$\frac{2\left(\frac{5}{2}\right)}{\frac{5}{2}-3} \le \frac{4}{\frac{5}{2}-3}.$$

For $(3, \infty)$, choose 4.

$$\frac{2(4)}{4-3} = 8 \text{ and } \frac{4}{4-3} = 4, \text{ so}$$

$$\frac{2(4)}{4-3} \not\le \frac{4}{4-3}.$$

The solution is $[2, 3)$.

46. $\dfrac{5}{p+1} > \dfrac{12}{p+1}$

Solve the equation $\dfrac{5}{p+1} = \dfrac{12}{p+1}$.

$$\frac{5}{p+1} = \frac{12}{p+1}$$

$$5 = 12$$

The equation has no solution.
Set the denominator equal to zero and solve for p.

$$p+1 = 0$$

$$p = -1$$

Intervals: $(-\infty, -1), (-1, \infty)$

For $(-\infty, -1)$, choose -2.

$$\frac{5}{-2+1} = -5 \text{ and } \frac{12}{-2+1} = -12, \text{ so}$$

$$\frac{5}{-2+1} > \frac{12}{-2+1}.$$

For $(-1, \infty)$, choose 0.

$$\frac{5}{0+1} = 5 \text{ and } \frac{12}{0+1} = 12, \text{ so}$$

$$\frac{5}{0+1} \not> \frac{12}{0+1}.$$

The solution is $(-\infty, -1)$.

47. $\dfrac{2x}{x^2-x-6} \ge 0$

Solve $\dfrac{2x}{x^2-x-6} = 0$.

$$\frac{2x}{x^2-x-6} = 0$$

$$2x = 0$$

$$x = 0$$

Set the denominator equal to 0 and solve for x.

$$x^2-x-6 = 0$$

$$(x+2)(x-3) = 0$$

$$x+2 = 0 \quad \text{or} \quad x-3 = 0$$

$$x = -2 \quad \text{or} \quad x = 3$$

Intervals: $(-\infty, -2), (-2, 0), (0, 3), (3, \infty)$

For $(-\infty, -2)$, choose -3.

$$\frac{2(-3)}{(-3)^2 - (-3) - 6} = -1 \not\geq 0$$

For $(-2, 0)$, choose -1.

$$\frac{2(-1)}{(-1)^2 - (-1) - 6} = \frac{1}{2} \geq 0$$

For $(0, 3)$, choose 2.

$$\frac{2(2)}{2^2 - 2 - 6} = -1 \not\geq 0$$

For $(3, \infty)$, choose 4.

$$\frac{2(4)}{4^2 - 4 - 6} = \frac{4}{3} \geq 0$$

The solution is $(-2, 0] \cup (3, \infty)$.

48. $\dfrac{8}{p^2 + 2p} > 1$

Solve the equation $\dfrac{8}{p^2 + 2p} = 1$.

$$\frac{8}{p^2 + 2p} = 1$$
$$8 = p^2 + 2p$$
$$0 = p^2 + 2p - 8$$
$$0 = (p + 4)(p - 2)$$
$$p + 4 = 0 \quad \text{or} \quad p - 2 = 0$$
$$p = -4 \quad \text{or} \quad p = 2$$

Set the denominator equal to zero and solve for p.

$$p^2 + 2p = 0$$
$$p(p + 2) = 0$$
$$p = 0 \quad \text{or} \quad p + 2 = 0$$
$$p = -2$$

Intervals: $(-\infty, -4), (-4, -2), (-2, 0),$ $(0, 2), (2, \infty)$

For $(-\infty, -4)$, choose -5.

$$\frac{8}{(-5)^2 + 2(-5)} = \frac{8}{15} \not> 1$$

For $(-4, -2)$, choose -3.

$$\frac{8}{(-3)^2 + 2(-3)} = \frac{8}{9 - 6} = \frac{8}{3} > 1$$

For $(-2, 0)$, choose -1.

$$\frac{8}{(-1)^2 + 2(-1)} = \frac{8}{-1} = -8 \not> 1$$

For $(0, 2)$, choose 1.

$$\frac{8}{(1)^2 + 2(1)} = \frac{8}{3} > 1$$

For $(2, \infty)$, choose 3.

$$\frac{8}{(3)^2 + (2)(3)} = \frac{8}{15} \not> 1$$

The solution is $(-4, -2) \cup (0, 2)$.

49. $\dfrac{z^2 + z}{z^2 - 1} \geq 3$

Solve

$$\frac{z^2 + z}{z^2 - 1} = 3.$$
$$z^2 + z = 3z^2 - 3$$
$$-2z^2 + z + 3 = 0$$
$$-1(2z^2 - z - 3) = 0$$
$$-1(z + 1)(2z - 3) = 0$$
$$z = -1 \quad \text{or} \quad z = \frac{3}{2}$$

Set $z^2 - 1 = 0$.

$$z^2 = 1$$
$$z = -1 \quad \text{or} \quad z = 1$$

Intervals: $(-\infty, -1), (-1, 1), \left(1, \frac{3}{2}\right), \left(\frac{3}{2}, \infty\right)$

For $(-\infty, -1)$, choose $x = -2$.

$$\frac{(-2)^2 + 3}{(-2)^2 - 1} = \frac{7}{3} \not\geq 3$$

For $(-1, 1)$, choose $x = 0$.

$$\frac{0^2 + 3}{0^2 - 1} = -3 \not\geq 3$$

For $\left(1, \frac{3}{2}\right)$, choose $x = \frac{3}{2}$.

$$\frac{\left(\frac{3}{2}\right)^2 + 3}{\left(\frac{3}{2}\right)^2 - 1} = \frac{21}{5} \geq 3$$

For $\left(\frac{3}{2}, \infty\right)$, choose $x = 2$.

$$\frac{2^2 + 3}{2^2 - 1} = \frac{7}{3} \not\geq 3$$

The solution is $\left(1, \frac{3}{2}\right]$.

50. $\dfrac{a^2 + 2a}{a^2 - 4} \leq 2$

Solve the equation $\dfrac{a^2 + 2a}{a^2 - 4} = 2$.

$$\dfrac{a^2 + 2a}{a^2 - 4} = 2$$
$$a^2 + 2a = 2(a^2 - 4)$$
$$a^2 + 2a = 2a^2 - 8$$
$$0 = a^2 - 2a - 8$$
$$0 = (a - 4)(a + 2)$$
$$a - 4 = 0 \quad \text{or} \quad a + 2 = 0$$
$$a = 4 \quad \text{or} \quad a = -2$$

But -2 is not a possible solution.
Set the denominator equal to zero and solve for a.

$$a^2 - 4 = 0$$
$$(a + 2)(a - 2) = 0$$
$$a + 2 = 0 \quad \text{or} \quad a - 2 = 0$$
$$a = -2 \quad \text{or} \quad a = 2$$

Intervals: $(-\infty, -2)$, $(-2, 2)$,
$(2, 4)$, $(4, \infty)$

For $(-\infty, -2)$, choose -3.

$$\dfrac{(-3)^2 + 2(-3)}{(-3)^2 - 4} = \dfrac{9 - 6}{9 - 4} = \dfrac{3}{5} \leq 2$$

For $(-2, 2)$, choose 0.

$$\dfrac{(0)^2 + 2(0)}{0 - 4} = \dfrac{0}{-4} = 0 \leq 2$$

For $(2, 4)$, choose 3.

$$\dfrac{(3)^2 + 2(3)}{(3)^2 - 4} = \dfrac{9 + 6}{9 - 5} = \dfrac{15}{4} \not\leq 2$$

For $(4, \infty)$, choose 5.

$$\dfrac{(5)^2 + 2(5)}{(5)^2 - 4} = \dfrac{25 + 10}{25 - 4} = \dfrac{35}{21} \leq 2$$

The value 4 will satisfy the original inequality, but the values -2 and 2 will not since they make the denominator zero. The solution is $(-\infty, -2) \cup (-2, 2) \cup [4, \infty)$.

R.6   Exponents

1. $8^{-2} = \dfrac{1}{8^2} = \dfrac{1}{64}$

2. $3^{-4} = \dfrac{1}{3^4} = \dfrac{1}{81}$

3. $5^0 = 1$, by definition.

4. $\left(-\dfrac{3}{4}\right)^0 = 1$, by definition.

5. $-(-3)^{-2} = -\dfrac{1}{(-3)^2} = -\dfrac{1}{9}$

6. $-(-3^{-2}) = -\left(-\dfrac{1}{3^2}\right) = -\left(-\dfrac{1}{9}\right) = \dfrac{1}{9}$

7. $\left(\dfrac{1}{6}\right)^{-2} = \dfrac{1}{\left(\frac{1}{6}\right)^2} = \dfrac{1}{\frac{1}{36}} = 36$

8. $\left(\dfrac{4}{3}\right)^{-3} = \dfrac{1}{\left(\frac{4}{3}\right)^3} = \dfrac{1}{\frac{64}{27}} = \dfrac{27}{64}$

9. $\dfrac{4^{-2}}{4} = 4^{-2-1} = 4^{-3} = \dfrac{1}{4^3} = \dfrac{1}{64}$

10. $\dfrac{8^9 \cdot 8^{-7}}{8^{-3}} = 8^{9+(-7)-(-3)} = 8^{9-7+3} = 8^5$

11. $\dfrac{10^8 \cdot 10^{-10}}{10^4 \cdot 10^2}$

$$= \dfrac{10^{8+(-10)}}{10^{4+2}} = \dfrac{10^{-2}}{10^6}$$
$$= 10^{-2-6} = 10^{-8}$$
$$= \dfrac{1}{10^8}$$

12. $\left(\dfrac{7^{-12} \cdot 7^3}{7^{-8}}\right)^{-1} = \left(7^{-12+3-(-8)}\right)^{-1}$

$$= \left(7^{-12+3+8}\right)^{-1} = \left(7^{-1}\right)^{-1}$$
$$= 7^{(-1)(-1)} = 7^1 = 7$$

13. $\dfrac{x^4 \cdot x^3}{x^5} = \dfrac{x^{4+3}}{x^5} = \dfrac{x^7}{x^5} = x^{7-5} = x^2$

14. $\dfrac{y^{10} \cdot y^{-4}}{y^6} = y^{10-4-6} = y^0 = 1$

15. $\dfrac{(4k^{-1})^2}{2k^{-5}} = \dfrac{4^2 k^{-2}}{2k^{-5}} = \dfrac{16k^{-2-(-5)}}{2}$

$$= 8k^{-2+5} = 8k^3$$
$$= 2^3 k^3$$

16. $\dfrac{(3z^2)^{-1}}{z^5} = \dfrac{3^{-1}(z^2)^{-1}}{z^5} = \dfrac{3^{-1}z^{2(-1)}}{z^5}$

$\qquad = \dfrac{3^{-1}z^{-2}}{z^5} = 3^{-1}z^{-2-5}$

$\qquad = 3^{-1}z^{-7} = \dfrac{1}{3} \cdot \dfrac{1}{z^7} = \dfrac{1}{3z^7}$

17. $\dfrac{3^{-1} \cdot x \cdot y^2}{x^{-4} \cdot y^5} = 3^{-1} \cdot x^{1-(-4)} \cdot y^{2-5}$

$\qquad = 3^{-1} \cdot x^{1+4} \cdot y^{-3}$

$\qquad = \dfrac{1}{3} \cdot x^5 \cdot \dfrac{1}{y^3}$

$\qquad = \dfrac{x^5}{3y^3}$

18. $\dfrac{5^{-2}m^2y^{-2}}{5^2 m^{-1}y^{-2}} = \dfrac{5^{-2}}{5^2} \cdot \dfrac{m^2}{m^{-1}} \cdot \dfrac{y^{-2}}{y^{-2}}$

$\qquad = 5^{-2-2}m^{2-(-1)}y^{-2-(-2)}$

$\qquad = 5^{-2-2}m^{2+1}y^{-2+2}$

$\qquad = 5^{-4}m^3y^0 = \dfrac{1}{5^4} \cdot m^3 \cdot 1$

$\qquad = \dfrac{m^3}{5^4}$

19. $\left(\dfrac{a^{-1}}{b^2}\right)^{-3} = \dfrac{(a^{-1})^{-3}}{(b^2)^{-3}} = \dfrac{a^{(-1)(-3)}}{b^{2(-3)}}$

$\qquad = \dfrac{a^3}{b^{-6}} = a^3b^6$

20. $\left(\dfrac{c^3}{7d^{-1/2}}\right)^{-2} = \dfrac{(c^3)^{-2}}{7^{-2}(d^{-1/2})^{-2}}$

$\qquad = \dfrac{c^{(3)(-2)}}{7^{-2}d^{(-1/2)(-2)}} = \dfrac{c^{-6}}{7^{-2}d^1}$

$\qquad = \dfrac{7^2}{c^6d} = \dfrac{49}{c^6d}$

21. $\left(\dfrac{x^6y^{-3}}{x^{-2}y^5}\right)^{1/2} = (x^{6-(-2)}y^{-3-5})^{1/2}$

$\qquad = (x^8y^{-8})^{1/2}$

$\qquad = (x^8)^{1/2}(y^{-8})^{1/2}$

$\qquad = x^4y^{-4}$

$\qquad = \dfrac{x^4}{y^4}$

22. $\left(\dfrac{a^{-7}b^{-1}}{b^{-4}a^2}\right)^{1/3} = \left(a^{-7-2}b^{-1-(-4)}\right)^{1/3}$

$\qquad = \left(a^{-9}b^3\right)^{1/3}$

$\qquad = \left(a^{-9}\right)^{1/3}\left(b^3\right)^{1/3}$

$\qquad = a^{-3}b^1$

$\qquad = \dfrac{b}{a^3}$

23. $a^{-1} + b^{-1} = \dfrac{1}{a} + \dfrac{1}{b}$

$\qquad = \left(\dfrac{b}{b}\right)\left(\dfrac{1}{a}\right) + \left(\dfrac{a}{a}\right)\left(\dfrac{1}{b}\right)$

$\qquad = \dfrac{b}{ab} + \dfrac{a}{ab}$

$\qquad = \dfrac{b+a}{ab}$

$\qquad = \dfrac{a+b}{ab}$

24. $b^{-2} - a = \dfrac{1}{b^2} - a$

$\qquad = \dfrac{1}{b^2} - a\left(\dfrac{b^2}{b^2}\right)$

$\qquad = \dfrac{1}{b^2} - \dfrac{ab^2}{b^2}$

$\qquad = \dfrac{1-ab^2}{b^2}$

25. $\dfrac{2n^{-1} - 2m^{-1}}{m+n^2} = \dfrac{\frac{2}{n} - \frac{2}{m}}{m+n^2}$

$\qquad = \dfrac{\frac{2}{n} \cdot \frac{m}{m} - \frac{2}{m} \cdot \frac{n}{n}}{(m+n^2)}$

$\qquad = \dfrac{2m-2n}{mn(m+n^2)} \quad \text{or} \quad \dfrac{2(m-n)}{mn(m+n^2)}$

26. $\left(\dfrac{m}{3}\right)^{-1} + \left(\dfrac{n}{2}\right)^{-2} = \left(\dfrac{3}{m}\right)^1 + \left(\dfrac{2}{n}\right)^2$

$\qquad = \dfrac{3}{m} + \dfrac{4}{n^2}$

$\qquad = \left(\dfrac{3}{m}\right)\left(\dfrac{n^2}{n^2}\right) + \left(\dfrac{4}{n^2}\right)\left(\dfrac{m}{m}\right)$

$\qquad = \dfrac{3n^2}{mn^2} + \dfrac{4m}{mn^2}$

$\qquad = \dfrac{3n^2+4m}{mn^2}$

27. $(x^{-1} - y^{-1})^{-1} = \dfrac{1}{\frac{1}{x} - \frac{1}{y}}$

$\qquad = \dfrac{1}{\frac{1}{x} \cdot \frac{y}{y} - \frac{1}{y} \cdot \frac{x}{x}}$

$\qquad = \dfrac{1}{\frac{y}{xy} - \frac{x}{xy}}$

$\qquad = \dfrac{1}{\frac{y-x}{xy}}$

$\qquad = \dfrac{xy}{y-x}$

28. $\left(x \cdot y^{-1} - y^{-2}\right)^{-2} = \left(\dfrac{x}{y} - \dfrac{1}{y^2}\right)^{-2}$

$$= \left[\left(\dfrac{x}{y}\right)\left(\dfrac{y}{y}\right) - \dfrac{1}{y^2}\right]^{-2}$$

$$= \left(\dfrac{xy}{y^2} - \dfrac{1}{y^2}\right)^{-2}$$

$$= \left(\dfrac{xy - 1}{y^2}\right)^{-2}$$

$$= \left(\dfrac{y^2}{xy - 1}\right)^{2}$$

$$= \dfrac{(y^2)^2}{(xy - 1)^2}$$

$$= \dfrac{y^4}{(xy - 1)^2}$$

29. $121^{1/2} = (11^2)^{1/2} = 11^{2(1/2)} = 11^1 = 11$

30. $27^{1/3} = \sqrt[3]{27} = 3$

31. $32^{2/5} = (32^{1/5})^2 = 2^2 = 4$

32. $-125^{2/3} = -(125^{1/3})^2 = -5^2 = -25$

33. $\left(\dfrac{36}{144}\right)^{1/2} = \dfrac{36^{1/2}}{144^{1/2}} = \dfrac{6}{12} = \dfrac{1}{2}$

This can also be solved by reducing the fraction first.

$$\left(\dfrac{36}{144}\right)^{1/2} = \left(\dfrac{1}{4}\right)^{1/2} = \dfrac{1^{1/2}}{4^{1/2}} = \dfrac{1}{2}$$

34. $\left(\dfrac{64}{27}\right)^{1/3} = \dfrac{64^{1/3}}{27^{1/3}} = \dfrac{4}{3}$

35. $8^{-4/3} = (8^{1/3})^{-4} = 2^{-4} = \dfrac{1}{2^4} = \dfrac{1}{16}$

36. $625^{-1/4} = \dfrac{1}{625^{1/4}} = \dfrac{1}{5}$

37. $\left(\dfrac{27}{64}\right)^{-1/3} = \dfrac{27^{-1/3}}{64^{-1/3}} = \dfrac{64^{1/3}}{27^{1/3}} = \dfrac{4}{3}$

38. $\left(\dfrac{121}{100}\right)^{-3/2} = \dfrac{1}{\left(\frac{121}{100}\right)^{3/2}} = \dfrac{1}{\left[\left(\frac{121}{100}\right)^{1/2}\right]^3}$

$$= \dfrac{1}{\left(\frac{11}{10}\right)^3} = \dfrac{1}{\frac{1331}{1000}} = \dfrac{1000}{1331}$$

39. $3^{2/3} \cdot 3^{4/3} = 3^{(2/3)+(4/3)} = 3^{6/3} = 3^2 = 9$

40. $27^{2/3} \cdot 27^{-1/3} = 27^{(2/3)+(-1/3)}$

$$= 27^{2/3 - 1/3}$$

$$= 27^{1/3}$$

41. $\dfrac{4^{9/4} \cdot 4^{-7/4}}{4^{-10/4}} = 4^{9/4 - 7/4 - (-10/4)}$

$$= 4^{12/4} = 4^3 = 64$$

42. $\dfrac{3^{-5/2} \cdot 3^{3/2}}{3^{7/2} \cdot 3^{-9/2}}$

$$= 3^{(-5/2)+(3/2)-(7/2)-(-9/2)}$$

$$= 3^{-5/2 + 3/2 - 7/2 + 9/2}$$

$$= 3^0 = 1$$

43. $\dfrac{7^{-1/3} \cdot 7r^{-3}}{7^{2/3} \cdot (r^{-2})^2}$

$$= \dfrac{7^{-1/3+1}r^{-3}}{7^{2/3} \cdot r^{-4}}$$

$$= 7^{-1/3+3/3-2/3}r^{-3-(-4)}$$

$$= 7^0 r^{-3+4} = 1 \cdot r^1 = r$$

44. $\dfrac{12^{3/4} \cdot 12^{5/4} \cdot y^{-2}}{12^{-1} \cdot (y^{-3})^{-2}}$

$$= \dfrac{12^{3/4+5/4} \cdot y^{-2}}{12^{-1} \cdot y^{(-3)(-2)}} = \dfrac{12^{8/4} \cdot y^{-2}}{12^{-1} \cdot y^6}$$

$$= \dfrac{12^2 \cdot y^{-2}}{12^{-1}y^6}$$

$$= 12^{2-(-1)} \cdot y^{-2-6} = 12^3 y^{-8}$$

$$= \dfrac{12^3}{y^8}$$

45. $\dfrac{3k^2 \cdot (4k^{-3})^{-1}}{4^{1/2} \cdot k^{7/2}}$

$$= \dfrac{3k^2 \cdot 4^{-1}k^3}{2 \cdot k^{7/2}}$$

$$= 3 \cdot 2^{-1} \cdot 4^{-1}k^{2+3-(7/2)}$$

$$= \dfrac{3}{8} \cdot k^{3/2}$$

$$= \dfrac{3k^{3/2}}{8}$$

46. $\dfrac{8p^{-3}(4p^2)^{-2}}{p^{-5}} = \dfrac{8p^{-3} \cdot 4^{-2}p^{(2)(-2)}}{p^{-5}}$

$\qquad\qquad = \dfrac{8p^{-3}4^{-2}p^{-4}}{p^{-5}}$

$\qquad\qquad = 8 \cdot 4^{-2}p^{(-3)+(-4)-(-5)}$

$\qquad\qquad = 8 \cdot 4^{-2}p^{-3-4+5}$

$\qquad\qquad = 8 \cdot 4^{-2}p^{-2}$

$\qquad\qquad = 8 \cdot \dfrac{1}{4^2} \cdot \dfrac{1}{p^2}$

$\qquad\qquad = 8 \cdot \dfrac{1}{16} \cdot \dfrac{1}{p^2}$

$\qquad\qquad = \dfrac{8}{16p^2} = \dfrac{1}{2p^2}$

47. $\dfrac{a^{4/3}}{a^{2/3}} \cdot \dfrac{b^{1/2}}{b^{-3/2}} = a^{4/3-2/3}b^{1/2-(-3/2)}$

$\qquad\qquad\qquad = a^{2/3}b^2$

48. $\dfrac{x^{3/2} \cdot y^{4/5} \cdot z^{-3/4}}{x^{5/3} \cdot y^{-6/5} \cdot z^{1/2}}$

$\qquad = x^{3/2-(5/3)} \cdot y^{4/5-(-6/5)} \cdot z^{-3/4-(1/2)}$

$\qquad = x^{-1/6} \cdot y^2 \cdot z^{-5/4}$

$\qquad = \dfrac{y^2}{x^{1/6}z^{5/4}}$

49. $\dfrac{k^{-3/5} \cdot h^{-1/3} \cdot t^{2/5}}{k^{-1/5} \cdot h^{-2/3} \cdot t^{1/5}}$

$\qquad = k^{-3/5-(-1/5)}h^{-1/3-(-2/3)}t^{2/5-1/5}$

$\qquad = k^{-3/5+1/5}h^{-1/3+2/3}t^{2/5-1/5}$

$\qquad = k^{-2/5}h^{1/3}t^{1/5}$

$\qquad = \dfrac{h^{1/3}t^{1/5}}{k^{2/5}}$

50. $\dfrac{m^{7/3} \cdot n^{-2/5} \cdot p^{3/8}}{m^{-2/3} \cdot n^{3/5} \cdot p^{-5/8}}$

$\qquad = m^{7/3-(-2/3)}n^{-2/5-(3/5)}p^{3/8-(-5/8)}$

$\qquad = m^{7/3+2/3}n^{-2/5-3/5}p^{3/8+5/8}$

$\qquad = m^{9/3}n^{-5/5}p^{8/8}$

$\qquad = m^3n^{-1}p^1$

$\qquad = \dfrac{m^3p}{n}$

51. $3x^3(x^2+3x)^2 - 15x(x^2+3x)^2$

$\qquad = 3x \cdot x^2(x^2+3x)^2 - 3x \cdot 5(x^2+3x)^2$

$\qquad = 3x(x^2+3x)^2(x^2-5)$

52. $6x(x^3+7)^2 - 6x^2(3x^2+5)(x^3+7)$

$\qquad = 6x(x^3+7)(x^3+7) - 6x(x)(3x^2+5)(x^3+7)$

$\qquad = 6x(x^3+7)[(x^3+7) - x(3x^2+5)]$

$\qquad = 6x(x^3+7)(x^3+7-3x^3-5x)$

$\qquad = 6x(x^3+7)(-2x^3-5x+7)$

53. $10x^3(x^2-1)^{-1/2} - 5x(x^2-1)^{1/2}$

$\qquad = 5x \cdot 2x^2(x^2-1)^{-1/2} - 5x(x^2-1)^{-1/2}(x^2-1)^1$

$\qquad = 5x(x^2-1)^{-1/2}[2x^2 - (x^2-1)]$

$\qquad = 5x(x^2-1)^{-1/2}(x^2+1)$

54. $9(6x+2)^{1/2} + 3(9x-1)(6x+2)^{-1/2}$

$\qquad = 3 \cdot 3(6x+2)^{-1/2}(6x+2)^1$

$\qquad\quad + 3(9x-1)(6x+2)^{-1/2}$

$\qquad = 3(6x+2)^{-1/2}[3(6x+2) + (9x-1)]$

$\qquad = 3(6x+2)^{-1/2}(18x+6+9x-1)$

$\qquad = 3(6x+2)^{-1/2}(27x+5)$

55. $x(2x+5)^2(x^2-4)^{-1/2} + 2(x^2-4)^{1/2}(2x+5)$

$\qquad = (2x+5)^2(x^2-4)^{-1/2}(x)$

$\qquad\quad + (x^2-4)^1(x^2-4)^{-1/2}(2)(2x+5)$

$\qquad = (2x+5)(x^2-4)^{-1/2}$

$\qquad\quad \cdot [(2x+5)(x) + (x^2-4)(2)]$

$\qquad = (2x+5)(x^2-4)^{-1/2}$

$\qquad\quad \cdot (2x^2+5x+2x^2-8)$

$\qquad = (2x+5)(x^2-4)^{-1/2}(4x^2+5x-8)$

56. $(4x^2+1)^2(2x-1)^{-1/2} + 16x(4x^2+1)(2x-1)^{1/2}$

$\qquad = (4x^2+1)(4x^2+1)(2x-1)^{-1/2}$

$\qquad\quad + 16x(4x^2+1)(2x-1)^{-1/2}(2x-1)$

$\qquad = (4x^2+1)(2x-1)^{-1/2}$

$\qquad\quad \cdot [(4x^2+1) + 16x(2x-1)]$

$\qquad = (4x^2+1)(2x-1)^{-1/2}(4x^2+1+32x^2-16x)$

$\qquad = (4x^2+1)(2x-1)^{-1/2}(36x^2-16x+1)$

R.7   Radicals

1. $\sqrt[3]{125} = 5$ because $5^3 = 125$.

2. $\sqrt[4]{1296} = \sqrt[4]{6^4} = 6$

3. $\sqrt[5]{-3125} = -5$ because $(-5)^5 = -3125$.

4. $\sqrt{50} = \sqrt{25 \cdot 2} = \sqrt{25}\sqrt{2} = 5\sqrt{2}$

5. $\sqrt{2000} = \sqrt{4 \cdot 100 \cdot 5}$
$= 2 \cdot 10\sqrt{5}$
$= 20\sqrt{5}$

6. $\sqrt{32y^5} = \sqrt{(16y^4)(2y)}$
$= \sqrt{16y^4}\sqrt{2y}$
$= 4y^2\sqrt{2y}$

7. $\sqrt{27} \cdot \sqrt{3} = \sqrt{27 \cdot 3} = \sqrt{81} = 9$

8. $\sqrt{2} \cdot \sqrt{32} = \sqrt{2 \cdot 32} = \sqrt{64} = 8$

9. $7\sqrt{2} - 8\sqrt{18} + 4\sqrt{72}$
$= 7\sqrt{2} - 8\sqrt{9 \cdot 2} + 4\sqrt{36 \cdot 2}$
$= 7\sqrt{2} - 8(3)\sqrt{2} + 4(6)\sqrt{2}$
$= 7\sqrt{2} - 24\sqrt{2} + 24\sqrt{2}$
$= 7\sqrt{2}$

10. $4\sqrt{3} - 5\sqrt{12} + 3\sqrt{75}$
$= 4\sqrt{3} - 5(\sqrt{4}\sqrt{3}) + 3(\sqrt{25}\sqrt{3})$
$= 4\sqrt{3} - 5(2\sqrt{3}) + 3(5\sqrt{3})$
$= 4\sqrt{3} - 10\sqrt{3} + 15\sqrt{3}$
$= (4 - 10 + 15)\sqrt{3} = 9\sqrt{3}$

11. $4\sqrt{7} - \sqrt{28} + \sqrt{343}$
$= 4\sqrt{7} - \sqrt{4}\sqrt{7} + \sqrt{49}\sqrt{7}$
$= 4\sqrt{7} - 2\sqrt{7} + 7\sqrt{7}$
$= (4 - 2 + 7)\sqrt{7}$
$= 9\sqrt{7}$

12. $3\sqrt{28} - 4\sqrt{63} + \sqrt{112}$
$= 3(\sqrt{4}\sqrt{7}) - 4(\sqrt{9}\sqrt{7}) + (\sqrt{16}\sqrt{7})$
$= 3(2\sqrt{7}) - 4(3\sqrt{7}) + (4\sqrt{7})$
$= 6\sqrt{7} - 12\sqrt{7} + 4\sqrt{7}$
$= (6 - 12 + 4)\sqrt{7}$
$= -2\sqrt{7}$

13. $\sqrt[3]{2} - \sqrt[3]{16} + 2\sqrt[3]{54}$
$= \sqrt[3]{2} - (\sqrt[3]{8 \cdot 2}) + 2(\sqrt[3]{27 \cdot 2})$
$= \sqrt[3]{2} - \sqrt[3]{8}\sqrt[3]{2} + 2(\sqrt[3]{27}\sqrt[3]{2})$
$= \sqrt[3]{2} - 2\sqrt[3]{2} + 2(3\sqrt[3]{2})$
$= \sqrt[3]{2} - 2\sqrt[3]{2} + 6\sqrt[3]{2}$
$= 5\sqrt[3]{2}$

14. $2\sqrt[3]{5} - 4\sqrt[3]{40} + 3\sqrt[3]{135}$
$= 2\sqrt[3]{5} - 4\sqrt[3]{8 \cdot 5} + 3\sqrt[3]{27 \cdot 5}$
$= 2\sqrt[3]{5} - 4(2)\sqrt[3]{5} + 3(3)\sqrt[3]{5}$
$= 2\sqrt[3]{5} - 8\sqrt[3]{5} + 9\sqrt[3]{5}$
$= 3\sqrt[3]{5}$

15. $\sqrt{2x^3y^2z^4} = \sqrt{x^2y^2z^4 \cdot 2x}$
$= xyz^2\sqrt{2x}$

16. $\sqrt{160r^7s^9t^{12}}$
$= \sqrt{(16 \cdot 10)(r^6 \cdot r)(s^8 \cdot s)(t^{12})}$
$= \sqrt{(16r^6s^8t^{12})(10rs)}$
$= \sqrt{16r^6s^8t^{12}}\sqrt{10rs}$
$= 4r^3s^4t^6\sqrt{10rs}$

17. $\sqrt[3]{128x^3y^8z^9} = \sqrt[3]{64x^3y^6z^9 \cdot 2y^2}$
$= \sqrt[3]{64x^3y^6z^9}\sqrt[3]{2y^2}$
$= 4xy^2z^3\sqrt[3]{2y^2}$

18. $\sqrt[4]{x^8y^7z^{11}} = \sqrt[4]{(x^8)(y^4 \cdot y^3)(z^8z^3)}$
$= \sqrt[4]{(x^8y^4z^8)(y^3z^3)}$
$= \sqrt[4]{x^8y^4z^8}\sqrt[4]{y^3z^3}$
$= x^2yz^2\sqrt[4]{y^3z^3}$

19. $\sqrt{a^3b^5} - 2\sqrt{a^7b^3} + \sqrt{a^3b^9}$
$= \sqrt{a^2b^4ab} - 2\sqrt{a^6b^2ab} + \sqrt{a^2b^8ab}$
$= ab^2\sqrt{ab} - 2a^3b\sqrt{ab} + ab^4\sqrt{ab}$
$= (ab^2 - 2a^3b + ab^4)\sqrt{ab}$
$= ab\sqrt{ab}(b - 2a^2 + b^3)$

20. $\sqrt{p^7q^3} - \sqrt{p^5q^9} + \sqrt{p^9q}$
$= \sqrt{(p^6p)(q^2q)} - \sqrt{(p^4p)(q^8q)}$
$\quad + \sqrt{(p^8p)q}$
$= \sqrt{(p^6q^2)(pq)} - \sqrt{(p^4q^8)(pq)}$
$\quad + \sqrt{(p^8)(pq)}$
$= \sqrt{p^6q^2}\sqrt{pq} - \sqrt{p^4q^8}\sqrt{pq} + \sqrt{p^8}\sqrt{pq}$
$= p^3q\sqrt{pq} - p^2q^4\sqrt{pq} + p^4\sqrt{pq}$
$= p^2pq\sqrt{pq} - p^2q^4\sqrt{pq} + p^2p^2\sqrt{pq}$
$= p^2\sqrt{pq}(pq - q^4 + p^2)$

21. $\sqrt{a} \cdot \sqrt[3]{a} = a^{1/2} \cdot a^{1/3} = a^{1/2+(1/3)} = a^{5/6} = \sqrt[6]{a^5}$

22. $\sqrt{b^3} \cdot \sqrt[4]{b^3} = b^{3/2} \cdot b^{3/4}$
$= b^{3/2+(3/4)} = b^{9/4}$
$= \sqrt[4]{b^9} = \sqrt[4]{b^8 \cdot b}$
$= \sqrt[4]{b^8}\sqrt[4]{b} = b^2\sqrt[4]{b}$

23. $\dfrac{5}{\sqrt{7}} = \dfrac{5}{\sqrt{7}} \cdot \dfrac{\sqrt{7}}{\sqrt{7}} = \dfrac{5\sqrt{7}}{7}$

24. $\dfrac{5}{\sqrt{10}} = \dfrac{5}{\sqrt{10}} \cdot \dfrac{\sqrt{10}}{\sqrt{10}} = \dfrac{5\sqrt{10}}{\sqrt{100}} = \dfrac{5\sqrt{10}}{10} = \dfrac{\sqrt{10}}{2}$

25. $\dfrac{-3}{\sqrt{12}} = \dfrac{-3}{\sqrt{4\cdot 3}}$

$\qquad = \dfrac{-3}{2\sqrt{3}} \cdot \dfrac{\sqrt{3}}{\sqrt{3}}$

$\qquad = \dfrac{-3\sqrt{3}}{6}$

$\qquad = -\dfrac{\sqrt{3}}{2}$

26. $\dfrac{4}{\sqrt{8}} = \dfrac{4}{\sqrt{8}} \cdot \dfrac{\sqrt{2}}{\sqrt{2}} = \dfrac{4\sqrt{2}}{\sqrt{16}} = \dfrac{4\sqrt{2}}{4} = \sqrt{2}$

27. $\dfrac{3}{1-\sqrt{2}} = \dfrac{3}{1-\sqrt{2}} \cdot \dfrac{1+\sqrt{2}}{1+\sqrt{2}}$

$\qquad = \dfrac{3(1+\sqrt{2})}{1-2}$

$\qquad = \dfrac{-3(1+\sqrt{2})}{4}$

28. $\dfrac{5}{2-\sqrt{6}} = \dfrac{5}{2-\sqrt{6}} \cdot \dfrac{2+\sqrt{6}}{2+\sqrt{6}}$

$\qquad = \dfrac{5(2+\sqrt{6})}{4+2\sqrt{6}-2\sqrt{6}-\sqrt{36}}$

$\qquad = \dfrac{5(2+\sqrt{6})}{4-\sqrt{36}}$

$\qquad = \dfrac{5(2+\sqrt{6})}{4-6}$

$\qquad = \dfrac{5(2+\sqrt{6})}{-2}$

$\qquad = -\dfrac{5(2+\sqrt{6})}{2}$

29. $\dfrac{6}{2+\sqrt{2}} = \dfrac{6}{2+\sqrt{2}} \cdot \dfrac{2-\sqrt{2}}{2-\sqrt{2}}$

$\qquad = \dfrac{6(2-\sqrt{2})}{4-2\sqrt{2}+2\sqrt{2}-\sqrt{4}}$

$\qquad = \dfrac{6(2-\sqrt{2})}{4-2}$

$\qquad = \dfrac{6(2-\sqrt{2})}{2}$

$\qquad = 3(2-\sqrt{2})$

30. $\dfrac{\sqrt{5}}{\sqrt{5}+\sqrt{2}} = \dfrac{\sqrt{5}}{\sqrt{5}+\sqrt{2}} \cdot \dfrac{\sqrt{5}-\sqrt{2}}{\sqrt{5}-\sqrt{2}}$

$\qquad = \dfrac{\sqrt{5}(\sqrt{5}-\sqrt{2})}{\sqrt{25}-\sqrt{10}+\sqrt{10}-\sqrt{4}}$

$\qquad = \dfrac{5-\sqrt{10}}{5-2}$

$\qquad = \dfrac{5-\sqrt{10}}{3}$

31. $\dfrac{1}{\sqrt{r}-\sqrt{3}} = \dfrac{1}{\sqrt{r}-\sqrt{3}} \cdot \dfrac{\sqrt{r}+\sqrt{3}}{\sqrt{r}+\sqrt{3}}$

$\qquad = \dfrac{\sqrt{r}+\sqrt{3}}{r-3}$

32. $\dfrac{5}{\sqrt{m}-\sqrt{5}} = \dfrac{5}{\sqrt{m}-\sqrt{5}} \cdot \dfrac{\sqrt{m}+\sqrt{5}}{\sqrt{m}+\sqrt{5}}$

$\qquad = \dfrac{5(\sqrt{m}+\sqrt{5})}{\sqrt{m^2}+\sqrt{5m}-\sqrt{5m}-\sqrt{25}}$

$\qquad = \dfrac{5(\sqrt{m}+\sqrt{5})}{\sqrt{m^2}-\sqrt{25}} = \dfrac{5(\sqrt{m}+\sqrt{5})}{m-5}$

33. $\dfrac{y-5}{\sqrt{y}-\sqrt{5}} = \dfrac{y-5}{\sqrt{y}-\sqrt{5}} \cdot \dfrac{\sqrt{y}+\sqrt{5}}{\sqrt{y}+\sqrt{5}}$

$\qquad = \dfrac{(y-5)(\sqrt{y}+\sqrt{5})}{y-5}$

$\qquad = \sqrt{y}+\sqrt{5}$

34. $\dfrac{\sqrt{z}-1}{\sqrt{z}-\sqrt{5}} = \dfrac{\sqrt{z}-1}{\sqrt{z}-\sqrt{5}} \cdot \dfrac{\sqrt{z}+\sqrt{5}}{\sqrt{z}+\sqrt{5}}$

$\qquad = \dfrac{\sqrt{z^2}+\sqrt{5z}-\sqrt{z}-\sqrt{5}}{\sqrt{z^2}+\sqrt{5z}-\sqrt{5z}-\sqrt{25}}$

$\qquad = \dfrac{z+\sqrt{5z}-\sqrt{z}-\sqrt{5}}{z-5}$

35. $\dfrac{\sqrt{x}+\sqrt{x+1}}{\sqrt{x}-\sqrt{x+1}} = \dfrac{\sqrt{x}+\sqrt{x+1}}{\sqrt{x}-\sqrt{x+1}} \cdot \dfrac{\sqrt{x}+\sqrt{x+1}}{\sqrt{x}+\sqrt{x+1}}$

$\qquad = \dfrac{x+2\sqrt{x(x+1)}+(x+1)}{x-(x+1)}$

$\qquad = \dfrac{2x+2\sqrt{x(x+1)}+1}{-1}$

$\qquad = -2x-2\sqrt{x(x+1)}-1$

36. $\dfrac{\sqrt{p}+\sqrt{p^2-1}}{\sqrt{p}-\sqrt{p^2-1}}$

$\qquad = \dfrac{\sqrt{p}+\sqrt{p^2-1}}{\sqrt{p}-\sqrt{p^2-1}} \cdot \dfrac{\sqrt{p}+\sqrt{p^2-1}}{\sqrt{p}+\sqrt{p^2-1}}$

$\qquad = \dfrac{(\sqrt{p})^2+2\sqrt{p}\sqrt{p^2-1}+(\sqrt{p^2-1})^2}{\sqrt{p^2}+\sqrt{p}\sqrt{p^2-1}-\sqrt{p}\sqrt{p^2-1}-(\sqrt{p^2-1})^2}$

$\qquad = \dfrac{p+2\sqrt{p}\sqrt{p^2-1}+(p^2-1)}{p-(p^2-1)}$

$\qquad = \dfrac{p^2+p+2\sqrt{p(p^2-1)}-1}{-p^2+p+1}$

37. $\dfrac{1+\sqrt{2}}{2} = \dfrac{(1+\sqrt{2})(1-\sqrt{2})}{2(1-\sqrt{2})}$

$\phantom{\dfrac{1+\sqrt{2}}{2}} = \dfrac{1-2}{2(1-\sqrt{2})}$

$\phantom{\dfrac{1+\sqrt{2}}{2}} = -\dfrac{1}{2(1-\sqrt{2})}$

38. $\dfrac{3-\sqrt{3}}{6} = \dfrac{3-\sqrt{3}}{6} \cdot \dfrac{3+\sqrt{3}}{3+\sqrt{3}}$

$\phantom{\dfrac{3-\sqrt{3}}{6}} = \dfrac{9+3\sqrt{3}-3\sqrt{3}-\sqrt{9}}{6(3+\sqrt{3})}$

$\phantom{\dfrac{3-\sqrt{3}}{6}} = \dfrac{9-3}{6(3+\sqrt{3})}$

$\phantom{\dfrac{3-\sqrt{3}}{6}} = \dfrac{6}{6(3+\sqrt{3})}$

$\phantom{\dfrac{3-\sqrt{3}}{6}} = \dfrac{1}{3+\sqrt{3}}$

39. $\dfrac{\sqrt{x}+\sqrt{x+1}}{\sqrt{x}-\sqrt{x+1}}$

$= \dfrac{\sqrt{x}+\sqrt{x+1}}{\sqrt{x}-\sqrt{x+1}} \cdot \dfrac{\sqrt{x}-\sqrt{x+1}}{\sqrt{x}-\sqrt{x+1}}$

$= \dfrac{x-(x+1)}{x-2\sqrt{x}\cdot\sqrt{x+1}+(x+1)}$

$= \dfrac{-1}{2x-2\sqrt{x(x+1)}+1}$

40. $\dfrac{\sqrt{p}-\sqrt{p-2}}{\sqrt{p}} = \dfrac{\sqrt{p}-\sqrt{p-2}}{\sqrt{p}} \cdot \dfrac{\sqrt{p}+\sqrt{p-2}}{\sqrt{p}+\sqrt{p-2}}$

$= \dfrac{\sqrt{p^2}+\sqrt{p}\sqrt{p-2}-\sqrt{p}\sqrt{p-2}-\sqrt{(p-2)^2}}{\sqrt{p^2}+\sqrt{p}\sqrt{p-2}}$

$= \dfrac{p-(p-2)}{p+\sqrt{p(p-2)}}$

$= \dfrac{2}{p+\sqrt{p(p-2)}}$

41. $\sqrt{16-8x+x^2}$

$= \sqrt{(4-x)^2}$

$= |4-x|$

Since $\sqrt{}$ denotes the nonnegative root, we must have $4-x \geq 0$.

42. $\sqrt{9y^2+30y+25} = \sqrt{(3y+5)^2} = |3y+5|$

Since $\sqrt{}$ denotes the nonnegative root, we must have $3y+5 \geq 0$.

43. $\sqrt{4-25z^2} = \sqrt{(2+5z)(2-5z)}$

This factorization does not produce a perfect square, so the expression $\sqrt{4-25z^2}$ cannot be simplified.

44. $\sqrt{9k^2+h^2}$

The expression $9k^2+h^2$ is the sum of two squares and cannot be factored. Therefore, $\sqrt{9k^2+h^2}$ cannot be simplified.

Chapter 1

LINEAR FUNCTIONS

1.1 Slopes and Equations of Lines

1. Find the slope of the line through $(4, 5)$ and $(-1, 2)$.

$$m = \frac{5 - 2}{4 - (-1)}$$

$$= \frac{3}{5}$$

3. Find the slope of the line through $(8, 4)$ and $(8, -7)$.

$$m = \frac{4 - (-7)}{8 - 8}$$

$$= \frac{11}{0}$$

The slope is undefined; the line is vertical.

5. $y = x$

Using the slope-intercept form, $y = mx + b$, we see that the slope is 1.

7. $5x - 9y = 11$

Rewrite the equation in slope-intercept form.

$$9y = 5x - 11$$

$$y = \frac{5}{9}x - \frac{11}{9}$$

The slope is $\frac{5}{9}$.

9. $x = 5$

This is a vertical line. The slope is undefined.

11. $y = 8$

This is a horizontal line, which has a slope of 0.

13. Find the slope of a line parallel to $6x - 3y = 12$.

Rewrite the equation in slope-intercept form.

$$-3y = -6x + 12$$

$$y = 2x - 4$$

The slope is 2, so a parallel line will also have slope 2.

15. The line goes through $(1, 3)$, with slope $m = -2$. Use point-slope form.

$$y - 3 = -2(x - 1)$$

$$y = -2x + 2 + 3$$

$$y = -2x + 5$$

17. The line goes through $(-5, -7)$ with slope $m = 0$. Use point-slope form.

$$y - (-7) = 0[x - (-5)]$$

$$y + 7 = 0$$

$$y = -7$$

19. The line goes through $(4, 2)$ and $(1, 3)$. Find the slope, then use point-slope form with either of the two given points.

$$m = \frac{3 - 2}{1 - 4}$$

$$= -\frac{1}{3}$$

$$y - 3 = -\frac{1}{3}(x - 1)$$

$$y = -\frac{1}{3}x + \frac{1}{3} + 3$$

$$y = -\frac{1}{3}x + \frac{10}{3}$$

21. The line goes through $\left(\frac{2}{3}, \frac{1}{2}\right)$ and $\left(\frac{1}{4}, -2\right)$.

$$m = \frac{-2 - \frac{1}{2}}{\frac{1}{4} - \frac{2}{3}} = \frac{-\frac{4}{2} - \frac{1}{2}}{\frac{3}{12} - \frac{8}{12}}$$

$$m = \frac{-\frac{5}{2}}{-\frac{5}{12}} = \frac{60}{10} = 6$$

$$y - (-2) = 6\left(x - \frac{1}{4}\right)$$

$$y + 2 = 6x - \frac{3}{2}$$

$$y = 6x - \frac{3}{2} - 2$$

$$y = 6x - \frac{3}{2} - \frac{4}{2}$$

$$y = 6x - \frac{7}{2}$$

23. The line goes through $(-8, 4)$ and $(-8, 6)$.

$$m = \frac{4 - 6}{-8 - (-8)} = \frac{-2}{0};$$

which is undefined.

This is a vertical line; the value of x is always -8. The equation of this line is $x = -8$.

25. The line has x-intercept -6 and y-intercept -3. Two points on the line are $(-6, 0)$ and $(0, -3)$. Find the slope; then use slope-intercept form.

$$m = \frac{-3 - 0}{0 - (-6)} = \frac{-3}{6} = -\frac{1}{2}$$

$$b = -3$$

$$y = -\frac{1}{2}x - 3$$

27. The vertical line through $(-6, 5)$ goes through the point $(-6, 0)$, so the equation is $x = -6$.

29. Write an equation of the line through $(-4, 6)$, parallel to $3x + 2y = 13$.

Rewrite the equation of the given line in slope-intercept form.

$$3x + 2y = 13$$

$$2y = -3x + 13$$

$$y = -\frac{3}{2}x + \frac{13}{2}$$

The slope is $-\frac{3}{2}$.

Use $m = -\frac{3}{2}$ and the point $(-4, 6)$ in the point-slope form.

$$y - 6 = -\frac{3}{2}[x - (-4)]$$

$$y = -\frac{3}{2}(x + 4) + 6$$

$$y = -\frac{3}{2}x - 6 + 6$$

$$y = -\frac{3}{2}x$$

31. Write an equation of the line through $(3, -4)$, perpendicular to $x + y = 4$.

Rewrite the equation of the given line as

$$y = -x + 4.$$

The slope of this line is -1. To find the slope of a perpendicular line, solve

$$-1m = -1.$$

$$m = 1$$

Use $m = 1$ and $(3, -4)$ in the point-slope form.

$$y - (-4) = 1(x - 3)$$

$$y = x - 3 - 4$$

$$y = x - 7$$

33. Write an equation of the line with y-intercept 4, perpendicular to $x + 5y = 7$.

Find the slope of the given line.

$$x + 5y = 7$$

$$5y = -x + 7$$

$$y = -\frac{1}{5}x + \frac{7}{5}$$

The slope is $-\frac{1}{5}$, so the slope of the perpendicular line will be 5. If the y-intercept is 4, then using the slope-intercept form we have

$$y = mx + b$$

$$y = 5x + 4.$$

35. Do the points $(4, 3), (2, 0)$, and $(-18, -12)$ lie on the same line?

Find the slope between $(4, 3)$ and $(2, 0)$.

$$m = \frac{0 - 3}{2 - 4} = \frac{-3}{-2} = \frac{3}{2}$$

Find the slope between $(4, 3)$ and $(-18, -12)$.

$$m = \frac{-12 - 3}{-18 - 4} = \frac{-15}{-22} = \frac{15}{22}$$

Since these slopes are not the same, the points do not lie on the same line.

37. A parallelogram has 4 sides, with opposite sides parallel. The slope of the line through $(1, 3)$ and $(2, 1)$ is

$$m = \frac{3 - 1}{1 - 2} = \frac{2}{-1} = -2.$$

The slope of the line through $\left(-\frac{5}{2}, 2\right)$ and $\left(-\frac{7}{2}, 4\right)$ is

$$m = \frac{2 - 4}{-\frac{5}{2} - \left(-\frac{7}{2}\right)} = \frac{-2}{1} = -2.$$

Since these slopes are equal, these two sides are parallel.

The slope of the line through $\left(-\frac{7}{2}, 4\right)$ and $(1, 3)$ is

$$m = \frac{4 - 3}{-\frac{7}{2} - 1} = \frac{1}{-\frac{9}{2}} = -\frac{2}{9}.$$

Slope of the line through $\left(-\frac{5}{2}, 2\right)$ and $(2, 1)$ is

$$m = \frac{2 - 1}{-\frac{5}{2} - 2} = \frac{1}{-\frac{9}{2}} = -\frac{2}{9}.$$

Since these slopes are equal, these two sides are parallel.

Since both pairs of opposite sides are parallel, the quadrilateral is a parallelogram.

39. The line goes through $(0, 2)$ and $(-2, 0)$

$$m = \frac{2 - 0}{0 - (-2)} = \frac{2}{2} = 1$$

The correct choice is (a).

41. The line appears to go through $(0, 0)$ and $(-1, 4)$.

$$m = \frac{4 - 0}{-1 - 0} = \frac{4}{-1} = -4$$

43. (a) See the figure in the textbook.
Segment MN is drawn perpendicular to segment PQ. Recall that MQ is the length of segment MQ.

$$m_1 = \frac{\triangle y}{\triangle x} = \frac{MQ}{PQ}$$

From the diagram, we know that $PQ = 1$. Thus, $m_1 = \frac{MQ}{1}$, so MQ has length m_1.

(b) $\quad m_2 = \dfrac{\triangle y}{\triangle x} = \dfrac{-QN}{PQ} = \dfrac{-QN}{1}$

$$QN = -m_2$$

(c) Triangles MPQ, PNQ, and MNP are right triangles by construction. In triangles MPQ and MNP,

$$\text{angle } M = \text{angle } M,$$

and in the right triangles PNQ and MNP,

$$\text{angle } N = \text{angle } N.$$

Since all right angles are equal, and since triangles with two equal angles are similar, triangle MPQ is similar to triangle MNP and triangle PNQ is similar to triangle MNP.

Therefore, triangles MPQ and PNQ are similar to each other.

(d) Since corresponding sides in similar triangles are proportional,

$$MQ = k \cdot PQ \quad \text{and} \quad PQ = k \cdot QN.$$

$$\frac{MQ}{PQ} = \frac{k \cdot PQ}{k \cdot QN}$$

$$\frac{MQ}{PQ} = \frac{PQ}{QN}$$

From the diagram, we know that $PQ = 1$.

$$MQ = \frac{1}{QN}$$

From (a) and (b), $m_1 = MQ$ and $-m_2 = QN$.

Substituting, we get

$$m_1 = \frac{1}{-m_2}.$$

Multiplying both sides by m_2, we have

$$m_1 m_2 = -1.$$

45. $y = 4x + 5$

Three ordered pairs that satisfy this equation are $(-2, -3)$, $(-1, 1)$, and $(0, 5)$. Plot these points and draw a line through them.

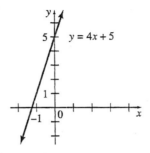

47. $y = -6x + 12$

Three ordered pairs that satisfy this equation are $(0, 12)$, $(1, 6)$, and $(2, 0)$. Plot these points and draw a line through them.

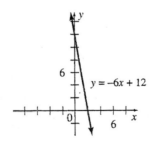

49. $3x - y = -9$

Find the intercepts.
If $y = 0$, then

$$3x - 0 = -9$$
$$3x = -9$$
$$x = -3$$

If $x = 0$, then

$$3(0) - y = -9$$
$$-y = -9$$
$$y = 9$$

so the y-intercept is 9.

Plot the ordered pairs $(-3, 0)$ and $(0, 9)$ and draw a line through these points. (A third point may be used as a check.)

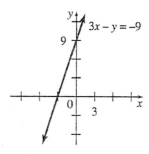

51. $5y + 6x = 11$

Find the intercepts.
If $y = 0$, then

$$5(0) + 6x = 11$$
$$6x = 11$$
$$x = \frac{11}{6}$$

so the x-intercept is $\frac{11}{6}$.
If $x = 0$, then

$$5y + 6(0) = 11$$
$$5y = 11$$
$$y = \frac{11}{5}$$

so the y-intercept is $\frac{11}{5}$.

Plot the ordered pairs $\left(\frac{11}{6}, 0\right)$ and $\left(0, \frac{11}{5}\right)$ and draw a line through these points. (A third point may be used as a check.)

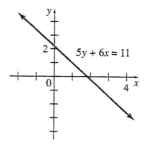

53. $x = 4$

For any value of y, the x-value is 4. Because all ordered pairs that satisfy this equation have the same first number, this equation does not represent a function. The graph is the vertical line with x-intercept 4.

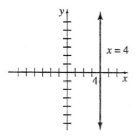

55. $y + 8 = 0$

This equation may be rewritten as $y = -8$, or, equivalently, $y = 0x + -8$. The y-value is -8 for any value of x. The graph is the horizontal line with y-intercept -8.

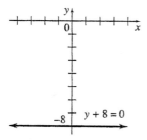

57. $y = -5x$

Three ordered pairs that satisfy this equation are $(0, 0)$, $(-1, 5)$, and $(1, -5)$. Use these points to draw the graph.

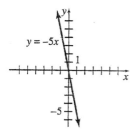

59. $3x - 5y = 0$

If $y = 0$, then $x = 0$, so the x-intercept is 0. If $x = 0$, then $y = 0$, so the y-intercept is 0. Both intercepts give the same ordered pair $(0, 0)$.

To get a second point, choose some other value of x (or y). For example, if $x = 5$, then

$$3x - 5y = 0$$
$$3(5) - 5y = 0$$
$$15 - 5y = 0$$
$$-5y = -15$$
$$y = 3$$

giving the ordered pair $(5, 3)$. Graph the line through $(0, 0)$ and $(5, 3)$.

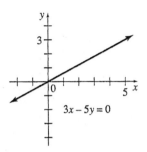

61. (a)

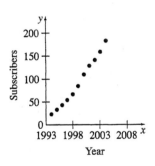

The number of subscribers is increasing and the data appear to be nearly linear.

(b) Find the slope using $(3, 44.04)$ and $(11, 182.14)$.

(c) $$m = \frac{182.14 - 44.04}{11 - 3} = \frac{138.1}{8} = 17.2625$$

$$y - 182.14 = 17.2625(x - 11)$$
$$y - 182.14 = 17.2625x - 189.89$$
$$y = 17.2625x - 7.75$$

Rounding the slope to the nearest hundredth, the equation is $y = 17.26x - 7.75$.

The year 2005 corresponds to $x = 2005 - 1993 = 12$.

$$y = 17.26(12) - 7.75$$
$$y = 199.37$$

The approximation using the equation is less than the actual number of subscribers.

63. (a) The line goes through $(0, 100)$ and $(24, 201.6)$.

$$m = \frac{201.6 - 100}{24 - 0} \approx 4.23$$
$$b = 100$$
$$y = 4.23x + 100$$

(b) The year 2000 corresponds to $x = 18$.

$$y = 4.23(18) + 100$$
$$y = 176.14$$

The estimate is more than, but close to, the actual CPI.

(c) It is increasing at a rate of 4.23 per year.

65. (a) Let $x =$ age.

$$u = 0.85(220 - x) = 187 - 0.85x$$
$$l = 0.7(220 - x) = 154 - 0.7x$$

(b) $u = 187 - 0.85(20) = 170$
$l = 154 - 0.7(20) = 140$

The target heart rate zone is 140 to 170 beats per minute.

(c) $u = 187 - 0.85(40) = 153$
$l = 154 - 0.7(40) = 126$

The target heart rate zone is 126 to 153 beats per minute.

(d) $154 - 0.7x = 187 - 0.85(x + 36)$
$154 - 0.7x = 187 - 0.85x - 30.6$
$154 - 0.7x = 156.4 - 0.85x$
$0.15x = 2.4$
$x = 16$

The younger woman is 16; the older woman is $16 + 36 = 52$. $l = 0.7(220 - 16) \approx 143$ beats per minute.

67. Let $x = 0$ correspond to 1900. Then the "life expectancy from birth" line contains the points $(0, 46)$ and $(104, 77.8)$.

$$m = \frac{77.8 - 46}{104 - 0} = \frac{31.3}{102} = 0.306$$

Since $(0, 46)$ is one of the points, the line is given by the equation

$$y = 0.306x + 46.$$

The "life expectancy from age 65" line contains the points $(0, 76)$ and $(104, 83.7)$.

$$m = \frac{83.7 - 76}{104 - 0} = \frac{7.7}{104} \approx 0.074$$

Since $(0, 76)$ is one of the points, the line is given by the equation

$$y = 0.074x + 76.$$

Set the two equations equal to determine where the lines intersect. At this point, life expectancy should increase no further.

$$0.306x + 46 = 0.074x + 76$$
$$0.232x = 30$$
$$x \approx 129$$

Determine the y-value when $x = 129$. Use the first equation.

$$y = 0.306(129) + 46$$
$$= 39.474 + 46$$
$$= 85.474$$

Thus, the maximum life expectancy for humans is about 86 years.

69. (a) $\quad m = \dfrac{27.4 - 22.8}{45 - 5} = \dfrac{4.6}{10} = 0.115$

$$y - 22.8 = 0.115(x - 5)$$
$$y - 22.8 = 0.115x - 0.575$$
$$y = 0.115x + 22.2$$

(b) $\quad m = \dfrac{25.8 - 20.6}{45 - 5} = \dfrac{5.2}{40} = 0.13$

$$y - 20.6 = 0.13(x - 5)$$
$$y - 20.6 = 0.13x - 0.65$$
$$y = 0.13x + 19.95$$

(c) Since $0.13 > 0.115$, women have the faster increase.

(d) Let $y = 30$ and use the equation from part (a) to solve for x.

$$30 = 0.115x + 22.2$$
$$7.8 = 0.115x$$
$$68 \approx x$$

68 years after 1960, or in the year 2028, men's median age at first marriage will reach 30.

(e) Let $x = 68$ and use the equation from part (b) to find y.

$$y = 0.13(68) + 19.95$$
$$y = 8.84 + 19.95$$
$$y = 28.79$$

The median age for women at first marriage will be about 28.8 years.

71. (a) The line goes through $(0, 1.59)$ and $(24, 5.08)$.

$$m = \frac{5.08 - 1.59}{24 - 0} = \frac{3.49}{24} \approx 0.145$$
$$b = 1.59$$
$$y = 0.145x + 1.59$$

(b) The year 2014 corresponds to $x = 30$.

$$y = 0.145(30) + 1.59$$
$$y = 5.94$$

In 2010, the number of cohabitating adults will be about 5.94 million.

73. (a) Plot the points $(15, 1600)$, $(200, 15{,}000)$, $(290, 24{,}000)$, and $(520, 40{,}000)$.

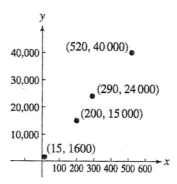

The points lie approximately on a line, so there appears to be a linear relationship between distance and time.

(b) The graph of any equation of the form $y = mx$ goes through the origin, so the line goes through $(520, 40{,}000)$ and $(0, 0)$.

$$m = \frac{40{,}000 - 0}{520 - 0} \approx 76.9$$
$$b = 0$$
$$y = 76.9x + 0$$
$$y = 76.9x$$

(c) Let $y = 60,000$; solve for x.

$$60,000 = 76.9x$$
$$780.23 \approx x$$

Hydra is about 780 megaparsecs from earth.

(d) $A = \dfrac{9.5 \times 10^{11}}{m}, m = 76.9$

$$A = \dfrac{9.5 \times 10^{11}}{76.9}$$
$$= 12.4 \text{ billion years}$$

75. (a)

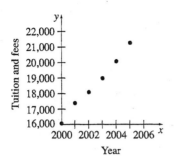

Yes, the data appear to lie roughtly along a straight line.

(b) $m = \dfrac{21,235 - 16,072}{5 - 0} = \dfrac{5163}{5} = 1032.6$

$$b = 16,072$$
$$y = 1032.6x + 16,072$$

The slope 1032.6 indicates that tuition and fees have increased approximately $1033 per year.

(c) The year 2025 is too far in the future to rely on this equation to predict costs; too many other factors may influence these costs by then.

1.2   Linear Functions and Applications

1. $f(2) = 7 - 5(2) = 7 - 10 = -3$

3. $f(-3) = 7 - 5(-3) = 7 + 15 = 22$

5. $g(1.5) = 2(1.5) - 3 = 3 - 3 = 0$

7. $g\left(-\dfrac{1}{2}\right) = 2\left(-\dfrac{1}{2}\right) - 3 = -1 - 3 = -4$

9. $f(t) = 7 - 5(t) = 7 - 5t$

11. This statement is true.
When we solve $y = f(x) = 0$, we are finding the value of x when $y = 0$, which is the x-intercept.

When we evaluate $f(0)$, we are finding the value of y when $x = 0$, which is the y-intercept.

13. This statement is true.
Only a vertical line has an undefined slope, but a vertical line is not the graph of a function. Therefore, the slope of a linear function cannot be defined.

15. The fixed cost is constant for a particular product and does not change as more items are made. The marginal cost is the rate of change of cost at a specific level of production and is equal to the slope of the cost function at that specific value; it approximates the cost of producing one additional item.

19. $10 is the fixed cost and $2.25 is the cost per hour.

Let $x = $ number of hours;
$R(x) = $ cost of renting a snowboard for x hours.

Thus,

$R(x) = $ fixed cost $+$ (cost per hour) $\cdot$ (number of hours)
$R(x) = 10 + (2.25)(x)$
$ = 2.25x + 10$

21. 50¢ is the fixed cost and 35¢ is the cost per half-hour.

Let $x = $ the number of half-hours;
$C(x) = $ the cost of parking a car for x half-hours.

Thus,

$$C(x) = 50 + 35x$$
$$= 35x + 50.$$

23. Fixed cost, $100; 50 items cost $1600 to produce.

Let $C(x) = $ cost of producing x items.
$C(x) = mx + b$, where b is the fixed cost.

$$C(x) = mx + 100$$

Now,

$C(x) = 1600$ when $x = 50$, so

$$1600 = m(50) + 100$$
$$1500 = 50m$$
$$30 = m.$$

Thus, $C(x) = 30x + 100$.

25. Marginal cost: $75; 50 items cost $4300.

$$C(x) = 75x + b$$

Now, $C(x) = 4300$ when $x = 50$.

$$4300 = 75(50) + b$$
$$4300 = 3750 + b$$
$$550 = b$$

Thus, $C(x) = 75x + 550$.

27. $D(q) = 16 - 1.25q$

(a) $D(0) = 16 - 1.25(0) = 16 - 0 = 16$

When 0 watches are demanded, the price is $16.

(b) $D(4) = 16 - 1.25(4) = 16 - 5 = 11$

When 400 watches are demanded, the price is $11.

(c) $D(8) = 16 - 1.25(8) = 16 - 10 = 6$

When 800 watches are demanded, the price is $6.

(d) Let $D(q) = 8$. Find q.

$$8 = 16 - 1.25q$$
$$\frac{5}{4}q = 8$$
$$q = 6.4$$

When the price is $8, 640 watches are demanded.

(e) Let $D(q) = 10$. Find q.

$$10 = 16 - 1.25q$$
$$\frac{5}{4}q = 6$$
$$q = 4.8$$

When the price is $10, 480 watches are demanded.

(f) Let $D(q) = 12$. Find q.

$$12 = 16 - 1.25q$$
$$\frac{5}{4}q = 4$$
$$q = 3.2$$

When the price is $12, 320 watches are demanded.

(g)

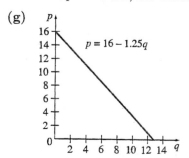

(h) $S(q) = 0.75q$

Let $S(q) = 0$. Find q.

$$0 = 0.75q$$
$$0 = q$$

When the price is $0, 0 watches are supplied.

(i) Let $S(q) = 10$. Find q.

$$10 = 0.75q$$
$$\frac{40}{3} = q$$
$$q = 13.\overline{3}$$

When the price is $10, about 1333 watches are supplied.

(j) Let $S(q) = 20$. Find q.

$$20 = 0.75q$$
$$\frac{80}{3} = q$$
$$q = 26.\overline{6}$$

When the price is $20, about 2667 watches are demanded.

(k)

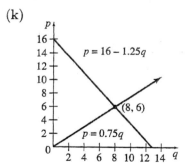

(l) $D(q) = S(q)$

$$16 - 1.25q = 0.75q$$
$$16 = 2q$$
$$8 = q$$
$$S(8) = 0.75(8) = 6$$

The equilibrium quantity is 800 watches, and the equilibrium price is $6.

29. $p = S(q) = \dfrac{2}{5}q$; $p = D(q) = 100 - \dfrac{2}{5}q$

(a)

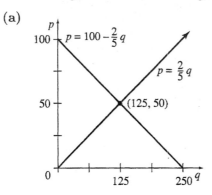

(b)
$$S(q) = D(q)$$
$$\frac{2}{5}q = 100 - \frac{2}{5}q$$
$$\frac{4}{5}q = 100$$
$$q = 125$$
$$S(125) = \frac{2}{5}(125) = 50$$

The equilibrium quantity is 125, the equilibrium price is $50

31. (a) $C(x) = mx + b$; $m = 3.50$; $C(60) = 300$

$$C(x) = 3.50x + b$$

Find b.
$$300 = 3.50(60) + b$$
$$300 = 210 + b$$
$$90 = b$$
$$C(x) = 3.50x + 90$$

(b) $R(x) = 9x$
 $C(x) = R(x)$

$$3.50x + 90 = 9x$$
$$90 = 5.5x$$
$$16.36 = x$$

Joanne must produce and sell 17 shirts.

(c) $P(x) = R(x) - C(x)$; $P(x) = 500$

$$500 = 9x - (3.50x + 90)$$
$$500 = 5.5x - 90$$
$$590 = 5.5x$$
$$107.27 = x$$

To make a profit of $500, Joanne must produce and sell 108 shirts.

33. (a) Using the points $(100, 11.02)$ and $(400, 40.12)$,

$$m = \frac{40.12 - 11.02}{400 - 100} = \frac{29.1}{300} = 0.097.$$
$$y - 11.02 = 0.097(x - 100)$$
$$y - 11.02 = 0.097x - 9.7$$
$$y = 0.097x + 1.32$$
$$C(x) = 0.097x + 1.32$$

(b) The fixed cost is given by the constant in $C(x)$. It is $1.32.

(c) $C(1000) = 0.097(1000) + 1.32 = 97 + 1.32$
$ = 98.32$

The total cost of producing 1000 cups is $98.32.

(d) $C(1001) = 0.097(1001) + 1.32 = 97.097 + 1.32$
$ = 98.417$

The total cost of producing 1001 cups is $98.417.

(e) Marginal cost $= 98.417 - 98.32$
$ = \$0.097$ or   9.7¢

(f) The marginal cost for *any* cup is the slope, $0.097 or 9.7¢. This means the cost of producing one additional cup of coffee would be 9.7¢.

35. (a) $(100{,}000)(50) = 5{,}000{,}000$

Sales in 1996 would be $100{,}000 + 5{,}000{,}000 = 5{,}100{,}000$.

(b) The ordered pairs are (1, 100,000) and (6, 5,100,000).

$$m = \frac{5{,}100{,}000 - 100{,}000}{6 - 1} = \frac{5{,}000{,}000}{5} = 1{,}000{,}000$$

$$y - 100{,}000 = 1{,}000{,}000(x - 1)$$
$$y - 100{,}000 = 1{,}000{,}000x - 1{,}000{,}000$$
$$y = 1{,}000{,}000x - 900{,}000$$
$$S(x) = 1{,}000{,}000x - 900{,}000$$

(c) Let $S(x) = 1{,}000{,}000{,}000$. Find x.

$$1{,}000{,}000{,}000 = 1{,}000{,}000x - 900{,}000$$
$$1{,}000{,}900{,}000 = 1{,}000{,}000x$$
$$x = 1000.9$$

Sales would reach $1 billion in about $1991 + 1000.9 = 2991.9$, or during the year 2991.
Sales would have to grow much faster than linearly to reach $1 billion by 2003.

(d) Use ordered pairs (13, 356,000,000) and (14, 479,000,000).

$$m = \frac{479,000,000 - 356,000,000}{14 - 13} = 123,000,000$$

$$S(x) - 356,000,000 = 123,000,000(x - 13)$$
$$S(x) - 356,000,000 = 123,000,000x - 1,599,000000$$
$$S(x) = 123,000,000x - 1,243,000,000$$

(e) The year 2005 corresponds to $x = 2005 - 1990 = 15$.

$$S(15) = 123,000,000(15) - 1,243,000,000$$
$$S(15) = 602,000,000$$

The estimated sales are \$602,000,000, which is less than the actual sales.

(f) Let $S(x) = 1,000,000,000$. Find x.

$$1,000,000,000 = 123,000,000x - 1,243,000,000$$
$$2,243,000,000 = 123,000,000x$$
$$x \approx 18.2$$

Sales would reach \$1 billion in about 1990 + 18.2 = 2008.2, or during the year 2009.

37. $C(x) = 12x + 39;\ R(x) = 25x$

(a) $C(x) = R(x)$
$$12x + 39 = 25x$$
$$39 = 13x$$
$$3 = x$$

The break-even quantity is 3 units.

(b) $P(x) = R(x) - C(x)$
$$P(x) = 25x - (12x + 39)$$
$$P(x) = 13x - 39$$
$$P(250) = 13(250) - 39$$
$$= 3250 - 39$$
$$= 3211$$

The profit from 250 units is \$3211.

(c) $P(x) = \$130$; find x.

$$130 = 13x - 39$$
$$169 = 13x$$
$$13 = x$$

For a profit of \$130, 13 units must be produced.

39. $C(x) = 105x + 6000$
$R(x) = 250x$

Set $C(x) = R(x)$ to find the break-even quantity.

$$105x + 6000 = 250x$$
$$6000 = 145x$$
$$41.38 \approx x$$

The break-even quantity is about 41 units, so you should decide to produce.

$$P(x) = R(x) - C(x)$$
$$= 250x - (105x + 6000)$$
$$= 145x - 6000$$

The profit function is $P(x) = 145x - 6000$.

41. $C(x) = 1000x + 5000$
$R(x) = 900x$

$$900x = 1000x + 5000$$
$$-5000 = 100x$$
$$-50 = x$$

It is impossible to make a profit when the break-even quantity is negative. Cost will always be greater than revenue.

$$P(x) = R(x) - C(x)$$
$$= 900x - (1000x + 5000)$$
$$= -100x - 5000$$

The profit function is $P(x) = -100x - 5000$ (always a loss).

43. Use the formula derived in Example 7 in this section of the textbook.

$$F = \frac{9}{5}C + 32$$

$$C = \frac{5}{9}(F - 32)$$

(a) $C = 37$; find F.

$$F = \frac{9}{5}(37) + 32$$

$$F = \frac{333}{5} + 32$$

$$F = 98.6$$

The Fahrenheit equivalent of 37°C is 98.6°F.

(b) $C = 36.5$; find F.

$$F = \frac{9}{5}(36.5) + 32$$

$$F = 65.7 + 32$$

$$F = 97.7$$

$C = 37.5$; find F.

$$F = \frac{9}{5}(37.5) + 32$$

$$= 67.5 + 32 = 99.5$$

The range is between 97.7°F and 99.5°F.

1.3 The Least Squares Line

3. (a)

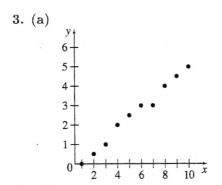

(b)

x	y	xy	x^2	y^2
1	0	0	1	0
2	0.5	1	4	0.25
3	1	3	9	1
4	2	8	16	4
5	2.5	12.5	25	6.25
6	3	18	36	9
7	3	21	49	9
8	4	32	64	16
9	4.5	40.5	81	20.25
10	5	50	100	25
55	25.5	186	385	90.75

$$r = \frac{n(\sum xy) - (\sum x)(\sum y)}{\sqrt{n(\sum x^2) - (\sum x)^2} \cdot \sqrt{n(\sum y^2) - (\sum y)^2}}$$

$$= \frac{10(186) - (55)(25.5)}{\sqrt{10(385) - (55)^2}\sqrt{10(90.75) - (25.5)^2}}$$

$$\approx 0.993$$

(c) The least squares line is of the form $Y = mx + b$. First solve for m.

$$m = \frac{n(\sum xy) - (\sum x)(\sum y)}{n(\sum x^2) - (\sum x)^2}$$

$$= \frac{10(186) - (55)(25.5)}{10(385) - (55)^2}$$

$$= 0.5545454545 \approx 0.55$$

Now find b.

$$b = \frac{\sum y - m(\sum x)}{n}$$

$$= \frac{25.5 - 0.5545454545(55)}{10}$$

$$= -0.5$$

Thus, $Y = 0.55x - 0.5$.

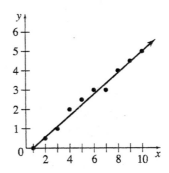

(d) Let $x = 11$. Find Y.

$$Y = 0.55(11) - 0.5 = 5.55$$

5.

$$nb + (\sum x)m = \sum y$$
$$(\sum x)b + (\sum x^2)m = \sum xy$$
$$nb + (\sum x)m = \sum y$$
$$nb = (\sum y) - (\sum x)m$$
$$b = \frac{\sum y - m(\sum x)}{n}$$

$$(\sum x)\left(\frac{\sum y - m(\sum x)}{n}\right) + (\sum x^2)m = \sum xy$$
$$(\sum x)[(\sum y) - m(\sum x)] + nm(\sum x^2) = n(\sum xy)$$
$$(\sum x)(\sum y) - m(\sum x)^2 + nm(\sum x^2) = n(\sum xy)$$
$$nm(\sum x^2) - m(\sum x)^2 = n(\sum xy) - (\sum x)(\sum y)$$
$$m\left[n(\sum x^2) - (\sum x)^2\right] = n(\sum xy) - (\sum x)(\sum y)$$

$$m = \frac{n(\sum xy) - (\sum x)(\sum y)}{n(\sum x^2) - (\sum x)^2}$$

7. (a) $m = \dfrac{n(\sum xy) - (\sum x)(\sum y)}{n(\sum x^2) - (\sum y)}$

$m = \dfrac{10(8501.39) - (995)(85.65)}{10(99,085) - 995^2}$

$m = -0.2519393939 \approx -0.2519$

$b = \dfrac{\sum y - m(\sum x)}{n}$

$b = \dfrac{85.65 - (-0.2519393939)(995)}{10} \approx 33.6330$

$Y = -0.2519x + 33.6330$

(b) The year 2010 corresponds to $x = 110$.

$$Y = -0.2519(110) + 33.6330 \approx 5.924 \text{ (in thousands)}$$

If the trend continues, there will be about 5924 banks in 2010.

(c) $r = \dfrac{10(8501.39) - (995)(85.65)}{\sqrt{10(99,085) - 995^2} \cdot \sqrt{10(739.08) - 85.65^2}} \approx -0.977$

This means that the least squares line fits the data points very well. The negative sign indicates that the number of banks is decreasing as the years increase.

9.

x	y	xy	x^2	y^2
97	6247	605,959	9409	39,025,009
98	6618	648,565	9604	43,797,924
99	7031	696,069	9801	49,434,961
100	7842	784,200	10,000	61,496,964
101	8234	831,634	10,201	67,798,756
102	8940	911,880	10,404	79,923,600
103	9205	948,115	10,609	84,732,025
104	9312	968,448	10,816	86,713,344
804	63,429	6,394,869	80,844	512,922,583

(a)

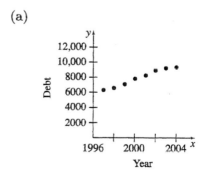

Yes, the pattern is linear.

(b) $m = \dfrac{n(\sum xy) - (\sum x)(\sum y)}{n(\sum x^2) - (\sum x)^2}$

$m = \dfrac{8(6{,}394{,}869) - (804)(63{,}429)}{8(80{,}844) - 804^2}$

$m = 482.25$

$b = \dfrac{\sum y - m(\sum x)}{n}$

$b = \dfrac{63{,}429 - 482.25(804)}{8} = -40{,}537.5$

$Y = 482.25x - 40{,}537.5$

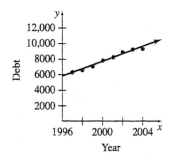

The least squares line seems to be a good fit.

(c) $r = \dfrac{8(6{,}394{,}869) - (804)(63{,}429)}{\sqrt{8(80{,}844) - 804^2} \cdot \sqrt{8(512{,}922{,}583) - 63{,}429^2}} \approx 0.987$

This confirms the least squares line is a good fit.

(d) Let $Y = 12{,}000$ and solve for x.

$$12{,}000 = 482.25x - 40{,}537.5$$
$$52{,}537.5 = 482.25x$$
$$x \approx 109$$

If the trend continues, credit card debt will reach $12,000 in $1900 + 109$, or the year 2009.

11. (a)

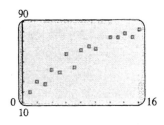

Yes, the points lie in a linear pattern.

(b) Using a calculator's STAT feature, the correlation coefficient is found to be $r \approx 0.959$. This indicates that the percentage of successful hunts does trend to increase with the size of the hunting party.

(c) $Y = 3.98x + 22.7$

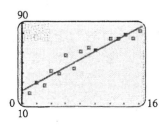

13. (a)

x	y	xy	x^2	y^2
88.6	20.0	1772	7849.96	400.0
71.6	16.0	1145.6	5126.56	256.0
93.3	19.8	1847.34	8704.89	392.04
84.3	18.4	1551.12	7106.49	338.56
80.6	17.1	1378.26	6496.36	292.41
75.2	15.5	1165.6	5655.04	240.25
69.7	14.7	1024.59	4858.09	216.09
82.0	17.1	1402.2	6724	292.41
69.4	15.4	1068.76	4816.36	237.16
83.3	16.2	1349.46	6938.89	262.44
79.6	15.0	1194	6336.16	225
82.6	17.2	1420.72	6822.76	295.84
80.6	16.0	1289.6	6496.36	256.0
83.5	17.0	1419.5	6972.25	289.0
76.3	14.4	1098.72	5821.69	207.36
1200.6	249.8	20,127.47	96,725.86	4200.56

$m = \dfrac{n(\sum xy) - (\sum x)(\sum y)}{n(\sum x^2) - (\sum x)^2}$

$\quad = \dfrac{15(20{,}127.47) - (1200.6)(249.8)}{15(96{,}725.86) - 1200.6^2}$

$\quad = 0.211925009 \approx 0.212$

$b = \dfrac{\sum y - m(\sum x)}{n}$

$\quad = \dfrac{249.8 - 0.212(1200.6)}{15}$

$\quad \approx -0.315$

$Y = 0.212x - 0.315$

(b) Let $x = 73$; find Y.

$Y = 0.212(73) - 0.315$

$\quad \approx 15.2$

If the temperature were $73°F$, you would expect to hear 15.2 chirps per second.

(c) Let $Y = 18$; find x.

$$18 = 0.212x - 0.315$$
$$18.315 = 0.212x$$
$$86.4 \approx x$$

When the crickets are chirping 18 times per second, the temperature is 86.4°F.

(d)

$$r = \frac{15(20,127) - (1200.6)(249.8)}{\sqrt{15(96,725.86) - (1200.6)^2} \cdot \sqrt{15(4200.56) - (249.8)^2}}$$

$$= 0.835$$

15. (a)

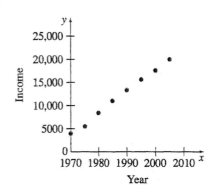

Yes, the data appear to lie along a straight line.

$$r = \frac{8(2159.635) - (140)(95.364)}{\sqrt{8(3500) - 140^2} \cdot \sqrt{8(1366.748) - 95.364^2}}$$

$$\approx 0.999$$

Yes, there is a strong positive linear correlation between the income and the year.

(c) $m = \dfrac{n(\sum xy) - (\sum x)(\sum y)}{n(\sum x^2) - (\sum x)^2}$

$$m = \frac{8(2159.635) - (140)(95.364)}{8(3500) - 140^2}$$

$$m = 0.4673952381 \approx 0.467$$

$$b = \frac{\sum y - m(\sum x)}{n}$$

$$b = \frac{95.364 - 0.4673952381(140)}{8}$$

$$\approx 3.74$$
$$Y = 0.467x + 3.74$$

(d) The year 2020 corresponds to $x = 50$.

$$Y = 0.467(50) + 3.74 = 27.09$$

The predicted poverty level in the year 2020 is $27,090.

17. (a)

x	y	xy	x^2	y^2
150	5000	750,000	22,500	25,000,000
175	5500	962,500	30,625	30,250,000
215	6000	1,290,000	46,225	36,000,000
250	6500	1,625,000	62,500	42,250,000
280	7000	1,960,000	78,400	49,000,000
310	7500	2,325,000	96,100	56,250,000
350	8000	2,800,000	122,500	64,000,000
370	8500	3,145,000	136,900	72,250,000
420	9000	3,780,000	176,400	81,000,000
450	9500	4,275,000	202,500	90,250,000
2970	72,500	22,912,500	974,650	546,250,000

$$m - \frac{n(\sum xy) - (\sum x)(\sum y)}{n(\sum x^2) - (\sum x)^2}$$

$$m = \frac{10(22,912,500) - (2970)(72,500)}{10(974,650) - 2970^2}$$

$$m = 14.90924806 \approx 14.9$$

$$b = \frac{\sum y - m(\sum x)}{n}$$

$$b = \frac{72,500 - 14.9(2970)}{10}$$

$$\approx 2820$$
$$Y = 14.9x + 2820$$

(b) Let $x = 150$; find Y.

$$Y = 14.9(150) + 2820$$
$$Y \approx 5060, \text{ compared to actual } 5000$$

Let $x = 280$; find Y.

$$Y = 14.9(280) + 2820$$
$$\approx 6990, \text{ compared to actual } 7000$$

Let $x = 420$; find Y.

$$Y = 14.9(420) + 2820$$
$$\approx 9080, \text{ compared to actual } 9000$$

(c) Let $x = 230$; find Y.

$$Y = 14.9(230) + 2820$$
$$\approx 6250$$

Adam would need to buy a 6500 BTU air conditioner.

19. (a) Use a calculator's statistical features to obtain the least squares line.

$$y = -0.1358x + 113.94$$

(b) $y = -0.3913x + 148.98$

(c) Set the two equations equal and solve for x.

$$-0.1358x + 113.94 = -0.3913x + 148.98$$
$$0.2555x = 35.04$$
$$x \approx 137$$

The women's record will catch up with the men's record in $1900 + 137$, or in the year 2037.

(d) $r_{men} \approx -0.9823$
$$r_{women} \approx -0.9487$$

Both sets of data points closely fit a line with negative slope.

(e)

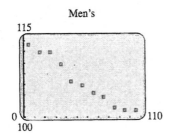

21. (a) $m = \dfrac{n(\sum xy) - (\sum x)(\sum y)}{n(\sum x^2) - (\sum x)^2}$

$$m = \dfrac{10(5496) - (110)(466)}{10(1540) - 110^2}$$

$$m = 1.121212121 \approx 1.12$$

$$b = \dfrac{\sum y - m(\sum x)}{n}$$

$$b = \dfrac{466 - 1.121212121(110)}{10}$$

$$\approx 34.27$$
$$Y = 1.12x + 34.27$$

(b) $= \dfrac{10(5496) - (110)(466)}{\sqrt{10(1540) - 110^2} \cdot \sqrt{10(22{,}232) - 466^2}}$

$$\approx 0.8963$$

Yes, the value indicates a good fit of the least squares line to the data.

(c) The year 2005 corresponds to $x = 25$.

$$Y = 1.12(25) + 34.27 = 62.27 \approx 62$$

The predicted length of a game in 2005 is 2 hours + 62 minutes, or 3:02.

Chapter 1 Review Exercises

3. Through $(-3, 7)$ and $(2, 12)$

$$m = \frac{12 - 7}{2 - (-3)} = \frac{5}{5} = 1$$

5. Through the origin and $(11, -2)$

$$m = \frac{-2 - 0}{11 - 0} = -\frac{2}{11}$$

7. $4x + 3y = 6$
$$3y = -4x + 6$$

$$y = -\frac{4}{3}x + 2$$

Therefore, the slope is $m = -\frac{4}{3}$.

9. $y + 4 = 9$
$$y = 5$$
$$y = 0x + 5$$
$$m = 0$$

11. $y = 5x + 4$
$$m = 5$$

13. Through $(5, -1)$; slope $\frac{2}{3}$

Use point-slope form.

$$y - (-1) = \frac{2}{3}(x - 5)$$

$$y + 1 = \frac{2}{3}(x - 5)$$

$$3(y + 1) = 2(x - 5)$$
$$3y + 3 = 2x - 10$$
$$3y = 2x - 13$$

$$y = \frac{2}{3}x - \frac{13}{3}$$

15. Through $(-6, 3)$ and $(2, -5)$

$$m = \frac{-5 - 3}{2 - (-6)} = \frac{-8}{8} = -1$$

Use point-slope form.

$$y - 3 = -1[x - (-6)]$$
$$y - 3 = -x - 6$$
$$y = -x - 3$$

17. Through $(-1, 4)$; undefined slope

Undefined slope means the line is vertical. The equation of the vertical line through $(-1, 4)$ is $x = -1$.

19. Through $(3, -4)$ parallel to $4x - 2y = 9$
Solve $4x - 2y = 9$ for y.

$$-2y = -4x + 9$$
$$y = 2x - \frac{9}{2}$$
$$m = 2$$

The desired line has the same slope. Use the point-slope form.

$$y - (-4) = 2(x - 3)$$
$$y + 4 = 2x - 6$$
$$y = 2x - 10$$

21. Through $(2, -10)$, perpendicular to a line with undefined slope
A line with undefined slope is a vertical line. A line perpendicular to a vertical line is a horizontal line with equation of the form $y = k$. The desired line passed through $(2, -10)$, so $k = -10$. Thus, an equation of the desired line is $y = -10$.

23. Through $(-3, 5)$, perpendicular to $y = -2$
The given line, $y = -2$, is a horizontal line. A line perpendicular to a horizontal line is a vertical line with equation of the form $x = h$.
The desired line passes through $(-3, 5)$, so $h = -3$. Thus, an equation of the desired line is $x = -3$.

25. $y = 6 - 2x$

Find the intercepts.
Let $x = 0$.

$$y = 6 - 2(0) = 6$$

The y-intercept is 6.
Let $y = 0$.

$$0 = 6 - 2x$$
$$2x = 6$$
$$x = 3$$

The x-intercept is 3.
Draw the line through $(0, 6)$ and $(3, 0)$.

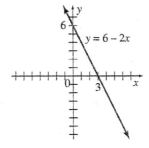

27. $4x + 6y = 12$

Find the intercepts.
When $x = 0$, $y = 2$, so the y-intercept is 2.
When $y = 0$, $x = 3$, so the x-intercept is 3.
Draw the line through $(0, 2)$ and $(3, 0)$.

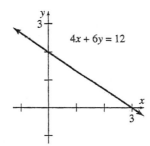

29. $y = 1$

This is the horizontal line passing through $(0, 1)$.

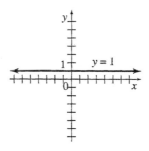

31. $x + 3y = 0$

When $x = 0$, $y = 0$.
When $x = 3$, $y = -1$.
Draw the line through $(0, 0)$ and $(3, -1)$.

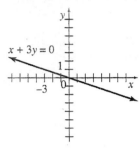

33. $S(q) = 6q + 3$; $D(q) = 19 - 2q$

(a) $S(q) = D(q) = 10$

$$10 = 6q + 3$$
$$7 = 6q$$
$$\frac{7}{6} = q \quad \text{(supply)}$$
$$10 = 19 - 2q$$
$$-9 = -2q$$
$$\frac{9}{2} = q \quad \text{(demand)}$$

When the price is \$10 per pound, the supply is $\frac{7}{6}$ pounds per day, and the demand is $\frac{9}{2}$ pounds per day.

(b) $S(q) = D(q) = 15$

$$15 = 6q + 3$$
$$12 = 6q$$
$$2 = q \quad \text{(supply)}$$
$$15 = 19 - 2q$$
$$-4 = -2q$$
$$2 = q \quad \text{(demand)}$$

When the price is \$15 per pound, the supply is 2 pounds per day, and the demand is 2 pounds per day.

(c) $S(q) = D(q) = \$18$

$$18 = 6q + 3$$
$$15 = 6q$$
$$\frac{5}{2} = q \quad \text{(supply)}$$
$$18 = 19 - 2q$$
$$-1 = -2q$$
$$\frac{1}{2} = q \quad \text{(demand)}$$

When the price is $\frac{5}{2}$ pounds per day, the demand is $\frac{1}{2}$ pound per day.

(d)

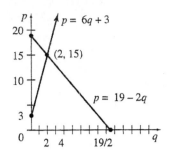

(e) The graph shows that the lines representing the supply and demand functions intersect at the point $(2, 15)$. The y-coordinate of this point gives the equilibrium price. Thus, the equilibrium price is \$15.

(f) The x-coordinate of the intersection point gives the equilibrium quantity. Thus, the equilibrium quantity is 2, representing 2 pounds of crabmeat per day.

35. Using the points $(50, 47.50)$ and $(80, 32.50)$,

$$m = \frac{47.50 - 32.50}{50 - 80} = \frac{15}{-30} = \frac{-1}{2} = -0.5.$$

$$p - 47.50 = -0.5(q - 50)$$
$$p - 47.50 = -0.5q + 25$$
$$p = -0.5q + 72.50$$
$$D(q) = -0.5q + 72.50$$

37. Eight units cost \$300; fixed cost is \$60.
The fixed cost is the cost if zero units are made.
$(8, 300)$ and $(0, 60)$ are points on the line.

$$m = \frac{60 - 300}{0 - 8} = 30$$

Use slope-intercept form.

$$y = 30x + 60$$
$$C(x) = 30x + 60$$

39. Twelve units cost \$445; 50 units cost \$1585. Points on the line are $(12, 445)$ and $(50, 1585)$.

$$m = \frac{1585 - 445}{50 - 12} = 30$$

Use point-slope form.

$$y - 445 = 30(x - 12)$$
$$y - 445 = 30x - 360$$
$$y = 30x + 85$$
$$C(x) = 30x + 85$$

41. $C(x) = 200x + 1000$
$R(x) = 400x$

  (a) $C(x) = R(x)$

$$200x + 1000 = 400x$$
$$1000 = 200x$$
$$5 = x$$

The break-even quantity is 5 cartons.

  (b) $R(5) = 400(5) = 2000$

The revenue from 5 cartons of CD's is $2000.

43. Let y represent imports from China in billions of dollars. Using the points $(1, 102)$ and $(5, 243)$,

$$m = \frac{243 - 102}{5 - 1} = \frac{141}{4} = 35.25$$
$$y - 102 = 35.25(x - 1)$$
$$y - 102 = 35.25x - 35.25$$
$$y = 35.25x + 66.75.$$

45. Using the points $(97, 44{,}883)$ and $(105, 46{,}326)$,

$$m = \frac{46{,}326 - 44{,}883}{105 - 97} = \frac{1443}{8} \approx 180.4$$

$$I - 44{,}883 = 180.4(x - 97)$$
$$I - 44{,}883 = 180.4x - 17{,}498.8$$
$$I(x) = 180.4x + 27{,}384.2.$$

Rounded to the nearest dollar,

$$I(x) = 180.4x + 27{,}384.$$

47. (a)

x	y
1960	43
2840	74
2060	54
3630	79
2420	63
3160	74
3220	78
2550	70
3140	80
3790	77

Using a graphing calculator, $r = 0.881$. Yes, the data seem to fit a straight line.

(b)

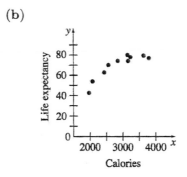

The data somewhat fit a straight line, but there is also a nonlinear trend.

(c) Using a graphing calculator, $Y = 0.0173x + 19.3$.

(d) Let $x = 3400$. Find Y.

$$Y = 0.0173(3400) + 19.3 \approx 78.1$$

The predicted life expectancy in the United Kingdom, with a daily calorie supply of 3400, is about 78.1 years. This agrees with the actual value of 78 years.

49. Using the points $(74, 142.3)$ and $(104, 118.4)$,

$$m = \frac{118.4 - 142.3}{104 - 74} = \frac{-23.9}{30} = -0.797$$

$$y - 142.3 = -0.797(x - 74)$$
$$y - 142.3 = -0.797x + 59$$
$$y = -0.797x + 201.3.$$

51. (a) Using a graphing calculator, $r = 0.749$.

The data seem to fit a line but the fit is not very good.

(b)

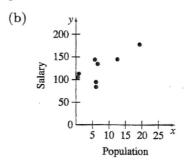

(c) Using a graphing calculator,

$$Y = 3.81x + 98.24$$

(d) The slope is 3.81 thousand (or 3810). On average, the governor's salary increases $3810 for each additional million in population.

Chapter 1 Test

[1.1]

Find the slope of each line that has a slope.

1. Through $(2, -5)$ and $(-1, 7)$

2. Through $(9, 5)$ and $(9, 2)$

3. $3x - 7y = 9$

4. Perpendicular to $2x + 5y = 7$

Find an equation in the form $ax + by = c$ for each line.

5. Through $(-1, 6)$ and $(5, -3)$

6. x–intercept 6, y–intercept -5

7. Through $(0, 3)$, parallel to $2x - 4y = 1$

8. Through $(1, -4)$, perpendicular to $3x + y = 1$

9. Through $(2, 5)$, perpendicular to $x = 5$

Graph each of the following.

10. $3x + 5y = 15$

11. $2x + y = 0$

12. $x - 3 = 0$

[1.2]

13. Let the supply and demand functions for a certain product be given by the following equations.

   Supply: $p = 0.20q - 5$ Demand: $p = 100 - 0.15q$,

   where p represents the price (in dollars) at a supply or demand, respectively, of q units.

   (a) Graph these equations on the same axes.
   (b) Find the equilibrium price.
   (c) Find the equilibrium quantity.

14. For a given product, eight units cost $450, while forty units cost $770.

   (a) Find the appropriate linear cost function.
   (b) What is the fixed cost?
   (c) What is the marginal cost per item?
   (d) Find the average cost function.

15. For a given product, the variable cost is $100, while 150 items cost $16,000 to produce.

   (a) Find the appropriate linear cost function.
   (b) What is the fixed cost?
   (c) What is the marginal cost per item?
   (d) Find the average cost function.

16. Producing x hundred units of widgets costs $C(x) = 4x + 16$; revenue is $12x$, where revenue and $C(x)$ are in thousands of dollars.

   (a) What is the break-even quantity?

   (b) What is the profit from 300 units?

   (c) How many units will produce a profit of $40,000?

[1.3]

17. An electronics firm was planning to expand its product line and wanted to get an idea of the salary picture for technicians it would hire in this field. The following data was collected.

$$n = 12 \qquad\qquad \sum x^2 = 3162$$
$$\sum x = 176 \qquad \sum y^2 = 10{,}870$$
$$\sum y = 356$$
$$\sum xy = 5629$$

   (a) Find an equation for the least squares line.

   (b) Find the coefficient of correlation.

18. An economist was interested in the production costs for companies supplying chemicals for use in fertilizers. The data below represents the relationship between the number of tons produced during a given year (x) and the production cost per ton (y) for seven companies.

Number of Tons (in thousands)	Cost per Ton (in dollars)
3.0	40
4.0	50
2.4	50
5.0	35
2.6	55
4.0	35
5.5	30

   (a) Find the equation for the least squares line.

   (b) Find the coefficient of correlation.

Chapter 1 Test Answers

1. -4

2. Undefined

3. $\frac{3}{7}$

4. $\frac{5}{2}$

5. $3x + 2y = 9$

6. $5x - 6y = 30$

7. $x - 2y = -6$

8. $x - 3y = 13$

9. $y = 5$

10.

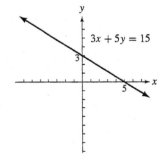

11.

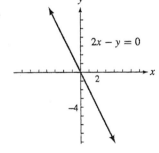

12.

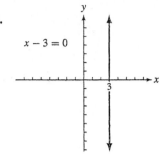

13. (a)

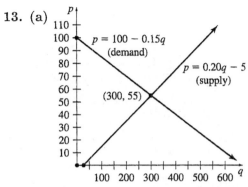

(b) \$55   (c) 300 units

14. (a) $y = 10x + 370$ (b) \$370
(c) \$10   (d) $\overline{C}(x) = 10 + \frac{370}{x}$

15. (a) $y = 100x + 1000$ (b) \$1000
(c) \$100   (d) $\overline{C}(x) = 100 + \frac{1000}{x}$

16. (a) 200 units   (b) \$8000   (c) 700 units

17. (a) $Y = 0.70x + 19.37$ (b) $r = 0.96$

18. (a) $Y = -6.37x + 66.24$ (b) $r = -0.79$

SYSTEMS OF LINEAR EQUATIONS AND MATRICES

2.1 Solution of Linear Systems by the Echelon Method

In Exercises 1-15 and 19-27, check each solution by substituting it in the original equation of the system.

1. $x + y = 5$ (1)
$2x - 2y = 2$ (2)

To eliminate x in equation (2), multiply equation (1) by -2 and add the result to equation (2). The new system is

$$x + y = 5 \quad (1)$$
$$-2R_1 + R_2 \rightarrow R_2 \qquad -4y = -8. \quad (3)$$

Now make the coefficient of the first term in each row equal 1. To accomplish this, multiply equation (3) by $-\frac{1}{4}$.

$$x + y = 5 \quad (1)$$
$$-\tfrac{1}{4}R_2 \rightarrow R_2 \qquad y = 2 \quad (4)$$

Substitute 2 for y in equation (1).

$$x + 2 = 5$$
$$y = 3$$

The solution is $(3, 2)$.

3. $3x - 2y = -3$ (1)
$5x - y = 2$ (2)

To eliminate x in equation (2), multiply equation (1) by -5 and equation (2) by 3. Add the results. The new system is

$$3x - 2y = -3 \quad (1)$$
$$-5R_1 + 3R_2 \rightarrow R_2 \qquad 7y = 21. \quad (3)$$

Now make the coefficient of the first term in each row equal 1. To accomplish this, multiply equation (1) by $\frac{1}{3}$ and equation (3) by $\frac{1}{7}$.

$$\tfrac{1}{3}R_1 \rightarrow R_1 \qquad x - \frac{2}{3}y = -1 \quad (4)$$
$$\tfrac{1}{7}R_2 \rightarrow R_2 \qquad y = 3 \quad (5)$$

Back-substitution of 3 for y in equation (4) gives

$$x - \frac{2}{3}(3) = -1$$
$$x - 2 = -1$$
$$x = 1.$$

The solution is $(1, 3)$.

5. $3x + 2y = -6$ (1)
$5x - 2y = -10$ (2)

Eliminate x in equation (2) to get the system

$$3x + 2y = -6 \quad (1)$$
$$5R_1 + (-3)R_2 \rightarrow R_2 \qquad 16y = 0. \quad (3)$$

Make the coefficient of the first term in each equation equal 1.

$$\tfrac{1}{3}R_1 \rightarrow R_1 \quad x + \frac{2}{3}y = -2 \quad (4)$$
$$\tfrac{1}{16}R_2 \rightarrow R_2 \qquad y = 0 \quad (5)$$

Substitute 0 for y in equation (4) to get $x = -2$. The solution is $(-2, 0)$.

7. $6x - 2y = -4$ (1)
$3x + 4y = 8$ (2)

Eliminate x in equation (2).

$$6x - 2y = -4 \quad (1)$$
$$-1R_1 + 2R_2 \rightarrow R_2 \qquad 10y = 20 \quad (3)$$

Make the coefficient of the first term in each row equal 1.

$$\tfrac{1}{6}R_1 \rightarrow R_1 \quad x - \frac{1}{3}y = -\frac{2}{3} \quad (4)$$
$$\tfrac{1}{10}R_2 \rightarrow R_2 \qquad y = 2 \quad (5)$$

Substitute 2 for y in equation (4) to get $x = 0$. The solution is $(0, 2)$.

9. $5p + 11q = -7$ (1)
  $3p - 8q = 25$ (2)

Eliminate p in equation (2).

$$5p + 11q = -7 \quad (1)$$

$-3R_1 + 5R_2 \rightarrow R_2 \qquad -73q = 146 \quad (3)$

Make the coefficient of the first term in each row equal 1.

$\frac{1}{5}R_1 \rightarrow R_1 \quad p + \frac{11}{5}q = -\frac{7}{5} \quad (4)$

$-\frac{1}{73}R_2 \rightarrow R_2 \qquad q = -1 \quad (5)$

Substitute -2 for q in equation (4) to get $p = 3$. The solution is $(3, -2)$.

11. $6x + 7y = -2$ (1)
  $7x - 6y = 26$ (2)

Eliminate x in equation (2).

$$6x + 7y = -2 \quad (1)$$

$7R_1 + (-6)R_2 \rightarrow R_2 \qquad 85y = -170 \quad (3)$

Make the coefficient of the first term in each equation equal 1.

$\frac{1}{6}R_1 \rightarrow R_1 \quad x + \frac{7}{6}y = -\frac{1}{3} \quad (4)$

$\frac{1}{85}R_2 \rightarrow R_2 \qquad y = -2 \quad (5)$

Substitute -2 for y in equation (4) to get $x = 2$. The solution is $(2, -2)$.

13. $3x + 2y = 5$ (1)
  $6x + 4y = 8$ (2)

Eliminate x in equation (2).

$$3x + 2y = 5 \quad (1)$$

$-2R_1 + R_2 \rightarrow R_2 \qquad 0 = -2 \quad (3)$

Equation (3) is a false statement.
The system is inconsistent and has no solution.

15. $3x - 2y = -4$ (1)
  $-6x + 4y = 8$ (2)

Eliminate x in equation (2).

$$3x - 2y = -4 \quad (1)$$

$2R_1 + R_2 \rightarrow R_2 \qquad 0 = 0 \quad (3)$

The true statement in equation (3) indicates that there are an infinite number of solutions for the system. Solve equation (1) for x.

$$3x - 2y = -4 \qquad (1)$$
$$3x = 2y - 4$$
$$x = \frac{2y - 4}{3} \quad (4)$$

For each value of y, equation (4) indicates that $x = \frac{2y-4}{3}$, and all ordered pairs of the form $\left(\frac{2y-4}{3}, y\right)$ are solutions.

17. An inconsistent system has *no* solutions.

19. $x - \frac{3y}{2} = \frac{5}{2}$ (1)
  $\frac{4x}{3} + \frac{2y}{3} = 6$ (2)

Rewrite the equations without fractions.

$2R_1 \rightarrow R_1 \quad 2x - 3y = 5 \quad (3)$
$3R_2 \rightarrow R_2 \quad 4x + 2y = 18 \quad (4)$

Eliminate x in equation (4).

$$2x - 3y = 5 \quad (3)$$

$-2R_1 + R_2 \rightarrow R_2 \qquad 8y = 8 \quad (5)$

Make the coefficient of the first term in each equation equal 1.

$\frac{1}{2}R_1 + R_1 \quad x - \frac{3}{2}y = \frac{5}{2} \quad (6)$

$\frac{1}{8}R_1 + R_2 \rightarrow R_2 \qquad y = 1 \quad (7)$

Substitute 1 for y in equation (6) to get $y = 4$. The solution is $(4, 1)$.

21. $\frac{x}{2} + y = \frac{3}{2}$ (1)
  $\frac{x}{3} + y = \frac{1}{3}$ (2)

Rewrite the equations without fractions.

$2R_1 \rightarrow R_1 \quad x + 2y = 3 \quad (3)$
$3R_2 \rightarrow R_2 \quad x + 3y = 1 \quad (4)$

Eliminate x in equation (4).

$$x + 2y = 3 \quad (3)$$

$-1R_1 + R_2 \rightarrow R_2 \qquad y = -2 \quad (5)$

Substitute -2 for y in equation (3) to get $x = 7$. The solution is $(7, -2)$.

23. $x + y + z = 2$ (1)
  $2x + y - z = 5$ (2)
  $x - y + z = -2$ (3)

Eliminate x in equations (2) and (3).

$$x + y + z = 2 \quad (1)$$

$-2R_1 + R_2 \rightarrow R_2 \qquad -y - 3z = 1 \quad (4)$
$-1R_1 + R_3 \rightarrow R_3 \qquad -2y = -4 \quad (5)$

Eliminate y in equation (5).

$$x + y + \ z = \ 2 \quad (1)$$
$$-y - 3z = \ 1 \quad (4)$$
$$-2R_2 + R_3 \rightarrow R_3 \qquad 6z = -6 \quad (6)$$

Make the coefficient of the first term in each equation equal 1.

$$x + y + \ z = \ 2 \quad (1)$$
$$-1R_2 \rightarrow R_2 \qquad y + 3z = -1 \quad (7)$$
$$\tfrac{1}{6}R_3 \rightarrow R_3 \qquad z = -1 \quad (8)$$

Substitute -1 for z in equation (7) to get $y = 2$. Finally, substitute -1 for z and 2 for y in equation (1) to get $x = 1$. The solution is $(1, 2, -1)$.

25. $\quad x + 3y + 4z = 14 \quad (1)$
$\quad\ 2x - 3y + 2z = 10 \quad (2)$
$\quad\ 3x - \ y + \ z = \ 9 \quad (3)$

Eliminate x in equations (2) and (3).

$$x + 3y + \ 4z = \ 14 \quad (1)$$
$$-2R_1 + R_2 \rightarrow R_2 \qquad -9y - \ 6z = -18 \quad (4)$$
$$-3R_1 + R_3 \rightarrow R_3 \qquad -10y - 11z = -33 \quad (5)$$

Eliminate y in equation (5).

$$x + 3y + \ 4z = \ 14 \quad (1)$$
$$-9y - \ 6z = -18 \quad (4)$$
$$10R_2 + (-9)R_3 \rightarrow R_3 \qquad 39z = 117 \quad (6)$$

Make the coefficient of the first term in each equation equal 1.

$$x + 3y + 4z = 14 \quad (1)$$
$$-\tfrac{1}{9}R_2 \rightarrow R_2 \qquad y + \tfrac{2}{3}z = \ 2 \quad (7)$$
$$\tfrac{1}{39}R_3 \rightarrow R_3 \qquad z = \ 3 \quad (8)$$

Substitute 3 for z in equation (2) to get $y = 0$. Finally, substitute 3 for z and 0 for y in equation (1) to get $x = 2$. The solution is $(2, 0, 3)$.

27. $\quad 2x + 5y + 4z = 10 \quad (1)$
$\quad\ 8x + 2y + 3z = 27 \quad (2)$
$\quad\ 4x + \ y + \ z = 13 \quad (3)$

Eliminate x in equations (2) and (3).

$$2x + \ 5y + \ 4z = \ 10 \quad (1)$$
$$-4R_1 + R_2 \rightarrow R_2 \qquad -18y - 13z = -13 \quad (4)$$
$$-2R_1 + R_3 \rightarrow R_3 \qquad -9y - \ 7z = \ -7 \quad (5)$$

Eliminate y in equation (5).

$$2x + \ 5y + \ 4z = \ 10 \quad (1)$$
$$-18y - 13z = -13 \quad (4)$$
$$R_2 + (-2)R_3 \rightarrow R_3 \qquad z = \ 1 \quad (6)$$

Make the coefficient of the first term in each row equal 1.

$$\tfrac{1}{2}R_1 \rightarrow R_1 \quad x + \tfrac{5}{2}y + 2z = \ 5 \quad (7)$$
$$-\tfrac{1}{18}R_2 \rightarrow R_2 \qquad y + \tfrac{13}{18}z = \tfrac{13}{18} \quad (8)$$
$$z = \ 1 \quad (6)$$

Substitute 1 for z in equation (8) to get $y = 0$. Finally, substitute 1 for z and 0 for y in equation (7) to get $x = 3$. The solution is $(3, 0, 1)$.

31. $3x + y - \ z = \ 0 \quad (1)$
$\quad\ 2x - y + 3z = -7 \quad (2)$

Eliminate x in equation (2).

$$3x + y - \ \ z = \ 0 \quad (1)$$
$$2R_1 + (-3)R_2 \rightarrow R_2 \qquad 5y - 11z = 21 \quad (3)$$

Make the coefficient of the first term in each equation equal 1.

$$\tfrac{1}{3}R_1 \rightarrow R_1 \quad x + \tfrac{1}{3}y - \tfrac{1}{3}z = \ 0 \quad (4)$$
$$\tfrac{1}{5}R_2 \rightarrow R_2 \qquad y - \tfrac{11}{5}z = \tfrac{21}{5} \quad (5)$$

Solve equation (5) for y in terms of z.

$$y = \tfrac{11}{5}z + \tfrac{21}{5}$$

Substitute this expression for y in equation (4), and solve the equation for x.

$$x + \tfrac{1}{3}\left(\tfrac{11}{5}z + \tfrac{21}{5}\right) - \tfrac{1}{3}z = 0$$
$$x + \tfrac{11}{15}z + \tfrac{7}{5} - \tfrac{1}{3}z = 0$$
$$x + \tfrac{2}{5}z = -\tfrac{7}{5}$$
$$x = -\tfrac{2}{5}z - \tfrac{7}{5}$$

The solution is

$$\left(-\tfrac{2}{5}z - \tfrac{7}{5}, \ \tfrac{11}{5}z + \tfrac{21}{5}, \ z\right) \ \text{or}$$
$$\left(\tfrac{-2z - 7}{5}, \ \tfrac{11z + 21}{5}, \ z\right).$$

33. $-x + y - z = -7$ (1)
$2x + 3y + z = 7$ (2)

Eliminate x in equation (2).

$$\quad -x + y - z = -7 \quad (1)$$
$$2R_1 + R_2 \to R_2 \qquad 5y - z = -7 \quad (3)$$

Make the coefficient of the first term in each equation equal 1.

$$-1R_1 \to R_1 \quad x - y + z = 7 \quad (4)$$
$$\tfrac{1}{5}R_2 \to R_2 \qquad y - \tfrac{1}{5}z = -\tfrac{7}{5} \quad (5)$$

Solve equation (5) for y in terms of z.

$$y = \tfrac{1}{5}z - \tfrac{7}{5}$$

Substitute this expression for y in equation (4), and solve the equation for x.

$$x - \left(\tfrac{1}{5}z - \tfrac{7}{5}\right) + z = 7$$
$$x - \tfrac{1}{5}z + \tfrac{7}{5} + z = 7$$
$$x + \tfrac{4}{5}z = \tfrac{28}{5}$$
$$x = -\tfrac{4}{5}z + \tfrac{28}{5}.$$

The solution of the system is

$$\left(-\tfrac{4}{5}z + \tfrac{28}{5}, \tfrac{1}{5}z - \tfrac{7}{5}, z\right) \text{ or}$$
$$\left(\frac{-4z + 28}{5}, \frac{z - 7}{5}, z\right).$$

37. Let $x =$ the number of skirts originally in the store, and
$y =$ the number of blouses originally in the store.

The system to be solved is

$$45x + 35y = 51,750 \quad (1)$$
$$45\left(\tfrac{1}{2}x\right) + 35\left(\tfrac{2}{3}y\right) = 30,600. \quad (2)$$

Simplify each equation. Multiply equation (1) by $\tfrac{1}{5}$ and equation (2) by $\tfrac{6}{5}$.

$$9x + 7y = 10,350 \quad (3)$$
$$27x + 28y = 36,720 \quad (4)$$

Eliminate x from equation (4).

$$\quad 9x + 7y = 10,350 \quad (3)$$
$$-3R_1 + R_2 \to R_2 \qquad 7y = 5670 \quad (5)$$

Make each leading coefficient equal 1.

$$\tfrac{1}{9}R_1 \to R_1 \quad x + \tfrac{7}{9}y = 1150 \quad (6)$$
$$\tfrac{1}{7}R_2 \to R_2 \qquad\quad y = 810 \quad (7)$$

Substitute 810 for y in equation (6).

$$x + \tfrac{7}{9}(810) = 1150$$
$$x + 630 = 1150$$
$$x = 520$$

Half of the skirts are sold, leaving half in the store, so

$$\tfrac{1}{2}x = \tfrac{1}{2}(520) = 260.$$

Two-thirds of the blouses are sold, leaving one-third in the store, so

$$\tfrac{1}{3}y = \tfrac{1}{3}(810) = 270.$$

There are 260 skirts and 270 blouses left in the store.

39. Let $x =$ the number of shares of Disney stock, and
$y =$ the number of shares of Intel stock.

$$30x + 70y = 16,000 \quad (1)$$
$$45x + 105y = 25,500 \quad (2)$$

Simplify each equation. Multiply equation (1) by $\tfrac{1}{10}$ and equation (2) by $\tfrac{1}{15}$.

$$3x + 7y = 1600 \quad (3)$$
$$3x + 7y = 1700 \quad (4)$$

Since $3x + 7y$ cannot equal both 1600 and 1700 for one point (x, y), we have an inconsistent system. Therefore, this situation is not possible.

41. Let $x =$ the number of fives,
$y =$ the number of tens, and
$z =$ the number of twenties.

Since the number of fives is three times the number of tens, $x = 3y$.

The system to be solved is

$$x + y + z = 70 \quad (1)$$
$$x - 3y = 0 \quad (2)$$
$$5x + 10y + 20z = 960. \quad (3)$$

Eliminate x in equations (2) and (3).

$$x + y + z = 70 \quad (1)$$
$$R_1 + (-1)R_2 \to R_2 \qquad 4y + z = 70 \quad (4)$$
$$-5R_1 + R_3 \to R_3 \qquad 5y + 15z = 610 \quad (5)$$

Eliminate y in equation (5).

$$x + y + z = 70 \quad (1)$$
$$4y + z = 70 \quad (4)$$
$$-5R_2 + 4R_3 \to R_3 \qquad 55z = 2090 \quad (6)$$

Make each leading coefficient equal 1.

$$x + y + z = 70 \quad (1)$$
$$\tfrac{1}{4}R_2 \to R_2 \qquad y + \frac{1}{4}z = \frac{35}{2} \quad (7)$$
$$\tfrac{1}{55}R_3 \to R_3 \qquad z = 38 \quad (8)$$

Substitute 38 for z in equation (7) to get $y = 8$. Finally, substitute 38 for z and 8 for y in equation (1) to get $x = 24$.

There are 24 fives, 8 tens, and 38 twenties.

43. Let $x =$ the number of buffets produced each week,

$y =$ the number of chairs produced each week, and

$z =$ the number of tables produced each week.

Make a table.

	Buffet	Chair	Table	Totals
Construction	30	10	10	350
Finishing	10	10	30	150

The system to be solved is

$$30x + 10y + 10z = 350 \quad (1)$$
$$10x + 10y + 30z = 150. \quad (2)$$

Make the coefficient of the first term in equation (1) equal 1.

$$\tfrac{1}{30}R_1 \to R_1 \qquad x + \frac{1}{3}y + \frac{1}{3}z = \frac{35}{3} \quad (3)$$
$$10x + 10y + 30z = 150 \quad (2)$$

Eliminate x from equation (2).

$$x + \frac{1}{3}y + \frac{1}{3}z = \frac{35}{3} \quad (3)$$
$$-10R_1 + R_2 \to R_2 \qquad \frac{20}{3}y + \frac{80}{3}z = \frac{100}{3} \quad (4)$$

Solve equation (4) for y. Multiply by 3.

$$20y + 80z = 100$$
$$y + 4z = 5$$
$$y = 5 - 4z$$

Substitute $5 - 4z$ for y in equation (1) and solve for x.

$$30x + 10(5 - 4z) + 10z = 350$$
$$30x + 50 - 40z + 10z = 350$$
$$30x = 300 + 30z$$
$$x = 10 + z$$

The solution is $(10 + z, 5 - 4z, z)$. All variables must be nonnegative integers. Therefore,

$$5 - 4z \geq 0$$
$$5 \geq 4z$$
$$z \leq \frac{5}{4},$$

so $z = 0$ or $z = 1$. (Any larger value of z would cause y to be negative, which would make no sense in the problem.) If $z = 0$, then the solution is $(10, 5, 0)$. If $z = 1$, then the solution is $(11, 1, 1)$.

Therefore, the company should make either 10 buffets, 5 chairs, and no tables or 11 buffets, 1 chair, and 1 table each week.

45. Let $x =$ the number of long-sleeve blouses,

$y =$ the number of short-sleeve blouses, and

$z =$ the number of sleeveless blouses.

Make a table.

	Long Sleeve	Short Sleeve	Sleeveless	Totals
Cutting	1.5	1	0.5	380
Sewing	1.2	0.9	0.6	330

The system to be solved is

$$1.5x + y + 0.5z = 380 \quad (1)$$
$$1.2x + 0.9y + 0.6z = 330. \quad (2)$$

Simplify the equations. Multiply equation (1) by 2 and equation (2) by $\frac{10}{3}$.

$$3x + 2y + z = 760 \quad (3)$$
$$4x + 3y + 2z = 1100 \quad (4)$$

Make the leading coefficient of equation (3) equal 1.

$$\tfrac{1}{3}R_1 \to R_1 \qquad x + \frac{2}{3}y + \frac{1}{3}z = \frac{760}{3} \quad (5)$$
$$4x + 3y + 2z = 1100 \quad (4)$$

Eliminate x from equation (4).

$$x + \frac{2}{3}y + \frac{1}{3}z = \frac{760}{3} \quad (5)$$
$$-4R_1 + R_2 \to R_2 \qquad \frac{1}{3}y + \frac{2}{3}z = \frac{260}{3} \quad (6)$$

Make the leading coefficient of equation (6) equal 1.

$$x + \frac{2}{3}y + \frac{1}{3}z = \frac{760}{3} \quad (5)$$

$$3R_2 \to R_2 \qquad y + 2z = 260 \quad (7)$$

From equation (7), $y = 260 - 2z$. Substitute this into equation (5).

$$x + \frac{2}{3}(260 - 2z) + \frac{1}{3}z = \frac{760}{3}$$

$$x + \frac{520}{3} - \frac{4}{3}z + \frac{1}{3}z = \frac{760}{3}$$

$$x - z = \frac{240}{3}$$

$$x = z + 80$$

The solution is $(z + 80, 260 - 2z, z)$. In this problem $x, y,$ and z must be nonnegative, so

$$260 - 2z \geq 0$$
$$-2z \geq -260$$
$$z \leq 130.$$

Therefore, the plant should make $z + 80$ long-sleeve blouses, $260 - 2z$ short-sleeve blouses, and z sleeveless blouses with $0 \leq z \leq 130$.

47. **(a)** For the first equation, the first sighting in 2000 was on day $y = 759 - 0.338(2000) = 83$, or during the eighty-third day of the year. Since 2000 was a leap year, the eighty-third day fell on March 23.

For the second equation, the first sighting in 2000 was on day $y = 1637 - 0.779(2000) = 79$, or during the seventy-ninth day of the year. Since 2000 was a leap year, the seventh-ninth day fell on March 19.

(b) $y = 759 - 0.338x \quad (1)$
$\quad\;\; y = 1637 - 0.779x \quad (2)$

Rewrite equations so that variables are on the left side and constant term is on the right side.

$$0.338x + y = 759 \quad (3)$$
$$0.779x + y = 1637 \quad (4)$$

Eliminate y from equation (4).

$$0.338x + y = 759 \quad (3)$$
$$-1R_1 + R_2 \to R_2 \qquad 0.441x = 878 \quad (5)$$

Make leading coefficient for equation (5) equal 1.

$$0.338x + y = 759 \qquad (3)$$
$$\tfrac{1}{0.441}R_2 \to R_2 \qquad x = \frac{878}{0.441} \quad (6)$$

The two estimates agree in the year closest to $x = \frac{878}{0.441} \approx 1990.93$, so they agree in 1991. The estimated number of days into the year when a robin can be expected is

$$0.338\left(\frac{878}{0.441}\right) + y = 759$$

$$y \approx 86.$$

49. Let $x =$ number of free throws, and
$\quad\;\; y =$ number of foul shots.

Then

$$x + y = \;\;64 \quad (1)$$
$$2x + y = 100 \quad (2).$$

Eliminate x in equation (2).

$$x + y = 64 \qquad (1)$$
$$-2R_1 + R_2 \to R_2 \quad -y = -28 \quad (3)$$

Make the coefficients of the first term of each equation equal 1.

$$x + y = 64 \quad (1)$$
$$-1R_2 \to R_2 \qquad y = 28 \quad (4)$$

Substitute 28 for y in equation (1) to get $x = 36$. Wilt Chamberlain made 36 free throws and 28 foul shots.

51. **(a)** Since 8 and 9 must be two of the four numbers combined using addition, subtraction, multiplication, and/or division to get 24, begin by finding two numbers to use with 8 and 9. One possibility is 8 and 3 since $(9 - 8) \cdot 8 \cdot 3 = 24$. If we can find values of x and y such that either $x + y = 8$ and $3x + 2y = 3$, or $x + y = 3$ and $3x + 2y = 8$, we will have found a solution. Solving the first system gives $x = -13$ and $y = 21$. This, however, does not satisfy the condition that x and y be single-digit positive integers. Solving the second system gives $x = 2$ and $y = 1$. Since both of these values are single-digit positive integers, we have one possible system. Thus, one system is

$$\begin{cases} x + y = 3 \\ 3x + 2y = 8 \end{cases}.$$

Its solution is $(2, 1)$. These values of x and y give the numbers 8, 9, 8, and 3 on the game card. These numbers can be combined as $(9 - 8) \cdot 8 \cdot 3$ to make 24.

2.2   Solution of Linear Systems by the Gauss-Jordan Method

1. $3x + y = 6$
  $2x + 5y = 15$

The equations are already in proper form. The augmented matrix obtained from the coefficients and the constants is

$$\begin{bmatrix} 3 & 1 & | & 6 \\ 2 & 5 & | & 15 \end{bmatrix}.$$

3. $2x + y + z = 3$
  $3x - 4y + 2z = -7$
  $x + y + z = 2$

leads to the augmented matrix

$$\begin{bmatrix} 2 & 1 & 1 & | & 3 \\ 3 & -4 & 2 & | & -7 \\ 1 & 1 & 1 & | & 2 \end{bmatrix}.$$

5. We are given the augmented matrix

$$\begin{bmatrix} 1 & 0 & | & 2 \\ 0 & 1 & | & 3 \end{bmatrix}.$$

This is equivalent to the system of equations

$$\begin{aligned} x \quad &= 2 \\ y &= 3, \end{aligned}$$

or $x = 2$, $y = 3$.

7. $\begin{bmatrix} 1 & 0 & 0 & | & 4 \\ 0 & 1 & 0 & | & -5 \\ 0 & 0 & 1 & | & 1 \end{bmatrix}$

The system associated with this matrix is

$$\begin{aligned} x \quad\quad &= 4 \\ y \quad &= -5 \\ z &= 1, \end{aligned}$$

or $x = 4$, $y = -5$, $z = 1$.

9. *Row operations* on a matrix correspond to transformations of a system of equations.

11. $\begin{bmatrix} 3 & 7 & 4 & | & 10 \\ 1 & 2 & 3 & | & 6 \\ 0 & 4 & 5 & | & 11 \end{bmatrix}$

Find $R_1 + (-3)R_2$.

In row 2, column 1,

$$3 + (-3)1 = 0.$$

In row 2, column 2,

$$7 + (-3)2 = 1.$$

In row 2, column 3,

$$4 + (-3)3 = -5.$$

In row 2, column 4,

$$10 + (-3)6 = -8.$$

Replace R_2 with these values. The new matrix is

$$\begin{bmatrix} 3 & 7 & 4 & | & 10 \\ 0 & 1 & -5 & | & -8 \\ 0 & 4 & 5 & | & 11 \end{bmatrix}.$$

13. $\begin{bmatrix} 1 & 6 & 4 & | & 7 \\ 0 & 3 & 2 & | & 5 \\ 0 & 5 & 3 & | & 7 \end{bmatrix}$

Find $(-2)R_2 + R_1 \to R_1$

$$\begin{bmatrix} (-2)0 + 1 & (-2)3 + 6 & (-2)2 + 4 & | & (-2)5 + 7 \\ 0 & 3 & 2 & | & 5 \\ 0 & 5 & 3 & | & 7 \end{bmatrix}$$

$$= \begin{bmatrix} 1 & 0 & 0 & | & -3 \\ 0 & 3 & 2 & | & 5 \\ 0 & 5 & 3 & | & 7 \end{bmatrix}$$

15. $\begin{bmatrix} 3 & 0 & 0 & | & 18 \\ 0 & 5 & 0 & | & 9 \\ 0 & 0 & 4 & | & 8 \end{bmatrix}$

$\frac{1}{3}R_1 \to R_1$

$$\begin{bmatrix} \frac{1}{3}(3) & \frac{1}{3}(0) & \frac{1}{3}(0) & | & \frac{1}{3}(18) \\ 0 & 5 & 0 & | & 9 \\ 0 & 0 & 4 & | & 8 \end{bmatrix} = \begin{bmatrix} 1 & 0 & 0 & | & 6 \\ 0 & 5 & 0 & | & 9 \\ 0 & 0 & 4 & | & 8 \end{bmatrix}$$

17. $x + y = 5$
$3x + 2y = 12$

Write the augmented matrix and use row operations.

$$\begin{bmatrix} 1 & 1 & | & 5 \\ 3 & 2 & | & 12 \end{bmatrix}$$

$-3R_1 + R_2 \to R_2$ $\begin{bmatrix} 1 & 1 & | & 5 \\ 0 & -1 & | & -3 \end{bmatrix}$

$-1R_2 \to R_2$ $\begin{bmatrix} 1 & 1 & | & 5 \\ 0 & 1 & | & 3 \end{bmatrix}$

$-1R_2 + R_1 \to R_1$ $\begin{bmatrix} 1 & 0 & | & 2 \\ 0 & 1 & | & 3 \end{bmatrix}$

The solution is $(2, 3)$.

19. $x + y = 7$
$4x + 3y = 22$

Write the augmented matrix and use row operations.

$$\begin{bmatrix} 1 & 1 & | & 7 \\ 4 & 3 & | & 22 \end{bmatrix}$$

$-4R_1 + R_2 \to R_2$ $\begin{bmatrix} 1 & 1 & | & 7 \\ 0 & -1 & | & -6 \end{bmatrix}$

$-1R_2 \to R_2$ $\begin{bmatrix} 1 & 1 & | & 7 \\ 0 & 1 & | & 6 \end{bmatrix}$

$-1R_2 + R_1 \to R_1$ $\begin{bmatrix} 1 & 0 & | & 1 \\ 0 & 1 & | & 6 \end{bmatrix}$

The solution is $(1, 6)$.

21. $2x - 3y = 2$
$4x - 6y = 1$

Write the augmented matrix and use row operations.

$$\begin{bmatrix} 2 & -3 & | & 2 \\ 4 & -6 & | & 1 \end{bmatrix}$$

$-2R_1 + R_2 \to R_2$ $\begin{bmatrix} 2 & -3 & | & 2 \\ 0 & 0 & | & -3 \end{bmatrix}$

The system associated with the last matrix is

$$2x - 3y = 2$$
$$0x + 0y = -3.$$

Since the second equation, $0 = -3$, is false, the system is inconsistent and therefore has no solution.

23. $6x - 3y = 1$
$-12x + 6y = -2$

Write the augmented matrix of the system and use row operations.

$$\begin{bmatrix} 6 & -3 & | & 1 \\ -12 & 6 & | & -2 \end{bmatrix}$$

$2R_1 + R_2 \to R_2$ $\begin{bmatrix} 6 & -3 & | & 1 \\ 0 & 0 & | & 0 \end{bmatrix}$

$\frac{1}{6}R_1 \to R_1$ $\begin{bmatrix} 1 & -\frac{1}{2} & | & \frac{1}{6} \\ 0 & 0 & | & 0 \end{bmatrix}$

This is as far as we can go with the Gauss-Jordan method. To complete the solution, write the equation that corresponds to the first row of the matrix.

$$x - \frac{1}{2}y = \frac{1}{6}$$

Solve this equation for x in terms of y.

$$x = \frac{1}{2}y + \frac{1}{6}$$
$$= \frac{3y + 1}{6}$$

The solution is $\left(\frac{3y+1}{6}, y\right)$, where y is any real number.

25. $y = x - 3$
$y = 1 + z$
$z = 4 - x$

First write the system in proper form.

$$\begin{aligned} -x + y &= -3 \\ y - z &= 1 \\ x + z &= 4 \end{aligned}$$

Write the augmented matrix and use row operations.

$$\begin{bmatrix} -1 & 1 & 0 & | & -3 \\ 0 & 1 & -1 & | & 1 \\ 1 & 0 & 1 & | & 4 \end{bmatrix}$$

$-1R_1 \rightarrow R_1$
$$\begin{bmatrix} 1 & -1 & 0 & | & 3 \\ 0 & 1 & -1 & | & 1 \\ 1 & 0 & 1 & | & 4 \end{bmatrix}$$

$-1R_1 + R_3 \rightarrow R_3$
$$\begin{bmatrix} 1 & -1 & 0 & | & 3 \\ 0 & 1 & -1 & | & 1 \\ 0 & 1 & 1 & | & 1 \end{bmatrix}$$

$R_2 + R_1 \rightarrow R_1$
$-1R_2 + R_3 \rightarrow R_3$
$$\begin{bmatrix} 1 & 0 & -1 & | & 4 \\ 0 & 1 & -1 & | & 1 \\ 0 & 0 & 2 & | & 0 \end{bmatrix}$$

$R_3 + 2R_1 \rightarrow R_1$
$R_3 + 2R_2 \rightarrow R_2$
$$\begin{bmatrix} 2 & 0 & 0 & | & 8 \\ 0 & 2 & 0 & | & 2 \\ 0 & 0 & 2 & | & 0 \end{bmatrix}$$

$\frac{1}{2}R_1 \rightarrow R_1$
$\frac{1}{2}R_2 \rightarrow R_2$
$\frac{1}{2}R_3 \rightarrow R_3$
$$\begin{bmatrix} 1 & 0 & 0 & | & 4 \\ 0 & 1 & 0 & | & 1 \\ 0 & 0 & 1 & | & 0 \end{bmatrix}$$

The solution is $(4, 1, 0)$.

27. $2x - 2y = -5$
$2y + z = 0$
$2x + z = -7$

Write the augmented matrix and use row operations.

$$\begin{bmatrix} 2 & -2 & 0 & | & -5 \\ 0 & 2 & 1 & | & 0 \\ 2 & 0 & 1 & | & -7 \end{bmatrix}$$

$-1R_1 + R_3 \rightarrow R_3$
$$\begin{bmatrix} 2 & -2 & 0 & | & -5 \\ 0 & 2 & 1 & | & 0 \\ 0 & 2 & 1 & | & -2 \end{bmatrix}$$

$R_2 + R_1 \rightarrow R_1$
$-1R_2 + R_3 \rightarrow R_3$
$$\begin{bmatrix} 2 & 0 & 1 & | & -5 \\ 0 & 2 & 1 & | & 0 \\ 0 & 0 & 0 & | & -2 \end{bmatrix}$$

This matrix corresponds to the system of equations

$$2x + z = -5$$
$$2y + z = 0$$
$$0 = -2.$$

This false statement $0 = -2$ indicates that the system is inconsistent and therefore has no solution.

29. $4x + 4y - 4z = 24$
$2x - y + z = -9$
$x - 2y + 3z = 1$

Write the augmented matrix and use row operations.

$$\begin{bmatrix} 4 & 4 & -4 & | & 24 \\ 2 & -1 & 1 & | & -9 \\ 1 & -2 & 3 & | & 1 \end{bmatrix}$$

$R_1 + (-2)R_2 \rightarrow R_2$
$R_1 + (-4)R_3 \rightarrow R_3$
$$\begin{bmatrix} 4 & 4 & -4 & | & 24 \\ 0 & 6 & -6 & | & 42 \\ 0 & 12 & -16 & | & 20 \end{bmatrix}$$

$2R_2 + (-3)R_1 \rightarrow R_1$
$-2R_2 + R_3 \rightarrow R_3$
$$\begin{bmatrix} -12 & 0 & 0 & | & 12 \\ 0 & 6 & -6 & | & 42 \\ 0 & 0 & -4 & | & -64 \end{bmatrix}$$

$-3R_3 + 2R_2 \rightarrow R_2$
$$\begin{bmatrix} -12 & 0 & 0 & | & 12 \\ 0 & 12 & 0 & | & 276 \\ 0 & 0 & -4 & | & -64 \end{bmatrix}$$

$-\frac{1}{12}R_1 \rightarrow R_1$
$-\frac{1}{12}R_2 \rightarrow R_2$
$-\frac{1}{4}R_3 \rightarrow R_3$
$$\begin{bmatrix} 1 & 0 & 0 & | & -1 \\ 0 & 1 & 0 & | & 23 \\ 0 & 0 & 1 & | & 16 \end{bmatrix}$$

The solution is $(-1, 23, 16)$.

31. $3x + 5y - z = 0$
$4x - y + 2z = 1$
$7x + 4y + z = 1$

Write the augmented matrix and use row operations.

$$\begin{bmatrix} 3 & 5 & -1 & | & 0 \\ 4 & -1 & 2 & | & 1 \\ 7 & 4 & 1 & | & 1 \end{bmatrix}$$

$4R_1 + (-3)R_2 \rightarrow R_2$
$7R_1 + (-3)R_3 \rightarrow R_3$
$$\begin{bmatrix} 3 & 5 & -1 & | & 0 \\ 0 & 23 & -10 & | & -3 \\ 0 & 23 & -10 & | & -3 \end{bmatrix}$$

$23R_1 + (-5)R_2 \rightarrow R_1$
$R_2 + (-1)R_3 \rightarrow R_3$
$$\begin{bmatrix} 69 & 0 & 27 & | & 15 \\ 0 & 23 & -10 & | & -3 \\ 0 & 0 & 0 & | & 0 \end{bmatrix}$$

$\frac{1}{69}R_1 \rightarrow R_1$
$\frac{1}{23}R_2 \rightarrow R_2$
$$\begin{bmatrix} 1 & 0 & \frac{9}{23} & | & \frac{5}{23} \\ 0 & 1 & -\frac{10}{23} & | & -\frac{3}{23} \\ 0 & 0 & 0 & | & 0 \end{bmatrix}$$

The row of zeros indicates dependent equations. Solve the first two equations respectively for x and y in terms of z to obtain

$$x = -\frac{9}{23}z + \frac{5}{23} = \frac{-9z + 5}{23}$$

and

$$y = \frac{10}{23}z - \frac{3}{23} = \frac{10x - 3}{23}.$$

The solution is $\left(\frac{-9z+5}{23}, \frac{10z-3}{23}, z\right)$.

33. $5x - 4y + 2z = 6$
    $5x + 3y - z = 11$
    $15x - 5y + 3z = 23$

Write the augmented matrix and use row operations.

$$\begin{bmatrix} 5 & -4 & 2 & | & 6 \\ 5 & 3 & -1 & | & 11 \\ 15 & -5 & 3 & | & 23 \end{bmatrix}$$

$$\begin{matrix} \\ -1R_1 + R_2 \rightarrow R_2 \\ -3R_1 + R_3 \rightarrow R_3 \end{matrix} \begin{bmatrix} 5 & -4 & 2 & | & 6 \\ 0 & 7 & -3 & | & 5 \\ 0 & 7 & -3 & | & 5 \end{bmatrix}$$

$$\begin{matrix} 4R_2 + 7R_1 \rightarrow R_1 \\ \\ -1R_2 + R_3 \rightarrow R_3 \end{matrix} \begin{bmatrix} 35 & 0 & 2 & | & 62 \\ 0 & 7 & -3 & | & 5 \\ 0 & 0 & 0 & | & 0 \end{bmatrix}$$

$$\begin{matrix} \frac{1}{35}R_1 \rightarrow R_1 \\ \\ \frac{1}{7}R_2 \rightarrow R_2 \\ \\ \end{matrix} \begin{bmatrix} 1 & 0 & \frac{2}{35} & | & \frac{62}{35} \\ 0 & 1 & -\frac{3}{7} & | & \frac{5}{7} \\ 0 & 0 & 0 & | & 0 \end{bmatrix}$$

The row of zeros indicates dependent equations. Solve the first two equations respectively for x and y in terms of z to obtain

$$x = -\frac{2}{35}z + \frac{62}{35} = \frac{-2z + 62}{35}$$

and

$$y = \frac{3}{7}z + \frac{5}{7} = \frac{3z + 5}{7}.$$

The solution is $\left(\frac{-2z+62}{35}, \frac{3z+5}{7}, z\right)$.

35. $2x + 3y + z = 9$
    $4x + 6y + 2z = 18$
    $-\frac{1}{2}x - \frac{3}{4}y - \frac{1}{4}z = -\frac{9}{4}$

Write the augmented matrix and use row operations.

$$\begin{bmatrix} 2 & 3 & 1 & | & 9 \\ 4 & 6 & 2 & | & 18 \\ -\frac{1}{2} & -\frac{3}{4} & -\frac{1}{4} & | & -\frac{9}{4} \end{bmatrix}$$

$$\begin{matrix} -2R_1 + R_2 \rightarrow R_2 \\ \frac{1}{4}R_1 + R_3 \rightarrow R_3 \end{matrix} \begin{bmatrix} 2 & 3 & 1 & | & 9 \\ 0 & 0 & 0 & | & 0 \\ 0 & 0 & 0 & | & 0 \end{bmatrix}$$

The rows of zeros indicate dependent equations. Since the equation involves $x, y,$ and z, let y and z be parameters. Solve the equation for x to obtain $x = \frac{9 - 3y - z}{2}$.

The solution is $\left(\frac{9-3y-z}{2}, y, z\right)$, where y and z are any real numbers.

37. $x + 2y \qquad - w = 3$
    $2x \qquad + 4z + 2w = -6$
    $x + 2y - z \qquad = 6$
    $2x - y + z + w = -3$

Write the augmented matrix and use row operations.

$$\begin{bmatrix} 1 & 2 & 0 & -1 & | & 3 \\ 2 & 0 & 4 & 2 & | & -6 \\ 1 & 2 & -1 & 0 & | & 6 \\ 2 & -1 & 1 & 1 & | & -3 \end{bmatrix}$$

$$\begin{matrix} \\ -2R_1 + R_2 \rightarrow R_2 \\ -1R_1 + R_3 \rightarrow R_3 \\ -2R_1 + R_4 \rightarrow R_4 \end{matrix} \begin{bmatrix} 1 & 2 & 0 & -1 & | & 3 \\ 0 & -4 & 4 & 4 & | & -12 \\ 0 & 0 & -1 & 1 & | & 3 \\ 0 & -5 & 1 & 3 & | & -9 \end{bmatrix}$$

$$\begin{matrix} R_2 + 2R_1 \rightarrow R_1 \\ \\ \\ -5R_2 + 4R_4 \rightarrow R_4 \end{matrix} \begin{bmatrix} 2 & 0 & 4 & 2 & | & -6 \\ 0 & -4 & 4 & 4 & | & -12 \\ 0 & 0 & -1 & 1 & | & 3 \\ 0 & 0 & -16 & -8 & | & 24 \end{bmatrix}$$

$$\begin{matrix} 4R_3 + R_1 \rightarrow R_1 \\ 4R_3 + R_2 \rightarrow R_2 \\ \\ 16R_3 + (-1)R_4 \rightarrow R_4 \end{matrix} \begin{bmatrix} 2 & 0 & 0 & 6 & | & 6 \\ 0 & -4 & 0 & 8 & | & 0 \\ 0 & 0 & -1 & 1 & | & 3 \\ 0 & 0 & 0 & 24 & | & 24 \end{bmatrix}$$

$$\begin{matrix} R_4 + (-4R_1) \rightarrow R_1 \\ R_4 + (-3R_2) \rightarrow R_2 \\ R_4 + (-24R_3) \rightarrow R_3 \\ \\ \end{matrix} \begin{bmatrix} -8 & 0 & 0 & 0 & | & 0 \\ 0 & 12 & 0 & 0 & | & 24 \\ 0 & 0 & 24 & 0 & | & -48 \\ 0 & 0 & 0 & 24 & | & 24 \end{bmatrix}$$

$$\begin{matrix} -\frac{1}{8}R_1 \rightarrow R_1 \\ \frac{1}{12}R_2 \rightarrow R_2 \\ \frac{1}{24}R_3 \rightarrow R_3 \\ \frac{1}{24}R_4 \rightarrow R_4 \end{matrix} \begin{bmatrix} 1 & 0 & 0 & 0 & | & 0 \\ 0 & 1 & 0 & 0 & | & 2 \\ 0 & 0 & 1 & 0 & | & -2 \\ 0 & 0 & 0 & 1 & | & 1 \end{bmatrix}$$

The solution is $x = 0$, $y = 2$, $z = -2$, $w = 1$, or $(0, 2, -2, 1)$.

39. $x + y - z + 2w = -20$
  $2x - y + z + w = 11$
  $3x - 2y + z - 2w = 27$

$$\begin{bmatrix} 1 & 1 & -1 & 2 & \bigm| & -20 \\ 2 & -1 & 1 & 1 & \bigm| & 11 \\ 3 & -2 & 1 & -2 & \bigm| & 27 \end{bmatrix}$$

$-2R_1 + R_2 \rightarrow R_2$
$-3R_1 + R_3 \rightarrow R_3$
$$\begin{bmatrix} 1 & 1 & -1 & 2 & \bigm| & -20 \\ 0 & -3 & 3 & -3 & \bigm| & 51 \\ 0 & -5 & 4 & -8 & \bigm| & 87 \end{bmatrix}$$

$-\frac{1}{3}R_2 \rightarrow R_2$
$$\begin{bmatrix} 1 & 1 & -1 & 2 & \bigm| & -20 \\ 0 & 1 & -1 & 1 & \bigm| & -17 \\ 0 & -5 & 4 & -8 & \bigm| & 87 \end{bmatrix}$$

$-1R_2 + R_1 \rightarrow R_1$
$5R_2 + R_3 \rightarrow R_3$
$$\begin{bmatrix} 1 & 0 & 0 & 1 & \bigm| & -3 \\ 0 & 1 & -1 & 1 & \bigm| & -17 \\ 0 & 0 & -1 & -3 & \bigm| & 2 \end{bmatrix}$$

$-1R_3 \rightarrow R_3$
$$\begin{bmatrix} 1 & 0 & 0 & 1 & \bigm| & -3 \\ 0 & 1 & -1 & 1 & \bigm| & -17 \\ 0 & 0 & 1 & 3 & \bigm| & -2 \end{bmatrix}$$

$R_3 + R_2 \rightarrow R_2$
$$\begin{bmatrix} 1 & 0 & 0 & 1 & \bigm| & -3 \\ 0 & 1 & 0 & 4 & \bigm| & -19 \\ 0 & 0 & 1 & 3 & \bigm| & -2 \end{bmatrix}$$

This is as far as we can go using row operations. To complete the solution, write the equations that correspond to the matrix.

$$\begin{aligned} x + w &= -3 \\ y + 4w &= -19 \\ z + 3w &= -2 \end{aligned}$$

Let w be the parameter and express x, y, and z in terms of w. From the equations above, $x = -w - 3$, $y = -4w - 19$, and $z = -3w - 2$.
The solution is $(-w - 3, -4w - 19, -3w - 2, w)$, where w is any real number.

41. $10.47x + 3.52y + 2.58z - 6.42w = 218.65$
  $8.62x - 4.93y - 1.75z + 2.83w = 157.03$
  $4.92x + 6.83y - 2.97z + 2.65w = 462.3$
  $2.86x + 19.10y - 6.24z - 8.73w = 398.4$

Write the augmented matrix of the system.

$$\begin{bmatrix} 10.47 & 3.52 & 2.58 & -6.42 & \bigm| & 218.65 \\ 8.62 & -4.93 & -1.75 & 2.83 & \bigm| & 157.03 \\ 4.92 & 6.83 & -2.97 & 2.65 & \bigm| & 462.3 \\ 2.86 & 19.10 & -6.24 & -8.73 & \bigm| & 398.4 \end{bmatrix}$$

This exercise should be solved by graphing calculator or computer methods. The solution, which may vary slightly, is $x \approx 28.9436$, $y \approx 36.6326$, $z \approx 9.6390$, and $w \approx 37.1036$, or

$$(28.9436, 36.6326, 9.6390, 37.1036).$$

43. Insert the given values, introduce variables, and the table is as follows.

$\frac{3}{8}$	a	b
c	d	$\frac{1}{4}$
e	f	g

From this, we obtain the following system of equations.

$$
\begin{aligned}
a + b \qquad\qquad\qquad\qquad + \tfrac{3}{8} &= 1 \\
c + d \qquad\qquad + \tfrac{1}{4} &= 1 \\
e \; + f \; + g \qquad\qquad &= 1 \\
c \qquad + e \qquad\quad + \tfrac{3}{8} &= 1 \\
a \qquad + d \qquad\quad + f \qquad &= 1 \\
b \qquad\qquad\qquad + g \; + \tfrac{1}{4} &= 1 \\
d \qquad\qquad + g \; + \tfrac{3}{8} &= 1 \\
b \; + d \; + e \qquad\qquad &= 1
\end{aligned}
$$

The augmented matrix and the final form after row operations are as follows.

$$
\begin{bmatrix}
1 & 1 & 0 & 0 & 0 & 0 & 0 & \frac{5}{8} \\
0 & 0 & 1 & 1 & 0 & 0 & 0 & \frac{3}{4} \\
0 & 0 & 0 & 0 & 1 & 1 & 1 & 1 \\
0 & 0 & 1 & 0 & 1 & 0 & 0 & \frac{5}{8} \\
1 & 0 & 0 & 1 & 0 & 1 & 0 & 1 \\
0 & 1 & 0 & 0 & 0 & 0 & 1 & \frac{3}{4} \\
0 & 0 & 0 & 1 & 0 & 0 & 1 & \frac{5}{8} \\
0 & 1 & 0 & 1 & 1 & 0 & 0 & 1
\end{bmatrix}
\rightarrow
\begin{bmatrix}
1 & 0 & 0 & 0 & 0 & 0 & 0 & \frac{1}{6} \\
0 & 1 & 0 & 0 & 0 & 0 & 0 & \frac{11}{24} \\
0 & 0 & 1 & 0 & 0 & 0 & 0 & \frac{5}{12} \\
0 & 0 & 0 & 1 & 0 & 0 & 0 & \frac{1}{3} \\
0 & 0 & 0 & 0 & 1 & 0 & 0 & \frac{5}{24} \\
0 & 0 & 0 & 0 & 0 & 1 & 0 & \frac{1}{2} \\
0 & 0 & 0 & 0 & 0 & 0 & 1 & \frac{7}{12} \\
0 & 0 & 0 & 0 & 0 & 0 & 0 & 0
\end{bmatrix}
$$

The solution to the system is read from the last column.

$$a = \tfrac{1}{6}, b = \tfrac{11}{24}, c = \tfrac{5}{12}, d = \tfrac{1}{3}, e = \tfrac{5}{24}, f = \tfrac{1}{2}, \text{ and } g = \tfrac{7}{24}$$

So the magic square is:

$\frac{3}{8}$	$\frac{1}{6}$	$\frac{11}{24}$
$\frac{5}{12}$	$\frac{1}{3}$	$\frac{1}{4}$
$\frac{5}{24}$	$\frac{1}{2}$	$\frac{7}{24}$

45. Let $x =$ the number of units ordered from Toronto,

$y =$ the number of units ordered from Montreal, and

$z =$ the number of units ordered from Ottawa.

The system to be solved is

$$\begin{aligned} x + y + z &= 100 \quad (1) \\ 80x + 50y + 65z &= 5990 \quad (2) \\ x &= z. \quad (3) \end{aligned}$$

Write the augmented matrix of the system.

$$\begin{bmatrix} 1 & 1 & 1 & 100 \\ 80 & 50 & 65 & 5990 \\ 1 & 0 & -1 & 0 \end{bmatrix}$$

$$\begin{array}{c} -80R_1 + R_2 \rightarrow R_2 \\ -1R_1 + R_3 \rightarrow R_3 \end{array} \begin{bmatrix} 1 & 1 & 1 & 100 \\ 0 & -30 & -15 & -2010 \\ 0 & -1 & -2 & -100 \end{bmatrix}$$

$$-\tfrac{1}{30}R_2 \rightarrow R_2 \begin{bmatrix} 1 & 1 & 1 & 100 \\ 0 & 1 & \tfrac{1}{2} & 67 \\ 0 & -1 & -2 & -100 \end{bmatrix}$$

$$\begin{array}{c} -1R_2 + R_1 \rightarrow R_1 \\ \\ R_2 + R_3 \rightarrow R_3 \end{array} \begin{bmatrix} 1 & 0 & \tfrac{1}{2} & 33 \\ 0 & 1 & \tfrac{1}{2} & 67 \\ 0 & 0 & -\tfrac{3}{2} & -33 \end{bmatrix}$$

$$-\tfrac{2}{3}R_3 \rightarrow R_3 \begin{bmatrix} 1 & 0 & \tfrac{1}{2} & 33 \\ 0 & 1 & \tfrac{1}{2} & 67 \\ 0 & 0 & 1 & 22 \end{bmatrix}$$

$$\begin{array}{c} -\tfrac{1}{2}R_3 + R_1 \rightarrow R_1 \\ -\tfrac{1}{2}R_3 + R_2 \rightarrow R_2 \end{array} \begin{bmatrix} 1 & 0 & 0 & 22 \\ 0 & 1 & 0 & 56 \\ 0 & 0 & 1 & 22 \end{bmatrix}$$

The solution is $(22, 56, 22)$. There were 22 units ordered from Toronto, 56 units ordered from Montreal, and 22 units ordered from Ottawa.

47. (a) Let $x =$ the number of deluxe models,

$y =$ the number of super-deluxe models, and

$z =$ the number of ultra models.

Make a table to organize the information.

	Deluxe	Super-Deluxe	Ultra	Totals
Electronic	2	1	3	100
Assembly	2	3	2	100
Finishing	1	1	2	65

The system to be solved is

$$\begin{aligned} 2x + y + 3z &= 100 \quad (1) \\ 2x + 3y + 2z &= 100 \quad (2) \\ x + y + 2z &= 65. \quad (3) \end{aligned}$$

Write the augmented matrix of the system.

$$\begin{bmatrix} 2 & 1 & 3 & 100 \\ 2 & 3 & 2 & 100 \\ 1 & 1 & 2 & 65 \end{bmatrix}$$

Interchange rows 1 and 3.

$$\begin{bmatrix} 1 & 1 & 2 & 65 \\ 2 & 3 & 2 & 100 \\ 2 & 1 & 3 & 100 \end{bmatrix}$$

$$\begin{array}{c} -2R_1 + R_2 \rightarrow R_2 \\ -2R_1 + R_3 \rightarrow R_3 \end{array} \begin{bmatrix} 1 & 1 & 2 & 65 \\ 0 & 1 & -2 & -30 \\ 0 & -1 & -1 & -30 \end{bmatrix}$$

$$\begin{array}{c} -1R_2 + R_1 \rightarrow R_1 \\ \\ R_2 + R_3 \rightarrow R_3 \end{array} \begin{bmatrix} 1 & 0 & 4 & 95 \\ 0 & 1 & -2 & -30 \\ 0 & 0 & -3 & -60 \end{bmatrix}$$

$$-\tfrac{1}{3}R_3 \rightarrow R_3 \begin{bmatrix} 1 & 0 & 4 & 95 \\ 0 & 1 & -2 & -30 \\ 0 & 0 & 1 & 20 \end{bmatrix}$$

$$\begin{array}{c} -4R_3 + R_1 \rightarrow R_1 \\ 2R_3 + R_2 \rightarrow R_2 \end{array} \begin{bmatrix} 1 & 0 & 0 & 15 \\ 0 & 1 & 0 & 10 \\ 0 & 0 & 1 & 20 \end{bmatrix}$$

The solution is $(15, 10, 20)$. Each week 15 deluxe models, 10 super-deluxe models, and 20 ultra models should be produced.

(b) The system to be solved is

$$\begin{aligned} 2x + y + 6z &= 100 \\ 2x + 3y + 2z &= 100 \\ x + y + 2z &= 65 \end{aligned}$$

Write the augmented matrix of the system.

$$\begin{bmatrix} 2 & 1 & 6 & 100 \\ 2 & 3 & 2 & 100 \\ 1 & 1 & 2 & 65 \end{bmatrix}$$

Interchange rows 1 and 3.

$$\begin{bmatrix} 1 & 1 & 2 & | & 65 \\ 2 & 3 & 2 & | & 100 \\ 2 & 1 & 6 & | & 100 \end{bmatrix}$$

$\begin{matrix} -2R_1 + R_2 \rightarrow R_2 \\ -2R_1 + R_3 \rightarrow R_3 \end{matrix}$ $\begin{bmatrix} 1 & 1 & 2 & | & 65 \\ 0 & 1 & -2 & | & -30 \\ 0 & -1 & 2 & | & -30 \end{bmatrix}$

$\begin{matrix} -1R_2 + R_1 \rightarrow R_1 \\ \\ R_2 + R_3 \rightarrow R_3 \end{matrix}$ $\begin{bmatrix} 1 & 0 & 4 & | & 95 \\ 0 & 1 & -2 & | & -30 \\ 0 & 0 & 0 & | & -60 \end{bmatrix}$

The last row indicates there is no solution to the system.

(c) The system to be solved is

$$\begin{aligned} 2x + \ y + 6z &= 160 \\ 2x + 3y + 2z &= 100 \\ x + \ y + 2z &= \ 65 \end{aligned}$$

Write the augmented matrix of the system.

$$\begin{bmatrix} 2 & 1 & 6 & | & 160 \\ 2 & 3 & 2 & | & 100 \\ 1 & 1 & 2 & | & 65 \end{bmatrix}$$

Interchange rows 1 and 3.

$$\begin{bmatrix} 1 & 1 & 2 & | & 65 \\ 2 & 3 & 2 & | & 100 \\ 2 & 1 & 6 & | & 160 \end{bmatrix}$$

$\begin{matrix} -2R_1 + R_2 \rightarrow R_2 \\ -2R_1 + R_3 \rightarrow R_3 \end{matrix}$ $\begin{bmatrix} 1 & 1 & 2 & | & 65 \\ 0 & 1 & -2 & | & -30 \\ 0 & -1 & 2 & | & -30 \end{bmatrix}$

$\begin{matrix} -1R_2 + R_1 \rightarrow R_1 \\ \\ R_2 + R_3 \rightarrow R_3 \end{matrix}$ $\begin{bmatrix} 1 & 0 & 4 & | & 95 \\ 0 & 1 & -2 & | & -30 \\ 0 & 0 & 0 & | & 0 \end{bmatrix}$

The system is dependent. Let z be the parameter and solve the first two equations for x and y, yielding

$$x = 95 - 4z \quad \text{and} \quad y = 2z - 30.$$

Since $x, y,$ and z must be nonnegative integers, $95 - 4z \geq 0$ and $2z - 30 \geq 0$.

$$\begin{aligned} 95 - 4z &\geq 0 \\ -4z &\geq -95 \\ z &\leq 23.75 \\ 2z - 30 &\geq 0 \\ 2z &\geq 30 \\ z &\geq 15 \end{aligned}$$

Thus, z in an integer such that $15 \geq z \leq 23.75$. There are 9 solutions.

49. Let $x =$ the number of vans to be purchased,
$y =$ the number of small trucks to be purchased, and
$z =$ the number of large trucks to be purchased.

The system to be solved is

$$\begin{aligned} x + y + z &= 200 \\ 35{,}000x + 30{,}000y + 50{,}000z &= 7{,}000{,}000 \\ x &= 2y. \end{aligned}$$

To simplify the system, divide the second equation by 1000. Write the system in proper form, obtain the augmented matrix, and use row operations to solve.

$$\begin{bmatrix} 1 & 1 & 1 & | & 200 \\ 35 & 30 & 50 & | & 7000 \\ 1 & -2 & 0 & | & 0 \end{bmatrix}$$

$\begin{matrix} -35R_1 + R_2 \rightarrow R_2 \\ -1R_1 + R_3 \rightarrow R_3 \end{matrix}$ $\begin{bmatrix} 1 & 1 & 1 & | & 200 \\ 0 & -5 & 15 & | & 0 \\ 0 & -3 & -1 & | & -200 \end{bmatrix}$

$\begin{matrix} R_2 + 5R_1 \rightarrow R_1 \\ \\ -3R_2 + 5R_3 \rightarrow R_3 \end{matrix}$ $\begin{bmatrix} 5 & 0 & 20 & | & 1000 \\ 0 & -5 & 15 & | & 0 \\ 0 & 0 & -50 & | & -1000 \end{bmatrix}$

$\begin{matrix} 2R_3 + 5R_1 \rightarrow R_1 \\ 3R_3 + 10R_2 \rightarrow R_2 \end{matrix}$ $\begin{bmatrix} 25 & 0 & 0 & | & 3000 \\ 0 & -50 & 0 & | & -3000 \\ 0 & 0 & -50 & | & -1000 \end{bmatrix}$

$\begin{matrix} \frac{1}{25}R_1 \rightarrow R_1 \\ \\ -\frac{1}{50}R_2 \rightarrow R_2 \\ \\ -\frac{1}{50}R_3 \rightarrow R_3 \end{matrix}$ $\begin{bmatrix} 1 & 0 & 0 & | & 120 \\ 0 & 1 & 0 & | & 60 \\ 0 & 0 & 1 & | & 20 \end{bmatrix}$

The solution is (120, 60, 20). U-Drive Rent-A-Truck should buy 120 vans, 60 small trucks, and 20 large trucks.

51. Let $x_1 =$ the number of cars sent from
  I to A,

 $x_2 =$ the number of cars sent from
  II to A,

 $x_3 =$ the number of cars sent from
  I to B, and

 $x_4 =$ the number of cars sent from
  II to B.

	A	B
I	x_1	x_3
II	x_2	x_4

Plant I has 28 cars, so

$$x_1 + x_3 = 28.$$

Plant II has 8 cars, so

$$x_2 + x_4 = 8.$$

Dealer A needs 20 cars, so

$$x_1 + x_2 = 20.$$

Dealer B needs 16 cars, so

$$x_3 + x_4 = 16.$$

The total transportation cost is $10,640, so

$$220x_1 + 400x_2 + 300x_3 + 180x_4 = 10,640.$$

The system to be solved is

$$
\begin{array}{rcrcrcrcr}
x_1 & & & + & x_3 & & & = & 28 \\
& & x_2 & & & + & x_4 & = & 8 \\
x_1 & + & x_2 & & & & & = & 20 \\
& & & & x_3 & + & x_4 & = & 16 \\
220x_1 & + & 400x_2 & + & 300x_3 & + & 180x_4 & = & 10,640.
\end{array}
$$

Write the augmented matrix and use row operations.

$$
\left[\begin{array}{cccc|c}
1 & 0 & 1 & 0 & 28 \\
0 & 1 & 0 & 1 & 8 \\
1 & 1 & 0 & 0 & 20 \\
0 & 0 & 1 & 1 & 16 \\
220 & 400 & 300 & 180 & 10,640
\end{array}\right]
$$

$$
\begin{array}{c}
\\
\\
-1R_1 + R_3 \rightarrow R_3 \\
\\
-220R_1 + R_5 \rightarrow R_5
\end{array}
\left[\begin{array}{cccc|c}
1 & 0 & 1 & 0 & 28 \\
0 & 1 & 0 & 1 & 8 \\
0 & 1 & -1 & 0 & -8 \\
0 & 0 & 1 & 1 & 16 \\
0 & 400 & 80 & 180 & 4480
\end{array}\right]
$$

$$
\begin{array}{c}
\\
\\
-1R_2 + R_3 \rightarrow R_3 \\
\\
-400R_2 + R_5 \rightarrow R_5
\end{array}
\left[\begin{array}{cccc|c}
1 & 0 & 1 & 0 & 28 \\
0 & 1 & 0 & 1 & 8 \\
0 & 0 & -1 & -1 & -16 \\
0 & 0 & 1 & 1 & 16 \\
0 & 0 & 80 & -220 & 1280
\end{array}\right]
$$

$$
\begin{array}{c}
R_1 + R_3 \rightarrow R_1 \\
\\
\\
R_3 + R_4 \rightarrow R_4 \\
80R_3 + R_5 \rightarrow R_5
\end{array}
\left[\begin{array}{cccc|c}
1 & 0 & 0 & -1 & 12 \\
0 & 1 & 0 & 1 & 8 \\
0 & 0 & -1 & -1 & -16 \\
0 & 0 & 0 & 0 & 0 \\
0 & 0 & 0 & -300 & 0
\end{array}\right]
$$

There is a 0 now in row 4, column 4, where we would like to get a 1. To proceed, interchange the fourth and fifth rows.

$$
\left[\begin{array}{cccc|c}
1 & 0 & 0 & -1 & 12 \\
0 & 1 & 0 & 1 & 8 \\
0 & 0 & -1 & -1 & -16 \\
0 & 0 & 0 & -300 & 0 \\
0 & 0 & 0 & 0 & 0
\end{array}\right]
$$

$$
\begin{array}{c}
-1R_4 + 300R_1 \rightarrow R_1 \\
R_4 + 300R_2 \rightarrow R_2 \\
-1R_4 + 300R_3 \rightarrow R_3 \\
\\
\end{array}
\left[\begin{array}{cccc|c}
300 & 0 & 0 & 0 & 3600 \\
0 & 300 & 0 & 0 & 2400 \\
0 & 0 & -300 & 0 & -4800 \\
0 & 0 & 0 & -300 & 0 \\
0 & 0 & 0 & 0 & 0
\end{array}\right]
$$

$$
\begin{array}{c}
\frac{1}{300}R_1 \rightarrow R_1 \\
\frac{1}{300}R_2 \rightarrow R_2 \\
-\frac{1}{300}R_3 \rightarrow R_3 \\
-\frac{1}{300}R_4 \rightarrow R_4 \\
\end{array}
\left[\begin{array}{cccc|c}
1 & 0 & 0 & 0 & 12 \\
0 & 1 & 0 & 0 & 8 \\
0 & 0 & 1 & 0 & 16 \\
0 & 0 & 0 & 1 & 0 \\
0 & 0 & 0 & 0 & 0
\end{array}\right]
$$

Each of the original variables has a value, so the last row of all zeros may be ignored. The solution of the system is $x_1 = 12$, $x_2 = 8$, $x_3 = 16$, $x_4 = 0$. Therefore, 12 cars should be sent from I to A, 8 cars from II to A, 16 cars from I to B, and no cars from II to B.

53. Let $x =$ the number of Italian style vegetable packages,

$y =$ the number of French style vegetable packages, and

$z =$ the number of Oriental style vegetable packages.

$$0.3x \qquad + 0.2z = 16{,}200$$
$$0.3x + 0.6y + 0.5z = 41{,}400$$
$$0.4x + 0.4y + 0.3z = 29{,}400$$

Expressing the amount of each vegetable in thousands, the system becomes

$$0.3x \qquad + 0.2z = 16.2$$
$$0.3x + 0.6y + 0.5z = 41.4$$
$$0.4x + 0.4y + 0.3z = 29.4.$$

Multiply each equation by 10, and write the augmented matrix.

$$\begin{bmatrix} 3 & 0 & 2 & | & 162 \\ 3 & 6 & 5 & | & 414 \\ 4 & 4 & 3 & | & 294 \end{bmatrix}$$

$\frac{1}{3}R_1 \to R_1$
$$\begin{bmatrix} 1 & 0 & \frac{2}{3} & | & 54 \\ 3 & 6 & 5 & | & 414 \\ 4 & 4 & 3 & | & 294 \end{bmatrix}$$

$-3R_1 + R_2 \to R_2$
$-4R_1 + R_3 \to R_3$
$$\begin{bmatrix} 1 & 0 & \frac{2}{3} & | & 54 \\ 0 & 6 & 3 & | & 252 \\ 0 & 4 & \frac{1}{3} & | & 78 \end{bmatrix}$$

$\frac{1}{6}R_2 \to R_2$
$$\begin{bmatrix} 1 & 0 & \frac{2}{3} & | & 54 \\ 0 & 1 & \frac{1}{2} & | & 42 \\ 0 & 4 & \frac{1}{3} & | & 78 \end{bmatrix}$$

$-4R_2 + R_3 \to R_3$
$$\begin{bmatrix} 1 & 0 & \frac{2}{3} & | & 54 \\ 0 & 1 & \frac{1}{2} & | & 42 \\ 0 & 0 & -\frac{5}{3} & | & -90 \end{bmatrix}$$

$-\frac{3}{5}R_3 \to R_3$
$$\begin{bmatrix} 1 & 0 & \frac{2}{3} & | & 54 \\ 0 & 1 & \frac{1}{2} & | & 42 \\ 0 & 0 & 1 & | & 54 \end{bmatrix}$$

$-\frac{2}{3}R_3 + R_1 \to R_1$
$-\frac{1}{2}R_3 + R_2 \to R_2$
$$\begin{bmatrix} 1 & 0 & 0 & | & 18 \\ 0 & 1 & 0 & | & 15 \\ 0 & 0 & 1 & | & 54 \end{bmatrix}$$

Therefore, the company should prepare 18 thousand or 18,000 packages of Italian style, 15 thousand or 15,000 of French style, and 54 thousand or 54,000 of Oriental style vegetables.

55. Let $x_1 =$ the number of cases of Brand A,

$x_2 =$ the number of cases of Brand B,

$x_3 =$ the number of cases of Brand C, and

$x_4 =$ the number of cases of Brand D.

$$25x_1 + 50x_2 + 75x_3 + 100x_4 = 1200$$
$$30x_1 + 30x_2 + 30x_3 + 60x_4 = 600$$
$$30x_1 + 20x_2 + 20x_3 + 30x_4 = 400$$

The augmented matrix of the system is

$$\begin{bmatrix} 25 & 50 & 75 & 100 & | & 1200 \\ 30 & 30 & 30 & 60 & | & 600 \\ 30 & 20 & 20 & 30 & | & 400 \end{bmatrix}.$$

$\frac{1}{5}R_1 \to R_1$
$\frac{1}{30}R_2 \to R_2$
$\frac{1}{10}R_3 \to R_3$
$$\begin{bmatrix} 5 & 10 & 15 & 20 & | & 240 \\ 1 & 1 & 1 & 2 & | & 20 \\ 3 & 2 & 2 & 3 & | & 40 \end{bmatrix}$$

Interchange rows 1 and 2.

$$\begin{bmatrix} 1 & 1 & 1 & 2 & | & 20 \\ 5 & 10 & 15 & 20 & | & 240 \\ 3 & 2 & 2 & 3 & | & 40 \end{bmatrix}$$

$-5R_1 + R_2 \to R_2$
$-3R_1 + R_3 \to R_3$
$$\begin{bmatrix} 1 & 1 & 1 & 2 & | & 20 \\ 0 & 5 & 10 & 10 & | & 140 \\ 0 & -1 & -1 & -3 & | & -20 \end{bmatrix}$$

$\frac{1}{5}R_2 \to R_2$
$$\begin{bmatrix} 1 & 1 & 1 & 2 & | & 20 \\ 0 & 1 & 2 & 2 & | & 28 \\ 0 & -1 & -1 & -3 & | & -20 \end{bmatrix}$$

$-1R_2 + R_1 \to R_1$
$R_2 + R_3 \to R_3$
$$\begin{bmatrix} 1 & 0 & -1 & 0 & | & -8 \\ 0 & 1 & 2 & 2 & | & 28 \\ 0 & 0 & 1 & -1 & | & 8 \end{bmatrix}$$

$R_3 + R_1 \to R_1$
$-2R_3 + R_2 \to R_2$
$$\begin{bmatrix} 1 & 0 & 0 & -1 & | & 0 \\ 0 & 1 & 0 & 4 & | & 12 \\ 0 & 0 & 1 & -1 & | & 8 \end{bmatrix}$$

We cannot change the values in column 4 further without changing the form of the other three columns. Therefore, let x_4 be arbitrary. This matrix gives the equations

$$x_1 - x_4 = 0 \quad \text{or} \quad x_1 = x_4,$$
$$x_2 + 4x_4 = 12 \quad \text{or} \quad x_2 = 12 - 4x_4,$$
$$x_3 - x_4 = 8 \quad \text{or} \quad x_3 = 8 + x_4.$$

The solution is $(x_4, 12 - 4x_4, 8 + x_4, x_4)$. Since all solutions must be nonnegative,

$$12 - 4x_4 \geq 0$$
$$x_4 \leq 3.$$

If $x_4 = 0$, then $x_1 = 0$, $x_2 = 12$, and $x_3 = 8$.
If $x_4 = 1$, then $x_1 = 1$, $x_2 = 8$, and $x_3 = 9$.

If $x_4 = 2$, then $x_1 = 2$, $x_2 = 4$, and $x_3 = 10$.
If $x_4 = 3$, then $x_1 = 3$, $x_2 = 0$, and $x_3 = 11$.

Therefore, there are four possible solutions. The breeder should mix

1. 0 cases of A, 12 cases of B, 8 cases of C, and 0 cases of D;

2. 1 case of A, 8 cases of B, 9 cases of C, and 1 case of D;

3. 2 cases of A, 4 cases of B, 10 cases of C, and 2 cases of D; or

4. 3 cases of A, 0 cases of B, 11 cases of C, and 3 cases of D.

57. Let $x = $ the number of the first species,
$y = $ the number of the second species, and
$z = $ the number of the third species.

$$1.3x + 1.1y + 8.1z = 16{,}000$$
$$1.3x + 2.4y + 2.9z = 28{,}000$$
$$2.3x + 3.7y + 5.1z = 44{,}000$$

Write the augmented matrix of the system.

$$\begin{bmatrix} 1.3 & 1.1 & 8.1 & | & 16{,}000 \\ 1.3 & 2.4 & 2.9 & | & 28{,}000 \\ 2.3 & 3.7 & 5.1 & | & 44{,}000 \end{bmatrix}$$

This exercise should be solved by graphing calculator or computer methods. The solution, which may vary slightly, is 2340 of the first species, 10,128 of the second species, and 224 of the third species. (All of these are rounded to the nearest whole number.)

59. Let $x = $ the number of acres for honeydews,
$y = $ the number of acres for yellow onions, and
$z = $ the number of acres for lettuce.

(a)
$$x + y + z = 220$$
$$120x + 150y + 180z = 29{,}100$$
$$180x + 80y + 80z = 32{,}600$$
$$4.97x + 4.45y + 4.65z = 480$$

Write the augmented matrix for this system.

$$\begin{bmatrix} 1 & 1 & 1 & | & 220 \\ 120 & 150 & 180 & | & 29{,}100 \\ 180 & 80 & 80 & | & 32{,}600 \\ 4.97 & 4.45 & 4.65 & | & 480 \end{bmatrix}$$

Using graphing calculator or computer methods, we obtain

$$\begin{bmatrix} 1 & 0 & 0 & | & 150 \\ 0 & 1 & 0 & | & 50 \\ 0 & 0 & 1 & | & 20 \\ 0 & 0 & 0 & | & -581 \end{bmatrix}.$$

There is no solution to the system. Therefore, it is not possible to utilize all resources completely.

(b) If 1061 hr of labor are available, the augmented matrix becomes,

$$\begin{bmatrix} 1 & 1 & 1 & | & 220 \\ 120 & 150 & 180 & | & 29{,}100 \\ 180 & 80 & 80 & | & 32{,}600 \\ 4.97 & 4.45 & 4.65 & | & 1061 \end{bmatrix}.$$

Again, using graphing calculator or computer methods we obtain

$$\begin{bmatrix} 1 & 0 & 0 & | & 150 \\ 0 & 1 & 0 & | & 50 \\ 0 & 0 & 1 & | & 20 \\ 0 & 0 & 0 & | & 0 \end{bmatrix}.$$

The solution is $(150, 50, 20)$. Therefore, allot 150 acres for honeydews, 50 acres for onions, and 20 acres for lettuce.

61. (a) In 1980, $t = 0$ and $R = 207.9$.

$$207.9 = a(0)^2 + b(0) + c$$
$$207.9 = c$$

In 1990, $t = 10$ and $R = 216.0$.

$$216.0 = a(10)^2 + b(10) + c$$
$$216.0 = 100a + 10b + c$$

In 2000, $t = 20$ and $R = 199.6$.

$$199.6 = a(20)^2 + b(20) + c$$
$$199.6 = 400a + 20b + c$$

The linear system to be solved is

$$400a + 20b + c = 199.6$$
$$100a + 10b + c = 216.0$$
$$c = 207.9$$

Write the augmented matrix and use row operations to solve.

$$\begin{bmatrix} 400 & 20 & 1 & | & 199.6 \\ 100 & 10 & 1 & | & 216.0 \\ 0 & 0 & 1 & | & 207.9 \end{bmatrix}$$

$-1R_1 + 4R_2 \rightarrow R_2$
$$\begin{bmatrix} 400 & 20 & -2 & | & 199.6 \\ 0 & 20 & 3 & | & 664.4 \\ 0 & 0 & 1 & | & 207.9 \end{bmatrix}$$

$-1R_2 + R_1 \rightarrow R_1$
$$\begin{bmatrix} 400 & 0 & -2 & | & -464.8 \\ 0 & 20 & 3 & | & 664.4 \\ 0 & 0 & 1 & | & 207.9 \end{bmatrix}$$

$2R_3 + R_1 \rightarrow R_1$
$-3R_3 + R_2 \rightarrow R_2$
$$\begin{bmatrix} 400 & 0 & 0 & | & -49 \\ 0 & 20 & 0 & | & 40.7 \\ 0 & 0 & 1 & | & 207.9 \end{bmatrix}$$

$\frac{1}{400}R_1 \rightarrow R_1$
$\frac{1}{20}R_2 \rightarrow R_2$
$$\begin{bmatrix} 1 & 0 & 0 & | & -0.1225 \\ 0 & 1 & 0 & | & 2.035 \\ 0 & 0 & 1 & | & 207.9 \end{bmatrix}$$

The solution is $a = -0.1225, b = 2.035$, and $c = 207.9$.

(b) In 2003, $t = 23$.

$$R = -0.1225t^2 + 2.035t + 207.9$$
$$= -0.1225(23)^2 + 2.035(23) + 207.9$$
$$\approx 189.9$$

The predicated value is close to the actual value, 194.0 deaths per 1,000,000.

(c) In 1980, $t = 0$ and $R = 207.9$.

$$207.9 = a(0)^3 + b(0)^2 + c(0) + d$$
$$207.9 = d$$

Similarly, we obtain the remaining three equations for $t = 10, t = 20$, and $t = 23$. The linear system to be solved is

$$12{,}167a + 529b + 23c + d = 190.1$$
$$8000a + 400b + 20c + d = 199.6$$
$$1000a + 100b + 10c + d = 216.0$$
$$d = 207.9.$$

The augmented matrix is

$$\begin{bmatrix} 12{,}167 & 529 & 23 & 1 & | & 190.1 \\ 8000 & 400 & 20 & 1 & | & 199.6 \\ 1000 & 100 & 10 & 1 & | & 216.0 \\ 0 & 0 & 0 & 1 & | & 207.9 \end{bmatrix}$$

Use a graphing calculator or computer methods to obtain the solution $a = 0.0002202, b = -0.1291$, $c = 2.079$, and $d = 207.9$.

63. (a) The other two equations are

$$x_2 + x_3 = 700$$
$$x_3 + x_4 = 600.$$

(b) The augmented matrix is

$$\begin{bmatrix} 1 & 0 & 0 & 1 & | & 1000 \\ 1 & 1 & 0 & 0 & | & 1100 \\ 0 & 1 & 1 & 0 & | & 700 \\ 0 & 0 & 1 & 1 & | & 600 \end{bmatrix}.$$

$-1R_1 + R_2 \rightarrow R_2$
$$\begin{bmatrix} 1 & 0 & 0 & 1 & | & 1000 \\ 0 & 1 & 0 & -1 & | & 100 \\ 0 & 1 & 1 & 0 & | & 700 \\ 0 & 0 & 1 & 1 & | & 600 \end{bmatrix}$$

$-1R_2 + R_3 \rightarrow R_3$
$$\begin{bmatrix} 1 & 0 & 0 & 1 & | & 1000 \\ 0 & 1 & 0 & -1 & | & 100 \\ 0 & 0 & 1 & 1 & | & 600 \\ 0 & 0 & 1 & 1 & | & 600 \end{bmatrix}$$

$-1R_3 + R_4 \rightarrow R_4$
$$\begin{bmatrix} 1 & 0 & 0 & 1 & | & 1000 \\ 0 & 1 & 0 & -1 & | & 100 \\ 0 & 0 & 1 & 1 & | & 600 \\ 0 & 0 & 0 & 0 & | & 0 \end{bmatrix}$$

Let x_4 be arbitrary. Solve the first three equations for x_1, x_2, and x_3.

$$x_1 = 1000 - x_4$$
$$x_2 = 100 + x_4$$
$$x_3 = 600 - x_4$$

The solution is $(1000 - x_4, 100 + x_4, 600 - x_4, x_4)$.

(c) For x_4, we see that $x_4 \geq 0$ and $x_4 \leq 600$ since $600 - x_4$ must be nonnegative. Therefore, $0 \leq x_4 \leq 600$.

(d) x_1:　If $x_4 = 0$, then $x_1 = 1000$.
　　　If $x_4 = 600$, then $x_1 = 1000 - 600 = 400$.

Therefore, $400 \leq x_1 \leq 1000$.

　x_2:　If $x_4 = 0$, then $x_2 = 100$.
　　　If $x_4 = 600$, then $x_2 = 100 + 600 = 700$.

Therefore, $100 \leq x_2 \leq 700$.

　x_3:　If $x_4 = 0$, then $x_3 = 600$.
　　　If $x_4 = 600$, then $x_3 = 600 - 600 = 0$.

Therefore, $0 \leq x_3 \leq 600$.

(e) If you know the number of cars entering or leaving three of the intersections, then the number entering or leaving the fourth is automatically determined because the number leaving must equal the number entering.

65. Let $x =$ the number of balls,
$y =$ the number of dolls, and
$z =$ the number of cars.

(a) The system to be solved is

$$\begin{aligned} x + \quad y \quad\;\; z &= 100 \\ 2x + \;\; 3y + \;\; 4z &= 295 \\ 12x + 16y + 18z &= 1542. \end{aligned}$$

Write the augmented matrix of the system.

$$\begin{bmatrix} 1 & 1 & 1 & 100 \\ 2 & 3 & 4 & 295 \\ 12 & 16 & 18 & 1542 \end{bmatrix}$$

$$\begin{matrix} -2R_1 + R_2 \rightarrow R_2 \\ -12R_1 + R_3 \rightarrow R_3 \end{matrix} \quad \begin{bmatrix} 1 & 1 & 1 & 100 \\ 0 & 1 & 2 & 95 \\ 0 & 4 & 6 & 342 \end{bmatrix}.$$

$$\begin{matrix} -1R_2 + R_1 \rightarrow R_1 \\ \\ -4R_2 + R_3 \rightarrow R_3 \end{matrix} \quad \begin{bmatrix} 1 & 0 & -1 & 5 \\ 0 & 1 & 2 & 95 \\ 0 & 0 & -2 & -38 \end{bmatrix}$$

$$-\tfrac{1}{2}R_3 \rightarrow R_3 \quad \begin{bmatrix} 1 & 0 & -1 & 5 \\ 0 & 1 & 2 & 95 \\ 0 & 0 & 1 & 19 \end{bmatrix}$$

$$\begin{matrix} R_3 + R_1 \rightarrow R_1 \\ -2R_3 + R_2 \rightarrow R_2 \end{matrix} \quad \begin{bmatrix} 1 & 0 & 0 & 24 \\ 0 & 1 & 0 & 57 \\ 0 & 0 & 1 & 19 \end{bmatrix}$$

The solution is $(24, 57, 19)$. There were 24 balls, 57 dolls, and 19 cars.

(b) The augmented matrix becomes

$$\begin{bmatrix} 1 & 1 & 1 & 100 \\ 2 & 3 & 4 & 295 \\ 11 & 15 & 19 & 1542 \end{bmatrix}.$$

$$\begin{matrix} -2R_1 + R_2 \rightarrow R_2 \\ -11R_1 + R_3 \rightarrow R_3 \end{matrix} \quad \begin{bmatrix} 1 & 1 & 1 & 100 \\ 0 & 1 & 2 & 95 \\ 0 & 4 & 8 & 442 \end{bmatrix}$$

$$\begin{matrix} -1R_2 + R_1 \rightarrow R_1 \\ \\ -4R_2 + R_3 \rightarrow R_3 \end{matrix} \quad \begin{bmatrix} 1 & 0 & -1 & 5 \\ 0 & 1 & 2 & 95 \\ 0 & 0 & 0 & 62 \end{bmatrix}$$

Since row 3 yields a false statement, $0 = 62$, there is no solution.

(c) The augmented matrix becomes

$$\begin{bmatrix} 1 & 1 & 1 & 100 \\ 2 & 3 & 4 & 295 \\ 11 & 15 & 19 & 1480 \end{bmatrix}.$$

$$\begin{matrix} -2R_1 + R_2 \rightarrow R_2 \\ -11R_1 + R_3 \rightarrow R_3 \end{matrix} \quad \begin{bmatrix} 1 & 1 & 1 & 100 \\ 0 & 1 & 2 & 95 \\ 0 & 4 & 8 & 380 \end{bmatrix}$$

$$\begin{matrix} -1R_2 + R_1 \rightarrow R_1 \\ \\ -4R_2 + R_3 \rightarrow R_3 \end{matrix} \quad \begin{bmatrix} 1 & 0 & -1 & 5 \\ 0 & 1 & 2 & 95 \\ 0 & 0 & 0 & 0 \end{bmatrix}$$

Since the last row is all zeros, there are infinitely many solutions. Let z be the parameter. The matrix gives

$$\begin{aligned} x - z &= \;\; 5 \\ y + 2z &= 95. \end{aligned}$$

Solving these equations for x and y, the solution is $(5 + z, 95 - 2z, z)$. The numbers in the solution must be nonnegative integers. Therefore,

$$\begin{aligned} 95 - 2z &\geq 0 \\ -2z &\geq -95 \\ z &\leq 47.5. \end{aligned}$$

Thus, $z \in \{0, 1, 2, 3, \ldots, 47\}$. There are 48 possible solutions.

(d) For the smallest number of cars, $z = 0$, the solution is $(5, 95, 0)$. This means 5 balls, 95 dolls, and no cars.

(e) For the largest number of cars, $z = 47$, the solution is $(52, 1, 47)$. This means 52 balls, 1 doll, and 47 cars.

67. Let $x =$ the number of singles,
$y =$ the number of doubles,
$z =$ the number of triples, and
$w =$ the number of home runs hit by Ichiro Suzuki.

The system to be solved is

$$\begin{aligned} x + y + z + w &= 262 \\ z &= w - 3 \\ y &= 3w \\ x &= 45z \end{aligned}$$

Write the equations in proper form, obtain the augmented matrix, and use row operations to solve.

$$\begin{bmatrix} 1 & 1 & 1 & 1 & | & 262 \\ 0 & 0 & 1 & -1 & | & -3 \\ 0 & 1 & 0 & -3 & | & 0 \\ 1 & 0 & -45 & 0 & | & 0 \end{bmatrix}$$

$$-1R_1 + R_4 \rightarrow R_4 \begin{bmatrix} 1 & 1 & 1 & 1 & | & 262 \\ 0 & 0 & 1 & -1 & | & -3 \\ 0 & 1 & 0 & -3 & | & 0 \\ 0 & -1 & -46 & -1 & | & -262 \end{bmatrix}$$

$$R_2 \longleftrightarrow R_3 \begin{bmatrix} 1 & 1 & 1 & 1 & | & 262 \\ 0 & 1 & 0 & -3 & | & 0 \\ 0 & 0 & 1 & -1 & | & -3 \\ 0 & -1 & -46 & -1 & | & -262 \end{bmatrix}$$

$$\begin{array}{c} -1R_2 + R_1 \rightarrow R_1 \\ \\ \\ R_2 + R_4 \rightarrow R_4 \end{array} \begin{bmatrix} 1 & 0 & 1 & 4 & | & 262 \\ 0 & 1 & 0 & -3 & | & 0 \\ 0 & 0 & 1 & -1 & | & -3 \\ 0 & 0 & -46 & -4 & | & -262 \end{bmatrix}$$

$$\begin{array}{c} -1R_3 + R_1 \rightarrow R_1 \\ \\ \\ 46R_3 + R_4 \rightarrow R_4 \end{array} \begin{bmatrix} 1 & 0 & 0 & 5 & | & 265 \\ 0 & 1 & 0 & -3 & | & 0 \\ 0 & 0 & 1 & -1 & | & -3 \\ 0 & 0 & 0 & -50 & | & -400 \end{bmatrix}$$

$$\begin{array}{c} R_4 + 10R_1 \rightarrow R_1 \\ -3R_4 + 50R_2 \rightarrow R_2 \\ -1R_4 + 50R_3 \rightarrow R_3 \end{array} \begin{bmatrix} 10 & 0 & 0 & 0 & | & 2250 \\ 0 & 50 & 0 & 0 & | & 1200 \\ 0 & 0 & 50 & 0 & | & 250 \\ 0 & 0 & 0 & -50 & | & -400 \end{bmatrix}$$

$$\begin{array}{c} \frac{1}{10}R_1 \rightarrow R_1 \\ \frac{1}{50}R_2 \rightarrow R_2 \\ \frac{1}{50}R_3 \rightarrow R_3 \\ -\frac{1}{50}R_4 \rightarrow R_4 \end{array} \begin{bmatrix} 1 & 0 & 0 & 0 & | & 225 \\ 0 & 1 & 0 & 0 & | & 24 \\ 0 & 0 & 1 & 0 & | & 5 \\ 0 & 0 & 0 & 1 & | & 8 \end{bmatrix}$$

Ichiro Suzuki hit 225 singles, 24 doubles, 5 triples, and 8 home runs during the 2004 season.

2.3   Addition and Subtraction of Matrices

1. $\begin{bmatrix} 1 & 3 \\ 5 & 7 \end{bmatrix} = \begin{bmatrix} 1 & 5 \\ 3 & 7 \end{bmatrix}$

This statement is false, since not all corresponding elements are equal.

3. $\begin{bmatrix} x \\ y \end{bmatrix} = \begin{bmatrix} -2 \\ 8 \end{bmatrix}$ if $x = -2$ and $y = 8$.

This statement is true. The matrices are the same size and corresponding elements are equal.

5. $\begin{bmatrix} 1 & 9 & -4 \\ 3 & 7 & 2 \\ -1 & 1 & 0 \end{bmatrix}$ is a square matrix.

This statement is true. The matrix has 3 rows and 3 columns.

7. $\begin{bmatrix} -4 & 8 \\ 2 & 3 \end{bmatrix}$ is a 2 × 2 square matrix.

Its additive inverse is $\begin{bmatrix} 4 & -8 \\ -2 & -3 \end{bmatrix}$.

9. $\begin{bmatrix} -6 & 8 & 0 & 0 \\ 4 & 1 & 9 & 2 \\ 3 & -5 & 7 & 1 \end{bmatrix}$ is a 3 × 4 matrix.

Its additive inverse is

$$\begin{bmatrix} 6 & -8 & 0 & 0 \\ -4 & -1 & -9 & -2 \\ -3 & 5 & -7 & -1 \end{bmatrix}.$$

11. $\begin{bmatrix} -7 \\ 5 \end{bmatrix}$ is a 2 × 1 column matrix.

Its additive inverse is

$$\begin{bmatrix} 7 \\ -5 \end{bmatrix}.$$

13. The sum of an $n \times m$ matrix and its additive inverse is the $n \times m$ zero matrix.

15. $\begin{bmatrix} 3 & 4 \\ -8 & 1 \end{bmatrix} = \begin{bmatrix} 3 & x \\ y & z \end{bmatrix}$

Corresponding elements must be equal for the matrices to be equal. Therefore, $x = 4, y = -8$, and $z = 1$.

17. $\begin{bmatrix} s-4 & t+2 \\ -5 & 7 \end{bmatrix} = \begin{bmatrix} 6 & 2 \\ -5 & r \end{bmatrix}$

Corresponding elements must be equal

$$\begin{array}{ccc} s - 4 = 6 & t + 2 = 2 & r = 7. \\ s = 10 & t = 0 & \end{array}$$

Thus, $s = 10, t = 0$, and $r = 7$.

19. $\begin{bmatrix} a+2 & 3b & 4c \\ d & 7f & 8 \end{bmatrix} + \begin{bmatrix} -7 & 2b & 6 \\ -3d & -6 & -2 \end{bmatrix} = \begin{bmatrix} 15 & 25 & 6 \\ -8 & 1 & 6 \end{bmatrix}$

Add the two matrices on the left side to obtain

$\begin{bmatrix} a+2 & 3b & 4c \\ d & 7f & 8 \end{bmatrix} + \begin{bmatrix} -7 & 2b & 6 \\ -3d & -6 & -2 \end{bmatrix}$

$= \begin{bmatrix} (a+2)+(-7) & 3b+2b & 4c+6 \\ d+(-3d) & 7f+(-6) & 8+(-2) \end{bmatrix}$

$= \begin{bmatrix} a-5 & 5b & 4c+6 \\ -2d & 7f-6 & 6 \end{bmatrix}$

Corresponding elements of this matrix and the matrix on the right side of the original equation must be equal.

$a-5=15 \qquad 5b=25 \qquad 4c+6=6$
$\quad a=20 \qquad\quad b=5 \qquad\quad\; c=0$

$-2d=-8 \qquad 7f-6=1$
$\quad\; d=4 \qquad\qquad f=1$

Thus, $a=20, b=5, c=0, d=4$, and $f=1$.

21. $\begin{bmatrix} 2 & 4 & 5 & -7 \\ 6 & -3 & 12 & 0 \end{bmatrix} + \begin{bmatrix} 8 & 0 & -10 & 1 \\ -2 & 8 & -9 & 11 \end{bmatrix}$

$= \begin{bmatrix} 2+8 & 4+0 & 5+(-10) & -7+1 \\ 6+(-2) & -3+8 & 12+(-9) & 0+11 \end{bmatrix}$

$= \begin{bmatrix} 10 & 4 & -5 & -6 \\ 4 & 5 & 3 & 11 \end{bmatrix}$

23. $\begin{bmatrix} 1 & 3 & -2 \\ 4 & 7 & 1 \end{bmatrix} + \begin{bmatrix} 3 & 0 \\ 6 & 4 \\ -5 & 2 \end{bmatrix}$

These matrices cannot be added since the first matrix has size 2×3, while the second has size 3×2. Only matrices that are the same size can be added.

25. The matrices have the same size, so the subtraction can be done. Let A and B represent the given matrices. Using the definition of subtraction, we have

$A - B = A + (-B)$

$= \begin{bmatrix} 2 & 8 & 12 & 0 \\ 7 & 4 & -1 & 5 \\ 1 & 2 & 0 & 10 \end{bmatrix} + \begin{bmatrix} -1 & -3 & -6 & -9 \\ -2 & 3 & 3 & -4 \\ -8 & 0 & 2 & -17 \end{bmatrix}$

$= \begin{bmatrix} 1 & 5 & 6 & -9 \\ 5 & 7 & 2 & 1 \\ -7 & 2 & 2 & -7 \end{bmatrix}$.

27. $\begin{bmatrix} 2 & 3 \\ -2 & 4 \end{bmatrix} + \begin{bmatrix} 4 & 3 \\ 7 & 8 \end{bmatrix} - \begin{bmatrix} 3 & 2 \\ 1 & 4 \end{bmatrix}$

$= \begin{bmatrix} 2+4-3 & 3+3-2 \\ -2+7-1 & 4+8-4 \end{bmatrix} = \begin{bmatrix} 3 & 4 \\ 4 & 8 \end{bmatrix}$

29. $\begin{bmatrix} 2 & -1 \\ 0 & 13 \end{bmatrix} - \begin{bmatrix} 4 & 8 \\ -5 & 7 \end{bmatrix} + \begin{bmatrix} 12 & 7 \\ 5 & 3 \end{bmatrix}$

$= \begin{bmatrix} 2 & -1 \\ 0 & 13 \end{bmatrix} + \begin{bmatrix} -4 & -8 \\ 5 & -7 \end{bmatrix} + \begin{bmatrix} 12 & 7 \\ 5 & 3 \end{bmatrix}$

$= \begin{bmatrix} 2+(-4)+12 & -1+(-8)+7 \\ 0+5+5 & 13+(-7)+3 \end{bmatrix}$

$= \begin{bmatrix} 10 & -2 \\ 10 & 9 \end{bmatrix}$

31. $\begin{bmatrix} -4x+2y & -3x+y \\ 6x-3y & 2x-5y \end{bmatrix} + \begin{bmatrix} -8x+6y & 2x \\ 3y-5x & 6x+4y \end{bmatrix}$

$= \begin{bmatrix} (-4x+2y)+(-8x+6y) & (-3x+y)+2x \\ (6x-3y)+(3y-5x) & (2x-5y)+(6x+4y) \end{bmatrix}$

$= \begin{bmatrix} -12x+8y & -x+y \\ x & 8x-y \end{bmatrix}$

33. The additive inverse of

$$X = \begin{bmatrix} x & y \\ z & w \end{bmatrix}$$

is

$$-X = \begin{bmatrix} -x & -y \\ -z & -w \end{bmatrix}.$$

35. Show that $X + (T + P) = (X + T) + P$.

On the left side, the sum $T + P$ is obtained first, and then

$$X + (T + P).$$

This gives the matrix

$$\begin{bmatrix} x+(r+m) & y+(s+n) \\ z+(t+p) & w+(u+q) \end{bmatrix}.$$

For the right side, first the sum $X + T$ is obtained, and then

$$(X + T) + P.$$

This gives the matrix

$$\begin{bmatrix} (x+r)+m & (y+s)+n \\ (z+t)+p & (w+u)+q \end{bmatrix}.$$

Comparing corresponding elements, we see that they are equal by the associative property of addition of real numbers. Thus,

$$X + (T + P) = (X + T) + P.$$

37. Show that $P + O = P$.

$$P + O = \begin{bmatrix} m & n \\ p & q \end{bmatrix} + \begin{bmatrix} 0 & 0 \\ 0 & 0 \end{bmatrix}$$

$$= \begin{bmatrix} m+0 & n+0 \\ p+0 & q+0 \end{bmatrix}$$

$$= \begin{bmatrix} m & n \\ p & q \end{bmatrix}$$

$$= P$$

Thus, $P + O = P$.

39. (a) The production cost matrix for Chicago is

	Phones	Calculators
Material	4.05	7.01
Labor	3.27	3.51

The production cost matrix for Seattle is

	Phones	Calculators
Material	4.40	6.90
Labor	3.54	3.76

(b) The new production cost matrix for Chicago is

	Phones	Calculators
Material	4.05 + 0.37	7.01 + 0.42
Labor	3.27 + 0.11	3.51 + 0.11

or $\begin{bmatrix} 4.42 & 7.43 \\ 3.38 & 3.62 \end{bmatrix}$.

41. (a) There are four food groups and three meals. To represent the data by a 3 × 4 matrix, we must use the rows to correspond to the meals, breakfast, lunch, and dinner, and the columns to correspond to the four food groups. Thus, we obtain the matrix

$$\begin{bmatrix} 2 & 1 & 2 & 1 \\ 3 & 2 & 2 & 1 \\ 4 & 3 & 2 & 1 \end{bmatrix}.$$

(b) There are four food groups. These will correspond to the four rows. There are three components in each food group: fat, carbohydrates, and protein. These will correspond to the three columns. The matrix is

$$\begin{bmatrix} 5 & 0 & 7 \\ 0 & 10 & 1 \\ 0 & 15 & 2 \\ 10 & 12 & 8 \end{bmatrix}.$$

(c) The matrix is

$$\begin{bmatrix} 8 \\ 4 \\ 5 \end{bmatrix}.$$

43.

	Obtained Pain Relief	
	Yes	No
Painfree	22	3
Placebo	8	17

(a) Of the 25 patients who took the placebo, 8 got relief.

(b) Of the 25 patients who took Painfree, 3 got no relief.

(c) $\begin{bmatrix} 22 & 3 \\ 8 & 17 \end{bmatrix} + \begin{bmatrix} 21 & 4 \\ 6 & 19 \end{bmatrix} + \begin{bmatrix} 19 & 6 \\ 10 & 15 \end{bmatrix} + \begin{bmatrix} 23 & 2 \\ 3 & 22 \end{bmatrix}$

$= \begin{bmatrix} 85 & 15 \\ 27 & 73 \end{bmatrix}$

(d) Yes, it appears that Painfree is effective. Of the 100 patients who took the medication, 85% got relief.

45. (a) The matrix for the life expectancy of African Americans is

	M	F
1970	60.0	68.3
1980	63.8	72.5
1990	64.5	73.6
2000	68.2	74.9

(b) The matrix for the life expectancy of White Americans is

	M	F
1970	68.0	75.6
1980	70.7	78.1
1990	72.7	79.4
2000	74.8	80.0

(c) The matrix showing the difference between the life expectancy between the two groups is

$$\begin{bmatrix} 60.0 & 68.3 \\ 63.8 & 72.5 \\ 64.5 & 73.6 \\ 68.2 & 74.9 \end{bmatrix} - \begin{bmatrix} 68.0 & 75.6 \\ 70.7 & 78.1 \\ 72.7 & 79.4 \\ 74.8 & 80.0 \end{bmatrix}$$

$$= \begin{bmatrix} 60.0 & 68.3 \\ 63.8 & 72.5 \\ 64.5 & 73.6 \\ 68.2 & 74.9 \end{bmatrix} + \begin{bmatrix} -68.0 & -75.6 \\ -70.7 & -78.1 \\ -72.7 & -79.4 \\ -74.8 & -80.0 \end{bmatrix}$$

$$= \begin{bmatrix} -8.0 & -7.3 \\ -6.9 & -5.6 \\ -8.2 & -5.8 \\ -6.6 & -5.1 \end{bmatrix}$$

47. (a) The matrix for the educational attainment of African Americans is

	Four Years of High School or More	Four Years of College or More
1980	51.4	7.9
1985	59.9	11.1
1990	66.2	11.3
1995	73.8	13.3
2000	78.9	16.6
2004	80.6	17.6

(b) The matrix for the educational attainment of Hispanic Americans is

	Four Years of High School or More	Four Years of College or More
1980	44.5	7.6
1985	47.9	8.5
1990	50.8	9.2
1995	53.4	9.3
2000	57.0	10.6
2004	58.4	12.1

(c) The matrix showing the difference in the educational attainment between African and Hispanic Americans is

$$\begin{bmatrix} 51.4 & 7.9 \\ 59.9 & 11.1 \\ 66.2 & 11.3 \\ 73.8 & 13.3 \\ 78.9 & 16.6 \\ 80.6 & 17.6 \end{bmatrix} - \begin{bmatrix} 44.5 & 7.6 \\ 47.9 & 8.5 \\ 50.8 & 9.2 \\ 53.4 & 9.3 \\ 57.0 & 10.6 \\ 58.4 & 12.1 \end{bmatrix}$$

$$= \begin{bmatrix} 51.4 & 7.9 \\ 59.9 & 11.1 \\ 66.2 & 11.3 \\ 73.8 & 13.3 \\ 78.9 & 16.6 \\ 80.6 & 17.6 \end{bmatrix} + \begin{bmatrix} -44.5 & -7.6 \\ -47.9 & -8.5 \\ -50.8 & -9.2 \\ -53.4 & -9.3 \\ -57.0 & -10.6 \\ -58.4 & -12.1 \end{bmatrix}$$

$$= \begin{bmatrix} 6.9 & 0.3 \\ 12.0 & 2.6 \\ 15.4 & 2.1 \\ 20.4 & 4.0 \\ 21.9 & 6.0 \\ 22.2 & 5.5 \end{bmatrix}$$

2.4 Multiplication of Matrices

In Exercises 1-5, let

$$A = \begin{bmatrix} -2 & 4 \\ 0 & 3 \end{bmatrix} \text{ and } B = \begin{bmatrix} -6 & 2 \\ 4 & 0 \end{bmatrix}.$$

1. $2A = 2\begin{bmatrix} -2 & 4 \\ 0 & 3 \end{bmatrix} = \begin{bmatrix} -4 & 8 \\ 0 & 6 \end{bmatrix}$

3. $-6A = -6\begin{bmatrix} -2 & 4 \\ 0 & 3 \end{bmatrix} = \begin{bmatrix} 12 & -24 \\ 0 & -18 \end{bmatrix}$

5. $-4A + 5B = -4\begin{bmatrix} -2 & 4 \\ 0 & 3 \end{bmatrix} + 5\begin{bmatrix} -6 & 2 \\ 4 & 0 \end{bmatrix}$

$$= \begin{bmatrix} 8 & -16 \\ 0 & -12 \end{bmatrix} + \begin{bmatrix} -30 & 10 \\ 20 & 0 \end{bmatrix}$$

$$= \begin{bmatrix} -22 & -6 \\ 20 & -12 \end{bmatrix}$$

7.   Matrix A size   Matrix B size

$2 \times \underline{2}$ $\underline{2} \times 2$

The number of columns of A is the same as the number of rows of B, so the product AB exists. The size of the matrix AB is 2×2.

Matrix B size   Matrix A size

$2 \times \underline{2}$ $\underline{2} \times 2$

Since the number of columns of B is the same as the number of rows of A, the product BA also exists and has size 2×2.

9.   Matrix A size   Matrix B size

$3 \times \underline{4}$ $\underline{4} \times 4$

Since matrix A has 4 columns and matrix B has 4 rows, the product AB exists and has size 3×4.

Matrix B size   Matrix A size

$4 \times \underline{4}$ $\underline{3} \times 4$

Since B has 4 columns and A has 3 rows, the product BA does not exist.

11.   Matrix A size   Matrix B size

$4 \times \underline{2}$ $\underline{3} \times 4$

The number of columns of A is not the same as the number of rows of B, so the product AB does not exist.

Matrix B size   Matrix A size

$3 \times \underline{4}$ $\underline{4} \times 2$

The number of columns of B is the same as the number of rows of A, so the product BA exists and has size 3×2.

13. To find the product matrix AB, the number of *columns* of A must be the same as the number of *rows* of B.

15. Call the first matrix A and the second matrix B. The product matrix AB will have size 2×1.

Step 1: Multiply the elements of the first row of A by the corresponding elements of the column of B and add.

$$\begin{bmatrix} 2 & -1 \\ 5 & 8 \end{bmatrix} \begin{bmatrix} 3 \\ -2 \end{bmatrix} \quad 2(3) + (-1)(-2) = 8$$

Therefore, 8 is the first row entry of the product matrix AB.

Step 2: Multiply the elements of the second row of A by the corresponding elements of the column of B and add.

$$\begin{bmatrix} 2 & -1 \\ 5 & 8 \end{bmatrix} \begin{bmatrix} 3 \\ -2 \end{bmatrix} \quad 5(3) + 8(-2) = -1$$

The second row entry of the product is -1.

Step 3: Write the product using the two entries found above.

$$AB = \begin{bmatrix} 2 & -1 \\ 5 & 8 \end{bmatrix} \begin{bmatrix} 3 \\ -2 \end{bmatrix} = \begin{bmatrix} 8 \\ -1 \end{bmatrix}$$

17. $\begin{bmatrix} 2 & -1 & 7 \\ -3 & 0 & -4 \end{bmatrix} \begin{bmatrix} 5 \\ 10 \\ 2 \end{bmatrix} = \begin{bmatrix} 2 \cdot 5 + (-1) \cdot 10 + 7 \cdot 2 \\ (-3) \cdot 5 + 0 \cdot 10 + (-4) \cdot 2 \end{bmatrix} = \begin{bmatrix} 14 \\ -23 \end{bmatrix}$

19. $\begin{bmatrix} 2 & -1 \\ 3 & 6 \end{bmatrix} \begin{bmatrix} -1 & 0 & 4 \\ 5 & -2 & 0 \end{bmatrix} = \begin{bmatrix} 2 \cdot (-1) + (-1) \cdot 5 & 2 \cdot 0 + (-1) \cdot (-2) & 2 \cdot 4 + (-1) \cdot 0 \\ 3 \cdot (-1) + 6 \cdot 5 & 3 \cdot 0 + 6 \cdot (-2) & 3 \cdot 4 + 6 \cdot 0 \end{bmatrix} = \begin{bmatrix} -7 & 2 & 8 \\ 27 & -12 & 12 \end{bmatrix}$

21. $\begin{bmatrix} 2 & 2 & -1 \\ 3 & 0 & 1 \end{bmatrix} \begin{bmatrix} 0 & 2 \\ -1 & 4 \\ 0 & 2 \end{bmatrix} = \begin{bmatrix} 2 \cdot 0 + 2(-1) + (-1)0 & 2 \cdot 2 + 2 \cdot 4 + (-1)2 \\ 3 \cdot 0 + 0(-1) + 1(0) & 3 \cdot 2 + 0 \cdot 4 + 1 \cdot 2 \end{bmatrix} = \begin{bmatrix} -2 & 10 \\ 0 & 8 \end{bmatrix}$

23. $\begin{bmatrix} 1 & 2 \\ 3 & 4 \end{bmatrix} \begin{bmatrix} -1 & 5 \\ 7 & 0 \end{bmatrix} = \begin{bmatrix} 1(-1) + 2 \cdot 7 & 1 \cdot 5 + 2 \cdot 0 \\ 3(-1) + 4 \cdot 7 & 3 \cdot 5 + 4 \cdot 0 \end{bmatrix} = \begin{bmatrix} 13 & 5 \\ 25 & 15 \end{bmatrix}$

25. $\begin{bmatrix} -2 & -3 & 7 \\ 1 & 5 & 6 \end{bmatrix} \begin{bmatrix} 1 \\ 2 \\ 3 \end{bmatrix} = \begin{bmatrix} -2(1) + (-3)2 + 7 \cdot 3 \\ 1 \cdot 1 + 5 \cdot 2 + 6 \cdot 3 \end{bmatrix} = \begin{bmatrix} 13 \\ 29 \end{bmatrix}$

27. $\left(\begin{bmatrix} 2 & 1 \\ -3 & -6 \\ 4 & 0 \end{bmatrix} \begin{bmatrix} 1 & -2 \\ 2 & -1 \end{bmatrix} \right) \begin{bmatrix} 3 \\ 1 \end{bmatrix} = \begin{bmatrix} 4 & -5 \\ -15 & 12 \\ 4 & -8 \end{bmatrix} \begin{bmatrix} 3 \\ 1 \end{bmatrix} = \begin{bmatrix} 7 \\ -33 \\ 4 \end{bmatrix}$

29. $\begin{bmatrix} 2 & -2 \\ 1 & -1 \end{bmatrix} \left(\begin{bmatrix} 4 & 3 \\ 1 & 2 \end{bmatrix} + \begin{bmatrix} 7 & 0 \\ -1 & 5 \end{bmatrix} \right) = \begin{bmatrix} 2 & -2 \\ 1 & -1 \end{bmatrix} \begin{bmatrix} 11 & 3 \\ 0 & 7 \end{bmatrix} = \begin{bmatrix} 22 & -8 \\ 11 & -4 \end{bmatrix}$

31. (a) $AB = \begin{bmatrix} -2 & 4 \\ 1 & 3 \end{bmatrix} \begin{bmatrix} -2 & 1 \\ 3 & 6 \end{bmatrix} = \begin{bmatrix} 16 & 22 \\ 7 & 19 \end{bmatrix}$

(b) $BA = \begin{bmatrix} -2 & 1 \\ 3 & 6 \end{bmatrix} \begin{bmatrix} -2 & 4 \\ 1 & 3 \end{bmatrix} = \begin{bmatrix} 5 & -5 \\ 0 & 30 \end{bmatrix}$

(c) No, AB and BA are not equal here.

(d) No, AB does not always equal BA.

33. Verify that $P(X + T) = PX + PT$.

Find $P(X + T)$ and $PX + PT$ separately and compare their values to see if they are the same.

$$P(X + T) = \begin{bmatrix} m & n \\ p & q \end{bmatrix} \left(\begin{bmatrix} x & y \\ z & w \end{bmatrix} + \begin{bmatrix} r & s \\ t & u \end{bmatrix} \right) = \begin{bmatrix} m & n \\ p & q \end{bmatrix} \left(\begin{bmatrix} x+r & y+s \\ z+t & w+u \end{bmatrix} \right)$$

$$= \begin{bmatrix} m(x+r) + n(z+t) & m(y+s) + n(w+u) \\ p(x+r) + q(z+t) & p(y+s) + q(w+u) \end{bmatrix} = \begin{bmatrix} mx + mr + nz + nt & my + ms + nw + nu \\ px + pr + qz + qt & py + ps + qw + qu \end{bmatrix}$$

$$PX + PT = \begin{bmatrix} m & n \\ p & q \end{bmatrix} \begin{bmatrix} x & y \\ z & w \end{bmatrix} + \begin{bmatrix} m & n \\ p & q \end{bmatrix} \begin{bmatrix} r & s \\ t & u \end{bmatrix} = \begin{bmatrix} mx + nz & my + nw \\ px + qz & py + qw \end{bmatrix} + \begin{bmatrix} mr + nt & ms + nu \\ pr + qt & ps + qu \end{bmatrix}$$

$$= \begin{bmatrix} (mx + nz) + (mr + nt) & (my + nw) + (ms + nu) \\ (px + qz) + (pr + qt) & (py + qw) + (ps + qu) \end{bmatrix} = \begin{bmatrix} mx + nz + mr + nt & my + nw + ms + nu \\ px + qz + pr + qt & py + qw + ps + qu \end{bmatrix}$$

$$= \begin{bmatrix} mx + mr + nz + nt & my + ms + nw + nu \\ px + pr + qz + qt & py + ps + qw + qu \end{bmatrix}$$

Observe that the two results are identical. Thus, $P(X + T) = PX + PT$.

35. Verify that $(k+h)P = kP+hP$ for any real numbers k and h.

$$(k+h)P = (k+h)\begin{bmatrix} m & n \\ p & q \end{bmatrix}$$

$$= \begin{bmatrix} (k+h)m & (k+h)n \\ (k+h)p & (k+h)q \end{bmatrix}$$

$$= \begin{bmatrix} km+hm & kn+hn \\ kp+hp & kq+hq \end{bmatrix}$$

$$= \begin{bmatrix} km & kn \\ kp & kq \end{bmatrix} + \begin{bmatrix} hm & hn \\ hp & hq \end{bmatrix}$$

$$= k\begin{bmatrix} m & n \\ p & q \end{bmatrix} + h\begin{bmatrix} m & n \\ p & q \end{bmatrix}$$

$$= kP + hP$$

Thus, $(k+h)P = kP + hP$ for any real numbers k and h.

37. $\begin{bmatrix} 2 & 3 & 1 \\ 1 & -4 & 5 \end{bmatrix} \begin{bmatrix} x_1 \\ x_2 \\ x_3 \end{bmatrix} = \begin{bmatrix} 2x_1 + 3x_2 + x_3 \\ x_1 - 4x_2 + 5x_3 \end{bmatrix}$,

and $\begin{bmatrix} 2x_1 + 3x_2 + x_3 \\ x_1 - 4x_2 + 5x_3 \end{bmatrix} = \begin{bmatrix} 5 \\ 8 \end{bmatrix}$.

This is equivalent to

$$2x_1 + 3x_2 + x_3 = 5$$
$$x_1 - 4x_2 + 5x_3 = 8$$

since corresponding elements of equal matrices must be equal. Reversing this, observe that the given system of linear equations can be written as the matrix equation

$$\begin{bmatrix} 2 & 3 & 1 \\ 1 & -4 & 5 \end{bmatrix} \begin{bmatrix} x_1 \\ x_2 \\ x_3 \end{bmatrix} = \begin{bmatrix} 5 \\ 8 \end{bmatrix}.$$

39. (a) Use a graphing calculator or a computer to find the product matrix. The answer is

$$AC = \begin{bmatrix} 6 & 106 & 158 & 222 & 28 \\ 120 & 139 & 64 & 75 & 115 \\ -146 & -2 & 184 & 144 & -129 \\ 106 & 94 & 24 & 116 & 110 \end{bmatrix}.$$

(b) CA does not exist.

(c) AC and CA are clearly not equal, since CA does not even exist.

41. Use a graphing calculator or computer to find the matrix products and sums. The answers are as follows.

(a) $C + D = \begin{bmatrix} -1 & 5 & 9 & 13 & -1 \\ 7 & 17 & 2 & -10 & 6 \\ 18 & 9 & -12 & 12 & 22 \\ 9 & 4 & 18 & 10 & -3 \\ 1 & 6 & 10 & 28 & 5 \end{bmatrix}$

(b) $(C + D)B = \begin{bmatrix} -2 & -9 & 90 & 77 \\ -42 & -63 & 127 & 62 \\ 413 & 76 & 180 & -56 \\ -29 & -44 & 198 & 85 \\ 137 & 20 & 162 & 103 \end{bmatrix}$

(c) $CB = \begin{bmatrix} -56 & -1 & 1 & 45 \\ -156 & -119 & 76 & 122 \\ 315 & 86 & 118 & -91 \\ -17 & -17 & 116 & 51 \\ 118 & 19 & 125 & 77 \end{bmatrix}$

(d) $DB = \begin{bmatrix} 54 & -8 & 89 & 32 \\ 114 & 56 & 51 & -60 \\ 98 & -10 & 62 & 35 \\ -12 & -27 & 82 & 34 \\ 19 & 1 & 37 & 26 \end{bmatrix}$

(e) $CB + DB = \begin{bmatrix} -2 & -9 & 90 & 77 \\ -42 & -63 & 127 & 62 \\ 413 & 76 & 180 & -56 \\ -29 & -44 & 198 & 85 \\ 137 & 20 & 162 & 103 \end{bmatrix}$

(f) Yes, $(C+D)B$ and $CB+DB$ are equal, as can be seen by observing that the answers to parts (b) and (e) are identical.

43. (a) $\begin{bmatrix} 10 & 4 & 3 & 5 & 6 \\ 7 & 2 & 2 & 3 & 8 \\ 4 & 5 & 1 & 0 & 10 \\ 0 & 3 & 4 & 5 & 5 \end{bmatrix} \begin{bmatrix} 2 & 3 \\ 1 & 1 \\ 4 & 3 \\ 3 & 3 \\ 1 & 2 \end{bmatrix}$

$$= \begin{array}{r} \\ \text{Dept. 1} \\ \text{Dept. 2} \\ \text{Dept. 3} \\ \text{Dept. 4} \end{array} \begin{matrix} A & B \\ \begin{bmatrix} 57 & 70 \\ 41 & 54 \\ 27 & 40 \\ 39 & 40 \end{bmatrix} \end{matrix}$$

(b) The total cost to buy from supplier A is $57 + 41 + 27 + 39 = \$164$, and the total cost to buy from supplier B is $70 + 54 + 40 + 40 = \$204$. The company should make the purchase from supplier A, since \$164 is a lower total cost than \$204.

45. (a) To find the average, add the matrices. Then multiply the resulting matrix by $\frac{1}{3}$. (Multiplying by $\frac{1}{3}$ is the same as dividing by 3.)

$$\frac{1}{3}\left(\begin{bmatrix} 4.27 & 6.94 \\ 3.45 & 3.65 \end{bmatrix} + \begin{bmatrix} 4.05 & 7.01 \\ 3.27 & 3.51 \end{bmatrix} + \begin{bmatrix} 4.40 & 6.90 \\ 3.54 & 3.76 \end{bmatrix}\right) = \frac{1}{3}\begin{bmatrix} 12.72 & 20.85 \\ 10.26 & 10.92 \end{bmatrix} = \begin{bmatrix} 4.24 & 6.95 \\ 3.42 & 3.64 \end{bmatrix}$$

(b) To find the new average, add the new matrix for the Chicago plant and the matrix for the Seattle plant. Since there are only two matrices now, multiply the resulting matrix by $\frac{1}{2}$ to get the average. (Multiplying by $\frac{1}{2}$ is the same as dividing by 2.)

$$\frac{1}{2}\left(\begin{bmatrix} 4.42 & 7.43 \\ 3.38 & 3.62 \end{bmatrix} + \begin{bmatrix} 4.40 & 6.90 \\ 3.54 & 3.76 \end{bmatrix}\right) = \frac{1}{2}\begin{bmatrix} 8.82 & 14.33 \\ 6.92 & 7.38 \end{bmatrix} = \begin{bmatrix} 4.41 & 7.17 \\ 3.46 & 3.69 \end{bmatrix}$$

47. (a)

$$P = \begin{matrix} \\ \text{Sal's} \\ \text{Fred's} \end{matrix} \begin{matrix} \text{Sh} & \text{Sa} & \text{B} \\ \begin{bmatrix} 80 & 40 & 120 \\ 60 & 30 & 150 \end{bmatrix} \end{matrix}$$

(b)

$$F = \begin{matrix} \\ \text{Sh} \\ \text{Sa} \\ \text{B} \end{matrix} \begin{matrix} \text{CA} & \text{AR} \\ \begin{bmatrix} \frac{1}{2} & \frac{1}{5} \\ \frac{1}{4} & \frac{1}{5} \\ \frac{1}{4} & \frac{3}{5} \end{bmatrix} \end{matrix}$$

(c) $PF = \begin{bmatrix} 80 & 40 & 120 \\ 60 & 30 & 150 \end{bmatrix}\begin{bmatrix} \frac{1}{2} & \frac{1}{5} \\ \frac{1}{4} & \frac{1}{5} \\ \frac{1}{4} & \frac{3}{5} \end{bmatrix} = \begin{bmatrix} 80\left(\frac{1}{2}\right)+40\left(\frac{1}{4}\right)+120\left(\frac{1}{4}\right) & 80\left(\frac{1}{5}\right)+40\left(\frac{1}{5}\right)+120\left(\frac{3}{5}\right) \\ 60\left(\frac{1}{2}\right)+30\left(\frac{1}{4}\right)+150\left(\frac{1}{4}\right) & 60\left(\frac{1}{5}\right)+30\left(\frac{1}{5}\right)+150\left(\frac{3}{5}\right) \end{bmatrix} = \begin{bmatrix} 80 & 96 \\ 75 & 108 \end{bmatrix}$

The rows give the average price per pair of footwear sold by each store, and the columns give the state.

49. (a) $XY = \begin{bmatrix} 2 & 1 & 2 & 1 \\ 3 & 2 & 2 & 1 \\ 4 & 3 & 2 & 1 \end{bmatrix}\begin{bmatrix} 5 & 0 & 7 \\ 0 & 10 & 1 \\ 0 & 15 & 2 \\ 10 & 12 & 8 \end{bmatrix} = \begin{bmatrix} 20 & 52 & 27 \\ 25 & 62 & 35 \\ 30 & 72 & 43 \end{bmatrix}$

The rows give the amounts of fat, carbohydrates, and protein, respectively, in each of the daily meals.

(b) $YZ = \begin{bmatrix} 5 & 0 & 7 \\ 0 & 10 & 1 \\ 0 & 15 & 2 \\ 10 & 12 & 8 \end{bmatrix}\begin{bmatrix} 8 \\ 4 \\ 5 \end{bmatrix} = \begin{bmatrix} 75 \\ 45 \\ 70 \\ 168 \end{bmatrix}$

The rows give the number of calories in one exchange of each of the food groups.

(c) Use the matrices found for XY and YZ from parts (a) and (b).

$$(XY)Z = \begin{bmatrix} 20 & 52 & 27 \\ 25 & 62 & 35 \\ 30 & 72 & 43 \end{bmatrix}\begin{bmatrix} 8 \\ 4 \\ 5 \end{bmatrix} = \begin{bmatrix} 503 \\ 623 \\ 743 \end{bmatrix}$$

$$X(YZ) = \begin{bmatrix} 2 & 1 & 2 & 1 \\ 3 & 2 & 2 & 1 \\ 4 & 3 & 2 & 1 \end{bmatrix}\begin{bmatrix} 75 \\ 45 \\ 70 \\ 168 \end{bmatrix} = \begin{bmatrix} 503 \\ 623 \\ 743 \end{bmatrix}$$

The rows give the number of calories in each meal.

51. $\dfrac{1}{6}\left(\begin{bmatrix} 60.0 & 68.3 \\ 63.8 & 72.5 \\ 64.5 & 73.6 \\ 68.2 & 74.9 \end{bmatrix} + 5\begin{bmatrix} 68.0 & 75.6 \\ 70.7 & 78.1 \\ 72.7 & 79.4 \\ 74.8 & 80.0 \end{bmatrix}\right) = \dfrac{1}{6}\left(\begin{bmatrix} 60.0 & 68.3 \\ 63.8 & 72.5 \\ 64.5 & 73.6 \\ 68.2 & 74.9 \end{bmatrix} + 5\begin{bmatrix} 68.0 & 75.6 \\ 70.7 & 78.1 \\ 72.7 & 79.4 \\ 74.8 & 80.0 \end{bmatrix}\right)$

$= \dfrac{1}{6}\left(\begin{bmatrix} 60.0 & 68.3 \\ 63.8 & 72.5 \\ 64.5 & 73.6 \\ 68.2 & 74.9 \end{bmatrix} + \begin{bmatrix} 340.0 & 378.0 \\ 353.5 & 390.5 \\ 363.5 & 397.0 \\ 374.0 & 400.0 \end{bmatrix}\right)$

$= \dfrac{1}{6}\begin{bmatrix} 400.0 & 446.3 \\ 417.3 & 463.0 \\ 428.0 & 470.6 \\ 442.2 & 474.9 \end{bmatrix}$

$= \begin{bmatrix} 66.7 & 74.4 \\ 69.6 & 77.2 \\ 71.3 & 78.4 \\ 73.7 & 79.2 \end{bmatrix}$

53. (a) The matrices are

$$A = \begin{bmatrix} 0.036 & 0.014 \\ 0.019 & 0.008 \\ 0.021 & 0.006 \\ 0.014 & 0.008 \\ 0.011 & 0.011 \end{bmatrix} \text{ and } B = \begin{bmatrix} 283 & 1628 & 218 & 199 & 425 \\ 361 & 2038 & 286 & 227 & 460 \\ 473 & 2494 & 362 & 252 & 484 \\ 627 & 2978 & 443 & 278 & 499 \\ 839 & 3518 & 539 & 320 & 513 \end{bmatrix}$$

(b) The total number of births and deaths each year is found by multiplying matrix B by matrix A.

$$BA = \begin{bmatrix} 283 & 1628 & 218 & 199 & 425 \\ 361 & 2038 & 286 & 227 & 460 \\ 473 & 2494 & 362 & 252 & 484 \\ 627 & 2978 & 443 & 278 & 499 \\ 839 & 3518 & 539 & 320 & 513 \end{bmatrix}\begin{bmatrix} 0.036 & 0.014 \\ 0.019 & 0.008 \\ 0.021 & 0.006 \\ 0.014 & 0.008 \\ 0.011 & 0.011 \end{bmatrix}$$

	Births	Deaths
1960	53.159	24.561
1970	65.962	29.950
= 1980	80.868	36.086
1990	97.838	42.973
2002	118.488	51.327

2.5 Matrix Inverses

1. $\begin{bmatrix} 2 & 1 \\ 5 & 3 \end{bmatrix}\begin{bmatrix} 3 & -1 \\ -5 & 2 \end{bmatrix} = \begin{bmatrix} 6-5 & -2+2 \\ 15-15 & -5+6 \end{bmatrix} = \begin{bmatrix} 1 & 0 \\ 0 & 1 \end{bmatrix} = I$

$\begin{bmatrix} 3 & -1 \\ -5 & 2 \end{bmatrix}\begin{bmatrix} 2 & 1 \\ 5 & 3 \end{bmatrix} = \begin{bmatrix} 6-5 & 3-3 \\ -10+10 & -5+6 \end{bmatrix} = \begin{bmatrix} 1 & 0 \\ 0 & 1 \end{bmatrix} = I$

Since the products obtained by multiplying the matrices in either order are both the 2×2 identity matrix, the given matrices are inverses of each other.

3. $\begin{bmatrix} 2 & 6 \\ 2 & 4 \end{bmatrix} \begin{bmatrix} -1 & 2 \\ 2 & -4 \end{bmatrix} = \begin{bmatrix} 10 & -20 \\ 6 & -12 \end{bmatrix} \neq I$

No, the matrices are not inverses of each other since their product matrix is not I.

5. $\begin{bmatrix} 2 & 0 & 1 \\ 1 & 1 & 2 \\ 0 & 1 & 0 \end{bmatrix} \begin{bmatrix} 1 & 1 & -1 \\ 0 & 1 & 0 \\ -1 & -2 & 2 \end{bmatrix}$

$= \begin{bmatrix} 2+0-1 & 2+0-2 & -2+0+2 \\ 1+0-2 & 1+1-4 & -1+0+4 \\ 0+0+0 & 0+1+0 & 0+0+0 \end{bmatrix}$

$= \begin{bmatrix} 1 & 0 & 0 \\ -1 & -2 & 3 \\ 0 & 1 & 0 \end{bmatrix} \neq I$

No, the matrices are not inverses of each other since their product matrix is not I.

7. $\begin{bmatrix} 1 & 3 & 3 \\ 1 & 4 & 3 \\ 1 & 3 & 4 \end{bmatrix} \begin{bmatrix} 7 & -3 & -3 \\ -1 & 1 & 0 \\ -1 & 0 & 1 \end{bmatrix} = \begin{bmatrix} 1 & 0 & 0 \\ 0 & 1 & 0 \\ 0 & 0 & 1 \end{bmatrix} = I$

$\begin{bmatrix} 7 & -3 & -3 \\ -1 & 1 & 0 \\ -1 & 0 & 1 \end{bmatrix} \begin{bmatrix} 1 & 3 & 3 \\ 1 & 4 & 3 \\ 1 & 3 & 4 \end{bmatrix} = \begin{bmatrix} 1 & 0 & 0 \\ 0 & 1 & 0 \\ 0 & 0 & 1 \end{bmatrix} = I$

Yes, these matrices are inverses of each other.

9. No, a matrix with a row of all zeros does not have an inverse; the row of all zeros makes it impossible to get all the 1's in the main diagonal of the identity matrix.

11. Let $A = \begin{bmatrix} 1 & -1 \\ 2 & 0 \end{bmatrix}$.

Form the augmented matrix $[A|I]$.

$[A|I] = \begin{bmatrix} 1 & -1 & 1 & 0 \\ 2 & 0 & 0 & 1 \end{bmatrix}$

Perform row operations on $[A|I]$ to get a matrix of the form $[I|B]$.

$\begin{bmatrix} 1 & -1 & 1 & 0 \\ 2 & 0 & 0 & 1 \end{bmatrix}$

$-2R_1 + R_2 \to R_2 \quad \begin{bmatrix} 1 & -1 & 1 & 0 \\ 0 & 2 & -2 & 1 \end{bmatrix}$

$2R_1 + R_2 \to R_1 \quad \begin{bmatrix} 2 & 0 & 0 & 1 \\ 0 & 2 & -2 & 1 \end{bmatrix}$

$\begin{array}{c} \frac{1}{2}R_1 \to R_1 \\ \frac{1}{2}R_2 \to R_2 \end{array} \begin{bmatrix} 1 & 0 & 0 & \frac{1}{2} \\ 0 & 1 & -1 & \frac{1}{2} \end{bmatrix} = [I|B]$

The matrix B in the last transformation is the desired multiplicative inverse.

$$A^{-1} = \begin{bmatrix} 0 & \frac{1}{2} \\ -1 & \frac{1}{2} \end{bmatrix}$$

This answer may be checked by showing that $AA^{-1} = I$ and $A^{-1}A = I$.

13. Let $A = \begin{bmatrix} 3 & -1 \\ -5 & 2 \end{bmatrix}$.

$[A|I] = \begin{bmatrix} 3 & -1 & 1 & 0 \\ -5 & 2 & 0 & 1 \end{bmatrix}$

$5R_1 + 3R_2 \to R_2 \quad \begin{bmatrix} 3 & -1 & 1 & 0 \\ 0 & 1 & 5 & 3 \end{bmatrix}$

$R_1 + R_2 \to R_1 \quad \begin{bmatrix} 3 & 0 & 6 & 3 \\ 0 & 1 & 5 & 3 \end{bmatrix}$

$\frac{1}{3}R_1 \to R_1 \quad \begin{bmatrix} 1 & 0 & 2 & 1 \\ 0 & 1 & 5 & 3 \end{bmatrix} = [I|B]$

The desired inverse is

$$A^{-1} = \begin{bmatrix} 2 & 1 \\ 5 & 3 \end{bmatrix}.$$

15. Let $A = \begin{bmatrix} 1 & -3 \\ -2 & 6 \end{bmatrix}$.

$[A|I] = \begin{bmatrix} 1 & -3 & 1 & 0 \\ -2 & 6 & 0 & 1 \end{bmatrix}$

$2R_1 + R_2 \to R_2 \quad \begin{bmatrix} 1 & -3 & 1 & 0 \\ 0 & 0 & 2 & 1 \end{bmatrix}$

Because the last row has all zeros to the left of the vertical bar, there is no way to complete the desired transformation. A has no inverse.

17. Let $A = \begin{bmatrix} 1 & 0 & 0 \\ 0 & -1 & 0 \\ 1 & 0 & 1 \end{bmatrix}$.

$[A|I] = \begin{bmatrix} 1 & 0 & 0 & 1 & 0 & 0 \\ 0 & -1 & 0 & 0 & 1 & 0 \\ 1 & 0 & 1 & 0 & 0 & 1 \end{bmatrix}$

$-1R_1 + R_3 \to R_3 \quad \begin{bmatrix} 1 & 0 & 0 & 1 & 0 & 0 \\ 0 & -1 & 0 & 0 & 1 & 0 \\ 0 & 0 & 1 & -1 & 0 & 1 \end{bmatrix}$

$-1R_2 \to R_2 \quad \begin{bmatrix} 1 & 0 & 0 & 1 & 0 & 0 \\ 0 & 1 & 0 & 0 & -1 & 0 \\ 0 & 0 & 1 & -1 & 0 & 1 \end{bmatrix}$

$A^{-1} = \begin{bmatrix} 1 & 0 & 0 \\ 0 & -1 & 0 \\ -1 & 0 & 1 \end{bmatrix}$

19. Let $A = \begin{bmatrix} -1 & -1 & -1 \\ 4 & 5 & 0 \\ 0 & 1 & -3 \end{bmatrix}$.

$$[A|I] = \begin{bmatrix} -1 & -1 & -1 & | & 1 & 0 & 0 \\ 4 & 5 & 0 & | & 0 & 1 & 0 \\ 0 & 1 & -3 & | & 0 & 0 & 1 \end{bmatrix}$$

$4R_1 + R_2 \rightarrow R_2$ $\begin{bmatrix} -1 & -1 & -1 & | & 1 & 0 & 0 \\ 0 & 1 & -4 & | & 4 & 1 & 0 \\ 0 & 1 & -3 & | & 0 & 0 & 1 \end{bmatrix}$

$R_2 + R_1 \rightarrow R_1$
$-1R_2 + R_3 \rightarrow R_3$ $\begin{bmatrix} -1 & 0 & -5 & | & 5 & 1 & 0 \\ 0 & 1 & -4 & | & 4 & 1 & 0 \\ 0 & 0 & 1 & | & -4 & -1 & 1 \end{bmatrix}$

$5R_3 + R_1 \rightarrow R_1$
$4R_3 + R_2 \rightarrow R_2$ $\begin{bmatrix} -1 & 0 & 0 & | & -15 & -4 & 5 \\ 0 & 1 & 0 & | & -12 & -3 & 4 \\ 0 & 0 & 1 & | & -4 & -1 & 1 \end{bmatrix}$

$-1R_1 \rightarrow R_1$ $\begin{bmatrix} 1 & 0 & 0 & | & 15 & 4 & -5 \\ 0 & 1 & 0 & | & -12 & -3 & 4 \\ 0 & 0 & 1 & | & -4 & -1 & 1 \end{bmatrix}$

$$A^{-1} = \begin{bmatrix} 15 & 4 & -5 \\ -12 & -3 & 4 \\ -4 & -1 & 1 \end{bmatrix}$$

21. Let $A = \begin{bmatrix} 1 & 2 & 3 \\ -3 & -2 & -1 \\ -1 & 0 & 1 \end{bmatrix}$.

$$[A|I] = \begin{bmatrix} 1 & 2 & 3 & | & 1 & 0 & 0 \\ -3 & -2 & -1 & | & 0 & 1 & 0 \\ -1 & 0 & 1 & | & 0 & 0 & 1 \end{bmatrix}$$

$3R_1 + R_2 \rightarrow R_2$
$R_1 + R_3 \rightarrow R_3$ $\begin{bmatrix} 1 & 2 & 3 & | & 1 & 0 & 0 \\ 0 & 4 & 8 & | & 3 & 1 & 0 \\ 0 & 2 & 4 & | & 1 & 0 & 1 \end{bmatrix}$

$R_2 + (-2R_1) \rightarrow R_1$
$R_2 + (-2R_3) \rightarrow R_3$ $\begin{bmatrix} -2 & 0 & 2 & | & 1 & 1 & 0 \\ 0 & 4 & 8 & | & 3 & 1 & 0 \\ 0 & 0 & 0 & | & 1 & 1 & -2 \end{bmatrix}$

Because the last row has all zeros to the left of the vertical bar, there is no way to complete the desired transformation. A has no inverse.

23. Find the inverse of $A = \begin{bmatrix} 1 & 3 & -2 \\ 2 & 7 & -3 \\ 3 & 8 & -5 \end{bmatrix}$, if it exists.

$$[A|I] = \begin{bmatrix} 1 & 3 & -2 & | & 1 & 0 & 0 \\ 2 & 7 & -3 & | & 0 & 1 & 0 \\ 3 & 8 & -5 & | & 0 & 0 & 1 \end{bmatrix}$$

$-2R_1 + R_2 \rightarrow R_2$
$-3R_1 + R_3 \rightarrow R_3$ $\begin{bmatrix} 1 & 3 & -2 & | & 1 & 0 & 0 \\ 0 & 1 & 1 & | & -2 & 1 & 0 \\ 0 & -1 & 1 & | & -3 & 0 & 1 \end{bmatrix}$

$-3R_2 + R_1 \rightarrow R_1$
$R_2 + R_3 \rightarrow R_3$ $\begin{bmatrix} 1 & 0 & -5 & | & 7 & -3 & 0 \\ 0 & 1 & 1 & | & -2 & 1 & 0 \\ 0 & 0 & 2 & | & -5 & 1 & 1 \end{bmatrix}$

$5R_3 + 2R_1 \rightarrow R_1$
$-1R_3 + 2R_2 \rightarrow R_2$ $\begin{bmatrix} 2 & 0 & 0 & | & -11 & -1 & 5 \\ 0 & 2 & 0 & | & 1 & 1 & -1 \\ 0 & 0 & 2 & | & -5 & 1 & 1 \end{bmatrix}$

$\frac{1}{2}R_1 \rightarrow R_1$
$\frac{1}{2}R_2 \rightarrow R_2$
$\frac{1}{2}R_3 \rightarrow R_3$ $\begin{bmatrix} 1 & 0 & 0 & | & -\frac{11}{2} & -\frac{1}{2} & \frac{5}{2} \\ 0 & 1 & 0 & | & \frac{1}{2} & \frac{1}{2} & -\frac{1}{2} \\ 0 & 0 & 1 & | & -\frac{5}{2} & \frac{1}{2} & \frac{1}{2} \end{bmatrix}$

$$A^{-1} = \begin{bmatrix} -\frac{11}{2} & -\frac{1}{2} & \frac{5}{2} \\ \frac{1}{2} & \frac{1}{2} & -\frac{1}{2} \\ -\frac{5}{2} & \frac{1}{2} & \frac{1}{2} \end{bmatrix}$$

25. Let $A = \begin{bmatrix} 1 & -2 & 3 & 0 \\ 0 & 1 & -1 & 1 \\ -2 & 2 & -2 & 4 \\ 0 & 2 & -3 & 1 \end{bmatrix}$.

$$[A|I] = \begin{bmatrix} 1 & -2 & 3 & 0 & | & 1 & 0 & 0 & 0 \\ 0 & 1 & -1 & 1 & | & 0 & 1 & 0 & 0 \\ -2 & 2 & -2 & 4 & | & 0 & 0 & 1 & 0 \\ 0 & 2 & -3 & 1 & | & 0 & 0 & 0 & 1 \end{bmatrix}$$

$2R_1 + R_3 \rightarrow R_3$ $\begin{bmatrix} 1 & -2 & 3 & 0 & | & 1 & 0 & 0 & 0 \\ 0 & 1 & -1 & 1 & | & 0 & 1 & 0 & 0 \\ 0 & -2 & 4 & 4 & | & 2 & 0 & 1 & 0 \\ 0 & 2 & -3 & 1 & | & 0 & 0 & 0 & 1 \end{bmatrix}$

$2R_2 + R_1 \rightarrow R_1$

$2R_2 + R_3 \rightarrow R_3$
$-2R_2 + R_4 \rightarrow R_4$ $\begin{bmatrix} 1 & 0 & 1 & 2 & | & 1 & 2 & 0 & 0 \\ 0 & 1 & -1 & 1 & | & 0 & 1 & 0 & 0 \\ 0 & 0 & 2 & 6 & | & 2 & 2 & 1 & 0 \\ 0 & 0 & -1 & -1 & | & 0 & -2 & 0 & 1 \end{bmatrix}$

$R_3 + (-2)R_1 \rightarrow R_1$
$R_3 + 2R_2 \rightarrow R_2$

$R_3 + 2R_4 \rightarrow R_4$ $\begin{bmatrix} -2 & 0 & 0 & 2 & | & 0 & -2 & 1 & 0 \\ 0 & 2 & 0 & 8 & | & 2 & 4 & 1 & 0 \\ 0 & 0 & 2 & 6 & | & 2 & 2 & 1 & 0 \\ 0 & 0 & 0 & 4 & | & 2 & -2 & 1 & 2 \end{bmatrix}$

$$
\begin{matrix}
-2R_1+R_4 \to R_1 \\
R_2+(-2)R_4 \to R_2 \\
2R_3+(-3)R_4 \to R_3
\end{matrix}
\left[\begin{array}{cccc|cccc}
4 & 0 & 0 & 0 & 2 & 2 & -1 & 2 \\
0 & 2 & 0 & 0 & -2 & 8 & -1 & -4 \\
0 & 0 & 4 & 0 & -2 & 10 & -1 & -6 \\
0 & 0 & 0 & 4 & 2 & -2 & 1 & 2
\end{array}\right]
$$

$$
\begin{matrix}
\frac{1}{4}R_1 \to R_1 \\
\frac{1}{2}R_2 \to R_2 \\
\frac{1}{4}R_3 \to R_3 \\
\frac{1}{4}R_4 \to R_4
\end{matrix}
\left[\begin{array}{cccc|cccc}
1 & 0 & 0 & 0 & \frac{1}{2} & \frac{1}{2} & -\frac{1}{4} & \frac{1}{2} \\
0 & 1 & 0 & 0 & -1 & 4 & -\frac{1}{2} & -2 \\
0 & 0 & 1 & 0 & -\frac{1}{2} & \frac{5}{2} & -\frac{1}{4} & -\frac{3}{2} \\
0 & 0 & 0 & 1 & \frac{1}{2} & -\frac{1}{2} & \frac{1}{4} & \frac{1}{2}
\end{array}\right]
$$

$$
A^{-1} = \begin{bmatrix}
\frac{1}{2} & \frac{1}{2} & -\frac{1}{4} & \frac{1}{2} \\
-1 & 4 & -\frac{1}{2} & -2 \\
-\frac{1}{2} & \frac{5}{2} & -\frac{1}{4} & -\frac{3}{2} \\
\frac{1}{2} & -\frac{1}{2} & \frac{1}{4} & \frac{1}{2}
\end{bmatrix}
$$

27. $2x + 5y = 15$
   $x + 4y = 9$

First, write the system in matrix form.

$$
\begin{bmatrix} 2 & 5 \\ 1 & 4 \end{bmatrix}
\begin{bmatrix} x \\ y \end{bmatrix} =
\begin{bmatrix} 15 \\ 9 \end{bmatrix}
$$

Let $A = \begin{bmatrix} 2 & 5 \\ 1 & 4 \end{bmatrix}$, $X = \begin{bmatrix} x \\ y \end{bmatrix}$, and $B = \begin{bmatrix} 15 \\ 9 \end{bmatrix}$.

The system in matrix form is $AX = B$. We wish to find $X = A^{-1}AX = A^{-1}B$. Use row operations to find A^{-1}.

$$
[A|I] = \left[\begin{array}{cc|cc} 2 & 5 & 1 & 0 \\ 1 & 4 & 0 & 1 \end{array}\right]
$$

$$
-1R_1 + 2R_2 \to R_2 \left[\begin{array}{cc|cc} 2 & 5 & 1 & 0 \\ 0 & 3 & -1 & 2 \end{array}\right]
$$

$$
-5R_2 + 3R_1 \to R_1 \left[\begin{array}{cc|cc} 6 & 0 & 8 & -10 \\ 0 & 3 & -1 & 2 \end{array}\right]
$$

$$
\begin{matrix} \frac{1}{6}R_1 \to R_1 \\ \frac{1}{3}R_2 \to R_2 \end{matrix}
\left[\begin{array}{cc|cc} 1 & 0 & \frac{4}{3} & -\frac{5}{3} \\ 0 & 1 & -\frac{1}{3} & \frac{2}{3} \end{array}\right]
$$

$$
A^{-1} = \begin{bmatrix} \frac{4}{3} & -\frac{5}{3} \\ -\frac{1}{3} & \frac{2}{3} \end{bmatrix} = \frac{1}{3}\begin{bmatrix} 4 & -5 \\ -1 & 2 \end{bmatrix}
$$

Next find the product $A^{-1}B$.

$$
X = A^{-1}B = \begin{bmatrix} \frac{4}{3} & -\frac{5}{3} \\ -\frac{1}{3} & \frac{2}{3} \end{bmatrix} \begin{bmatrix} 15 \\ 9 \end{bmatrix}
$$

$$
= \frac{1}{3}\begin{bmatrix} 4 & -5 \\ -1 & 2 \end{bmatrix} \begin{bmatrix} 15 \\ 9 \end{bmatrix}
$$

$$
= \frac{1}{3}\begin{bmatrix} 15 \\ 3 \end{bmatrix} = \begin{bmatrix} 5 \\ 1 \end{bmatrix}
$$

Thus, the solution is $(5, 1)$.

29. $2x + y = 5$
   $5x + 3y = 13$

Let $A = \begin{bmatrix} 2 & 1 \\ 5 & 3 \end{bmatrix}$, $X = \begin{bmatrix} x \\ y \end{bmatrix}$, $B = \begin{bmatrix} 5 \\ 13 \end{bmatrix}$.

Use row operations to obtain

$$
A^{-1} = \begin{bmatrix} 3 & -1 \\ -5 & 2 \end{bmatrix}.
$$

$$
X = A^{-1}B = \begin{bmatrix} 3 & -1 \\ -5 & 2 \end{bmatrix}\begin{bmatrix} 5 \\ 13 \end{bmatrix} = \begin{bmatrix} 2 \\ 1 \end{bmatrix}
$$

The solution is $(2, 1)$.

31. $3x - 2y = 3$
   $7x - 5y = 0$

First, write the system in matrix form.

$$
\begin{bmatrix} 3 & -2 \\ 7 & -5 \end{bmatrix}\begin{bmatrix} x \\ y \end{bmatrix} = \begin{bmatrix} 3 \\ 0 \end{bmatrix}
$$

Let $A = \begin{bmatrix} 3 & -2 \\ 7 & -5 \end{bmatrix}$, $X = \begin{bmatrix} x \\ y \end{bmatrix}$, and $B = \begin{bmatrix} 3 \\ 0 \end{bmatrix}$.

The system is in matrix form $AX = B$. We wish to find $X = A^{-1}AX = A^{-1}B$. Use row operations to find A^{-1}.

$$
[A|I] = \left[\begin{array}{cc|cc} 3 & -2 & 1 & 0 \\ 7 & -5 & 0 & 1 \end{array}\right]
$$

$$
-7R_1 + 3R_2 \to R_2 \left[\begin{array}{cc|cc} 3 & -2 & 1 & 0 \\ 0 & -1 & -7 & 3 \end{array}\right]
$$

$$
-2R_2 + R_1 \to R_1 \left[\begin{array}{cc|cc} 3 & 0 & 15 & -6 \\ 0 & -1 & -7 & 3 \end{array}\right]
$$

$$
\begin{matrix} \frac{1}{3}R_1 \to R_1 \\ -1R_2 \to R_2 \end{matrix}
\left[\begin{array}{cc|cc} 1 & 0 & 5 & -2 \\ 0 & 1 & 7 & -3 \end{array}\right]
$$

$$
A^{-1} = \begin{bmatrix} 5 & -2 \\ 7 & -3 \end{bmatrix}
$$

Next find the product $A^{-1}B$.

$$
X = A^{-1}B = \begin{bmatrix} 5 & -2 \\ 7 & -3 \end{bmatrix}\begin{bmatrix} 3 \\ 0 \end{bmatrix} = \begin{bmatrix} 15 \\ 21 \end{bmatrix}
$$

Thus, the solution is $(15, 21)$.

33. $-x - 8y = 12$
$3x + 24y = -36$

Let $A = \begin{bmatrix} -1 & -8 \\ 3 & 24 \end{bmatrix}$, $X = \begin{bmatrix} x \\ y \end{bmatrix}$, $B = \begin{bmatrix} 12 \\ -36 \end{bmatrix}$.

Using row operations on $[A|I]$ leads to the matrix

$$\left[\begin{array}{cc|cc} 1 & 8 & -1 & 0 \\ 0 & 0 & 3 & 1 \end{array} \right],$$

but the zeros in the second row indicate that matrix A does not have an inverse. We cannot complete the solution by this method.

Since the second equation is a multiple of the first, the equations are dependent. Solve the first equation of the system for x.

$$-x - 8y = 12$$
$$-x = 8y + 12$$
$$x = -8y - 12$$

The solution is $(-8y - 12, y)$, where y is any real number.

35. $-x - y - z = 1$
$4x + 5y = -2$
$ y - 3z = 3$

has coefficient matrix

$$A = \begin{bmatrix} -1 & -1 & -1 \\ 4 & 5 & 0 \\ 0 & 1 & -3 \end{bmatrix}.$$

In Exercise 19, it was found that

$$A^{-1} = \begin{bmatrix} -1 & -1 & -1 \\ 4 & 5 & 0 \\ 0 & 1 & 3 \end{bmatrix}^{-1}$$

$$= \begin{bmatrix} 15 & 4 & -5 \\ -12 & -3 & 4 \\ -4 & -1 & 1 \end{bmatrix}.$$

Since $X = A^{-1}B$,

$$\begin{bmatrix} x \\ y \\ z \end{bmatrix} = \begin{bmatrix} 15 & 4 & -5 \\ -12 & -3 & 4 \\ -4 & -1 & 1 \end{bmatrix} \begin{bmatrix} 1 \\ -2 \\ 3 \end{bmatrix} = \begin{bmatrix} -8 \\ 6 \\ 1 \end{bmatrix}.$$

The solution is $(-8, 6, 1)$.

37. $x + 3y - 2z = 4$
$2x + 7y - 3z = 8$
$3x + 8y - 5z = -4$

has coefficient matrix

$$A = \begin{bmatrix} 1 & 3 & -2 \\ 2 & 7 & -3 \\ 3 & 8 & -5 \end{bmatrix}.$$

In Exercise 23, it was calculated that

$$A^{-1} = \begin{bmatrix} 1 & 3 & -2 \\ 2 & 7 & -3 \\ 3 & 8 & -5 \end{bmatrix}^{-1} = \begin{bmatrix} -\frac{11}{2} & -\frac{1}{2} & \frac{5}{2} \\ \frac{1}{2} & \frac{1}{2} & -\frac{1}{2} \\ -\frac{5}{2} & \frac{1}{2} & \frac{1}{2} \end{bmatrix}$$

$$= \frac{1}{2} \begin{bmatrix} -11 & -1 & 5 \\ 1 & 1 & -1 \\ -5 & 1 & 1 \end{bmatrix}.$$

Since $X = A^{-1}B$.

$$\begin{bmatrix} x \\ y \\ z \end{bmatrix} = \frac{1}{2} \begin{bmatrix} -11 & -1 & 5 \\ 1 & 1 & -1 \\ -5 & 1 & 1 \end{bmatrix} \begin{bmatrix} 4 \\ 8 \\ -4 \end{bmatrix}$$

$$= \frac{1}{2} \begin{bmatrix} -72 \\ 16 \\ -16 \end{bmatrix} = \begin{bmatrix} -36 \\ 8 \\ -8 \end{bmatrix}.$$

Thus, the solution is $(-36, 8, -8)$.

39. $2x - 2y = 5$
$ 4y + 8z = 7$
$x + 2z = 1$

has coefficient matrix

$$A = \begin{bmatrix} 2 & -2 & 0 \\ 0 & 4 & 8 \\ 1 & 0 & 2 \end{bmatrix}.$$

However, using row operations on $[A|I]$ shows that A does not have an inverse, so another method must be used.

Try the Gauss-Jordan method. The augmented matrix is

$$\left[\begin{array}{ccc|c} 2 & -2 & 0 & 5 \\ 0 & 4 & 8 & 7 \\ 1 & 0 & 2 & 1 \end{array} \right].$$

After several row operations, we obtain the matrix

$$\left[\begin{array}{ccc|c} 1 & 0 & 2 & \frac{17}{4} \\ 0 & 1 & 2 & \frac{7}{4} \\ 0 & 0 & 0 & 13 \end{array} \right].$$

The bottom row of this matrix shows that the system has no solution, since $0 = 13$ is a false statement.

41.
$$\begin{aligned}
x - 2y + 3z \quad\quad &= 4 \\
y - z + w &= -8 \\
-2x + 2y - 2z + 4w &= 12 \\
2y - 3z + w &= -4
\end{aligned}$$

has coefficient matrix

$$A = \begin{bmatrix} 1 & -2 & 3 & 0 \\ 0 & 1 & -1 & 1 \\ -2 & 2 & -2 & 4 \\ 0 & 2 & -3 & 1 \end{bmatrix}.$$

In Exercise 25, it was found that

$$A^{-1} = \begin{bmatrix} \frac{1}{2} & \frac{1}{2} & -\frac{1}{4} & \frac{1}{2} \\ -1 & 4 & -\frac{1}{2} & -2 \\ -\frac{1}{2} & \frac{5}{2} & -\frac{1}{4} & -\frac{3}{2} \\ \frac{1}{2} & -\frac{1}{2} & \frac{1}{4} & \frac{1}{2} \end{bmatrix}.$$

Since $X = A^{-1}B$,

$$\begin{bmatrix} x \\ y \\ z \\ w \end{bmatrix} = \begin{bmatrix} \frac{1}{2} & \frac{1}{2} & -\frac{1}{4} & \frac{1}{2} \\ -1 & 4 & -\frac{1}{2} & -2 \\ -\frac{1}{2} & \frac{5}{2} & -\frac{1}{4} & -\frac{3}{2} \\ \frac{1}{2} & -\frac{1}{2} & \frac{1}{4} & \frac{1}{2} \end{bmatrix} \begin{bmatrix} 4 \\ -8 \\ 12 \\ -4 \end{bmatrix} = \begin{bmatrix} -7 \\ -34 \\ -19 \\ 7 \end{bmatrix}.$$

The solution is $(-7, -34, -19, 7)$.

In Exercises 43–47, let $A = \begin{bmatrix} a & b \\ c & d \end{bmatrix}$.

43. $IA = \begin{bmatrix} 1 & 0 \\ 0 & 1 \end{bmatrix} \begin{bmatrix} a & b \\ c & d \end{bmatrix} = \begin{bmatrix} a & b \\ c & d \end{bmatrix} = A$

Thus, $IA = A$.

45. $A \cdot 0 = \begin{bmatrix} a & b \\ c & d \end{bmatrix} \begin{bmatrix} 0 & 0 \\ 0 & 0 \end{bmatrix} = \begin{bmatrix} 0 & 0 \\ 0 & 0 \end{bmatrix} = 0$

Thus, $A \cdot 0 = 0$.

47. In Exercise 46, it was found that

$$A^{-1} = \frac{1}{ad - bc} \begin{bmatrix} d & -b \\ -c & a \end{bmatrix}.$$

$$A^{-1}A = \left(\frac{1}{ad - bc} \begin{bmatrix} d & -b \\ -c & a \end{bmatrix} \right) \begin{bmatrix} a & b \\ c & d \end{bmatrix} = \frac{1}{ad - bc} \left(\begin{bmatrix} d & -b \\ -c & a \end{bmatrix} \begin{bmatrix} a & b \\ c & d \end{bmatrix} \right)$$

$$= \frac{1}{ad - bc} \begin{bmatrix} ad - bc & 0 \\ 0 & ad - bc \end{bmatrix} = \begin{bmatrix} 1 & 0 \\ 0 & 1 \end{bmatrix} = I$$

Thus, $A^{-1}A = I$.

49.
$$AB = O$$
$$A^{-1}(AB) = A^{-1} \cdot O$$
$$(A^{-1}A)B = O$$
$$I \cdot B = O$$
$$B = O$$

Thus, if $AB = O$ and A^{-1} exists, then $B = O$.

51. This exercise should be solved by graphing calculator or computer methods. The solution, which may vary slightly, is

$$C^{-1} = \begin{bmatrix} -0.0477 & -0.0230 & 0.0292 & 0.0895 & -0.0402 \\ 0.0921 & 0.0150 & 0.0321 & 0.0209 & -0.0276 \\ -0.0678 & 0.0315 & -0.0404 & 0.0326 & 0.0373 \\ 0.0171 & -0.0248 & 0.0069 & -0.0003 & 0.0246 \\ -0.0208 & 0.0740 & 0.0096 & -0.1018 & 0.0646 \end{bmatrix}.$$

(Entries are rounded to 4 places.)

53. This exercise should be solved by graphing calculator or computer methods. The solution, which may vary slightly, is

$$D^{-1} = \begin{bmatrix} 0.0394 & 0.0880 & 0.0033 & 0.0530 & -0.1499 \\ -0.1492 & 0.0289 & 0.0187 & 0.1033 & 0.1668 \\ -0.1330 & -0.0543 & 0.0356 & 0.1768 & 0.1055 \\ 0.1407 & 0.0175 & -0.0453 & -0.1344 & 0.0655 \\ 0.0102 & -0.0653 & 0.0993 & 0.0085 & -0.0388 \end{bmatrix}.$$

(Entries are rounded to 4 places.)

55. This exercise should be solved by graphing calculator or computer methods. The solution may vary slightly.

The answer is, yes, $D^{-1}C^{-1} = (CD)^{-1}$.

57. This exercise should be solved by graphing calculator or computer methods. The solution, which may vary slightly, is

$$\begin{bmatrix} 1.51482 \\ 0.053479 \\ -0.637242 \\ 0.462629 \end{bmatrix}.$$

59. (a) The matrix is $B = \begin{bmatrix} 72 \\ 48 \\ 60 \end{bmatrix}$.

(b) The matrix equation is

$$\begin{bmatrix} 2 & 4 & 2 \\ 2 & 1 & 2 \\ 2 & 1 & 3 \end{bmatrix} \begin{bmatrix} x_1 \\ x_2 \\ x_3 \end{bmatrix} = \begin{bmatrix} 72 \\ 48 \\ 60 \end{bmatrix}.$$

(c) To solve the system, begin by using row operations to find A^{-1}.

$$[A|I] = \begin{bmatrix} 2 & 4 & 2 & | & 1 & 0 & 0 \\ 2 & 1 & 2 & | & 0 & 1 & 0 \\ 2 & 1 & 3 & | & 0 & 0 & 1 \end{bmatrix}$$

$$\begin{matrix} \\ R_1 - 1R_2 \rightarrow R_2 \\ R_1 - 1R_3 \rightarrow R_3 \end{matrix} \begin{bmatrix} 2 & 4 & 2 & | & 1 & 0 & 0 \\ 0 & 3 & 0 & | & 1 & -1 & 0 \\ 0 & 3 & -1 & | & 1 & 0 & -1 \end{bmatrix}$$

$$\begin{matrix} -4R_2 + 3R_1 \rightarrow R_1 \\ \\ R_2 - 1R_3 \rightarrow R_3 \end{matrix} \begin{bmatrix} 6 & 0 & 6 & | & -1 & 4 & 0 \\ 0 & 3 & 0 & | & 1 & -1 & 0 \\ 0 & 0 & 1 & | & 0 & -1 & 1 \end{bmatrix}$$

$$\begin{matrix} \\ -6R_3 + R_1 \rightarrow R_1 \\ \\ \end{matrix} \begin{bmatrix} 6 & 0 & 0 & | & -1 & 10 & -6 \\ 0 & 3 & 0 & | & 1 & -1 & 0 \\ 0 & 0 & 1 & | & 0 & -1 & 1 \end{bmatrix}$$

$$\begin{matrix} \frac{1}{6}R_1 \rightarrow R_1 \\ \frac{1}{3}R_2 \rightarrow R_2 \\ \\ \end{matrix} \begin{bmatrix} 1 & 0 & 0 & | & -\frac{1}{6} & \frac{5}{3} & -1 \\ 0 & 1 & 0 & | & \frac{1}{3} & -\frac{1}{3} & 0 \\ 0 & 0 & 1 & | & 0 & -1 & 1 \end{bmatrix}$$

The inverse matrix is

$$A^{-1} = \begin{bmatrix} -\frac{1}{6} & \frac{5}{3} & -1 \\ \frac{1}{3} & -\frac{1}{3} & 0 \\ 0 & -1 & 1 \end{bmatrix}.$$

Since $X = A^{-1}B$,

$$\begin{bmatrix} x_1 \\ x_2 \\ x_3 \end{bmatrix} = \begin{bmatrix} -\frac{1}{6} & \frac{5}{3} & -1 \\ \frac{1}{3} & -\frac{1}{3} & 0 \\ 0 & -1 & 1 \end{bmatrix} \begin{bmatrix} 72 \\ 48 \\ 60 \end{bmatrix} = \begin{bmatrix} 8 \\ 8 \\ 12 \end{bmatrix}.$$

There are 8 daily orders for type I, 8 for type II, and 12 for type III.

61. Let $x =$ the amount invested in AAA bonds,
   $y =$ the amount invested in A bonds, and
   $z =$ amount invested in B bonds.

(a) The total investment is $x + y + z = 25{,}000$.
   The annual return is
   $0.06x = 0.065y + 0.08z = 1650$.
   Since twice as much is invested in AAA
   bonds as in B bonds, $x = 2z$.
   The system to be solved is

$$\begin{aligned} x + \quad\;\; y + \quad\;\;\;\; z &= 25{,}000 \\ 0.06x + 0.065y + 0.08z &= \;\;1650 \\ x \qquad\qquad\quad - \quad 2z &= \qquad 0 \end{aligned}$$

Let $A = \begin{bmatrix} 1 & 1 & 1 \\ 0.06 & 0.065 & 0.08 \\ 1 & 0 & -2 \end{bmatrix}$, $B = \begin{bmatrix} 25{,}000 \\ 1650 \\ 0 \end{bmatrix}$, and $X = \begin{bmatrix} x \\ y \\ z \end{bmatrix}$.

Use a graphing calculator to obtain

$$A^{-1} \begin{bmatrix} -26 & 400 & 3 \\ 40 & -600 & -4 \\ -13 & 200 & 1 \end{bmatrix}.$$

Use a graphing calculator again to solve the matrix equation $X = A^{-1}B$.

$$\begin{bmatrix} x \\ y \\ z \end{bmatrix} = \begin{bmatrix} -26 & 400 & 3 \\ 40 & -600 & -4 \\ -13 & 200 & 1 \end{bmatrix} \begin{bmatrix} 25{,}000 \\ 1650 \\ 0 \end{bmatrix} = \begin{bmatrix} 10{,}000 \\ 10{,}000 \\ 5000 \end{bmatrix}$$

$10,000 should be invested at 6% in AAA bonds, $10,000 at 6.5% in A bonds, and $5000 at 8% in B bonds.

(b) The matrix of constants is changed to

$$B = \begin{bmatrix} 30{,}000 \\ 1985 \\ 0 \end{bmatrix}.$$

$$\begin{bmatrix} x \\ y \\ z \end{bmatrix} = \begin{bmatrix} -26 & 400 & 3 \\ 40 & -600 & -4 \\ -13 & 200 & 1 \end{bmatrix} \begin{bmatrix} 30{,}000 \\ 1985 \\ 0 \end{bmatrix} = \begin{bmatrix} 14{,}000 \\ 9000 \\ 7000 \end{bmatrix}$$

$14,000 should be invested at 6% in AAA bonds, $9000 at 6.5% in A bonds, and $7000 at 8% in B bonds.

(c) The matrix of constants is changed to $B = \begin{bmatrix} 40{,}000 \\ 2660 \\ 0 \end{bmatrix}.$

$$\begin{bmatrix} x \\ y \\ z \end{bmatrix} = \begin{bmatrix} -26 & 400 & 3 \\ 40 & -600 & -4 \\ -13 & 200 & 1 \end{bmatrix} \begin{bmatrix} 40{,}000 \\ 2660 \\ 0 \end{bmatrix} = \begin{bmatrix} 24{,}000 \\ 4000 \\ 12{,}000 \end{bmatrix}$$

$24,000 should be invested at 6% in AAA bonds, $4000 at 6.5% in A bonds, and $12,000 at 8% in B bonds.

63. Let $x =$ the number of Super Vim tablets,
$\quad\quad y =$ the number of Multitab tablets, and
$\quad\quad z =$ the number of Mighty Mix tablets.

The total number of vitamins is

$$x + y + z.$$

The total amount of niacin is

$$15x + 20y + 25z.$$

The total amount of Vitamin E is

$$12x + 15y + 35z.$$

(a) The system to be solved is

$$\begin{array}{rcl} x + y + z &=& 225 \\ 15x + 20y + 25z &=& 4750 \\ 12x + 15y + 35z &=& 5225. \end{array}$$

Let $A = \begin{bmatrix} 1 & 1 & 1 \\ 15 & 20 & 25 \\ 12 & 15 & 35 \end{bmatrix}$, $X = \begin{bmatrix} x \\ y \\ z \end{bmatrix}$, $B = \begin{bmatrix} 225 \\ 4750 \\ 5225 \end{bmatrix}.$

Thus, $AX = B$ and

$$\begin{bmatrix} 1 & 1 & 1 \\ 15 & 20 & 25 \\ 12 & 15 & 35 \end{bmatrix} \begin{bmatrix} x \\ y \\ z \end{bmatrix} = \begin{bmatrix} 225 \\ 4750 \\ 5225 \end{bmatrix}.$$

Use row operations to obtain the inverse of the coefficient matrix.

$$A^{-1} = \begin{bmatrix} \frac{65}{17} & -\frac{4}{17} & \frac{1}{17} \\ -\frac{45}{17} & \frac{23}{85} & -\frac{2}{17} \\ -\frac{3}{17} & -\frac{3}{85} & \frac{1}{17} \end{bmatrix}$$

Since $X = A^{-1}B$,

$$\begin{bmatrix} x \\ y \\ z \end{bmatrix} = \begin{bmatrix} \frac{65}{17} & -\frac{4}{17} & \frac{1}{17} \\ -\frac{45}{17} & \frac{23}{85} & -\frac{2}{17} \\ -\frac{3}{17} & -\frac{3}{85} & \frac{1}{17} \end{bmatrix} \begin{bmatrix} 225 \\ 4750 \\ 5225 \end{bmatrix} = \begin{bmatrix} 50 \\ 75 \\ 100 \end{bmatrix}.$$

There are 50 Super Vim tablets, 75 Multitab tablets, and 100 Mighty Mix tablets.

(b) The matrix of constants is changed to

$$B = \begin{bmatrix} 185 \\ 3625 \\ 3750 \end{bmatrix}.$$

$$\begin{bmatrix} x \\ y \\ z \end{bmatrix} = \begin{bmatrix} \frac{65}{17} & -\frac{4}{17} & \frac{1}{17} \\ -\frac{45}{17} & \frac{23}{85} & -\frac{2}{17} \\ -\frac{3}{17} & -\frac{3}{85} & \frac{1}{17} \end{bmatrix} \begin{bmatrix} 185 \\ 3625 \\ 3750 \end{bmatrix} = \begin{bmatrix} 75 \\ 50 \\ 60 \end{bmatrix}$$

There are 75 Super Vim tablets, 50 Multitab tablets, and 60 Mighty Mix tablets.

(c) The matrix of constants is changed to

$$B = \begin{bmatrix} 230 \\ 4450 \\ 4210 \end{bmatrix}.$$

$$\begin{bmatrix} x \\ y \\ z \end{bmatrix} = \begin{bmatrix} \frac{65}{17} & -\frac{4}{17} & \frac{1}{17} \\ -\frac{45}{17} & \frac{23}{85} & -\frac{2}{17} \\ -\frac{3}{17} & -\frac{3}{85} & \frac{1}{17} \end{bmatrix} \begin{bmatrix} 230 \\ 4450 \\ 4210 \end{bmatrix} = \begin{bmatrix} 80 \\ 100 \\ 50 \end{bmatrix}$$

There are 80 Super Vim tablets, 100 Multitab tablets, and 50 Mighty Mix tablets.

65. (a) First, divide the letters and spaces of the message into groups of three, writing each group as a column vector.

$$\begin{bmatrix} T \\ o \\ \text{(space)} \end{bmatrix}, \begin{bmatrix} b \\ e \\ \text{(space)} \end{bmatrix}, \begin{bmatrix} o \\ r \\ \text{(space)} \end{bmatrix}, \begin{bmatrix} n \\ o \\ t \end{bmatrix}, \begin{bmatrix} \text{(space)} \\ t \\ o \end{bmatrix}, \begin{bmatrix} \text{(space)} \\ b \\ e \end{bmatrix}$$

Next, convert each letter into a number, assigning 1 to A, 2 to B, and so on, with the number 27 used to represent each space between words.

$$\begin{bmatrix} 20 \\ 15 \\ 27 \end{bmatrix}, \begin{bmatrix} 2 \\ 5 \\ 27 \end{bmatrix}, \begin{bmatrix} 15 \\ 18 \\ 27 \end{bmatrix}, \begin{bmatrix} 14 \\ 15 \\ 20 \end{bmatrix}, \begin{bmatrix} 27 \\ 20 \\ 15 \end{bmatrix}, \begin{bmatrix} 27 \\ 2 \\ 5 \end{bmatrix}$$

Now, find the product of the coding matrix B and each column vector. This produces a new set of vectors, which represents the coded message.

$$\begin{bmatrix} 262 \\ -161 \\ -12 \end{bmatrix}, \begin{bmatrix} 186 \\ -103 \\ -22 \end{bmatrix}, \begin{bmatrix} 264 \\ -168 \\ -9 \end{bmatrix}, \begin{bmatrix} 208 \\ -134 \\ -5 \end{bmatrix}, \begin{bmatrix} 224 \\ -152 \\ 5 \end{bmatrix}, \begin{bmatrix} 92 \\ -50 \\ -3 \end{bmatrix}$$

This message will be transmitted as 262, -161, -12, 186, -103, -22, 264, -168, -9, 208, -134, -5, 224, -152, 5, 92, -50, -3.

(b) Use row operations or a graphing calculator to find the inverse of the coding matrix B.

$$B^{-1} = \begin{bmatrix} 1.75 & 2.5 & 3 \\ -0.25 & -0.5 & 0 \\ -0.25 & -0.5 & -1 \end{bmatrix}$$

(c) First, divide the coded message into groups of three numbers and form each group into a column vector.

$$\begin{bmatrix} 116 \\ -60 \\ -15 \end{bmatrix}, \begin{bmatrix} 294 \\ -197 \\ -2 \end{bmatrix}, \begin{bmatrix} 148 \\ -92 \\ -9 \end{bmatrix}, \begin{bmatrix} 96 \\ -64 \\ 4 \end{bmatrix}, \begin{bmatrix} 264 \\ -182 \\ -2 \end{bmatrix}$$

Next, find the product of the decoding matric B^{-1} and each of the column vectors. This produces a new set of vectors, which represents the decoded message.

$$\begin{bmatrix} 8 \\ 1 \\ 16 \end{bmatrix}, \begin{bmatrix} 16 \\ 25 \\ 27 \end{bmatrix}, \begin{bmatrix} 2 \\ 9 \\ 18 \end{bmatrix}, \begin{bmatrix} 20 \\ 8 \\ 4 \end{bmatrix}, \begin{bmatrix} 1 \\ 25 \\ 27 \end{bmatrix}$$

Last, convert each number into a letter, assigning A to 1, B to 2, and so on, with the number 27 used to represent a space between words. The decoded message is HAPPY BIRTHDAY.

2.6   Input-Output Models

1. $A = \begin{bmatrix} 0.8 & 0.2 \\ 0.2 & 0.7 \end{bmatrix}$, $D = \begin{bmatrix} 2 \\ 3 \end{bmatrix}$

To find the production matrix, first calculate $I - A$.

$$I - A = \begin{bmatrix} 1 & 0 \\ 0 & 1 \end{bmatrix} - \begin{bmatrix} 0.8 & 0.2 \\ 0.2 & 0.7 \end{bmatrix} = \begin{bmatrix} 0.2 & -0.2 \\ -0.2 & 0.3 \end{bmatrix}$$

Using row operations, find the inverse of $I - A$.

$$[I - A \mid I] = \begin{bmatrix} 0.2 & -0.2 & 1 & 0 \\ -0.2 & 0.3 & 0 & 1 \end{bmatrix}$$

$$\begin{matrix} 10R_1 \rightarrow R_1 \\ 10R_2 \rightarrow R_2 \end{matrix} \begin{bmatrix} 2 & -2 & 10 & 0 \\ -2 & 3 & 0 & 10 \end{bmatrix}$$

$$R_1 + R_2 \rightarrow R_2 \begin{bmatrix} 2 & -2 & 10 & 0 \\ 0 & 1 & 10 & 10 \end{bmatrix}$$

$$2R_2 + R_1 \rightarrow R_1 \begin{bmatrix} 2 & 0 & 30 & 20 \\ 0 & 1 & 10 & 10 \end{bmatrix}$$

$$\tfrac{1}{2}R_1 \rightarrow R_1 \begin{bmatrix} 1 & 0 & 15 & 10 \\ 0 & 1 & 10 & 10 \end{bmatrix}$$

$$(I - A)^{-1} = \begin{bmatrix} 15 & 10 \\ 10 & 10 \end{bmatrix}$$

Since $X = (I - A)^{-1}D$, the product matrix is

$$X = \begin{bmatrix} 15 & 10 \\ 10 & 10 \end{bmatrix} \begin{bmatrix} 2 \\ 3 \end{bmatrix} = \begin{bmatrix} 60 \\ 50 \end{bmatrix}.$$

3. $A = \begin{bmatrix} 0.1 & 0.03 \\ 0.07 & 0.6 \end{bmatrix}$, $D = \begin{bmatrix} 5 \\ 10 \end{bmatrix}$

First, calculate $I - A$.

$$I - A = \begin{bmatrix} 0.9 & -0.03 \\ -0.07 & 0.4 \end{bmatrix}$$

Use row operations to find the inverse of $I - A$, which is

$$(I - A)^{-1} \approx \begin{bmatrix} 1.118 & 0.084 \\ 0.196 & 2.515 \end{bmatrix}.$$

Since $X = (I - A)^{-1}D$, the production matrix is

$$X = \begin{bmatrix} 1.118 & 0.084 \\ 0.196 & 2.515 \end{bmatrix} \begin{bmatrix} 5 \\ 10 \end{bmatrix}$$

$$= \begin{bmatrix} 6.43 \\ 26.12 \end{bmatrix}.$$

5. $A = \begin{bmatrix} 0.8 & 0 & 0.1 \\ 0.1 & 0.5 & 0.2 \\ 0 & 0 & 0.7 \end{bmatrix}$, $D = \begin{bmatrix} 1 \\ 6 \\ 3 \end{bmatrix}$

To find the production matrix, first calculate $I - A$.

$$I - A = \begin{bmatrix} 1 & 0 & 0 \\ 0 & 1 & 0 \\ 0 & 0 & 1 \end{bmatrix} - \begin{bmatrix} 0.8 & 0 & 0.1 \\ 0.1 & 0.5 & 0.2 \\ 0 & 0 & 0.7 \end{bmatrix}$$

$$= \begin{bmatrix} 0.2 & 0 & -0.1 \\ -0.1 & 0.5 & -0.2 \\ 0 & 0 & 0.3 \end{bmatrix}$$

Using row operations, find the inverse of $I - A$.

$$[I - A \mid I] = \begin{bmatrix} 0.2 & 0 & -0.1 & 1 & 0 & 0 \\ -0.1 & 0.5 & -0.2 & 0 & 1 & 0 \\ 0 & 0 & 0.3 & 0 & 0 & 1 \end{bmatrix}$$

$$\begin{matrix} 10R_1 \to R_1 \\ 10R_2 \to R_2 \\ 10R_3 \to R_3 \end{matrix} \begin{bmatrix} 2 & 0 & -1 & 10 & 0 & 0 \\ -1 & 5 & -2 & 0 & 10 & 0 \\ 0 & 0 & 3 & 0 & 0 & 10 \end{bmatrix}$$

$$R_1 + 2R_2 \to R_2 \begin{bmatrix} 2 & 0 & -1 & 10 & 0 & 0 \\ 0 & 10 & -5 & 10 & 20 & 0 \\ 0 & 0 & 3 & 0 & 0 & 10 \end{bmatrix}$$

$$\begin{matrix} R_3 + 3R_1 \to R_1 \\ 5R_3 + 3R_2 \to R_2 \end{matrix} \begin{bmatrix} 6 & 0 & 0 & 30 & 0 & 10 \\ 0 & 30 & 0 & 30 & 60 & 50 \\ 0 & 0 & 3 & 0 & 0 & 10 \end{bmatrix}$$

$$\begin{matrix} \frac{1}{6}R_1 \to R_1 \\ \frac{1}{30}R_2 \to R_2 \\ \frac{1}{3}R_3 \to R_3 \end{matrix} \begin{bmatrix} 1 & 0 & 0 & 5 & 0 & \frac{5}{3} \\ 0 & 1 & 0 & 1 & 2 & \frac{5}{3} \\ 0 & 0 & 1 & 0 & 0 & \frac{10}{3} \end{bmatrix}$$

$$(I - A)^{-1} = \begin{bmatrix} 5 & 0 & \frac{5}{3} \\ 1 & 2 & \frac{5}{3} \\ 0 & 0 & \frac{10}{3} \end{bmatrix}$$

Since $X = (I - A)^{-1}D$, the product matrix is

$$X = \begin{bmatrix} 5 & 0 & \frac{5}{3} \\ 1 & 2 & \frac{5}{3} \\ 0 & 0 & \frac{10}{3} \end{bmatrix} \begin{bmatrix} 1 \\ 6 \\ 3 \end{bmatrix} = \begin{bmatrix} 10 \\ 18 \\ 10 \end{bmatrix}.$$

7.
$$\begin{array}{c} \\ A \\ B \\ C \end{array} \begin{array}{ccc} A & B & C \\ \begin{bmatrix} 0.3 & 0.1 & 0.8 \\ 0.5 & 0.6 & 0.1 \\ 0.2 & 0.3 & 0.1 \end{bmatrix} & = A \end{array}$$

$$I - A = \begin{bmatrix} 0.7 & -0.1 & -0.8 \\ -0.5 & 0.4 & -0.1 \\ -0.2 & -0.3 & 0.9 \end{bmatrix}$$

Set $(I - A)X = O$ to obtain the following.

$$\begin{bmatrix} 0.7 & -0.1 & -0.8 \\ -0.5 & 0.4 & -0.1 \\ -0.2 & -0.3 & 0.9 \end{bmatrix} \begin{bmatrix} x_1 \\ x_2 \\ x_3 \end{bmatrix} = \begin{bmatrix} 0 \\ 0 \\ 0 \end{bmatrix}$$

$$\begin{bmatrix} 0.7x_1 - 0.1x_2 - 0.8x_3 \\ -0.5x_1 + 0.4x_2 - 0.1x_3 \\ -0.2x_1 - 0.3x_2 + 0.9x_3 \end{bmatrix} = \begin{bmatrix} 0 \\ 0 \\ 0 \end{bmatrix}$$

Rewrite this matrix equation as a system of equations.

$$0.7x_1 - 0.1x_2 - 0.8x_3 = 0$$
$$-0.5x_1 + 0.4x_2 - 0.1x_3 = 0$$
$$-0.2x_1 - 0.3x_2 + 0.9x_3 = 0$$

Rewrite the equations without decimals.

$$7x_1 - x_2 - 8x_3 = 0 \quad (1)$$
$$-5x_1 + 4x_2 - x_3 = 0 \quad (2)$$
$$-2x_1 - 3x_2 + 9x_3 = 0 \quad (3)$$

Use row operations to solve this system of equations. Begin by eliminating x_1 in equations (2) and (3).

$$7x_1 - x_2 - 8x_3 = 0 \quad (1)$$
$$5R_1 + 7R_2 \rightarrow R_2 \qquad 23x_2 - 47x_3 = 0 \quad (4)$$
$$2R_1 + 7R_3 \rightarrow R_3 \qquad -23x_2 + 47x_3 = 0 \quad (5)$$

Eliminate x_2 in equations (1) and (5).

$$23R_1 + R_2 \rightarrow R_1 \quad 161x_1 \qquad - 231x_3 = 0 \quad (6)$$
$$23x_2 - 47x_3 = 0 \quad (4)$$
$$R_2 + R_3 \rightarrow R_3 \qquad\qquad\qquad 0 = 0 \quad (7)$$

The true statement in equation (7) indicates that the equations are dependent. Solve equation (6) for x_1 and equation (4) for x_2, each in terms of x_3.

$$x_1 = \frac{231}{161}x_3 = \frac{33}{23}x_3$$
$$x_2 = \frac{47}{23}x_3$$

The solution of the system is

$$\left(\frac{33}{23}x_3, \frac{47}{23}x_3, x_3 \right).$$

If $x_3 = 23$, then $x_1 = 33$ and $x_2 = 47$, so the production of the three commodities should be in the ratio 33:47:23.

9. Use a graphing calculator or a computer to find the production matrix $X = (I - A)^{-1}D$. The answer is

$$X = \begin{bmatrix} 7697 \\ 4205 \\ 6345 \\ 4106 \end{bmatrix}.$$

Values have been rounded.

11. In Example 4, it was found that

$$(I - A)^{-1} \approx \begin{bmatrix} 1.3882 & 0.1248 \\ 0.5147 & 1.1699 \end{bmatrix}.$$

Since $X = (I - A)^{-1}D$, the production matrix is

$$X = \begin{bmatrix} 1.3882 & 0.1248 \\ 0.5147 & 1.1699 \end{bmatrix} \begin{bmatrix} 925 \\ 1250 \end{bmatrix} = \begin{bmatrix} 1440.085 \\ 1938.473 \end{bmatrix}.$$

Thus, about 1440.1 metric tons of wheat and 1938.5 metric tons of oil should be produced.

13. In Example 3, it was found that

$$(I - A)^{-1} \approx \begin{bmatrix} 1.40 & 0.50 & 0.59 \\ 0.84 & 1.36 & 0.62 \\ 0.56 & 0.47 & 1.30 \end{bmatrix}.$$

Since $X = (I - A)^{-1}D$, the production matrix is

$$X = \begin{bmatrix} 1.40 & 0.50 & 0.59 \\ 0.84 & 1.36 & 0.62 \\ 0.56 & 0.47 & 1.30 \end{bmatrix} \begin{bmatrix} 607 \\ 607 \\ 607 \end{bmatrix} = \begin{bmatrix} 1511.43 \\ 1711.74 \\ 1414.31 \end{bmatrix}.$$

Thus, about 1511.4 units of agriculture, 1711.7 units of manufacturing, and 1414.3 units of transportation should be produced.

15. From the given data, we get the input-output matrix

$$A = \begin{bmatrix} 0 & \frac{1}{2} & \frac{1}{4} \\ \frac{1}{4} & 0 & \frac{1}{4} \\ \frac{1}{2} & \frac{1}{4} & 0 \end{bmatrix}.$$

$$I - A = \begin{bmatrix} 1 & -\frac{1}{2} & -\frac{1}{4} \\ -\frac{1}{4} & 1 & -\frac{1}{4} \\ -\frac{1}{2} & -\frac{1}{4} & 1 \end{bmatrix}.$$

Use row operations to find the inverse of $I - A$, which is

$$(I - A)^{-1} \approx \begin{bmatrix} 1.538 & 0.923 & 0.615 \\ 0.615 & 1.436 & 0.513 \\ 0.923 & 0.821 & 1.436 \end{bmatrix}.$$

Since $X = (I - A)^{-1}D$, the production matrix is

$$X = \begin{bmatrix} 1.538 & 0.923 & 0.615 \\ 0.615 & 1.436 & 0.513 \\ 0.923 & 0.821 & 1.436 \end{bmatrix} \begin{bmatrix} 1000 \\ 1000 \\ 1000 \end{bmatrix}$$

$$\approx \begin{bmatrix} 3077 \\ 2564 \\ 3179 \end{bmatrix}.$$

Thus, the production should be about 3077 units of agriculture, 2564 units of manufacturing, and 3179 units of transportation.

17. From the given data, we get the input-output matrix

$$A = \begin{bmatrix} \frac{1}{4} & \frac{1}{6} \\ \frac{1}{2} & 0 \end{bmatrix}.$$

$$I - A = \begin{bmatrix} \frac{3}{4} & -\frac{1}{6} \\ -\frac{1}{2} & 1 \end{bmatrix}$$

Use row operations to find the inverse of $I - A$, which is

$$(I - A)^{-1} = \begin{bmatrix} \frac{3}{2} & \frac{1}{4} \\ \frac{3}{4} & \frac{9}{8} \end{bmatrix}.$$

(a) The production matrix is

$$X = (I - A)^{-1}D = \begin{bmatrix} \frac{3}{2} & \frac{1}{4} \\ \frac{3}{4} & \frac{9}{8} \end{bmatrix} \begin{bmatrix} 1 \\ 1 \end{bmatrix} = \begin{bmatrix} \frac{7}{4} \\ \frac{15}{8} \end{bmatrix}.$$

Thus, $\frac{7}{4}$ bushels of yams and $\frac{15}{8} \approx 2$ pigs should be produced.

(b) The production matrix is

$$X = (I - A)^{-1}D = \begin{bmatrix} \frac{3}{2} & \frac{1}{4} \\ \frac{3}{4} & \frac{9}{8} \end{bmatrix} \begin{bmatrix} 100 \\ 70 \end{bmatrix} = \begin{bmatrix} 167.5 \\ 153.75 \end{bmatrix}.$$

Thus, 167.5 bushels of yams and $153.75 \approx 154$ pigs should be produced.

19. Use a graphing calculator or a computer to find the production matrix $X = (I - A)^{-1}D$. The answer is

$$\begin{bmatrix} 848 \\ 516 \\ 2970 \end{bmatrix}.$$

Values have been rounded.

Produce 848 units of agriculture, 516 units of manufacturing, and 2970 units of households.

21. Use a graphing calculator or a computer to find the production matrix $X = (I - A)^{-1}D$. The answer is

$$\begin{bmatrix} 195,492 \\ 25,933 \\ 13,580 \end{bmatrix}.$$

Values have been rounded. Change from thousands of pounds to millions of pounds.

Produce about 195 million lb of agriculture, 26 million lb of manufacturing, and 13.6 million lb of energy.

23. Use a graphing calculator or a computer to find the production matrix $X = (I - A)^{-1}D$. The answer is

$$\begin{bmatrix} 532 \\ 481 \\ 805 \\ 1185 \end{bmatrix}.$$

Values have been rounded.

Produce about 532 units of natural resources, 481 manufacturing units, 805 trade and service units, and 1185 personal consumption units.

25. (a) Use a graphing calculator or a computer to find the matrix $(I - A)^{-1}$. The answer is

$$\begin{bmatrix} 1.67 & 0.56 & 0.56 \\ 0.19 & 1.17 & 0.06 \\ 3.15 & 3.27 & 4.38 \end{bmatrix}.$$

Values have been rounded.

(b) These multipliers imply that if the demand for one community's output increases by \$1 then the output in the other community will increase by the amount in the row and column of this matrix. For example, if the demand for Hermitage's output increases by \$1, then output from Sharon will increase \$0.56, from Farrell by \$0.06, and from Hermitage by \$4.38.

27. Calculate $I - A$, and then set $(I - A)X = O$ to find X.

$$(I - A)X = \left(\begin{bmatrix} 1 & 0 \\ 0 & 1 \end{bmatrix} - \begin{bmatrix} \frac{3}{4} & \frac{1}{3} \\ \frac{1}{4} & \frac{2}{3} \end{bmatrix} \right) \begin{bmatrix} x_1 \\ x_2 \end{bmatrix}$$

$$= \begin{bmatrix} \frac{1}{4} & -\frac{1}{3} \\ -\frac{1}{4} & \frac{1}{3} \end{bmatrix} \begin{bmatrix} x_1 \\ x_2 \end{bmatrix}$$

$$= \begin{bmatrix} \frac{1}{4}x_1 - \frac{1}{3}x_2 \\ -\frac{1}{4}x_1 + \frac{1}{3}x_2 \end{bmatrix} \begin{bmatrix} 0 \\ 0 \end{bmatrix}$$

Thus,

$$\frac{1}{4}x_1 - \frac{1}{3}x_2 = 0$$

$$\frac{1}{4}x_1 = \frac{1}{3}x_2$$

$$x_1 = \frac{4}{3}x_2$$

If $x_2 = 3$, $x_1 = 4$. Therefore, produce 4 units of steel for every 3 units of coal.

29. For this economy,

$$A = \begin{bmatrix} \frac{1}{5} & \frac{3}{5} & 0 \\ \frac{2}{5} & \frac{1}{5} & \frac{4}{5} \\ \frac{2}{5} & \frac{1}{5} & \frac{1}{5} \end{bmatrix}.$$

Find the value of $I - A$, then set $(I - A)X = O$.

$$(I - A)X = \left(\begin{bmatrix} 1 & 0 & 0 \\ 0 & 1 & 0 \\ 0 & 0 & 1 \end{bmatrix} - \begin{bmatrix} \frac{1}{5} & \frac{3}{5} & 0 \\ \frac{2}{5} & \frac{1}{5} & \frac{4}{5} \\ \frac{2}{5} & \frac{1}{5} & \frac{1}{5} \end{bmatrix} \right) \begin{bmatrix} x_1 \\ x_2 \\ x_3 \end{bmatrix}$$

$$= \begin{bmatrix} \frac{4}{5} & -\frac{3}{5} & 0 \\ -\frac{2}{5} & \frac{4}{5} & -\frac{4}{5} \\ -\frac{2}{5} & -\frac{1}{5} & \frac{4}{5} \end{bmatrix} \begin{bmatrix} x_1 \\ x_2 \\ x_3 \end{bmatrix}$$

$$= \begin{bmatrix} \frac{4}{5}x_1 - \frac{3}{5}x_2 \\ -\frac{2}{5}x_1 + \frac{4}{5}x_2 - \frac{4}{5}x_3 \\ -\frac{2}{5}x_1 - \frac{1}{5}x_2 + \frac{4}{5}x_3 \end{bmatrix}$$

$$= \begin{bmatrix} 0 \\ 0 \\ 0 \end{bmatrix}$$

The system to be solved is

$$\frac{4}{5}x_1 - \frac{3}{5}x_2 \qquad = 0$$

$$-\frac{2}{5}x_1 + \frac{4}{5}x_2 - \frac{4}{5}x_3 = 0$$

$$-\frac{2}{5}x_1 - \frac{1}{5}x_2 + \frac{4}{5}x_3 = 0.$$

Write the augmented matrix of the system.

$$\begin{bmatrix} \frac{4}{5} & -\frac{3}{5} & 0 & | & 0 \\ -\frac{2}{5} & \frac{4}{5} & -\frac{4}{5} & | & 0 \\ -\frac{2}{5} & -\frac{1}{5} & \frac{4}{5} & | & 0 \end{bmatrix}$$

$$\begin{array}{c} \frac{5}{4}R_1 \to R_1 \\ 5R_2 \to R_2 \\ 5R_3 \to R_3 \end{array} \begin{bmatrix} 1 & -\frac{3}{4} & 0 & | & 0 \\ -2 & 4 & -4 & | & 0 \\ -2 & -1 & 4 & | & 0 \end{bmatrix}$$

$$\begin{array}{c} 2R_1 + R_2 \to R_2 \\ 2R_1 + R_3 \to R_3 \end{array} \begin{bmatrix} 1 & -\frac{3}{4} & 0 & | & 0 \\ 0 & \frac{5}{2} & -4 & | & 0 \\ 0 & -\frac{5}{2} & 4 & | & 0 \end{bmatrix}$$

$$\frac{2}{5}R_2 \to R_2 \begin{bmatrix} 1 & -\frac{3}{4} & 0 & | & 0 \\ 0 & 1 & -\frac{8}{5} & | & 0 \\ 0 & -\frac{5}{2} & 4 & | & 0 \end{bmatrix}$$

$$\begin{array}{c} \frac{3}{4}R_2 + R_1 \to R_1 \\ \frac{5}{2}R_2 + R_3 \to R_3 \end{array} \begin{bmatrix} 1 & 0 & -\frac{6}{5} & | & 0 \\ 0 & 1 & -\frac{8}{5} & | & 0 \\ 0 & 0 & 0 & | & 0 \end{bmatrix}$$

Use x_3 as the parameter. Therefore, $x_1 = \frac{6}{5}x_3$ and $x_2 = \frac{8}{5}x_3$, and the solution is $\left(\frac{6}{5}x_3, \frac{8}{5}x_3, x_3\right)$. If $x_3 = 5$, then $x_1 = 6$ and $x_2 = 8$.

Produce 6 units of mining for every 8 units of manufacturing and 5 units of communication.

Chapter 2 Review Exercises

3. $2x - 3y = 14$ (1)
       $3x + 2y = -5$ (2)

Eliminate x in equation (2).

$$\begin{array}{cc} & 2x - 3y = 14 \quad (1) \\ -3R_1 + 2R_2 \to R_2 & 13y = -52 \quad (3) \end{array}$$

Make each leading coefficient equal 1.

$$\begin{array}{cc} \frac{1}{2}R_1 \to R_1 & x - \frac{3}{2}y = 7 \quad (4) \\ \frac{1}{13}R_2 \to R_2 & y = -4 \quad (5) \end{array}$$

Substitute -4 for y in equation (4) to get $x = 1$. The solution is $(1, -4)$.

5. $2x - 3y + z = -5$ (1)
$\quad x + 4y + 2z = 13$ (2)
$\quad 5x + 5y + 3z = 14$ (3)

Eliminate x in equations (2) and (3).

$$
\begin{array}{ll}
 & 2x - 3y + z = -5 \quad (1) \\
-2R_2 + R_1 \rightarrow R_2 & -11y - 3z = -31 \quad (4) \\
5R_1 + (-2)R_3 \rightarrow R_3 & -25y - z = -53 \quad (5)
\end{array}
$$

Eliminate y in equation (5).

$$
\begin{array}{ll}
 & 2x - 3y + z = -5 \quad (1) \\
 & -11y - 3z = -31 \quad (4) \\
-25R_2 + 11R_3 \rightarrow R_3 & 64z = 192 \quad (6)
\end{array}
$$

Make each leading coefficient equal 1.

$$
\begin{array}{ll}
\frac{1}{2}R_1 \rightarrow R_1 & x - \frac{3}{2}y + \frac{1}{2}z = -\frac{5}{2} \quad (7) \\
-\frac{1}{11}R_2 \rightarrow R_2 & y + \frac{3}{11}z = \frac{31}{11} \quad (8) \\
\frac{1}{64}R_3 \rightarrow R_3 & z = 3 \quad (9)
\end{array}
$$

Substitute 3 for z in equation (8) to get $y = 2$. Substitute 3 for z and 2 for y in equation (7) to get $x = -1$.

The solution is $(-1, 2, 3)$.

7. $2x + 4y = -6$
$\quad -3x - 5y = 12$

Write the augmented matrix and use row operations.

$$
\begin{bmatrix} 2 & 4 & | & -6 \\ -3 & -5 & | & 12 \end{bmatrix}
$$

$$
3R_1 + 2R_2 \rightarrow R_2 \quad \begin{bmatrix} 2 & 4 & | & -6 \\ 0 & 2 & | & 6 \end{bmatrix}
$$

$$
-2R_2 + R_1 \rightarrow R_1 \quad \begin{bmatrix} 2 & 0 & | & -18 \\ 0 & 2 & | & 6 \end{bmatrix}
$$

$$
\begin{array}{l} \frac{1}{2}R_1 \rightarrow R_1 \\ \frac{1}{2}R_2 \rightarrow R_2 \end{array} \quad \begin{bmatrix} 1 & 0 & | & -9 \\ 0 & 1 & | & 3 \end{bmatrix}
$$

The solution is $(-9, 3)$.

9. $x - y + 3z = 13$
$\quad 4x + y + 2z = 17$
$\quad 3x + 2y + 2z = 1$

Write the augmented matrix and use row operations.

$$
\begin{bmatrix} 1 & -1 & 3 & | & 13 \\ 4 & 1 & 2 & | & 17 \\ 3 & 2 & 2 & | & 1 \end{bmatrix}
$$

$$
\begin{array}{l} -4R_1 + R_2 \rightarrow R_2 \\ -3R_1 + R_3 \rightarrow R_3 \end{array} \quad \begin{bmatrix} 1 & -1 & 3 & | & 13 \\ 0 & 5 & -10 & | & -35 \\ 0 & 5 & -7 & | & -38 \end{bmatrix}
$$

$$
\begin{array}{l} R_2 + 5R_1 \rightarrow R_1 \\ -1R_2 + R_3 \rightarrow R_3 \end{array} \quad \begin{bmatrix} 5 & 0 & 5 & | & 30 \\ 0 & 5 & -10 & | & -35 \\ 0 & 0 & 3 & | & -3 \end{bmatrix}
$$

$$
\begin{array}{l} 5R_3 + (-3R_1) \rightarrow R_1 \\ 10R_3 + 3R_2 \rightarrow R_2 \end{array} \quad \begin{bmatrix} -15 & 0 & 0 & | & -105 \\ 0 & 15 & 0 & | & -135 \\ 0 & 0 & 3 & | & -3 \end{bmatrix}
$$

$$
\begin{array}{l} -\frac{1}{15}R_1 \rightarrow R_1 \\ \frac{1}{15}R_2 \rightarrow R_2 \\ \frac{1}{3}R_3 \rightarrow R_3 \end{array} \quad \begin{bmatrix} 1 & 0 & 0 & | & 7 \\ 0 & 1 & 0 & | & -9 \\ 0 & 0 & 1 & | & -1 \end{bmatrix}
$$

The solution is $(7, -9, -1)$.

11. $3x - 6y + 9z = 12$
$\quad -x + 2y - 3z = -4$
$\quad x + y + 2z = 7$

Write the augmented matrix and use row operations.

$$
\begin{bmatrix} 3 & -6 & 9 & | & 12 \\ -1 & 2 & -3 & | & -4 \\ 1 & 1 & 2 & | & 7 \end{bmatrix}
$$

$$
\begin{array}{l} R_1 + 3R_2 \rightarrow R_2 \\ -1R_1 + 3R_3 \rightarrow R_3 \end{array} \quad \begin{bmatrix} 3 & -6 & 9 & | & 12 \\ 0 & 0 & 0 & | & 0 \\ 0 & 9 & -3 & | & 9 \end{bmatrix}
$$

The zero in row 2, column 2 is an obstacle. To proceed, interchange the second and third rows.

$$
\begin{bmatrix} 3 & -6 & 9 & | & 12 \\ 0 & 9 & -3 & | & 9 \\ 0 & 0 & 0 & | & 0 \end{bmatrix}
$$

$$
3R_1 + 2R_2 \rightarrow R_1 \quad \begin{bmatrix} 9 & 0 & 21 & | & 54 \\ 0 & 9 & -3 & | & 9 \\ 0 & 0 & 0 & | & 0 \end{bmatrix}
$$

$$
\begin{array}{l} \frac{1}{9}R_1 \rightarrow R_1 \\ \frac{1}{9}R_2 \rightarrow R_2 \end{array} \quad \begin{bmatrix} 1 & 0 & \frac{7}{3} & | & 6 \\ 0 & 1 & -\frac{1}{3} & | & 1 \\ 0 & 0 & 0 & | & 0 \end{bmatrix}
$$

The row of zeros indicates dependent equations.

Solve the first two equations respectively for x and y in terms of z to obtain

$$x = 6 - \frac{7}{3}z \quad \text{and} \quad y = 1 + \frac{1}{3}z.$$

The solution of the system is

$$\left(6 - \frac{7}{3}z, 1 + \frac{1}{3}z, z \right),$$

where z is any real number.

In Exercises 13 – 15, corresponding elements must be equal.

13. $\begin{bmatrix} 2 & x \\ y & 6 \\ 5 & z \end{bmatrix} = \begin{bmatrix} a & -1 \\ 4 & 6 \\ p & 7 \end{bmatrix}$

The size of these matrices is 3×2. For matrices to be equal, corresponding elements must be equal, so $a = 2$, $x = -1$, $y = 4$, $p = 5$, and $z = 7$.

15. $\begin{bmatrix} a+5 & 3b & 6 \\ 4c & 2+d & -3 \\ -1 & 4p & q-1 \end{bmatrix} = \begin{bmatrix} -7 & b+2 & 2k-3 \\ 3 & 2d-1 & 4\ell \\ m & 12 & 8 \end{bmatrix}$

These are 3×3 square matrices. Since corresponding elements must be equal,

$a + 5 = -7$, so $\qquad a = -12$;

$3b = b + 2$, so $\qquad b = 1$;

$6 = 2k - 3$, so $\qquad k = \frac{9}{2}$;

$4c = 3$, so $\qquad c = \frac{3}{4}$;

$2 + d = 2d - 1$, so $\quad d = 3$;

$-3 = 4\ell$, so $\qquad \ell = -\frac{3}{4}$;

$\qquad\qquad\qquad\qquad m = -1$;

$4p = 12$, so $\qquad p = 3$; and

$q - 1 = 8$, so $\qquad q = 9$.

17. $2G - 4F = 2\begin{bmatrix} -2 & 0 \\ 1 & 5 \end{bmatrix} - 4\begin{bmatrix} -1 & 4 \\ 3 & 7 \end{bmatrix}$

$$= \begin{bmatrix} -4 & 0 \\ 2 & 10 \end{bmatrix} + \begin{bmatrix} 4 & -16 \\ -12 & -28 \end{bmatrix}$$

$$= \begin{bmatrix} 0 & -16 \\ -10 & -18 \end{bmatrix}$$

19. Since B is a 3×3 matrix, and C is a 3×2 matrix, the calculation of $B - C$ is not possible.

21. A has size 3×2 and G has 2×2, so AG will have size 3×2.

$$AG = \begin{bmatrix} 4 & 10 \\ -2 & -3 \\ 6 & 9 \end{bmatrix} \begin{bmatrix} -2 & 0 \\ 1 & 5 \end{bmatrix} = \begin{bmatrix} 2 & 50 \\ 1 & -15 \\ -3 & 45 \end{bmatrix}$$

23. D has size 3×1 and E has size 1×3, so DE will have size 3×3.

$$DE = \begin{bmatrix} 6 \\ 1 \\ 0 \end{bmatrix} \begin{bmatrix} 1 & 3 & -4 \end{bmatrix} = \begin{bmatrix} 6 & 18 & -24 \\ 1 & 3 & -4 \\ 0 & 0 & 0 \end{bmatrix}$$

25. B has size 3×3 and D has size 3×1, so BD will have size 3×1.

$$BD = \begin{bmatrix} 2 & 3 & -2 \\ 2 & 4 & 0 \\ 0 & 1 & 2 \end{bmatrix} \begin{bmatrix} 6 \\ 1 \\ 0 \end{bmatrix} = \begin{bmatrix} 15 \\ 16 \\ 1 \end{bmatrix}$$

27. $\qquad\qquad F = \begin{bmatrix} -1 & 4 \\ 3 & 7 \end{bmatrix}$

$$[F|I] = \begin{bmatrix} -1 & 4 & 1 & 0 \\ 3 & 7 & 0 & 1 \end{bmatrix}$$

$3R_1 + R_2 \to R_2 \quad \begin{bmatrix} -1 & 4 & 1 & 0 \\ 0 & 19 & 3 & 1 \end{bmatrix}$

$4R_2 + (-19R_1) \to R_1 \quad \begin{bmatrix} 19 & 0 & -7 & 4 \\ 0 & 19 & 3 & 1 \end{bmatrix}$

$\frac{1}{19}R_1 \to R_1 \quad \begin{bmatrix} 1 & 0 & -\frac{7}{19} & \frac{4}{19} \\ 0 & 1 & \frac{3}{19} & \frac{1}{19} \end{bmatrix}$
$\frac{1}{19}R_2 \to R_2$

$$F^{-1} = \begin{bmatrix} -\frac{7}{19} & \frac{4}{19} \\ \frac{3}{19} & \frac{1}{19} \end{bmatrix}$$

29. A and C are 3×2 matrices, so their sum $A + C$ is a 3×2 matrix. Only square matrices have inverses. Therefore, $(A + C)^{-1}$ does not exist.

31. Let $\qquad\qquad A = \begin{bmatrix} -4 & 2 \\ 0 & 3 \end{bmatrix}.$

$$[A|I] = \begin{bmatrix} -4 & 2 & 1 & 0 \\ 0 & 3 & 0 & 1 \end{bmatrix}$$

$2R_2 + (-3R_1) \to R_1 \quad \begin{bmatrix} 12 & 0 & -3 & 2 \\ 0 & 3 & 0 & 1 \end{bmatrix}$

$\frac{1}{12}R_1 \to R_1 \quad \begin{bmatrix} 1 & 0 & -\frac{1}{4} & \frac{1}{6} \\ 0 & 1 & 0 & \frac{1}{3} \end{bmatrix}$
$\frac{1}{3}R_2 \to R_2$

$$A^{-1} = \begin{bmatrix} -\frac{1}{4} & \frac{1}{6} \\ 0 & \frac{1}{3} \end{bmatrix}$$

33. Let

$$A = \begin{bmatrix} 6 & 4 \\ 3 & 2 \end{bmatrix}.$$

$$[A|I] = \begin{bmatrix} 6 & 4 & | & 1 & 0 \\ 3 & 2 & | & 0 & 1 \end{bmatrix}$$

$$R_1 + (-2)R_2 \rightarrow R_2 \quad \begin{bmatrix} 6 & 4 & | & 1 & 0 \\ 0 & 0 & | & 1 & -2 \end{bmatrix}$$

The zeros in the second row indicate that the original matrix has no inverse.

35. Let

$$A = \begin{bmatrix} 2 & 0 & 4 \\ 1 & -1 & 0 \\ 0 & 1 & -2 \end{bmatrix}.$$

$$[A|I] = \begin{bmatrix} 2 & 0 & 4 & | & 1 & 0 & 0 \\ 1 & -1 & 0 & | & 0 & 1 & 0 \\ 0 & 1 & -2 & | & 0 & 0 & 1 \end{bmatrix}$$

$$-2R_2 + R_1 \rightarrow R_2 \quad \begin{bmatrix} 2 & 0 & 4 & | & 1 & 0 & 0 \\ 0 & 2 & 4 & | & 1 & -2 & 0 \\ 0 & 1 & -2 & | & 0 & 0 & 1 \end{bmatrix}$$

$$-2R_3 + R_2 \rightarrow R_3 \quad \begin{bmatrix} 2 & 0 & 4 & | & 1 & 0 & 0 \\ 0 & 2 & 4 & | & 1 & -2 & 0 \\ 0 & 0 & 8 & | & 1 & -2 & -2 \end{bmatrix}$$

$$\begin{matrix} -1R_3 + 2R_1 \rightarrow R_1 \\ -1R_3 + 2R_2 \rightarrow R_2 \end{matrix} \quad \begin{bmatrix} 4 & 0 & 0 & | & 1 & 2 & 2 \\ 0 & 4 & 0 & | & 1 & -2 & 2 \\ 0 & 0 & 8 & | & 1 & -2 & -2 \end{bmatrix}$$

$$\begin{matrix} \frac{1}{4}R_1 \rightarrow R_1 \\ \frac{1}{4}R_2 \rightarrow R_2 \\ \frac{1}{8}R_3 \rightarrow R_3 \end{matrix} \quad \begin{bmatrix} 1 & 0 & 0 & | & \frac{1}{4} & \frac{1}{2} & \frac{1}{2} \\ 0 & 1 & 0 & | & \frac{1}{4} & -\frac{1}{2} & \frac{1}{2} \\ 0 & 0 & 1 & | & \frac{1}{8} & -\frac{1}{4} & -\frac{1}{4} \end{bmatrix}$$

$$A^{-1} = \begin{bmatrix} \frac{1}{4} & \frac{1}{2} & \frac{1}{2} \\ \frac{1}{4} & -\frac{1}{2} & \frac{1}{2} \\ \frac{1}{8} & -\frac{1}{4} & -\frac{1}{4} \end{bmatrix}.$$

37. Find the inverse of $A = \begin{bmatrix} 2 & -3 & 4 \\ 1 & 5 & 7 \\ -4 & 6 & -8 \end{bmatrix}$, if it exists.

$$[A|I] = \begin{bmatrix} 2 & -3 & 4 & | & 1 & 0 & 0 \\ 1 & 5 & 7 & | & 0 & 1 & 0 \\ -4 & 6 & -8 & | & 0 & 0 & 1 \end{bmatrix}$$

$$\begin{matrix} -1R_1 + 2R_2 \rightarrow R_2 \\ 2R_1 + R_3 \rightarrow R_3 \end{matrix} \quad \begin{bmatrix} 2 & -3 & 4 & | & 1 & 0 & 0 \\ 0 & 13 & 10 & | & -1 & 2 & 0 \\ 0 & 0 & 0 & | & 2 & 0 & 1 \end{bmatrix}$$

The zeros in the third row to the left of the vertical bar indicate that the original matrix has no inverse.

39. $A = \begin{bmatrix} 1 & 2 \\ 2 & 4 \end{bmatrix}, B = \begin{bmatrix} 5 \\ 10 \end{bmatrix}$

Row operations may be used to see that matrix A has no inverse. The matrix equation $AX = B$ may be written as the system of equations

$$\begin{aligned} x + 2y &= 5 \quad (1) \\ 2x + 4y &= 10. \quad (2) \end{aligned}$$

Use the elimination method to solve this system. Begin by eliminating x in equation (2).

$$\begin{aligned} & & x + 2y &= 5 \quad (1) \\ -2R_1 + R_2 \rightarrow R_2 & & 0 &= 0 \quad (3) \end{aligned}$$

The true statement in equation (3) indicates that the equations are dependent. Solve equation (1) for x in terms of y.

$$x = -2y + 5$$

The solution is $(-2y + 5, y)$, where y is any real number.

41. $A = \begin{bmatrix} 2 & 4 & 0 \\ 1 & -2 & 0 \\ 0 & 0 & 3 \end{bmatrix}, B = \begin{bmatrix} 72 \\ -24 \\ 48 \end{bmatrix}$

Use row operations to find the inverse of A, which is

$$A^{-1} = \begin{bmatrix} \frac{1}{4} & \frac{1}{2} & 0 \\ \frac{1}{8} & -\frac{1}{4} & 0 \\ 0 & 0 & \frac{1}{3} \end{bmatrix}.$$

Since $X = A^{-1}B$,

$$X = \begin{bmatrix} \frac{1}{4} & \frac{1}{2} & 0 \\ \frac{1}{8} & -\frac{1}{4} & 0 \\ 0 & 0 & \frac{1}{3} \end{bmatrix} \begin{bmatrix} 72 \\ -24 \\ 48 \end{bmatrix} = \begin{bmatrix} 6 \\ 15 \\ 16 \end{bmatrix}.$$

43. $\begin{aligned} 5x + 10y &= 80 \\ 3x - 2y &= 120 \end{aligned}$

Let $A = \begin{bmatrix} 5 & 10 \\ 3 & -2 \end{bmatrix}, X = \begin{bmatrix} x \\ y \end{bmatrix}, B = \begin{bmatrix} 80 \\ 120 \end{bmatrix}.$

Use row operations to find the inverse of A, which is

$$A^{-1} = \begin{bmatrix} \frac{1}{20} & \frac{1}{4} \\ \frac{3}{40} & -\frac{1}{8} \end{bmatrix}.$$

Since $X = A^{-1}B$,

$$\begin{bmatrix} x \\ y \end{bmatrix} = \begin{bmatrix} \frac{1}{20} & \frac{1}{4} \\ \frac{3}{40} & -\frac{1}{8} \end{bmatrix} \begin{bmatrix} 80 \\ 120 \end{bmatrix} = \begin{bmatrix} 34 \\ -9 \end{bmatrix}.$$

The solution is $(34, -9)$.

45.
$$x - 4y + 2z = -1$$
$$-2x + y - 3z = -9$$
$$3x + 5y - 2z = 7$$

Let $A = \begin{bmatrix} 1 & -4 & 2 \\ -2 & 1 & -3 \\ 3 & 5 & -2 \end{bmatrix}$, $X = \begin{bmatrix} x \\ y \\ z \end{bmatrix}$,

$B = \begin{bmatrix} -1 \\ -9 \\ 7 \end{bmatrix}$.

Use row operations to find the inverse of A, which is

$$A^{-1} = \begin{bmatrix} \frac{1}{3} & \frac{2}{39} & \frac{10}{39} \\ -\frac{1}{3} & -\frac{8}{39} & -\frac{1}{39} \\ -\frac{1}{3} & -\frac{17}{39} & -\frac{7}{39} \end{bmatrix} = \frac{1}{39} \begin{bmatrix} 13 & 2 & 10 \\ -13 & -8 & -1 \\ -13 & -17 & -7 \end{bmatrix}.$$

Since $X = A^{-1}B$,

$$\begin{bmatrix} x \\ y \\ z \end{bmatrix} = \frac{1}{39} \begin{bmatrix} 13 & 2 & 10 \\ -13 & -8 & -1 \\ -13 & -17 & -7 \end{bmatrix} \begin{bmatrix} -1 \\ -9 \\ 7 \end{bmatrix}$$

$$= \frac{1}{39} \begin{bmatrix} 39 \\ 78 \\ 117 \end{bmatrix}$$

$$= \begin{bmatrix} 1 \\ 2 \\ 3 \end{bmatrix}.$$

The solution is $(1, 2, 3)$.

47. $A = \begin{bmatrix} 0.2 & 0.1 & 0.3 \\ 0.1 & 0 & 0.2 \\ 0 & 0 & 0.4 \end{bmatrix}$, $D = \begin{bmatrix} 500 \\ 200 \\ 100 \end{bmatrix}$

$X = (I - A)^{-1}D$

$$I - A = \begin{bmatrix} 0.8 & -0.1 & -0.3 \\ -0.1 & 1 & -0.2 \\ 0 & 0 & 0.6 \end{bmatrix}$$

$$(I - A)^{-1} \approx \begin{bmatrix} 1.266 & 0.1266 & 0.6751 \\ 0.1266 & 1.0127 & 0.40084 \\ 0 & 0 & 1.6667 \end{bmatrix}$$

Since $X = (I - A)^{-1}D$,

$$X = \begin{bmatrix} 1.266 & 0.1266 & 0.6751 \\ 0.1266 & 1.0127 & 0.40084 \\ 0 & 0 & 1.6667 \end{bmatrix} \begin{bmatrix} 500 \\ 200 \\ 100 \end{bmatrix}$$

$$= \begin{bmatrix} 725.7 \\ 305.9 \\ 166.7 \end{bmatrix}.$$

49. Use a table to organize the information.

	Standard	Extra Large	Time Available
Hours Cutting	$\frac{1}{4}$	$\frac{1}{3}$	4
Hours Shaping	$\frac{1}{2}$	$\frac{1}{3}$	6

Let $x =$ the number of standard paper clips (in thousands),

and $y =$ the number of extra large paper clips (in thousands).

The given information leads to the system

$$\tfrac{1}{4}x + \tfrac{1}{3}y = 4$$
$$\tfrac{1}{2}x + \tfrac{1}{3}y = 6.$$

Solve this system by any method to get $x = 8$, $y = 6$. The manufacturer can make 8 thousand (8000) standard and 6 thousand (6000) extra large paper clips.

51. Let $x =$ Tulsa's number of gallons,

$y =$ New Orleans' number of gallons, and

$z =$ Ardmore's number of gallons.

The system that may be written is

$$0.5x + 0.4y + 0.3z = 219,000 \quad \textit{Chicago}$$
$$0.2x + 0.4y + 0.4z = 192,000 \quad \textit{Dallas}$$
$$0.3x + 0.2y + 0.3z = 144,000. \quad \textit{Atlanta}$$

The augmented matrix is

$$\begin{bmatrix} 0.5 & 0.4 & 0.3 & | & 219,000 \\ 0.2 & 0.4 & 0.4 & | & 192,000 \\ 0.3 & 0.2 & 0.3 & | & 144,000 \end{bmatrix}.$$

$$\begin{matrix} \\ 2R_1 + (-5)R_2 \to R_2 \\ 3R_1 + (-5)R_3 \to R_3 \end{matrix} \begin{bmatrix} 0.5 & 0.4 & 0.3 & | & 219,000 \\ 0 & -1.2 & -1.4 & | & -522,000 \\ 0 & 0.2 & -0.6 & | & -63,000 \end{bmatrix}$$

$$\begin{matrix} -2R_3 + R_1 \to R_1 \\ \\ R_2 + 6R_3 \to R_3 \end{matrix} \begin{bmatrix} 0.5 & 0 & 1.5 & | & 345,000 \\ 0 & -1.2 & -1.4 & | & -522,000 \\ 0 & 0 & -5 & | & -900,000 \end{bmatrix}$$

$$\begin{matrix} 0.3R_3 + R_1 \to R_1 \\ \\ -14R_3 + 50R_2 \to R_2 \end{matrix} \begin{bmatrix} 0.5 & 0 & 0 & | & 75,000 \\ 0 & -60 & 0 & | & -13,500,000 \\ 0 & 0 & -5 & | & -900,000 \end{bmatrix}$$

$$\begin{matrix} 2R_1 \to R_1 \\ -\frac{1}{60}R_2 \to R_2 \\ -\frac{1}{5}R_3 \to R_3 \end{matrix} \begin{bmatrix} 1 & 0 & 0 & | & 150,000 \\ 0 & 1 & 0 & | & 225,000 \\ 0 & 0 & 1 & | & 180,000 \end{bmatrix}$$

Thus, 150,000 gal were produced at Tulsa, 225,000 gal at New Orleans, and 180,000 gal at Ardmore.

53. (a)
$$\begin{array}{c}\text{High}\\\text{Medium}\\\text{Coated}\end{array}\begin{bmatrix}3170\\2360\\1800\end{bmatrix}$$

(b)
$$\begin{bmatrix}x\\y\\z\end{bmatrix}$$

(c)
$$\begin{bmatrix}10 & 5 & 8\\12 & 0 & 4\\0 & 10 & 5\end{bmatrix}\begin{bmatrix}x\\y\\z\end{bmatrix}=\begin{bmatrix}3170\\2360\\1800\end{bmatrix}$$

(d)
$$\begin{bmatrix}x\\y\\z\end{bmatrix}=\begin{bmatrix}10 & 5 & 8\\12 & 0 & 4\\0 & 10 & 5\end{bmatrix}^{-1}\begin{bmatrix}3170\\2360\\1800\end{bmatrix}$$
$$=\begin{bmatrix}-0.154 & 0.212 & 0.0769\\-0.231 & 0.192 & 0.2154\\0.462 & -0.385 & -0.231\end{bmatrix}\begin{bmatrix}3170\\2360\\1800\end{bmatrix}$$
$$=\begin{bmatrix}150\\110\\140\end{bmatrix}$$

55. (a) Use a graphing calculator or a computer to find $(I-A)^{-1}$. The solution, which may vary slightly, is

$$\begin{bmatrix}1.30 & 0.045 & 0.567 & 0.012 & 0.068 & 0.020\\0.204 & 1.03 & 0.183 & 0.004 & 0.022 & 0.006\\0.155 & 0.038 & 1.12 & 0.020 & 0.114 & 0.034\\0.018 & 0.021 & 0.028 & 1.08 & 0.016 & 0.033\\0.537 & 0.525 & 0.483 & 0.279 & 1.74 & 0.419\\0.573 & 0.346 & 0.497 & 0.536 & 0.087 & 1.94\end{bmatrix}$$

Values have been rounded.

The value in row 2, column 1 of this matrix, 0.204, indicates that every \$1 of increased demand for livestock will result in an increase of production demand of \$0.204 in crops.

(b) Use a graphing calculator or computer to find $(I-A)^{-1}D$. The solution, which may vary slightly, is

$$\begin{bmatrix}3855\\1476\\2726\\1338\\8439\\10{,}256\end{bmatrix}.$$

Values have been rounded.

In millions of dollars, produce \$3855 in livestock, \$1476 in crops, \$2726 in food products, \$1338 in mining and manufacturing, \$8439 in households, and \$10,256 in other business sectors.

57. (a) The X-ray passes through cells B and C, so the attenuation value for beam 3 is $b+c$.

(b) Beam 1: $a+b=0.8$
Beam 2: $a+c=0.55$
Beam 3: $b+c=0.65$

$$\begin{bmatrix}1 & 1 & 0\\1 & 0 & 1\\0 & 1 & 1\end{bmatrix}\begin{bmatrix}a\\b\\c\end{bmatrix}=\begin{bmatrix}0.8\\0.55\\0.65\end{bmatrix}$$

$$\begin{bmatrix}a\\b\\c\end{bmatrix}=\begin{bmatrix}1 & 1 & 0\\1 & 0 & 1\\0 & 1 & 1\end{bmatrix}^{-1}\begin{bmatrix}0.8\\0.55\\0.65\end{bmatrix}$$

$$=\begin{bmatrix}\frac{1}{2} & \frac{1}{2} & -\frac{1}{2}\\\frac{1}{2} & -\frac{1}{2} & \frac{1}{2}\\-\frac{1}{2} & \frac{1}{2} & \frac{1}{2}\end{bmatrix}\begin{bmatrix}0.8\\0.55\\0.65\end{bmatrix}$$

$$=\begin{bmatrix}0.35\\0.45\\0.2\end{bmatrix}$$

The solution is $(0.35, 0.45, 0.2)$, so A is tumorous, B is bone, and C is healthy.

(c) For patient X,

$$\begin{bmatrix}a\\b\\c\end{bmatrix}=\begin{bmatrix}\frac{1}{2} & \frac{1}{2} & -\frac{1}{2}\\\frac{1}{2} & -\frac{1}{2} & \frac{1}{2}\\-\frac{1}{2} & \frac{1}{2} & \frac{1}{2}\end{bmatrix}\begin{bmatrix}0.54\\0.40\\0.52\end{bmatrix}=\begin{bmatrix}0.21\\0.33\\0.19\end{bmatrix}.$$

A and C are healthy; B is tumorous.

For patient Y,

$$\begin{bmatrix}a\\b\\c\end{bmatrix}=\begin{bmatrix}\frac{1}{2} & \frac{1}{2} & -\frac{1}{2}\\\frac{1}{2} & -\frac{1}{2} & \frac{1}{2}\\-\frac{1}{2} & \frac{1}{2} & \frac{1}{2}\end{bmatrix}\begin{bmatrix}0.65\\0.80\\0.75\end{bmatrix}=\begin{bmatrix}0.35\\0.3\\0.45\end{bmatrix}.$$

A and B are tumorous; C is bone.

For patient Z,

$$\begin{bmatrix}a\\b\\c\end{bmatrix}=\begin{bmatrix}\frac{1}{2} & \frac{1}{2} & -\frac{1}{2}\\\frac{1}{2} & -\frac{1}{2} & \frac{1}{2}\\-\frac{1}{2} & \frac{1}{2} & \frac{1}{2}\end{bmatrix}\begin{bmatrix}0.51\\0.49\\0.44\end{bmatrix}=\begin{bmatrix}0.28\\0.23\\0.21\end{bmatrix}.$$

A could be healthy or tumorous; B and C are healthy.

59. The matrix representing the rates per 1000 athlete-exposures for specific injuries that caused a player wearing either shield to miss one or more events is

$$\begin{bmatrix} 3.54 & 1.41 \\ 1.53 & 1.57 \\ 0.34 & 0.29 \\ 7.53 & 6.21 \end{bmatrix}.$$

Since an equal number of players wear each type of shield and the total number of athlete-exposures for the league in a season is 8000, each type of shield is worn by 4000 players. Since the rates are given per 1000 athletic-exposures, the matrix representing the number of 1000 athlete-exposures for each type of shield is

$$\begin{bmatrix} 4 \\ 4 \end{bmatrix}.$$

The product of these matrices is

$$\begin{bmatrix} 20 \\ 12 \\ 3 \\ 55 \end{bmatrix}.$$

Values have been rounded.

There would be about 20 head and face injuries, 12 concussions, 3 neck injuries, and 55 other injuries.

61. $\dfrac{1}{2}W_1 + \dfrac{\sqrt{2}}{2}W_2 = 150$ (1)

$\dfrac{\sqrt{3}}{2}W_1 - \dfrac{\sqrt{2}}{2}W_2 = 0$ (2)

Adding equations (1) and (2) gives

$$\left(\dfrac{1}{2} + \dfrac{\sqrt{3}}{2} \right) W_1 = 150.$$

Multiply by 2.

$$(1 + \sqrt{3})W_1 = 300$$

$$W_1 = \dfrac{300}{1 + \sqrt{3}} \approx 110$$

From equation (2),

$$\dfrac{\sqrt{3}}{2}W_1 = \dfrac{\sqrt{2}}{2}W_2$$

$$W_2 = \dfrac{\sqrt{3}}{\sqrt{2}}W_1.$$

Substitute $\dfrac{300}{1+\sqrt{3}}$ from above for W_1.

$$W_2 = \dfrac{\sqrt{3}}{\sqrt{2}} \cdot \dfrac{300}{1 + \sqrt{3}} = \dfrac{300\sqrt{3}}{(1 + \sqrt{3})\sqrt{2}} \approx 134$$

Therefore, $W_1 \approx 110$ lb and $W_2 \approx 134$ lb.

63. (a) $\begin{bmatrix} 1 \\ 1 \end{bmatrix} x + \begin{bmatrix} 0 \\ 2 \end{bmatrix} y = \begin{bmatrix} 1 \\ 2 \end{bmatrix}$

$$\begin{bmatrix} x \\ x \end{bmatrix} + \begin{bmatrix} 0 \\ 2y \end{bmatrix} = \begin{bmatrix} 1 \\ 2 \end{bmatrix}$$

$$\begin{bmatrix} x \\ x + 2y \end{bmatrix} = \begin{bmatrix} 1 \\ 2 \end{bmatrix}$$

Since corresponding elements must be equal, $x = 1$ and $x + 2y = 2$. Substituting $x = 1$ in the second equation gives $y = \frac{1}{2}$. Note that $x = 1$ and $y = \frac{1}{2}$ are the values that balance the equation.

(b) $\qquad x\text{CO}_2 + y\text{H}_2 + z\text{CO} = \text{H}_2\text{O}$

$$\begin{bmatrix} 1 \\ 0 \\ 2 \end{bmatrix} x + \begin{bmatrix} 0 \\ 2 \\ 0 \end{bmatrix} y + \begin{bmatrix} 1 \\ 0 \\ 1 \end{bmatrix} z = \begin{bmatrix} 0 \\ 2 \\ 1 \end{bmatrix}$$

$$\begin{bmatrix} x \\ 0 \\ 2x \end{bmatrix} + \begin{bmatrix} 0 \\ 2y \\ 0 \end{bmatrix} + \begin{bmatrix} z \\ 0 \\ z \end{bmatrix} = \begin{bmatrix} 0 \\ 2 \\ 1 \end{bmatrix}$$

$$\begin{bmatrix} x + z \\ 2y \\ 2x + z \end{bmatrix} = \begin{bmatrix} 0 \\ 2 \\ 1 \end{bmatrix}$$

Since corresponding elements must be equal, $x + z = 0$, $2y = 2$, and $2x + z = 1$. Solving $2y = 2$ gives $y = 1$. Solving the system $\begin{cases} x + z = 0 \\ 2x + z = 1 \end{cases}$ gives $x = 1$ and $z = -1$. Thus, the values that balance the equation are $x = 1$, $y = 1$, and $z = -1$.

65. Let $x =$ the number of singles,
$\quad y =$ the number of doubles,
$\quad z =$ the number of triples, and
$\quad w =$ the number of home runs
$\qquad$ hit by Barry Bonds.

If he had a total of 135 hits, then $x + y + z + w = 135$.

If he hit 15 times as many home runs as triples, then $w = 15z$.

If he hit 50% more home runs than doubles and triples, then $w = 1.5(y + z)$.

If he hit twice as many singles as doubles and triples, then $x = 2(y + z)$.

Write the equations in proper form, obtain the augmented matrix, and use row operations to solve.

$$\begin{bmatrix} 1 & 1 & 1 & 1 & 135 \\ 0 & 0 & 15 & -1 & 0 \\ 0 & 1.5 & 1.5 & -1 & 0 \\ 1 & -2 & -2 & 0 & 0 \end{bmatrix}$$

$R_2 \longleftrightarrow R_3$
$$\begin{bmatrix} 1 & 1 & 1 & 1 & 135 \\ 0 & 1.5 & 1.5 & -1 & 0 \\ 0 & 0 & 15 & -1 & 0 \\ 1 & -2 & -2 & 0 & 0 \end{bmatrix}$$

$-1R_1 + R_4 \rightarrow R_4$
$$\begin{bmatrix} 1 & 1 & 1 & 1 & 135 \\ 0 & 1.5 & 1.5 & -1 & 0 \\ 0 & 0 & 15 & -1 & 0 \\ 0 & -3 & -3 & -1 & -135 \end{bmatrix}$$

$-10R_2 + 15R_1 \rightarrow R_1$

$2R_2 + R_4 \rightarrow R_4$
$$\begin{bmatrix} 15 & 0 & 0 & 25 & 2025 \\ 0 & 1.5 & 1.5 & -1 & 0 \\ 0 & 0 & 15 & -1 & 0 \\ 0 & 0 & 0 & -3 & -135 \end{bmatrix}$$

$-1R_3 + 10R_2 \rightarrow R_2$
$$\begin{bmatrix} 15 & 0 & 0 & 25 & 2025 \\ 0 & 15 & 0 & -9 & 0 \\ 0 & 0 & 15 & -1 & 0 \\ 0 & 0 & 0 & -3 & -135 \end{bmatrix}$$

$25R_4 + 3R_1 \rightarrow R_1$
$-3R_4 + R_2 \rightarrow R_2$
$-1R_4 + 3R_3 \rightarrow R_3$
$$\begin{bmatrix} 45 & 0 & 0 & 0 & 2700 \\ 0 & 15 & 0 & 0 & 405 \\ 0 & 0 & 45 & 0 & 135 \\ 0 & 0 & 0 & -3 & -135 \end{bmatrix}$$

$\frac{1}{45}R_1 \rightarrow R_1$
$\frac{1}{15}R_2 \rightarrow R_2$
$\frac{1}{45}R_3 \rightarrow R_3$
$-\frac{1}{3}R_4 \rightarrow R_4$
$$\begin{bmatrix} 1 & 0 & 0 & 0 & 60 \\ 0 & 1 & 0 & 0 & 27 \\ 0 & 0 & 1 & 0 & 3 \\ 0 & 0 & 0 & 1 & 45 \end{bmatrix}$$

Barry Bonds hit 60 singles, 27 doubles, 3 triples, and 45 home runs.

Chapter 2 Test

[2.1]

1. Solve the system using the echelon method.

$$3x + y = 11$$
$$x - 2y = -8$$

[2.2]

2. Solve the system using the Gauss-Jordan method.

$$x + 2y + 3z = 5$$
$$2x - y + z = 5$$
$$x + y + z = 2$$

3. Use the Gauss-Jordan method to find all solutions of the following system, given that all variables must be nonnegative integers.

$$2x + 2y - z = 30$$
$$3x + 2y - 2z = 41$$
$$x + 4y + z = 27$$

[2.1–2.2]

4. Use a system of equations to solve the following problem. Solve the system by the method of your choice.

   An investor has $80,000 and she would like to earn 7.675% per year by investing it in mutual funds, bonds, and certificates of deposit. She wants to invest $4000 more in bonds than the total investment in mutual funds and certificates of deposit. Mutual funds pay 10% per year, bonds pay 7% per year, and certificates of deposit pay 5% per year. How much should she invest in each of the three?

[2.3]

5. Find the values of the variables.

$$\begin{bmatrix} 6 - x & 2y + 1 \\ 3m & 5p + 2 \end{bmatrix} = \begin{bmatrix} 8 & 10 \\ -5 & 3p - 1 \end{bmatrix}$$

[2.3–2.4]

6. Given the following matrices, perform the indicated operations, if possible.

$$A = \begin{bmatrix} 1 & 2 & -1 \\ 0 & 1 & 1 \\ 1 & 0 & 1 \end{bmatrix} \qquad B = \begin{bmatrix} 1 & -2 \\ 1 & 1 \\ 0 & 1 \end{bmatrix} \qquad C = \begin{bmatrix} 2 & 1 & 3 \\ 0 & 4 & 1 \\ 1 & 1 & 1 \end{bmatrix}$$

   (a) AB (b) $2A - C$ (c) $A + 2B$

[2.5]

Find the inverse of each matrix which has an inverse.

7. $A = \begin{bmatrix} 1 & 0 & -1 \\ 2 & 1 & 1 \\ 1 & 1 & 5 \end{bmatrix}$

8. $B = \begin{bmatrix} 2 & -1 & 1 \\ 0 & 2 & 4 \\ 2 & 1 & 5 \end{bmatrix}$

9. For $A = \begin{bmatrix} 1 & 0 & 1 \\ 1 & 1 & 1 \\ 2 & 1 & 3 \end{bmatrix}$, $A^{-1} = \begin{bmatrix} 2 & 1 & -1 \\ -1 & 1 & 0 \\ -1 & -1 & 1 \end{bmatrix}$.

Use this inverse to solve the equation $AX = B$, where $B = \begin{bmatrix} 1 \\ 2 \\ -1 \end{bmatrix}$.

10. Solve the system using the inverse of the coefficient matrix.

$$\begin{aligned} x - 2y &= 3 \\ x + 3y &= 5 \end{aligned}$$

11. Solve the system by using the inverse of the coefficient matrix. Use a graphing calculator.

$$\begin{aligned} 0.103x - 0.247y + 0.489z &= 0.936 \\ -0.218x + 0.379y + 0.702z &= 0.863 \\ 0.315x - 0.742y - 0.913z &= -0.768 \end{aligned}$$

[2.6]

12. Find the production matrix, given the following input-output and demand matrices.

$$A = \begin{bmatrix} 0.2 & 0.1 & 0.3 \\ 0.1 & 0 & 0.2 \\ 0 & 0 & 0.4 \end{bmatrix}, \qquad D = \begin{bmatrix} 1000 \\ 2000 \\ 5000 \end{bmatrix}$$

Chapter 2 Test Answers

1. $(2, 5)$

2. $(1, -1, 2)$

3. $(11, 4, 0)$, $(13, 3, 2)$, $(15, 2, 4)$, $(17, 1, 6)$, $(19, 0, 8)$

4. $26,000 in mutual funds; $42,000 in bonds, $12,000 in certificates of deposit

5. $x = -2$, $y = \frac{9}{2}$, $m = -\frac{5}{3}$, $p = -\frac{3}{2}$

6. (a) $\begin{bmatrix} 3 & -1 \\ 1 & 2 \\ 1 & -1 \end{bmatrix}$ **(b)** $\begin{bmatrix} 0 & 3 & -5 \\ 0 & -2 & 1 \\ 1 & -1 & 1 \end{bmatrix}$ **(c)** Not possible

7. $A^{-1} = \begin{bmatrix} \frac{4}{3} & -\frac{1}{3} & \frac{1}{3} \\ -3 & 2 & -1 \\ \frac{1}{3} & -\frac{1}{3} & \frac{1}{3} \end{bmatrix}$

8. B^{-1} does not exist.

9. $X = \begin{bmatrix} 5 \\ 1 \\ -4 \end{bmatrix}$

10. $\left(\frac{19}{5}, \frac{2}{5} \right)$

11. $(-2.061, -1.683, 1.498)$

12. $\begin{bmatrix} 4895 \\ 4156 \\ 8333 \end{bmatrix}$

LINEAR PROGRAMMING: THE GRAPHICAL METHOD

3.1 Graphing Linear Inequalities

1. $x + y \leq 2$

First graph the boundary line $x + y = 2$ using the points $(2, 0)$ and $(0, 2)$. Since the points on this line satisfy $x + y \leq 2$, draw a solid line. To find the correct region to shade, choose any point not on the line. If $(0, 0)$ is used as the test point, we have

$$x + y \leq 2$$
$$0 + 0 \leq 2$$
$$0 \leq 2,$$

which is a true statement. Shade the half-plane containing $(0, 0)$, or all points below the line.

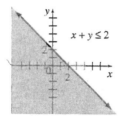

3. $x \geq 2 - y$

First graph the boundary line $x = 2 - y$ using the points $(0, 2)$ and $(2, 0)$. This will be a solid line. Choose $(0, 0)$ as a test point.

$$x \geq 2 - y$$
$$0 \geq 2 - 0$$
$$0 \geq 2,$$

which is a false statement. Shade the half-plane that does not contain $(0, 0)$, or all points below the line.

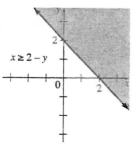

5. $4x - y < 6$

Graph $4x - y = 6$ as a dashed line, since the points on the line are not part of the solution; the line passes through the points $(0, -6)$ and $\left(\frac{3}{2}, 0\right)$. Using the test point $(0, 0)$, we have $0 - 0 < 6$ or $0 < 6$, a true statement. Shade the half-plane containing $(0, 0)$, or all points above the line.

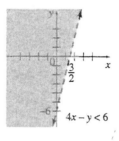

7. $4x + y < 8$

Graph $4x + y = 8$ as a dashed line through $(2, 0)$ and $(0, 8)$. Using the test point $(0, 0)$, we get $4(0) + 0 < 8$ or $0 < 8$, a true statement. Shade the half-plane containing $(0, 0)$, or all points below the line.

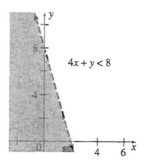

9. $x + 3y \geq -2$

The graph includes the line $x + 3y = -2$, whose intercepts are the points $\left(0, -\frac{2}{3}\right)$ and $(-2, 0)$. Graph $x + 3y = -2$ as a solid line and use the origin as a test point. Since $0 + 3(0) \geq -2$ is true, shade the half-plane containing $(0, 0)$, or all points above the line.

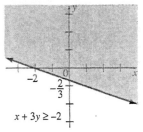

11. $x \leq 3y$

Graph $x = 3y$ as a solid line through the points $(0, 0)$ and $(3, 1)$. Since this line contains the origin, some point other than $(0, 0)$ must be used as a test point. If we use the point $(1, 2)$, we obtain $1 \leq 3(2)$ or $1 \leq 6$, a true statement. Shade the half-plane containing $(1, 2)$, or all points above the line.

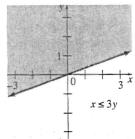

13. $x + y \leq 0$

Graph $x + y = 0$ as a solid line through the points $(0, 0)$ and $(1, -1)$. This line contains $(0, 0)$. If we use $(-1, 0)$ as a test point, we obtain $-1 + 0 \leq 0$ or $-1 \leq 0$, a true statement. Shade the half-plane containing $(-1, 0)$, or all points below the line.

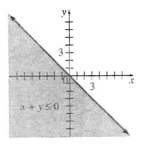

15. $y < x$

Graph $y = x$ as a dashed line through the points $(0, 0)$ and $(1, 1)$. Since this line contains the origin, choose a point other than $(0, 0)$ as a test point. If we use $(2, 3)$, we obtain $3 < 2$, which is false. Shade the half-plane that does not contain $(2, 3)$, or all points below the line.

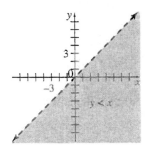

17. $x < 4$

Graph $x = 4$ as a dashed line. This is the vertical line crossing the x-axis at the point $(4, 0)$. Using $(0, 0)$ as a test point, we obtain $0 < 4$, which is true. Shade the half-plane containing $(0, 0)$, or all points to the left of the line.

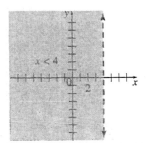

19. $y \leq -2$

Graph $y = -2$ as a solid horizontal line through the point $(0, -2)$. Using the origin as a test point, we obtain $0 \leq -2$, which is false. Shade the half-plane that does not contain $(0, 0)$, or all points below the line.

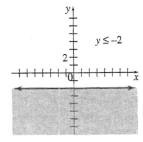

21. $x + y \le 1$
$\quad x - y \ge 2$

Graph the solid lines

$$x + y = 1 \text{ and}$$
$$x - y = 2.$$

$0 + 0 \le 1$ is true, and $0 - 0 \ge 2$ is false. In each case, the graph is the region below the line. Shade the overlapping part of these two half-planes, which is the region below both lines. The shaded region is the feasible region for this system.

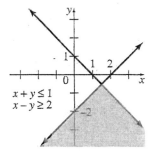

23. $x + 3y \le 6$
$\quad 2x + 4y \ge 7$

Graph the solid lines $x + 3y = 6$ and $2x + 4y = 7$. Use $(0, 0)$ as a test point. $0 + 0 \le 6$ is true, and $0 + 0 \ge 7$ is false. Shade all points below $x + 3y = 6$ and above $2x + 4y = 7$. The feasible region is the overlap of the two half-planes.

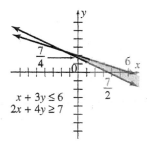

25. $x + y \le 7$
$\quad x - y \le -4$
$\quad 4x + y \ge 0$

The graph of $x + y \le 7$ consists of the solid line $x + y = 7$ and all the points below it. The graph of $x - y \le -4$ consists of the solid line $x - y = -4$ and all the points above it. The graph of $4x + y \ge 0$ consists of the solid line $4x + y = 0$ and all the points above it. The feasible region is the overlapping part of these three half-planes.

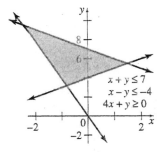

27. $-2 < x < 3$
$\quad -1 \le y \le 5$
$\quad 2x + y < 6$

The graph of $-2 < x < 3$ is the region between the vertical lines $x = -2$ and $x = 3$, but not including the lines themselves (so the two vertical boundaries are drawn as dashed lines). The graph of $-1 \le y \le 5$ is the region between the horizontal lines $y = -1$ and $y = 5$, including the lines (so the two horizontal boundaries are drawn as solid lines). The graph of $2x + y < 6$ is the region below the line $2x + y = 6$ (so the boundary is drawn as a dashed line). Shade the region common to all three graphs to show the feasible region.

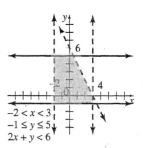

29. $y - 2x \leq 4$

$\quad\quad y \geq 2 - x$

$\quad\quad x \geq 0$

$\quad\quad y \geq 0$

The graph of $y - 2x \leq 4$ consists of the boundary line $y - 2x = 4$ and the region below it. The graph of $y \geq 2 - x$ consists of the boundary line $y = 2 - x$ and the region above it. The inequalities $x \geq 0$ and $y \geq 0$ restrict the feasible region to the first quadrant. Shade the region in the first quadrant where the first two graphs overlap to show the feasible region.

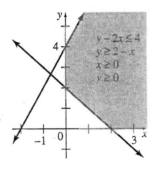

31. $3x + 4y > 12$

$\quad\quad 2x - 3y < 6$

$\quad\quad 0 \leq y \leq 2$

$\quad\quad\quad x \geq 0$

$3x + 4y > 12$ is the set of points above the dashed line $3x + 4y = 12$; $2x - 3y < 6$ is the set of points above the dashed line $2x - 3y = 6$; $0 \leq y \leq 2$ is the set of points lying on or between the horizontal lines $y = 0$ and $y = 2$; and $x \geq 0$ consists of all the points on or to the right of the y-axis. Shade the feasible region, which is the triangular region satisfying all of the inequalities.

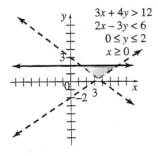

33. $2x - 6y > 12$

Use a graphing calculator. The boundary line is the graph of $2x - 6y = 12$. Solve this equation for y.

$$-6y = -2x + 12$$

$$y = \frac{-2}{-6}x + \frac{12}{-6}$$

$$y = \frac{1}{3}x - 2$$

Enter $y_1 = \frac{1}{3}x - 2$ and graph it. Using the origin as a test point, we obtain $0 > 12$, which is false. Shade the region that does not contain the origin.

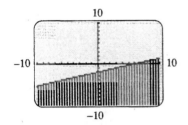

35. $3x - 4y < \; 6$

$\quad\quad 2x + 5y > 15$

Use a graphing calculator. One boundary line is the graph of $3x - 4y = 6$. Solve this equation for y.

$$-4y = -3x + 6$$

$$y = \frac{-3}{-4}x + \frac{6}{-4}$$

$$y = \frac{3}{4}x - \frac{3}{2}$$

Enter $y_1 = \frac{3}{4}x - \frac{3}{2}$ and graph it. Using the origin as a test point, we obtain $0 < 6$, which is true. Shade the region that contains the origin.

The other boundary line is the graph of $2x + 5y = 15$. Solve this equation for y.

$$5y = -2x + 15$$

$$y = -\frac{2}{5}x + 3$$

Enter $y_2 = -\frac{2}{5}x + 3$ and graph it. Using the origin as a test point, we obtain $0 > 15$, which is false. Shade the region that does not contain the origin. The overlap of the two graphs is the feasible region.

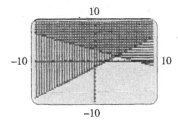

37. The region B is described by the inequalities

$$x + 3y \leq 6$$
$$x + y \leq 3$$
$$x - 2y \leq 2$$
$$x \geq 0$$
$$y \geq 0.$$

The region C is described by the inequalities

$$x + 3y \geq 6$$
$$x + y \geq 3$$
$$x - 2y \leq 2$$
$$x \geq 0$$
$$y \geq 0.$$

The region D is described by the inequalities

$$x + 3y \leq 6$$
$$x + y \geq 3$$
$$x - 2y \leq 2$$
$$x \geq 0$$
$$y \geq 0.$$

The region E is described by the inequalities

$$x + 3y \leq 6$$
$$x + y \leq 3$$
$$x - 2y \geq 2$$
$$x \geq 0$$
$$y \geq 0.$$

The region F is described by the inequalities

$$x + 3y \leq 6$$
$$x + y \geq 3$$
$$x - 2y \geq 2$$
$$x \geq 0$$
$$y \geq 0.$$

The region G is described by the inequalities

$$x + 3y \geq 6$$
$$x + y \geq 3$$
$$x - 2y \geq 2$$
$$x \geq 0$$
$$y \geq 0.$$

39. (a)

	Shawls	Afghans	Total
Number Made	x	y	
Spinning Time	1	2	≤ 8
Dyeing Time	1	1	≤ 6
Weaving Time	1	4	≤ 14

(b) $x + 2y \leq 8$ *Spinning inequality*
   $x + y \leq 6$ *Dyeing inequality*
   $x + 4y \leq 14$ *Weaving inequality*
   $x \geq 0$ *Ensures a nonnegative*
   $y \geq 0$ *number of each*

Graph the solid lines $x + 2y = 8$, $x + y = 6$, $x + 4y = 14$, $x = 0$, and $y = 0$, and shade the appropriate half-planes to get the feasible region.

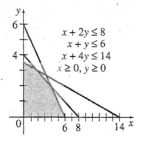

(c) Yes, 3 shawls and 2 afghans can be made because this corresponds to the point $(3, 2)$, which is in the feasible region.

No, 4 shawls and 3 afghans cannot be made because this corresponds to the point $(4, 3)$, which is not in the feasible region.

41. (a) The first sentence of the problem tells us that a total of \$30 million or $x + y \leq 30$ has been set aside for loans. The second sentence of the problem gives $x \geq 4y$. The third and fourth sentences give

$$0.06x + 0.08y \geq 1.6$$

Also, $x \geq 0$ and $y \geq 0$ ensure nonnegative numbers. Thus,

$$x + y \leq 30$$
$$x \geq 4y$$
$$0.06x + 0.08y \geq 1.6$$
$$x \geq 0$$
$$y \geq 0$$

(b) Using the above system, graph solid lines and shade appropriate half-planes to get the feasible region.

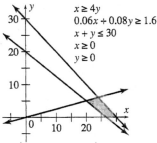

43. (a) The second sentence of the problem tells us that the number of M3 Power^{TM} razors is never more than half the number of Fusion Power^{TM} razors or $x \leq \frac{1}{2}y$. The third sentence tells us that a total of at most 800 razors can be produced per week or $x + y \leq 800$. The inequalities $x \geq 0$ and $y \geq 0$ ensure nonnegative numbers. Thus,

$$x \leq \frac{1}{2}y$$
$$x + y \leq 800$$
$$x \geq 0$$
$$y \geq 0.$$

(b)

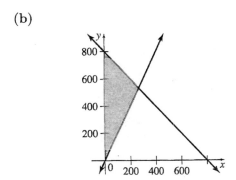

45. (a) The problem tells us that each ounce of fruit supplies 1 unit of protein and each ounce of nuts supplies 1 unit of protein. Thus, $1x + 1y$ is the number of units of protein per package. Since each package must provide at least 7 units of protein, we get the inequality $x + y \geq 7$. Similarly, we get the inequalities $2x + y \geq 10$ for carbohydrates and $x + y \leq 9$ for fat. The inequalities $x \geq 0$ and $y \geq 0$ ensure nonnegative numbers. Thus,

$$x + y \geq 7$$
$$2x + y \geq 10$$
$$x + y \leq 9$$
$$x \geq 0$$
$$y \geq 0.$$

(b)

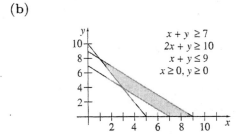

3.2   Solving Linear Programming Problems Graphically

1. (a)

Corner Point	Value of $z = 3x + 2y$
$(0, 5)$	$3(0) + 2(5) = 10$ Minimum
$(3, 8)$	$3(3) + 2(8) = 25$
$(7, 4)$	$3(7) + 2(4) = 29$ Maximum
$(4, 1)$	$3(4) + 2(1) = 14$

The maximum value of 29 occurs at $(7, 4)$. The minimum value of 10 occurs at $(0, 5)$.

(b)

Corner Point	Value of $z = x + 4y$
$(0, 5)$	$0 + 4(5) = 20$
$(3, 8)$	$3 + 4(8) = 35$ Maximum
$(7, 4)$	$7 + 4(4) = 23$
$(4, 1)$	$4 + 4(1) = 8$ Minimum

The maximum value of 35 occurs at $(3, 8)$. The minimum value of 8 occurs at $(4, 1)$.

3. (a)

Corner Point	Value of $z = 0.40x + 0.75y$	
$(0,0)$	$0.40(0) + 0.75(0) = 0$	Minimum
$(0,12)$	$0.40(0) + 0.75(12) = 9$	Maximum
$(4,8)$	$0.40(4) + 0.75(8) = 7.6$	
$(7,3)$	$0.40(7) + 0.75(3) = 5.05$	
$(8,0)$	$0.40(8) + 0.75(0) = 3.2$	

The maximum value of 9 occurs at $(0,12)$. The minimum value of 0 occurs at $(0,0)$.

(b)

Corner Point	Value of $z = 1.50x + 0.25y$	
$(0,0)$	$1.50(0) + 0.25(0) = 0$	Minimum
$(0,12)$	$1.50(0) + 0.25(12) = 3$	
$(4,8)$	$1.50(4) + 0.25(8) = 8$	
$(7,3)$	$1.50(7) + 0.25(3) = 11.25$	
$(8,0)$	$1.50(8) + 0.25(0) = 12$	Maximum

The maximum value of 12 occurs at $(8,0)$. The minimum value of 0 occurs at $(0,0)$.

5.

(a)

Corner Point	Value of $z = 4x + 2y$	
$(0,8)$	$4(0) + 2(8) = 16$	Minimum
$(3,4)$	$4(3) + 2(4) = 20$	
$\left(\frac{13}{2},2\right)$	$4\left(\frac{13}{2}\right) + 2(2) = 30$	
$(12,0)$	$4(12) + 2(0) = 48$	

The minimum value is 16 at $(0,8)$. Since the feasible region is unbounded, there is no maximum value.

(b)

Corner Point	Value of $z = 2x + 3y$	
$(0,8)$	$2(0) + 3(8) = 24$	
$(3,4)$	$2(3) + 3(4) = 18$	Minimum
$\left(\frac{13}{2},2\right)$	$2\left(\frac{13}{2}\right) + 3(2) = 19$	
$(12,0)$	$2(12) + 3(0) = 24$	

The minimum value is 18 at $(3,4)$; there is no maximum value since the feasible region is unbounded.

(c)

Corner Point	Value of $z = 2x + 4y$	
$(0,8)$	$2(0) + 4(8) = 32$	
$(3,4)$	$2(3) + 4(4) = 22$	
$\left(\frac{13}{2},2\right)$	$2\left(\frac{13}{2}\right) + 4(2) = 21$	Minimum
$(12,0)$	$2(12) + 4(0) = 24$	

The minimum value is 21 at $\left(\frac{13}{2},2\right)$; there is no maximum value since the feasible region is unbounded.

(d)

Corner Point	Value of $z = x + 4y$	
$(0,8)$	$0 + 4(8) = 32$	
$(3,4)$	$3 + 4(4) = 19$	
$\left(\frac{13}{2},2\right)$	$\frac{13}{2} + 4(2) = \frac{29}{2}$	
$(12,0)$	$12 + 4(0) = 12$	Minimum

The minimum value is 12 at $(12,0)$; there is no maximum value since the feasible region is unbounded.

7. Minimize $z = 4x + 7y$

subject to: $x - y \geq 1$
$3x + 2y \geq 18$
$x \geq 0$
$y \geq 0.$

Sketch the feasible region.

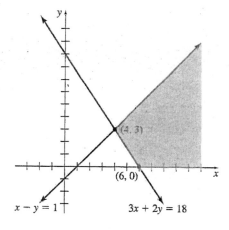

The sketch shows that the feasible region is unbounded. The corner points are $(4,3)$ and $(6,0)$. The corner point $(4,3)$ can be found by solving the system

$$x - y = 1$$
$$3x + 2y = 18.$$

Use the corner points to find the minimum value of the objective function.

Corner Point	Value of $z = 4x + 7y$
$(4,3)$	$4(4) + 7(3) = 37$
$(6,0)$	$4(6) + 7(0) = 24$ Minimum

The minimum value is 24 when $x = 6$ and $y = 0$.

9. Maximize $z = 5x + 2y$

subject to: $4x - y \leq 16$
$2x + y \geq 11$
$x \geq 3$
$y \leq 8.$

Sketch the feasible region.

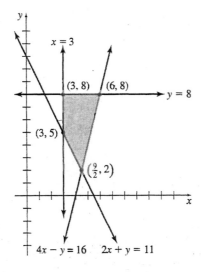

The sketch shows that the feasible region is bounded. The corner points are: $(3,5)$, the intersection of $2x + y = 11$ and $x = 3$; $(3,8)$, the intersection of $x = 3$ and $y = 8$; $(6,8)$, the intersection of $y = 8$ and $4x - y = 16$; and $\left(\frac{9}{2}, 2\right)$, the intersection of $4x - y = 16$ and $2x + y = 11$. Use the corner points to find the maximum value of the objective function.

Corner Point	Value of $z = 5x + 2y$
$(3,5)$	$5(3) + 2(5) = 25$
$(3,8)$	$5(3) + 2(8) = 31$
$(6,8)$	$5(6) + 2(8) = 46$ Maximum
$\left(\frac{9}{2}, 2\right)$	$5\left(\frac{9}{2}\right) + 2(2) = 26.5$

The maximum value is 46 when $x = 6$ and $y = 8$.

11. Maximize $z = 10x + 10y$

subject to: $5x + 8y \geq 200$
$25x - 10y \geq 250$
$x + y \leq 150$
$x \geq 0$
$y \geq 0.$

Sketch the feasible region.

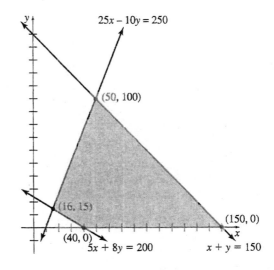

The sketch shows that the feasible region is bounded. The corner points are: $(16,15)$, the intersection of $5x + 8y = 200$ and $25x - 10y = 250$; $(50,100)$, the intersection of $25x - 10y = 250$ and $x + y = 150$; $(150,0)$; and $(40,0)$. Use the corner points to find the maximum value of the objective function.

Corner Point	Value of $z = 10x + 10y$
$(16,15)$	$10(16) + 10(15) = 310$
$(50,100)$	$10(50) + 10(100) = 1500$ Maximum
$(150,0)$	$10(150) + 10(0) = 1500$ Maximum
$(40,0)$	$10(40) + 10(0) = 400$

The maximum value is 1500 when $x = 50$ and $y = 100$, as well as when $x = 150$ and $y = 0$ and all points on the line between.

13. Maximize $z = 3x + 6y$

subject to: $2x - 3y \leq 12$

$\qquad x + y \geq 5$

$\qquad 3x + 4y \geq 24$

$\qquad\qquad x \geq 0$

$\qquad\qquad y \geq 0.$

Sketch the feasible region.

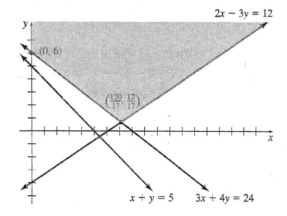

The graph shows that the feasible region is unbounded. Therefore, there is no maximum value of the objective function on the feasible region, hence, no solution.

15. Maximize $z = 10x + 12y$ subject to the following sets of constraints, with $x \geq 0$ and $y \geq 0$.

(a) $\quad x + y \leq 20$

$\qquad x + 3y \leq 24$

Sketch the feasible region in the first quadrant, and identify the corner points at $(0,0)$, $(0,8)$, $(18,2)$, which is the intersection of $x+y = 20$ and $x+3y = 24$, and $(20,0)$.

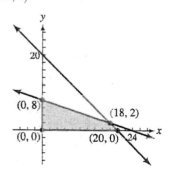

Corner Point	Value of $z = 10x + 12y$
$(0,0)$	$10(0) + 12(0) = 0$
$(0,8)$	$10(0) + 12(8) = 96$
$(18,2)$	$10(18) + 12(2) = 204$ Maximum
$(20,0)$	$10(20) + 12(0) = 200$

The maximum value of 204 occurs when $x = 18$ and $y = 2$.

(b) $3x + y \leq 15$

$\quad x + 2y \leq 18$

Sketch the feasible region in the first quadrant, and identify the corner points. The corner point $\left(\frac{12}{5}, \frac{39}{5}\right)$ can be found by solving the system

$$3x + \ y = 15$$
$$x + 2y = 18.$$

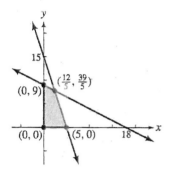

Corner Point	Value of $z = 10x + 12y$
$(0,0)$	$10(0) + 12(0) = 0$
$(0,9)$	$10(0) + 12(9) = 108$
$\left(\frac{12}{5}, \frac{39}{5}\right)$	$10\left(\frac{12}{5}\right) + 12\left(\frac{39}{5}\right) = \frac{588}{5} = 117\frac{3}{5}$
	Maximum
$(5,0)$	$10(5) + 12(0) = 50$

The maximum value of $\frac{588}{5}$ occurs when $x = \frac{12}{5}$ and $y = \frac{39}{5}$.

(c) $2x + 5y \geq 22$

$\quad 4x + 3y \leq 28$

$\quad 2x + 2y \leq 17$

Sketch the feasible region in the first quadrant, and identify the corner points. The corner point $\left(\frac{5}{2}, 6\right)$ can be found by solving the system

$$4x + 3y = 28$$
$$2x + 2y = 17,$$

and the corner point $\left(\frac{37}{7}, \frac{16}{7}\right)$ can be found by solving the system

$$2x + 5y = 22$$
$$4x + 3y = 28.$$

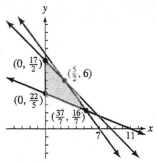

Corner Point	Value of $z = 10x + 12y$
$\left(0, \frac{22}{5}\right)$	$10(0) + 12\left(\frac{22}{5}\right) = \frac{264}{5} = 52.8$
$\left(0, \frac{17}{2}\right)$	$10(0) + 12\left(\frac{17}{2}\right) = 102$ Maximum
$\left(\frac{5}{2}, 6\right)$	$10\left(\frac{5}{2}\right) + 12(6) = 97$
$\left(\frac{37}{7}, \frac{16}{7}\right)$	$10\left(\frac{37}{7}\right) + 12\left(\frac{16}{7}\right) = \frac{562}{7} \approx 80.3$

The maximum value of 102 occurs when $x = 0$ and $y = \frac{17}{2}$.

17. Maximize $z = c_1 x_1 + c_2 x_2$

subject to: $2x_1 + x_2 \le 11$
$-x_1 + 2x_2 \le 2$
$x_1 \ge 0,\ x_2 \ge 0.$

Sketch the feasible region.

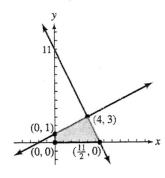

The region is bounded, with corner points $(0,0)$, $(0,1)$, $(4,3)$, and $\left(\frac{11}{2}, 0\right)$.

Corner Point	Value of $z = c_1 x_1 + c_2 x_2$
$(0,0)$	$c_1(0) + c_2(0) = 0$
$(0,1)$	$c_1(0) + c_2(1) = c_2$
$(4,3)$	$c_1(4) + c_2(3) = 4c_1 + 3c_2$
$\left(\frac{11}{2}, 0\right)$	$c_1\left(\frac{11}{2}\right) + c_2(0) = \frac{11}{2}c_1$

If we are to have $(x_1, x_2) = (4,3)$ as an optimal solution, then it must be true that both $4c_1 + 3c_2 \ge c_2$ and $4c_1 + 3c_2 \ge \frac{11}{2}c_1$, because the value of z at

$(4,3)$ cannot be smaller than the other values of z in the table. Manipulate the symbols in these two inequalities in order to isolate $\frac{c_1}{c_2}$ in each; keep in mind the given information that $c_2 > 0$ when performing division by c_2. First,

$$4c_1 + 3c_2 \ge c_2$$
$$4c_1 \ge -2c_2$$
$$\frac{4c_1}{4c_2} \ge \frac{-2c_2}{4c_2}$$
$$\frac{c_1}{c_2} \ge -\frac{1}{2}.$$

Then,

$$4c_1 + 3c_2 \ge \frac{11}{2}c_1$$
$$-\frac{3}{2}c_1 + 3c_2 \ge 0$$
$$3c_1 - 6c_2 \le 0$$
$$3c_1 \le 6c_2$$
$$\frac{3c_1}{3c_2} \le \frac{6c_2}{3c_2}$$
$$\frac{c_1}{c_2} \le 2.$$

Since $\frac{c_1}{c_2} \ge -\frac{1}{2}$ and $\frac{c_1}{c_2} \le 2$, the desired range for $\frac{c_1}{c_2}$ is $\left[-\frac{1}{2}, 2\right]$, which corresponds to choice (b).

3.3 Applications of Linear Programming

1. Let x represent the number of product A made and y represent the number of product B. Each item of A uses 3 hr on the machine, so $2x$ represents the total hours required for x items of product A. Similarly, $5y$ represents the total hours used for product B. There are only 60 hr available, so

$$3x + 5y \le 60.$$

3. Let x = the amount of calcium carbonate supplement
and y = the amount of calcimum citrate supplement.

Then $600x$ represents the number of units of calcium provided by the calcium carbonate supplement and $250y$ represents the number of units provided by the calcium citrate supplement. Since at least 1500 units are needed per day,

$$600x + 250y \ge 1500.$$

5. Let x represent the number of pounds of $8 coffee and y represent the number of pounds of $10 coffee. Since the mixture must weigh at least 40 lb,

$$x + y \geq 40.$$

(Notice that the price per pound is not used in setting up this inequality.)

7. Let $x =$ the number of engines to ship to
    plant I
and $y =$ the number of engines to ship to
    plant II.

Minimize $z = 30x + 40y$

subject to: $x \geq 45$
                    $y \geq 32$
                $x + y \leq 90$
            $20x + 15y \geq 1200$
                    $x \geq 0.$
                    $y \geq 0.$

Sketch the feasible region in quadrant I, and identify the corner points.

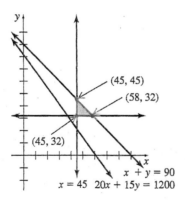

The corner points are: $(45, 32)$, the intersection of $x = 45$ and $y = 32$; $(45, 45)$, the intersection of $x = 45$ and $x + y = 90$; and $(58, 32)$, the intersection of $x + y = 90$ and $y = 32$. Use the corner points to find the minimum value of the objective function.

Corner Point	Value of $z = 30x + 40y$	
$(45, 32)$	$30(45) + 40(32) = 2630$	Minimum
$(45, 45)$	$30(45) + 40(45) = 3150$	
$(58, 32)$	$30(58) + 40(32) = 3020$	

The minimum value is $2630, which occurs when 45 engines are shipped to plant I and 32 engines are shipped to plant II.

9. Let $x =$ the number of units of policy A and $y =$ the number of units of policy B.

(a) Minimize $z = 50x + 40y$

subject to:

$$10,000x + \;\;15,000y \geq 300,000$$
$$180,000x + 120,000y \geq 3,000,000$$
$$x \geq 0$$
$$y \geq 0.$$

Sketch the feasible region in quadrant I, and identify the corner points. The corner point $(6, 16)$ can be found by solving the system

$$10,000x + \;\;15,000y = 300,000$$
$$180,000x + 120,000y = 3,000,000,$$

which can be simplified as

$$2x + 3y = 60$$
$$3x + 2y = 50.$$

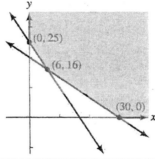

Corner Point	Value of $z = 50x + 40y$	
$(0, 25)$	$50(0) + 40(25) = 1000$	
$(6, 16)$	$50(6) + 40(16) = 940$	Minimum
$(30, 0)$	$50(30) + 40(0) = 1500$	

The minimum cost is $940, which occurs when 6 units of policy A and 16 units of policy B are purchased.

(b) The objective function changes to $z = 25x + 40y$, but the constraints remain the same. Use the same corner points as in part (a).

Corner Point	Value of $z = 25x + 40y$	
$(0, 25)$	$25(0) + 40(25) = 1000$	
$(6, 16)$	$25(6) + 40(16) = 790$	
$(30, 0)$	$25(30) + 40(0) = 750$	Minimum

The minimum cost is $750, which occurs when 30 units of policy A and no units of policy B are purchased.

11. Let $x =$ the number of type I bolts
and $y =$ the number of type II bolts.

Maximize $z = 0.15x + 0.20y$
subject to: $0.2x + 0.2y \le 300$
$0.6x + 0.2y \le 720$
$0.04x + 0.08y \le 100$
$x \ge 0$
$y \ge 0.$

Graph the feasible region and identify the corner points.

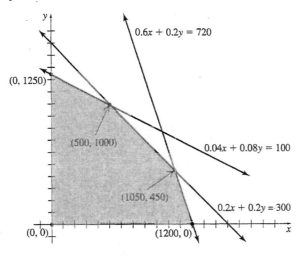

The corner points are $(0,0)$; $(0,1250)$; $(500,1000)$, the intersection of $0.04x + 0.08y = 100$ and $0.2x + 0.2y = 300$; $(1050,450)$, the intersection of $0.2x + 0.2y = 300$ and $0.6x + 0.2y = 720$; and $(1200,0)$. Use the corner points to find the maximum value of the objective function.

Corner Point	Value of $z = 0.15x + 0.90y$
$(0,0)$	$0.15(0) + 0.20(0) = 0$
$(0,1250)$	$0.15(0) + 0.20(1250) = 250$
$(500,1000)$	$0.15(500) + 0.20(1000) = 275$
	Maximum
$(1050,450)$	$0.15(1050) + 0.20(450) = 247.5$
$(1200,0)$	$0.15(1200) + 0.20(0) = 180$

The shop should manufacture 500 type I bolts and 1000 type II bolts to maximize revenue.

(b) The maximum revenue is $275.

(c) Notice that each one-cent increase in the selling price of type I bolts increases the value of z at $(500,1000)$ by $5 and at $(1050,450)$ by $10.50. (At $(0,1200)$, there would be no increase in z.)

Therefore, the corner point $(1050,450)$ will begin to maximize the revenue when n, the number of one-cent increases, is larger than the solution to the following equation.

$$(0.15 + 0.01n)(1050) + 0.20(450) = (0.15 + 0.01n)(500) + 0.20(1000)$$
$$(0.15 + 0.01n)(550) = 110$$
$$5.5n = 27.5$$
$$n = 5$$

Therefore, the selling price of the type I bolts can increase up to $0.15 + 0.01(5) = 0.20$, or 20¢, before a different number of each type of bolts should be produced to maximize the revenue. If the price of the type I bolts exceeds 20¢, then it is more profitable to produce 1050 type I bolts and 450 type II bolts.

13. (a) Let $x =$ the number of kg of the half-and-half mixture
and $y =$ the number of kg of the second mixture.

Maximize $z = 7x + 9.5y$

subject to: $\dfrac{1}{2}x + \dfrac{3}{4}y \le 150$

$\dfrac{1}{2}x + \dfrac{1}{4}y \le 90$

$x \ge 0$
$y \ge 0.$

Sketch the feasible region and identify the corner points.

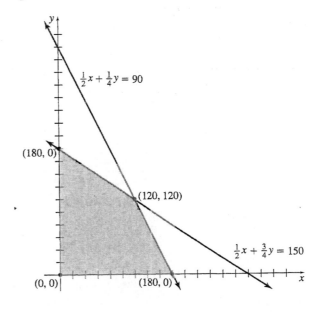

The corner points are $(0,0)$; $(0,200)$; $(120,120)$, the intersection of $\frac{1}{2}x + \frac{3}{4}y = 150$ and $\frac{1}{2}x + \frac{1}{4}y = 90$; and $(180,0)$. Use the corner points to find the maximum value of the objective function.

Corner Point	Value of $z = 7x + 9.5y$
$(0,0)$	$7(0) + 9.5(0) = 0$
$(0,200)$	$7(0) + 9.5(200) = 1900$
$(120,120)$	$7(120) + 9.5(120) = 1980$
	Maximum
$(180,0)$	$7(180) + 9.5(0) = 1260$

The candy company should prepare 120 kg of the half-and-half mixture and 120 kg of the second mixture for a maximum revenue of $1980.

(b) The objective function to be maximized is now $z = 7x + 11y$. The corner points remain the same.

Corner Point	Value of $z = 7x + 11y$
$(0,0)$	$7(0) + 11(0) = 0$
$(0,200)$	$7(0) + 11(200) = 2200$
$(120,120)$	Maximum
	$7(120) + 11(120) = 2160$
$(180,0)$	$7(180) + 11(0) = 1260$

In order to maximize the revenue under the altered conditions, the candy company should prepare 0 kg of the half-and-half mixture and 200 kg of the second mixture for a maximum revenue of $2200.

15. (a) Let $x =$ the number of gallons from dairy I and $y =$ the number of gallons from dairy II.

Maximize $z = 0.037x + 0.032y$

subject to: $0.60x + 0.20y \le 36$
$x \le 50$
$y \le 80$
$x + y \le 100$
$x \ge 0$
$y \ge 0.$

Sketch the feasible region, and identify the corner points.

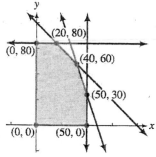

Corner Point	Value of $z = 0.037x + 0.032y$
$(0,0)$	0
$(0,80)$	2.56
$(20,80)$	3.30
$(40,60)$	3.40 Maximum
$(50,30)$	2.81
$(50,0)$	1.85

The maximum amount of butterfat is 3.4%, which occurs when 40 gal are purchased from dairy I and 60 gal are purchased from dairy II.

(b) In the solution to part (a), Mostpure uses 40 gallons from dairy I with a capacity for 50 gallons. Therefore, there is an excess capacity of 10 gallons from dairy I. Similarly, there is an excess capacity of $80 - 60$ or 20 gallons from dairy II.

17. Let $x =$ the amount (in millions) invested in U.S. Treasury bonds
and $y =$ the amount (in millions) invested in mutual funds.

Maximize $z = 0.04x + 0.08y$

subject to: $x + y \le 30$
$x \ge 5$
$y \ge 10$
$100x + 200y \ge 5000$
$x \ge 0$
$y \ge 0.$

Sketch the feasible region, and identify the corner points.

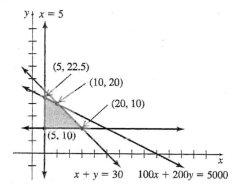

The corner points are $(5,10)$; $(5,22.5)$, the intersection of $x = 5$ and $100x + 200y = 5000$; $(10,20)$, the intersection of $100x + 200y = 5000$ and $x + y = 30$; and $(20,10)$. Use the corner points to find the maximum value of the objective function.

Corner Point	Value of $z = 0.04x + 0.08y$	
$(5, 10)$	$0.04(5) + 0.08(10) = 1$	
$(5, 22.5)$	$0.04(5) + 0.08(22.5) = 2$	Maximum
$(10, 20)$	$0.04(10) + 0.08(20) = 2$	Maximum
$(20, 10)$	$0.04(20) + 0.08(10) = 1.6$	

The maximum annual interest of $2 million can be achieved by investing $5 million in U.S. Treasurey bonds and $22.5 million in mutual funds, or $10 million in bonds and $20 million in mutual funds (or in any solution on the line between those two points).

19. Beta is limited to 400 units per day, so Beta ≤ 400. The correct answer is choice (a).

21. (a) Let $x =$ the number of pill 1 and $y =$ the number of pill 2.

Minimize $z = 0.15x + 0.30y$

subject to: $8x + 2y \geq 16$
$$x + y \geq 5$$
$$2x + 7y \geq 20$$
$$x \geq 0$$
$$y \geq 0.$$

Sketch the feasible region in quadrant I.

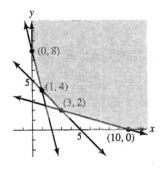

The corner points $(0, 8)$ and $(10, 0)$ can be identified from the graph. The coordinates of the corner point $(1, 4)$ can be found by solving the system

$$8x + 2y = 16$$
$$x + y = 5.$$

The coordinates of the corner point $(3, 2)$ can be found by solving the system

$$2x + 7y = 20$$
$$x + y = 5.$$

Corner Point	Value of $z = 0.15x + 0.30y$	
$(1, 4)$	1.35	
$(3, 2)$	1.05	Minimum
$(0, 8)$	2.40	
$(10, 0)$	1.50	

A minimum daily cost of $1.05 is incurred by taking three of pill 1 and two of pill 2.

(b) In the solution to part (a), Mark receives $8(3) + 2(2)$ or 28 units of vitamin A. This is a surplus of $28 - 16$ or 12 units of vitamin A.

23. Let $x =$ the number of ounces of fruit and $y =$ the number of ounces of nuts.

Minimize $z = 20x + 30y$

subject to: $3y \geq 6$
$$2x + y \geq 10$$
$$x + 2y \leq 9$$
$$x \geq 0$$
$$y \geq 0.$$

Sketch the feasible region, and identify the corner points.

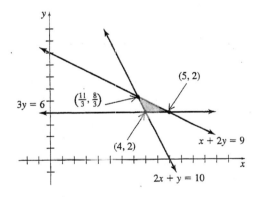

The corner points are: $(4, 2)$, the intersection of $y = 2$ and $2x + y = 10$; $(\frac{11}{3}, \frac{8}{3})$, the intersection of $2x + y = 10$ and $x + 2y = 9$; and $(5, 2)$, the intersection of $x + 2y = 9$ and $y = 2$. Use the corner points to find the minimum value of the objective function.

Corner Point	Value of $z = 20x + 30y$	
$(4, 2)$	$20(4) + 30(2) = 140$	Minimum
$(\frac{11}{3}, \frac{8}{3})$	$20(\frac{11}{3}) + 30(\frac{8}{3}) = \frac{460}{3} \approx 153$	
$(5, 2)$	$20(5) + 30(2) = 160$	

The dietician should use 4 ounces of fruit and 2 ounces of nuts for a minimum of 140 calories.

25. Let x = the number of units of plants and y = the number of animals.

Minimize $z = 30x + 15y$
subject to: $30x + 20y \geq 360$
$10x + 25y \geq 300$
$y \geq 8$
$0 \leq x \leq 25$
$0 \leq y \leq 25.$

Sketch the feasible region in quadrant I.

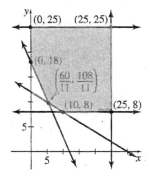

The corner points $(0, 18)$, $(0, 25)$, $(25, 25)$, and $(25, 8)$ can be determined from the graph. The corner point $\left(\frac{60}{11}, \frac{108}{11}\right)$ can be found by solving the system

$$30x + 20y = 360$$
$$10x + 25y = 300.$$

The corner point $(10, 8)$ can be found by solving the system

$$10x + 25y = 300$$
$$y = 8.$$

Corner Point	Value of $z = 30x + 15y$	
$(0, 18)$	270	Minimum
$(0, 25)$	375	
$(25, 25)$	1125	
$(25, 8)$	870	
$\left(\frac{60}{11}, \frac{108}{11}\right)$	$\frac{3420}{11} \approx 310.91$	
$(10, 8)$	420	

The minimum labor is 270 hours and is achieved when 0 units of plants and 18 animals are collected.

Chapter 3 Review Exercises

3. $y \geq 2x + 3$

Graph $y = 2x + 3$ as a solid line, using the intercepts $(0, 3)$ and $\left(-\frac{3}{2}, 0\right)$. Using the origin as a test point, we get $0 \geq 2(0) + 3$ or $0 \geq 3$, which is false. Shade the region that does not contain the origin, that is, the half-plane above the line.

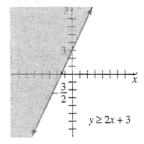

5. $2x + 6y \leq 8$

Graph $2x + 6y = 8$ as a solid line, using the intercepts $\left(0, \frac{4}{3}\right)$ and $(4, 0)$. Using the origin as a test point, we get $0 \leq 8$, which is true. Shade the region that contains the origin, that is, the half-plane below the line.

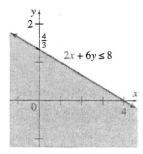

7. $y \geq x$

Graph $y = x$ as a solid line. Since this line contains the origin, choose a point other than $(0, 0)$ as a test point. If we use $(1, 4)$, we get $4 \geq 1$, which is true. Shade the region that contains the test point, that is, the half-plane above the line.

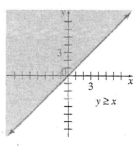

9. $x + y \leq 6$
  $2x - y \geq 3$

$x + y \leq 6$ is the half-plane on or below the line $x + y = 6$; $2x - y \geq 3$ is the half-plane on or below the line $2x - y = 3$. Shade the overlapping part of these two half-planes, which is the region below both lines. The only corner point is the intersection of the two boundary lines, the point $(3, 3)$.

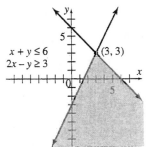

11. $-4 \leq x \leq 2$
  $-1 \leq y \leq 3$
  $x + y \leq 4$

$-4 \leq x \leq 2$ is the rectangular region lying on or between the two vertical lines, $x = -4$ and $x = 2$; $-1 \leq y \leq 3$ is the rectangular region lying on or between the two horizontal lines, $y = -1$ and $y = 3$; $x + y \leq 4$ is the half-plane lying on or below the line $x + y = 4$. Shade the overlapping part of these three regions. The corner points are $(-4, -1)$, $(-4, 3)$, $(1, 3)$, $(2, 2)$, and $(2, -1)$.

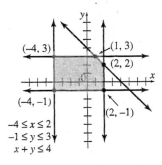

13. $x + 2y \leq 4$
  $5x - 6y \leq 12$
  $x \geq 0$
  $y \geq 0$

$x + 2y \leq 4$ is the half-plane on or below the line $x + 2y = 4$; $5x - 6y \leq 12$ is the half-plane on or above the line $5x - 6y = 12$; $x \geq 0$ and $y \geq 0$ together restrict the graph to the first quadrant. Shade the portion of the first quadrant where the half-planes overlap. The corner points are $(0, 0)$, $(0, 2)$, $\left(\frac{12}{5}, 0\right)$, and $\left(3, \frac{1}{2}\right)$, which can be found by solving the system

$$x + 2y = 4$$
$$5x - 6y = 12.$$

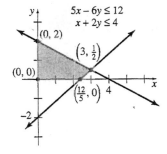

15. Evaluate the objective function $z = 2x + 4y$ at each corner point.

Corner Point	Value of $z = 2x + 4y$
$(0, 0)$	$2(0) + 4(0) = 0$ Minimum
$(0, 4)$	$2(0) + 4(4) = 16$
$(3, 4)$	$2(3) + 4(4) = 22$ Maximum
$(6, 2)$	$2(6) + 4(2) = 20$
$(4, 0)$	$2(4) + 4(0) = 8$

The maximum value 22 occurs at $(3, 4)$, and the minimum value of 0 occurs at $(0, 0)$.

17. Maximize $z = 2x + 4y$

subject to: $3x + 2y \leq 12$
  $5x + y \geq 5$
  $x \geq 0$
  $y \geq 0.$

Sketch the feasible region in quadrant I.

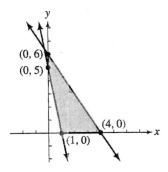

The corner points are $(0, 5)$, $(0, 6)$, $(4, 0)$, and $(1, 0)$.

Corner Point	Value of $z = 2x + 4y$	
$(0, 5)$	20	
$(0, 6)$	24	Maximum
$(4, 0)$	8	
$(1, 0)$	2	

The maximum value is 24 at $(0, 6)$.

19. Minimize $z = 4x + 2y$

subject to: $x + y \le 50$
$2x + y \ge 20$
$x + 2y \ge 30$
$x \ge 0$
$y \ge 0.$

Sketch the feasible region.

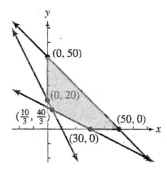

The corner points are $(0, 20)$, $\left(\frac{10}{3}, \frac{40}{3}\right)$, $(30, 0)$, $(50, 0)$, and $(0, 50)$. The corner point $\left(\frac{10}{3}, \frac{40}{3}\right)$ can be found by solving the system

$$2x + y = 20$$
$$x + 2y = 30.$$

Corner Point	Value of $z = 4x + 2y$	
$(0, 20)$	40	Minimum
$\left(\frac{10}{3}, \frac{40}{3}\right)$	40	Minimum
$(30, 0)$	120	
$(50, 0)$	200	
$(0, 50)$	100	

Thus, the minimum value is 40 and occurs at every point on the line segment joining $(0, 20)$ and $\left(\frac{10}{3}, \frac{40}{3}\right)$.

23. Maximize $z = 2x + 5y$

subject to: $3x + 2y \le 6$
$-x + 2y \le 4$
$x \ge 0$
$y \ge 0.$

(a) Sketch the feasible region. All corner points except one can be read from the graph. Solving the system

$$3x + 2y = 6$$
$$-x + 2y = 4$$

gives the final corner point, $\left(\frac{1}{2}, \frac{9}{4}\right)$.

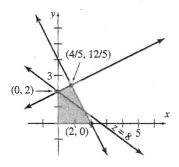

(b)

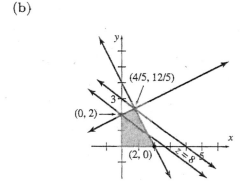

25. Let $x =$ the number of batches of cakes and $y =$ the number of batches of cookies.

Then we have the following inequalities.

$$2x + \frac{3}{2}y \le 15 \quad \text{(oven time)}$$

$$3x + \frac{2}{3}y \le 13 \quad \text{(decorating)}$$

$$x \ge 0$$
$$y \ge 0$$

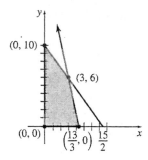

27. (a) From the graph for Exercise 25, the corner points are $(0, 10)$, $(3, 6)$, $\left(\frac{13}{3}, 0\right)$, and $(0, 0)$. Since x was the number of batches of cakes and y the number of batches of cookies, the revenue function is

$$z = 30x + 20y.$$

Evaluate this objective function at each corner point.

Corner Point	Value of $z = 30x + 20y$
$(0, 10)$	200
$(3, 6)$	210 Maximum
$\left(\frac{13}{3}, 0\right)$	130
$(0, 0)$	0

Therefore, 3 batches of cakes and 6 batches of cookies should be made to produce a maximum profit of \$210.

(b) Note that each \$1 increase in the price of cookies increases the profit by \$10 at $(0, 10)$ and by \$6 at $(3, 6)$. (The increase has no effect at the other corner points since, for those, $y = 0$.) Therefore, the corner point $(0, 10)$ will begin to maximize the profit when x, the number of one-dollar increases, is larger than the solution to the following equation.

$$30(0) + (20 + 1x)(10) = 30(3) + (20 + 1x)(6)$$
$$(20 + x)(4) = 90$$
$$4x = 10$$
$$x = 2.5$$

If the profit per batch of cookies increases by more than \$2.50 (to \$22.50), then it will be more profitable to make 10 batches of cookies and no batches of cake.

29. Let $x =$ number of packages of gardening mixture and $y =$ number of packages of potting mixture.

Maximize $z = 3x + 5y$

subject to: $2x + y \le 16$
$x + 2y \le 11$
$x + 3y \le 15$
$x \ge 0$
$y \ge 0$.

Sketch the feasible region in quadrant I.

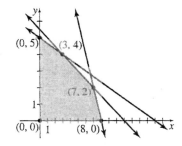

The corner points $(0, 0), (0, 5)$, and $(8, 0)$ can be identified from the graph. The corner point $(3, 4)$ can be found by solving the system

$$x + 2y = 11$$
$$x + 3y = 15.$$

The corner point $(7, 2)$ can be found by solving the system

$$2x + y = 16$$
$$x + 2y = 11.$$

Corner Point	Value of $z = 3x + 5y$
$(0, 0)$	0
$(0, 5)$	25
$(3, 4)$	29
$(7, 2)$	31 Maximum
$(8, 0)$	24

A maximum income of \$31 can be achieved by preparing 7 packages of gardening mixture and 2 packages of potting mixture.

31. Let $x =$ number of runs of type I
and $y =$ number of runs of type II.

Minimize $z = 15,000x + 6000y$
subject to: $3000x + 3000y \geq 18,000$
$\qquad 2000x + 1000y \geq 7000$
$\qquad 2000x + 3000y \geq 14,000$
$\qquad\qquad\qquad x \geq 0$
$\qquad\qquad\qquad y \geq 0.$

Sketch the feasible region in quadrant I.

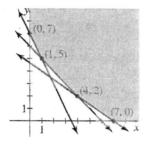

The corner points $(0, 7)$ and $(7, 0)$ can be identified from the graph. The corner point $(1, 5)$ can be found by solving the system

$$3000x + 3000y = 18,000$$
$$2000x + 1000y = 7000.$$

The corner point $(4, 2)$ can be found by solving the system

$$3000x + 3000y = 18,000$$
$$2000x + 3000y = 14,000.$$

Corner Point	Value of $z = 15,000x + 6000y$
$(0, 7)$	42,000 Minimum
$(1, 5)$	45,000
$(4, 2)$	72,000
$(7, 0)$	105,000

The company should produce 0 runs of type I and 7 runs of type II for a minimum cost of $42,000.

33. Let $x =$ the number of acres devoted to millet
and $y =$ the number of acres devoted to wheat.

Maximize $z = 400x + 800y$
subject to: $36x + 8y \leq 48$
$\qquad\qquad\quad x + y \leq 2$
$\qquad\qquad\qquad x \geq 0$
$\qquad\qquad\qquad y \geq 0.$

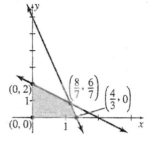

The corner points $(0, 0)$ and $(0, 2)$ can be identified from the graph. The corner point $(\frac{4}{3}, 0)$ can be found by solving the system

$$36x + 8y = 48$$
$$y = 0.$$

The corner point $(\frac{8}{7}, \frac{6}{7})$ can be found by solving the system

$$36x + 8y = 48$$
$$x + y = 2.$$

Corner Point	Value of $z = 400x + 800y$
$(0, 0)$	0
$(0, 2)$	1600 Maximum
$(\frac{8}{7}, \frac{6}{7})$	$\frac{8000}{7}$
$(\frac{4}{3}, 0)$	$\frac{1600}{3}$

The maximum amount of grain is 1600 pounds and can be obtained by planting 2 acres of wheat and no millet.

Chapter 3 Test

[3.1]

1. Graph the following linear inequality.

$$2x + y \leq 4$$

[3.1–3.2]

2. Graph the feasible region for the following system of inequalities. Find all corner points.

$$
\begin{aligned}
x + y &\leq 4 \\
2x + y &\geq 6 \\
x &\geq 0 \\
y &\geq 0
\end{aligned}
$$

[3.2]

3. Find the maximum and minimum values of the objective function $z = 3x + 2y$ for the region sketched below.

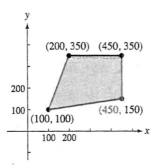

Use the graphical method to solve the following linear programming problems.

4. Maximize $z = 4x + 3y$

   subject to: $3x + y \leq 12$
   $x + y \geq 3$
   $x \geq 0$
   $y \geq 0.$

5. Minimize $z = x + 10y$

   subject to: $5x + 2y \geq 20$
   $x + y \geq 7$
   $x + 6y \geq 27$
   $y \geq 3$
   $x \geq 0.$

[3.3]

6. The Gigantic Zipper Company manufactures two kinds of zippers. Type I zippers require 2 minutes on machine A and 3 minutes on machine B. Type II zippers require 1 minute on machine A and 4 minutes on machine B. Machine A is available for 20 minutes, while machine B is available for 12 minutes. The profit on each type I zipper is $0.30 and on each type II zipper is $0.20. How many of each type of zipper should be manufactured to ensure the maximum profit?

Chapter 3 Test Answers

1.

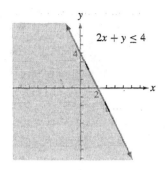

$2x + y \leq 4$

2.

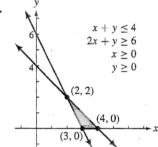

$x + y \leq 4$
$2x + y \geq 6$
$x \geq 0$
$y \geq 0$

Corner points: $(3, 0)$, $(4, 0)$, $(2, 2)$

3. Maximum of 2050 at $(450, 350)$; minimum of 500 at $(100, 100)$

4. Maximum of 36 at $(0, 12)$

5. Minimum of 39 at $(9, 3)$

6. 4 type I zippers and 0 type II zippers

LINEAR PROGRAMMING: THE SIMPLEX METHOD

4.1 Slack Variables and the Pivot

1. $x_1 + 2x_2 \leq 6$

Add s_1 to the given inequality to obtain

$$x_1 + 2x_2 + s_1 = 6.$$

3. $2.3x_1 + 5.7x_2 + 1.8x_3 \leq 17$

Add s_1 to the given inequality to obtain

$$2.3x_1 + 5.7x_2 + 1.8x_3 + s_1 = 17.$$

5. (a) Since there are three constraints to be converted into equations, we need three slack variables.

(b) We use s_1, s_2, and s_3 for the slack variables.

(c) The equations are

$$\begin{aligned}
2x_1 + 3x_2 + s_1 \qquad\quad &= 15 \\
4x_1 + 5x_2 \qquad + s_2 \quad &= 35 \\
x_1 + 6x_2 \qquad\qquad +s_3 &= 20.
\end{aligned}$$

7. (a) There are two constraints to be converted into equations, so we must introduce two slack variables.

(b) Call the slack variables s_1 and s_2.

(c) The equations are

$$\begin{aligned}
7x_1 + 6x_2 + 8x_3 + s_1 \qquad &= 118 \\
4x_1 + 5x_2 + 10x_3 \qquad + s_2 &= 220.
\end{aligned}$$

9.

x_1	x_2	x_3	s_1	s_2	z	
1	0	4	5	1	0	8
3	1	1	2	0	0	4
−2	0	2	3	0	1	28

The variables x_2 and s_2 are basic variables, because the columns for these variables have all zeros except for one nonzero entry. If the remaining variables x_1, x_3, and s_1 are zero, then $x_2 = 4$ and $s_2 = 8$. From the bottom row, $z = 28$. The basic feasible solution is $x_1 = 0, x_2 = 4, x_3 = 0, s_1 = 0, s_2 = 8$, and $z = 28$.

11.

x_1	x_2	x_3	s_1	s_2	s_3	z	
6	2	2	3	0	0	0	16
2	2	0	1	0	5	0	35
2	1	0	3	1	0	0	6
−3	−2	0	2	0	0	3	36

The basic variables are x_3, s_2, and s_3. If x_1, x_2, and s_1 are zero, then $2x_3 = 16$, so $x_3 = 8$. Similarly, $s_2 = 6$ and $5s_3 = 35$, so $s_3 = 7$. From the bottom row, $3z = 36$, so $z = 12$. The basic feasible solution is $x_1 = 0, x_2 = 0, x_3 = 8, s_1 = 0, s_2 = 6, s_3 = 7$, and $z = 12$.

13.

x_1	x_2	x_3	s_1	s_2	z	
1	2	4	1	0	0	56
2	$\boxed{2}$	1	0	1	0	40
−1	−3	−2	0	0	1	0

Clear the x_2 column.

$-R_2 + R_1 \rightarrow R_1$

x_1	x_2	x_3	s_1	s_2	z	
−1	0	3	1	−1	0	16
2	$\boxed{2}$	1	0	1	0	40
−1	−3	−2	0	0	1	0

$3R_2 + 2R_3 \rightarrow R_3$

x_1	x_2	x_3	s_1	s_2	z	
−1	0	3	1	−1	0	16
2	2	1	0	1	0	40
4	0	−1	0	3	2	120

x_2 and s_1 are now basic. The solution is $x_1 = 0, x_2 = 20, x_3 = 0, s_1 = 16, s_2 = 0$, and $z = 60$.

15.

x_1	x_2	x_3	s_1	s_2	s_3	z	
2	2	$\boxed{1}$	1	0	0	0	12
1	2	3	0	1	0	0	45
3	1	1	0	0	1	0	20
−2	−1	−3	0	0	0	1	0

Clear the x_3 column.

$-3R_1 + R_2 \rightarrow R_2$
$-R_1 + R_3 \rightarrow R_3$
$3R_1 + R_4 \rightarrow R_4$

x_1	x_2	x_3	s_1	s_2	s_3	z	
2	2	1	1	0	0	0	12
−5	−4	0	−3	1	0	0	9
1	−1	0	−1	0	1	0	8
4	5	0	3	0	0	1	36

x_3, s_2, and s_3 are now basic. The solution is $x_1 = 0, x_2 = 0, x_3 = 12, s_1 = 0, s_2 = 9, s_3 = 8$, and $z = 36$.

17.

$$\begin{array}{ccccccc} x_1 & x_2 & x_3 & s_1 & s_2 & s_3 & z \\ \left[\begin{array}{ccccccc|c} 2 & \boxed{2} & 3 & 1 & 0 & 0 & 0 & 500 \\ 4 & 1 & 1 & 0 & 1 & 0 & 0 & 300 \\ 7 & 2 & 4 & 0 & 0 & 1 & 0 & 700 \\ \hline -3 & -4 & -2 & 0 & 0 & 0 & 1 & 0 \end{array}\right] \end{array}$$

Clear the x_2 column.

$$\begin{array}{c} \\ -R_1 + 2R_2 \to R_2 \\ -R_1 + R_3 \to R_3 \\ 2R_1 + R_4 \to R_4 \end{array} \begin{array}{ccccccc} x_1 & x_2 & x_3 & s_1 & s_2 & s_3 & z \\ \left[\begin{array}{ccccccc|c} 2 & \boxed{2} & 3 & 1 & 0 & 0 & 0 & 500 \\ 6 & 0 & -1 & -1 & 2 & 0 & 0 & 100 \\ 5 & 0 & 1 & -1 & 0 & 1 & 0 & 200 \\ \hline 1 & 0 & 4 & 2 & 0 & 0 & 1 & 1000 \end{array}\right] \end{array}$$

x_2, s_2, and s_3 are now basic. Thus, the solution is $x_1 = 0, x_2 = 250, x_3 = 0, s_1 = 0, s_2 = 50, s_3 = 200$, and $z = 1000$.

19. Find $x_1 \geq 0$ and $x_2 \geq 0$ such that

$$4x_1 + 2x_2 \leq 5$$
$$x_1 + 2x_2 \leq 4$$

and $z = 7x_1 + x_2$ is maximized.

We need two slack variables, s_1 and s_2. Then the problem can be restated as:

Find $x_1 \geq 0, x_2 \geq 0, s_1 \geq 0$, and $s_2 \geq 0$ such that

$$4x_1 + 2x_2 + s_1 \qquad = 5$$
$$x_1 + 2x_2 \qquad + s_2 = 4.$$

and $z = 7x_1 + x_2$ is maximized.

Rewrite the objective function as

$$-7x_1 - x_2 + z = 0.$$

The initial simplex tableau is

$$\begin{array}{ccccc} x_1 & x_2 & s_1 & s_2 & z \\ \left[\begin{array}{ccccc|c} 4 & 2 & 1 & 0 & 0 & 5 \\ 1 & 2 & 0 & 1 & 0 & 4 \\ \hline -7 & -1 & 0 & 0 & 1 & 0 \end{array}\right]. \end{array}$$

21. Find $x_1 \geq 0$ and $x_2 \geq 0$ such that

$$x_1 + x_2 \leq 10$$
$$5x_1 + 2x_2 \leq 20$$
$$x_1 + 2x_2 \leq 36$$

and $z = x_1 + 3x_2$ is maximized.

Using slack variables s_1, s_2, and s_3, the problem can be restated as:

Find $x_1 \geq 0, x_2 \geq 0, s_1 \geq 0, s_2 \geq 0$, and $s_3 \geq 0$ such that

$$x_1 + x_2 + s_1 \qquad\qquad = 10$$
$$5x_1 + 2x_2 \qquad + s_2 \qquad = 20$$
$$x_1 + 2x_2 \qquad\qquad + s_3 = 36$$

and $z = x_1 + 3x_2$ is maximized.

Rewrite the objective function as

$$-x_1 - 3x_2 + z = 0.$$

The initial simplex tableau is

$$\begin{array}{cccccc} x_1 & x_2 & s_1 & s_2 & s_3 & z \\ \left[\begin{array}{cccccc|c} 1 & 1 & 1 & 0 & 0 & 0 & 10 \\ 5 & 2 & 0 & 1 & 0 & 0 & 20 \\ 1 & 2 & 0 & 0 & 1 & 0 & 36 \\ \hline -1 & -3 & 0 & 0 & 0 & 1 & 0 \end{array}\right]. \end{array}$$

23. Find $x_1 \geq 0$ and $x_2 \geq 0$ such that

$$3x_1 + x_2 \leq 12$$
$$x_1 + x_2 \leq 15$$

and $z = 2x_1 + x_2$ is maximized.

Using slack variables s_1 and s_2, the problem can be restated as:

Find $x_1 \geq 0, x_2 \geq 0, s_1 \geq 0$, and $s_2 \geq 0$ such that

$$3x_1 + x_2 + s_1 \qquad = 12$$
$$x_1 + x_2 \qquad + s_2 = 15$$

and $z = 2x_1 + x_2$ is maximized.

Rewrite the objective function as

$$-2x_1 - x_2 + z = 0.$$

The initial simplex tableau is

$$\begin{array}{ccccc} x_1 & x_2 & s_1 & s_2 & z \\ \left[\begin{array}{ccccc|c} 3 & 1 & 1 & 0 & 0 & 12 \\ 1 & 1 & 0 & 1 & 0 & 15 \\ \hline -2 & -1 & 0 & 0 & 1 & 0 \end{array}\right]. \end{array}$$

25. Let x_1 represent the number of simple figures, x_2 the number of figures with additions, and x_3 the number of computer-drawn sketches. Organize the information in a table.

	Simple Figures	Figures with Additions	Computer-Drawn Sketches	Maximum Allowed
Cost	20	35	60	2200
Royalties	95	200	325	

The cost constraint is

$$20x_1 + 35x_2 + 60x_3 \leq 2200.$$

The limit of 400 figures leads to the constraint

$$x_1 + x_2 + x_3 \leq 400.$$

The other stated constraints are

$$x_3 \leq x_1 + x_2 \text{ and } x_1 \geq 2x_2,$$

and these can be rewritten in standard form as

$$-x_1 - x_2 + x_3 \leq 0 \text{ and } -x_1 + 2x_2 \leq 0$$

respectively. The problem may be stated as:

Find $x_1 \geq 0, x_2 \geq 0$, and $x_3 \geq 0$ such that

$$\begin{aligned} 20x_1 + 35x_2 + 60x_3 &\leq 2200 \\ x_1 + x_2 + x_3 &\leq 400 \\ -x_1 - x_2 + x_3 &\leq 0 \\ -x_1 + 2x_2 \quad\quad &\leq 0 \end{aligned}$$

and $z = 95x_1 + 200x_2 + 325x_3$ is maximized.

Introduce slack variables s_1, s_2, s_3, and s_4, and the problem can be restated as:

Find $x_1 \geq 0, x_2 \geq 0, x_3 \geq 0, s_1 \geq 0, s_2 \geq 0, s_3 \geq 0$, and $s_4 \geq 0$ such that

$$\begin{aligned} 20x_1 + 35x_2 + 60x_3 + s_1 \quad\quad\quad\quad &= 2200 \\ x_1 + x_2 + x_3 \quad + s_2 \quad\quad\quad &= 400 \\ -x_1 - x_2 + x_3 \quad\quad + s_3 \quad\quad &= 0 \\ -x_1 + 2x_2 \quad\quad\quad\quad\quad + s_4 &= 0 \end{aligned}$$

and $z = 95x_1 + 200x_2 + 325x_3$ is maximized.

Rewrite the objective function as

$$-95x_1 - 200x_2 - 325x_3 + z = 0.$$

The initial simplex tableau is

$$\left[\begin{array}{cccccccc|c} x_1 & x_2 & x_3 & s_1 & s_2 & s_3 & s_4 & z & \\ 20 & 35 & 60 & 1 & 0 & 0 & 0 & 0 & 2200 \\ 1 & 1 & 1 & 0 & 1 & 0 & 0 & 0 & 400 \\ -1 & -1 & 1 & 0 & 0 & 1 & 0 & 0 & 0 \\ -1 & 2 & 0 & 0 & 0 & 0 & 1 & 0 & 0 \\ \hline -95 & -200 & -325 & 0 & 0 & 0 & 0 & 1 & 0 \end{array}\right].$$

27. Let x_1 represent the number of redwood tables, x_2 the number of stained Douglas fir tables, and x_3 the number of stained white spruce tables. Organize the information in a table.

	Redwood	Douglas Fir	White Spruce	Maximum Available
Assembly Time	8	7	8	90 8-hr days = 720 hr
Staining Time	0	2	2	60 8-hr days = 480 hr
Cost	$159	$138.85	$129.35	$15,000

The limit of 720 hr for carpenters leads to the constraint

$$8x_1 + 7x_2 + 8x_3 \leq 720.$$

The limit of 480 hr for staining leads to the constraint

$$2x_2 + 2x_3 \leq 480.$$

The cost constraint is

$$159x_1 + 138.85x_2 + 129.35x_3 \leq 15,000.$$

The problem may be stated as:

Find $x_1 \geq 0, x_2 \geq 0$, and $x_3 \geq 0$ such that

$$\begin{aligned} 8x_1 + 7x_2 + 8x_3 &\leq 720 \\ 2x_2 + 2x_3 &\leq 480 \\ 159x_1 + 138.85x_2 + 129.35x_3 &\leq 15,000 \end{aligned}$$

and $z = x_1 + x_2 + x_3$ is maximized.

Introduce slack variables s_1, s_2, and s_3, and the problem can be restated as:

Find $x_1 \geq 0, x_2 \geq 0, x_3 \geq 0, s_1 \geq 0, s_2 \geq 0$, and $s_3 \geq 0$ such that

$$\begin{aligned} 8x_1 + 7x_2 + 8x_3 + s_1 \quad\quad\quad &= 720 \\ 2x_2 + 2x_3 \quad + s_2 \quad &= 480 \\ 159x_1 + 138.85x_2 + 129.35x_3 \quad\quad + s_3 &= 15,000 \end{aligned}$$

and $z = x_1 + x_2 + x_3$ is maximized.

Rewrite the objective function as

$$-x_1 - x_2 - x_3 + z = 0.$$

The initial simplex tableau is

$$\begin{bmatrix}
x_1 & x_2 & x_3 & s_1 & s_2 & s_3 & z & \\
8 & 7 & 8 & 1 & 0 & 0 & 0 & 720 \\
0 & 2 & 2 & 0 & 1 & 0 & 0 & 480 \\
159 & 138.85 & 129.35 & 0 & 0 & 1 & 0 & 15{,}000 \\
\hline
-1 & -1 & -1 & 0 & 0 & 0 & 1 & 0
\end{bmatrix}.$$

29. Let $x_1 =$ the number of newspaper ads,

$x_2 =$ the number of internet banners,

and $x_3 =$ the number of TV ads.

Organize the information in a table.

	Newspaper Ads	Internet Banners	TV Ads
Cost per Ad	400	20	2000
Maximum Number	30	60	10
Women Seeing Ad	4000	3000	10,000

The cost constraint is

$$400x_1 + 20x_2 + 2000x_3 \le 8000$$

The constraints on the numbers of ads is

$$x_1 \le 30$$
$$x_2 \le 60$$
$$x_3 \le 10.$$

The problem may be stated as:

Find $x_1 \ge 0, x_2 \ge 0,$ and $x_3 \ge 0$ such that

$$\begin{aligned}
400x_1 + 20x_2 + 2000x_3 &\le 8000 \\
x_1 &\le 30 \\
x_2 &\le 60 \\
x_3 &\le 10
\end{aligned}$$

and $z = 4000x_1 + 3000x_2 + 10{,}000x_3$ is maximized.

Introduce slack variables s_1, s_2, s_3 and s_4, and the problem can be restated as:

Find $x_1 \ge 0, x_2 \ge 0, x_3 \ge 0, s_1 \ge 0, s_2 \ge 0, s_3 \ge 0,$ and $s_4 \ge 0$ such that

$$\begin{aligned}
400x_1 + 20x_2 + 2000x_3 + s_1 &= 8000 \\
x_1 + s_2 &= 30 \\
x_2 + s_3 &= 60 \\
x_3 + s_4 &= 10
\end{aligned}$$

and $z = 4000x_1 + 3000x_2 + 10{,}000x_3$ is maximized.

Rewrite the objective function as

$$-4000x_1 - 3000x_2 - 10{,}000x_3 + z = 0.$$

The initial simplex tableau is

$$\begin{bmatrix}
x_1 & x_2 & x_3 & s_1 & s_2 & s_3 & s_4 & z & \\
400 & 20 & 2000 & 1 & 0 & 0 & 0 & 0 & 8000 \\
1 & 0 & 0 & 0 & 1 & 0 & 0 & 0 & 30 \\
0 & 1 & 0 & 0 & 0 & 1 & 0 & 0 & 60 \\
0 & 0 & 1 & 0 & 0 & 0 & 1 & 0 & 10 \\
\hline
-4000 & -3000 & -10{,}000 & 0 & 0 & 0 & 0 & 1 & 0
\end{bmatrix}$$

4.2 Maximization Problems

1.
$$\begin{bmatrix}
x_1 & x_2 & x_3 & s_1 & s_2 & z & \\
1 & 4 & 4 & 1 & 0 & 0 & 16 \\
2 & 1 & 5 & 0 & 1 & 0 & 20 \\
\hline
-3 & -1 & -2 & 0 & 0 & 1 & 0
\end{bmatrix}$$

The most negative indicator is -3, in the first column. Find the quotients $\frac{16}{1} = 16$ and $\frac{20}{2} = 10$; since 10 is the smaller quotient, 2 in row 2, column 1 is the pivot.

$$\begin{array}{c} \frac{16}{1} = 16 \\ \\ \frac{20}{2} = 10 \end{array}
\begin{bmatrix}
x_1 & x_2 & x_3 & s_1 & s_2 & z & \\
1 & 4 & 4 & 1 & 0 & 0 & 16 \\
\boxed{2} & 1 & 5 & 0 & 1 & 0 & 20 \\
\hline
-3 & -1 & -2 & 0 & 0 & 1 & 0
\end{bmatrix}$$

Performing row transformations, we get the following tableau.

$$\begin{array}{c} -R_2 + 2R_1 \to R_1 \\ \\ 3R_2 + 2R_3 \to R_3 \end{array}
\begin{bmatrix}
x_1 & x_2 & x_3 & s_1 & s_2 & z & \\
0 & 7 & 3 & 2 & -1 & 0 & 12 \\
2 & 1 & 5 & 0 & 1 & 0 & 20 \\
0 & 1 & 11 & 0 & 3 & 2 & 60
\end{bmatrix}$$

All of the numbers in the last row are nonnegative, so we are finished pivoting. Create a 1 in the columns corresponding to x_1, s_1 and z.

$$\begin{array}{c} \frac{1}{2}R_1 \to R_1 \\ \frac{1}{2}R_2 \to R_2 \\ \frac{1}{2}R_3 \to R_3 \end{array}
\begin{bmatrix}
x_1 & x_2 & x_3 & s_1 & s_2 & z & \\
0 & \frac{7}{2} & \frac{3}{2} & 1 & -\frac{1}{2} & 0 & 6 \\
1 & \frac{1}{2} & \frac{5}{2} & 0 & \frac{1}{2} & 0 & 10 \\
\hline
0 & \frac{1}{2} & \frac{11}{2} & 0 & \frac{3}{2} & 1 & 30
\end{bmatrix}$$

The maximum value is 30 and occurs when $x_1 = 10, x_2 = 0, x_3 = 0, s_1 = 6,$ and $s_2 = 0.$

3.

$$\begin{array}{cccccc|c} x_1 & x_2 & s_1 & s_2 & s_3 & z & \\ 1 & 3 & 1 & 0 & 0 & 0 & 12 \\ 2 & 1 & 0 & 1 & 0 & 0 & 10 \\ 1 & 1 & 0 & 0 & 1 & 0 & 4 \\ \hline -2 & -1 & 0 & 0 & 0 & 1 & 0 \end{array}$$

The most negative indicator is -2, in the first column. Find the quotients $\frac{12}{1} = 12$, $\frac{10}{2} = 5$, and $\frac{4}{1} = 4$; since 4 is the smallest quotient, 1 in row 3, column 1 is the pivot.

$$\begin{array}{cccccc|c} x_1 & x_2 & s_1 & s_2 & s_3 & z & \\ 1 & 3 & 1 & 0 & 0 & 0 & 12 \\ 2 & 1 & 0 & 1 & 0 & 0 & 10 \\ \boxed{1} & 1 & 0 & 0 & 1 & 0 & 4 \\ \hline -2 & -1 & 0 & 0 & 0 & 1 & 0 \end{array}$$

$$\begin{array}{c} \\ -R_3 + R_1 \to R_1 \\ -2R_3 + R_2 \to R_2 \\ \\ 2R_3 + R_4 \to R_4 \end{array} \begin{array}{cccccc|c} x_1 & x_2 & s_1 & s_2 & s_3 & z & \\ 0 & 2 & 1 & 0 & -1 & 0 & 8 \\ 0 & -1 & 0 & 1 & -2 & 0 & 2 \\ 1 & 1 & 0 & 0 & 1 & 0 & 4 \\ \hline 0 & 1 & 0 & 0 & 2 & 1 & 8 \end{array}$$

This is a final tableau since all of the numbers in the last row are nonnegative. The maximum value is 8 when $x_1 = 4, x_2 = 0, s_1 = 8, s_2 = 2$, and $s_3 = 0$.

5.

$$\begin{array}{ccccccc|c} x_1 & x_2 & x_3 & s_1 & s_2 & s_3 & z & \\ 2 & 2 & 8 & 1 & 0 & 0 & 0 & 40 \\ 4 & -5 & 6 & 0 & 1 & 0 & 0 & 60 \\ 2 & -2 & 6 & 0 & 0 & 1 & 0 & 24 \\ \hline -14 & -10 & -12 & 0 & 0 & 0 & 1 & 0 \end{array}$$

The most negative indicator is -14, in the first column. Find the quotients $\frac{40}{2} = 20$, $\frac{60}{4} = 15$, and $\frac{24}{2} = 12$; since 12 is the smallest quotient, 2 in row 3, column 1 is the pivot.

$$\begin{array}{ccccccc|c} x_1 & x_2 & x_3 & s_1 & s_2 & s_3 & z & \\ 2 & 2 & 8 & 1 & 0 & 0 & 0 & 40 \\ 4 & -5 & 6 & 0 & 1 & 0 & 0 & 60 \\ \boxed{2} & -2 & 6 & 0 & 0 & 1 & 0 & 24 \\ \hline -14 & -10 & -12 & 0 & 0 & 0 & 1 & 0 \end{array}$$

Performing row transformations, we get the following tableau.

$$\begin{array}{c} -R_3 + R_1 \to R_1 \\ \\ -2R_3 + R_2 \to R_2 \\ \\ \\ 7R_3 + R_4 \to R_4 \end{array} \begin{array}{ccccccc|c} x_1 & x_2 & x_3 & s_1 & s_2 & s_3 & z & \\ 0 & \boxed{4} & 2 & 1 & 0 & -1 & 0 & 16 \\ 0 & -1 & -6 & 0 & 1 & -2 & 0 & 12 \\ 2 & -2 & 6 & 0 & 0 & 1 & 0 & 24 \\ \hline 0 & -24 & 30 & 0 & 0 & 7 & 1 & 168 \end{array}$$

Since there is still a negative indicator, we must repeat the process. The second pivot is the 4 in column 2, since $\frac{16}{4}$ is the only nonnegative quotient in the only column with a negative indicator. Performing row transformations again, we get the following tableau.

$$\begin{array}{c} \\ R_1 + 4R_2 \to R_2 \\ R_1 + 2R_3 \to R_3 \\ 6R_1 + R_4 \to R_4 \end{array} \begin{array}{ccccccc|c} x_1 & x_2 & x_3 & s_1 & s_2 & s_3 & z & \\ 0 & 4 & 2 & 1 & 0 & -1 & 0 & 16 \\ 0 & 0 & -22 & 1 & 4 & -9 & 0 & 64 \\ 4 & 0 & 14 & 1 & 0 & 1 & 0 & 64 \\ \hline 0 & 0 & 42 & 6 & 0 & 1 & 1 & 264 \end{array}$$

All of the numbers in the last row are nonnegative, so we are finished pivoting. Create a 1 in the columns corresponding to x_1, x_2, and s_2.

$$\begin{array}{c} \frac{1}{4}R_1 \to R_1 \\ \\ \frac{1}{4}R_2 \to R_2 \\ \frac{1}{4}R_3 \to R_3 \\ \\ \end{array} \begin{array}{ccccccc|c} x_1 & x_2 & x_3 & s_1 & s_2 & s_3 & z & \\ 0 & 1 & \frac{1}{2} & \frac{1}{4} & 0 & -\frac{1}{4} & 0 & 4 \\ 0 & 0 & -\frac{11}{2} & \frac{1}{4} & 1 & -\frac{9}{4} & 0 & 16 \\ 1 & 0 & \frac{7}{2} & \frac{1}{4} & 0 & \frac{1}{4} & 0 & 16 \\ \hline 0 & 0 & 42 & 6 & 0 & 1 & 1 & 264 \end{array}$$

The maximum value is 264 and occurs when $x_1 = 16, x_2 = 4, x_3 = 0, s_1 = 0, s_2 = 16$, and $s_3 = 0$.

7. Maximize $z = 3x_1 + 5x_2$
subject to: $4x_1 + x_2 \le 25$
$2x_1 + 3x_2 \le 15$
with $x_1 \ge 0, x_2 \ge 0$.

Two slack variables, s_1 and s_2, need to be introduced. The problem can be restated as:

Maximize $z = 3x_1 + 5x_2$
subject to: $4x_1 + x_2 + s_1 = 25$
$2x_1 + 3x_2 + s_2 = 15$
with $x_1 \ge 0, x_2 \ge 0, s_1 \ge 0, s_2 \ge 0$.

Rewrite the objective function as

$$-3x_1 - 5x_2 + z = 0.$$

The initial simplex tableau follows.

$$\begin{array}{ccccc|c} x_1 & x_2 & s_1 & s_2 & z & \\ 4 & 1 & 1 & 0 & 0 & 25 \\ 2 & 3 & 0 & 1 & 0 & 15 \\ \hline -3 & -5 & 0 & 0 & 1 & 0 \end{array}$$

The most negative indicator is -5, in the second column. To select the pivot from column 2, find the quotients $\frac{25}{1} = 25$ and $\frac{15}{3} = 5$. The smaller quotient is 5, so 3 is the pivot.

$$\begin{array}{ccccc} x_1 & x_2 & s_1 & s_2 & z \\ \left[\begin{array}{ccccc|c} 4 & 1 & 1 & 0 & 0 & 25 \\ 2 & \boxed{3} & 0 & 1 & 0 & 15 \\ \hline -3 & -5 & 0 & 0 & 1 & 0 \end{array}\right] \end{array}$$

$$\begin{array}{c} \\ -R_2 + 3R_1 \rightarrow R_1 \\ \\ 5R_2 + 3R_3 \rightarrow R_3 \end{array} \begin{array}{ccccc} x_1 & x_2 & s_1 & s_2 & z \\ \left[\begin{array}{ccccc|c} 10 & 0 & 3 & -1 & 0 & 60 \\ 2 & 3 & 0 & 1 & 0 & 15 \\ 1 & 0 & 0 & 5 & 3 & 75 \end{array}\right] \end{array}$$

All of the indicators are nonnegative. Create a 1 in the columns corresponding to $x_2, s_1,$ and z.

$$\begin{array}{c} \frac{1}{3}R_1 \rightarrow R_1 \\ \frac{1}{3}R_2 \rightarrow R_2 \\ \frac{1}{3}R_3 \rightarrow R_3 \end{array} \begin{array}{ccccc} x_1 & x_2 & s_1 & s_2 & z \\ \left[\begin{array}{ccccc|c} \frac{10}{3} & 0 & 1 & -\frac{1}{3} & 0 & 20 \\ \frac{2}{3} & 1 & 0 & \frac{1}{3} & 0 & 5 \\ \frac{1}{3} & 0 & 0 & \frac{5}{3} & 1 & 25 \end{array}\right] \end{array}$$

The maximum value is 25 when $x_1 = 0$, $x_2 = 5, s_1 = 20,$ and $s_2 = 0$.

9. Maximize $\quad z = 10x_1 + 12x_2$

subject to: $\quad 4x_1 + 2x_2 \leq 20$
$\qquad\qquad\quad 5x_1 + \ x_2 \leq 50$
$\qquad\qquad\quad 2x_1 + 2x_2 \leq 24$

with $\qquad\quad x_1 \geq 0, x_2 \geq 0.$

Three slack variables, $s_1, s_2,$ and s_3, need to be introduced. The initial tableau is as follows.

$$\begin{array}{cccccc} x_1 & x_2 & s_1 & s_2 & s_3 & z \\ \left[\begin{array}{cccccc|c} 4 & 2 & 1 & 0 & 0 & 0 & 20 \\ 5 & 1 & 0 & 1 & 0 & 0 & 50 \\ 2 & 2 & 0 & 0 & 1 & 0 & 24 \\ \hline -10 & -12 & 0 & 0 & 0 & 1 & 0 \end{array}\right] \end{array}$$

The most negative indicator is -12, in column 2. The quotients are $\frac{20}{2} = 10, \frac{50}{1} = 50,$ and $\frac{24}{2} = 12$; the smallest is 10, so 2 in row 1, column 2 is the pivot.

$$\begin{array}{cccccc} x_1 & x_2 & s_1 & s_2 & s_3 & z \\ \left[\begin{array}{cccccc|c} 4 & \boxed{2} & 1 & 0 & 0 & 0 & 20 \\ 5 & 1 & 0 & 1 & 0 & 0 & 50 \\ 2 & 2 & 0 & 0 & 1 & 0 & 24 \\ \hline -10 & -12 & 0 & 0 & 0 & 1 & 0 \end{array}\right] \end{array}$$

$$\begin{array}{c} \\ -R_1 + 2R_2 \rightarrow R_2 \\ -R_1 + \ R_3 \rightarrow R_3 \\ 6R_1 + \ R_4 \rightarrow R_4 \end{array} \begin{array}{cccccc} x_1 & x_2 & s_1 & s_2 & s_3 & z \\ \left[\begin{array}{cccccc|c} 4 & 2 & 1 & 0 & 0 & 0 & 20 \\ 6 & 0 & -1 & 2 & 0 & 0 & 80 \\ -2 & 0 & -1 & 0 & 1 & 0 & 4 \\ \hline 14 & 0 & 6 & 0 & 0 & 1 & 120 \end{array}\right] \end{array}$$

All of the indicators are nonnegative, so we are finished pivoting. Create a 1 in the columns corresponding to x_2 and s_2.

$$\begin{array}{c} \\ \cdot\frac{1}{2}R_1 \rightarrow R_1 \\ \frac{1}{2}R_2 \rightarrow R_2 \\ \\ \end{array} \begin{array}{cccccc} x_1 & x_2 & s_1 & s_2 & s_3 & z \\ \left[\begin{array}{cccccc|c} 2 & 1 & \frac{1}{2} & 0 & 0 & 0 & 10 \\ 3 & 0 & -\frac{1}{2} & 1 & 0 & 0 & 40 \\ -2 & 0 & -1 & 0 & 1 & 0 & 4 \\ \hline 14 & 0 & 6 & 0 & 0 & 1 & 120 \end{array}\right] \end{array}$$

The maximum value is 120 when $x_1 = 0$, $x_2 = 10, s_1 = 0, s_2 = 40,$ and $s_3 = 4$.

11. Maximize $\quad z = 8x_1 + 3x_2 + x_3$

subject to: $\quad x_1 + 6x_2 + \ 8x_3 \leq 118$
$\qquad\qquad\quad x_1 + 5x_2 + 10x_3 \leq 220$

with $\qquad\quad x_1 \geq 0, x_2 \geq 0, x_3 \geq 0.$

Two slack variables, s_1 and s_2, need to be introduced. The initial simplex tableau is as follows.

$$\begin{array}{cccccc} x_1 & x_2 & x_3 & s_1 & s_2 & z \\ \left[\begin{array}{cccccc|c} \boxed{1} & 6 & 8 & 1 & 0 & 0 & 118 \\ 1 & 5 & 10 & 0 & 1 & 0 & 220 \\ \hline -8 & -3 & -1 & 0 & 0 & 1 & 0 \end{array}\right] \end{array}$$

The most negative indicator is -8, in the first column. The quotients are $\frac{118}{1} = 118$ and $\frac{220}{1} = 220$; since 118 is the smaller, 1 in row 1, column 1 is the pivot. Performing row transformations, we get the following tableau.

$$\begin{array}{c} \\ -R_1 + R_2 \rightarrow R_2 \\ 8R_1 + R_3 \rightarrow R_3 \end{array} \begin{array}{cccccc} x_1 & x_2 & x_3 & s_1 & s_2 & z \\ \left[\begin{array}{cccccc|c} 1 & 6 & 8 & 1 & 0 & 0 & 118 \\ 0 & -1 & 2 & -1 & 1 & 0 & 102 \\ 0 & 45 & 63 & 8 & 0 & 1 & 944 \end{array}\right] \end{array}$$

All of the indicators are nonnegative, so we are finished pivoting. The maximum value is 944 when $x_1 = 118, x_2 = 0, x_3 = 0, s_1 = 0,$ and $s_2 = 102$.

13. Maximize $z = 10x_1 + 15x_2 + 10x_3 + 5x_4$

   subject to: $x_1 + x_2 + x_3 + x_4 \le 300$

   $x_1 + 2x_2 + 3x_3 + x_4 \le 360$

   with $x_1 \ge 0, x_2 \ge 0, x_3 \ge 0, x_4 \ge 0.$

The initial tableau is as follows.

$$
\begin{array}{ccccccc}
x_1 & x_2 & x_3 & x_4 & s_1 & s_2 & z \\
\end{array}
$$

$$
\left[
\begin{array}{ccccccc|c}
1 & 1 & 1 & 1 & 1 & 0 & 0 & 300 \\
1 & \boxed{2} & 3 & 1 & 0 & 1 & 0 & 360 \\
\hline
-10 & -15 & -10 & -5 & 0 & 0 & 1 & 0 \\
\end{array}
\right]
$$

In the column with the most negative indicator, -15, the quotients are $\frac{300}{1} = 300$ and $\frac{360}{2} = 180$. The smaller quotient is 180, so the 2 in row 2, column 2, is the pivot.

$$
\begin{array}{ccccccc}
x_1 & x_2 & x_3 & x_4 & s_1 & s_2 & z \\
\end{array}
$$

$$
\begin{array}{l}
-R_2 + 2R_1 \to R_1 \\
\\
15R_2 + 2R_3 \to R_3
\end{array}
\left[
\begin{array}{ccccccc|c}
\boxed{1} & 0 & -1 & 1 & 2 & -1 & 0 & 240 \\
1 & 2 & 3 & 1 & 0 & 1 & 0 & 360 \\
\hline
-5 & 0 & 25 & 5 & 0 & 15 & 2 & 5400 \\
\end{array}
\right]
$$

Pivot on the 1 in row 1, column 1.

$$
\begin{array}{ccccccc}
x_1 & x_2 & x_3 & x_4 & s_1 & s_2 & z \\
\end{array}
$$

$$
\begin{array}{l}
\\
-R_1 + R_2 \to R_2 \\
5R_1 + R_3 \to R_3
\end{array}
\left[
\begin{array}{ccccccc|c}
1 & 0 & -1 & 1 & 2 & -1 & 0 & 240 \\
0 & 2 & 4 & 0 & -2 & 2 & 0 & 120 \\
\hline
0 & 0 & 20 & 10 & 10 & 10 & 2 & 6600 \\
\end{array}
\right]
$$

Create a 1 in the columns corresponding to x_2 and z.

$$
\begin{array}{ccccccc}
x_1 & x_2 & x_3 & x_4 & s_1 & s_2 & z \\
\end{array}
$$

$$
\begin{array}{l}
\\
\frac{1}{2}R_2 \to R_2 \\
\frac{1}{2}R_3 \to R_3
\end{array}
\left[
\begin{array}{ccccccc|c}
1 & 0 & -1 & 1 & 2 & -1 & 0 & 240 \\
0 & 1 & 2 & 0 & -1 & 1 & 0 & 60 \\
\hline
0 & 0 & 10 & 5 & 5 & 5 & 1 & 3300 \\
\end{array}
\right]
$$

The maximum value is 3300 when $x_1 = 240, x_2 = 60, x_3 = 0, x_4 = 0, s_1 = 0,$ and $s_2 = 0.$

15. Maximize $z = 4x_1 + 6x_2$

   subject to: $x_1 - 5x_2 \le 25$

   $4x_1 - 3x_2 \le 12$

   with $x_1 \ge 0, x_2 \ge 0.$

$$
\begin{array}{ccccc}
x_1 & x_2 & s_1 & s_2 & z \\
\end{array}
$$

$$
\left[
\begin{array}{ccccc|c}
1 & -5 & 1 & 0 & 0 & 25 \\
4 & -3 & 0 & 1 & 0 & 12 \\
\hline
-4 & -6 & 0 & 0 & 1 & 0 \\
\end{array}
\right]
$$

The most negative indicator is -6. The negative quotients $25/(-5)$ and $12/(-3)$ indicate an unbounded feasible region, so there is no unique optimum solution.

17. Maximize $z = 37x_1 + 34x_2 + 36x_3 + 30x_4 + 35x_5$

subject to: $16x_1 + 19x_2 + 23x_3 + 15x_4 + 21x_5 \leq 42{,}000$

$15x_1 + 10x_2 + 19x_3 + 23x_4 + 10x_5 \leq 25{,}000$

$9x_1 + 16x_2 + 14x_3 + 12x_4 + 11x_5 \leq 23{,}000$

$18x_1 + 20x_2 + 15x_3 + 17x_4 + 19x_5 \leq 36{,}000$

with $x_1 \geq 0, x_2 \geq 0, x_3 \geq 0, x_4 \geq 0, x_5 \geq 0.$

Four slack variables, $s_1, s_2, s_3,$ and s_4, need to be introduced. The initial simplex tableau follows.

x_1	x_2	x_3	x_4	x_5	s_1	s_2	s_3	s_4	z	
16	19	23	15	21	1	0	0	0	0	42,000
15	10	19	23	10	0	1	0	0	0	25,000
9	16	14	12	11	0	0	1	0	0	23,000
18	20	15	17	19	0	0	0	1	0	36,000
−37	−34	−36	−30	−35	0	0	0	0	1	0

Using a graphing calculator or computer program, the maximum value is found to be 70,818.18 when $x_1 = 181.82$, $x_2 = 0, x_3 = 454.55, x_4 = 0, x_5 = 1363.64, s_1 = 0, s_2 = 0, s_3 = 0,$ and $s_4 = 0$.

21. Organize the information in a table.

	Church Group	Labor Union	Maximum Time Available
Letter Writing	2	2	16
Follow-up	1	.3	12
Money Raised	$100	$200	

Let x_1 and x_2 be the number of church groups and labor unions contacted respectively. We need two slack variables, s_1 and s_2.

Maximize $z = 100x_1 + 200x_2$

subject to: $2x_1 + 2x_2 + s_1 \quad\quad = 16$

$x_1 + 3x_2 \quad\quad + s_2 = 12$

with. $x_1 \geq 0, x_2 \geq 0, s_1 \geq 0, s_2 \geq 0.$

The initial simplex tableau is as follows.

x_1	x_2	s_1	s_2	z	
2	2	1	0	0	16
1	3	0	1	0	12
−100	−200	0	0	1	0

Pivot on the 3 in row 2, column 2.

$-2R_2 + 3R_1 \rightarrow R_1$

$200R_2 + 3R_3 \rightarrow R_3$

x_1	x_2	s_1	s_2	z	
4	0	3	−2	0	24
1	3	0	1	0	12
−100	0	0	200	3	2400

Pivot on the 4 in row 1, column 1.

$-R_1 + 4R_2 \rightarrow R_2$

$25R_1 + R_3 \rightarrow R_3$

x_1	x_2	s_1	s_2	z	
4	0	3	−2	0	24
0	12	−3	6	0	24
0	0	75	150	3	3000

This is a final tableau, since all of the indicators are nonnegative. Create a 1 in the columns corresponding to x_1, x_2, and z.

$$
\begin{array}{c}
\frac{1}{4}R_1 \to R_1 \\
\frac{1}{12}R_2 \to R_2 \\
\\
\frac{1}{3}R_3 \to R_3
\end{array}
\begin{array}{c}
\begin{array}{cccccc}
x_1 & x_2 & s_1 & s_2 & z & \\
\end{array} \\
\left[
\begin{array}{ccccc|c}
1 & 0 & \frac{3}{4} & -\frac{1}{2} & 0 & 6 \\
0 & 1 & -\frac{1}{4} & \frac{1}{2} & 0 & 2 \\
\hline
0 & 0 & 25 & 50 & 1 & 1000
\end{array}
\right]
\end{array}
$$

The maximum amount of money raised is \$1000/mo when $x_1 = 6$ and $x_2 = 2$, that is, when 6 churches and 2 labor unions are contacted.

23. (a) Let x_1 be the number of Royal Flush poker sets, x_2 be the number of Deluxe Diamond sets, and x_3 be the number of Full House sets. The problem can be stated as follows.

Maximize $z = 38x_1 + 22x_2 + 12x_3$

subject to:
$$
\begin{aligned}
1000x_1 + 600x_2 + 300x_3 &\le 2{,}800{,}000 \\
4x_1 + 2x_2 + 2x_3 &\le 10{,}000 \\
10x_1 + 5x_2 + 5x_3 &\le 25{,}000 \\
2x_1 + x_2 + x_3 &\le 6000
\end{aligned}
$$

with $x_1 \ge 0, x_2 \ge 0, x_3 \ge 0$.

Since there are four constraints, introduce slack variables, s_1, s_2, s_3, and s_4 and set up the initial simplex tableau.

$$
\begin{array}{cccccccc}
x_1 & x_2 & x_3 & s_1 & s_2 & s_3 & s_4 & z \\
\end{array}
$$
$$
\left[
\begin{array}{cccccccc|c}
1000 & 600 & 300 & 1 & 0 & 0 & 0 & 0 & 2{,}800{,}000 \\
4 & 2 & 2 & 0 & 1 & 0 & 0 & 0 & 10{,}000 \\
10 & 5 & 5 & 0 & 0 & 1 & 0 & 0 & 25{,}000 \\
2 & 1 & 1 & 0 & 0 & 0 & 1 & 0 & 6000 \\
\hline
-38 & -22 & -12 & 0 & 0 & 0 & 0 & 1 & 0
\end{array}
\right]
$$

Using a graphing calculator or computer program, the maximum profit is \$104,000 and is obtained when 1000 Royal Flush poker sets, 3000 Deluxe Diamond poker sets, and no Full House poker sets are assembled.

(b) According to the poker chip constraint:

$$1000(1000) + 600(3000) + 300(0) + s_2 = 2{,}800{,}000$$
$$s_1 = 0.$$

So all of the poker chips are used. Checking the card constraint:

$$4(1000) + 2(3000) + 2(0) + s_2 = 10{,}000$$
$$s_2 = 0.$$

So all of the poker chips are used. Checking the dice constraint:

$$10(1000) + 5(3000) + 5(0) + s_3 = 25{,}000$$
$$s_3 = 0.$$

So all of the dice are used. Finally, checking the dealer button constraint:

$$2(1000) + 3000 + 0 + s_4 = 6000$$
$$s_4 = 1000.$$

This means there are 1000 unused dealer buttons.

25. (a) Let x_1 represent the number of racing bicycles, x_2 the number of touring bicycles, and x_3 the number of mountain bicycles.

From Exercise 26 in Section 4.1, the initial simplex tableau is as follows.

$$
\begin{array}{ccccccc}
x_1 & x_2 & x_3 & s_1 & s_2 & z & \\
\end{array}
$$

$$
\left[\begin{array}{cccccc|c}
17 & 27 & \boxed{34} & 1 & 0 & 0 & 91{,}800 \\
12 & 21 & 15 & 0 & 1 & 0 & 42{,}000 \\
\hline
-8 & -12 & -22 & 0 & 0 & 1 & 0
\end{array}\right]
$$

Pivot on the 34 in row 1, column 3.

$$
\begin{array}{c}
\\
-15R_1 + 34R_2 \rightarrow R_2 \\
11R_1 + 17R_3 \rightarrow R_3
\end{array}
\left[\begin{array}{cccccc|c}
17 & 27 & 34 & 1 & 0 & 0 & 91{,}800 \\
153 & 309 & 0 & -15 & 34 & 0 & 51{,}000 \\
51 & 93 & 0 & 11 & 0 & 17 & 1{,}009{,}800
\end{array}\right]
$$

This is a final tableau, since all of the indicators are nonnegative. Create a 1 in the columns corresponding to x_3, s_2, and z.

$$
\begin{array}{c}
\frac{1}{34}R_1 \rightarrow R_1 \\
\frac{1}{34}R_2 \rightarrow R_2 \\
\frac{1}{17}R_3 \rightarrow R_3
\end{array}
\left[\begin{array}{cccccc|c}
\frac{1}{2} & \frac{27}{34} & 1 & \frac{1}{34} & 0 & 0 & 2700 \\
\frac{9}{2} & \frac{309}{34} & 0 & -\frac{15}{34} & 1 & 0 & 1500 \\
3 & \frac{93}{17} & 0 & \frac{11}{17} & 0 & 1 & 59{,}400
\end{array}\right]
$$

From the tableau, $x_1 = 0$, $x_2 = 0$, and $x_3 = 2700$. The company should make no racing or touring bicycles and 2700 mountain bicycles.

(b) From the third row of the final tableau, the maximum profit is \$59,400.

(c) When $x_1 = 0, x_2 = 0$, and $x_3 = 2700$, the number of units of steel used is

$$17(0) + 27(0) + 34(2700) = 91{,}800$$

which is all the steel available. The number of units of aluminum used is

$$12(0) + 21(0) + 15(2700) = 40{,}500$$

which leaves $42{,}000 - 40{,}500 = 1500$ units of aluminum unused.

Checking the second constraint:

$$
\begin{aligned}
12x_1 + 21x_2 + 15x_3 + s_2 &= 42{,}000 \\
12(0) + 21(0) + 15(2700) + s_2 &= 42{,}000 \\
s_2 &= 1500.
\end{aligned}
$$

27. (a) Let x_1 be the number of newspaper ads, x_2 be the number of Internet banner ads, and x_3 be the number of TV ads. From

$$
\begin{array}{cccccccc}
x_1 & x_2 & x_3 & s_1 & s_2 & s_3 & s_4 & z \\
\end{array}
$$

$$
\left[\begin{array}{ccccccc|c}
400 & 20 & \boxed{2000} & 1 & 0 & 0 & 0 & 8000 \\
1 & 0 & 0 & 0 & 1 & 0 & 0 & 30 \\
0 & 1 & 0 & 0 & 0 & 1 & 0 & 60 \\
0 & 0 & 1 & 0 & 0 & 0 & 1 & 10 \\
\hline
-4000 & -3000 & -10{,}000 & 0 & 0 & 0 & 0 & 1 & 0
\end{array}\right]
$$

Pivot on the 2000 in row 1, column 3.

$$
\begin{array}{r}
\\
\\
\\
-R_1 + 2000R_4 \to R_4 \\
5R_1 + R_5 \to R_5
\end{array}
\begin{array}{c}
x_1 \quad x_2 \quad x_3 \quad s_1 \; s_2 \; s_3 \quad s_4 \quad z \\
\left[\begin{array}{ccccccccc|c}
400 & 20 & 2000 & 1 & 0 & 0 & 0 & 0 & 8000 \\
1 & 0 & 0 & 0 & 1 & 0 & 0 & 0 & 30 \\
0 & \boxed{1} & 0 & 0 & 0 & 1 & 0 & 0 & 60 \\
-400 & -20 & 0 & -1 & 0 & 0 & 2000 & 0 & 12{,}000 \\
-2000 & -2900 & 0 & 5 & 0 & 0 & 0 & 1 & 40{,}000
\end{array}\right]
\end{array}
$$

Pivot on the 1 in row 3, column 2.

$$
\begin{array}{r}
-20R_3 + R_1 \to R_1 \\
\\
\\
20R_3 + R_4 \to R_4 \\
2900R_3 + R_5 \to R_5
\end{array}
\begin{array}{c}
x_1 \quad x_2 \; x_3 \quad s_1 \; s_2 \quad s_3 \quad s_4 \quad z \\
\left[\begin{array}{ccccccccc|c}
\boxed{400} & 0 & 2000 & 1 & 0 & -20 & 0 & 0 & 6800 \\
1 & 0 & 0 & 0 & 1 & 0 & 0 & 0 & 30 \\
0 & 1 & 0 & 0 & 0 & 1 & 0 & 0 & 60 \\
-400 & 0 & 0 & -1 & 0 & 20 & 2000 & 0 & 13{,}200 \\
-2000 & 0 & 0 & 5 & 0 & 2900 & 0 & 1 & 214{,}000
\end{array}\right]
\end{array}
$$

Pivot on the 400 in row 1, column 1.

$$
\begin{array}{r}
\\
-R_1 + 400R_2 \to R_2 \\
\\
R_1 + R_4 \to R_4 \\
5R_1 + R_5 \to R_5
\end{array}
\begin{array}{c}
x_1 \; x_2 \quad x_3 \quad s_1 \; s_2 \quad s_3 \quad s_4 \quad z \\
\left[\begin{array}{ccccccccc|c}
400 & 0 & 2000 & 1 & 0 & -20 & 0 & 0 & 6800 \\
0 & 0 & -2000 & -1 & 400 & 20 & 0 & 0 & 5200 \\
0 & 1 & 0 & 0 & 0 & 1 & 0 & 0 & 60 \\
0 & 0 & 2000 & 0 & 0 & 0 & 2000 & 0 & 20{,}000 \\
0 & 0 & 10{,}000 & 10 & 0 & 2800 & 0 & 1 & 248{,}000
\end{array}\right]
\end{array}
$$

Create a 1 in the columns corresponding to $x_1, s_2,$ and s_4.

$$
\begin{array}{r}
\frac{1}{400}R_1 \to R_1 \\
\frac{1}{400}R_2 \to R_2 \\
\\
\frac{1}{2000}R_4 \to R_4
\end{array}
\begin{array}{c}
x_1 \; x_2 \quad x_3 \quad s_1 \quad s_2 \quad s_3 \; s_4 \; z \\
\left[\begin{array}{ccccccccc|c}
1 & 0 & 5 & \frac{1}{400} & 0 & -\frac{1}{20} & 0 & 0 & 17 \\
0 & 0 & -5 & -\frac{1}{400} & 1 & \frac{1}{20} & 0 & 0 & 13 \\
0 & 1 & 0 & 0 & 0 & 1 & 0 & 0 & 60 \\
0 & 0 & 1 & 0 & 0 & 0 & 1 & 0 & 10 \\
0 & 0 & 10{,}000 & 10 & 0 & 2800 & 0 & 1 & 248{,}000
\end{array}\right]
\end{array}
$$

This is the final tableau. The maximum exposure is 248,000 women when 17 newspaper ads, 60 Internet banner ads, and no TV ads are used.

29. (a) The coefficients of the objective function are the profit coefficients from the table: 5, 4, and 3; choice (3) is correct.

(b) The constraints are the available man-hours for the 2 departments, 400 and 600; choice (4) is correct.

(c) $2X_1 + 3X_2 + 1X_3 \le 400$ is the constraint on department 1; choice (3) is correct.

31. Maximize $z = 100x + 200y$

subject to:
$$2x + 2y \leq 16$$
$$x + 3y \leq 12$$
with
$$x \geq 0, y \geq 0.$$

Using Excel, we enter the variables x and y in cells A1 and B1, respectively. Enter the x- and y-coordinates of the initial corner point of the feasible region, $(0,0)$, in cells A2 and B2, respectively, and NAME these cells x and y, respectively. In cells C2, C4, C5, C6, and C7, enter the formula for the function to maximize and each of the constraints: $100x + 200y, 2x + 2y, x + 3y, x$, and y. Since x and y have been set to 0, all the cells containing formulas should also show the value 0, as below.

	A	B	C
1	x	y	
2	0	0	0
3			
4			0
5			0
6			0
7			0

Using the SOLVER, ask Excel to maximize the value in cell C2 subject to the constraints $C4 \leq 16$, $C5 \leq 12$, $C6 \geq 0$, $C7 \geq 0$. Make sure you have checked off the box *Assume Linear Model* in SOLVER OPTIONS.

Excel returns the following values and allows you to choose a report.

	A	B	C
1	x	y	
2	6	2	1000
3			
4			16
5			12
6			6
7			2

Select the sensitivity report. The report will appear on a new sheet of the spread sheet.

Adjustable Cells

Cell	Name	Final Value	Reduced Cost	Objective Coefficient	Allowable Increase	Allowable Decrease
A2	x	6	0	100	100	33.33333333
B2	y	2	0	200	100	100

Constraints

Cell	Name	Final Value	Shadow Price	Constraint R.H. Side	Allowable Increase	Allowable Decrease
C4		16	25	16	8	8
C5		12	50	12	12	4
C6		6	0	0	6	1E+30
C7		2	0	0	2	1E+30

The church group's allowable increase is \$100 and the allowable decrease is \$33.33. So their contribution can be as high as \$100 + \$100 = \$200 or as low as \$100 − \$33.33 = \$66.67 and the original solution is still optimal. The unions' allowable increase is \$100 and the allowable decrease is \$100. So their contribution can be as high as \$200 + \$100 = \$300 or as low as \$200 − \$100 = \$100 and the original solution is still optimal.

33. Let x_1 = number of hours running, x_2 be the number of hours biking, and x_3 be the number hours walking. The problem can be stated as follows.

$$\text{Maximize} \quad z = 531x_1 + 472x_2 + 354x_3$$
$$\text{subject to:} \quad
\begin{aligned}
x_1 + x_2 + x_3 &\le 15 \\
x_1 &\le 3 \\
2x_2 - x_3 &\le 0
\end{aligned}$$
$$\text{with} \quad x_1 \ge 0, x_2 \ge 0, x_3 \ge 0.$$

We need three slack variables, $s_1, s_2,$ and s_3. The initial simplex tableau as follows.

$$\begin{array}{ccccccc|c}
x_1 & x_2 & x_3 & s_1 & s_2 & s_3 & z & \\
1 & 1 & 1 & 1 & 0 & 0 & 0 & 15 \\
\boxed{1} & 0 & 0 & 0 & 1 & 0 & 0 & 3 \\
0 & 2 & -1 & 0 & 0 & 1 & 0 & 0 \\
\hline
-531 & -472 & -354 & 0 & 0 & 0 & 1 & 0
\end{array}$$

Pivot on the 1 in row 2, column 1.

$$\begin{array}{l}
\\
-R_2 + R_1 \to R_1 \\
\\
\\
531R_2 + R_4 \to R_4
\end{array}
\begin{array}{ccccccc|c}
x_1 & x_2 & x_3 & s_1 & s_2 & s_3 & z & \\
0 & 1 & 1 & 1 & -1 & 0 & 0 & 12 \\
1 & 0 & 0 & 0 & 1 & 0 & 0 & 3 \\
0 & \boxed{2} & -1 & 0 & 0 & 1 & 0 & 0 \\
\hline
0 & -472 & -354 & 0 & 351 & 0 & 1 & 1593
\end{array}$$

Pivot on the 2 in row 3, column 2.

$$\begin{array}{l}
\\
-R_3 + 2R_1 \to R_1 \\
\\
\\
236R_3 + R_4 \to R_4
\end{array}
\begin{array}{ccccccc|c}
x_1 & x_2 & x_3 & s_1 & s_2 & s_3 & z & \\
0 & 0 & \boxed{3} & 2 & -2 & -1 & 0 & 24 \\
1 & 0 & 0 & 0 & 1 & 0 & 0 & 3 \\
0 & 2 & -1 & 0 & 0 & 1 & 0 & 0 \\
\hline
0 & 0 & -590 & 0 & 531 & 236 & 1 & 1593
\end{array}$$

Finally pivot on the 3 in row 1, column 3.

$$\begin{array}{l}
\\
\\
R_1 + 3R_3 \to R_3 \\
590R_1 + 3R_4 \to R_4
\end{array}
\begin{array}{ccccccc|c}
x_1 & x_2 & x_3 & s_1 & s_2 & s_3 & z & \\
0 & 1 & 3 & 2 & -2 & -1 & 0 & 24 \\
1 & 0 & 0 & 0 & 1 & 0 & 0 & 3 \\
0 & 6 & 0 & 2 & -2 & 2 & 0 & 24 \\
\hline
0 & 0 & 0 & 1180 & 413 & 118 & 3 & 18,939
\end{array}$$

Create a 1 in the columns corresponding to $x_2, x_3,$ and z.

$$\begin{array}{l}
\tfrac{1}{3}R_1 \to R_1 \\
\\
\tfrac{1}{6}R_3 \to R_3 \\
\\
\tfrac{1}{3}R_4 \to R_4
\end{array}
\begin{array}{ccccccc|c}
x_1 & x_2 & x_3 & s_1 & s_2 & s_3 & z & \\
0 & 0 & 1 & \tfrac{2}{3} & -\tfrac{2}{3} & -\tfrac{1}{3} & 0 & 8 \\
1 & 0 & 0 & 0 & 1 & 0 & 0 & 3 \\
0 & 1 & 0 & \tfrac{1}{3} & -\tfrac{1}{3} & \tfrac{1}{3} & 0 & 4 \\
\hline
0 & 0 & 0 & \tfrac{1180}{3} & \tfrac{413}{3} & \tfrac{118}{3} & 1 & 6313
\end{array}$$

Rachel should run 3 hours, bike 4 hours, and walk 8 hours for a maximum calorie expenditure of 6313 calories.

35. (a) Let $x_1 =$ amount of P, $x_2 =$ amount of Q, $x_3 =$ amount of R, and $x_4 =$ amount of S (all in kilograms).

We desire to maximize

$$z = 90x_1 + 70x_2 + 60x_3 + 50x_4$$

subject to:

$$0.375x_3 + 0.625x_4 \le 500$$
$$0.75x_2 + \quad 0.5x_3 + 0.375x_4 \le 600$$
$$x_1 + 0.25x_2 + 0.125x_3 \qquad\qquad \le 300$$

with $x_1 \ge 0, x_2 \ge 0, x_3 \ge 0, x_4 \ge 0.$

If we rewrite the constraints as

$$\frac{3}{8}x_3 + \frac{5}{8}x_4 \le 500$$

$$\frac{3}{4}x_2 + \frac{1}{2}x_3 + \frac{3}{8}x_4 \le 600$$

$$x_1 + \frac{1}{4}x_2 + \frac{1}{8}x_3 \qquad \le 300,$$

and then multiply each inequality by the least common denominator, 8, we get a set of constraints without fractions.

$$3x_3 + 5x_4 \le 4000$$
$$6x_2 + 4x_3 + 3x_4 \le 4800$$
$$8x_1 + 2x_2 + \quad x_3 \qquad \le 2400$$

We need three slack variables. The initial simplex tableau is as follows.

$$
\begin{array}{cccccccc|c}
x_1 & x_2 & x_3 & x_4 & s_1 & s_2 & s_3 & z & \\
\hline
0 & 0 & 3 & 5 & 1 & 0 & 0 & 0 & 4000 \\
0 & 6 & 4 & 3 & 0 & 1 & 0 & 0 & 4800 \\
\boxed{8} & 2 & 1 & 0 & 0 & 0 & 1 & 0 & 2400 \\
\hline
-90 & -70 & -60 & -50 & 0 & 0 & 0 & 1 & 0
\end{array}
$$

The first pivot is the 8 in row 3, column 1.

$$
\begin{array}{cccccccc|c}
x_1 & x_2 & x_3 & x_4 & s_1 & s_2 & s_3 & z & \\
\hline
0 & 0 & 3 & \boxed{5} & 1 & 0 & 0 & 0 & 4000 \\
0 & 6 & 4 & 3 & 0 & 1 & 0 & 0 & 4800 \\
8 & 2 & 1 & 0 & 0 & 0 & 1 & 0 & 2400 \\
\hline
0 & -190 & -195 & -200 & 0 & 0 & 45 & 4 & 108{,}000
\end{array}
$$
$45R_3 + 4R_4 \rightarrow R_4$

Pivot on the 5 in row 1, column 4.

$$
\begin{array}{cccccccc|c}
x_1 & x_2 & x_3 & x_4 & s_1 & s_2 & s_3 & z & \\
\hline
0 & 0 & 3 & 5 & 1 & 0 & 0 & 0 & 4000 \\
0 & \boxed{30} & 11 & 0 & -3 & 5 & 0 & 0 & 12{,}000 \\
8 & 2 & 1 & 0 & 0 & 0 & 1 & 0 & 2400 \\
\hline
0 & -190 & -75 & 0 & 40 & 0 & 45 & 4 & 268{,}000
\end{array}
$$
$-3R_1 + 5R_2 \rightarrow R_2$

$40R_1 + R_4 \rightarrow R_4$

Pivot on the 30 in row 2, column 2.

$$
\begin{array}{cccccccc|c}
x_1 & x_2 & x_3 & x_4 & s_1 & s_2 & s_3 & z & \\
\hline
0 & 0 & 3 & 5 & 1 & 0 & 0 & 0 & 4000 \\
0 & 30 & \boxed{11} & 0 & -3 & 5 & 0 & 0 & 12{,}000 \\
120 & 0 & 4 & 0 & 3 & -5 & 15 & 0 & 24{,}000 \\
\hline
0 & 0 & -16 & 0 & 63 & 95 & 135 & 12 & 1{,}032{,}000
\end{array}
$$
$-R_2 + 15R_3 \rightarrow R_3$

$19R_2 + 3R_4 \rightarrow R_4$

Pivot on the 11 in row 2, column 3.

$$
\begin{array}{c}
\\
-3R_2 + 11R_1 \rightarrow R_1 \\
\\
-4R_2 + 11R_3 \rightarrow R_3 \\
16R_2 + 11R_4 \rightarrow R_4
\end{array}
\begin{array}{ccccccccc}
x_1 & x_2 & x_3 & x_4 & s_1 & s_2 & s_3 & z & \\
\hline
0 & -90 & 0 & 55 & 20 & -15 & 0 & 0 & 8000 \\
0 & 30 & 11 & 0 & -3 & 5 & 0 & 0 & 12{,}000 \\
1320 & -120 & 0 & 0 & 45 & -75 & 165 & 0 & 216{,}000 \\
\hline
0 & 480 & 0 & 0 & 645 & 1125 & 1485 & 132 & 11{,}544{,}000
\end{array}
$$

$$
\begin{array}{c}
\frac{1}{55}R_1 \rightarrow R_1 \\[4pt]
\frac{1}{11}R_2 \rightarrow R_2 \\[4pt]
\frac{1}{1320}R_3 \rightarrow R_3 \\[6pt]
\frac{1}{132}R_4 \rightarrow R_4
\end{array}
\begin{array}{ccccccccc}
x_1 & x_2 & x_3 & x_4 & s_1 & s_2 & s_3 & z & \\
\hline
0 & -\frac{18}{11} & 0 & 1 & \frac{4}{11} & -\frac{3}{11} & 0 & 0 & \frac{1600}{11} \\[4pt]
0 & \frac{30}{11} & 1 & 0 & -\frac{3}{11} & \frac{5}{11} & 0 & 0 & \frac{12{,}000}{11} \\[4pt]
1 & -\frac{1}{11} & 0 & 0 & \frac{3}{88} & -\frac{15}{264} & \frac{1}{8} & 0 & \frac{1800}{11} \\[6pt]
\hline
0 & \frac{40}{11} & 0 & 0 & \frac{215}{44} & \frac{1125}{132} & \frac{45}{4} & 1 & \frac{962{,}000}{11}
\end{array}
$$

This final tableau gives the solution $x_1 = \frac{1800}{11} \approx 163.6, x_2 = 0, x_3 = \frac{12{,}000}{11} \approx 1090.9, x_4 = \frac{1600}{11} \approx 145.5$, and $z = \frac{962{,}000}{11} \approx 87{,}454.5$. Produce 163.6 kg of food P, none of food Q, 1090.9 kg of R, and 145.5 kg of S.

(b) The maximum total growth value is read from the bottom row of the final tableau: $\frac{962{,}000}{11} \approx 87{,}454.5$.

(c) When $x_1 = \frac{1800}{11}, x_2 = 0, x_3 = \frac{12{,}000}{11}$, and $x_4 = \frac{1600}{11}$, the number of units of nutrient A used is

$$0.375 \left(\frac{12{,}000}{11} \right) + 0.625 \left(\frac{1600}{11} \right) = 500$$

which is the total amount of nutrient A available. The number of units of nutrient B used is

$$0.75(0) + 0.5 \left(\frac{12{,}000}{11} \right) + 0.375 \left(\frac{1600}{11} \right) = 600$$

which is all the units of nutrient B. The amount of nutrient C used is

$$\left(\frac{1800}{11} \right) + 0.25(0) + 0.125 \left(\frac{12{,}000}{11} \right) = 300$$

which is all of the nutrient C. So none of the nutrients are left over.

37. Let x_1 represent the number of minutes for the senator, x_2 the number of minutes for the congresswoman, and x_3 the number of minutes for the governor.

Of the half-hour show's time, at most only $30 - 3 = 27$ min are available to be allotted to the politicians. The given information leads to the inequality

$$x_1 + x_2 + x_3 \le 27$$

and the inequalities

$$x_1 \ge 2x_3 \quad \text{and} \quad x_1 + x_3 \ge 2x_2,$$

and we are to maximize the objective function

$$z = 35x_1 + 40x_2 + 45x_3.$$

Rewrite the equation as

$$x_3 \le 27 - x_1 - x_2$$

and the inequalities as

$$-x_1 + 2x_3 \le 0 \quad \text{and} \quad -x_1 + 2x_2 - x_3 \le 0.$$

Substitute $27 - x_1 - x_2$ for x_3 in the objective function and the inequalities, and the problem is as follows.

Maximize $z = 35x_1 + 40x_2 + 45x_3$

subject to: $-x_1 \qquad + 2x_3 \le 0$

$\qquad\qquad -x_1 + 2x_2 - \ x_3 \le 0$

$\qquad\qquad\ x_1 + \ x_2 + \ x_3 \le 27$

with $x_1 \ge 0, x_2 \ge 0, x_3 \ge 0.$

We need three slack variables. The initial simplex tableau is as follows.

$$
\begin{array}{ccccccc|c}
x_1 & x_2 & x_3 & s_1 & s_2 & s_3 & z & \\
-1 & 0 & 2 & 1 & 0 & 0 & 0 & 0 \\
-1 & \boxed{2} & -1 & 0 & 1 & 0 & 0 & 0 \\
1 & 1 & 1 & 0 & 0 & 1 & 0 & 27 \\
\hline
-35 & -40 & -45 & 0 & 0 & 0 & 1 & 0
\end{array}
$$

Pivot on the 2 in row 2, column 2.

$$
\begin{array}{ccccccc|c}
x_1 & x_2 & x_3 & s_1 & s_2 & s_3 & z & \\
-1 & 0 & \boxed{2} & 1 & 0 & 0 & 0 & 0 \\
-1 & 2 & -1 & 0 & 1 & 0 & 0 & 0 \\
3 & 0 & 3 & 0 & -1 & 2 & 0 & 54 \\
\hline
-55 & 0 & -65 & 0 & 20 & 0 & 1 & 0
\end{array}
$$

$-R_2 + 2R_3 \to R_3$

$20R_2 + R_4 \to R_4$

Pivot on the 2 in row 1, column 3.

$R_1 + 2R_2 \to R_2$

$-3R_1 + 2R_3 \to R_3$

$65R_1 + 2R_4 \to R_4$

$$
\begin{array}{ccccccc|c}
x_1 & x_2 & x_3 & s_1 & s_2 & s_3 & z & \\
-1 & 0 & 2 & 1 & 0 & 0 & 0 & 0 \\
-3 & 4 & 0 & 1 & 2 & 0 & 0 & 0 \\
\boxed{9} & 0 & 0 & -3 & -2 & 4 & 0 & 108 \\
\hline
-175 & 0 & 0 & 65 & 40 & 0 & 2 & 0
\end{array}
$$

Pivot on the 9 in row 3, column 1.

$R_3 + 9R_1 \to R_1$

$R_3 + 3R_2 \to R_2$

$175R_3 + 9R_4 \to R_4$

$$
\begin{array}{ccccccc|c}
x_1 & x_2 & x_3 & s_1 & s_2 & s_3 & z & \\
0 & 0 & 18 & 6 & -2 & 4 & 0 & 108 \\
0 & 12 & 0 & 0 & 4 & 4 & 0 & 108 \\
9 & 0 & 0 & -3 & -2 & 4 & 0 & 108 \\
\hline
0 & 0 & 0 & 60 & 10 & 700 & 18 & 18{,}900
\end{array}
$$

Create a 1 in the columns corresponding to $x_1, x_2, x_3,$ and z.

$\frac{1}{18} R_1 \to R_1$

$\frac{1}{12} R_2 \to R_2$

$\frac{1}{9} R_3 \to R_3$

$\frac{1}{18} R_4 \to R_4$

$$
\begin{array}{ccccccc|c}
x_1 & x_2 & x_3 & s_1 & s_2 & s_3 & z & \\
0 & 0 & 1 & \frac{1}{3} & -\frac{1}{9} & \frac{2}{9} & 0 & 6 \\
0 & 1 & 0 & 0 & \frac{1}{3} & \frac{1}{3} & 0 & 9 \\
1 & 0 & 0 & -\frac{1}{3} & -\frac{2}{9} & \frac{4}{9} & 0 & 12 \\
\hline
0 & 0 & 0 & \frac{10}{3} & \frac{5}{9} & \frac{350}{9} & 1 & 1050
\end{array}
$$

The maximum value of z is 1050 when $x_1 = 12, x_2 = 9,$ and $x_3 = 6$. That is, for a maximum of 1,050,000 viewers, the time allotments should be 12 minutes for the senator, 9 minutes for the congresswoman, and 6 minutes for the governor.

4.3   Minimization Problems; Duality

1. To form the transpose of a matrix, the rows of the original matrix are written as the columns of the transpose. The transpose of

$$\begin{bmatrix} 1 & 2 & 3 \\ 3 & 2 & 1 \\ 1 & 10 & 0 \end{bmatrix}$$

is

$$\begin{bmatrix} 1 & 3 & 1 \\ 2 & 2 & 10 \\ 3 & 1 & 0 \end{bmatrix}.$$

3. The transpose of

$$\begin{bmatrix} 4 & 5 & -3 & 15 \\ 7 & 14 & 20 & -8 \\ 5 & 0 & -2 & 23 \end{bmatrix}$$

is

$$\begin{bmatrix} 4 & 7 & 5 \\ 5 & 14 & 0 \\ -3 & 20 & -2 \\ 15 & -8 & 23 \end{bmatrix}.$$

5. Maximize $z = 4x_1 + 3x_2 + 2x_3$
   subject to: $x_1 + x_2 + x_3 \le 5$
   $x_1 + x_2 \qquad \le 4$
   $2x_1 + x_2 + 3x_3 \le 15$
   with $x_1 \ge 0, x_2 \ge 0, x_3 \ge 0.$

To form the dual, first write the augmented matrix for the given problem.

$$\begin{bmatrix} 1 & 1 & 1 & 5 \\ 1 & 1 & 0 & 4 \\ 2 & 1 & 3 & 15 \\ \hline 4 & 3 & 2 & 0 \end{bmatrix}$$

Then form the transpose of this matrix.

$$\begin{bmatrix} 1 & 1 & 2 & 4 \\ 1 & 1 & 1 & 3 \\ 1 & 0 & 3 & 2 \\ \hline 5 & 4 & 15 & 0 \end{bmatrix}$$

The dual problem is stated from this second matrix (using y instead of x).

Minimize $w = 5y_1 + 4y_2 + 15y_3$
subject to: $y_1 + y_2 + 2y_3 \ge 4$
$y_1 + y_2 + y_3 \ge 3$
$y_1 \qquad + 3y_3 \ge 2$
with $y_1 \ge 0, y_2 \ge 0, y_3 \ge 0.$

7. Minimize $w = 3y_1 + 6y_2 + 4y_3 + y_4$
   subject to: $y_1 + y_2 + y_3 + y_4 \ge 150$
   $2y_1 + 2y_2 + 3y_3 + 4y_4 \ge 275$
   with $y_1 \ge 0, y_2 \ge 0, y_3 \ge 0, y_4 \ge 0.$

To find the dual problem, first write the augmented matrix for the problem.

$$\begin{bmatrix} 1 & 1 & 1 & 1 & 150 \\ 2 & 2 & 3 & 4 & 275 \\ \hline 3 & 6 & 4 & 1 & 0 \end{bmatrix}$$

Then form the transpose of this matrix.

$$\begin{bmatrix} 1 & 2 & 3 \\ 1 & 2 & 6 \\ 1 & 3 & 4 \\ 1 & 4 & 1 \\ \hline 150 & 275 & 0 \end{bmatrix}$$

The dual problem is

Maximize $z = 150x_1 + 275x_2$
subject to: $x_1 + 2x_2 \le 3$
$x_1 + 2x_2 \le 6$
$x_1 + 3x_2 \le 4$
$x_1 + 4x_2 \le 1$
with $x_1 \ge 0, x_2 \ge 0.$

9. Find $y_1 \ge 0$ and $y_2 \ge 0$ such that

$$2y_1 + 3y_2 \ge 6$$
$$2y_1 + y_2 \ge 7$$

and $w = 5y_1 + 2y_2$ is minimized.

Write the augmented matrix for this problem.

$$\begin{bmatrix} 2 & 3 & 6 \\ 2 & 1 & 7 \\ \hline 5 & 2 & 0 \end{bmatrix}$$

Form the transpose of this matrix.

$$\begin{bmatrix} 2 & 2 & 5 \\ 3 & 1 & 2 \\ \hline 6 & 7 & 0 \end{bmatrix}$$

Use this matrix to write the dual problem.

Find $x_1 \geq 0$ and $x_2 \geq 0$ such that

$$2x_1 + 2x_2 \leq 5$$
$$3x_1 + x_2 \leq 2$$

and $z = 6x_1 + 7x_2$ is maximized.

Introduce slack variables s_1 and s_2. The initial tableau is as follows.

$$\begin{array}{ccccc} x_1 & x_2 & s_1 & s_2 & z \\ \left[\begin{array}{ccccc|c} 2 & 2 & 1 & 0 & 0 & 5 \\ 3 & \boxed{1} & 0 & 1 & 0 & 2 \\ \hline -6 & -7 & 0 & 0 & 1 & 0 \end{array}\right] \end{array}$$

Pivot on the 1 in row 2, column 2, since that column has the most negative indicator and that row has the smallest nonnegative quotient.

$$\begin{array}{c} \\ -2R_2 + R_1 \rightarrow R_1 \\ \\ 7R_2 + R_3 \rightarrow R_3 \end{array} \begin{array}{ccccc} x_1 & x_2 & s_1 & s_2 & z \\ \left[\begin{array}{ccccc|c} -4 & 0 & 1 & -2 & 0 & 1 \\ 3 & 1 & 0 & 1 & 0 & 2 \\ \hline 15 & 0 & 0 & 7 & 1 & 14 \end{array}\right] \end{array}$$

The minimum value of w is the same as the maximum value of z. The minimum value of w is 14 when $y_1 = 0$ and $y_2 = 7$. (Note that the values of y_1 and y_2 are given by the entries in the bottom row of the columns corresponding to the slack variables in the final tableau.)

11. Find $y_1 \geq 0$ and $y_2 \geq 0$ such that

$$10y_1 + 5y_2 \geq 100$$
$$20y_1 + 10y_2 \geq 150$$

and $w = 4y_1 + 5y_2$ is minimized.

Write the augmented matrix for this problem.

$$\left[\begin{array}{cc|c} 10 & 5 & 100 \\ 20 & 10 & 150 \\ \hline 4 & 5 & 0 \end{array}\right]$$

Form the transpose of this matrix.

$$\left[\begin{array}{cc|c} 10 & 20 & 4 \\ 5 & 10 & 5 \\ \hline 100 & 150 & 0 \end{array}\right]$$

Write the dual problem from this matrix.

Find $x_1 \geq 0$ and $x_2 \geq 0$ such that

$$10x_1 + 20x_2 \leq 4$$
$$5x_1 + 10x_2 \leq 5$$

and $z = 100x_1 + 150x_2$ is maximized.

The initial simplex tableau is as follows.

$$\begin{array}{ccccc} x_1 & x_2 & s_1 & s_2 & z \\ \left[\begin{array}{ccccc|c} 10 & \boxed{20} & 1 & 0 & 0 & 4 \\ 5 & 10 & 0 & 1 & 0 & 5 \\ \hline -100 & -150 & 0 & 0 & 1 & 0 \end{array}\right] \end{array}$$

Pivot on the 20 in row 1, column 2.

$$\begin{array}{c} \\ \\ -R_1 + 2R_1 \rightarrow R_2 \\ 15R_1 + 2R_3 \rightarrow R_3 \end{array} \begin{array}{ccccc} x_1 & x_2 & s_1 & s_2 & z \\ \left[\begin{array}{ccccc|c} \boxed{10} & 20 & 1 & 0 & 0 & 4 \\ 0 & 0 & -1 & 2 & 0 & 6 \\ \hline -50 & 0 & 15 & 0 & 2 & 60 \end{array}\right] \end{array}$$

Pivot on the 10 in row 1, column 1.

$$\begin{array}{c} \\ \\ 5R_1 + R_3 \rightarrow R_3 \end{array} \begin{array}{ccccc} x_1 & x_2 & s_1 & s_2 & z \\ \left[\begin{array}{ccccc|c} 10 & 20 & 1 & 0 & 0 & 4 \\ 0 & 0 & -1 & 2 & 0 & 6 \\ \hline 0 & 100 & 20 & 0 & 2 & 80 \end{array}\right] \end{array}$$

Create a 1 in the columns corresponding to x_1, s_2, and z.

$$\begin{array}{c} \frac{1}{10}R_1 \rightarrow R_1 \\ \frac{1}{2}R_2 \rightarrow R_2 \\ \frac{1}{2}R_3 \rightarrow R_3 \end{array} \begin{array}{ccccc} x_1 & x_2 & s_1 & s_2 & z \\ \left[\begin{array}{ccccc|c} 1 & 2 & \frac{1}{10} & 0 & 0 & \frac{2}{5} \\ 0 & 0 & -\frac{1}{2} & 1 & 0 & 3 \\ \hline 0 & 50 & 10 & 0 & 1 & 40 \end{array}\right] \end{array}$$

The minimum value of w is 40 when $y_1 = 10$ and $y_2 = 0$. (These values of y_1 and y_2 are read from the last row of the columns corresponding to s_1 and s_2 in the final tableau.)

13. Minimize $w = 2y_1 + y_2 + 3y_3$
   subject to: $y_1 + y_2 + y_3 \geq 100$
   $2y_1 + y_2 \geq 50$
   with $y_1 \geq 0, y_2 \geq 0, y_3 \geq 0.$

Write the augmented matrix.

$$\left[\begin{array}{ccc|c} 1 & 1 & 1 & 100 \\ 2 & 1 & 0 & 50 \\ \hline 2 & 1 & 3 & 0 \end{array}\right]$$

Form the transpose of this matrix.

$$\left[\begin{array}{cc|c} 1 & 2 & 2 \\ 1 & 1 & 1 \\ 1 & 0 & 3 \\ \hline 100 & 50 & 0 \end{array}\right]$$

The dual problem is as follows.

Maximize $z = 100x_1 + 50x_2$
subject to: $x_1 + 2x_2 \leq 2$
$$x_1 + x_2 \leq 1$$
$$x_1 \leq 3$$
with $x_1 \geq 0, x_2 \geq 0.$

The initial simplex tableau is as follows.

$$
\begin{array}{cccccc|c}
x_1 & x_2 & s_1 & s_2 & s_3 & z & \\
1 & 2 & 1 & 0 & 0 & 0 & 2 \\
\boxed{1} & 1 & 0 & 1 & 0 & 0 & 1 \\
1 & 0 & 0 & 0 & 1 & 0 & 3 \\
\hline
-100 & -50 & 0 & 0 & 0 & 1 & 0
\end{array}
$$

Pivot on the 1 in row 2, column 1.

$$
\begin{array}{l}
-R_2 + R_1 \to R_1 \\
\\
-R_2 + R_3 \to R_3 \\
100R_2 + R_4 \to R_4
\end{array}
\begin{array}{cccccc|c}
x_1 & x_2 & s_1 & s_2 & s_3 & z & \\
0 & 1 & 1 & -1 & 0 & 0 & 1 \\
1 & 1 & 0 & 1 & 0 & 0 & 1 \\
0 & -1 & 0 & -1 & 1 & 0 & 2 \\
0 & 50 & 0 & 100 & 0 & 1 & 100
\end{array}
$$

The minimum value of w is 100 when $y_1 = 0, y_2 = 100$, and $y_3 = 0$.

15. Minimize $z = x_1 + 2x_2$
subject to: $-2x_1 + x_2 \geq 1$
$$x_1 - 2x_2 \geq 1$$
with $x_1 \geq 0, x_2 \geq 0.$

A quick sketch of the constraints $-2x_1 + x_2 \geq 1$ and $x_1 - 2x_2 \geq 1$ will verify that the two corresponding half planes do not overlap in the first quadrant of the x_1x_2-plane. Therefore, this problem (P) has no feasible solution. The dual of the given problem is as follows:

Maximize $w = y_1 + y_2$
subject to: $-2y_1 + y_2 \leq 1$
$$y_1 - 2y_2 \leq 2$$
with $y_1 \geq 0, y_2 \geq 0.$

A quick sketch here will verify that there is a feasible region in the y_1y_2-plane, and it is unbounded. Therefore, there is no maximum value of w in this problem (D).

(P) has no feasible solution and the objective function of (D) is unbounded; this is choice (a).

17. (a) Let $y_1 =$ the number of small test tubes
and $y_2 =$ the number of large test tubes.

Minimize $w = 18y_1 + 15y_2$
Subject to: $y_1 \geq 900$
$$y_2 \geq 600$$
$$y_1 + y_2 \geq 2700$$
$$y_1 \geq 2y_2$$
with $y_1 \geq 0, y_2 \geq 0.$

The last constraint can be written as

$$y_1 - 2y_2 \geq 0.$$

Write the augmented matrix for this problem.

$$\left[\begin{array}{cc|c} 1 & 0 & 900 \\ 0 & 1 & 600 \\ 1 & 1 & 2700 \\ 1 & -2 & 0 \\ \hline 18 & 15 & 0 \end{array}\right]$$

Transpose to get the matrix for the dual problem.

$$\left[\begin{array}{cccc|c} 1 & 0 & 1 & 1 & 18 \\ 0 & 1 & 1 & -2 & 15 \\ \hline 900 & 600 & 2700 & 0 & 0 \end{array}\right]$$

Write the dual problem.

Maximize $z = 900x_1 + 600x_2 + 2700x_3$
Subject to: $x_1 \qquad + x_3 + x_4 \ \le 18$
$\qquad\qquad x_2 + x_3 - 2x_4 \le 15$
with $\qquad x_1 \ge 0, x_2 \ge 0, x_3 \ge 0, x_4 \ge 0$

Write the initial simplex tableau.

$$\begin{array}{ccccccc} x_1 & x_2 & x_3 & x_4 & s_1 & s_2 & z \end{array}$$
$$\left[\begin{array}{ccccccc|c} 1 & 0 & 1 & 1 & 1 & 0 & 0 & 18 \\ 0 & 1 & \boxed{1} & -2 & 0 & 1 & 0 & 15 \\ \hline -900 & -600 & -2700 & 0 & 0 & 0 & 1 & 0 \end{array}\right]$$

Pivot on the 1 in row 2, column 3.

$$\begin{array}{ccccccc} x_1 & x_2 & x_3 & x_4 & s_1 & s_2 & z \end{array}$$
$$\begin{array}{r} -R_2+R_1 \rightarrow R_1 \\ \\ 2700R_2+R_3 \rightarrow R_3 \end{array} \left[\begin{array}{ccccccc|c} 1 & -1 & 0 & \boxed{3} & 1 & -1 & 0 & 3 \\ 0 & 1 & 1 & -2 & 0 & 1 & 0 & 15 \\ \hline -900 & 2100 & 0 & -5400 & 0 & 2700 & 1 & 40{,}500 \end{array}\right]$$

Pivot on the 3 in row 1, column 4.

$$\begin{array}{ccccccc} x_1 & x_2 & x_3 & x_4 & s_1 & s_2 & z \end{array}$$
$$\begin{array}{r} \\ 2R_1+3R_2 \rightarrow R_2 \\ 1800R_1+R_3 \rightarrow R_3 \end{array} \left[\begin{array}{ccccccc|c} 1 & -1 & 0 & 3 & 1 & -1 & 0 & 3 \\ 2 & 1 & 3 & 0 & 2 & 1 & 0 & 51 \\ 900 & 300 & 0 & 0 & 1800 & 900 & 1 & 45{,}900 \end{array}\right]$$

Create a 1 in the columns corresponding to x_3 and x_4.

$$\begin{array}{ccccccc} x_1 & x_2 & x_3 & x_4 & s_1 & s_2 & z \end{array}$$
$$\begin{array}{r} \frac{1}{3}R_1 \rightarrow R_1 \\ \frac{1}{3}R_2 \rightarrow R_2 \\ \\ \end{array} \left[\begin{array}{ccccccc|c} \frac{1}{3} & -\frac{1}{3} & 0 & 1 & \frac{1}{3} & -\frac{1}{3} & 0 & 1 \\ \frac{2}{3} & \frac{1}{3} & 1 & 0 & \frac{2}{3} & \frac{1}{3} & 0 & 17 \\ \hline 900 & 300 & 0 & 0 & 1800 & 900 & 1 & 45{,}900 \end{array}\right]$$

The minimum cost is 45,900¢, or \$459, when 1800 small test tubes and 900 test tubes are ordered.

(b) The shadow cost for the test tubes is \$0.17. An increase in the minimum number of test tubes by $(3000 - 2700) = 300$ will increase the cost to

$$\$459 + \$0.17(300) = \$510.$$

19. (a) Maximize $x_1 + 1.5x_2 = z$

subject to: $x_1 + 2x_2 \leq 200$

$4x_1 + 3x_2 \leq 600$

$0 \leq x_2 \leq 90$

with $x_1 \geq 0.$

(b) Write the initial tableau.

$$
\begin{array}{cccccc|c}
x_1 & x_2 & s_1 & s_2 & s_3 & z & \\
\hline
1 & 2 & 1 & 0 & 0 & 0 & 200 \\
4 & 3 & 0 & 1 & 0 & 0 & 600 \\
0 & \boxed{1} & 0 & 0 & 1 & 0 & 90 \\
\hline
-1 & -1.5 & 0 & 0 & 0 & 1 & 0
\end{array}
$$

Pivot on the 1 in row 3, column 2.

$$
\begin{array}{l}
-2R_3 + R_1 \rightarrow R_1 \\
-3R_3 + R_2 \rightarrow R_2 \\
\\
1.5R_3 + R_4 \rightarrow R_4
\end{array}
\begin{array}{cccccc|c}
x_1 & x_2 & s_1 & s_2 & s_3 & z & \\
\boxed{1} & 0 & 1 & 0 & -2 & 0 & 20 \\
4 & 0 & 0 & 1 & -3 & 0 & 330 \\
0 & 1 & 0 & 0 & 1 & 0 & 90 \\
\hline
-1 & 0 & 0 & 0 & 1.5 & 1 & 135
\end{array}
$$

Pivot on the 1 in row 1, column 1.

$$
\begin{array}{l}
\\
-4R_1 + R_2 \rightarrow R_2 \\
\\
R_1 + R_4 \rightarrow R_4
\end{array}
\begin{array}{cccccc|c}
x_1 & x_2 & s_1 & s_2 & s_3 & z & \\
1 & 0 & 1 & 0 & -2 & 0 & 20 \\
0 & 0 & -4 & 1 & \boxed{5} & 0 & 250 \\
0 & 1 & 0 & 0 & 1 & 0 & 90 \\
\hline
0 & 0 & 1 & 0 & -0.5 & 1 & 155
\end{array}
$$

Pivot on the 5 in row 2, column 5.

$$
\begin{array}{l}
\frac{2}{5}R_2 + R_1 \rightarrow R_1 \\
\frac{1}{5}R_2 \rightarrow R_2 \\
\\
-\frac{1}{5}R_2 + R_3 \rightarrow R_3 \\
\\
\frac{1}{10}R_2 + R_4 \rightarrow R_4
\end{array}
\begin{array}{cccccc|c}
x_1 & x_2 & s_1 & s_2 & s_3 & z & \\
1 & 0 & -\frac{3}{5} & \frac{2}{5} & 0 & 0 & 120 \\
0 & 0 & -\frac{4}{5} & \frac{1}{5} & 1 & 0 & 50 \\
0 & 1 & \frac{4}{5} & -\frac{1}{5} & 0 & 0 & 40 \\
\hline
0 & 0 & 0.6 & 0.1 & 0 & 1 & 180
\end{array}
$$

The maximum profit is $180 when $x_1 = 120$ and $x_2 = 40$, that is, when 120 bears and 40 monkeys are produced.

(c) The corresponding dual problem is as follows:

Minimize $w = 200y_1 + 600y_2 + 90y_3$

subject to: $y_1 + 4y_2 \qquad \geq 1$

$2y_1 + 3y_2 + y_3 \geq 1.5$

with $y_1 \geq 0, y_2 \geq 0, y_3 \geq 0.$

(d) From the given final tableau, the optimal solution to the dual problem is $y_1 = 0.6, y_2 = 0.1, y_3 = 0$, and $w = 180$.

(e) The shadow value for felt is 0.6; an increase in supply of 10 units of felt will increase profit to

$$\$180 + 0.6(10) = \$186.$$

(f) The shadow values are 0.1 for stuffing and 0 for trim. If stuffing and trim are each decreased by 10 units, the profit will be

$$\$180 - 0.1(10) - 0(10) = \$179.$$

21. (a) Let y_1 = the number of grams of soybean meal,

y_2 = the number of grams of meat byproducts,

and y_3 = the number of grams of grain.

Minimize $w = 8y_1 + 9y_2 + 10y_3$

subject to: $2.5y_1 + 4.5y_2 + 5y_3 \geq 54$

$5y_1 + 3y_2 + 10y_3 \geq 60$

with $y_1 \geq 0, y_2 \geq 0, y_3 \geq 0.$

Write the augmented matrix for this problem.

$$\begin{bmatrix} 2.5 & 4.5 & 5 & | & 54 \\ 5 & 3 & 10 & | & 60 \\ \hline 8 & 9 & 10 & | & 0 \end{bmatrix}$$

Transpose to get the matrix for the dual problem.

$$\begin{bmatrix} 2.5 & 5 & | & 8 \\ 4.5 & 3 & | & 9 \\ 5 & 10 & | & 10 \\ \hline 54 & 60 & | & 0 \end{bmatrix}$$

Write the dual problem.

Maximize $z = 54x_1 + 60x_2$

subject to: $2.5x_1 + 5x_2 \leq 8$

$4.5x_1 + 3x_2 \leq 9$

$5x_1 + 10x_2 \leq 10$

with $x_1 \geq 0, x_2 \geq 0.$

Write the initial tableau.

$$\begin{array}{cccccc} x_1 & x_2 & s_1 & s_2 & s_3 & z \\ \end{array}$$
$$\begin{bmatrix} 2.5 & 5 & 1 & 0 & 0 & 0 & | & 8 \\ 4.5 & 3 & 0 & 1 & 0 & 0 & | & 9 \\ 5 & \boxed{10} & 0 & 0 & 1 & 0 & | & 10 \\ \hline -54 & -60 & 0 & 0 & 0 & 1 & | & 0 \end{bmatrix}$$

To eliminate the decimal entries, multiply rows 1 and 2 by 2.

$$\begin{array}{ccccccc} x_1 & x_2 & s_1 & s_2 & s_3 & z \\ \end{array}$$
$$\begin{bmatrix} 5 & 10 & 2 & 0 & 0 & 0 & | & 16 \\ 9 & 6 & 0 & 2 & 0 & 0 & | & 18 \\ 5 & \boxed{10} & 0 & 0 & 1 & 0 & | & 10 \\ \hline -54 & -60 & 0 & 0 & 0 & 1 & | & 0 \end{bmatrix}$$

Pivot on the 10 in row 3, column 2.

$$\begin{array}{l} -R_3 + R_1 \to R_1 \\ -3R_3 + 5R_2 \to R_2 \\ \\ 6R_3 + R_4 \to R_4 \end{array} \begin{array}{cccccc} x_1 & x_2 & s_1 & s_2 & s_3 & z \\ \end{array}$$
$$\begin{bmatrix} 0 & 0 & 2 & 0 & -1 & 0 & | & 6 \\ \boxed{30} & 0 & 0 & 10 & -3 & 0 & | & 60 \\ 5 & 10 & 0 & 0 & 1 & 0 & | & 10 \\ \hline -24 & 0 & 0 & 0 & 6 & 1 & | & 60 \end{bmatrix}$$

Pivot on the 30 in row 2, column 1.

$$\begin{array}{cccccc} x_1 & x_2 & s_1 & s_2 & s_3 & z \\ \end{array}$$
$$\begin{array}{l} \\ \\ -R_2 + 6R_3 \to R_3 \\ 4R_2 + 5R_4 \to R_4 \end{array} \begin{bmatrix} 0 & 0 & 2 & 0 & -1 & 0 & | & 6 \\ 30 & 0 & 0 & 10 & -3 & 0 & | & 60 \\ 0 & 60 & 0 & -10 & 9 & 0 & | & 0 \\ \hline 0 & 0 & 0 & 40 & 18 & 5 & | & 540 \end{bmatrix}$$

Create a 1 in the columns representing $x_1, x_2, s_1,$ and z.

$$\begin{array}{cccccc} x_1 & x_2 & s_1 & s_2 & s_3 & z \\ \end{array}$$
$$\begin{array}{l} \frac{1}{2}R_1 \to R_1 \\ \\ \frac{1}{30}R_2 \to R_2 \\ \\ \frac{1}{60}R_3 \to R_3 \\ \\ \frac{1}{5}R_4 \to R_4 \end{array} \begin{bmatrix} 0 & 0 & 1 & 0 & -\frac{1}{2} & 0 & | & 3 \\ 1 & 0 & 0 & \frac{1}{3} & -\frac{1}{10} & 0 & | & 2 \\ 0 & 1 & 0 & -\frac{1}{6} & \frac{3}{20} & 0 & | & 0 \\ \hline 0 & 0 & 0 & 8 & 3.6 & 1 & | & 108 \end{bmatrix}$$

The minimum cost is obtained when 0g of soybean meal, 8g of meat by products, and 3.6g of grain are used, or 0g of soybean meal, 0g of meat by products and 10.8g of grain are used.

(b) The minimum cost is $1.08.

(c) After the initial pivot, the tableau is

$$\begin{array}{cccccc} x_1 & x_2 & s_1 & s_2 & s_3 & z \\ \end{array}$$
$$\begin{bmatrix} 0 & 0 & 2 & 0 & -1 & 0 & | & 6 \\ 30 & 0 & 0 & 10 & -3 & 0 & | & 60 \\ \boxed{5} & 10 & 0 & 0 & 1 & 0 & | & 10 \\ \hline -24 & 0 & 0 & 0 & 6 & 1 & | & 60 \end{bmatrix}$$

Now pivot on the 5 in row 3, column 1.

$$\begin{array}{cccccc} x_1 & x_2 & s_1 & s_2 & s_3 & z \\ \end{array}$$
$$\begin{array}{l} \\ -6R_3 + R_2 \to R_2 \\ \\ 24R_3 + 5R_4 \to R_4 \end{array} \begin{bmatrix} 0 & 0 & 2 & 0 & -1 & 0 & | & 6 \\ 0 & -60 & 0 & 10 & -9 & 0 & | & 0 \\ 5 & 10 & 0 & 0 & 1 & 0 & | & 10 \\ \hline 0 & 240 & 0 & 0 & 54 & 5 & | & 540 \end{bmatrix}$$

$$\begin{array}{cccccc} x_1 & x_2 & s_1 & s_2 & s_3 & z \\ \end{array}$$
$$\begin{array}{l} \\ \\ \\ \frac{1}{5}R_4 \to R_4 \end{array} \begin{bmatrix} 0 & 0 & 2 & 0 & -1 & 0 & | & 6 \\ 0 & -60 & 0 & 10 & -9 & 0 & | & 0 \\ 5 & 10 & 0 & 0 & 1 & 0 & | & 10 \\ \hline 0 & 48 & 0 & 0 & 10.8 & 1 & | & 108 \end{bmatrix}$$

The minimum cost is $108 when $y_1 = 0, y_2 = 8$ and $y_3 = 10.8$, that is, when 0 grams of soybean meal, 0 grams of meat by-products, and 10.8 grams of grain are mixed.

23. Let $y_1 =$ the number of large bowls.

$y_2 =$ the number of small bowls.

$y_3 =$ the number of pots for plants.

Minimize $w = 5y_1 + 6y_2 + 4y_3$

subject to: $3y_1 + 2y_2 + 4y_3 \geq 72$

$6y_1 + 6y_2 + 2y_3 \geq 108$

with $y_1 \geq 0, y_2 \geq 0, y_3 \geq 0.$

Write the augmented matrix for this problem.

$$\begin{bmatrix} 3 & 2 & 4 & 72 \\ 6 & 6 & 2 & 108 \\ \hline 5 & 6 & 4 & 0 \end{bmatrix}$$

Transpose to get the matrix for the dual problem.

$$\begin{bmatrix} 3 & 6 & 5 \\ 2 & 6 & 6 \\ 4 & 2 & 4 \\ \hline 72 & 108 & 0 \end{bmatrix}$$

Write the dual problem.

Maximize $z = 72x_1 + 108x_2$

subject to: $3x_1 + 6x_2 \leq 5$

$2x_1 + 6x_2 \leq 6$

$4x_1 + 2x_2 \leq 4$

with $x_1 \geq 0, x_2 \geq 0.$

Write the initial tableau.

$$\begin{array}{cccccc} x_1 & x_2 & s_1 & s_2 & s_3 & z \\ \end{array}$$
$$\begin{bmatrix} 3 & \boxed{6} & 1 & 0 & 0 & 0 & 5 \\ 2 & 6 & 0 & 1 & 0 & 0 & 6 \\ 4 & 2 & 0 & 0 & 1 & 0 & 4 \\ \hline -72 & -108 & 0 & 0 & 0 & 1 & 0 \end{bmatrix}$$

Pivot on the 6 in row 1, column 2.

$$\begin{array}{c} \frac{1}{6}R_1 \rightarrow R_1 \\ -R_1 + R_2 \rightarrow R_2 \\ -\frac{1}{3}R_1 + R_3 \rightarrow R_3 \\ 18R_1 + R_4 \rightarrow R_4 \end{array} \begin{array}{ccccccc} x_1 & x_2 & s_1 & s_2 & s_3 & z \\ \frac{1}{2} & 1 & \frac{1}{6} & 0 & 0 & 0 & \frac{5}{6} \\ -1 & 0 & -1 & 1 & 0 & 0 & 1 \\ \boxed{3} & 0 & -\frac{1}{3} & 0 & 1 & 0 & \frac{7}{3} \\ \hline -18 & 0 & 18 & 0 & 0 & 1 & 90 \end{array}$$

Pivot on the 3 in row 3, column 1.

$$\begin{array}{c} -\frac{1}{6}R_3 + R_1 \rightarrow R_1 \\ \frac{1}{3}R_3 + R_2 \rightarrow R_2 \\ \frac{1}{3}R_3 \rightarrow R_3 \\ 6R_3 + R_4 \rightarrow R_4 \end{array} \begin{array}{ccccccc} x_1 & x_2 & s_1 & s_2 & s_3 & z \\ 0 & 1 & \frac{2}{9} & 0 & -\frac{1}{6} & 0 & \frac{4}{9} \\ 0 & 0 & -\frac{10}{9} & 1 & \frac{1}{3} & 0 & \frac{16}{9} \\ 1 & 0 & -\frac{1}{9} & 0 & \frac{1}{3} & 0 & \frac{7}{9} \\ \hline 0 & 0 & 16 & 0 & 6 & 1 & 104 \end{array}$$

The minimum time is 104 hours when $y_1 = 16$, $y_2 = 0$, and $y_3 = 6$, that is, when 16 large bowls, 0 small bowls, and 6 pots for flowers are made.

25. Let $y_1 =$ the number of #1 pills

and $y_2 =$ the number of #2 pills.

Organize the given information in a table.

	Vitamin A	Vitamin B_1	Vitamin C	Cost
#1	8	1	2	$0.10
#2	2	1	7	$0.20
Total Needed	16	5	20	

The problem is:

Minimize $w = 0.1y_1 + 0.2y_2$

subject to: $8y_1 + 2y_2 \geq 16$

$y_1 + y_2 \geq 5$

$2y_1 + 7y_2 \geq 20$

with $y_1 \geq 0, y_2 \geq 0 .$

The dual problem is as follows.

Maximize $z = 16x_1 + 5x_2 + 20x_3$

subject to: $8x_1 + x_2 + 2x_3 \leq 0.1$

$2x_1 + x_2 + 7x_3 \leq 0.2$

with $x_1 \geq 0, x_2 \geq 0.$

The initial tableau is as follows.

$$\begin{array}{cccccc} x_1 & x_2 & x_3 & s_1 & s_2 & z \\ \end{array}$$
$$\begin{bmatrix} 8 & 1 & 2 & 1 & 0 & 0 & 0.1 \\ 2 & 1 & \boxed{7} & 0 & 1 & 0 & 0.2 \\ \hline -16 & -5 & -20 & 0 & 0 & 1 & 0 \end{bmatrix}$$

Pivot as indicated.

$$\begin{array}{c} -2R_2 + 7R_1 \rightarrow R_1 \\ \\ 20R_2 + 7R_3 \rightarrow R_3 \end{array} \begin{array}{cccccc} x_1 & x_2 & x_3 & s_1 & s_2 & z \\ \boxed{52} & 5 & 0 & 7 & -2 & 0 & 0.3 \\ 2 & 1 & 7 & 0 & 1 & 0 & 0.2 \\ \hline -72 & -15 & 0 & 0 & 20 & 7 & 4 \end{array}$$

$$\begin{array}{c} \\ -R_1 + 26R_2 \rightarrow R_2 \\ 18R_1 + 13R_3 \rightarrow R_3 \end{array} \begin{array}{cccccc} x_1 & x_2 & x_3 & s_1 & s_2 & z \\ 52 & \boxed{5} & 0 & 7 & -2 & 0 & 0.3 \\ 0 & 21 & 182 & -7 & 28 & 0 & 4.9 \\ 0 & -105 & 0 & 126 & 224 & 91 & 57.4 \end{array}$$

$$\begin{array}{cc} & \begin{array}{cccccc} x_1 & x_2 & x_3 & s_1 & s_2 & z \end{array} \\ \begin{array}{c} \\ -21R_1 + 5R_2 \to R_2 \\ 21R_1 + R_3 \to R_3 \end{array} & \left[\begin{array}{cccccc|c} 52 & 5 & 0 & 7 & -2 & 0 & 0.3 \\ -1092 & 0 & 910 & -182 & 182 & 0 & 18.2 \\ \hline 1092 & 0 & 0 & 273 & 182 & 91 & 63.7 \end{array} \right] \end{array}$$

Create a 1 in the columns corresponding to $x_2, x_3,$ and z.

$$\begin{array}{cc} & \begin{array}{cccccc} x_1 & x_2 & x_3 & s_1 & s_2 & z \end{array} \\ \begin{array}{c} \frac{1}{5}R_1 \to R_1 \\[4pt] \frac{1}{910}R_2 \to R_2 \\[4pt] \frac{1}{91}R_3 \to R_3 \end{array} & \left[\begin{array}{cccccc|c} \frac{52}{5} & 1 & 0 & \frac{7}{5} & -\frac{2}{5} & 0 & 0.06 \\[4pt] -\frac{6}{5} & 0 & 1 & -\frac{1}{5} & \frac{1}{5} & 0 & 0.02 \\[4pt] \hline 12 & 0 & 0 & 3 & 2 & 1 & 0.7 \end{array} \right] \end{array}$$

From the last row, the minimum value is 0.7 when $y_1 = 3$ and $y_2 = 2$. Mark should buy 3 of pill #1 and 2 of pill #2 for a minimum cost of 70¢.

4.4   Nonstandard Problems

1. $2x_1 + 3x_2 \le 8$
   $x_1 + 4x_2 \ge 7$

Introduce the slack variable s_1 and the surplus variable s_2 to obtain the following equations:

$$\begin{aligned} 2x_1 + 3x_2 + s_1 \quad\ \ &= 8 \\ x_1 + 4x_2 \quad\ \ - s_2 &= 7. \end{aligned}$$

3. $2x_1 + x_2 + 2x_3 \le 50$
   $x_1 + 3x_2 + x_3 \ge 35$
   $x_1 + 2x_2 \ge 15$

Introduce the slack variable s_1 and the surplus variables s_2 and s_3 to obtain the following equations:

$$\begin{aligned} 2x_1 + x_2 + 2x_3 + s_1 \qquad\qquad\ &= 50 \\ x_1 + 3x_2 + x_3 \qquad - s_2 \qquad &= 35 \\ x_1 + 2x_2 \qquad\qquad\quad - s_3 &= 15. \end{aligned}$$

5. Minimize $w = 3y_1 + 4y_2 + 5y_3$
   subject to: $y_1 + 2y_2 + 3y_3 \ge 9$
   $y_2 + 2y_3 \ge 8$
   $2y_1 + y_2 + 2y_3 \ge 6$
   with $y_1 \ge 0, y_2 \ge 0, y_3 \ge 0.$

Change this to a maximization problem by letting $z = -w$. The problem can now be stated equivalently as follows:

Maximize $z = -3y_1 - 4y_2 - 5y_3$
subject to: $y_1 + 2y_2 + 3y_3 \ge 9$
$y_2 + 2y_3 \ge 8$
$2y_1 + y_2 + 2y_3 \ge 6$
with $y_1 \ge 0, y_2 \ge 0, y_3 \ge 0.$

7. Minimize $\quad w = y_1 + 2y_2 + y_3 + 5y_4$
subject to: $\quad y_1 + y_2 + \; y_3 + y_4 \geq \; 50$
$\qquad\qquad 3y_1 + y_2 + 2y_3 + y_4 \geq 100$
with $\qquad\quad y_1 \geq 0, y_2 \geq 0, y_3 \geq 0, y_4 \geq 0.$

Change this to a maximization problem by letting $z = -w$. The problem can now be stated equivalently as follows:

Maximize $\quad z = -y_1 - 2y_2 - y_3 - 5y_4$
subject to: $\quad y_1 + y_2 + \; y_3 + y_4 \geq \; 50$
$\qquad\qquad 3y_1 + y_2 + 2y_3 + y_4 \geq 100$
with $\qquad\quad y_1 \geq 0, y_2 \geq 0, y_3 \geq 0, y_4 \geq 0.$

9. Find $x_1 \geq 0$ and $x_2 \geq 0$ such that

$$x_1 + 2x_2 \geq 24$$
$$x_1 + \; x_2 \leq 40$$

and $z = 12x_1 + 10x_2$ is maximized.

Subtracting the surplus variable s_1 and adding the slack variable s_2 leads to the equations

$$x_1 + 2x_2 - s_1 \qquad = 24$$
$$x_1 + \; x_2 \qquad + s_2 = 40.$$

The initial simplex tableau is as follows.

$$\begin{array}{ccccc} x_1 & x_2 & s_1 & s_2 & z \\ \end{array}$$
$$\left[\begin{array}{ccccc|c} \boxed{1} & 2 & -1 & 0 & 0 & 24 \\ 1 & 1 & 0 & 1 & 0 & 40 \\ \hline -12 & -10 & 0 & 0 & 1 & 0 \end{array}\right]$$

The initial basic solution is not feasible since $s_1 = -24$ is negative, so row transformations must be used. Pivot on the 1 in row 1, column 1, since it is the positive entry that is farthest to the left in the first row (the row containing the -1) and since, in the first column, $\frac{24}{1} = 24$ is a smaller quotient than $\frac{40}{1} = 40$. After row transformations, we obtain the following tableau.

$$\begin{array}{ccccc} & x_1 & x_2 & s_1 & s_2 & z \\ \end{array}$$
$$\begin{array}{c} \\ -R_1 + R_2 \to R_2 \\ \\ 12R_1 + R_3 \to R_3 \end{array} \left[\begin{array}{ccccc|c} 1 & 2 & -1 & 0 & 0 & 24 \\ 0 & -1 & \boxed{1} & 1 & 0 & 16 \\ \hline 0 & 14 & -12 & 0 & 1 & 288 \end{array}\right]$$

The basic solution is now feasible, but the problem is not yet finished since there is a negative indicator. Continue in the usual way. The 1 in column 3 is the next pivot. After row transformations, we get the following tableau.

$$\begin{array}{ccccc} & x_1 & x_2 & s_1 & s_2 & z \\ \end{array}$$
$$\begin{array}{c} R_1 + R_2 \to R_1 \\ \\ 12R_2 + R_3 \to R_3 \end{array} \left[\begin{array}{ccccc|c} 1 & 1 & 0 & 1 & 0 & 40 \\ 0 & -1 & 1 & 1 & 0 & 16 \\ \hline 0 & 2 & 0 & 12 & 1 & 480 \end{array}\right]$$

This is a final tableau since the entries in the last row are all nonnegative. The maximum value is 480 when $x_1 = 40$ and $x_2 = 0$.

11. Find $x_1 \geq 0, x_2 \geq 0$, and $x_3 \geq 0$ such that

$$x_1 + x_2 + x_3 \leq 150$$
$$x_1 + x_2 + x_3 \geq 100$$

and $z = 2x_1 + 5x_2 + 3x_3$ is maximized.

The initial tableau is as follows.

$$\begin{array}{cccccc} x_1 & x_2 & x_3 & s_1 & s_2 & z \\ \end{array}$$
$$\left[\begin{array}{cccccc|c} 1 & 1 & 1 & 1 & 0 & 0 & 150 \\ \boxed{1} & 1 & 1 & 0 & -1 & 0 & 100 \\ \hline -2 & -5 & -3 & 0 & 0 & 1 & 0 \end{array}\right]$$

Note that s_1 is a slack variable, while s_2 is a surplus variable. The initial basic solution is not feasible, since $s_2 = -100$ is negative. Pivot on the 1 in row 2, column 1.

$$\begin{array}{cccccc} & x_1 & x_2 & x_3 & s_1 & s_2 & z \\ \end{array}$$
$$\begin{array}{c} -R_2 + R_1 \to R_1 \\ \\ 2R_2 + R_3 \to R_3 \end{array} \left[\begin{array}{cccccc|c} 0 & 0 & 0 & 1 & 1 & 0 & 50 \\ 1 & \boxed{1} & 1 & 0 & -1 & 0 & 100 \\ \hline 0 & -3 & -1 & 0 & -2 & 1 & 200 \end{array}\right]$$

Pivot on the 1 in row 2, column 2.

$$\begin{array}{cccccc} & x_1 & x_2 & x_3 & s_1 & s_2 & z \\ \end{array}$$
$$\begin{array}{c} \\ \\ 3R_2 + R_3 \to R_3 \end{array} \left[\begin{array}{cccccc|c} 0 & 0 & 0 & 1 & \boxed{1} & 0 & 50 \\ 1 & 1 & 1 & 0 & -1 & 0 & 100 \\ \hline 3 & 0 & 2 & 0 & -5 & 1 & 500 \end{array}\right]$$

Pivot on the 1 in row 1, column 5.

$$\begin{array}{cccccc} & x_1 & x_2 & x_3 & s_1 & s_2 & z \\ \end{array}$$
$$\begin{array}{c} \\ R_1 + R_2 \to R_2 \\ 5R_1 + R_3 \to R_3 \end{array} \left[\begin{array}{cccccc|c} 0 & 0 & 0 & 1 & 1 & 0 & 50 \\ 1 & 1 & 1 & 1 & 0 & 0 & 150 \\ \hline 3 & 0 & 2 & 5 & 0 & 1 & 750 \end{array}\right]$$

This is a final tableau. The maximum value is 750 when $x_1 = 0, x_2 = 150$, and $x_3 = 0$.

13. Find $x_1 \geq 0$ and $x_2 \geq 0$ such that

$$
\begin{aligned}
x_1 + \ x_2 &\leq 100 \\
2x_1 + 3x_2 &\leq \ 75 \\
x_1 + 4x_2 &\geq \ 50
\end{aligned}
$$

and $z = 5x_1 - 3x_2$ is maximized.

The initial simplex tableau is

$$
\begin{array}{cccccc}
x_1 & x_2 & s_1 & s_2 & s_3 & z \\
\end{array}
$$

$$
\left[
\begin{array}{cccccc|c}
1 & 1 & 1 & 0 & 0 & 0 & 100 \\
\boxed{2} & 3 & 0 & 1 & 0 & 0 & 75 \\
1 & 4 & 0 & 0 & -1 & 0 & 50 \\
\hline
-5 & 3 & 0 & 0 & 0 & 1 & 0
\end{array}
\right]
$$

The initial basic solution is not feasible since $s_3 = -50$. Pivot on the 2 in row 2, column 1.

$$
\begin{array}{cccccc}
 & x_1 & x_2 & x_3 & s_1 & s_2 & z \\
\end{array}
$$

$$
\begin{array}{c}
-R_2 + 2R_1 \to R_1 \\
\\
-R_2 + 2R_3 \to R_3 \\
\\
5R_2 + 2R_4 \to R_4
\end{array}
\left[
\begin{array}{cccccc|c}
0 & -1 & 2 & -1 & 0 & 0 & 125 \\
2 & 3 & 0 & 1 & 0 & 0 & 75 \\
0 & \boxed{5} & 0 & -1 & -2 & 0 & 25 \\
\hline
0 & 21 & 0 & 5 & 0 & 2 & 375
\end{array}
\right]
$$

This solution is still not feasible since $s_3 = -\frac{25}{2}$. Pivot on the 5 in row 3, column 2.

$$
\begin{array}{cccccc}
 & x_1 & x_2 & x_3 & s_1 & s_2 & z \\
\end{array}
$$

$$
\begin{array}{c}
R_3 + 5R_1 \to R_1 \\
-3R_3 + 5R_2 \to R_2 \\
\\
-21R_3 + 5R_4 \to R_4
\end{array}
\left[
\begin{array}{cccccc|c}
0 & 0 & 10 & -6 & -2 & 0 & 650 \\
10 & 0 & 0 & 8 & 6 & 0 & 300 \\
0 & \boxed{5} & 0 & -1 & -2 & 0 & 25 \\
\hline
0 & 0 & 0 & 46 & 42 & 10 & 1350
\end{array}
\right]
$$

Create a 1 in the columns corresponding to x_1, x_2, s_1, and z.

$$
\begin{array}{cccccc}
 & x_1 & x_2 & x_3 & s_1 & s_2 & z \\
\end{array}
$$

$$
\begin{array}{c}
\frac{1}{10}R_1 \to R_1 \\
\frac{1}{10}R_2 \to R_2 \\
\frac{1}{5}R_3 \to R_3 \\
\frac{1}{10}R_4 \to R_4
\end{array}
\left[
\begin{array}{cccccc|c}
0 & 0 & 1 & -\frac{3}{5} & -\frac{1}{5} & 0 & 65 \\
1 & 0 & 0 & \frac{4}{5} & \frac{3}{5} & 0 & 30 \\
0 & 1 & 0 & -\frac{1}{5} & -\frac{2}{5} & 0 & 5 \\
\hline
0 & 0 & 0 & \frac{23}{5} & \frac{21}{5} & 1 & 135
\end{array}
\right]
$$

This is a final tableau. The maximum is 135 when $x_1 = 30, x_2 = 5$.

15. Find $y_1 \geq 0, y_2 \geq 0$, and $y_3 \geq 0$ such that

$$
\begin{aligned}
5y_1 + \ 3y_2 + 2y_3 &\leq 150 \\
5y_1 + 10y_2 + 3y_3 &\geq \ 90
\end{aligned}
$$

and $w = 10y_1 + 12y_2 + 10y_3$ is minimized. Let $z = -w = -10y - 12y_2 - 10y_3$. Maximize z. The initial simplex tableau is

$$
\begin{array}{cccccc}
y_1 & y_2 & y_3 & s_1 & s_2 & z \\
\end{array}
$$

$$
\left[
\begin{array}{cccccc|c}
5 & 3 & 2 & 1 & 0 & 0 & 150 \\
\boxed{5} & 10 & 3 & 0 & -1 & 0 & 90 \\
\hline
10 & 12 & 10 & 0 & 0 & 1 & 0
\end{array}
\right]
$$

The initial basic solution is not feasible since $s_2 = -90$. Pivot on the 5 in row 2, column 1.

$$
\begin{array}{cccccc}
 & y_1 & y_2 & y_3 & s_1 & s_2 & z \\
\end{array}
$$

$$
\begin{array}{c}
-R_2 + R_1 \to R_1 \\
\\
-2R_2 + R_3 \to R_3
\end{array}
\left[
\begin{array}{cccccc|c}
0 & -7 & -1 & 1 & 1 & 0 & 60 \\
5 & \boxed{10} & 3 & 0 & -1 & 0 & 90 \\
\hline
0 & -8 & 4 & 0 & 2 & 1 & -180
\end{array}
\right]
$$

Pivot on the 10 in row 2, column 2.

$$
\begin{array}{cccccc}
 & y_1 & y_2 & y_3 & s_1 & s_2 & z \\
\end{array}
$$

$$
\begin{array}{c}
7R_2 + 10R_1 \to R_1 \\
\\
8R_2 + 10R_3 \to R_3
\end{array}
\left[
\begin{array}{cccccc|c}
35 & 0 & 11 & 10 & 3 & 0 & 1230 \\
5 & 10 & 3 & 0 & -1 & 0 & 90 \\
\hline
40 & 0 & 64 & 0 & 12 & 10 & -1080
\end{array}
\right]
$$

Create a 1 in the columns corresponding to y_2, s_1, and z.

$$
\begin{array}{cccccc}
 & x_1 & x_2 & s_1 & s_2 & s_3 & z \\
\end{array}
$$

$$
\begin{array}{c}
\frac{1}{10}R_1 \to R_1 \\
\frac{1}{10}R_2 \to R_2 \\
\frac{1}{10}R_3 \to R_3
\end{array}
\left[
\begin{array}{cccccc|c}
\frac{7}{2} & 0 & \frac{11}{10} & 1 & \frac{3}{10} & 0 & 123 \\
\frac{1}{2} & 1 & \frac{3}{10} & 0 & -\frac{1}{10} & 0 & 9 \\
\hline
4 & 0 & \frac{32}{5} & 0 & \frac{6}{5} & 1 & -108
\end{array}
\right]
$$

This is a final tableau. The minimum is 108 when $y_1 = 0, y_2 = 9$, and $y_3 = 0$.

17. Maximize $z = 3x_1 + 2x_2$
  subject to: $x_1 + x_2 = 50$
$$4x_1 + 2x_2 \geq 120$$
$$5x_1 + 2x_2 \leq 200$$
  with $x_1 \geq 0, x_2 \geq 0.$

The artificial variable a_1 is used to rewrite $x_1 + x_2 = 50$ as $x_1 + x_2 + a_1 = 50$; note that a_1 must equal 0 for this equation to be a true statement. Also the surplus variable s_1 and the slack variable s_2 are needed. The initial tableau is as follows.

$$
\begin{array}{cccccc|c}
x_1 & x_2 & a_1 & s_1 & s_2 & z & \\
\hline
1 & 1 & 1 & 0 & 0 & 0 & 50 \\
4 & 2 & 0 & -1 & 0 & 0 & 120 \\
5 & 2 & 0 & 0 & 1 & 0 & 200 \\
\hline
-3 & -2 & 0 & 0 & 0 & 1 & 0
\end{array}
$$

The initial basic solution is not feasible. Pivot on the 4 in row 2, column 1.

$$
\begin{array}{l}
-R_2 + 4R_1 \to R_1 \\
\\
-5R_2 + 4R_3 \to R_3 \\
\\
3R_2 + 4R_4 \to R_4
\end{array}
\begin{array}{cccccc|c}
x_1 & x_2 & a_1 & s_1 & s_2 & z & \\
\hline
0 & 2 & 4 & 1 & 0 & 0 & 80 \\
4 & 2 & 0 & -1 & 0 & 0 & 120 \\
0 & -2 & 0 & 5 & 4 & 0 & 200 \\
\hline
0 & -2 & 0 & -3 & 0 & 4 & 360
\end{array}
$$

The basic solution is now feasible, but there are negative indicators. Pivot on the 5 in row 3, column 4 (which is the column with the most negative indicator and the row with the smallest nonnegative quotient).

$$
\begin{array}{l}
-R_3 + 5R_1 \to R_1 \\
R_3 + 5R_2 \to R_2 \\
\\
3R_3 + 5R_4 \to R_4
\end{array}
\begin{array}{cccccc|c}
x_1 & x_2 & a_1 & s_1 & s_2 & z & \\
\hline
0 & 12 & 20 & 0 & -4 & 0 & 200 \\
20 & 8 & 0 & 0 & 4 & 0 & 800 \\
0 & -2 & 0 & 5 & 4 & 0 & 200 \\
\hline
0 & -16 & 0 & 0 & 12 & 20 & 2400
\end{array}
$$

Pivot on the 12 in row 1, column 2.

$$
\begin{array}{l}
\\
-2R_1 + 3R_2 \to R_2 \\
R_1 + 6R_3 \to R_3 \\
4R_1 + 3R_4 \to R_4
\end{array}
\begin{array}{cccccc|c}
x_1 & x_2 & a_1 & s_1 & s_2 & z & \\
\hline
0 & 12 & 20 & 0 & -4 & 0 & 200 \\
60 & 0 & -40 & 0 & 20 & 0 & 2000 \\
0 & 0 & 20 & 30 & 20 & 0 & 1400 \\
\hline
0 & 0 & 80 & 0 & 20 & 60 & 8000
\end{array}
$$

We now have $a_1 = 0$, so drop the a_1 column.

$$
\begin{array}{ccccc|c}
x_1 & x_2 & s_1 & s_2 & z & \\
\hline
0 & 12 & 0 & -4 & 0 & 200 \\
60 & 0 & 0 & 20 & 0 & 2000 \\
0 & 0 & 30 & 20 & 0 & 1400 \\
\hline
0 & 0 & 0 & 20 & 60 & 8000
\end{array}
$$

We are finished pivoting. Create a 1 in the columns corresponding to x_1, x_2, s_1, and z.

$$
\begin{array}{c}
\frac{1}{12}R_1 \to R_1 \\
\frac{1}{60}R_2 \to R_2 \\
\frac{1}{30}R_3 \to R_3 \\
\\
\frac{1}{60}R_4 \to R_4
\end{array}
\left[
\begin{array}{ccccc|c}
x_1 & x_2 & s_1 & s_2 & z & \\
0 & 1 & 0 & -\frac{1}{3} & 0 & \frac{50}{3} \\
1 & 0 & 0 & \frac{1}{3} & 0 & \frac{100}{3} \\
0 & 0 & 1 & \frac{2}{3} & 0 & \frac{140}{3} \\
\hline
0 & 0 & 0 & \frac{1}{3} & 1 & \frac{400}{3}
\end{array}
\right]
$$

The maximum value is $\frac{400}{3}$ when $x_1 = \frac{100}{3}$ and $x_2 = \frac{50}{3}$.

19. Minimize $w = 32y_1 + 40y_2 + 48y_3$
    subject to: $20y_1 + 10y_2 + 5y_3 = 200$
                  $25y_1 + 40y_2 + 50y_3 \le 500$
                  $18y_1 + 24y_2 + 12y_3 \ge 300$
    with $y_1 \ge 0, y_2 \ge 0, y_3 \ge 0$

With artificial, slack, and surplus variables, this problem becomes

Maximize $z = -32y_1 - 40y_2 - 48y_3$
subject to: $20y_1 + 10y_2 + 5y_3 + a_1 \qquad\qquad = 200$
               $25y_1 + 40y_2 + 50y_3 + \qquad s_1 \qquad = 500$
               $18y_1 + 24y_2 + 12y_3 \qquad\qquad - s_2 = 300.$

The initial tableau is as follows.

$$
\left[
\begin{array}{ccccccc|c}
y_1 & y_2 & y_3 & a_1 & s_1 & s_2 & z & \\
\boxed{20} & 10 & 5 & 1 & 0 & 0 & 0 & 200 \\
25 & 40 & 50 & 0 & 1 & 0 & 0 & 500 \\
18 & 24 & 12 & 0 & 0 & -1 & 0 & 300 \\
\hline
32 & 40 & 48 & 0 & 0 & 0 & 1 & 0
\end{array}
\right]
$$

The initial basic tableau is not feasible. Pivot on the 20 in row 1, column 1.

$$
\begin{array}{c}
\\
-5R_1 + 4R_2 \to R_2 \\
-9R_1 + 10R_3 \to R_3 \\
-8R_1 + 5R_4 \to R_4
\end{array}
\left[
\begin{array}{ccccccc|c}
y_1 & y_2 & y_3 & a_1 & s_1 & s_2 & z & \\
20 & 10 & 5 & 1 & 0 & 0 & 0 & 200 \\
0 & 110 & 175 & -5 & 4 & 0 & 0 & 1000 \\
0 & 150 & 75 & -9 & 0 & -10 & 0 & 1200 \\
\hline
0 & 120 & 200 & -8 & 0 & 0 & 5 & -1600
\end{array}
\right]
$$

Eliminate the a_1 column.

$$
\left[
\begin{array}{cccccc|c}
y_1 & y_2 & y_3 & s_1 & s_2 & z & \\
20 & 10 & 5 & 0 & 0 & 0 & 200 \\
0 & 110 & 175 & 4 & 0 & 0 & 1000 \\
0 & \boxed{150} & 75 & 0 & -10 & 0 & 1200 \\
\hline
0 & 120 & 200 & 0 & 0 & 5 & -1600
\end{array}
\right]
$$

Pivot on the 150 in row 3, column 2.

$$
\begin{array}{c}
-R_3 + 15R_1 \to R_1 \\
11R_3 + 15R_2 \to R_2 \\
\\
-4R_3 + 5R_4 \to R_4
\end{array}
\left[
\begin{array}{cccccc|c}
y_1 & y_2 & y_3 & s_1 & s_2 & z & \\
300 & 0 & 0 & 0 & 10 & 0 & 1800 \\
0 & 0 & 1800 & 60 & 110 & 0 & 1800 \\
0 & 150 & 75 & 0 & -10 & 0 & 1200 \\
\hline
0 & 0 & 700 & 0 & 40 & 25 & -12,800
\end{array}
\right]
$$

Create ones in the columns corresponding to $y_1, y_2, s_1,$ and z.

$$
\begin{array}{c}
\\
\frac{1}{300}R_1 \to R_1 \\
\frac{1}{60}R_2 \to R_2 \\
\frac{1}{150}R_3 \to R_3 \\
\\
\frac{1}{25}R_4 \to R_4
\end{array}
\begin{array}{c}
\begin{array}{cccccc}
y_1 & y_2 & y_3 & s_1 & s_2 & z \\
\end{array} \\
\left[\begin{array}{cccccc|c}
1 & 0 & 0 & 0 & \frac{1}{30} & 0 & 6 \\
0 & 0 & 30 & 1 & \frac{11}{6} & 0 & 30 \\
0 & 1 & \frac{1}{2} & 0 & -\frac{1}{15} & 0 & 8 \\ \hline
0 & 0 & 28 & 0 & \frac{8}{5} & 1 & -512
\end{array}\right]
\end{array}
$$

This is a final tableau. The minimum value is 512 when $y_1 = 6$, $y_2 = 8$, and $y_3 = 0$.

23. (a)   Let $y_1 =$ amount shipped from S_1 to D_1,
$y_2 =$ amount shipped from S_1 to D_2,
$y_3 =$ amount shipped from S_2 to D_1,
and $y_4 =$ amount shipped from S_2 to D_2.

Minimize $w = 30y_1 + 20y_2 + 25y_3 + 22y_4$

subject to:
$$
\begin{aligned}
y_1 + y_3 &\geq 3000 \\
y_2 + y_4 &\geq 5000 \\
y_1 + y_2 &\leq 5000 \\
y_3 + y_4 &\leq 5000 \\
2y_1 + 6y_2 + 5y_3 + 4y_4 &\leq 40{,}000
\end{aligned}
$$
with $\quad y_1 \geq 0, y_2 \geq 0, y_3 \geq 0, y_4 \geq 0.$

Maximize $z = -w = -30y_1 - 20y_2 - 25y_3 - 22y_4.$

$$
\begin{array}{c}
\begin{array}{ccccccccccc}
y_1 & y_2 & y_3 & y_4 & s_1 & s_2 & s_3 & s_4 & s_5 & z & \\
\end{array} \\
\left[\begin{array}{cccccccccc|c}
\boxed{1} & 0 & 1 & 0 & -1 & 0 & 0 & 0 & 0 & 0 & 3000 \\
0 & 1 & 0 & 1 & 0 & -1 & 0 & 0 & 0 & 0 & 5000 \\
1 & 1 & 0 & 0 & 0 & 0 & 1 & 0 & 0 & 0 & 5000 \\
0 & 0 & 1 & 1 & 0 & 0 & 0 & 1 & 0 & 0 & 5000 \\
2 & 6 & 5 & 4 & 0 & 0 & 0 & 0 & 1 & 0 & 40{,}000 \\ \hline
30 & 20 & 25 & 22 & 0 & 0 & 0 & 0 & 0 & 1 & 0
\end{array}\right]
\end{array}
$$

Pivot on the 1 in row 1, column 1 since the feasible solution has a negative value, $s_1 = -3000$.

$$
\begin{array}{c}
\\
\\
\\
-R_1 + R_3 \to R_3 \\
\\
-2R_1 + R_5 \to R_5 \\
-30R_1 + R_6 \to R_6
\end{array}
\begin{array}{c}
\begin{array}{ccccccccccc}
y_1 & y_2 & y_3 & y_4 & s_1 & s_2 & s_3 & s_4 & s_5 & z & \\
\end{array} \\
\left[\begin{array}{cccccccccc|c}
1 & 0 & 1 & 0 & -1 & 0 & 0 & 0 & 0 & 0 & 3000 \\
0 & 1 & 0 & 1 & 0 & -1 & 0 & 0 & 0 & 0 & 5000 \\
0 & \boxed{1} & -1 & 0 & 1 & 0 & 1 & 0 & 0 & 0 & 2000 \\
0 & 0 & 1 & 1 & 0 & 0 & 0 & 1 & 0 & 0 & 5000 \\
0 & 6 & 3 & 4 & 2 & 0 & 0 & 0 & 1 & 0 & 34{,}000 \\ \hline
0 & 20 & -5 & 22 & 30 & 0 & 0 & 0 & 0 & 1 & -90{,}000
\end{array}\right]
\end{array}
$$

Since the feasible solution has a negative value ($s_2 = -5000$), pivot on the 1 in row 3, column 2.

$$
\begin{array}{c}
\\
-R_3 + R_2 \to R_2 \\
\\
\\
-6R_3 + R_5 \to R_5 \\
\\
-20R_3 + R_6 \to R_6
\end{array}
\begin{array}{c}
\begin{array}{ccccccccccc}
y_1 & y_2 & y_3 & y_4 & s_1 & s_2 & s_3 & s_4 & s_5 & z & \\
\end{array} \\
\left[\begin{array}{cccccccccc|c}
1 & 0 & 1 & 0 & -1 & 0 & 0 & 0 & 0 & 0 & 3000 \\
0 & 0 & 1 & 1 & -1 & -1 & -1 & 0 & 0 & 0 & 3000 \\
0 & 1 & -1 & 0 & 1 & 0 & 1 & 0 & 0 & 0 & 2000 \\
0 & 0 & 1 & 1 & 0 & 0 & 0 & 1 & 0 & 0 & 5000 \\
0 & 0 & \boxed{9} & 4 & -4 & 0 & -6 & 0 & 1 & 0 & 22{,}000 \\ \hline
0 & 0 & 15 & 22 & 10 & 0 & -20 & 0 & 0 & 1 & -130{,}000
\end{array}\right]
\end{array}
$$

Since the feasible solution has a negative value ($s_2 = -3000$), pivot on the 9 in row 5, column 3.

$$
\begin{array}{c}
-R_5 + 9R_1 \rightarrow R_1 \\
-R_5 + 9R_2 \rightarrow R_2 \\
R_5 + 9R_3 \rightarrow R_3 \\
-R_5 + 9R_4 \rightarrow R_4 \\
\\
-5R_5 + 3R_6 \rightarrow R_6
\end{array}
$$

	y_1	y_2	y_3	y_4	s_1	s_2	s_3	s_4	s_5	z	
	9	0	0	-4	-5	0	6	0	-1	0	5000
	0	0	0	$\boxed{5}$	-5	-9	-3	0	-1	0	5000
	0	9	0	4	5	0	3	0	1	0	40,000
	0	0	0	5	4	0	6	9	-1	0	23,000
	0	0	9	4	-4	0	-6	0	1	0	22,000
	0	0	0	46	50	0	-30	0	-5	3	-500,000

Pivot on the 5 in row 2, column 4.

$$
\begin{array}{c}
4R_2 + 5R_1 \rightarrow R_1 \\
\\
-4R_2 + 5R_3 \rightarrow R_3 \\
-R_2 + R_4 \rightarrow R_4 \\
-4R_2 + 5R_5 \rightarrow R_5 \\
-46R_2 + 5R_6 \rightarrow R_6
\end{array}
$$

	y_1	y_2	y_3	y_4	s_1	s_2	s_3	s_4	s_5	z	
	45	0	0	0	-45	-36	18	0	-9	0	45,000
	0	0	0	5	-5	-9	-3	0	-1	0	5000
	0	45	0	0	45	36	27	0	9	0	180,000
	0	0	0	0	9	9	$\boxed{9}$	9	0	0	18,000
	0	0	45	0	0	36	-18	0	9	0	90,000
	0	0	0	0	480	414	-12	0	21	15	-2,730,000

Pivot on the 9 in row 4, column 7.

$$
\begin{array}{c}
-2R_4 + R_1 \rightarrow R_1 \\
R_4 + 3R_2 \rightarrow R_2 \\
-3R_4 + R_3 \rightarrow R_3 \\
\\
2R_4 + R_5 \rightarrow R_5 \\
4R_4 + 3R_6 \rightarrow R_6
\end{array}
$$

	y_1	y_2	y_3	y_4	s_1	s_2	s_3	s_4	s_5	z	
	45	0	0	0	-63	-54	0	-18	-9	0	9000
	0	0	0	15	-6	-18	0	9	-3	0	33,000
	0	45	0	0	18	9	0	-27	9	0	126,000
	0	0	0	0	9	9	9	9	0	0	18,000
	0	0	45	0	18	54	0	18	9	0	126,000
	0	0	0	0	1476	450	0	36	63	45	-8,118,000

Create a 1 in the columns corresponding to y_1, y_2, y_3, y_4, and z.

$$
\begin{array}{c}
\frac{1}{45}R_1 \rightarrow R_1 \\
\frac{1}{15}R_2 \rightarrow R_2 \\
\frac{1}{45}R_3 \rightarrow R_3 \\
\\
\frac{1}{45}R_5 \rightarrow R_5 \\
\\
\frac{1}{45}R_6 \rightarrow R_6
\end{array}
$$

	y_1	y_2	y_3	y_4	s_1	s_2	s_3	s_4	s_5	z	
	1	0	0	0	$-\frac{7}{5}$	$-\frac{6}{5}$	0	$-\frac{2}{5}$	$-\frac{1}{5}$	0	200
	0	0	0	1	$-\frac{2}{5}$	$-\frac{6}{5}$	0	$\frac{3}{5}$	$-\frac{1}{5}$	0	2200
	0	1	0	0	$\frac{2}{5}$	$\frac{1}{5}$	0	$-\frac{3}{5}$	$\frac{1}{5}$	0	2800
	0	0	0	0	9	9	9	9	0	0	18,000
	0	0	1	0	$\frac{2}{5}$	$\frac{6}{5}$	0	$\frac{2}{5}$	$\frac{1}{5}$	0	2800
	0	0	0	0	$\frac{164}{5}$	10	0	$\frac{4}{5}$	$\frac{7}{5}$	1	-180,400

Here, $y_1 = 200, y_2 = 2800, y_3 = 2800, y_4 = 2200$, and $-z = w = 180,400$. So, ship 200 barrels of oil from supplier S_1 to distributor D_1. Ship 2800 barrels of oil from supplier S_1 to distributor D_2. Ship 2800 barrels of oil from supplier S_2 to distributor D_1. Ship 2200 barrels of oil from supplier S_2 to distributor D_2. The minimum cost is $180,400.

(b) From the final tableau, $9s_3 = 18,000$, so $s_3 = 2000$. Therefore, S_1 could furnish 2000 more barrels of oil.

25. Let $x_1 =$ the number of million dollars for home loans

and $x_2 =$ the number of million dollars for commercial loans.

$$
\begin{array}{lll}
\text{Maximize} & z = 0.12x_1 + 0.10x_2 \\
\text{subject to:} & x_1 \qquad\ \geq 4x_2 \text{ or } x_1 - 4x_2 \geq 0 \\
& x_1 + \ x_2 \geq 10 \\
& 3x_1 + 2x_2 \leq 72 \\
& x_1 + \ x_2 \leq 25 \\
\text{with} & x_1 \geq 0, x_2 \geq 0.
\end{array}
$$

$$
\begin{array}{ccccccc}
x_1 & x_2 & s_1 & s_2 & s_3 & s_4 & z \\
\end{array}
$$

$$
\left[\begin{array}{ccccccc|c}
1 & -4 & -1 & 0 & 0 & 0 & 0 & 0 \\
1 & 1 & 0 & -1 & 0 & 0 & 0 & 10 \\
3 & 2 & 0 & 0 & 1 & 0 & 0 & 72 \\
1 & 1 & 0 & 0 & 0 & 1 & 0 & 25 \\
\hline
-0.12 & -0.10 & 0 & 0 & 0 & 0 & 1 & 0
\end{array}\right]
$$

Eliminate the decimals in the last row by multiplying by 100

$$
\begin{array}{ccccccc}
x_1 & x_2 & s_1 & s_2 & s_3 & s_4 & z \\
\end{array}
$$

$$
\left[\begin{array}{ccccccc|c}
1 & -4 & -1 & 0 & 0 & 0 & 0 & 0 \\
\boxed{1} & 1 & 0 & -1 & 0 & 0 & 0 & 10 \\
3 & 2 & 0 & 0 & 1 & 0 & 0 & 72 \\
1 & 1 & 0 & 0 & 0 & 1 & 0 & 25 \\
\hline
-12 & -10 & 0 & 0 & 0 & 0 & 100 & 0
\end{array}\right]
$$

Pivot on the 1 in row 2, column 1.

$$
\begin{array}{ccccccc}
& x_1 & x_2 & s_1 & s_2 & s_3 & s_4 & z \\
\end{array}
$$

$$
\begin{array}{l}
-R_2 + R_1 \rightarrow R_1 \\
\\
-3R_2 + R_3 \rightarrow R_3 \\
-R_2 + R_4 \rightarrow R_4 \\
12R_2 + R_5 \rightarrow R_5
\end{array}
\left[\begin{array}{ccccccc|c}
0 & -5 & -1 & 1 & 0 & 0 & 0 & -10 \\
1 & 1 & 0 & -1 & 0 & 0 & 0 & 10 \\
0 & -1 & 0 & \boxed{3} & 1 & 0 & 0 & 42 \\
0 & 0 & 0 & 1 & 0 & 1 & 0 & 15 \\
\hline
0 & 2 & 0 & -12 & 0 & 0 & 100 & 120
\end{array}\right]
$$

Pivot on the 3 in row 3, column 4.

$$
\begin{array}{ccccccc}
& x_1 & x_2 & s_1 & s_2 & s_3 & s_4 & z \\
\end{array}
$$

$$
\begin{array}{l}
-R_3 + 3R_1 \rightarrow R_1 \\
R_3 + 3R_2 \rightarrow R_2 \\
\\
-R_3 + 3R_4 \rightarrow R_4 \\
\\
4R_3 + R_5 \rightarrow R_5
\end{array}
\left[\begin{array}{ccccccc|c}
0 & -14 & -3 & 0 & -1 & 0 & 0 & -72 \\
3 & 2 & 0 & 0 & -2 & 0 & 0 & 72 \\
0 & -1 & 0 & 3 & 1 & 0 & 0 & 42 \\
0 & \boxed{1} & 0 & 0 & -1 & 3 & 0 & 3 \\
\hline
0 & -2 & 0 & 0 & 4 & 0 & 100 & 288
\end{array}\right]
$$

Pivot on the 1 in row 4, column 2.

$$
\begin{array}{ccccccc}
& x_1 & x_2 & s_1 & s_2 & s_3 & s_4 & z \\
\end{array}
$$

$$
\begin{array}{l}
14R_4 + R_1 \rightarrow R_1 \\
-2R_4 + R_2 \rightarrow R_2 \\
R_4 + R_3 \rightarrow R_3 \\
\\
2R_4 + R_5 \rightarrow R_5
\end{array}
\left[\begin{array}{ccccccc|c}
0 & 0 & -3 & 0 & -15 & 42 & 0 & -30 \\
3 & 0 & 0 & 0 & 0 & -6 & 0 & 66 \\
0 & 0 & 0 & 3 & 0 & 3 & 0 & 45 \\
0 & 1 & 0 & 0 & -1 & 3 & 0 & 3 \\
\hline
0 & 0 & 0 & 0 & 2 & 6 & 100 & 294
\end{array}\right]
$$

Create a 1 in the columns corresponding to x_1 and z.

$$
\begin{array}{c}
\\
\\
\frac{1}{3}R_2 \to R_2 \\
\\
\\
\frac{1}{100}R_5 \to R_5
\end{array}
\begin{array}{c}
\begin{array}{ccccccc}
x_1 & x_2 & s_1 & s_2 & s_3 & s_4 & z
\end{array} \\
\left[
\begin{array}{ccccccc|c}
0 & 0 & -3 & 0 & -15 & 42 & 0 & -30 \\
1 & 0 & 0 & 0 & 0 & -2 & 0 & 22 \\
0 & 0 & 0 & 3 & 0 & 3 & 0 & 45 \\
0 & 1 & 0 & 0 & -1 & 3 & 0 & 3 \\
\hline
0 & 0 & 0 & 0 & 0.02 & 0.06 & 1 & 2.94
\end{array}
\right]
\end{array}
$$

Here, $x_1 = 22, x_2 = 3$, and $z = 2.94$. Make \$22 million (\$22,000,000) in home loans and \$3 million (\$3,000,000) in commercial loans for a maximum return of \$2.94 million, or \$2,940,000.

27. Let $x_1 = $ the number of pounds of bluegrass seed,

$\quad\quad x_2 = $ the number of pounds of rye seed,

and $x_3 = $ the number of pounds of Bermuda seed.

If each batch must contain at least 25% bluegrass seed, then

$$y_1 \geq 0.25(y_1 + y_2 + y_3)$$
$$0.75y_1 - 0.25y_2 - 0.25y_3 \geq 0.$$

And if the amount of Bermuda must be no more than $\frac{2}{3}$ the amount of rye, then

$$y_3 \leq \frac{2}{3}y_2$$

$$-2y_2 + 3y_3 = 0.$$

Using these forms for our constraints, we can now state the problem as follows.

Minimize $w = 16y_1 + 14y_2 + 12y_3$

subject to: $0.75y_1 - 0.25y_2 - 0.25y_3 \geq 0$

$\quad\quad\quad\quad\quad -\; 2y_2 +\quad 3y_3 \leq 0$

$\quad\quad\quad y_1 +\quad y_2 +\quad y_3 \geq 6000$

with $y_1 \geq 0,\; y_2 \geq 0,\; y_3 \geq 0.$

The initial simplex tableau is

$$
\begin{array}{ccccccc}
y_1 & y_2 & y_3 & s_1 & s_2 & a & z
\end{array}
$$
$$
\left[
\begin{array}{ccccccc|c}
0.75 & -0.25 & -0.25 & -1 & 0 & 0 & 0 & 0 \\
0 & -2 & 3 & 0 & 1 & 0 & 0 & 0 \\
\boxed{1} & 1 & 1 & 0 & 0 & 1 & 0 & 6000 \\
\hline
16 & 14 & 12 & 0 & 0 & 0 & 1 & 0
\end{array}
\right]
$$

First eliminate the artificial variable a. Pivot on the 1 in row 3, column 1.

$$
\begin{array}{c}
\\
0.75R_3 - R_1 \to R_1 \\
\\
\\
-16R_3 + R_4 \to R_4
\end{array}
\begin{array}{c}
\begin{array}{ccccccc}
y_1 & y_2 & y_3 & s_1 & s_2 & a & z
\end{array} \\
\left[
\begin{array}{ccccccc|c}
0 & 1 & 1 & 1 & 0 & 0.75 & 0 & 4500 \\
0 & -2 & 3 & 0 & 1 & 0 & 0 & 0 \\
1 & 1 & 1 & 0 & 0 & 1 & 0 & 6000 \\
\hline
0 & -2 & -4 & 0 & 0 & -16 & 1 & -96{,}000
\end{array}
\right]
\end{array}
$$

Since $a = 0$, we can drop the a column.

$$\begin{array}{cccccc} y_1 & y_2 & y_3 & s_1 & s_2 & z \\ \left[\begin{array}{cccccc|c} 0 & 1 & 1 & 1 & 0 & 0 & 4500 \\ 0 & -2 & \boxed{3} & 0 & 1 & 0 & 0 \\ 1 & 1 & 1 & 0 & 0 & 0 & 6000 \\ \hline 0 & -2 & -4 & 0 & 0 & 1 & -96{,}000 \end{array}\right] \end{array}$$

Pivot on the 3 in row 2, column 3.

$$\begin{array}{c} \\ -R_2+3R_1\rightarrow R_1 \\ \\ -R_2+3R_3\rightarrow R_3 \\ 4R_2+3R_4\rightarrow R_4 \end{array} \begin{array}{cccccc} y_1 & y_2 & y_3 & s_1 & s_2 & z \\ \left[\begin{array}{cccccc|c} 0 & \boxed{5} & 0 & 3 & -1 & 0 & 13{,}500 \\ 0 & -2 & 3 & 0 & 1 & 0 & 0 \\ 3 & 5 & 0 & 0 & -1 & 0 & 18{,}000 \\ 0 & -14 & 0 & 0 & 4 & 3 & -288{,}000 \end{array}\right] \end{array}$$

Pivot on the 5 in row 1, column 2.

$$\begin{array}{c} \\ 2R_1+5R_2\rightarrow R_2 \\ -R_1+R_3\rightarrow R_3 \\ 14R_1+5R_4\rightarrow R_4 \end{array} \begin{array}{cccccc} y_1 & y_2 & y_3 & s_1 & s_2 & z \\ \left[\begin{array}{cccccc|c} 0 & 5 & 0 & 3 & -1 & 0 & 13{,}500 \\ 0 & 0 & 15 & 6 & 3 & 0 & 27{,}000 \\ 3 & 0 & 0 & -3 & 0 & 0 & 4500 \\ 0 & 0 & 0 & 42 & 6 & 15 & -1{,}251{,}000 \end{array}\right] \end{array}$$

Create a 1 in the columns corresponding to $y_1, y_2, y_3,$ and z.

$$\begin{array}{c} \\ \frac{1}{5}R_1\rightarrow R_1 \\ \frac{1}{15}R_2\rightarrow R_2 \\ \frac{1}{3}R_3\rightarrow R_3 \\ \frac{1}{15}R_4\rightarrow R_4 \end{array} \begin{array}{cccccc} x_1 & x_2 & x_3 & s_1 & s_2 & z \\ \left[\begin{array}{cccccc|c} 0 & 1 & 0 & 0.6 & -0.2 & 0 & 2700 \\ 0 & 0 & 1 & 0.4 & 0.2 & 0 & 1800 \\ 1 & 0 & 0 & -1 & 0 & 0 & 1500 \\ \hline 0 & 0 & 0 & 2.8 & 0.4 & 1 & -83{,}400 \end{array}\right] \end{array}$$

Here, $y_1 = 1500$, $y_2 = 2700$, $y_3 = 1800$, and $z = -w = 83{,}400$. Therefore, use 1500 lb of bluegrass, 2700 lb of rye, and 1800 lb of Bermuda for a minimum cost of \$834.

29. (a)   Let $x_1 =$ the number of computers shipped from W_1 to D_1,
$x_2 =$ the number of computers shipped from W_1 to D_2,
$x_3 =$ the number of computers shipped from W_2 to D_1,
and $x_4 =$ the number of computers shipped from W_2 to D_2.

Minimize $w = 14x_1 + 12x_2 + 12x_3 + 10x_4$
subject to: $x_1 + x_3 \geq 32$
$x_2 + x_4 \geq 20$
$x_1 + x_2 \leq 25$
$x_3 + x_4 \leq 30$
with $x_1 \geq 0, x_2 \geq 0, x_3 \geq 0, x_4 \geq 0.$

Maximize $z = -w = -14x_1 - 12x_2 - 12x_3 - 10x_4.$

$$\begin{array}{ccccccccc} x_1 & x_2 & x_3 & x_4 & s_1 & s_2 & s_3 & s_4 & z \\ \left[\begin{array}{ccccccccc|c} \boxed{1} & 0 & 1 & 0 & -1 & 0 & 0 & 0 & 0 & 32 \\ 0 & 1 & 0 & 1 & 0 & -1 & 0 & 0 & 0 & 20 \\ 1 & 1 & 0 & 0 & 0 & 0 & 1 & 0 & 0 & 25 \\ 0 & 0 & 1 & 1 & 0 & 0 & 0 & 1 & 0 & 30 \\ \hline 14 & 12 & 12 & 10 & 0 & 0 & 0 & 0 & 1 & 0 \end{array}\right] \end{array}$$

Since the basic solution is not feasible ($s_1 = -32$), pivot on the 1 in row 1, column 1.

$$
\begin{array}{c}
\\
\\
R_1 - R_3 \to R_3 \\
\\
-14R_1 + R_5 \to R_5
\end{array}
\begin{array}{ccccccccc|c}
x_1 & x_2 & x_3 & x_4 & s_1 & s_2 & s_3 & s_4 & z & \\
1 & 0 & 1 & 0 & -1 & 0 & 0 & 0 & 0 & 32 \\
0 & \boxed{1} & 0 & 1 & 0 & -1 & 0 & 0 & 0 & 20 \\
0 & -1 & 1 & 0 & -1 & 0 & -1 & 0 & 0 & 7 \\
0 & 0 & 1 & 1 & 0 & 0 & 0 & 1 & 0 & 30 \\
\hline
0 & 12 & -2 & 10 & 14 & 0 & 0 & 0 & 1 & -448
\end{array}
$$

Pivot on the 1 in row 2, column 2 since $s_2 = -20$.

$$
\begin{array}{c}
\\
\\
R_2 + R_3 \to R_3 \\
\\
-12R_2 + R_5 \to R_5
\end{array}
\begin{array}{ccccccccc|c}
x_1 & x_2 & x_3 & x_4 & s_1 & s_2 & s_3 & s_4 & z & \\
1 & 0 & 1 & 0 & -1 & 0 & 0 & 0 & 0 & 32 \\
0 & 1 & 0 & 1 & 0 & -1 & 0 & 0 & 0 & 20 \\
0 & 0 & \boxed{1} & 1 & -1 & -1 & -1 & 0 & 0 & 27 \\
0 & 0 & 1 & 1 & 0 & 0 & 0 & 1 & 0 & 30 \\
\hline
0 & 0 & -2 & -2 & 14 & 12 & 0 & 0 & 1 & -688
\end{array}
$$

Pivot on the 1 in row 3, column 3 since $s_3 = -27$.

$$
\begin{array}{c}
-R_3 + R_1 \to R_1 \\
\\
\\
-R_3 + R_4 \to R_4 \\
2R_3 + R_5 \to R_5
\end{array}
\begin{array}{ccccccccc|c}
x_1 & x_2 & x_3 & x_4 & s_1 & s_2 & s_3 & s_4 & z & \\
1 & 0 & 0 & -1 & 0 & 1 & 1 & 0 & 0 & 5 \\
0 & 1 & 0 & 1 & 0 & -1 & 0 & 0 & 0 & 20 \\
0 & 0 & 1 & 1 & -1 & -1 & -1 & 0 & 0 & 27 \\
0 & 0 & 0 & 0 & 1 & 1 & \boxed{1} & 1 & 0 & 3 \\
\hline
0 & 0 & 0 & 0 & 12 & 10 & -2 & 0 & 1 & -634
\end{array}
$$

Pivot on the 1 in row 4, column 7.

$$
\begin{array}{c}
-R_4 + R_1 \to R_1 \\
\\
R_4 + R_3 \to R_3 \\
\\
2R_4 + R_5 \to R_5
\end{array}
\begin{array}{ccccccccc|c}
x_1 & x_2 & x_3 & x_4 & s_1 & s_2 & s_3 & s_4 & z & \\
1 & 0 & 0 & -1 & -1 & 0 & 0 & -1 & 0 & 2 \\
0 & 1 & 0 & 1 & 0 & -1 & 0 & 0 & 0 & 20 \\
0 & 0 & 1 & 1 & 0 & 0 & 0 & 1 & 0 & 30 \\
0 & 0 & 0 & 0 & 1 & 1 & 1 & 1 & 0 & 3 \\
\hline
0 & 0 & 0 & 0 & 14 & 12 & 0 & 2 & 1 & -628
\end{array}
$$

Here, $x_1 = 2, x_2 = 20, x_3 = 30, x_4 = 0$, and $z = -w = 628$. Therefore, ship 2 computers from W_1 to D_1, 20 computers from W_1 to D_2, 30 computers from W_2 to D_1, and 0 computers from W_2 to D_2 for a minimum cost of \$628.

(b) From the final tableau, $s_3 = 3$. Therefore, warehouse W_1 has three more computers that it could ship.

31. Let $y_1 =$ the number of ounces of ingredient I,
      $y_2 =$ the number of ounces of ingredient II,
and $y_3 =$ the number of ounces of ingredient III.

The problem is:

$$
\begin{array}{ll}
\text{Minimize} & w = 0.30y_1 + 0.09y_2 + 0.27y_3 \\
\text{subject to:} & y_1 + y_2 + y_3 \geq 10 \\
& y_1 + y_2 + y_3 \leq 15 \\
& y_1 \geq \tfrac{1}{4}y_2 \\
& y_3 \geq y_1 \\
\text{with} & y_1 \geq 0, y_2 \geq 0, y_3 \geq 0.
\end{array}
$$

Use a graphing calculator or computer to find that the minimum is $w = 1.55$ when $y_1 = \frac{5}{3}, y_2 = \frac{20}{3}$, and $y_3 = \frac{5}{3}$. Therefore, the additive should consist of $\frac{5}{3}$ oz of ingredient I, $\frac{20}{3}$ oz of ingredient II, and $\frac{5}{3}$ oz of ingredient III, for a minimum cost of \$1.55/gal. The amount of additive that should be used per gallon of gasoline is $\frac{5}{3} + \frac{20}{3} + \frac{5}{3} = 10$ oz.

33. (a) Let $x_1 =$ the number of hours spent doing calisthenics,

$x_2 =$ the number of hours spent swimming,

and $x_3 =$ the number of hours spent playing the drums.

The problem can be stated as follows.

Maximize $z = 388x_1 + 518x_2 + 345x_3$

subject to: $x_1 + x_2 + x_3 \leq 10$

$x_1 - 2x_2 + x_3 \geq 0$

$x_3 \leq 4$

$x_3 \geq 1.$

with $x_1 \geq 0, x_2 \geq 0, x_3 \geq 0.$

The initial simplex tableau is

$$
\begin{array}{cccccccc|c}
x_1 & x_2 & x_3 & s_1 & s_2 & s_3 & s_4 & z & \\
\hline
1 & 1 & 1 & 1 & 0 & 0 & -1 & 0 & 10 \\
1 & -2 & 1 & 0 & -1 & 0 & 0 & 0 & 0 \\
0 & 0 & 1 & 0 & 0 & 1 & 0 & 0 & 4 \\
0 & 0 & \boxed{1} & 0 & 0 & 0 & -1 & 0 & 1 \\
\hline
-388 & -518 & -345 & 0 & 0 & 0 & 0 & 1 & 0
\end{array}
$$

Since $s_4 = 1$, the initial basic solution is not feasible. So pivot on the 1 in row 4, column 3.

$$
\begin{array}{l}
-R_4 + R_1 \rightarrow R_1 \\
R_4 - R_2 \rightarrow R_2 \\
-R_4 + R_3 \rightarrow R_3 \\
\\
345R_4 + R_5 \rightarrow R_5
\end{array}
\begin{array}{cccccccc|c}
x_1 & x_2 & x_3 & s_1 & s_2 & s_3 & s_4 & z & \\
\hline
1 & 1 & 0 & 1 & 0 & 0 & 0 & 0 & 9 \\
-1 & \boxed{2} & 0 & 0 & 1 & 0 & -1 & 0 & 1 \\
0 & 0 & 0 & 0 & 0 & 1 & 1 & 0 & 3 \\
0 & 0 & 1 & 0 & 0 & 0 & -1 & 0 & 1 \\
\hline
-388 & -518 & 0 & 0 & 0 & 0 & -345 & 1 & 345
\end{array}
$$

Pivot on the 2 in row 2, column 2.

$$
\begin{array}{l}
-R_2 + 2R_1 \rightarrow R_1 \\
\\
\\
\\
259R_2 + R_5 \rightarrow R_5
\end{array}
\begin{array}{cccccccc|c}
x_1 & x_2 & x_3 & s_1 & s_2 & s_3 & s_4 & z & \\
\hline
\boxed{3} & 0 & 0 & 2 & -1 & 0 & 1 & 0 & 17 \\
-1 & 2 & 0 & 0 & 1 & 0 & -1 & 0 & 1 \\
0 & 0 & 0 & 0 & 0 & 1 & 1 & 0 & 3 \\
0 & 0 & 1 & 0 & 0 & 0 & -1 & 0 & 1 \\
\hline
-647 & 0 & 0 & 0 & 259 & 0 & -604 & 1 & 604
\end{array}
$$

Pivot on the 3 in row 1, column 1.

$$
\begin{array}{l}
\\
R_1 + 3R_2 \rightarrow R_2 \\
\\
\\
647R_1 + 3R_5 \rightarrow R_5
\end{array}
\begin{array}{cccccccc|c}
x_1 & x_2 & x_3 & s_1 & s_2 & s_3 & s_4 & z & \\
\hline
3 & 0 & 0 & 2 & -1 & 0 & 1 & 0 & 17 \\
0 & 6 & 0 & 2 & 2 & 0 & -2 & 0 & 20 \\
0 & 0 & 0 & 0 & 0 & 1 & 1 & 0 & 3 \\
0 & 0 & 1 & 0 & 0 & 0 & -1 & 0 & 1 \\
\hline
0 & 0 & 0 & 1294 & 130 & 0 & -1165 & 3 & 12{,}811
\end{array}
$$

Create a 1 in the columns corresponding to $x_1, x_2,$ and z.

$$
\begin{array}{c}
\begin{array}{c}\\ \frac{1}{3}R_1 \to R_1 \\ \frac{1}{6}R_2 \to R_2 \\ \\ \\ \frac{1}{3}R_5 \to R_5\end{array}
\begin{array}{c}
\quad x_1 \quad x_2 \quad x_3 \quad s_1 \quad s_2 \quad s_3 \quad s_4 \quad z \\
\left[\begin{array}{ccccccc|c}
1 & 0 & 0 & \frac{2}{3} & -\frac{1}{3} & 0 & \frac{1}{3} & 0 & \frac{17}{3} \\
0 & 1 & 0 & \frac{1}{3} & \frac{1}{3} & 0 & -\frac{1}{3} & 0 & \frac{10}{3} \\
0 & 0 & 0 & 0 & 0 & 1 & 1 & 0 & 3 \\
0 & 0 & 1 & 0 & 0 & 0 & -1 & 0 & 1 \\
\hline
0 & 0 & 0 & \frac{1294}{3} & \frac{130}{3} & 0 & -\frac{1165}{3} & 1 & \frac{12{,}811}{3}
\end{array}\right]
\end{array}
\end{array}
$$

Joe should spend $\frac{17}{3}$ hours doing calisthenics, $\frac{10}{3}$ hours swimming, and 1 hour playing drums for a maximum calorie expenditure of $\frac{12{,}811}{3}$ or $4270\frac{1}{3}$ calories.

Chapter 4 Review Exercises

1. The simplex method should be used for problems with more than two variables or problems with two variables and many constants.

3. (a) Maximize $z = 2x_1 + 7x_2$

    subject to: $4x_1 + 6x_2 \le 60$
    $$3x_1 + x_2 \le 18$$
    $$2x_1 + 5x_2 \le 20$$
    $$x_1 + x_2 \le 15$$

    with $x_1 \ge 0,\ x_2 \ge 0.$

Adding slack variables $s_1, s_2, s_3,$ and s_4, we obtain the following equations.

$$
\begin{aligned}
4x_1 + 6x_2 + s_1 &= 60 \\
3x_1 + x_2 + s_2 &= 18 \\
2x_1 + 5x_2 + s_3 &= 20 \\
x_1 + x_2 + s_4 &= 15.
\end{aligned}
$$

(b) The initial simplex tableau is as follows.

$$
\begin{array}{c}
\quad x_1 \quad x_2 \quad s_1 \quad s_2 \quad s_3 \quad s_4 \quad z \\
\left[\begin{array}{ccccccc|c}
4 & 6 & 1 & 0 & 0 & 0 & 0 & 60 \\
3 & 1 & 0 & 1 & 0 & 0 & 0 & 18 \\
2 & 5 & 0 & 0 & 1 & 0 & 0 & 20 \\
1 & 1 & 0 & 0 & 0 & 1 & 0 & 15 \\
\hline
-2 & -7 & 0 & 0 & 0 & 0 & 1 & 0
\end{array}\right]
\end{array}.
$$

5. Maximize $z = 5x_1 + 8x_2 + 6x_3$
    subject to: $x_1 + x_2 + x_3 \le 90$
    $$2x_1 + 5x_2 + x_3 \le 120$$
    $$x_1 + 3x_2 \ge 80$$
    with $x_1 \ge 0, x_2 \ge 0, x_3 > 0.$

(a) Adding the slack variables s_1 and s_2 and
subtracting the surplus variable s_3, we obtain the following equations:

$$
\begin{aligned}
x_1 + x_2 + x_3 + s_1 &= 90 \\
2x_1 + 5x_2 + x_3 + s_2 &= 120 \\
x_1 + 3x_2 - s_3 &= 80.
\end{aligned}
$$

(b) The initial tableau is

$$
\begin{array}{ccccccc|c}
x_1 & x_2 & x_3 & s_1 & s_2 & s_3 & z & \\
\hline
1 & 1 & 1 & 1 & 0 & 0 & 0 & 90 \\
2 & 5 & 1 & 0 & 1 & 0 & 0 & 120 \\
1 & 3 & 0 & 0 & 0 & -1 & 0 & 80 \\
\hline
-5 & -8 & -6 & 0 & 0 & 0 & 1 & 0
\end{array}
$$

7.

$$
\begin{array}{cccccc|c}
x_1 & x_2 & x_3 & s_1 & s_2 & z & \\
\hline
4 & 5 & \boxed{2} & 1 & 0 & 0 & 18 \\
2 & 8 & \boxed{6} & 0 & 1 & 0 & 24 \\
\hline
-5 & -3 & -6 & 0 & 0 & 1 & 0
\end{array}
$$

The most negative entry in the last row is -6, and the smaller of the two quotients is $\frac{24}{6} = 4$. Hence, the 6 in row 2, column 3, is the first pivot. Performing row transformations leads to the following tableau.

$$
\begin{array}{l}
-R_2 + 3R_1 \to R_1 \\
\\
R_2 + R_3 \to R_3
\end{array}
\qquad
\begin{array}{cccccc|c}
x_1 & x_2 & x_3 & s_1 & s_2 & z & \\
\hline
\boxed{10} & 7 & 0 & 3 & -1 & 0 & 30 \\
2 & 8 & 6 & 0 & 1 & 0 & 24 \\
\hline
-3 & 5 & 0 & 0 & 1 & 1 & 24
\end{array}
$$

Pivot on the 10 in row 1, column 1.

$$
\begin{array}{l}
\\
-R_1 + 5R_2 \to R_2 \\
3R_1 + 10R_3 \to R_3
\end{array}
\qquad
\begin{array}{cccccc|c}
x_1 & x_2 & x_3 & s_1 & s_2 & z & \\
\hline
10 & 7 & 0 & 3 & -1 & 0 & 30 \\
0 & 33 & 30 & -3 & 6 & 0 & 90 \\
\hline
0 & 71 & 0 & 9 & 7 & 10 & 330
\end{array}
$$

Create a 1 in the columns corresponding to x_1, x_3, and z.

$$
\begin{array}{l}
\frac{1}{10}R_1 \to R_1 \\
\frac{1}{30}R_2 \to R_2 \\
\\
\frac{1}{10}R_3 \to R_3
\end{array}
\qquad
\begin{array}{cccccc|c}
x_1 & x_2 & x_3 & s_1 & s_2 & z & \\
\hline
1 & \frac{7}{10} & 0 & \frac{3}{10} & -\frac{1}{10} & 0 & 3 \\
0 & \frac{11}{10} & 1 & -\frac{1}{10} & \frac{1}{5} & 0 & 3 \\
\hline
0 & \frac{71}{10} & 0 & \frac{9}{10} & \frac{7}{10} & 1 & 33
\end{array}
$$

The maximum value is 33 when $x_1 = 3, x_2 = 0, x_3 = 3, s_1 = 0$, and $s_2 = 0$.

9.

$$
\begin{array}{ccccccc|c}
x_1 & x_2 & x_3 & s_1 & s_2 & s_3 & z & \\
\hline
1 & 2 & 2 & 1 & 0 & 0 & 0 & 50 \\
\boxed{3} & 1 & 0 & 0 & 1 & 0 & 0 & 20 \\
1 & 0 & 2 & 0 & 0 & -1 & 0 & 15 \\
\hline
-5 & -3 & -2 & 0 & 0 & 0 & 1 & 0
\end{array}
$$

The initial basic solution is not feasible since $s_3 = -15$. In the third row where the negative coefficient appears, the nonnegative entry that appears farthest to the left is the 1 in the first column. In the first column, the smallest nonnegative quotient is $\frac{20}{3}$. Pivot on the 3 in row 2, column 1.

$$
\begin{array}{l}
-R_2 + 3R_1 \to R_1 \\
\\
\\
-R_2 + 3R_3 \to R_3 \\
\\
5R_2 + 3R_4 \to R_4
\end{array}
\qquad
\begin{array}{ccccccc|c}
x_1 & x_2 & x_3 & s_1 & s_2 & s_3 & z & \\
\hline
0 & 5 & 6 & 3 & -1 & 0 & 0 & 130 \\
3 & 1 & 0 & 0 & 1 & 0 & 0 & 20 \\
0 & -1 & \boxed{6} & 0 & -1 & -3 & 0 & 25 \\
\hline
0 & -4 & -6 & 0 & 5 & 0 & 3 & 100
\end{array}
$$

Continue by pivoting on each circled entry.

$$
\begin{array}{c}
 \\
-R_3+R_2\to R_1 \\
 \\
 \\
R_3+R_4\to R_4
\end{array}
\begin{array}{cccccccc}
x_1 & x_2 & x_3 & s_1 & s_2 & s_3 & z & \\
\left[\begin{array}{ccccccc|c}
0 & \boxed{6} & 0 & 3 & 0 & 3 & 0 & 105 \\
3 & 1 & 0 & 0 & 1 & 0 & 0 & 20 \\
0 & -1 & 6 & 0 & -1 & -3 & 0 & 25 \\
\hline
0 & -5 & 0 & 0 & 4 & -3 & 3 & 125
\end{array}\right]
\end{array}
$$

The basic solution is now feasible, but there are negative indicators.
Continue pivoting.

$$
\begin{array}{c}
 \\
-R_1+6R_2\to R_2 \\
R_1+6R_3\to R_3 \\
5R_1+6R_4\to R_4
\end{array}
\begin{array}{cccccccc}
x_1 & x_2 & x_3 & s_1 & s_2 & s_3 & z & \\
\left[\begin{array}{ccccccc|c}
0 & 6 & 0 & 3 & 0 & \boxed{3} & 0 & 105 \\
18 & 0 & 0 & -3 & 6 & -3 & 0 & 15 \\
0 & 0 & 36 & 3 & 0 & -15 & 0 & 255 \\
\hline
0 & 0 & 0 & 15 & 24 & -3 & 18 & 1275
\end{array}\right]
\end{array}
$$

$$
\begin{array}{c}
 \\
R_1+R_2\to R_2 \\
5R_1+R_3\to R_3 \\
R_1+R_4\to R_4
\end{array}
\begin{array}{cccccccc}
x_1 & x_2 & x_3 & s_1 & s_2 & s_3 & z & \\
\left[\begin{array}{ccccccc|c}
0 & 6 & 0 & 3 & 0 & 3 & 0 & 105 \\
18 & 6 & 0 & 0 & 6 & 0 & 0 & 120 \\
0 & 30 & 36 & 18 & 0 & 0 & 0 & 780 \\
\hline
0 & 6 & 0 & 18 & 24 & 0 & 18 & 1380
\end{array}\right]
\end{array}
$$

Create a 1 in the columns corresponding to x_1, x_3, s_3, and z.

$$
\begin{array}{c}
\frac{1}{3}R_1\to R_1 \\
\frac{1}{18}R_2\to R_2 \\
\frac{1}{36}R_3\to R_3 \\
\frac{1}{18}R_4\to R_4
\end{array}
\begin{array}{cccccccc}
x_1 & x_2 & x_3 & s_1 & s_2 & s_3 & z & \\
\left[\begin{array}{ccccccc|c}
0 & 2 & 0 & 1 & 0 & 1 & 0 & 35 \\
1 & .33 & 0 & 0 & .33 & 0 & 0 & 6.67 \\
0 & .83 & 1 & .5 & 0 & 0 & 0 & 21.67 \\
\hline
0 & .33 & 0 & 1 & 1.33 & 0 & 1 & 76.67
\end{array}\right]
\end{array}
$$

The maximum value is about 76.67 when $x_1 \approx 6.67, x_2 = 0, x_3 \approx 21.67, s_1 = 0, s_2 = 0$, and $s_3 = 35$.

11. Minimize $\quad w = 10y_1 + 15y_2$
   subject to: $\quad y_1 + y_2 \geq 17$
   $\qquad\qquad\quad 5y_1 + 8y_2 \geq 42$
   with $\qquad\quad y_1 \geq 0, y_2 \geq 0.$

Using the dual method:

To form the dual, write the augmented matrix for the given problem.

$$
\left[\begin{array}{cc|c}
1 & 1 & 17 \\
5 & 8 & 42 \\
\hline
10 & 15 & 0
\end{array}\right]
$$

Form the transpose of this matrix.

$$
\left[\begin{array}{cc|c}
1 & 5 & 10 \\
1 & 8 & 15 \\
\hline
17 & 42 & 0
\end{array}\right]
$$

Write the dual problem.

   Maximize $\quad z = 17x_1 + 42x_2$
   subject to: $\quad x_1 + 5x_2 \leq 10$
   $\qquad\qquad\quad x_1 + 8x_2 \leq 15$
   with $\qquad\quad x_1 \geq 0, x_2 \geq 0.$

The initial simplex tableau is as follows.

$$
\begin{array}{ccccc}
x_1 & x_2 & s_1 & s_2 & z \\
\end{array}
$$

$$
\left[
\begin{array}{ccccc|c}
1 & 5 & 1 & 0 & 0 & 10 \\
1 & \boxed{8} & 0 & 1 & 0 & 15 \\
\hline
-17 & -42 & 0 & 0 & 1 & 0 \\
\end{array}
\right]
$$

Pivot on the 8 in row 2 column 2.

$$
\begin{array}{ccccc}
 & x_1 & x_2 & s_1 & s_2 & z \\
\end{array}
$$

$$
\begin{array}{l}
-5R_2 + 8R_1 \rightarrow R_1 \\
\\
21R_2 + 4R_3 \rightarrow R_3
\end{array}
\left[
\begin{array}{ccccc|c}
\boxed{3} & 0 & 8 & -5 & 0 & 5 \\
1 & 8 & 0 & 1 & 0 & 15 \\
\hline
-47 & 0 & 0 & 21 & 4 & 315 \\
\end{array}
\right]
$$

Pivot on the 3 in row 1, column 1.

$$
\begin{array}{ccccc}
x_1 & x_2 & s_1 & s_2 & z \\
\end{array}
$$

$$
\begin{array}{l}
\\
-R_1 + 3R_2 \rightarrow R_2 \\
\\
47R_1 + 3R_3 \rightarrow R_3
\end{array}
\left[
\begin{array}{ccccc|c}
3 & 0 & 8 & -5 & 0 & 5 \\
0 & 24 & -8 & \boxed{8} & 0 & 40 \\
\hline
0 & 0 & 376 & -172 & 12 & 1180 \\
\end{array}
\right]
$$

Pivot on the 8 in row 2, column 4.

$$
\begin{array}{ccccc}
x_1 & x_2 & s_1 & s_2 & z \\
\end{array}
$$

$$
\begin{array}{l}
5R_2 + 8R_1 \rightarrow R_1 \\
\\
43R_2 + 2R_3 \rightarrow R_3
\end{array}
\left[
\begin{array}{ccccc|c}
24 & 120 & 24 & 0 & 0 & 240 \\
0 & 24 & -8 & 8 & 0 & 40 \\
\hline
0 & 1032 & 408 & 0 & 24 & 4080 \\
\end{array}
\right]
$$

Create a 1 in the columns corresponding to x_1, x_2, and z.

$$
\begin{array}{ccccc}
x_1 & x_2 & s_1 & s_2 & z \\
\end{array}
$$

$$
\begin{array}{l}
\frac{1}{24}R_1 \rightarrow R_1 \\
\frac{1}{8}R_2 \rightarrow R_2 \\
\frac{1}{24}R_3 \rightarrow R_3
\end{array}
\left[
\begin{array}{ccccc|c}
1 & 5 & 1 & 0 & 0 & 10 \\
0 & 3 & -1 & 1 & 0 & 5 \\
\hline
0 & 43 & 17 & 0 & 1 & 170 \\
\end{array}
\right]
$$

The minimum value is 170 when $y_1 = 17$ and $y_2 = 0$.

Using the method of 4.4:

Change the objective function to

$$\text{Maximize } z = -w = -10y_1 - 15y_2.$$

The constraints are not changed.

The initial simplex tableau is as follows.

$$
\begin{array}{ccccc}
y_1 & y_2 & s_1 & s_2 & z \\
\end{array}
$$

$$
\left[
\begin{array}{ccccc|c}
1 & 1 & -1 & 0 & 0 & 17 \\
\boxed{5} & 8 & 0 & -1 & 0 & 42 \\
\hline
10 & 15 & 0 & 0 & 1 & 0 \\
\end{array}
\right]
$$

The solution is not feasible since $s_1 = -17$ and $s_2 = -42$. Pivot on the 5 in row 2, column 1.

$$\begin{array}{c} -R_2 + 5R_1 \rightarrow R_1 \\ \\ -2R_2 + R_3 \rightarrow R_3 \end{array} \quad \begin{array}{ccccc} y_1 & y_2 & s_1 & s_2 & z \\ \left[\begin{array}{ccccc|c} 0 & -3 & -5 & 1 & 0 & 43 \\ 5 & 8 & 0 & -1 & 0 & 42 \\ \hline 0 & -1 & 0 & 2 & 1 & -84 \end{array}\right] \end{array}$$

The solution is still not feasible since $s_1 = -\frac{43}{5}$. But there are no positive entries to the left of the -5 in column 3 so it is not possible to choose a pivot element. The method of 4.4 fails to provide a solution in this case.

13. Minimize $\quad w = 7y_1 + 2y_2 + 3y_3$

subject to: $\quad y_1 + y_2 + 2y_3 \geq 48$

$\qquad\qquad y_1 + y_2 \qquad \geq 12$

$\qquad\qquad\qquad\quad y_3 \geq 10$

$\qquad\quad 3y_1 \qquad + y_3 \geq 30$

with $\qquad y_1 \geq 0, _2 \geq 0, y_3 \geq 0.$

Using the dual method:

To form the dual, write the augmented matrix for the given problem.

$$\left[\begin{array}{ccc|c} 1 & 1 & 2 & 48 \\ 1 & 1 & 0 & 12 \\ 0 & 0 & 1 & 10 \\ 3 & 0 & 1 & 30 \\ \hline 7 & 2 & 3 & 0 \end{array}\right]$$

Form the transpose of this matrix.

$$\left[\begin{array}{cccc|c} 1 & 1 & 0 & 3 & 7 \\ 1 & 1 & 0 & 0 & 2 \\ 2 & 0 & 1 & 1 & 3 \\ \hline 48 & 12 & 10 & 30 & 0 \end{array}\right]$$

Write the dual problem.

Maximize $\quad z = 48x_1 + 12x_2 + 10x_3 + 30x_4$

subject to: $\quad x_1 + x_2 \qquad + 3x_4 \leq 7$

$\qquad\qquad x_1 + x_2 \qquad\qquad \leq 2$

$\qquad\qquad 2x_1 \qquad + x_3 + x_4 \leq 3$

with $\qquad x_1 \geq 0, x_2 \geq 0, x_3 \geq 0, x_4 \geq 0.$

The initial simplex tableau is as follows.

$$\begin{array}{cccccccc} x_1 & x_2 & x_3 & x_4 & s_1 & s_2 & s_3 & z \\ \left[\begin{array}{cccccccc|c} 1 & 1 & 0 & 3 & 1 & 0 & 0 & 0 & 7 \\ 1 & 1 & 0 & 0 & 0 & 1 & 0 & 0 & 2 \\ \boxed{2} & 0 & 1 & 1 & 0 & 0 & 1 & 0 & 3 \\ \hline -48 & -12 & -10 & -30 & 0 & 0 & 0 & 1 & 0 \end{array}\right] \end{array}$$

Pivot on the 2 in row 3, column 1.

$$\begin{array}{c} -R_3 + 2R_1 \rightarrow R_1 \\ -R_3 + 2R_2 \rightarrow R_2 \\ \\ 24R_3 + R_4 \rightarrow R_4 \end{array} \quad \begin{array}{cccccccc} x_1 & x_2 & x_3 & x_4 & s_1 & s_2 & s_3 & z \\ \left[\begin{array}{cccccccc|c} 0 & 2 & -1 & 5 & 2 & 0 & -1 & 0 & 11 \\ 0 & \boxed{2} & -1 & -1 & 0 & 2 & -1 & 0 & 1 \\ 2 & 0 & 1 & 1 & 0 & 0 & 1 & 0 & 3 \\ \hline 0 & -12 & 14 & -6 & 0 & 0 & 24 & 1 & 72 \end{array}\right] \end{array}$$

Pivot on the 2 in row 2, column 2.

	x_1	x_2	x_3	x_4	s_1	s_2	s_3	z	
$-R_2+R_1 \rightarrow R_1$	0	0	0	6	2	-2	0	0	10
	0	2	-1	-1	0	2	-1	0	1
	2	0	1	1	0	0	1	0	3
$6R_2+R_4 \rightarrow R_4$	0	0	8	-12	0	12	18	1	78

Pivot on the 6 in row 1, column 4.

	x_1	x_2	x_3	x_4	s_1	s_2	s_3	z	
	0	0	0	6	2	-2	0	0	10
$R_1+6R_2 \rightarrow R_2$	0	12	-6	0	2	10	-6	0	16
$-R_1+6R_3 \rightarrow R_3$	12	0	6	0	-2	2	6	0	8
$2R_1+ R_4 \rightarrow R_4$	0	0	8	0	4	8	18	1	98

Create a 1 in the columns corresponding to x_1, x_2, and x_4.

	x_1	x_2	x_3	x_4	s_1	s_2	s_3	z	
$\frac{1}{6}R_1 \rightarrow R_1$	0	0	0	1	$\frac{1}{3}$	$-\frac{1}{3}$	0	0	$\frac{5}{3}$
$\frac{1}{12}R_2 \rightarrow R_2$	0	1	$-\frac{1}{2}$	0	$\frac{1}{6}$	$\frac{5}{6}$	$-\frac{1}{2}$	0	$\frac{4}{3}$
$\frac{1}{12}R_3 \rightarrow R_3$	1	0	$\frac{1}{2}$	0	$-\frac{1}{6}$	$\frac{1}{6}$	$\frac{1}{2}$	0	$\frac{2}{3}$
	0	0	8	0	4	8	18	1	98

The minimum value is 98 when $y_1 = 4, y_2 = 8$, and $y_3 = 18$.

Using the method of 4.4:

Change the objective function to

$$\text{Maximize } z = -w = -7y_1 - 2y_2 - 3y_3.$$

The constraints are not changed.

The initial simplex tableau is as follows.

y_1	y_2	y_3	s_1	s_2	s_3	s_4	z	
1	1	2	-1	0	0	0	0	48
1	1	0	0	-1	0	0	0	12
0	0	1	0	0	-1	0	0	10
3	0	1	0	0	0	-1	0	30
7	2	3	0	0	0	0	1	0

The solution is not feasible since $s_1 = -48, s_2 = -12, s_3 = -10$, and $s_4 = -30$. Pivot on the 3 in row 4, column 1.

	y_1	y_2	y_3	s_1	s_2	s_3	s_4	z	
$-R_4+3R_1 \rightarrow R_1$	0	3	5	-3	0	0	1	0	114
$-R_4+3R_2 \rightarrow R_2$	0	3	-1	0	-3	0	1	0	6
	0	0	1	0	0	-1	0	0	10
	3	0	1	0	0	0	-1	0	30
$-7R_4+3R_5 \rightarrow R_5$	0	6	2	0	0	0	7	3	-210

The solution is still not feasible since $s_1 = -38, s_2 = -2$, and $s_3 = -10$. Pivot on the 3 in row 1, column 2.

$$R_1 - R_2 \rightarrow R_2$$
$$-2R + R_5 \rightarrow R_5$$

	y_1	y_2	y_3	s_1	s_2	s_3	s_4	z	
	0	3	5	-3	0	0	1	0	114
	0	0	6	-3	3	0	0	0	108
	0	0	1	0	0	-1	0	0	10
	3	0	1	0	0	0	-1	0	30
	0	0	-8	6	0	0	5	3	-438

Again, the solution is not feasible since $s_3 = -10$ and $s_4 = -30$. Pivot on the 1 in row 3, column 3.

$$-5R_3 + R_1 \rightarrow R_1$$
$$-6R_3 + R_2 \rightarrow R_2$$
$$-R_3 + R_4 \rightarrow R_4$$
$$8R_3 + R_5 \rightarrow R_5$$

	y_1	y_2	y_3	s_1	s_2	s_3	s_4	z	
	0	3	0	-3	0	5	1	0	64
	0	0	0	-3	3	6	0	0	48
	0	0	1	0	0	-1	0	0	10
	3	0	0	0	0	1	-1	0	20
	0	0	0	6	0	-8	5	3	-358

The solution is feasible because all variables are nonnegative. But it is still not optimal. Pivot on the 6 in row 2, column 6.

$$-5R_2 + 6R_1 \rightarrow R_1$$
$$R_2 + 6R_3 \rightarrow R_3$$
$$-R_2 + 6R_4 \rightarrow R_4$$
$$4R_2 + 3R_5 \rightarrow R_5$$

	y_1	y_2	y_3	s_1	s_2	s_3	s_4	z	
	0	18	0	-3	-15	0	6	0	144
	0	0	0	-3	3	6	0	0	48
	0	0	6	-3	3	0	0	0	108
	18	0	0	3	-3	0	-6	0	72
	0	0	0	6	12	0	15	9	-882

Create a 1 in the columns corresponding to y_1, y_2, y_3, s_3 and z.

$$\tfrac{1}{18}R_1 \rightarrow R_1$$
$$\tfrac{1}{6}R_2 \rightarrow R_2$$
$$\tfrac{1}{6}R_3 \rightarrow R_3$$
$$\tfrac{1}{18}R_4 \rightarrow R_4$$
$$\tfrac{1}{9}R_5 \rightarrow R_5$$

	y_1	y_2	y_3	s_1	s_2	s_3	s_4	z	
	0	1	0	$-\tfrac{1}{6}$	$-\tfrac{5}{6}$	0	$\tfrac{1}{3}$	0	8
	0	0	0	$-\tfrac{1}{2}$	$\tfrac{1}{2}$	1	0	0	8
	0	0	1	$-\tfrac{1}{2}$	$\tfrac{1}{2}$	0	0	0	18
	1	0	0	$\tfrac{1}{6}$	$-\tfrac{1}{6}$	0	$-\tfrac{1}{3}$	0	4
	0	0	0	$\tfrac{2}{3}$	$\tfrac{4}{3}$	0	$\tfrac{5}{3}$	1	-98

Since $z = -w = -98$, the minimum value is 98 when $y_1 = 4, y_2 = 8$, and $y_3 = 18$.

15.

	y_1	y_2	s_1	s_2	s_3	s_4	z	
	0	0	3	0	1	1	0	2
	1	0	-2	0	2	0	0	8
	0	1	7	0	0	0	0	12
	0	0	1	1	-4	0	0	1
	0	0	5	0	8	0	1	-62

From this final tableau, read that the maximum value of $z = -w$ is -62 when $y_1 = 8, y_2 = 12, s_1 = 0, s_2 = 1, s_3 = 0$, and $s_4 = 2$. Therefore, the minimum value of w is 62 when $y_1 = 8, y_2 = 12, s_1 = 0, s_2 = 1, s_3 = 0$, and $s_4 = 2$.

17.

$$
\begin{array}{ccccc}
x_1 & x_2 & s_1 & s_2 & z \\
\end{array}
$$

$$
\left[\begin{array}{ccccc|c}
5 & 10 & 1 & 0 & 0 & 120 \\
\boxed{10} & 15 & 0 & -1 & 0 & 200 \\
\hline
-20 & -30 & 0 & 0 & 1 & 0
\end{array}\right]
$$

The initial tableau is not feasible. Pivot on the 10 in row 2, column 1.

$$
\begin{array}{ccccc}
x_1 & x_2 & s_1 & s_2 & z \\
\end{array}
$$

$$
\begin{array}{l}
-R_2 + 2R_1 \rightarrow R_1 \\
\\
2R_2 + R_3 \rightarrow R_3
\end{array}
\left[\begin{array}{ccccc|c}
0 & 5 & 2 & \boxed{1} & 0 & 40 \\
10 & 15 & 0 & -1 & 0 & 200 \\
\hline
0 & 0 & 0 & -2 & 1 & 400
\end{array}\right]
$$

The basic solution is feasible, but there are negative indicators. Pivot on the 1 in row 1, column 4.

$$
\begin{array}{ccccc}
x_1 & x_2 & s_1 & s_2 & z \\
\end{array}
$$

$$
\begin{array}{l}
\\
R_2 + 2R_1 \rightarrow R_2 \\
2R_1 + R_3 \rightarrow R_3
\end{array}
\left[\begin{array}{ccccc|c}
0 & 5 & 2 & 1 & 0 & 40 \\
10 & 20 & 2 & 0 & 0 & 240 \\
\hline
0 & 10 & 4 & 0 & 1 & 480
\end{array}\right]
$$

Create a one in the column corresponding to x_1.

$$
\begin{array}{ccccc}
x_1 & x_2 & s_1 & s_2 & z \\
\end{array}
$$

$$
\begin{array}{l}
\\
\frac{1}{10}R_2 \rightarrow R_2 \\
\\
\end{array}
\left[\begin{array}{ccccc|c}
0 & 5 & 2 & 1 & 0 & 40 \\
1 & 2 & \frac{1}{5} & 0 & 0 & 24 \\
\hline
0 & 10 & 4 & 0 & 1 & 480
\end{array}\right]
$$

The maximum value is $z = 480$ when $x_1 = 24$ and $x_2 = 0$.

19. Maximize $z = 10x_1 + 12x_2$

subject to: $2x_1 + 2x_2 = 17$

$\qquad\qquad 2x_1 + 5x_2 \geq 22$

$\qquad\qquad 4x_1 + 3x_2 \leq 28$

with $x_1 \geq 0, x_2 \geq 0.$

Introduce artificial variable a, surplus variable s_1, and slack variable s_2. The initial simplex tableau as follows.

$$
\begin{array}{cccccc}
x_1 & x_2 & a & s_1 & s_2 & z \\
\end{array}
$$

$$
\left[\begin{array}{cccccc|c}
\boxed{2} & 2 & 1 & 0 & 0 & 0 & 17 \\
2 & 5 & 0 & -1 & 0 & 0 & 22 \\
4 & 3 & 0 & 0 & 1 & 0 & 28 \\
\hline
-10 & -12 & 0 & 0 & 0 & 1 & 0
\end{array}\right]
$$

First, eliminate the artificial variable a. Pivot on the 2 in row 1, column 1.

$$
\begin{array}{cccccc}
x_1 & x_2 & a & s_1 & s_2 & z \\
\end{array}
$$

$$
\begin{array}{l}
\\
-R_1 + R_2 \rightarrow R_2 \\
2R_1 - R_3 \rightarrow R_3 \\
5R_1 + R_4 \rightarrow R_4
\end{array}
\left[\begin{array}{cccccc|c}
2 & 2 & 1 & 0 & 0 & 0 & 17 \\
0 & 3 & -1 & -1 & 0 & 0 & 5 \\
0 & 1 & 2 & 0 & -1 & 0 & 6 \\
\hline
0 & -2 & 5 & 0 & 0 & 1 & 85
\end{array}\right]
$$

Now $a = 0$, so we can drop the a column.

$$\begin{bmatrix} x_1 & x_2 & s_1 & s_2 & z & \\ 2 & 2 & 0 & 0 & 0 & 17 \\ 0 & \boxed{3} & -1 & 0 & 0 & 5 \\ 0 & 1 & 0 & -1 & 0 & 6 \\ \hline 0 & -2 & 0 & 0 & 1 & 85 \end{bmatrix}$$

Because $s_1 = -5$, we choose the 3 in row 2, column 2, as the next pivot.

$$\begin{array}{l} -2R_2 + 3R_1 \rightarrow R_1 \\ \\ \\ -R_2 + 3R_3 \rightarrow R_3 \\ 2R_2 + 3R_4 \rightarrow R_4 \end{array} \begin{bmatrix} x_1 & x_2 & s_1 & s_2 & z & \\ 6 & 0 & 2 & 0 & 0 & 41 \\ 0 & 3 & -1 & 0 & 0 & 5 \\ 0 & 0 & \boxed{1} & -3 & 0 & 13 \\ \hline 0 & 0 & -2 & 0 & 3 & 265 \end{bmatrix}$$

The solution is still not feasible since $s_2 = -\frac{13}{3}$. Pivot on the 1 in row 3, column 3.

$$\begin{array}{l} -2R_3 + R_1 \rightarrow R_1 \\ R_3 + R_2 \rightarrow R_2 \\ \\ 2R_3 + R_4 \rightarrow R_4 \end{array} \begin{bmatrix} x_1 & x_2 & s_1 & s_2 & z & \\ 6 & 0 & 0 & \boxed{6} & 0 & 15 \\ 0 & 3 & 0 & -3 & 0 & 18 \\ 0 & 0 & 1 & -3 & 0 & 13 \\ \hline 0 & 0 & 0 & -6 & 3 & 291 \end{bmatrix}$$

The solution is now feasible but is not yet optimal. Pivot on the 6 in row 1, column 4.

$$\begin{array}{l} \\ R_1 + 2R_2 \rightarrow R_2 \\ R_1 + 2R_3 \rightarrow R_3 \\ R_1 + R_4 \rightarrow R_4 \end{array} \begin{bmatrix} x_1 & x_2 & s_1 & s_2 & z & \\ 6 & 0 & 0 & 6 & 0 & 15 \\ 6 & 6 & 0 & 0 & 0 & 51 \\ 6 & 0 & 2 & 0 & 0 & 41 \\ 6 & 0 & 0 & 0 & 3 & 306 \end{bmatrix}$$

Create a 1 in the columns corresponding to x_2, s_1, s_2, and z.

$$\begin{array}{l} \frac{1}{6}R_1 \rightarrow R_1 \\ \frac{1}{6}R_2 \rightarrow R_2 \\ \frac{1}{2}R_3 \rightarrow R_3 \\ \frac{1}{3}R_4 \rightarrow R_4 \end{array} \begin{bmatrix} x_1 & x_2 & s_1 & s_2 & z & \\ 1 & 0 & 0 & 1 & 0 & \frac{5}{2} \\ 1 & 1 & 0 & 0 & 0 & \frac{17}{2} \\ 3 & 0 & 1 & 0 & 0 & \frac{41}{2} \\ 2 & 0 & 0 & 0 & 1 & 102 \end{bmatrix}$$

The maximum is 102 when $x_1 = 0$ and $x_2 = \frac{17}{2}$.

21. Any maximizing or minimizing problems can be solved using slack, surplus, and artificial variables. Slack variables are used in problems involving "$\leq$" constraints. Surplus variables are used in problems involving "$\geq$" constraints. Artificial variables are used in problems involving "$=$" constraints.

23.

$$\left[\begin{array}{cccccc|c} 4 & 2 & 3 & 1 & 0 & 0 & 9 \\ 5 & 4 & 1 & 0 & 1 & 0 & 10 \\ \hline -6 & -7 & -5 & 0 & 0 & 1 & 0 \end{array}\right]$$

(a) The 1 in column 4 and the 1 in column 5 indicate that the constraints involve $\leq$. The problem being solved with this tableau is:

Maximize $z = 6x_1 + 7x_2 + 5x_3$

subject to: $4x_1 + 2x_2 + 3x_3 \leq 9$

 $5x_1 + 4x_2 + x_3 \leq 10$

with $x_1 \geq 0, x_2 \geq 0, x_3 \geq 0.$

(b)   If the 1 in row 1, column 4 was -1 rather than 1, then the first constraint would have a surplus variable rather than a slack variable, which means the first constraint would be $4x_1 + 2x_2 + 3x_3 \geq 9$ instead of $4x_1 + 2x_2 + 3x_3 \leq 9$.

(c)

	x_1	x_2	x_3	s_1	s_2	z	
	3	0	5	2	−1	0	8
	11	10	0	−1	3	0	21
	47	0	0	13	11	10	227

From this tableau, the solution is $x_1 = 0$,

$x_2 = \frac{21}{10} = 2.1, x_3 = \frac{8}{5} = 1.6,$ and

$z = \frac{227}{10} = 22.7.$

(d) The dual of the original problem is as follows:

Minimize $w = 9y_1 + 10y_2$

subject to: $4y_1 + 5y_2 \geq 6$

 $2y_1 + 4y_2 \geq 7$

 $3y_1 + y_2 \geq 5$

with $y_1 \geq 0, y_2 \geq 0.$

(e) From the tableau in part (c), the solution of the dual in part (d) is $y_1 = \frac{13}{10} = 1.3, y_2 = \frac{11}{10} = 1.1,$ and $w = \frac{227}{10} = 22.7.$

25. (a)   Let $x_1 =$ the number of cake plates,

$x_2 =$ the number of bread plates,

and $x_3 =$ the number of dinner plates.

(b) The objective function to maximize is

$z = 15x_1 + 12x_2 + 5x_3.$

(c) The constraints are

$$15x_1 + 10x_2 + 8x_3 \leq 1500$$
$$5x_1 + 4x_2 + 4x_3 \leq 2700$$
$$6x_1 + 5x_2 + 5x_3 \leq 1200.$$

27. (a) Let $x_1 = $ number of gallons of fruity wine
and $x_2 = $ number of gallons of crystal wine.

(b) The profit function is

$$z = 12x_1 + 15x_2.$$

(c) The ingredients available are the limitations; the constraints are

$$2x_1 + x_2 \leq 110$$
$$2x_1 + 3x_2 \leq 125$$
$$2x_1 + x_2 \leq 90.$$

29. Maximize $z = 15x_1 + 12x_2 + 5x_3$

subject to: $15x_1 + 10x_2 + 8x_3 \leq 1500$
$$5x_1 + 4x_2 + 4x_3 \leq 2700$$
$$6x_1 + 5x_2 + 5x_3 \leq 1200$$
with $\qquad x_1 \geq 0, x_2 \geq 0, x_3 \geq 0.$

The initial tableau is as follows.

$$
\begin{array}{ccccccc|c}
x_1 & x_2 & x_3 & s_1 & s_2 & s_3 & z & \\
\hline
\boxed{15} & 10 & 8 & 1 & 0 & 0 & 0 & 1500 \\
5 & 4 & 4 & 0 & 1 & 0 & 0 & 2700 \\
6 & 5 & 5 & 0 & 0 & 1 & 0 & 1200 \\
\hline
-15 & -12 & -5 & 0 & 0 & 0 & 1 & 0
\end{array}
$$

Pivot on the 15 in row 1, column 1.

$$
\begin{array}{c}
\\
-R_1 + 3R_2 \to R_2 \\
-2R_1 + 5R_3 \to R_3 \\
R_1 + R_4 \to R_4
\end{array}
\begin{array}{ccccccc|c}
x_1 & x_2 & x_3 & s_1 & s_2 & s_3 & z & \\
\hline
15 & \boxed{10} & 8 & 1 & 0 & 0 & 0 & 1500 \\
0 & 2 & 4 & -1 & 3 & 0 & 0 & 6600 \\
0 & 5 & 9 & -2 & 0 & 5 & 0 & 3000 \\
\hline
0 & -2 & 3 & 1 & 0 & 0 & 1 & 1500
\end{array}
$$

Pivot on the 10 in row 1, column 2.

$$
\begin{array}{c}
\\
-R_1 + 5R_2 \to R_2 \\
-R_1 + 2R_3 \to R_3 \\
R_1 + 5R_4 \to R_4
\end{array}
\begin{array}{ccccccc|c}
x_1 & x_2 & x_3 & s_1 & s_2 & s_3 & z & \\
\hline
15 & 10 & 8 & 1 & 0 & 0 & 0 & 1500 \\
-15 & 0 & 12 & -6 & 15 & 0 & 0 & 31{,}500 \\
-15 & 0 & 10 & -5 & 0 & 10 & 0 & 4500 \\
\hline
15 & 0 & 23 & 6 & 0 & 0 & 5 & 9000
\end{array}
$$

Create a 1 in the columns corresponding to $x_2, s_2, s_3,$ and z.

$$
\begin{array}{c}
\frac{1}{10}R_1 \to R_1 \\
\frac{1}{15}R_2 \to R_2 \\
\frac{1}{10}R_3 \to R_3 \\
\\
\frac{1}{5}R_4 \to R_4
\end{array}
\begin{array}{ccccccc|c}
x_1 & x_2 & x_3 & s_1 & s_2 & s_3 & z & \\
\hline
\frac{3}{2} & 1 & \frac{4}{5} & \frac{1}{10} & 0 & 0 & 0 & 150 \\
-1 & 0 & \frac{4}{5} & -\frac{2}{5} & 1 & 0 & 0 & 2100 \\
-\frac{3}{2} & 0 & 1 & -\frac{1}{2} & 0 & 1 & 0 & 450 \\
\hline
3 & 0 & \frac{23}{5} & \frac{6}{5} & 0 & 0 & 1 & 1800
\end{array}
$$

The maximum profit of $1800 when no cake plates, 150 bread plates, and no dinner places are produced.

31. Based on Exercise 27, the initial tableau is

$$\begin{array}{cccccc} x_1 & x_2 & s_1 & s_2 & s_3 & z \\ \end{array}$$

$$\left[\begin{array}{cccccc|c} 2 & 1 & 1 & 0 & 0 & 0 & 110 \\ 2 & \boxed{3} & 0 & 1 & 0 & 0 & 125 \\ 2 & 1 & 0 & 0 & 1 & 0 & 90 \\ \hline -12 & -15 & 0 & 0 & 0 & 1 & 0 \end{array}\right]$$

Locating the first pivot in the usual way, it is found to be the 3 in row 2, column 2. After row transformations, we get the next tableau.

$$\begin{array}{cccccc} x_1 & x_2 & s_1 & s_2 & s_3 & z \\ \end{array}$$

$$\begin{array}{l} -R_2 + 3R_1 \to R_1 \\ \\ -R_2 + 3R_3 \to R_3 \\ \\ 5R_2 + \ R_4 \to R_4 \end{array} \left[\begin{array}{cccccc|c} 4 & 0 & 3 & -1 & 0 & 0 & 205 \\ 2 & 3 & 0 & 1 & 0 & 0 & 125 \\ \boxed{4} & 0 & 0 & -1 & 3 & 0 & 145 \\ \hline -2 & 0 & 0 & 5 & 0 & 1 & 625 \end{array}\right]$$

Pivot on the 4 in row 3, column 1.

$$\begin{array}{cccccc} x_1 & x_2 & s_1 & s_2 & s_3 & z \\ \end{array}$$

$$\begin{array}{l} -R_3 + \ R_1 \to R_1 \\ -R_3 + 2R_2 \to R_2 \\ \\ R_3 + 2R_4 \to R_4 \end{array} \left[\begin{array}{cccccc|c} 0 & 0 & 3 & 0 & -3 & 0 & 60 \\ 0 & 6 & 0 & 3 & -3 & 0 & 105 \\ 4 & 0 & 0 & -1 & 3 & 0 & 145 \\ \hline 0 & 0 & 0 & 9 & 3 & 2 & 1395 \end{array}\right]$$

$$\begin{array}{cccccc} x_1 & x_2 & s_1 & s_2 & s_3 & z \\ \end{array}$$

$$\begin{array}{l} \frac{1}{3}R_1 \to R_1 \\ \\ \frac{1}{6}R_2 \to R_2 \\ \\ \frac{1}{4}R_3 \to R_3 \\ \\ \frac{1}{2}R_4 \to R_4 \end{array} \left[\begin{array}{cccccc|c} 0 & 0 & 1 & 0 & -1 & 0 & 20 \\ 0 & 1 & 0 & \frac{1}{2} & -\frac{1}{2} & 0 & \frac{35}{2} \\ 1 & 0 & 0 & -\frac{1}{4} & \frac{3}{4} & 0 & \frac{145}{4} \\ \hline 0 & 0 & 0 & \frac{9}{2} & \frac{3}{2} & 1 & \frac{1395}{2} \end{array}\right]$$

The final tableau gives the solution $x_1 = \frac{145}{4}$, $x_2 = \frac{35}{2}$, and $z = \frac{1395}{2} = 697.5$. 36.25 gal of fruity wine and 17.5 gal of crystal wine should be produced for a maximum profit of $697.50.

33. (a) Let $y_1 =$ the number of cases of corn,
$\quad\quad\quad\quad y_2 =$ the number of cases of beans,
$\quad$ and $y_3 =$ the number of cases of carrots.

Minimize $\quad w = 10y_1 + 15y_2 + 25y_3$
subject to: $\quad y_1 + y_2 + y_3 \geq 1000$
$$\quad\quad\quad\quad\quad y_1 \geq 2y_2$$
$$\quad\quad\quad\quad\quad y_3 \geq 340$$
with $\quad\quad y_1 \geq 0, y_2 \geq 0.$

The second constraint can be rewritten as $y_1 - 2y_2 \geq 0$. Change this to a maximization problem by letting $z = -w = -10y_1 - 15y_2 - 25y_3$. Now maximize $z = -10y_1 - 15y_2 - 25y_3$ subject to the constraints above. Begin by inserting surplus variables to set up the first tableau.

$$\begin{array}{ccccccc} y_1 & y_2 & y_3 & s_1 & s_2 & s_3 & z \\ \end{array}$$

$$\left[\begin{array}{ccccccc|c} \boxed{1} & 1 & 1 & -1 & 0 & 0 & 0 & 1000 \\ 1 & -2 & 0 & 0 & -1 & 0 & 0 & 0 \\ 0 & 0 & 1 & 0 & 0 & -1 & 0 & 340 \\ \hline 10 & 15 & 25 & 0 & 0 & 0 & 1 & 0 \end{array}\right]$$

Multiply row 2 by -1 so that s_2 is positive.

$$-R_2 \to R_2 \quad \begin{array}{cccccccc} & y_1 & y_2 & y_3 & s_1 & s_2 & s_3 & z & \\ \left[\begin{array}{ccccccc|c} \boxed{1} & 1 & 1 & -1 & 0 & 0 & 0 & 1000 \\ -1 & 2 & 0 & 0 & 1 & 0 & 0 & 0 \\ 0 & 0 & 1 & 0 & 0 & -1 & 0 & 340 \\ \hline 10 & 15 & 25 & 0 & 0 & 0 & 1 & 0 \end{array}\right] \end{array}$$

Pivot on the 1 in row 1, column 1.

$$\begin{array}{c} \\ R_1 + R_2 \to R_2 \\ \\ -10R_1 + R_4 \to R_4 \end{array} \quad \begin{array}{cccccccc} y_1 & y_2 & y_3 & s_1 & s_2 & s_3 & z & \\ \left[\begin{array}{ccccccc|c} 1 & 1 & 1 & -1 & 0 & 0 & 0 & 1000 \\ 0 & 3 & 1 & -1 & 1 & 0 & 0 & 1000 \\ 0 & 0 & \boxed{1} & 0 & 0 & -1 & 0 & 340 \\ \hline 0 & 5 & 15 & 10 & 0 & 0 & 1 & -10{,}000 \end{array}\right] \end{array}$$

Pivot on the 1 in row 3, column 3.

$$\begin{array}{c} -R_3 + R_1 \to R_1 \\ -R_3 + R_2 \to R_2 \\ \\ -15R_3 + R_4 \to R_4 \end{array} \quad \begin{array}{cccccccc} y_1 & y_2 & y_3 & s_1 & s_2 & s_3 & z & \\ \left[\begin{array}{ccccccc|c} 1 & 1 & 0 & -1 & 0 & 1 & 0 & 660 \\ 0 & 3 & 0 & -1 & 1 & 1 & 0 & 660 \\ 0 & 0 & 1 & 0 & 0 & -1 & 0 & 340 \\ \hline 0 & 5 & 0 & 10 & 0 & 15 & 1 & -15{,}100 \end{array}\right] \end{array}$$

The maximum value of z is $-15{,}100$ when $y_1 = 660, y_2 = 0$, and $y_3 = 340$. Hence the minimum value of w is $15{,}100$ when $y_1 = 660, y_2 = 0$, and $y_3 = 340$.

(b) The dual problem is as follows.

Maximize $\quad z = 1000x_1 + 340x_3$

subject to: $\quad x_1 + \ x_2 \le 10$

$\qquad\qquad\ x_1 - 2x_2 \le 15$

$\qquad\qquad\ x_1 + x_3 \le 25$

with $\qquad\ x_1 \ge 0, x_2 \ge 0, x_3 \ge 0.$

The initial simplex tableau is as follows.

$$\begin{array}{cccccccc} x_1 & x_2 & x_3 & s_1 & s_2 & s_3 & z & \\ \left[\begin{array}{ccccccc|c} \boxed{1} & 1 & 0 & 1 & 0 & 0 & 0 & 10 \\ 1 & -2 & 0 & 0 & 1 & 0 & 0 & 15 \\ 1 & 0 & 1 & 0 & 0 & 1 & 0 & 25 \\ \hline -1000 & 0 & -340 & 0 & 0 & 0 & 1 & 0 \end{array}\right] \end{array}$$

Pivot on the 1 in row 1, column 1.

$$\begin{array}{c} \\ -R_1 + R_2 \to R_2 \\ -R_1 + R_3 \to R_3 \\ 1000R_1 + R_4 \to R_4 \end{array} \quad \begin{array}{cccccccc} x_1 & x_2 & x_3 & s_1 & s_2 & s_3 & z & \\ \left[\begin{array}{ccccccc|c} 1 & 1 & 0 & 1 & 0 & 0 & 0 & 10 \\ 0 & -3 & 0 & -1 & 1 & 0 & 0 & 5 \\ 0 & -1 & \boxed{1} & -1 & 0 & 1 & 0 & 15 \\ \hline 0 & 1000 & -340 & 1000 & 0 & 0 & 1 & 10{,}000 \end{array}\right] \end{array}$$

Pivot on the 1 in row 3, column 3.

$$\begin{array}{c} \\ \\ \\ 340R_3 + R_4 \to R_4 \end{array} \quad \begin{array}{cccccccc} x_1 & x_2 & x_3 & s_1 & s_2 & s_3 & z & \\ \left[\begin{array}{ccccccc|c} 1 & 1 & 0 & 1 & 0 & 0 & 0 & 10 \\ 0 & -3 & 0 & -1 & 1 & 0 & 0 & 5 \\ 0 & -1 & 1 & -1 & 0 & 1 & 0 & 15 \\ \hline 0 & 660 & 0 & 660 & 0 & 340 & 1 & 15{,}100 \end{array}\right] \end{array}$$

The minimum value of w is 15,100 when $y_1 = 660$, $y_2 = 0$, and $y_3 = 340$, that is, 660 cases of corn, 0 cases of beans, and 340 cases of carrots should be produced to minimize costs, and the minimum cost is \$15,100.

35. Let $x_1 =$ the number of hours doing tai chi,
$\quad\quad x_2 =$ the number of hours riding a unicycle,
and $x_3 =$ the number of hours fencing.

If Ginger wants the total time doing tai chi to be at least twice as long as she rides a unicycle, then

$$x_1 \geq 2x_2$$
$$\text{or} \quad -x_1 + 2x_2 \leq 0.$$

The problem can be stated as follows.

Maximize $z = 236x_1 + 295x_2 + 354x_3$
subject to: $\quad x_1 + \ x_2 + \ x_3 \leq 10$
$$\quad\quad\quad\quad\quad\quad\quad\quad x_3 \leq \ 2$$
$$\quad\quad -x_1 + \ 2x_2 \quad\quad \leq \ 0$$
with $\quad\quad x_1 \geq 0, x_2 \geq 0, x_3 \geq 0.$

The initial simplex tableau is as follows.

$$\begin{bmatrix}
x_1 & x_2 & x_3 & s_1 & s_2 & s_3 & z & \\
1 & 1 & 1 & 1 & 0 & 0 & 0 & 10 \\
0 & 0 & \boxed{1} & 0 & 1 & 0 & 0 & 2 \\
-1 & 2 & 0 & 0 & 0 & 1 & 0 & 0 \\
\hline
-236 & -295 & -354 & 0 & 0 & 0 & 1 & 0
\end{bmatrix}$$

Pivot on the 1 in row 2, column 3.

$$\begin{array}{c}
\\
-R_2 + R_1 \rightarrow R_1 \\
\\
\\
354R_2 + R_4 \rightarrow R_4
\end{array}
\begin{bmatrix}
x_1 & x_2 & x_3 & s_1 & s_2 & s_3 & z & \\
1 & 1 & 0 & 1 & -1 & 0 & 0 & 8 \\
0 & 0 & 1 & 0 & 1 & 0 & 0 & 2 \\
-1 & \boxed{2} & 0 & 0 & 0 & 1 & 0 & 0 \\
\hline
-236 & -295 & 0 & 0 & 354 & 0 & 1 & 708
\end{bmatrix}$$

Pivot on the 2 in row 3, column 2.

$$\begin{array}{c}
-R_3 + 2R_1 \rightarrow R_1 \\
\\
\\
295R_3 + 2R_4 \rightarrow R_4
\end{array}
\begin{bmatrix}
x_1 & x_2 & x_3 & s_1 & s_2 & s_3 & z & \\
\boxed{3} & 0 & 0 & 2 & -2 & -1 & 0 & 16 \\
0 & 0 & 1 & 0 & 1 & 0 & 0 & 2 \\
-1 & 2 & 0 & 0 & 0 & 1 & 0 & 0 \\
\hline
-767 & 0 & 0 & 0 & 708 & 295 & 2 & 1416
\end{bmatrix}$$

Pivot on the 3 in row 1, column 1.

$$\begin{array}{c}
\\
\\
R_1 + 3R_3 \rightarrow R_3 \\
767R_1 + 3R_4 \rightarrow R_4
\end{array}
\begin{bmatrix}
x_1 & x_2 & x_3 & s_1 & s_2 & s_3 & z & \\
3 & 0 & 0 & 2 & -2 & -1 & 0 & 16 \\
0 & 0 & 1 & 0 & 1 & 0 & 0 & 2 \\
0 & 6 & 0 & 2 & -2 & 2 & 0 & 16 \\
\hline
0 & 0 & 0 & 1534 & 590 & 118 & 6 & 16{,}520
\end{bmatrix}$$

Create a 1 in the columns corresponding to $x_1, x_2,$ and z.

$$
\begin{array}{c}
\\
\tfrac{1}{3}R_1 \to R_1 \\
\\
\\
\tfrac{1}{6}R_3 \to R_3 \\
\\
\tfrac{1}{6}R_4 \to R_4
\end{array}
\begin{array}{ccccccc}
x_1 & x_2 & x_3 & s_1 & s_2 & s_3 & z \\
\end{array}
\left[
\begin{array}{ccccccc|c}
1 & 0 & 0 & \tfrac{2}{3} & -\tfrac{2}{3} & -\tfrac{1}{3} & 0 & \tfrac{16}{3} \\
0 & 0 & 1 & 0 & 1 & 0 & 0 & 2 \\
0 & 1 & 0 & \tfrac{1}{3} & -\tfrac{1}{3} & \tfrac{1}{3} & 0 & \tfrac{8}{3} \\
\hline
0 & 0 & 0 & \tfrac{767}{3} & \tfrac{295}{3} & \tfrac{59}{3} & 1 & \tfrac{8260}{3}
\end{array}
\right]
$$

Ginger will burn a maximum of $2753\tfrac{1}{3}$ calories if she does $\tfrac{16}{3}$ hours of tai chi, $\tfrac{8}{3}$ hours riding a unicycle, and 2 hours fencing.

Chapter 4 Test

[4.1]

1. For the following maximization problem,

    (a) determine the number of slack variables needed;

    (b) convert each constraint into a linear equation.

   $$\text{Maximize } z = 50x_1 + 80x_2$$

   $$\begin{aligned}
   \text{subject to:} \quad x_1 + 2x_2 &\leq 32 \\
   3x_1 + 4x_2 &\leq 84 \\
   x_2 &\leq 12
   \end{aligned}$$

   $$\text{with} \qquad x_1 \geq 0, \, x_2 \geq 0.$$

2. For the following maximization problem,

    (a) set up the initial simplex tableau;

    (b) determine the initial basic solution;

    (c) find the first pivot element and justify your choice.

   $$\text{Maximize } z = 10x_1 + 5x_2$$

   $$\begin{aligned}
   \text{subject to:} \quad 6x_1 + 2x_2 &\leq 36 \\
   2x_1 + 4x_2 &\leq 32
   \end{aligned}$$

   $$\text{with} \qquad x_1 \geq 0, \, x_2 \geq 0.$$

3. Determine the basic feasible solution from the following tableau.

x_1	x_2	x_3	s_1	s_2	s_3	z	
1	0	-2	1	0	2	0	6
0	1	4	1	0	-1	0	14
0	0	3	0	1	3	0	12
0	0	5	7	0	6	1	40

[4.2]

4. Solve the problem with the given initial tableau.

x_1	x_2	x_3	s_1	s_2	s_3	z	
1	1	1	1	0	0	0	1000
40	20	30	0	1	0	0	3200
1	2	1	0	0	1	0	160
-100	-300	-200	0	0	0	1	0

Use the simplex method to solve each linear programming problem.

5. Maximize $z = 6x_1 + 9x_2 + 6x_3$

    subject to: $\quad 2x_1 + 3x_2 + 3x_3 \leq 30$

    $\qquad\qquad\quad 2x_1 + 2x_2 + \quad x_3 \leq 20$

    $\qquad\qquad\quad 2x_1 + 5x_2 + \quad x_3 \leq 40$

    with $\qquad\qquad x_1 \geq 0,\ x_2 \geq 0,\ x_3 \geq 0.$

6. Mammoth Micros markets computers with single-sided and double-sided disk drives. They obtain these drives from Large Disks, Inc. and Double Drives Are Us. Large Disk charges \$250 for a single-sided and \$350 for a double-sided disk. Double drives charges \$290 and \$320 for single-sided and double-sided disks, respectively. Each month Large Disks can supply at most 1000 drives in all. Double Drives can supply at most 2000. Mammoth needs at least 1200 single and 1600 double drives. How many of each type should they buy from each company to minimize their total costs? What is the minimum cost?

[4.3]

7. Find the transpose of the following matrix.

    $$\begin{bmatrix} 2 & 1 & 7 & 6 & 3 \\ 5 & 9 & 0 & 4 & 2 \\ 6 & 8 & 5 & 1 & 4 \end{bmatrix}$$

8. For the following minimization problem,

    (a) state the dual problem;

    (b) solve the problem using the simplex method.

    Minimize $w = 5y_1 + 7y_2$

    subject to: $\quad y_1 \qquad\quad \geq 4$

    $\qquad\qquad\quad y_1 + \quad y_2 \geq 8$

    $\qquad\qquad\quad y_1 + 2y_2 \geq 10$

    with $\qquad\qquad y_1 \geq 0,\ y_2 \geq 0.$

9. State the dual problem for the following minimization problem.

    Minimize $w = 3y_1 + 4y_2$

    subject to: $\quad y_1 + 2y_2 \geq 8$

    $\qquad\qquad\quad 2y_1 + 2y_2 \geq 10$

    $\qquad\qquad\quad y_1 + 4y_2 \geq 12$

    with $\qquad\qquad y_1 \geq 0,\ y_2 \geq 0.$

[4.4]

10. Rewrite the following system of inequalities, adding slack variables or subtracting surplus variables as necessary.

    $$x_1 + 3x_2 + 2x_3 \leq 42$$
    $$3x_1 \qquad\quad + \quad x_3 \geq 20$$
    $$2x_1 + \quad x_2 + 2x_3 \geq 13$$

11. Convert the following problem into a maximization problem.

   Minimize $w = 8y_1 + 5y_2 + 3y_3$

   subject to:
   $$y_1 + 3y_2 + 2y_3 \geq 90$$
   $$3y_1 + y_2 + y_3 \geq 75$$
   $$2y_1 \qquad + 3y_3 \geq 60$$

   with $\qquad y_1 \geq 0,\ y_2 \geq 0,\ y_3 \geq 0.$

12. Use the simplex method to solve the following problem.

   Maximize $z = 6x_1 - 2x_2$

   subject to:
   $$x_1 + x_2 \leq 10$$
   $$3x_1 + 2x_2 \geq 24$$

   with $\qquad x_1 \geq 0,\ x_2 \geq 0.$

Chapter 4 Test Answers

1. **(a)** 3 slack variables are needed.   **(b)**
$$
\begin{aligned}
x_1 + 2x_2 + s_1 \quad\quad\quad\quad &= 32 \\
3x_1 + 4x_2 \quad\quad + s_2 \quad\quad &= 84 \\
x_2 \quad\quad\quad\quad + s_3 &= 12
\end{aligned}
$$

2. **(a)**

x_1	x_2	s_1	s_2	z	
6	2	1	0	0	36
2	4	0	1	0	32
−10	−5	0	1	1	0

(b) $s_1 = 36$, $s_2 = 32$, $z = 0$

(c) Column one has the most negative indicator. The smallest nonnegative quotient occurs in row 1, since $\frac{36}{6}$ is smaller than $\frac{32}{2}$. Pivot on the 6.

3. $x_1 = 6$, $x_2 = 14$, $x_3 = 0$, $s_1 = 0$, $s_2 = 12$, $s_3 = 0$, $z = 40$

4. The maximum value is 28,000 when $x_1 = 0$, $x_2 = 40$, and $x_3 = 80$.

5. The maximum value is 85 when $x_1 = \frac{5}{2}$, $x_2 = \frac{20}{3}$, and $x_3 = \frac{5}{3}$.

6. They should buy 1000 single-sided from Large Disks, and they should buy 200 single-sided and 1600 double-sided from Doubles Are Us for a minimum cost of $820,000.

7.
$$
\begin{bmatrix}
2 & 5 & 6 \\
1 & 9 & 8 \\
7 & 0 & 5 \\
6 & 4 & 1 \\
3 & 2 & 4
\end{bmatrix}
$$

8. **(a)**   Maximize $z = 4x_1 + 8x_2 + 10x_3$
   $$
   \begin{aligned}
   \text{subject to:} \quad x_1 + x_2 + \ x_3 &\le 5 \\
   x_2 + 2x_3 &\le 7 \\
   \text{with} \quad x_1 \ge 0, \ x_2 \ge 0, \ x_3 &\ge 0.
   \end{aligned}
   $$

   (b) The minimum value is 44 when $y_1 = 6$ and $y_2 = 2$.

9.   Maximize $z = 8x_1 + 10x_2 + 12x_3$
   $$
   \begin{aligned}
   \text{subject to:} \quad x_1 + 2x_2 + \ x_3 &\le 3 \\
   2x_1 + 2x_2 + 4x_3 &\le 4 \\
   \text{with} \quad x_1 \ge 0, \ x_2 \ge 0, \ x_3 &\ge 0.
   \end{aligned}
   $$

10.
$$
\begin{aligned}
x_1 + 3x_2 + 2x_3 + s_1 \quad\quad\quad\quad &= 42 \\
3x_1 \quad\quad + \ x_3 \quad\quad - s_2 \quad\quad &= 20 \\
2x_1 + \ x_2 + 2x_3 \quad\quad\quad\quad - s_3 &= 13
\end{aligned}
$$

11.   Maximize $z = -w = -8y_1 - 5y_2 - 3y_3$

    subject to: $y_1 + 3y_2 + 2y_3 \geq 90$

                      $3y_1 + y_2 + y_3 \geq 75$

                      $2y_1 \qquad + 3y_3 \geq 60$

    with $y_1 \geq 0, \ y_2 \geq 0, \ y_3 \geq 0.$

12. The maximum value is 60 when $x_1 = 10$ and $x_2 = 0$.

Chapter 5

MATHEMATICS OF FINANCE

5.1 Simple and Compound Interest

5. $25,000 at 3% for 9 mo

Use the formula for simple interest.

$$I = Prt$$
$$= 25,000(0.03)\left(\frac{9}{12}\right)$$
$$= 562.50$$

The simple interest is $562.50.

7. $1974 at 6.3% for 25 wk

Use the formula for simple interest.

$$I = Prt$$
$$= 1974(0.063)\left(\frac{25}{52}\right)$$
$$\approx 59.79$$

The simple interest is $59.79.

9. $8192.17 at 3.1% for 72 days

Use the formula for simple interest.

$$I = Prt$$
$$= 8192.17(0.031)\left(\frac{72}{360}\right)$$
$$\approx 50.79$$

The simple interest is $50.79.

11. Use the formula for future value for simple interest.

$$A = P(1 + rt)$$
$$= 3125\left[1 + 0.0285\left(\frac{7}{12}\right)\right]$$
$$\approx 3176.95$$

The maturity value is $3176.95. The interest earned is $3176.95 - 3125 = $51.95.

13. Use the formula for simple interest.

$$I = Prt$$
$$56.25 = 1500r\left(\frac{6}{12}\right)$$
$$r = 0.075$$

The interest rate was 7.5%.

19. Use the formula for compound amount with $P = 1000$, $i = 0.06$, and $n = 8$.

$$A = P(1 + i)^n$$
$$= 1000(1 + 0.06)^8$$
$$\approx 1593.85$$

The compound amount is $1593.85. The interest earned is $1593.85 - 1000 = $593.85.

21. Use the formula for compound amount with $P = 470, i = \frac{0.054}{2} = 0.027$, and $n = 12(2) = 24$.

$$A = P(1 + i)^n$$
$$= 470(1 + 0.027)^{24}$$
$$\approx 890.82$$

The compound amount is $890.82. The interest earned is $890.82 - 470 = $420.82.

23. Use the formula for compound amount with $P = 8500, i = \frac{0.08}{4} = 0.02$, and $n = 5(4) = 20$.

$$A = P(1 + i)^n$$
$$= 8500(1 + 0.02)^{20}$$
$$\approx 12,630.55$$

The compound amount is $12,630.55. The interest earned is $12,630.55 - 8500 = $4130.55.

25. Use the formula for present value for compound interest with $A = 12,820.77, i = 0.048$, and $n = 6$.

$$P = \frac{A}{(1 + r)^n}$$
$$= \frac{12,820.77}{(1 + 0.048)^6}$$
$$\approx 9677.13$$

The present value $9677.13.

27. Use the formula for present value for compound interest with $A = 2000, i = \frac{0.06}{2} = 0.03$, and $n = 8(2) = 16$.

$$P = \frac{A}{(1+r)^n}$$

$$= \frac{2000}{(1+0.03)^{16}}$$

$$\approx 1246.33$$

The present value $1246.33.

29. Use the formula for present value for compound interest with $A = 8800, i = \frac{0.05}{4} = 0.0125$, and $n = 5(4) = 20$.

$$P = \frac{A}{(1+r)^n}$$

$$= \frac{8800}{(1+0.0125)^{20}}$$

$$\approx 6864.08$$

The present value $6864.08.

33. 4% compounded quarterly.

Use the formula for effective rate with $r = 0.04$ and $m = 4$.

$$r_e = \left(1 + \frac{r}{m}\right)^m - 1$$

$$= \left(1 + \frac{0.04}{4}\right)^4 - 1$$

$$\approx 0.04060$$

The effective rate is about 4.06%.

35. 7.25% compounded semiannually.

Use the formula for effective rate with $r = 0.0725$ and $m = 2$.

$$r_e = \left(1 + \frac{r}{m}\right)^m - 1$$

$$= \left(1 + \frac{0.0725}{2}\right)^2 - 1$$

$$\approx 0.07381$$

The effective rate is about 7.381%, or rounding to two decimal places, 7.38%.

37. Start by finding the total amount repaid. Use the formula for future value for simple interest, with $P = 2700, r = 0.062$, and $t = \frac{9}{12}$.

$$A = P(1 + rt)$$

$$= 7200\left[1 + 0.062\left(\frac{9}{12}\right)\right]$$

$$= 7534.80$$

Amy repaid her father $7534.80. To find the amount of this which was interest, subtract the original loan amount from the repayment amount.

$$7534.80 - 7200 = 334.80$$

Of the amount repaid, $334.80 was interest.

39. The interest earned was

$$\$1521.25 - \$1500 = \$21.25$$

Use the formula for simple interest, with $I = 21.25$, $P = 1500$, and $t = \frac{75}{360}$.

$$I = Prt$$

$$21.25 = 1500r\left(\frac{75}{360}\right)$$

$$0.068 = r$$

The interest rate was 6.8%.

41. The total cost of the 8 computers is $2309(8) = $18,472. We want to find the present value of 18,472 dollars compounded at interest rate $i = \frac{0.0479}{12}$ per month for $n = 6$ months.

$$P = \frac{A}{(1+i)^n}$$

$$= \frac{18,472}{\left(1 + \frac{0.0479}{12}\right)^6}$$

$$\approx 18,035.71$$

The department should deposit $18,035.71 now.

43. Use the formula for compound amount with $P = 50,000$, $i = \frac{r}{m} = \frac{0.08}{12} = 0.00\overline{6}$, and $n = mt = 12(6) = 72$.

$$A = P(1+i)^n$$
$$= 50,000(1 + 0.00\overline{6})^{72}$$
$$\approx 80,675.11$$

The amount of interest is found by subtracting.

$$I = A - P$$
$$= 80,675.11 - 50,000$$
$$= 30,675.11$$

The interest earned was $30,675.11.

45. (a) Use the Rule of 72 to find the doubling time.

$$\text{Doubling time} = \frac{72}{4.5}$$
$$= 16$$

It takes about 16 years for the trust fund to double in size, from $10,000 to $20,000. The grandchild will be 16 years old.

(b) Use the formula for compound amount with $P = 10,000$, $i = \frac{0.045}{12}$, and $n = 16(12) = 192$.

$$A = P(1+i)^n$$
$$= 10,000\left(1 + \frac{0.045}{12}\right)^{192}$$
$$\approx 20,516.69$$

The actual amount in the trust fund after 16 years is $20,516.69. Obviously, it actually takes slightly less than 16 years for the fund to reach $20,000.

47. Use the formula for present value for compound interest with $A = 30,000$, $i = \frac{0.055}{4} = 0.01375$, and $n = 5(4) = 20$.

$$P = \frac{A}{(1+r)^n}$$
$$= \frac{30,000}{(1 + 0.01375)^{20}}$$
$$\approx 22,829.89$$

The present value is $22,829.89, or rounding up to the nearest cent (to make sure that the investment really grows to $30,000), $22,829.90. That is how much of the inherited $25,000 Phyllis should invest in order to have $30,000 for a down payment in 5 years.

49. (a) Use the formula for compound amount to find the value of $1000 in 5 yr.

$$A = P(1+i)^n$$
$$= 1000(1.06)^5$$
$$\approx 1338.23$$

In 5 yr, $1000 will be worth $1338.23. Since this is larger than the $1210 one would receive in 5 yr, it would be more profitable to take the $1000 now.

51. The yield is the effective rate. Find the corresponding nominal rate by using the formula for effective rate. Use the formula for effective rate, with $r = 0.0546$ and $m = 12$.

$$r_e = \left(1 + \frac{r}{m}\right)^m - 1$$
$$0.0546 = \left(1 + \frac{r}{12}\right)^{12} - 1$$
$$(1 + 0.0546)^{1/12} = 1 + \frac{r}{12}$$
$$0.05328 \approx r$$

The effective rate is about 5.33%.

53. Use 8% compounded quarterly for 20 yr. Then $i = \frac{0.08}{4} = 0.02$ and $n = 20(4) = 80$.

$$A = P(1+i)^n = P(1.02)^{80}$$

For $10,000,

$$A = 10,000(1.02)^{80} \approx 48,754.39,$$

that is, $48,754.39.

For $149,000,

$$A = 149,000(1.02)^{80} \approx 726,440.43,$$

that is, $726,440.43.

For $1,000,000,

$$A = 1,000,000(1.02)^{80} \approx 4,875,439.16,$$

that is, $4,875,439.16.

55. $2 = (1 + 0.05)^n$
$2 = (1.05)^n$

Try various values for n.

$$(1.05)^{14} \approx 1.979932 \approx 2$$

Thus, $n \approx 14$. It would take about 14 yr for the general level of prices in the economy to double at the annual inflation rate of 5%.

57. Find n such that

$$2 = (1.02)^n$$

By trying various values of n, we see that $n \approx 35$ is approximately correct because

$$(1.02)^{35} \approx 1.999890 \approx 2.$$

It will take about 35 yr before the utilities will need to double their generating capacity.

59. Let $P = 150,000$, $i = -0.024$, and $n = 8$.

$$\begin{aligned} A &= P(1+i)^n \\ &= 150,000(1-0.024)^8 \\ &= 150,000(0.976)^8 \\ &= 123,506.50 \end{aligned}$$

After 8 yr, the amount on deposit will be $123,506.50.

61. Use the formula

$$A = P(1+i)^n$$

with $P = \frac{2}{8}$ cent $= \$0.0025$ and $r = 0.04$ compounded quarterly for 2000 yr.

$$\begin{aligned} A &= 0.0025 \left(1 + \frac{0.04}{4}\right)^{4(2000)} \\ &= 0.0025(1.01)^{8000} \\ &\approx 9.31 \times 10^{31} \end{aligned}$$

The money would be worth 9.31×10^{31} 2000 yr later.

63. Use the formula

$$A = P(1+i)^n$$

with $P = 10,000$ and $r = 0.05$ for 10 yr.

(a) If interest is compounding annually,

$$\begin{aligned} A &= 10,000(1+0.05)^{10} \\ &\approx 16,288.95. \end{aligned}$$

The future value is $16,288.95.

(b) If interest is compounding quarterly,

$$\begin{aligned} A &= 10,000 \left(1 + \frac{0.05}{4}\right)^{40} \\ &\approx 16,436.19. \end{aligned}$$

The future value is $16,436.19.

(c) If interest is compounding monthly,

$$\begin{aligned} A &= 10,000 \left(1 + \frac{0.05}{12}\right)^{120} \\ &\approx 16,470.09. \end{aligned}$$

The future value is $16,470.09.

(d) If interest is compounding daily,

$$\begin{aligned} A &= 10,000 \left(1 + \frac{0.05}{365}\right)^{3650} \\ &\approx 16,486.65. \end{aligned}$$

The future value is $16,486.65.

65. First consider the case of earning interest at a rate of k per annum compounded quarterly for all 8 yr and earning $2203.76 on the $1000 investment.

$$2203.76 = 1000 \left(1 + \frac{k}{4}\right)^{8(4)}$$

$$2.20376 = \left(1 + \frac{k}{4}\right)^{32}$$

Use a calculator to raise both sides to the power $\frac{1}{32}$.

$$1.025 = 1 + \frac{k}{4}$$

$$0.025 = \frac{k}{4}$$

$$0.1 = k$$

Next consider the actual investments. The $1000 was invested for the first 5 yr at a rate of j per annum compounded semiannually.

$$A = 1000 \left(1 + \frac{j}{2}\right)^{5(2)}$$

$$A = 1000 \left(1 + \frac{j}{2}\right)^{10}$$

This amount was then invested for the remaining 3 yr at $k = .1$ per annum compounded quarterly for a final compound amount of $1990.76.

$$1990.76 = A \left(1 + \frac{0.1}{4}\right)^{3(4)}$$

$$1990.76 = A(1.025)^{12}$$

$$1480.24 \approx A$$

Recall that $A = 1000\left(1+\dfrac{j}{2}\right)^{10}$ and substitute this value into the above equation.

$$1480.24 = 1000\left(1+\frac{j}{2}\right)^{10}$$

$$1.48024 = \left(1+\frac{j}{2}\right)^{10}$$

Use a calculator to raise both sides to the power $\frac{1}{10}$.

$$1.04 \approx 1+\frac{j}{2}$$

$$0.04 = \frac{j}{2}$$

$$0.08 = j$$

The ratio of k to j is

$$\frac{k}{j} = \frac{0.1}{0.08} = \frac{10}{8} = \frac{5}{4}.$$

5.2   Future Value of an Annuity

1. $a = 3;\ r = 2$

The first five terms are

$$3,\ 3(2),\ 3(2)^2,\ 3(2)^3,\ 3(2)^4$$

or

$$3,\ 6,\ 12,\ 24,\ 48.$$

The fifth term is 48.

Or, use the formula $a_n = ar^{n-1}$ with $n = 5$.

$$a_5 = ar^{5-1} = 3(2)^4 = 3(16) = 48$$

3. $a = -8;\ r = 3;\ n = 5$

$$a_5 = ar^{5-1} = -8(3)^4 = -8(81) = -648$$

The fifth term is -648.

5. $a = 1;\ r = -3;\ n = 5$

$$a_5 = ar^{5-1} = 1(-3)^4 = 81$$

The fifth term is 81.

7. $a = 256;\ r = \frac{1}{4};\ n = 5$

$$a_5 = ar^{5-1} = 256\left(\frac{1}{4}\right)^4 = 256\left(\frac{1}{256}\right) = 1$$

The fifth term is 1.

9. $a = 1;\ r = 2;\ n = 4$

To find the sum of the first 4 terms, S_4, use the formula for the sum of the first n terms of a geometric sequence.

$$S_n = \frac{a(r^n - 1)}{r - 1}$$

$$S_4 = \frac{1(2^4 - 1)}{2 - 1} = \frac{16 - 1}{1} = 15$$

11. $a = 5;\ r = \frac{1}{5};\ n = 4$

$$S_n = \frac{a(r^n - 1)}{r - 1}$$

$$S_4 = \frac{5\left[\left(\frac{1}{5}\right)^4 - 1\right]}{\frac{1}{5} - 1} = \frac{5\left(-\frac{624}{625}\right)}{-\frac{4}{5}}$$

$$= \frac{-\frac{624}{125}}{-\frac{4}{5}} = \left(-\frac{624}{125}\right)\left(-\frac{5}{4}\right) = \frac{156}{25}$$

13. $a = 128;\ r = -\frac{3}{2};\ n = 4$

$$S_n = \frac{a(r^n - 1)}{r - 1}$$

$$S_4 = \frac{128\left[\left(-\frac{3}{2}\right)^4 - 1\right]}{-\frac{3}{2} - 1} = \frac{128\left(\frac{65}{16}\right)}{-\frac{5}{2}}$$

$$= -208$$

15. $s_{\overline{n}|i} = \dfrac{(1+i)^n - 1}{i}$

$$s_{\overline{12}|0.05} = \frac{(1+0.05)^{12} - 1}{0.05} \approx 15.91713$$

17. $s_{\overline{n}|i} = \dfrac{(1+i)^n - 1}{i}$

$$s_{\overline{10}|0.052} = \frac{(1+0.052)^{10} - 1}{0.052} \approx 12.69593$$

21. $R = 100;\ i = 0.06;\ n = 4$

Use the formula for the future value of an ordinary annuity.

$$S = R\left[\frac{(1+i)^n - 1}{i}\right]$$

$$= 100\left[\frac{(1.06)^4 - 1}{0.06}\right]$$

$$= 100\left(\frac{1.262477 - 1}{0.06}\right)$$

$$\approx 437.46$$

The future value is $437.46.

23. $R = 25,000$; $i = 0.045$; $n = 36$

$$S = R\left[\frac{(1+i)^n - 1}{i}\right]$$

$$= 25,000\left[\frac{(1+0.045)^{36} - 1}{0.045}\right]$$

$$\approx 2,154,099.15$$

The future value is $2,154,099.15.

25. $R = 9200$; 10% interest compounded semiannually for 7 yr

Interest of $\frac{10\%}{2} = 5\%$ is earned semiannually, so $i = 0.05$. In 7 yr, there are $7(2) = 14$ semiannual periods, so $n = 14$.

$$S = R\left[\frac{(1+i)^n - 1}{i}\right]$$

$$= 9200\left[\frac{(1.05)^{14} - 1}{0.05}\right]$$

$$\approx 180,307.41$$

The future value is $180,307.41.
$9200 is contributed in each of 14 periods. The total contribution is

$$\$9200(14) = \$128,800.$$

The amount from interest is

$$\$180,307.41 - 128,800 = \$51,507.41$$

27. $R = 800$; 6.51% interest compounded semiannually for 12 yr

Interest of $\frac{6.51\%}{2}$ is earned semiannually, so $i = \frac{0.0651}{2} = 0.03255$. In 12 yr, there are $12(2) = 24$ semiannual periods, so $n = 24$.

$$S = R\left[\frac{(1+i)^n - 1}{i}\right]$$

$$= 800\left[\frac{(1+0.03255)^{24} - 1}{0.03255}\right]$$

$$\approx 28,438.21$$

The future value is $28,438.21.

$800 is contributed in each of 24 periods. The total contribution is

$$\$800(24) = \$19,200.$$

The amount from interest is

$$\$28,438.21 - 19,200 = \$9238.21.$$

29. $R = 12,000$; $i = \frac{0.048}{2} = 0.012$; $n = 16(4) = 64$

$$S = R\left[\frac{(1+i)^n - 1}{i}\right]$$

$$= 12,000\left[\frac{(1+0.012)^{64} - 1}{0.012}\right]$$

$$\approx 1,145,619.96$$

The future value is $1,145,619.96.

$12,000 is contributed in each of 64 periods. The total contribution is

$$\$12,000(64) = \$768,000.$$

The amount from interest is

$$\$1,145,619.96 - 768,000 = \$377,619.96.$$

31. $R = 600$; $i = 0.06$; $n = 8$

To find the future value of an annuity due, use the formula for the future value of an ordinary annuity, but include one additional time period and subtract the amount of one payment.

$$S = R\left[\frac{(1+i)^{n+1} - 1}{i}\right] - R$$

$$= 600\left[\frac{(1+0.06)^9 - 1}{0.06}\right] - 600$$

$$\approx 6294.79$$

The future value is $6294.79.

33. $R = 16,000$; $i = 0.05$; $n = 7$

$$S = R\left[\frac{(1+i)^{n+1} - 1}{i}\right] - R$$

$$= 16,000\left[\frac{(1+0.05)^8 - 1}{0.05}\right] - 16,000$$

$$\approx 136,785.74$$

The future value is $136,785.74.

35. $R = 1000$; $i = \frac{0.0815}{2} = 0.04075$; $n = 9(2) = 18$

$$S = R\left[\frac{(1+i)^{n+1} - 1}{i}\right] - R$$

$$= 1000\left[\frac{(1+0.04075)^{19} - 1}{0.04075}\right] - 1000$$

$$\approx 26,874.97$$

The future value is $26,874.97.

$1000 is contributed in each of 18 periods. The total contribution is

$$\$1000(18) = \$18,000.$$

The amount from interest is

$$\$26,874.97 - 18,000 = \$8874.97.$$

37. $R = 250;\ i = \frac{0.042}{2} = 0.0105;\ n = 12(4) = 48$

$$S = R\left[\frac{(1+i)^{n+1}-1}{i}\right] - R$$

$$= 250\left[\frac{(1+0.0105)^{49}-1}{0.0105}\right] - 250$$

$$\approx 15{,}662.40$$

The future value is $15,662.40.

$250 is contributed in each of 48 periods. The total contribution is

$$\$250(48) = \$12,000.$$

The amount from interest is

$$15{,}662.40 - 12{,}000 = \$3662.40.$$

39. $S = \$10,000$; interest is 5% compounded annually; payments are made at the end of each year for 12 yr.

This is a sinking fund. Use the formula for an ordinary annuity with $S = 10,000$, $i = 0.05$, and $n = 12$ to find the value of R, the amount of each payment.

$$10{,}000 = Rs_{\overline{12}|0.05}$$

$$R = \frac{10{,}000}{s_{\overline{12}|0.05}} = \frac{10{,}000}{\frac{(1+0.05)^{12}-1}{0.05}} \approx 628.25$$

The required periodic payment is $628.25.

43. $2750; money earns 5% compounded annually; 5 annual payments

Let R be the amount of each payment.

$$2750 = Rs_{\overline{5}|0.05}$$

$$R = \frac{2750}{s_{\overline{5}|0.05}}$$

$$= \frac{2750}{\frac{(1+0.05)^5-1}{0.05}}$$

$$\approx 497.68$$

The amount of each payment is $497.68.

45. $25,000; money earns 5.7% compounded quarterly for $3\frac{1}{2}$ yr.

Thus, $i = \frac{0.057}{4} = 0.01425$ and $n = \left(3\frac{1}{2}\right)4 = 14$.

$$R = \frac{25{,}000}{s_{\overline{14}|0.01425}}$$

$$= \frac{25{,}000}{\frac{(1+0.01425)^{14}-1}{0.01425}}$$

$$\approx 1626.16$$

The amount of each payment is $1626.16.

47. $9000; money earns 4.8% compounded monthly for $2\frac{1}{2}$ years

Thus, $i = \frac{0.048}{12} = 0.004$ and $n = \left(2\frac{1}{2}\right)12 = 30$.

$$R = \frac{9000}{s_{\overline{30}|0.004}} = \frac{9000}{\frac{(1+0.004)^{30}-1}{0.004}} \approx 282.96$$

The amount of each payment is $282.96.

49. Use the formula for the future value of an ordinary annuity with $R = 100$, $i = \frac{0.0225}{12} = 0.001875$, and $n = 12(2) = 24$.

$$S = R\left[\frac{(1+i)^{n+1}-1}{i}\right]$$

$$= 100\left[\frac{(1+.001875)^{24}-1}{0.001875}\right]$$

$$\approx 2452.47$$

The amount in the account after 2 years is $2452.47.

Tom deposited $100 in each of 24 periods. The total amount deposited was

$$\$100(24) = \$2400.$$

The amount of interest earned was

$$\$2452.47 - 2400 = \$52.47.$$

51. $130.50 is invested each month at 4.8% compounded monthly for 40 yr. Thus, $i = \frac{0.048}{12} = 0.004$ and $n = 40(12) = 480$. Use the formula for the future value of an ordinary annuity.

$$S = R\left[\frac{(1+i)^n-1}{i}\right]$$

$$= 130.50\left[\frac{(1+0.004)^{480}-1}{0.004}\right]$$

$$\approx 189{,}058.14$$

The account would be worth $189,058.14.

53. For the first 15 yr, we have an ordinary annuity with $R = 2500$, $i = \frac{0.06}{4} = 0.015$, and $n = 15(4) = 60$. The amount on deposit after 15 yr is

$$S = R\left[\frac{(1+i)^n - 1}{i}\right]$$

$$= 2500\left[\frac{(1 + 0.015)^{60} - 1}{0.015}\right]$$

$$\approx 240{,}536.63$$

For the remaining 5 yr, this amount earns compound interest at 6% compounded quarterly. To find the final amount on deposit, use the formula for the compound amount with $P = 240{,}536.63$, $i = \frac{0.06}{4} = 0.015$, and $n = 5(4) = 20$.

$$A = P(1 + i)^n$$

$$= 240{,}536.63(1.015)^{20}$$

$$\approx 323{,}967.96$$

The man will have about $323,967.96 in the account when he retires.

55. For the first 8 yr, we have an annuity due with $R = 2435$, $i = \frac{0.06}{2} = 0.03$, and $n = 8(2) = 16$. The amount on deposit after 8 yr is

$$S = R\left[\frac{(1+i)^{n+1} - 1}{i}\right] - R$$

$$= 2435\left[\frac{(1 + 0.03)^{17} - 1}{0.03}\right] - 2435$$

$$\approx 50{,}554.47.$$

For the remaining 5 yr, this amount, $50,554.47, earns compound interest at 6% compounded semi-annually. To find the final amount on deposit, use the formula for the compound amount with $P = 50{,}554.47$, $i = \frac{0.06}{2} = 0.03$, and $n = 5(2) = 10$.

$$A = P(1 + i)^n$$

$$= 50{,}554.47(1.03)^{10}$$

$$\approx 67{,}940.98$$

The final amount on deposit will be about $67,940.98.

57. **(a)** This is a sinking fund with $S = 10{,}000$, $i = \frac{0.08}{4} = 0.02$, and $n = 8(4) = 32$. Let R represent the amount of each payment.

$$S = Rs_{\overline{n}|i}$$

$$10{,}000 = Rs_{\overline{32}|0.02}$$

$$R = \frac{10{,}000}{s_{\overline{32}|0.02}}$$

$$= \frac{10{,}000(0.02)}{(1 + 0.02)^{32} - 1}$$

$$\approx 226.11$$

If the money is deposited at 8% compounded quarterly, Greg's quarterly deposit will need to be about $226.11.

(b) Here $S = 10{,}000$, $i = \frac{0.06}{4} = 0.015$, and $n = 8(4) = 32$. Let R represent the amount of each payment.

$$S = Rs_{\overline{n}|i}$$

$$10{,}000 = Rs_{\overline{32}|0.015}$$

$$R = \frac{10{,}000}{s_{\overline{32}|0.015}}$$

$$= \frac{10{,}000(0.015)}{(1 + 0.015)^{32} - 1}$$

$$\approx 245.77$$

If the money is deposited at 6% compounded quarterly, Greg's quarterly deposit will need to be about $245.77.

59. $S = 20{,}000$, $i = \frac{0.032}{4} = 0.008$, $n = 6(4) = 24$

Let R represent the amount of each payment.

$$S = Rs_{\overline{n}|i}$$

$$20{,}000 = Rs_{\overline{24}|0.008}$$

$$R = \frac{20{,}000}{Rs_{\overline{24}|0.008}}$$

$$= \frac{20{,}000(0.008)}{(1 + 0.008)^{24} - 1}$$

$$\approx 759.21$$

She must deposit about $759.21 at the end of each quarter.

61. $R = 1000$, $i = \frac{0.08}{4} = 0.02$, and $n = 25(4) = 100.$

$$S = R\left[\frac{(1+i)^n - 1}{i}\right]$$

$$= 1000\left[\frac{(1 + 0.02)^{100} - 1}{0.02}\right]$$

$$\approx 312{,}232.31$$

There will be about \$312,232.31 in the IRA.

The total amount deposited was $\$1000(100) = \$100{,}000.$

Thus, the amount of interest earned was

$$\$312{,}232.31 - 100{,}000 = \$212{,}232.31.$$

63. $R = 1000$, $i = \frac{0.10}{4} = 0.025$, and $n = 100.$

$$S = R\left[\frac{(1+i)^n - 1}{i}\right]$$

$$= 1000\left[\frac{(1 + 0.025)^{100} - 1}{0.025}\right]$$

$$\approx 432{,}548.65$$

There will be about \$432,548.65 in the IRA.

The total amount deposited was \$100,000.

Thus, the amount of interest earned was

$$\$432{,}548.65 - 100{,}000 = \$332{,}548.65.$$

65. Let $x =$ the annual interest rate.

$n = 30(12) = 360$

Graph $y_1 = 330{,}000$ and

$$y_2 = 250\left[\frac{\left(1 + \frac{x}{12}\right)^{360} - 1}{\frac{x}{12}}\right].$$

The x-coordinate of the point of intersection is 0.0739706. Thus, she would need to earn an annual interest rate of about 7.397%.

67. This exercise should be solved by graphing calculator or computer methods. The answers, which may vary slightly, are as follows.

(a) The buyer's quarterly interest payment will be

$$I = Prt$$

$$= \$60{,}000(0.08)\frac{1}{4}$$

$$= \$1200.$$

(b) The buyer's semiannual payments into the sinking fund will be \$3511.58 for each of the first 13 payments and \$3511.59 for the last payment. A table showing the amount in the sinking fund after each deposit is as follows.

Payment Number	Amount of Deposit	Interest Earned	Total
1	\$3511.58	\$0	\$3511.58
2	\$3511.58	\$105.35	\$7128.51
3	\$3511.58	\$213.86	\$10,853.94
4	\$3511.58	\$325.62	\$14,691.14
5	\$3511.58	\$440.73	\$18,643.46
6	\$3511.58	\$559.30	\$22,714.34
7	\$3511.58	\$681.43	\$26,907.35
8	\$3511.58	\$807.22	\$31,226.15
9	\$3511.58	\$936.78	\$35,674.51
10	\$3511.58	\$1070.24	\$40,256.33
11	\$3511.58	\$1207.69	\$44,975.60
12	\$3511.58	\$1349.27	\$49,836.45
13	\$3511.58	\$1495.09	\$54,843.12
14	\$3511.59	\$1645.29	\$60,000.00

69. Using the compound amount formula the future value of the down payment, D, is given by

$$A = D(1 + i)^n$$

The rest of the payments form an ordinary annuity with future value given by the formula

$$S = R\left[\frac{(1+i)^n - 1}{i}\right]$$

The future value of the loan, including the down payment, is the sum of the future value of the down payment and the future value of the annuity, or

$$S = D(1 + i)^n + R\left[\frac{(1+i)^n - 1}{i}\right]$$

5.3   Present Value of an Annuity; Amortization

1. $\dfrac{1 - (1 + i)^{-n}}{i}$

is represented by $a_{\overline{n}|i}$, and it is choice (c).

3. $a_{\overline{n}|i} = \dfrac{1 - (1 + i)^{-n}}{i}$

$a_{\overline{15}|0.065} = \dfrac{1 - (1 + 0.065)^{-15}}{0.065}$

$$\approx 9.40267$$

5. $a_{\overline{n}|i} = \dfrac{1 - (1+i)^{-n}}{i}$

$a_{\overline{18}|0.055} = \dfrac{1 - (1 + 0.055)^{-18}}{0.055}$

≈ 11.24607

9. Payments of \$890 each year for 16 years at 6% compounded annually

Use the formula for present value of an annuity with $R = 890$, $i = 0.06$, and $n = 16$.

$$P = R \left[\dfrac{1 - (1+i)^{-n}}{i} \right]$$

$$= 890 \left[\dfrac{1 - (1 + 0.06)^{-16}}{0.06} \right]$$

$$\approx 8994.25$$

The present value is \$8994.25.

11. Payments of \$10,000 semiannually for 15 years at 5% compounded semiannually

Use the formula for present value of an annuity with $R = 10,000$, $i = \frac{0.05}{2} = 0.025$, and $n = 15(2) = 30$.

$$P = R \left[\dfrac{1 - (1+i)^{-n}}{i} \right]$$

$$= 10,000 \left[\dfrac{1 - (1 + 0.025)^{-30}}{0.025} \right]$$

$$\approx 209,302.93$$

The present value is \$209,302.93.

13. Payments of \$15,806 quarterly for 3 years at 6.8 compounded quarterly

Use the formula for present value of an annuity with $R = 15,806$, $i = \frac{0.068}{4} = 0.017$, and $n = 3(4) = 12$.

$$P = R \left[\dfrac{1 - (1+i)^{-n}}{i} \right]$$

$$= 15,806 \left[\dfrac{1 - (1 + 0.017)^{-12}}{0.017} \right]$$

$$\approx 170,275.47$$

The present value is \$170,275.47.

15. 4% compounded annually

We want the present value, P, of an annuity with $R = 10,000$, $i = 0.04$, and $n = 15$.

$$P = R \left[\dfrac{1 - (1+i)^{-n}}{i} \right]$$

$$= 10,000 \left[\dfrac{1 - (1.04)^{-15}}{0.04} \right]$$

$$\approx 111,183.87$$

The required lump sum is \$111,183.87.

17. $P = 2500$, $i = \frac{0.06}{4} = 0.015$; $n = 6$

To find the payment amount, use the formula for amortization payments.

$$R = \dfrac{Pi}{1 - (1+i)^{-n}}$$

$$R = \dfrac{2500(0.015)}{1 - (1 + 0.015)^{-6}}$$

$$\approx 438.81$$

Each payment is \$438.81.

To find the total payments, multiply the amount of one payment by $n = 6$.

$$438.81(6) = 2632.86$$

The total payments come out to \$2632.86.

To find the total amount of interest paid, subtract the original loan amount from the total payments.

$$2632.86 - 2500 = 132.86$$

The total amount of interest paid is \$132.86.

19. $P = 90,000$; $i = 0.06$; $n = 12$

To find the payment amount, use the formula for amortization payments.

$$R = \dfrac{Pi}{1 - (1+i)^{-n}}$$

$$R = \dfrac{90,000(0.06)}{1 - (1 + 0.06)^{-12}}$$

$$\approx 10,734.93$$

Each payment is \$10,734.93.

To find the total payments, multiply the amount of one payment by $n = 12$.

$$10734.93(12) = 128,819.16$$

The total payments come out to \$128,819.16.

To find the total amount of interest paid, subtract the original loan amount from the total payments.

$$128{,}819.16 - 90{,}000 = 38{,}819.16$$

The total amount of interest paid is \$38,819.16.

21. $P = 7400$; $i = \frac{0.062}{2} = 0.031$; $n = 18$

To find the payment amount, use the formula for amortization payments.

$$R = \frac{Pi}{1 - (1+i)^{-n}}$$

$$R = \frac{7400(0.031)}{1 - (1 + 0.031)^{-18}} \approx 542.60$$

Each payment is \$542.60.

To find the total payments, multiply the amount of one payment by $n = 18$.

$$542.60(18) = 9766.80$$

The total payments come out to \$9766.80.

To find the total amount of interest paid, subtract the original loan amount from the total payments

$$9766.80 - 7400 = 2366.80$$

The total amount of interest paid is \$2366.80.

23. Look at the entry for payment number 4 under the heading "Interest for Period." The amount of interest included in the fourth payment is \$7.61.

25. To find the amount of interest paid in the first 4 mo of the loan, add the entries for payments 1, 2, 3, and 4 under the heading "Interest for Period."

$$\$10.00 + 9.21 + 8.42 + 7.61 = \$35.24$$

In the first 4 mo of the loan, \$35.24 of interest is paid.

27. First, find the value of the annuity at the end of 8 yr. Use the formula for future value of an ordinary annuity.

$$S = R\left[\frac{(1+i)^n - 1}{i}\right]$$

$$= 1000\left[\frac{(1 + 0.06)^8 - 1}{0.06}\right]$$

$$\approx 9897.47$$

The future value of the annuity is \$9897.47.

Now find the present value of \$9897.47 at 5% compounded annually for 8 yr. Use the formula for present value for compound interest.

$$P = \frac{A}{(1+i)^n} = \frac{9897.47}{(1.05)^8} \approx 6699.00$$

The required amount is \$6699.

29. $P = 199{,}000$; $i = \frac{0.0701}{12}$; $n = 25(12) = 300$

To find the payment amount, use the formula for amorization payments.

$$R = \frac{Pi}{1 - (1+i)^{-n}}$$

$$R = \frac{199{,}000\left(\frac{0.0701}{12}\right)}{1 - \left(1 + \frac{0.0701}{12}\right)^{-300}}$$

$$\approx 1407.76$$

Each payment is \$1407.76.

To find the total payments, multiply the amount of one payment by $n = 300$.

$$1407.76(300) = 422{,}328$$

The total payments come out to \$422,328.

To find the total amount of interest paid, subtract the original loan amount from the total payments.

$$422{,}328 - 199{,}000 = 223{,}328$$

The total amount of interest paid is \$223,328.

31. $P = 253{,}000$, $i = \frac{0.0645}{12}$, $n = 30(12) = 360$

To find the payment amount, use the formula for amorization payments.

$$R = \frac{Pi}{1 - (1+i)^{-n}}$$

$$R = \frac{253{,}000\left(\frac{0.0645}{12}\right)}{1 - \left(1 + \frac{0.0645}{12}\right)^{-360}}$$

$$\approx 1590.82$$

Each payment is \$1590.82.

To find the total payments, multiply the amount of one payment by $n = 360$.

$$1590.82(360) = 572{,}695.20$$

The total payments come out to \$572,695.20.

To find the total amount of interest paid, subtract the original loan amount from the total payments.

$$572{,}695.20 - 253{,}000 = 319{,}695.20$$

The total amount of interest paid is \$319,695.20.

33. From Example 3, $P = 220,000$ and $i = \frac{0.06}{12} = 0.005$. For a 15-year loan, use $n = 15(12) = 180$.

$$R = \frac{Pi}{1 - (1+i)^{-n}}$$
$$= \frac{220,000(0.005)}{1 - (1 + 0.005)^{-180}}$$
$$\approx 1856.49$$

The monthly payments would be $1856.49. The family makes 180 payments of $1856.49 each, for a total of $334,168.20. Since the amount of the loan was $220,000, the total interest paid is

$$334,168.20 - 220,000 = 114,168.20.$$

The total amount of interest paid is $114,168.20.

The payments for the 15-year loan are

$$1856.49 - \$1319.01 = \$537.48$$

more than those for the 30-year loan in Example 3. However, the total interest paid is

$$254,843.60 - \$114,168.20 = \$140,675.40$$

less than for the 30-year loan in Example 3.

35. (a) $P = 14,000$, $i = \frac{0.07}{12}$, $n = 4(12) = 48$

$$R = \frac{Pi}{1 - (1+i)^{-n}}$$
$$= \frac{14,000(\frac{0.07}{12})}{1 - (1 + \frac{0.07}{12})^{-48}}$$
$$\approx 335.25$$

The amount of each payment is $335.25.

(b) 48 payments of $335.25 are made, and $48(\$335.25) = \$16,092$. The total amount of interest Le will pay is $16,092 - \$14,000 = \$2092$.

37. (a) $P = 20,000$, $i = \frac{0.019}{12}$, $n = 36$

$$R = \frac{Pi}{1 - (1+i)^{-n}}$$
$$= \frac{20,000(\frac{0.019}{12})}{1 - (1 + \frac{0.019}{12})^{-36}}$$
$$\approx 571.98$$

The amount of each payment is $571.98. Since 36 payments are made, the total amount paid will be $36(\$571.98) = \$20,591.28$.

(b) $P = 15,000$, $i = \frac{0.0693}{12}$, $n = 4(12) = 48$

$$R = \frac{Pi}{1 - (1+i)^{-n}}$$
$$= \frac{20,000(\frac{0.0693}{12})}{1 - (1 + \frac{0.0693}{12})^{-48}}$$
$$\approx 358.71$$

The amount of each payment will be $358.71. Since 48 payments are made, the total amount paid will be $48(\$358.71) = \$17,218.08$.

39. For parts (a) and (b), if $1 million is divided into 20 equal payments, each payment is $50,000.

(a) $i = 0.05$, $n = 20$

$$P = R\left[\frac{1 - (1+i)^{-n}}{i}\right]$$
$$= 50,000\left[\frac{1 - (1 + 0.05)^{-20}}{0.05}\right]$$
$$\approx 623,110.52$$

The present value is $623,110.52.

(b) $i = 0.09$, $n = 20$

$$P = R\left[\frac{1 - (1+i)^{-n}}{i}\right]$$
$$= 50,000\left[\frac{1 - (1 + 0.09)^{-20}}{0.09}\right]$$
$$\approx 456,427.28$$

The present value is $456,427.28.

For parts (c) and (d), if $1 million is divided into 25 equal payments, each payment is $40,000.

(c) $i = 0.05$, $n = 25$

$$P = R\left[\frac{1 - (1+i)^{-n}}{i}\right]$$
$$= 40,000\left[\frac{1 - (1 + 0.05)^{-25}}{0.05}\right]$$
$$\approx 563,757.78$$

The present value is $563,757.78.

(d) $i = 0.09$, $n = 25$

$$P = R\left[\frac{1 - (1+i)^{-n}}{i}\right]$$
$$= 40,000\left[\frac{1 - (1 + 0.09)^{-25}}{0.09}\right]$$
$$\approx 392,903.18$$

The present value is $392,903.18.

41. $P = 35,000$ at 7.43% compounded monthly for 20 yr. Thus, $i = \frac{0.0743}{12} = 0.006191\overline{6}$ and $n = 20(12) = 240$.

$$R = \frac{Pi}{1 - (1+i)^{-n}}$$

$$= \frac{35,000(0.006191\overline{6})}{1 - (1 + 0.006191\overline{6})^{-240}}$$

$$\approx 280.46$$

The monthly payment is $280.46. The total interest is given by

$$240(280.46) - 35,000 = 32,310.40.$$

The total interest is $32,310.40.

43. $P = 110,000, i = \frac{0.08}{2} = 0.04, n = 9$

$$R = \frac{110,000}{a_{\overline{9}|0.04}} \approx \$14,794.23$$

is the amount of each payment.

Of the first payment, the company owes interest of

$$I = Prt = 110,000(0.08)\left(\tfrac{1}{2}\right) = \$4400.$$

Therefore, from the first payment, $4400 goes to interest, and the balance.

$$\$14,794.23 - 4400 = \$10,394.23$$

,goes to principal. The principal at the end of this period is

$$\$110,000 - 10,394.23 = \$99,605.77.$$

The interest for the second payment is

$$I = Prt = 99,605.77(0.08)\left(\tfrac{1}{2}\right) \approx \$3984.23$$

Of the second payment, $3984.23 goes to interest and

$$\$14,794.23 - 3984.23 = \$10,810.00$$

goes to principal. Continue in this fashion to complete the amorization schedule for the first four payments.

Payment Number	Amount of Payment	Interest for Period	Portion to Principal	Principal at End of Period
0	—	—	—	$110,000.00
1	$14,794.23	$4400.00	$10,394.23	$99,605.77
2	$14,794.23	$3984.23	$10,810.00	$88,795.77
3	$14,794.23	$3551.83	$11,242.40	$77,553.37
4	$14,794.23	$3102.13	$11,692.10	$65,861.27

45. $150,000 is the future value of an annuity over 79 yr compounded quarterly. So, there are $79(4) = 316$ payment periods.

(a) The interest per quarter is $\frac{5.25\%}{4} = 1.3125\%$. Thus, $S = 150,000, n = 316, i = 0.013125$, and we must find the quarterly payment R in the formula

$$S = R\left[\frac{(1+i)^n - 1}{i}\right]$$

$$150,000 = R\left[\frac{(1.013125)^{316} - 1}{0.013125}\right]$$

$$R \approx 32.4923796$$

She would have to put $32.49 into her savings at the end of every three months.

(b) For a 2% interest rate, the interest per quarter is $\frac{2\%}{4} = 0.5\%$. Thus, $S = 150,000, n = 316, i = 0.005$, and we must find the quarterly payment R in the formula

$$S = R\left[\frac{(1+i)^n - 1}{i}\right]$$

$$150,000 = R\left[\frac{(1.005)^{316} - 1}{0.005}\right]$$

$$R \approx 195.5222794$$

She would have to put $195.52 into her savings at the end of every three months.

For a 7% interest rate, the interest per quarter is $\frac{7\%}{4} = 1.75\%$. Thus, $S = 150,000, n = 316, i = 0.0175$, and we must find the quarterly payment R in the formula

$$S = R\left[\frac{(1+i)^n - 1}{i}\right]$$

$$150,000 = R\left[\frac{(1.0175)^{316} - 1}{0.0175}\right]$$

$$R \approx 10.9663932$$

She would have to put $10.97 into her savings at the end of every three months.

47. Throughout this exercise, $i = \frac{0.065}{12}$ and $P = $ the total amount financed, which is

$$\$285,000 - 60,000 = \$225,000.$$

(a) $n = 15(12) = 180$

$$R = \frac{Pi}{1 - (1+i)^{-n}}$$
$$= \frac{225,000\left(\frac{0.065}{12}\right)}{1 - \left(1 + \frac{0.065}{12}\right)^{-180}}$$
$$\approx 1959.99$$

The monthly payment is $1959.99.

Total payments $= 180(\$1959.99) = \$352,798.20$

Total interest $= \$352,798.20 - 225,000$
$$= \$127,798.20$$

(b) $n = 20(12) = 240$

$$R = \frac{Pi}{1 - (1+i)^{-n}}$$
$$= \frac{225,000\left(\frac{0.065}{12}\right)}{1 - \left(1 + \frac{0.065}{12}\right)^{-240}}$$
$$\approx 1677.54$$

The monthly payment is $1677.54.

Total payments $= 240(\$1677.54) = \$402,609.60$

Total interest $= \$402,609.60 - 225,000$
$$= \$177,609.60$$

(c) $n = 25(12) = 300$

$$R = \frac{Pi}{1 - (1+i)^{-n}}$$
$$= \frac{225,000\left(\frac{0.065}{12}\right)}{1 - \left(1 + \frac{0.065}{12}\right)^{-300}}$$
$$\approx 1519.22$$

The monthly payment is $1519.22.

Total payments $= 300(\$1519.22) = \$455,766$

Total interest $= \$455,766 - 225,000$
$$= \$230,766$$

(d) Graph

$$y_1 = 1677.54\left[\frac{1 - \left(1 + \frac{0.065}{12}\right)^{-(240-x)}}{\frac{0.065}{12}}\right] \text{ and}$$

$$y_2 = \frac{285,000 - 60,000}{2}.$$

The x-coordinate of the point of intersection is 156.44167, which rounds up to 157. Half the loan will be paid after 157 payments.

49. $P = 150,000$, $i = \frac{0.082}{12}$, and $n = 30(12) = 360$.

$$R = \frac{Pi}{1 - (1+i)^{-n}}$$
$$= \frac{150,000\left(\frac{0.082}{12}\right)}{1 - \left(1 + \frac{0.082}{12}\right)^{-360}}$$
$$\approx 1121.63$$

The monthly payment is $1121.63.

Total payments $= 360(\$1121.63) = \$403,786.80$

Total interest $= \$403,786.80 - 150,000$
$$= \$253,786.80$$

(b) 15 years of payments means $15(12) = 180$ payments.

$$y_{15} = 1121.63\left[\frac{1 - \left(1 + \frac{0.082}{12}\right)^{-(360-180)}}{\frac{0.082}{12}}\right]$$
$$\approx 115,962.66$$

The unpaid balance after 15 years is approximately $115,962.66.

The total of the remaining 180 payments is

$$180(\$1121.63) = \$201,893.40.$$

(c) The unpaid balance from part (b) is the new loan amount. Now $P = 115,962.66$, $i = \frac{0.065}{12}$, and again $n = 30(12) = 360$.

$$R = \frac{Pi}{1 - (1+i)^{-n}}$$
$$= \frac{115,962.66\left(\frac{0.065}{12}\right)}{1 - \left(1 + \frac{0.065}{12}\right)^{-360}}$$
$$\approx 732.96$$

The new monthly payment would be $732.96.

Total payments $= 360(\$732.96) + \$3400 = \$267,265.60$

(d) Again the unpaid balance from part (b) is the new loan amount. Again $P = 115,962.66$ and $i = \frac{0.065}{12}$, and this time $n = 15(12) = 180$.

$$R = \frac{Pi}{1 - (1+i)^{-n}}$$
$$= \frac{115,962.66\left(\frac{0.065}{12}\right)}{1 - \left(1 + \frac{0.065}{12}\right)^{-180}}$$
$$\approx 1010.16$$

The new monthly payment would be $1010.16.

Total payments $= 180(\$1010.16) + \$4500 = \$186,328.80$

51. This is just like a sinking fund in reverse.

(a) $P = 150,000$, $i = \frac{0.06}{2} = 0.03$, $n = 2(5) = 10$

$$R = \frac{Pi}{1 - (1+i)^{-n}}$$
$$= \frac{150,000(.03)}{1 - (1 + 0.03)^{-10}}$$
$$\approx 17,584.58$$

The amount of each withdrawal is $17,584.58.

(b) $P = 150,000$, $i = \frac{0.06}{2} = 0.03$, $n = 2(6) = 12$

$$R = \frac{Pi}{1 - (1+i)^{-n}}$$
$$= \frac{150,000(0.03)}{1 - (1 + 0.03)^{-12}}$$
$$\approx 15,069.31$$

If the money must last 6 yr, the amount of each withdrawal is $15,069.31.

53. This exercise should be solved by graphing calculator or computer methods. The amortization schedule, which may vary slightly, is as follows.

Payment Number	Amount of Payment	Interest for Period	Portion to Principal	Principal at End of Period
0	—	—	—	$4836.00
1	$585.16	$175.31	$409.85	$4426.15
2	$585.16	$160.45	$424.71	$4001.43
3	$585.16	$145.05	$440.11	$3561.32
4	$585.16	$129.10	$456.06	$3105.26
5	$585.16	$112.57	$472.59	$2632.67
6	$585.16	$95.43	$489.73	$2142.94
7	$585.16	$77.68	$507.48	$1635.46
8	$585.16	$59.29	$525.87	$1109.59
9	$585.16	$40.22	$544.94	$564.65
10	$585.12	$20.47	$564.65	$0.00

55. (a) Here $R = 1000$ and $i = 0.04$ and we have

$$P = \frac{R}{i} = \frac{100}{0.04} = 25,000$$

Therefore, the present value of the perpetuity is $25,000.

(b) Here $R = 600$ and $i = \frac{0.06}{4} = 0.015$ and we have

$$P = \frac{R}{i} = \frac{600}{0.015} = 40,000$$

Therefore, the present value of the perpetuity is $40,000.

Chapter 5 Review Exercises

1. $I = Prt$

$$= 15,903(0.06)\left(\frac{8}{12}\right)$$
$$= 636.12$$

The simple interest is $636.12.

3. $I = Prt$

$$= 42,368(0.0522)\left(\frac{7}{12}\right)$$
$$\approx 1290.11$$

The simple interest is $1290.11.

5. For a given amount of money at a given interest rate for a given time period greater than 1, compound interest produces more interest than simple interest.

7. $19,456.11 at 8% compounded semiannually for 7 yr

Use the formula for compound amount with $P = 19,456.11$, $i = \frac{0.08}{2} = 0.04$, and $n = 7(2) = 14$.

$$A = P(1+i)^n$$
$$= 19,456.11(1.04)^{14}$$
$$\approx 33,691.69$$

The compound amount is $33,691.69.

9. $57,809.34 at 6% compounded quarterly for 5 yr

Use the formula for compound amount with $P = 57,809.34$, $i = \frac{0.06}{4} = 0.015$, and $n = 5(4) = 20$.

$$A = P(1+i)^n$$
$$= 57,809.34(1.015)^{20}$$
$$\approx 77,860.80$$

The compound amount is $77,860.80.

11. $12,699.36 at 5% compounded semiannually for 7 yr

Here $P = 12,699.36$, $i = \frac{0.05}{2} = 0.025$, and $n = 7(2) = 14$. First find the compound amount.

$$A = P(1+i)^n$$
$$= 12,699.36(1.025)^{14}$$
$$\approx 17,943.86$$

The compound amount is $17,943.86.

To find the amount of interest earned, subtract the initial deposit from the compound amount. The interest earned is

$$\$\ 17,943.86 - 12,699.36 = \$5244.50.$$

13. $34,677.23 at 4.8% compounded monthly for 32 mo

Here $P = 34,677.23$, $i = \frac{0.048}{12} = 0.004$, and $n = 32$.

$$A = P(1+i)^n$$
$$= 34,677.23(1.004)^{32}$$
$$\approx 39,402.45$$

The compound amount is $39,402.45

The interest earned is

$$\$39,402.45 - 34,677.23 = \$4725.22.$$

15. $42,000 in 7 yr, 6% compounded monthly

Use the formula for present value for compound interest with $A = 42,000$, $i = \frac{0.06}{12} = 0.005$, and $n = 7(12) = 84$.

$$P = \frac{A}{(1+i)^n} = \frac{42,000}{(1.005)^{84}} \approx 27,624.86$$

The present value is $27,624.86.

17. $1347.89 in 3.5 yr, 6.77% compounded semiannually

Use the formula for present value for compound interest with $A = 1347.89$, $i = \frac{0.0677}{2} = 0.03385$, and $n = 3.5(2) = 7$.

$$P = \frac{A}{(1+i)^n} = \frac{1347.89}{(1.03385)^7} \approx 1067.71$$

The present value is $1067.71.

19. $a = 2$; $r = 3$

The first five terms are

$$2, \; 2(3), \; 2(3)^2, \; 2(3)^3, \text{ and } 2(3)^4,$$

or

$$2, 6, 18, 54, \text{ and } 162.$$

21. $a = -3$; $r = 2$

To find the sixth term, use the formula $a_n = ar^{n-1}$ with $a = -3$, $r = 2$, and $n = 6$.

$$a_6 = ar^{6-1} = -3(2)^5 = -3(32) = -96$$

23. $a = -3$; $r = 3$

To find the sum of the first 4 terms of this geometric sequence, use the formula $S_n = \frac{a(r^n - 1)}{r-1}$ with $n = 4$.

$$S_4 = \frac{-3(3^4 - 1)}{3-1} = \frac{-3(80)}{2} = \frac{-240}{2} = -120$$

25. $s_{\overline{n}|i} = \frac{(1+i)^n - 1}{i}$

$$s_{\overline{30}|0.02} = \frac{(1.02)^{30} - 1}{0.02} \approx 40.56808$$

29. $R = 1288, i = 0.04, n = 14$

This is an ordinary annuity.

$$S = Rs_{\overline{n}|i}$$
$$S = 1288 s_{\overline{14}|0.04}$$
$$= 1288 \left[\frac{(1+0.04)^{14} - 1}{0.04} \right]$$
$$\approx 23,559.98$$

The future value is $23,559.98.

The total amount deposited is $1288(14) = \$18,032$. Thus, the amount of interest is

$$\$23,559.98 - 18,032 = \$5527.98.$$

31. $R = 233, i = \frac{0.048}{12} = 0.004, n = 4(12) = 48$

This is an ordinary annuity.

$$S = R \left[\frac{(1+i)^n - 1}{i} \right]$$
$$S = 233 \left[\frac{(1.004)^{48} - 1}{0.004} \right] \approx 12,302.78$$

The future value is $12,302.78.

The total amount deposited is $233(48) = \$11,184.$ Thus, the amount of interest is

$$\$12,302.78 - 11,184 = \$1118.78.$$

33. $R = 11,900, i = \frac{0.06}{12} = 0.005, n = 13$

This is an annuity due, so we use the formula for future value of an ordinary annuity, but include one additional time period and subtract the amount of one payment.

$$S = R\left[\frac{(1+i)^{n+1} - 1}{i}\right] - R$$

$$= 11,900\left[\frac{(1.005)^{14} - 1}{0.005}\right] - 11,900$$

$$\approx 160,224.29$$

The future value is $\$160,224.29.$

The total amount deposited is $\$11,900(13) = \$154,700.$ Thus, the amount of interest is

$$\$160,224.29 - 154,700 = \$5524.29.$$

35. $6500; money earns 5% compounded annually; 6 annual payments

$$S = 6500, i = 0.05, n = 6$$

Let R be the amount of each payment.

$$S = Rs_{\overline{n}|i}$$

$$R = \frac{6500}{s_{\overline{6}|0.05}}$$

$$= \frac{6500(0.05)}{(1.05)^6 - 1}$$

$$\approx 955.61$$

The amount of each payment is $955.61.

37. $233,188; money earns 5.2% compounded quarterly for $7\frac{3}{4}$

$$S = 233,188, i = \frac{0.052}{4} = 0.013, n = \left(7\frac{3}{4}\right)(4) = 31$$

Let R be the amount of each payment.

$$S = Rs_{\overline{n}|i}$$

$$R = \frac{233,188}{s_{\overline{31}|0.013}}$$

$$= \frac{233,188(0.013)}{(1.013)^{31} - 1}$$

$$\approx 6156.14$$

The amount of each payment is $6156.14.

39. Deposits of $850 annually for 4 years at 6% compounded annually

Use the formula for the present value of an annuity with $R = 850, i = 0.06,$ and $n = 4.$

$$P = R\left[\frac{1 - (1+i)^{-n}}{i}\right]$$

$$= 850\left[\frac{1 - (1+0.06)^{-4}}{0.06}\right]$$

$$\approx 2945.34$$

The present value is $2945.34.

41. Deposits of $4210 semiannually for 8 years at 4.2% compounded annually

Use the formula for the present value of an annuity with $R = 4210, i = \frac{0.042}{2} = 0.021, n = 8(2) = 16.$

$$P = R\left[\frac{1 - (1+i)^{-n}}{i}\right]$$

$$= 4210\left[\frac{1 - (1.021)^{-16}}{0.021}\right]$$

$$\approx 56,711.93$$

The present value is $56,711.93.

43. Two types of loans that are commonly amortized are home loans and auto loans.

45. $P = 3200, i = \frac{0.08}{4} = 0.02, n = 12$

$$R = \frac{Pi}{1 - (1+i)^{-n}} = \frac{3200(0.02)}{1 - (1.02)^{-12}} \approx 302.59$$

The amount of each payment is $302.59.

The total amount paid is $302.59(12) = \$3631.08.$ Thus, the total interest paid is

$$\$3631.08 - 3200 = \$431.08.$$

47. $P = 51,607, i = \frac{0.08}{12} = 0.00\overline{6}, n = 32$

$$R = \frac{Pi}{1 - (1+i)^{-n}}$$

$$= \frac{51,607(0.00\overline{6})}{1 - (1.00\overline{6})^{-32}}$$

$$\approx 1796.20$$

The amount of each payment is $1796.20.

The total amount paid is $1796.20(32) = \$57,478.40.$ Thus, the total interest paid is

$$\$57,478.40 - 51,607 = \$5871.40.$$

49. $P = 177,110$, $i = \frac{0.0668}{12} = 0.0055\overline{6}$, $n = 30(12) = 360$

$$R = \frac{Pi}{1 - (1+i)^{-n}}$$

$$= \frac{177,110(0.0055\overline{6})}{1 - (1.0055\overline{6})^{-360}}$$

$$\approx 1140.50$$

The amount of each payment is $1140.50.

The total amount paid is $1140.50(360) = \$410,580$. Thus, the total interest paid is

$$\$410,580 - 177,110 = \$233,470.$$

51. The answer can be found in the table under payment number 12 in the column labeled "Portion to Principal." The amount of principal repayment included in the fifth payment is $132.99.

53. The last entry in the column "Principal at End of Period," $125,464.39, shows the debt remaining at the end of the first year (after 12 payments). Since the original debt (loan principal) was $127,000, the amount by which the debt has been reduced at the end of the first year is

$$\$127,000 - 125,464.39 = \$1535.61.$$

55. Here $P = 9820$, $r = 6.7\% = 0.067$, and $t = \frac{7}{12}$.

$$I = Prt$$

$$= 9820(0.067)\left(\frac{7}{12}\right)$$

$$\approx 383.80$$

The interest he will pay is $383.80. The total amount he will owe in 7 mo is

$$\$9820 + 383.80 = \$10,203.80.$$

57. $P = 84,720$, $t = \frac{7}{12}$, $I = 4055.46$

Substitute these values into the formula for simple interest to find the value of r.

$$I = Prt$$

$$4055.46 = 84,720r\left(\frac{7}{12}\right)$$

$$4055.46 = 49,420r$$
$$0.0821 \approx r$$

The interest rate is 8.21%.

59. In both cases use the formula for compound amount with $P = 500$ and $i = \frac{0.05}{4} = 0.0125$. For the investment at age 23 use $n = 42(4) = 168$.

$$A = P(1+i)^n$$
$$= 500(1 + 0.0125)^{168}$$
$$\approx 4030.28$$

For the investment at age 40 use $n = 25(4) = 100$.

$$A = P(1+i)^n$$
$$= 500(1 + 0.0125)^{100}$$
$$\approx 1731.70$$

The increased amount of money Tom will have if he invests now is

$$\$4030.28 - 1731.70 = \$2298.58.$$

61. $R = 5000$, $i = \frac{0.10}{2} = 0.05$, $n = 7\frac{1}{2}(2) = 15$

This is an ordinary annuity.

$$S = R\left[\frac{(1+i)^n - 1}{i}\right]$$

$$S = 5000\left[\frac{(1 + 0.05)^{15} - 1}{0.05}\right]$$

$$\approx 107,892.82$$

The future value is $107,892.82. The amount of interest earned is

$$\$107,892.82 - 15(5000) = \$32,892.82.$$

63. Use the formula for amortization payments with $P = 48,000$, $i = 0.065$, and $n = 7$.

$$R = \frac{Pi}{1 - (1+i)^{-n}}$$

$$= \frac{48,000(0.065)}{1 - (1.065)^{-7}}$$

$$\approx 8751.91$$

The owner should deposit $8751.91 at the end of each year.

The total amount deposited is $8751.91(7) = \$61,263.37$. Thus, the total interest paid is

$$\$61,263.37 - 48,000 \doteq \$13,263.37.$$

65. To find the effective rates use the formula
$r_e = \left(1 + \frac{r}{m}\right)^m - 1$.

First Community Bank:

$$r_e = \left(1 + \frac{0.0515}{4}\right)^4 - 1$$

$$\approx 0.052503$$

$$\approx 5.250\%$$

UFB Direct.com:

$$r_e = \left(1 + \frac{0.0513}{12}\right)^{12} - 1$$

$$\approx 0.052524$$

$$\approx 5.252\%$$

UFB Direct.com has the higher effective rate, even though it has a lower stated rate.

67. **(a)** There is no interest with 0% financing, so the monthly payment is $20,000 \div 36 \approx \$555.56$.

The total payments will be $20,000.

(b) 1.9% financing for 48 months:

$P = 20,000$, $i = \frac{0.019}{12} = 0.0015\overline{83}$, $n = 48$

$$P = R\left[\frac{1 - (1+i)^{-n}}{i}\right]$$

$$20,000 = R\left[\frac{1 - (1 + 0.0015\overline{83})^{-48}}{0.0015\overline{83}}\right]$$

$$R \approx 433.03$$

The monthly payments are $433.03.

The total amount of the payments is $433.03(48) = \$20,785.44$.

2.9% financing for 60 months:

$P = 20,000$, $i = \frac{0.029}{12} = 0.00241\overline{6}$, $n = 60$

$$P = R\left[\frac{1 - (1+i)^{-n}}{i}\right]$$

$$20,000 = R\left[\frac{1 - (1 + 0.00241\overline{6})^{-60}}{0.00241\overline{6},}\right]$$

$$R \approx 358.49$$

The monthly payments are $358.49.

The total amount of the payments is $358.49(60) = \$21,509.40$.

(c) $P = 18,000$, $i = \frac{0.0635}{12} = 0.0052916$, $n = 48$

$$P = R\left[\frac{1 - (1+i)^{-n}}{i}\right]$$

$$18,000 = R\left[\frac{1 - (1 + 0.0052916)^{-48}}{0.0052916}\right]$$

$$R \approx 425.62$$

The monthly payments are $425.62.

The total amount of the payments is $425.62(48) = \$20,429.76$.

69. Amount of loan $= \$191,000 - 40,000$
$= \$151,000$

(a) Use the formula for amortization payments with $P = 151,000$, $i = \frac{0.065}{12} = 0.005541\overline{6}$, and $n = 30(12) = 360$.

$$R = \frac{Pi}{1 - (1+i)^{-n}}$$

$$= \frac{151,000(0.005541\overline{6})}{1 - (1.005541\overline{6})^{-360}}$$

$$\approx 954.42$$

The monthly payment for this mortgage is $954.42.

(b) To find the amount of the first payment that goes to interest, use $I = Prt$ with $P = 151,000$, $i = 0.005541\overline{6}$, and $t = 1$.

$$I = 151,000(0.065)\left(\tfrac{1}{12}\right) = 817.92$$

Of the first payment, $817.92 is interest.

(c) Using method 1, since 180 of 360 payments were made, there are 180 remaining payments. The present value is

$$954.42\left[\frac{1 - (1.005541\overline{6})^{-180}}{0.005541\overline{6}}\right] \approx 109,563.99,$$

so the remaining balance is $109,563.99.

Using method 2, since 180 payments were already made, we have

$$954.42\left[\frac{1 - (1.005541\overline{6})^{-180}}{0.005541\overline{6}}\right] \approx 109,563.99.$$

She still owes

$$\$151,000 - 109,563.99 = \$41,436.01.$$

Furthermore, she owes the interest on this amount for 180 mo, for a total remaining balance of

$$41,436.01(1.005541\overline{6})^{180} = 109,565.13.$$

(d) Closing costs $= 3700 + 0.025(238,000)$

$$= 3700 + 5950$$

$$= 9650$$

Closing costs are $9650.

(e) Amount of money received

$$\doteq \text{Selling price} - \text{Closing costs}$$
$$- \text{Current mortgage balance}$$

Using method 1, the amount received is

$238,000 - 9650 - 109,563.99 = \$118,786.01$.

Using method 2, the amount received is

$238,000 - 9650 - 109,565.13 = \$118,784.87$.

71. (a) Use the formula for effective rate with $r_e = 0.10$ and $m = 12$.

$$r_e = \left(1 + \frac{r}{m}\right)^m - 1$$

$$0.10 = \left(1 + \frac{r}{12}\right)^{12} - 1$$

$$1.10 = \left(1 + \frac{r}{12}\right)^{12}$$

$$(1.10)^{1/12} = 1 + \frac{r}{12}$$

$$1.007974 \approx 1 + \frac{r}{12}$$

$$0.007974 \approx \frac{r}{12}$$

$$0.095688 \approx r$$

The annual interest rate is 9.569%.

(b) Use the formula for amortization payments with $P = 140,000$, $i = \frac{0.06625}{12}$, and $n = 30(12) = 360$.

$$R = \frac{Pi}{1 - (1+i)^{-n}}$$

$$= \frac{140,000 \left(\frac{0.06625}{12}\right)}{1 - \left(1 + \frac{0.06625}{12}\right)^{-360}}$$

$$\approx 896.44$$

Her monthly payment is $896.44.

(c) This investment is an annuity with $R = 1200 - 896.44 = 303.56$, $i = \frac{0.09569}{12}$, and $n = 30(12) = 360$. The future value is

$$S = R\left[\frac{(1+i)^n - 1}{i}\right]$$

$$= 303.56 \left[\frac{\left(1 + \frac{0.09569}{12}\right)^{360} - 1}{\frac{0.09569}{12}}\right]$$

$$\approx 626,200.88$$

In 30 yr she will have $626,200.88 in the fund.

(d) Use the formula for amortization payments with $P = 140,000$, $i = \frac{0.0625}{12}$, and $n = 15(12) = 180$.

$$R = \frac{Pi}{1 - (1+i)^{-n}}$$

$$= \frac{140,000 \left(\frac{0.0625}{12}\right)}{1 - \left(1 + \frac{0.0625}{12}\right)^{-180}}$$

$$\approx 1200.39$$

His monthly payment is $1200.39.

(e) This investment is an annuity with $R = 1200$, $i = \frac{0.09569}{12}$, and $n = 15(12) = 180$. The future value is

$$S = R\left[\frac{(1+i)^n - 1}{i}\right]$$

$$= 1200 \left[\frac{\left(1 + \frac{0.09569}{12}\right)^{180} - 1}{\frac{0.09569}{12}}\right]$$

$$\approx 478,134.14$$

In 30 yr he will have $478,134.14.

(f) Sue is ahead by

$$\$626,200.88 - 478,134.14 = \$148,066.74.$$

Chapter 5 Test

[5.1]

1. Find the simple interest on $1252 at 5% for 11 months.

2. Using a 360 day year, find the simple interest on $12,000 at 6.25% for 170 days.

3. Find the compound amount if $7000 is deposited for 8 years in an account paying 6% per year compounded quarterly.

4. Find the amount that should be invested now to accumulate $8000 at 8.1% compounded monthly for 2 years.

5. Find the effective rate (to the nearest hundredth of a percent) corresponding to a nominal rate of 6.5% compounded monthly.

6. To the nearest tenth of a year, how long will it take for an investment to double if it is invested at 5.86% compounded monthly?

[5.2]

7. Find the fifth term of the geometric sequence with $a = 4.7$, $r = 2$.

8. Find the sum of the first eight terms of the geometric sequence with $a = 14$ and $r = \frac{1}{2}$.

9. Find the future value of an ordinary annuity in which $750 is deposited at the end of each quarter for 6 years, with 12% interest compounded quarterly.

10. Find the future value of an annuity due in which $1375 is deposited at the beginning of each six months for 12 years, with 8% interest compounded semiannually.

11. Ralph deposits $100 at the end of each month for 3 years in an account paying 6% compounded monthly. He then uses the money from this account to buy a certificate of deposit which pays 7.5% compounded annually. If he redeems the certificate 4 years later, how much money will he receive?

12. What amount must be deposited into a sinking fund at the end of each quarter at 8% compounded quarterly to have $10,000 in 10 years?

[5.3]

13. Find the present value of an ordinary annuity with payments of $800 per month at 6% compounded monthly for 4 years.

14. Ms. Morroco borrows $12,000 at 8% compounded quarterly, to be paid off with equal quarterly payments over 2 years.

    (a) What quarterly payment is needed to amortize this loan?
    (b) Prepare an amortization table for the first 4 payments.

Chapter 5 Test Answers

1. $57.38

2. $354.17

3. $11,272.27

4. $6807.23

5. 6.70%

6. 11.9 years

7. 75.2

8. $\frac{1785}{64} = 27.890625$

9. $25,819.85

10. $55,888.12

11. $5253.21

12. $165.56

13. $34,064.25

14. (a) $1638.12

(b)

Payment Number	Amount of Payment	Interest for Period	Portion to Principal	Principal at End of Period
0	——	——	——	$12,000.00
1	$1638.12	$240.00	$1398.12	$10,601.88
2	$1638.12	$212.04	$1426.08	$9175.80
3	$1638.12	$183.52	$1454.60	$7721.20
4	$1638.12	$154.42	$1483.70	$6237.50

Chapter 6

LOGIC

6.1 Statements and Quantifiers

1. Because the declarative sentence "Montevideo is the capital of Uruguay" has the property of being true or false, it is considered a statement. It is not compound.

3. "Don't cry for me Argentina" is not a declarative sentence and does not have the property of being true or false. Hence, it is not considered a statement.

5. "$2 + 2 = 5$ and $3 + 3 = 7$" is a declarative sentence that is true and, therefore, is considered a statement. It is compound.

7. "Where's the beef?" is a question, not a declarative sentence, and, therefore, is not considered a statement.

9. "I am not a crook" is a compound statement because it contains the logical connective "not."

11. "She enjoys the comedy team of Penn and Teller" is not compound because only one assertion is being made.

13. "If ever I would leave you, it wouldn't be in summer" is a compound statement because it consists of two simple statements combined by the connective "if . . . then."

15. The negation of "My favorite flavor is chocolate" is "My favorite flavor is not chocolate."

17. A negation for "$y > 12$" (without using a slash sign) would be "$y \leq 12$."

19. A negation for "$q \geq 5$" would be "$q < 5$."

23. A translation of "$\sim b$" is "I'm not getting better."

25. A translation of "$\sim b \vee d$" is "I'm not getting better or my parrot is dead."

27. A translation of "$\sim (b \wedge \sim d)$" is "It is not the case that both I'm getting better and my parrot is not dead."

29. If q is false, then $(p \wedge \sim q) \wedge q$ is false, since both parts of the conjunction must be true for the compound statement to be true.

31. If the conjunction $p \wedge q$ is true, then both p and q must be true. Thus, q must be true.

33. If $\sim (p \vee q)$ is true, then $p \vee q$ must be false, since a statement and its negation have opposite truth values. In order for the disjunction $p \vee q$ to be false, both component statements must be false. Thus, p and q are both false.

35. Since p is false, $\sim p$ is true, since a statement and its negation have opposite truth values.

37. Since p is false and q is true, we may consider the statement $p \vee q$ as

$$F \vee T,$$

which is true by the *or* truth table. That is, $p \vee q$ is true.

39. Since p is false and q is true, we may consider $p \vee \sim q$ as

$$F \vee \sim T$$
$$F \vee F$$
$$F.$$

That is, $p \vee \sim q$ is false.

41. With the given truth values for p and q, we may consider $\sim p \vee \sim q$ as

$$\sim F \vee \sim T$$
$$T \vee F$$
$$T.$$

Thus, $\sim p \vee \sim q$ is true.

43. Replacing p and q with the given truth values, we have

$$\sim (F \wedge \sim T)$$
$$\sim (F \wedge F)$$
$$\sim F$$
$$T.$$

Thus, the compound statement $\sim (p \wedge \sim q)$ is true.

45. Replacing p and q with the given truth values, we have

$$\sim[\sim F \land (\sim T \lor F)]$$
$$\sim[T \land (F \lor F)]$$
$$\sim[T \land F]$$
$$\sim F$$
$$T.$$

Thus, the compound statement $\sim[\sim p \land (\sim q \lor p)]$ is true.

47. The statement $3 \geq 1$ is a disjunction since it means "$3 > 1$" or "$3 = 1$."

49. Replacing p, q, and r with the given truth values, we have

$$(T \land F) \lor \sim F$$
$$F \lor T$$
$$T.$$

Thus, the compound statement $(p \land r) \lor \sim q$ is true.

51. Replacing p, q, and r with the given truth values, we have

$$T \land (F \lor F)$$
$$T \land F$$
$$F.$$

Thus, the compound statement $p \land (q \lor r)$ is false.

53. Replacing p, q, and r with the given truth values, we have

$$\sim(T \land F) \land (F \lor \sim F)$$
$$\sim F \land (F \lor T)$$
$$T \land T$$
$$T.$$

Thus, the compound statement $\sim(p \land q) \land (r \lor \sim q)$ is true.

55. Replacing p, q, and r with the given truth values, we have

$$\sim[(\sim T \land F) \lor F]$$
$$\sim[(F \land F) \lor F]$$
$$\sim[F \lor F]$$
$$\sim F$$
$$T.$$

Thus, the compound statement $\sim[(p \land q) \lor r]$ is true.

57. Since p is false and r is true, we have

$$F \land T$$
$$F.$$

The compound statement $p \land r$ is false.

59. Since q is false and r is true, we have

$$\sim F \lor \sim T$$
$$T \lor F$$
$$T.$$

The compound statement $\sim q \lor \sim r$ is true.

61. Since p and q are false and r is true, we have

$$(F \land F) \lor T$$
$$F \lor T$$
$$T.$$

The compound statement $(p \land q) \lor r$ is true.

63. Since p and q are false and r is true, we have

$$(\sim T \land F) \lor \sim F$$
$$(F \land F) \lor T$$
$$F \lor T$$
$$T.$$

The compound statement $(\sim r \land q) \lor \sim p$ is true.

65. **a**, **c**, and **d** are declarative sentences that are true or false and are therefore statements.

b is a command, not a declarative sentence, and is therefore not a statement.

67. The negation is formed by adding "not": "We may not charge a fee of $35 in each billing period the New Balance on your statement exceeds your credit line."

69. **c** is a compound statement, formed using "Tax rates are lower for a head of household than for a person filing as single" and "the standard deduction is higher."

d is a compound statement, formed by negating the statement "You reduce the exemptions because of the shorter tax table."

b is a simple statement, not a compound one.

71. $p \land q$, where p is the statement "Tax rates are lower for a head of household than for a person filing as single" and q is the statement "the standard deduction is higher."

73. (a) This is a question, not a declarative sentence, and therefore not a statement.

(b), (d) These are compound statements that are conjunctions.

(c) This is a compound statement in the "If . . . then" form.

(e) This is a simple statement.

75. (a), (b) These are commands, not declarative sentences, and therefore not statements.

(c), (d), (e) These are declarative sentences that are true or false and therefore statements.

77. One choice: "Most legal problems are not matters of civil law."

81. "New England did not win the Super Bowl or Tom Brady is not the best quarterback" may be symbolized as $\sim n \vee \sim b$.

83. "New England did not win the Super Bowl but Tom Brady is the best quarterback" may be symbolized as $\sim n \wedge b$.

85. "Either New England won the Super Bowl or Tom Brady is the best quarterback" and it is not the case that both New England won the Super Bowl and Tom Brady is the best quarterback" may be symbolized as $(n \vee b) \wedge [\sim(n \wedge b)]$.

87. Assume that n is false and that b is false. Under these conditions, the statements in Exercises 80-85 have the following truth values.

80. $n \wedge \sim b$: False, because n is false and $\sim b$ is true.

81. $\sim n \vee \sim b$: True, because $\sim n$ is true and $\sim b$ is true.

82. $\sim n \vee b$: True, because $\sim n$ is true and $\sim b$ is false.

83. $\sim n \wedge b$: False, because $\sim n$ is true and b is false.

84. $\sim n \wedge \sim b$: True, because $\sim n$ is true and $\sim b$ is true.

85. $(n \vee b) \wedge [\sim(n \wedge b)]$: False, since $(n \vee b)$ is false but $[\sim(n \wedge b)]$ is true.

Therefore, the answer is 81, 82, and 84 are true statements.

6.2   Truth Tables and Equivalent Statements

1. Since there are two simple statements (p and r), we have $2^2 = 4$ rows in the truth table.

3. Since there are four simple statements (p, q, r, and s), we have $2^4 = 16$ rows in the truth table.

5. Since there are seven simple statements (p, q, r, s, t, u, and v), we have $2^7 = 128$ rows in the truth table.

7. If the truth table for a certain compound statement has 64 rows, then there must be six distinct component statements since $2^6 = 64$.

9. $\sim p \wedge q$

p	q	$\sim p$	$\sim p \wedge q$
T	T	F	F
T	F	F	F
F	T	T	T
F	F	T	F

11. $\sim(p \wedge q)$

p	q	$p \wedge q$	$\sim(p \wedge q)$
T	T	T	F
T	F	F	T
F	T	F	T
F	F	F	T

13. $(q \vee \sim p) \vee \sim q$

p	q	$\sim p$	$\sim q$	$q \vee \sim p$	$(q \vee \sim p) \vee \sim q$
T	T	F	F	T	T
T	F	F	T	F	T
F	T	T	F	T	T
F	F	T	T	T	T

In Exercises 15–23 to save space we are using the alternative method, filling in columns in the order indicated by the numbers. Observe that columns with the same number are combined (by the logical definition of the connective) to get the next numbered column. Note that this is different from the way the numbered columns are used in the textbook. Remember that the last column (highest numbered column) completed yields the truth values for the complete compound statement. Be sure to align truth values under the appropriate logical connective or simple statement.

15. $\sim q \wedge (\sim p \vee q)$

p	q	$\sim q$	$\wedge$	$(\sim p$	$\vee$	$q)$
T	T	F	F	F	T	T
T	F	T	F	F	F	F
F	T	F	F	T	T	T
F	F	T	T	T	T	F
		1	4	2	3	2

17. $(p \lor \sim q) \land (p \land q)$

p	q	$(p$	$\lor$	$\sim q)$	$\land$	$(p$	$\land$	$q)$
T	T	T	T	F	T	T	T	T
T	F	T	T	T	F	T	F	F
F	T	F	F	F	F	F	F	T
F	F	F	T	T	F	F	F	F
		1	2	1	5	3	4	3

19. $(\sim p \land q) \land r$

p	q	r	$(\sim p$	$\land$	$q)$	$\land$	r
T	T	T	F	F	T	F	T
T	T	F	F	F	T	F	F
T	F	T	F	F	F	F	T
T	F	F	F	F	F	F	F
F	T	T	T	T	T	T	T
F	T	F	T	T	T	F	F
F	F	T	T	F	F	F	T
F	F	F	T	F	F	F	F
			1	2	1	4	3

21. $(\sim p \land \sim q) \lor (\sim r \lor \sim p)$

p	q	r	$(\sim p$	$\land$	$\sim q)$	$\lor$	$(\sim r$	$\lor$	$\sim p)$
T	T	T	F	F	F	F	F	F	F
T	T	F	F	F	F	T	T	T	F
T	F	T	F	F	T	F	F	F	F
T	F	F	F	F	T	T	T	T	F
F	T	T	T	F	F	T	F	T	T
F	T	F	T	F	F	T	T	T	T
F	F	T	T	T	T	T	F	T	T
F	F	F	T	T	T	T	T	T	T
			1	2	1	5	3	4	3

23. $\sim (\sim p \land \sim q) \lor (\sim r \lor \sim s)$

p	q	r	s	$\sim$	$(\sim p$	$\land$	$\sim q)$	$\lor$	$(\sim r$	$\lor$	$\sim s)$
T	T	T	T	T	F	F	F	T	F	F	F
T	T	T	F	T	F	F	F	T	F	T	T
T	T	F	T	T	F	F	F	T	T	T	F
T	T	F	F	T	F	F	F	T	T	T	T
T	F	T	T	T	F	F	T	T	F	F	F
T	F	T	F	T	F	F	T	T	F	T	T
T	F	F	T	T	F	F	T	T	T	T	F
T	F	F	F	T	F	F	T	T	T	T	T
F	T	T	T	T	T	F	F	T	F	F	F
F	T	T	F	T	T	F	F	T	F	T	T
F	T	F	T	T	T	F	F	T	T	T	F
F	T	F	F	T	T	F	F	T	T	T	T
F	F	T	T	F	T	T	T	F	F	F	F
F	F	T	F	F	T	T	T	T	F	T	T
F	F	F	T	F	T	T	T	T	T	T	F
F	F	F	F	F	T	T	T	T	T	T	T
				3	1	2	1	6	4	5	4

25. "It's summertime and the living is easy" has the symbolic form $p \land q$. The negation, $\sim (p \land q)$, is equivalent, by one of DeMorgan's laws, to $\sim p \lor \sim q$. The corresponding word statement is "It's not summertime, or the living is not easy."

27. "Either the door was unlocked or the thief broke a window" has the symbolic form $p \vee q$. The negation, $\sim (p \vee q)$, is equivalent, by one of DeMorgan's laws, to $\sim p \wedge \sim q$. The corresponding word statement is "The door was locked and the thief didn't break a window."

29. "I'm ready to go, but Emily Portwood isn't" has the symbolic form $p \wedge \sim q$. (The connective "but" is logically equivalent to "and.") The negation, $\sim (p \wedge \sim q)$, is equivalent, by one of DeMorgan's laws, to $\sim p \vee q$. The corresponding word statement is "I'm not ready to go, or Emily Portwood is."

31. "$12 > 4$ or $8 = 9$" has the symbolic form $p \vee q$. The negation, $\sim (p \vee q)$, is equivalent, by one of DeMorgan's laws, to $\sim p \wedge \sim q$. The corresponding statement is "$12 \leq 4$ and $8 \neq 9$." (Note that the inequality "$\leq$" is logically equivalent to "$\not>$.")

33. "Larry or Moe is out sick today" has the symbolic form $p \vee q$. The negation, $\sim (p \vee q)$, is equivalent, by one of De Morgan's laws, to $\sim p \wedge \sim q$. The corresponding word statement is "Neither Larry nor Moe is out sick today."

35. $p \veebar q$

p	q	$p \veebar q$
T	T	F
T	F	T
F	T	T
F	F	F

Observe that it is only the first line in the truth table that changes for "exclusive disjunction" since the component statements can not both be true at the same time.

37. "$3 + 1 = 4 \veebar 2 + 5 = 9$" is <u>true</u> since the first component statement is true and the second is false.

39. Store the truth values of the statements $p, q,$ and s as P, Q, and S, respectively.

Use the stored values of P, Q, and S to find the truth values of each of the compound statements.

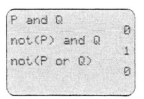   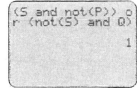

(a) P and Q returns 0, meaning $p \wedge q$ is false.

(b) not (P) and Q returns 1, meaning $\sim p \wedge q$ is true.

(c) not (P or Q) returns 0, meaning $\sim (p \vee q)$ is false.

(d) (S and not (P)) or (not (S) and Q) returns 1, meaning $(s \wedge \sim p) \vee (\sim s \wedge q)$ is true.

43. Letting s represent "You will be completely satisfied," r represent "We will refund your money," and q represent "We will ask you questions," the guarantee translates into symbols as $s \vee (r \wedge \sim q)$. The truth table follows.

s	r	q	s	$\vee$	$(r$	$\wedge$	$\sim q)$
T	T	T	T	T	T	F	F
T	T	F	T	T	T	T	T
T	F	T	T	T	F	F	F
T	F	F	T	T	F	F	T
F	T	T	F	F	T	F	F
F	T	F	F	T	T	T	T
F	F	T	F	F	F	F	F
F	F	F	F	F	F	F	T
			1	1	4	2	3	2

The guarantee would be false in the three cases indicated as "F" in the column labeled "6." These would be if you are not completely satisfied, and they either don't refund your money or they ask you questions.

45. p: The Pennsylvania Fish and Boat Commission is sensitive to the needs of the physically challenged.
q: The Pennsylvania Fish and Boat Commission works to make our facilities accessible.
The statement is $p \wedge q$. The negation is $\sim p \vee \sim q$.
Negation: Either the Pennsylvania Fish and Boat Commission is not sensitive to the needs of the physically challenged or it does not work to make our facilities accessible.

47. p: The court won't do it for you.

q: Hiring an attorney is usually not cost effective. The statement is $p \wedge q$. The negation is $\sim p \vee \sim q$.

Negation: Either the court will do it for you or hiring an attorney is often cost effective.

49. Letting c represent "I will cut taxes," e represent "I will eliminate the deficit," and r represent "I will run for reelection," the guarantee translates into symbols as $(c \wedge e) \vee \sim r$. The truth table follows.

c	e	r	$(c$	$\wedge$	$e)$	$\vee$	$\sim r$
T	T	T	T	T	T	T	F
T	T	F	T	T	T	T	T
T	F	T	T	F	F	F	F
T	F	F	T	F	F	T	T
F	T	T	F	F	T	F	F
F	T	F	F	F	T	T	T
F	F	T	F	F	F	F	F
F	F	F	F	F	F	T	T
			1	2	1	4	3

The promise would be false in the three cases indicated as "F" in the column labeled "6." These would be if the senator runs for reelection and either doesn't cut taxes or doesn't eliminate the deficit.

51. Since only one sign is true and the sign on Door 2 is true, the sign on Door 1 must be false. Thus, either the lady is not behind Door 1 or the tiger is not behind Door 2. Since the lady and the tiger are not behind the same door, the lady must be behind Door 2 and the tiger behind Door 1.

6.3   The Conditional and Circuits

1. The statement "If the antecedent of a conditional statement is false, the conditional statement is true" is <u>true</u>, since a false antecedent will always yield a true conditional statement.

3. The statement "If q is true, then $(p \wedge q) \to q$ is true" is <u>true</u>, since with a true consequent the conditional statement is always true (even though the antecedent may be false)."

5. "Given that $\sim p$ is true and q is false, the conditional $p \to q$ is true" is a <u>true</u> statement since the antecedent, p, must be false.

9. "$F \to (4 \neq 7)$" is a <u>true</u> statement, since a false antecedent always yields a conditional statement which is true.

11. "$(4 = 11 - 7) \to (8 > 0)$" is <u>true</u> since the antecedent and the consequent are both true.

13. $d \to (e \wedge s)$ expressed in words becomes "If she dances tonight, then I'm leaving early and he sings loudly."

15. $\sim s \to (d \vee \sim e)$ expressed in words becomes "If he doesn't sing loudly, then she dances tonight or I'm not leaving early."

17. The statement "My dog ate my homework, or if I receive a failing grade then I'll run for governor" can be symbolized as $d \vee (f \to g)$.

19. The statement "I'll run for governor if I don't receive a failing grade" can be symbolized as $\sim f \to g$.

21. Replacing r and q with the given truth values, we have

$$\sim F \to T$$
$$T \to T$$
$$T.$$

Thus, the statement $\sim r \to q$ is true.

23. Replacing r and p with the given truth values, we have

$$\sim F \to F$$
$$T \to F$$
$$F.$$

Thus, the statement $\sim r \to p$ is false.

25. Replacing p, q, and r with the given truth values, we have

$$\sim F \to (T \wedge F)$$
$$T \to F$$
$$F.$$

Thus, the statement $\sim p \to (q \wedge r)$ is false.

27. Replacing p, q, and r with the given truth values, we have

$$\sim T \to (F \wedge F)$$
$$F \to F$$
$$T.$$

Thus, the statement $\sim q \to (p \wedge r)$ is true.

29. Replacing p, q, and r with the given truth values, we have

$$(\sim F \rightarrow \ \sim T) \rightarrow (\sim F \wedge \ \sim F)$$
$$(T \rightarrow F) \rightarrow (T \wedge T)$$
$$F \rightarrow T$$
$$T.$$

Thus, the statement $(p \rightarrow \sim q) \rightarrow (\sim p \wedge \sim r)$ is true.

33. According to Equivalent Statement 8, $p \rightarrow q \equiv \sim p \rightarrow q$. Therefore,

$$\sim p \rightarrow q \equiv \sim (\sim p) \vee \sim q \qquad \text{by Equivalent Statement 8}$$
$$\sim (\sim p) \vee \sim q \equiv p \vee \sim q \qquad \text{by Equivalent Statement 7.}$$

35. $p \rightarrow \sim q$

p	q	p	$\rightarrow$	$\sim q$
T	T	T	F	F
T	F	T	T	T
F	T	F	T	F
F	F	F	T	T
		1	2	1

37. $(\sim q \rightarrow \sim p) \rightarrow \sim q$

p	q	$(\sim q \rightarrow \sim p)$			$\rightarrow$	$\sim q$
T	T	F	T	F	F	F
T	F	T	F	F	T	T
F	T	F	T	T	F	F
F	F	T	T	T	T	T
		1	2	1	4	3

39.

p	q	$(p$	$\wedge$	$\sim q)$	$\wedge$	$(p$	$\rightarrow$	$q)$
T	T	T	F	F	F	T	T	T
T	F	T	T	T	F	T	F	F
F	T	F	F	F	F	F	T	T
F	F	F	F	T	F	F	T	F
		1	2	1	4	3	4	3

Since this statement is always false (The truth values in column 4 are all false.), it is a contradiction.

41. $r \rightarrow (p \wedge \sim q)$

p	q	r	r	$\rightarrow$	$(p$	$\wedge$	$\sim q)$
T	T	T	T	F	T	F	F
T	T	F	F	T	T	F	F
T	F	T	T	T	T	T	T
T	F	F	F	T	T	T	T
F	T	T	T	F	F	F	F
F	T	F	F	T	F	F	F
F	F	T	T	F	F	F	T
F	F	F	F	T	F	F	T
			1	4	2	3	2

43. $(\sim r \rightarrow s) \vee (p \rightarrow \sim q)$

p	q	r	s	$(\sim$	r	$\rightarrow$	$s)$	$\vee$	$(p$	$\rightarrow$	$\sim q)$
T	T	T	T	F	T	T	T	T	T	F	F
T	T	T	F	F	T	F	T	T	T	F	F
T	T	F	T	T	T	T	T	T	T	F	F
T	T	F	F	T	F	F	F	T	T	F	F
T	F	T	T	F	T	T	T	T	T	T	T
T	F	T	F	F	T	F	T	T	T	T	T
T	F	F	T	T	T	T	T	T	T	T	T
T	F	F	F	T	F	F	T	T	T	T	T
F	T	T	T	F	T	T	T	T	F	T	F
F	T	T	F	F	T	F	T	T	F	T	F
F	T	F	T	T	T	T	T	T	F	T	F
F	T	F	F	T	F	F	T	T	F	T	F
F	F	T	T	F	T	T	T	T	F	T	T
F	F	T	F	F	T	F	T	T	F	T	T
F	F	F	T	T	T	T	T	T	F	T	T
F	F	F	F	T	F	F	T	T	F	T	T
				1	2	1	5	3	4	3	

45. The statement is not a tautology if at least <u>one</u> F appears in the final column of a truth table, since a tautology requires all T's in the final column.

47. Let p represent "they can see me now" and q represent "they'd never believe it." The statement has the form $p \rightarrow q$. In words, the equivalent form $\sim p \vee q$ becomes "They cannot see me now or they'd never believe it."

49. Let p represent "I am you" and q represent "I would watch out." The statement has the form $p \rightarrow q$. In words, the equivalent form $\sim p \vee q$ becomes "I am not you or I would watch out."

51. "If I can make it there, I'll make it anywhere" has the form $p \rightarrow q$. The negation has the form $p \wedge \sim q$, which translates as "I can make it there and I won't make it anywhere."

53. "If he's my brother, then he's not heavy" has the form $p \rightarrow \sim q$. The negation has the form $p \wedge q$, which translates as "He's my brother and he's heavy."

55.

p	q	$\sim$	$(p$	$\rightarrow$	$q)$	p	$\wedge$	$\sim q$
T	T	F	T	T	T	T	F	F
T	F	T	T	F	F	T	T	T
F	T	F	F	T	T	F	F	F
F	F	F	F	T	F	F	F	T
		3	1	2	1	1	2	1

Since the truth values in the final columns for each statement are the same, the statements are equivalent.

57.

p	q	q	$\rightarrow$	p	$\sim p$	$\rightarrow$	$\sim q$
T	T	T	T	T	F	T	F
T	F	F	T	T	F	T	T
F	T	T	F	F	T	F	F
F	F	F	T	F	T	T	T
		1	2	1	1	2	1

Since the truth values in the final columns for each statement are the same, the statements are equivalent.

59.

p	q	p	$\rightarrow$	q	$\sim q$	$\rightarrow$	$\sim p$
T	T	T	T	T	F	T	F
T	F	T	F	F	T	F	F
F	T	F	T	T	F	T	T
F	F	F	T	F	T	T	T
		1	2	1	1	2	1

Since the truth values in the final columns for each statement are the same, the statements are equivalent.

61.

p	q	$\sim p$	$\wedge$	q	$\sim p$	$\rightarrow$	q
T	T	F	F	T	F	T	T
T	F	F	F	F	F	T	F
F	T	T	T	T	T	T	T
F	F	T	F	F	T	F	F
		1	2	1	1	2	1

Since the truth values in the final columns for each statement are not the same, the statements are not equivalent.

63.

p	q	p	$\vee$	q	$\sim p$	$\rightarrow$	q
T	T	T	T	T	F	T	T
T	F	T	T	F	F	T	F
F	T	F	T	T	T	T	T
F	F	F	F	F	T	F	F
		1	2	1	3	4	3

The columns labeled 2 and 4 are identical.

65.

p	q	p	$\wedge$	q	q	$\wedge$	p
T	T	T	T	T	T	T	T
T	F	T	F	F	F	F	T
F	T	F	F	T	T	F	F
F	F	F	F	F	F	F	F
		1	2	1	3	4	3

The columns labeled 2 and 4 are identical.

67.

p	q	r	$(p$	$\wedge$	$q)$	$\wedge$	r	p	$\wedge$	$(q$	$\wedge$	$r)$
T	T	T	T	T	T	T	T	T	T	T	T	T
T	T	F	T	T	T	F	F	T	F	T	F	F
T	F	T	T	F	F	F	T	T	F	F	F	T
T	F	F	T	F	F	F	F	T	F	F	F	F
F	T	T	F	F	T	F	T	F	F	T	T	T
F	T	F	F	F	T	F	F	F	F	T	F	F
F	F	T	F	F	F	F	T	F	F	F	F	T
F	F	F	F	F	F	F	F	F	F	F	F	F
			1	2	1	4	3	5	8	6	7	6

The columns labeled 4 and 8 are identical.

69.

p	q	r	p	$\wedge$	$(q$	$\vee$	$r)$	$(p$	$\wedge$	$q)$	$\vee$	$(p$	$\wedge$	$r)$
T	T	T	T	T	T	T	T	T	T	T	T	T	T	T
T	T	F	T	T	T	T	F	T	T	T	T	T	F	F
T	F	T	T	T	F	T	T	T	F	F	T	T	T	T
T	F	F	T	F	F	F	F	T	F	F	F	T	F	F
F	T	T	F	F	T	T	T	F	F	T	F	F	F	T
F	T	F	F	F	T	T	F	F	F	T	F	F	F	F
F	F	T	F	F	F	T	T	F	F	F	F	F	F	T
F	F	F	F	F	F	F	F	F	F	F	F	F	F	F
			1	4	2	3	2	5	6	5	9	7	8	7

The columns labeled 4 and 9 are identical.

71.

p	q	$(p$	$\vee$	$q)$	$\wedge$	p
T	T	T	T	T	T	T
T	F	T	T	F	T	T
F	T	F	T	T	F	F
F	F	F	F	F	F	F
		1	2	1	4	3

The p column and the column labeled 4 are identical.

73. In the diagram, a parallel circuit is shown, which corresponds to $r \vee q$. This circuit, in turn, is in series with p. Thus, the logical statement is

$$p \wedge (r \vee q).$$

75. The diagram shows q in parallel with a series circuit consisting of p and the parallel circuit involving q and $\sim p$. Thus, the logical statement is

$$q \vee [p \wedge (q \vee \sim p)].$$

One pair of equivalent statements listed in the text includes

$$p \wedge (q \vee \sim p) \equiv (p \wedge q) \vee (p \wedge \sim p).$$

Since $(p \wedge \sim p)$ is never true, $p \wedge (q \vee \sim p)$ simplifies to

$$(p \wedge q) \vee F \equiv (p \wedge q).$$

Thus, $q \vee [p \wedge (q \vee \sim p)] \equiv q \vee (p \wedge q)$
$$\equiv (q \vee p) \wedge (q \wedge q)$$
$$\equiv (q \vee p) \wedge q$$
$$\equiv q.$$

77. The diagram shows two parallel circuits, $\sim p \vee q$ and $\sim p \vee \sim q$, which are parallel to each other. Thus, the total circuit can be represented as

$$(\sim p \vee q) \vee (\sim p \vee \sim q).$$

This circuit can be simplified using the following equivalencies.

$$(\sim p \vee q) \vee (\sim p \vee \sim q) \equiv \sim p \vee q \vee \sim p \vee \sim q$$
$$\equiv \sim p \vee q \vee \sim q$$
$$\equiv \sim p \vee (q \vee \sim q)$$
$$\equiv \sim p \vee T$$
$$\equiv T$$

79. In the diagram, series circuit $\sim p \wedge \sim q$ is parallel to the parallel circuit $p \vee q$. This entire circuit is in series with the parallel circuit $p \vee q$ and p. The logical statement is

$$\{[(\sim p \wedge \sim q) \vee (p \vee q)] \wedge (p \vee q)\} \wedge p.$$

This statement simplifies to p as follows:

$$\begin{aligned}
\{[(\sim p \wedge \sim q) &\vee (p \vee q)] \wedge (p \vee q)\} \wedge p \\
&\equiv \{[(\sim (p \vee q) \vee (p \vee q)] \wedge (p \vee q)\} \wedge p \\
&\equiv [T \wedge (p \vee q)] \wedge p \\
&\equiv (p \vee q) \wedge p \\
&\equiv p.
\end{aligned}$$

81. The logical statement $(\sim p \wedge \sim q) \wedge \sim r$ can be represented by the following circuit.

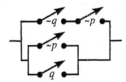

83. The logical statement $(\sim q \wedge \sim p) \vee (\sim p \vee q)$ can be represented by the following circuit.

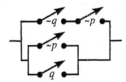

The statement $(\sim q \wedge \sim p) \vee (\sim p \vee q)$ simplifies to $\sim p \vee q$ as follows:

$$\begin{aligned}
(\sim q \wedge \sim p) &\vee (\sim p \vee q) \\
&\equiv [\sim q \vee (\sim p \vee q)] \wedge [\sim p \vee (\sim p \vee q)] \\
&\equiv [\sim q \vee \sim p \vee q] \wedge [\sim p \vee \sim p \vee q] \\
&\equiv [(\sim q \vee q) \vee \sim p] \wedge [(\sim p \vee \sim p) \vee q] \\
&\equiv (T \vee \sim p) \wedge (\sim p \vee q) \\
&\equiv T \wedge (\sim p \vee q) \\
&\equiv \sim p \vee q.
\end{aligned}$$

85. The statement $[(\sim p \wedge \sim r) \vee \sim q] \wedge (\sim p \wedge r)$ can be represented by the following circuit.

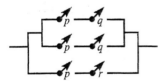

The statement $[(\sim p \wedge \sim r) \vee \sim q] \wedge (\sim p \wedge r)$ simplifies to $(\sim p \wedge r) \wedge \sim q$ in the following manner. Both $[(\sim p \wedge \sim r) \vee \sim q]$ and $(\sim p \wedge r)$ must be true. But if $(\sim p \wedge r)$ is true, then $(\sim p \wedge \sim r)$ is false. If $(\sim p \wedge \sim r)$ is false, then $\sim q$ must be true for the original disjunction to be true. Thus,

$$\begin{aligned}
[(\sim p \wedge \sim r) &\vee \sim q] \wedge (\sim p \wedge r) \\
&\equiv (F \vee \sim q) \wedge (\sim p \wedge r) \\
&\equiv \sim q \wedge (\sim p \wedge r) \\
&\equiv (\sim p \wedge r) \wedge \sim q.
\end{aligned}$$

87. The logical statement $\sim p \rightarrow (\sim p \vee \sim q)$ can be represented by the following circuit.

The statement $\sim p \rightarrow (\sim p \vee \sim q)$ simplifies to T as follows:

$$\begin{aligned}
\sim p \rightarrow (\sim p \vee \sim q) &\equiv p \vee (\sim p \vee \sim q) \\
&\equiv p \vee \sim p \vee \sim q \\
&\equiv (p \vee \sim p) \vee \sim q \\
&\equiv T \vee \sim q \\
&\equiv T.
\end{aligned}$$

89. The logical statement $[(p \wedge q) \vee (p \wedge q)] \vee (p \wedge r)$ can be represented by the following circuit.

The statement simplifies to $p \wedge (q \vee r)$ as follows:

$$\begin{aligned}
[(p \wedge q) \vee (p \wedge q)] \vee (p \wedge r) &\equiv (p \wedge q) \vee (p \wedge r) \\
&\equiv p \wedge (q \vee r).
\end{aligned}$$

91. Referring to Figures 10 and 11 of Example 8 in the text:

Cost per year of the circuit in Figure 10
= number of switches × \$0.03 × 24 hrs × 365 days
= 4 × 0.03 × 24 × 365
= \$1051.20.

Cost per year of the circuit in Figure 11
= number of switches × \$0.03 × 24 hrs × 365 days
= 3 × 0.03 × 24 × 365
= \$788.40

The savings is \$1051.20 − \$788.40 = \$262.80.

93. The form of the statements is $p \rightarrow q$.

An equivalent form is $\sim p \vee q$, which translates as "You are not wheezing persistently or you should see your doctor."

The negation is $p \wedge \sim q$, which translates as "You are wheezing persistently and you should not see your doctor."

6.4   More on the Conditional

Wording may vary in the answers to Exercises 1–27.

1. *The direct statement*: If the exit is ahead, then I don't see it.

  (a) *Converse:* If I don't see it, then the exit is ahead.

  (b) *Inverse:* If the exit is not ahead, then I see it.

  (c) *Contrapositive:* If I see it, then the exit is not ahead.

3. *The direct statement*: If I knew you were coming, I'd have baked a cake.

  (a) *Converse:* If I baked a cake, then I knew you were coming.

  (b) *Inverse:* If I didn't know you were coming, I wouldn't have baked a cake.

  (c) *Contrapositive:* If I didn't bake a cake, then I didn't know you were coming.

5. *It is helpful to reword the given statement.*

The direct statement: If a man is dead, then he doesn't wear plaid.

  (a) *Converse:* If a man doesn't wear plaid, then he's dead.

  (b) *Inverse:* If a man is not dead, then he wears plaid.

  (c) *Contrapositive:* If a man wears plaid, then he's not dead.

7. The direct statement: $p \rightarrow \sim q$.

  (a) *Converse:* $\sim q \rightarrow p$.

  (b) *Inverse:* $\sim p \rightarrow q$.

  (c) *Contrapositive:* $q \rightarrow \sim p$.

9. The direct statement: $p \rightarrow (q \vee r)$.

  (a) *Converse:* $(q \vee r) \rightarrow p$.

  (b) *Inverse:* $\sim p \rightarrow \sim(q \vee r)$ or $\sim p \rightarrow (\sim q \wedge \sim r)$.

  (c) *Contrapositive:* $(\sim q \wedge \sim r) \rightarrow \sim p$.

13. The statement "Your signature implies that you accept the conditions" becomes "If you sign, then you accept the conditions."

15. The statement "You can take this course pass/fail only if you have prior permission" becomes "If you can take this course pass/fail, then you have prior permission."

17. The statement "You can skate on the pond when the temperature is below 10°" becomes "If the temperature is below 10°, then you can skate on the pond."

19. The statement "Eating ten hot dogs is sufficient to make someone sick" becomes "If someone eats ten hot dogs, then he or she will get sick."

21. The statement "A valid passport is necessary for travel to France" becomes "If you travel to France, then you have a valid passport."

23. The statement "For a number to have a real square root, it is necessary that it be nonnegative" becomes "If a number has a real square root, then it is nonnegative."

25. The statement "All brides are beautiful" becomes "If someone is a bride, then she is beautiful."

27. The statement "A number is divisible by 3 if the sum of its digits is divisible by 3" becomes "If the sum of a number's digits is divisible by 3, then it is divisible by 3."

29. Option d is the answer since "r is necessary for s" represents the converse, $s \rightarrow r$, of all of the other statements.

33. The statement "$5 = 9 - 4$ if and only if $8 + 2 = 10$" is <u>true</u>, since this is a biconditional composed of two true statements.

35. The statement "$8 + 7 \neq 15$ if and only if $3 \times 5 \neq 9$" is <u>false</u>, since this is a biconditional consisting of one false statement and one true statement.

37. The statement "China is in Asia if and only if Mexico is in Europe" is <u>false</u>, since it is a biconditional consisting of a true statement and a false statement.

39.

p	q	$(\sim p$	$\wedge$	$q)$	$\leftrightarrow$	$(p$	$\rightarrow$	$q)$
T	T	F	F	T	F	T	T	T
T	F	F	F	F	T	T	F	F
F	T	T	T	T	T	F	T	T
F	F	T	F	F	F	F	T	F
		1	2	1	5	3	4	3

41. (a) "Your employer must tell you if the award qualifies for full or partial tax-free treatment" becomes "If the award qualifies for full or partial tax-free treatment, then your employer must tell you."

(b) "Medical expenses are 'qualified' only if incurred *after* the HSA has been established" becomes "If medical expenses are 'qualified,' then they are incurred *after* the HSA has been established."

(c) "You can avoid this interest deduction limitation if you elect to report the market discount annually as interest income" becomes "If you elect to report the market discount annually as interest income, you can avoid this interest deduction limitation."

43. For $p \rightarrow q$, the converse is $q \rightarrow p$, the inverse is $\sim p \rightarrow \sim q$, and the contrapositive is $\sim q \rightarrow \sim p$.

Converse: If you may avoid paying the annual fee billed on this statement, then you close your account within 30 days from the date this statement was mailed.
Inverse: If you do not close your account within 30 days from the date this statement was mailed, you may not avoid paying the annual fee billed on this statement.
Contrapositive: If you may not avoid paying the annual fee billed on this statement, then you do not close your account within 30 days from the date this statement was mailed.

The converse and the inverse are equivalent, and the contrapositive and the original statement are equivalent.

45. (a) Let p represent "there are triplets," let q represent "the most persistent stands to gain an extra meal," and let r represent "it may eat at the expense of another." Then the statement can be written as $p \rightarrow (q \wedge r)$.

(b) The contrapositive is $\sim (q \wedge r) \rightarrow p$, which is equivalent to $(\sim q \vee \sim r) \rightarrow p$: If the most persistent does not stand to gain an extra meal or it does not eat at the expense of another, then there are not triplets.

47. If liberty and equality are not best attained when all persons share alike in the government to the utmost, then they are not, as is thought by some, chiefly to be found in democracy.

49. The statement can be written as

$$(d \rightarrow l) \wedge \sim (l \rightarrow d)$$

51. If there is an R.P.F. alliance, there there is a Modéré incumbent. *Converse:* If there is a Modéré incumbent, then there is an R.P.F. alliance. *Inverse:* If there is not an R.P.F. alliance, then there is not a Modéré incumbent. *Contrapositive:* If there is not a Modéré incumbent, then there is not an R.P.F. alliance.
The contrapositive is equivalent to the original.

53. "Worked on the weekend": Must be turned over to see whether the employee got a day off.
"Did not work on the weekend": Need not be turned over, since it does not describe an employee who worked on the weekend.
"Did get a day off": Need not be turned over, since it cannot describe an employee who worked on the weekend without getting a day off.
"Did not get a day off": Must be turned over to see whether the other side says "worked on the weekend."

6.5   Analyzing Arguments and Proofs

1. Let p represent "she weighs the same as a duck," q represent "she's made of wood," and r represent "she's a witch." The argument is then represented symbolically by:

$$\frac{\begin{array}{l} p \to q \\ q \to r \end{array}}{p \to r.}$$

This is the <u>valid</u> argument form Reasoning by Transitivity.

3. Let p represent "I had a hammer" and q represent "I'd hammer in the morning." The argument is then represented symbolically by:

$$\frac{\begin{array}{l} p \to q \\ p \end{array}}{q.}$$

This is the <u>valid</u> argument form Modus Ponens.

5. Let p represent "you want to make trouble" and q represent "the door is that way." The argument is then represented symbolically by:

$$\frac{\begin{array}{l} p \to q \\ q \end{array}}{p.}$$

Since this is the form Fallacy of the Converse, it is invalid and considered a <u>fallacy</u>.

7. Let p represent "Martin Broudeur plays" and q represent "the opponent gets shut out." The argument is then represented symbolically by:

$$\frac{\begin{array}{l} p \to q \\ \sim q \end{array}}{\sim p.}$$

This is the <u>valid</u> argument form Modus Tollens.

9. Let p represent "we evolved a race of Isaac Newtons" and q represent "that would not be progress." The argument is then represented symbolically by:

$$\frac{\begin{array}{l} p \to q \\ \sim p \end{array}}{\sim q.}$$

Note: that since we let q represent "that <u>would not</u> be progress," then $\sim q$ represents "that <u>is</u> progress." Since this is the form Fallacy of the Inverse, it is <u>invalid</u> and considered a fallacy.

11. Let p represent "Something is rotten in the state of Denmark" and q represent "my name isn't Hamlet." The argument is then represented symbolically by:

$$\frac{\begin{array}{l} p \vee q \ (\text{or } q \vee p) \\ \sim q \end{array}}{\sim p.}$$

Since this is the form Disjunctive Syllogism, it is a <u>valid</u> argument.

To show validity for the arguments in the following exercises we must show that the conjunction of the premises implies the conclusion. That is, the conditional statement $[P_1 \wedge P_2 \wedge \ldots \wedge P_n] \to C$ must be a tautology.

13. 1. $p \vee q$ T
          2. $\dfrac{p}{\sim q}$ $\begin{array}{l} \text{T} \\ \text{F} \end{array}$

The argument is <u>invalid</u>. When $p = $ T and $q = $ T, the premises are true but the conclusion is false.

15. 1. $p \to q$ T
          2. $\dfrac{q \to p}{p \wedge q}$ $\begin{array}{l} \text{T} \\ \text{F} \end{array}$

The argument is <u>invalid</u>. When $p = $ F and $q = $ F, the premises are true but the conclusion is false.

17. The argument is <u>valid</u>.

1.	$\sim p \to \sim q$	Premise
2.	q	Premise
3.	p	1, 2, Modus Tollens

19. The argument is <u>valid</u>.

1.	$p \to q$	Premise
2.	$\sim q$	Premise
3.	$\sim p \to r$	Premise
4.	$\sim p$	1, 2, Modus Tollens
5.	r	3, 4, Modus Ponens

21. The argument is <u>valid</u>.

1.	$p \to q$	Premise
2.	$q \to r$	Premise
3.	$\sim r$	Premise
4.	$p \to r$	1, 2, Transitivity
5.	$\sim p$	3, 4, Modus Tollens

23. The argument is <u>valid</u>.

    1. $p \rightarrow q$ Premise
    2. $q \rightarrow \sim r$ Premise
    3. p Premise
    4. $r \vee s$ Premise
    5. q 1, 3, Modus Ponens
    6. $\sim r$ 2, 5, Modus Ponens
    7. s 4, 6, Disjunctive Syllogism

25. Make a truth table for the statement $(p \wedge q) \rightarrow p$.

p	q	$(p$	$\wedge$	$q)$	$\rightarrow$	p
T	T	T	T	T	T	T
T	F	T	F	F	T	T
F	T	F	F	T	T	F
F	F	F	F	F	T	F
		1	2	1	3	2

Since the final column, 3, indicates that the conditional statement that represents the argument is true for all possible truth values of p and q, the statement is a tautology.

27. Make a truth table for the statement $(p \wedge q) \rightarrow (p \wedge q)$.

p	q	$(p$	$\wedge$	$q)$	$\rightarrow$	$(p$	$\wedge$	$q)$
T	T	T	T	T	T	T	T	T
T	F	T	F	F	T	T	F	F
F	T	F	F	T	T	F	F	T
F	F	F	F	F	T	F	F	F
		1	2	1	3	1	2	1

Since the final column, 3, indicates that the conditional statement that represents the argument is true for all possible truth values of p and q, the statement is a tautology.

29. Let c represent "my computer is working," f represent "the power supply is faulty," and s represent "my stereo works." The argument is then presented symbolically by:

$$c$$
$$f \rightarrow \sim c$$
$$\sim f \rightarrow s$$
$$s.$$

The argument is <u>valid</u>.

    1. c Premise
    2. $f \rightarrow \sim c$ Premise
    3. $\sim f \rightarrow s$ Premise
    4. $\sim f$ 1, 2, Modus Tollens
    5. s 3, 4, Modus Ponens

31. Let s represent "you have strep throat," f represent "you have a fever," and c represent "you have a serious cough." The argument is then represented symbolically by:

$$s \rightarrow f$$
$$c \vee f$$
$$\sim c$$
$$\sim s.$$

The argument is <u>invalid</u>. Suppose s is T, f is T, and c is F. Then the premises are true but the conclusion is false as seen in the table below.

    1. $s \rightarrow f$ T
    2. $c \vee f$ T
    3. $\sim c$ T
    4. $\sim s$ F

33. Let y represent "the Yankees will be in the World Series," m represent "the Marlins will be in the World Series," and n represent "the National League wins." The argument is then represented symbolically by:

$$y \vee \sim m$$
$$\sim m \rightarrow \sim n$$
$$n$$
$$y.$$

The argument is <u>valid</u>.

    1. $y \vee \sim m$ Premise
    2. $\sim m \rightarrow \sim n$ Premise
    3. n Premise
    4. m 2, 3, Modus Tollens
    5. y 1,4, Disjunctive Syllogism

35. Let p represent "I am your woman," q represent "you are my man," and r represent "I stop loving you." The argument is then represented symbolically by:

$$(p \wedge q) \rightarrow \sim r$$
$$r$$
$$\sim p \vee \sim q.$$

The argument is <u>valid</u>.

1. $(p \wedge q) \rightarrow \sim r$ Premise
2. r Premise
3. $\sim (p \wedge q)$ 1, 2, Modus Tollens
4. $\sim p \vee \sim q$ 3, De Morgan's Law

37. We apply reasoning by repeated transitivity to the six premises. A conclusion from this reasoning, which makes the argument valid, is reached by linking the first antecedent to the last consequent. This conclusion is "If I tell you the time, then my life will be miserable."

39. (a) $s \rightarrow l$ or $\sim l \rightarrow s$ **(b)** $\sim s \rightarrow \sim j$ **(c)** $y \rightarrow \sim l$ **(d)** $y \rightarrow \sim j$, *Conclusion:* If he is your son, then he is not fit to serve on a jury. In Lewis Carroll's words, "None of our sons are fit to serve on a jury."

41. (a) $a \rightarrow s$ or $\sim s \rightarrow \sim a$ **(b)** $g \rightarrow i$ **(c)** $i \rightarrow \sim s$ **(d)** $g \rightarrow \sim a$, *Conclusion:* If it is a guinea pig, then it does not appreciate Beethoven. In Lewis Carroll's words, "Guinea pigs never really appreciate Beethoven."

43. (a) $p \rightarrow b$ or $\sim b \rightarrow \sim p$ **(b)** $\sim t \rightarrow \sim l$ **(c)** $o \rightarrow \sim s$ **(d)** $b \rightarrow l$ **(e)** $w \rightarrow p$ **(f)** $\sim s \rightarrow \sim t$ **(g)** $o \rightarrow \sim w$, *Conclusion:* If he is an opium eater, then he doesn'tt wear white gloves. In Lewis Carroll's words, "Opium eaters never wear white kid gloves."

6.6   Analyzing Arguments with Quantifiers

1. Let $b(x)$ represent "x is a book" and $s(x)$ represent "x is a bestseller."

   (a) $\exists x[b(x) \wedge s(x)]$

   (b) $\forall x[b(x) \rightarrow \sim s(x)]$

   (c) No books are bestsellers.

3. Let $c(x)$ represent "x is a CEO" and $s(x)$ represent "x sleeps well at night."

   (a) $\forall x[c(x) \rightarrow \sim s(x)]$

   (b) $\exists x[c(x) \wedge s(x)]$

   (c) There is a CEO who sleeps well at night.

5. Let $l(x)$ represent "x is a leaf" and $b(x)$ represent "x is brown."

   (a) $\forall x[l(x) \rightarrow b(x)]$

   (b) $\exists x[l(x) \wedge \sim b(x)]$

   (c) There is a leaf that's not brown.

7. **(a)** Let $g(x)$ represent "x is a girl" and $f(x)$ represent "x wants to have fun." Let t represent Teri Lovelace. We can represent the argument symbolically as follows.

$$\forall x[g(x) \rightarrow f(x)]$$
$$\underline{g(t)}$$
$$f(t)$$

   (b) Draw an Euler diagram where the region representing "girls" must be inside the region representing "people who want to have fun" so that the first premise is true.

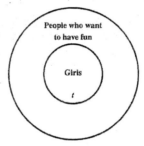

By the second premise, t must lie in the "girls" region. Since this forces the conclusion to be true, the argument is <u>valid</u>.

9. **(a)** Let $p(x)$ represent "x is a professor" and $c(x)$ represent "x is covered with chalk dust." Let o represent Otis Taylor. We can represent the argument symbolically as follows.

$$\forall x[p(x) \rightarrow c(x)]$$
$$\underline{c(o)}$$
$$p(o)$$

   (b) Draw an Euler diagram where the region representing "professors" must be inside the region representing "those who are covered with chalk dust" so that the first premise is true.

By the second premise, o must lie in the "those who are covered with chalk dust" region. Thus, o could be inside or outside the inner region "professors." Since this allows for a false conclusion (Otis doesn't have to be a professor to be covered with chalk dust), the argument is <u>invalid</u>.

11. (a) Let $c(x)$ represent "x is an accountant" and $p(x)$ represent "x uses a spreadsheet." Let n represent Nancy Hart. We can represent the argument symbolically as follows.

$$\forall x[c(x) \rightarrow p(x)]$$
$$\underline{\sim p(n)}$$
$$\sim c(n)$$

(b) Draw an Euler diagram where the region representing "accountants" must be inside the region representing "those who use spreadsheets" so that the first premise is true.

By the second premise, n must lie outside the region representing "those who use spreadsheets." Since this forces the conclusion to be true, the argument is <u>valid</u>.

13. (a) Let $t(x)$ represent "x is turned down for a mortgage," $s(x)$ represent "x has a second income," and $b(x)$ represent "x needs a mortgage broker." We can represent the argument symbolically as follows.

$$\exists x[t(x) \wedge s(x)]$$
$$\underline{\forall x[t(x) \rightarrow b(x)]}$$
$$\exists x[s(x) \wedge b(x)]$$

(b) Draw an Euler diagram where the region representing "Those who are turned down for a mortgage" intersects the region representing "Those with a 2nd income." This keeps the first premise true.

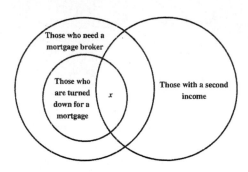

By the second premise the region representing "Those who are turned down for a mortgage" must be inside the region "Those who need a mortgage broker." Since x lies in both regions "These with a 2nd income" and "Those who need a mortgage broker," the conclusion is true and so the argument is <u>valid</u>.

15. (a) Let $w(x)$ represent "x wanders" and $l(x)$ represent "x is lost." Let m represent Martha MacDonald We can represent the argument symbolically as follows.

$$\exists x[w(x) \wedge l(x)]$$
$$\underline{w(m)}$$
$$l(m)$$

(b) Draw an Euler diagram where the region representing "those who wander" intersects the region representing "those who are lost." This keeps the first premise true.

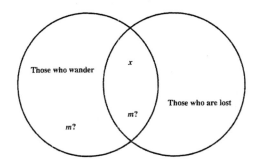

By the second premise, m must lie in the region representing "those who wander." But, m could be inside or outside the region representing "those who are lost." Since this allows for a false conclusion, the argument is <u>invalid</u>.

17. (a) Let $p(x)$ represent "x is a psychologist," $u(x)$ represent "x is a university professor," and $r(x)$ represent "x has a private practice." We can represent the argument symbolically as follows.

$$\exists x[p(x) \wedge u(x)]$$
$$\underline{\exists x[p(x) \wedge r(x)]}$$
$$\exists x[u(x) \wedge r(x)]$$

(b) Draw an Euler diagram where the region representing "psychologists" and "university professors" intersect each other to keep the first premise true. Then add a region representing "those with a private practice" intersecting the region "university professors" to keep the second premise true. In the most general case, this region should also intersect the region representing "psychologists."

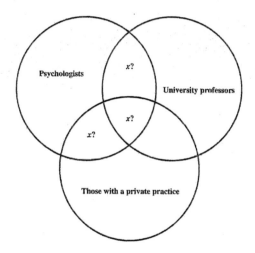

By the first premise, x must lie in the region shared by "psychologists" and "university professors"; by the second premise, x must lie in the region shared by "psychologists" and "those with a private practice." But x may not lie in the region shared by all three regions. Since this diagram shows true premises but a false conclusion, the argument is invalid.

19. (a) Let $a(x)$ represent "x is a saint" and $i(x)$ represent "x is a sinner." We can represent the argument symbolically as follows.

$$\forall x[a(x) \wedge i(x)]$$
$$\underline{\exists x[\sim a(x)}$$
$$\exists x[i(x)]$$

(b) Draw an Euler diagram where the region representing "saints" intersects the region representing "sinners" to keep the first premise true.

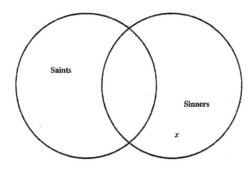

By the second premise, x must lie outside the region representing "saints." But that means x must lie in the part of the "sinners" region not shared by the "saints" region." Hence, the conclusion is true and so the argument is valid.

21. Interchanging the second premise and the conclusion of Example 4 yields the following argument.

All well-run businesses generate profits.
Monsters, Inc. is a well-run business.
Monsters, Inc. generates profits.

Draw an Euler diagram where the region representing "well-run businesses" must be inside the region representing "things that generate profits" so that the first premise is true.

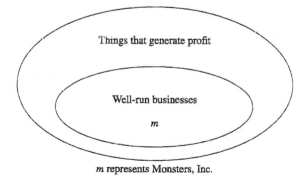

m represents Monsters, Inc.

Let m represent Monsters, Inc. By the second premise, m must lie inside the region representing "well-run businesses." Since this forces the conclusion to be true, the argument is valid, which makes the answer to the question yes.

23. Since the region representing "major league baseball players" lies entirely inside the region representing "people who earn at least $300,000 a year," a possible first premise is

All major league baseball players earn at least $300,000 a year.

And if r represents Ryan Howard and r is inside the region representing "major league baseball players," then a possible second premise is

Ryan Howard is a major league baseball player.

A valid conclusion drawn from these two premises is that

Ryan Howard earns at least $300,000 a year.

Therefore, a valid argument based on the Euler diagram is as follows.

All major league baseball players earn at least $300,000 a year.
Ryan Howard is a major league baseball player.

Ryan Howard earns at least $300,000 a year.

25. The following diagram yields true premises. It also forces the conclusion to be true.

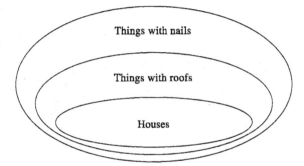

Thus, the argument is <u>valid</u>. Observe that the diagram is the only way to show true premises.

27. The following represents one way to diagram the premises so that they are true but does not lead to a true conclusion.

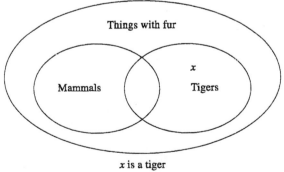

x is a tiger

If we let x be a tiger, according to the premises, x could also be a tiger but not a mammal. Thus, the argument is <u>invalid</u>

29. The following Euler diagram illustrates that the conclusion is not forced to be true.

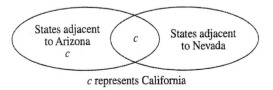

c represents California

The argument is <u>invalid</u> even though the conclusion is true.

31. The following Euler diagram represents true premises.

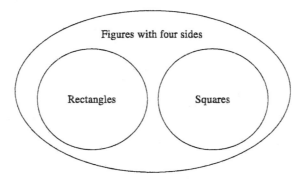

According to the premises, the region representing squares may lie entirely outside of the region representing rectangles. Thus, the argument is <u>invalid</u>, even though the conclusion is true.

37. (a) Let $l(x)$ represent "x is a legislative power herein granted" and $g(x)$ represent "x is vested in a Congress of the United States." We can represent the passage symbolically as follows.

$$\forall x[l(x) \rightarrow g(x)]$$

(b) We make the following argument.

> All legislative power herein granted shall be vested in a Congress of the United States.
> The power to collect taxes is a legislative power herein granted.
> _____
> The power to collect taxes shall be vested in a Congress of the United States.

The conclusion is true. Thus, the argument is <u>valid</u>.

(c) Draw an Euler diagram where the region representing "legislative power herein granted" must be inside the region representing "things that are vested in a Congress of the United States" so that the first premise is true. Let c represent "the power to collect taxes." By the second premise, c must lie in the region representing "legislative powers herein granted."

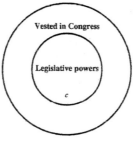

39. (a) Let $b(x)$ represent "x is a bill of attainder," $e(x)$ represent "x is an ex post facto law," and $p(x)$ represent "x has been passed." We can represent the passage symbolically as follows.

$$\forall x\{[b(x) \vee e(x)] \rightarrow\, \sim p(x)\}$$

(b) We make the following argument.

> No bill of attainder or ex post facto law shall be passed.
> The law forbidding members of the Communist Party to serve as an officer or as an employee of a labor union was a bill of attainder.
> _____
> The law forbidding members of the Communist Party to serve as an officer or as an employee of a labor union shall not be passed.

The conclusion is true. Thus, the argument is <u>valid</u>.

(c) Draw an Euler diagram where the region representing "bill of attainder" and the region representing "ex post facto law" intersect but lie outside of the region representing "bills and laws passed." Let l represent "the law forbidding members of the Communist Party to serve as an officer or as an employee of a labor union." By the second premise, l must lie in the region representing "bill of attainder" and so will lie outside of the region representing "bills and laws passed."

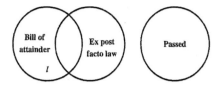

41. (a) Since the region corresponding to "Tourette syndrome" is entirely contained inside the region corresponding to "Chronic tics," the conclusion is valid.

(b) Since the region corresponding to "Tourette syndrome" is not entirely contained inside the region corresponding to "OCD," the conclusion is invalid.

(c) Since some of the region corresponding to "Chronic tics" also lies in the region corresponding to "OCD," the conclusion is valid.

(d) Since some of the region corresponding to "OCD" lies outside the region corresponding to "Tourette syndrome," the conclusion is valid.

(e) Since some of the region corresponding to "Chronic tics" lies outside the region corresponding to "Tourette syndrome," the conclusion is invalid.

Thus, the answer is a, c, and d.

In Exercises 43–47, the premises marked A, B, and C are followed by several possible conclusions. Take each conclusions in turn, and check the resulting argument as valid or invalid.

    A.   *All kittens are cute animals.*
    B.   *All cute animals are admired by animal lovers.*
    C.   *Some dangerous animals are admired by animal lovers.*

Diagram the first two premises to be true. Then, notice that premise C is correctly represented by Case I, Case II, or Case III in the diagram.

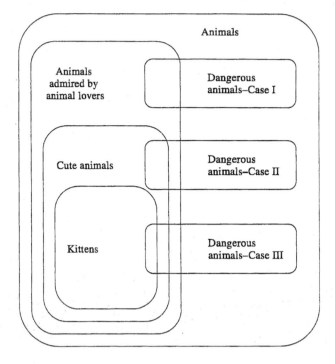

43. We are not forced into the conclusion "Some kittens are dangerous animals" since Case I and Case II represent true premises where this conclusion is false. Thus, the argument is <u>invalid</u>.

45. We are not forced into the conclusion "Some dangerous animals are cute" since Case I represents true premises where this conclusion is false. Thus, the argument is <u>invalid</u>.

47. The conclusion "All kittens are admired by animal lovers" yields a <u>valid</u> argument since premises A and B force the conclusion to be true.

Chapter 6 Review Exercises

1. The negation of "If she doesn't pay me, I won't have enough cash" is "She pays me and I have enough cash."

3. The symbolic form of "He loses the election, but he wins the hearts of the voters" is $l \wedge w$.

5. The symbolic form of "He loses the election only if he doesn't win the hearts of the voters" is
$l \rightarrow\ \sim w$.

7. Writing the symbolic form $\sim l \wedge w$ in words, we get "He doesn't lose the election and he wins the hearts of the voters."

9. Replacing q and r with the given truth values, we have

$$\sim F \wedge \sim F$$
$$T \wedge T$$
$$T$$

The compound statement $\sim q \wedge \sim r$ is true.

11. Replacing r with the given truth value (s not known), we have

$$F \rightarrow (s \wedge F)$$
$$T$$

since a conditional statement with a false antecedent is true.
The compound statement $r \rightarrow (s \vee r)$ is true.

15.

p	q	p	$\wedge$	$(\sim p$	$\vee$	$q)$
T	T	T	T	F	T	T
T	F	T	F	F	F	F
F	T	F	F	T	T	T
F	F	F	F	T	T	F
		1	4	2	3	2

The statement is not a tautology.

17. "All mathematicians are lovable" can be restated as "If someone is a mathematician, then that person is lovable."

19. "Having at least as many equations as unknowns is necessary for a system to have a unique solution" can be restated as "If a system has a unique solution, then it has a least as many equations as unknowns."

21. *The direct statement*: If the proposed regulations have been approved, then we need to change the way we do business.

(a) *Converse*: If we need to change the way we do business, then the proposed regulations have been approved.

(b) *Inverse*: If the proposed regulations have not been approved, then we do not need to change the way we do business.

(c) *Contrapositive:* If we do not need to change the way we do business, then the proposed regulations have not been approved.

23. In the diagram, a series circuit corresponding to $p \wedge p$ is followed in series by a parallel circuit represented by $\sim p \vee q$. The logical statement is $(p \wedge p) \wedge (\sim p \vee q)$. This statement is equivalent to $p \wedge q$.

p	q	$(p$	$\wedge$	$p)$	$\wedge$	$(\sim p$	$\vee$	$q)$	p	$\wedge$	q
T	T	T	T	T	T	F	T	T	T	T	T
T	F	T	T	T	F	F	F	F	T	F	F
F	T	F	F	F	F	T	T	T	F	F	T
F	F	F	F	F	F	T	T	F	F	F	F
		1	2	1	6	3	5	4	7	9	8

Columns 6 and 9 are identical.

25. The logical statement $(p \wedge q) \vee (p \wedge p)$ can be represented by the following circuit.

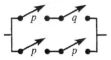

The statement simplifies to p as follows:

$$(p \wedge q) \vee (p \wedge p) = p \wedge (q \wedge p)$$
$$= p$$

27.

p	q	$(p$	$\veebar$	$q)$	$(p$	$\vee$	$q)$	$\wedge$	$\sim$	$(p$	$\wedge$	$q)$
T	T	T	F	T	T	T	T	F	F	T	T	T
T	F	T	T	F	T	T	F	T	T	T	F	F
F	T	F	T	T	F	T	T	T	T	F	F	T
F	F	F	F	F	F	F	F	F	T	F	F	F
		1	2	1	3	4	3	8	7	5	6	5

The columns labeled 2 and 8 are identical.

29. (a) Yes, the statement is true because "this year is 2002" is false and "$1 + 1 = 3$" is false and $F \rightarrow F$ is true.

(b) No, the statement was not true in 2002 because then the statement had the value $T \rightarrow F$, which is false.

31. Let l represent "you're late one more time" and d represent "you'll be docked." The argument is then presented symbolically as follows.

$$l \rightarrow d$$
$$\underline{l \qquad\quad}$$
$$d.$$

The argument is valid by Modus Ponens.

33. Let l represent "the instructor is late" and w represent "my watch is wrong." The argument is then presented symbolically as follows.

$$l \vee w$$
$$\underline{\sim w \quad}$$
$$l.$$

The argument is valid by Disjunctive Syllogism.

35. Let p represent "you play that song one more time" and n represent "I'm going nuts." The argument is then presented symbolically as follows.

$$p \rightarrow n$$
$$\underline{n \qquad\quad}$$
$$p.$$

The argument is invalid by Fallacy of the Converse.

37. Let h represent "we hire a new person," t represent "we'll spend more on training," and r represent "we rewrite the manual." The argument is then presented symbolically as follows.

$$h \rightarrow t$$
$$r \rightarrow \sim t$$
$$\underline{r \qquad\quad}$$
$$\sim h.$$

The argument is valid.

$$
\begin{array}{lll}
1. & h \rightarrow t & \text{Premise} \\
2. & r \rightarrow \sim t & \text{Premise} \\
3. & r & \text{Premise} \\
4. & \sim t & 2,\ 3,\ \text{Modus Ponens} \\
5. & \sim h & 1,\ 4,\ \text{Modus Tollens}
\end{array}
$$

39.

$$
\begin{array}{ll}
1. \sim p \rightarrow \sim q & \text{T} \\
2. \underline{\quad q \rightarrow \ p \quad} & \text{T} \\
\quad\ p \ \lor \ q & \text{F}
\end{array}
$$

The argument is <u>invalid</u>. When $p = \text{F}$ and $q = \text{F}$, the premises are true but the conclusion is false.

41. Let $d(x)$ represent "x is a dog" and $h(x)$ represent "x goes to heaven."

(a) $\forall x[d(x) \rightarrow h(x)]$

(b) $\exists x[d(x) \land \sim h(x)]$

(c) There is a dog that doesn't go to heaven.

43. (a) Let $f(x)$ represent "x is a member of that fraternity," $w(x)$ represent "x does well academically," and j represent John Cross. We can represent the argument symbolically as follows.

$$
\begin{array}{l}
\forall x[f(x) \rightarrow w(x)] \\
\underline{f(j) \quad\quad\quad\quad\quad\quad} \\
w(j)
\end{array}
$$

(b) Because of the first premise, the region representing "members of that fraternity" must be inside the region representing "those who do well academically." And j must be within the region representing "members of that fraternity" because of the second premise. Complete the Euler diagram as follows.

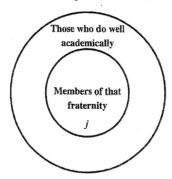

Since, when the premises are diagrammed as being true, we are forced into a true conclusion, the argument is <u>valid</u>.

45.

p	q	r	p	$\rightarrow$	$(q$	$\rightarrow$	$r)$	$(p$	$\rightarrow$	$q)$	$\rightarrow$	r
T	T	T	T	T	T	T	T	T	T	T	T	T
T	T	F	T	F	T	F	F	T	T	T	F	F
T	F	T	T	T	F	T	T	T	F	F	T	T
T	F	F	T	T	F	T	F	T	F	F	T	F
F	T	T	F	T	T	T	T	F	T	T	T	T
F	T	F	F	T	T	F	F	F	T	T	F	F
F	F	T	F	T	F	T	T	F	T	F	T	T
F	F	F	F	T	F	T	F	F	T	F	F	F
			1	4	2	3	2	5	6	5	8	7

To determine if the statements are equivalent, compare columns 4 and 8. Since they are not identical, the statements are not equivalent.

47. (a)

p	q	$(p$	$\wedge$	$\sim p)$	$\rightarrow$	q
T	T	T	F	F	T	T
T	F	T	F	F	T	F
F	T	F	F	T	T	T
F	F	F	F	T	T	F
		1	2	1	4	3

49. a, b, and **c** are compound statements.

51. The contrapositive of "The Schedule D Tax Worksheet in the Schedule D instructions is used only if you have a net 28% rate gain or unrecaptured Section 1250 gain " is "If you do not have a net 28% rate gain or unrecaptured Section 1250 gain, then the Schedule D Tax Worksheet in the Schedule D instructions is not used."

53. The negation of $p \wedge q$ is $\sim p \vee \sim q$.

(a) Regulations do not have both costs and benefits or rules that are passed to solve a problem cannot make it worse.

(b) Shooters do not overwhelmingly have problems with alcoholism or they do not have long criminal histories, particularly arrests for violent acts.

(c) They are not disproportionately involved in automobile crashes or they are not much more likely to have their driver's license suspended or revoked.

55. (a) Since "real things are not things we can deal with," the regions representing "real things" and "things we can deal with" do not intersect. Since "only things we can measure are things we can deal with," the region representing "things we can deal with" should lie entirely within the region representing "things we can measure." And since "only things we can measure are real things," the region representing "real things" should lie entirely within the region representing "things we can measure." The Euler diagram is as follows.

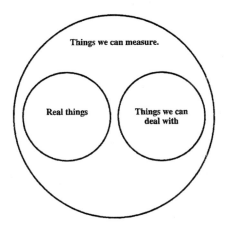

57. (a), (b), (d) These are not statements.

(c) This is a compound statement, formed using "Someone finally runs out of money" and "The player with the most cash wins this colorful, fast-paced junior version of the world's most popular board game."

59. (a) $h \to o$

(b) $a \to m$

(c) $p \to \sim o \equiv o \to \sim p$

(d) $m \to h$

(e) If the bird is in this aviary, it does not live on mince pies. In Lewis Carroll's words, "No bird in this aviary lives on mince pies."

61. (a) $u \to c$

(b) $p \to r$

(c) $s \to h$

(d) $\sim u \to \sim r \equiv r \to u$

(e) $h \to p$

(f) If the writer is Shakespeare, the writer is clever. In Lewis Carroll's words, "Shakespeare was clever."

Chapter 6 Test

[6.1]

Write a negation for each of the following statements.

1. It is raining in Spain.

2. Everybody loves a winner.

3. If it quacks like a duck, it's a duck.

4. King Edward is just, and I am treacherous.

[6.2]

Let p represent "Xanthippe loves Socrates" and let q represent "Socrates drinks hemlock." Write each of the following in symbols.

5. If Xanthippe does not love Socrates, then Socrates drinks hemlock.

6. Socrates does not drink hemlock unless Xanthippe loves him.

7. Socrates drinks hemlock if and only if Xanthippe does not love him.

[6.3]

In Exercises 8-9, assume that p and q are false and that r is true. Find the truth value of each statement.

8. $r \vee \sim q$

9. $(q \wedge \sim r) \rightarrow \sim p$

10. Construct a truth table for $\sim q \vee (p \rightarrow \sim q)$.

[6.4]

11. In the following statement, assume that p and q are false and that r is true. Find the truth value of $(p \longleftrightarrow (p \vee r)) \wedge r$.

12. Construct a truth table for $\sim p \longleftrightarrow (r \wedge \sim q)$.

13. Write "Whenever it rains, it pours." in *if...then* form.

14. Write the contrapositive of "If interest rates fall, stocks will recover."

[6.5]

Determine whether each argument is valid or invalid. If it is valid, give a proof. If it is invalid, give an assignment of truth values to the variables that makes the premises true and the conclusion false.

15. $\sim p \to q$
    $\sim p$
    _____
    $\sim p$

16. $p \vee q$
    $q \to \sim r$
    r
    _____
    p

17. If Maria does not buy lunch, Roberta will go to the theater. But John can't go to school if Roberta goes to the theater. Therefore, if John goes to school, Maria will buy lunch.

[6.6]

Decide whether each statement is true or false.

18. Any brother is a sibling.

19. Some twin is an orphan.

20. Determine whether the following argument is *valid* or *invalid*.

> All nasturtiums are red.
> Some flower is not red.
> _____
> Some flower is not a nasturtium

Chapter 6 Test Answers

1. It is not raining in Spain.

2. Somebody does not love a winner.

3. It quacks like a duck and is not a duck.

4. King Edward is not just or I am not treacherous.

5. $\sim p \rightarrow q$

6. $\sim p \rightarrow \sim q$ (or $q \rightarrow p$)

7. $q \longleftrightarrow \sim p$

8. True

9. True

10.

p	q	$\sim q$	$\vee$	$(p$	$\rightarrow$	$\sim q)$
T	T	F	F	T	F	F
T	F	T	T	T	T	T
F	T	F	T	F	T	F
F	F	T	T	F	T	T
		1	4	2	3	2

11. False

12.

p	q	r	$\sim p$	$\longleftrightarrow$	$(r$	$\wedge$	$\sim q)$
T	T	T	F	T	T	F	F
T	T	F	F	T	F	F	F
T	F	T	F	F	T	T	T
T	F	F	F	T	F	F	T
F	T	T	T	F	T	F	F
F	T	F	T	F	F	F	F
F	F	T	T	T	T	T	T
F	F	F	T	F	F	F	T
			1	4	2	3	2

13. If it rains, then it pours.

14. If stocks will not recover, interest rates do not fall.

15. The argument is <u>valid</u>.

1.	$\sim p \rightarrow q$	Premise
2.	$\sim q$	Premise
3.	p	1, 2, Modus Tollens

16. The argument is <u>valid</u>.

1.	$p \vee q$	Premise
2.	$q \rightarrow \sim r$	Premise
3.	r	Premise
4.	$r \rightarrow \sim q$	2, De Morgan's law
5.	$\sim q$	3, 4, Modus Ponens
6.	p	1, 5, Disjunctive Syllogism

17. The argument is <u>valid</u>.

1.	$\sim m \rightarrow r$	Premise
2.	$r \rightarrow \sim j$	Premise
3.	$j \rightarrow \sim r$	2, Contrapositive
4.	$\sim r \rightarrow m$	1, Contrapositive
5.	$j \rightarrow m$	4, Transitivity

18. True

19. True

20. Valid

Chapter 7

SETS AND PROBABILITY

7.1 Sets

1. $3 \in \{2, 5, 7, 9, 10\}$

The number 3 is not an element of the set, so the statement is false.

3. $9 \notin \{2, 1, 5, 8\}$

Since 9 is not an element of the set, the statement is true.

5. $\{2, 5, 8, 9\} = \{2, 5, 9, 8\}$

The sets contain exactly the same elements, so they are equal. The statement is true.

7. {All whole numbers greater than 7 and less than $10\} = \{8, 9\}$

Since 8 and 9 are the only such numbers, the statement is true.

9. $0 \in \emptyset$

The empty set has no elements. The statement is false.

In Exercises 11-21,

$$A = \{2, 4, 6, 8, 10, 12\},$$
$$B = \{2, 4, 8, 10\},$$
$$C = \{4, 8, 12\},$$
$$D = \{2, 10\},$$
$$E = \{6\},$$
$$\text{and} \quad U = \{2, 4, 6, 8, 10, 12, 14\}.$$

11. Since every element of A is also an element of U, A is a subset of U, written $A \subseteq U$.

13. A contains elements that do not belong to E, namely 2, 4, 8, 10, and 12, so A is not a subset of E, written $A \nsubseteq E$.

15. The empty set is a subset of every set, so $\emptyset \subseteq A$.

17. Every element of D is also an element of B, so D is a subset of B, $D \subseteq B$.

19. Since every element of A is also an element of U, and $A \neq U$, $A \boxed{\subset} U$.
Since every element of E is also an element of A, and $E \neq A$, $E \boxed{\subset} A$.
Since every element of A is not also an element of E, $A \boxed{\not\subset} E$.
Since every element of B is not also an element of C, $B \boxed{\not\subset} C$.
Since ϕ is a subset of every set, and $\neq A$, $\phi \boxed{\subset} A$.
Since every element of $\{0, 2\}$ is not also an element of D, $\{0, 2\} \boxed{\not\subset} D$.
Since every element of D is also an element of B, and $D \neq B$, $D \boxed{\subset} B$.
Since every element of A is not also an element of C, $A \boxed{\not\subset} C$.

21. Since B has 4 elements, it has 2^4 or 16 subsets. There are exactly 16 subsets of B.

23. Since D has 2 elements, it has 2^2 or 4 subsets. There are exactly 4 subsets of D.

25. Since $\{7, 9\}$ is the set of elements belonging to both sets, which is the intersection of the two sets, we write

$$\{5, 7, 9, 19\} \cap \{7, 9, 11, 15\} = \{7, 9\}.$$

27. Since $\{1\}$ is the set of elements belonging to both sets, we write

$$\{2, 1, 7\} \cap \{1, 5, 9\} = \{1\}.$$

29. Since $\emptyset$ contains no elements, there are no elements belonging to both sets. Thus, the intersection is the empty set, and we write

$$\{3, 5, 9, 10\} \cap \emptyset = \emptyset.$$

31. $\{1, 2, 4\}$ is the set of elements belonging to both sets, and $\{1, 2, 4\}$ is also the set of elements in the first set or in the second set or possibly both. Thus,

$$\{1, 2, 4\} \cap \{1, 2, 4\} = \{1, 2, 4\}$$

and

$$\{1, 2, 4\} \cup \{1, 2, 4\} = \{1, 2, 4\}$$

are both true statements.

In Exercises 33-41,

$$U = \{1, 2, 3, 4, 5, 6, 7, 8, 9\},$$
$$X = \{2, 4, 6, 8\},$$
$$Y = \{2, 3, 4, 5, 6\},$$
$$\text{and} \quad Z = \{1, 2, 3, 8, 9\}.$$

33. $X \cap Y$, the intersection of X and Y, is the set of elements belonging to both X and Y. Thus,

$$X \cap Y = \{2, 4, 6, 8\} \cap \{2, 3, 4, 5, 6\} = \{2, 4, 6\}.$$

35. X', the complement of X, consists of those elements of U that are not in X. Thus,

$$X' = \{1, 3, 5, 7, 9\}.$$

37. From Exercise 35, $X' = \{1, 3, 5, 7, 9\}$; from Exercise 36, $Y' = \{1, 7, 8, 9\}$. There are no elements common to both X' and Y' so

$$X' \cap Y' = \{1, 7, 9\}.$$

39. First find $X \cup Z$.

$$X \cup Z = \{2, 4, 6, 8\} \cup \{1, 2, 3, 8, 9\} = \{1, 2, 3, 4, 6, 8, 9\}$$

Now find $Y \cap (X \cup Z)$.

$$Y \cap (X \cup Z) = \{2, 3, 4, 5, 6\} \cap \{1, 2, 3, 4, 6, 8, 9\} = \{2, 3, 4, 6\}$$

41. From Exercise 36, $Y' = \{1, 7, 8, 9\}$; $Z' = \{4, 5, 6, 7\}$.

$$(X \cap Y') \cup Z') = (\{2, 4, 6, 8\} \cup \{1, 7, 8, 9\}) \cup \{4, 5, 6, 7\} = \{8\} \cup \{4, 5, 6, 7\} = \{4, 5, 6, 7, 8\}$$

43. M' consists of all students in U who are not in M, so M' consists of all students in this school not taking this course.

45. $N \cap P$ is the set of all students in this school taking both accounting and zoology.

47. $A = \{2, 4, 6, 8, 10, 12\}$,
$B = \{2, 4, 8, 10\}$,
$C = \{4, 8, 12\}$,
$D = \{2, 10\}$,
$E = \{6\}$,
$U = \{2, 4, 6, 8, 10, 12, 14\}$

A pair of sets is disjoint if the two sets have no elements in common. The pairs of these sets that are disjoint are B and E, C and E, D and E, and C and D.

49. B' is the set of all stocks with a closing price below \$26 or above \$30. Therefore,

$$B' = \{\text{AT\&T, CocaCola, Office Max Inc., Texas Instruments}\}.$$

51. $(A \cap B)'$ is the set of all stocks that do not have both a high price greater than \$34 and a closing price between \$26 and \$30;

$$(A \cap B)' = \{\text{AT\&T, CocaCola, Disney, Office Max Inc., Texas Instruments}\}.$$

53. $A = \{1, 2, 3, \{3\}, \{1, 4, 7\}\}$

    (a) $1 \in A$ is true.

    (b) $\{3\} \in A$ is true.

    (c) $\{2\} \in A$ is false. $(\{2\} \subseteq A)$

    (d) $4 \in A$ is false. $(4 \in \{1, 4, 7\})$

    (e) $\{\{3\}\} \subset A$ is true.

    (f) $\{1, 4, 7\} \in A$ is true.

    (g) $\{1, 4, 7\} \subseteq A$ is false. $(\{1, 4, 7\} \in A)$

55. $V \cap J = \{$General Electric Co., ExxonMobil Corp., Citigroup, Inc., Microsoft Corp., Proctor & Gamble$\}$
            $\cap \{$Boeing Co., Proctor & Gamble, Yahoo!, Inc., UnitedHealth Group, Microsoft Corp.$\}$
      $= \{$MicrosoftCorp., Proctor & Gamble$\}$

57. $(J \cup F)' = (\{$Boeing Co., Procter & Gamble, Yahoo!, Inc., UnitedHealth Group, Microsoft Corp.$\}$
            $\cup \{$Nokia Corp., UnitedHealth Group, Schlumberger Ltd., Google Inc., General Electric Co.$\}$
      $= (\{$Boeing Co., Procter & Gamble, Yahoo!, Inc., UnitedHealth Group, Microsoft Corp.,
         Nokia Corp., Schlumberger Ltd., Google Inc., General Electric Co.,$\})'$
      $= \{$ExxonMobil Corp., Citigroup Inc., American International Group.$\}$

59. $U = \{s, d, c, g, i, m, h\}$ and $O = \{i, m, h, g\}$, so $O' = \{s, d, c\}$.

61. $N \cap O = \{s, d, c, g\} \cap \{i, m, h, g\} = \{g\}$

63. $N \cap O' = \{s, d, c, g\} \cap \{s, d, c\} = \{s, d, c\}$

65. The total number of subsets is $2^4 = 16$. The number of subsets with no elements is 1, and the number of subsets with one element is 4. Therefore, the number of subsets with at least two elements is $16 - (1 + 4)$ or 11.

67. $F = \{$The Disney Channel, Showtime, HBO, Encore$\}$

69. $H = \{$Encore, Starz$\}$

71. $G \cup H = \{$Showtime, HBO$\} \cup \{$Encore, Starz$\} = \{$Showtime, HBO, Encore, Starz$\}$

73. Joe should always first choose the complement of what Dorothy chose. This will leave only two sets to choose from, and Joe will get the last choice.

75. **(a)** $(A \cup B)' \cap C$

    $A \cup B$ is the set of states whose name contains the letter e or has a population over 4,000,000. Therefore, $(A \cup B)'$ is the set of states who are not among those whose name contains the letter e or has a population over 4,000,000. As a result, $(A \cup B)' \cap C$ is the set of states who are not among those whose name contains the letter e or has a population over 4,000,000 who also have an area over 40,000 miles.

    (b) $(A \cup B)' = \{($Kentucky, Maine, Nebraska, New Jersey$\} \cup \{$Alabama, Colorado, Florida, Indiana, Kentucky, New Jersey$\})'$
             $= \{$Alabama, Colorado, Florida, Indiana, Kentucky, Maine, Nebraska, New Jersey$\}'$
             $= \{$Alaska, Hawaii$\}$
    $(A \cup B)' \cap C = \{$Alaska, Hawaii$\} \cap \{$Alabama, Alaska, Colorado, Florida, Kentucky, Nebraska$\}$
             $= \{$Alaska$\}$

7.2   Applications of Venn Diagrams

1. $B \cap A'$ is the set of all elements in B *and* not in A.

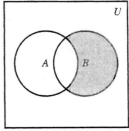

$B \cap A'$

3. $A' \cup B$ is the set of all elements that do not belong to A *or* that do belong to B, or both.

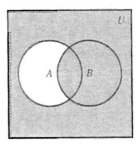

$A' \cup B$

5. $B' \cup (A' \cap B')$

First find $A' \cap B'$, the set of elements not in A *and* not in B.

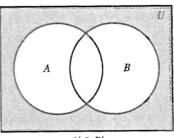

$A' \cap B'$

For the union, we want those elements in B' *or* $(A' \cap B')$, or both.

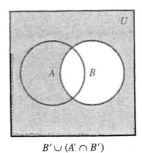

$B' \cup (A' \cap B')$

7. U' is the empty set $\emptyset$.

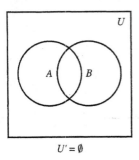

$U' = \emptyset$

9. Three sets divide the universal set into at most 8 regions. (Examples of this situation will be seen in Exercises 11-17.)

11. $(A \cap B) \cap C$

First form the intersection of A with B.

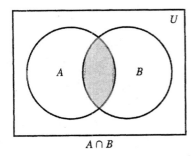

$A \cap B$

Now form the intersection of $A \cap B$ with C. The result will be the set of all elements that belong to all three sets.

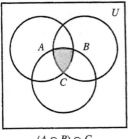

$(A \cap B) \cap C$

13. $A \cap (B \cup C')$

C' is the set of all elements in U that are not elements of C.

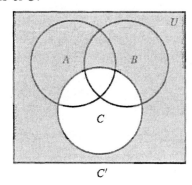

C'

Now form the union of C' with B.

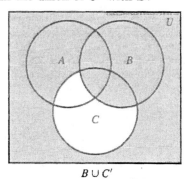

$B \cup C'$

Finally, find the intersection of this region with A.

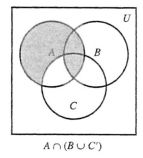

$A \cap (B \cup C')$

15. $(A' \cap B') \cap C$

$A' \cap B'$ is the region of the universal set not in A *and* not in B.

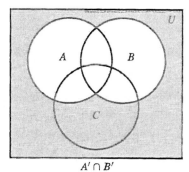

$A' \cap B'$

Now intersect this region with C.

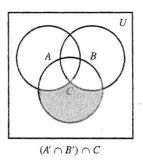

$(A' \cap B') \cap C$

17. $(A \cap B') \cup C$

First find $A \cap B'$, the region in A and not in B.

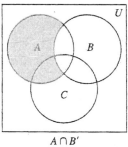

$A \cap B'$

For the union, we want the region in $(A \cap B')$ or in C, or both.

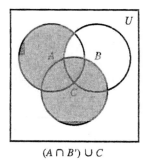

$(A \cap B') \cup C$

19. $(A \cup B') \cap C$

First find $A \cup B'$, the region in A or B' or both.

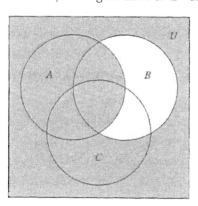

Intersect this with C.

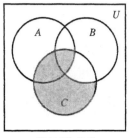

$$(A \cup B') \cap C$$

21. $n(A \cup B) = n(A) + n(B) - n(A \cap B)$
$$= 5 + 12 - 4$$
$$= 13$$

23. $n(A \cup B) = n(A) + n(B) - n(A \cap B)$
$$22 = n(A) + 9 - 5$$
$$22 = n(A) + 4$$
$$18 = n(A)$$

25. $n(U) = 41$
$n(A) = 16$
$n(A \cap B) = 12$
$n(B') = 20$

First put 12 in $A \cap B$. Since $n(A) = 16$, and 12 are in $A \cap B$, there must be 4 elements in A that are not in $A \cap B$. $n(B') = 20$, so there are 20 not in B. We already have 4 not in B (but in A), so there must be another 16 outside B *and* outside A. So far we have accounted for 32, and $n(U) = 41$, so 9 must be in B but not in any region yet identified. Thus $n(A' \cap B) = 9$.

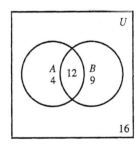

27. $n(A \cup B) = 24$
$n(A \cap B) = 6$
$n(A) = 11$
$n(A' \cup B') = 25$

Start with $n(A \cap B) = 6$. Since $n(A) = 11$, there must be 5 more in A not in B. $n(A \cup B) = 24$; we already have 11, so 13 more must be in B not yet counted. $A' \cup B'$ consists of all the region not in $A \cap B$, where we have 6. So far $5 + 13 = 18$

are in this region, so another $25 - 18 = 7$ must be outside both A and B.

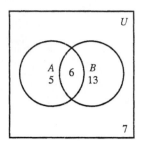

29. $n(A) = 28$
$n(B) = 34$
$n(C) = 25$
$n(A \cap B) = 14$
$n(B \cap C) = 15$
$n(A \cap C) = 11$
$n(A \cap B \cap C) = 9$
$n(U) = 59$

We start with $n(A \cap B \cap C) = 9$. If $n(A \cap B) = 14$, an additional 5 are in $A \cap B$ but not in $A \cap B \cap C$. Similarly, $n(B \cap C) = 15$, so $15 - 9 = 6$ are in $B \cap C$ but not in $A \cap B \cap C$. Also, $n(A \cap C) = 11$, so $11 - 9 = 2$ are in $A \cap C$ but not in $A \cap B \cap C$.

Now we turn our attention to $n(A) = 28$. So far we have $2 + 9 + 5 = 16$ in A; there must be another $28 - 16 = 12$ in A not yet counted. Similarly, $n(B) = 34$; we have $5 + 9 + 6 = 20$ so far, and $34 - 20 = 14$ more must be put in B. For C, $n(C) = 25$; we have $2 + 9 + 6 = 17$ counted so far. Then there must be 8 more in C not yet counted. The count now stands at 56, and $n(U) = 59$, so 3 must be outside the three sets.

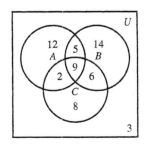

31. $n(A \cap B) = 6$
$n(A \cap B \cap C) = 4$
$n(A \cap C) = 7$
$n(B \cap C) = 4$
$n(A \cap C') = 11$
$n(B \cap C') = 8$
$n(C) = 15$
$n(A' \cap B' \cap C') = 5$

Start with $n(A \cap B) = 6$ and $n(A \cap B \cap C) = 4$ to get $6 - 4 = 2$ in that portion of $A \cap B$ outside of C. From $n(B \cap C) = 4$, there are $4 - 4 = 0$ elements in that portion of $B \cap C$ outside of A. Use $n(A \cap C) = 7$ to get $7 - 4 = 3$ elements in that portion of $A \cap C$ outside of B.
Since $n(A \cap C') = 11$, there are $11 - 2 = 9$ elements in that part of A outside of B and C. Use $n(B \cap C') = 8$ to get $8 - 2 = 6$ elements in that part of B outside of A and C. Since $n(C) = 15$, there are $15 - 3 - 4 - 0 = 8$ elements in C outside of A and B. Finally, 5 must be outside all three sets, since $n(A' \cap B' \cap C') = 5$.

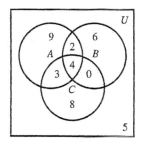

33. $(A \cup B)' = A' \cap B'$

For $(A \cup B)'$, first find $A \cup B$.

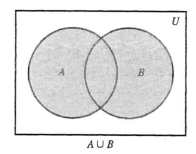

$A \cup B$

Now find $(A \cup B)'$, the region outside $A \cup B$.

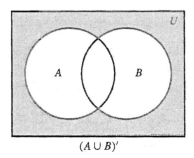

$(A \cup B)'$

For $A' \cap B'$, first find A' and B' individually.

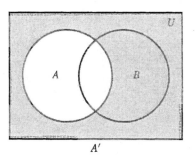

A'

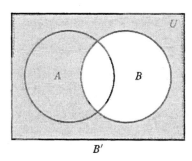

B'

Then $A' \cap B'$ is the region where A' and B' overlap, which is the entire region outside $A \cup B$ (the same result as in the second diagram). Therefore,

$$(A \cup B)' = A' \cap B'.$$

35. $A \cap (B \cup C) = (A \cap B) \cup (A \cap C)$

First find A and $B \cup C$ individually.

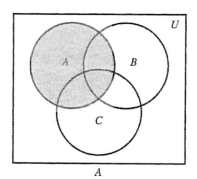

A

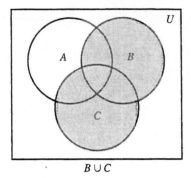

$B \cup C$

Then $A \cap (B \cup C)$ is the region where the above two diagrams overlap.

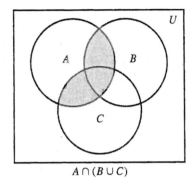

$A \cap (B \cup C)$

Next find $A \cap B$ and $A \cap C$ individually.

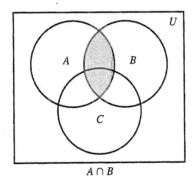

$A \cap B$

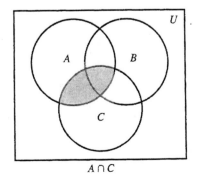

$A \cap C$

Then $(A \cap B) \cup (A \cap C)$ is the union of the above two diagrams.

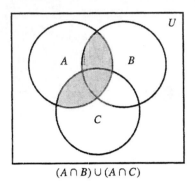

$(A \cap B) \cup (A \cap C)$

The Venn diagram for $A \cap (B \cup C)$ is identical to the Venn diagram for $(A \cap B) \cup (A \cap C)$, so conclude that

$$A \cap (B \cup C) = (A \cap B) \cup (A \cap C).$$

37. Prove

$$n(A \cup B \cup C)$$
$$= n(A) + n(B) + n(C) - n(A \cap B) - n(A \cap C)$$
$$- n(B \cap C) + n(A \cap B \cap C)$$

$$n(A \cup B \cup C)$$
$$= n[A \cup (B \cup C)]$$
$$= n(A) + n(B \cup C) - n[A \cap (B \cup C)]$$
$$= n(A) + n(B) + n(C) - n(B \cap C)$$
$$- n[(A \cap B) \cup (A \cap C)]$$
$$= n(A) + n(B) + n(C) - n(B \cap C)$$
$$- \{n(A \cap B) + n(A \cap C)$$
$$- n[(A \cap B) \cap (A \cap C)]\}$$
$$= n(A) + n(B) + n(C) - n(B \cap C) - n(A \cap B)$$
$$- n(A \cap C) + n(A \cap B \cap C)$$

39. Let A be the set of trucks that carried early peaches, B be the set of trucks that carried late peaches, and C be the set of trucks that carried extra late peaches. We are given the following information.

$$n(A) = 34$$
$$n(B) = 61$$
$$n(C) = 50$$
$$n(A \cap B) = 25$$
$$n(B \cap C) = 30$$
$$n(A \cap C) = 8$$
$$n(A \cap B \cap C) = 6$$
$$n(A' \cap B' \cap C') = 9$$

Start with $A \cap B \cap C$.
We know that $n(A \cap B \cap C) = 6$.

Since $n(A \cap B) = 25$, the number in $A \cap B$ but not in C is $25 - 6 = 19$.

Since $n(B \cap C) = 30$, the number in $B \cap C$ but not in A is $30 - 6 = 24$.

Since $n(A \cap C) = 8$, the number in $A \cap C$ but not in B is $8 - 6 = 2$.

Since $n(A) = 34$, the number in A but not in B or C is $34 - (19 + 6 + 2) = 7$.

Since $n(B) = 61$, the number in B but not in A or C is $61 - (19 + 6 + 24) = 12$.

Since $n(C) = 50$, the number in C but not in A or B is $50 - (24 + 6 + 2) = 18$.

Since $n(A' \cap B' \cap C') = 9$, the number outside $A \cup B \cup C$ is 9.

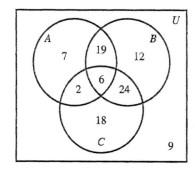

(a) From the Venn diagram, 12 trucks carried only late peaches.

(b) From the Venn diagram, 18 trucks carried only extra late peaches.

(c) From the Venn diagram, $7 + 12 + 18 = 37$ trucks carried only one type of peach.

(d) From the Venn diagram, $6 + 2 + 19 + 24 + 7 + 12 + 18 + 9 = 97$ trucks went out during the week.

41. (a) $n(Y \cap B) = 2$ since 2 is the number in the table where the Y row and the B column meet.

(b) $n(M \cup A) = n(M) + n(A) - n(M \cap A)$
$$= 33 + 41 - 14 = 60$$

(c) $n[Y \cap (S \cup B)] = 6 + 2 = 8$

(d) $n[O' \cup (S \cup A)]$
$$= n(O') + n(S \cup A) - n[O' \cap (S \cup A)]$$
$$= (23 + 33) + (52 + 41) - (6 + 14 + 15 + 14)$$
$$= 100$$

(e) Since $M' \cup O'$ is the entire set, $(M' \cup O') \cap B = B$. Therefore,

$$n[(M' \cup O') \cap B] = n(B) = 27.$$

(f) $Y \cap (S \cup B)$ is the set of all bank customers who are of age 18-29 and who invest in stocks or bonds.

43. Let T be the set of all tall pea plants, G be the set of plants with green peas, and S be the set of plants with smooth peas. We are given the following information.

$$n(U) = 50$$
$$n(T) = 22$$
$$n(G) = 25$$
$$n(S) = 39$$
$$n(T \cap G) = 9$$
$$n(G \cap S) = 20$$
$$n(T \cap G \cap S) = 6$$
$$n(T' \cap G' \cap S') = 4$$

Start by filling in the Venn Diagram with the numbers for the last two regions, $T \cap G \cap S$ and $T' \cap G' \cap S'$, as shown below. With $n(T \cap G) = 9$, this leaves $n(T \cap G \cap S') = 9 - 6 = 3$. With $n(G \cap S) = 20$, this leaves $n(T' \cap G \cap S) = 20 - 6 = 14$. Since $n(G) = 25$, $n(T' \cap G \cap S') = 25 - 3 - 6 - 14 = 2$. With no other regions that we can calculate, denote by x the number in $T \cap G' \cap S'$. Then $n(T \cap G' \cap S') = 22 - 3 - 6 - x = 13 - x$, and $n(T' \cap G' \cap S) = 39 - 6 - 14 - x = 19 - x$, as shown. Summing the values for all eight regions,

$$(13 - x) + 3 + 2 + x + 6 + 14 + (19 - x) + 4 = 50$$
$$61 - x = 50$$
$$x = 11$$

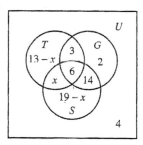

(a) $n(T \cap S) = 11 + 6 = 17$

(b) $n(T \cap G' \cap S') = 13 - x = 13 - 11 = 2$

(c) $n(T' \cap G \cap S) = 14$

45. First fill in the Venn diagram, starting with the region common to all three sets.

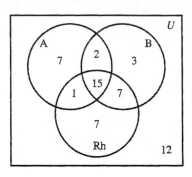

(a) The total of these numbers in the diagram is 54.

(b) $7 + 3 + 7 = 17$ had only one antigen.

(c) $1 + 2 + 7 = 10$ had exactly two antigens.

(d) A person with O-positive blood has only the Rh antigen, so this number is 7.

(e) A person with AB-positive blood has all three antigens, so this number is 15.

(f) A person with B-negative blood has only the B antigen, so this number is 3.

(g) A person with O-negative blood has none of the antigens. There are 12 such people.

(h) A person with A-positive blood has the A and Rh antigens, but not the B-antigen. The number is 1.

47. Extend the table to include totals for each row and each column.

	H	F	Total
A	95	34	129
B	41	38	79
C	9	7	16
D	202	150	352
Total	347	229	576

(a) $n(A \cap F)$ is the entry in the table that is in both row A and column F. Thus, there are 34 players in the set $A \cap F$.

(b) Since all players in the set C are either in set H or set F, $C \cap (H \cup F) = C$. Thus, $n(C \cap (H \cup F)) = n(C) = 16$, the total for row C. There are 16 players in the set $C \cap (H \cup F)$.

(c) $n(D \cup F) = n(D) + n(F) - n(D \cap F)$
$$= 352 + 229 - 150$$
$$= 431$$

(d) $B' \cap C'$ is the set of players who are both *not* in B and *not* in C. Thus, $B' \cap C' = A \cup D$, and since A and D are disjoints $n(A \cup D) = n(A) + n(D) = 129 + 352 = 481$. There are 481 players in the set $B' \cap C'$.

49. Reading directly from the table, $n(A \cap B) = 110.6$. Thus, there are 110.6 million people in the set $A \cap B$.

51. $n(G \cup (C \cap H)) = n(G) + n(C \cap H) = 80.4 + 5.0 = 85.4$

There are 85.4 million people in the set $G \cup (C \cap H)$.

53. $n(H \cup D) = n(H) + n(D) - n(H \cap D) = 53.6 + 19.6 - 2.2 = 71.0$

There are 71.0 million people in the set $H \cup D$.

For Exercises 55−57, extend the table to include totals for each row and each column.

	W	B	H	A	Total
N	49,101	11,783	9862	3181	73,927
M	104,689	9279	14,239	5594	133,801
I	11,754	1730	918	418	14,820
D	21,372	4250	2917	607	29,146
Total	186,916	27,042	27,936	9800	251,694

55. $n(N \cap (B \cup H)) = n(N \cap B) + n(N \cap H) = 11{,}783 + 9862 = 21{,}645$

There are 21,645 thousand people in the set $N \cap (B \cup H)$.

57. First, notice that $A' = W \cup B \cup H$. Therefore, the set $(D \cup W) \cap A'$ is just $D \cup W$ restricted to the sets, $W, B,$ or H. In other words,

$$
\begin{aligned}
n((D \cup W) \cap A') &= n(D \cap A') + n(W \cap A') - n((D \cap W) \cap A') \\
&= n(D \cap A') + n(W) - n(D \cap W) \\
&= [n(D \cap W) + n(D \cap B) + n(D \cap H)] + n(W) - n(D \cap W) \\
&= n(D \cap B) + n(D \cap H) + n(W) \\
&= 4250 + 2917 + 186{,}916 \\
&= 194{,}083
\end{aligned}
$$

There are 194,083 thousand people in the set $(D \cup W) \cap A'$.

59. Let W be the set of women, C be the set of those who speak Cantonese, and F be the set of those who set off firecrackers. We are given the following information.

$n(W) = 120$
$n(C) = 150$
$n(F) = 170$
$n(W' \cap C) = 108$
$n(W' \cap F') = 100$
$n(W \cap C' \cap F) = 18$
$n(W' \cap C' \cap F') = 78$
$n(W \cap C \cap F) = 30$

Note that

$n(W' \cap C \cap F') = n(W' \cap F') - n(W' \cap C' \cap F') = 100 - 78 = 22.$

Furthermore,

$n(W' \cap C \cap F) = n(W' \cap C) - n(W' \cap C \cap F') = 108 - 22 = 86.$

We now have

$n(W \cap C \cap F') = n(C) - n(W' \cap C \cap F) - n(W \cap C \cap F) - n(W' \cap C \cap F') = 150 - 86 - 30 - 22 = 12.$

With all of the overlaps of W, C, and F determined, we can now compute $n(W \cap C' \cap F') = 60$ and $n(W' \cap C' \cap F) = 36$.

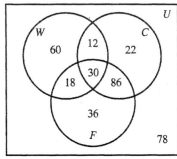

(a) Adding up the disjoint components, we find the total attendance to be

$$60 + 12 + 18 + 30 + 22 + 86 + 36 + 78 = 342.$$

(b) $n(C') = 342 - n(C) = 342 - 150 = 192$

(c) $n(W \cap F') = 60 + 12 = 72$

(d) $n(W' \cap C \cap F) = 86$

61. Let F be the set of fat chickens (so F' is the set of thin chickens), R be the set of red chickens (so R' is the set of brown chickens), and M be the set of male chickens, or roosters (so M' is the set of female chickens, or hens). We are given the following information.

$n(F \cap R \cap M) = 9$
$n(F' \cap R' \cap M') = 13$
$n(R \cap M) = 15$
$n(F' \cap R) = 11$
$n(R \cap M') = 17$
$n(F) = 56$
$n(M) = 41$
$n(M') = 48$

First, note that $n(M) + n(M') = n(U) = 89$, the total number of chickens.
Since $n(R \cap M) = 15$, $n(F' \cap R \cap M) = 15 - 9 = 6$.
Since $n(F' \cap R) = 11$, $n(F' \cap R \cap M') = 11 - 6 = 5$.
Since $n(R \cap M') = 17$, $n(F \cap R \cap M') = 17 - 5 = 12$.
Since $n(M') = 48$, $n(F \cap R' \cap M') = 48 - (12 + 5 + 13) = 18$.
Since $n(F) = 56$, $n(F \cap R' \cap M) = 56 - (18 + 12 + 9) = 17$.
And, finally, since $n(M) = 41$, $n(F' \cap R' \cap M) = 41 - (17 + 9 + 6) = 9$.

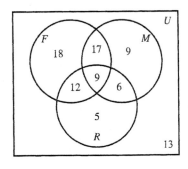

(a) $n(F) = 56$

(b) $n(R) = n(R \cap M) + n(R \cap M')$
$$= 15 + 17 = 32$$

(c) $n(F \cap M) = n(F \cap R \cap M) + n(F \cap R' \cap M)$
$$= 17 + 9 = 26$$

(d) $n(F \cap M') = n(F) - n(F \cap M) = 56 - 26$
$$= 30$$

(e) $n(F' \cap R') = n(F' \cap R' \cap M) + n(F' \cap R' \cap M')$
$$= 9 + 13 = 22$$

(f) $n(F \cap R) = n(F \cap R \cap M) + n(F \cap R \cap M')$
$$= 9 + 12 = 21$$

7.3   Introduction to Probability

3. The sample space is the set of the twelve months, {January, February, March, ..., December}.

5. The possible number of points earned could be any whole number from 0 to 80. The sample space is the set
$$\{0, 1, 2, 3, \ldots, 79, 80\}.$$

7. The possible decisions are to go ahead with a new oil shale plant or to cancel it. The sample space is the set {go ahead, cancel}.

9. Let h = heads and t = tails for the coin; the die can display 6 different numbers. There are 12 possible outcomes in the sample space, which is the set
$$\{(h, 1),\ (t, 1),\ (h, 2),\ (t, 2),\ (h, 3),\ (t, 3),$$
$$(h, 4),\ (t, 4),\ (h, 5),\ (t, 5),\ (h, 6),\ (t, 6)\}.$$

13. Use the first letter of each name. The sample space is the set
$$S = \{AB,\ AC,\ AD,\ AE,\ BC,\ BD,\ BE,\ CD,\ CE,\ DE\}.$$
$n(S) = 10$. Assuming the committee is selected at random, the outcomes are equally likely.
(a) One of the committee members must be Chinn. This event is {AC, BC, CD, CE}.

(b) Alam, Bartolini, and Chinn may be on any committee; Dickson and Ellsberg may not be on the same committee. This event is
$$\{AB,\ AC,\ AD,\ AE,\ BC,\ BD,\ BE,\ CD,\ CE\}.$$

(c) Both Alam and Chinn are on the committee. This event is {AC}.

15. Each outcome consists of two of the numbers 1, 2, 3, 4, and 5, without regard for order. For example, let (2, 5) represent the outcome that the slips of paper marked with 2 and 5 are drawn. There are ten equally likely outcomes in this sample space, which is
$$S = \{(1, 2),\ (1, 3),\ (1, 4),\ (1, 5),\ (2, 3),$$
$$(2, 4),\ (2, 5),\ (3, 4),\ (3, 5),\ (4, 5)\}.$$

(a) Both numbers in the outcome pair are even. This event is $\{(2, 4)\}$, which is called a simple event since it consists of only one outcome.

(b) One number in the pair is even and the other number is odd. This event is
$$\{(1, 2),\ (1, 4), (2, 3), (2, 5), (3, 4), (4, 5)\}.$$

(c) Each slip of paper has a different number written on it, so it is not possible to draw two slips marked with the same number. This event is $\emptyset$, which is called an impossible event since it contains no outcomes.

17. $S = \{HH,\ THH,\ HTH,\ TTHH,\ THTH,\ HTTH,$
$$TTTH,\ TTHT,\ THTT,\ HTTT,\ TTTT\}$$
$n(S) = 11$. The outcomes are not equally likely.

(a) The coin is tossed four times. This event is written $\{TTHH,\ THTH,\ HTTH,\ TTTH,\ TTHT,$ $THTT,\ HTTT,\ TTTT\}$.

(b) Exactly two heads are tossed. This event is written $\{HH,\ THH,\ HTH,\ TTHH,\ THTH,\ HTTH\}$.

(c) No heads are tossed. This event is written $\{TTTT\}$.

For Exercises 19−23, use the sample space
$$S = \{1, 2, 3, 4, 5, 6\}.$$

19. "Getting a 2" is the event $E = \{2\}$, so $n(E) = 1$ and $n(S) = 6$.

If all the outcomes in a sample space S are equally likely, then the probability of an event E is
$$P(E) = \frac{n(E)}{n(S)}.$$

In this problem,
$$P(E) = \frac{n(E)}{n(S)} = \frac{1}{6}.$$

21. "Getting a number less than 5" is the event $E = \{1, 2, 3, 4\}$, so $n(E) = 4$.

$$P(E) = \frac{4}{6} = \frac{2}{3}.$$

23. "Getting a 3 or a 4" is the event $E = \{3, 4\}$, so $n(E) = 2$.

$$P(E) = \frac{2}{6} = \frac{1}{3}.$$

For Exercises 25–33, the sample space contains all 52 cards in the deck, so $n(S) = 52$.

25. Let E be the event "a 9 is drawn." There are four 9's in the deck, so $n(E) = 4$.

$$P(9) = P(E) = \frac{n(E)}{n(S)} = \frac{4}{52} = \frac{1}{13}$$

27. Let F be the event "a black 9 is drawn." There are two black 9's in the deck, so $n(F) = 2$.

$$P(\text{black } 9) = P(F) = \frac{n(F)}{n(S)} = \frac{2}{52} = \frac{1}{26}$$

29. Let G be the event "a 9 of hearts is drawn." There is only one 9 of hearts in a deck of 52 cards, so $n(G) = 1$.

$$P(9 \text{ of hearts}) = P(G) = \frac{n(G)}{n(S)} = \frac{1}{52}$$

31. Let H be the event "a 2 or a queen is drawn." There are four 2's and four queens in the deck, so $n(H) = 8$.

$$P(2 \text{ or queen}) = P(H) = \frac{n(H)}{n(S)} = \frac{8}{52} = \frac{2}{13}$$

33. Let E be the event "a red card or a ten is drawn." There are 26 red cards and 4 tens in the deck. But 2 tens are red cards and are counted twice. Use the result from the previous section.

$$n(E) = n(\text{red cards}) + n(\text{tens}) - n(\text{red tens})$$
$$= 26 + 4 - 2$$
$$= 28$$

Now calculate the probability of E.

$$P(\text{red card or ten}) = \frac{n(E)}{n(S)}$$
$$= \frac{28}{52}$$
$$= \frac{7}{13}$$

For Exercises 35-39, the sample space consists of all the marbles in the jar. There are $3 + 4 + 5 + 8 = 20$ marbles, so $n(S) = 20$.

35. 3 of the marbles are white, so

$$P(\text{white}) = \frac{3}{20}.$$

37. 5 of the marbles are yellow, so

$$P(\text{yellow}) = \frac{5}{20} = \frac{1}{4}.$$

39. $3 + 4 + 5 = 12$ of the marbles are not black, so

$$P(\text{not black}) = \frac{12}{20} = \frac{3}{5}.$$

41. The outcomes are not equally likely.

43. E: worker is female
F: worker has worked less than 5 yr
G: worker contributes to a voluntary retirement plan

(a) E' occurs when E does not, so E' is the event "worker is male."

(b) $E \cap F$ occurs when both E and F occur, so $E \cap F$ is the event "worker is female and has worked less than 5 yr."

(c) $E \cup G'$ is the event "worker is female or does not contribute to a voluntary retirement plan."

(d) F' occurs when F does not, so F' is the event "worker has worked 5 yr or more."

(e) $F \cup G$ occurs when F or G occurs or both, so $F \cup G$ is the event "worker has worked less than 5 yr or contributes to a voluntary retirement plan."

(f) $F' \cap G'$ occurs when F does not and G does not, so $F' \cap G'$ is the event "worker has worked 5 yr or more and does not contribute to a voluntary retirement plan."

45. (a) $P(\text{invested in stocks and bonds}) = \frac{80}{150}$
$$= \frac{8}{15}$$

(b) $P(\text{invested in stocks and bonds and CDs})$
$$= \frac{80}{150}$$
$$= \frac{8}{15}$$

47. E: person smokes

F: person has a family history of heart disease

G: person is overweight

(a) $E \cup F$ occurs when E or F or both occur, so $E \cup F$ is the event "person smokes or has a family history of heart disease, or both."

(b) $E' \cap F$ occurs when E does not occur and F does occur, so $E' \cap F$ is the event "person does not smoke and has a family history of heart disease."

(c) $F' \cup G'$ is the event "person does not have a family history of heart disease or is not overweight, or both."

49. The total population for 2020 is 322,742, and the total for 2050 is 393,931.

(a) $P(\text{Hispanic in } 2020) = \dfrac{52{,}652}{322{,}742}$

≈ 0.1631

(b) $P(\text{Hispanic in } 2050) = \dfrac{96{,}508}{393{,}931}$

≈ 0.2450

(c) $P(\text{Black in } 2020) = \dfrac{41{,}538}{322{,}742}$

≈ 0.1287

(d) $P(\text{Black in } 2050) = \dfrac{53{,}555}{393{,}931}$

≈ 0.1360

51. (a) $P(\text{III Corps}) = \dfrac{22{,}083}{70{,}076} \approx 0.3151$

(b) $P(\text{lost in battle}) = \dfrac{22{,}557}{70{,}076} \approx 0.3219$

(c) $P(\text{I Corps lost in battle}) = \dfrac{7661}{20{,}706}$

≈ 0.3670

(d) $P(\text{I Corps not lost in battle}) = \dfrac{20{,}706 - 7661}{20{,}706}$

≈ 0.6300

$P(\text{II Corps not lost in battle}) = \dfrac{20{,}666 - 6603}{20{,}666}$

≈ 0.6805

$P(\text{III Corps not lost in battle}) = \dfrac{22{,}083 - 8007}{22{,}083}$

≈ 0.6374

$P(\text{Calvary not lost in battle}) = \dfrac{6621 - 286}{6621}$

≈ 0.9568

The Calvary had the highest probability of not being lost in battle.

(e) $P(\text{I Corps loss}) = \dfrac{7661}{20{,}706} \approx 0.3700$

$P(\text{II Corps loss}) = \dfrac{6603}{20{,}666} \approx 0.3195$

$P(\text{III Corps loss}) = \dfrac{8007}{22{,}083} \approx 0.3626$

$P(\text{Calvary loss}) = \dfrac{286}{6621} \approx 0.0432$

I Corps had the highest probability of loss.

53. There were 342 in attendance.

(a) $P(\text{speaks Cantonese}) = \dfrac{150}{342}$

$= \dfrac{25}{57}$

(b) $P(\text{does not speak Cantonese}) = \dfrac{192}{342}$

$= \dfrac{32}{57}$

(c) $P(\text{woman who did not light firecracker}).$

$= \dfrac{72}{342} = \dfrac{4}{19}$

7.4   Basic Concepts of Probability

3. A person can be from Texas and be a business major at the same time. No, these events are not mutually exclusive.

5. A person cannot be a teenager and be 70 years old at the same time. Yes, these events are mutually exclusive.

7. A person can be male and be a nurse at the same time. No, these events are not mutually exclusive.

9. When the two dice are rolled, there are 36 equally likely outcomes. Let 5-3 represent the outcome "the first die shows a 5 and the second die shows a 3," and so on.

(a) Rolling a sum of 8 occurs when the outcome is 2-6, 3-5, 4-4, 5-3, or 6-2. Therefore, since there are five such outcomes, the probability of this event is

$$P(\text{sum is 8}) = \frac{5}{36}.$$

(b) A sum of 9 occurs when the outcome is 3-6, 4-5, 5-4, or 6-3, so

$$P(\text{sum is 9}) = \frac{4}{36} = \frac{1}{9}.$$

(c) A sum of 10 occurs when the outcome is 4-6, 5-5, or 6-4, so

$$P(\text{sum is 10}) = \frac{3}{36} = \frac{1}{12}.$$

(d) A sum of 13 does not occur in any of the 36 outcomes, so

$$P(\text{sum is 13}) = \frac{0}{36} = 0.$$

11. Again, when two dice are rolled, there are 36 equally likely outcomes.

(a) $P(\text{sum is not more than 5})$
$$= P(2) + P(3) + P(4) + P(5)$$
$$= \frac{1}{36} + \frac{2}{36} + \frac{3}{36} + \frac{4}{36}$$
$$= \frac{10}{36}$$
$$= \frac{5}{18}$$

(b) $P(\text{sum is not less than 8})$
$$= P(8) + P(9) + P(10) + P(11) + P(12)$$
$$= \frac{5}{36} + \frac{4}{36} + \frac{3}{36} + \frac{2}{36} + \frac{1}{36}$$
$$= \frac{15}{36}$$
$$= \frac{5}{12}$$

(c) $P(\text{sum is between 3 and 7})$
$$= P(4) + P(5) + P(6)$$
$$= \frac{3}{36} + \frac{4}{36} + \frac{5}{36}$$
$$= \frac{12}{36}$$
$$= \frac{1}{3}$$

13. $P(\text{second die is 5 or the sum is 10})$
$$= P(\text{second die is 5}) + P(\text{sum is 10})$$
$$\quad - P(\text{second die is 5 and sum is 10})$$
$$= \frac{6}{36} + \frac{3}{36} - \frac{1}{36}$$
$$= \frac{8}{36}$$
$$= \frac{2}{9}$$

15. (a) The events E, "9 is drawn," and F, "10 is drawn," are mutually exclusive, so $P(E \cap F) = 0$.

Using the union rule,

$$P(9 \text{ or } 10) = P(9) + P(10)$$
$$= \frac{4}{52} + \frac{4}{52}$$
$$= \frac{8}{52}$$
$$= \frac{2}{13}.$$

(b) $P(\text{red or 3}) = P(\text{red}) + P(3) - P(\text{red and 3})$
$$= \frac{26}{52} + \frac{4}{52} - \frac{2}{52}$$
$$= \frac{28}{52}$$
$$= \frac{7}{13}$$

(c) Since these events are mutually exclusive,

$$P(9 \text{ or black 10}) = P(9) + P(\text{black 10})$$
$$= \frac{4}{52} + \frac{2}{52}$$
$$= \frac{6}{52}$$
$$= \frac{3}{26}.$$

(d) $P(\text{heart or black}) = P(\text{heart}) + P(\text{black})$

$$- P(\text{heart and black})$$

$$= \frac{13}{52} + \frac{26}{52} - \frac{0}{52}$$

$$= \frac{39}{52}$$

$$= \frac{3}{4}.$$

(e) $P(\text{face card or diamond})$
$$= P(\text{face card}) + P(\text{diamond})$$
$$- P(\text{face card and diamond})$$

$$= \frac{12}{52} + \frac{13}{52} - \frac{3}{52}$$

$$= \frac{22}{52}$$

$$= \frac{11}{26}.$$

17. (a) Since these events are mutually exclusive,

$$P(\text{brother or uncle}) = P(\text{brother}) + P(\text{uncle})$$

$$= \frac{2}{13} + \frac{3}{13}$$

$$= \frac{5}{13}.$$

(b) Since these events are mutually exclusive,

$$P(\text{brother or cousin}) = P(\text{brother}) + P(\text{cousin})$$

$$= \frac{2}{13} + \frac{5}{13}$$

$$= \frac{7}{13}.$$

(c) Since these events are mutually exclusive,

$$P(\text{brother or mother}) = P(\text{brother}) + P(\text{mother})$$

$$= \frac{2}{13} + \frac{1}{13}$$

$$= \frac{3}{13}.$$

19. (a) There are 5 possible numbers on the first slip drawn, and for each of these, 4 possible numbers on the second, so the sample space contains $5 \cdot 4 = 20$ ordered pairs. Two of these ordered pairs have a sum of 9: $(4, 5)$ and $(5, 4)$. Thus,

$$P(\text{sum is 9}) = \frac{2}{20} = \frac{1}{10}.$$

(b) The outcomes for which the sum is 5 or less are $(1, 2), (1, 3), (1, 4), (2, 1), (2, 3), (3, 1), (3, 2),$ and $(4, 1)$. Thus,

$$P(\text{sum is 5 or less}) = \frac{8}{20} = \frac{2}{5}.$$

(c) Let A be the event "the first number is 2" and B the event "the sum is 6." Use the union rule.

$$P(A \cup B) = P(A) + P(B) - P(A \cap B)$$

$$= \frac{4}{20} + \frac{4}{20} - \frac{1}{20}$$

$$= \frac{7}{20}$$

21. Since $P(E \cap F) = 0.16$, the overlapping region $E \cap F$ is assigned the probability 0.16 in the diagram. Since $P(F) = 0.26$ and $P(E \cap F) = 0.16$, the region in E but not F is given the label 0.10. Similarly, the remaining regions are labeled.

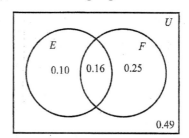

(a) $P(E \cup F) = 0.10 + 0.16 + 0.25$
$$= 0.51$$

Consequently, the part of U outside $E \cup F$ receives the label

$$1 - 0.51 = 0.49.$$

(b) $P(E' \cap F) = P(\text{in } F \text{ but not in } E)$
$$= 0.25$$

(c) The region $E \cap F'$ is that part of E which is not in F. Thus,

$$P(E \cap F') = 0.10.$$

(d) $P(E' \cup F') = P(E') + P(F') - P(E' \cap F')$
$$= 0.74 + 0.59 - 0.49$$
$$= 0.84$$

25. Let E be the event "4, 5, or 6 is rolled." Here $E = \{4, 5, 6\}$, so $P(E) = \frac{3}{6} = \frac{1}{2}$, and $P(E') = \frac{1}{2}$. The odds in favor of rolling 4, 5 or 6 are

$$\frac{P(E)}{P(E')} = \frac{\frac{1}{2}}{\frac{1}{2}} = \frac{1}{1}.$$

which is written "1 to 1."

27. Let F be the event "some number less than 6 is rolled." Here $F = \{1, 2, 3, 4, 5\}$, so $P(F) = \frac{5}{6}$ and $P(F') = \frac{1}{6}$. The odds in favor of rolling a number less than 2 are

$$\frac{P(F)}{P(F')} = \frac{\frac{5}{6}}{\frac{1}{6}} = \frac{5}{1},$$

which is written "5 to 1."

29. Let E be the event "draw a white marble." Then $P(E) = \frac{2}{9}$ and $P(E') = \frac{7}{9}$. The odds of not drawing a white marble are

$$\frac{P(E')}{P(E)} = \frac{\frac{7}{9}}{\frac{2}{9}} = \frac{7}{2},$$

which is written "7 to 2."

31. The statement is not correct.

Assume two dice are rolled.

$P(\text{you win}) = P(\text{first die is greater than second})$

$$= \frac{15}{36}$$

$$= \frac{5}{12}$$

$P(\text{other player wins}) = 1 - P(\text{you win})$

$$= 1 - \frac{5}{12}$$

$$= \frac{7}{12}$$

33. $P(\text{You will eat out today.}) = \dfrac{1}{1+2} = \dfrac{1}{3}$

$P(\text{Bottled water will be tap water.}) = \dfrac{1}{1+4} = \dfrac{1}{5}$

$P(\text{Earth will be struck by meteor.}) = \dfrac{1}{1+9000} = \dfrac{1}{9001}$

$P(\text{You will go to Disney World.}) = \dfrac{1}{1+9} = \dfrac{1}{10}$

$P(\text{You'll regain weight.}) = \dfrac{9}{9+10} = \dfrac{9}{19}$

35. This is empirical; only a survey could determine the probability.

37. This is not empirical; a formula can compute the probability exactly.

39. This is empirical, based on experience and conditions rather than probability theory.

41. This is empirical, based on experience rather than probability theory.

43. Each of the probabilities is between 0 and 1 and the sum of all the probabilities is

$$0.09 + 0.32 + 0.21 + 0.25 + 0.13 = 1,$$

so this assignment is possible.

45. The sum of the probabilities

$$\frac{1}{3} + \frac{1}{4} + \frac{1}{6} + \frac{1}{8} + \frac{1}{10} = \frac{117}{120} < 1,$$

so this assignment is not possible.

47. This assignment is not possible because one of the probabilities is -0.08, which is not between 0 and 1. A probability cannot be negative.

49. The answers that are given are theoretical. Using the Monte Carlo method with at least 50 repetitions on a graphing calculator should give values close to these.

(a) 0.2778   **(b)** 0.4167

51. The answers that are given are theoretical. Using the Monte Carlo method with at least 100 repetitions should give values close to these.

(a) 0.0463   **(b)** 0.2963

55. Let C be the event "the calculator has a good case," and let B be the event "the calculator has good batteries."

$P(C \cap B)$
   $= 1 - P[(C \cap B)']$
   $= 1 - P(C' \cup B')$
   $= 1 - [P(C') + P(B') - P(C' \cap B')]$
   $= 1 - (0.08 + 0.11 - 0.03)$
   $= 0.84$

Thus, the probability that the calculator has a good case and good batteries is 0.84.

57. (a) $P(\text{less than } \$25) = 0.02 + 0.05 = 0.07$

(b) $P(\text{more than } \$24.99) = 1 - P(\text{less than } \$25)$
$$= 1 - 0.07 = 0.93$$

(c) $P(\$50 \text{ to } \$199.99) = P(\$50 - \$74.99)$
$$+ P(\$75 - \$99.99)$$
$$+ P(\$100 - \$199.99)$$
$$= 0.13 + 0.14 + 0.22 = 0.49$$

59. We are given that $P(\text{profit}) = 0.74$, so

$$P(\text{no profit}) = 1 - 0.74 = 0.26.$$

The odds against the company's making a profit are

$$\frac{P(\text{no profit})}{P(\text{profit})} = \frac{0.26}{0.74} = \frac{13}{37},$$

or 13 to 37.

61. $P(C) = 0.039,\ P(M \cap C) = 0.035,$
$P(M \cup C) = 0.491$

Place the given information in a Venn diagram by starting with 0.035 in the intersection of the regions for M and C.

Since $P(C) = 0.039$,

$$0.039 - 0.035 = 0.004$$

goes inside region C, but outside the intersection of C and M. Thus,

$$P(C \cap M') = 0.004.$$

Since $P(M \cup C) = 0.491$,

$$0.491 - 0.035 - 0.004 = 0.452$$

goes inside region M, but outside the intersection of C and M. Thus, $P(M \cap C') = 0.452$. The labeled regions have probability

$$0.452 + 0.035 + 0.004 = 0.491.$$

Since the entire region of the Venn diagram must have probability 1, the region outside M and C, or $M' \cap C'$, has probability

$$1 - 0.491 = 0.509.$$

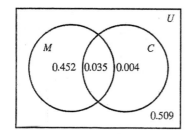

(a) $P(C') = 1 - P(C)$
$= 1 - 0.039$
$= 0.961$

(b) $P(M) = 0.452 + 0.035$
$= 0.487$

(c) $P(M') = 1 - P(M)$
$= 1 - 0.487$
$= 0.513$

(d) $P(M' \cap C') = 0.509$

(e) $P(C \cap M') = 0.004$

(f) $P(C \cup M')$
$= P(C) + P(M') - P(C \cap M')$
$= 0.039 + 0.513 - 0.004$
$= 0.548$

63. (a) Now red is no longer dominant, and RW or WR results in pink, so

$$P(\text{red}) = P(RR) = \frac{1}{4}.$$

(b) $P(\text{pink}) = P(RW) + P(WR)$
$$= \frac{1}{4} + \frac{1}{4} = \frac{1}{2}.$$

(c) $P(\text{white}) = P(WW) = \dfrac{1}{4}$

65. Let L be the event "visit results in lab work" and R be the event "visit results in referral to specialist." We are given the probability a visit results in neither is 35%, so $P((L \cup R)') = 0.35$. Since $P(L \cup R) = 1 - P((L \cup R)')$, we have

$$P(L \cup R) = 1 - 0.35 = 0.65.$$

We are also given $P(L) = 0.40$ and $P(R) = 0.30$. Using the union rule for probability,

$$\begin{aligned}
P(L \cup R) &= P(L) + P(R) - P(L \cap R) \\
0.65 &= 0.40 + 0.30 - P(L \cap R) \\
0.65 &= 0.50 - P(L \cap R) \\
P(L \cap R) &= 0.05
\end{aligned}$$

The correct answer choice is **a.**

67. Let $x = P(A \cap B)$,
$\quad\quad\quad y = P(B \cap C)$,
$\quad\quad\quad z = P(A \cap C)$,
and $\quad w = P((A \cup B \cup C)')$.

If an employee must choose exactly two or none of the supplementary coverages A, B, and C, then $P(A \cap B \cap C) = 0$ and the probabilities of the region representing a single choice of coverages A, B, or C are also 0. We can represent the choices and probabilities with the following Venn diagram.

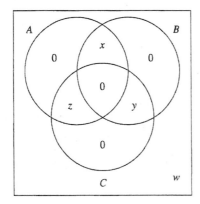

The information given leads to the following system of equations.

$$x + y + z + w = 1$$
$$x \quad\quad + z \quad\quad = \frac{1}{4}$$
$$x + y \quad\quad\quad = \frac{1}{3}$$
$$\quad\quad y + z \quad\quad = \frac{5}{12}$$

Using a graphing calculator or computer program, the solution to the system is $x = 1/2, y = 1/4, z = 1/6, w = 1/2$. Since the probability that a randomly chosen employee will choose no supplementary coverage is w, the correct answer choice is **c.**

69. Since 55 of the workers were women, $130 - 55 = 75$ were men. Since 3 of the women earned more than $40,000, $55 - 3 = 52$ of them earned $40,000 or less. Since 62 of the men earned $40,000 or less, $75 - 62 = 13$ earned more than $40,000.

These data for the 130 workers can be summarized in the following table.

	Men	Women
$40,000 or less	62	52
Over $40,000	13	3

(a) P(a woman earning $40,000 or less)

$$= \frac{52}{130} = 0.4$$

(b) P(a man earning more than $40,000)

$$= \frac{13}{130} = 0.1$$

(c) P(a man or is earning more than $40,000)

$$= \frac{62 + 13 + 3}{130}$$
$$= \frac{78}{130} = 0.6$$

(d) P(a woman or is earning $40,000 or less)

$$= \frac{52 + 3 + 62}{130}$$
$$= \frac{117}{130} = 0.9$$

71. Let A be the set of refugees who came to escape abject poverty and B be the set of refugees who came to escape political oppression. Then $P(A) = 0.80$, $P(B) = 0.90$, and $P(A \cap B) = 0.70$.

$$P(A \cup B) = P(A) + P(B) - P(A \cap B)$$
$$= 0.80 + 0.90 - 0.70 = 1$$
$$P(A' \cap B') = 1 - P(A \cap B)$$
$$= 1 - 1 = 0$$

The probability that a refugee in the camp was neither poor nor seeking political asylum is 0.

73. The odds of winning are 3 to 2; this means there are 3 ways to win and 2 ways to lose, out of a total of $2 + 3 = 5$ ways altogether. Hence, the probability of losing is $\frac{2}{5}$.

75. (a) P(somewhat or extremely intolerant of Facists)
$\quad = P$(somewhat intolerant of Fascists)
$\quad\quad + P$(extremely intolerant of Facists)

$$= \frac{27.1}{100} + \frac{59.5}{100} = \frac{86.6}{100} = 0.866$$

(b) P(completely tolerant of Communists)
$\quad = P$(no intolerance at all of Communists)

$$= \frac{47.8}{100} = 0.478$$

77. Since the odds are 4 to 7, the probability of rain is

$$\frac{4}{4 + 7} = \frac{4}{11}.$$

7.5 Conditional Probability; Independent Events

1. Let A be the event "the number is 2" and B be the event "the number is odd."

   The problem seeks the conditional probability $P(A|B)$. Use the definition

   $$P(A|B) = \frac{P(A \cap B)}{P(B)}.$$

   Here, $P(A \cap B) = 0$ and $P(B) = \frac{1}{2}$. Thus,

   $$P(A|B) = \frac{0}{\frac{1}{2}} = 0.$$

3. Let A be the event "the number is even" and B be the event "the number is 6." Then

   $$P(A|B) = \frac{P(A \cap B)}{P(B)} = \frac{\frac{1}{6}}{\frac{1}{6}} = 1.$$

5. Let A be the event "sum of 6" and B be the event "double." 6 of the 36 ordered pairs are doubles, so $P(B) = \frac{6}{36} = \frac{1}{6}$. There is only one outcome, 3-3, in $A \cap B$ (that is, a double with a sum of 6), so $P(A \cap B) = \frac{1}{36}$. Thus,

   $$P(A|B) = \frac{\frac{1}{36}}{\frac{1}{6}} = \frac{6}{36} = \frac{1}{6}.$$

7. Use a reduced sample space. After the first card drawn is a heart, there remain 51 cards, of which 12 are hearts. Thus,

   $$P(\text{heart on 2nd}|\text{heart on 1st}) = \frac{12}{51} = \frac{4}{17}.$$

9. Use a reduced sample space. After the first card drawn is a jack, there remain 51 cards, of which 11 are face cards. Thus,

   $$P(\text{face card on 2nd}|\text{jack on 1st}) = \frac{11}{51}.$$

11. $P(\text{a jack and a 10})$
    $= P(\text{jack followed by 10})$
    $\quad + P(10 \text{ followed by jack})$

    $$= \frac{4}{52} \cdot \frac{4}{51} + \frac{4}{52} \cdot \frac{4}{51}$$

    $$= \frac{16}{2652} + \frac{16}{2652}$$

    $$= \frac{32}{2652} = \frac{8}{663}.$$

13. $P(\text{two black cards})$
    $= P(\text{black on 1st})$
    $\quad \cdot P(\text{black on 2nd}|\text{black on 1st})$

    $$= \frac{26}{52} \cdot \frac{25}{51}$$

    $$= \frac{650}{2652} = \frac{25}{102}.$$

17. The knowledge that "it rains more than 10 days" affects the knowledge that "it rains more than 15 days." (For instance, if it hasn't rained more than 10 days, it couldn't possibly have rained more than 15 days.) Since $P(D|C) \neq P(D)$, the events are dependent.

19. First not that $P(G) = 1/7$ and $P(H) = 1/2$. A tree diagram of all the possible outcomes of the experiment would have 14 branches, exactly one of which would correspond to the outcome "today is Tuesday and the coin comes up heads," or $G \cap H$. Therefore, $P(G \cap H) = 1/14$. But notice that

    $$P(G \cap H) = \frac{1}{14} = \frac{1}{7} \cdot \frac{1}{2} = P(G) \cdot P(H).$$

    Thus, the events are independent.

21. (a) Given that the first number is 5, there are four possible sums: $5 + 1 = 6$, $5 + 2 = 7$, $5 + 3 = 8$, and $5 + 4 = 9$. One of the four possible outcomes corresponds to the sum 8. Therefore,

    $$P(\text{sum is 8}|\text{first number is 5}) = \frac{1}{4}.$$

    (b) Given that the first number is 4, there are four possible sums: $4 + 1 = 5$, $4 + 2 = 6$, $4 + 3 = 7$, and $4 + 5 = 9$. None of these corresponds to the sum 8. Therefore,

    $$P(\text{sum is 8}|\text{first number is 4}) = \frac{0}{4} = 0.$$

27. (a) Let C be the event "the coin comes up heads" and D be the event "6 is rolled on the die." We have

    $$P(C) = \frac{1}{2},\ P(D) = \frac{1}{6},\ P(C \cap D) = \frac{1}{12}.$$

    So the probability of winning—of the event $C \cap D$—is $1/12$.

What is the probability that either the head or the 6 occurred—in other words, what is $P(C \cup D)$? Use the union for probability.

$$P(C \cup D) = P(C) + P(D) - P(C \cap D)$$

$$= \frac{1}{2} + \frac{1}{6} - \frac{1}{12}$$

$$= \frac{6}{12} + \frac{2}{12} - \frac{1}{12}$$

$$= \frac{7}{12}$$

We want to find the probability of "head and 6 occured" given that "head or 6 occurred"—that is, $P((C \cap D)|(C \cup D))$. Use the formula for the conditional probability.

$$P((C \cap D)|(C \cup D)) = \frac{P((C \cap D)|(C \cup D))}{P(C \cup D)}$$

$$= \frac{P(C \cap D)}{P(C \cup D)}$$

$$= \frac{\frac{1}{12}}{\frac{7}{12}}$$

$$= \frac{1}{7}$$

The probability of winning the original game with the additional information is now 1/7. On the other hand, the probability of winning the new game, of the event "6 is rolled," is simply $P(D) = 1/6$. Since $1/6 > 1/7$, it is better to switch.

29. For a two-child family, the sample space is

$$M-M \quad M-F \quad F-M \quad F-F.$$

$$P(\text{same sex}) = \frac{2}{4} = \frac{1}{2}$$

$$P(\text{same sex}|\text{at most one male}) = \frac{1}{3}$$

The events are not independent.
For a three-child family, the same space is

$$M-M-M \quad M-M-F \quad M-F-M \quad M-F-F$$
$$F-M-M \quad F-M-F \quad F-F-M \quad F-F-F.$$

$$P(\text{same sex}) = \frac{2}{8} = \frac{1}{4}$$

$$P(\text{same sex}|\text{at most one male}) = \frac{1}{4}$$

The events are independent.

31. (a) If A and B are mutually exclusive, then

$$P(B) = P(A \cup B) - P(A)$$
$$= 0.7 - 0.5 = 0.2$$

(b) If A and B are independent, then

$$\frac{P(A \cap B)}{P(B)} = P(A|B) = P(A) = 0.5.$$

Thus,

$$P(A \cap B) = 0.5P(B).$$

Solving

$$P(A \cup B) = P(A) + P(B) - P(A \cap B)$$

for

$$P(A \cap B)$$

and substituting into the previous equations we get

$$P(A) + P(B) - P(A \cup B) = 0.5P(B)$$
$$0.5 + P(B) - 0.7 = 0.5P(B)$$
$$-0.2 = -0.5P(B)$$
$$0.4 = P(B).$$

$$P(B) = 0.4$$

33. $P(C|D)$ is the probability that a customer cashes a check, given that the customer made a deposit.

$$P(C|D) = \frac{P(C \cap D)}{P(D)}$$

$$= \frac{n(C \cap D)}{n(D)}$$

$$= \frac{60}{80}$$

$$= \frac{3}{4}$$

35. $P(C'|D')$ is the probability that a customer does not cash a check, given that the customer did not make a deposit.

$$P(C'|D') = \frac{P(C' \cap D')}{P(D')}$$

$$= \frac{n(C' \cap D')}{n(D')}$$

$$= \frac{10}{40}$$

$$= \frac{1}{4}$$

37. $P[(C \cap D)']$ is the probability that a customer does not both cash a check and make a deposit.

$$P[(C \cap D)'] = 1 - P(C \cap D)$$
$$= 1 - \frac{60}{120}$$
$$= \frac{60}{120}$$
$$= \frac{1}{2}$$

39. Let M represent the event "the main computer fails" and B the event "the backup computer fails." Since these events are assumed to be independent,

$$P(M \cap B) = P(M) \cdot P(B)$$
$$= 0.003(0.005)$$
$$= 0.000015.$$

This is the probability that both computers fail, which means that the company will not have computer service. The fraction of the time it will have service is

$$1 - 0.000015 = 0.999985.$$

Independence is a fairly realistic assumption. Situations such as floods or electric surges might cause both computers to fail at the same time.

41. Since 60% of production comes off assembly line A, $P(A) = 0.60$. Also $P(\text{pass inspection}|A) = 0.95$, so $P(\text{not pass}|A) = 0.05$. Therefore,

$$P(A \cap \text{not pass}) = P(A) \cdot P(\text{not pass}|A)$$
$$= 0.60(0.05)$$
$$= 0.03.$$

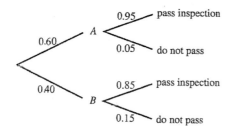

43. $P(\text{bike did not pass inspection})$

$$= 0.60(0.05) + 0.40(0.15)$$
$$= 0.03 + 0.06$$
$$= 0.09$$

Use the following tree diagram for Exercises 45 through 49

1st child	2nd child	3rd child	Branch	Probability
		1/2 B	1	1/8
	1/2 B	1/2 G	2	1/8
1/2 B	1/2 G	1/2 B	3	1/8
		1/2 G	4	1/8
	1/2 B	1/2 B	5	1/8
1/2 G		1/2 G	6	1/8
	1/2 G	1/2 B	7	1/8
		1/2 G	8	1/8

45. $P(\text{all girls}|\text{first is a girl})$

$$= \frac{P(\text{all girls and first is a girl})}{P(\text{first is a girl})}$$
$$= \frac{n(\text{all girls and first is a girl})}{n(\text{first is a girl})}$$
$$= \frac{1}{4}$$

47. $P(\text{all girls}|\text{second is a girl})$

$$= \frac{P(\text{all girls and second is a girl})}{P(\text{second is a girl})}$$
$$= \frac{n(\text{all girls and second is a girl})}{n(\text{second is a girl})}$$
$$= \frac{1}{4}$$

49. $P(\text{all girls}|\text{at least 1 girl})$

$$= \frac{P(\text{all girls and at least 1 girl})}{P(\text{at least 1 girl})}$$
$$= \frac{n(\text{all girls and at least 1 girl})}{n(\text{at least 1 girl})}$$
$$= \frac{1}{7}$$

51. By the product rule for independent events, two events are independent if the product of their probabilities is the probability of their intersection.

$$0.75(0.4) = 0.3$$

Therefore, the given events are independent.

53. $P(C) = 0.039$, the total of the C row.

55. $P(M \cup C) = P(M) + P(C) - P(M \cap C)$
$$= 0.487 + 0.039 - 0.035$$
$$= 0.491$$

57. $P(C|M) = \dfrac{P(C \cap M)}{P(M)}$

$$= \dfrac{0.035}{0.487}$$

$$\approx 0.072$$

59. By the definition of independent events, C and M are independent if

$$P(C|M) = P(C).$$

From Exercises 44 and 48,

$$P(C) = 0.039$$
$$\text{and}\quad P(C|M) = 0.072.$$

Since $P(C|M) \neq P(C)$, events C and M are not independent, so we say that they are dependent. This means that red-green color blindness does not occur equally among men and women.

61. First draw a tree diagram.

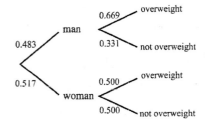

(a) $P(\text{overweight man}) = 0.483(0.669) \approx 0.323$

(b) $P(\text{overweight}) = 0.323 + 0.517(0.500)$
$$= 0.323 + 0.25$$
$$\approx 0.582$$

(c) The two events "man" and "overweight" are independent if

$$P(\text{overweight man}) = P(\text{overweight}) \cdot P(\text{man}).$$

We are given $P(\text{man}) = 0.483$ so that

$$P(\text{overweight}) \cdot P(\text{man}) = 0.582 \cdot 0.483 \approx 0.281$$

but we found $P(\text{overweight man}) = 0.323$ in part (a). Therefore, the events are not independent.

63. Let　H be the event "patient has high blood pressure,"
　　　　N be the event "patient has normal blood pressure,"
　　　　L be the event "patient has low blood pressure,"
　　　　R be the event "patient has a regular heartbeat,"
and I be the event "patient has an irregular heartbeat."

We wish to determine $P(R \cap L)$.

Statement (i) tells us $P(H) = 0.14$ and statement (ii) tells us $P(L) = 0.22$. Therefore,

$$P(H) + P(N) + P(L) = 1$$
$$0.14 + P(N) + 0.22 = 1$$
$$P(N) = 0.64.$$

Statement (iii) tells us $P(I) = 0.15$. This and statement (iv) lead to

$$P(I \cap H) = \frac{1}{3}P(I) = \frac{1}{3}(0.15) = 0.05.$$

Statement (v) tells us

$$P(N \cap I) = \frac{1}{8}P(N) = \frac{1}{3}(0.64) = 0.08.$$

Make a table and fill in the data just found.

	H	N	L	Totals
R	—	—	—	—
I	0.05	0.08	—	0.15
Totals	0.14	0.64	0.22	1.00

To determine $P(R \cap L)$, we need to find only $P(I \cap L)$.

$$P(I) = P(I \cap H) + P(I \cap N) + P(I \cap L)$$
$$0.15 = 0.05 + 0.08 + P(I \cap L)$$
$$0.15 = 0.13 + P(I \cap L)$$
$$P(I \cap L) = 0.02$$

Now calculate $P(R \cap L)$.

$$P(L) = P(R \cap L) + P(I \cap L)$$
$$0.22 = P(R \cap L) + 0.02$$
$$P(I \cap L) = 0.20$$

The correct answer choice is **e**.

65. (a) Since only 1 of the 2000 substances results in a marketable drug,

$P(\text{compound survives and becomes marketable})$

$$= \frac{1}{2000} = 0.0005.$$

(b) 1999 out of every 2000 compounds fail to get to market, so the probability of failure is

$$\frac{1999}{2000} = 0.9995.$$

(c) The probability that any one compound does *not* produce a marketable drug is $\frac{1999}{2000}$. If there are "a" compounds and we assume independence,

$$P(\text{none produces a marketable drug}) = \left(\frac{1999}{2000}\right)^{a}.$$

(d) This is the complement of the event in part (c). The probability that at least one of the drugs will prove marketable is

$$1 - \left(\frac{1999}{2000}\right)^{a}.$$

(e) For each scientist, $P(\text{none of } c \text{ compounds produces marketable drug}) = \left(\frac{1999}{2000}\right)^{c}.$

However, there are N scientists. Thus,

$$P(\text{no marketable drug will be discovered in a year}) = \left[\left(\frac{1999}{2000}\right)^{c}\right]^{N} = \left(\frac{1999}{2000}\right)^{Nc}.$$

(f) This is the complement of the event in part (e), so the probability that at least one marketable drug will be discovered is

$$1 - \left(\frac{1999}{2000}\right)^{Nc},$$

where Nc is the number of drugs tested.

67. $P(C|F) = \dfrac{n(C \cap F)}{n(F)} = \dfrac{7}{229}$

69. $P(B'|H') = P((A \cup C \cup D)|F) = \dfrac{n((A \cup C \cup D) \cap F)}{n(F)} = \dfrac{34 + 7 + 150}{229} = \dfrac{191}{229}$

71. Draw the tree diagram.

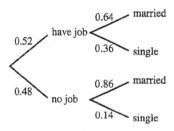

(a) $P(\text{married}) = P(\text{job and married}) + P(\text{no job and married}) = 0.52(0.64) + 0.48(0.86)$
$$= 0.3328 + 0.4128 = 0.7456$$

(b) $P(\text{job and single}) = 0.52(0.36) = 0.1872$

73. (a) In this exercise, it is easier to work with complementary events. Let E be the event "at least one of the faults erupts." Then the complementary event E' is "none of the faults erupts," and we can use $P(E) = 1 - P(E')$. Consider the event E': "none of the faults erupts." This means "the first fault does not erupt and the second fault does not erupt and...and the seventh fault does not erupt." Letting F_i denote the event "the ith fault erupts," we wish to find

$$P(E') = P(F_1' \cap F_2' \cap F_3' \cap F_4' \cap F_5' \cap F_6' \cap F_7').$$

Since we are assuming the events are independent, we have

$$P(E') = P(F_1' \cap F_2' \cap F_3' \cap F_4' \cap F_5' \cap F_6' \cap F_7') = P(F_1') \cdot P(F_2') \cdot P(F_3') \cdot P(F_4') \cdot P(F_5') \cdot P(F_6') \cdot P(F_7')$$

Now use $P(F_i') = 1 - P(F_i)$ and perform the calculations.

$$P(E') = P(F_1') \cdot P(F_2') \cdot P(F_3') \cdot P(F_4') \cdot P(F_5') \cdot P(F_6') \cdot P(F_7')$$
$$= (1 - 0.27) \cdot (1 - 0.21) \cdot \ldots \cdot (1 - 0.03)$$
$$= (0.73)(0.79)(0.89)(0.90)(0.96)(0.97)(0.97)$$
$$\approx 0.42$$

Therefore,

$$P(E) = 1 - P(E') \approx 1 - 0.42 \approx 0.58.$$

75. (a) $P(\text{second class}) = \dfrac{357}{1316} \approx 0.2713$

(b) $P(\text{surviving}) = \dfrac{499}{1316} \approx 0.3792$

(c) $P(\text{surviving|first class}) = \dfrac{203}{325} \approx 0.6246$

(d) $P(\text{surviving|child and third class}) = \dfrac{27}{79} \approx 0.3418$

(e) $P(\text{woman|first class and survived}) = \dfrac{140}{203} \approx 0.6897$

(f) $P(\text{third class|man and survived}) = \dfrac{75}{146} \approx 0.5137$

(g) $P(\text{survived|man}) = \dfrac{146}{805} \approx 0.1814$

$P(\text{survived|man and third class}) = \dfrac{75}{462} \approx 0.1623$

No, the events are not independent.

77. First draw the tree diagram.

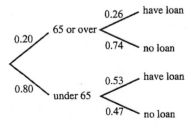

(a) $P(\text{person is 65 or over and has a loan}) = P(\text{65 or over}) \cdot P(\text{has loan|65 or over}) = 0.20(0.26) = 0.052$

(b) $P(\text{person has a loan}) = P(\text{65 or over and has loan}) + P(\text{under 65 and has loan})$
$$= 0.20(0.26) + 0.80(0.53) = 0.052 + 0.424 = 0.476$$

79.

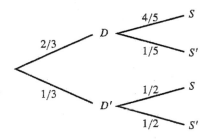

From the tree diagram, we see that the probability that a person

(a) drinks diet soft drinks is

$$\frac{2}{3}\left(\frac{4}{5}\right) + \frac{1}{3}\left(\frac{1}{2}\right) = \frac{8}{15} + \frac{1}{6} = \frac{21}{30} = \frac{7}{10};$$

(b) diets, but does not drink diet soft drinks is

$$\frac{2}{3}\left(\frac{1}{5}\right) = \frac{2}{15}.$$

81. Let F_i be the event "the ith burner fails." The event "all four burners fail" is equivalent to the event "the first burner fails <u>and</u> the second burner fails <u>and</u> the third burner fails <u>and</u> the fourth burner fails"—that is, the event $F_1 \cap F_2 \cap F_3 \cap F_4$. We are told that the burners are independent. Therefore

$$P(F_1 \cap F_2 \cap F_3 \cap F_4) = P(F_1) \cdot P(F_2) \cdot P(F_3) \cdot P(F_4) = (0.001)(0.001)(0.001)(0.001)$$
$$= 0.000000000001 = 10^{-12}.$$

83. $P(\text{luxury car}) = 0.04$ and $P(\text{luxury car}|\text{CPA}) = 0.17$

Use the formal definition of independent events. Since these probabilities are not equal, the events are not independent.

85. (a)

<table>
<tr><td>1-point kick
after first TD</td><td>2-point
conversion
after second TD</td><td>Outcome</td></tr>
</table>

```
                        2-point
      1-point kick      conversion
      after first TD    after second TD    Outcome

                              r      2      win
                      1  <
              k              1 - r   0      lose
         <
              1 - k          r       2      tie
                      0  <
                             1 - r   0      lose
```

From the tree diagram,

$$P(\text{win}) = kr,$$
$$P(\text{tie}) = (1 - k)r,$$
$$P(\text{lose}) = k(1 - r) + (1 - k)(1 - r) = k - kr + 1 - r - k + kr = 1 - r.$$

(b)

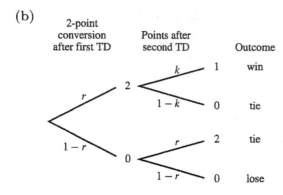

From the tree diagram,

$$P(\text{win}) = rk,$$
$$P(\text{tie}) = r(1-k) + (1-r)r = r - rk + r - r^2 = 2r - rk - r^2 = r(2 - k - r),$$
$$P(\text{lose}) = (1-r)(1-r) = (1-r)^2.$$

(c) $P(\text{win})$ is the same under both strategies.

(d) If $r < 1$, $(1-r) > (1-r)^2$. The probability of losing is smaller for the 2-point first strategy.

7.6 Bayes' Theorem

1. Use Bayes' theorem with two possibilities M and M'.

$$P(M|N) = \frac{P(M) \cdot P(N|M)}{P(M) \cdot P(N|M) + P(M') \cdot P(N|M')} = \frac{0.4(0.3)}{0.4(0.3) + 0.6(0.4)}$$
$$= \frac{0.12}{0.12 + 0.24} = \frac{0.12}{0.36} = \frac{12}{36} = \frac{1}{3}$$

3. Using Bayes' theorem,

$$P(R_1|Q) = \frac{P(R_1) \cdot P(Q|R_1)}{P(R_1) \cdot P(Q|R_1) + P(R_2) \cdot P(Q|R_2) + P(R_3) \cdot P(Q|R_3)}$$
$$= \frac{0.15(0.40)}{(0.15)(0.40) + 0.55(0.20) + 0.30(0.70)}$$
$$= \frac{0.06}{0.38} = \frac{6}{38} = \frac{3}{19}.$$

5. Using Bayes' theorem,

$$P(R_3|Q) = \frac{P(R_3) \cdot P(Q|R_3)}{P(R_1) \cdot P(Q|R_1) + P(R_2) \cdot P(Q|R_2) + P(R_3) \cdot P(Q|R_3)}$$
$$= \frac{0.30(0.70)}{(0.15)(0.40) + 0.55(0.20) + 0.30(0.70)}$$
$$= \frac{0.21}{0.38} = \frac{21}{38}.$$

7. We first draw the tree diagram and determine the probabilities as indicated below.

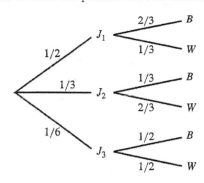

We want to determine the probability that if a white ball is drawn, it came from the second jar. This is $P(J_2|W)$. Use Bayes' theorem.

$$P(J_2|W) = \frac{P(J_2) \cdot P(W|J_2)}{P(J_2) \cdot P(W|J_2) + P(J_1) \cdot P(W|J_1) + P(J_3) \cdot P(W|J_3)} = \frac{\frac{1}{3} \cdot \frac{2}{3}}{\frac{1}{3} \cdot \frac{2}{3} + \frac{1}{2} \cdot \frac{1}{3} + \frac{1}{6} \cdot \frac{1}{2}}$$

$$= \frac{\frac{2}{9}}{\frac{2}{9} + \frac{1}{6} + \frac{1}{12}} = \frac{\frac{2}{9}}{\frac{17}{36}} = \frac{8}{17}$$

9. Let G represent "good worker," B represent "bad worker," S represent "pass the test," and F represent "fail the test." The given information if $P(G) = 0.70$, $P(B) = P(G') = 0.30$, $P(S|G) = 0.85$ (and therefore $P(F|G) = 0.15$), and $P(S|B) = 0.35$ (and therefore $P(F|B) = 0.65$). If passing the test is made a requirement for employment, then the percent of the new hires that will turn out to be good workers is

$$P(G|S) = \frac{P(G) \cdot P(S|G)}{P(G) \cdot P(S|G) + P(B) \cdot P(S|B)} = \frac{0.70(0.85)}{0.70(0.85) + 0.30(0.35)} = \frac{0.595}{0.700} = 0.85.$$

85% of new hires become good workers.

11. Let Q represent "qualified" and A represent "approved by the manager." Set up the tree diagram.

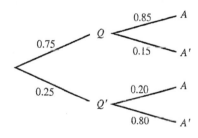

$$P(Q'|A) = \frac{P(Q') \cdot P(A|Q')}{P(Q) \cdot P(A|Q) + P(Q') \cdot P(A|Q')} = \frac{0.25(0.20)}{0.75(0.85) + 0.25(0.20)} = \frac{0.05}{0.6875} = \frac{4}{55} \approx 0.0727$$

13. Let D represent "damaged," A represent "from supplier A," and B represent "from supplier B." Set up the tree diagram.

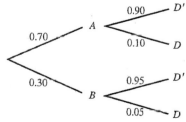

$$P(B|D) = \frac{P(B) \cdot P(D|B)}{P(B) \cdot P(D|B) + P(A) \cdot P(D|A)} = \frac{0.30(0.05)}{0.30(0.05) + 0.70(0.10)} = \frac{0.015}{0.015 + 0.07} = \frac{0.015}{0.085} \approx 0.1765$$

15. Start with the tree diagram, where the first state refers to the companies and the second to a defective appliance.

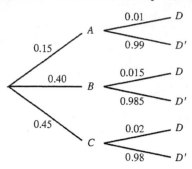

$$P(B|D) = \frac{P(B) \cdot P(D|B)}{P(A) \cdot P(D|A) + P(B) \cdot P(D|B) + P(C) \cdot P(D|C)}$$

$$= \frac{0.40(0.015)}{0.15(0.01) + 0.40(0.015) + 0.45(0.02)} = \frac{0.0060}{0.0165} \approx 0.3636$$

17. Let H represent "high rating," F_1 represent "sponsors college game," F_2 represent "sponsors baseball game," and F_3 represent "sponsors pro football game."

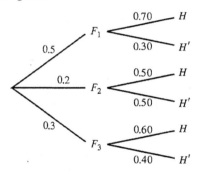

$$P(F_3|H) = \frac{P(F_3) \cdot P(H|F_3)}{P(F_1) \cdot P(H|F_1) + P(F_2) \cdot P(H|F_2) + P(F_3) \cdot P(H|F_3)} = \frac{0.3(0.60)}{0.5(0.70) + 0.2(0.50) + 0.3(0.60)}$$

$$= \frac{0.18}{0.35 + 0.10 + 0.18} = \frac{0.18}{0.63} = \frac{18}{63} = \frac{2}{7}$$

19. Using the given information, construct a table similar to the one in the previous exercise.

Category of Policyholder	Portion of Policyholders	Probability of Dying in the Next Year
Standard	0.50	0.010
Preferred	0.40	0.005
Ultra-preferred	0.10	0.001

Let S represent "standard policyholder,"
  R represent "preferred policyholder,"
  U represent "ultra-preferred policyholder,"
and D represent "policyholder dies in the next year."

We wish to find $P(U|D)$.

$$P(U|D) = \frac{P(U) \cdot P(D|U)}{P(S) \cdot P(D|S) + P(R) \cdot P(D|R) + P(U) \cdot P(D|U)} = \frac{0.10(0.001)}{0.50(0.010) + 0.40(0.005) + 0.10(0.001)}$$

$$= \frac{0.0001}{0.0071} \approx 0.141$$

The correct answer choice is **d**.

21. Let L be the event "the object was shipped by land," A be the event "the object was shipped by air," S be the event "the object was shipped by sea," and E be the event "an error occurred."

$$P(L|E) = \frac{P(L) \cdot P(E|L)}{P(L) \cdot P(E|L) + P(A) \cdot P(E|A) + P(S) \cdot P(E|S)}$$

$$= \frac{0.50(0.02)}{0.50(0.02) + 0.40(0.04) + 0.10(0.14)}$$

$$= \frac{0.0100}{0.0400} = 0.25$$

The correct response is **c**.

23. Let E represent the event "hemoccult test is positive," and let F represent the event "has colorectal cancer." We are given

$$P(F) = 0.003, P(E|F) = 0.5, \text{ and } P(E|F') = 0.03$$

and we want to find $P(F|E)$. Since $P(F) = 0.003$, $P(F') = 0.997$. Therefore,

$$P(F|E) = \frac{P(F) \cdot P(E|F)}{P(F) \cdot P(E|F) + P(F') \cdot P(E|F')} = \frac{0.003 \cdot 0.5}{0.003 \cdot 0.5 + 0.997 \cdot 0.03} \approx 0.0478.$$

25. Draw a tree diagram.

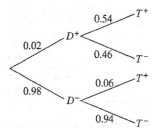

(a) $P(D^+|T^+) = \dfrac{0.02(0.54)}{0.02(0.54) + 0.98(0.06)} = \dfrac{0.0108}{0.0696} \approx 0.16$

(b) $P(D^-|T^-) = \dfrac{0.98(0.94)}{0.02(0.46) + 0.98(0.94)} = \dfrac{0.9212}{0.9304} \approx 0.99$

(c) $P(D^- \cap T^+) = 0.98(0.06) = 0.0588$
$0.0588(1000) = 58.8$

We would expect about 59 false positives in 1000 examinations.

27. (a) Let H represent "heavy smoker," L be "light smoker," N be "nonsmoker," and D be "person died." Let $x = P(D|N)$, that is, let x be the probability that a nonsmoker died. Then $P(D|L) = 2x$ and $P(D|H) = 4x$. Create a table.

Level of Smoking	Probability of Level	Probability of Death for Level
H	0.2	$4x$
L	0.3	$2x$
N	0.5	x

We wish to find $P(H|D)$.

$$P(H|D) = \frac{P(H) \cdot P(D|H)}{P(H) \cdot P(D|H) + P(L) \cdot P(D|L) + P(N) \cdot P(D|N)}$$

$$= \frac{0.2(4x)}{0.2(4x) + 0.3(2x) + 0.5(x)} = \frac{0.8x}{1.9x} \approx 0.42$$

The correct answer choice is **d**.

29. Let H represent "person has the disease" and R be "test indicates presence of the disease." We wish to determine $P(H|R)$.

Construct a table as before.

Category of Person	Portion of Population	Probability of Presence of Disease
H	0.01	0.950
H'	0.99	0.005

$$P(H|R) = \frac{P(H) \cdot P(R|H)}{P(H) \cdot P(R|H) + P(H') \cdot P(R|H')} = \frac{0.01(0.950)}{0.01(0.950) + 0.99(0.005)} = \frac{0.00950}{0.01445} \approx 0.657$$

The correct answer choice is **b**.

31. Start with the tree diagram.

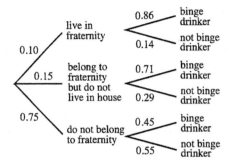

(a) $P(\text{binge drinker}) = 0.10(0.86) + 0.15(0.71) + 0.75(0.45) = 0.53$

(b) $P(\text{lives in fraternity}|\text{binge drinker}) = \dfrac{0.10(0.86)}{0.53} \approx 0.1623$

33. $P(\text{between 35 and 44}|\text{never married}) = \dfrac{0.212(0.195)}{0.135(0.0895) + 0.191(0.434) + 0.212(0.195) + 0.320(0.088) + 0.142(0.074)}$

$$= \frac{0.041340}{0.283727} \approx 0.1457$$

35. $P(\text{between 45 and 64}|\text{never married}) = \dfrac{0.316(0.075)}{0.123(0.806) + 0.179(0.311) + 0.203(0.132) + 0.316(0.075) + 0.179(0.037)}$

$$= \frac{0.023700}{0.211926} \approx 0.1118$$

37. Draw a tree diagram.

$$P(\text{not wearing seat belt}|\text{unharmed}) = \frac{0.51(0.09)}{0.49(0.29) + 0.51(0.09)} \approx 0.2441$$

39.

Category	Proportion of Population	Probability of Being Picked Up
Has terrorist ties	$\frac{1}{1,000,000}$	0.99
Does not have terrorists ties	$\frac{999,999}{1,000,000}$	0.01

$$P(\text{Has terrorist ties}|\text{Picked up}) = \frac{\frac{1}{1,000,000}(0.99)}{\frac{1}{1,000,000}(0.99) + \frac{999,999}{1,000,000}(0.01)}$$

$$= \frac{\frac{1}{1,000,000}(0.99)}{\frac{1}{1,000,000}(0.99) + \frac{999,999}{1,000,000}(0.01)} \cdot \frac{1,000,000}{1,000,000}$$

$$= \frac{0.99}{10,000.98}$$

$$\approx 9.9 \times 10^{-5}$$

Chapter 7 Review Exercises

1. $9 \in \{8, 4, -3, -9, 6\}$

Since 9 is not an element of the set, this statement is false.

3. $2 \notin \{0, 1, 2, 3, 4\}$

Since 2 is an element of the set, this statement is false.

5. $\{3, 4, 5\} \subseteq \{2, 3, 4, 5, 6\}$

Every element of $\{3, 4, 5\}$ is an element of $\{2, 3, 4, 5, 6\}$, so this statement is true.

7. $\{3, 6, 9, 10\} \subseteq \{3, 9, 11, 13\}$

10 is an element of $\{3, 6, 9, 10\}$, but 10 is not an element of $\{3, 9, 11, 13\}$. Therefore, $\{3, 6, 9, 10\}$ is not a subset of $\{3, 9, 11, 13\}$. The statement is false.

9. $\{2, 8\} \nsubseteq \{2, 4, 6, 8\}$

Since both 2 and 8 are elements of $\{2, 4, 6, 8\}$, $\{2, 8\}$ is a subset of $\{2, 4, 6, 8\}$. This statement is false.

In Exercises 11−19,

$$U = \{a, b, c, d, e, f, g, h\},$$
$$K = \{c, d, e, f, h\},$$
$$\text{and} \quad R = \{a, c, d, g\}.$$

11. K has 5 elements, so it has $2^5 = 32$ subsets.

13. K' (the complement of K) is the set of all elements of U that do *not* belong to K.

$$K' = \{a, b, g\}$$

15. $K \cap R$ (the intersection of K and R) is the set of all elements belonging to both set K and set R.

$$K \cap R = \{c, d\}$$

17. $(K \cap R)' = \{a, b, e, f, g, h\}$ since these elements are in U but not in $K \cap R$. (See Exercise 15.)

19. $\emptyset' = U$

21. $A \cap C$ is the set of all female employees in the K.O. Brown Company who are in the accounting department.

23. $A \cup D$ is the set of all employees in the K.O. Brown Company who are in the accounting department *or* have MBA degrees.

25. $B' \cap C'$ is the set of all male employees who are not in the sales department.

27. $A \cup B'$ is the set of all elements which belong to A or do not belong to B, or both.

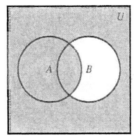

$A \cup B'$

29. $(A \cap B) \cup C$

First find $A \cap B$.

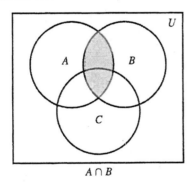

$A \cap B$

Now find the union of this region with C.

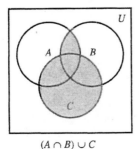

$(A \cap B) \cup C$

31. The sample space for rolling a die is

$$S = \{1, 2, 3, 4, 5, 6\}.$$

33. The sample space of the possible weights is

$$S = \{0, 0.5, 1, 1.5, 2, \ldots, 299.5, 300\}.$$

35. The sample space consists of all ordered pairs (a, b) where a can be 3, 5, 7, 9, or 11, and b is either R (red) or G(green). Thus,

$$S = \{(3,R), (3,G), (5,R), (5,G), (7,R),$$
$$(7,G), (9,R), (9,G), (11,R), (11,G)\}.$$

37. The event F that the second ball is green is

$$F = \{(3,G), (5,G), (7,G), (9,G), (11,G)\}.$$

39. There are 13 hearts out of 52 cards in a deck. Thus,

$$P(\text{heart}) = \frac{13}{52} = \frac{1}{4}.$$

41. There are 3 face cards in each suit (jack, queen, and king) and there are 4 suits, so there are $3 \cdot 4 = 12$ face cards out of the 52 cards. Thus,

$$P(\text{face card}) = \frac{12}{52} = \frac{3}{13}.$$

43. There are 4 queens of which 2 are red, so

$$P(\text{red}|\text{queen}) = \frac{n(\text{red and queen})}{n(\text{queen})} = \frac{2}{4} = \frac{1}{2}.$$

45. There are 4 kings of which all 4 are face cards. Thus,

$$P(\text{face card}|\text{king}) = \frac{n(\text{face card and king})}{n(\text{king})} = \frac{4}{4} = 1.$$

51. Marilyn vos Savant's answer is that the contestant should switch doors. To understand why, recall that the puzzle begins with the contestant choosing door 1 and then the host opening door 3 to reveal a goat. When the host opens door 3 and shows the goat, that does not affect the probability of the car being behind door 1; the contestant had a $\frac{1}{3}$ probability of being correct to begin with, and he still has a $\frac{1}{3}$ probability after the host opens door 3.

The contestant knew that the host would open another door regardless of what was behind door 1, so opening either other door gives no new information about door 1. The probability of the car being behind door 1 is still $\frac{1}{3}$; with the goat behind door 3, the only other place the car could be is behind door 2, so the probability that the car is behind door 2 is now $\frac{2}{3}$. By switching to door 2, the contestant can double his chances of winning the car.

53. Let E represent the event "draw a black jack." $P(E) = \frac{2}{52} = \frac{1}{26}$ and then $P(E') = \frac{25}{26}$. The odds in favor of drawing a black jack are

$$\frac{P(E)}{P(E')} = \frac{\frac{1}{26}}{\frac{25}{26}} = \frac{1}{25},$$

or 1 to 25.

55. The sum is 8 for each of the 5 outcomes 2-6, 3-5, 4-4, 5-3, and 6-2. There are 36 outcomes in all in the sample space.

$$P(\text{sum is 8}) = \frac{5}{36}$$

57. $P(\text{sum is at least 10})$
$$= P(\text{sum is 10}) + P(\text{sum is 11})$$
$$+ P(\text{sum is 12})$$
$$= \frac{3}{36} + \frac{2}{36} + \frac{1}{36}$$
$$= \frac{6}{36} = \frac{1}{6}$$

59. The sum can be 9 or 11. $P(\text{sum is 9}) = \frac{4}{36}$ and $P(\text{sum is 11}) = \frac{2}{36}$.

$P(\text{sum is odd number greater than 8})$
$$= \frac{4}{36} + \frac{2}{36}$$
$$= \frac{6}{36} = \frac{1}{6}$$

61. Consider the reduced sample space of the 11 outcomes in which at least one die is a four. Of these, 2 have a sum of 7, 3-4 and 4-3. Therefore,

$P(\text{sum is 7}|\text{at least one die is a 4})$
$$= \frac{2}{11}.$$

63. $P(E) = 0.51$, $P(F) = 0.37$, $P(E \cap F) = 0.22$

(a) $P(E \cup F) = P(E) + P(F) - P(E \cap F)$
$$= 0.51 + 0.37 - 0.22$$
$$= 0.66$$

(b) Draw a Venn diagram.

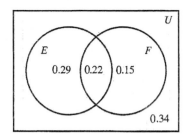

$E \cap F'$ is the portion of the diagram that is inside E and outside F.

$$P(E \cap F') = 0.29$$

(c) $E' \cup F$ is outside E or inside F, or both.

$$P(E' \cup F) = 0.22 + 0.15 + 0.34 = 0.71.$$

(d) $E' \cap F'$ is outside E and outside F.

$$P(E' \cap F') = 0.34$$

65. The probability that the ball came from box B, given that it is red, is

$P(B|\text{red})$

$$= \frac{P(B) \cdot P(\text{red}|B)}{P(B) \cdot P(\text{red}|B) + P(A) \cdot P(\text{red}|A)}$$

$$= \frac{\frac{5}{8}\left(\frac{2}{5}\right)}{\frac{5}{8}\left(\frac{2}{5}\right) + \frac{3}{8}\left(\frac{5}{6}\right)}$$

$$= \frac{4}{9}.$$

67. First make a tree diagram letting C represent "a competent shop" and R represent "an appliance is repaired correctly."

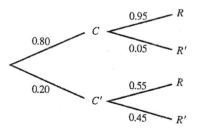

To obtain $P(C|R)$, use Bayes' theorem.

$$P(C|R) = \frac{P(C) \cdot P(R|C)}{P(C) \cdot P(R|C) + P(C') \cdot P(R|C')}$$

$$= \frac{0.80(0.95)}{0.80(0.95) + 0.20(0.55)}$$

$$= \frac{0.76}{0.87} \approx 0.8736$$

69. Refer to the tree diagram for Exercise 67. Use Bayes' theorem.

$$P(C|R') = \frac{P(C) \cdot P(R'|C)}{P(C) \cdot P(R'|C) + P(C') \cdot P(R'|C')}$$

$$= \frac{0.80(0.05)}{0.80(0.05) + 0.20(0.45)}$$

$$= \frac{0.04}{0.13} \approx 0.3077$$

71. To find $P(R)$, use

$$P(R) = P(C) \cdot P(R|C) + P(C') \cdot P(R|C')$$
$$= 0.80(0.95) + 0.20(0.55) = 0.87.$$

73. **(a)** "A customer buys neither machine" may be written $(E \cup F)'$ or $E' \cap F'$.

(b) "A customer buys at least one of the machines" is written $E \cup F$.

75. Use Bayes' theorem to find the required probabilities.

(a) Let D be the event "item is defective" and E_k be the event "item came from supplier k," $k = 1, 2, 3, 4$.

$$P(D) = P(E_1) \cdot P(D|E_1) + P(E_2) \cdot P(D|E_2)$$
$$+ P(E_3) \cdot P(D|E_3) + P(E_4) \cdot P(D|E_4)$$
$$= 0.17(0.01) + 0.39(0.02) + 0.35(0.05)$$
$$+ 0.09(0.03)$$
$$= 0.0297$$

(b) Find $P(E_4|D)$. Using Bayes' theorem, the numerator is

$$P(E_4) \cdot P(D|E_4) = 0.09(0.03) = 0.0027.$$

The denominator is $P(E_1) \cdot P(D|E_1) + P(E_2) \cdot P(D|E_2) + P(E_3) \cdot P(D|E_3) + P(E_4) \cdot P(D|E_4)$, which, from part (a), equals 0.0297.

Therefore,

$$P(E_4|D) = \frac{0.0027}{0.0297} \approx 0.0909.$$

(c) Find $P(E_2|D)$. Using Bayes' theorem with the same denominator as in part (a),

$$P(E_2|D) = \frac{P(E_2) \cdot P(D|E_2)}{0.0418}$$
$$= \frac{0.39(0.02)}{0.0297}$$
$$= \frac{0.0078}{0.0297}$$
$$\approx 0.2626.$$

(d) Since $P(D) = 0.0297$ and $P(D|E_4) = 0.03$,

$$P(D) \neq P(D|E_4)$$

Therefore, the events are not independent.

77. Let E represent "customer insures exactly one car" and S represent "customer insures a sports car." Let x be the probability that a customer who insures exactly one car insures a sports car, or $P(S|E)$. Make a tree diagram.

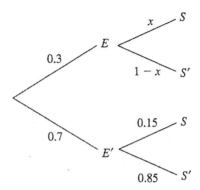

We are told that 20% of the customers insure a sports car, or $P(S) = 0.20$.

$$P(S) = P(E) \cdot P(S|E) + P(E') \cdot P(S|E')$$
$$0.20 = 0.30(x) + 0.70(0.15)$$
$$0.20 = 0.3x + 0.105$$
$$0.3x = 0.095$$
$$x \approx 0.316666667$$

Therefore, the probability that a customer insures a car other than a sports car is

$$P(S') = 1 - P(S)$$
$$\approx 1 - 0.316666667$$
$$= 0.683333333.$$

Finally, the probability that a randomly selected customer insures exactly one car and that car is not a sports car is

$$P(E \cap S') = P(E) \cdot P(S')$$
$$\approx 0.3 \cdot 0.683333333$$
$$\approx 0.21.$$

The correct answer choice is **b**.

79. Let C represent "the automobile owner purchases collision coverage" and D represent "the automobile owner purchases disability coverage." We want to find $P(C' \cap D') = P[(C \cup D)'] = 1 - P(C \cup D)$. We are given that $P(C) = 2 \cdot P(D)$ and that $P(C \cap D) = 0.15$. Let $x = P(D)$.

$$P(C \cap D) = P(C) \cdot P(D)$$
$$0.15 = 2x \cdot x$$
$$0.075 = x^2$$
$$x = \sqrt{0.075}$$
$$x \approx 0.2739$$

So $P(D) = x \approx 0.27$ and $P(C) = 2x \approx 0.55$. Using the union rule for probability,

$$P(C \cup D) = P(C) + P(D) - P(C \cap D) = 0.55 + 0.27 - 0.15 = 0.67.$$

Finally, $P(C' \cap D') = 1 - 0.67 = 0.33$. The correct answer choice is **b**.

81. (a)

	N_2	T_2
N_1	$N_1 N_2$	$N_1 T_2$
T_1	$T_1 N_2$	$T_1 T_2$

Since the four combinations are equally likely, each has probability $\frac{1}{4}$.

(b) $P(\text{two trait cells}) = P(T_1 T_2) = \frac{1}{4}$

(c) $P(\text{one normal cell and one trait cell}) = P(N_1 T_2) + P(T_1 N_2) = \frac{1}{4} + \frac{1}{4} = \frac{1}{2}$

(d) $P(\text{not a carrier and does not have the disease}) = P(N_1 N_2) = \frac{1}{4}$

83. Let D represent "man died from causes related to heart desease" and E represent "at least one parent suffered from heart disease." We want to find $P(D|E')$. We are given that $n(U) = 937, n(D) = 210, n(E) = 312,$ and $n(U \cap E) = 102$. Construct a table and calculate the missing information.

	E	E'	Totals
D	102	108	210
$D\prime$	210	517	727
Totals	312	625	937

Therefore,

$$P(D|E') = \frac{P(D \cap E')}{P(E')} = \frac{\frac{108}{937}}{\frac{625}{937}} = \frac{108}{625} \approx 0.173.$$

The correct answer choice is **b**.

85. Let B represent "the person voted for Bush,"

K represent "the person voted for Kerry,"

O represent "the person voted for another candidate,"

and M represent "the person was male."

Use a tree diagram and fill in the missing information.

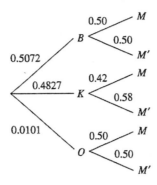

(a) $P(M) = P(B) \cdot P(M|B) + P(K) \cdot P(M|K) + P(O) \cdot P(M|O)$

$\qquad = 0.5072(0.50) + 0.4827(0.42) + 0.0101(0.50)$

$\qquad \approx 0.4614$

About 46.14% of the voters were male.

(b) $P(B|M) = \dfrac{P(B) \cdot P(M|B)}{P(B) \cdot P(M|B) + P(K) \cdot P(M|K) + P(O) \cdot P(M|O)}$

$\qquad = \dfrac{P(B) \cdot P(M|B)}{P(M)} \approx \dfrac{0.5072(0.50)}{0.4614} \approx 0.5497$

(c) First, note that $P(\text{female}) = P(M') = 1 - P(M) = 1 - 0.4614 = 0.5386$. Therefore,

$$P(B|M') = \frac{P(B) \cdot P(M'|B)}{P(M')} \approx \frac{0.5072(0.50)}{0.5386} \approx 0.4708.$$

87. (a) $P(\text{answer yes}) = P(\text{answer } B) \cdot P(\text{answer yes}|\text{answer } B) + P(\text{answer } A) \cdot P(\text{answer yes}|\text{answer } A)$

Divide by $P(\text{answer } B)$.

$\dfrac{P(\text{answer yes})}{P(\text{answer } B)} = P(\text{answer yes}|\text{answer } B) + \dfrac{P(\text{answer } A) \cdot P(\text{answer yes}|\text{answer } A)}{P(\text{answer } B)}$

Solve for $P(\text{answer yes}|\text{answer } B)$.

$P(\text{answer yes}|\text{answer } B) = \dfrac{P(\text{answer yes}) - P(\text{answer } A) \cdot P(\text{answer yes}|\text{answer } A)}{P(\text{answer } B)}$

(b) Using the formula from part (a),

$$\frac{0.6 - \frac{1}{2}\left(\frac{1}{2}\right)}{\frac{1}{2}} = \frac{7}{10}.$$

89. In calculating the probability of two babies in a family would die of SIDS is $(1/8543)^2$, he assumed that the events that either infant died of SIDS are independent. There may be a genetic factor, in which case the events are dependent.

91. (a) $P(\text{making a 1st down with } n \text{ yards to go}) = \dfrac{\text{number of successes}}{\text{number of trials}}$

n	Trials	Successes	Probability of Making First Down with n Yards to Go
1	543	388	$\frac{388}{543} \approx 0.7145$
2	327	186	$\frac{186}{327} \approx 0.5688$
3	356	146	$\frac{146}{356} \approx 0.4101$
4	302	97	$\frac{97}{302} \approx 0.3212$
5	336	91	$\frac{91}{336} \approx 0.2708$

93. Let L be the set of songs about love,

          P be the set of songs about prison,

and T be the set of songs about trucks.

We are given the following information.

$$n(L \cap P \cap T) = 12$$
$$n(L \cap P) = 13$$
$$n(L) = 28$$
$$n(L \cap T) = 18$$
$$n(L' \cap P) = 5$$
$$n(P) = 18$$
$$n(P \cap T) = 15$$
$$n(P' \cap T) = 16$$

Start with $L \cap P \cap T$: $n(L \cap P \cap T) = 12$.
Since $n(L \cap P) = 13$, $n(L \cap P \cap T') = 1$.
Since $n(L \cap T) = 18$, $n(L \cap P' \cap T) = 6$.
Since $n(L) = 28$, $n(L \cap P' \cap T') = 28 - 12 - 1 - 6 = 9$.
Since $n(P \cap T) = 15$, $n(L' \cap P \cap T) = 3$.
Since $n(P) = 18$, $n(L' \cap P \cap T') = 18 - 1 - 12 - 3 = 2$.
Since $n(P' \cap T) = 16$, $n(L' \cap P' \cap T) = 16 - 6 = 10$.
Finally, $n(L' \cap P' \cap T') = 8$.

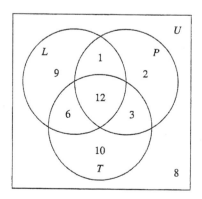

(a) $n(U) = 9 + 1 + 2 + 6 + 12 + 3 + 10 + 8 = 51$

(b) $n(T) = 6 + 12 + 3 + 10 = 31$

(c) $n(P) = 1 + 2 + 12 + 3 = 18$

(d) $n(T \cap P) = 12 + 3 = 15$

(e) $n(P') = n(U) - n(P) = 51 - 18 = 33$

(f) $n(L') = n(U) - n(L) = 51 - 28 = 23$

95. (a) $P(\text{double miss}) = 0.05(0.05) = 0.0025$

(b) $P(\text{specific silo destroyed}) = 1 - P(\text{double miss}) = 1 - 0.0025 = 0.9975$

(c) $P(\text{all ten destroyed}) = (0.9975)^{10} \approx 0.9753$

(d) $P(\text{at least one survived})$
$$= 1 - P(\text{none survived}) = 1 - P(\text{all ten destroyed})$$
$$= 1 - 0.9753 = 0.0247 \quad \text{or} \quad 2.47\%$$

This does not agree with the quote of a 5% chance that at least one would survive.

(e) The events that each of the two bombs hit their targets are assumed to be independent. The events that each silo is destroyed are assumed to be independent.

Chapter 7 Test

[7.1]

1. Write *true* or *false* for each statement.

(a) $3 \in \{1, 5, 7, 9\}$ (b) $\{1, 3\} \not\subset \{0, 1, 2, 3, 4\}$ (c) $\emptyset \subset \{2\}$

(d) A set of 6 distinct elements has exactly 64 subsets.

2. Let $U = \{1, 2, 3, 4, 5, 6, 7, 8, 9\}$, $A = \{1, 3, 4, 5\}$, $B = \{2, 4, 5\}$, and $C = \{1, 3, 5, 7\}$.
List the members of each of the following sets, using set braces.

(a) $A \cap B'$ (b) $A \cap (B \cup C')$

[7.2]

3. Draw a Venn diagram and shade the region that represents $A \cap (B \cup C')$.

4. Draw a Venn diagram and fill in the number of elements in each region given
that $n(U) = 25, n(A) = 11, n(B \cap A') = 9$, and $n(A \cap B) = 6$.

5. A survey of 70 children obtained the following results:

> 32 play soccer;
> 29 play basketball;
> 13 play tennis only;
> 6 play all three sports;
> 18 play soccer and basketball;
> 15 play soccer and tennis;
> 10 play basketball and tennis;
> 14 play none of the three sports.

Use a Venn diagram to answer the following questions.

(a) How many children play basketball only?

(b) How many children play tennis and basketball, but not soccer?

(c) How many children play soccer and tennis, but not basketball?

(d) How many children play exactly one of the three sports?

[7.3]

6. A single fair die is rolled. Find the probabilities of the following events.

(a) Getting a 4

(b) Getting an even number

(c) Getting a number less than 5

(d) Getting a number greater than 1

(e) Getting any number except 4 or 5

(f) Getting a number less than 8

[7.4]

7. Suppose that for events A and B, $P(A) = 0.4$, $P(B) = 0.3$ and $P(A \cup B) = 0.68$. Find each of the following probabilities.

   (a) $P(A \cap B)$ (b) $P(A')$

   (c) $P(A \cap B')$ (d) $P(A' \cap B')$

8. An urn contains 4 red, 3 blue, and 2 yellow marbles. A single marble is drawn.

   (a) Find the odds in favor of drawing a red marble.

   (b) Find the probability that a red or a blue marble is drawn.

[7.5]

9. For events E and F, $P(E) = 0.4$, $P(F) = 0.5$, and $P(E \cup F) = 0.8$. Find

   (a) $P(E|F)$ (b) $P(F|E')$.

10. Three cards are drawn without replacement from a standard deck of 52.

    (a) What is the probability that all three are spades?

    (b) What is the probability that all three are spades, given that the first card drawn is a spade?

11. The probability of passing the University of Waterloo's physical fitness test is 0.3. If you fail the first time, your chances of passing on the second try drop to 0.1. Draw a tree diagram and compute the probability that a person will pass on the first or second try.

[7.6]

12. The Magnum Opus Publishing Company uses three printers to put out its lengthy tomes. Printer A produces 40% of their books with a 20% failure rate. Printer B produces 25% with a 10% failure rate. Printer C produces the remainder with a 40% failure rate. Given that a book is badly printed, what is the probability that it was printed by Printer C?

Chapter 7 Test Answers

1. (a) False   (b) False   (c) True   (d) True

2. (a) $\{1,3\}$ (b) $\{4,5\}$

3.

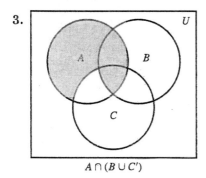

$A \cap (B \cup C')$

4.

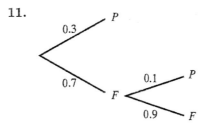

5. (a) 7   (b) 4   (c) 9   (d) 25

6. (a) $\frac{1}{6}$ (b) $\frac{1}{2}$ (c) $\frac{2}{3}$ (d) $\frac{5}{6}$ (e) $\frac{2}{3}$ (f) 1

7. (a) 0.02   (b) 0.6   (c) 0.38   (d) 0.32

8. (a) 4 to 5   (b) $\frac{7}{9}$

9. (a) $\frac{1}{5}$ (b) $\frac{2}{3}$

10. (a) $\left(\frac{13}{52}\right)\left(\frac{12}{51}\right)\left(\frac{11}{50}\right) \approx 0.013$

   (b) $\left(\frac{12}{51}\right)\left(\frac{11}{50}\right) \approx 0.052$

11.

$P(\text{pass}) = 0.3 + 0.07 = 0.37$

12. $\dfrac{(0.35)(0.4)}{(0.35)(0.4)+(0.4)(0.2)+(0.25)(0.1)} \approx 0.57$

Chapter 8

COUNTING PRINCIPLES; FURTHER PROBABILITY TOPICS

8.1   The Multiplication Principle; Permutations

1. $6! = 6 \cdot 5 \cdot 4 \cdot 3 \cdot 2 \cdot 1 = 720$

3. $15! = 15 \cdot 14 \cdot 13 \cdot 12 \cdot 11 \cdot 10 \cdot 9 \cdot 8 \cdot 7$
$\cdot 6 \cdot 5 \cdot 4 \cdot 3 \cdot 2 \cdot 1$
$\approx 1.308 \cdot 10^{12}$

5. $P(13, 2) = \dfrac{13!}{(13 - 2)!} = \dfrac{13!}{11!}$

$\qquad = \dfrac{13 \cdot 12 \cdot 11!}{11!}$

$\qquad = 156$

7. $P(38, 17) = \dfrac{38!}{(38 - 17)!} = \dfrac{38!}{21!}$

$\qquad\qquad \approx 1.024 \cdot 10^{25}$

9. $P(n, 0) = \dfrac{n!}{(n - 0)!} = \dfrac{n!}{n!} = 1$

11. $P(n, 1) = \dfrac{n!}{(n - 1)!} = \dfrac{n(n - 1)!}{(n - 1)!} = n$

13. By the multiplication principle, there will be $6 \cdot 3 \cdot 2 = 36$ different home types available.

15. There are 4 choices for the first name and 5 choices for the middle name, so, by the multiplication principle, there are $4 \cdot 5 = 20$ possible arrangements.

19. In Example 7, there are only 3 unordered 2-letter subsets of letters A, B, and C. They are AB, AC, and BC.

21. Use the formula for distinguishable permutations. The number of different "words" is

$$\frac{n!}{n_1! n_2! n_3! n_4!} = \frac{13!}{5! 4! 2! 2!} = 540,540.$$

23. (a) Since there are 14 distinguishable objects to be arranged, use permutations. The number of arrangements is

$$P(14, 14) = 14! = 87,178,291,000$$
$$\text{or } 8.7178291 \times 10^{10}.$$

(b) There are 3! ways to arrange the pyramids among themselves, 4! ways to arrange the cubes, and 7! ways to arrange the spheres. We must also consider the number of ways to arrange the order of the three groups of shapes. This can be done in 3! ways. Using the multiplication principle, the number of arrangements is

$$3! 4! 7! 3! = 6 \cdot 24 \cdot 5040 \cdot 6$$
$$= 4,354,560.$$

(c) In this case, all of the objects that are the same shape are indistinguishable. Use the formula for distinguishable permutations. The number of distinguishable arrangements is

$$\frac{n!}{n_1! n_2! n_3!} = \frac{14!}{3! 4! 7!} = 120,120.$$

(d) There are 3 choices for the pyramid, 4 for the cube, and 7 for the sphere. The total number of ways is

$$3 \cdot 4 \cdot 7 = 84.$$

(e) From part (d) there are 84 ways if the order does not matter. There are 3! ways to choose the order. Using the multiplication principle, the number of possible ways is

$$84 \cdot 3! = 84 \cdot 6 = 504.$$

25. $10! = 10 \cdot 9!$

To find the value of 10!, multiply the value of 9! by 10.

27. (a) The number 13! has 2 factors of five so there must be 2 ending zeros in the answer.

(b) The number 27! has 6 factors of five (one each in 5, 10, 15, and 20 and two factors in 25), so there must be 6 ending zeros in the answer.

(c) The number 75! has $15 + 3 = 18$ factors of five (one each in 5, 10,..., 75 and two factors each in 25, 50, and 75), so there must be 18 ending zeros in the answer.

29. Use the multiplication principle. There are

$$8 \cdot 7 \cdot 4 \cdot 5 = 1120$$

varieties of automobile available.

31. If each species were to be assigned 3 initials, since there are 26 different letters in the alphabet, there could be $26^3 = 17,576$ different 3-letter designations. This would not be enough. If 4 initials were used, the biologist could represent $26^4 = 456,976$ different species, which is more than enough. Therefore, the biologist should use at least 4 initials.

33. The number of ways to seat the people is

$$P(6,6) = \frac{6!}{0!} = \frac{6!}{1}$$
$$= 6 \cdot 5 \cdot 4 \cdot 3 \cdot 2 \cdot 1$$
$$= 720.$$

35. The number of ways to arrange a schedule of 3 classes is

$$P(6,3) = \frac{6!}{(6-3)!} = \frac{6!}{3!}$$
$$= \frac{6 \cdot 5 \cdot 4 \cdot 3!}{3!}$$
$$= 120.$$

37. The number of possible batting orders is

$$P(19,9) = \frac{19!}{(19-9)!} = \frac{19!}{10!}$$
$$= 33{,}522{,}128{,}640$$
$$\approx 3.352 \times 10^{10}.$$

39. **(a)** The number of ways 5 works can be arranged is

$$P(5,5) = 5! = 120.$$

(b) If one of the 2 overtures must be chosen first, followed by arrangements of the 4 remaining pieces, then

$$P(2,1) \cdot P(4,4) = 2 \cdot 24 = 48$$

is the number of ways the program can be arranged.

41. By the multiplication principle, a person could schedule the evening of television viewing in

$$8 \cdot 5 \cdot 7 = 280$$

different ways.

43. **(a)** There are 5 odd digits: 1, 3, 5, 7, and 9. There are 7 decisions to be made, one for each digit; there are 5 choices for each digit. Thus, $5^7 = 78,125$ phone numbers are possible.

(b) The first digit has 9 possibilities, since 0 is not allowed; the middle 5 digits each have 10 choices; the last digit must be 0. Thus, there are

$$9 \cdot 10^5 \cdot 1 = 900,000$$

possible phone numbers.

(c) Solve as in part (b), except that the last *two* digits must be 0; therefore there are

$$9 \cdot 10^4 \cdot 1 \cdot 1 = 90,000$$

possible phone numbers.

(d) There are no choices for the first three digits; thus,
$$1^3 \cdot 10^4 = 10,000$$

phone numbers are possible.

(e) The first digit cannot be 0; in the absence of repetitions there are 9 choices for the second digit, and the choices decrease by one for each subsequent digit. The result is

$$9 \cdot 9 \cdot 8 \cdot 7 \cdot 6 \cdot 5 \cdot 4 = 544,320$$

phone numbers.

45. There are 8 choices for the first digit, since it cannot be 0 or 1. Since restrictions are eliminated for the second digit, there are 10 possibilities for each of the second and third digits. Thus, the total number of area codes would be

$$8 \cdot 10 \cdot 10 = 800.$$

47. Since a social security number has 9 digits with no restrictions, there are

$$10^9 = 1,000,000,000 \text{ (1 billion)}$$

different social security numbers. Yes, this is enough for every one of the 281 million people in the United States to have a social security number.

49. Since a zip code has nine digits with no restrictions, there are

$$10^9 = 1,000,000,000$$

different 9-digit zip codes.

51. Since a 20-sided die is rolled 12 times, the number of possible games is

$$20^{12} \quad \text{or} \quad 4.096 \cdot 10^{15} \text{ games.}$$

53. **(a)** The number of different circuits is $P(9,9)$ since we do not count the city he is starting in.

$$P(9,9) = 9! = 362,880$$

is the number of different circuits.

(b) He must check half of the circuits since, for each circuit, there is a corresponding one in the reverse order. Therefore,

$$\frac{1}{2}(362,880) = 181,440$$

circuits should be checked.

(c) No, it would not be feasible.

8.2   Combinations

3. $\dbinom{12}{5} = \dfrac{12!}{(12-5)!5!} = \dfrac{12!}{7!5!}$

$\qquad = \dfrac{12 \cdot 11 \cdot 10 \cdot 9 \cdot 8 \cdot 7!}{7! \cdot 5 \cdot 4 \cdot 3 \cdot 2 \cdot 1}$

$\qquad = 792$

5. $\dbinom{40}{18} = \dfrac{40!}{(40-18)!18!}$

$\qquad = \dfrac{40!}{22!18!}$

$\qquad \approx 1.134 \cdot 10^{11}$

7. $\dbinom{n}{n} = \dfrac{n!}{(n-n)!n!} = \dfrac{n!}{0!n!} = \dfrac{n!}{1 \cdot n!} = 1$

9. $\dbinom{n}{n-1} = \dfrac{n!}{[n-(n-1)]!(n-1)!}$

$\qquad = \dfrac{n!}{(n-n+1)!(n-1)!}$

$\qquad = \dfrac{n(n-1)!}{1!(n-1)!}$

$\qquad = n$

11. **(a)** There are

$$\binom{5}{2} = \frac{5!}{3!2!} = \frac{5 \cdot 4 \cdot 3!}{3! \cdot 2 \cdot 1} = 10$$

different 2-card combinations possible.

(b) The 10 possible hands are

$$\{1,2\}, \{2,3\}, \{3,4\}, \{4,5\}, \{1,3\},$$
$$\{2,4\}, \{3,5\}, \{1,4\}, \{2,5\}, \{1,5\}.$$

Of these, 7 contain a card numbered less than 3.

13. Choose 2 letters from $\{L, M, N\}$; order is important.

(a)

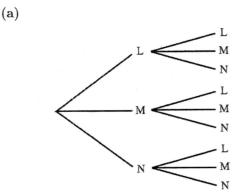

There are 9 ways to choose 2 letters if repetition is allowed.

(b)

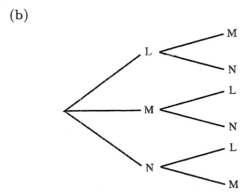

There are 6 ways to choose 2 letters if no repeats are allowed.

(c) The number of 3 elements taken 2 at a time is

$$\binom{3}{2} = \frac{3!}{1!2!} = 3.$$

This answer differs from both parts (a) and (b).

17. Order does not matter in choosing members of a committee, so use combinations rather than permutations.

(a) The number of committees whose members are all men is

$$\binom{9}{5} = \frac{9!}{4!5!} = \frac{9 \cdot 8 \cdot 7 \cdot 6 \cdot 5!}{4 \cdot 3 \cdot 2 \cdot 1 \cdot 5!} = 126.$$

(b) The number of committees whose members are all women is

$$\binom{11}{5} = \frac{11!}{6!5!} = \frac{11 \cdot 10 \cdot 9 \cdot 8 \cdot 7 \cdot 6!}{6! \cdot 5 \cdot 4 \cdot 3 \cdot 2 \cdot 1} = 462.$$

(c) The 3 men can be chosen in

$$\binom{9}{3} = \frac{9!}{6!3!} = \frac{9 \cdot 8 \cdot 7 \cdot 6!}{6! \cdot 3 \cdot 2 \cdot 1} = 84 \text{ ways.}$$

The 2 women can be chosen in

$$\binom{11}{2} = \frac{11!}{9!2!} = \frac{11 \cdot 10 \cdot 9!}{9! \cdot 2 \cdot 1} = 55 \text{ ways.}$$

Using the multiplication principle, a committee of 3 men and 2 women can be chosen in

$$84 \cdot 55 = 4620 \text{ ways.}$$

19. Order is important, so use permutations. The number of ways in which the children can find seats is

$$P(12, 11) = \frac{12!}{(12 - 11)!} = \frac{12!}{1!}$$
$$= 12!$$
$$= 479,001,600.$$

21. Since order does not matter, the answers are combinations.

(a) $\binom{16}{2} = \frac{16!}{14!2!} = \frac{16 \cdot 15 \cdot 14!}{14! \cdot 2 \cdot 1} = 120$

120 samples of 2 marbles can be drawn.

(b) $\binom{16}{4} = 1820$

1820 samples of 4 marbles can be drawn.

(c) Since there are 9 blue marbles in the bag, the number of samples containing 2 blue marbles is

$$\binom{9}{2} = 36.$$

23. Since order does not matter, use combinations.

(a) $\binom{5}{3} = \frac{5!}{2!3!} = \frac{5 \cdot 4 \cdot 3!}{2 \cdot 1 \cdot 3!} = 10$

There are 10 possible samples with all black jelly beans.

(b) There is only 1 red jelly bean, so there are no samples in which all 3 are red.

(c) $\binom{3}{3} = 1$

There is 1 sample with all yellow.

(d) $\binom{5}{2}\binom{1}{1} = 10 \cdot 1 = 10$

There are 10 samples with 2 black and 1 red.

(e) $\binom{5}{2}\binom{3}{1} = 10 \cdot 3 = 30$

There are 30 samples with 2 black and 1 yellow.

(f) $\binom{3}{2}\binom{5}{1} = 3 \cdot 5 = 15$

There are 15 samples with 2 yellow and 1 black.

(g) There is only 1 red jelly bean, so there are no samples containing 2 red jelly beans.

25. Show that $\binom{n}{r} = \binom{n}{n-r}$.

Work with each side of the equation separately.

$$\binom{n}{r} = \frac{n!}{r!(n-r)!}$$
$$\binom{n}{n-r} = \frac{n!}{(n-r)![n-(n-r)]!}$$
$$= \frac{n!}{(n-r)!r!}$$

Since both results are the same, we have shown that

$$\binom{n}{r} = \binom{n}{n-r}.$$

27. There are 7 digits. The number of cases with the same number of dots on both sides is $\binom{7}{1} = 7$. The number of cases with a different number of dots on each side is $\binom{7}{2} = \frac{7!}{5!2!} = 21$. The total number of dominoes that can be formed is $\binom{7}{1} + \binom{7}{2} = 7 + 21 = 28$.

29. Order is important in arranging a schedule, so use permutations.

(a) $P(6,6) = \dfrac{6!}{0!} = 6! = 720$

She can arrange her schedule in 720 ways if she calls on all 6 prospects.

(b) $P(6,4) = \dfrac{6!}{2!} = 360$

She can arrange her schedule in 360 ways if she calls on only 4 of the 6 prospects.

31. There are 2 types of meat and 6 types of extras. Order does not matter here, so use combinations.

(a) There are $\binom{2}{1}$ ways to choose one type of meat and $\binom{6}{3}$ ways to choose exactly three extras. By the multiplication principle, there are

$$\binom{2}{1}\binom{6}{3} = 2 \cdot 20 = 40$$

different ways to order a hamburger with exactly three extras.

(b) There are

$$\binom{6}{3} = 20$$

different ways to choose exactly three extras.

(c) "At least five extras" means "5 extras or 6 extras." There are $\binom{6}{5}$ different ways to choose exactly 5 extras and $\binom{6}{6}$ ways to choose exactly 6 extras, so there are

$$\binom{6}{5} + \binom{6}{6} = 6 + 1 = 7$$

different ways to choose at least five extras.

33. Select 8 of the 16 smokers and 8 of the 22 non-smokers; order does not matter in the group, so use combinations. There are

$$\binom{16}{8}\binom{22}{8} = 4{,}115{,}439{,}900$$

different ways to select the study group.

35. Order does not matter in choosing a delegation, so use combinations. This committee has $5 + 4 = 9$ members.

(a) There are

$$\binom{9}{3} = \frac{9!}{6!3!}$$

$$= \frac{9 \cdot 8 \cdot 7 \cdot 6!}{6! \cdot 3 \cdot 2 \cdot 1}$$

$$= 84 \text{ possible delegations.}$$

(b) To have all Democrats, the number of possible delegations is

$$\binom{5}{3} = 10.$$

(c) To have 2 Democrats and 1 Republican, the number of possible delegations is

$$\binom{5}{2}\binom{4}{1} = 10 \cdot 4 = 40.$$

(d) We have previously calculated that there are 84 possible delegations, of which 10 consist of all Democrats. Those 10 delegations are the only ones with no Republicans, so the remaining $84 - 10 = 74$ delegations include at least one Republican.

37. Order does not matter in choosing the panel, so use combinations.

$$\binom{45}{3} = \frac{45!}{42!3!} = \frac{45 \cdot 44 \cdot 43 \cdot 42!}{3 \cdot 2 \cdot 1 \cdot 42!} = 14{,}190$$

The publisher was wrong. There are 14,190 possible three judge panels.

39. Since the cards are chosen at random, that is, order does not matter, the answers are combinations.

(a) There are 4 queens and 48 cards that are not queens. The total number of hands is

$$\binom{4}{4}\binom{48}{1} = 1 \cdot 48 = 48.$$

(b) Since there are 12 face cards (3 in each suit), there are 40 nonface cards. The number of ways to choose no face cards (all 5 nonface cards) is

$$\binom{40}{5} = \frac{40!}{35!5!} = 658{,}008.$$

(c) If there are exactly 2 face cards, there will be 3 nonface cards. The number of ways in which the face cards can be chosen is $\binom{12}{2}$, while the number of ways in which the nonface cards can be chosen is $\binom{40}{3}$. Using the multiplication principle, the number of ways to get this result is

$$\binom{12}{2}\binom{40}{3} = 66 \cdot 9880 = 652{,}080.$$

(d) If there are at least 2 face cards, there must be either 2 face cards and 3 nonface cards, 3 face cards and 2 nonface cards, 4 face cards and 1 nonface card, or 5 face cards. Use the multiplication principle as in part (c) to find the number of ways to obtain each of these possibilities. Then add these numbers. The total number of ways to get at least 2 face cards is

$$\binom{12}{2}\binom{40}{3} + \binom{12}{3}\binom{40}{2} + \binom{12}{4}\binom{40}{1} + \binom{12}{5}$$

$$= 66 \cdot 9880 + 220 \cdot 780 + 495 \cdot 40 + 792$$
$$= 652{,}080 + 171{,}600 + 19{,}800 + 792$$
$$= 844{,}272.$$

(e) The number of ways to choose 1 heart is $\binom{13}{1}$, the number of ways to choose 2 diamonds is $\binom{13}{2}$, and the number of ways to choose 2 clubs is $\binom{13}{2}$. Using the multiplication principle, the number of ways to get this result is

$$\binom{13}{1}\binom{13}{2}\binom{13}{2} = 13 \cdot 78 \cdot 78 = 79{,}092.$$

41. Since order does not matter, use combinations.

2 good hitters: $\binom{5}{2}\binom{4}{1} = 10 \cdot 4 = 40$

3 good hitters: $\binom{5}{3}\binom{4}{0} = 10 \cdot 1 = 10$

The total number of ways is $40 + 10 = 50$.

43. Since order does not matter, use combinations.

(a) There are

$$\binom{20}{5} = 15{,}504$$

different ways to select 5 of the orchids.

(b) If 2 special orchids must be included in the show, that leaves 18 orchids from which the other 3 orchids for the show must be chosen. This can be done in

$$\binom{18}{3} = 816$$

different ways.

45. In the lottery, 6 different numbers are to be chosen from the 99 numbers.

(a) There are

$$\binom{99}{6} = \frac{99!}{93!6!} = 1{,}120{,}529{,}256$$

different ways to choose 6 numbers if order is not important.

(b) There are

$$P(99, 6) = \frac{99!}{93!} = 806{,}781{,}064{,}320$$

different ways to choose 6 numbers if order matters.

47. (a) There can be 5, 4, 3, 2, 1, or no toppings. The total number of possibilities for the first pizza is

$$\binom{11}{5} + \binom{11}{4} + \binom{11}{3} + \binom{11}{2} + \binom{11}{1} + \binom{11}{0}$$
$$= 462 + 330 + 165 + 55 + 11 + 1$$
$$= 1024.$$

The total number of possibilities for the toppings on two pizzas is

$$1024 \cdot 1024 = 1{,}048{,}576.$$

(b) In part (a), we found that if the order of the two pizzas matters, there are

$$1024^2 = 1{,}048{,}576$$

possibilities. If we had a list of all of these possibilities and if the order of the pizzas doesn't matter, we must eliminate all of the possibilities that involve the same two pizzas. There are 1024 such items on the list, one of each of the possibilities for one pizza. Therefore, the number of items on the list that have a duplicate is

$$1{,}048{,}576 - 1024 = 1{,}047{,}552.$$

To eliminate duplicates, we eliminate the second listing of each of these, that is,

$$\frac{1{,}047{,}552}{2} = 523{,}776.$$

Subtracting this from the number of possibilities on the list, we see that if the order of the two pizzas doesn't matter, the number of possibilities is

$$1{,}048{,}576 - 523{,}776 = 524{,}800.$$

49. (a) $\binom{8}{1} + \binom{8}{2} + \binom{8}{3} + \binom{8}{4} + \binom{8}{5} + \binom{8}{6}$

$$+ \binom{8}{7} + \binom{8}{8}$$

$$= 8 + 28 + 56 + 70 + 56 + 28 + 8 + 1$$
$$= 255$$

There are 255 breakfasts that can be made.

(b) She has 2 choices. For the first choice she has 4 items. For the second choice she has 4 items.

$$4 \cdot 4 = 16$$

She can make 16 breakfasts.

(c) She has $\binom{4}{2}$ choices of cereal mix and $\binom{4}{3}$ choices of add-in mix. Her total number of choices is

$$\binom{4}{2}\binom{4}{3} = 6 \cdot 4 = 24.$$

(d) He has

$$\binom{4}{1} + \binom{4}{2} + \binom{4}{3} + \binom{4}{4} = 4 + 6 + 4 + 1 = 15$$

choices of cereal mix and

$$\binom{4}{1} + \binom{4}{2} + \binom{4}{3} + \binom{4}{4} = 4 + 6 + 4 + 1 = 15$$

choices of add-in mix. His total number of breakfasts is

$$15 \cdot 15 = 225.$$

(e) $\binom{7}{0} + \binom{7}{1} + \binom{7}{2} + \binom{7}{3} + \binom{7}{4} + \binom{7}{5}$

$$+ \binom{7}{6} + \binom{7}{7}$$

$$= 1 + 7 + 21 + 35 + 35 + 21 + 7 + 1$$
$$= 128$$

She has 128 different cereals.

51. (a) The number of ways the names can be arranged is

$$18! \approx 6.402 \times 10^{15}.$$

(b) 4 lines consist of a 3 syllable name repeated, followed by a 2 syllable name and then a 4 syllable name. Including order, the number of arrangements is

$$10 \cdot 4 \cdot 4 \cdot 9 \cdot 3 \cdot 3 \cdot 8 \cdot 2 \cdot 2 \cdot 7 \cdot 1 \cdot 1$$
$$= 2{,}903{,}040.$$

2 lines consist of a 3 syllable name repeated, followed by two more 3 syllable names. Including order, the number of arrangements is

$$6 \cdot 5 \cdot 4 \cdot 3 \cdot 2 \cdot 1 = 720.$$

The number of ways the similar 4 lines can be arranged among the 6 total lines is

$$\binom{6}{4} = 15.$$

The number of arrangements that fit the pattern is

$$2{,}903{,}040 \cdot 720 \cdot 15 \approx 3.135 \times 10^{10}.$$

53. (a) The number of different committees possible is

$$\binom{5}{2} + \binom{5}{3} + \binom{5}{4} + \binom{5}{5} = 10 + 10 + 5 + 1 = 26.$$

(b) The total number of subsets is

$$2^5 = 32.$$

The number of different committees possible is

$$2^5 - \binom{5}{1} - \binom{5}{0} = 32 - 5 - 1 = 26.$$

8.3   Probability Applications of Counting Principles

1. There are $\binom{11}{3}$ samples of 3 apples.

$$\binom{11}{3} = \frac{11 \cdot 10 \cdot 9}{3 \cdot 2 \cdot 1} = 165$$

There are $\binom{7}{3}$ samples of 3 red apples.

$$\binom{7}{3} = \frac{7 \cdot 6 \cdot 5}{3 \cdot 2 \cdot 1} = 35$$

Thus,

$$P(\text{all red apples}) = \frac{35}{165} = \frac{7}{33}.$$

3. There are $\binom{4}{2}$ samples of 2 yellow apples.

$$\binom{4}{2} = \frac{4 \cdot 3}{2 \cdot 1} = 6$$

There are $\binom{7}{1} = 7$ samples of 1 red apple. Thus, there are $6 \cdot 7 = 42$ samples of 3 in which 2 are yellow and 1 red. Thus,

$$P(\text{2 yellow and 1 red apple}) = \frac{42}{165} = \frac{14}{55}.$$

5. The number of 2-card hands is

$$\binom{52}{2} = \frac{52 \cdot 51}{2 \cdot 1} = 1326.$$

7. There are $\binom{52}{2} = 1326$ different 2-card hands. The number of 2-card hands with exactly one ace is

$$\binom{4}{1}\binom{48}{2} = 4 \cdot 48 = 192.$$

The number of 2-card hands with two aces is

$$\binom{4}{2} = 6.$$

Thus there are 198 hands with at least one ace. Therefore,

P(the 2-card hand contains an ace)

$$= \frac{198}{1326} = \frac{33}{221} \approx 0.149.$$

9. There are $\binom{52}{2} = 1326$ different 2-card hands. There are $\binom{13}{2} = 78$ ways to get a 2-card hand where both cards are of a single named suit, but there are 4 suits to choose from. Thus,

P(two cards of same suit)

$$= \frac{4 \cdot \binom{13}{2}}{\binom{52}{2}} = \frac{312}{1326} = \frac{52}{221} \approx 0.235.$$

11. There are $\binom{52}{2} = 1326$ different 2-card hands. There are 12 face cards in a deck, so there are 40 cards that are not face cards. Thus,

P(no face cards)

$$= \frac{\binom{40}{2}}{\binom{52}{2}} = \frac{780}{1326} = \frac{130}{221} \approx 0.588.$$

13. There are 26 choices for each slip pulled out, and there are 5 slips pulled out, so there are

$$26^5 = 11,881,376$$

different "words" that can be formed from the letters. If the "word" must be "chuck," there is only one choice for each of the 5 letters (the first slip must contain a "c," the second an "h," and so on). Thus,

P(word is "chuck")

$$= \frac{1^5}{26^5} = \left(\frac{1}{26}\right)^5 \approx 8.417 \times 10^{-8}.$$

15. There are $26^5 = 11,881,376$ different "words" that can be formed. If the "word" is to have no repetition of letters, then there are 26 choices for the first letter, but only 25 choices for the second (since the letters must all be different), 24 choices for the third, and so on. Thus,

P(all different letters)

$$= \frac{26 \cdot 25 \cdot 24 \cdot 23 \cdot 22}{26^5}$$

$$= \frac{1 \cdot 25 \cdot 24 \cdot 23 \cdot 22}{26^4}$$

$$= \frac{303,600}{456,976}$$

$$= \frac{18,975}{28,561} \approx 0.664.$$

19. P(at least 2 presidents have the same birthday)
$$= 1 - P(\text{no 2 presidents have the same birthday})$$

The number of ways that 42 people can have the same or different birthdays is $(365)^{42}$. The number of ways that 42 people can have all different birthdays is the number of permutations of 365 things taken 42 at a time or $P(365, 42)$. Thus,

P(at least 2 presidents have the same birthday)

$$= 1 - \frac{P(365, 42)}{365^{42}}.$$

(Be careful to realize that the symbol P is sometimes used to indicate permutations and sometimes used to indicate probability; in this solution, the symbol is used both ways.)

21. Since there are 435 members of the House of Representatives, and there are only 365 days in a year, it is a certain event that at least 2 people will have the same birthday. Thus,

P(at least 2 members have the same birthday) $= 1$.

23. Each of the 4 people can choose to get off at any one of the 7 floors, so there are 7^4 ways the four people can leave the elevator. The number of ways the people can leave at different floors is the number of permutations of 7 things (floors) taken 4 at a time or

$$P(7, 4) = 7 \cdot 6 \cdot 5 \cdot 4 = 840.$$

The probability that no 2 passengers leave at the same floor is

$$\frac{P(7, 4)}{7^4} = \frac{840}{2401} \approx 0.3499.$$

Thus, the probability that at least 2 passengers leave at the same floor is

$$1 - 0.3499 = 0.6501.$$

(Note the similarity of this problem and the "birthday problem.")

25. $P(\text{at least one \$100-bill})$

$$= P(1 \ \$100\text{-bill}) + P(2 \ \$100\text{-bills})$$

$$= \frac{\binom{2}{1}\binom{4}{1}}{\binom{6}{2}} + \frac{\binom{2}{2}\binom{4}{0}}{\binom{6}{2}}$$

$$= \frac{8}{15} + \frac{1}{15}$$

$$= \frac{9}{15} = \frac{3}{5}$$

$$P(\text{no \$100-bill}) = \frac{\binom{2}{0}\binom{4}{2}}{\binom{6}{2}} = \frac{6}{15} = \frac{2}{5}$$

It is more likely to get at least one \$100-bill.

27. The number of orders of the three types of birds is $P(3,3)$. The number of arrangements of the crows is $P(3,3)$, of the bluejays is $P(4,4)$, and of the starlings is $P(5,5)$. The total number of arrangements of all the birds is $P(12,12)$.

$P(\text{all birds of same type are sitting together})$

$$= \frac{P(3,3) \cdot P(3,3) \cdot P(4,4) \cdot P(5,5)}{P(12,12)}$$

$$\approx 2.165 \times 10^{-4}$$

29. There are 11 letters so the number of possible spellings (counting duplicates) is $11! = 39,916,800$. Since the letter i is repeated 4 times, the letter s is repeated 4 times, and the letter p is repeated 2 times, the spelling Mississippi will occur $4!4!2! = 1152$ times. The probability that Mississippi will be spelled is

$$\frac{1152}{39,916,800} \approx 0.0000289.$$

31. There are $\binom{9}{2}$ possible ways to choose 2 nondefective typewriters out of the $\binom{11}{2}$ possible ways of choosing any 2. Thus,

$$P(\text{no defective}) = \frac{\binom{9}{2}}{\binom{11}{2}} = \frac{36}{55}.$$

33. There are $\binom{9}{4}$ possible ways to choose 4 nondefective typewriters out of the $\binom{11}{4}$ possible ways of choosing any 4. Thus,

$$P(\text{no defective}) = \frac{\binom{9}{4}}{\binom{11}{4}} = \frac{126}{330} = \frac{21}{55}.$$

35. There are $\binom{12}{5} = 792$ ways to pick a sample of 5. It will be shipped if all 5 are good. There are $\binom{10}{5} = 252$ ways to pick 5 good ones, so

$$P(\text{all good}) = \frac{252}{792} = \frac{7}{22} \approx 0.318.$$

37. There are 20 people in all, so the number of possible 5-person committees is $\binom{20}{5} = 15,504$. Thus, in parts (a)-(g), $n(S) = 15,504$.

(a) There are $\binom{10}{3}$ ways to choose the 3 men and $\binom{10}{2}$ ways to choose the 2 women. Thus,

$P(3 \text{ men and } 2 \text{ women})$

$$= \frac{\binom{10}{3}\binom{10}{2}}{\binom{20}{5}} = \frac{120 \cdot 45}{15,504} = \frac{225}{646} \approx 0.348.$$

(b) There are $\binom{6}{3}$ ways to choose the 3 Miwoks and $\binom{9}{2}$ ways to choose the 2 Pomos. Thus,

$P(\text{exactly } 3 \text{ Miwoks and } 2 \text{ Pomos})$

$$= \frac{\binom{6}{3}\binom{9}{2}}{\binom{20}{5}} = \frac{20 \cdot 36}{15,504} = \frac{15}{323} \approx 0.046.$$

(c) Choose 2 of the 6 Miwoks, 2 of the 5 Hoopas, and 1 of the 9 Pomos. Thus,

$P(2 \text{ Miwoks, } 2 \text{ Hoopas, and a Pomo})$

$$= \frac{\binom{6}{2}\binom{5}{2}\binom{9}{1}}{\binom{20}{5}} = \frac{15 \cdot 10 \cdot 9}{15,504} = \frac{225}{2584} \approx 0.087.$$

(d) There cannot be 2 Miwoks, 2 Hoopas, and 2 Pomos, since only 5 people are to be selected. Thus,

$P(2 \text{ Miwoks, } 2 \text{ Hoopas, and } 2 \text{ Pomos}) = 0.$

(e) Since there are more women then men, there must be 3, 4, or 5 women.

$P(\text{more women than men})$

$$= \frac{\binom{10}{3}\binom{10}{2} + \binom{10}{4}\binom{10}{1} + \binom{10}{5}\binom{10}{0}}{\binom{20}{5}}$$

$$= \frac{7752}{15,504} = \frac{1}{2}$$

(f) Choose 3 of 5 Hoopas and any 2 of the 15 non-Hoopas.

$P(\text{exactly 3 Hoopas})$

$$= \frac{\binom{5}{3}\binom{15}{2}}{\binom{20}{5}} = \frac{175}{2584} \approx 0.068$$

(g) There can be 2 to 5 Pomos, the rest chosen from the 11 nonPomos.

$P(\text{at least 2 Pomos})$

$$= \frac{\binom{9}{2}\binom{11}{3} + \binom{9}{3}\binom{11}{2} + \binom{9}{4}\binom{11}{1} + \binom{9}{5}\binom{11}{0}}{\binom{20}{5}}$$

$$= \frac{503}{646} \approx 0.779$$

39. There are $\binom{52}{5}$ different 5-card poker hands. There are 4 royal flushes, one for each suit. Thus,

$P(\text{royal flush})$

$$= \frac{4}{\binom{52}{5}} = \frac{4}{2,598,960} = \frac{1}{649,740}$$

$$\approx 1.539 \times 10^{-6}.$$

41. The four of a kind can be chosen in 13 ways and then is matched with 1 of the remaining 48 cards to make a 5-card hand containing four of a kind. Thus, there are $13 \cdot 48 = 624$ poker hands with four of a kind. It follows that

$P(\text{four of a kind})$

$$= \frac{624}{\binom{52}{5}} = \frac{624}{2,598,960} = \frac{1}{4165}$$

$$\approx 2.401 \times 10^{-4}.$$

43. There are 13 different values with 4 cards of each value. The total number of possible three of a kind is then $13 \cdot \binom{4}{3}$. The other 2 cards must be chosen from the remaining 48 cards of different value. However, these 2 cards must be different. Thus, for the last 2 cards, there are 48 cards to choose from, but the cards must not have the same value. The number of possibilities for the last 2 cards is $\binom{48}{2} - 12 \cdot \binom{4}{2}$, and

$$P(\text{three of a kind}) = \frac{13 \cdot \binom{4}{3}\left[\binom{48}{2} - 12 \cdot \binom{4}{2}\right]}{\binom{52}{5}}$$

$$\approx 0.0211.$$

45. There are 13 different values with 4 cards of each value. The total number of possible pairs is $13 \cdot \binom{4}{2}$. The remaining 3 cards must be chosen from the 48 cards of different value. However, among these 3 we cannot have 3 of a kind nor can we have 2 of a kind.

$P(\text{one pair})$

$$= \frac{13 \cdot \binom{4}{2}\left[\binom{48}{3} - 12 \cdot \binom{4}{3} - 12\binom{4}{2} \cdot \binom{44}{1}\right]}{\binom{52}{5}}$$

$$\approx 0.4226$$

47. The hand can have exactly 3 aces or 4 aces. There are $\binom{4}{3} = 4$ ways to pick exactly 3 aces, and there are $\binom{48}{10}$ ways to pick the other 10 cards. Also, there is only $\binom{4}{4} = 1$ way to pick 4 aces, and there are $\binom{48}{9}$ ways to pick the other 9 cards. Hence,

$$P(\text{3 aces}) = \frac{\binom{4}{3}\binom{48}{10} + \binom{4}{4}\binom{48}{9}}{\binom{52}{13}} \approx 0.0438.$$

49. The number of ways of choosing 3 suits is $P(4,3)$. The number of ways of choosing 6 of one suit is $\binom{13}{6}$, 4 of another is $\binom{13}{4}$, and 3 of another is $\binom{13}{3}$. Thus,

$P(\text{6 of one suit, 4 of another, and 3 of another})$

$$= \frac{P(4,3)\binom{13}{6}\binom{13}{4}\binom{13}{3}}{\binom{52}{13}}$$

$$\approx 0.0133.$$

Order is important in this problem because 6 spades, 4 hearts, and 3 clubs would be different than 6 hearts, 4 clubs, and 3 spades.

51. There are $\binom{99}{6} = 1,120,529,256$ different ways to pick 6 numbers from 1 to 99, but there is only 1 way to win; the 6 numbers you pick must exactly match the 6 winning numbers, without regard to order. Thus,

$P(\text{win the big prize})$

$$= \frac{1}{1,120,529,256} \approx 8.924 \times 10^{-10}.$$

53. Let A be the event of drawing four royal flushes in a row all in spades, and B be the event of meeting four strangers all with the same birthday. Then,

$$P(A) = \left[\frac{1}{\binom{52}{5}}\right]^4 \cdot$$

For four people, the number of possible birthdays is 365^4. Of these there are 365 which are the same (that is, all birthdays January 1 or January 2 or January 3, etc.).

$$P(B) = \frac{365}{365^4} = \frac{1}{365^3}$$

Therefore,

$$P(A \cap B) = \left[\frac{1}{\binom{52}{5}}\right]^4 \frac{1}{365^3} \approx 4.507 \times 10^{-34}.$$

No, this probability is much smaller than that of winning the lottery.

55. (a) The number of ways to select 5 numbers between 1 and 55 is $\binom{55}{5} = 3{,}478{,}761$ and there are 42 ways to select the bonus number.

$$P(\text{winning jackpot}) = \frac{1}{3{,}478{,}761 \cdot 42}$$
$$= \frac{1}{146{,}107{,}962}$$

(b) The number of selections you would make over 138 years is

$$138 \cdot 365 \cdot 24 \cdot 60 = 72{,}582{,}480.$$

The probability that none of the selections win is

$$\left(\frac{146{,}107{,}961}{146{,}107{,}962}\right)^{72{,}482{,}480} \approx 0.6085.$$

Therefore, the probability of winning is $1 - 0.6085 = 0.3915$.

57. $P(\text{saying "Math class is tough."})$

$$= \frac{\binom{1}{1}\binom{269}{3}}{\binom{270}{4}} \approx 0.0148$$

No, it is not correct. The correct figure is 1.48%.

59. (a) There are only 4 ways to win in just 4 calls: the 2 diagonals, the center column, and the center row. There are $\binom{75}{4}$ combinations of 4 numbers that can occur. The probability that a person will win bingo after just 4 numbers are called is $\frac{4}{\binom{75}{4}} \approx 3.291 \times 10^{-6}$.

(b) There is only 1 way to get an L. It can occur in as few as 9 calls. There are $\binom{75}{9}$ combinations of 9 numbers that can occur in 9 calls is $\frac{1}{\binom{75}{9}} \approx 7.962 \times 10^{-12}$.

(c) There is only 1 way to get an X-out. It can ocur in as few as 8 calls. There are $\binom{75}{8}$ combinations of 8 numbers that can occur. The probability that an X-out occurs in 8 calls is $\frac{1}{\binom{75}{8}} \approx 5.927 \times 10^{-11}$.

(d) Four columns contain a permutation of 15 numbers taken 5 at a time. One column contains a permutation of 15 numbers taken 4 at a time. The number of distinct cards is $P(15,5)^4 \cdot P(15,4) \approx 5.524 \times 10^{26}$.

8.4 Binomial Probability

1. This is a Bernoulli trial problem with $P(\text{success}) = P(\text{girl}) = \frac{1}{2}$. The probability of exactly x successes in n trials is

$$\binom{n}{x}p^x(1-p)^{n-x},$$

where p is the probability of success in a single trial. We have $n = 5$, $x = 2$, and $p = \frac{1}{2}$. Note that

$$1 - p = 1 - \frac{1}{2} = \frac{1}{2}.$$

$$P(\text{exactly 2 girls and 3 boys}) = \binom{5}{2}\left(\frac{1}{2}\right)^2\left(\frac{1}{2}\right)^3$$
$$= \frac{10}{32} = \frac{5}{16} \approx 0.313$$

3. We have $n = 5$, $x = 0$, $p = \frac{1}{2}$, and $1 - p = \frac{1}{2}$.

$$P(\text{no girls}) = \binom{5}{0}\left(\frac{1}{2}\right)^0\left(\frac{1}{2}\right)^5 = \frac{1}{32} \approx 0.031$$

5. "At least 4 girls" means either 4 or 5 girls.

P(at least 4 girls)

$$= \binom{5}{4}\left(\frac{1}{2}\right)^4\left(\frac{1}{2}\right)^1 + \binom{5}{5}\left(\frac{1}{2}\right)^5\left(\frac{1}{2}\right)^0$$

$$= \frac{5}{32} + \frac{1}{32} = \frac{6}{32} = \frac{3}{16} \approx 0.188$$

7. P(no more than 3 boys)
$$= 1 - P(\text{at least 4 boys})$$
$$= 1 - P(4 \text{ boys or } 5 \text{ boys})$$
$$= 1 - [P(4 \text{ boys}) + P(5 \text{ boys})]$$

$$= 1 - \left(\frac{5}{32} + \frac{1}{32}\right)$$

$$= 1 - \frac{6}{32}$$

$$= 1 - \frac{3}{16} = \frac{13}{16} \approx 0.813$$

9. On one roll, $P(1) = \frac{1}{6}$. We have $n = 12$, $x = 12$, and $p = \frac{1}{6}$. Note that $1 - p = \frac{5}{6}$. Thus,

$$P(\text{exactly 12 ones}) = \binom{12}{12}\left(\frac{1}{6}\right)^{12}\left(\frac{5}{6}\right)^0$$

$$\approx 4.594 \times 10^{-10}.$$

11. $P(\text{exactly 1 one }) = \binom{12}{1}\left(\frac{1}{6}\right)^1\left(\frac{5}{6}\right)^{11} \approx 0.2692$

13. "No more than 3 ones" means 0, 1, 2, or 3 ones. Thus,

P(no more than 3 ones)
$$= P(0 \text{ ones}) + P(1 \text{ one}) + P(2 \text{ ones})$$
$$+ P(3 \text{ ones})$$

$$= \binom{12}{0}\left(\frac{1}{6}\right)^0\left(\frac{5}{6}\right)^{12} + \binom{12}{1}\left(\frac{1}{6}\right)^1\left(\frac{5}{6}\right)^{11}$$

$$+ \binom{12}{2}\left(\frac{1}{6}\right)^2\left(\frac{5}{6}\right)^{10} + \binom{12}{3}\left(\frac{1}{6}\right)^3\left(\frac{5}{6}\right)^9$$

$$\approx 0.8748.$$

15. Each time the coin is tossed, $P(\text{head}) = \frac{1}{2}$. We have $n = 6$, $x = 6$, $p = \frac{1}{2}$, and $1 - p = \frac{1}{2}$. Thus,

$$P(\text{all heads}) = \binom{6}{6}\left(\frac{1}{2}\right)^6\left(\frac{1}{2}\right)^0$$

$$= \frac{1}{64} \approx 0.016.$$

17. P(no more than 3 heads)
$$= P(0 \text{ heads}) + P(1 \text{ head}) + P(2 \text{ heads})$$
$$+ P(3 \text{ heads})$$

$$= \binom{6}{0}\left(\frac{1}{2}\right)^0\left(\frac{1}{2}\right)^6 + \binom{6}{1}\left(\frac{1}{2}\right)^1\left(\frac{1}{2}\right)^5$$

$$+ \binom{6}{2}\left(\frac{1}{2}\right)^2\left(\frac{1}{2}\right)^4 + \binom{6}{3}\left(\frac{1}{2}\right)^3\left(\frac{1}{2}\right)^3$$

$$= \frac{42}{64} = \frac{21}{32} \approx 0.656$$

21. $\binom{n}{r} + \binom{n}{r+1}$

$$= \frac{n!}{r!(n-r)!} + \frac{n!}{(r+1)![n-(r+1)]!}$$

$$= \frac{n!(r+1)}{r!(r+1)(n-r)!}$$

$$+ \frac{n!(n-r)}{(r+1)![n-(r+1)]!(n-r)}$$

$$= \frac{rn! + n!}{(r+1)!(n-r)!} + \frac{n(n!) - rn!}{(r+1)!(n-r)!}$$

$$= \frac{rn! + n! + n(n!) - rn!}{(r+1)!(n-r)!}$$

$$= \frac{n!(n+1)}{(r+1)!(n-r)!}$$

$$= \frac{(n+1)!}{(r+1)![(n+1)-(r+1)]!}$$

$$= \binom{n+1}{r+1}$$

23. Since the potential callers are not likely to have birthdates that are distributed evenly throughout the twentieth century, the use of binomial probabilities is not applicable and thus, the probabilities that are computed are not correct.

25. We define a success to be the event that a customer overpays. In this situation, $n = 15$, $x = 0$, $p = \frac{1}{10}$, and $1 - p = \frac{9}{10}$.

P(customer does not overpay for any item)

$$= \binom{15}{0}\left(\frac{1}{10}\right)^0\left(\frac{9}{10}\right)^{15}$$

$$\approx 0.2059$$

27. In Exercise 25, we defined a success to be the event that a customer overpays. In this situation, $n = 15$; $x = 2, 3, 4, \ldots, 15$; $p = \frac{1}{10}$; and $1 - p = \frac{9}{10}$.

P(a customer overpays on at least 2 items)
   $= 1 - P$(a customer overpays on 0 or 1 item)

$$= 1 - \binom{15}{0}\left(\frac{1}{10}\right)^0 \left(\frac{9}{10}\right)^{15} - \binom{15}{1}\left(\frac{1}{10}\right)^1 \left(\frac{9}{10}\right)^{14}$$

$$\approx 0.4510$$

29. $n = 20, x = 6, p = 0.256$

$$P(6) = \binom{20}{6}(0.256)^6 (0.744)^{14}$$

$$\approx 0.1737$$

31. $n = 20, p = 0.256$

P(at least 4)

$$= 1 - \binom{20}{0}(0.256)^0 (0.744)^{20} - \binom{20}{1}(0.256)^1 (0.744)^{19}$$

$$- \binom{20}{2}(0.256)^2 (0.744)^{18} - \binom{20}{3}(0.256)^3 (0.744)^{17}$$

$$\approx 0.7925$$

33. We have $n = 6$, $x = 2$, $p = \frac{1}{5}$, and $1 - p = \frac{4}{5}$. Thus,

$$P(\text{exactly 2 correct}) = \binom{6}{2}\left(\frac{1}{5}\right)^2 \left(\frac{4}{5}\right)^4 \approx 0.2458.$$

35. We have

P(at least 4 correct)
   $= P(4 \text{ correct}) + P(5 \text{ correct}) + P(6 \text{ correct})$

$$= \binom{6}{4}\left(\frac{1}{5}\right)^4 \left(\frac{4}{5}\right)^2 + \binom{6}{5}\left(\frac{1}{5}\right)^5 \left(\frac{4}{5}\right)^1$$

$$+ \binom{6}{6}\left(\frac{1}{5}\right)^6 \left(\frac{4}{5}\right)^0$$

$$\approx 0.0170.$$

37. $n = 20, p = 0.05, x = 0$

P(0 defective transistors)

$$= \binom{20}{0}(0.05)^0 (0.95)^{20} \approx 0.3585$$

39. Let success mean producing a defective item. Then we have $n = 75$, $p = 0.05$, and $1 - p = 0.95$.

(a) If there are exactly 5 defective items, then $x = 5$. Thus,

$$P(\text{exactly 5 defective}) = \binom{75}{5}(0.05)^5 (0.95)^{70}$$

$$\approx 0.1488.$$

(b) If there are no defective items, then $x = 0$. Thus,

$$P(\text{none defective}) = \binom{75}{0}(0.05)^0 (0.95)^{75}$$

$$\approx 0.0213.$$

(c) If there is at least 1 defective item, then we are interested in $x \geq 1$. We have

$$P(\text{at least one defective}) = 1 - P(x = 0)$$
$$\approx 1 - 0.021$$
$$= 0.9787.$$

41. (a) Since 80% of the "good nuts" are good, 20% of the "good nuts" are bad. Let's let success represent "getting a bad nut." Then 0.2 is the probability of success in a single trial. The probability of 8 successes in 20 trials is

$$\binom{20}{8}(0.2)^8 (1 - 0.2)^{20-8} = \binom{20}{8}(0.2)^8 (0.8)^{12}$$
$$\approx 0.0222$$

(b) Since 60% of the "blowouts" are good, 40% of the "blowouts" are bad. Let's let success represent "getting a bad nut." Then 0.4 is the probability of success in a single trial. The probability of 8 successes in 20 trials is

$$\binom{20}{8}(0.4)^8 (1 - 0.4)^{20-8} = \binom{20}{8}(0.4)^8 (0.6)^{12}$$
$$\approx 0.1797$$

(c) The probability that the nuts are "blowouts" is

$$\frac{\text{Probability of "Blowouts"}}{\text{Probability of "Good Nuts" or "Blowouts"}}$$
having 8 bad nuts out of 20 / having 8 bad nuts out of 20

$$= \frac{0.3\left[\binom{20}{8}(0.4)^8 (0.6)^{12}\right]}{0.7\left[\binom{20}{8}(0.2)^8 (0.8)^{12}\right] + 0.3\left[\binom{20}{8}(0.4)^8 (0.6)^{12}\right]}$$

$$\approx 0.7766.$$

43. $n = 15, p = 0.85$

$$P(\text{all } 15) = \binom{15}{15}(0.85)^{15}(0.15)^{0} \approx 0.0874$$

45. $n = 15, p = 0.85$

$$P(\text{not all}) = 1 - P(\text{all } 15)$$

$$= 1 - \binom{15}{15}(0.85)^{15}(0.15)^{0}$$

$$\approx 0.9126$$

47. $n = 100, \ p = 0.012, \ x = 2$

$$P(\text{exactly 2 sets of twins})$$

$$= \binom{100}{2}(0.012)^{2}(0.988)^{98} \approx 0.2183$$

49. We have $n = 10{,}000, \ p = 2.5 \cdot 10^{-7} = 0.00000025$, and $1 - p = 0.99999975$. Thus,

$$P(\text{at least 1 mutation occurs})$$
$$= 1 - P(\text{none occurs})$$

$$= 1 - \binom{10{,}000}{0}p^{0}(1-p)^{10{,}000}$$

$$= 1 - (0.99999975)^{10{,}000}$$
$$\approx 0.0025.$$

51. $n = 53, \ p = 0.042$

(a) The probability that exactly 5 men are color-blind is

$$P(5) = \binom{53}{5}(0.042)^{5}(0.958)^{48} \approx 0.0478.$$

(b) The probability that no more than 5 men are color-blind is

$$P(\text{no more than 5 men are color-blind})$$

$$= \binom{53}{0}(0.042)^{0}(0.958)^{53} + \binom{53}{1}(0.042)^{1}(0.958)^{52}$$

$$+ \binom{53}{2}(0.042)^{2}(0.958)^{51} + \binom{53}{3}(0.042)^{3}(0.958)^{50}$$

$$+ \binom{53}{4}(0.042)^{4}(0.958)^{49} + \binom{53}{5}(0.042)^{5}(0.958)^{48}$$

$$\approx 0.9767.$$

(c) The probability that at least 1 man is color-blind is

$$1 - P(0 \text{ men are color-blind})$$
$$= 1 - \binom{53}{0}(0.042)^{0}(0.958)^{53} \approx 0.8971.$$

53. (a) Since the probability of a particular band matching is 1 in 4 or $\frac{1}{4}$, the probability that 5 bands match is $\left(\frac{1}{4}\right)^{5} = \frac{1}{1024}$ or 1 chance in 1024.

(b) The probability that 20 bands match is

$\left(\frac{1}{4}\right)^{20} \approx \frac{1}{1.1 \times 10^{12}}$ or about 1 chance in 1.1×10^{12}.

(c) If 20 bands are compared, the probability that 16 or more bands match is

$$P(\text{at least } 16)$$
$$= P(16) + P(17) + P(18) + P(19) + P(20)$$

$$= \binom{20}{16}\left(\frac{1}{4}\right)^{16}\left(\frac{3}{4}\right)^{4} + \binom{20}{17}\left(\frac{1}{4}\right)^{17}\left(\frac{3}{4}\right)^{3}$$

$$+ \binom{20}{18}\left(\frac{1}{4}\right)^{18}\left(\frac{3}{4}\right)^{2} + \binom{20}{19}\left(\frac{1}{4}\right)^{19}\left(\frac{3}{4}\right)^{1}$$

$$+ \binom{20}{20}\left(\frac{1}{4}\right)^{20}\left(\frac{3}{4}\right)^{0}$$

$$= 4845\left(\frac{1}{4}\right)^{16}\left(\frac{3}{4}\right)^{4} + 1140\left(\frac{1}{4}\right)^{17}\left(\frac{3}{4}\right)^{3}$$

$$+ 190\left(\frac{1}{4}\right)^{18}\left(\frac{3}{4}\right)^{2} + 20\left(\frac{1}{4}\right)^{19}\left(\frac{3}{4}\right)^{1}$$

$$+ \left(\frac{1}{4}\right)^{20} \cdot 1$$

$$= \left(\frac{1}{4}\right)^{16}\left[4845\left(\frac{81}{256}\right) + 1140\left(\frac{1}{4}\right)\left(\frac{27}{64}\right)\right.$$

$$\left. + 190\left(\frac{1}{16}\right)\left(\frac{9}{16}\right) + 20\left(\frac{1}{64}\right)\left(\frac{3}{4}\right) + \frac{1}{256}\right]$$

$$= \left(\frac{1}{4}\right)^{16}\left(\frac{392{,}445 + 30{,}780 + 1710 + 60 + 1}{256}\right)$$

$$= \frac{424{,}996}{4^{20}}$$

$$= \frac{1}{\frac{4^{20}}{424{,}996}}$$

$$\approx \frac{1}{2{,}587{,}110}$$

or about 1 chance in 2.587×10^{6}.

55. $n = 4800, p = 0.001$

$$P(\text{more than } 1) = 1 - P(1) - P(0)1 - \binom{4800}{1}(0.001)^1(0.999)^{4799} - \binom{4800}{0}(0.001)^0(0.999)^{4800} \approx 0.9523$$

57. First, find the probability that one group of ten has at least 9 participants complete the study.
$n = 10, p = 0.8$,

$$P(\text{at least } 9 \text{ complete}) = P(9) + P(10) = \binom{10}{9}(0.8)^9(0.2)^1 + \binom{10}{10}(0.8)^{10}(0.2)^0 \approx 0.3758$$

The probability that 2 or more drop out in one group is $1 - 0.3758 = 0.6242$. Thus, the probability that at least 9 participants complete the study in one of the two groups, but not in both groups, is $(0.3758)(0.6242) + (0.6242)(0.3758) \approx 0.469$. The answer is e.

59. $n = 12, x = 7, p = 0.83$

$$P(7) = \binom{12}{7}(0.83)^7(0.17)^5 \approx 0.0305$$

61. $n = 12, p = 0.83$

$P(\text{at least } 9)$
$= P(9) + P(10) + P(11) + P(12)$

$$= \binom{12}{9}(0.83)^9(0.17)^3 + \binom{12}{10}(0.83)^{10}(0.17)^2 + \binom{12}{11}(0.83)^{11}(0.17)^1 + \binom{12}{12}(0.83)^{12}(0.17)^0$$

≈ 0.8676

63. $n = 10, \; p = 0.33, \; 1 - p = 0.67$

(a) The probability that exactly 2 belong to an ethnic minority is

$$\binom{10}{2}(0.33)^2(0.67)^8 \approx 0.1990.$$

(b) The probability that 3 or fewer belong to an ethnic minority is

$$\binom{10}{0}(0.33)^0(0.67)^{10} + \binom{10}{1}(0.33)^1(0.67)^9 + \binom{10}{2}(0.33)^2(0.67)^8 + \binom{10}{3}(0.33)^3(0.67)^7 \approx 0.5684.$$

(c) The probability that one person does not belong to an ethnic minority is 0.67. The probability that exactly 5 do not belong to an ethnic minority is

$$\binom{10}{5}(0.67)^5(0.33)^5 \approx 0.1332.$$

(d) The probability that 6 or more do not belong to an ethnic minority is

$$\binom{10}{6}(0.67)^6(0.33)^4 + \binom{10}{7}(0.67)^7(0.33)^3 + \binom{10}{8}(0.67)^8(0.33)^2 + \binom{10}{9}(0.67)^9(0.33)^1 + \binom{10}{10}(0.67)^{10}(0.33)^0$$

$\approx 0.7936.$

65. (a) The probability that less than 15 will graduate from high school by age 19 is

$$P(\text{less than } 15) = P(0) + P(1) + P(2) + P(3) + \ldots + P(14)$$

$$= \binom{40}{0}(0.152)^0(0.848)^{40} + \binom{40}{1}(0.152)^1(0.848)^{39} + \binom{40}{2}(0.152)^2(0.848)^{38}$$

$$+ \binom{40}{3}(0.152)^3(0.848)^{37} + \binom{40}{4}(0.152)^4(0.848)^{36} + \binom{40}{5}(0.152)^5(0.848)^{35}$$

$$+ \binom{40}{6}(0.152)^6(0.848)^{34} + \binom{40}{7}(0.152)^7(0.848)^{33} + \binom{40}{8}(0.152)^8(0.848)^{32}$$

$$+ \binom{40}{9}(0.152)^9(0.848)^{31} + \binom{40}{10}(0.152)^{10}(0.848)^{30} + \binom{40}{11}(0.152)^{11}(0.848)^{29}$$

$$+ \binom{40}{12}(0.152)^{12}(0.848)^{28} + \binom{40}{13}(0.152)^{13}(0.848)^{27} + \binom{40}{14}(0.152)^{14}(0.848)^{26}$$

$$\approx 0.00137 + 0.00980 + 0.03426 + 0.07779 + 0.12898$$
$$+ 0.16646 + 0.17405 + 0.15153 + 0.11204 + 0.07140$$
$$+ 0.03968 + 0.01940 + 0.00840 + 0.00324 + 0.00112$$
$$\approx 0.9995$$

67. (a) Suppose the National League wins the series in four games. Then they must win all four games and $P = \binom{4}{4}(0.5)^4(0.5)^0 = 0.0625$. Since the probability that the American League wins the series in four games is equally likely, the probability the series lasts four games is $2(0.0625) = 0.125$.

Suppose the National League wins the series in five games. Then they must win exactly three of the previous four games and $P = \binom{4}{3}(0.5)^3(0.5)^1 \cdot (0.5) = 0.125$. Since the probability that the American League wins the series in five games is equally likely, the probability the series lasts five games is $2(0.125) = 0.25$. Suppose the National League wins the series in six games. Then they must win exactly three of the previous five games and $P = \binom{5}{3}(0.5)^3(0.5)^2 \cdot (0.5) = 0.15625$. Since the probability that the American League wins the series in six games is equally likely, the probability the series lasts six games is $2(0.15625) = 0.3125$.

Suppose the National League wins the series in seven games. Then they must win exactly three of the previous six games and $P = \binom{6}{3}(0.5)^3(0.5)^3 \cdot (0.5) = 0.15625$. Since the probability that the American League wins the series in seven games is equally likely, the probability the series last seven games is $2(0.15625) = 0.3125$.

(b) Suppose the better team wins the series in four games. Then they must win all four games and $P = \binom{4}{4}(0.73)^4(0.27)^0 \approx 0.2840$. Suppose the other team wins the series in four games. Then they must win all four games and

$P = \binom{4}{4}(0.27)^4(0.73)^0 \approx 0.0053$. The probability the series lasts four games is the sum of two probabilities, 0.2893.

Suppose the better team wins the series in five games. Then they must win exactly three of the previous four games and $P = \binom{4}{3}(0.73)^3(0.27)^1 \cdot (0.73) \approx 0.3067$. Suppose the other team wins the series in five games. Then they must win exactly three of the previous four games and $P = \binom{4}{3}(0.27)^3(0.73)^1 \cdot (0.27) \approx 0.0155$. The probability the series lasts five games is the sum of the two probabilities, 0.3222.

Suppose the better team wins the series in six games. Then they must win exactly three of the previous five games and $P = \binom{5}{3}(0.73)^3(0.27)^2 \cdot (0.73) \approx 0.2070$. Suppose the other team wins the series in six games. Then they must win exactly three of the previous five games and $P = \binom{5}{3}(0.27)^3(0.73)^2 \cdot (0.27) \approx 0.0283$. The probability the series lasts six games is the sum of the two probabilities, 0.2353.

Suppose the better team wins the series in seven games. Then they must win exactly three of the previous six games and $P = \binom{6}{3}(0.73)^3(0.27)^3 \cdot (0.73) \approx 0.1118$. Suppose the other team wins the series in seven games. Then they must win exactly three of the previous six games and $P = \binom{6}{3}(0.27)^3(0.73)^3 \cdot (0.27) \approx 0.0413$. The probability the series lasts seven games is the sum of the two probabilties, 0.1531.

8.5   Probability Distributions; Expected Value

1. Let x denote the number of heads observed. Then x can take on 0, 1, 2, 3, or 4 as values. The probabilities are as follows.

$$P(x=0) = \binom{4}{0}\left(\frac{1}{2}\right)^0\left(\frac{1}{2}\right)^4 = \frac{1}{16}$$

$$P(x=1) = \binom{4}{1}\left(\frac{1}{2}\right)^1\left(\frac{1}{2}\right)^3 = \frac{4}{16} = \frac{1}{4}$$

$$P(x=2) = \binom{4}{2}\left(\frac{1}{2}\right)^2\left(\frac{1}{2}\right)^2 = \frac{6}{16} = \frac{3}{8}$$

$$P(x=3) = \binom{4}{3}\left(\frac{1}{2}\right)^3\left(\frac{1}{2}\right)^1 = \frac{4}{16} = \frac{1}{4}$$

$$P(x=4) = \binom{4}{4}\left(\frac{1}{2}\right)^4\left(\frac{1}{2}\right)^0 = \frac{1}{16}$$

Therefore, the probability distribution is as follows.

Number of Heads	0	1	2	3	4
Probability	$\frac{1}{16}$	$\frac{1}{4}$	$\frac{3}{8}$	$\frac{1}{4}$	$\frac{1}{16}$

3. Let x denote the number of aces drawn. Then x can take on values 0, 1, 2, or 3. The probabilities are as follows.

$$P(x=0) = \binom{3}{0}\left(\frac{48}{52}\right)\left(\frac{47}{51}\right)\left(\frac{46}{50}\right) \approx 0.7826$$

$$P(x=1) = \binom{3}{1}\left(\frac{4}{52}\right)\left(\frac{48}{51}\right)\left(\frac{47}{50}\right) \approx 0.2042$$

$$P(x=2) = \binom{3}{2}\left(\frac{4}{52}\right)\left(\frac{3}{51}\right)\left(\frac{48}{50}\right) \approx 0.0130$$

$$P(x=3) = \binom{3}{3}\left(\frac{4}{52}\right)\left(\frac{3}{51}\right)\left(\frac{2}{50}\right) \approx 0.0002$$

Therefore, the probability distribution is as follows.

Number of Aces	0	1	2	3
Probability	0.7826	0.2042	0.0130	0.0002

5. Use the probabilities that were calculated in Exercise 1. Draw a histogram with 5 rectangles, corresponding to $x=0$, $x=1$, $x=2$, $x=3$, and $x=4$. $P(x \le 2)$ corresponds to

$$P(x=0) + P(x=1) + P(x=2),$$

so shade the first 3 rectangles in the histogram.

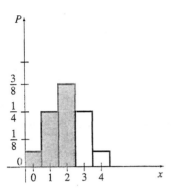

7. Use the probabilities that were calculated in Exercise 3. Draw a histogram with 4 rectangles, corresponding to $x=0$, $x=1$, $x=2$, and $x=3$. $P(\text{at least one ace}) = P(x \ge 1)$ corresponds to

$$P(x=1) + P(x=2) + P(x=3),$$

so shade the last 3 rectangles.

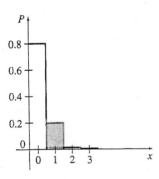

9. $E(x) = 2(0.1) + 3(0.4) + 4(0.3) + 5(0.2)$
$= 3.6$

11. $E(z) = 9(0.14) + 12(0.22) + 15(0.38) + 18(0.19)$
$+ 21(0.07)$
$= 14.49$

13. It is possible (but not necessary) to begin by writing the histogram's data as a probability distribution, which would look as follows.

x	1	2	3	4
$P(x)$	0.2	0.3	0.1	0.4

The expected value of x is

$$E(x) = 1(0.2) + 2(0.3) + 3(0.1) + 4(0.4)$$
$$= 2.7.$$

15. The expected value of x is

$$E(x) = 6(0.1) + 12(0.2) + 18(0.4) + 24(0.2)$$
$$+ 30(0.1)$$
$$= 18.$$

17. Using the data from Example 4, the expected winnings for Mary are

$$E(x) = -1.2 \left(\frac{1}{4}\right) + 1.2 \left(\frac{1}{4}\right) + 1.2 \left(\frac{1}{4}\right) + (-1.2) \left(\frac{1}{4}\right)$$
$$= 0.$$

Yes, it is still a fair game if Mary tosses and Donna calls.

19. (a)

Number of Yellow Marbles	Probability
0	$\dfrac{\binom{3}{0}\binom{4}{3}}{\binom{7}{3}} = \dfrac{4}{35}$
1	$\dfrac{\binom{3}{1}\binom{4}{2}}{\binom{7}{3}} = \dfrac{18}{35}$
2	$\dfrac{\binom{3}{2}\binom{4}{1}}{\binom{7}{3}} = \dfrac{12}{35}$
3	$\dfrac{\binom{3}{3}\binom{4}{0}}{\binom{7}{3}} = \dfrac{1}{35}$

Draw a histogram with four rectangles corresponding to $x = 0, 1, 2,$ and 3.

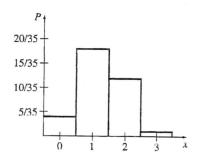

(b) Expected number of yellow marbles

$$= 0 \left(\frac{4}{35}\right) + 1 \left(\frac{18}{35}\right) + 2 \left(\frac{12}{35}\right) + 3 \left(\frac{1}{35}\right)$$
$$= \frac{45}{35} = \frac{9}{7} \approx 1.286$$

21. (a) Let x be the number of times 1 is rolled. Since the probability of getting a 1 on any single roll is $\frac{1}{6}$, the probability of any other outcome is $\frac{5}{6}$. Use combinations since the order of outcomes is not important.

$$P(x = 0) = \binom{4}{0} \left(\frac{1}{6}\right)^0 \left(\frac{5}{6}\right)^4 = \frac{625}{1296}$$

$$P(x = 1) = \binom{4}{1} \left(\frac{1}{6}\right)^1 \left(\frac{5}{6}\right)^3 = \frac{125}{324}$$

$$P(x = 2) = \binom{4}{2} \left(\frac{1}{6}\right)^2 \left(\frac{5}{6}\right)^2 = \frac{25}{216}$$

$$P(x = 3) = \binom{4}{3} \left(\frac{1}{6}\right)^3 \left(\frac{5}{6}\right)^1 = \frac{5}{324}$$

$$P(x = 4) = \binom{4}{4} \left(\frac{1}{6}\right)^4 \left(\frac{5}{6}\right)^0 = \frac{1}{1296}$$

x	0	1	2	3	4
$P(x)$	$\frac{625}{1296}$	$\frac{125}{324}$	$\frac{25}{216}$	$\frac{5}{324}$	$\frac{1}{1296}$

(b) $E(x) = 0 \left(\dfrac{625}{1296}\right) + 1 \left(\dfrac{125}{324}\right) + 2 \left(\dfrac{25}{216}\right)$
$$+ 3 \left(\frac{5}{324}\right) + 4 \left(\frac{1}{1296}\right)$$
$$= \frac{2}{3}$$

23. Set up the probability distribution.

Number of Women	0	1	2
Probability	$\dfrac{\binom{3}{0}\binom{5}{2}}{\binom{8}{2}}$	$\dfrac{\binom{3}{1}\binom{5}{1}}{\binom{8}{2}}$	$\dfrac{\binom{3}{2}\binom{5}{0}}{\binom{8}{2}}$
Simplified	$\frac{5}{14}$	$\frac{15}{28}$	$\frac{3}{28}$

$$E(x) = 0 \left(\frac{5}{14}\right) + 1 \left(\frac{15}{28}\right) + 2 \left(\frac{3}{28}\right)$$
$$= \frac{21}{28} = \frac{3}{4} = 0.75$$

25. Set up the probability distribution as in Exercise 20.

Number of Diamonds	0	1	2
Probability	$\dfrac{\binom{13}{0}\binom{39}{2}}{\binom{52}{2}}$	$\dfrac{\binom{13}{1}\binom{39}{1}}{\binom{52}{2}}$	$\dfrac{\binom{13}{2}\binom{39}{0}}{\binom{52}{2}}$
Simplified	$\dfrac{741}{1326}$	$\dfrac{507}{1326}$	$\dfrac{78}{1326}$

$$E(x) = 0\left(\frac{741}{1326}\right) + 1\left(\frac{507}{1326}\right) + 2\left(\frac{78}{1326}\right)$$

$$= \frac{663}{1326} = \frac{1}{2}$$

29. (a) First list the possible sums, 5, 6, 7, 8, and 9, and find the probabilities for each. The total possible number of results are $4 \cdot 3 = 12$. There are two ways to draw a sum of 5 (2 then 3, and 3 then 2). The probability of 5 is $\frac{2}{12} = \frac{1}{6}$. There are two ways to draw a sum of 6 (2 then 4, and 4 then 2). The probability of 6 is $\frac{2}{12} = \frac{1}{6}$. There are four ways to draw a sum of 7 (2 then 5, 3 then 4, 4 then 3, and 5 then 2). The probability of 7 is $\frac{4}{12} = \frac{1}{3}$. There are two ways to draw a sum of 8 (3 then 5, and 5 then 3). The probability of 7 is $\frac{2}{12} = \frac{1}{6}$. There are two ways to draw a sum of 9 (4 then 5, and 5 then 4). The probability of 9 is $\frac{2}{12} = \frac{1}{6}$. The distribution is as follows.

Sum	5	6	7	8	9
Probability	$\frac{1}{6}$	$\frac{1}{6}$	$\frac{1}{3}$	$\frac{1}{6}$	$\frac{1}{6}$

(b)

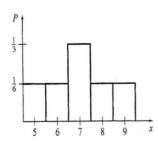

(c) The probability that the sum is even is $\frac{1}{6} + \frac{1}{6} = \frac{1}{3}$. Thus the odds are 1 to 2.

(d) $E(x) = \frac{1}{6}(5) + \frac{1}{6}(6) + \frac{1}{3}(7) + \frac{1}{6}(8) + \frac{1}{6}(9) = 7$

31. We first compute the amount of money the company can expect to pay out for each kind of policy. The sum of these amounts will be the total amount the company can expect to pay out. For a single $100,000 policy, we have the following probability distribution.

Outcome	Pay $100,000	Don't Pay $100,000
Probability	0.0012	0.9998

$$E(\text{payoff}) = 100{,}000(0.0012) + 0(0.9998)$$
$$= \$120$$

For all 100 such policies, the company can expect to pay out

$$100(120) = \$12{,}000.$$

For a single $50,000 policy,

$$E(\text{payoff}) = 50{,}000(0.0012) + 0(0.9998)$$
$$= \$60.$$

For all 500 such policies, the company can expect to pay out

$$500(60) = \$30{,}000.$$

Similarly, for all 1000 policies of $10,000, the company can expect to pay out

$$1000(12) = \$12{,}000.$$

Thus, the total amount the company can expect to pay out is

$$\$12{,}000 + \$30{,}000 + \$12{,}000 = \$54{,}000.$$

33. (a) Expected number of good nuts in 50 "blow outs" is

$$E(x) = 50(0.60) = 30.$$

(b) Since 80% of the "good nuts" are good, 20% are bad. Expected number of bad nuts in 50 "good nuts" is

$$E(x) = 50(0.20) = 10.$$

35. The tour operator earns $1050 if 1 or more tourists do not show up. The tour operator earns $950 if all tourists show up. The probability that all tourists show up is $(0.98)^{21} \approx 0.6543$. The expected revenue is $1050(0.3457) + 950(0.6543) = 984.57$

The answer is **e**.

37. (a) Expected cost of Amoxicillin:

$$E(x) = 0.75(\$59.30) + 0.25(\$96.15)$$
$$= \$68.51$$

Expected cost of Cefaclor:

$$E(x) = 0.90(\$69.15) + 0.10(\$106.00)$$
$$= \$72.84$$

(b) Amoxicillin should be used to minimize total expected cost.

39. $E(x) = 250(0.152) = 38$

We would expect 38 low-birth-weight babies to graduate from high school.

41. (a) Using binomial probability, $n = 48, x = 0$, $p = 0.0976$.

$$P(0) = \binom{48}{0}(0.0976)^0(0.9024)^{48} \approx 0.007230$$

(b) Using combinations, the probability is

$$\frac{\binom{74}{48}}{\binom{82}{48}} \approx 5.094 \times 10^{-4}.$$

(c) Using binomial probability, $n = 6, x = 5$, $p = 0.1$.

$$P(0) = \binom{6}{5}(0.1)^5(0.9)^1 + (0.1)^6 = 5.5 \times 10^{-5}$$

(d) Using binomial probability, $n = 6, p = 0.1$.

$$P(\text{at least } 2) = 1 - \binom{6}{0}(0.1)^0(0.9)^6 - \binom{6}{1}(0.1)^1(0.9)^5$$

$$\approx 0.1143$$

43. (a) We define a success to be a cat sitting in the chair with Kimberly. For this situation, $n = 4$; $x = 0, 1, 2, 3$, or 4; $p = 0.3$; and $1 - p = 0.7$.

Number of Cats	Probability
0	$\binom{4}{0}(0.3)^0(0.7)^4 = 0.2401$
1	$\binom{4}{1}(0.3)^1(0.7)^3 = 0.4116$
2	$\binom{4}{2}(0.3)^2(0.7)^2 = 0.2646$
3	$\binom{4}{3}(0.3)^3(0.7)^1 = 0.0756$
4	$\binom{4}{4}(0.3)^4(0.7)^0 = 0.0081$

(b) Expected number of cats
$$= 0(0.2401) + 1(0.4116) + 2(0.2646)$$
$$+ 3(0.0756) + 4(0.0081)$$
$$= 1.2$$

(c) Expected number of cats
$$= np = 4(0.3) = 1.2$$

45. Below is the probability distribution of x, which stands for the person's payback.

x	$\$398$	$\$78$	$-\$2$
$P(x)$	$\frac{1}{500} = 0.002$	$\frac{3}{500} = 0.006$	$\frac{497}{500} = 0.994$

The expected value of the person's winnings is

$$E(x) = 398(0.002) + 78(0.006) + (-2)(0.994)$$
$$\approx -\$0.72 \quad \text{or} \quad -72\cancel{c}.$$

Since the expected value of the payback is not 0, this is not a fair game.

47. There are 13 possible outcomes for each suit. That would make $13^4 = 28,561$ total possible outcomes. In one case, you win $5000 (minus the $1 cost to play the game). In the other 28,560 cases, you lose your dollar.

$$E(x) = 4999\left(\frac{1}{28,561}\right) + (-1)\left(\frac{28,560}{28,561}\right)$$
$$= -82\cancel{c}$$

49. There are $18 + 20 = 38$ possible outcomes. In 18 cases you win a dollar and in 20 you lose a dollar; hence,

$$E(x) = 1\left(\frac{18}{38}\right) + (-1)\left(\frac{20}{38}\right)$$
$$= -\frac{1}{19}, \text{ or about } -5.3\cancel{c}.$$

51. You have one chance in a thousand of winning $500 on a $1 bet for a net return of $499. In the 999 other outcomes, you lose your dollar.

$$E(x) = 499\left(\frac{1}{1000}\right) + (-1)\left(\frac{999}{1000}\right)$$
$$= \frac{-500}{1000} = -50\cancel{c}$$

53. Let x represent the payback. The probability distribution is as follows.

x	$P(x)$
100,000	$\frac{1}{2,000,000}$
40,000	$\frac{2}{2,000,000}$
10,000	$\frac{2}{2,000,000}$
0	$\frac{1,999,995}{2,000,000}$

The expected value is

$$E(x) = 100,000 \left(\frac{1}{2,000,000}\right) + 40,000 \left(\frac{2}{2,000,000}\right)$$

$$+ 10,000 \left(\frac{2}{2,000,000}\right) + 0 \left(\frac{1,999,995}{2,000,000}\right)$$

$$= 0.05 + 0.04 + 0.01 + 0$$

$$= \$0.10 = 10\cent.$$

Since the expected payback is $10\cent$ and if entering the contest costs $50\cent$, then it would not be worth it to enter.

55. **(a)** The possible scores are 0, 2, 3, 4, 5, 6. Each score has a probability of $\frac{1}{6}$.

$$E(x) = 0 \left(\frac{1}{6}\right) + 2 \left(\frac{1}{6}\right) + 3 \left(\frac{1}{6}\right) + 4 \left(\frac{1}{6}\right)$$

$$+ 5 \left(\frac{1}{6}\right) + 6 \left(\frac{1}{6}\right)$$

$$= \frac{1}{6}(20) = \frac{10}{3}$$

(b) The possible scores are

0 which has a probability of $\frac{11}{36}$,

4 which has a probability of $\frac{1}{36}$,

5 which has a probability of $\frac{2}{36}$,

6 which has a probability of $\frac{3}{36}$,

7 which has a probability of $\frac{4}{36}$,

8 which has a probability of $\frac{5}{36}$,

9 which has a probability of $\frac{4}{36}$,

10 which has a probability of $\frac{3}{36}$,

11 which has a probability of $\frac{2}{36}$,

12 which has a probability of $\frac{1}{36}$.

$$E(x) = 0 \left(\frac{11}{36}\right) + 4 \left(\frac{1}{36}\right) + 5 \left(\frac{2}{36}\right) + 6 \left(\frac{3}{36}\right)$$

$$+ 7 \left(\frac{4}{36}\right) + 8 \left(\frac{5}{36}\right) + 9 \left(\frac{4}{36}\right)$$

$$+ 10 \left(\frac{3}{36}\right) + 11 \left(\frac{2}{36}\right) + 12 \left(\frac{1}{36}\right)$$

$$= \frac{4}{36} + \frac{10}{36} + \frac{18}{36} + \frac{28}{36} + \frac{40}{36} + \frac{36}{36}$$

$$+ \frac{30}{36} + \frac{22}{36} + \frac{12}{36} = \frac{200}{36} = \frac{50}{9}$$

(c) If a single die does not result in a score of zero, the possible scores are 2, 3, 4, 5, 6 with each of these having a probability of $\frac{1}{5}$.

$$E(x) = 2 \left(\frac{1}{5}\right) + 3 \left(\frac{1}{5}\right) + 4 \left(\frac{1}{5}\right)$$

$$+ 5 \left(\frac{1}{5}\right) + 6 \left(\frac{1}{5}\right)$$

$$= \frac{1}{5}(20)$$

$$= 4$$

Thus, if a player rolls n dice the expected average score is

$$n \cdot E(x) = n \cdot 4 = 4n.$$

(d) If a player rolls n dice, a nonzero score will occur whenever each die rolls a number other than 1. For each die, there are 5 possibilities so the possible scoring ways for n dice is 5^n. When rolling die there are 6 possibilities so the possible outcomes for n die is 6^n. The probability of rolling a scoring set of die is $\frac{5^n}{6^n}$; thus, the expected value of the player's score when rolling n dice is $E(x) = \frac{5^n(4n)}{6^n}$.

57. **(a)** Let x be the number of hits. Since the probability of getting a hit on any one time at bat is 0.335, the probability of not getting a hit is 0.765. Use combinations since the order of hits is not important.

$$P(x = 0) = \binom{4}{0} (0.335)^0 (0.765)^4 \approx 0.1956$$

$$P(x = 1) = \binom{4}{1} (0.335)^1 (0.765)^3 \approx 0.3941$$

$$P(x = 2) = \binom{4}{2} (0.335)^2 (0.765)^2 \approx 0.2978$$

$$P(x = 3) = \binom{4}{3} (0.335)^3 (0.765)^1 \approx 0.1000$$

$$P(x = 4) = \binom{4}{4} (0.335)^4 (0.765)^0 \approx 0.0126$$

x	0	1	2	3	4
$P(x)$	0.1956	0.3941	0.2978	0.1000	0.0126

(b) $E(x) = 0(0.1956) + 1(0.3941) + 2(0.2978)$
$$+ 3(0.1000) + 4(0.0126)$$
$$\approx 1.34$$

Chapter 8 Review Exercises

1. 6 shuttle vans can line up at the airport in
$$P(6,6) = 6! = 720$$
different ways.

3. 3 oranges can be taken from a bag of 12 in
$$\binom{12}{3} = \frac{12!}{9!3!} = \frac{12 \cdot 11 \cdot 10}{3 \cdot 2 \cdot 1} = 220$$
different ways.

5. 2 pictures from a group of 5 different pictures can be arranged in
$$P(5,2) = 5 \cdot 4 = 20$$
different ways.

7. (a) There are 2! ways to arrange the landscapes, 3! ways to arrange the puppies, and 2 choices whether landscapes or puppies come first. Thus, the pictures can be arranged in
$$2!3! \cdot 2 = 24$$
different ways.

(b) The pictures must be arranged puppy, landscape, puppy, landscape, puppy. Arrange the puppies in 3! or 6 ways. Arrange the landscapes in 2! or 2 ways. In this scheme, the pictures can be arranged in $6 \cdot 2 = 12$ different ways.

9. (a) There are $7 \cdot 5 \cdot 4 = 140$ different groups of 3 representatives possible.

(b) $7 \cdot 5 \cdot 4 = 140$ is the number of groups with 3 representatives. For 2 representatives, the number of groups is
$$7 \cdot 5 + 7 \cdot 4 + 5 \cdot 4 = 83.$$
For 1 representative, the number of groups is
$$7 + 5 + 4 = 16.$$

The total number of these groups is
$$140 + 83 + 16 = 239$$
groups.

13. It is impossible to draw 3 blue balls, since there are only 2 blue balls in the basket; hence,
$$P(\text{all blue balls}) = 0.$$

15. $P(\text{exactly 2 black balls})$
$$= \frac{\binom{4}{2}\binom{9}{1}}{\binom{13}{3}} = \frac{54}{286} = \frac{27}{143} \approx 0.1888$$

17. $P(\text{2 green balls and 1 blue ball})$
$$= \frac{\binom{7}{2}\binom{2}{1}}{\binom{13}{3}} = \frac{42}{286} = \frac{21}{143} \approx 0.1469$$

19. Let x represent the number of girls. We have $n = 6$, $x = 6$, $p = \frac{1}{2}$, and $1 - p = \frac{1}{2}$, so
$$P(\text{all girls}) = \binom{6}{6}\left(\frac{1}{2}\right)^6\left(\frac{1}{2}\right)^0 = \frac{1}{64} \approx 0.016.$$

21. Let x represent the number of boys, and then $p = \frac{1}{2}$ and $1 - p = \frac{1}{2}$. We have

$P(\text{no more than 2 boys})$
$$= P(x \leq 2)$$
$$= P(x = 0) + P(x = 1) + P(x = 2)$$
$$= \binom{6}{0}\left(\frac{1}{2}\right)^0\left(\frac{1}{2}\right)^6 + \binom{6}{1}\left(\frac{1}{2}\right)^1\left(\frac{1}{2}\right)^5$$
$$+ \binom{6}{2}\left(\frac{1}{2}\right)^2\left(\frac{1}{2}\right)^4$$
$$= \frac{11}{32} \approx 0.344.$$

23. $P(\text{2 spades}) = \frac{\binom{13}{2}}{\binom{52}{2}} = \frac{78}{1326} = \frac{1}{17} \approx 0.059$

25. $P(\text{exactly 1 face card})$
$$= \frac{\binom{12}{1}\binom{40}{1}}{\binom{52}{2}} = \frac{480}{1326} = \frac{80}{221} \approx 0.3620$$

27. $P(\text{at most 1 queen})$
$$= P(\text{0 queens}) + P(\text{1 queen})$$
$$= \frac{\binom{48}{2}}{\binom{52}{2}} + \frac{\binom{4}{1}\binom{48}{1}}{\binom{52}{2}}$$
$$= \frac{1128}{1326} + \frac{192}{1326}$$
$$= \frac{1320}{1326} = \frac{220}{221} \approx 0.9955$$

29. (a) There are $n = 36$ possible outcomes. Let x represent the sum of the dice, and note that the possible values of x are the whole numbers from 2 to 12. The probability distribution is as follows.

x	2	3	4	5	6
$P(x)$	$\frac{1}{36}$	$\frac{2}{36} = \frac{1}{18}$	$\frac{3}{36} = \frac{1}{12}$	$\frac{4}{36} = \frac{1}{9}$	$\frac{5}{36}$

x	7	8	9	10	11	12
$P(x)$	$\frac{6}{36} = \frac{1}{6}$	$\frac{5}{36}$	$\frac{4}{36} = \frac{1}{9}$	$\frac{3}{36} = \frac{1}{12}$	$\frac{2}{36} = \frac{1}{18}$	$\frac{1}{36}$

(b) The histogram consists of 11 rectangles.

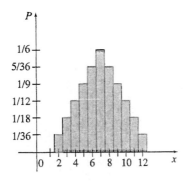

(c) The expected value is

$$E(x) = 2\left(\frac{1}{36}\right) + 3\left(\frac{2}{36}\right) + 4\left(\frac{3}{36}\right) + 5\left(\frac{4}{36}\right)$$
$$+ 6\left(\frac{5}{36}\right) + 7\left(\frac{6}{36}\right) + 8\left(\frac{5}{36}\right) + 9\left(\frac{4}{36}\right)$$
$$+ 10\left(\frac{3}{36}\right) + 11\left(\frac{2}{36}\right) + 12\left(\frac{1}{36}\right)$$
$$= \frac{252}{36} = 7.$$

31. The probability that corresponds to the shaded region of the histogram is the total of the shaded areas, that is,

$$1(0.1) + 1(0.3) + 1(0.2) = 0.6.$$

33. Let x represent the number of girls. The probability distribution is as follows.

x	0	1	2	3	4	5
$P(x)$	$\frac{1}{32}$	$\frac{5}{32}$	$\frac{10}{32}$	$\frac{10}{32}$	$\frac{5}{32}$	$\frac{1}{32}$

The expected value is

$$E(x) = 0\left(\frac{1}{32}\right) + 1\left(\frac{5}{32}\right) + 2\left(\frac{10}{32}\right) + 3\left(\frac{10}{32}\right)$$
$$+ 4\left(\frac{5}{32}\right) + 5\left(\frac{1}{32}\right)$$
$$= \frac{80}{32} = 2.5 \text{ girls.}$$

35. $P(3 \text{ clubs}) = \dfrac{\binom{13}{3}}{\binom{52}{3}} = \dfrac{286}{22,100} \approx 0.0129$

Thus,

$P(\text{win}) = 0.0129$ and
$P(\text{lose}) = 1 - 0.0129 = 0.9871.$

Let x represent the amount you should pay. Your net winnings are $100 - x$ if you win and $-x$ if you lose. If it is a fair game, your expected winnings will be 0. Thus, $E(x) = 0$ becomes

$$0.0129(100 - x) + 0.9871(-x) = 0$$
$$1.29 - 0.0129x - 0.9871x = 0$$
$$1.29 - x = 0$$
$$x = 1.29.$$

You should pay $1.29.

37. (a) Given a set with n elements, the number of subsets of size

$$0 \text{ is } \binom{n}{0} = 1,$$

$$1 \text{ is } \binom{n}{1} = n,$$

$$2 \text{ is } \binom{n}{2} = \frac{n(n-1)}{2}, \text{ and}$$

$$n \text{ is } \binom{n}{n} = 1.$$

(b) The total number of subsets is

$$\binom{n}{0} + \binom{n}{1} + \binom{n}{2} + \cdots + \binom{n}{n}.$$

(d) Let $n = 4$.

$$\binom{4}{0} + \binom{4}{1} + \binom{4}{2} + \binom{4}{3} + \binom{4}{4}$$
$$= 1 + 4 + 6 + 4 + 1$$
$$= 16$$
$$= 2^4 = 2^n$$

Let $n = 5$.

$$\binom{5}{0} + \binom{5}{1} + \binom{5}{2} + \binom{5}{3} + \binom{5}{4} + \binom{5}{5}$$
$$= 1 + 5 + 10 + 10 + 5 + 1$$
$$= 32$$
$$= 2^5 = 2^n$$

(e) The sum of the elements in row n of Pascal's triangle is 2^n.

39. $n = 12, x = 0, p = \dfrac{1}{6}$

$$P(0) = \binom{12}{0}\left(\frac{1}{6}\right)^0\left(\frac{5}{6}\right)^{12} \approx 0.1122$$

41. $n = 12, x = 10, p = \dfrac{1}{6}$

$$P(10) = \binom{12}{10}\left(\frac{1}{6}\right)^{10}\left(\frac{5}{6}\right)^{2} \approx 7.580 \times 10^{-7}$$

43. $n = 12, p = \dfrac{1}{6}$

$$P(\text{at least } 2) = 1 - P(\text{at most } 1)$$
$$= 1 - P(0) - P(1)$$
$$= 1 - \binom{12}{0}\left(\frac{1}{6}\right)^0\left(\frac{5}{6}\right)^{12}$$
$$- \binom{12}{1}\left(\frac{1}{6}\right)^1\left(\frac{5}{6}\right)^{11}$$
$$\approx 0.6187$$

45. The expected value is $\frac{1}{6}(12) = 2$.

47. Observe that for $a+b = 7$, $P(a)P(b) = \left(\frac{1}{2^{a+1}}\right)\left(\frac{1}{2^{b+1}}\right) = \frac{1}{2^{a+b+2}} = \frac{1}{2^9}$. The probability that exactly seven claims will be received during a given two-week period is

$$P(0)P(7) + P(1)P(6) + P(2)P(5) + P(3)P(4)$$
$$+ P(4)P(3) + P(5)P(2)$$
$$+ P(6)P(1) + P(7)P(0)$$
$$= 8\left(\frac{1}{2^9}\right) = \frac{1}{64}.$$

The answer is **d**.

49. Denote by S the event that a product is successful.
Denote by U the event that a product is unsuccessful.
Denote by Q the event of passing quality control. We must calculate the conditional probabilities $P(S|Q)$ and $P(U|Q)$ using Bayes' Theorem in order to calculate the expected net profit (in millions).

$E = 40P(S|Q) - 15P(U|Q)$.
$P(S) = P(U) = 0.5$
$P(Q|S) = 0.8, P(Q|U) = 0.25$

$$P(S|Q) = \frac{P(S) \cdot P(Q|S)}{P(S) \cdot P(Q|S) + P(U) \cdot P(Q|U)}$$
$$= \frac{0.5(0.8)}{0.5(0.8) + 0.5(0.25)}$$
$$= \frac{0.4}{0.4 + 0.125} = 0.762$$

$$P(U|Q) = \frac{P(U) \cdot P(Q|U)}{P(U) \cdot P(Q|U) + P(S) \cdot P(Q|S)}$$
$$= \frac{0.125}{0.525} = 0.238$$

Therefore,

$$E = 40P(S|Q) - 15P(U|Q)$$
$$= 40(0.762) - 15(0.238)$$
$$\approx 27.$$

So the expected net profit is $27 million, or the correct answer is **e**.

51. Let $I(x)$ represent the airline's net income if x people show up.

$I(0) = 0$
$I(1) = 400$
$I(2) = 2(400) = 800$
$I(3) = 3(400) = 1200$
$I(4) = 3(400) - 400 = 800$
$I(5) = 3(400) - 2(400) = 400$
$I(6) = 3(400) - 3(400) = 0$

Let $P(x)$ represent the probability that x people will show up. Use the binomial probability formula to find the values of $P(x)$.

$$P(0) = \binom{6}{0}(0.6)^0(0.4)^6 = 0.0041$$

$$P(1) = \binom{6}{1}(0.6)^1(0.4)^5 = 0.0369$$

$$P(2) = \binom{6}{2}(0.6)^2(0.4)^4 = 0.1382$$

$$P(3) = \binom{6}{3}(0.6)^3(0.4)^3 = 0.2765$$

$$P(4) = \binom{6}{4}(0.6)^4(0.4)^2 = 0.3110$$

$$P(5) = \binom{6}{5}(0.6)^5(0.4)^1 = 0.1866$$

$$P(6) = \binom{6}{6}(0.6)^6(0.4)^0 = 0.0467$$

(a) $E(I) = 0(0.0041) + 400(0.0369) + 800(0.1382) + 1200(0.2765) + 800(0.3110) + 400(0.1866) + 0(0.0467)$
$\qquad = \$780.56$

(b) $n = 3$

x	0	1	2	3
Income	0	100	200	300
$P(x)$	0.064	0.288	0.432	0.216

$E(I) = 0(0.064) + 400(0.288) + 800(0.432) + 1200(0.216) = \$720$

On the basis of all the calculations, the table given in the exercise is completed as follows.

x	0	1	2	3	4	5	6
Income	0	400	800	1200	800	400	0
$P(x)$	0.004	0.037	0.138	0.276	0.311	0.187	0.047

$n = 4$

x	1	1	2	3	4
Income	0	400	800	1200	800
$P(x)$	0.0256	0.1536	0.3456	0.3456	0.1296

$E(I) = 0(0.0256) + 400(0.1536) + 800(0.3456) + 1200(0.3456) + 800(0.1296) = \$856.32$

$n = 5$

x	0	1	2	3	4	5
Income	0	400	800	1200	800	400
$P(x)$	0.01024	0.0768	0.2304	0.3456	0.2592	0.07776

$E(I) = 0(0.01024) + 400(0.0768) + 800(0.2304) + 1200(0.3456) + 800(0.2592) + 400(0.07776) = \$868.22$

Since $E(I)$ is greatest when $n = 5$, the airlines should book 5 reservations to maximize revenue.

53. $P(\text{exactly } 5) = \binom{40}{5}\left(\frac{1}{8}\right)^5 \left(\frac{7}{8}\right)^{35} \approx 0.1875$

The probability that exactly 5 will choose a green piece of paper is about 0.1875.

55. (a) We define a success to be the event that a woman athlete is selected. In this situation, $n = 5$; $x = 0, 1, 2, 3, 4,$ or 5; $p = 0.4$; and $1 - p = 0.6$.

Number of Women	Probability
0	$\binom{5}{0}(0.4)^0(0.6)^5 = 0.0778$
1	$\binom{5}{1}(0.4)^1(0.6)^4 = 0.2592$
2	$\binom{5}{2}(0.4)^2(0.6)^3 = 0.3456$
3	$\binom{5}{3}(0.4)^3(0.6)^2 = 0.2304$
4	$\binom{5}{4}(0.4)^4(0.6)^1 = 0.0768$
5	$\binom{5}{5}(0.4)^5(0.6)^0 = 0.01021$

(b) Sketch the histogram with 6 rectangles.

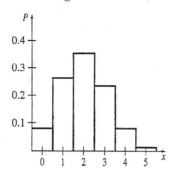

(c) Expected value = $np = 5(0.4) = 2$

57. (a)

Number Who Did Not Do Homework	Probability
0	$\dfrac{\binom{3}{0}\binom{7}{5}}{\binom{10}{5}} = \dfrac{21}{252} = \dfrac{1}{12}$
1	$\dfrac{\binom{3}{1}\binom{7}{4}}{\binom{10}{5}} = \dfrac{105}{252} = \dfrac{5}{12}$
2	$\dfrac{\binom{3}{2}\binom{7}{3}}{\binom{10}{5}} = \dfrac{105}{252} = \dfrac{5}{12}$
3	$\dfrac{\binom{3}{3}\binom{7}{2}}{\binom{10}{5}} = \dfrac{21}{252} = \dfrac{1}{12}$

(b) Draw a histogram with four rectangles.

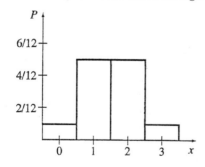

(c) Expected number who did not do homework

$$= 0\left(\frac{1}{12}\right) + 1\left(\frac{5}{12}\right) + 2\left(\frac{5}{12}\right) + 3\left(\frac{1}{12}\right)$$

$$= \frac{18}{12} = \frac{3}{2}$$

59. It costs $2(0.41 + 0.04) = 0.90$ to play the game.

x	\$1999.18	−\$0.90
$P(x)$	$\frac{1}{8000}$	$\frac{7999}{8000}$

$$E(x) = \$1999.18\left(\frac{1}{8000}\right) - \$0.90\left(\frac{7999}{8000}\right)$$

$$= -\$0.65$$

61. If the game was played 365 times a year for 26 years, it was played 9490 times. About $\frac{1}{1000}$ of those times, any one outcome–specifically, 000–would result. So, 000 would result $\frac{1}{1000}(9490) = 9.49$, or about 9.5 times.

63. (a) (i) When 5 socks are selected, we could get 1 matching pair and 3 odd socks or 2 matching pairs and 1 odd sock.

First consider 1 matching pair and 3 odd socks. The number of ways this could be done is

$$\binom{10}{1}\left[\binom{18}{3} - \binom{9}{1}\binom{16}{1}\right] = 6720.$$

$\binom{10}{1}$ gives the number of ways for 1 pair, while $\left[\binom{18}{3} - \binom{9}{1}\binom{16}{1}\right]$ gives the number of ways for the remaining 3 socks from the 18 socks left. We must subtract the number of ways the last 3 socks could contain a pair from the 9 pairs remaining.

Next consider 2 matching pairs and 1 odd sock. The number of ways this could be done is

$$\binom{10}{2}\binom{16}{1} = 720.$$

$\binom{10}{2}$ gives the number of ways for 2 pairs, while $\binom{16}{1}$ gives the number of ways for the 1 odd sock.

The total number of ways is

$$6720 + 720 = 7440.$$

Then

$$P(\text{matching pair}) = \frac{7440}{\binom{20}{5}} \approx 0.4799.$$

(ii) When 6 socks are selected, we could get 3 matching pairs and no odd socks or 2 matching pairs and 2 odd socks or 1 matching pair and 4 odd socks. The number of ways of obtaining 3 matching pairs is $\binom{10}{3} = 120$. The number of ways of obtaining 2 matching pairs and 2 odd socks is

$$\binom{10}{2}\left[\binom{16}{2} - \binom{8}{1}\right] = 5040.$$

The 2 odd socks must come from the 16 socks remaining but cannot be one of the 8 remaining pairs.

The number of ways of obtaining 1 matching pair and 4 odd socks is

$$\binom{10}{1}\left[\binom{18}{4} - \binom{9}{2} - \binom{9}{1}\left[\binom{16}{2} - 8\right]\right] = 20{,}160.$$

The 4 odd socks must come from the 18 socks remaining but can be 2 pairs and cannot be 1 pair and 2 odd socks.

The total number of ways is

$$120 + 5040 + 20{,}160 = 25{,}320.$$

Thus,

$$P(\text{matching pair}) = \frac{25{,}320}{\binom{20}{6}} \approx 0.6533.$$

(c) Suppose 6 socks are lost at random. The worst case is they are 6 odd socks. The best case is they are 3 matching pairs.

First find the number of ways of selecting 6 odd socks. This is

$$\binom{10}{6}\binom{2}{1}\binom{2}{1}\binom{2}{1}\binom{2}{1}\binom{2}{1}\binom{2}{1} = 13{,}440.$$

The $\binom{10}{6}$ gives the number of ways of choosing 6 different socks from the 10 pairs. But with each pair, $\binom{2}{1}$ gives the number of ways of selecting 1 sock. Then

$$P(\text{6 odd socks}) = \frac{13{,}440}{\binom{20}{6}} \approx 0.3467.$$

Next find the number of ways of selecting three matching pairs. This is $\binom{10}{3} = 120$. Then

$$P(\text{3 matching pairs}) = \frac{120}{\binom{20}{6}} \approx 0.003096.$$

Chapter 8 Test

[8.1]

1. The 24 members of the 3rd grade Mitey Mites hockey team must select a head basher, a second basher, and a designated tripper for their team. How many ways can this be done?

2. In Ohio, most license plates consist of three letters followed by three digits. How many different plates are possible in the following situations?

   (a) Repeats of letters and digits are allowed.

   (b) Repeats of letters, but not digits are allowed.

   (c) The first two letters must be AR, and repeats are allowed.

[8.1–8.2]

3. Evaluate each of the following.

   (a) $8!$ (b) $P(6,4)$ (c) $\binom{10}{4}$ (d) $\binom{17}{17}$

[8.2]

4. A basketball team consists of 7 good shooters and 5 poor shooters.

   (a) In how many ways can a 5-person team be selected?

   (b) In how many ways can a team consisting of only good shooters be selected?

   (c) In how many ways can a team consisting of 3 good and 2 poor shooters be selected?

5. A bag contains 4 red, 5 blue, and 8 white marbles. A sample of 6 marbles is drawn from the bag. In how many ways is it possible to get the following results?

   (a) 3 blue and 3 white marbles

   (b) 5 blue marbles and 1 white marble

   (c) 2 red, 2 blue, and 2 white marbles

[8.3]

6. A three-card hand is drawn from a standard deck of 52 cards. Set up the probability of each of the following events. (Do not calculate the answers.)

   (a) The hand contains exactly 2 hearts.

   (b) The hand contains fewer than 2 hearts.

   (c) The hand contains exactly 2 queens.

7. A mathematics class has 15 female students and 8 male students. Five students are randomly selected. Find the probability that there will be at least one male in the group. (Round your answer to four decimal places.)

[8.4]

8. A shipment of bolts has 20% of the bolts defective.

   (a) What is the probability of finding 2 or fewer defective bolts in a sample of 15?
   (b) What is the probability that if 9 bolts are sampled, 2 defectives will be found?
   (c) How many bolts must be sampled to ensure a probability of at least 0.6 that a defective bolt is found?

[8.5]

9. A nickel, a dime, and a quarter are tossed simultaneously. Let the random variable x denote the number of heads observed in the experiment, and prepare a probability distribution for this experiment.

10. Two dice are rolled, and the total number of points is recorded. Find the expected value.

11. There is a game called Double or Nothing, and it costs $1 to play. If you draw the ace of spades, you are paid $2, but you are paid nothing if you draw any other card. Is this a fair game?

12. At a large university, 62% of the students enrolled are female. A sample of 3 students are selected and the number of female students is noted.

   (a) Give the probability distribution for the number of females.
   (b) Sketch its histogram.
   (c) Find the expected value.

Chapter 8 Test Answers

1. 12,144

2. (a) 17,576,000   (b) 12,654,720   (c) 26,000

3. (a) 40,320   (b) 360   (c) 210   (d) 1

4. (a) 792   (b) 21   (c) 350

5. (a) 560   (b) 8   (c) 1680

6. (a) $\dfrac{\binom{13}{2}\binom{39}{1}}{\binom{52}{3}}$ (b) $\dfrac{\binom{13}{0}\binom{39}{3}+\binom{13}{1}\binom{39}{2}}{\binom{52}{3}}$ (c) $\dfrac{\binom{4}{2}\binom{48}{1}}{\binom{52}{3}}$

7. 0.9108

8. (a) $\binom{15}{0}(0.2)^0(0.8)^{15}+\binom{15}{1}(0.2)^1(0.8)^{14}+\binom{15}{2}(0.2)^2(0.8)^{13}\approx 0.3980$ (b) $\binom{9}{2}(0.2)^2(0.8)^7\approx 0.3020$ (c) 5

9.

x	0	1	2	3
$P(x)$	$\frac{1}{8}$	$\frac{3}{8}$	$\frac{3}{8}$	$\frac{1}{8}$

10. 7

11. No, it is not a fair game.

12. (a)

Number of Females	0	1	2	3
Probability	0.0549	0.2686	0.4382	0.2383

(b)

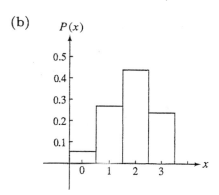

(c) 1.86

STATISTICS

9.1 Frequency Distributions; Measures of Central Tendency

1. **(a)-(b)** Since 0-24 is to be the first interval and there are 25 numbers between 0 and 24 inclusive, we will let all six intervals be of size 25. The other five intervals are 25-49, 50-74, 75-99, 100-124, and 125-149. Making a tally of how many data values lie in each interval leads to the following frequency distribution.

Interval	Frequency
0-24	4
25-49	8
50-74	5
75-99	10
100-124	4
125-149	5

(c) Draw the histogram. It consists of 6 bars of equal width having heights as determined by the frequency of each interval. See the histogram in part (d).

(d) To construct the frequency polygon, join consecutive midpoints of the tops of the histogram bars with line segments.

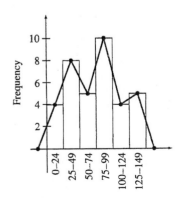

3. **(a)-(b)** There are eight intervals starting with 0-19. Making a tally of how many data values lie in each interval leads to the following frequency distribution.

Interval	Frequency
0-19	4
20-39	5
40-59	4
60-79	5
80-99	9
100-119	3
120-139	4
140-159	2

(c) Draw the histogram. It consists of 8 rectangles of equal width having heights as determined by the frequency of each interval. See the histogram in part (d).

(d) To construct the frequency polygon, join consecutive midpoints of the tops of the histogram bars with line segments.

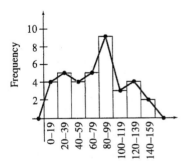

7. $\bar{x} = \dfrac{\sum x}{n}$

$= \dfrac{8 + 10 + 16 + 21 + 25}{5}$

$= \dfrac{80}{5} = 16$

9. $\sum x = 30{,}200 + 23{,}700 + 33{,}320 + 29{,}410 + 24{,}600$
$\qquad\qquad + 27{,}750 + 27{,}300 + 32{,}680$
$\qquad = 228{,}960$

The mean of the 8 numbers is

$$\bar{x} = \frac{\sum x}{n} = \frac{228{,}960}{8} = 28{,}620.$$

11. $\sum x = 9.4 + 11.3 + 10.5 + 7.4 + 9.1$
$\qquad\qquad + 8.4 + 9.7 + 5.2 + 1.1 + 4.7$
$\qquad = 76.8$

The mean of the 10 numbers is

$$\bar{x} = \frac{\sum x}{n} = \frac{76.8}{10} = 7.68.$$

13. Add to the frequency distribution a new column, "Value × Frequency."

Value	Frequency	Value × Frequency
4	6	$4 \cdot 6 =$ 24
6	1	$6 \cdot 1 =$ 6
9	3	$9 \cdot 3 =$ 27
15	2	$15 \cdot 2 =$ 30
Totals:	12	87

The mean is

$$\bar{x} = \frac{\sum xf}{n} = \frac{87}{12} = 7.25.$$

15. 27, 35, 39, 42, 47, 51, 54

The numbers are already arranged in numerical order, from smallest to largest. The median is the middle number, 42.

17. 100, 114, 125, 135, 150, 172

The median is the mean of the two middle numbers, which is

$$\frac{125 + 135}{2} = \frac{260}{2} = 130.$$

19. Arrange the numbers in numerical order, from smallest to largest.

$$3.4, 9.1, 27.6, 28.4, 29.8, 32.1, 47.6, 59.8$$

There are eight numbers here; the median is the mean of the two middle numbers, which is

$$\frac{28.4 + 29.8}{2} = \frac{58.2}{2} = 29.1.$$

21. Using a graphing calculator, $\bar{x} \approx 73.861$ and the median is 80.5.

23. 4, 9, 8, 6, 9, 2, 1, 3

The mode is the number that occurs most often. Here, the mode is 9.

25. 55, 62, 62, 71, 62, 55, 73, 55, 71

The mode is the number that occurs most often. Here, there are two modes, 55 and 62, since they both appear three times.

27. 6.8, 6.3, 6.3, 6.9, 6.7, 6.4, 6.1, 6.0

The mode is 6.3.

31.

Interval	Midpoint, x	Frequency, f	Product, xf
0-24	12	4	48
25-49	37	8	296
50-74	62	5	310
75-99	87	10	870
100-124	112	4	448
125-149	137	5	685
Totals:		36	2657

The mean of this collection of grouped data is

$$\bar{x} = \frac{\sum xf}{n} = \frac{2657}{36} \approx 73.8.$$

The interval 75-99 contains the most data values, 10, so it is the modal class.

35. Find the mean of the numbers in the "Production" column.

$$\bar{x} = \frac{\sum x}{n} = \frac{21{,}990}{10} = 2199$$

The mean production is 2199 million bushels. To find the median, list the ten values in the "Production" column from smallest to largest.

$$1606, 1947, 2105, 2158, 2228,$$
$$2277, 2296, 2345, 2481, 2547$$

The median is the mean of the two middle entries.

$$\frac{2228 + 2277}{2} = 2252.5$$

The median production is 2252.5 million bushels.

37. Find the mean for the grouped data.

$$\overline{x} = \frac{\sum xf}{n} = \frac{(10{,}000)(4741) + (30{,}000)(3893) + \ldots + (225{,}000)(58)}{4741 + 3893 + \ldots + 58} = \frac{541{,}265{,}000}{13{,}635} \approx \$39{,}696.74$$

The estimated mean household income for African Americans in 2004 is \$39,696.74.

39. (a) Find the mean of the numbers in the "Complaints" column.

$$\overline{x} = \frac{\sum x}{n} = \frac{2639}{12} \approx 219.92$$

The mean number of complaints is 219.92 complaints per airline.
To find the median, list the 12 values in the "Complaints" column from smallest to largest.

$$28,\ 49,\ 61,\ 69,\ 83,\ 93,\ 211,\ 227,\ 378,\ 403,\ 457,\ 580$$

The median is the mean of the two middle numbers.

$$\text{median} = \frac{93 + 211}{2} = 152.$$

The median number of complaints per airline is 152.

(b) The average found is not meaningful because not all airlines carry the same number of passengers.

(c) Find the mean of the numbers in the "Complaints per 100,000 Passengers Boarding" column.

$$\overline{x} = \frac{\sum x}{n} = \frac{10.08}{12} = 0.84$$

The mean number of complaints per 100,000 passengers boarding is 0.84.

To find the median, list the 12 values in the "Complaints per 100,000 Passengers Boarding" column from smallest to largest.

$$0.18,\ 0.32,\ 0.59,\ 0.62,\ 0.73,\ 0.84,\ 0.91,\ 1.00,\ 1.02,\ 1.17,\ 1.32,\ 1.38$$

The median is the mean of the two middle numbers.

$$\text{median} = \frac{0.84 + 0.91}{2} = 0.875$$

The median number of complaints per 100,000 passengers boarding is 0.875.

41. Find the mean.

$$\overline{x} = \frac{\sum x}{n} = \frac{16 + 12 + \ldots + 2}{13} = \frac{96}{13} \approx 7.38$$

The mean number of recognized blood types is 7.38.
Find the median.
The values are listed in order.
Since there are 13 values, the median is the seventh value, 7.
The median number of recognized blood types is 7.
The values 7, 5, and 4 each occur the greatest number of times, 2.
The modes are 7, 5, and 4.

43. (a) The height of the bar representing age group 20-29 is 12.

Therefore, 12% of the population is estimated to be in the age groups 20-29.

(b) The height of the bar representing age group 70+ is 13.

Therefore, 13% of the population is estimated to be in the age group 70+.

(c) The shortest bar represents age groups 50-59. Therefore, the age group 50-59 will have the smallest percent of the population.

(d) In Exercise 41, the percentages for various age ranges vary from 7% to 15%, while the estimated percentages for 2025 range from 11% to 13%. So, it seems the U.S. population is becoming uniform for all age groups.

45. (a) Find the mean of the numbers in the maximum temperature column.

$$\bar{x} = \frac{\sum x}{n} = \frac{666}{12} = 55.5$$

The mean of the maximum temperatures is 55.5°F. To find the median, list the 12 maximum temperatures from smallest to largest.

$$39, 39, 40, 44, 47, 50, 51, 60, 69, 70, 78, 79$$

The median is the mean of the two middle values.

$$\frac{50 + 51}{2} = 50.5°F$$

(b) Find the mean of the numbers in the minimum temperature column.

$$\bar{x} = \frac{\sum x}{n} = \frac{347}{12} \approx 28.9$$

The mean of the minimum temperatures is about 28.9°F.

To find the median, list the 12 minimum temperatures from smallest to largest.

$$16, 18, 20, 21, 24, 26, 31, 32, 37, 37, 42, 43$$

The median is the mean of the two middle values.

$$\frac{26 + 31}{2} = 28.5°F$$

47. (a) $\dfrac{5,700,000,000 \cdot 1 + 100 \cdot 80,000}{1 + 80,000}$

$$= \frac{5,708,000,000}{80,001}$$

$$\approx 71,349$$

The average worth of a citizen of Chukotka is $71,349.

49. (a) Find the mean:

$$\bar{x} = \frac{\sum x}{n} = \frac{96,487,919}{27} \approx 3,573,626.63$$

The mean salary is $3,573,627.

Find the median: The values are listed in order. Therefore, the median salary is the middle number, $1,750,000.

Find the mode: The mode is the most frequent entry in the list. Since the values $345,000 and $7,000,000 each occur twice, there are two modes. The modal salaries are $345,000 and $7,000,000.

(b) Since most of the team earned far below the mean salary of $3,573,627 and most of the team earned between the two modes of $345,000 and $7,000,000, the median salary of $1,750,000 best describes this data.

9.2   Measures of Variation

1. The standard deviation of a sample of numbers is the square root of the variance of the sample.

3. The range is the difference of the highest and lowest numbers in the list, or $85 - 52 = 33$.

To find the standard deviation, first find the mean.

$$\bar{x} = \frac{72 + 61 + 57 + 83 + 52 + 66 + 85}{7}$$

$$= \frac{476}{7} = 68$$

To prepare for calculating the standard deviation, construct a table.

x	x^2
72	5184
61	3721
57	3249
83	6889
52	2704
66	4356
85	7225
Total:	33,328

The variance is

$$s^2 = \frac{\sum x^2 - n\bar{x}^2}{n-1}$$

$$= \frac{33{,}328 - 7(68)^2}{7-1}$$

$$= \frac{33{,}328 - 32{,}368}{6}$$

$$= 160$$

and the standard deviation is

$$s = \sqrt{160} \approx 12.6.$$

5. The range is $287 - 241 = 46$. The mean is

$$\bar{x} = \frac{241 + 248 + 251 + 257 + 252 + 287}{6} = 256.$$

x	x^2
241	58,081
248	61,504
251	63,001
257	66,049
252	63,504
287	82,369
Total:	394,508

The standard deviation is

$$s = \sqrt{\frac{\sum x^2 - n\bar{x}^2}{n-1}}$$

$$= \sqrt{\frac{394{,}508 - 6(256)^2}{5}}$$

$$= \sqrt{258.4} \approx 16.1.$$

7. The range is $27 - 3 = 24$. The mean is

$$\bar{x} = \frac{\sum x}{n} = \frac{140}{10} = 14.$$

x	x^2
3	9
7	49
4	16
12	144
15	225
18	324
19	361
27	729
24	576
11	121
Total:	2554

The standard deviation is

$$s = \sqrt{\frac{\sum x^2 - n\bar{x}^2}{n-1}}$$

$$= \sqrt{\frac{2554 - 10(14)^2}{9}}$$

$$= \sqrt{66} \approx 8.1.$$

9. Using a graphing calculator, enter the 36 numbers into a list. Using the 1-Var Stats feature of a TI-83/84 Plus calculator, the standard deviation is found to be $Sx \approx 40.04793754$, or 40.05.

11. Expand the table to include columns for the midpoint x of each interval for xf, x^2, and fx^2.

Interval	f	x	xf	x^2	fx^2
30-39	4	12	48	144	576
40-49	8	37	296	1369	10,952
50-59	5	62	310	3844	19,220
60-69	10	87	870	7569	75,690
70-79	4	112	448	12,544	50,176
80-89	5	137	685	18,769	93,845
Totals:	36		2657		250,459

The mean of the grouped data is

$$\bar{x} = \frac{\sum xf}{n} = \frac{2657}{36} \approx 73.8.$$

The standard deviation for the grouped data is

$$s = \sqrt{\frac{\sum fx^2 - n\bar{x}^2}{n-1}}$$

$$= \sqrt{\frac{250{,}459 - 36(73.8)^2}{35}}$$

$$\approx \sqrt{1554}$$

$$\approx 39.4.$$

13. Use $k = 3$ in Chebyshev's theorem.

$$1 - \frac{1}{k^2} = 1 - \frac{1}{3^2} = \frac{8}{9},$$

so at least $\frac{8}{9}$ of the distribution is within 3 standard deviations of the mean.

15. Use $k = 5$ in Chebyshev's theorem.

$$1 - \frac{1}{k^2} = 1 - \frac{1}{5^2} = \frac{24}{25},$$

so at least $\frac{24}{25}$ of the distribution is within 5 standard deviations of the mean.

17. We have $48 = 60 - (3/2) \cdot 8 = \overline{x} - (3/2)s$ and $72 = 60 + (3/2) \cdot 8 = \overline{x} + (3/2)s$, so Chebyshev's theorem applies with $k = 3/2$. Hence, at least

$$1 - \frac{1}{k^2} = 1 - \frac{4}{9} = \frac{5}{9}$$

of the numbers lie between 48 and 72.

19. The answer here is the complement of the answer to Exercise 17. It was found there that at least 5/9 of the distribution of the numbers are between 48 and 72, so at most $1 - 5/9 = 4/9$ of the numbers are less than 48 or more than 72.

23. $15, 18, 19, 23, 25, 25, 28, 30, 34, 38$

(a) $\overline{x} = \dfrac{1}{10}(15 + 18 + 19 + 23 + 25 + 25 + 28$

$+\ 30 + 34 + 38)$

$= \dfrac{1}{10}(255) = 25.5$

The mean life of the sample of Brand X batteries is 25.5 hr.

x	x^2
15	225
18	324
19	361
23	529
25	625
25	625
28	784
30	900
34	1156
38	1444
Total:	6973

$$s = \sqrt{\frac{\sum x^2 - n\overline{x}^2}{n-1}}$$

$$= \sqrt{\frac{6973 - 10(25.5)^2}{9}}$$

$$\approx \sqrt{52.28} \approx 7.2$$

The standard deviation of the Brand X lives is 7.2 hr.

(b) Forever Power has a smaller standard deviation (4.1 hr, as opposed to 7.2 hr for Brand X), which indicates a more uniform life.

(c) Forever Power has a higher mean (26.2 hr, as opposed to 25.5 hr for Brand X), which indicates a longer average life.

25. (a)   (b)

Sample Number	$\overline{x}$	s
1	$\frac{1}{3}$	2.1
2	2	2.6
3	$-\frac{1}{3}$	1.5
4	0	2.6
5	$\frac{5}{3}$	2.5
6	$\frac{7}{3}$	0.6
7	1	1.0
8	$\frac{4}{3}$	2.1
9	$\frac{7}{3}$	0.6
10	$\frac{2}{3}$	1.2

(c) $\overline{X} = \dfrac{\sum \overline{x}}{n} \approx \dfrac{11.3}{10} = 1.13$

(d) $\overline{s} = \dfrac{\sum s}{n} = \dfrac{16.8}{10} = 1.68$

(e) The upper control limit for the sample means is

$$\overline{X} + k_1 \overline{s} = 1.13 + 1.954(1.68) \approx 4.41.$$

The lower control limit for the sample means is

$$\overline{X} - k_1 \overline{s} = 1.13 - 1.954(1.68) \approx -2.15.$$

(f) The upper control limit for the sample standard deviations is

$$k_2 \overline{s} = 2.568(1.68) \approx 4.31.$$

The lower control limit for the sample standard deviations is

$$k_3 \overline{s} = 0(1.68) = 0.$$

27. This exercise should be solved using a calculator with a standard deviation key. The answers are $\overline{x} = 1.8158$ mm and $s = 0.4451$ mm.

29. (a) This exercise should be solved using a calculator with a standard deviation key. The answers are $\bar{x} = 7.3571$ and $s = 0.1326$.

(b) $\bar{x} + 2s = 7.3571 + 2(0.1326) = 7.6223$
$\bar{x} - 2s = 7.3571 - 2(0.1326) = 7.0919$

All the data, or 100%, are within these two values, that is, within 2 standard deviations of the mean.

31. (a) Find the mean.

$$\bar{x} = \frac{\sum x}{n} = \frac{84 + 91 + \ldots + 164}{7} = \frac{894}{7} = 127.71$$

The mean is 127.71 days.
Find the standard deviation with a graphing calculator or spreadsheet.

$$s = 30.16$$

The standard deviation is 30.16 days.

(b) $\bar{x} + 2s = 127.71 + 2(30.16) = 188.03$
$\bar{x} - 2s = 127.71 - 2(30.16) = 67.39$

All seven of these cancers have doubling times that are within two standard deviations of the mean.

33. (a) Using a graphing calculator, the standard deviation is $s \approx 4,233,387$.

(b) The mean is $\bar{x} \approx 3,573,627$.

$\bar{x} + 3s = 3,573,627 + 3(4,233,387) = 16,273,788$
$\bar{x} - 3s = 3,573,627 - 3(4,233,387) = -9,126,534$

1 of the 27 players (about 4%), has a salary ($19,331,470) that is beyond 3 standard deviations of the mean.

9.3   The Normal Distribution

1. The peak in a normal curve occurs directly above *the mean*.

3. For normal distributions where $\mu \neq 0$ or $\sigma \neq 1$, z-scores are found by using the formula

$$z = \frac{x - \mu}{\sigma}.$$

5. Use the table, "Area Under a Normal Curve to the Left of z", in the Appendix. To find the percent of the area under a normal curve between the mean and 1.70 standard deviations from the mean, subtract the table entry for $z = 0$ (representing the mean) from the table entry for $z = 1.7$.

$$0.9554 - 0.5000 = 0.4554$$

Therefore, 45.54% of the area lies between μ and $\mu + 1.7\sigma$.

7. Subtract the table entry for $z = -2.31$ from the table entry for $z = 0$.

$$0.5000 - 0.0104 = 0.4896$$

48.96% of the area lies between μ and $\mu - 2.31\sigma$.

9. $P(0.32 \leq z \leq 3.18)$
$= P(z \leq 3.18) - P(z \leq 0.32)$
$=$ (area to the left of 3.18)
$-$ (area to the left of 0.32)
$= 0.9993 - 0.6255$
$= 0.3738$ or 37.38%

11. $P(-1.83 \leq z \leq -0.91)$
$= P(z \leq -0.91) - P(z \leq -1.83)$
$= 0.1814 - 0.0336$
$= 0.1478$ or 14.78%

13. $P(-2.95 \leq z \leq 2.03)$
$= P(z \leq 2.03) - P(z \leq -2.95)$
$= 0.9788 - 0.0016$
$= 0.9772$ or 97.72%

15. 5% of the total area is to the left of z.

Use the table backwards. Look in the body of the table for an area of 0.05, and find the corresponding z using the left column and top column of the table.

The closest values to 0.05 in the body of the table are 0.0505, which corresponds to $z = -1.64$, and 0.0495, which corresponds to $z = -1.65$.

17. 10% of the total area is to the right of z.

If 10% of the area is to the right of z, then 90% of the area is to the left of z. The closest value to 0.90 in the body of the table is 0.8997, which corresponds to $z = 1.28$.

19. For any normal distribution, the value of $P(x \leq \mu)$ is 0.5 since half of the distribution is less than the mean. Similarly, $P(x \geq \mu)$ is 0.5 since half of the distribution is greater than the mean.

21. According to Chebyshev's theorem, the probability that a number will lie within 3 standard deviations of the mean of a probability distribution is at least

$$1 - \frac{1}{3^2} = 1 - \frac{1}{9} = \frac{8}{9} \approx 0.8889.$$

Using the normal distribution, the probability that a number will lie within 3 standard deviations of the mean is 0.9974.

These values are not contradictory since "at least 0.8889" means 0.8889 or more. For the normal distribution, the value is more.

In Exercises 23-27, let x represent the life of a light bulb.

23. Less than 500 hr

$$z = \frac{x - \mu}{\sigma} = \frac{500 - 500}{100} = 0, \text{ so}$$

$$\begin{aligned} P(x < 500) &= P(z < 0) \\ &= \text{area to the left of } z = 0 \\ &= 0.5000. \end{aligned}$$

Hence, $0.5000(10,000) = 5000$ bulbs can be expected to last less than 500 hr.

25. Between 350 and 550 hr

For $x = 350$,

$$z = \frac{350 - 500}{100} = -1.5,$$

and for $x = 550$,

$$z = \frac{550 - 500}{100} = 0.5.$$

Then

$$\begin{aligned} P(350 < x < 550) &= P(-1.5 < z < 0.5) \\ &= \text{area between } z = -1.5 \\ &\quad \text{and } z = 0.5 \\ &= 0.6915 - 0.0668 \\ &= 0.6247. \end{aligned}$$

Hence, $0.6247(10,000) = 6247$ bulbs should last between 350 and 550 hr.

27. More than 440 hr

For $x = 440$,

$$z = \frac{440 - 500}{100} = -0.6.$$

Then

$$\begin{aligned} P(x > 440) &= P(z > -0.6) \\ &= \text{area to the right of } z = -0.6 \\ &= 1 - 0.2743 \\ &= 0.7257. \end{aligned}$$

Hence, $0.7257(10,000) = 7257$ bulbs should last more than 440 hr.

In Exercises 29—31, let x represent the weight of a package.

29. Here, $\mu = 16.5$, $\sigma = 0.5$.

For $x = 16$,

$$z = \frac{16 - 16.5}{0.5} = -1.$$

$$P(x < 16) = P(z < -1) = 0.1587$$

The fraction of the boxes that are underweight is 0.1587.

31. Here, $\mu = 16.5$, $\sigma = 0.2$.

For $x = 16$,

$$z = \frac{16 - 16.5}{0.2} = -2.5.$$

$$P(x < 16) = P(z < -2.5) = 0.0062$$

The fraction of the boxes that are underweight is 0.0062.

In Exercises 33—37, let x represent the weight of a chicken.

33. More than 1700 g means $x > 1700$.

For $x = 1700$,

$$z = \frac{1700 - 1850}{150} = -1.0.$$

$$\begin{aligned} P(x > 1700) &= 1 - P(x \le 1700) \\ &= 1 - P(z \le -1.0) \\ &= 1 - 0.1587 \\ &= 0.8413 \end{aligned}$$

Thus, 84.13% of the chickens will weigh more than 1700 g.

35. Between 1750 and 1900 g means $1750 \le x \le 1900$.

For $x = 1750$,

$$z = \frac{1750 - 1850}{150} = -0.67.$$

For $x = 1900$,

$$z = \frac{1900 - 1850}{150} = 0.33.$$

$$\begin{aligned} P(1750 \le x \le 1900) &= P(-0.67 \le z \le 0.33) \\ &= P(z \le 0.33) - P(z \le -0.67) \\ &= 0.6293 - 0.2514 \\ &= 0.3779 \end{aligned}$$

Thus, 37.79% of the chickens will weigh between 1750 and 1900 g.

37. More than 2100 g or less than 1550 g

$$P(x < 1550 \text{ or } x > 2100) = 1 - P(1550 \leq x \leq 2100).$$

For $x = 1550$,

$$z = \frac{1550 - 1850}{150} = -2.00.$$

For $x = 2100$,

$$z = \frac{2100 - 1850}{150} = 1.67.$$

$$\begin{aligned}
P(x &< 1550 \text{ or } x > 2100) \\
&= P(z \leq -2.00) + [1 - P(z \leq 1.67)] \\
&= 0.0228 + (1 - 0.9525) \\
&= 0.0228 + 0.0475 \\
&= 0.0703
\end{aligned}$$

Thus, 7.03% of chickens will weigh more than 2100 g or less than 1550 g.

39. Let x represent the bolt diameter.

$$\mu = 0.25, \ \sigma = 0.02$$

First, find the probability that a bolt has a diameter less than or equal to 0.3 in, that is, $P(x \leq 0.3)$. The z-score corresponding to $x = 0.3$ is

$$z = \frac{x - \mu}{\sigma} = \frac{0.3 - 0.25}{0.02} = 2.5.$$

Using the table, find the area to the left of $z = 2.5$. This gives

$$P(x \leq 0.3) = P(z \leq 2.5) = 0.9938.$$

Then

$$\begin{aligned}
P(x > 0.3) &= 1 - P(x \leq 0.3) \\
&= 1 - 0.9938 \\
&= 0.0062.
\end{aligned}$$

41. Let x represent the amount of a grocery bill. We are given

$$\mu = 74.50 \text{ and } \sigma = 24.30.$$

The middle 50% of the grocery bills have cutoffs at 25% below the mean and 25% above the mean. At 25% below the mean, the area to the left is 0.2500, which corresponds to about $z = -0.67$. At 25% above the mean, the area to the left is 0.7500, which corresponds to about $z = 0.67$. Find the x-value that corresponds to each z-score.

The largest amount of a grocery bill corresponds to $z = 0.67$.

$$\frac{x - 74.50}{24.30} = 0.67$$
$$x - 74.50 = 16.281$$
$$x \approx 90.78$$

The smallest amount of a grocery bill corresponds to $z = -0.67$.

$$\frac{x - 74.50}{24.30} = -0.67$$
$$x - 74.50 = -16.281$$
$$x \approx 58.22$$

The middle 50% of customers spend between \$58.22 and \$90.78.

43. Let x represent the amount of vitamins a person needs. Then

$$\begin{aligned}
P(x \leq \mu + 2.5\sigma) &= P(z \leq 2.5) \\
&= 0.9938.
\end{aligned}$$

99.38% of the people will receive adequate amounts of vitamins.

45. The Recommended Daily Allowance is

$$\begin{aligned}
\mu + 2.5\sigma &= 159 + 2.5(12) \\
&= 189 \text{ units.}
\end{aligned}$$

47. Let x represent an individual's blood clotting time (in seconds).

$$\mu = 7.45, \ \sigma = 3.6$$

For $x = 7$,

$$z = \frac{x - \mu}{\sigma} = \frac{7 - 7.45}{3.6} \approx -0.13,$$

and for $x = 8$,

$$z = \frac{8 - 7.45}{3.6} \approx 0.15.$$

Then

$$\begin{aligned}
P(x &< 7) + P(x > 8) \\
&= P(z < -0.13) + P(z > 0.15) \\
&= (\text{area to the left of } z = -0.13) \\
&\quad + (\text{area to the right of } z = 0.15) \\
&= 0.4483 + 0.4404 \\
&= 0.8887.
\end{aligned}$$

49. Let x represent a driving speed.

$$\mu = 52, \ \sigma = 8$$

At the 85th percentile, the area to the left is 0.8500, which corresponds to about $z = 1.04$. Find the x-value that corresponds to this z-score.

$$z = \frac{x - \mu}{\sigma}$$
$$1.04 = \frac{x - 52}{8}$$
$$8.32 = x - 52$$
$$60.32 = x$$

The 85th percentile speed for this road is 60.32 mph.

51. $P\left(x \geq \mu + \frac{3}{2}\sigma\right) = P(z \geq 1.5)$
$$= 1 - P(z \leq 1.5)$$
$$= 1 - 0.9332 = 0.0668$$

Thus, 6.68% of the students receive A's.

53. $P\left(\mu - \frac{1}{2}\sigma \leq x \leq \mu + \frac{1}{2}\sigma\right)$

$$= P(-0.5 \leq z \leq 0.5)$$
$$= P(z \leq 0.5) - P(z \leq -0.5)$$
$$= 0.6915 - 0.3085$$
$$= 0.383$$

Thus, 38.3% of the students receive C's.

In Exercises 55–57, let x represent a student's test score.

55. Since the top 8% get A's, we want to find the number a for which

$$P(x \geq a) = 0.08,$$
$$\text{or} \quad P(x \leq a) = 0.92.$$

Read the table backwards to find the z-score for an area of 0.92, which is 1.41. Find the value of x that corresponds to $z = 1.41$.

$$z = \frac{x - \mu}{\sigma}$$
$$1.41 = \frac{x - 76}{8}$$
$$11.28 = x - 76$$
$$87.28 = x$$

The bottom cutoff score for an A is 87.

57. 28% of the students will receive D's and F's, so to find the bottom cutoff score for a C we need to find the number c for which

$$P(x \leq c) = 0.28.$$

Read the table backwards to find the z-score for an area of 0.28, which is -0.58. Find the value of x that corresponds to $z = -0.58$.

$$-0.58 = \frac{x - 76}{8}$$
$$-4.64 = x - 76$$
$$71.36 = x$$

The bottom cutoff score for a C is 71.

59. (a) The area above the 55th percentile is equal to the area below the 45th percentile.

$$2P(z > 0.55) = 2[1 - P(z \leq 0.55)]$$
$$= 2(1 - 0.7088)$$
$$= 2(0.2912)$$
$$= 0.5824$$

$0.58 = 58\%$

(b) The area above the 60th percentile is equal to the area below the 40th percentile.

$$2P(z > 0.6) = 2[1 - P(z \leq 0.6)]$$
$$= 2(1 - 0.7257)$$
$$= 2(0.2743)$$
$$= 0.5486$$

$0.55 = 55\%$

The probability that the student will be above the 60th percentile or below the 40th percentile is 55%.

61. (a) $\mu = 93, \sigma = 16$

For $x = 130.5$,

$$z = \frac{130.5 - 93}{16} = 2.34.$$

Then,

$$P(x \geq 130.5) = P(z \geq 2.34)$$
$$= \text{area to the right of } 2.34$$
$$= 1 - 0.9904$$
$$= 0.0096$$

The probability is about 0.01 that a person from this time period would have a lead level of 130.5 ppm or higher. Yes, this provides evidence that Jackson suffered from lead poisoning during this time period.

(b) $\mu = 10, \sigma = 5$

For $x = 130.5$,

$$z = \frac{130.5 - 10}{5} = 24.1.$$

Then,

$P(x \geq 130.5) = P(z \geq 24.1) = $ area to the right of $24.1 \approx 0$

The probability is essentially 0 by these standards. From this we can conclude that Andrew Jackson had lead poisioning.

63.

	Reference	Models

a. Head size: $z = \dfrac{55 - 55.3}{2.0} = -0.15$ $z = \dfrac{55 - 50.0}{2.4} = 2.08$

$$P(x \geq 55) = P(z \geq -0.15)$$
$$= 1 - 0.4404$$
$$= 0.5596$$

$$P(x \geq 55) = P(z \geq 2.08)$$
$$= 1 - 0.9812$$
$$= 0.0188$$

b. Neck size: $z = \dfrac{23.9 - 32.7}{1.4} = -6.29$ $z = \dfrac{23.9 - 31.0}{1.0} = -7.1$

$$P(x \leq 23.9) = P(z \leq -6.29)$$
$$\approx 0$$

$$P(x \leq 23.9) = P(z \leq -7.1)$$
$$\approx 0$$

c. Bust size: $z = \dfrac{82.3 - 90.3}{5.5} = -1.45$ $z = \dfrac{82.3 - 87.4}{3.0} = -1.70$

$$P(x \geq 82.3) = P(z \geq -1.45)$$
$$= 1 - 0.0735$$
$$= 0.9265$$

$$P(x \geq 82.3) = P(z \geq -1.70)$$
$$= 1 - 0.0446$$
$$= 0.9554$$

d. Wrist size: $z = \dfrac{10.6 - 16.1}{0.8} = -6.88$ $z = \dfrac{10.6 - 15.0}{0.6} = -7.33$

$$P(x \leq 10.6) = P(z \leq -6.88)$$
$$\approx 0$$

$$P(x \leq 10.6) = P(z \leq -7.33)$$
$$\approx 0$$

e. Waist size: $z = \dfrac{40.7 - 69.8}{4.7} = -6.19$ $z = \dfrac{40.7 - 65.7}{3.5} = -7.14$

$$P(x \leq 40.7) = P(z \leq -6.19)$$
$$\approx 0$$

$$P(x \leq 40.7) = P(z \leq -7.14)$$
$$\approx 0$$

9.4   Normal Approximation to the Binomial Distribution

1. In order to find the mean and standard deviation of a binomial distribution, you must know the number of trials and the probability of a success on each trial.

3. Let x represent the number of heads tossed. For this experiment, $n = 16$, $x = 4$, and $p = \frac{1}{2}$.

   (a) $P(x = 4) = \binom{16}{4}\left(\frac{1}{2}\right)^4\left(1 - \frac{1}{2}\right)^4$

   ≈ 0.0278

   (b) $\mu = np = 16\left(\frac{1}{2}\right) = 8$

   $\sigma = \sqrt{np(1 - p)}$

   $= \sqrt{16\left(\frac{1}{2}\right)\left(\frac{1}{2}\right)}$

   $= \sqrt{4} = 2$

   For $x = 3.5$,

   $$z = \frac{3.5 - 8}{2} = -2.25.$$

   For $x = 4.5$,

   $$z = \frac{4.5 - 8}{2} = -1.75.$$

   $P(z < -1.75) - P(z < -2.25)$
   $= 0.0401 - 0.0122 = 0.0279$

5. Let x represent the number of tails tossed. For this experiment, $n = 16$; $x = 13$, 14, 15, or 16; and $p = \frac{1}{2}$.

   (a)
   $P(x = 13, 14, 15,\text{ or } 16)$

   $= \binom{16}{13}\left(\frac{1}{2}\right)^{13}\left(1 - \frac{1}{2}\right)^3 + \binom{16}{14}\left(\frac{1}{2}\right)^{14}\left(1 - \frac{1}{2}\right)^2$

   $+ \binom{16}{15}\left(\frac{1}{2}\right)^{15}\left(1 - \frac{1}{2}\right)^1 + \binom{16}{16}\left(\frac{1}{2}\right)^{16}\left(1 - \frac{1}{2}\right)^0$

   $\approx 0.00854 + 0.00183 + 0.00024 + 0.00001$
   ≈ 0.0106

(b) $\mu = np = 16\left(\frac{1}{2}\right) = 8$

   $\sigma = \sqrt{np(1 - p)} = \sqrt{16\left(\frac{1}{2}\right)\left(\frac{1}{2}\right)} = \sqrt{4} = 2$

For $x = 12.5$,

$$z = \frac{12.5 - 8}{2} = 2.25.$$

$P(z > 2.25) = 1 - P(z \le 2.25)$
$= 1 - 0.9878$
$= 0.0122$

In Exercises 7–9, let x represent the number of heads tossed. Since $n = 1000$ and $p = \frac{1}{2}$,

$$\mu = np = 1000\left(\frac{1}{2}\right) = 500$$

and

$$\sigma = \sqrt{np(1 - p)} = \sqrt{1000\left(\frac{1}{2}\right)\left(\frac{1}{2}\right)}$$
$$= \sqrt{250}$$
$$\approx 15.8.$$

7. To find $P(\text{exactly 500 heads})$, find the z-scores for $x = 499.5$ and $x = 500.5$.

   For $x = 499.5$,

   $$z = \frac{499.5 - 500}{15.8} \approx -0.03.$$

   For $x = 500.5$,

   $$z = \frac{500.5 - 500}{15.8} \approx 0.03.$$

   Using the table,

   $P(\text{exactly 500 heads}) = 0.5120 - 0.4880$
   $= 0.0240.$

9. Since we want 475 heads or more, we need to find the area to the right of $x = 474.5$. This will be $1 -$ (the area to the left of $x = 474.5$). Find the z-score for $x = 474.5$.

   $$z = \frac{474.5 - 500}{15.8} \approx -1.61$$

   The area to the left of 474.5 is 0.0537, so

   $P(480 \text{ heads or more}) = 1 - 0.0537$
   $= 0.9463.$

11. Let x represent the number of 5's tossed.

$n = 120$, $p = \frac{1}{6}$

$$\mu = np = 120 \left(\frac{1}{6} \right) = 20$$

$$\sigma = \sqrt{np(1-p)} = \sqrt{120 \left(\frac{1}{6} \right) \left(\frac{5}{6} \right)} \approx 4.08$$

Since we want the probability of getting exactly twenty 5's, we need to find the area between $x = 19.5$ and $x = 20.5$. Find the corresponding z-scores.

For $x = 19.5$,

$$z = \frac{19.5 - 20}{4.08} \approx -0.12.$$

For $x = 20.5$,

$$z = \frac{20.5 - 20}{4.08} \approx 0.12.$$

Using values from the table,

$$P(\text{exactly twenty 5's}) = 0.5478 - 0.4522$$
$$= 0.0956.$$

13. Let x represent the number of 3's tossed.

$n = 120$, $p = \frac{1}{6}$

$\mu = 20$, $\sigma \approx 4.08$

(These values for μ and σ are calculated in the solution for Exercise 11.)

Since

$P(\text{more than fifteen 3's})$
$\quad = 1 - P(\text{fifteen 3's or less}),$

find the z-score for $x = 15.5$.

$$z = \frac{15.5 - 20}{4.08} \approx -1.10$$

From the table, $P(z < -1.10) = 0.1357$.

Thus,

$$P(\text{more than fifteen 3's}) = 1 - 0.1357$$
$$= 0.8643.$$

15. Let x represent the number of times the chosen number appears.

$n = 130$; $x = 26, 27, 28, \ldots, 130$; and $p = \frac{1}{6}$

$$\mu = np = 130 \left(\frac{1}{6} \right) = \frac{65}{3}$$

$$\sigma = \sqrt{np(1-p)} = \sqrt{130 \left(\frac{1}{6} \right) \left(\frac{5}{6} \right)} = \frac{5}{6} \sqrt{26}$$

For $x = 25.5$,

$$z = \frac{25.5 - \frac{65}{3}}{\frac{5}{6}\sqrt{26}} \approx 0.90.$$

$$P(z > 0.90) = 1 - P(z \le 0.90) = 1 - 0.8159 = 0.1841$$

17. Let x represent the number of defective items.

$n = 75$, $p = 0.05$

$$\mu = np = 75(0.05) = 3.75$$
$$\sigma = \sqrt{np(1-p)} = \sqrt{75(0.05)(0.95)} = \sqrt{3.5625}$$

(a) To find $P(\text{exactly 5 defectives})$, find the z-scores for $x = 4.5$ and $x = 5.5$.

For $x = 4.5$,

$$z = \frac{4.5 - 3.75}{\sqrt{3.5625}} \approx 0.40.$$

For $x = 5.5$,

$$z = \frac{5.5 - 3.75}{\sqrt{3.5625}} \approx 0.93.$$

$$P(0.40 < z < 0.93) = P(z < 0.93) - P(z < 0.40)$$
$$= 0.8238 - 0.6554$$
$$= 0.1684$$

(b) To find $P(\text{no defectives})$, find the z-scores for $x = -0.5$ and $x = 0.5$.

For $x = -0.5$,

$$z = \frac{-0.5 - 3.75}{\sqrt{3.5625}} \approx -2.25.$$

For $x = 0.5$,

$$z = \frac{0.5 - 3.75}{\sqrt{3.5625}} \approx -1.72.$$

$$P(-2.25 < z < -1.72)$$
$$= P(z < -1.72) - P(z < -2.25)$$
$$= 0.0427 - 0.0122$$
$$= 0.0305$$

(c) To find P(at least 1 defective), find the z-score for $x = 0.5$.

$$z = \frac{0.5 - 3.75}{\sqrt{3.5625}} \approx -1.72$$

$$P(z > -1.72) = 1 - P(z < -1.72)$$
$$= 1 - 0.0427$$
$$= 0.9573$$

19. Let x represent the number of minimum wage earners who are 16 to 24 years old. $n = 600, p = 0.513, x = 341, 342, \ldots, 600$

$$\mu = np = 600(0.513) = 307.8$$
$$\sigma = \sqrt{np(1-p)}$$
$$= \sqrt{600(0.513)(0.487)}$$
$$= \sqrt{149.8986}$$
$$\approx 12.24$$

To find $P(x > 340)$, find the z-score for 340.5.

$$z = \frac{340.5 - 307.8}{12.24} = 2.67$$

$$P(z > 2.67) = 1 - P(z < 2.67)$$
$$= 1 - 0.9962$$
$$= 0.0038$$

21. (a) Let x represent the number of units an animal consumes.

$n = 120, x = 80, p = 0.6$

$$\mu = np = 120(0.6) = 72$$
$$\sigma = \sqrt{np(1-p)} = \sqrt{120(0.6)(0.4)} = \sqrt{28.8}$$

To find P(80 units consumed), find the z-scores for $x = 79.5$ and $x = 80.5$.

For $x = 79.5$,

$$z = \frac{79.5 - 72}{\sqrt{28.8}} \approx 1.40.$$

For $x = 80.5$,

$$z = \frac{80.5 - 72}{\sqrt{28.8}} \approx 1.58.$$

$$P(1.40 < z < 1.58) = P(z < 1.58) - P(z < 1.40)$$
$$= 0.9429 - 0.9192$$
$$= 0.0237$$

(b) To find P(at least 70 units), find the z-score for $x = 69.5$.

$$z = \frac{69.5 - 72}{\sqrt{28.8}} \approx -0.47$$

$$P(z > -0.47) = 1 - P(z < -0.47)$$
$$= 1 - 0.3192$$
$$= 0.6808$$

23. (a) Let x represent the number of people cured.

$n = 25, p = 0.80$

$$\mu = np = 25(0.80) = 20$$
$$\sigma = \sqrt{np(1-p)} = \sqrt{25(0.80)(0.20)} = 2$$

To find

$$P(\text{exactly 20 cured}) = P(19.5 \leq x \leq 20.5),$$

find the z-scores for $x = 19.5$ and $x = 20.5$.

For $x = 19.5$,

$$z = \frac{19.5 - 20}{2} = -0.25.$$

For $x = 20.5$,

$$z = \frac{20.5 - 20}{2} = 0.25.$$

Using the table,

$$P(\text{exactly 20 cured}) = 0.5987 - 0.4013$$
$$= 0.1974.$$

(b) Let x represent the number of people who are cured.

$n = 25, p = 0.8$

$$\mu = np = 25(0.8) = 20$$
$$\sigma = \sqrt{np(1-p)} = \sqrt{25(0.8)(0.2)} = \sqrt{4} = 2$$

To find P(all are cured) $= P$(25 are cured), find the z-scores for $x = 24.5$ and $x = 25.5$.

$$z = \frac{24.5 - 20}{2} = 2.25 \qquad z = \frac{25.5 - 20}{2} = 2.75.$$

Using the table,

$$P(\text{all are cured}) = 0.9970 - 0.9878$$
$$= 0.0092.$$

(c) $P(x = 0) = \binom{25}{0}(0.80)^0(0.20)^{25}$

$$= (0.20)^{25}$$
$$= 3.36 \times 10^{-18}$$
$$\approx 0$$

(d) From parts a and b, $\mu = 20$ and $\sigma = 2$.

To find P(12 or fewer are cured) $= P(x \leq 12)$, find the z-score for $x = 12.5$.

$$z = \frac{12.5 - 20}{2} = -3.75$$

Since the table does not go out to $z = -3.75$, we must extrapolate, that is, read beyond the values in the table.

$$P(x \leq 12) \approx 0.0001$$

25. (a) Let x represent the number that are AB−.

$n = 1000; x = 10, 11, 12, \ldots, 1000; p = 0.006$

$\mu = np = 1000(0.006) = 6$

$$\sigma = \sqrt{np(1-p)}$$
$$= \sqrt{1000(0.006)(0.994)}$$
$$\approx 2.442$$

To find $P(10 \text{ or more})$, find the z-score for $x = 9.5$.

$$z = \frac{9.5 - 6}{2.442} = 1.43$$

$P(z > 1.43) = 1 - 0.9236 = 0.0764$

(b) Let x represent the number that are B−.

$n = 1000; x = 20, 21, \ldots, 39, 40; p = 0.015$

$\mu = np = 1000(0.015) = 15$

$$\sigma = \sqrt{np(1-p)}$$
$$= \sqrt{1000(0.015)(0.985)}$$
$$\approx 3.844$$

To find $P(\text{between 20 to 40 inclusive})$, find the z-scores for $x = 19.5$ and $x = 40.5$.

$$z = \frac{19.5 - 15}{3.844} = 1.17, z = \frac{40.5 - 15}{3.844} = 6.63$$

$P(1.17 < z < 6.63) = P(z < 6.63) - P(z < 1.17)$
$$= 1 - 0.8790$$
$$= 0.1210$$

(c) $\mu = np = 500(0.015) = 7.5$

$$\sigma = \sqrt{np(1-p)}$$
$$= \sqrt{500(0.015)(0.985)}$$
$$\approx 2.718$$

To find P(15 or more donors being B−), find the z-score for $x = 14.5$.

$$z = \frac{14.5 - 7.5}{2.718} = 2.57$$

$$P(z > 2.57) = 1 - 0.9949 = 0.0051$$

The probability that 15 or more donors are B− is only 0.0051. It is very unlikely that this town has a higher than normal number of donors who are B−.

27. $n = 1400, p = 0.55$

$\mu = np = 1400(0.55) = 770$

$\sigma = \sqrt{np(1-p)} = \sqrt{1400(0.55)(0.45)} \approx 18.6$

To find $P(\text{at least 750 people}) = P(x \geq 749.5)$, find the z-score for $x = 749.5$.

$$z = \frac{749.5 - 770}{18.6} \approx -1.10$$

$$P(z \geq -1.10) = 1 - P(z \leq -1.10)$$
$$= 1 - 0.1357$$
$$= 0.8643$$

29. Let x represent the number of high school students who have carried a weapon. $n = 1200, p = 0.185, x = 201, 202, \ldots, 249$

$\mu = np = 1200(0.185) = 222$
$\sigma = \sqrt{np(1-p)}$
$= \sqrt{1200(0.185)(0.815)}$
$= \sqrt{180.93}$
≈ 13.45

To find $P(200 < x < 250)$, find the z-scores for 200.5 and 249.5.

$$z_1 = \frac{200.5 - 222}{13.45} = -1.60 \qquad z_2 = \frac{249.5 - 222}{13.45} = 2.04$$

$P(-1.60 < z < 2.04) = P(z < 2.04) - P(z < -1.60)$
$$= 0.9793 - 0.0548$$
$$= 0.9245$$

31. Let x represent the number of parents who require homework before TV watching.

$n = 51; x = 0, 1, 2, 3, 4, \text{ or } 5; p = \frac{1}{12}$

$$\mu = np = 51\left(\frac{1}{12}\right) = \frac{17}{4}$$

$$\sigma = \sqrt{np(1-p)} = \sqrt{51\left(\tfrac{1}{12}\right)\left(\tfrac{11}{12}\right)} = \frac{\sqrt{561}}{12}$$

To find $P(5 \text{ or fewer})$, find the z-score for $x = 5.5$.

$$z = \frac{5.5 - \frac{17}{4}}{\frac{\sqrt{561}}{12}} \approx 0.63$$

$$P(z < 0.63) = 0.7357$$

33. (a) $1 - P(x = 0, 1, 2, 3)$

$$= 1 - \left[\binom{156}{0}\left(\frac{1}{3709}\right)^0\left(1 - \frac{1}{3709}\right)^{156}\right.$$

$$+ \binom{156}{1}\left(\frac{1}{3709}\right)^1\left(1 - \frac{1}{3709}\right)^{155}$$

$$+ \binom{156}{2}\left(\frac{1}{3709}\right)^2\left(1 - \frac{1}{3709}\right)^{154}$$

$$+ \left.\binom{156}{3}\left(\frac{1}{3709}\right)^3\left(1 - \frac{1}{3709}\right)^{153}\right]$$

$$= 1.2139 \times 10^{-7}$$

(b) $n = 156, p = \dfrac{1}{3709}$

$$\mu = np = 156\left(\dfrac{1}{3709}\right) = 0.042$$

$$\sigma = \sqrt{np(1-p)}$$

$$= \sqrt{156\left(\tfrac{1}{3709}\right)\left(\tfrac{3708}{3709}\right)}$$

$$\approx 0.205$$

$$P(x \geq 4) = P\left(z \geq \dfrac{3.5 - 0.042}{0.205}\right)$$

$$= P(z \geq 16.9)$$

$$\approx 0$$

(c) $n = 20{,}000, p = \dfrac{1}{3709}$

$$\mu = np = 20{,}000\left(\dfrac{1}{3709}\right) = 5.39$$

$$\sigma = \sqrt{np(1-p)}$$

$$= \sqrt{20{,}000\left(\tfrac{1}{3709}\right)\left(\tfrac{3708}{3709}\right)}$$

$$\approx 2.32$$

$$P(x \geq 4) = P\left(z \geq \dfrac{3.5 - 5.39}{2.32}\right)$$

$$= P(z \geq -0.81)$$
$$= 1 - P(z < -0.81)$$
$$= 1 - 0.2090$$
$$= 0.7910$$

Chapter 9 Review Exercises

3. (a) Since 450-474 is to be the first interval, let all the intervals be of size 25. The largest data value is 566, so the last interval that will be needed is 550-574. The frequency distribution is as follows.

Interval	Frequency
450-474	5
475-499	6
500-524	5
525-549	2
550-574	2

(b) Draw the histogram. It consists of 5 bars of equal width having heights as determined by the frequency of each interval. See the histogram in part (c).

(c) Construct the frequency polygon by joining consecutive midpoints of the tops of the histogram bars with line segments.

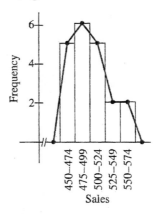

5. $\sum x = 30 + 24 + 34 + 30 + 29 + 28 + 30 + 29$
   $= 234$

The mean of the 8 numbers is

$$\bar{x} = \dfrac{\sum x}{n} = \dfrac{234}{8} = 29.25.$$

7.

Interval	Midpoint, x	Frequency, f	Product, xf
10-19	14.5	6	87
20-29	24.5	12	294
30-39	34.5	14	483
40-49	44.5	10	445
50-59	54.5	8	436
Totals:		50	1745

The mean of this collection of grouped data is

$$\bar{x} = \dfrac{\sum xf}{n} = \dfrac{1745}{50} = 34.9.$$

11. Arrange the numbers in numerical order, from smallest to largest.

$$35, 36, 36, 38, 38, 42, 44, 48$$

There are 8 numbers here; the median is the mean of the two middle numbers, which is

$$\dfrac{38 + 38}{2} = \dfrac{76}{2} = 38.$$

The mode is the number that occurs most often. Here, there are two modes, 36 and 38, since they both appear twice.

13. The modal class for the distribution of Exercise 8 is the interval 55-59, since it contains more data values than any of the other intervals.

17. The range is $93 - 26 = 67$, the difference of the highest and lowest numbers in the distribution.

The mean is

$$\bar{x} = \frac{\sum x}{n} = \frac{520}{10} = 52.$$

Construct a table with the values of x and x^2.

x	x^2
26	676
43	1849
51	2601
29	841
37	1369
56	3136
29	841
82	6724
74	5476
93	8649
Total:	32,162

The standard deviation is

$$s = \sqrt{\frac{\sum x^2 - n\bar{x}^2}{n - 1}}$$

$$= \sqrt{\frac{32,162 - 10(52)^2}{9}}$$

$$\approx \sqrt{569.1} \approx 23.9.$$

19. Start with the frequency distribution that was the answer to Exercise 8, and expand the table to include columns for the midpoint x of each interval and for xf, x^2, and fx^2.

Interval	f	x	xf	x^2	fx^2
40-44	3	42	126	1764	5292
45-49	6	47	282	2209	13,254
50-54	7	52	364	2704	18,928
55-59	14	57	798	3249	45,486
60-64	3	62	186	3844	11,532
65-69	2	67	134	4489	8978
Totals:	35		1890		103,470

The mean of the grouped data is

$$\bar{x} = \frac{\sum x}{n} = \frac{1890}{35} = 54.$$

The standard deviation for the grouped data is

$$s = \sqrt{\frac{\sum fx^2 - n\bar{x}^2}{n - 1}}$$

$$= \sqrt{\frac{103,470 - 35(54)^2}{34}}$$

$$\approx \sqrt{41.47}$$

$$\approx 6.4.$$

21. A skewed distribution has the largest frequency at one end rather than in the middle.

23. To the left of $z = 0.84$

Using the standard normal curve table,

$$P(z < 0.84) = 0.7995.$$

25. Between $z = 1.53$ and $z = 2.82$

$P(1.53 \leq z \leq 2.82)$
$= P(z \leq 2.82) - P(z \leq 1.53)$
$= 0.9976 - 0.9370$
$= 0.0606$

27. The normal distribution is not a good approximation of a binomial distribution that has a value of p close to 0 or 1 because the histogram of such a binomial distribution is skewed and therefore not close to the shape of a normal distribution.

29.

Number of Heads, x	Frequency, f	xf	fx^2
0	1	0	0
1	5	5	5
2	7	14	28
3	5	15	45
4	2	8	32
Totals:	20	42	110

(a) $\bar{x} = \dfrac{\sum xf}{n} = \dfrac{42}{20} = 2.1$

$$s = \sqrt{\frac{\sum fx^2 - n\bar{x}^2}{n - 1}} = \sqrt{\frac{110 - 20(2.1)^2}{20 - 1}} \approx 1.07$$

(b) For this binomial experiment,

$$\mu = np = 4\left(\frac{1}{2}\right) = 2,$$

and

$$\sigma = \sqrt{np(1 - p)} = \sqrt{4\left(\frac{1}{2}\right)\left(\frac{1}{2}\right)} = \sqrt{1} = 1.$$

(c) The answers to parts (a) and (b) should be close.

31. (a) For Stock I,

$$\bar{x} = \frac{11 + (-1) + 14}{3} = 8,$$

so, the mean (average return) is 8%.

$$s = \sqrt{\frac{\sum x^2 - n\bar{x}^2}{n-1}} = \sqrt{\frac{318 - 3(8)^2}{2}} = \sqrt{63} \approx 7.9$$

so the standard deviation is 7.9%.

For Stock II,

$$\bar{x} = \frac{9 + 5 + 10}{3} = 8,$$

so the mean is also 8%.

$$s = \sqrt{\frac{\sum x^2 - n\bar{x}^2}{n-1}} = \sqrt{\frac{206 - 3(8)^2}{2}} = \sqrt{7} \approx 2.6,$$

so the standard deviation is 2.6%.

(b) Both stocks offer an average (mean) return of 8%. The smaller standard deviation for Stock II indicates a more stable return and thus greater security.

33. Let x represent the number of overstuffed frankfurters.

$n = 500, p = 0.04, 1 - p = 0.96$

We also need the following results.

$$\mu = np = 500(0.04) = 20$$
$$\sigma = \sqrt{np(1-p)}$$
$$= \sqrt{500(0.04)(0.96)}$$
$$= \sqrt{19.2}$$
$$\approx 4.38$$

(a) P(twenty-five or fewer) or, equivalently, $P(x \leq 25)$

First, using the binomial probability formula:

$$P(x \leq 25) = \binom{500}{0}(0.04)^0(0.96)^{500}$$

$$+ \binom{500}{1}(0.04)^1(0.96)^{499}$$

$$+ \ldots + \binom{500}{25}(0.04)^{25}(0.96)^{475}$$

$$= (1.4 \times 10^{-9}) + (2.8 \times 10^{-8})$$
$$+ \ldots + 0.0446$$
$$\approx 0.8924$$

(To evaulate the sum, use a calculator or computer program. For example, using a TI-83/84 Plus calculator, enter the following:

sum(seq((500nCrX)(0.04^X)(0.96^(500−X)), X, 0, 25, 1))

The displayed result is 0.8923644609.)

Second, using the normal approximation:

To find $P(x \leq 25)$, first find the z-score for 25.5.

$$z = \frac{25.5 - 20}{4.38}$$

$$\approx 1.26$$
$$P(z < 1.26) = 0.8962$$

(b) P(exactly twenty-five) or $P(x = 25)$

Using the binomial probability formula:

$$P(x = 25) = \binom{500}{25}(0.04)^{25}(0.96)^{475}$$

$$\approx 0.0446$$

Using the normal approximation:

P(exactly twenty-five) corresponds to the area under the normal curve between $x = 24.5$ and $x = 25.5$. The corresponding z-scores are found as follows.

$$z = \frac{24.5 - 20}{4.38} \approx 1.03 \text{ and } z = \frac{25.5 - 20}{4.38} \approx 1.26$$

$$\begin{aligned} P(x = 25) &= P(24.5 < x < 25.5) \\ &= P(1.03 < z < 1.26) \\ &= P(z < 1.26) - P(z < 1.03) \\ &= 0.8962 - 0.8485 \\ &= 0.0477 \end{aligned}$$

(c) P(at least 30), or equivalently, $P(x \geq 30)$

Using the binomial probability formula:

This is the complementary event to "less than 30," which requires fewer calculations. We can use the results from Part (a) to reduce the amount of work even more.

$$\begin{aligned} P(x < 30) &= P(x \leq 25) + P(x = 26) \\ &\quad + P(x = 27) + P(x = 28) \\ &\quad + P(x = 29) \end{aligned}$$

$$= 0.8924 + \binom{500}{26}(0.04)^{26}(0.96)^{474}$$

$$+ \ldots + \binom{500}{29}(0.04)^{29}(0.96)^{471}$$

$$\approx 0.9804$$

Therefore,

$$P(x \geq 30) = 1 - P(x < 30)$$
$$= 1 - 0.9804$$
$$= 0.0196.$$

Using the normal approximation:

Again, use the complementary event. To find $P(x < 30)$, find the z-score for 29.5.

$$z = \frac{29.5 - 20}{4.38} \approx 2.17$$

$$P(z < 2.17) = 0.9850$$

Therefore,

$$P(x \geq 30) = 1 - P(x < 30)$$
$$= 1 - 0.9850$$
$$= 0.0150.$$

35. The table below records the mean and standard deviation for diet A and for diet B.

	$\bar{x}$	s
Diet A	2.7	2.26
Diet B	1.3	0.95

(a) Diet A had the greater mean gain, since the mean for diet A is larger.

(b) Diet B had a more consistent gain, since diet B has a smaller standard deviation.

37. Let x represent the number of flies that are killed.

$n = 1000$; $x = 0, 1, 2, \ldots, 986$; $p = 0.98$

$\mu = np = 1000(0.98) = 980$

$\sigma = \sqrt{np(1-p)} = \sqrt{1000(0.98)(0.02)} = \sqrt{19.6}$

To find P(no more than 986), find the z-score for $x = 986.5$.

$$z = \frac{986.5 - 980}{\sqrt{19.6}} \approx 1.47$$

$$P(z < 1.47) = 0.9292$$

39. Again, let x represent the number of flies that are killed.

$n = 1000$; $x = 973, 974, 975, \ldots, 993$; $p = 0.98$

As in Exercise 37, $\mu = 980$ and $\sigma = \sqrt{19.6}$. To find P(between 973 and 993), find the z-scores for $x = 972.5$ and $x = 993.5$.

For $x = 972.5$,

$$z = \frac{972.5 - 980}{\sqrt{19.6}} \approx -1.69.$$

For $x = 993.5$,

$$z = \frac{993.5 - 980}{\sqrt{19.6}} \approx 3.05.$$

$$P(-1.69 \leq z \leq 3.05) = P(z \leq 3.05) - P(z \leq -1.69)$$
$$= 0.9989 - 0.0455$$
$$= 0.9534$$

41. No more than 40 min/day

$\mu = 42$, $\sigma = 12$

Find the z-score for $x = 35$.

$$z = \frac{40 - 42}{12} \approx -0.17$$

$$P(x \leq 40) = P(z \leq -0.17) = 0.4325$$

43.25% of the residents commute no more than 40 min/day.

43. Between 38 and 60 min/day

$\mu = 42$, $\sigma = 12$

Find the z-scores for $x = 38$ and $x = 60$.

For $x = 38$,

$$z = \frac{38 - 42}{12} \approx -0.33.$$

For $x = 60$,

$$z = \frac{60 - 42}{12} = 1.5.$$

$$P(38 \leq x \leq 60) = P(-0.33 \leq z \leq 1.5)$$
$$= P(z \leq 1.5) - P(z \leq -0.33)$$
$$= 0.9332 - 0.3707$$
$$= 0.5625$$

56.25% of the residents commute between 38 and 60 min/day.

45. (a) The mean is found as follows.

$$\bar{x} = \frac{\sum x}{n} = \frac{1647}{12} = 137.25$$

To find the median, arrange the data in order from smallest to largest.

80, 83, 92, 93, 112, 121, 140,
159, 169, 197, 199, 202

Because there are an even number of items, the median is the mean of the two middle entries.

$$\text{median} = \frac{121 + 140}{2} = 130.5$$

Each number occurs only once in the list so there is no mode.

(b) To find the standard deviation, construct a table.

x	x^2
83	6889
93	8649
112	12,544
121	14,641
92	8464
80	6400
140	19,600
159	25,281
169	28,561
197	38,809
199	39,601
202	40,804
Total:	250,243

The standard deviation is

$$s = \sqrt{\frac{\sum x^2 - n\bar{x}^2}{n - 1}}$$

$$= \sqrt{\frac{250{,}243 - 12(137.25)^2}{12 - 1}}$$

$$\approx \sqrt{2199.30} \approx 46.9.$$

(c) $\bar{x} + s = 137.25 + 46.90 = 184.15$
$\bar{x} - s = 137.25 + 46.90 = 90.35$

Seven of the 12 years fall between these two values. Thus,

$$\frac{7}{12} \approx 0.583 \text{ or } 58.3\%$$

of the data is within one standard deviation of the mean.

(d) $\bar{x} + s = 137.25 + 2(46.90) = 231.05$
$\bar{x} - s = 137.25 - 2(46.90) = 43.45$

All of the data fall between these two values. Thus, 100% of the data is within two standard deviations of the mean.

47. (a) $P(x \geq 35) = P\left(z \geq \dfrac{35 - 28.0}{5.5}\right)$

$= P(z \geq 1.27)$
$= 1 - P(z < 1.27)$
$= 1 - 0.8980$
$= 0.1020$

(b) $P(x \geq 35) = P\left(z \geq \dfrac{35 - 32.2}{8.4}\right)$

$= P(z \geq 0.33)$
$= 1 - P(z < 0.33)$
$\doteq 1 - 0.6293$
$= 0.3707$

(d) $P(x \geq 1.4) = P\left(z \geq \dfrac{1.4 - 1.64}{0.08}\right)$

$= P(z \geq -3)$
$= 1 - P(z < -3)$
$= 1 - 0.0013$
$= 0.9987$

(e) $P(z < -1.5) + P(z > 1.5)$
$= 2P(z < -1.5)$
$= 2(0.0668)$
$= 0.1336$

Chapter 9 Test

[9.1]

1. Refer to the following data, which give the number of pairs of socks sold at the Sock and Accessory Shop during the past 20 weeks.

125	155	148	110	162	128	132	119	150	129
132	168	124	115	153	143	148	143	128	138

(a) Construct a grouped frequency distribution using six intervals, the first one being 110-119.

(b) Construct a histogram.

(c) Construct a frequency polygon.

2. Find the mean for the data shown in the following frequency distribution.

Value	Frequency
10	3
11	9
12	18
13	25
14	17
15	14
16	7

3. Find the mean, median, and mode (or modes) for the following list of numbers. Round the mean to the nearest tenth.

68, 72, 25, 49, 97, 58, 91, 25

[9.2]

4. Find the range, the mean, and the standard deviation for the following set of numbers.

3, 3, 4, 4, 4, 5, 6, 6, 6, 6

5. Find the standard deviation for the following grouped data. (Round your answer to the nearest tenth.)

Interval	Frequency
10-19	5
20-29	7
30-39	3
40-49	10
50-59	12

6. A distribution of 100 incomes has $\mu = \$10,500$ and $\sigma = \$1000$. Use Chebyshev's theorem to find:

   (a) A range of incomes which would include at least 75 of the 100 incomes.

   (b) The minimum number of incomes we would expect to find in the interval from $7500 to $13,500.

[9.3]

7. Using the normal curve table, find the following areas under the standard normal curve.

   (a) The area between $z = -1.13$ and $z = 2.14$

   (b) The area to the right of $z = 2.1$

8. If a normal distribution has mean 60 and standard deviation 6, find the following z-scores.

   (a) The z-score for $x = 72$

   (b) The z-score for $x = 51$

   (c) The z-score for $x = 43.5$

9. A survey of students enrolled at a community college finds the students have an average age of 26.5 years, with a standard deviation of 5.5 years. The ages are closely approximated by a normal curve.

   Using a normal curve table, find the percent of students whose ages are in the following ranges. (Round your answers to the nearest percent.)

   (a) Younger than 24   (b) Between 24 and 29   (c) Over 35

[9.4]

10. Find the mean and standard deviation of a binomial distribution with $n = 56$ and $p = 0.4$.

11. Al's Quick Photo store develops an average roll of film in 2.3 minutes. The standard deviation in time is 0.6 minutes. Assume a normal distribution.

    (a) Out of 1000 rolls of film, how many will be developed in less than 3.5 minutes?

    (b) What percentage of film is developed in between 2 and 3 minutes?

12. A loaded coin with $P(\text{head}) = 0.6$ is tossed 100 times. Using the normal curve approximation to the binomial distribution, find the probability of getting each of the following:

    (a) at least 55 heads;   (b) exactly 61 heads;

    (c) between 55 and 65 heads (inclusive).

Chapter 9 Test Answers

1. (a)

Interval	Frequency
110-119	3
120-129	5
130-139	3
140-149	4
150-159	3
160-169	2

(b)-(c)

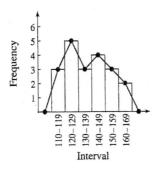

2. 13.2

3. Mean: 60.6; median: 63; mode: 25

4. Range: 3; mean: 4.7; standard deviation: 1.25

5. 14.6

6. (a) $8500 to $12,500   (b) 88

7. (a) 0.8546   (b) 0.0179

8. (a) 2   (b) -1.5 (c) -2.75

9. (a) 33%   (b) 35%   (c) 6%

10. $\mu = 22.4$, $\sigma \approx 3.67$

11. (a) 977   (b) 57%

12. (a) 0.8686   (b) 0.0819   (c) 0.7372

NONLINEAR FUNCTIONS

10.1   Properties of Functions

1. The x-value of 82 corresponds to two y-values, 93 and 14. In a function, each value of x must correspond to exactly one value of y.
The rule is not a function.

3. Each x-value corresponds to exactly one y-value.
The rule is a function.

5. $y = x^3 + 2$

Each x-value corresponds to exactly one y-value.
The rule is a function.

7. $x = |y|$

Each value of x (except 0) corresponds to two y-values.
The rule is not a function.

9. $y = 2x + 3$

x	-2	-1	0	1	2	3
y	-1	1	3	5	7	9

Pairs: $(-2,-1)$, $(-1,1)$, $(0,3)$,
$(1,5)$, $(2,7)$, $(3,9)$

Range: $\{-1,\ 1,\ 3,\ 5,\ 7,\ 9\}$

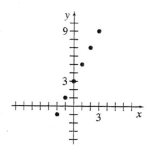

11. $2y - x = 5$
$\qquad 2y = 5 + x$
$\qquad y = \dfrac{1}{2}x + \dfrac{5}{2}$

x	-2	-1	0	1	2	3
y	$\frac{3}{2}$	2	$\frac{5}{2}$	3	$\frac{7}{2}$	4

Pairs:$(-2, \frac{3}{2})$, $(-1, 2)$, $(0, \frac{5}{2})$, $(1, 3)$, $(2, \frac{7}{2})$, $(3, 4)$

Range: $\{\frac{3}{2},\ 2,\ \frac{5}{2}\ 3,\ \frac{7}{2},\ 4\}$

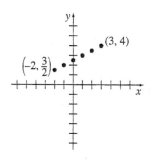

13. $y = x(x + 2)$

x	-2	-1	0	1	2	3
y	0	-1	0	3	8	15

Pairs: $(-2, 0)$, $(-1, -1)$, $(0, 0)$,
$(1, 3)$, $(2, 8)$, $(3, 15)$

Range: $\{-1,\ 0,\ 3,\ 8,\ 15\}$

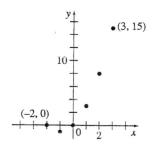

15. $y = x^2$

x	-2	-1	0	1	2	3
y	4	1	0	1	4	9

Pairs: $(-2, 4)$, $(-1, 1)$, $(0, 0)$,
        $(1, 1)$, $(2, 4)$, $(3, 9)$

Range: $\{0, 1, 4, 9\}$

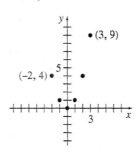

17. $y = \dfrac{1}{x + 3}$

x	-2	-1	0	1	2	3
y	1	$\frac{1}{2}$	$\frac{1}{3}$	$\frac{1}{4}$	$\frac{1}{5}$	$\frac{1}{6}$

Pairs: $(-2, 1), (-1, \frac{1}{2}), (0, \frac{1}{3}), (1, \frac{1}{4}), (2, \frac{1}{5}), (3, \frac{1}{6})$

Range: $\{1, \frac{1}{2}, \frac{1}{3}, \frac{1}{4}, \frac{1}{5}, \frac{1}{6}\}$

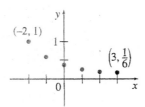

19. $y = \dfrac{2x - 2}{x + 4}$

x	-2	-1	0	1	2	3
y	-3	$-\frac{4}{3}$	$-\frac{1}{2}$	0	$\frac{1}{3}$	$\frac{4}{7}$

Pairs: $(-2, -3),\ (-1, -\frac{4}{3}),\ (0, -\frac{1}{2}), (1, 0), (2, \frac{1}{3}), (3, \frac{4}{7})$

Range: $\{-3, -\frac{4}{3}, -\frac{1}{2}, 0, \frac{1}{3}, \frac{4}{7}\}$

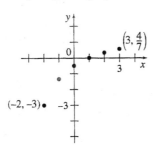

21. $f(x) = 2x$

x can take on any value, so the domain is the set of real numbers, $(-\infty, \infty)$.

23. $f(x) = x^4$

x can take on any value, so the domain is the set of real numbers, $(-\infty, \infty)$.

25. $f(x) = \sqrt{4 - x^2}$

For $f(x)$ to be a real number, $4 - x^2 \geq 0$.
Solve $4 - x^2 = 0$.

$$(2 - x)(2 + x) = 0$$
$$x = 2 \quad \text{or} \quad x = -2$$

The numbers form the intervals $(-\infty, -2)$, $(-2, 2)$, and $(2, \infty)$.
Only values in the interval $(-2, 2)$ satisfy the inequality. The domain is $[-2, 2]$.

27. $f(x) = (x - 3)^{1/2} = \sqrt{x - 3}$

For $f(x)$ to be a real number,

$$x - 3 \geq 0$$
$$x \geq 3.$$

The domain is $[3, \infty)$.

29. $f(x) = \dfrac{2}{1 - x^2} = \dfrac{2}{(1 - x)(1 + x)}$

Since division by zero is not defined,
$(1 - x) \cdot (1 + x) \neq 0$.
When $(1 - x)(1 + x) = 0$,

$$1 - x = 0 \quad \text{or} \quad 1 + x = 0$$
$$x = 1 \quad \text{or} \quad x = -1.$$

Thus, x can be any real number except ± 1.
The domain is

$$(-\infty, -1) \cup (-1, 1) \cup (1, \infty).$$

31. $f(x) = -\sqrt{\dfrac{2}{x^2 - 16}} = -\sqrt{\dfrac{2}{(x - 4)(x + 4)}}.$

$(x-4) \cdot (x+4) > 0$, since $(x-4) \cdot (x+4) < 0$ would produce a negative radicand and $(x-4) \cdot (x+4) = 0$ would lead to division by zero.

Solve $(x - 4) \cdot (x + 4) = 0$.
$x - 4 = 0 \quad \text{or} \quad x + 4 = 0$
$x = 4 \quad \text{or} \quad x = -4$

Use the values -4 and 4 to divide the number line into 3 intervals, $(-\infty, -4), (-4, 4)$ and $(4, \infty)$.
Only the values in the intervals $(-\infty, -4)$ and $(4, \infty)$ satisfy the inequality.
The domain is

$$(-\infty, -4) \cup (4, \infty).$$

33. $f(x) = \sqrt{x^2 - 4x - 5} = \sqrt{(x-5)(x+1)}$

See the method used in Exercise 25.

$$(x-5)(x+1) \geq 0$$

when $x \geq 5$ and when $x \leq -1$.
The domain is $(-\infty, -1] \cup [5, \infty)$.

35. $f(x) = \dfrac{1}{\sqrt{3x^2 + 2x - 1}} = \dfrac{1}{\sqrt{(3x-1)(x+1)}}$

$(3-2)(x+1) > 0$, since the radicand cannot be negative and the denominator of the function cannot be zero.
Solve $(3-1)(x+1) = 0$.

$$3 - 1 = 0 \quad \text{or} \quad x + 1 = 0$$
$$x = \tfrac{1}{3} \quad \text{or} \quad x = -1$$

Use the values -1 and $\frac{1}{3}$ to divide the number line into 3 intervals, $(-\infty, -1)$, $(-1, 4)$ and $(\frac{1}{3}, \infty)$. Only the values in the intervals $(-\infty, -\frac{1}{3})$ and $(\frac{1}{3}, \infty)$ satisfy the inequality.
The domain is $(-\infty, -1) \cup (\frac{1}{3}, \infty)$.

37. By reading the graph, the domain is all numbers greater than or equal to -5 and less than 4. The range is all numbers greater than or equal to -2 and less than or equal to 6.
Domain: $[-5, 4)$; range: $[-2, 6]$

39. By reading the graph, x can take on any value, but y is less than or equal to 12.
Domain: $(-\infty, \infty)$; range: $(-\infty, 12]$

41. $f(x) = 3x^2 - 4x + 1$

(a) $f(4) = 3(4)^2 - 4(4) + 1$
$\quad = 48 - 16 + 1$
$\quad = 33$

(b) $f\left(-\dfrac{1}{2}\right) = 3\left(-\dfrac{1}{2}\right)^2 - 4\left(-\dfrac{1}{2}\right) + 1$

$\quad = \dfrac{3}{4} + 2 + 1$

$\quad = \dfrac{15}{4}$

(c) $f(a) = 3(a)^2 - 4(a) + 1$
$\quad = 3a^2 - 4a + 1$

(d) $f\left(\dfrac{2}{m}\right) = 3\left(\dfrac{2}{m}\right)^2 - 4\left(\dfrac{2}{m}\right) + 1$

$\quad = \dfrac{12}{m^2} - \dfrac{8}{m} + 1$

$\quad \text{or} \quad \dfrac{12 - 8m + m^2}{m^2}$

(e) $\qquad f(x) = 1$
$$3x^2 - 4x + 1 = 1$$
$$3x^2 - 4x = 0$$
$$x(3x - 4) = 0$$
$$x = 0 \quad \text{or} \quad x = \dfrac{4}{3}$$

43. $f(x) = \dfrac{2x + 1}{x - 2}$

(a) $f(4) = \dfrac{2(4) + 1}{4 - 2} = \dfrac{9}{2}$

(b) $f\left(-\dfrac{1}{2}\right) = \dfrac{2(-\frac{1}{2}) + 1}{-\frac{1}{2} - 2}$

$\quad = \dfrac{-1 + 1}{\frac{5}{2}}$

$\quad = \dfrac{0}{\frac{5}{2}} = 0$

(c) $f(a) = \dfrac{2(a) + 1}{(a) - 2} = \dfrac{2a + 1}{a - 2}$

(d) $f\left(\dfrac{2}{m}\right) = \dfrac{2\left(\frac{2}{m}\right) + 1}{\frac{2}{m} - 2}$

$\quad = \dfrac{\frac{4}{m} + \frac{m}{m}}{\frac{2}{m} - \frac{2m}{m}}$

$\quad = \dfrac{\frac{4+m}{m}}{\frac{2-2m}{m}}$

$\quad = \dfrac{4 + m}{m} \cdot \dfrac{m}{2 - 2m}$

$\quad = \dfrac{4 + m}{2 - 2m}$

(e) $\qquad f(x) = 1$
$$\dfrac{2x + 1}{x - 2} = 1$$
$$2x + 1 = x - 2$$
$$x = -3$$

45. The domain is all real numbers between the endpoints of the curve, or $[-2, 4]$.
The range is all real numbers between the minimum and maximum values of the function or $[0, 4]$.

(a) $f(-2) = 0$

(b) $f(0) = 4$

(c) $f\left(\dfrac{1}{2}\right) = 3$

(d) From the graph, $f(x) = 1$ when $x = -1.5$, 1.5, or 2.5.

47. The domain is all real numbers between the endpoints of the curve, or $[-2, 4]$.

The range is all real numbers between the minimum and maximum values of the function or $[-3, 2]$.

(a) $f(-2) = -3$

(b) $f(0) = -2$

(c) $f\left(\dfrac{1}{2}\right) = -1$

(d) From the graph, $f(x) = 1$ when $x = 2.5$.

49. $f(x) = 6x^2 - 2$

$$f(t+1) = 6(t+1)^2 - 2$$
$$= 6(t^2 + 2t + 1) - 2$$
$$= 6t^2 + 12t + 6 - 2$$
$$= 6t^2 + 12t + 4$$

51. $g(r + h)$
$$= (r+h)^2 - 2(r+h) + 5$$
$$= r^2 + 2hr + h^2 - 2r - 2h + 5$$

53. $g\left(\dfrac{3}{q}\right) = \left(\dfrac{3}{q}\right)^2 - 2\left(\dfrac{3}{q}\right) + 5$

$$= \dfrac{9}{q^2} - \dfrac{6}{q} + 5$$

$$\text{or} \quad \dfrac{9 - 6q + 5q^2}{q^2}$$

55. A vertical line drawn anywhere through the graph will intersect the graph in only one place. The graph represents a function.

57. A vertical line drawn through the graph may intersect the graph in two places. The graph does not represent a function.

59. A vertical line drawn anywhere through the graph will intersect the graph in only one place. The graph represents a function.

61. $f(x) = 2x + 1$

(a) $f(x + h) = 2(x + h) + 1$
$$= 2x + 2h + 1$$

(b) $f(x + h) - f(x)$

$$= \dfrac{2x + 2h + 1}{2x + 1}$$

$$= 2x + 2h + 1 - 2x - 1$$

$$= 2h$$

(c) $\dfrac{f(x + h) - f(x)}{h}$

$$= \dfrac{\frac{2x+2h+1}{2x+1}}{h}$$

$$= \dfrac{2x + 2h + 1 - 2x - 1}{h}$$

$$= \dfrac{2h}{h}$$

$$= 2$$

63. $f(x) = 2x^2 - 4x - 5$

(a) $f(x + h)$
$$= 2(x + h)^2 - 4(x + h) - 5$$
$$= 2(x^2 + 2hx + h^2) - 4x - 4h - 5$$
$$= 2x^2 + 4hx + 2h^2 - 4x - 4h - 5$$

(b) $f(x + h) - f(x)$
$$= 2x^2 + 4hx + 2h^2 - 4x - 4h - 5$$
$$\quad - (2x^2 - 4x - 5)$$
$$= 2x^2 + 4hx + 2h^2 - 4x - 4h - 5$$
$$\quad - 2x^2 + 4x + 5$$
$$= 4hx + 2h^2 - 4h$$

(c) $\dfrac{f(x + h) - f(x)}{h}$

$$= \dfrac{4hx + 2h^2 - 4h}{h}$$

$$= \dfrac{h(4x + 2h - 4)}{h}$$

$$= 4x + 2h - 4$$

65. $f(x) = \dfrac{1}{x}$

(a) $f(x + h) = \dfrac{1}{x + h}$

(b) $f(x + h) - f(x)$

$$= \dfrac{1}{x + h} - \dfrac{1}{x}$$

$$= \left(\dfrac{x}{x}\right)\dfrac{1}{x + h} - \dfrac{1}{x}\left(\dfrac{x + h}{x + h}\right)$$

$$= \dfrac{x - (x + h)}{x(x + h)}$$

$$= \dfrac{-h}{x(x + h)}$$

(c) $\dfrac{f(x+h) - f(x)}{h}$

$= \dfrac{\frac{1}{x+h} - \frac{1}{x}}{h}$

$= \dfrac{\frac{1}{x+h}\left(\frac{x}{x}\right) - \frac{1}{x}\left(\frac{x+h}{x+h}\right)}{h}$

$= \dfrac{\frac{x-(x+h)}{(x+h)x}}{h}$

$= \dfrac{1}{h}\left[\dfrac{x-x-h}{(x+h)x}\right]$

$= \dfrac{1}{h}\left[\dfrac{-h}{(x+h)x}\right]$

$= \dfrac{-1}{x(x+h)}$

67. $f(x) = 3x$
$f(-x) = 3(-x)$
$\quad\quad = -(3x)$
$\quad\quad = -f(x)$

The function is odd.

69. $f(x) = 2x^2$
$f(-x) = 2(-x)^2$
$\quad\quad = 2x^2$
$\quad\quad = f(x)$

The function is even.

71. $f(x) = \dfrac{1}{x^2 + 4}$

$f(-x) = \dfrac{1}{(-x)^2 + 4}$

$\quad\quad = \dfrac{1}{x^2 + 4}$

$\quad\quad = f(x)$

The function is even.

73. $f(x) = \dfrac{x}{x^2 - 9}$

$f(-x) = \dfrac{-x}{(-x)^2 - 9}$

$\quad\quad = -\dfrac{x}{x^2 - 9}$

$\quad\quad = -f(x)$

The function is odd.

75. (a) The independent variable is the years.

(b) The dependent variable is the number of Internet users.

(c) $f(2003) = 719$ million users.

(d) The domain is $1995 \le x \le 2006$.
The range is $16{,}000{,}000 \le y \le 1{,}043{,}000{,}000$.

77. If x is a whole number of days, the cost of renting a car is given by

$$C(x) = 54x + 44.$$

For x in whole days plus a fraction of a day, substitute the next whole number for x in $54x + 44$ because a fraction of a day is charged as a whole day.

(a) $C\left(\frac{3}{4}\right) = C(1)$
$\quad\quad\quad = 54(1) + 44$
$\quad\quad\quad = \$98$

(b) $C\left(\frac{9}{10}\right) = C(1)$
$\quad\quad\quad = \$98$

(c) $C(1) = \$98$

(d) $C\left(1\frac{5}{8}\right) = C(2)$
$\quad\quad\quad\quad = 54(2) + 44$
$\quad\quad\quad\quad = \$152$

(e) $C(2.4) = C(3)$
$\quad\quad\quad = 54(3) + 44$
$\quad\quad\quad = \$206$

(f)
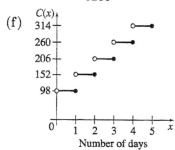

(g) Yes, C is a function.

(h) No, C is not a linear function.

79. (a)   (i) $y = f(5) = 19.7(5)^{0.753}$
$\quad\quad\quad\quad\quad\quad \approx 66$ kcal/day
(ii) $y = f(25) = 19.7(25)^{0.753}$
$\quad\quad\quad\quad\quad\quad \approx 222$ kcal/day

(b) Since 1 pound equals 0.454 kg, then $x = g(z) = 0.454z$ is the number of kilograms equal in weight to z pounds.

81. (a) In the graph, the curves representing wood and coal intersect approximately at the point $(1880, 50)$. So, in 1880 use of wood and coal were both about 50% of the global energy consumption.

(b) In the graph, the curves representing oil and coal intersect approximately at the point $(1965, 35)$. So, in 1965 use of oil and coal were both about 35% of the global energy consumption.

83. (a) Let $w =$ the width of the field;
$\qquad l =$ the length.

The perimeter of the field is 6000 ft, so

$$2l + 2w = 6000$$
$$l + w = 3000$$
$$l = 3000 - w.$$

Thus, the area of the field is given by

$$A = lw$$
$$A = (3000 - w)w.$$

(b) Since $l = 3000 - w$ and w cannot be negative, $0 \leq w \leq 3000$.
The domain of A is $0 \leq w \leq 3000$.

(c)

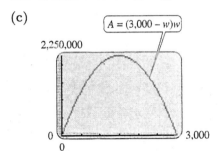

10.2 Quadratic Functions; Translation and Reflection

3. The graph of $y = x^2 - 3$ is the graph of $y = x^2$ translated 3 units downward.
This is graph d.

5. The graph of $y = (x - 3)^2 + 2$ is the graph of $y = x^2$ translated 3 units to the right and 2 units upward.
This is graph a.

7. The graph of $y = -(3 - x)^2 + 2$ is the same as the graph of $y = -(x - 3)^2 + 2$. This is the graph of $y = x^2$ reflected in the x-axis, translated 3 units to the right, and translated 2 units upward.
This is graph c.

9. $y = x^2 + 5x + 6$
$\quad y = (x + 3)(x + 2)$

Set $y = 0$ to find the x-intercepts.

$$0 = (x + 3)(x + 2)$$
$$x = -3, \ x = -2$$

The x-intercepts are -3 and -2.
Set $x = 0$ to find the y-intercept.

$$y = 0^2 + 5(0) + 6$$
$$y = 6$$

The y-intercept is 6.
The x-coordinate of the vertex is

$$x = \frac{-b}{2a} = \frac{-5}{2} = -\frac{5}{2}.$$

Substitute to find the y-coordinate.

$$y = (-\frac{5}{2})^2 + 5(-\frac{5}{2}) + 6 = \frac{25}{4} - \frac{25}{2} + 6 = -\frac{1}{4}$$

The vertex is $(-\frac{5}{2}, -\frac{1}{4})$.
The axis is $x = -\frac{5}{2}$, the vertical line through the vertex.

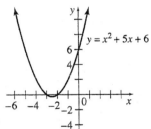

11. $y = -2x^2 - 12x - 16$
$\qquad = -2(x^2 + 6x + 8)$
$\qquad = -2(x + 4)(x + 2)$

Let $y = 0$.

$$0 = -2(x + 4)(x + 2)$$
$$x = -4, \ x = -2$$

-4 and -2 are the x-intercepts.
Let $x = 0$.

$$y = -2(0)^2 + 12(0) - 16$$

-16 is the y-intercept.

Vertex: $x = \dfrac{-b}{2a} = \dfrac{12}{-4} = -3$

$$y = -2(-3)^2 - 12(-3) - 16$$
$$= -18 + 36 - 16 = 2$$

The vertex is $(-3, 2)$.

The axis is $x = -3$, the vertical line through the vertex.

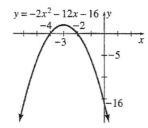

13. $y = 2x^2 + 8x - 8$

Let $y = 0$.

$$2x^2 + 8x - 8 = 0$$
$$x^2 + 4x - 4 = 0$$

$$x = \frac{-4 \pm \sqrt{4^2 - 4(1)(-4)}}{2(1)}$$
$$= \frac{-4 \pm \sqrt{32}}{2} = \frac{-4 \pm 4\sqrt{2}}{2}$$
$$= -2 \pm 2\sqrt{2}$$

The x-intercepts are $-2 \pm 2\sqrt{2} \approx 0.83$ or -4.83.
Let $x = 0$.

$$y = 2(0)^2 + 8(0) - 8 = -8$$

The y-intercept is -8.

The x-coordinate of the vertex is

$$x = \frac{-b}{2a} = -\frac{8}{4} = -2.$$

If $x = -2$,

$$y = 2(-2)^2 + 8(-2) - 8 = 8 - 16 - 8 = -16.$$

The vertex is $(-2, -16)$.
The axis is $x = -2$.

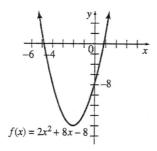

15. $f(x) = 2x^2 - 4x + 5$

Let $f(x) = 0$.

$$0 = 2x^2 - 4x + 5$$

$$x = \frac{-(-4) \pm \sqrt{(-4)^2 - 4(2)(5)}}{2(2)}$$
$$= \frac{4 \pm \sqrt{16 - 40}}{4}$$
$$= \frac{4 \pm \sqrt{-24}}{4}$$

Since the radicand is negative, there are no x-intercepts.

Let $x = 0$.

$$y = 2(0)^2 - 4(0) + 5$$
$$y = 5$$

5 is the y-intercept.

Vertex: $x = \dfrac{-b}{2a} = \dfrac{-(-4)}{2(2)} = \dfrac{4}{4} = 1$

$$y = 2(1)^2 - 4(1) + 5 = 2 - 4 + 5 = 3$$

The vertex is $(1, 3)$.
The axis is $x = 1$.

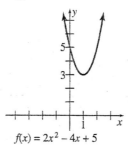

17. $f(x) = -2x^2 + 16x - 21$

Let $f(x) = 0$
Use the quadratic formula.

$$x = \frac{-16 \pm \sqrt{16^2 - 4(-2)(-21)}}{2(-2)}$$
$$= \frac{-16 \pm \sqrt{88}}{-4}$$
$$= \frac{-16 \pm 2\sqrt{22}}{-4}$$
$$= 4 \pm \frac{\sqrt{22}}{2}$$

The x-intercepts are $4 + \frac{\sqrt{22}}{2} \approx 6.35$ and

$4 - \frac{\sqrt{22}}{2} \approx 1.65$.

Let $x = 0$.

$$y = -2(0)^2 + 16(0) - 21$$

$$y = -21$$

-21 is the y-intercept.

Vertex: $x = \dfrac{-b}{2a} = \dfrac{-16}{2(-2)} = \dfrac{-16}{-4} = 4$

$$y = -2(4)^2 + 16(4) - 21$$

$$= -32 + 64 - 21 = 11$$

The vertex is $(4, 11)$.
The axis is $x = 4$.

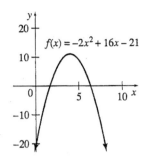

19. $y = \dfrac{1}{3}x^2 - \dfrac{8}{3}x + \dfrac{1}{3}$

Let $y = 0$.

$$0 = \dfrac{1}{3}x^2 - \dfrac{8}{3}x + \dfrac{1}{3}$$

Multiply by 3.

$$0 = x^2 - 8x + 1$$

$$x = \dfrac{-(-8) \pm \sqrt{(-8)^2 - 4(1)(1)}}{2(1)}$$

$$= \dfrac{8 \pm \sqrt{64 - 4}}{2} = \dfrac{8 \pm \sqrt{60}}{2}$$

$$= \dfrac{8 \pm 2\sqrt{15}}{2} = 4 \pm \sqrt{15}$$

The x-intercepts are $4 + \sqrt{15} \approx 7.87$ and $4 - \sqrt{15} \approx 0.13$.

Let $x = 0$.

$$y = \dfrac{1}{3}(0)^2 - \dfrac{8}{3}(0) + \dfrac{1}{3}$$

$\dfrac{1}{3}$ is the y-intercept.

Vertex: $x = \dfrac{-b}{2a} = \dfrac{-\left(-\frac{8}{3}\right)}{2\left(\frac{1}{3}\right)} = \dfrac{\frac{8}{3}}{\frac{2}{3}} = 4$

$$y = \dfrac{1}{3}(4)^2 - \dfrac{8}{3}(4) + \dfrac{1}{3}$$

$$= \dfrac{16}{3} - \dfrac{32}{3} + \dfrac{1}{3} = -\dfrac{15}{3} = -5$$

The vertex is $(4, -5)$.
The axis is $x = 4$.

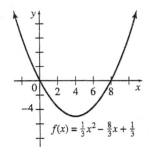

21. The graph of $y = \sqrt{x + 2} - 4$ is the graph of $y = \sqrt{x}$ translated 2 units to the left and 4 units downward.
This is graph d.

23. The graph of $y = \sqrt{-x + 2} - 4$ is the graph of $y = \sqrt{-(x - 2)} - 4$, which is the graph of $y = \sqrt{x}$ reflected in the y-axis, translated 2 units to the right, and translated 4 units downward.
This is graph c.

25. The graph of $y = -\sqrt{x + 2} - 4$ is the graph of $y = \sqrt{x}$ reflected in the x-axis, translated 2 units to the left, and translated 4 units downward.
This is graph e.

27. The graph of $y = -f(x)$ is the graph of $y = f(x)$ reflected in the x-axis.

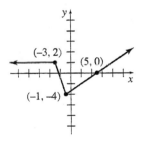

29. The graph of $y = f(-x)$ is the graph of $y = f(x)$ reflected in the y-axis.

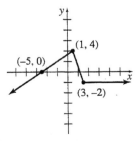

31. $f(x) = \sqrt{x - 2} + 2$

Translate the graph of $f(x) = \sqrt{x}$ 2 units right and 2 units up.

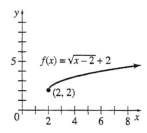

33. $f(x) = -\sqrt{2-x} - 2$
$$= -\sqrt{-(x-2)} - 2$$

Reflect the graph of $f(x)$ vertically and horizontally.

Translate the graph 2 units right and 2 units down.

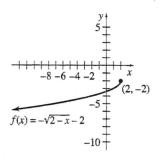

35. If $0 < a < 1$, the graph of $f(ax)$ will be flatter and wider than the graph of $f(x)$.

Multiplying x by a fraction makes the y-values less than the original y-values.

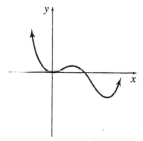

37. If $-1 < a < 0$, the graph of $f(ax)$ will be reflected horizontally, since a is negative. It will be flatter because multiplying x by a fraction decreases the corresponding y-values.

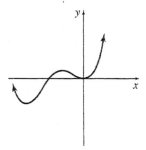

39. If $0 < a < 1$, the graph of $af(x)$ will be flatter and wider than the graph of $f(x)$. Each y-value is only a fraction of the height of the original y-values.

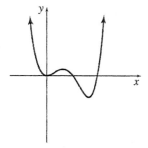

41. If $-1 < a < 0$, the graph will be reflected vertically, since a will be negative. Also, because a is a fraction, the graph will be flatter because each y-value will only be a fraction of its original height.

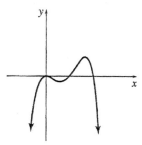

43. **(a)** Since the graph of $y = f(x)$ is reflected vertically to obtain the graph of $y = -f(x)$, the x-intercept is unchanged. The x-intercept of the graph of $y = f(x)$ is r.

(b) Since the graph of $y = f(x)$ is reflected horizontally to obtain the graph of $y = f(-x)$, the x-intercept of the graph of $y = f(-x)$ is $-r$.

(c) Since the graph of $y = f(x)$ is reflected both horizontally and vertically to obtain the graph of $y = -f(-x)$, the x-intercept of the graph of $y = -f(-x)$ is $-r$.

45. (a)

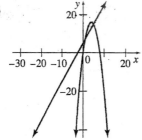

(b) Break-even quantities are values of $x =$ number of widgets for which revenue and cost are equal. Set $R(x) = C(x)$ and solve for x.

$$-x^2 + 8x = 2x + 5$$
$$x^2 - 6x + 5 = 0$$
$$(x - 5)(x - 1) = 0$$
$$x - 5 = 0 \quad \text{or} \quad x - 1 = 0$$
$$x = 5 \quad \text{or} \quad x = 1$$

So, the break-even quantities are 1 and 5. The minimum break-even quantity is $x = 1$.

(c) The maximum revenue occurs at the vertex of R. Since $R(x) = -x^2 + 8x$, then the x-coordinate of the vertex is

$$x = -\frac{b}{2a} = -\frac{8}{2(-1)} = 4.$$

So, the maximum revenue is

$$R(4) = -4^2 + 8(4) = 16.$$

(d) The maximum profit is the maximum difference $R(x) - C(x)$. Since

$$P(x) = R(x) - C(x)$$
$$= -x^2 + 8x - (2x + 5)$$
$$= -x^2 + 6x - 5$$

is a quadratic function, we can find the maximum profit by finding the vertex of P. This occurs at

$$x = -\frac{b}{2a} = \frac{-6}{2(-1)} = 3.$$

Therefore, the maximum profit is

$$P(3) = -(3)^2 + 6(3) - 5 = 4.$$

47. (a)

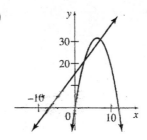

(b) Break-even quantities are values of $x =$ number of widgets for which revenue equals cost. Set $R(x) = C(x)$ and solve for x.

$$-\frac{4}{5}x^2 + 10x = 2x + 15$$
$$\frac{4}{5}x^2 - 8x + 15 = 0$$
$$4x^2 - 40x + 75 = 0$$
$$4x^2 - 10x - 30x + 75 = 0$$
$$2x(2x - 5) - 15(2x - 5) = 0$$
$$(2x - 5)(2x - 15) = 0$$
$$2x - 5 = 0 \quad \text{or} \quad 2x - 15 = 0$$
$$x = 2.5 \quad \text{or} \quad x = 7.5$$

So, the break-even quantities are 2.5 and 7.5 with $x = 2.5$ the minimum break-even quantity.

(c) The maximum revenue occurs at the vertex of R. Since $R(x) = -\frac{4}{5}x^2 + 10x$, then the x-coordinate of the vertex is

$$x = -\frac{b}{2a} = -\frac{10}{2\left(\frac{-4}{5}\right)} = 6.25.$$

So, the maximum revenue is

$$R(6.25) = 31.25.$$

(d) The maximum profit is the maximum difference $R(x) - C(x)$. Since

$$P(x) = R(x) - C(x)$$
$$= -\frac{4}{5}x^2 + 10x - (2x + 15)$$
$$= -\frac{4}{5}x^2 + 8x - 15$$

is a quadratic function, we can find the maximum profit by finding the vertex of P. This occurs at

$$x = -\frac{b}{2a} = -\frac{8}{2\left(\frac{-4}{5}\right)} = 5.$$

Therefore, the maximum profit is

$$P(5) = -\frac{4}{5}5^2 + 8(5) - 15 = 5.$$

49. $R(x) = 8000 + 70x - x^2$
$$= -x^2 + 70x + 8000$$

The maximum revenue occurs at the vertex.

$$x = \frac{-b}{2a} = \frac{-70}{2(-1)} = 35$$

$$y = 8000 + 70(35) - (35)^2$$
$$= 8000 + 2450 - 1225$$
$$= 9225$$

The vertex is $(35, 9225)$.
The maximum revenue of $9225 is realized when 35 seats are left unsold.

51. $p = 500 - x$

(a) The revenue is

$$R(x) = px$$
$$= (500 - x)(x)$$
$$= 500x - x^2.$$

(b)

(c) From the graph, the vertex is halfway between $x = 0$ and $x = 500$, so $x = 250$ units corresponds to maximum revenue. Then the price is

$$p = 500 - x$$
$$= 500 - 250 = \$250.$$

Note that price, p, cannot be read directly from the graph of

$$R(x) = 500x - x^2.$$

(d) $R(x) = 500x - x^2$
$$= -x^2 + 500x$$

Find the vertex.

$$x = \frac{-b}{2a} = \frac{-500}{2(-1)} = 250$$

$$y = -(250)^2 + 500(250)$$
$$= 62,500$$

The vertex is $(250, 62,500)$.
The maximum revenue is $62,500.

53. Let $x =$ the number of $25 increases.

(a) Rent per apartment: $800 + 25x$

(b) Number of apartments rented: $80 - x$

(c) Revenue:
$$R(x) = (\text{number of apartments rented})$$
$$\times (\text{rent per apartment})$$
$$= (80 - x)(800 + 25x)$$
$$= -25x^2 + 1200x + 64,000$$

(d) Find the vertex:

$$x = \frac{-b}{2a} = \frac{-1200}{2(-25)} = 24$$

$$y = -25(24)^2 + 1200(24) + 64,000$$
$$= 78,400$$

The vertex is $(24, 78,400)$. The maximum revenue occurs when $x = 24$.

(e) The maximum revenue is the y-coordinate of the vertex, or $78,400.

55. $S(x) = 1 - 0.058x - 0.076x^2$

(a) $$0.50 = 1 - 0.058x - 0.076x^2$$
$$0.076x^2 + 0.058x - 0.50 = 0$$
$$76x^2 + 58x - 500 = 0$$
$$38x^2 + 29x - 250 = 0$$

$$x = \frac{-29 \pm \sqrt{(29)^2 - 4(38)(-250)}}{2(38)}$$

$$= \frac{-29 \pm \sqrt{38,841}}{76}$$

$\frac{-29 - \sqrt{38,841}}{76} \approx -2.97$ and $\frac{-29 + \sqrt{38,841}}{76} \approx 2.21$

We ignore the negative value.
The value $x = 2.2$ represents 2.2 decades or 22 years, and 22 years after 65 is 87.
The median length of life is 87 years.

(b) If nobody lives, $S(x) = 0$.

$$1 - 0.058x - 0.076x^2 = 0$$
$$76x^2 + 58x - 1000 = 0$$
$$38x^2 + 29x - 500 = 0$$

$$x = \frac{-29 \pm \sqrt{(29)^2 - 4(38)(-500)}}{2(38)}$$

$$= \frac{-29 \pm \sqrt{76,841}}{76}$$

$\frac{-29 - \sqrt{76,841}}{76} \approx -4.03$ and $\frac{-29 + \sqrt{76,841}}{76} \approx 3.27$

We ignore the negative value.

The value $x = 3.3$ represents 3.3 decades or 33 years, and 33 years after 65 is 98.
Virtually nobody lives beyond 98 years.

57. **(a)** The vertex of the quadratic function $y = 0.057x - 0.001x^2$ is at

$$x = -\frac{b}{2a} = -\frac{0.057}{2(-0.001)} = 28.5.$$

Since the coefficient of the leading term, -0.001, is negative, then the graph of the function opens downward, so a maximum is reached at 28.5 weeks of gestation.

(b) The maximum splenic artery resistance reached at the vertex is

$$y = 0.057(28.5) - 0.001(28.5)^2$$
$$\approx 0.81.$$

(c) The splenic artery resistance equals 0, when $y = 0$.

$$0.057x - 0.001x^2 = 0 \quad \text{Substitute in the expression in } x \text{ for } y.$$
$$x(0.057 - 0.001x) = 0 \quad \text{Factor.}$$
$$x = 0 \text{ or } 0.057 - 0.001x = 0 \quad \text{Set each factor equal to 0.}$$
$$x = \frac{0.057}{0.001} = 57$$

So, the splenic artery resistance equals 0 at 0 weeks or 57 weeks of gestation.
No, this is not reasonable because at $x = 0$ or 57 weeks, the fetus does not exist.

59. **(a)**

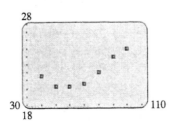

(c) $y = 0.002726x^2 - 0.3113x + 29.33$

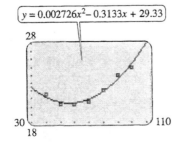

(d) Given that $(h, k) = (60, 20.3)$, the equation has the form

$$y = a(x - 60)^2 + 20.3.$$

Since $(100, 25.1)$ is also on the curve.

$$25.1 = a(100 - 60)^2 + 20.3$$
$$4.8 = 1600a$$
$$a = 0.003$$

(e)

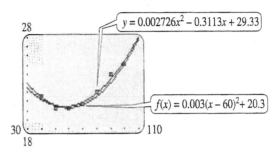

The two graphs are very close.

61. $f(x) = 60.0 - 2.28x + 0.0232x^2$

$$\frac{-b}{2a} = -\frac{-2.28}{2(0.0232)} \approx 49.1$$

The minimum value occurs when $x \approx 49.1$. The age at which the accident rate is a minimum is 49 years. The minimum rate is

$$f(49.1) = 60.0 - 2.28(49.1) + 0.0232(49.1)^2$$
$$= 60.0 - 111.948 + 55.930792$$
$$\approx 3.98.$$

63. $y = 0.056057x^2 + 1.06657x$

(a) If $x = 25$ mph,

$$y = 0.056057(25)^2 + 1.06657(25)$$
$$y \approx 61.7.$$

At 25 mph, the stopping distance is approximately 61.7 ft.

(b) $0.056057x^2 + 1.06657x = 150$
$$0.056057x^2 + 1.06657x - 150 = 0$$

$$x = \frac{-1.06657 \pm \sqrt{(1.06657)^2 - 4(0.056057)(-150)}}{2(0.056057)}$$

$$x \approx 43.08 \text{ or } x \approx -62.11$$

We ignore the negative value.
To stop within 150 ft, the fastest speed you can drive is 43 mph.

65. Let $x =$ the length of the lot and
$y =$ the width of the lot.

The perimeter is given by

$$P = 2x + 2y$$
$$380 = 2x + 2y$$
$$190 = x + y$$
$$190 - x = y.$$

Area $= xy$ (quantity to be maximized)

$$A = x(190 - x)$$
$$= 190x - x^2$$
$$= -x^2 + 190x$$

Find the vertex: $\dfrac{-b}{2a} = \dfrac{-190}{-2} = 95$

$$y = -(95)^2 + 190(95)$$
$$= 9025$$

This is a parabola with vertex $(95, 9025)$ that opens downward. The maximum area is the value of A at the vertex, or 9025 sq ft.

67. Sketch the culvert on the xy-axes as a parabola that opens upward with vertex at $(0, 0)$.

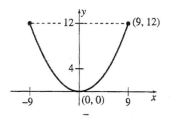

The equation is of the form $y = ax^2$. Since the culvert is 18 ft wide at 12 ft from its vertex, the points $(9, 12)$ and $(-9, 12)$ are on the parabola. Use $(9, 12)$ as one point on the parabola.

$$12 = a(9)^2$$
$$12 = 81a$$
$$\frac{12}{81} = a$$
$$\frac{4}{27} = a$$

Thus,

$$y = \frac{4}{27}x^2.$$

To find the width 8 feet from the top, find the points with

$$y\text{-value} = 12 - 8 = 4.$$

Thus,

$$4 = \frac{4}{27}x^2$$
$$108 = 4x^2$$
$$27 = x^2$$
$$x^2 = 27$$
$$x = \pm\sqrt{27}$$
$$x = \pm 3\sqrt{3}.$$

The width of the culvert is $3\sqrt{3} + \left|-3\sqrt{3}\right|$
$= 6\sqrt{3}$ ft ≈ 10.39 ft.

10.3   Polynomial and Rational Functions

3. The graph of $f(x) = (x - 2)^3 + 3$ is the graph of $y = x^3$ translated 2 units to the right and 3 units upward.

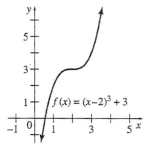

5. The graph of $f(x) = -(x + 3)^4 + 1$ is the graph of $y = x^4$ reflected horizontally, translated 3 units to the left, and translated 1 unit upward.

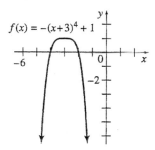

7. The graph of $y = x^3 - 7x - 9$ has the right end up, the left end down, at most two turning points, and a y-intercept of -9.
This is graph d.

9. The graph of $y = -x^3 - 4x^2 + x + 6$ has the right end down, the left end up, at most two turning points, and a y-intercept of 6.
This is graph e.

11. The graph of $y = x^4 - 5x^2 + 7$ has both ends up, at most three turning points, and a y-intercept of 7.
This is graph i.

13. The graph of $y = -x^4 + 2x^3 + 10x + 15$ has both ends down, at most three turning points, and a y-intercept of 15.
This is graph g.

15. The graph of $y = -x^5 + 4x^4 + x^3 - 16x^2 + 12x + 5$ has the right end down, the left end up, at most four turning points, and a y-intercept of 5.
This is graph a.

17. The graph of $y = \frac{2x^2 + 3}{x^2 + 1}$ has no vertical asymptote, the line with equation $y = 2$ as a horizontal asymptote, and a y-intercept of 3.
This is graph d.

19. The graph $y = \frac{-2x^2 - 3}{x^2 + 1}$ has no vertical asymptote, the line with equation $y = -2$ as a horizontal asymptote, and a y-intercept of -3.
This is graph e.

21. The right end is up and the left end is up. There are three turning points.
The degree is an even integer equal to 4 or more. The x^n term has a $+$ sign.

23. The right end is up and the left end is down. There are four turning points. The degree is an odd integer equal to 5 or more. The x^n term has a $+$ sign.

25. The right end is down and the left end is up. There are six turning points. The degree is an odd integer equal to 7 or more. The x^n term has a $-$ sign.

27. $y = \dfrac{-4}{x + 2}$

The function is undefined for $x = -2$, so the line $x = -2$ is a vertical asymptote.

x	-102	-12	-7	-5	-3	-1	8	98
$x + 2$	-100	-10	-5	-3	-1	1	10	100
y	0.04	0.4	0.8	1.3	4	-4	-0.4	-0.04

The graph approaches $y = 0$, so the line $y = 0$ (the x-axis) is a horizontal asymptote.
Asymptotes: $y = 0$, $x = -2$

x-intercept:
none, because the x-axis is an asymptote

y-intercept:
-2, the value when $x = 0$

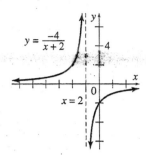

29. $y = \dfrac{2}{3 + 2x}$

$3 + 2x = 0$ when $2x = -3$ or $x = -\frac{3}{2}$, so the line $x = -\frac{3}{2}$ is a vertical asymptote.

x	-51.5	-6.5	-2	-1	3.5	48.5
$3 + 2x$	-100	-10	-1	1	10	100
y	-0.02	-0.2	-2	2	0.2	0.02

The graph approaches $y = 0$, so the line $y = 0$ (the x-axis) is a horizontal asymptote.
Asymptote: $y = 0$, $x = -\frac{3}{2}$

x-intercept:
none, since the x-axis is an asymptote

y-intercept:
$\frac{2}{3}$, the value when $x = 0$

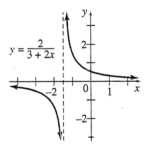

31. $y = \dfrac{2x}{x - 3}$

$x - 3 = 0$ when $x = 3$, so the line $x = 3$ is a vertical asymptote.

x	-97	-7	-1	1	2	2.5
$2x$	-194	-14	-2	2	4	5
$x - 3$	-100	-10	-4	-2	-1	-0.5
y	1.94	1.4	0.5	-1	-4	-10

x	3.5	4	5	7	11	103
$2x$	7	8	10	14	22	206
$x - 3$	0.5	1	2	4	8	100
y	14	8	5	3.5	2.75	2.06

As x gets larger,

$$\frac{2x}{x-3} \approx \frac{2x}{x} = 2.$$

Thus, $y = 2$ is a horizontal asymptote.
Asymptotes: $y = 2$, $x = 3$

x-intercept:

0, the value when $y = 0$

y-intercept:

0, the value when $x = 0$

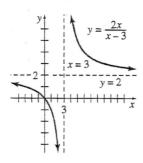

33. $y = \dfrac{x+1}{x-4}$

$x - 4 = 0$ when $x = 4$, so $x = 4$ is a vertical asymptote.

x	-96	-6	-1	0	3
$x+1$	-95	-5	0	1	4
$x-4$	-100	-10	-5	-4	-1
y	0.95	0.5	0	-0.25	-4

x	3.5	4.5	5	14	104
$x+1$	4.5	5.5	6	15	105
$x-4$	-0.5	0.5	1	10	100
y	-9	11	6	1.5	1.05

As x gets larger,

$$\frac{x+1}{x-4} \approx \frac{x}{x} = 1.$$

Thus, $y = 1$ is a horizontal asymptote.
Asymptotes: $y = 1$, $x = 4$

x-intercept:
-1, the value when $y = 0$

y-intercept:
$-\frac{1}{4}$, the value when $x = 0$

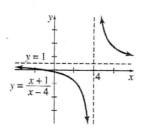

35. $y = \dfrac{3-2x}{4x+20}$

$4x + 20 = 0$ when $4x = -20$ or $x = -5$, so the line $x = -5$ is a vertical asymptote.

x	-8	-7	-6	-4	-3	-2
$3-2x$	-26	-23	-20	-14	-11	-8
$4x+20$	-12	-8	-4	4	8	12
y	2.17	2.88	5	-3.5	-1.38	-0.67

As x gets larger,

$$\frac{3-2x}{4x+20} \approx \frac{-2x}{4x} = -\frac{1}{2}.$$

Thus, the line $y = -\frac{1}{2}$ is a horizontal asymptote.

Asymptotes: $x = -5$, $y = -\frac{1}{2}$

x-intercept:

$\frac{3}{2}$, the value when $y = 0$

y-intercept:

$\frac{3}{20}$, the value when $x = 0$

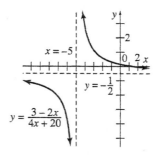

37. $y = \dfrac{-x-4}{3x+6}$

$3x + 6 = 0$ when $3x = -6$ or $x = -2$, so the line $x = -2$ is a vertical asymptote.

x	-5	-4	-3	-1	0	1
$-x-4$	1	0	-1	-3	-4	-5
$3x+6$	-9	-6	-3	3	6	9
y	-0.11	0	0.33	-1	-0.67	-0.56

As x gets larger,

$$\frac{-x-4}{3x+6} \approx \frac{-x}{3x} = -\frac{1}{3}.$$

The line $y = -\frac{1}{3}$ is a horizontal asymptote.
Asymptotes: $y = -\frac{1}{3}$, $x = -2$

x-intercept:
-4, the value when $y = 0$

y-intercept:

$-\frac{2}{3}$, the value when $x = 0$

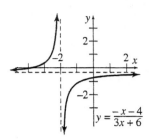

39. $y = \dfrac{x^2 + 7x + 12}{x + 4}$

$= \dfrac{(x + 3)(x + 4)}{x + 4}$

$= x + 3, \ x \neq -4$

There are no asymptotes, but there is a hole at $x = -4$.

x-intercept: -3, the value when $y = 0$.

y-intercept: 3, the value when $x = 0$.

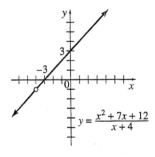

41. For a vertical asymptote at $x = 1$, put $x - 1$ in the denominator. For a horizontal asymptote at $y = 2$, the degree of the numerator must equal the degree of the denominator and the quotient of their leading terms must equal 2. So, $2x$ in the numerator would cause y to approach 2 as x gets larger.

So, one possible answer is $y = \dfrac{2x}{x - 1}$.

43. $f(x) = (x - 1)(x - 2)(x + 3)$,
$g(x) = x^3 + 2x^2 - x - 2$,
$h(x) = 3x^3 + 6x^2 - 3x - 6$

(a) $f(1) = (0)(-1)(4) = 0$

(b) $f(x)$ is zero when $x = 2$ and when $x = -3$.

(c) $g(-1) = (-1)^3 + 2(-1)^2 - (-1) - 2$
$= -1 + 2 + 1 - 2 = 0$
$g(1) = (1)^3 + 2(1)^2 - (1) - 2$
$= 1 + 2 - 1 - 2$
$= 0$
$g(-2) = (-2)^3 + 2(-2)^2 - (-2) - 2$
$= -8 + 8 + 2 - 2$
$= 0$

(d) $g(x) = [x - (-1)](x - 1)[x - (-2)]$
$g(x) = (x + 1)(x - 1)(x + 2)$

(e) $h(x) = 3g(x)$
$= 3(x + 1)(x - 1)(x + 2)$

(f) If f is a polynomial and $f(a) = 0$ for some number a, then one factor of the polynomial is $x - a$.

45. $f(x) = \dfrac{1}{x^5 - 2x^3 - 3x^2 + 6}$

(a) Two vertical asymptotes appear, one at $x = -1.4$ and one at $x = 1.4$.

(b) Three vertical asymptotes appear, one at $x = -1.414$, one at $x = 1.414$, and one at $x = 1.442$.

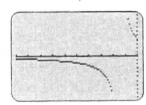

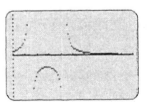

47. $\overline{C}(x) = \dfrac{220{,}000}{x + 475}$

(a) If $x = 25$,

$$\overline{C}(25) = \frac{220{,}000}{25 + 475} = \frac{220{,}000}{500} = 440.$$

If $x = 50$,

$$\overline{C}(50) = \frac{220{,}000}{50 + 475} = \frac{220{,}000}{525} \approx 419.$$

If $x = 100$,

$$\overline{C}(100) = \frac{220,000}{100 + 475} = \frac{220,000}{575} \approx 383.$$

If $x = 200$,

$$\overline{C}(200) = \frac{220,000}{200 + 475} = \frac{220,000}{675} \approx 326.$$

If $x = 300$,

$$\overline{C}(300) = \frac{220,000}{300 + 475} = \frac{220,000}{775} \approx 284.$$

If $x = 400$,

$$\overline{C}(400) = \frac{220,000}{400 + 475} = \frac{220,000}{875} \approx 251.$$

(b) A vertical asymptote occurs when the denominator is 0.

$$x + 475 = 0$$
$$x = -475$$

A horizontal asymptote occurs when $\overline{C}(x)$ approaches a value as x gets larger. In this case, $\overline{C}(x)$ approaches 0.
The asymptotes are $x = -475$ and $y = 0$.

(c) x-intercepts:

$$0 = \frac{220,000}{x + 475}; \text{ no such } x, \text{ so no } x\text{-intercepts}$$

y-intercepts:

$$\overline{C}(0) = \frac{220,000}{0 + 475} \approx 463.2$$

(d) Use the following ordered pairs:
$(25, 440)$, $(50, 419)$, $(100, 383)$,
$(200, 326)$, $(300, 284)$, $(400, 251)$.

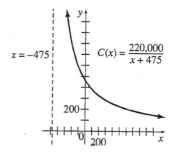

49. Quadratic functions with roots at $x = 0$ and $x = 100$ are of the form $f(x) = ax(100 - x)$.
$f_1(x)$ has a maximum of 100, which occurs at the vertex. The x-coordinate of the vertex lies between the two roots.
The vertex is $(50, 100)$.

$$100 = a(50)(100 - 50)$$
$$100 = a(50)(50)$$
$$\frac{100}{2500} = a$$
$$\frac{1}{25} = a$$

$$f_1(x) = \frac{1}{25}x(100 - x) \text{ or } \frac{x(100 - x)}{25}$$

$f_2(x)$ has a maximum of 250, occurring at $(50, 250)$.

$$250 = a(50)(100 - 50)$$
$$250 = a(50)(50)$$
$$\frac{250}{2500} = a$$
$$\frac{1}{10} = a$$

$$f_2(x) = \frac{1}{10}x(100 - x) \text{ or } \frac{x(100 - x)}{10}$$

$$f_1(x) \cdot f_2(x) = \left[\frac{x(100 - x)}{25}\right] \cdot \left[\frac{x(100 - x)}{10}\right]$$
$$= \frac{x^2(100 - x)^2}{250}$$

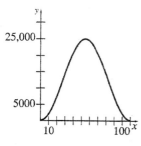

$$f(x) = \frac{x^2(100 - x)^2}{250}$$

51. $y = \dfrac{6.7x}{100 - x}$,

Let $x =$ percent of pollutant;
$\quad y =$ cost in thousands.

(a) $x = 50$

$$y = \frac{6.7(50)}{100 - 50} = 6.7$$

The cost is $6700.

$x = 70$

$$y = \frac{6.7(70)}{100 - 70} \approx 15.6$$

The cost is $15,600.

$x = 80$

$$y = \frac{6.7(80)}{100 - 80} = 26.8$$

The cost is $26,800.

$x = 90$

$$y = \frac{6.7(90)}{100 - 90} = 60.3$$

The cost is $60,300.

$x = 95$

$$y = \frac{6.7(95)}{100 - 95}$$

The cost is $127,300.

$x = 98$

$$y = \frac{6.7(98)}{100 - 98} = 328.3$$

The cost is $328,300.

$x = 99$

$$y = \frac{6.7(99)}{100 - 99} = 663.3$$

The cost is $663,300.

(b) No, because $x = 100$ makes the denominator zero, so $x = 100$ is a vertical asymptote.

(c)

Cost (in thousands of dollars) vs Percent removed; $y = \frac{6.7x}{100 - x}$; $x = 100$

---

53. (a) $a = \dfrac{k}{d}$

$k = ad$

d	a	$k = ad$
36.000	9.37	337.32
36.125	9.34	337.4075
36.250	9.31	337.4875
36.375	9.27	337.19625
36.500	9.24	337.26
36.625	9.21	337.31625
36.750	9.18	337.365
36.875	9.15	337.40625
37.000	9.12	337.44

We find the average of the nine values of k by adding them and dividing by 9. This gives 337.35, or, rounding to the nearest integer, $k = 337$. Therefore,

$$a = \frac{337}{d}.$$

(b) When $d = 40.50$,

$$a = \frac{337}{40.50} \approx 8.32.$$

The strength for 40.50 diopter lenses is 8.32 mm of arc.

55. $D(x) = -0.125x^5 + 3.125x^4 + 4000$

(a)

x	0	5	10	15
$D(x)$	4000	5563	22,750	67,281

x	20	25
$D(x)$	104,000	4000

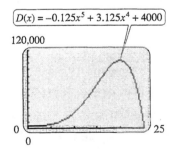

$D(x) = -0.125x^5 + 3.125x^4 + 4000$

120,000

0   25

(b) $D(x)$ increases from $x = 0$ to $x = 20$. This corresponds to an increasing population from 1905 to 1925. $D(x)$ does not change much from $x = 0$ to $x = 5$. This corresponds to a relatively stable population from 1905 to 1910.

$D(x)$ decreases from $x = 20$ to $x = 25$. This corresponds to a decreasing population from 1925 to 1930.

57. $f(t) = -0.0014t^3 + 0.092t^2 - 0.67t + 11.89$

(a)

x	9	18	27	36	45
$A(x)$	12.29	21.47	33.31	41.68	40.47

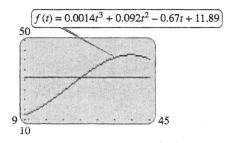

(b) The graph of $f(t)$ intersects the graph of $g(t) = 31$ at about $t = 25$, which corresponds to the year 1985.

59. (a) A reasonable domain for the function is $[0, \infty)$. Populations are not measured using negative numbers, and they may get extremely large.

(b) $f(x) = \dfrac{Kx}{A + x}$

When $K = 5$ and $A = 2$,

$$f(x) = \frac{5x}{2 + x}.$$

The graph has a horizontal asymptote at $y = 5$ since

$$\frac{5x}{2 + x} \approx \frac{5x}{x} = 5$$

as x gets larger.

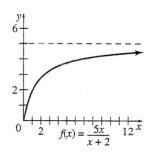

(c) $f(x) = \dfrac{Kx}{A + x}$

As x gets larger,

$$\frac{Kx}{A + x} \approx \frac{Kx}{x} = K.$$

Thus, $y = K$ will always be a horizontal asymptote for this function.

(d) K represents the maximum growth rate. The function approaches this value asymptotically, showing that although the growth rate can get very close to K, it can never reach the maximum, K.

(e) $f(x) = \dfrac{Kx}{A + x}$

Let $A = x$, the quantity of food present.

$$f(x) = \frac{Kx}{A + x} = \frac{Kx}{2x} = \frac{K}{2}$$

K is the maximum growth rate, so $\frac{K}{2}$ is half the maximum. Thus, A represents the quantity of food for which the growth rate is half of its maximum.

61. (a)

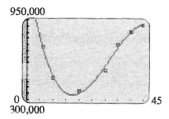

(b) $y = 890.37x^2 - 36,370x + 830,144$

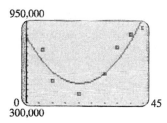

(c) $y = -52.954x^3 + 5017.88x^2 - 127,714x + 1,322,606$

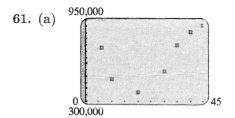

63. (a)

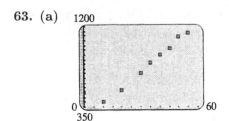

(b) $f(x) = 0.19327x^2 + 3.0039x + 431.30$

(c)

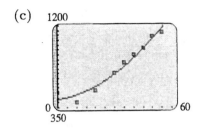

(d) $f(x) = -0.010883x^3 + 1.1079x^2 - 16.432x + 485.45$

(e)

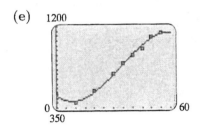

10.4   Exponential Functions

1.

number of folds	1	2	3	4	5 ...		10 ...		50
layers of paper	2	4	8	16	32 ...		1024 ...		2^{50}

$2^{50} = 1.125899907 \times 10^{15}$

3. The graph of $y = 3^x$ is the graph of an exponential function $y = a^x$ with $a > 1$.
This is graph e.

5. The graph of $y = \left(\frac{1}{3}\right)^{1-x}$ is the graph of $y = (3^{-1})^{1-x}$ or $y = 3^{x-1}$. This is the graph of $y = 3^x$ translated 1 unit to the right.
This is graph c.

7. The graph of $y = 3(3)^x$ is the same as the graph of $y = 3^{x+1}$. This is the graph of $y = 3^x$ translated 1 unit to the left.
This is graph f.

9. The graph of $y = 2 - 3^{-x}$ is the same as the graph of $y = -3^{-x} + 2$. This is the graph of $y = 3^x$ reflected in the x-axis, reflected in the y-axis, and translated up 2 units.
This is graph a.

11. The graph of $y = 3^{x-1}$ is the graph of $y = 3^x$ translated 1 unit to the right.
This is graph c.

13. $2^x = 32$

$2^x = 2^5$

$x = 5$

15. $3^x = \dfrac{1}{81}$

$3^x = \dfrac{1}{3^4}$

$3^x = 3^{-4}$

$x = -4$

17. $4^x = 8^{x+1}$

$(2^2)^x = (2^3)^{x+1}$

$2^{2x} = 2^{3x+3}$

$2x = 3x + 3$

$-x = 3$

$x = -3$

19. $16^{x+3} = 64^{2x-5}$

$(2^4)^{x+3} = (2^6)^{2x-5}$

$2^{4x+12} = 2^{12x-30}$

$4x + 12 = 12x - 30$

$42 = 8x$

$\dfrac{21}{4} = x$

21. $e^{-x} = (e^4)^{x+3}$

$e^{-x} = e^{4x+12}$

$-x = 4x + 12$

$-5x = 12$

$x = -\dfrac{12}{5}$

23. $5^{-|x|} = \dfrac{1}{25}$

$5^{-|x|} = 5^{-2}$

$-|x| = -2$

$|x| = 2$

$x = 2 \quad \text{or} \quad x = -2$

25. $5^{x^2+x} = 1$

$5^{x^2+x} = 5^0$

$x^2 + x = 0$

$x(x+1) = 0$

$x = 0 \quad \text{or} \quad x + 1 = 0$

$x = 0 \quad \text{or} \qquad x = -1$

27. $27^x = 9^{x^2+x}$

$(3^3)^x = (3^2)^{x^2+x}$

$3^{3x} = 3^{2x^2+2x}$

$3x = 2x^2 + 2x$

$0 = 2x^2 - x$

$0 = x(2x - 1)$

$x = 0 \quad \text{or} \quad 2x - 1 = 0$

$x = 0 \quad \text{or} \qquad x = \dfrac{1}{2}$

33. $A = P\left(1 + \dfrac{r}{m}\right)^{tm}$, $P = 10{,}000$, $r = 0.04$, $t = 5$

(a) annually, $m = 1$

$A = 10{,}000\left(1 + \dfrac{0.04}{1}\right)^{5(1)}$

$= 10{,}000(1.04)^5$

$= \$12{,}166.53$

Interest $= \$12{,}166.53 \ - \ \$10{,}000$

$= \$2166.53$

(b) semiannually, $m = 2$

$A = 10{,}000\left(1 + \dfrac{0.04}{2}\right)^{5(2)}$

$= 10{,}000(1.02)^{10}$

$= \$12{,}189.94$

Interest $= \$12{,}189.94 \ - \ \$10{,}000$

$= \$2189.94$

(c) quarterly, $m = 4$

$A = 10{,}000\left(1 + \dfrac{0.04}{4}\right)^{5(4)}$

$= 10{,}000(1.01)^{20}$

$= \$12{,}201.90$

Interest $= \$12{,}201.90 \ - \ \$10{,}000$

$= \$2201.90$

(d) monthly, $m = 12$

$A = 10{,}000\left(1 + \dfrac{0.04}{12}\right)^{5(12)}$

$= 10{,}000(1.00\overline{3})^{60}$

$= \$12{,}209.97$

Interest $= \$12{,}209.97 \ - \ \$10{,}000$

$= \$2209.97$

35. For 6% compounded annually for 2 years,

$$A = 18{,}000(1 + 0.06)^2$$
$$= 18{,}000(1.06)^2$$
$$= 20{,}224.80$$

For 5.9% compounded monthly for 2 years,

$$A = 18{,}000\left(1 + \frac{0.059}{12}\right)^{12(2)}$$
$$= 18{,}000\left(\frac{12.059}{12}\right)^{24}$$
$$= 20{,}248.54$$

The 5.9% investment is better. The additional interest is

$$\$20{,}248.54 - \$20{,}224.80 = \$23.74.$$

37. $A = Pe^{rt}$

(a) $r = 3\%$

$A = 10e^{0.03(3)} = \$10.94$

(b) $r = 4\%$

$A = 10e^{0.04(3)} = \$11.27$

(c) $r = 5\%$

$A = 10e^{0.05(3)} = \$11.62$

39. $1200 = 500\left(1 + \dfrac{r}{4}\right)^{(14)(4)}$

$\dfrac{1200}{500} = \left(1 + \dfrac{r}{4}\right)^{56}$

$2.4 = \left(1 + \dfrac{r}{4}\right)^{56}$

$1 + \dfrac{r}{4} = (2.4)^{1/56}$

$4 + r = 4(2.4)^{1/56}$

$r = 4(2.4)^{1/56} - 4$

$r \approx 0.0630$

The required interest rate is 6.30%.

41. $y = (0.92)^t$

(a)

t	y
0	$(0.92)^0 = 1$
1	$(0.92)^1 = 0.92$
2	$(0.92)^2 \approx 0.85$
3	$(0.92)^3 \approx 0.78$
4	$(0.92)^4 \approx 0.72$
5	$(0.92)^5 \approx 0.66$
6	$(0.92)^6 \approx 0.61$
7	$(0.92)^7 \approx 0.56$
8	$(0.92)^8 \approx 0.51$
9	$(0.92)^9 \approx 0.47$
10	$(0.92)^{10} \approx 0.43$

(b)

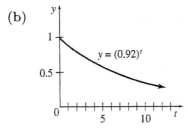

(c) Let $x =$ the cost of the house in 10 years.

Then, $0.43x = 165{,}000$
$$x \approx 383{,}721.$$

In 10 years, the house will cost about $384,000.

(d) Let $x =$ the cost of the book in 8 years.

Then, $0.51x = 50$
$$x \approx 98$$

In 8 years, the textbook will cost about $98.

43. $A = P\left(1 + \dfrac{r}{m}\right)^{tm}$

$$A = 1000\left(1 + \frac{j}{2}\right)^{5(2)} = 1000\left(1 + \frac{j}{2}\right)^{10}$$

This represents the amount in Bank X on January 1, 1985.

$$A = P\left(1 + \frac{r}{m}\right)^{tm}$$

$$= \left[1000\left(1 + \frac{j}{2}\right)^{10}\right]\left(1 + \frac{k}{4}\right)^{3(4)}$$

$$= 1000\left(1 + \frac{j}{2}\right)^{10}\left(1 + \frac{k}{4}\right)^{12}$$

This represents the amount in Bank Y on January 1, 1988, $1990.76.

$$A = P\left(1 + \frac{r}{m}\right)^{tm} = 1000\left(1 + \frac{k}{4}\right)^{8 \cdot 4}$$

$$= 1000\left(1 + \frac{k}{4}\right)^{32}$$

This represents the amount he could have had from January 1, 1980, to January 1, 1988, at a rate of k per annum compounded quarterly, $2203.76.

So,

$$1000\left(1 + \frac{j}{2}\right)^{10}\left(1 + \frac{k}{4}\right)^{12} = 1990.76$$

and $\qquad 1000\left(1 + \dfrac{k}{4}\right)^{32} = 2203.76.$

$$\left(1 + \frac{k}{4}\right)^{32} = 2.20376$$

$$1 + \frac{k}{4} = (2.20376)^{1/32}$$

$$1 + \frac{k}{4} = 1.025$$

$$\frac{k}{4} = 0.025$$

$$k = 0.1 \quad \text{or} \quad 10\%$$

Substituting, we have

$$1000\left(1 + \frac{j}{2}\right)^{10}\left(1 + \frac{1.0}{4}\right)^{12} = 1990.76$$

$$1000\left(1 + \frac{j}{2}\right)^{10}(1.025)^{12} = 1990.76$$

$$\left(1 + \frac{j}{2}\right)^{10} = 1.480$$

$$1 + \frac{j}{2} = (1.480)^{1/10}$$

$$1 + \frac{j}{2} = 1.04$$

$$\frac{j}{2} = 0.04$$

$$j = 0.08 \quad \text{or} \quad 8\%.$$

The ratio $\dfrac{k}{j} = \dfrac{0.1}{0.08} = 1.25$, is choice (a).

45. $f(x) = 500 \cdot 2^{3x}$

(a) After 1 hour:

$$f(1) = 500 \cdot 2^{3(1)} = 500 \cdot 8 = 4000 \text{ bacteria}$$

(b) initially:

$$f(0) = 500 \cdot 2^{3(0)} = 500 \cdot 1 = 500 \text{ bacteria}$$

(c) The bacteria double every $3x = 1$ hour, or every $\frac{1}{3}$ hour.

(d) When does $f(x) = 32,000$?

$$32,000 = 500 \cdot 2^{3x}$$
$$64 = 2^{3x}$$
$$2^6 = 2^{3x}$$
$$6 = 3x$$
$$x = 2$$

The number of bacteria will increase to 32,000 in 2 hours.

47. (a)

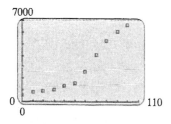

The emissions appear to grow exponentially.

(b) $f(x) = f_0 a^x$
$$f_0 = 534$$

Use the point $(100, 6672)$ to find a.

$$6672 = 534a^{100}$$
$$a^{100} = \frac{6672}{534}$$
$$a = \sqrt[100]{\frac{6672}{534}}$$
$$\approx 1.026$$
$$f(x) = 534(1.026)^x$$

(c) $1.026 - 1 = 0.026 = 2.6\%$

(d) Double the 2000 value is $2(6672) = 13,344$.

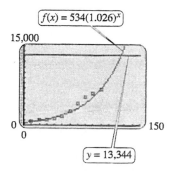

The doubling point is reached when $x \approx 125.4$. The first year in which emissions equal or exceed that threshold is 2026.

49. (a) When $x = 0$, $P = 1013$.
When $x = 10,000$, $P = 265$.
First we fit $P = ae^{kx}$.

$$1013 = ae^0$$
$$a = 1013$$
$$P = 1013e^{kx}$$
$$265 = 1013e^{k(10,000)}$$
$$\frac{265}{1013} = e^{10,000k}$$
$$10,000k = \ln\left(\frac{265}{1013}\right)$$
$$k = \frac{\ln\left(\frac{265}{1013}\right)}{10,000} \approx -1.34 \times 10^{-4}$$

Therefore $P = 1013e^{(-1.34 \times 10^{-4})x}$.

Next we fit $P = mx + b$.
We use the points $(0, 1013)$ and $(10,000, 265)$.

$$m = \frac{265 - 1013}{10,000 - 0} = -0.0748$$
$$b = 1013$$

Therefore $P = -0.0748x + 1013$.

Finally, we fit $P = \frac{1}{ax+b}$.

$$1013 = \frac{1}{a(0) + b}$$
$$b = \frac{1}{1013} \approx 9.87 \times 10^{-4}$$
$$P = \frac{1}{ax + \frac{1}{1013}}$$
$$265 = \frac{1}{10,000a + \frac{1}{1013}}$$
$$\frac{1}{265} = 10,000a + \frac{1}{1013}$$
$$10,000a = \frac{1}{265} - \frac{1}{1013}$$
$$a = \frac{\frac{1}{265} - \frac{1}{1013}}{10,000} \approx 2.79 \times 10^{-7}$$

Therefore,

$$P = \frac{1}{(2.79 \times 10^{-7})x + (9.87 \times 10^{-4})}.$$

(b)

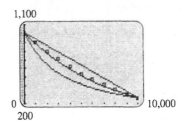

$P = 1013e^{(-1.34\times10^{-4})x}$ is the best fit.

(c) $P(1500) = 1013e^{-1.34\times10^{-4}(1500)} \approx 829$

$P(11,000) = 1013e^{-1.34\times10^{-4}(11,000)} \approx 232$

We predict that the pressure at 1500 meters will be 829 millibars, and at 11,000 meters will be 232 millibars.

(d) Using exponential regression, we obtain $P = 1038(0.99998661)^x$ which differs slightly from the function found in part (b) which can be rewritten as

$$P = 1013(0.99998660)^x.$$

10.5 Logarithmic Functions

1. $5^3 = 125$

Since $a^y = x$ means $y = \log_a x$, the equation in logarithmic form is

$$\log_5 125 = 3.$$

3. $3^4 = 81$

The equation in logarithmic form is

$$\log_3 81 = 4.$$

5. $3^{-2} = \dfrac{1}{9}$

The equation in logarithmic form is

$$\log_3 \frac{1}{9} = -2.$$

7. $\log_2 32 = 5$

Since $y = \log_a x$ means $a^y = x$, the equation in exponential form is

$$2^5 = 32.$$

9. $\ln \dfrac{1}{e} = -1$

The equation in exponential form is

$$e^{-1} = \frac{1}{e}.$$

11. $\log 100,000 = 5$
      $\log_{10} 100,000 = 5$
         $10^5 = 100,000$

When no base is written, $\log_{10}$ is understood.

13. Let $\log_8 64 = x$.

Then, $8^x = 64$
      $8^x = 8^2$
       $x = 2.$

Thus, $\log_8 64 = 2$.

15. $\log_4 64 = x$
      $4^x = 64$
      $4^x = 4^3$
       $x = 3$

17. $\log_2 \dfrac{1}{16} = x$

$2^x = \dfrac{1}{16}$

$2^x = 2^{-4}$

$x = -4$

19. $\log_2 \sqrt[3]{\dfrac{1}{4}} = x$

$2^x = \left(\dfrac{1}{4}\right)^{1/3}$

$2^x = \left(\dfrac{1}{2^2}\right)^{1/3}$

$2^x = 2^{-2/3}$

$x = -\dfrac{2}{3}$

21. $\ln e = x$

Recall that ln means $\log_e$.

$e^x = e$
$x = 1$

23. $\ln e^{5/3} = x$

$e^x = e^{5/3}$

$x = \dfrac{5}{3}$

25. The logarithm to the base 3 of 4 is written $\log_3 4$. The subscript denotes the base.

27. $\log_5 (3k) = \log_5 3 + \log_5 k$

29. $\log_3 \dfrac{3p}{5k}$

$= \log_3 3p - \log_3 5k$

$= (\log_3 3 + \log_3 p) - (\log_3 5 + \log_3 k)$

$= 1 + \log_3 p - \log_3 5 - \log_3 k$

31. $\ln \dfrac{3\sqrt{5}}{\sqrt[3]{6}}$

$= \ln 3\sqrt{5} - \ln \sqrt[3]{6}$

$= \ln 3 \cdot 5^{1/2} - \ln 6^{1/3}$

$= \ln 3 + \ln 5^{1/2} - \ln 6^{1/3}$

$= \ln 3 + \dfrac{1}{2} \ln 5 - \dfrac{1}{3} \ln 6$

33. $\log_b 32 = \log_b 2^5$

$= 5 \log_b 2$

$= 5a$

35. $\log_b 72b = \log_b 72 + \log_b b$

$= \log_b 72 + 1$

$= \log_b 2^3 \cdot 3^3 + 1$

$= \log_b 2^3 + \log_b 3^2 + 1$

$= 3 \log_b 2 + 2 \log_b 3 + 1$

$= 3a + 2c + 1$

37. $\log_5 30 = \dfrac{\ln 30}{\ln 5}$

$\approx \dfrac{3.4012}{1.6094}$

≈ 2.113

39. $\log_{1.2} 0.95 = \dfrac{\ln 0.95}{\ln 1.2}$

≈ -0.281

41. $\log_x 36 = -2$

$x^{-2} = 36$

$(x^{-2})^{-1/2} = 36^{-1/2}$

$x = \dfrac{1}{6}$

43. $\log_8 16 = z$

$8^z = 16$

$(2^3)^z = 2^4$

$2^{3z} = 2^4$

$3z = 4$

$z = \dfrac{4}{3}$

45. $\log_r 5 = \dfrac{1}{2}$

$r^{1/2} = 5$

$(r^{1/2})^2 = 5^2$

$r = 25$

47. $\log_5 (9x - 4) = 1$

$5^1 = 9x - 4$

$9 = 9x$

$1 = x$

49. $\log_9 m - \log_9 (m - 4) = -2$

$\log_9 \dfrac{m}{m-4} = -2$

$9^{-2} = \dfrac{m}{m-4}$

$\dfrac{1}{81} = \dfrac{m}{m-4}$

$m - 4 = 81m$

$-4 = 80m$

$-0.05 = m$

This value is not possible since $\log_9 (-0.05)$ does not exist.

Thus, there is no solution to the original equation.

51. $\log_3 (x - 2) + \log_3 (x + 6) = 2$

$\log_3 [(x - 2)(x + 6)] = 2$

$(x - 2)(x + 6) = 3^2$

$x^2 + 4x - 12 = 9$

$x^2 + 4x - 21 = 0$

$(x + 7)(x - 3) = 0$

$x = -7 \quad \text{or} \quad x = 3$

$x = -7$ does not check in the original equation. The only solution is 3.

53. $\log_2(x^2 - 1) - \log_2(x + 1) = 2$

$\log_2 \dfrac{x^2 - 1}{x + 1} = 2$

$2^2 = \dfrac{x^2 - 1}{x + 1}$

$4 = \dfrac{(x - 1)(x + 1)}{x + 1}$

$4 = x - 1$

$x = 5$

55. $5^x = 12$

$x \log 5 = \log 12$

$x = \dfrac{\log 12}{\log 5}$

≈ 1.544

57.
$$e^{2y} = 15$$
$$\ln e^{2y} = \ln 15$$
$$2y \ln e = \ln 15$$
$$2y(1) = \ln 15$$
$$y = \frac{\ln 15}{2}$$
$$\approx 1.354$$

59.
$$10e^{3z-7} = 100$$
$$\ln 10e^{3z-7} = \ln 100$$
$$\ln 10 + \ln e^{3z-7} = \ln 100$$
$$\ln 10 + (3z - 7) \ln e = \ln 100$$
$$3z - 7 = \ln 100 - \ln 10$$
$$3z = \ln 100 - \ln 10 + 7$$
$$z = \frac{\ln 100 - \ln 10 + 7}{3}$$
$$\approx 3.101$$

61.
$$1.5(1.05)^x = 2(1.01)^x$$
$$\ln[1.5(1.05)^x] = \ln[2(1.01)^x]$$
$$\ln 1.5 + x \ln 1.05 = \ln 2 + x \ln 1.01$$
$$x(\ln 1.05 - \ln 1.01) = \ln 2 - \ln 1.5$$
$$x = \frac{\ln 2 - \ln 1.5}{\ln 1.05 - \ln 1.01}$$
$$\approx 7.407$$

63. $f(x) = \ln(x^2 - 9)$

Since the domain of $f(x) = \ln x$ is $(0, \infty)$, the domain of $f(x) = \ln(x^2 - 9)$ is the set of all real numbers x for which

$$x^2 - 9 > 0.$$

To solve this quadratic inequality, first solve the corresponding quadratic equation.

$$x^2 - 9 = 0$$
$$(x + 3)(x - 3) = 0$$
$$x + 3 = 0 \quad \text{or} \quad x - 3 = 0$$
$$x = -3 \quad \text{or} \quad x = 3$$

These two solutions determine three intervals on the number line: $(-\infty, -3), (-3, 3)$, and $(3, \infty)$.

If $x = -4$, $(-4 + 2)(-4 - 2) > 0$.
If $x = 0$, $(0 + 2)(0 - 2) \not> 0$.
If $x = 4$, $(4 + 2)(4 - 2) > 0$.

The domain is $x < -3$ or $x > 3$, which is written in interval notation as $(-\infty, -3) \cup (3, \infty)$.

65. Let $m = \log_a \dfrac{x}{y}$, $n = \log_a x$, and $p = \log_a y$.

Then $a^m = \dfrac{x}{y}$, $a^n = x$, and $a^p = y$.

Substituting gives

$$a^m = \frac{x}{y} = \frac{a^n}{a^p} = a^{n-p}.$$

So $m = n - p$.
Therefore,

$$\log_a \frac{x}{y} = \log_a x - \log_a y.$$

67. From Example 7, the doubling time t in years when $m = 1$ is given by

$$t = \frac{\ln 2}{\ln(1 + r)}.$$

(a) Let $r = 0.03$.

$$t = \frac{\ln 2}{\ln 1.03}$$
$$= 23.4 \text{ years}$$

(b) Let $r = 0.06$.

$$t = \frac{\ln 2}{\ln 1.06}$$
$$= 11.9 \text{ years}$$

(c) Let $r = 0.08$.

$$t = \frac{\ln 2}{\ln 1.08}$$
$$= 9.0 \text{ years}$$

(d) Since $0.001 \leq 0.03 \leq 0.05$, for $r = 0.03$, we use the rule of 70.

$$\frac{70}{100r} = \frac{70}{100(0.03)} = 23.3 \text{ years}$$

Since $0.05 \leq 0.06 \leq 0.12$, for $r = 0.06$, we use the rule of 72.

$$\frac{72}{100r} = \frac{72}{100(0.06)} = 12 \text{ years}$$

For $r = 0.08$, we use the rule of 72.

$$\frac{72}{100(0.08)} = 9 \text{ years}$$

69. $A = Pe^{rt}$

$$1200 = 500e^{r \cdot 14}$$
$$2.4 = e^{14r}$$
$$\ln(2.4) = \ln e^{14r}$$
$$\ln(2.4) = 14r$$
$$\frac{\ln(2.4)}{14} = r$$
$$0.0625 \approx r$$

The interest rate should be 6.25%.

71. After x years at Humongous Enterprises, your salary would be $45{,}000 \, (1 + 0.04)^x$ or $45{,}000 \, (1.04)^x$.
After x years at Crabapple Inc., your salary would be $30{,}000 \, (1 + 0.06)^x$ or $30{,}000 \, (1.06)^x$.
First we find when the salaries would be equal.

$$45{,}000(1.04)^x = 30{,}000(1.06)^x$$
$$\frac{(1.04)^x}{(1.06)^x} = \frac{30{,}000}{45{,}000}$$
$$\left(\frac{1.04}{1.06}\right)^x = \frac{2}{3}$$
$$\log\left(\frac{1.04}{1.06}\right)^x = \log\left(\frac{2}{3}\right)$$
$$x \log\left(\frac{1.04}{1.06}\right) = \log\left(\frac{2}{3}\right)$$

$$x = \frac{\log\left(\frac{2}{3}\right)}{\log\left(\frac{1.04}{1.06}\right)}$$

$$x \approx 21.29$$
$$2009 + 21.29 = 2030.29$$

Therefore, on July 1, 2031, the job at Crabapple, Inc., will pay more.

73. (a) The total number of individuals in the community is $50 + 50$, or 100.

Let $P_1 = \dfrac{50}{100} = 0.5$, $P_2 = 0.5$.
$$H = -1[P_1 \ln P_1 + P_2 \ln P_2]$$
$$= -1[0.5 \ln 0.5 + 0.5 \ln 0.5]$$
$$\approx 0.693$$

(b) For 2 species, the maximum diversity is $\ln 2$.

(c) Yes, $\ln 2 \approx 0.693$.

75. (a) 3 species, $\frac{1}{3}$ each:

$$P_1 = P_2 = P_3 = \frac{1}{3}$$
$$H = -(P_1 \ln P_1 + P_2 \ln P_2 + P_3 \ln P_3)$$
$$= -3\left(\frac{1}{3} \ln \frac{1}{3}\right)$$
$$= -\ln \frac{1}{3}$$
$$\approx 1.099$$

(b) 4 species, $\frac{1}{4}$ each:

$$P_1 = P_2 = P_3 = P_4 = \frac{1}{4}$$
$$H = (P_1 \ln P_1 + P_2 \ln P_2 + P_3 \ln P_3 + P_4 \ln P_4)$$
$$= -4\left(\frac{1}{4} \ln \frac{1}{4}\right)$$
$$= -\ln \frac{1}{4}$$
$$\approx 1.386$$

(c) Notice that

$$-\ln \frac{1}{3} = \ln(3^{-1})^{-1} = \ln 3 \approx 1.099$$

and

$$-\ln \frac{1}{4} = \ln(4^{-1})^{-1} = \ln 4 \approx 1.386$$

by Property c of logarithms, so the populations are at a maximum index of diversity.

77. $C(t) = C_0 e^{-kt}$

When $t = 0$, $C(t) = 2$, and when $t = 3$, $C(t) = 1$.
$$2 = C_0 e^{-k(0)}$$
$$C_0 = 2$$
$$1 = 2e^{-3k}$$
$$\frac{1}{2} = e^{-3k}$$
$$-3k = \ln \frac{1}{2} = \ln 2^{-1} = -\ln 2$$
$$k = \frac{\ln 2}{3}$$
$$T = \frac{1}{k} \ln \frac{C_2}{C_1}$$
$$T = \frac{1}{\frac{\ln 2}{3}} \ln \frac{5\,C_1}{C_1}$$
$$T = \frac{3 \ln 5}{\ln 2}$$
$$T \approx 7.0$$

The drug should be given about every 7 hours.

79. (a) $h(t) = 37.79(1.021)^t$

Double the 2005 population is $2(42.69) = 85.38$ million

$$85.38 = 37.79(1.021)^t$$

$$\frac{85.38}{37.79} = (1.021)^t$$

$$\log_{1.021}\left(\frac{85.38}{37.79}\right) = t$$

$$t = \frac{\ln\left(\frac{85.38}{37.79}\right)}{\ln 1.021}$$

$$\approx 39.22$$

The Hispanic population is estimated to double their 2005 population in 2039.

(b) $h(t) = 11.14(1.023)^t$

Double the 2005 population is $2(12.69) = 25.38$ million

$$25.38 = 11.14(1.023)^t$$

$$\frac{25.38}{11.14} = (1.023)^t$$

$$\log_{1.023}\left(\frac{25.38}{11.14}\right) = t$$

$$t = \frac{\ln\left(\frac{25.38}{11.14}\right)}{\ln 1.023}$$

$$\approx 36.21$$

The Asian population is estimated to double their 2005 population in 2036.

81. $\qquad C = B\log_2\left(\frac{s}{n} + 1\right)$

$$\frac{C}{B} = \log_2\left(\frac{s}{n} + 1\right)$$

$$2^{C/B} = \frac{s}{n} + 1$$

$$\frac{s}{n} = 2^{C/B} - 1$$

83. Let I_1 be the intensity of the sound whose decibel rating is 85.

(a) $\qquad 10\log\frac{I_1}{I_0} = 85$

$$\log\frac{I_1}{I_0} = 8.5$$

$$\log I_1 - \log I_0 = 8.5$$

$$\log I_1 = 8.5 + \log I_0$$

Let I_2 be the intensity of the sound whose decimal rating is 75.

$$10\log\frac{I_2}{I_0} = 75$$

$$\log\frac{I_2}{I_0} = 7.5$$

$$\log I_2 - \log I_0 = 7.5$$

$$\log I_0 = \log I_2 - 7.5$$

Substitute for I_0 in the equation for $\log I_1$.

$$\log I_1 = 8.5 + \log I_0$$
$$= 8.5 + \log I_2 - 7.5$$
$$= 1 + \log I_2$$

$$\log I_1 - \log I_2 = 1$$

$$\log\frac{I_1}{I_2} = 1$$

Then $\frac{I_1}{I_2} = 10$, so $I_2 = \frac{1}{10}I_1$. This means the intensity of the sound that had a rating of 75 decibels is $\frac{1}{10}$ as intense as the sound that had a rating of 85 decibels.

85. $\text{pH} = -\log\,[\text{H}^+]$

(a) For pure water:

$$7 = -\log\,[\text{H}^+]$$
$$-7 = \log\,[\text{H}^+]$$
$$10^{-7} = [\text{H}^+]$$

For acid rain:

$$4 = -\log\,[\text{H}^+]$$
$$-4 = \log\,[\text{H}^+]$$
$$10^{-4} = [\text{H}^+]$$

$$\frac{10^{-4}}{10^{-7}} = 10^3 = 1000$$

The acid rain has a hydrogen ion concentration 1000 times greater than pure water.

(b) For laundry solution:

$$11 = -\log\,[\text{H}^+]$$
$$10^{-11} = [\text{H}^+]$$

For black coffee:

$$5 = -\log\,[\text{H}^+]$$
$$10^{-5} = [\text{H}^+]$$

$$\frac{10^{-5}}{10^{-11}} = 10^6 = 1,000,000$$

The coffee has a hydrogen ion concentration 1,000,000 times greater than the laundry mixture.

10.6 Applications: Growth and Decay; Mathematics of Finance

5. Assume that $y = y_0 e^{kt}$ represents the amount remaining of a radioactive substance decaying with a half-life of T. Since $y = y_0$ is the amount of the substance at time $t = 0$, then $y = \frac{y_0}{2}$ is the amount at time $t = T$. Therefore, $\frac{y_0}{2} = y_0 e^{kT}$, and solving for k yields

$$\frac{1}{2} = e^{kT}$$

$$\ln\left(\frac{1}{2}\right) = kT$$

$$k = \frac{\ln\left(\frac{1}{2}\right)}{T}$$

$$= \frac{\ln(2^{-1})}{T}$$

$$= -\frac{\ln 2}{T}.$$

7. $r = 4\%$ compounded quarterly, $m = 4$

$$r_E = \left(1 + \frac{r}{m}\right)^m - 1$$

$$= \left(1 + \frac{0.04}{4}\right)^4 - 1$$

$$\approx 0.0406$$

$$\approx 4.06\%$$

9. $r = 8\%$ compounded continuously

$$r_E = e^r - 1$$
$$= e^{0.08} - 1$$
$$= 0.0833$$
$$= 8.33\%$$

11. $A = \$10,000$, $r = 6\%$, $m = 4$, $t = 8$

$$P = A\left(1 + \frac{r}{m}\right)^{-tm}$$

$$= 10,000\left(1 + \frac{0.06}{4}\right)^{-8(4)}$$

$$\approx \$6209.93$$

13. $A = \$7300$, $r = 5\%$ compounded continuously, $t = 3$

$$A = Pe^{rt}$$

$$P = \frac{A}{e^{rt}}$$

$$= \frac{7300}{e^{0.5(3)}}$$

$$\approx \$6283.17$$

15. $r = 9\%$ compounded semiannually

$$r_E = \left(1 + \frac{0.09}{2}\right)^2 - 1$$

$$\approx 0.0920$$

$$\approx 9.20\%$$

17. $r = 6\%$ compounded monthly

$$r_E = \left(1 + \frac{0.06}{12}\right)^{12} - 1$$

$$\approx 0.0617$$

$$\approx 6.17\%$$

19. (a) $A = \$307,000$, $t = 3$, $r = 6\%$, $m = 2$

$$A = P\left(1 + \frac{r}{m}\right)^{mt}$$

$$307,000 = P\left(1 + \frac{0.06}{2}\right)^{3(2)}$$

$$307,000 = P(1.03)^6$$

$$\frac{307,000}{(1.03)^6} = P$$

$$\$257,107.67 = P$$

(b) Interest $= 307,000 - 257,107.67$
$$= \$49,892.33$$

(c) $P = \$200,000$
$$A = 200,000(1.03)^6$$
$$= 238,810.46$$

The additional amount needed is

$$307,000 - 238,810.46$$
$$= \$68,189.54.$$

21. $P = \$60,000$

(a) $r = 8\%$ compounded quarterly:

$$A = P \left(1 + \frac{r}{m}\right)^{tm}$$

$$= 60,000 \left(1 + \frac{0.08}{4}\right)^{5(4)}$$

$$\approx \$89,156.84$$

$r = 7.75\%$ compounded continuously

$$A = Pe^{rt}$$
$$= 60,000e^{0.0775(5)}$$
$$\approx \$88,397.58$$

Linda will earn more money at 8% compounded quarterly.

(b) She will earn $759.26 more.

(c) $r = 8\%$, $m = 4$:

$$r_E = \left(1 + \frac{r}{m}\right)^m - 1$$

$$= \left(1 + \frac{0.08}{4}\right)^4 - 1$$

$$\approx 0.0824$$
$$= 8.24\%$$

$r = 7.75\%$ compounded continuously:

$$r_E = e^r - 1$$
$$= e^{0.0775} - 1$$
$$\approx 0.0806$$
$$= 8.06\%$$

(d) $A = \$80,000$

$$A = Pe^{rt}$$
$$80,000 = 60,000e^{0.0775t}$$

$$\frac{4}{3} = e^{0.0775t}$$

$$\ln \frac{4}{3} = \ln e^{0.0775t}$$

$$\ln 4 - \ln 3 = 0.0775t$$

$$\frac{\ln 4 - \ln 3}{0.0775} = t$$

$$3.71 = t$$

$60,000 will grow to $80,000 in about 3.71 years.

(e) $60,000 \left(1 + \dfrac{0.08}{4}\right)^{4x} \geq 80,000$

$$(1.02)^{4x} \geq \frac{80,000}{60,000}$$

$$(1.02)^{4x} \geq \frac{4}{3}$$

$$\log (1.02)^{4x} \geq \log \left(\frac{4}{3}\right)$$

$$4x \log (1.02) \geq \log \left(\frac{4}{3}\right)$$

$$x \geq \frac{\log \left(\frac{4}{3}\right)}{4 \log (1.02)} \approx 3.63$$

It will take about 3.63 years.

23. The figure is not correct.

$$(1 + 0.09)(1 + 0.08)(1 + 0.07) = 1.2596$$

This is a 25.96% increase.

25. $S(x) = 1000 - 800e^{-x}$

(a) $S(0) = 1000 - 800e^0$
$$= 1000 - 800$$
$$= 200$$

(b) $S(x) = 500$
$$500 = 1000 - 800e^{-x}$$
$$-500 = -800e^{-x}$$

$$\frac{5}{8} = e^{-x}$$

$$\ln \frac{5}{8} = \ln e^{-x}$$

$$-\ln \frac{5}{8} = x$$

$$0.47 \approx x$$

Sales reach 500 in about $\frac{1}{2}$ year.

(c) Since $800e^{-x}$ will never actually be zero, $S(x) = 1000 - 800e^{-x}$ will never be 1000.

(d) Graphing the function $y = S(x)$ on a graphing calculator will show that there is a horizontal asymptote at $y = 1000$. This indicates that the limit on sales is 1000 units.

27. (a) $P = P_0 e^{kt}$

When $t = 1650$, $P = 470$.
When $t = 2005$, $P = 6451$.

$$470 = P_0 e^{1650k}$$
$$6451 = P_0 e^{2005k}$$

$$\frac{6451}{470} = \frac{P_0 e^{2005k}}{P_0 e^{1650k}}$$

$$\frac{6451}{470} = e^{355k}$$

$$355k = \ln\left(\frac{6451}{470}\right)$$

$$k = \frac{\ln\left(\frac{6451}{470}\right)}{355}$$

$$k \approx 0.007378$$

Substitute this value into $470 = P_0 e^{1650k}$ to find P_0.

$$470 = P_0 e^{1650(0.007378)}$$

$$P_0 = \frac{470}{e^{1650(0.007378)}}$$

$$P_0 \approx 0.002427$$

Therefore, $P(t) = 0.002427 e^{0.007378t}$.

(b) $P(1) = 0.002427 e^{0.007378} \approx 0.002445$ million or 2445

The exponential equation gives a world population of only 2445 in the year 1.

(c) No, the answer in part (b) is too small. Exponential growth does not accurately describe population growth for the world over a long period of time.

29. From 1960 to 2005 is an interval of $t = 45$ years.

$$P = P_0 e^{rt}$$
$$1500 = 59 e^{45r}$$

$$\frac{1500}{59} = e^{45r}$$

$$45r = \ln\frac{1500}{59}$$

$$r = \frac{1}{45}\ln\frac{1500}{59}$$

$$\approx 0.0719$$

This is an annual increase of 7.19%.

31. $y = y_0 e^{kt}$

(a) $y = 20{,}000$, $y_0 = 50{,}000$, $t = 9$

$$20{,}000 = 50{,}000 e^{9k}$$
$$0.4 = e^{9k}$$
$$\ln 0.4 = 9k$$
$$-0.102 = k$$

The equation is

$$y = 50{,}000 e^{-0.102t}.$$

(b) $\frac{1}{2}(50{,}000) = 25{,}000$

$$25{,}000 = 50{,}000 e^{-0.102t}$$
$$0.5 = e^{-0.102t}$$
$$\ln 0.5 = -0.102t$$
$$6.8 = t$$

Half the bacteria remain after about 6.8 hours.

33. Use $y = y_0 e^{-kt}$.

When $t = 5$, $y = 0.37 y_0$.

$$0.37 y_0 = y_0 e^{-5k}$$
$$0.37 = e^{-5k}$$
$$-5k = \ln(0.37)$$
$$k = \frac{\ln(0.37)}{-5}$$
$$k \approx 0.1989$$

35.
$$A(t) = A_0 e^{kt}$$
$$0.60\, A_0 = A_0 e^{(-\ln 2/5600)t}$$
$$0.60 = e^{(-\ln 2/5600)t}$$
$$\ln 0.60 = -\frac{\ln 2}{5600}t$$
$$\frac{5600(\ln 0.60)}{-\ln 2} = t$$
$$4127 \approx t$$

The sample was about 4100 years old.

37.
$$\frac{1}{2} A_0 = A_0 e^{-0.00043t}$$
$$\frac{1}{2} = e^{-0.00043t}$$
$$\ln \frac{1}{2} = -0.00043t$$
$$\ln 1 - \ln 2 = -0.00043t$$
$$\frac{0 - \ln 2}{-0.00043} = t$$
$$1612 \approx t$$

The half-life of radium 226 is about 1600 years.

39. (a) $A(t) = A_0 \left(\dfrac{1}{2}\right)^{t/1620}$

$A(100) = 4.0 \left(\dfrac{1}{2}\right)^{100/1620}$

$A(100) \approx 3.8$

After 100 years, about 3.8 grams will remain.

(b) $0.1 = 4.0 \left(\dfrac{1}{2}\right)^{t/1620}$

$\dfrac{0.1}{4} = \left(\dfrac{1}{2}\right)^{t/1620}$

$\ln 0.025 = \dfrac{t}{1620} \ln \dfrac{1}{2}$

$t = \dfrac{1620 \ln 0.025}{\ln \left(\frac{1}{2}\right)}$

$t \approx 8600$

The half-life is about 8600 years.

41. (a) $y = y_0 e^{kt}$

When $t = 0$, $y = 25.0$, so $y_0 = 25.0$.
When $t = 50$, $y = 19.5$.

$19.5 = 25.0 e^{50k}$

$\dfrac{19.5}{25.0} = e^{50k}$

$50k = \ln \left(\dfrac{19.5}{25.0}\right)$

$k = \dfrac{\ln \left(\frac{19.5}{25.0}\right)}{50}$

$k \approx -0.00497$

$y = 25.0 e^{-0.00497t}$

(b) $\dfrac{1}{2} y_0 = y_0 e^{-0.00497t}$

$\dfrac{1}{2} = e^{-0.00497t}$

$-0.00497t = \ln \left(\dfrac{1}{2}\right)$

$t = \dfrac{\ln \left(\frac{1}{2}\right)}{-0.00497}$

$t \approx 139$

The half-life is about 139 days.

43. $A(t) = A_0 \left(\dfrac{1}{2}\right)^{t/5600}$

$A(43{,}000) = A_0 \left(\dfrac{1}{2}\right)^{43{,}000/5600}$

$\approx 0.005 A_0$

About 0.5% of the original carbon 14 was present.

45. (a)   Let $t =$ the number of degrees Celsius.

$y = y_0 \cdot e^{kt}$
$y_0 = 10$ when $t = 0°$.
To find k, let $y = 11$ when $t = 10°$.

$11 = 10 e^{10k}$

$e^{10k} = \dfrac{11}{10}$

$10k = \ln 1.1$

$k = \dfrac{\ln 1.1}{10}$

≈ 0.0095

The equation is

$y = 10 e^{0.0095t}.$

(b) Let $y = 15$; solve for t.

$15 = 10 e^{0.0095t}$
$\ln 1.5 = 0.0095t$

$t = \dfrac{\ln 1.5}{0.0095}$

≈ 42.7

15 grams will dissolve at 42.7°C.

47. $f(t) = T_0 + Ce^{-kt}$

$25 = 20 + 100 e^{-0.1t}$
$5 = 100 e^{-0.1t}$
$e^{-0.1t} = 0.05$
$-0.1t = \ln 0.05$

$t = \dfrac{\ln 0.05}{-0.1}$

≈ 30

It will take about 30 min.

Chapter 10 Review Exercises

5. $y = (2x - 1)(x + 1)$
   $= 2x^2 + x - 1$

x	-3	-2	-1	0	1	2	3
y	14	5	0	-1	2	9	20

Pairs: $(-3, 14), (-2, 5), (-1, 0), (0, -1), (1, 2),$
$(2, 9), (3,\ 20)$
Range: $\{-1, 0, 2, 5, 9, 14, 20\}$

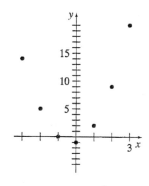

7. $f(x) = 5x^2 - 3$ and $g(x) = -x^2 + 4x + 1$

(a) $f(-2) = 5(-2)^2 - 3 = 17$

(b) $g(3) = -(3)^2 + 4(3) + 1 = 4$

(c) $f(-k) = 5(-k)^2 - 3 = 5k^2 - 3$

(d) $g(3m) = -(3m)^2 + 4(3m) + 1$
   $= -9m^2 + 12m + 1$

(e) $f(x + h) = 5(x + h)^2 - 3$
   $= 5(x^2 + 2xh + h^2) - 3$
   $= 5x^2 + 10xh + 5h^2 - 3$

(f) $g(x + h) = -(x + h)^2 + 4(x + h) + 1$
   $= -(x^2 + 2xh + h^2) + 4x + 4h + 1$
   $= -x^2 - 2xh - h^2 + 4x + 4h + 1$

(g) $\dfrac{f(x + h) - f(x)}{h}$

   $= \dfrac{5(x + h)^2 - 3 - (5x^2 - 3)}{h}$

   $= \dfrac{5(x^2 + 2hx + h^2) - 3 - 5x^2 + 3}{h}$

   $= \dfrac{5x^2 + 10hx + 5h^2 - 5x^2}{h}$

   $= \dfrac{10hx + 5h^2}{h}$

   $= 10x + 5h$

(h) $\dfrac{g(x + h) - g(x)}{h}$

   $= \dfrac{-(x + h)^2 + 4(x + h) + 1 - (-x^2 + 4x + 1)}{h}$

   $= \dfrac{-(x^2 + 2xh + h^2) + 4x + 4h + 1 + x^2 - 4x - 1}{h}$

   $= \dfrac{-x^2 - 2xh - h^2 + 4h + x^2}{h}$

   $= \dfrac{-2xh - h^2 + 4h}{h}$

   $= -2x - h + 4$

9. $y = \ln(x + 7)$

$$x + 7 > 0$$
$$x > -7$$

Domain: $(-7, \infty)$.

11. $y = \dfrac{3x - 4}{x}$

$x \neq 0$

Domain: $(-\infty, 0) \cup (0, \infty)$

13. $y = 2x^2 + 3x - 1$

The graph is a parabola.
Let $y = 0$.

$$0 = 2x^2 + 3x - 1$$

$$x = \frac{-3 \pm \sqrt{3^2 - 4(2)(-1)}}{2(2)}$$

$$= \frac{-3 \pm \sqrt{9 + 8}}{4}$$

$$= \frac{-3 \pm \sqrt{17}}{4}$$

The x-intercepts are $\frac{-3 + \sqrt{17}}{4} \approx 0.28$ and
$\frac{-3 - \sqrt{17}}{4} \approx -1.48$.

Let $x = 0$.

$$y = 2(0)^2 + 3(0) - 1$$

-1 is the y-intercept.

Vertex: $x = \dfrac{-b}{2a} = \dfrac{-3}{2(2)} = -\dfrac{3}{4}$

$$y = 2\left(-\frac{3}{4}\right)^2 + 3\left(-\frac{3}{4}\right) - 1$$

$$= \frac{9}{8} - \frac{9}{4} - 1$$

$$= -\frac{17}{8}$$

The vertex is $\left(-\frac{3}{4}, -\frac{17}{8}\right)$.

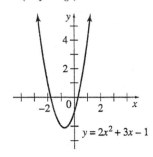

15. $y = -x^2 + 4x + 2$

Let $y = 0$.

$$0 = -x^2 + 4x + 2$$

$$x = \frac{-4 \pm \sqrt{4^2 - 4(-1)(2)}}{2(-1)}$$

$$= \frac{-4 \pm \sqrt{24}}{-2}$$

$$= 2 \pm \sqrt{6}$$

The x-intercepts are $2 + \sqrt{6} \approx 4.45$ and $2 - \sqrt{6} \approx -0.45$.

Let $x = 0$.

$$y = -0^2 + 4(0) + 2$$

2 is the y-intercept.

Vertex: $x = \frac{-b}{2a} = \frac{-4}{2(-1)} = \frac{-4}{-2} = 2$

$$y = -2^2 + 4(2) + 2 = 6$$

The vertex is $(2, 6)$.

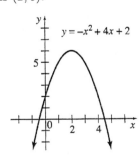

17. $f(x) = x^3 - 3$

Translate the graph of $f(x) = x^3$ 3 units down.

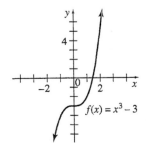

19. $y = -(x - 1)^4 + 4$

Translate the graph of $y = x^4$ 1 unit to the right and reflect vertically. Translate 4 units upward.

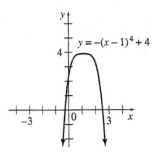

21. $f(x) = \frac{8}{x}$

Vertical asymptote: $x = 0$

Horizontal asymptote:

$\frac{8}{x}$ approaches zero as x gets larger.

$y = 0$ is an asymptote.

x	-4	-3	-2	-1	1	2	3	4
y	-2	-2.7	-4	-8	8	4	2.7	2

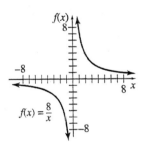

23. $f(x) = \frac{4x - 2}{3x + 1}$

Vertical asymptote:

$$3x + 1 = 0$$

$$x = -\frac{1}{3}$$

Horizontal asymptote:

As x gets larger,

$$\frac{4x - 2}{3x - 1} \approx \frac{4x}{3x} = \frac{4}{3}.$$

$y = \frac{4}{3}$ is an asymptote.

x	-3	-2	-1	0	1	2	3
y	1.75	2	3	-2	0.5	0.86	1

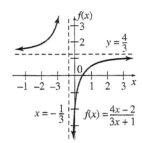

25. $y = 4^x$

x	-2	-1	0	1	2
y	$\frac{1}{16}$	$\frac{1}{4}$	1	4	16

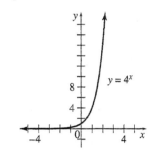

27. $y = \left(\dfrac{1}{5}\right)^{2x-3}$

x	0	1	2
y	125	5	$\frac{1}{5}$

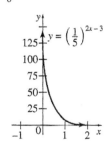

29. $y = \log_2 (x - 1)$

$2^y = x - 1$

$x = 1 + 2^y$

x	2	3	5	9
y	0	1	2	3

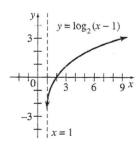

31.

$$y = -\ln(x+3)$$
$$-y = \ln(x+3)$$
$$e^{-y} = x+3$$
$$e^{-y} - 3 = x$$

x	-2.63	-2	-0.28	4.39
y	1	0	-1	-2

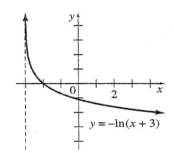

33. $2^{x+2} = \dfrac{1}{8}$

$$2^{x+2} = \frac{1}{2^3}$$
$$2^{x+2} = 2^{-3}$$
$$x + 2 = -3$$
$$x = -5$$

35. $9^{2y+3} = 27^y$

$$(3^2)^{2y+3} = (3^3)^y$$
$$3^{4y+6} = 3^{3y}$$
$$4y + 6 = 3y$$
$$y = -6$$

37. $3^5 = 243$

The equation in logarithmic form is

$$\log_3 243 = 5.$$

39. $e^{0.8} = 2.22554$

The equation in logarithmic form is

$$\ln 2.22554 = 0.8.$$

41. $\log_2 32 = 5$

The equation in exponential form is

$$2^5 = 32.$$

43. $\ln 82.9 = 4.41763$

The equation in exponential form is

$$e^{4.41763} = 82.9.$$

45. $\log_3 81 = x$
$$3^x = 81$$
$$3^x = 3^4$$
$$x = 4$$

47. $\log_4 8 = x$
$$4^x = 8$$
$$(2^2)^x = 2^3$$
$$2x = 3$$
$$x = \frac{3}{2}$$

49. $\log_5 3k + \log_5 7k^3$
$$= \log_5 3k(7k^3)$$
$$= \log_5 (21k^4)$$

51. $4 \log_3 y - 2 \log_3 x$
$$= \log_3 y^4 - \log_3 x^2$$
$$= \log_3 \left(\frac{y^4}{x^2}\right)$$

53. $6^p = 17$
$$\ln 6^p = \ln 17$$
$$p \ln 6 = \ln 17$$
$$p = \frac{\ln 17}{\ln 6}$$
$$\approx 1.581$$

55. $2^{1-m} = 7$
$$\ln 2^{1-m} = \ln 7$$
$$(1 - m) \ln 2 = \ln 7$$
$$1 - m = \frac{\ln 7}{\ln 2}$$
$$-m = \frac{\ln 7}{\ln 2} - 1$$
$$m = 1 - \frac{\ln 7}{\ln 2}$$
$$\approx -1.807$$

57. $e^{-5-2x} = 5$
$$\ln e^{-5-2x} = \ln 5$$
$$(-5 - 2x) \ln e = \ln 5$$
$$(-5 - 2x) \cdot 1 = \ln 5$$
$$-2x = \ln 5 + 5$$
$$x = \frac{\ln 5 + 5}{-2}$$
$$\approx -3.305$$

59. $\left(1 + \dfrac{m}{3}\right)^5 = 15$
$$\left[\left(1 + \frac{m}{3}\right)^5\right]^{1/5} = 15^{1/5}$$
$$1 + \frac{m}{3} = 15^{1/5}$$
$$\frac{m}{3} = 15^{1/5} - 1$$
$$m = 3(15^{1/5} - 1)$$
$$\approx 2.156$$

61. $\log_k 64 = 6$
$$k^6 = 64$$
$$k^6 = 2^6$$
$$k = 2$$

63. $\log(4p + 1) + \log p = \log 3$
$$\log[p(4p + 1)] = \log 3$$
$$\log(4p^2 + p) = \log 3$$
$$4p^2 + p = 3$$
$$4p^2 + p - 3 = 0$$
$$(4p - 3)(p + 1) = 0$$
$$4p - 3 = 0 \quad \text{or} \quad p + 1 = 0$$
$$p = \frac{3}{4} \qquad \qquad p = -1$$

p cannot be negative, so $p = \frac{3}{4}$.

65. $f(x) = a^x; a > 0, a \neq 1$

(a) The domain is $(-\infty, \infty)$.

(b) The range is $(0, \infty)$.

(c) The y-intercept is 1.

(d) The graph has no discontinuities.

(e) The x-axis, $y = 0$, is a horizontal asymptote.

(f) The function is increasing if $a > 1$.

(g) The function is decreasing if $0 < a < 1$.

69. $y = \dfrac{7x}{100 - x}$

(a) $y = \dfrac{7(80)}{100 - 80} = \dfrac{560}{20} = 28$

The cost is $28,000.

(b) $y = \dfrac{7(50)}{100 - 50} = \dfrac{350}{50} = 7$

The cost is $7000.

(c) $\dfrac{7(90)}{100 - 90} = \dfrac{630}{10} = 63$

The cost is $63,000.

(d) Plot the points $(80, 28)$, $(50, 7)$, and $(90, 63)$.

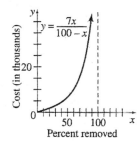

(e) No, because all of the pollutant would be removed when $x = 100$, at which point the denominator of the function would be zero.

71. $P = \$2781.36$, $r = 4.8\%$, $t = 6$, $m = 4$

$$A = P\left(1 + \frac{r}{m}\right)^{tm}$$

$$A = 2781.36\left(1 + \frac{0.048}{4}\right)^{(6)(4)}$$

$$= 2781.36(1.012)^{24}$$
$$= \$3703.31$$

$$\text{Interest} = \$3703.31 - \$2781.36$$
$$= \$921.95$$

73. \$2100 deposited at 4% compounded quarterly.

$$A = P\left(1 + \frac{r}{m}\right)^{tm}$$

To double:

$$2(2100) = 2100\left(1 + \frac{0.04}{4}\right)^{t \cdot 4}$$

$$2 = 1.01^{4t}$$
$$\ln 2 = 4t \ln 1.01$$

$$t = \frac{\ln 2}{4 \ln 1.01}$$

$$\approx 17.4$$

Because interest is compounded quarterly, round the result up to the nearest quarter, which is 17.5 years or 70 quarters.

To triple:

$$3(2100) = 2100\left(1 + \frac{0.04}{4}\right)^{t \cdot 4}$$

$$3 = 1.01^{4t}$$
$$\ln 3 = 4t \ln 1.01$$

$$t = \frac{\ln 3}{4 \ln 1.01}$$

$$\approx 27.6$$

Because interest is compounded quarterly, round the result up to the nearest quarter, which is 27.75 years or 111 quarters.

75. $P = \$12{,}104$, $r = 6.2\%$, $t = 4$

$$A = Pe^{rt}$$
$$A = 12{,}104e^{0.062(4)}$$
$$= 12{,}104e^{0.248}$$
$$= \$15{,}510.79$$

77. $P = \$12{,}000$, $r = 0.05$, $t = 8$

$$A = 12{,}000e^{0.05(8)}$$
$$= 12{,}000e^{0.40}$$
$$= \$17{,}901.90$$

79. $r = 6\%$, $m = 12$

$$r_E = \left(1 + \frac{r}{m}\right)^m - 1$$

$$= \left(1 + \frac{0.06}{12}\right)^{12} - 1$$

$$= 0.0617 = 6.17\%$$

81. $A = \$2000$, $r = 6\%$, $t = 5$, $m = 1$

$$P = A\left(1 + \frac{r}{m}\right)^{-tm}$$

$$= 2000\left(1 + \frac{0.06}{1}\right)^{-5(1)}$$

$$= 2000(1.06)^{-5}$$
$$= \$1494.52$$

83. $r = 7\%$, $t = 8$, $m = 2$, $P = 10{,}000$

$$A = P\left(1 + \frac{r}{m}\right)^{tm}$$

$$= 10{,}000\left(1 + \frac{0.07}{2}\right)^{8(2)}$$

$$= 10{,}000(1.035)^{16}$$
$$= \$17{,}339.86$$

85. $P = \$6000$, $A = \$8000$, $t = 3$

$$A = Pe^{rt}$$
$$8000 = 6000e^{3r}$$
$$\frac{4}{3} = e^{3r}$$

$$\ln 4 - \ln 3 = 3r$$
$$r = \frac{\ln 4 - \ln 3}{3}$$

$$r \approx 0.0959 \text{ or about } 9.59\%$$

87. (a) $n = 1000 - (p - 50)(10), \ p \geq 50$

$= 1000 - 10p + 500$

$= 1500 - 10p$

(b) $R = pn$

$R = p(1500 - 10p)$

(c) $p \geq 50$

Since n cannot be negative,

$$1500 - 10p \geq 0$$
$$-10p \geq -1500$$
$$p \leq 150.$$

Therefore, $50 \leq p \leq 150$.

(d) Since $n = 1500 - 10p$,

$$10p = 1500 - n$$
$$p = 150 - \frac{n}{10}.$$

$$R = pn$$
$$R = \left(150 - \frac{n}{10}\right) n$$

(e) Since she can sell at most 1000 tickets, $0 \leq n \leq 1000$.

(f) $R = -10p^2 + 1500p$

$$\frac{-b}{2a} = \frac{-1500}{2(-10)} = 75$$

The price producing maximum revenue is $75.

(g) $R = -\frac{1}{10}n^2 + 150n$

$$\frac{-b}{2a} = \frac{-150}{2\left(-\frac{1}{10}\right)} = 750$$

The number of tickets producing maximum revenue is 750.

(h) $R(p) = -10p^2 + 1500p$

$R(75) = -10(75)^2 + 1500(75)$

$= -56{,}250 + 112{,}500$

$= 56{,}250$

The maximum revenue is $56,250.

(i)

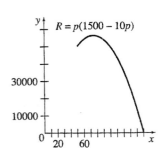

(j) The revenue starts at $50,000 when the price is $50, rises to a maximum of $56,250 when the price is $75, and falls to 0 when the price is $150.

89. $C(x) = x^2 + 4x + 7$

(a)

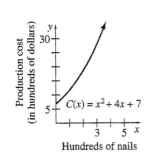

(b) $C(x+1) - C(x)$

$= (x+1)^2 + 4(x+1) + 7$

$\quad - (x^2 + 4x + 7)$

$= x^2 + 2x + 1 + 4x + 4 + 7$

$\quad - x^2 - 4x - 7$

$= 2x + 5$

(c) $A(x) = \dfrac{C(x)}{x} = \dfrac{x^2 + 4x + 7}{x}$

$$= x + 4 + \frac{7}{x}$$

(d) $A(x+1) - A(x)$

$= (x+1) + 4 + \dfrac{7}{x+1}$

$\quad - \left(x + 4 + \dfrac{7}{x}\right)$

$= x + 1 + 4 + \dfrac{7}{x+1} - x - 4 - \dfrac{7}{x}$

$= 1 + \dfrac{7}{x+1} - \dfrac{7}{x}$

$= 1 + \dfrac{7x - 7(x+1)}{x(x+1)}$

$= 1 + \dfrac{7x - 7x - 7}{x(x+1)}$

$= 1 - \dfrac{7}{x(x+1)}$

91. $F(x) = -\dfrac{2}{3}x^2 + \dfrac{14}{3}x + 96$

The maximum fever occurs at the vertex of the parabola.

$$x = \frac{-b}{2a} = \frac{-\frac{14}{3}}{-\frac{4}{3}} = \frac{7}{2}$$

$$y = -\frac{2}{3}\left(\frac{7}{2}\right)^2 + \frac{14}{3}\left(\frac{7}{2}\right) + 96$$

$$= -\frac{2}{3}\left(\frac{49}{4}\right) + \frac{49}{3} + 96$$

$$= -\frac{49}{6} + \frac{49}{3} + 96$$

$$= -\frac{49}{6} + \frac{98}{6} + \frac{576}{6} = \frac{625}{6} \approx 104.2$$

The maximum fever occurs on the third day. It is about 104.2°F.

93. (a)

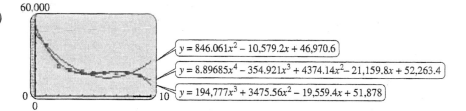

(b) Quadratic:

$$y = 846.061x^2 - 10{,}579.2x + 46{,}970.6$$

Cubic:

$$y = -194.777x^3 + 3475.56x^2 - 19{,}558.4x + 51{,}879$$

Quartic:

$$y = 8.89685x^4 - 354.921x^3 + 4374.14x^2 - 21{,}159.8x + 52{,}263.4$$

(c)

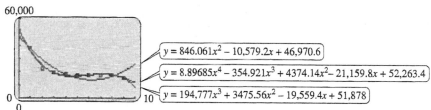

(d) Use the functions found in part (b), with $x = 11$.

Quadratic: 32,973
Cubic: −1969
Quartic: 6635

95.

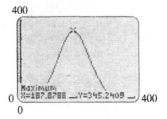

This function has a maximum value at $x \approx 187.9$. At $x \approx 187.9, y \approx 345$. The largest girth for which this formula gives a reasonable answer is 187.9 cm. The predicted mass of a polar bear with this girth is 345 kg.

97. $p(t) = \dfrac{1.79 \cdot 10^{11}}{(2026.87 - t)^{0.99}}$

(a) $p(2005) \approx 8.441$ billion

This is about 1.964 billion more than the estimate of 6.477 billion.

(b) $p(2020) \approx 26.56$ billion
$p(2025) \approx 96.32$ billion

99. Graph

$$y = c(t) = e^{-t} - e^{-2t}$$

on a graphing calculator and locate the maximum point. A calculator shows that the x-coordinate of the maximum point is about 0.69, and the y-coordinate is exactly 0.25. Thus, the maximum concentration of 0.25 occurs at about 0.69 minutes.

101. $y = y_0 e^{-kt}$

(a) $100{,}000 = 128{,}000 e^{-k(5)}$
$128{,}000 = 100{,}000 e^{5k}$

$$\frac{128}{100} = e^{5k}$$

$$\ln\left(\frac{128}{100}\right) = 5k$$

$$0.05 \approx k$$

$y = 100{,}000 e^{-0.05t}$

(b) $70{,}000 = 100{,}000 e^{-0.05t}$

$$\frac{7}{10} = e^{-0.05t}$$

$$\ln \frac{7}{10} = -0.05t$$

$$7.1 \approx t$$

It will take about 7.1 years.

103. (a) Since the speed in one direction is $v + w$ and in the other direction is $v - w$, the time in one direction is $\frac{d}{v+w}$ and in the other direction is $\frac{d}{v-w}$. So the total time is $\frac{d}{v+w} + \frac{d}{v-w}$.

(b) The average speed is the total distance divided by the total time. So

$$v_{aver} = \frac{2d}{\dfrac{d}{v+w} + \dfrac{d}{v-w}}.$$

(c) $\dfrac{2d}{\dfrac{d}{v+w} + \dfrac{d}{v-w}}$

$$= \frac{2d}{\dfrac{d}{v+w} + \dfrac{d}{v-w}} \cdot \frac{(v+w)(v-w)}{(v+w)(v-w)}$$

$$= \frac{2d(v^2 - w^2)}{d(v-w) + d(v+w)}$$

$$= \frac{2d(v^2 - w^2)}{dv - dw + dv + dw}$$

$$= \frac{2d(v^2 - w^2)}{2dv}$$

$$= \frac{v^2 - w^2}{v} = v - \frac{w^2}{v}$$

(d) $v_{aver} = v - \dfrac{w^2}{v}$

v_{aver} will be greatest when $w = 0$.

105. (a) $P = kD^1$
$164.8 = k(30.1)$

$$k = \frac{164.8}{30.1} \approx 5.48$$

For $n = 1$, $P = 5.48D$.

$P = kD^{1.5}$
$164.8 = k(30.1)^{1.5}$

$$k = \frac{164.8}{(30.1)^{1.5}} \approx 1.00$$

For $n = 1.5$, $P = 1.00D^{1.5}$.

$P = kD^2$
$164.8 = k(30.1)^2$

$$k = \frac{164.8}{(30.1)^2} \approx 0.182$$

For $n = 2$, $P = 0.182D^2$.

(b)

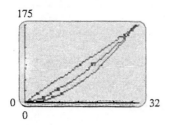

$P = 1.00D^{1.5}$ appears to be the best fit.

(c) $P = 1.00(39.5)^{1.5} \approx 248.3$ years

(d) We obtain

$$P = 1.00D^{1.5}.$$

This is the same as the function found in part (b).

Chapter 10 Test

[10.1]

1. List the ordered pairs obtained from each equation, given $\{-3, -2, -1, 0, 1, 2, 3\}$ as the domain. Give the range.

   (a) $2x + 3y = 6$ (b) $y = \dfrac{1}{x^2 - 2}$

2. Let $f(x) = 3x - 4$ and $g(x) = -x^2 + 5x$. Find each of the following.

   (a) $f(-3)$ (b) $g(-2)$ (c) $f(2m)$ (d) $g(k-1)$ (e) $f(x+h)$

[10.2]

3. Graph the parabola

   $$y = 2x^2 - 4x - 2$$

   and give its vertex, axis, x–intercept, and y–intercept.

4. The manufacturer of a certain product has determined that his profit in dollars for making x units of this product is given by the equation

   $$p = -2x^2 + 120x + 3000.$$

   (a) Find the number of units that will maximize the profit.

   (b) What is the maximum profit for making this product?

Use translations and reflections to graph the following functions.

5. $y = 4 - x^2$ 6. $f(x) = \dfrac{1}{2}x^3 + 1$

[10.3]

Graph each function.

7. $f(x) = x^3 - 2x^2 - x + 2$ 8. $f(x) = \dfrac{3}{2x - 1}$ 9. $f(x) = \dfrac{x - 1}{3x + 6}$

10. Find the horizontal and vertical asymptotes and $x-$ and y–intercepts, if any, for the following function.

   $$f(x) = \frac{3x - 5}{5x + 10}$$

11. Suppose a cost-benefit model is given by the equation

$$y = \frac{20x}{110 - x},$$

where y is the cost in thousands of dollars of removing x percent of a certain pollutant. Find the cost in thousands of dollars of removing each of the following percents of pollution.

(a) 70%   (b) 90%   (c) 100%

[10.4]

12. Solve the equation $8^{2y-1} = 4^{y+1}$.   13. Graph the function $f(x) = 4^{x-1}$.

14. \$315 is deposited in an account paying 6% compounded quarterly for 3 years. Find the following.

(a) The amount in the account after 3 years.

(b) The amount of interest earned by this deposit.

[10.5]

15. Evaluate each logarithm without using a calculator.

(a) $\log_8 16$ (b) $\ln e^{-3/4}$

16. Use properties of logarithms to simplify the following.

(a) $\log_2 3k + \log_2 4k^2$ (b) $4 \log_3 r - 3 \log_3 m$

17. Solve each equation. Round to the nearest thousandth if necessary.

(a) $2^{x+1} = 10$ (b) $\log_2 (x+3) + \log_2 (x-3) = 4$

[10.6]

18. Find the interest rate needed for \$5000 to grow to \$10,000 in 8 years with continuous compounding.

19. How long will it take for \$1 to triple at an average rate of 7% compounded continuously?

20. Suppose sales of a certain item are given by $S(x) = 2500 - 1500e^{-2x}$, where x represents the number of years that the item has been on the market and $S(x)$ represents sales in thousands. Find the limit on sales.

21. Find the effective rate for the account described in Problem 14.

22. The population of Smalltown has grown exponentially from 14,000 in 1994 to 16,500 in 1997. At this rate, in what year will the population reach 17,400?

23. Potassium 42 decays exponentially. A sample which contained 1000 grams 5 hours ago has decreased to 758 grams at present.

    (a) Write an exponential equation to express the amount, y, present after t hours.

    (b) What is the half-life of potassium 42?

24. Find the present value of $15,000 at 6% compounded quarterly for 4 years.

25. Mr. Jones needs $20,000 for a down payment on a house in 5 years. How much must he deposit now at 5.8% compounded quarterly in order to have $20,000 in 5 years?

Chapter 10 Test Answers

1. (a) $(-3,4)$, $\left(-2,\frac{10}{3}\right)$, $\left(-1,\frac{8}{3}\right)$, $(0,2)$, $\left(1,\frac{4}{3}\right)$, $\left(2,\frac{2}{3}\right)$, $(3,0)$; range: $\left\{0,\frac{2}{3},\frac{4}{3},2,\frac{8}{3},\frac{10}{3},4\right\}$

   (b) $\left(-3,\frac{1}{7}\right)$, $\left(-2,\frac{1}{2}\right)$, $(-1,-1)$, $\left(0,-\frac{1}{2}\right)$, $(1,-1)$, $\left(2,\frac{1}{2}\right)$, $\left(3,\frac{1}{7}\right)$; range: $\left\{-1,-\frac{1}{2},\frac{1}{7},\frac{1}{2}\right\}$

2. (a) -13 (b) -14 (c) $6m-4$
   (d) $-k^2+7k-6$ (e) $3x+3h-4$

3. Vertex: $(1,-4)$; axis: $x=1$;
   x–intercepts: $1-\sqrt{2}\approx -0.414$,
   $1+\sqrt{2}\approx 2.414$; y–intercept: -2

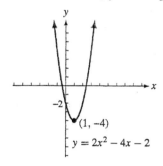

4. (a) 30   (b) $4800

5.

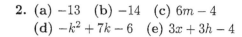

6.

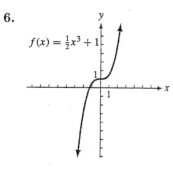

7.

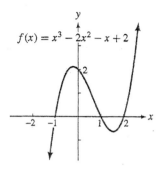

8.

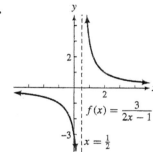

9.

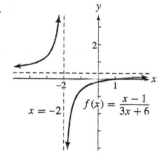

10. Horizontal: $y=\frac{3}{5}$; vertical: $x=-2$;

   x–intercept: $\frac{5}{3}$; y–intercept: $-\frac{1}{2}$

11. (a) $35,000   (b) $90,000   (c) $200,000

12. $\frac{5}{4}$

13.

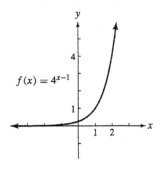

$f(x) = 4^{x-1}$

14. (a) $376.62   (b) $61.62

15. (a) $\frac{4}{3}$ (b) $-\frac{3}{4}$

16. (a) $\log_2 12k^3$ (b) $\log_3 \frac{r^4}{m^3}$

17. (a) 2.322   (b) 5

18. 8.66%

19. About 15.7 years

20. 2,500,000

21. 6.14%

22. 1998

23. (a) $y = 1000e^{-0.0554t}$ (b) About 12.5 hours

24. $11,820.47

25. $14,996.47

Chapter 11

THE DERIVATIVE

11.1 Limits

1. Since $\lim_{x\to 2^-} f(x)$ does not equal $\lim_{x\to 2^+} f(x)$, $\lim_{x\to 2} f(x)$ does not exist. The answer is c.

3. Since $\lim_{x\to 4^-} f(x) = \lim_{x\to 4^+} f(x) = 6$,

$\lim_{x\to 4} f(x) = 6$. The answer is b.

5. (a) By reading the graph, as x gets closer to 3 from the left or right, $f(x)$ gets closer to 3.

$$\lim_{x\to 3} f(x) = 3$$

(b) By reading the graph, as x gets closer to 0 from the left or right, $f(x)$ gets closer to 1.

$$\lim_{x\to 0} f(x) = 1$$

7. (a) By reading the graph, as x gets closer to 0 from the left or right, $f(x)$ gets closer to 0.

$$\lim_{x\to 0} f(x) = 0$$

(b) By reading the graph, as x gets closer to 2 from the left, $f(x)$ gets closer to -2, but as x gets closer to 2 from the right, $f(x)$ gets closer to 1.

$$\lim_{x\to 2} f(x) \text{ does not exist.}$$

9. (a) (i) By reading the graph, as x gets closer to -2 from the left, $f(x)$ gets closer to -1.

$$\lim_{x\to -2^-} f(x) = -1$$

(ii) By reading the graph, as x gets closer to -2 from the right, $f(x)$ gets closer to $-\frac{1}{2}$.

$$\lim_{x\to -2^+} f(x) = -\frac{1}{2}$$

(iii) Since $\lim_{x\to -2^-} f(x) = -1$ and $\lim_{x\to -2^+} f(x) = -\frac{1}{2}$, $\lim_{x\to -2} f(x)$ does not exist.

(iv) $f(-2)$ does not exist since there is no point on the graph with an x-coordinate of -2.

(b) (i) By reading the graph, as x gets closer to -1 from the left, $f(x)$ gets closer to $-\frac{1}{2}$.

$$\lim_{x\to -1^-} f(x) = -\frac{1}{2}$$

(ii) By reading the graph, as x gets closer to -1 from the right, $f(x)$ gets closer to $-\frac{1}{2}$.

$$\lim_{x\to -1^+} f(x) = -\frac{1}{2}$$

(iii) Since $\lim_{x\to -1^-} f(x) = -\frac{1}{2}$ and $\lim_{x\to -1^+} f(x) = -\frac{1}{2}$, $\lim_{x\to -1} f(x) = -\frac{1}{2}$.

(iv) $f(-1) = -\frac{1}{2}$ since $\left(-1, -\frac{1}{2}\right)$ is a point of the graph.

11. By reading the graph, as x moves further to the right, $f(x)$ gets closer to 3. Therefore, $\lim_{x\to\infty} f(x) = 3$.

13. $\lim_{x\to 2} F(x)$ in Exercise 6 exists because $\lim_{x\to 2^-} F(x) = 4$ and $\lim_{x\to 2^+} F(x) = 4$. $\lim_{x\to -2} f(x)$ in Exercise 9 does not exist since $\lim_{x\to -2^-} f(x) = -1$, but $\lim_{x\to -2^+} f(x) = -\frac{1}{2}$.

15. From the table, as x approaches 1 from the left or the right, $f(x)$ approaches 4.

$$\lim_{x\to 1} f(x) = 4$$

17. $k(x) = \dfrac{x^3 - 2x - 4}{x - 2}$; find $\lim_{x\to 2} k(x)$.

x	1.9	1.99	1.999
$k(x)$	9.41	9.9401	9.9941

x	2.001	2.01	2.1
$k(x)$	10.006	10.0601	10.61

As x approaches 2 from the left or the right, $k(x)$ approaches 10.

$$\lim_{x\to 2} k(x) = 10$$

19. $h(x) = \dfrac{\sqrt{x}-2}{x-1}$; find $\lim\limits_{x\to 1} h(x)$.

x	0.9	0.99	0.999
$h(x)$	10.51317	100.50126	1000.50013

x	1.001	1.01	1.1
$h(x)$	−999.50012	−99.50124	−9.51191

$\lim\limits_{x\to 1^-} = -\infty$

$\lim\limits_{x\to 1^+} = -\infty$

Thus, $\lim\limits_{x\to 1} h(x)$ does not exist.

21. $\lim\limits_{x\to 4} [f(x) - g(x)] = \lim\limits_{x\to 4} f(x) - \lim\limits_{x\to 4} g(x)$
$$= 9 - 27 = -18$$

23. $\lim\limits_{x\to 4} \dfrac{f(x)}{g(x)} = \dfrac{\lim\limits_{x\to 4} f(x)}{\lim\limits_{x\to 4} g(x)} = \dfrac{9}{27} = \dfrac{1}{3}$

25. $\lim\limits_{x\to 4} \sqrt{f(x)} = \lim\limits_{x\to 4} [f(x)]^{1/2}$
$$= [\lim\limits_{x\to 4} f(x)]^{1/2}$$
$$= 9^{1/2} = 3$$

27. $\lim\limits_{x\to 4} 2^{f(x)} = 2^{\lim\limits_{x\to 4} f(x)}$
$$= 2^9$$
$$= 512$$

29. $\lim\limits_{x\to 4} \dfrac{f(x) + g(x)}{2g(x)}$
$$= \dfrac{\lim\limits_{x\to 4} [f(x) + g(x)]}{\lim\limits_{x\to 4} 2g(x)}$$
$$= \dfrac{\lim\limits_{x\to 4} f(x) + \lim\limits_{x\to 4} g(x)}{2 \lim\limits_{x\to 4} g(x)}$$
$$= \dfrac{9 + 27}{2(27)} = \dfrac{36}{54} = \dfrac{2}{3}$$

31. $\lim\limits_{x\to 3} \dfrac{x^2 - 9}{x - 3} = \lim\limits_{x\to 3} \dfrac{(x-3)(x+3)}{x-3}$
$$= \lim\limits_{x\to 3} (x + 3)$$
$$= \lim\limits_{x\to 3} x + \lim\limits_{x\to 3} 3$$
$$= 3 + 3$$
$$= 6$$

33. $\lim\limits_{x\to 1} \dfrac{5x^2 - 7x + 2}{x^2 - 1} = \lim\limits_{x\to 1} \dfrac{(5x-2)(x-1)}{(x+1)(x-1)}$
$$= \lim\limits_{x\to 1} \dfrac{5x - 2}{x + 1}$$
$$= \dfrac{5 - 2}{2}$$
$$= \dfrac{3}{2}$$

35. $\lim\limits_{x\to -2} \dfrac{x^2 - x - 6}{x + 2} = \lim\limits_{x\to -2} \dfrac{(x-3)(x+2)}{x+2}$
$$= \lim\limits_{x\to -2} (x - 3)$$
$$= \lim\limits_{x\to -2} x + \lim\limits_{x\to -2} (-3)$$
$$= -2 - 3$$
$$= -5$$

37. $\lim\limits_{x\to 0} \dfrac{\frac{1}{x+3} - \frac{1}{3}}{x}$
$$= \lim\limits_{x\to 0} \left(\dfrac{1}{x+3} - \dfrac{1}{3} \right) \left(\dfrac{1}{x} \right)$$
$$= \lim\limits_{x\to 0} \left[\dfrac{3}{3(x+3)} - \dfrac{x+3}{3(x+3)} \right] \left(\dfrac{1}{x} \right)$$
$$= \lim\limits_{x\to 0} \dfrac{3 - x - 3}{3(x+3)(x)}$$
$$= \lim\limits_{x\to 0} \dfrac{-x}{3(x+3)x}$$
$$= \lim\limits_{x\to 0} \dfrac{-1}{3(x+3)}$$
$$= \dfrac{-1}{3(0+3)}$$
$$= -\dfrac{1}{9}$$

39. $\lim\limits_{x\to 25} \dfrac{\sqrt{x} - 5}{x - 25}$
$$= \lim\limits_{x\to 25} \dfrac{\sqrt{x} - 5}{x - 25} \cdot \dfrac{\sqrt{x} + 5}{\sqrt{x} + 5}$$
$$= \lim\limits_{x\to 25} \dfrac{x - 25}{(x - 25)(\sqrt{x} + 5)}$$
$$= \lim\limits_{x\to 25} \dfrac{1}{\sqrt{x} + 5}$$
$$= \dfrac{1}{\sqrt{25} + 5}$$
$$= \dfrac{1}{10}$$

41. $\lim\limits_{h \to 0} \dfrac{(x+h)^2 - x^2}{h}$

$= \lim\limits_{h \to 0} \dfrac{x^2 + 2hx + h^2 - x^2}{h}$

$= \lim\limits_{h \to 0} \dfrac{2hx + h^2}{h}$

$= \lim\limits_{h \to 0} \dfrac{h(2x + h)}{h}$

$= \lim\limits_{h \to 0} (2x + h)$

$= 2x + 0 = 2x$

43. $\lim\limits_{x \to \infty} \dfrac{3x}{7x - 1} = \lim\limits_{x \to \infty} \dfrac{\frac{3x}{x}}{\frac{7x}{x} - \frac{1}{x}}$

$= \lim\limits_{x \to \infty} \dfrac{3}{7 - \frac{1}{x}}$

$= \dfrac{3}{7 - 0} = \dfrac{3}{7}$

45. $\lim\limits_{x \to -\infty} \dfrac{-x^2 + 2x}{2x^2 - 2x + 1}$

$= \lim\limits_{x \to -\infty} \dfrac{\frac{3x^2}{x^2} + \frac{2x}{x^2}}{\frac{2x^2}{x^2} - \frac{2x}{x^2} + \frac{1}{x^2}}$

$= \lim\limits_{x \to -\infty} \dfrac{3 + \frac{2}{x}}{2 - \frac{2}{x} + \frac{1}{x^2}}$

$= \dfrac{3 - 0}{2 + 0 + 0} = \dfrac{3}{2}$

47. $\lim\limits_{x \to \infty} \dfrac{3x^3 + 2x - 1}{2x^4 - 3x^3 - 2}$

$= \lim\limits_{x \to \infty} \dfrac{\frac{3x^3}{x^4} + \frac{2x}{x^4} - \frac{1}{x^4}}{\frac{2x^4}{x^4} - \frac{3x^3}{x^4} - \frac{2}{x^4}}$

$= \lim\limits_{x \to \infty} \dfrac{\frac{3}{x} + \frac{2}{x^3} - \frac{1}{x^4}}{2 - \frac{3}{x} - \frac{2}{x^4}}$

$= \dfrac{0 + 0 - 0}{2 - 0 - 0} = 0$

49. $\lim\limits_{x \to \infty} \dfrac{2x^3 - x - 3}{6x^2 - x - 1}$

$= \lim\limits_{x \to \infty} \dfrac{\frac{2x^3}{x^3} - \frac{x}{x^3} - \frac{3}{x^3}}{\frac{6x^2}{x^3} - \frac{x}{x^3} - \frac{1}{x^3}}$

$= \lim\limits_{x \to \infty} \dfrac{2 - \frac{1}{x^2} - \frac{3}{x^3}}{\frac{6}{x} - \frac{1}{x^2} - \frac{1}{x^3}}$

$= \dfrac{2 - 0 - 0}{0 - 0 - 0} = \dfrac{2}{0} = \infty$

Therefore $\lim\limits_{x \to \infty} \dfrac{2x^3 - x - 3}{6x^2 - x - 1}$ does not exist.

51. $\lim\limits_{x \to \infty} \dfrac{2x^2 - 7x^4}{9x^2 + 5x - 6} = \lim\limits_{x \to \infty} \dfrac{\frac{2x^2}{x^2} - \frac{7x^4}{x^2}}{\frac{9x^2}{x^2} + \frac{5x}{x^2} - \frac{6}{x^2}}$

$= \lim\limits_{x \to \infty} \dfrac{2 - 7x^2}{9 + \frac{5}{x} - \frac{6}{x^2}}$

The denominator approaches 9, while the numerator becomes a negative number that is larger and larger in magnitude, so

$$\lim\limits_{x \to \infty} \dfrac{2x^2 - 7x^4}{9x^2 + 5x - 6} = -\infty.$$

53. Find $\lim\limits_{x \to 3} f(x)$, where $f(x) = \frac{x^2 - 9}{x - 3}$.

x	2.9	2.99	2.999	3.001	3.01	3.1
$f(x)$	5.9	5.99	5.999	6.001	6.01	6.1

$\lim\limits_{x \to 3} f(x) = \lim\limits_{x \to 3} \dfrac{x^2 - 9}{x - 3} = 6.$

55. Find $\lim\limits_{x \to 1} f(x)$, where $f(x) = \frac{5x^2 - 7x + 2}{x^2 - 1}$.

x	0.9	0.99	0.999	1.001	1.01	1.1
$f(x)$	1.316	1.482	1.498	1.502	1.517	1.667

$\lim\limits_{x \to 1} f(x) = \lim\limits_{x \to 1} \dfrac{5x^2 - 7x + 2}{x^2 - 1} = 1.5 = \dfrac{3}{2}.$

57. (a) $\lim\limits_{x \to -2} \frac{3x}{(x+2)^3}$ does not exist since

$\lim\limits_{x \to -2+} \frac{3x}{(x+2)^3} = -\infty$ and $\lim\limits_{x \to -2-} \frac{3x}{(x+2)^3} = \infty$.

(b) Since $(x + 2)^3 = 0$ when $x = -2$, $x = -2$ is the vertical asymptote of the graph of $F(x)$.

(c) The two answers are related. Since $x = -2$ is a vertical asymptote, we know that $\lim\limits_{x \to -2} F(x)$ does not exist.

61. (a) $\lim\limits_{x \to -\infty} e^x = 0$ since, as the graph goes further to the left, e^x gets closer to 0.

(b) The graph of e^x has a horizontal asymptote at $y = 0$ since $\lim\limits_{x \to -\infty} e^x = 0$.

63. (a) $\lim\limits_{x \to 0+} \ln x = -\infty$ since, as the graph gets closer to $x = 0$, the value of $\ln x$ get smaller.

(b) The graph of $y = \ln x$ has a vertical asymptote at $x = 0$ since $\lim\limits_{x \to 0+} \ln x = -\infty$.

67. $\lim\limits_{x\to 1}\dfrac{x^4+4x^3-9x^2+7x-3}{x-1}$

(a)

x	1.01	1.001	1.0001	0.99	0.999	0.9999
$f(x)$	5.0908	5.009	5.0009	4.9108	4.991	4.9991

As $x\to 1^-$ and as $x\to 1^+$, we see that $f(x)\to 5$.

(b) Graph

$$y=\frac{x^4+4x^3-9x^2+7x-3}{x-1}$$

on a graphing calculator. One suitable choice for the viewing window is $[-6,6]$ by $[-10,40]$ with Xscl = 1, Yscl = 10.

Because $x-1=0$ when $x=1$, we know that the function is undefined at this x-value. The graph does not show an asymptote at $x=1$. This indicates that the rational expression that defines this function is not written in lowest terms, and that the graph should have an open circle to show a "hole" in the graph at $x=1$. The graphing calculator doesn't show the hole, but if we try to find the value of the function at $x=1$, we see that it is undefined. (Using the TABLE feature on a TI-83, we see that for $x=1$, the y-value is listed as "ERROR.")

By viewing the function near $x=1$ and using the ZOOM feature, we see that as x gets close to 1 from the left or the right, y gets close to 5, suggesting that

$$\lim_{x\to 1}\frac{x^4+4x^3-9x^2+7x-3}{x-1}=5.$$

69. $\lim\limits_{x\to -1}\dfrac{x^{1/3}+1}{x+1}$

(a)

x	-1.01	-1.001	-1.0001
$f(x)$	0.33223	0.33322	0.33332

x	-0.99	-0.999	-0.9999
$f(x)$	0.33445	0.33344	0.33334

We see that as $x\to -1^-$ and as $x\to -1^+$,

$f(x)\to 0.3333$ or $\frac{1}{3}$.

(b) Graph

$$y=\frac{x^{1/3}+1}{x+1}.$$

One suitable choice for the viewing window is $[-5,5]$ by $[-2,2]$.

Because $x+1=0$ when $x=-1$, we know that the function is undefined at this x-value. The graph does not show an asymptote at $x=-1$. This indicates that the rational expression that defined this function is not written lowest terms, and that the graph should have an open circle to show a "hole" in the graph at $x=-1$. The graphing calculator doesn't show the hole, but if we try to find the value of the function at $x=-1$, we see that it is undefined. (Using the TABLE feature on a TI-83, we see that for $x=-1$, the y-value is listed as "ERROR.")

By viewing the function near $x=-1$ and using the ZOOM feature, we see that as x gets close to -1 from the left or right, y gets close to 0.3333, suggesting that

$$\lim_{x\to -1}\frac{x^{1/3}+1}{x+1}=0.3333\ \text{or}\ \frac{1}{3}.$$

71. $\lim\limits_{x\to\infty}\dfrac{\sqrt{9x^2+5}}{2x}$

Graph the functions on a graphing calculator. A good choice for the viewing window is $[-10,10]$ by $[-5,5]$.

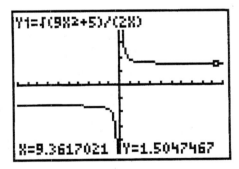

(a) The graph appears to have horizontal asymptotes at $y=\pm 1.5$. We see that as $x\to\infty$, $y\to 1.5$, so we determine that

$$\lim_{x\to\infty}\frac{\sqrt{9x^2+5}}{2x}=1.5.$$

(b) As $x\to\infty$,

$$\sqrt{9x^2+5}\to\sqrt{9x^2}=3\,|x|,$$

and

$$\frac{\sqrt{9x^2+5}}{2x}\to\frac{3\,|x|}{2x}.$$

Since $x>0$, $|x|=x$, so

$$\frac{3\,|x|}{2x}=\frac{3x}{2x}=\frac{3}{2}.$$

Thus,

$$\lim_{x\to\infty}\frac{\sqrt{9x^2+5}}{2x}=\frac{3}{2}\text{ or }1.5.$$

73. $\lim\limits_{x\to-\infty}\dfrac{\sqrt{36x^2+2x+7}}{3x}$

Graph this function on a graphing calculator. A good choice for the viewing window is $[-10,10]$ by $[-5,5]$.

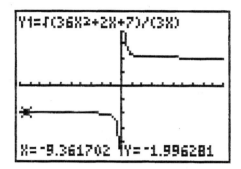

(a) The graph appears to have horizontal asymptotes at $y=\pm2$. We see that as $x\to-\infty$, $y\to-2$, so we determine that

$$\lim_{x\to-\infty}\frac{\sqrt{36x^2+2x+7}}{3x}=-2.$$

(b) As $x\to-\infty$,

$$\sqrt{36x^2+2x+7}\to\sqrt{36x^2}=6\,|x|$$

and

$$\frac{\sqrt{36x^2+2x+7}}{3x}\to\frac{6\,|x|}{3x}.$$

Since $x<0$, $|x|=-x$, so

$$\frac{6\,|x|}{3x}=\frac{6(-x)}{3x}=-2.$$

Thus,

$$\lim_{x\to-\infty}\frac{\sqrt{36x^2+2x+7}}{3x}=-2.$$

75. $\lim\limits_{x\to\infty}\dfrac{\left(1+5x^{1/3}+2x^{5/3}\right)^3}{x^5}$

Graph this function on a graphing calculator. A good choice for the viewing window is $[-20,20]$ by $[0,20]$ with Xscl $=5$, Yscl $=5$.

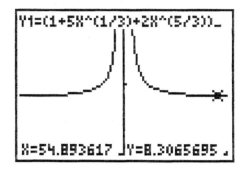

(a) The graph appears to have a horizontal asymptote at $y=8$. We see that as $x\to\infty$, $y\to8$, so we determine that

$$\lim_{x\to\infty}\frac{\left(1+5x^{1/3}+2x^{5/3}\right)^3}{x^5}=8.$$

(b) As $x\to\infty$, the highest power term dominates in the numerator, so

$$(1+5x^{1/3}+2x^{5/3})^3\to(2x^{5/3})^3=2^3x^5$$
$$=8x^5.$$

and

$$\frac{\left(1+5x^{1/3}+2x^{5/3}\right)^3}{x^5}\to\frac{8x^5}{x^5}=8.$$

Thus,

$$\lim_{x\to\infty}\frac{\left(1+5x^{1/3}+2x^{5/3}\right)^3}{x^5}=8.$$

79. (a) $\lim\limits_{x\to98}T(x)=7.25$ cents because $T(x)$ is constant at 7.25 cents as x approaches 98 from the left or the right.

(b) $\lim\limits_{x\to02^-}T(x)=7$ cents because as x approaches 02 from the left, $T(x)$ is constant at 7 cents.

(c) $\lim\limits_{x\to02^+}T(x)=7.25$ cents because as x approaches 02 from the right, $T(x)$ is constant at 7.25 cents.

(d) $\lim\limits_{x\to02}T(x)$ does not exist because

$$\lim_{x\to02^-}T(x)\neq\lim_{x\to02^+}T(x).$$

(e) $T(02)=7.25$ cents since $(02,7.25)$ is a point on the graph.

81. $\overline{C}(x) = \dfrac{C(x)}{x}$

$\qquad = \dfrac{0.0738x + 111.83}{x}$

$\qquad = 0.0738 + \dfrac{111.83}{x}$

$\lim\limits_{x\to\infty} \overline{C}(x) = 0.0738$

The average cost approaches \$0.0738 per mile as the number of miles becomes very large.

83. $\lim\limits_{n\to\infty}\left[R\left[\dfrac{1-(1+i)^{-n}}{i}\right]\right]$

$\qquad = \dfrac{R}{i}\lim\limits_{n\to\infty}\left[1-(1+i)^{-n}\right]$

$\qquad = \dfrac{R}{i}\left[\lim\limits_{n\to\infty}1 - \lim\limits_{n\to\infty}(1+i)^{-n}\right]$

$\qquad = \dfrac{R}{i}[1-0] = \dfrac{R}{i}$

85. (a) $N(65) = 71.8e^{-8.96e^{(-0.0685(65))}}$

$\qquad \approx 64.68$

To the nearest whole number, this species of alligator has approximately 65 teeth after 65 days of incubation by this formula.

(b) Since $\lim\limits_{t\to\infty}(-8.96e^{-0.0685t}) = -8.96\cdot 0 = 0$, it follows that

$$\lim\limits_{t\to\infty}71.8e^{-8.96e^{(-0.0685t)}} = 71.8e^{0}$$
$$= 71.8\cdot 1$$
$$= 71.8$$

So, to the nearest whole number, $\lim\limits_{t\to\infty}N(t) \approx 72$. Therefore, by this model a newborn alligator of this species will have about 72 teeth.

87. $A(h) = \dfrac{0.17h}{h^2 + 2}$

$\lim\limits_{x\to\infty}A(h) = \lim\limits_{x\to\infty}\dfrac{0.17h}{h^2 + 2}$

$\qquad = \lim\limits_{x\to\infty}\dfrac{\frac{0.17h}{h^2}}{\frac{h^2}{h^2} + \frac{2}{h^2}}$

$\qquad = \lim\limits_{x\to\infty}\dfrac{\frac{0.17}{h}}{1 + \frac{2}{h^2}}$

$\qquad = \dfrac{0}{1 + 0} = 0$

This means that the concentration of the drug in the bloodstream approaches 0 as the number of hours after injection increases.

11.2   Continuity

1. Discontinuous at $x = -1$

(a) $\lim\limits_{x\to -1^-}f(x) = \dfrac{1}{2}$

(b) $\lim\limits_{x\to -1^+}f(x) = \dfrac{1}{2}$

(c) $\lim\limits_{x\to -1}f(x) = \dfrac{1}{2}$ (since (a) and (b) have the same answers)

(d) $f(-1)$ does not exist.

(e) $f(-1)$ does not exist.

3. Discontinuous at $x = 1$

(a) $\lim\limits_{x\to 1^-}f(x) = -2$

(b) $\lim\limits_{x\to 1^+}f(x) = -2$

(c) $\lim\limits_{x\to 1}f(x) = -2$ (since (a) and (b) have the same answers)

(d) $f(1) = 2$

(e) $\lim\limits_{x\to 1}f(x) \neq f(1)$

5. Discontinuous at $x = -5$ and $x = 0$

(a) $\lim\limits_{x\to -5^-}f(x) = \infty$ (limit does not exist)

$\qquad\qquad = \lim\limits_{x\to 0^-}f(x) = 0$

(b) $\lim\limits_{x\to -5^+}f(x) = -\infty$ (limit does not exist)

$\qquad\qquad = \lim\limits_{x\to 0^+}f(x) = 0$

(c) $\lim\limits_{x\to -5}f(x)$ does not exist, since the answers to (a) and (b) are different.

$\lim\limits_{x\to 0}f(x) = 0$, since the answers to (a) and (b) are the same.

(d) $f(-5)$ does not exist. $f(0)$ does not exist.

(e) $f(-5)$ does not exist and $\lim\limits_{x\to -5}f(x)$ does not exist. $f(0)$ does not exist.

7. $f(x) = \dfrac{5+x}{x(x-2)}$

$f(x)$ is discontinuous at $x = 0$ and $x = 2$ since the denominator equals 0 at these two values.

$\lim\limits_{x \to 0} f(x)$ does not exist since $\lim\limits_{x \to 0^-} f(x) = \infty$ and $\lim\limits_{x \to 0^+} f(x) = -\infty$.

$\lim\limits_{x \to 2} f(x)$ does not exist since $\lim\limits_{x \to 2^-} f(x) = -\infty$ and $\lim\limits_{x \to 2^+} f(x) = \infty$.

9. $f(x) = \dfrac{x^2 - 4}{x - 2}$

$f(x)$ is discontinuous at $x = 2$ since the denominator equals zero at that value.

Since

$$\frac{x^2 - 4}{x - 2} = \frac{(x+2)(x-2)}{x - 2} = x + 2,$$

$\lim\limits_{x \to 2} f(x) = 2 + 2 = 4.$

11. $p(x) = x^2 - 4x + 11$

Since $p(x)$ is a polynomial function, it is continuous everywhere and thus discontinuous nowhere.

13. $p(x) = \dfrac{|x + 2|}{x + 2}$

$p(x)$ is discontinuous at $x = -2$ since the denominator is undefined at that value.

Since $\lim\limits_{x \to -2^-} p(x) = -1$ and $\lim\limits_{x \to -2^+} p(x) = 1$, $\lim\limits_{x \to -2} p(x)$ does not exist.

15. $k(x) = e^{\sqrt{x-1}}$

The function is undefined for $x < 1$, so the function is discontinuous for $a < 1$. The limit as x approaches any $a < 1$ does not exist because the function is undefined for $x < 1$.

17. As x approaches 0 from the left or the right, $\left|\frac{x}{x-1}\right|$ approaches 0 and $r(x) = \ln\left|\frac{x}{x-1}\right|$ goes to $-\infty$. So $\lim\limits_{x \to 0} r(x)$ does not exist. As x approaches 1 from the left or the right, $\left|\frac{x}{x-1}\right|$ goes to ∞ and so does $r(x) = \ln\left|\frac{x}{x-1}\right|$. So $\lim\limits_{x \to 1} r(x)$ does not exist.

19. $f(x) = \begin{cases} 1 & \text{if} \quad x < 2 \\ x + 3 & \text{if} \quad 2 \le x \le 4 \\ 7 & \text{if} \quad x > 4 \end{cases}$

(a)

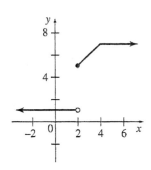

(b) $f(x)$ is discontinuous at $x = 2$.

(c) $\lim\limits_{x \to 2^-} f(x) = 1$ $\qquad\qquad$ $\lim\limits_{x \to 2^+} f(x) = 5$

21. $g(x) = \begin{cases} 11 & \text{if} \quad x < -1 \\ x^2 + 2 & \text{if} \quad -1 \le x \le 3 \\ 11 & \text{if} \quad x > 3 \end{cases}$

(a)

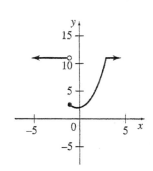

(b) $g(x)$ is discontinuous at $x = -1$.

(c) $\lim\limits_{x \to -1^-} g(x) = 11$

$\lim\limits_{x \to -1^+} g(x) = (-1)^2 + 2 = 3$

23. $h(x) = \begin{cases} 4x + 4 & \text{if} \quad x \le 0 \\ x^2 - 4x + 4 & \text{if} \quad x > 0 \end{cases}$

(a)

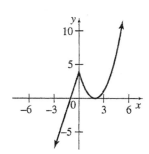

(b) There are no points of discontinuity.

25. Find k so that $kx^2 = x + k$ for $x = 2$.

$$k(2)^2 = 2 + k$$
$$4k = 2 + k$$
$$3k = 2$$
$$k = \frac{2}{3}$$

27. $\dfrac{2x^2 - x - 15}{x - 3} = \dfrac{(2x + 5)(x - 3)}{x - 3} = 2x + 5$

Find k so that $2x + 5 = kx - 1$ for $x = 3$.

$$2(3) + 5 = k(3) - 1$$
$$6 + 5 = 3k - 1$$
$$11 = 3k - 1$$
$$12 = 3k$$
$$4 = k$$

31. $f(x) = \dfrac{x^2 + x + 2}{x^3 - 0.9x^2 + 4.14x - 5.4} = \dfrac{P(x)}{Q(x)}$

(a) Graph

$$Y_1 = \frac{P(x)}{Q(x)} = \frac{x^2 + x + 2}{x^3 - 0.9x^2 + 4.14x - 5.4}$$

on a graphing calculator. A good choice for the viewing window is $[-3, 3]$ by $[-10, 10]$.

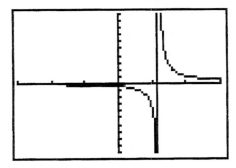

The graph has a vertical asymptote at $x = 1.2$, which indicates that f is discontinuous at $x = 1.2$.

(b) Graph

$$Y_2 = Q(x) = x^3 - 0.9x^2 + 4.14x - 5.4$$

using the same viewing window.

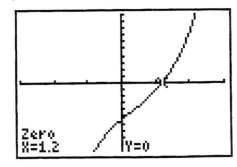

We see that this graph has one x-intercept, 1.2. This indicates that 1.2 is the only real solution of the equation $Q(x) = 0$.

This result verifies our answer from part (a) because a rational function of the form

$$f(x) = \frac{P(x)}{Q(x)}$$

will be discontinuous wherever $Q(x) = 0$.

33. $g(x) = \dfrac{x + 4}{x^2 + 2x - 8}$

$$= \frac{x + 4}{(x - 2)(x + 4)}$$

$$= \frac{1}{x - 2}, \; x \neq -4$$

If $g(x)$ is defined so that $g(-4) = \dfrac{1}{-4 - 2} = -\dfrac{1}{6}$, then the function becomes continuous at -4. It cannot be made continuous at 2. The correct answer is (a).

35. In dollars,

$$C(x) = 4x \text{ if } 0 < x \leq 150$$
$$C(x) = 3x \text{ if } 150 < x \leq 400$$
$$C(x) = 2.5x \text{ if } 400 < x.$$

(a) $C(130) = 4(130) = \$520$

(b) $C(150) = 4(150) = \$600$

(c) $C(210) = 3(210) = \$630$

(d) $C(400) = 3(400) = \$1200$

(e) $C(500) = 2.5(500) = \$1250$

(f) C is discontinuous at $x = 150$ and $x = 400$ because those represent points of price change.

37. In dollars,

$$C(t) = 36t \text{ if } 0 < t \leq 5$$
$$C(t) = 36(5) = 180 \text{ if } t = 6 \text{ or } t = 7$$
$$C(t) = 180 + 36(t - 7) \text{ if } 7 < t \leq 12.$$

The average cost per day is

$$A(t) = \frac{C(t)}{t}.$$

(a) $A(4) = \dfrac{36(4)}{4} = \$36$

(b) $A(5) = \dfrac{36(5)}{5} = \$36$

(c) $A(6) = \dfrac{180}{6} = \$30$

(d) $A(7) = \dfrac{180}{7} \approx \$25.71$

(e) $A(8) = \dfrac{180 + 36(8 - 7)}{8}$

$= \dfrac{216}{8} = \$27$

(f) $\lim\limits_{t \to 5^-} A(t) = 36$ because as t approaches 5 from the left, $A(t)$ approaches 36 (think of the graph for $t = 1, 2, ..., 5$).

(g) $\lim\limits_{t \to 5^+} A(t) = 30$ because as t approaches 5 from the right, $A(t)$ approaches 30. 1, 2, 3, 4, 7, 8, 9, 10

(h) A is discontinuous at $t = 11$, and because the average cost will differ for each different rental length.

39. (a) Since $t = 0$ weeks the woman weighs 120 lbs. and at $t = 40$ weeks she weighs 147 lbs., graph the line beginning at coordinate $(0, 120)$ and ending at $(40, 147)$, with closed circles at these points. Since immediately after giving birth, she loses 14 lbs. and continues to lose 13 more lbs. over the following 20 weeks, graph the line between the points $(40, 133)$ and $(60, 120)$ with an open circle at $(40, 133)$ and a closed circle at $(60, 120)$.

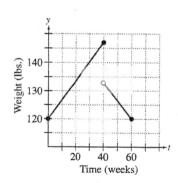

(b) From the graph, we see that

$$\lim\limits_{t \to 40^-} w(t) = 147 \neq 133$$

$$= \lim\limits_{t \to 40^+} w(t),$$

where $w(t)$ is the weight in pounds t weeks after conception. Therefore, w is discontinuous at $t = 40$.

11.3   Rates of Change

1. $y = x^2 + 2x = f(x)$ between $x = 1$ and $x = 3$

Average rate of change

$= \dfrac{f(3) - f(1)}{3 - 1}$

$= \dfrac{15 - 3}{2}$

$= 6$

3. $y = -3x^3 + 2x^2 - 4x + 1 = f(x)$ between $x = -2$ and $x = 1$

Average rate of change $= \dfrac{f(1) - f(-2)}{1 - (-2)}$

$= \dfrac{(-4) - (41)}{1 - (-2)}$

$= \dfrac{-45}{3} = -15$

5. $y = \sqrt{x} = f(x)$ between $x = 1$ and $x = 4$

Average rate of change

$= \dfrac{f(4) - f(1)}{4 - 1}$

$= \dfrac{2 - 1}{3}$

$= \dfrac{1}{3}$

7. $y = e^x = f(x)$ between $x = -2$ and $x = 0$

Average rate of change

$= \dfrac{f(0) - f(-2)}{0 - (-2)}$

$= \dfrac{1 - e^{-2}}{2}$

≈ 0.4323

9. $\lim\limits_{h \to 0} \dfrac{s(6 + h) - s(6)}{h}$

$= \lim\limits_{h \to 0} \dfrac{(6+h)^2 + 5(6+h) + 2 - [6^2 + 5(6) + 2]}{h}$

$= \lim\limits_{h \to 0} \dfrac{h^2 + 17h + 68 - 68}{h} = \lim\limits_{h \to 0} \dfrac{h^2 + 17h}{h}$

$= \lim\limits_{h \to 0} \dfrac{h(h + 17)}{h} = \lim\limits_{h \to 0} (h + 17) = 17$

The instantaneous velocity at $t = 6$ is 17.

11. $s(t) = 5t^2 - 2t - 7$

$$\lim_{h \to 0} \frac{s(2+h) - s(2)}{h}$$

$$= \lim_{h \to 0} \frac{[5(2+h)^2 - 2(2+h) - 7] - [5(2)^2 - 2(2) - 7]}{h}$$

$$= \lim_{h \to 0} \frac{[20 + 20h + 5h^2 - 4 - 2h - 7] - [20 - 4 - 7]}{h}$$

$$= \lim_{h \to 0} \frac{9 + 18h + 5h^2 - 9}{h}$$

$$= \lim_{h \to 0} \frac{18h + 5h^2}{h}$$

$$= \lim_{h \to 0} \frac{h(18 + 5h)}{h}$$

$$= \lim_{h \to 0} (18 + 5h) = 18$$

The instantaneous velocity at $t = 2$ is 18.

13. $s(t) = t^3 + 2t + 9$

$$\lim_{h \to 0} \frac{s(1+h) - s(1)}{h}$$

$$= \lim_{h \to 0} \frac{[(1+h)^3 + 2(1+h) + 9] - [(1)^3 + 2(1) + 9]}{h}$$

$$= \lim_{h \to 0} \frac{[1 + 3h + 3h^2 + h^3 + 2 + 2h + 9] - [1 + 2 + 9]}{h}$$

$$= \lim_{h \to 0} \frac{h^3 + 3h^2 + 5h + 12 - 12}{h}$$

$$= \lim_{h \to 0} \frac{h^3 + 3h^2 + 5h}{h} = \lim_{h \to 0} \frac{h(h^2 + 3h + 5)}{h}$$

$$= \lim_{h \to 0} (h^2 + 3h + 5) = 5$$

The instantaneous velocity at $t = 1$ is 5.

15. $f(x) = x^2 + 2x$ at $x = 0$

$$\lim_{h \to 0} \frac{f(0+h) - f(0)}{h}$$

$$= \lim_{h \to 0} \frac{(0+h)^2 + 2(0+h) - [0^2 + 2(0)]}{h}$$

$$= \lim_{h \to 0} \frac{h^2 + 2h}{h} = \lim_{h \to 0} \frac{h(h + 2)}{h}$$

$$= \lim_{h \to 0} h + 2 = 2$$

The instantaneous rate of change at $x = 0$ is 2.

17. $g(t) = 1 - t^2$ at $t = -1$

$$\lim_{h \to 0} \frac{g(-1+h) - g(-1)}{h}$$

$$= \lim_{h \to 0} \frac{1 - (-1+h)^2 - [1 - (-1)^2]}{h}$$

$$= \lim_{h \to 0} \frac{1 - (1 - 2h + h^2) - 1 + 1}{h}$$

$$= \lim_{h \to 0} \frac{2h - h^2}{h} = \lim_{h \to 0} \frac{h(2 - h)}{h}$$

$$= \lim_{h \to 0} (2 - h) = 2$$

The instantaneous rate of change at $t = -1$ is 2.

19. $f(x) = x^x$ at $x = 2$

h	
0.01	$\dfrac{f(2 + 0.01) - f(2)}{0.01}$
	$= \dfrac{2.01^{2.01} - 2^2}{0.01}$
	$= 6.84$
0.001	$\dfrac{f(2 + 0.001) - f(2)}{0.001}$
	$= \dfrac{2.001^{2.001} - 2^2}{0.001}$
	$= 6.779$
0.0001	$\dfrac{f(2 + 0.0001) - f(2)}{0.0001}$
	$= \dfrac{2.0001^{2.0001} - 2^2}{0.0001}$
	$= 6.773$
0.00001	$\dfrac{f(2 + 0.00001) - f(2)}{0.00001}$
	$= \dfrac{2.00001^{2.00001} - 2^2}{0.00001}$
	$= 6.7727$
0.000001	$\dfrac{f(2 + 0.000001) - f(2)}{0.000001}$
	$= \dfrac{2.000001^{2.000001} - 2^2}{0.000001}$
	$= 6.7726$

The instantaneous rate of change at $x = 2$ is 6.7726.

21. $f(x) = x^{\ln x}$ at $x = 2$

h	
0.01	$\dfrac{f(2 + 0.01) - f(2)}{0.01}$ $= \dfrac{2.01^{\ln 2.01} - 2^{\ln 2}}{0.01}$ $= 1.1258$
0.001	$\dfrac{f(2 + 0.001) - f(2)}{0.001}$ $= \dfrac{2.001^{\ln 2.001} - 22^{\ln 2}}{0.001}$ $= 1.1212$
0.0001	$\dfrac{f(2 + 0.0001) - f(2)}{0.0001}$ $= \dfrac{2.0001^{\ln 2.0001} - 2^{\ln 2}}{0.0001}$ $= 1.1207$
0.00001	$\dfrac{f(2 + 0.00001) - f(2)}{0.00001}$ $= \dfrac{2.00001^{\ln 2.00001} - 2^{\ln 2}}{0.00001}$ $= 1.1207$

The instantaneous rate of change at $x = 2$ is 1.1207.

25. Let $B(t) =$ the amount in the Medicare Trust Fund for year t.

(a) $B(1994) = 152$
$B(1998) = 125$

Average change in fund

$= \dfrac{125 - 152}{1998 - 1994}$

$= \dfrac{-27}{4}$

$= -6.75$

On average, the amount in the fund decreased approximately \$6.75 billion per year from 1994 − 1998.

(b) $B(1998) = 125$
$B(2010) = 300$

Average change in fund

$= \dfrac{300 - 125}{2010 - 1998}$

$= \dfrac{175}{12}$

≈ 14.58

On average, the amount in the fund increases approximately \$14.58 billion per year from 1998 − 2010.

(c) $B(1990) = 125$
$B(1998) = 125$

Average change in fund

$= \dfrac{125 - 125}{1998 - 1990}$

$= \dfrac{0}{8}$

$= 0$

On average, the amount in the fund changes approximately \$0 per year from 1990 to 1998.

27. (a) From June 2005 to December 2005 is 6 months. During that time, the number of single family housing starts went from 1,724,000 to 1,633,000—a drop of 91,000. The average monthly rate of change is

$$\dfrac{-91,000}{6} = -15,167 \text{ starts per month}$$

(b) From December 2005 to June 2006 is 6 months. During that time, the number of single family housing starts went from 1,633,000 to 1,486,000—a drop of 147,000. The average monthly rate of change is

$$\dfrac{-147,000}{6} = -24,500 \text{ starts per month}$$

(c) From June 2005 to June 2006 is 12 months. During that time, the number of single family housing starts went from 1,724,000 to 1,486,000—a drop of 238,000. The average monthly rate of change is

$$\dfrac{-238,000}{12} = -19,833 \text{ starts per month}$$

(d) $\dfrac{-15{,}167 + (-24{,}500)}{2} = -19{,}833$ (allowing for rounding)

They are equal. This will not be true for all time periods (only for time periods of equal length).

29. $P(x) = 2x^2 - 5x + 6$

(a) $P(4) = 18$
$P(2) = 4$

Average rate of change of profit

$= \dfrac{P(4) - P(2)}{4 - 2}$

$= \dfrac{18 - 4}{2}$

$= \dfrac{14}{2} = 7,$

which is $700 per item.

(b) $P(3) = 9$
$P(2) = 4$

Average rate of change of profit

$= \dfrac{P(3) - P(2)}{3 - 2} = \dfrac{9 - 4}{1} = 5$

which is $500 per item.

(c) $\displaystyle\lim_{h \to 0} \dfrac{P(2 + h) - P(2)}{h}$

$= \displaystyle\lim_{h \to 0} \dfrac{2(2 + h)^2 - 5(2 + h) + 6 - 4}{h}$

$= \displaystyle\lim_{h \to 0} \dfrac{8 + 8h + 2h^2 - 10 - 5h + 2}{h}$

$= \displaystyle\lim_{h \to 0} \dfrac{2h^2 + 3h}{h}$

$= \displaystyle\lim_{h \to 0} \dfrac{h(2h + 3)}{h}$

$= \displaystyle\lim_{h \to 0} (2h + 3) = 3,$

which is $300 per item.

(d) $\displaystyle\lim_{h \to 0} \dfrac{P(4 + h) - P(4)}{h}$

$= \displaystyle\lim_{h \to 0} \dfrac{2(4 + h)^2 - 5(4 + h) + 6 - 18}{h}$

$= \displaystyle\lim_{h \to 0} \dfrac{32 + 16h + 2h^2 - 20 - 5h - 12}{h}$

$= \displaystyle\lim_{h \to 0} \dfrac{2h^2 + 11h}{h}$

$= \displaystyle\lim_{h \to 0} \dfrac{h(2h + 11)}{h}$

$= \displaystyle\lim_{h \to 0} 2h + 11 = 11,$

which is $1100 per item.

31. $N(p) = 80 - 5p^2,\ 1 \le p \le 4$

(a) Average rate of change of demand is

$\dfrac{N(3) - N(2)}{3 - 2} = \dfrac{35 - 60}{1}$

$= -25$ boxes per dollar.

(b) Instantaneous rate of change when p is 2 is

$\displaystyle\lim_{h \to 0} \dfrac{N(2 + h) - N(2)}{h}$

$= \displaystyle\lim_{h \to 0} \dfrac{80 - 5(2 + h)^2 - [80 - 5(2)^2]}{h}$

$= \displaystyle\lim_{h \to 0} \dfrac{80 - 20 - 20h - 5h^2 - (80 - 20)}{h}$

$= \displaystyle\lim_{h \to 0} \dfrac{-5h^2 - 20h}{h}$

$= -20$ boxes per dollar. Around the $2 point, a $1 price increase (say, from $1.50 to $2.50) causes a drop in demand of about 20 boxes.

(c) Instantaneous rate of change when p is 3 is

$\displaystyle\lim_{h \to 0} \dfrac{80 - 5(3 + h)^2 - [80 - 5(3)^2]}{h}$

$= \displaystyle\lim_{h \to 0} \dfrac{80 - 45 - 30h - 5h^2 - 80 + 45}{h}$

$= \displaystyle\lim_{h \to 0} \dfrac{-30h - 5h^2}{h}$

$= -30$ boxes per dollar.

(d) As the price increases, the demand decreases; this is an expected change.

33. Let $P(t)$ = world population estimated in billions for year t.

(a) $P(1990) = 5.3$

If replacement-level fertility is reached in 2010, $P(2050) = 8.6$.

Average rate of change $= \dfrac{P(2050) - P(1990)}{2050 - 1990}$

$= \dfrac{8.6 - 5.3}{60}$

$= 0.055$

On average, the population will increase 55 million per year.

If replacement-level fertility is reached in 2030, $P(2050) = 9.2$.

Average rate of change $= \dfrac{P(2050) - P(1990)}{2050 - 1990}$

$= \dfrac{9.2 - 5.3}{60}$

$= 0.065$

On average, the population will increase 65 million per year.

If replacement-level fertility is reached in 2050, $P(2050) = 9.8$.

Average rate of change $= \dfrac{P(2050) - P(1990)}{2050 - 1990}$

$= \dfrac{9.8 - 5.3}{60}$

$= 0.075$

On average, the population will increase 75 million per year.

The projection for replacement-level fertility by 2010 predicts the smallest rate of change in world population.

(b) If replacement-level fertility is reached in 2010

$$P(2090) = 9.3$$
$$P(2130) = 9.6$$

Average rate of change $= \dfrac{P(2130) - P(2090)}{2130 - 2090}$

$= \dfrac{9.6 - 9.3}{40}$

$= 0.0075$

On average, the population will increase 7.5 million per year.

If replacement-level fertility is reached in 2030,

$$P(2090) = 10.3$$
$$P(2130) = 10.6$$

Average rate of change $= \dfrac{P(2130) - P(2090)}{2130 - 2090}$

$= \dfrac{10.6 - 10.3}{40}$

$= 0.0075$

On average, the population will increase 7.5 million per year.

If replacement-level fertility is reached in 2050,

$$P(2090) = 11.35$$
$$P(2130) = 11.75$$

Average rate of change $= \dfrac{P(2130) - P(2090)}{2130 - 2090}$

$= \dfrac{11.75 - 11.35}{40}$

$= 0.01$

On average, the population will increase 10 million per year.

From 2090 – 2130 the three projections show almost the same rate of change in world population.

35. $L(t) = -0.01t^2 + 0.788t - 7.048$

(a) $\dfrac{L(28) - L(22)}{28 - 22} = \dfrac{7.176 - 5.448}{6}$

$= 0.288$

The average rate of growth during weeks 22 through 28 is 0.288 mm per week.

(b) $\displaystyle\lim_{h \to 0} \dfrac{L(t+h) - L(t)}{h}$

$= \displaystyle\lim_{h \to 0} \dfrac{L(22+h) - L(22)}{h}$

$= \displaystyle\lim_{h \to 0} \dfrac{[-0.01(22+h)^2 + 0.788(22+h) - 7.048] - 5.448}{h}$

$= \displaystyle\lim_{h \to 0} \dfrac{-0.01(h^2 + 44h + 484) + 17.336 + 0.788h - 12.496}{h}$

$= \displaystyle\lim_{h \to 0} \dfrac{-0.01h^2 + 0.348h}{h}$

$= \displaystyle\lim_{h \to 0} (-0.01h + 0.348)$

$= 0.348$

The instantaneous rate of growth at exactly 22 weeks is 0.348 mm per week.

(c)

37. (a) The average rate of change of $M(t)$ on the interval $[105, 115]$ is

$$\frac{M(115) - M(105)}{115 - 105} = \frac{0.8}{10} = 0.08$$

kilograms per day.

(b) Calculate $\lim\limits_{h \to 0} \dfrac{M(105 + h) - M(105)}{h}$

$$\begin{aligned}
M(105 + h) &= 27.5 + 0.3(105 + h) \\
&\quad - 0.001(105 + h)^2 \\
&= 27.5 + 31.5 + 0.3h \\
&\quad - (11.025 + 0.21h + 0.001h^2) \\
&= 47.975 + 0.09h - 0.001h^2 \\
M(105) &= 47.975
\end{aligned}$$

So, the instantaneous rate of change of $M(t)$ at $t = 105$ is

$$\lim_{h \to 0} \left(\frac{47.975 + 0.09h - 0.001h^2 - 47.975}{h} \right)$$

$$= \lim_{h \to 0} \left(\frac{0.09h - 0.001h^2}{h} \right)$$

$$= \lim_{h \to 0} (0.09 - 0.001h)$$

$$= 0.09 \text{ kilograms per day.}$$

(c)

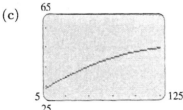

39. (a) Let $D(t)$ represent percent of kids who have used drugs by grade 8 in year t.

$$\frac{D(2001) - D(1998)}{2001 - 1998} = \frac{26.8 - 29}{3}$$

$$\approx -0.73$$

The average rate of change from 1998 to 2001 is -0.73 percent per year.

$$\frac{D(2005) - D(2002)}{2005 - 2002} = \frac{21.4 - 24.5}{3}$$

$$\approx -1.03$$

The average rate of change from 2002 to 2005 is -1.03 percent per year.

(b) Let $D(t)$ represent percent of kids who have used drugs by grade 10 in year t.

$$\frac{D(2001) - D(1998)}{2001 - 1998} = \frac{45.6 - 44.9}{3}$$

$$\approx 0.23$$

The average rate of change from 1998 to 2001 is 0.23 percent per year.

$$\frac{D(2005) - D(2002)}{2005 - 2002} = \frac{38.2 - 44.6}{3}$$

$$\approx -2.13$$

The average rate of change from 2002 to 2005 is -2.13 percent per year.

(c) Let $D(t)$ represent percent of kids who have used drugs by grade 12 in year t.

$$\frac{D(2001) - D(1998)}{2001 - 1998} = \frac{53.9 - 54.1}{3}$$

$$\approx -0.07$$

The average rate of change from 1998 to 2001 is -0.07 percent per year.

$$\frac{D(2005) - D(2002)}{2005 - 2002} = \frac{50.4 - 53}{3}$$

$$\approx -0.87$$

The average rate of change from 2002 to 2005 is -0.87 percent per year.

41. (a) $\dfrac{s(2) - s(0)}{2 - 0} = \dfrac{10 - 0}{2} = 5$ ft/sec

(b) $\dfrac{s(4) - s(2)}{4 - 2} = \dfrac{14 - 10}{2} = 2$ ft/sec

(c) $\dfrac{s(6) - s(4)}{6 - 4} = \dfrac{20 - 14}{2} = 3$ ft/sec

(d) $\dfrac{s(8) - s(6)}{8 - 6} = \dfrac{30 - 20}{2} = 5$ ft/sec

(e) (i) $\dfrac{f(x_0 + h) - f(x_0 - h)}{2h}$

$$= \frac{f(4 + 2) - f(4 - 2)}{(2)(2)}$$

$$= \frac{f(6) - f(2)}{4}$$

$$= \frac{20 - 10}{4}$$

$$= \frac{10}{4} = 2.5 \text{ ft/sec}$$

(ii) $\dfrac{2 + 3}{2} = 2.5$ ft/sec

(f) (i)
$$\frac{f(x_0 + h) - f(x_0 - h)}{2h}$$
$$= \frac{f(6 + 2) - f(6 - 2)}{(2)(2)}$$
$$= \frac{f(8) - f(4)}{4}$$
$$= \frac{30 - 14}{4}$$
$$= \frac{16}{4} = 4 \text{ ft/sec}$$

(ii) $\frac{3 + 5}{2} = 4 \text{ ft/sec}$

43. $s(t) = t^2 + 5t + 2$

(a) Average velocity $= \frac{s(6) - s(4)}{6 - 4}$
$$= \frac{68 - 38}{6 - 4}$$
$$= \frac{30}{2} = 15 \text{ ft/sec}$$

(b) Average velocity $= \frac{s(5) - s(4)}{5 - 4}$
$$= \frac{52 - 38}{5 - 4}$$
$$= \frac{14}{1} = 14 ft/sec$$

(c) $\lim_{h \to 0} \frac{s(4 + h) - s(4)}{h}$
$$= \lim_{h \to 0} \frac{(4 + h)^2 + 5(4 + h) + 2 - 38}{h}$$
$$= \lim_{h \to 0} \frac{16 + 8h + h^2 + 20 + 5h + 2 - 38}{h}$$
$$= \lim_{h \to 0} \frac{h^2 + 13h}{h} = \lim_{h \to 0} \frac{h(h + 13)}{h}$$
$$= \lim_{h \to 0} (h + 13) = 13 \text{ ft/sec}$$

11.4   Definition of the Derivative

1. (a) $f(x) = 5$ is a horizontal line and has slope 0; the derivative is 0.

(b) $f(x) = x$ has slope 1; the derivative is 1.

(c) $f(x) = -x$ has slope of -1; the derivative is -1.

(d) $x = 3$ is vertical and has undefined slope; the derivative does not exist.

(e) $y = mx + b$ has slope m; the derivative is m.

3. $f(x) = \frac{x^2 - 1}{x + 2}$ is not differentiable when $x + 2 = 0$ or $x = -2$ because the function is undefined and a vertical asymptote occurs there.

5. Using the points $(5, 3)$ and $(6, 5)$, we have
$$m = \frac{5 - 3}{6 - 5} = \frac{2}{1} = 2.$$

7. Using the points $(-2, 2)$ and $(2, 3)$, we have
$$m = \frac{3 - 2}{2 - (-2)} = \frac{1}{4}.$$

9. Using the points $(-3, -3)$ and $(0, -3)$, we have
$$m = \frac{-3 - (-3)}{0 - 3} = \frac{0}{-3} = 0.$$

11. $f(x) = -4x^2 + 9x + 2$

Step 1 $f(x + h) = -4(x + h)^2 + 9(x + h) + 2$
$$= -4(x^2 + 2xh + h^2) + 9x + 9h + 2$$
$$= -4x^2 - 8xh - 4h^2 + 9x + 9h + 2$$

Step 2 $f(x + h) - f(x)$
$$= -4x^2 - 8xh - 4h^2 + 9x + 9h + 2$$
$$- (-4x^2 + 9x + 2)$$
$$= -8xh - 4h^2 + 9h$$
$$= h(-8x - 4h + 9)$$

Step 3 $\frac{f(x + h) - f(x)}{h} = \frac{h(-8x - 4h + 9)}{h}$
$$= -8x - 4h + 9$$

Step 4 $f'(x) = \lim_{h \to 0} \frac{f(x + h) - f(x)}{h}$
$$= \lim_{h \to 0} (-8x - 4h + 9)$$
$$= -8x + 9$$

$$f'(-2) = -8(-2) + 9 = 25$$
$$f'(0) = -8(0) + 9 = 9$$
$$f'(3) = -8(3) + 9 = -15$$

13. $f(x) = \frac{12}{x}$

$$f(x + h) = \frac{12}{x + h}$$

$$f(x + h) - f(x) = \frac{12}{x + h} - \frac{12}{x}$$
$$= \frac{12x - 12(x + h)}{x(x + h)}$$
$$= \frac{12x - 12x - 12h}{x(x + h)}$$
$$= \frac{-12h}{x(x + h)}$$

$$\frac{f(x+h)-f(x)}{h} = \frac{-12h}{hx(x+h)}$$

$$= \frac{-12}{x(x+h)}$$

$$= \frac{-12}{x^2+xh}$$

$$f'(x) = \lim_{h \to 0} \frac{f(x+h)-f(x)}{h}$$

$$= \lim_{h \to 0} \frac{-12}{x^2+xh}$$

$$= \frac{-12}{x^2}$$

$$f'(-2) = \frac{-12}{(-2)^2} = \frac{-12}{4} = -3$$

$f'(0) = \frac{-12}{0^2}$ which is undefined so $f'(0)$ does not exist.

$$f'(3) = \frac{-12}{3^2} = \frac{-12}{9} = -\frac{4}{3}$$

15. $f(x) = \sqrt{x}$

Steps 1-3 are combined.

$$\frac{f(x+h)-f(x)}{h}$$

$$= \frac{\sqrt{x+h}-\sqrt{x}}{h}$$

$$= \frac{\sqrt{x+h}-\sqrt{x}}{h} \cdot \frac{\sqrt{x+h}+\sqrt{x}}{\sqrt{x+h}+\sqrt{x}}$$

$$= \frac{x+h-x}{h(\sqrt{x+h}+\sqrt{x})}$$

$$= \frac{1}{\sqrt{x+h}+\sqrt{x}}$$

$$f'(x) = \lim_{h \to 0} \frac{f(x+h)-f(x)}{h}$$

$$= \lim_{h \to 0} \frac{1}{\sqrt{x+h}+\sqrt{x}}$$

$$= \frac{1}{2\sqrt{x}}$$

$f'(-2) = \dfrac{1}{2\sqrt{-2}}$ which is undefined so $f'(-2)$ does not exist.

$f'(0) = \dfrac{1}{2\sqrt{0}} = \dfrac{1}{0}$ which is undefined so $f'(0)$ does not exist.

$$f'(3) = \frac{1}{2\sqrt{3}}$$

17. $f(x) = 2x^3 + 5$

Steps 1-3 are combined.

$$\frac{f(x+h)-f(x)}{h}$$

$$= \frac{2(x+h)^3+5-(2x^3+5)}{h}$$

$$= \frac{2(x^3+3x^2h+3xh^2+h^3)+5-2x^3-5}{h}$$

$$= \frac{2x^3+6x^2h+6xh^2+2h^3+5-2x^3-5}{h}$$

$$= \frac{6x^2h+6xh^2+2h^3}{h}$$

$$= \frac{h(6x^2+6xh+2h^2)}{h}$$

$$= 6x^2+6xh+2h^2$$

$$f'(x) = \lim_{h \to 0} (6x^2+6xh+2h^2)$$

$$= 6x^2$$

$$f'(-2) = 6(-2)^2 = 24$$

$$f'(0) = 6(0)^2 = 0$$

$$f'(3) = 6(3)^2 = 54$$

19. (a) $f(x) = x^2 + 2x; x = 3, x = 5$

Slope of secant line $= \dfrac{f(5)-f(3)}{5-3}$

$$= \frac{(5)^2+2(5)-[(3)^2+2(3)]}{2}$$

$$= \frac{35-15}{2}$$

$$= 10$$

Now use $m = 10$ and $(3, f(3)) = (3, 15)$ in the point-slope form.

$$y - 15 = 10(x-3)$$
$$y - 15 = 10x - 30$$
$$y = 10x - 30 + 15$$
$$y = 10x - 15$$

(b) $f(x) = x^2 + 2x;\ x = 3$

$$\frac{f(x+h)-f(x)}{h}$$

$$= \frac{[(x+h)^2+2(x+h)]-(x^2+2x)}{h}$$

$$= \frac{(x^2+2hx+h^2+2x+2h)-(x^2+2x)}{h}$$

$$= \frac{2hx+h^2+2h}{h} = 2x+h+2$$

$$f'(x) = \lim_{h \to 0} (2x+h+2) = 2x+2$$

$f'(3) = 2(3) + 2 = 8$ is the slope of the tangent line at $x = 3$.

Use $m = 8$ and $(3, 15)$ in the point-slope form.

$$y - 15 = 8(x - 3)$$
$$y = 8x - 9$$

21. (a) $f(x) = \frac{5}{x}; x = 2, x = 5$

Slope of secant line $= \dfrac{f(5) - f(2)}{5 - 2}$

$$= \frac{\frac{5}{5} - \frac{5}{2}}{3}$$

$$= \frac{1 - \frac{5}{2}}{3}$$

$$= -\frac{1}{2}$$

Now use $m = -\frac{1}{2}$ and $(5, f(5)) = (5, 1)$ in the point-slope form.

$$y - 1 = -\frac{1}{2}[x - 5]$$

$$y - 1 = -\frac{1}{2}x + \frac{5}{2}$$

$$y = -\frac{1}{2}x + \frac{5}{2} + 1$$

$$y = -\frac{1}{2}x + \frac{7}{2}$$

(b) $f(x) = \dfrac{5}{x}; \; x = 2$

$$\frac{f(x+h) - f(x)}{h} = \frac{\frac{5}{x+h} - \frac{5}{x}}{h}$$

$$= \frac{\frac{5x - 5(x+h)}{(x+h)x}}{h}$$

$$= \frac{5x - 5x - 5h}{h(x+h)(x)}$$

$$= \frac{-5h}{h(x+h)x}$$

$$= \frac{-5}{(x+h)x}$$

$$f'(x) = \lim_{h \to 0} \frac{-5}{(x+h)(x)} = -\frac{5}{x^2}$$

$f'(2) = \frac{-5}{2^2} = -\frac{5}{4}$ is the slope of the tangent line at $x = 2$.

Now use $m = -\frac{5}{4}$ and $\left(2, \frac{5}{2}\right)$ in the point-slope form.

$$y - \frac{5}{2} = -\frac{5}{4}(x - 2)$$

$$y - \frac{5}{2} = -\frac{5}{4}x + \frac{10}{4}$$

$$y = -\frac{5}{4}x + 5$$

$$5x + 4y = 20$$

23. (a) $f(x) = 4\sqrt{x}; x = 9, x = 16$

Slope of secant line $= \dfrac{f(16) - f(9)}{16 - 9}$

$$= \frac{4\sqrt{16} - 4\sqrt{9}}{7}$$

$$= \frac{16 - 12}{7}$$

$$= \frac{4}{7}$$

Now use $m = \frac{4}{7}$ and $(9, f(9)) = (9, 12)$ in the point-slope form.

$$y - 12 = \frac{4}{7}(x - 9)$$

$$y - 12 = \frac{4}{7}x - \frac{36}{7}$$

$$y = \frac{4}{7}x - \frac{36}{7} + 12$$

$$y = \frac{4}{7}x + \frac{48}{7}$$

(b) $f(x) = 4\sqrt{x}; \; x = 9$

$$\frac{f(x+h) - f(x)}{h}$$

$$= \frac{4\sqrt{x+h} - 4\sqrt{x}}{h} \cdot \frac{4\sqrt{x+h} + 4\sqrt{x}}{4\sqrt{x+h} + 4\sqrt{x}}$$

$$= \frac{16(x+h) - 16x}{h(4\sqrt{x+h} + 4\sqrt{x})}$$

$$f'(x) = \lim_{h \to 0} \frac{16(x+h) - 16x}{h(4\sqrt{x+h} + 4\sqrt{x})}$$

$$= \lim_{h \to 0} \frac{16h}{h(4\sqrt{x+h} + 4\sqrt{x})}$$

$$= \lim_{h \to 0} \frac{4}{(\sqrt{x+h} + \sqrt{x})} = \frac{4}{2\sqrt{x}}$$

$$= \frac{2}{\sqrt{x}}$$

$f'(9) = \frac{2}{\sqrt{9}} = \frac{2}{3}$ is the slope of the tangent line at $x = 9$.

Use $m = \frac{2}{3}$ and $(9, 12)$ in the point-slope form.

$$y - 12 = \frac{2}{3}(x - 9)$$

$$y = \frac{2}{3}x + 6$$

$$3y = 2x + 18$$

25. $f(x) = -4x^2 + 11x$

$$\frac{f(x + h) - f(x)}{h}$$

$$= \frac{-4(x + h)^2 + 11(x + h) - (-4x^2 + 11x)}{h}$$

$$= \frac{-8xh - 4h^2 + 11h}{h}$$

$$f'(x) = \lim_{h \to 0} (-8x - 4h + 11)$$
$$= -8x + 11$$
$$f'(2) = -8(2) + 11 = -5$$
$$f'(16) = -8(16) + 11 = -117$$
$$f'(-3) = -8(-3) + 11 = 35$$

27. $f(x) = e^x$

$$\frac{f(x + h) - f(x)}{h} = \frac{e^{x+h} - e^x}{h}$$

$$f'(x) = \lim_{h \to 0} \frac{e^{x+h} - e^x}{h}$$

$$f'(2) \approx 7.3891; f'(16) \approx 8,886,111; f'(-3) \approx 0.0498$$

29. $f(x) = -\frac{2}{x}$

$$\frac{f(x + h) - f(x)}{h} = \frac{\frac{-2}{x + h} - \left(\frac{-2}{x}\right)}{h}$$

$$= \frac{\frac{-2x + 2(x + h)}{(x + h)x}}{h}$$

$$= \frac{2h}{h(x + h)x}$$

$$= \frac{2}{(x + h)x}$$

$$f'(x) = \lim_{h \to 0} \frac{2}{(x + h)x}$$

$$= \frac{2}{x^2}$$

$$f'(2) = \frac{2}{2^2} = \frac{1}{2}$$

$$f'(16) = \frac{2}{16^2}$$

$$= \frac{2}{256}$$

$$\frac{1}{128}.$$

$$f'(-3) = \frac{2}{(-3)^2}$$

$$= \frac{2}{9}$$

31. $f(x) = \sqrt{x}$

$$\frac{f(x + h) - f(x)}{h}$$

$$= \frac{\sqrt{x + h} - \sqrt{x}}{h} \cdot \frac{\sqrt{x + h} + \sqrt{x}}{\sqrt{x + h} + \sqrt{x}}$$

$$= \frac{(x + h) - x}{h(\sqrt{x + h} + \sqrt{x})}$$

$$= \frac{h}{h(\sqrt{x + h} + \sqrt{x})}$$

$$= \frac{1}{\sqrt{x + h} + \sqrt{x}}$$

$$f'(x) = \lim_{h \to 0} \frac{1}{\sqrt{x + h} + \sqrt{x}} = \frac{1}{2\sqrt{x}}$$

$$f'(2) = \frac{1}{2\sqrt{2}}$$

$$f'(16) = \frac{1}{2\sqrt{16}} = \frac{1}{8}$$

$$f'(-3) = \frac{1}{2\sqrt{-3}}$$ is not a real number, so

$f'(-3)$ does not exist.

33. At $x = 0$, the graph of $f(x)$ has a sharp point. Therefore, there is no derivative for $x = 0$.

35. For $x = -3$ and $x = 0$, the tangent to the graph of $f(x)$ is vertical. For $x = -1$, there is a gap in the graph $f(x)$. For $x = 2$, the function $f(x)$ does not exist. For $x = 3$ and $x = 5$, the graph $f(x)$ has sharp points. Therefore, no derivative exists for $x = -3$, $x = -1$, $x = 0$, $x = 2$, $x = 3$, and $x = 5$.

37. **(a)** The rate of change of $f(x)$ is positive when $f(x)$ is increasing, that is, on $(a, 0)$ and (b, c).

(b) The rate of change of $f(x)$ is negative when $f(x)$ is decreasing, that is, on $(0, b)$.

(c) The rate of change is zero when the tangent to the graph is horizontal, that is, at $x = 0$ and $x = b$.

39. The zeros of graph (b) correspond to the turning points of graph (a), the points where the derivative is zero. Graph (a) gives the distance, while graph (b) gives the velocity.

41. $f(x) = x^x, a = 3$

(a)

h	
0.01	$\dfrac{f(3+0.01) - f(3)}{0.01}$ $= \dfrac{3.01^{3.01} - 3^3}{0.01}$ $= 57.3072$
0.001	$\dfrac{f(3+0.001) - f(3)}{0.001}$ $= \dfrac{3.001^{3.001} - 3^3}{0.001}$ $= 56.7265$
0.00001	$\dfrac{f(3+.00001) - f(3)}{0.00001}$ $= \dfrac{3.00001^{3.00001} - 3^3}{0.00001}$ $= 56.6632$
0.000001	$\dfrac{f(3+0.000001) - f(3)}{0.000001}$ $= \dfrac{3.000001^{3.000001} - 3^3}{0.000001}$ $= 56.6626$
0.0000001	$\dfrac{f(3+0.0000001) - f(3)}{0.0000001}$ $= \dfrac{3.0000001^{3.0000001} - 3^3}{0.0000001}$ $= 56.6625$

It appears that $f'(3) = 56.6625$.

(b) Graph the function on a graphing calculator and move the cursor to an x-value near $x = 3$. A good choice for the initial viewing window is $[0, 4]$ by $[0, 60]$ with Xscl $= 1$, Yscl $= 10$.

Now zoom in on the function several times. Each time you zoom in, the graph will look less like a curve and more like a straight line. Use the TRACE feature to select two points on the graph, and record their coordinates. Use these two points to compute the slope. The result will be close to

the most accurate value found in part (a), which is 56.6625.

Note: In this exercise, the method used in part (a) gives more accurate results than the method used in part (b).

43. $f(x) = x^{1/x}, a = 3$

(a)

h	
0.01	$\dfrac{f(3+0.01) - f(3)}{0.01}$ $= \dfrac{3.01^{1/3.01} - 3^{1/3}}{0.01}$ $= -0.0160$
0.001	$\dfrac{f(3+0.001) - f(3)}{0.001}$ $= \dfrac{3.001^{1/3.001} - 3^{1/3}}{0.001}$ $= -0.0158$
0.0001	$\dfrac{f(3+0.0001) - f(3)}{0.0001}$ $= \dfrac{3.0001^{1/3.0001} - 3^{1/3}}{0.0001}$ $= -0.0158$

It appears that $f'(3) = -0.0158$.

(b) Graph the function on a graphing calculator and move the cursor to an x-value near $x = 3$. A good choice for the initial viewing window is $[0, 5]$ by $[0, 3]$.

Follow the procedure outlined in the solution for Exercise 39, part (b). Note that near $x = 3$, the graph is very close to a horizontal line, so we expect that it slope will be close to 0. The final result will be close to the value found in part (a) of this exercise, which is -0.0158.

47. $D(p) = -2p^2 - 4p + 300$

D is demand; p is price.

(a) Given that $D'(p) = -4p - 4$, the rate of change of demand with respect to price is $-4p - 4$, the derivative of the function $D(p)$.

(b) $D'(10) = -4(10) - 4$
$\qquad\qquad = -44$

The demand is decreasing at the rate of about 44 items for each increase in price of $1.

49. $R(x) = 20x - \dfrac{x^2}{500}$

(a) $R'(x) = 20 - \dfrac{1}{250}x$

At $y = 1000$,

$$R'(1000) = 20 - \frac{1}{250}(1000)$$

$$= \$16 \text{ per table.}$$

(b) The marginal revenue for the 1001st table is approximately $R'(1000)$. From (a), this is about $16.

(c) The actual revenue is

$$R(1001) - R(1000) = 20(1001) - \frac{1001^2}{500}$$

$$- \left[20(1000) - \frac{1000^2}{500} \right]$$

$$= 18,015.998 - 18,000$$

$$= \$15.998 \text{ or } \$16.$$

(d) The marginal revenue gives a good approximation of the actual revenue from the sale of the 1001st table.

51. (a) $f(x) = -0.0142x^4 + 0.6698x^3 - 6.113x^2$
$\qquad\qquad + 84.05x + 203.9$

$\quad f(10) = 961$
$\quad f(20) = 2526$
$\quad f(30) = 3806$

(b) $Y_1 = -0.0142x^4 + 0.6698x^3 - 6.113x^2$
$\qquad\quad + 84.05x + 203.9$

$$nDeriv(Y_1, x, 10) \approx 106$$
$$nDeriv(Y_1, x, 20) \approx 189$$
$$nDeriv(Y_1, x, 30) \approx -8$$
$$nDeriv(Y_1, x, 35) \approx -318$$

53. The derivative at $(2, 4000)$ can be approximated by the slope of the line through $(0, 2000)$ and $(2, 4000)$. The derivative is approximately

$$\frac{4000 - 2000}{2 - 0} = \frac{2000}{2} = 1000.$$

Thus the shellfish population is increasing at a rate of 1000 shellfish per unit time.

The derivative at about $(10, 10,300)$ can be approximated by the slope of the line through $(10, 10,300)$ and $(13, 12,000)$. The derivative is approximately

$$\frac{12,000 - 10,300}{13 - 10} = \frac{1700}{3} \approx 570.$$

The shellfish population is increasing at a rate of about 583 shellfish per unit time. The derivative at about $(13, 11,250)$ can be approximated by the slope of the line through $(13, 11,250)$ and $(16, 12,000)$. The derivative is approximately

$$\frac{12,000 - 11,250}{16 - 13} = \frac{750}{3} = 250.$$

The shellfish population is increasing at a rate of 250 shellfish per unit time.

55. (a) Set $M(v) = 150$ and solve for v.

$$0.0312443v^2 - 101.39v + 82,264 = 150$$
$$0.0312443v^2 - 101.39v + 82,114 = 0$$

Solve using the quadratic formula. Let D equal the discriminant.

$$D = b^2 - 4ac$$
$$= (-101.39)^2 - 4(0.0312443)(82,114)$$
$$\approx 17.55$$

$$v = \frac{101.39 \pm \sqrt{D}}{2(0.0312443)}$$

$v \approx 1690$ meter per second or
$v \approx 1560$ meters per second.
Since the functions is defined only for $v \geq 1620$, the only solution is 1690 meters per second.

(b) Calculate $\lim\limits_{h \to 0} \dfrac{M(1700 + h) - M(1700)}{h}$

$M(1700 + h)$
$\quad = 0.0312443(1700 + h)^2 - 101.39(1700 + h)$
$\qquad + 82,264$
$\quad = 90,296.027 + 106.23062h + 0.0312443h^2$
$\qquad - 172,363 - 101.39h + 82,264$
$\quad = 0.01312443h^2 + 4.84062h + 197.027$

$M(1700) = 197.027$, so the derivative of $M(v)$ at $v = 1700$ is

$$\lim_{h \to 0} \left(\frac{0.0312443h^2 + 4.84062h + 197.027 - 197.027}{h} \right)$$

$$= \lim_{h \to 0} \left(\frac{0.0312443h^2 + 4.84062h}{h} \right)$$

$$= \lim_{h \to 0} (0.0312443h + 4.84062)$$

$$= 4.84062$$

$$\approx 4.84 \text{ days per meter per second}$$

The increase in velocity for this cheese from 1700 m/s to 1701 m/s indicates that the approximate age of the cheese has increased by 4.84 days.

57. (a) The derivative does not exist at the two "corners" or "sharp points." The x-values of these points are 0.75 and 3.

(b) To find $T'(0.5)$, calculate the slope of the line segment with positive slope, since this portion of the graph includes the value $x = 0.5$ (marked as $\frac{1}{2}$ hr on the graph.) To find this slope, use the points $(0, 100)$ (the starting point) and $(0.75, 875)$ (the beginning point of the cleaning cycle).

$$m = T'(0.5) = \frac{875 - 100}{0.75 - 0}$$
$$= \frac{775}{0.75}$$
$$\approx 1033$$

The oven temperature is increasing at 1033° per hour.

(c) To find $T'(2)$, find the slope of the line segment containing the value $x = 2$. This segment is horizontal, so

$$m = T'(2) = 0.$$

The oven temperature is not changing.

(d) To find $T'(3.5)$, calculate the slope of the line segment with negative slope, since this portion of the graph includes the value $x = 3.5$. To find this slope, use the points $(3, 875)$ (the end of the cleaning cycle) and $(3.75, 100)$ (the stopping point).

$$m = T'(3.5) = \frac{100 - 875}{3.75 - 3}$$
$$= \frac{-775}{0.75}$$
$$\approx -1033$$

The oven temperature is decreasing at 1033° per hour.

59. The velocities are equal at approximately $t = 0.13$. The acceleration for the hands is approximately 0 mph per sec. The acceleration for the bat is approximately 640 mph per sec.

11.5   Graphical Differentiation

3. Since the x-intercepts of the graph of f' occur whenever the graph of f has a horizontal tangent line, Y_1 is the derivative of Y_2. Notice that Y_1 has 2 x-intercepts; each occurs at an x-value where the tangent line to Y_2 is horizontal.

Note also that Y_1 is positive whenever Y_2 is increasing, and that Y_1 is negative whenever Y_2 is decreasing.

5. Since the x-intercepts of the graph of f' occur whenever the graph of f has a horizontal tangent line, Y_2 is the derivative of Y_1. Notice that Y_2 has 1 x-intercept which occurs at the x-value where the tangent line to Y_1 is horizontal. Also notice that the range on which Y_1 is increasing, Y_2 is positive and the range on which it is decreasing, Y_2 is negative.

7. To graph f', observe the intervals where the slopes of tangent lines are positive and where they are negative to determine where the derivative is positive and where it is negative. Also, whenever f has a horizontal tangent, f' will be 0, so the graph of f' will have an x-intercept. The x-values of the three turning point on the graph of f become the three x-intercepts of the graph of f.

Estimate the magnitude of the slope at several points by drawing tangents to the graph of f.

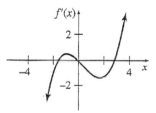

9. On the interval $(-\infty, -2)$, the graph of f is a horizontal line, so its slope is 0. Thus, on this interval, the graph of f' is $y = 0$ on $(-\infty, -2)$. On the interval $(-2, 0)$, the graph of f is a straight line, so its slope is constant. To find this slope, use the points $(-2, 2)$ and $(0, 0)$.

$$m = \frac{2 - 0}{-2 - 0} = \frac{2}{-2} = -1$$

On the interval $(0, 1)$, the slope is also constant. To find this slope, use the points $(0, 0)$ and $(1, 1)$.

$$m = \frac{1 - 0}{1 - 0} = 1$$

On the interval $(1, \infty)$, the graph is again a horizontal line, so $m = 0$. The graph of f' will be made up of portions of the y-axis and the lines $y = -1$ and $y = 1$.

Because the graph of f has "sharp points" or "corners" at $x = -2, x = 0$, and $x = 1$, we know that $f'(-2), f'(0)$, and $f'(1)$ do not exist. We show this on the graph of f' by using open circles at the endpoints of the portions of the graph.

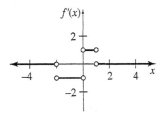

11. On the interval $(-\infty, -2)$, the graph of f is a straight line, so its slope is constant. To find this slope, use the points $(-4, 2)$ and $(-2, 0)$.

$$m = \frac{0 - 2}{-2 - (-4)} = \frac{-2}{2} = -1$$

On the interval $(2, \infty)$, the slope of f is also constant. To find this slope, use the points $(2, 0)$ and $(3, 2)$.

$$m = \frac{2 - 0}{3 - 2} = \frac{2}{1} = 2$$

Thus, we have $f'(x) = -1$ on $(-\infty, -2)$ and $f'(x) = 2$ on $(2, \infty)$.

Because f is discontinuous at $x = -2$ and $x = 2$, we know that $f'(-2)$ and $f'(2)$ do not exist, which we indicate with open circles at $(-2, -1)$ and $(2, 2)$ on the graph of f'.

On the interval $(-2, 2)$, all tangent lines have positive slopes, so the graph of f' will be above the y-axis. Notice that the slope of f (and thus the y-value of f') decreases on $(-2, 0)$ and increases on $(0, 2)$, with a minimum value on this interval of 1 at $x = 0$.

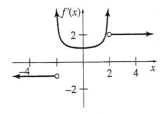

13. We observe that the slopes of tangent lines are positive on the interval $(-\infty, 0)$ and negative on the interval $(0, \infty)$, so the value of f' will be positive on $(-\infty, 0)$ and negative on $(0, \infty)$. Since f is undefined at $x = 0$, $f'(0)$ does not exist.

Notice that the graph of f becomes very flat when $|x| \to \infty$. The *value* of f approaches 0 and also the *slope* approaches 0. Thus, $y = 0$ (the x-axis) is a horizontal asymptote for both the graph of f and the graph of f'.

As $x \to 0^-$ and $x \to 0^+$, the graph of f gets very steep, so $|f'(x)| \to \infty$. Thus, $x = 0$ (the y-axis) is a vertical asymptote for both the graph of f and the graph of f'.

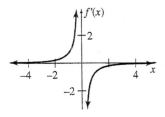

15. The slope of $f(x)$ is undefined at $x = -2, -1, 0, 1$, and 2, and the graph approaches vertical (unbounded slope) as x approaches those values. Accordingly, the graph of $f'(x)$ has vertical asymptotes at $x = -2, -1, 0, 1$, and 2. $f(x)$ has turning points (zero slope) at $x = -1.5, -0.5, 0.5$, and 1.5, so the graph of $f'(x)$ crosses the x-axis at those values. Elsewhere, the graph of $f'(x)$ is negative where $f(x)$ is decreasing and positive where $f(x)$ is increasing.

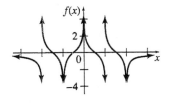

17. The graph rises steadily, with varying degrees of steepness. The graph is steepest around 1976 and nearly flat around 1950 and 1980. Accordingly, the rate of change is always positive, with a maximum value around 1976 and values near zero around 1950 and 1980.

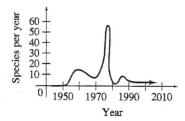

19. The growth rate of the function $y = f(x)$ is given by the derivative of this function $y' = f(x)$. We use the graph of f to sketch the graph of f'.

First, notice as x increase, y increases throughout the domain of f, but at a slower and slower rate. The slope of f is positive but always decreasing, and approaches 0 as t gets large. Thus, y' will always be positive and decreasing. It will approach but never reach 0.

To plot point on the graph of f', we need to estimate the slope of f at several points. From the graph of f, we obtain the values given in the following table.

t	y'
2	1000
10	700
13	250

Use these points to sketch the graph.

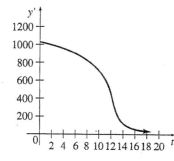

21.

About 9 cm; about 2.6 cm less per year

Chapter 11 Review Exercises

5. (a) $\lim\limits_{x \to -3^-} = 4$

   (b) $\lim\limits_{x \to -3^+} = 4$

   (c) $\lim\limits_{x \to -3} = 4$ (since parts (a) and (b) have the same answer)

   (d) $f(-3) = 4$, since $(-3, 4)$ is a point of the graph.

7. (a) $\lim\limits_{x \to 4^-} f(x) = \infty$

   (b) $\lim\limits_{x \to 4^+} f(x) = -\infty$

   (c) $\lim\limits_{x \to 4} f(x)$ does not exist since the answers to parts (a) and (b) are different.

   (d) $f(4)$ does not exist since the graph has no point with an x-value of 4.

9. $\lim\limits_{x \to -\infty} g(x) = \infty$ since the y-value gets very large as the x-value gets very small.

11. $\lim\limits_{x \to 6} \dfrac{2x + 7}{x + 3} = \dfrac{2(6) + 7}{6 + 3}$

    $= \dfrac{19}{9}$

13. $\lim\limits_{x \to 4} \dfrac{x^2 - 16}{x - 4} = \lim\limits_{x \to 4} \dfrac{(x - 4)(x + 4)}{x - 4}$

    $= \lim\limits_{x \to 4} (x + 4)$

    $= 4 + 4$

    $= 8$

15. $\lim\limits_{x \to -4} \dfrac{2x^2 + 3x - 20}{x + 4} = \lim\limits_{x \to -4} \dfrac{(2x - 5)(x + 4)}{x + 4}$

    $= \lim\limits_{x \to -4} (2x - 5)$

    $= 2(-4) - 5$

    $= -13$

17. $\lim\limits_{x \to 9} \dfrac{\sqrt{x} - 3}{x - 9} = \lim\limits_{x \to 9} \dfrac{\sqrt{x} - 3}{x - 9} \cdot \dfrac{\sqrt{x} + 3}{\sqrt{x} + 3}$

    $= \lim\limits_{x \to 9} \dfrac{x - 9}{(x - 9)(\sqrt{x} + 3)}$

    $= \lim\limits_{x \to 9} \dfrac{1}{\sqrt{x} + 3}$

    $= \dfrac{1}{\sqrt{9} + 3}$

    $= \dfrac{1}{6}$

19. $\lim\limits_{x\to\infty} \dfrac{2x^2+5}{5x^2-1} = \lim\limits_{x\to\infty} \dfrac{\frac{2x^2}{x^2}+\frac{5}{x^2}}{\frac{5x^2}{x^2}-\frac{1}{x^2}}$

$$= \lim_{x\to\infty} \frac{2+\frac{5}{x^2}}{5-\frac{1}{x^2}}$$

$$= \frac{2+0}{5-0}$$

$$= \frac{2}{5}$$

21. $\lim\limits_{x\to-\infty}\left(\dfrac{3}{8}+\dfrac{3}{x}-\dfrac{6}{x^2}\right)$

$$= \lim_{x\to-\infty}\frac{3}{8} + \lim_{x\to-\infty}\frac{3}{x} - \lim_{x\to-\infty}\frac{6}{x^2}$$

$$= \frac{3}{8}+0-0$$

$$= \frac{3}{8}$$

23. As shown on the graph, $f(x)$ is discontinuous at x_2 and x_4.

25. $f(x)$ is discontinuous at $x=0$ and $x=-\frac{1}{3}$ since that is where the denominator of $f(x)$ equals 0. $f(0)$ and $f\left(-\frac{1}{3}\right)$ do not exist.
$\lim\limits_{x\to 0} f(x)$ does not exist since $\lim\limits_{x\to 0^+} f(x) = -\infty$, but $\lim\limits_{x\to 0^-} f(x) = \infty$. $\lim\limits_{x\to-\frac{1}{3}} f(x)$ does not exist since $\lim\limits_{x\to-\frac{1}{3}^-} = -\infty$, but $\lim\limits_{x\to-\frac{1}{3}^+} f(x) = \infty$.

27. $f(x)$ is discontinuous at $x=-5$ since that is where the denominator of $f(x)$ equals 0.
$f(-5)$ does not exist.
$\lim\limits_{x\to-5} f(x)$ does not exist since $\lim\limits_{x\to-5^-} f(x) = \infty$, but $\lim\limits_{x\to-5^+} f(x) = -\infty$.

29. $f(x) = x^2 + 3x - 4$ is continuous everywhere since f is a polynomial function.

31. (a)

(b) The graph is discontinuous at $x=1$.

(c) $\lim\limits_{x\to 1^-} f(x) = 0$; $\lim\limits_{x\to 1^+} f(x) = 2$

33. $f(x) = \dfrac{x^4+2x^3+2x^2-10x+5}{x^2-1}$

(a) Find the values of $f(x)$ when x is close to 1.

x	y
1.1	2.6005
1.01	2.06
1.001	2.006
1.0001	2.0006
0.99	1.94
0.999	1.994
0.9999	1.9994

It appears that $\lim\limits_{x\to 1} f(x) = 2$.

(b) Graph

$$y = \frac{x^4+2x^3+2x^2-10x+5}{x^2-1}$$

on a graphing calculator. One suitable choice for the viewing window is $[-2,6]$ by $[-10,10]$.

Because $x^2-1=0$ when $x=-1$ or $x=1$, this function is discontinuous at these two x-values. The graph shows a vertical asymptote at $x=-1$ but not at $x=1$. The graph should have an open circle to show a "hole" in the graph at $x=1$. The graphing calculator doesn't show the hole, but trying to find the value of the function of $x=1$ will show that this value is undefined.

By viewing the function near $x=1$ and using the ZOOM feature, we see that as x gets close to 1 from the left or the right, y gets close to 2, suggesting that

$$\lim_{x\to 1}\frac{x^4+2x^3+2x^2-10x+5}{x^2-1} = 2.$$

35. $y = 6x^3+2 = f(x)$; from $x=1$ to $x=4$

$$f(4) = 6(4)^3+2 = 386$$
$$f(1) = 6(1)^3+2 = 8$$

Average rate of change:

$$= \frac{386-8}{4-1} = \frac{378}{3} = 126$$

$y' = 18x$

Instantaneous rate of change at $x=1$:

$$f'(1) = 18(1) = 18$$

37. $y = \dfrac{-6}{3x-5} = f(x)$; from $x = 4$ to $x = 9$

$$f(9) = \frac{-6}{3(9)-5} = \frac{-6}{22} = -\frac{3}{11}$$

$$f(4) = \frac{-6}{3(4)-5} = -\frac{6}{7}$$

Average rate of change:

$$= \frac{\frac{-3}{11} - \left(-\frac{6}{7}\right)}{9-4}$$

$$= \frac{\frac{-21+66}{77}}{5}$$

$$= \frac{45}{5(77)} = \frac{9}{77}$$

$$y' = \frac{(3x-5)(0)-(-6)(3)}{(3x-5)^2}$$

$$= \frac{18}{(3x-5)^2}$$

Instantaneous rate of change at $x = 4$:

$$f'(4) = \frac{18}{(3\cdot 4-5)^2} = \frac{18}{7^2} = \frac{18}{49}$$

39. (a) $f(x) = 3x^2 - 5x + 7$; $x = 2, x = 4$

Slope of secant line

$$= \frac{f(4)-f(2)}{4-2}$$

$$= \frac{[3(4)^2-5(4)+7]-[3(2)^2-5(2)+7]}{2}$$

$$= \frac{35-9}{2}$$

$$= 13$$

Now use $m = 13$ and $(2, f(2)) = (2,9)$ in the point-slope form.

$$y - 9 = 13(x-2)$$
$$y - 9 = 13x - 26$$
$$y = 13x - 26 + 9$$
$$y = 13x - 17$$

(b) $f(x) = 3x^2 - 5x + 7$; $x = 2$

$$\frac{f(x+h)-f(x)}{h}$$

$$= \frac{[3(x+h)^2-5(x+h)+7]-[3x^2-5x+7]}{h}$$

$$= \frac{3x^2+6xh+3h^2-5x-5h+7-3x^2+5x-7}{h}$$

$$= \frac{6xh+3h^2-5h}{h}$$

$$= 6x + 3h - 5$$

$$f'(x) = \lim_{h\to 0} 6x + 3h - 5$$
$$= 6x - 5$$
$$f'(2) = 6(2) - 5$$
$$= 7$$

Now use $m = 7$ and $(2, f(2)) = (2, 9)$ in the point-slope form.

$$y - 9 = 7(x-2)$$
$$y - 9 = 7x - 14$$
$$y = 7x - 14 + 9$$
$$y = 7x - 5$$

41. (a) $f(x) = \dfrac{12}{x-1}$; $x = 3, x = 7$

Slope of secant line $= \dfrac{f(7)-f(3)}{7-3}$

$$= \frac{\frac{12}{7-1} - \frac{12}{3-1}}{4}$$

$$= \frac{2-6}{4}$$

$$= -1$$

Now use $m = -1$ and $(3, f(x)) = (3, 6)$ in the point-slope form.

$$y - 6 = -1(x-3)$$
$$y - 6 = -x + 3$$
$$y = -x + 3 + 6$$
$$y = -x + 9$$

(b) $f(x) = \dfrac{12}{x-1}$; $x = 3$

$$\frac{f(x+h)-f(x)}{h} = \frac{\frac{12}{x+h-1} - \frac{12}{x-1}}{h}$$

$$= \frac{12(x-1)-12(x+h-1)}{h(x-1)(x+h-1)}$$

$$= \frac{-12h}{h(x-1)(x+h-1)}$$

$$= -\frac{12}{(x-1)(x+h-1)}$$

$$f'(x) = \lim_{h \to 0} -\frac{12}{(x-1)(x+h-1)} = -\frac{12}{(x-1)^2}$$

$$f'(3) = -\frac{12}{(3-1)^2} = -3$$

Now use $m = -3$ and $(3, f(x)) = (3, 6)$ in the point-slope form.

$$y - 6 = -3(x - 3)$$
$$y - 6 = -3x + 9$$
$$y = -3x + 9 + 6$$
$$y = -3x + 15$$

43. $y = 4x^2 + 3x - 2 = f(x)$

$$y' = \lim_{h \to 0} \frac{f(x+h) - f(x)}{h} = \lim_{h \to 0} \frac{[4(x+h)^2 + 3(x+h) - 2] - [4x^2 + 3x - 2]}{h}$$

$$= \lim_{h \to 0} \frac{4(x^2 + 2xh + h^2) + 3x + 3h - 2 - 4x^2 - 3x + 2}{h} = \lim_{h \to 0} \frac{4x^2 + 8xh + 4h^2 + 3x + 3h - 2 - 4x^2 - 3x + 2}{h}$$

$$= \lim_{h \to 0} \frac{8xh + 4h^2 + 3h}{h} = \lim_{h \to 0} \frac{h(8x + 4h + 3)}{h} = \lim_{h \to 0} (8x + 4h + 3) = 8x + 3$$

45. $f(x) = (\ln x)^x, x_0 = 3$

(a)

h	
0.01	$\dfrac{f(3 + 0.01) - f(3)}{0.01}$ $= \dfrac{(\ln 3.01)^{3.01} - (\ln 3)3}{0.01}$ $= 1.3385$
0.001	$\dfrac{f(3 + 0.001) - f(3)}{0.001}$ $= \dfrac{(\ln 3.001)^{3.001} - (\ln 3)^3}{0.001}$ $= 1.3323$
0.0001	$\dfrac{f(3 + 0.0001) - f(3)}{0.0001}$ $= \dfrac{(\ln 3.0001)^{3.0001} - (\ln 3)^3}{0.0001}$ $= 1.3317$
0.00001	$\dfrac{f(3 + 0.00001) - f(3)}{0.00001}$ $= \dfrac{(\ln 3.00001)^{3.00001} - (\ln 3)^3}{0.00001}$ $= 1.3317$

(b) Using a graphing calculator will confirm this result.

47. On the interval $(-\infty, 0)$, the graph of f is a straight line, so its slope is constant. To find this slope, use the points $(-2, 2)$ and $(0, 0)$.

$$m = \frac{0 - 2}{0 - (-2)} = \frac{-2}{2} = -1$$

Thus, the value of f' will be -1 on this interval. The graph of f has a sharp point at 0, so $f'(0)$ does not exist. To show this, we use an open circle on the graph of f' at $(0, -1)$.

We also observe that the slope of f is positive but decreasing from $x = 0$ to about $x = 1$, and then negative from there on. As $x \to \infty$, $f(x) \to 0$ and also $f'(x) = 0$.

Use this information to complete the graph of f'.

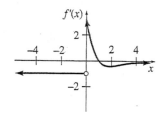

49. $\displaystyle \lim_{x \to \infty} \frac{cf(x) - dg(x)}{f(x) - g(x)} = \frac{\displaystyle \lim_{x \to \infty} [cf(x) - dg(x)]}{\displaystyle \lim_{x \to \infty} [f(x) - g(x)]}$

$$= \frac{\displaystyle \lim_{x \to \infty} [cf(x)] - \lim_{x \to \infty} [dg(x)]}{\displaystyle \lim_{x \to \infty} [f(x)] - \lim_{x \to \infty} [g(x)]}$$

$$= \frac{\displaystyle c \lim_{x \to \infty} [f(x)] - d \lim_{x \to \infty} [g(x)]}{\displaystyle \lim_{x \to \infty} [f(x)] - \lim_{x \to \infty} [g(x)]}$$

$$= \frac{c \cdot c - d \cdot d}{c - d}$$

$$= \frac{(c + d)(c - d)}{c - d}$$

$$= c + d$$

The answer is (e).

51. $R(x) = 5000 + 16x - 3x^2$

(a) $R'(x) = 16 - 6x$

(b) Since x is in hundreds of dollars, $1000 corresponds to $x = 10$.

$$R'(10) = 16 - 6(10)$$
$$= 16 - 60 = -44$$

An increase of $100 spent on advertising when advertising expenditures are $1000 will result in the revenue decreasing by $44.

53. $P(x) = 15x + 25x^2$

(a) $P(6) = 15(6) + 25(6)^2$
$\quad = 90 + 900 = 990$
$P(7) = 15(7) + 25(7)^2$
$\quad = 105 + 1225 = 1330$

Average rate of change:

$$= \frac{P(7) - P(6)}{7 - 6} = \frac{1330 - 990}{1}$$

$$= 340 \text{ cents or } \$3.40$$

(b) $P(6) = 990$
$P(6.5) = 15(6.5) + 25(6.5)^2$
$\quad = 97.5 + 1056.25$
$\quad = 1153.75$

Average rate of change:

$$= \frac{P(6.5) - P(6)}{6.5 - 6}$$

$$= \frac{1153.75 - 990}{0.5}$$

$$= 327.5 \text{ cents or } \$3.28$$

(c) $P(6) = 990$
$P(6.1) = 15(6.1) + 25(6.1)^2$
$\quad = 91.5 + 930.25$
$\quad = 1021.75$

Average rate of change:

$$= \frac{P(6.1) - P(6)}{6.1 - 6}$$

$$= \frac{1021.75 - 990}{0.1}$$

$$= 317.5 \text{ cents or } \$3.18$$

(d) $P'(x) = 15 + 50x$
$P'(6) = 15 + 50(6)$
$\quad = 15 + 300$
$\quad = 315 \text{ cents or } \$3.15$

(e) $P'(20) = 15 + 50(20)$
$\quad = 1015 \text{ cents or } \$10.15$

(f) $P'(30) = 15 + 50(30)$
$\quad = 1515 \text{ cents or } \$15.15$

(g) The domain of x is $[0, \infty)$ since pounds cannot be measured with negative numbers.

(h) Since $P'(x) = 15 + 50x$ gives the marginal profit, and $x \geq 0$, $P'(x)$ can never be negative.

(i) $\overline{P}(x) = \dfrac{P(x)}{x}$

$\qquad = \dfrac{15x + 25x^2}{x}$

$\qquad = 15 + 25x$

(j) $\overline{P}'(x) = 25$

(k) The marginal average profit cannot change since $\overline{P}'(x)$ is constant. The profit per pound never changes, no matter now many pounds are sold.

55. (a) $\displaystyle\lim_{x \to 29,300^-} T(x) = (29,300)(0.15)$

$\qquad\qquad = \$4395$

(b) $\displaystyle\lim_{x \to 29,300^+} T(x) = 4350 + (0.27)(29,300 - 29,300)$

$\qquad\qquad = \$4350$

(c) $\displaystyle\lim_{x \to 29,300} T(x)$ does not exist since parts (a) and (b) have different answers.

(d)

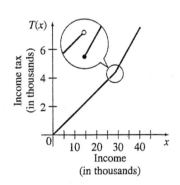

(e) The graph is discontinuous at $x = 29,300$.

(f) For $0 \le x \le 29,300$,

$$A(x) = \dfrac{T(x)}{x} = \dfrac{0.15x}{x} = 0.15.$$

For $x > 29,300$,

$$A(x) = \dfrac{T(x)}{x}$$

$$= \dfrac{4350 + (0.27)(x - 29,300)}{x}$$

$$= \dfrac{0.27x - 3561}{x}$$

$$= 0.27 - \dfrac{3561}{x}.$$

(g) $\displaystyle\lim_{x \to 29,300^-} A(x) = 0.15$

(h) $\displaystyle\lim_{x \to 29,300^+} A(x) = 0.27 - \dfrac{3561}{29,300} = 0.14846$

(i) $\displaystyle\lim_{x \to 29,300} A(x)$ does not exist since parts (g) and (h) have different answers.

(j) $\displaystyle\lim_{x \to \infty} A(x) = 0.27 - 0 = 0.27$

(k)

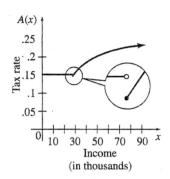

57. $V(t) = -t^2 + 6t - 4$

(a)

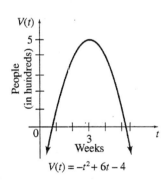

(b) The x-intercepts of the parabola are 0.8 and 5.2, so a reasonable domain would be $[0.8, 5.2]$, which represents the time period from 0.8 to 5.2 weeks.

(c) The number of cases reaches a maximum at the vertex;

$$x = \dfrac{-b}{2a} = \dfrac{-6}{-2} = 3$$

$$V(3) = -3^2 + 6(3) - 4 = 5$$

The vertex of the parabola is $(3, 5)$. This represents a maximum at 3 weeks of 500 cases.

(d) The rate of change function is

$$V'(t) = -2t + 6.$$

(e) The rate of change in the number of cases at the maximum is

$$V'(3) = -2(3) + 6 = 0.$$

(f) The sign of the rate of change up to the maximum is + because the function is increasing. The sign of the rate of change after the maximum is − because the function is decreasing.

59. (a) The curve slants downward up to about age 4, where it turns and begins to rise. There is a slight decline in steepness between ages 10.5 and 18. Correspondingly, the graph of the rate of change lies below the horizontal axis to the left of 4 years and above the horizontal axis to the right of that point. The graph of the rate of change is declining between ages 10.5 and 18.

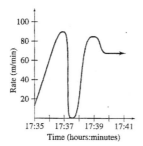

(b) The curve slants downward up to about age 5.25, where it turns and begins to rise. There is a slight decline in steepness between ages 11 and 20. Correspondingly, the graph of the rate of change lies below the horizontal axis to the left of 5.25 years and above the horizontal axis to the right of that point. The graph of the rate of change is declining between ages 11 and 20.

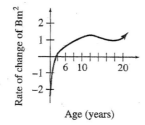

61. (a) The slope of the tangent line at $x = 100$ is 1. This means that the ball is rising 1 ft for each foot it travels horizontally.

(b) The slope of the tangent line at $x = 200$ is −2.7. This means that the ball is dropping 2.7 ft for each foot it travels horizontally.

Chapter 11 Test

[11.1]

Decide whether each limit exists. If a limit exists, find its value.

1. $\lim\limits_{x \to 1} f(x)$

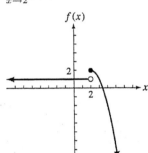

2. $\lim\limits_{x \to -1} f(x)$

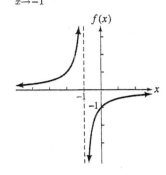

3. $\lim\limits_{x \to 2} f(x)$

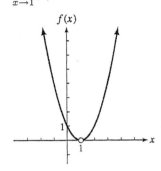

4. $\lim\limits_{x \to -2} f(x)$

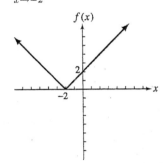

Use the properties of limits to help decide whether the following limits exist. If a limit exists, find its value.

5. $\lim\limits_{x \to 2} \left(\dfrac{1}{x} + 1 \right)(3x - 2)$

6. $\lim\limits_{x \to 0} \dfrac{3x + 5}{4x}$

7. $\lim\limits_{x \to 2} \dfrac{x^2 - 5x + 6}{x - 2}$

8. $\lim\limits_{x \to 1} \dfrac{x - 1}{\sqrt{x} - 1}$

[11.2]

Find all points $x = a$ where the function is discontinuous. For each point of discontinuity, give $\lim\limits_{x \to a} f(x)$ if it exists.

9. $f(x) = \dfrac{x^2 - 16}{x + 4}$

10. $g(x) = \dfrac{3 + x}{(2x + 3)(x - 5)}$

11. $h(x) = \dfrac{3x^4 + 2x^2 - 7}{4}$

12. Is $f(x) = \frac{x-2}{x(3-x)(x+4)}$ continuous at the given values of x?

(a) $x = 2$ (b) $x = 0$ (c) $x = 5$ (d) $x = 3$

13. Use the graph to answer the following questions.

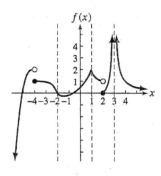

(a) On which of the following intervals is the graph continuous?
$(-5, -2)$, $(-3, 2)$, $(1, 4)$

(b) Where is the function discontinuous?

(c) Find $\lim\limits_{x \to -4^-} f(x)$.   (d) Find $\lim\limits_{x \to -4^+} f(x)$.

14. Consider the following function.

$$f(x) = \begin{cases} -3 & \text{if } x < -1 \\ 2x - 1 & \text{if } -1 \le x \le 2 \\ 6 & \text{if } x > 2 \end{cases}$$

(a) Graph this function.   (b) Find any points of discontinuity.

(c) Find the limit from the left and from the right at any point(s) of discontinuity.

[11.3]

15. Find the average rate of change of $f(x) = x^3 - 5x$ between $x = 1$ and $x = 5$.

16. Use the graph to find the average rate of change of f on the given intervals.

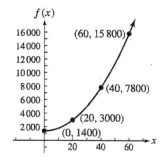

(a) From $x = 0$ to $x = 20$ (b) From $x = 20$ to $x = 60$

(c) From $x = 0$ to $x = 40$ (d) From $x = 0$ to $x = 60$

Find the instantaneous rate of change for each function at the given value.

17. $f(x) = 3x^2 - 7$ at $x = 1$

18. $g(t) = 4 - 2t^2$ at $t = -3$

19. Suppose the total profit in thousands of dollars from selling x units is given by

$$P(x) = 2x^2 - 6x + 9.$$

(a) Find the average rate of change of profit as x increases from 2 to 4.

(b) Find the marginal profit when 10 units are sold.

(c) Find the average rate of change of profit when sales are increased from 100 to 200 units.

[11.4]

Use the definition of the derivative to find the derivative of each function.

20. $y = x^3 - 5x^2$

21. $y = \dfrac{3}{x}$

Find $f'(x)$ and use it to find $f'(3)$, $f'(0)$, $f'(-1)$.

22. $f(x) = \dfrac{-3}{x}$

23. $f(x) = 3x^2 + 4x$

24. $f(x) = -\dfrac{1}{2}\sqrt{x}$

[11.5]

Sketch the graph of the derivative for each function shown.

25.

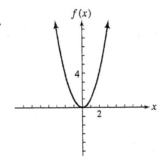

26.

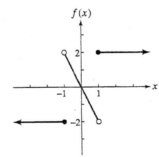

Chapter 11 Test Answers

1. 0

2. Does not exist

3. Does not exist

4. 0

5. 6

6. Does not exist

7. -1

8. 2

9. $a = -4,\ \lim\limits_{x \to -4} f(x) = -8$

10. $a = -\frac{3}{2}$, limit does not exist; $a = 5$, limit does not exist.

11. Discontinuous nowhere

12. (a) Yes (b) No (c) Yes (d) No

13. (a) $(-3, 2)$ (b) At $x = -4$, $x = 2$, and $x = 3$
(c) 2 (d) 1

14. (a)

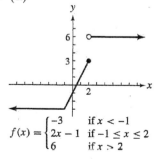

$$f(x) = \begin{cases} -3 & \text{if } x < -1 \\ 2x - 1 & \text{if } -1 \le x \le 2 \\ 6 & \text{if } x > 2 \end{cases}$$

(b) $x = 2$
(c) From the left: 3;
from the right: 6

15. 26

16. (a) 80 (b) 320 (c) 160 (d) 240

17. 6

18. 12

19. (a) \$6000 per unit (b) \$34,000 per unit
(c) \$594,000 per unit

20. $y' = 3x^2 - 10x$

21. $y' = -\frac{3}{x^2}$

22. $f'(x) = \frac{3}{x^2}$, $\frac{1}{3}$, undefined, 3

23. $f'(x) = 6x + 4$, 22, 4, -2

24. $f'(x) = \frac{-1}{4\sqrt{x}}$, $\frac{-1}{4\sqrt{3}}$ or $\frac{-\sqrt{3}}{12}$, undefined, undefined

25.

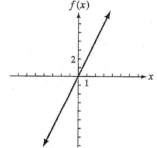

26.

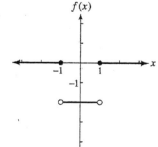

CALCULATING THE DERIVATIVE

12.1 Techniques for Finding Derivatives

1. $y = 12x^3 - 8x^2 + 7x + 5$

$$\frac{dy}{dx} = 12(3x^{3-1}) - 8(2x^{2-1}) + 7x^{1-1} + 0$$

$$= 36x^2 - 16x + 7$$

3. $y = 3x^4 - 6x^3 + \frac{x^2}{8} + 5$

$$\frac{dy}{dx} = 3(4x^{4-1}) - 6(3x^{3-1}) + \frac{1}{8}(2x^{2-1}) + 0$$

$$= 12x^3 - 18x^2 + \frac{1}{4}x$$

5. $y = 6x^{3.5} - 10x^{0.5}$

$$\frac{dy}{dx} = 6(3.5x^{3.5-1}) - 10(0.5x^{0.5-1})$$

$$= 21x^{2.5} - 5x^{-0.5} \text{ or } 21x^{2.5} - \frac{5}{x^{0.5}}$$

7. $y = 8\sqrt{x} + 6x^{3/4}$
$\quad = 8x^{1/2} + 6x^{3/4}$

$$\frac{dy}{dx} = 8\left(\frac{1}{2}x^{1/2-1}\right) + 6\left(\frac{3}{4}x^{3/4-1}\right)$$

$$= 4x^{-1/2} + \frac{9}{2}x^{-1/4}$$

$$\text{or} \quad \frac{4}{x^{1/2}} + \frac{9}{2x^{1/4}}$$

9. $g(x) = 6x^{-5} - x^{-1}$
$\quad g'(x) = 6(-5)x^{-5-1} - (-1)x^{-1-1}$
$\quad \quad = -30x^{-6} + x^{-2}$

$$\text{or} \quad \frac{-30}{x^6} + \frac{1}{x^2}$$

11. $y = 5x^{-5} - 6x^{-2} + 13x^{-1}$

$$\frac{dy}{dx} = 5(-5x^{-5-1}) - 6(-2x^{-2-1}) + 13(-1x^{-1-1})$$

$$= -25x^{-6} + 12x^{-3} - 13x^{-2}$$

$$\text{or} \quad \frac{-25}{x^6} + \frac{12}{x^3} - \frac{13}{x^2}$$

13. $f(t) = \frac{14}{t} + \frac{12}{t^4} + \sqrt{2}$

$$= 14t^{-1} + 12t^{-4} + \sqrt{2}$$
$$f'(t) = 14(-1t^{-1-1}) + 12(-4t^{-4-1}) + 0$$
$$= -14t^{-2} - 48t^{-5} \text{ or } \frac{-14}{t^2} - \frac{48}{t^5}$$

15. $y = \frac{3}{x^6} + \frac{1}{x^5} - \frac{7}{x^2}$

$$= 3x^{-6} + x^{-5} - 7x^{-2}$$

$$\frac{dy}{dx} = 3(-6x^{-7}) + (-5x^{-6}) - 7(-2x^{-3})$$

$$= -18x^{-7} - 5x^{-6} + 14x^{-3}$$

$$\text{or} \quad \frac{-18}{x^7} - \frac{5}{x^6} + \frac{14}{x^3}$$

17. $h(x) = x^{-1/2} - 14x^{-3/2}$

$$h'(x) = -\frac{1}{2}x^{-3/2} - 14\left(-\frac{3}{2}x^{-5/2}\right)$$

$$= \frac{-x^{-3/2}}{2} + 21x^{-5/2}$$

$$\text{or} \quad \frac{-1}{2x^{3/2}} + \frac{21}{x^{5/2}}$$

19. $y = \frac{-2}{\sqrt[3]{x}}$

$$= \frac{-2}{x^{1/3}} = -2x^{-1/3}$$

$$\frac{dy}{dx} = -2\left(-\frac{1}{3}x^{-4/3}\right)$$

$$= \frac{2x^{-4/3}}{3} \text{ or } \frac{2}{3x^{4/3}}$$

21. $g(x) = \frac{x^3 - 4x}{\sqrt{x}}$

$$= \frac{x^3 - 4x}{x^{1/2}}$$

$$= x^{5/2} - 4x^{1/2}$$

$$g'(x) = \frac{5}{2}x^{5/2-1} - 4\left(\frac{1}{2}x^{1/2-1}\right)$$

$$= \frac{5}{2}x^{3/2} - 2x^{-1/2}$$

$$\text{or} \quad \frac{5}{2}x^{3/2} - \frac{2}{\sqrt{x}}$$

23. $h(x) = (x^2 - 1)^3$
$$= x^6 - 3x^4 + 3x^2 - 1$$
$$h'(x) = 6x^{6-1} - 3(4x^{4-1}) + 3(2x^{2-1}) - 0$$
$$= 6x^5 - 12x^3 + 6x$$

27. $D_x \left[9x^{-1/2} + \dfrac{2}{x^{3/2}} \right]$

$$= D_x[9x^{-1/2} + 2x^{-3/2}]$$

$$= 9\left(-\frac{1}{2}x^{-3/2}\right) + 2\left(-\frac{3}{2}x^{-5/2}\right)$$

$$= -\frac{9}{2}x^{-3/2} - 3x^{-5/2}$$

or $\dfrac{-9}{2x^{3/2}} - \dfrac{3}{x^{5/2}}$

29. $f(x) = \dfrac{x^4}{6} - 3x$

$$= \frac{1}{6}x^4 - 3x$$

$$f'(x) = \frac{1}{6}(4x^3) - 3$$

$$= \frac{2}{3}x^3 - 3$$

$$f'(-2) = \frac{2}{3}(-2)^3 - 3$$

$$= -\frac{16}{3} - 3$$

$$= -\frac{25}{3}$$

31. $y = x^4 - 5x^3 + 2; \ x = 2$
$$y' = 4x^3 - 15x^2$$
$$y'(2) = 4(2)^3 - 15(2)^2$$
$$= -28$$

The slope of tangent line at $x = 2$ is -28.
Use $m = -28$ and $(x_1, y_1) = (2, -22)$ to obtain the equation.

$$y - (-22) = -28(x - 2)$$
$$y = -28x + 34$$

33. $y = -2x^{1/2} + x^{3/2}$
$$y' = -2\left(\frac{1}{2}x^{-1/2}\right) + \frac{3}{2}x^{1/2}$$

$$= -x^{-1/2} + \frac{3}{2}x^{1/2}$$

$$= -\frac{1}{x^{1/2}} + \frac{3x^{1/2}}{2}$$

$$y'(9) = -\frac{1}{(9)^{1/2}} + \frac{3(9)^{1/2}}{2}$$

$$= -\frac{1}{3} + \frac{9}{2}$$

$$= \frac{25}{6}$$

The slope of the tangent line at $x = 9$ is $\frac{25}{6}$.

35. $f(x) = 9x^2 - 8x + 4$
$$f'(x) = 18x - 8$$

Let $f'(x) = 0$ to find the point where the slope of the tangent line is zero.

$$18x - 8 = 0$$
$$18x = 8$$
$$x = \frac{8}{18} = \frac{4}{9}$$

Find the y-coordinate.

$$f(x) = 9x^2 - 8x + 4$$

$$f\left(\frac{4}{9}\right) = 9\left(\frac{4}{9}\right)^2 - 8\left(\frac{4}{9}\right) + 4$$

$$= 9\left(\frac{16}{81}\right) - \frac{32}{9} + 4$$

$$= \frac{16}{9} - \frac{32}{9} + \frac{36}{9} = \frac{20}{9}$$

The slope of the tangent line is zero at one point, $\left(\frac{4}{9}, \frac{20}{9}\right)$.

37. $f(x) = 2x^3 + 9x^2 - 60x + 4$
$$f'(x) = 6x^2 + 18x - 60$$

If the tangent line is horizontal, then its slope is zero and $f'(x) = 0$.

$$6x^2 + 18x - 60 = 0$$
$$6(x^2 + 3x - 10) = 0$$
$$6(x + 5)(x - 2) = 0$$
$$x = -5 \quad \text{or} \quad x = 2$$

Thus, the tangent line is horizontal at $x = -5$ and $x = 2$.

39. $f(x) = x^3 - 4x^2 - 7x + 8$
$f'(x) = 3x^2 - 8x - 7$

If the tangent line is horizontal, then its slope is zero and $f'(x) = 0$.

$$3x^2 - 8x - 7 = 0$$

$$x = \frac{8 \pm \sqrt{64 + 84}}{6}$$

$$x = \frac{8 \pm \sqrt{148}}{6}$$

$$x = \frac{8 \pm 2\sqrt{37}}{6}$$

$$x = \frac{2(4 \pm \sqrt{37})}{6}$$

$$x = \frac{4 \pm \sqrt{37}}{3}$$

Thus, the tangent line is horizontal at $x = \frac{4 \pm \sqrt{37}}{3}$.

41. $f(x) = 6x^2 + 4x - 9$
$f'(x) = 12x + 4$

If the slope of the tangent line is -2, $f'(x) = -2$.

$$12x + 4 = -2$$
$$12x = -6$$

$$x = -\frac{1}{2}$$

$$f\left(-\frac{1}{2}\right) = -\frac{19}{2}$$

The slope of the tangent line is -2 at $\left(-\frac{1}{2}, -\frac{19}{2}\right)$.

43. $f(x) = x^3 + 6x^2 + 21x + 2$
$f'(x) = 3x^2 + 12x + 21$

If the slope of the tangent line is 9, $f'(x) = 9$.

$$3x^2 + 12x + 21 = 9$$
$$3x^2 + 12x + 12 = 0$$
$$3(x^2 + 4x + 4) = 0$$
$$3(x + 2)^2 = 0$$
$$x = -2$$
$$f(-2) = -24$$

The slope of the tangent line is 9 at $(-2, -24)$.

45. $f(x) = \frac{1}{2}g(x) + \frac{1}{4}h(x)$

$f'(x) = \frac{1}{2}g'(x) + \frac{1}{4}h'(x)$

$f'(2) = \frac{1}{2}g'(2) + \frac{1}{4}h'(2)$

$= \frac{1}{2}(7) + \frac{1}{4}(14) = 7$

49. $\dfrac{f(x)}{k} = \dfrac{1}{k} \cdot f(x)$

Use the rule for the derivative of a constant times a function.

$$\frac{d}{dx}\left[\frac{f(x)}{k}\right] = \frac{d}{dx}\left[\frac{1}{k} \cdot f(x)\right]$$

$$= \frac{1}{k}f'(x)$$

$$= \frac{f'(x)}{k}$$

51. The demand is given by $q = 5000 - 100p$.

Solve for p.

$$p = \frac{5000 - q}{100}$$

$$R(q) = q\left(\frac{5000 - q}{100}\right)$$

$$= \frac{5000q - q^2}{100}$$

$$R'(q) = \frac{5000 - 2q}{100}$$

(a) $R'(1000) = \dfrac{5000 - 2(1000)}{100}$

$= 30$

(b) $R'(2500) = \dfrac{5000 - 2(2500)}{100}$

$= 0$

(c) $R'(3000) = \dfrac{5000 - 2(3000)}{100}$

$= -10$

53. $S(t) = 100 - 100t^{-1}$
$S'(t) = -100(-1t^{-2})$
$= 100t^{-2}$

$= \dfrac{100}{t^2}$

(a) $S'(1) = \dfrac{100}{(1)^2} = \dfrac{100}{1} = 100$

(b) $S'(10) = \dfrac{100}{(10)^2} = \dfrac{100}{100} = 1$

55. Profit = Revenue − Cost

$$P(q) = qp(q) - C(q)$$

$$P(q) = q\left(\frac{1000}{q^2} + 1000\right) - (0.2q^2 + 6q + 50)$$

$$= \frac{1000}{q} + 1000q - 0.2q^2 - 6q - 50$$

$$= 1000q^{-1} + 994q - 0.2x^2 - 50$$

$$P'(q) = -1000q^{-2} + 994 - 0.4q$$

$$= 994 - 0.4q - \frac{1000}{q^2}$$

$$P'(10) = 994 - 0.4(10) - \frac{1000}{(10)^2}$$

$$= 994 - 4 - 10$$

$$= 980$$

The marginal profit is $980.

57. (a) 1982 when $t = 50$:

$$C(50) = 0.00875(50)^2 - 0.108(50) + 1.42$$
$$= 17.895 \approx 17.9 \text{ cents}$$

2002 when $t = 70$:

$$C(70) = 0.00875(70)^2 - 0.108(70) + 1.42$$
$$= 36.735 \approx 36.7 \text{ cents}$$

(b) $C'(t) = 0.00875(2t) - 0.108(1)$
$$= 0.0175t - 0.108$$

1982 when $t = 50$:

$$C'(50) = 0.0175(50) - 0.108$$
$$= 0.767 \text{ cents/year}$$

2002 when $t = 70$:

$$C'(70) = 0.0175(70) - 0.108$$
$$\approx 1.12 \text{ cents/year}$$

(c) Using a graphing calculator, a cubic function that models the data is

$$C(t) = (-1.790 \times 10^{-4})t^3 + 0.02947t^2$$
$$- 0.7105t + 3.291.$$

Using the values from the calculator, the rate of change in 1982 is $C'(50) \approx 0.894$ cents/year. Using the values from the calculator, the rate of change in 2002 is $C'(70) \approx 0.784$ cents/year.

59. $N(t) = 0.00437t^{3.2}$
$N'(t) = 0.013984t^{2.2}$

(a) $N'(5) \approx 0.4824$

(b) $N'(10) \approx 2.216$

61. $V(t) = -2159 + 1313t - 60.82t^2$

(a) $V(3) = -2159 + 1313(3) - 60.82(3)^2$
$$= 1232.62 \text{ cm}^3$$

(b) $V'(t) = 1313 - 121.64t$
$V'(3) = 1313 - 121.64(3)$
$$= 948.08 \text{ cm}^3/\text{yr}$$

63. $v = 2.69l^{1.86}$

$$\frac{dv}{dl} = (1.86)2.69l^{1.86-1} \approx 5.00l^{0.86}$$

65. $t = 0.0588s^{1.125}$

(a) When $s = 1609$, $t \approx 238.1$ seconds, or 3 minutes, 58.1 seconds.

(b) $\dfrac{dt}{ds} = 0.0588(1.125s^{1.125-1})$
$$= 0.06615s^{0.125}$$

When $s = 100$, $\frac{dt}{ds} \approx 0.118$ sec/m. At 100 meters, the fastest possible time increases by 0.118 seconds for each additional meter.

(c) Yes, they have been surpassed. In 2000, the world record in the mile stood at 3:43.13. (Ref: www.runnersworld.com)

67. BMI $= \dfrac{703w}{h^2}$

(a) $6'2'' = 74$ in.

$$\text{BMI} = \frac{703(220)}{74^2} \approx 28$$

(b) BMI $= \dfrac{703w}{74^2} = 24.9$ implies

$$w = \frac{24.9(74)^2}{703} \approx 194.$$

A 220-lb person needs to lose 26 pounds to get down to 194 lbs.

(c) If $f(h) = \dfrac{703(125)}{h^2} = 87{,}875h^{-2}$, then

$$f'(h) = 87{,}875(-2h^{-2-1})$$
$$= -175{,}750h^{-3} = -\frac{175{,}750}{h^3}$$

(d) $f'(65) = -\dfrac{175{,}750}{65^3} \approx -0.64$

For a 125-lb female with a height of 65 in. (5′5″), the BMI decreases by 0.64 for each additional inch of height.

(e) Sample Chart

ht/wt	140	160	180	200
60	27	31	35	39
65	23	27	30	33
70	20	23	26	29
75	17	20	22	25

69. $s(t) = 18t^2 - 13t + 8$

(a) $v(t) = s'(t) = 18(2t) - 13 + 0$
$$= 36t - 13$$

(b) $v(0) = 36(0) - 13 = -13$
$v(5) = 36(5) - 13 = 167$
$v(10) = 36(10) - 13 = 347$

71. $s(t) = -3t^3 + 4t^2 - 10t + 5$

(a) $v(t) = s'(t) = -3(3t^2) + 4(2t) - 10 + 0$
$$= -9t^2 + 8t - 10$$

(b) $v(0) = -9(0)^2 + 8(0) - 10 = -10$
$v(5) = -9(5)^2 + 8(5) - 10$
$$= -225 + 40 - 10 = -195$$
$v(10) = -9(10)^2 + 8(10) - 10$
$$= -900 + 80 - 10 = -830$$

73. $s(t) = -16t^2 + 64t$

(a) $v(t) = s'(t) = -16(2t) + 64$
$$= -32t + 64$$

$v(2) = -32(2) + 64 = -64 + 64 = 0$

$v(3) = -32(3) + 64 = -96 + 64 = -32$

The ball's velocity is 0 ft/sec after 2 seconds and −32 ft/sec after 3 seconds.

(b) As the ball travels upward, its speed decreases because of the force of gravity until, at maximum height, its speed is 0 ft/sec.
In part (a), we found that $v(2) = 0$.

It takes 2 seconds for the ball to reach its maximum height.

(c) $s(2) = -16(2)^2 + 64(2)$
$$= -16(4) + 128$$
$$= -64 + 128$$
$$= 64$$

It will go 64 ft high.

75. $y_1 = 4.13x + 14.63$
$y_2 = -0.033x^2 + 4.647x + 13.347$

(a) When $x = 5, y_1 \approx 35$ and $y_2 \approx 36$.

(b) $\dfrac{dy_1}{dx} = 4.13$

$\dfrac{dy_2}{dx} = 0.033(2x) + 4.647$
$$= -0.066x + 4.647$$

When $x = 5, \frac{dy_1}{dx} = 4.13$ and $\frac{dy_2}{dx} \approx 4.32$. These values are fairly close and represent the rate of change of four years for a dog for one year of a human, for a dog that is actually 5 years old.

(c) With the first three points eliminated, the dog age increases in 2-year steps and the human age increases in 8-year steps, for a slope of 4. The equation has the form $y = 4x + b$. A value of 16 for b makes the numbers come out right. $y = 4x + b$. For a dog of age $x = 5$ years or more, the equivalent human age is given by $y = 4x + 16$.

12.2 Derivatives of Products and Quotients

1. $y = (3x^2 + 2)(2x - 1)$

$\dfrac{dy}{dx} = (3x^2 + 2)(2) + (2x - 1)(6x)$
$$= 6x^2 + 4 + 12x^2 - 6x$$
$$= 18x^2 - 6x + 4$$

3. $y = (2x - 5)^2$

$$= (2x - 5)(2x - 5)$$
$\dfrac{dy}{dx} = (2x - 5)(2) + (2x - 5)(2)$
$$= 4x - 10 + 4x - 10$$
$$= 8x - 20$$

5. $k(t) = (t^2 - 1)^2 = (t^2 - 1)(t^2 - 1)$
$k'(t) = (t^2 - 1)(2t) + (t^2 - 1)(2t)$
$$= 2t^3 - 2t + 2t^3 - 2t$$
$$= 4t^3 - 4t$$

7. $y = (x+1)(\sqrt{x}+2)$
$\quad = (x+1)(x^{1/2}+2)$

$\dfrac{dy}{dx} = (x+1)\left(\dfrac{1}{2}x^{-1/2}\right) + (x^{1/2}+2)(1)$

$\quad = \dfrac{1}{2}x^{1/2} + \dfrac{1}{2}x^{-1/2} + x^{1/2} + 2$

$\quad = \dfrac{3}{2}x^{1/2} + \dfrac{1}{2}x^{-1/2} + 2$

$\quad$ or $\quad \dfrac{3x^{1/2}}{2} + \dfrac{1}{2x^{1/2}} + 2$

9. $p(y) = (y^{-1} + y^{-2})(2y^{-3} - 5y^{-4})$
$p'(y) = (y^{-1} + y^{-2})(-6y^{-4} + 20y^{-5})$
$\quad\quad + (-y^{2} - 2y^{-3})(2y^{-3} - 5y^{-4})$
$\quad = -6y^{-5} + 20y^{-6} - 6y^{-6} + 20y^{-7}$
$\quad\quad - 2y^{-5} + 5y^{-6} - 4y^{-6} + 10y^{-7}$
$\quad = -8y^{-5} + 15y^{-6} + 30y^{-7}$

11. $f(x) = \dfrac{6x+1}{3x+10}$

$f'(x) = \dfrac{(3x+10)(6) - (6x+1)(3)}{(3x+10)^2}$

$\quad = \dfrac{18x + 60 - 18x - 3}{(3x+10)^2}$

$\quad = \dfrac{57}{(3x+10)^2}$

13. $y = \dfrac{5-3t}{4+t}$

$\dfrac{dy}{dx} = \dfrac{(4+t)(-3) - (5-3t)(1)}{(4+t)^2}$

$\quad = \dfrac{-12 - 3t - 5 + 3t}{(4+t)^2}$

$\quad = \dfrac{-17}{(4+t)^2}$

15. $y = \dfrac{x^2+x}{x-1}$

$\dfrac{dy}{dx} = \dfrac{(x-1)(2x+1) - (x^2+x)(1)}{(x-1)^2}$

$\quad = \dfrac{2x^2 + x - 2x - 1 - x^2 - x}{(x-1)^2}$

$\quad = \dfrac{x^2 - 2x - 1}{(x-1)^2}$

17. $f(t) = \dfrac{4t^2+11}{t^2+3}$

$f'(t) = \dfrac{(t^2+3)(8t) - (4t^2+11)(2t)}{(t^2+3)^2}$

$\quad = \dfrac{8t^3 + 24t - 8t^3 - 22t}{(t^2+3)^2}$

$\quad = \dfrac{2t}{(t^2+3)^2}$

19. $g(x) = \dfrac{x^2 - 4x + 2}{x^2+3}$

$g'(x) = \dfrac{(x^2+3)(2x-4) - (x^2-4x+2)(2x)}{(x^2+3)^2}$

$\quad = \dfrac{2x^3 - 4x^2 + 6x - 12 - 2x^3 + 8x^2 - 4x}{(x^2+3)^2}$

$\quad = \dfrac{4x^2 + 2x - 12}{(x^2+3)^2}$

21. $p(t) = \dfrac{\sqrt{t}}{t-1}$

$\quad = \dfrac{t^{1/2}}{t-1}$

$p'(t) = \dfrac{(t-1)\left(\frac{1}{2}t^{-1/2}\right) - t^{1/2}(1)}{(t-1)^2}$

$\quad = \dfrac{\frac{1}{2}t^{1/2} - \frac{1}{2}t^{-1/2} - t^{1/2}}{(t-1)^2}$

$\quad = \dfrac{-\frac{1}{2}t^{1/2} - \frac{1}{2}t^{-1/2}}{(t-1)^2}$

$\quad = \dfrac{-\frac{\sqrt{t}}{2} - \frac{1}{2\sqrt{t}}}{(t-1)^2}$ $\quad$ or $\quad \dfrac{-t-1}{2\sqrt{t}(t-1)^2}$

23. $y = \dfrac{5x+6}{\sqrt{x}} = \dfrac{5x+6}{x^{1/2}}$

$\dfrac{dy}{dx} = \dfrac{(x^{1/2})(5) - (5x+6)\left(\frac{1}{2}x^{-1/2}\right)}{(x^{1/2})^2}$

$\quad = \dfrac{5x^{1/2} - \frac{5}{2}x^{1/2} - 3x^{-1/2}}{x}$

$\quad = \dfrac{\frac{5}{2}x^{1/2} - 3x^{-1/2}}{x}$

$\quad = \dfrac{\frac{5\sqrt{x}}{2} - \frac{3}{\sqrt{x}}}{x}$ $\quad$ or $\quad \dfrac{5x-6}{2x\sqrt{x}}$

25. $g(y) = \dfrac{y^{1.4}+1}{y^{2.5}+2}$

$g'(y) = \dfrac{(y^{2.5}+2)(1.4y^{0.4}) - (y^{1.4}+1)(2.5y^{1.5})}{(y^{2.5}+2)^2} = \dfrac{1.4y^{2.9}+2.8y^{0.4} - 2.5y^{2.9} - 2.5y^{1.5}}{(y^{2.5}+2)^2}$

$\quad = \dfrac{-1.1y^{2.9} - 2.5y^{1.5} + 2.8y^{0.4}}{(y^{2.5}+2)^2}$

27. $g(x) = \dfrac{(2x^2+3)(5x+2)}{6x-7}$

$g'(x) = \dfrac{(6x-7)[(2x^2+3)(5) + (4x)(5x+2)] - (2x^2+3)(5x+2)(6)}{(6x-7)^2}$

$\quad = \dfrac{(6x-7)(30x^2+8x+15) - (2x^2+3)(30x+12)}{(6x-7)^2}$

$\quad = \dfrac{180x^3 + 48x^2 + 90x - 210x^2 - 56x - 105 - 60x^3 - 24x^2 - 90x - 36}{(6x-7)^2}$

$\quad = \dfrac{120x^3 - 186x^2 - 56x - 141}{(6x-7)^2}$

29. $h(x) = \dfrac{f(x)}{g(x)}$

$h'(x) = \dfrac{g(x)f'(x) - f(x)g'(x)}{[g(x)]^2}$

$h'(3) = \dfrac{g(3)f'(3) - f(3)g'(3)}{[g(3)]^2} = \dfrac{4(8) - 9(5)}{4^2} = -\dfrac{13}{16}$

31. In the first step, the denominator, $(x^3)^2 = x^6$, was omitted. The correct work follows.

$D_x\left(\dfrac{x^2-4}{x^3}\right) = \dfrac{x^3(2x) - (x^2-4)(3x^2)}{(x^3)^2} = \dfrac{2x^4 - 3x^4 + 12x^2}{x^6} = \dfrac{-x^4 + 12x^2}{x^6} = \dfrac{x^2(-x^2+12)}{x^2(x^4)} = \dfrac{-x^2+12}{x^4}$

33. **(a)** $f(x) = \dfrac{3x^3+6}{x^{2/3}}$

$f'(x) = \dfrac{(x^{2/3})(9x^2) - (3x^3+6)(\frac{2}{3}x^{-1/3})}{(x^{2/3})^2} = \dfrac{9x^{8/3} - 2x^{8/3} - 4x^{-1/3}}{x^{4/3}} = \dfrac{7x^{8/3} - \frac{4}{x^{1/3}}}{x^{4/3}} = \dfrac{7x^3 - 4}{x^{5/3}}$

(b) $f(x) = 3x^{7/3} + 6x^{-2/3}$

$f'(x) = 3\left(\dfrac{7}{3}x^{4/3}\right) + 6\left(-\dfrac{2}{3}x^{-5/3}\right) = 7x^{4/3} - 4x^{-5/3}$

(c) The derivatives are equivalent.

35. $f(x) = \dfrac{u(x)}{v(x)}$

$f'(x) = \lim_{h \to 0} \dfrac{f(x+h) - f(x)}{h} = \lim_{h \to 0} \dfrac{\frac{u(x+h)}{v(x+h)} - \frac{u(x)}{v(x)}}{h} = \lim_{h \to 0} \dfrac{u(x+h)v(x) - u(x)v(x+h)}{hv(x+h)v(x)}$

$\quad = \lim_{h \to 0} \dfrac{u(x+h)v(x) - u(x)v(x) + u(x)v(x) - u(x)v(x+h)}{hv(x+h)v(x)}$

$\quad = \lim_{h \to 0} \dfrac{v(x)[u(x+h) - u(x)] - u(x)[v(x+h) - v(x)]}{hv(x+h)v(x)}$

$\quad = \lim_{h \to 0} \dfrac{v(x)\frac{u(x+h)-u(x)}{h} - u(x)\frac{v(x+h)-v(x)}{h}}{v(x+h)v(x)} = \dfrac{v(x) \cdot u'(x) - u(x)v'(x)}{[v(x)]^2}$

37. Graph the numerical derivative of $f(x) = (x^2 - 2)(x^2 - \sqrt{2})$ for x ranging from -2 to 2. The derivative crosses the x-axis at 0 and at approximately -1.307 and 1.307.

39. $C(x) = \dfrac{3x + 2}{x + 4}$

$\overline{C}(x) = \dfrac{C(x)}{x} = \dfrac{3x + 2}{x^2 + 4x}$

(a) $\overline{C}(10) = \dfrac{3(10) + 2}{10^2 + 4(10)} = \dfrac{32}{140} \approx 0.2286$ hundreds of dollars or $22.86 per unit

(b) $\overline{C}(20) = \dfrac{3(20) + 2}{(20)^2 + 4(20)} = \dfrac{62}{480} \approx 0.1292$ hundreds of dollars or $12.92 per unit

(c) $\overline{C}(x) = \dfrac{3x + 2}{x^2 + 4x}$ per unit

(d) $\overline{C}'(x) = \dfrac{(x^2 + 4x)(3) - (3x + 2)(2x + 4)}{(x^2 + 4x)^2} = \dfrac{3x^2 + 12x - 6x^2 - 12x - 4x - 8}{(x^2 + 4x)^2} = \dfrac{-3x^2 - 4x - 8}{(x^2 + 4x)^2}$

41. $M(d) = \dfrac{100d^2}{3d^2 + 10}$

(a) $M'(d) = \dfrac{(3d^2 + 10)(200d) - (100d^2)(6d)}{(3d^2 + 10)^2} = \dfrac{600d^3 + 2000d - 600d^3}{(3d^2 + 10)^2} = \dfrac{2000d}{(3d^2 + 10)^2}$

(b) $M'(2) = \dfrac{2000(2)}{[3(2)^2 + 10]^2} = \dfrac{4000}{484} \approx 8.3$

This means the new employee can assemble about 8.3 additional bicycles per day after 2 days of training.

$$M'(5) = \dfrac{2000(5)}{[3(5)^2 + 10]^2} = \dfrac{10,000}{7225} \approx 1.4$$

This means the new employee can assemble about 1.4 additional bicycles per day after 5 days of training.

43. $\overline{C}(x) = \dfrac{C(x)}{x}$

Let $u(x) = C(x)$, with $u'(x) = C'(x)$.
Let $v(x) = x$ with $v'(x) = 1$. Then, by the quotient rule,

$$\overline{C}(x) = \dfrac{v(x) \cdot u'(x) - u(x) \cdot v'(x)}{[v(x)]^2} = \dfrac{x \cdot C'(x) - C(x) \cdot 1}{x^2} = \dfrac{xC'(x) - C(x)}{x^2}$$

45. Let $C(t)$ be the cost as a function of time and $q(t)$ be the quantity as a function of time. Then $\overline{C}(t) = \frac{C(t)}{q(t)}$ is the revenue as a function of time. Let $t = t_1$ represent last month.

$$\overline{C}'(t) = \dfrac{q(t)C'(t) - C(t)q'(t)}{[g(t)]^2}$$

$$\overline{C}'(t_1) = \dfrac{q(t_1)C'(t_1) - C(t_1)q'(t_1)}{[g(t_1)]^2} = \dfrac{(12,500)(1200) - (27,000)(350)}{(12,500)^2} = 0.03552$$

The average cost is increasing at a rate of $0.03552 per gallon per month.

47. $f(x) = \dfrac{Kx}{A+x}$

(a) $f'(x) = \dfrac{(A+x)K - Kx(1)}{(A+x)^2}$

  $f'(x) = \dfrac{AK}{(A+x)^2}$

(b) $f'(A) = \dfrac{AK}{(A+A)^2} = \dfrac{AK}{4A^2} = \dfrac{K}{4A}$

49. $R(w) = \dfrac{30(w-4)}{w-1.5}$

(a) $R(5) = \dfrac{30(5-4)}{5-1.5} \approx 8.57$ min

(b) $R(7) = \dfrac{30(7-4)}{7-1.5} \approx 16.36$ min

(c) $R'(w) = \dfrac{(w-1.5)(30) - 30(w-4)(1)}{(w-1.5)^2}$

  $= \dfrac{30w - 45 - 30w + 120}{(w-1.5)^2}$

  $= \dfrac{75}{(w-1.5)^2}$

  $R'(5) = \dfrac{75}{(5-1.5)^2} \approx 6.12\dfrac{\text{min}^2}{\text{kcal}}$

  $R'(7) = \dfrac{75}{(7-1.5)^2} \approx 2.48\dfrac{\text{min}^2}{\text{kcal}}$

51. $f(t) = \dfrac{90t}{99t - 90}$

  $f'(t) = \dfrac{(99t-90)(90) - (90t)(99)}{(99t-90)^2} = \dfrac{-8100}{(99t-90)^2}$

(a) $f'(1) = \dfrac{-8100}{(99-90)^2} = \dfrac{-8100}{9^2} = \dfrac{-8100}{81} = -100$

(b) $f'(10) = \dfrac{-8100}{[99(10)-90]^2}$

  $= \dfrac{-8100}{(900)^2}$

  $= \dfrac{-8100}{810,000}$

  $= -\dfrac{1}{100}$ or -0.01

12.3   The Chain Rule

In Exercises 1-5, $f(x) = 5x^2 - 2x$ and $g(x) = 8x + 3$.

1. $g(2) = 8(2) + 3 = 19$
  $f[g(2)] = f[19]$
    $= 5(19)^2 - 2(19)$
    $= 1805 - 38 = 1767$

3. $f(2) = 5(2)^2 - 2(2)$
    $= 20 - 4 = 16$
  $g[f(2)] = g[16]$
    $= 8(16) + 3$
    $= 128 + 3 = 131$

5. $g(k) = 8k + 3$
  $f[g(k)] = f[8k + 3]$
    $= 5(8k + 3)^2 - 2(8k + 3)$
    $= 5(64k^2 + 48k + 9) - 16k - 6$
    $= 320k^2 + 224k + 39$

7. $f(x) = \dfrac{x}{8} + 7;\ g(x) = 6x - 1$

  $f[g(x)] = \dfrac{6x-1}{8} + 7$

    $= \dfrac{6x-1}{8} + \dfrac{56}{8}$

    $= \dfrac{6x+55}{8}$

  $g[f(x)] = 6\left[\dfrac{x}{8} + 7\right] - 1$

    $= \dfrac{6x}{8} + 42 - 1$

    $= \dfrac{3x}{4} + 41$

    $= \dfrac{3x}{4} + \dfrac{164}{4}$

    $= \dfrac{3x + 164}{4}$

9. $f(x) = \dfrac{1}{x};\ g(x) = x^2$

  $f[g(x)] = \dfrac{1}{x^2}$

  $g[f(x)] = \left(\dfrac{1}{x}\right)^2$

    $= \dfrac{1}{x^2}$

11. $f(x) = \sqrt{x+2};\ g(x) = 8x^2 - 6$

  $f[g(x)] = \sqrt{(8x^2-6)+2}$
    $= \sqrt{8x^2 - 4}$
  $g[f(x)] = 8(\sqrt{x+2})^2 - 6$
    $= 8x + 16 - 6$
    $= 8x + 10$

13. $f(x) = \sqrt{x + 1}$; $g(x) \doteq \dfrac{-1}{x}$

$$f[g(x)] = \sqrt{\dfrac{-1}{x} + 1}$$

$$= \sqrt{\dfrac{x - 1}{x}}$$

$$g[f(x)] = \dfrac{-1}{\sqrt{x + 1}}$$

17. $y = (5 - x^2)^{3/5}$

If $f(x) = x^{3/5}$ and $g(x) = 5 - x^2$, then

$$y = f[g(x)] = (5 - x^2)^{3/5}.$$

19. $y = -\sqrt{13 + 7x}$

If $f(x) = -\sqrt{x}$ and
$g(x) = 13 + 7x$,

then $y = f[g(x)] = -\sqrt{13 + 7x}$.

21. $y = (x^2 + 5x)^{1/3} - 2(x^2 + 5x)^{2/3} + 7$

If $f(x) = x^{1/3} - 2x^{2/3} + 7$ and
$g(x) = x^2 + 5x$,

then

$$y = f[g(x)] = (x^2 + 5x)^{1/3}$$
$$- 2(x^2 + 5x)^{2/3} + 7.$$

23. $y = (8x^4 - 5x^2 + 1)^4$

Let $f(x) = x^4$ and $g(x) = 8x^4 - 5x^2 + 1$. Then

$$(8x^4 - 5x^2 + 1)^4 = f[g(x)].$$

Use the alternate form of the chain rule.

$$\dfrac{dy}{dx} = f'[g(x)] \cdot g'(x)$$

$$f'(x) = 4x^3$$

$$f'[g(x)] = 4[g(x)]^3$$
$$= 4(8x^4 - 5x^2 + 1)^3$$
$$g'(x) = 32x^3 - 10x$$

$$\dfrac{dy}{dx} = 4(8x^4 - 5x^2 + 1)^3(32x^3 - 10x)$$

25. $f(x) = -2(12x^2 + 5)^{-6}$

Use the generalized power rule with
$u = 12x^2 + 5, n = -6$, and $u' = 24x$.

$$f'(x) = -2[-6(12x^2 + 5)^{-6-1} \cdot 24x]$$
$$= -2[-144x(12x^2 + 5)^{-7}]$$
$$= 288x(12x^2 + 5)^{-7}$$

27. $s(t) = 45(3t^3 - 8)^{3/2}$

Use the generalized power rule with
$u = 3t^3 - 8$, $n = \frac{3}{2}$, and $u' = 9t^2$.

$$s'(t) = 45\left[\dfrac{3}{2}(3t^3 - 8)^{1/2} \cdot 9t^2\right]$$

$$= 45\left[\dfrac{27}{2}t^2(3t^3 - 8)^{1/2}\right]$$

$$= \dfrac{1215}{2}t^2(3t^3 - 8)^{1/2}$$

29. $g(t) = -3\sqrt{7t^3 - 1}$
$\quad\quad = -3(7t^3 - 1)^{1/2}$

Use generalized power rule with
$u = 7t^3 - 1$, $n = \frac{1}{2}$, and $u' = 21t^2$.

$$g'(t) = -3\left[\dfrac{1}{2}(7t^3 - 1)^{-1/2} \cdot 21t^2\right]$$

$$= -3\left[\dfrac{21}{2}t^2(7t^3 - 1)^{-1/2}\right]$$

$$= \dfrac{-63}{2}t^2 \cdot \dfrac{1}{(7t^3 - 1)^{1/2}}$$

$$= \dfrac{-63t^2}{2\sqrt{7t^3 - 1}}$$

31. $m(t) = -6t(5t^4 - 1)^4$

Use the product rule and the power rule.

$$m'(t) = -6t[4(5t^4 - 1)^3 \cdot 20t^3] + (5t^4 - 1)^4(-6)$$
$$= -480t^4(5t^4 - 1)^3 - 6(5t^4 - 1)^4$$
$$= -6(5t^4 - 1)^3[80t^4 + (5t^4 - 1)]$$
$$= -6(5t^4 - 1)^3(85t^4 - 1)$$

33. $y = (3x^4 + 1)^4(x^3 + 4)$

Use the product rule and the power rule.

$$\dfrac{dy}{dx} = (3x^4 + 1)^4(3x^2) + (x^3 + 4)[4(3x^4 + 1)^3$$
$$\cdot 12x^3]$$
$$= 3x^2(3x^4 + 1)^4 + 48x^3(x^3 + 4)(3x^4 + 1)^3$$
$$= 3x^2(3x^4 + 1)^3[3x^4 + 1 + 16x(x^3 + 4)]$$
$$= 3x^2(3x^4 + 1)^3(3x^4 + 1 + 16x^4 + 64x)$$
$$= 3x^2(3x^4 + 1)^3(19x^4 + 64x + 1)$$

35. $q(y) - 4y^2(y^2 + 1)^{5/4}$

Use the product rule and the power rule.

$$q'(y) = 4y^2 \cdot \dfrac{5}{4}(y^2 + 1)^{1/4}(2y) + 8y(y^2 + 1)^{5/4}$$

$$= 10y^3(y^2 + 1)^{1/4} + 8y(y^2 + 1)^{5/4}$$
$$= 2y(y^2 + 1)^{1/4}[5y^2 + 4(y^2 + 1)^{4/4}]$$
$$= 2y(y^2 + 1)^{1/4}(9y^2 + 4)$$

37. $y = \dfrac{-5}{(2x^3 + 1)^2} = -5(2x^3 + 1)^{-2}$

$$\frac{dy}{dx} = -5[-2(2x^3 + 1)^{-3} \cdot 6x^2]$$
$$= -5[-12x^2(2x^3 + 1)^{-3}]$$
$$= 60x^2(2x^3 + 1)^{-3}$$

$$= \frac{60x^2}{(2x^3 + 1)^3}$$

39. $r(t) = \dfrac{(5t - 6)^4}{3t^2 + 4}$

$r'(t)$
$$= \frac{(3t^2 + 4)[4(5t - 6)^3 \cdot 5] - (5t - 6)^4(6t)}{(3t^2 + 4)^2}$$

$$= \frac{20(3t^2 + 4)(5t - 6)^3 - 6t(5t - 6)^4}{(3t^2 + 4)^2}$$

$$= \frac{2(5t - 6)^3[10(3t^2 + 4) - 3t(5t - 6)]}{(3t^2 + 4)^2}$$

$$= \frac{2(5t - 6)^3(30t^2 + 40 - 15t^2 + 18t)}{(3t^2 + 4)^2}$$

$$= \frac{2(5t - 6)^3(15t^2 + 18t + 40)}{(3t^2 + 4)^2}$$

41. $y = \dfrac{3x^2 - x}{(2x - 1)^5}$

$$\frac{dy}{dx} = \frac{(2x - 1)^5(6x - 1) - (3x^2 - x)[5(2x - 1)^4 \cdot 2]}{[(2x - 1)^5]^2}$$

$$= \frac{(2x - 1)^5(6x - 1) - 10(3x^2 - x)(2x - 1)^4}{(2x - 1)^{10}}$$

$$= \frac{(2x - 1)^4[(2x - 1)(6x - 1) - 10(3x^2 - x)]}{(2x - 1)^{10}}$$

$$= \frac{12x^2 - 2x - 6x + 1 - 30x^2 + 10x}{(2x - 1)^6}$$

$$= \frac{-18x^2 + 2x + 1}{(2x - 1)^6}$$

43. (a) $D_x(f[g(x)])$ at $x = 1$

$$= f'[g(1)] \cdot g'(1)$$

$$= f'(2) \cdot \left(\frac{2}{7}\right)$$

$$= -7\left(\frac{2}{7}\right)$$

$$= -2$$

(b) $D_x(f[g(x)])$ at $x = 2$

$$= f'[g(2)] \cdot g'(2)$$

$$= f'(3) \cdot \left(\frac{3}{7}\right)$$

$$= -8\left(\frac{3}{7}\right)$$

$$= -\frac{24}{7}$$

45. $f(x) = \sqrt{x^2 + 16}; x = 3$
$f(x) = (x^2 + 16)^{1/2}$

$$f'(x) = \frac{1}{2}(x^2 + 16)^{-1/2}(2x)$$

$$f'(x) = \frac{x}{\sqrt{x^2 + 16}}$$

$$f'(3) = \frac{3}{\sqrt{3^2 + 16}} = \frac{3}{5}$$

$$f(3) = \sqrt{3^2 + 16} = 5$$

We use $m = \frac{3}{5}$ and the point $P(3, 5)$ in the point-slope form.

$$y - 5 = \frac{3}{5}(x - 3)$$
$$y - 5 = \frac{3}{5}x - \frac{9}{5}$$
$$y = \frac{3}{5}x + \frac{16}{5}$$

47. $f(x) = x(x^2 - 4x + 5)^4; x = 2$
$f'(x) = x \cdot 4(x^2 - 4x + 5)^3 \cdot (2x - 4)$
$\qquad + 1 \cdot (x^2 - 4x + 5)^4$
$\qquad = (x^2 - 4x + 5)^3$
$\qquad \cdot [4x(2x - 4) + (x^2 - 4x + 5)]$
$\qquad = (x^2 - 4x + 5)^3(9x^2 - 20x + 5)$
$f'(2) = (1)^3(1) = 1$
$f(2) = 2(1)^4 = 2$

We use $m = 1$ and the point $P(2, 2)$.

$$y - 2 = 1(x - 2)$$
$$y - 2 = x - 2$$
$$y = x$$

49. $f(x) = \sqrt{x^3 - 6x^2 + 9x + 1}$

$f(x) = (x^3 - 6x^2 + 9x + 1)^{1/2}$

$f'(x) = \dfrac{1}{2}(x^3 - 6x^2 + 9x + 1)^{-1/2}$
$\cdot (3x^2 - 12x + 9)$

$f'(x) = \dfrac{3(x^2 - 4x + 3)}{2\sqrt{x^3 - 6x^2 + 9x + 1}}$

If the tangent line is horizontal, its slope is zero and $f'(x) = 0$.

$$\dfrac{3(x^2 - 4x + 3)}{2\sqrt{x^3 - 6x^2 + 9x + 1}} = 0$$

$$3(x^2 - 4x + 3) = 0$$
$$3(x - 1)(x - 3) = 0$$
$$x = 1 \quad \text{or} \quad x = 3$$

The tangent line is horizontal at $x = 1$ and $x = 3$.

53. $D(p) = \dfrac{-p^2}{100} + 500; \quad p(c) = 2c - 10$

The demand in terms of the cost is

$D(c) = D[p(c)]$

$= \dfrac{-(2c - 10)^2}{100} + 500$

$= \dfrac{-4(c - 5)^2}{100} + 500$

$= \dfrac{-c^2 + 10c - 25}{25} + 500$

$= \dfrac{-c^2 + 10c - 25 + 12{,}500}{25}$

$= \dfrac{-c^2 + 10c + 12{,}475}{25}.$

55. $A = 1500\left(1 + \dfrac{r}{36{,}500}\right)^{1825}$

$\frac{dA}{dr}$ is the rate of change of A with respect to r.

$\dfrac{dA}{dr} = 1500(1825)\left(1 + \dfrac{r}{36{,}500}\right)^{1824}\left(\dfrac{1}{36{,}500}\right)$

$= 75\left(1 + \dfrac{r}{36{,}500}\right)^{1824}$

(a) For $r = 6\%$,

$\dfrac{dA}{dr} = 75\left(1 + \dfrac{6}{36{,}500}\right)^{1824} = \$101.22.$

(b) For $r = 8\%$,

$\dfrac{dA}{dr} = 75\left(1 + \dfrac{8}{36{,}500}\right)^{1824} = \$111.86.$

(c) For $r = 9\%$,

$\dfrac{dA}{dr} = 75\left(1 + \dfrac{9}{36{,}500}\right)^{1824} = \$117.59.$

57. $V = \dfrac{60{,}000}{1 + 0.3t + 0.1t^2}$

The rate of change of the value is

$V'(t)$

$= \dfrac{(1 + 0.3t + 0.1t^2)(0) - 60{,}000(0.3 + 0.2t)}{(1 + 0.3t + 0.1t^2)^2}$

$= \dfrac{-60{,}000(0.3 + 0.2t)}{(1 + 0.3t + 0.1t^2)^2}.$

(a) 2 years after purchase, the rate of change in the value is

$V'(2) = \dfrac{-60{,}000[0.3 + 0.2(2)]}{[1 + 0.3(2) + 0.1(2)^2]^2}$

$= \dfrac{-60{,}000(0.3 + 0.4)}{(1 + 0.6 + 0.4)^2}$

$= \dfrac{-42{,}000}{4}$

$= -\$10{,}500.$

(b) 4 years after purchase, the rate of change in the value is

$V'(4) = \dfrac{-60{,}000[0.3 + 0.2(4)]}{[1 + 0.3(4) + 0.1(4)^2]^2}$

$= \dfrac{-66{,}000}{14.44}$

$= -\$4570.64.$

59. $P(x) = 2x^2 + 1; \quad x = f(a) = 3a + 2$

$P[f(a)] = 2(3a + 2)^2 + 1$
$= 2(9a^2 + 12a + 4) + 1$
$= 18a^2 + 24a + 9$

61. (a) $r(t) = 2t; \quad A(r) = \pi r^2$

$A[r(t)] = \pi(2t)^2$
$= 4\pi t^2$

$A = 4\pi t^2$ gives the area of the pollution in terms of the time since the pollutants were first emitted.

(b) $D_t A[r(t)] = 8\pi t$
$D_t A[r(4)] = 8\pi(4) = 32\pi$

At 12 P.M., the area of pollution is changing at the rate of 32π mi^2/hr.

63. $C(t) = \frac{1}{2}(2t + 1)^{-1/2}$

$$C'(t) = \frac{1}{2}\left(-\frac{1}{2}\right)(2t + 1)^{-3/2}(2)$$

$$= -\frac{1}{2}(2t + 1)^{-3/2}$$

(a) $C'(0) = -\frac{1}{2}[2(0) + 1]^{-3/2}$

$$= -\frac{1}{2}$$

$$= -0.5$$

(b) $C'(4) = -\frac{1}{2}[2(4) + 1]^{-3/2}$

$$= -\frac{1}{2}(9)^{-3/2}$$

$$= \frac{-1}{2} \cdot \frac{1}{(\sqrt{9})^3}$$

$$= -\frac{1}{54}$$

$$\approx -0.02$$

(c) $C'(7.5) = -\frac{1}{2}[2(7.5) + 1]^{-3/2}$

$$= -\frac{1}{2}(16)^{-3/2}$$

$$= -\frac{1}{2}\left(\frac{1}{(\sqrt{16})^3}\right)$$

$$= -\frac{1}{128}$$

$$\approx -0.008$$

(d) C is always decreasing because

$$C' = -\frac{1}{2}(2t + 1)^{-3/2}$$

is always negative for $t \geq 0$.
(The amount of calcium in the bloodstream will continue to decrease over time.)

65. $V(r) = \frac{4}{3}\pi r^3, S(r) = 4\pi r^2, r(t) = 6 - \frac{3}{17}t$

(a) $r(t) = 0$ when $6 - \frac{3}{17}t = 0$;

$$t = \frac{17(6)}{3} = 34 \text{ min.}$$

(b) $\frac{dV}{dr} = 4\pi r^2, \frac{dS}{dr} = 8\pi r, \frac{dr}{dt} = -\frac{3}{17}$

$$\frac{dV}{dt} = \frac{dV}{dr} \cdot \frac{dr}{dt} = -\frac{12}{17}\pi r^2$$

$$= -\frac{12}{17}\pi\left(6 - \frac{3}{17}t\right)^2$$

$$\frac{dS}{dt} = \frac{dS}{dr} \cdot \frac{dr}{dt} = -\frac{24}{17}\pi r$$

$$= -\frac{24}{17}\pi\left(6 - \frac{3}{17}t\right)$$

When $t = 17$,

$$\frac{dV}{dt} = -\frac{12}{17}\pi\left[6 - \frac{3}{17}(17)\right]^2$$

$$= -\frac{108}{17}\pi \text{ mm}^3/\text{min}$$

$$\frac{dS}{dt} = -\frac{24}{17}\pi\left[6 - \frac{3}{17}(17)\right]$$

$$= -\frac{72}{17}\pi \text{ mm}^2/\text{min}$$

At $t = 17$ minutes, the volume is decreasing by $\frac{108}{17}\pi$ mm^3 per minute and the surface area is decreasing by $\frac{72}{17}\pi$ mm^2 per minute.

12.4   Derivatives of Exponential Functions

1. $\qquad y = e^{4x}$

Let $\quad g(x) = 4x$,
with $\quad g'(x) = 4$.

$$\frac{dy}{dx} = 4e^{4x}$$

3. $\quad y = -8e^{3x}$

$$\frac{dy}{dx} = -8(3e^{3x}) = -24e^{3x}$$

5. $\qquad y = -16e^{2x+1}$
$\qquad g(x) = 2x + 1$
$\qquad g'(x) = 2$

$$\frac{dy}{dx} = -16(2e^{2x+1}) = -32e^{2x+1}$$

7. $\qquad y = e^{x^2}$
$\qquad g(x) = x^2$
$\qquad g'(x) = 2x$

$$\frac{dy}{dx} = 2xe^{x^2}$$

9. $y = 3e^{2x^2}$

$g(x) = 2x^2$

$g'(x) = 4x$

$\dfrac{dy}{dx} = 3(4xe^{2x^2})$

$= 12xe^{2x^2}$

11. $y = 4e^{2x^2-4}$

$g(x) = 2x^2 - 4$

$g'(x) = 4x$

$\dfrac{dy}{dx} = 4[(4x)e^{2x^2-4}]$

$= 16xe^{2x^2-4}$

13. $y = xe^x$

Use the product rule.

$\dfrac{dy}{dx} = xe^x + e^x \cdot 1$

$= e^x(x+1)$

15. $y = (x+3)^2 e^{4x}$

Use the product rule.

$\dfrac{dy}{dx} = (x+3)^2(4)e^{4x} + e^{4x} \cdot 2(x+3)$

$= 4(x+3)^2 e^{4x} + 2(x+3)e^{4x}$

$= 2(x+3)e^{4x}[2(x+3)+1]$

$= 2(x+3)(2x+7)e^{4x}$

17. $y = \dfrac{x^2}{e^x}$

Use the quotient rule.

$\dfrac{dy}{dx} = \dfrac{e^x(2x) - x^2 e^x}{(e^x)^2}$

$= \dfrac{xe^x(2-x)}{e^{2x}}$

$= \dfrac{x(2-x)}{e^x}$

19. $y = \dfrac{e^x + e^{-x}}{x}$

$\dfrac{dy}{dx} = \dfrac{x(e^x - e^{-x}) - (e^x + e^{-x})}{x^2}$

21. $p = \dfrac{10,000}{9 + 4e^{-0.2t}}$

$\dfrac{dp}{dt} = \dfrac{(9 + 4e^{-0.2t}) \cdot 0 - 10,000[0 + 4(-0.2)e^{-0.2t}]}{(9 + 4e^{-0.2t})^2}$

$= \dfrac{8000e^{-0.2t}}{(9 + 4e^{-0.2t})^2}$

23. $f(z) = (2z + e^{-z^2})^2$

$f'(z) = 2(2z + e^{-z^2})^1(2 - 2ze^{-z^2})$

$= 4(2z + e^{-z^2})(1 - ze^{-z^2})$

25. $y = 4^{-5x+2}$

Let $g(x) = -5x + 2$, with $g'(x) = -5$. Then

$\dfrac{dy}{dx} = (\ln 4)(4^{-5x+2}) \cdot (-5)$

$= -5(\ln 4)4^{-5x+2}$

27. $y = -10^{3x^2-4}$

Let $g(x) = 3x^2 - 4$, with $g'(x) = 6x$.

$\dfrac{dy}{dx} = -(\ln 10)10^{3x^2-4} \cdot 6x$

$= -6x(10^{3x^2-4})\ln 10$

29. $s = 5 \cdot 2^{\sqrt{t-2}}$

Let $g(t) = t - 2$, with $g'(t) = \dfrac{1}{2\sqrt{t-2}}$. Then

$\dfrac{ds}{dt} = 5(\ln 2)(2^{\sqrt{t-2}}) \cdot \dfrac{1}{2\sqrt{t-2}}$

$= \dfrac{(5\ln 2)2^{\sqrt{t-2}}}{2\sqrt{t-2}}$

31. $y = \dfrac{t^2 e^{2t}}{t + e^{3t}}$

Use the quotient rule and product rule.

$\dfrac{dy}{dt} = \dfrac{(t + e^{3t})(2te^{2t} + t^2 \cdot 2e^{2t}) - t^2 e^{2t}(1 + 3e^{3t})}{(t + e^{3t})^2}$

$= \dfrac{(t + e^{3t})(2te^{2t} + 2t^2 e^{2t}) - t^2 e^{2t}(1 + 3e^{3t})}{(t + e^{3t})^2}$

$= \dfrac{(2t^2 e^{2t} + 2t^3 e^{2t} + 2te^{5t} + 2t^2 e^{5t}) - (t^2 e^{2t} + 3t^2 e^{5t})}{(t + e^{3t})^2}$

$= \dfrac{t^2 e^{2t} + 2t^3 e^{2t} + 2te^{5t} - t^2 e^{5t}}{(t + e^{3t})^2}$

$= \dfrac{(2t^3 + t^2)e^{2t} + (2t - t^2)e^{5t}}{(t + e^{3t})^2}$

33. $f(x) = e^{x^2/(x^3+2)}$

Let $g(x) = \dfrac{x^2}{x^3 + 2}$

$g'(x) = \dfrac{(x^3 + 2)(2x) - x^2(3x^2)}{(x^3 + 2)^2}$

$= \dfrac{2x^4 + 4x - 3x^4}{(x^3 + 2)^2}$

$= \dfrac{4x - x^4}{(x^3 + 2)^2}$

$= \dfrac{x(4 - x^3)}{(x^3 + 2)^2}$

$$f'(x) = e^{x^2/(x^3+2)} \cdot \left[\frac{x(4-x^3)}{(x^3+2)^2}\right]$$

$$= \frac{x(4-x^3)e^{x^2/(x^3+2)}}{(x^3+2)^2}$$

35. Graph

$$y = \frac{e^{x+0.0001} - e^x}{0.0001}$$

on a graphing calculator. A good choice for the viewing window is $[-1, 4]$ by $[-1, 16]$ with Xscl = 1, Yscl = 2.

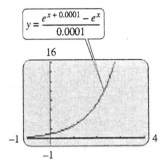

If we graph $y = e^x$ on the same screen, we see that the two graphs coincide. They are close enough to being identical that they are indistinguishable. By the definition of the derivative, if $f(x) = e^x$,

$$f'(x) = \lim_{h \to 0} \frac{f(x+h) - f(x)}{h} = \lim_{h \to 0} \frac{e^{x+h} - e^x}{h},$$

and $h = 0.0001$ is very close to 0. Comparing the two graphs provides graphical evidence that

$$f'(x) = e^x.$$

37. $S(t) = 100 - 90e^{-0.3t}$
$\ S'(t) = -90(-0.3)e^{-0.3t}$
$\ = 27e^{-0.3t}$

(a) $S'(1) = 27e^{-0.3(1)}$
$\ = 27e^{-0.3}$
$\ \approx 20$

(b) $S'(5) = 27e^{-0.3(5)}$
$\ = 27e^{-1.5}$
$\ \approx 6$

(c) As time goes on, the rate of change of sales is decreasing.

(d) $S'(t) = 27e^{-0.3t} \neq 0$, but

$$\lim_{t \to \infty} S'(t) = \lim_{t \to \infty} 27e^{-0.3t} = 0.$$

Although the rate of change of sales never equals zero, it gets closer and closer to zero as t increases.

39. $A(t) = 10t^2\, 2^{-t}$
$\ A'(t) = 10t^2(\ln 2)2^{-t}(-1) + 20t\, 2^{-t}$
$\ A'(t) = 10t\, 2^{-t}(-t \ln 2 + 2)$

(a) $A'(2) = 10(2)(2^{-2})(-2 \ln 2 + 2)$
$\ \approx 3.07$

(b) $A'(4) = 10(4)(2^{-4})(-4 \ln 2 + 2)$
$\ \approx -1.93$

(c) Public awareness increased at first and then decreased.

41. $y = 100e^{-0.03045t}$

(a) For $t = 0$,

$$y = 100e^{-0.03045(0)}$$
$$= 100e^0$$
$$= 100\%.$$

(b) For $t = 2$,

$$y = 100e^{-0.03045(2)}$$
$$= 100e^{-0.0609}$$
$$\approx 94\%.$$

(c) For $t = 4$,

$$y = 100e^{-0.03045(4)}$$
$$\approx 89\%.$$

(d) For $t = 6$,

$$y' = 100e^{-0.03045(6)}$$
$$\approx 83\%.$$

(e) $y' = 100(-0.03045)e^{-0.03045t}$
$\ = -3.045e^{-0.03045t}$

For $t = 0$,

$$y' = -3.045e^{-0.03045(0)}$$
$$= -3.045.$$

(f) For $t = 2$,

$$y' = -3.045e^{-0.03045(2)}$$
$$\approx -2.865.$$

(g) The percent of these cars on the road is decreasing, but at a slower rate as they age.

43. (a) $G_0 = 0.7, m = 10.3$

$$G(t) = \frac{10.3}{1 + \left(\frac{10.3}{0.7} - 1\right)e^{-0.03036(10.3)t}}$$

$$= \frac{10.3}{1 + 13.71e^{-0.3127t}}$$

(b) $G'(t) = -10.3(1 + 13.71e^{-0.3127t})^{-2}$
$$\cdot 13.71e^{-0.3127t}(-0.3127)$$

$$= \frac{44.1573051e^{-0.3127t}}{(1 + 13.71e^{-0.3127t})^2}$$

1990 when $t = 5$:

$$G(5) = \frac{10.3}{1 + 13.71e^{-0.3127(5)}}$$
$$\approx 2.66$$

$$G'(5) = \frac{44.1573051e^{-0.3127(5)}}{[1 + 13.71e^{-0.3127(5)}]^2}$$
$$\approx 0.617$$

The population in 1990 is 2.66 million and the growth rate is 0.617 million per year.

(c) 1995 when $t = 10$:

$$G(10) = \frac{10.3}{1 + 13.71e^{-0.3127(10)}}$$
$$\approx 6.43$$

$$G'(10) = \frac{44.1573051e^{-0.3127(10)}}{[1 + 13.71e^{-0.3127(10)}]^2}$$
$$\approx 0.755$$

The population in 1995 is 6.43 million and the growth rate is 0.755 million per year.

(d) 2000 when $t = 15$:

$$G(15) = \frac{10.3}{1 + 13.71e^{-0.3127(15)}}$$
$$\approx 9.15$$

$$G'(15) = \frac{44.1573051e^{-0.3127(15)}}{[1 + 13.71e^{-0.3127(15)}]^2}$$
$$\approx 0.320$$

The population in 2000 is 9.15 million and the growth rate is 0.320 million per year.

(e) The rate of growth over time increases for a while and then gradually decreases to 0.

45. $p(t) = 9.865(1.025)^t$
$p'(t) = 9.865(\ln 1.025)(1.025)^t$

(a) For 1998, $t = 18$.

$p'(18) = 9.865(\ln 1.025)(1.025)^{18}$
$\quad = 0.380$

The instantaneous rate of growth is 380,000 people per year.

(b) For 2006, $t = 26$.

$p'(26) = 9.865(\ln 1.025)(1.025)^{26}$
$\quad = 0.463$

The instantaneous rate of growth is 463,000 people per year.

47. $G(t) = \dfrac{mG_o}{G_o + (m - G_o)e^{-kmt}}$, where $G_o = 400$;

$m = 5200$; and $k = 0.0001$.

(a) $G(t) = \dfrac{(5200)(400)}{400 + (5200 - 400)e^{(-0.0001)(5200)t}}$

$$= \frac{(400)(5200)}{400 + 4800e^{-0.52t}}$$

$$= \frac{5200}{1 + 12e^{-0.52t}}$$

(b) $G(t) = 5200(1 + 12e^{-0.52t})^{-1}$
$G'(t) = -5200(1 + 12e^{-0.52t})^{-2}(-6.24e^{-0.52t})$

$$= \frac{32,448e^{-0.52t}}{(1 + 12e^{-0.52t})^2}$$

$$G(1) = \frac{5200}{1 + 12e^{-0.52}} \approx 639$$

$$G'(1) = \frac{32,448e^{-0.52}}{(1 + 12e^{-0.52})^2} \approx 292$$

(c) $G(4) = \dfrac{5200}{1 + 12e^{-2.08}} \approx 2081$

$$G'(4) = \frac{32,448e^{-2.08}}{(1 + 12e^{-2.08})^2} \approx 649$$

(d) $G(10) = \dfrac{5200}{1 + 12e^{-5.2}} \approx 4877$

$$G'(10) = \frac{34,448e^{-5.2}}{(1 + 12e^{-5.2})^2} \approx 167$$

(e) It increases for a while and then gradually decreases to 0.

49. $V(t) = 1100[1023e^{-0.02415t} + 1]^{-4}$

(a) $V(240) = 1100[1023e^{-0.02415(240)} + 1]^{-4}$
$\approx 3.857 \text{ cm}^3$

(b) $V = \dfrac{4}{3}\pi r^3$, so $r(V) = \sqrt[3]{\dfrac{3V}{4\pi}}$

$r(3.857) = \sqrt[3]{\dfrac{3(3.857)}{4\pi}} \approx 0.973 \text{ cm}$

(c) $V(t) = 1100[1023e^{-0.02415t} + 1]^{-4} = 0.5$

$[1023e^{-0.02415t} + 1]^{-4} = \dfrac{1}{2200}$

$(1023e^{-0.02415t} + 1)^4 = 2200$

$1023e^{-0.02415t} + 1 = 2200^{1/4}$

$1023e^{-0.02415t} = 2200^{1/4} - 1$

$e^{-0.02415t} = \dfrac{2200^{1/4} - 1}{1023}$

$-0.02415t = \ln\left(\dfrac{2200^{1/4} - 1}{1023}\right)$

$t = \dfrac{1}{-0.02415}\ln\left(\dfrac{2200^{1/4} - 1}{1023}\right) \approx 214 \text{ months}$

The tumor has been growing for almost 18 years.

(d) As t goes to infinity, $e^{-0.02415t}$ goes to zero, and $V(t) = 1100[1023e^{-0.02415t}+1]^{-4}$ goes to 1100 cm³, which corresponds to a sphere with a radius of $\sqrt[3]{\dfrac{3(1100)}{4\pi}} \approx 6.4$ cm. It makes sense that a tumor growing in a person's body reaches a maximum volume of this size.

(e) By the chain rule,

$\dfrac{dV}{dt} = 1100(-4)[1023e^{-0.02415t} + 1]^{-5}$
$\cdot (1023)(e^{-0.02415t})(-0.02415)$
$= 108{,}703.98[1023e^{-0.02415t} + 1]^{-5}e^{-0.02415t}$

At $t = 240$, $\dfrac{dV}{dt} \approx 0.282$.

At 240 months old, the tumor is increasing in volume at the instantaneous rate of 0.282 cm³/month.

51. $URR = 1 - \left\{(0.96)^{0.14t-1} + \dfrac{8t}{126t+900}[1-(0.96)^{0.14t-1}]\right\}$

(a) When $t = 180$, $URR \approx 0.589$. The patient has not received adequate dialysis.

(b) When $t = 240$, $URR \approx 0.690$. The patient has received adequate dialysis.

(c) $D_t URR$
$= -\left\{(\ln 0.96)(0.96)^{0.14t-1}(0.14)\right.$
$+ \dfrac{8t}{126t + 900}(-\ln 0.96)(0.96)^{0.14t-1}(0.14)$
$\left. + \dfrac{(126t + 900)(8) - 8t(126)}{(126t + 900)^2}[1 - (0.96)^{0.14t-1}]\right\}$

When $t = 240$, $D_t URR \approx 0.001$. The URR is increasing instantaneously by 0.001 units per minute when $t = 240$ minutes.

The rate of increase is low, and it will take a significant increase in time on dialysis to increase URR significantly.

53. $M(t) = 3102e^{-e^{-0.022(t-56)}}$

(a) $M(200) = 3102e^{-e^{-0.022(200-56)}} \approx 2974.15$ grams, or about 3 kilograms.

(b) As t gets very large, $-e^{-0.022(t-56)}$ goes to zero, $e^{-e^{-0.022(t-56)}}$ goes to 1, and $M(t)$ approaches 3102 grams or about 3.1 kilograms.

(c) 80% of 3102 is 2481.6.

$2481.6 = 3102e^{-e^{-0.022(t-56)}}$

$-\ln\dfrac{2481.6}{3102} = e^{-0.022(t-56)}$

$\ln\left(\ln\dfrac{3102}{2481.6}\right) = -0.022(t - 56)$

$t = -\dfrac{1}{0.022}\ln\left(\ln\dfrac{3102}{2481.6}\right) + 56$

$\approx 124 \text{ days}$

(d) $D_t M(t) = 3102e^{-e^{-0.022(t-56)}}D_t\left(-e^{-0.022(t-56)}\right)$
$= 3102e^{-e^{-0.022(t-56)}}\left(-e^{-0.022(t-56)}\right)(-0.022)$
$= 68.244e^{-e^{-0.022(t-56)}}e^{-0.022(t-56)}$

When $t = 200$, $D_t M(t) \approx 2.75$ g/day.

(e)

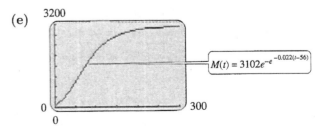

Growth is initially rapid, then tapers off.

(f)

Day	Weight	Rate
50	991	24.88
100	2122	17.73
150	2734	7.60
200	2974	2.75
250	3059	0.94
300	3088	0.32

55. $W_1(t) = 509.7(1 - 0.941e^{-0.00181t})$
$W_2(t) = 498.4(1 - 0.889e^{-0.00219t})^{1.25}$

(a) Both W_1 and W_2 are strictly increasing functions, so they approach their maximum values as t approaches ∞.

$$\lim_{t \to \infty} W_1(t) = \lim_{t \to \infty} 509.7(1 - 0.941e^{-0.00181t})$$
$$= 509.7(1 - 0) = 509.7$$

$$\lim_{t \to \infty} W_2(t) = \lim_{t \to \infty} 498.4(1 - 0.889e^{-0.00219t})^{1.25}$$
$$= 498.4(1 - 0)^{1.25} = 498.4$$

So, the maximum values of W_1 and W_2 are 509.7 kg and 498.4 kg respectively.

(b) $0.9(509.7) = 509.7(1 - 0.941e^{-0.00181t})$
$$0.9 = 1 - 0.941e^{-0.00181t}$$

$$\frac{0.1}{0.941} = e^{-0.00181t}$$

$$1239 \approx t$$

$$0.9(498.4) = 498.4(1 - 0.889e^{-0.00219t})^{1.25}$$
$$0.9 = (1 - 0.889e^{-0.00219t})^{1.25}$$

$$\frac{1 - 0.9^{0.8}}{0.889} = e^{-0.00219t}$$

$$1095 \approx t$$

Respectively, it will take the average beef cow about 1239 days or 1095 days to reach 90% of its maximum.

(c) $W_1'(t) = (509.7)(-0.941)(-0.00181)e^{-0.00181t}$
$$\approx 0.868126e^{-0.00181t}$$
$$W_1'(750) \approx 0.868126e^{-0.00181(750)}$$
$$\approx 0.22 \text{ kg/day}$$
$$W_2'(t) = (498.4)(1.25)(1 - 0.889e^{-0.00219t})^{0.25}$$
$$\cdot (-0.889)(-0.00219)e^{-0.00219t}$$
$$\approx 1.21292e^{-0.00219t}(1 - 0.889e^{-0.00219t})^{0.25}$$
$$W_2'(750) \approx 1.12192e^{-0.00219(750)}$$
$$\cdot (1 - 0.889e^{-0.00219(750)})^{0.25}$$
$$\approx 0.22 \text{ kg/day}$$

Both functions yield a rate of change of about 0.22 kg per day.

(d) Looking at the graph, the growth patterns of the two functions are very similar.

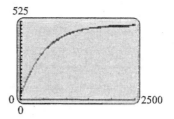

(e) The graphs of the rates of change of the two functions are also very similar.

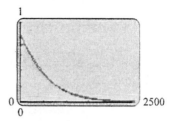

57. (a) $G_0 = 0.00369, m = 1, k = 3.5$

$$G(t) = \frac{1}{1 + \left(\frac{1}{0.00369} - 1\right)e^{-3.5(1)t}}$$

$$= \frac{1}{1 + 270e^{-3.5t}}$$

(b) $G'(t) = -(1 + 270e^{-3.5t})^{-2} \cdot 270e^{-3.5t}(-3.5)$

$$= \frac{945e^{-3.5t}}{(1 + 270e^{-3.5t})^2}$$

$$G(1) = \frac{1}{1 + 270e^{-3.5(1)}}$$

$$\approx 0.109$$

$$G'(1) = \frac{945e^{-3.5(1)}}{[1 + 270e^{-3.5(1)}]^2}$$

$$\approx 0.341$$

The proportion is 0.109 and the rate of growth is 0.341 per century.

(c) $G(2) = \dfrac{1}{1 + 270e^{-3.5(2)}}$

$$\approx 0.802$$

$$G'(2) = \frac{945e^{-3.5(2)}}{[1 + 270e^{-3.5(2)}]^2}$$

$$\approx 0.555$$

The proportion is 0.802 and the rate of growth is 0.555 per century.

(d) $G(3) = \dfrac{1}{1 + 270e^{-3.5(3)}}$

≈ 0.993

$G'(3) = \dfrac{945e^{-3.5(2)}}{[1 + 270e^{-3.5(2)}]^2}$
≈ 0.0256

The proportion is 0.993 and the rate of growth is 0.0256 per century.

(e) The rate of growth increases for a while and then gradually decreases to 0.

59. $P(t) = 37.79(1.012)^t$

(a) $P(10) = 37.79(1.021)^{10} \approx 46.5$

So, the U.S. Latino-American population in 2010 was approximately 46,500,000.

(b) $P'(t) = 37.79(\ln 1.021)(1.021)^t$
$\approx 0.7854(1.021)^t$
$P'(10) = 0.7854(1.021)^{10}$
≈ 0.967

The Latino-American population was increasing at the rate of 0.967 million/year at the end of the year 2010.

61. $Q(t) = CV(1 - e^{-t/RC})$

(a) $I_c = \dfrac{dQ}{dt} = CV\left[0 - e^{-t/RC}\left(-\dfrac{1}{RC}\right)\right]$

$= CV\left(\dfrac{1}{RC}\right)e^{-t/RC}$

$= \dfrac{V}{R}e^{-t/RC}$

(b) When $C = 10^{-5}$ farads, $R = 10^7$ ohms, and $V = 10$ volts, after 200 seconds

$I_c = \frac{10}{10^7}e^{-200/(10^7 \cdot 10^{-5})} \approx 1.35 \times 10^{-7}$ amps

63. $T(h) = 80e^{-0.000065h}$

$\dfrac{dT}{dt} = 80e^{-0.000065h}\left(-0.000065\dfrac{dh}{dt}\right)$

$= -0.0052e^{-0.000065h}\dfrac{dh}{dt}$

If $h = 1000$ and $\dfrac{dh}{dt} = 800$, then

$\dfrac{dT}{dt} = -0.0052e^{-0.000065(1000)}(800)$
≈ -3.90

The temperature is decreasing at 3.90 degrees/hr.

12.5   Derivatives of Logarithmic Functions

1. $y = \ln(8x)$

$\dfrac{dy}{dx} = \dfrac{d}{dx}(\ln 8x)$

$= \dfrac{d}{dx}(\ln 8 + \ln x)$

$= \dfrac{d}{dx}(\ln 8) + \dfrac{d}{dx}(\ln x)$

$= 0 + \dfrac{1}{x}$

$= \dfrac{1}{x}$

3. $y = \ln(8 - 3x)$
$g(x) = 8 - 3x$
$g'(x) = -3$

$\dfrac{dy}{dx} = \dfrac{g'(x)}{g(x)} = \dfrac{-3}{8 - 3x}$ or $\dfrac{3}{3x - 8}$

5. $y = \ln\left|4x^2 - 9x\right|$
$g(x) = 4x^2 - 9x$
$g'(x) = 8x - 9$

$\dfrac{dy}{dx} = \dfrac{g'(x)}{g(x)} = \dfrac{8x - 9}{4x^2 - 9x}$

7. $y = \ln\sqrt{x + 5}$
$g(x) = \sqrt{x + 5}$
$= (x + 5)^{1/2}$

$g'(x) = \dfrac{1}{2}(x + 5)^{-1/2}$

$\dfrac{dy}{dx} = \dfrac{\frac{1}{2}(x + 5)^{-1/2}}{(x + 5)^{1/2}}$

$= \dfrac{1}{2(x + 5)}$

9. $y = \ln (x^4 + 5x^2)^{3/2}$

$\quad = \dfrac{3}{2} \ln (x^4 + 5x^2)$

$\dfrac{dy}{dx} = \dfrac{3}{2} D_x \left[\ln (x^4 + 5x^2) \right]$

$g(x) = x^4 + 5x^2$

$g'(x) = 4x^3 + 10x$

$\dfrac{dy}{dx} = \dfrac{3}{2} \left(\dfrac{4x^3 + 10x}{x^4 + 5x^2} \right)$

$\quad = \dfrac{3}{2} \left[\dfrac{2x(2x^2 + 5)}{x^2(x^2 + 5)} \right]$

$\quad = \dfrac{3(2x^2 + 5)}{x(x^2 + 5)}$

11. $y = -5x \ln(3x + 2)$

Use the product rule.

$\dfrac{dy}{dx} = -5x \left[\dfrac{d}{dx} \ln(3x + 2) \right]$

$\quad\quad + \ln(3x + 2) \left[\dfrac{d}{dx} (-5x) \right]$

$\quad = -5x \left(\dfrac{3}{3x + 2} \right) + [\ln(3x + 2)](-5)$

$\quad = -\dfrac{15x}{3x + 2} - 5 \ln(3x + 2)$

13. $s = t^2 \ln |t|$

$\dfrac{ds}{dt} = t^2 \cdot \dfrac{1}{t} + 2t \ln |t|$

$\quad = t + 2t \ln |t|$

$\quad = t(1 + 2 \ln |t|)$

15. $y = \dfrac{2 \ln (x + 3)}{x^2}$

Use the quotient rule.

$\dfrac{dy}{dx} = \dfrac{x^2 \left(\frac{2}{x+3} \right) - 2 \ln (x + 3) \cdot 2x}{(x^2)^2}$

$\quad = \dfrac{\frac{2x^2}{x+3} - 4x \ln (x + 3)}{x^4}$

$\quad = \dfrac{2x^2 - 4x(x + 3) \ln (x + 3)}{x^4(x + 3)}$

$\quad = \dfrac{x[2x - 4(x + 3) \ln (x + 3)]}{x^4(x + 3)}$

$\quad = \dfrac{2x - 4(x + 3) \ln (x + 3)}{x^3(x + 3)}$

17. $y = \dfrac{\ln x}{4x + 7}$

Use the quotient rule.

$\dfrac{dy}{dx} = \dfrac{(4x + 7) \left(\frac{1}{x} \right) - (\ln x)(4)}{(4x + 7)^2}$

$\quad = \dfrac{\frac{4x+7}{x} - 4 \ln x}{(4x + 7)^2}$

$\quad = \dfrac{4x + 7 - 4x \ln x}{x(4x + 7)^2}$

19. $y = \dfrac{3x^2}{\ln x}$

$\dfrac{dy}{dx} = \dfrac{(\ln x)(6x) - 3x^2 \left(\frac{1}{x} \right)}{(\ln x)^2}$

$\quad = \dfrac{6x \ln x - 3x}{(\ln x)^2}$

21. $y = (\ln |x + 1|)^4$

$\dfrac{dy}{dx} = 4(\ln |x + 1|)^3 \left(\dfrac{1}{x + 1} \right)$

$\quad = \dfrac{4(\ln |x + 1|)^3}{x + 1}$

23. $y = \ln |\ln x|$

$g(x) = \ln x$

$g'(x) = \dfrac{1}{x}$

$\dfrac{dy}{dx} = \dfrac{g'(x)}{g(x)}$

$\quad = \dfrac{\frac{1}{x}}{\ln x}$

$\quad = \dfrac{1}{x \ln x}$

25. $y = e^{x^2} \ln x, \; x > 0$

$\dfrac{dy}{dx} = e^{x^2} \left(\dfrac{1}{x} \right) + (\ln x)(2x)e^{x^2}$

$\quad = \dfrac{e^{x^2}}{x} + 2xe^{x^2} \ln x$

27. $y = \dfrac{e^x}{\ln x}, \; x > 0$

Use the quotient rule.

$\dfrac{dy}{dx} = \dfrac{(\ln x)e^x - e^x \left(\frac{1}{x} \right)}{(\ln x)^2} \cdot \dfrac{x}{x}$

$\quad = \dfrac{xe^x \ln x - e^x}{x (\ln x)^2}$

29. $g(z) = (e^{2z} + \ln z)^3$

$$g'(z) = 3(e^{2z} + \ln z)^2 \left(e^{2z} \cdot 2 + \frac{1}{z} \right)$$

$$= 3(e^{2z} + \ln z)^2 \left(\frac{2ze^{2z} + 1}{z} \right)$$

31. $y = \log(4x - 3)$

$g(x) = 4x - 3$
$g'(x) = 4$

$$\frac{dy}{dx} = \frac{1}{\ln 10} \cdot \frac{4}{4x - 3}$$

$$= \frac{4}{(\ln 10)(4x - 3)}$$

33. $y = \log |3x|$

$g(x) = 3x$ and $g'(x) = 3$.

$$\frac{dy}{dx} = \frac{1}{\ln 10} \cdot \frac{3}{3x}$$

$$= \frac{1}{x \ln 10}$$

35. $y = \log_7 \sqrt{4x - 3}$

$g(x) = \sqrt{4x - 3}$
$$g'(x) = \frac{4}{2\sqrt{4x - 3}} = \frac{2}{\sqrt{4x - 3}}$$

$$\frac{dy}{dx} = \frac{1}{\ln 7} \cdot \frac{\frac{2}{\sqrt{4x-3}}}{\sqrt{4x - 3}}$$

$$= \frac{2}{(\ln 7)(4x - 3)}$$

37. $y = \log_2 (2x^2 - x)^{5/2}$

$g(x) = (2x^2 - x)^{5/2}$ and

$$g'(x) = \frac{5}{2}(2x^2 - x)^{3/2} \cdot (4x - 1).$$

$$\frac{dy}{dx} = \frac{1}{\ln 2} \cdot \frac{\frac{5}{2}(2x^2 - x)^{3/2} \cdot (4x - 1)}{(2x^2 - x)^{5/2}}$$

$$= \frac{5(4x - 1)}{(2 \ln 2)(2x^2 - x)}$$

39. $z = 10^y \log y$

$g(y) = 10^y$ and $g'(y) = (\ln 10)10^y$.

$$\frac{dz}{dy} = 10^y \cdot \frac{1}{(\ln 10)y} + \log y \cdot (\ln 10)10^y$$

$$= \frac{10^y}{(\ln 10)y} + (\log y)(\ln 10)10^y$$

41. $f(x) = \ln(xe^{\sqrt{x}} + 2)$

$g(x) = xe^{\sqrt{x}} + 2$

$$g'(x) = x \left[e^{\sqrt{x}} \left(\frac{1}{2\sqrt{x}} \right) \right] + e^{\sqrt{x}}(1)$$

$$= \frac{e^{\sqrt{x}}\sqrt{x}}{2} + e^{\sqrt{x}}$$

$$= \frac{e^{\sqrt{x}}}{2}(\sqrt{x} + 2)$$

$$f'(x) = \frac{g'(x)}{g(x)}$$

$$= \frac{\frac{e^{\sqrt{x}}}{2}(\sqrt{x} + 2)}{xe^{\sqrt{x}} + 2}$$

$$= \frac{e^{\sqrt{x}}(\sqrt{x} + 2)}{2(xe^{\sqrt{x}} + 2)}$$

43. $f(t) = \dfrac{2t^{3/2}}{\ln(2t^{3/2} + 1)}$

Use the quotient rule.

$u(t) = 2t^{3/2}, u'(t) = 3t^{1/2}$

$$v(t) = \ln(2t^{3/2} + 1), v'(t) = \frac{3t^{1/2}}{2t^{3/2} + 1}$$

$$f'(t) = \frac{\ln(2t^{3/2} + 1)(3t^{1/2}) - 2t^{3/2}\left[\frac{3t^{1/2}}{2t^{3/2}+1} \right]}{[\ln(2t^{3/2} + 1)]^2}$$

$$= \frac{(3t^{1/2})\ln(2t^{3/2} + 1) - \frac{6t^2}{2t^{3/2}+1}}{[\ln(2t^{3/2} + 1)]^2}$$

$$= \frac{(6t^2 + 3t^{1/2})\ln(2t^{3/2} + 1) - 6t^2}{(2t^{3/2} + 1)[\ln(2t^{3/2} + 1)]^2}$$

45. Note that a is a constant.

$$\frac{d}{dx} \ln |ax| = \frac{d}{dx}(\ln |a| + \ln |x|)$$

$$= \frac{d}{dx} \ln |a| + \frac{d}{dx} \ln x$$

$$= 0 + \frac{d}{dx} \ln |x|$$

$$= \frac{d}{dx} \ln |x|$$

Therefore,

$$\frac{d}{dx} \ln |ax| = \frac{d}{dx} \ln |x|.$$

47. Graph
$$y = \frac{\ln|x + 0.0001| - \ln|x|}{0.0001}$$
on a graphing calculator. A good choice for the viewing window is $[-3, 3]$.

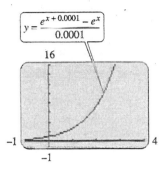

If we graph $y = \frac{1}{x}$ on the same screen, we see that the two graphs coincide.

By the definition of the derivative, if $f(x) = \ln|x|$,
$$f'(x) = \lim_{h \to 0} \frac{f(x+h) - f(x)}{h} = \lim_{h \to 0} \frac{\ln|x+h| - \ln|x|}{h},$$

and $h = 0.0001$ is very close to 0.

Comparing the two graphs provides graphical evidence that
$$f'(x) = \frac{1}{x}.$$

49. Use the derivative of $\ln x$.

$$\frac{d \ln \frac{u(x)}{v(x)}}{dx} = \frac{1}{\frac{u(x)}{v(x)}} \cdot \frac{d\left[\frac{u(x)}{v(x)}\right]}{dx} = \frac{v(x)}{u(x)} \cdot \frac{d\left[\frac{u(x)}{v(x)}\right]}{dx}$$

$$d \ln u(x) = \frac{1}{u(x)} \cdot \frac{d[u(x)]}{dx}$$

$$d \ln v(x) = \frac{1}{v(x)} \cdot \frac{d[v(x)]}{dx}$$

Then, since $\ln \frac{u(x)}{v(x)} = \ln u(x) - \ln v(x)$,

$$\frac{v(x)}{u(x)} \cdot \frac{d\left[\frac{u(x)}{v(x)}\right]}{dx} = \frac{1}{u(x)} \cdot \frac{d[u(x)]}{dx} - \frac{1}{v(x)} \cdot \frac{d[v(x)]}{dx}.$$

Multiply both sides of this equation by $\frac{u(x)}{v(x)}$. Then

$$\frac{d\left[\frac{u(x)}{v(x)}\right]}{dx} = \frac{1}{v(x)} \cdot \frac{d[u(x)]}{dx} - \frac{u(x)}{[v(x)]^2} \cdot \frac{d[v(x)]}{dx}$$

$$= \frac{v(x)}{[v(x)]^2} \cdot \frac{d[u(x)]}{dx} - \frac{u(x)}{[v(x)]^2} \cdot \frac{d[v(x)]}{dx}$$

$$= \frac{v(x) \cdot \frac{d[u(x)]}{dx} - u(x) \cdot \frac{d[v(x)]}{dx}}{[v(x)]^2}$$

This is the quotient rule.

51. The change-of-base theorem for logarithms states $\log_a x = \frac{\ln x}{\ln a}$. Find the derivative of each side.

$$\frac{d \log_a x}{dx} = \frac{\ln a \cdot \frac{d \ln x}{dx} - \ln x \cdot \frac{d \ln a}{dx}}{(\ln a)^2}$$

$$= \frac{\ln a \cdot \frac{1}{x} - \ln x \cdot 0}{(\ln a)^2}$$

$$= \frac{\frac{1}{x}}{\ln a}$$

$$= \frac{\frac{1}{x}}{x \ln a}$$

53. $h(x) = x^x$
$$u(x) = x, u'(x) = 1$$
$$v(x) = x, v'(x) = 1$$

$$h'(x) = x^x \left[\frac{x(1)}{x} + (\ln x)(1)\right]$$

$$= x^x(1 + \ln x)$$

55. $R(x) = 30 \ln(2x + 1)$

$$C(x) = \frac{x}{2}$$

$$P(x) = R(x) - C(x) = 30 \ln(2x + 1) - \frac{x}{2}$$

The profit will be a maximum when the derivative of the profit function is equal to 0.

$$P'(x) = 30\left(\frac{1}{2x+1}\right)(2) - \frac{1}{2} = \frac{60}{2x+1} - \frac{1}{2}$$

Now, $P'(x) = \frac{60}{2x+1} - \frac{1}{2} = 0$

when

$$\frac{60}{2x+1} = \frac{1}{2}$$
$$120 = 2x + 1$$
$$\frac{119}{2} = x.$$

Thus, a maximum profit occurs when $x = \frac{119}{2}$ or, in a practical sense, when 59 or 60 items are manufactured. (Both 59 and 60 give the same profit.)

57. $C(q) = 100q + 100$

(a) The marginal cost is given by $C'(q)$.

$$C'(q) = 100$$

(b) $P(q) = R(q) - C(q)$

$$= q\left(100 + \frac{50}{\ln q}\right)$$

$$- (100q + 100)$$

$$= 100q + \frac{50q}{\ln q} - 100q - 100$$

$$= \frac{50q}{\ln q} - 100$$

(c) The profit from one more unit is is $\frac{dP}{dq}$ for $q = 8$.

$$\frac{dP}{dq} = \frac{(\ln q)(50) - 50q\left(\frac{1}{q}\right)}{(\ln q)^2}$$

$$= \frac{50 \ln q - 50}{(\ln q)^2} = \frac{50(\ln q - 10)}{(\ln q)^2}$$

When $q = 8$, the profit from one more unit is

$$\frac{50(\ln 8 - 1)}{(\ln 8)^2} = \$12.48.$$

(d) The manager can use the information from part (c) to decide whether it is profitable to make and sell additional items.

59. $A(w) = 4.688 w^{0.8168 - 0.0154 \log_{10} w}$

(a) $A(4000) = 4.688(4000)^{0.8168 - 0.0154 \log_{10} 4000}$

$$\approx 2590 \text{ cm}^2$$

(b) $\dfrac{A(w)}{4.688} = w^{0.8166 - 0.0154 \log_{10} w}$

$$\ln A(w) - \ln 4.688 = (\ln w)\left(0.8168 - 0.0154 \frac{\ln w}{\ln 10}\right)$$

$$\frac{A'(w)}{A(w)} = \frac{1}{w}\left(0.8168 - 0.0154 \frac{\ln w}{\ln 10}\right)$$

$$+ \ln w\left(\frac{-0.0154}{\ln 10}\right)\frac{1}{w}$$

$$= \frac{0.8168}{w} - \frac{0.0308}{\ln 10}\frac{\ln w}{w}$$

$$= \frac{1}{w}\left(0.8168 - \frac{0.0308}{\ln 10}\ln w\right)$$

$$A'(w) = \frac{1}{w}\left(0.8168 - \frac{0.0308}{\ln 10}\ln w\right)$$

$$\cdot \left(4.688 w^{0.8168 - 0.0154 \log_{10} w}\right)$$

$$A'(4000) \approx 0.4571 \approx 0.46 \text{ g/cm}^2$$

When the infant weighs 4000 g, it is gaining 0.46 square centimeters per gram of weight increase.

(c)

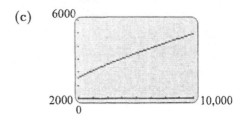

61. $F(x) = 0.774 + 0.727 \log(x)$

(a) $F(25,000) = 0.774 + 0.727 \log(25,000)$

$$= 3.9713\ldots$$

$$\approx 4 \text{ kJ/day}$$

(b) $F'(x) = 0.727 \dfrac{1}{x \ln 10}$

$$= \frac{0.727}{\ln 10} x^{-1}$$

$$F'(25,000) = \frac{0.727}{\ln 10} 25,000^{-1}$$

$$\approx 0.000012629\ldots$$

$$\approx 1.3 \times 10^{-5}$$

When a fawn is 25 kg in size, the rate of change of the energy expenditure of the fawn is about 1.3×10^{-5} kJ/day per gram.

(c)

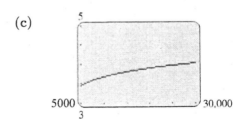

63. $M(t) = (0.1t + 1) \ln \sqrt{t}$

(a) $M(15) = [0.1(15) + 1] \ln \sqrt{15}$

$$\approx 3.385$$

When the temperature is 15°C, the number of matings is about 3.

(b) $M(25) = [0.1(25) + 1] \ln \sqrt{25}$

$$\approx 5.663$$

When the temperature is 25°C, the number of matings is about 6.

(c) $M(t) = (0.1t + 1) \ln \sqrt{t}$
$$= (0.1t + 1) \ln t^{1/2}$$

$$M'(t) = (0.1t + 1)\left(\frac{1}{2} \cdot \frac{1}{t}\right)$$
$$+ (\ln t^{1/2})(0.1)$$

$$= 0.1 \ln \sqrt{t} + \frac{1}{2t}(0.1t + 1)$$

$$M'(15) = 0.1 \ln \sqrt{15} + \frac{1}{2 \cdot 15}[(0.1)(15) + 1]$$

$$\approx 0.22$$

When the temperature is 15°C, the rate of change of the number of matings is about 0.22.

65. $M = \dfrac{2}{3} \log \dfrac{E}{0.007}$

(a) $8.9 = \dfrac{2}{3} \log \dfrac{E}{0.007}$

$$13.35 = \log \frac{E}{0.007}$$

$$10^{13.35} = \frac{E}{0.007}$$

$$E = 0.007(10^{13.35})$$
$$\approx 1.567 \times 10^{11} \text{ kWh}$$

(b) $10{,}000{,}000 \times 247$ kWh/month
$$= 2{,}470{,}000{,}000 \text{ kWh/month}$$

$$\frac{1.567 \times 10^{11} \text{ kWh}}{2{,}470{,}000{,}000 \text{ kWh/month}} \approx 63.4 \text{ months}$$

(c) $M = \dfrac{2}{3} \log E - \dfrac{2}{3} \log 0.007$

$$\frac{dM}{dE} = \frac{2}{3}\left(\frac{1}{(\ln 10)E}\right)$$

$$= \frac{2}{(3 \ln 10)E}$$

When $E = 70{,}000$,

$$\frac{dM}{dE} = \frac{2}{(3 \ln 10)70{,}000}$$

$$\approx 4.14 \times 10^{-6}$$

(d) $\frac{dM}{dE}$ varies inversely with E, so as E increases, $\frac{dM}{dE}$ decreases and approaches zero.

Chapter 12 Review Exercises

1. $y = 5x^3 - 7x^2 - 9x + \sqrt{5}$

$$\frac{dy}{dx} = 5(3x^2) - 7(2x) - 9 + 0$$
$$= 15x^2 - 14x - 9$$

3. $y = 9x^{8/3}$

$$\frac{dy}{dx} = 9\left(\frac{8}{3}x^{5/3}\right)$$

$$= 24x^{5/3}$$

5. $f(x) = 3x^{-4} + 6\sqrt{x}$
$$= 3x^{-4} + 6x^{1/2}$$

$$f'(x) = 3(-4x^{-5}) + 6\left(\frac{1}{2}x^{-1/2}\right)$$

$$= -12x^{-5} + 3x^{-1/2} \text{ or } -\frac{12}{x^5} + \frac{3}{x^{1/2}}$$

7. $k(x) = \dfrac{3x}{4x + 7}$

$$k'(x) = \frac{(4x + 7)(3) - (3x)(4)}{(4x + 7)^2}$$

$$= \frac{12x + 21 - 12x}{(4x + 7)^2}$$

$$= \frac{21}{(4x + 7)^2}$$

9. $y = \dfrac{x^2 - x + 1}{x - 1}$

$$\frac{dy}{dx} = \frac{(x - 1)(2x - 1) - (x^2 - x + 1)(1)}{(x - 1)^2}$$

$$= \frac{2x^2 - 3x + 1 - x^2 + x - 1}{(x - 1)^2}$$

$$= \frac{x^2 - 2x}{(x - 1)^2}$$

11. $f(x) = (3x^2 - 2)^4$
$$f'(x) = 4(3x^2 - 2)^3[3(2x)]$$
$$= 24x(3x^2 - 2)^3$$

13. $y = \sqrt{2t^7 - 5}$
$$= (2t^7 - 5)^{1/2}$$

$$\frac{dy}{dx} = \frac{1}{2}(2t^7 - 5)^{-1/2}[2(7t^6)]$$

$$= 7t^6(2t^7 - 5)^{-1/2} \text{ or } \frac{7t^6}{(2t^7 - 5)^{1/2}}$$

15. $y = 3x(2x + 1)^3$

$$\frac{dy}{dx} = 3x(3)(2x + 1)^2(2) + (2x + 1)^3(3)$$
$$= (18x)(2x + 1)^2 + 3(2x + 1)^3$$
$$= 3(2x + 1)^2[6x + (2x + 1)]$$
$$= 3(2x + 1)^2(8x + 1)$$

17. $r(t) = \dfrac{5t^2 - 7t}{(3t+1)^3}$

$r'(t) = \dfrac{(3t+1)^3(10t-7) - (5t^2-7t)(3)(3t+1)^2(3)}{[(3t+1)^3]^2}$

$= \dfrac{(3t+1)^3(10t-7) - 9(5t^2-7t)(3t+1)^2}{(3t+1)^6}$

$= \dfrac{(3t+1)(10t-7) - 9(5t^2-7t)}{(3t+1)^4}$

$= \dfrac{30t^2 - 11t - 7 - 45t^2 + 63t}{(3t+1)^4}$

$= \dfrac{-15t^2 + 52t - 7}{(3t+1)^4}$

19. $p(t) = t^2(t^2+1)^{5/2}$

$p'(t) = t^2 \cdot \dfrac{5}{2}(t^2+1)^{3/2} \cdot 2t + 2t(t^2+1)^{5/2}$

$= 5t^3(t^2+1)^{3/2} + 2t(t^2+1)^{5/2}$

$= t(t^2+1)^{3/2}[5t^2 + 2(t^2+1)^1]$

$= t(t^2+1)^{3/2}(7t^2+2)$

21. $y = -6e^{2x}$

$\dfrac{dy}{dx} = -6(2e^{2x}) = -12e^{2x}$

23. $y = e^{-2x^3}$

$g(x) = -2x^3$

$g'(x) = -6x^2$

$y' = -6x^2 e^{-2x^3}$

25. $y = 5x \cdot e^{2x}$

Use the product rule.

$\dfrac{dy}{dx} = 5x(2e^{2x}) + e^{2x}(5)$

$= 10xe^{2x} + 5e^{2x}$

$= 5e^{2x}(2x+1)$

27. $y = \ln(2+x^2)$

$g(x) = 2 + x^2$

$g'(x) = 2x$

$\dfrac{dy}{dx} = \dfrac{2x}{2+x^2}$

29. $y = \dfrac{\ln|3x|}{x-3}$

$\dfrac{dy}{dx} = \dfrac{(x-3)\left(\frac{1}{3x}\right)(3) - (\ln|3x|)(1)}{(x-3)^2}$

$= \dfrac{\frac{x-3}{x} - \ln|3x|}{(x-3)^2} \cdot \dfrac{x}{x}$

$= \dfrac{x - 3 - x\ln|3x|}{x(x-3)^2}$

31. $y = \dfrac{xe^x}{\ln(x^2-1)}$

$\dfrac{dy}{dx} = \dfrac{\ln(x^2-1)[xe^x + e^x] - xe^x\left(\frac{1}{x^2-1}\right)(2x)}{[\ln(x^2-1)]^2}$

$= \dfrac{e^x(x+1)\ln(x^2-1) - \frac{2x^2e^x}{x^2-1}}{[\ln(x^2-1)]^2} \cdot \dfrac{x^2-1}{x^2-1}$

$= \dfrac{e^x(x+1)(x^2-1)\ln(x^2-1) - 2x^2e^x}{(x^2-1)[\ln(x^2-1)]^2}$

33. $s = (t^2 + e^t)^2$

$s' = 2(t^2 + e^t)(2t + e^t)$

35. $y = 3 \cdot 10^{-x^2}$

$\dfrac{dy}{dx} = 3 \cdot (\ln 10)10^{-x^2}(-2x)$

$= -6x(\ln 10) \cdot 10^{-x^2}$

37. $g(z) = \log_2(z^3 + z + 1)$

$g'(z) = \dfrac{1}{\ln 2} \cdot \dfrac{3z^2+1}{z^3+z+1}$

$= \dfrac{3z^2+1}{(\ln 2)(z^3+z+1)}$

39. $f(x) = e^{2x}\ln(xe^x + 1)$

Use the product rule.

$f'(x) = e^{2x}\left(\dfrac{e^x + xe^x}{xe^x + 1}\right) + [\ln(xe^x+1)](2e^{2x})$

$= \dfrac{(1+x)e^{3x}}{xe^x + 1} + 2e^{2x}\ln(xe^x + 1)$

41. (a) $D_x(f[g(x)])$ at $x = 2$

$= f'[g(2)]g'(2)$

$= f'(1)\left(\dfrac{3}{10}\right)$

$= -5\left(\dfrac{3}{10}\right)$

$= -\dfrac{3}{2}$

(b) $D_x(f[g(x)])$ at $x = 3$

$$= f'[g(3)]g'(3)$$

$$= f'(2)\left(\frac{4}{11}\right)$$

$$= -6\left(\frac{4}{11}\right)$$

$$= -\frac{24}{11}$$

45. $y = 8 - x^2$; $x = 1$

$$y = 8 - x^2$$

$$\frac{dy}{dx} = -2x$$

slope $= y'(1) = -2(1) = -2$

Use $(1, 7)$ and $m = -2$ in the point-slope form.

$$y - 7 = -2(x - 1)$$
$$y - 7 = -2x + 2$$
$$2x + y = 9$$
$$y = -2x + 9$$

47. $y = \dfrac{x}{x^2 - 1}$; $x = 2$

$$\frac{dy}{dx} = \frac{(x^2 - 1) \cdot 1 - x(2x)}{(x^2 - 1)^2}$$

$$= \frac{-x^2 - 1}{(x^2 - 1)^2}$$

The value of $\frac{dy}{dx}$ when $x = 2$ is the slope.

$$m = \frac{-(2^2) - 1}{(2^2 - 1)^2} = \frac{-5}{9} = -\frac{5}{9}$$

When $x = 2$,

$$y = \frac{2}{4 - 1} = \frac{2}{3}.$$

Use $m = -\frac{5}{9}$ with $P\left(2, \frac{2}{3}\right)$.

$$y - \frac{2}{3} - -\frac{5}{9}(x - 2)$$

$$y - \frac{6}{9} = -\frac{5}{9}x + \frac{10}{9}$$

$$y = -\frac{5}{9}x + \frac{16}{9}$$

49. $y = -\sqrt{8x + 1}$; $x = 3$

$$y = -(8x + 1)^{1/2}$$

$$\frac{dy}{dx} = -\frac{1}{2}(8x + 1)^{-1/2}(8)$$

$$\frac{dy}{dx} = -\frac{4}{(8x + 1)^{1/2}}$$

The value of $\frac{dy}{dx}$ when $x = 3$ is the slope.

$$m = -\frac{4}{(24 + 1)^{1/2}} = -\frac{4}{5}$$

When $x = 3$,

$$y = -\sqrt{24 + 1} = -5.$$

Use $m = -\frac{4}{5}$ with $P(3, -5)$.

$$y + 5 = -\frac{4}{5}(x - 3)$$

$$y + \frac{25}{5} = -\frac{4}{5}x + \frac{12}{5}$$

$$y = -\frac{4}{5}x - \frac{13}{5}$$

51. $y = xe^x$; $x = 1$

$$\frac{dy}{dx} = xe^x + 1 \cdot e^x$$

$$= e^x(x + 1)$$

The value of $\frac{dy}{dx}$ when $x = 1$ is the slope.

$$m = e^1(1 + 1) = 2e$$

When $x = 1$, $y = 1e^1 = e$. Use $m = 2e$ with $P(1, e)$.

$$y - e = 2e(x - 1)$$
$$y = 2ex - e$$

53. $y = x \ln x$; $x = e$

$$\frac{dy}{dx} = x \cdot \frac{1}{x} + 1 \cdot \ln x$$

$$= 1 + \ln x$$

The value of $\frac{dy}{dx}$ when $x = e$ is the slope

$$m = 1 + \ln e = 1 + 1 - 2.$$

When $x = e$, $y = e \ln e = e \cdot 1 = e$. Use $m = 2$ with $P(e, e)$.

$$y - e = 2(x - e)$$
$$y = 2x - e$$

55. (a) Use the chain rule.

Let $g(x) = \ln x$. Then $g'(x) = \dfrac{1}{x}$.

Let $y = g[f(x)]$. Then $\dfrac{dy}{dx} = g'[f(x)] \cdot f'(x)$.

so

$$\frac{d \ln f(x)}{dx} = \frac{1}{f(x)} \cdot f'(x) = \frac{f'(x)}{f(x)}.$$

(b) $\hat{f} = \dfrac{f'(x)}{f(x)}, \hat{g} = f\dfrac{g'(x)}{g(x)}$

$\hat{fg} = \dfrac{(fg)'(x)}{(fg)(x)}$

$= \dfrac{f(x) \cdot g'(x) + g(x) \cdot f'(x)}{f(x)g(x)}$

$= \dfrac{f(x) \cdot g'(x)}{f(x) \cdot g(x)} + \dfrac{g(x) \cdot f'(x)}{f(x) \cdot g(x)}$

$= \dfrac{g'(x)}{g(x)} + \dfrac{f'(x)}{f(x)}$

$= \hat{g} + \hat{f}$ or $\hat{f} + \hat{g}$

57. $C(x) = \sqrt{x+1}$

$\overline{C}(x) = \dfrac{C(x)}{x} = \dfrac{\sqrt{x+1}}{x}$

$= \dfrac{(x+1)^{1/2}}{x}$

$\overline{C}'(x) = \dfrac{x\left[\frac{1}{2}(x+1)^{-1/2}\right] - (x+1)^{1/2}(1)}{x^2}$

$= \dfrac{\frac{1}{2}x(x+1)^{-1/2} - (x+1)^{1/2}}{x^2}$

$= \dfrac{x(x+1)^{-1/2} - 2(x+1)^{1/2}}{2x^2}$

$= \dfrac{(x+1)^{-1/2}[x - 2(x+1)]}{2x^2}$

$= \dfrac{(x+1)^{-1/2}(-x-2)}{2x^2}$

$= \dfrac{-x-2}{2x^2(x+1)^{1/2}}$

59. $C(x) = (x^2+3)^3$

$\overline{C}(x) = \dfrac{C(x)}{x} = \dfrac{(x^2+3)^3}{x}$

$\overline{C}'(x) = \dfrac{x[3(x^2+3)^2(2x)] - (x^2+3)^3(1)}{x^2}$

$= \dfrac{6x^2(x^2+3)^2 - (x^2+3)^3}{x^2}$

$= \dfrac{(x^2+3)^2[6x^2 - (x^2+3)]}{x^2}$

$= \dfrac{(x^2+3)^2(5x^2-3)}{x^2}$

61. $C(x) = 10 - e^{-x}$

$\overline{C}(x) = \dfrac{C(x)}{x}$

$\overline{C}(x) = \dfrac{10 - e^{-x}}{x}$

$\overline{C}'(x) = \dfrac{x(e^{-x}) - (10 - e^{-x}) \cdot 1}{x^2}$

$= \dfrac{e^{-x}(x+1) - 10}{x^2}$

63. $S(x) = 1000 + 60\sqrt{x} + 12x$
$\qquad = 1000 + 60x^{1/2} + 12x$

$\dfrac{dS}{dx} = 60\left(\dfrac{1}{2}x^{-1/2}\right) + 12$

$\qquad = 30x^{-1/2} + 12 = \dfrac{30}{\sqrt{x}} + 12$

(a) $\dfrac{dS}{dx}(9) = \dfrac{30}{\sqrt{9}} + 12 = \dfrac{30}{3} + 12 = 22$

Sales will increase by $22 million when $1000 more is spent on research.

(b) $\dfrac{dS}{dx}(16) = \dfrac{30}{\sqrt{16}} + 12 = \dfrac{30}{4} + 12 = 19.5$

Sales will increase by $19.5 million when $1000 more is spent on research.

(c) $\dfrac{dS}{dx}(25) = \dfrac{30}{\sqrt{25}} + 12 = \dfrac{30}{5} + 12 = 18$

Sales will increase by $18 million when $1000 more is spent on research.

(d) As more money is spent on research, the increase in sales is decreasing.

65. $T(x) = \dfrac{1000 + 60x}{4x + 5}$

$$T'(x) = \frac{(4x+5)(60) - (1000 + 60x)(4)}{(4x+5)^2} = \frac{240x + 300 - 4000 - 240x}{(4x+5)^2} = \frac{-3700}{(4x+5)^2}$$

(a) $T'(9) = \dfrac{-3700}{[4(9)+5]^2} = \dfrac{-3700}{1681} \approx -2.201$

Costs will decrease $2201 for the next $100 spent on training.

(b) $T'(19) = \dfrac{-3700}{[4(19)+5]^2} = \dfrac{-3700}{6561} \approx -0.564$

Costs will decrease $564 for the next $100 spent on training.

(c) Costs will always decrease because

$$T'(x) = \frac{-3700}{(4x+5)^2}$$

will always be negative.

67. $A(r) = 1000e^{12r/100}$

$$A'(r) = 1000e^{12r/100} \cdot \frac{12}{100} = 120e^{12r/100}$$
$$A'(5) = 120e^{0.6} \approx 218.65$$

The balance increases by approximately $218.65 for every 1% increase in the interest rate when the rate is 5%.

69. $f(t) = 1.5207t^4 - 19.166t^3 + 62.91t^2 + 6.0726t + 1026$
$f'(t) = 1.5207(4t^3) - 19.166(3t^2) + 62.91(2t) + 6.0726$
$\quad = 6.0828t^3 - 57.498t^2 + 125.82$

$t = 7$ corresponds to the beginning of 2005.

$f'(7) = 6.0828(7)^3 - 57.498(7)^2 + 125.82(7) + 6.0726$
$\quad \approx 156$

Rents were increasing at the rate of $156 per month per year.

71. (a) Using the regression feature on a graphing calculator, a cubic function that models the data is

$$y = (2.458 \times 10^{-5})t^3 - (6.767 \times 10^{-4})t^2 - 0.02561t + 2.031.$$

Using the regression feature on a graphing calculator, a quartic function that models the data is

$$y = (-1.314 \times 10^{-6})t^4 + (3.363 \times 10^{-4})t^3 - 0.02565t^2 + 0.7410t - 5.070.$$

(b) Using the cubic function, $\frac{dy}{dx}$ at $x = 95$ is about 0.51 dollar per year. Using the quartic function, $\frac{dy}{dx}$ at $x = 95$ is about 0.47 dollar per year.

73. $G(t) = \dfrac{m\,G_o}{G_o + (m - G_o)e^{-kmt}}$, where $m = 30{,}000$, $G_o = 2000$, and $k = 5 \cdot 10^{-6}$.

(a) $G(t) = \dfrac{(30{,}000)(2000)}{2000 + (30{,}000 - 2000)e^{-5 \cdot 10^{-6}(30{,}000)t}} = \dfrac{30{,}000}{1 + 14e^{-0.15t}}$

(b) $G(t) = 30{,}000(1 + 14e^{-0.15t})^{-1}$
$G(6) = 30{,}000(1 + 14e^{-0.90})^{-1} \approx 4483$

$G'(t) = -30{,}000(1 + 14e^{-0.15t})^{-2}(-2.1e^{-0.15t}) = \dfrac{63{,}000e^{-0.15t}}{(1 + 14e^{-0.15t})^2}$

$G'(6) = \dfrac{63{,}000e^{-0.90}}{(1 + 14e^{-0.90})^2} \approx 572$

The population is 4483, and the rate of growth is 572.

75. $M(t) = 3583e^{-e^{-0.020(t-66)}}$

(a) $M(250) = 3583e^{-e^{-0.020(250-66)}}$
$$\approx 3493.76 \text{ grams},$$
or about 3.5 kilograms

(b) As $t \to \infty$, $-e^{-0.020(t-66)} \to 0$, $e^{-e^{-0.020(t-66)}} \to 1$,
and $M(t) \to 3583$ grams or about 3.6 kilograms.

(c) 50% of 3583 is 1791.5.

$$1791.5 = 3583e^{-e^{-0.020(t-66)}}$$

$$\ln\left(\frac{1791.5}{3583}\right) = -e^{-0.020(t-66)}$$

$$\ln\left(\ln\frac{3583}{1791.5}\right) = -0.020(t-66)$$

$$t = -\frac{1}{0.020}\ln\left(\ln\frac{3583}{1791.5}\right) + 66$$

$$\approx 84 \text{ days}$$

(d) $D_t M(t)$
$$= 3583e^{-e^{-0.020(t-66)}} D_t\left(-e^{-0.020(t-66)}\right)$$
$$= 3583e^{-e^{-0.020(t-66)}}\left(-e^{-0.020(t-66)}\right)(-0.020)$$
$$= 71.66e^{-e^{-0.020(t-66)}}\left(e^{-0.020(t-66)}\right)$$

when $t = 250$, $D_t M(t) \approx 1.76$ g/day.

(e)

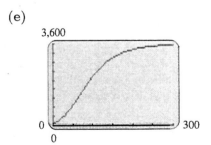

Growth is initially rapid, then tapers off.

(f)

Day	Weight	Rate
50	904	24.90
100	2159	21.87
150	2974	11.08
200	3346	4.59
250	3494	1.76
300	3550	0.66

77. $f(t) = \dfrac{8}{t+1} + \dfrac{20}{t^2+1}$

(a) The average velocity from $t = 1$ to $t = 3$ is given by

$$\text{average velocity} = \frac{f(3) - f(1)}{3 - 1}$$

$$= \frac{\left(\frac{8}{4} + \frac{20}{10}\right) - \left(\frac{8}{2} + \frac{20}{2}\right)}{2}$$

$$= \frac{4 - 14}{2}$$

$$= -5$$

Belmar's average velocity between 1 sec and 3 sec is -5 ft/sec.

(b) $f(t) = 8(t+1)^{-1} + 20(t^2+1)^{-1}$
$f'(t) = -8(t+1)^{-2} \cdot 1 - 20(t^2+1)^{-2} \cdot 2t$

$$= -\frac{8}{(t+1)^2} - \frac{40t}{(t^2+1)^2}$$

$$f'(3) = -\frac{8}{16} - \frac{120}{100}$$

$$= -0.5 - 1.2$$

$$= -1.7$$

Belmar's instantaneous velocity at 3 sec is -1.7 ft/sec.

79. (a) $N(t) = N_0 e^{-0.217t}$, where $t = 1$ and $N_0 = 210$

$$N(1) = 210e^{-0.217(1)}$$
$$\approx 169$$

The number of words predicted to be in use in 1950 is 169, and the actual number in use was 167.

(b) $N(2) = 210e^{-0.217(2)}$
$$\approx 136$$

In 2050 the will be about 136 words still being used.

(c) $N(t) = 210e^{-0.217t}$
$N'(t) = 210e^{-0.217t} \cdot (-0.217)$
$$= -45.57e^{-0.217t}$$
$N'(2) = -45.57e^{-0.217(2)}$
$$\approx -30$$

In the year 2050 the number of words in use will be decreasing by 30 words per millenium.

Chapter 12 Test

[12.1]

Find the derivative of each function.

1. $y = 2x^4 - 3x^3 + 4x^2 - x + 1$
2. $f(x) = x^{3/4} - x^{-2/3} + x^{-1}$
3. $f(x) = -2x^{-2} - 3\sqrt{x}$

4. Find the slope of the tangent line to $y = -2x + \frac{1}{x} + \sqrt{x}$ at $x = 1$. Find the equation of the tangent line.

5. For the function

$$f(x) = x^3 + \frac{3}{2}x^2 - 60x + 18,$$

   find all values of x where the tangent line is horizontal.

6. At the beginning of an experiment, a culture is determined to have 2×10^4 bacteria. Thereafter, the number of bacteria observed at time t (in hours) is given by the equation

$$B(t) = 10^4 \left(2 - 3\sqrt{t} + 2t + t^2\right).$$

   How fast is the population growing at the end of 4 hours?

[12.2]

Find the derivative of each function.

7. $f(x) = \left(2x^2 - 5x\right)\left(3x^2 + 1\right)$
8. $y = \dfrac{x^2 + x}{x^3 - 1}$

9. Management has determined that the cost in thousands of dollars for producing x units is given by the equation

$$C(x) = \frac{3x^2}{x^2 + 1} + 200.$$

   Find and interpret the marginal cost when $x = 3$.

[12.3]

10. Find $f[g(x)]$ and $g[f(x)]$ for the following pair of functions.

$$f(x) = \frac{3}{x^2}; \ g(x) = \frac{1}{x}$$

Find each of the following.

11. $\dfrac{dy}{dx}$ if $y = \dfrac{\sqrt{x} - 2}{x + 1}$
12. $D_x \left[\sqrt{\left(3x^2 - 1\right)^3}\right]$
13. $f'(2)$ if $f(t) = \dfrac{2t - 1}{\sqrt{t + 2}}$

Find the derivative of each function.

14. $y = -2x\sqrt{3x - 1}$

15. $y = \left(5x^3 - 2x\right)^5$

16. Find the equation of the tangent line to the graph of the function $f(x) = \sqrt{x^2 - 9}$ at $x = 5$. (Write the equation in slope-intercept form.)

[12.4]

Find the derivative of each function.

17. $y = 3x^2 e^{2x}$

18. $y = \left(e^{2x} - \ln |x|\right)^3$

19. The concentration of a certain drug in the bloodstream at time t in minutes is given by

$$c(t) = e^{-t} - e^{-3t}.$$

Find the rate of change in the concentration at each of the following times. Round to the nearest thousandth.

    (a) $t = 0$ **(b)** $t = 1$ **(c)** $t = 2$

[12.5]

Find the derivative of each function.

20. $y = \ln \left(2x^3 + 5\right)^{2/3}$

21. $y = x^2 \ln \left(x^2 + 1\right)$

22. $y = \dfrac{\ln |3x - 1|}{x - 2}.$

Chapter 12 Test Answers

1. $\frac{dy}{dx} = 8x^3 - 9x^2 + 8x - 1$

2. $f'(x) = \frac{3}{4x^{1/4}} + \frac{2}{3x^{5/3}} - \frac{1}{x^2}$

3. $f'(x) = \frac{4}{x^3} - \frac{3}{2\sqrt{x}}$

4. $-\frac{5}{2}; \ 5x + 2y = 5$

5. $x = -5, \ x = 4$

6. 9.25×10^4 bacteria per hour

7. $f'(x) = 24x^3 - 45x^2 + 4x - 5$

8. $\frac{dy}{dx} = \frac{-x^4 - 2x^3 - 2x - 1}{(x^3 - 1)^2}$

9. $C'(3) = 0.18$; after three units have been produced, the cost to produce one more unit will be approximately $0.18(1000)$ or \$180.

10. $f[g(x)] = 3x^2; \ g[f(x)] = \frac{x^2}{3}$

11. $\frac{5-x}{2\sqrt{x-2}(x+1)^2}$

12. $9x\sqrt{3x^2 - 1}$

13. $\frac{13}{16}$

14. $\frac{dy}{dx} = \frac{2 - 9x}{\sqrt{3x-1}}$

15. $\frac{dy}{dx} = 5\left(5x^3 - 2x\right)^4 \left(15x^2 - 2\right)$

16. $y = \frac{5}{4}x - \frac{9}{4}$

17. $6x^2 e^{2x} + 6xe^{2x}$

18. $\frac{3\left(e^{2x} - \ln|x|\right)^2 \left(2xe^{2x} - 1\right)}{x}$

19. (a) 2 (b) -0.219 (c) -0.128

20. $\frac{dy}{dx} = \frac{4x^2}{2x^3 + 5}$

21. $\frac{dy}{dx} = 2x \ln(x^2 + 1) + \frac{2x^3}{x^2 + 1}$

22. $\frac{3x - 6 - (3x - 1)\ln|3x - 1|}{(3x - 1)(x - 2)^2}$

GRAPHS AND THE DERIVATIVE

13.1 Increasing and Decreasing Functions

1. By reading the graph, f is

   (a) increasing on $(1, \infty)$ and
   (b) decreasing on $(-\infty, 1)$.

3. By reading the graph, g is

   (a) increasing on $(-\infty, -2)$ and
   (b) decreasing on $(-2, \infty)$.

5. By reading the graph, h is

   (a) increasing on $(-\infty, -4)$ and
   $(-2, \infty)$ and
   (b) decreasing on $(-4, -2)$.

7. By reading the graph, f is

   (a) increasing on $(-7, -4)$ and
   $(-2, \infty)$ and
   (b) decreasing on $(-\infty, -7)$ and
   $(-4, -2)$.

9. (a) Since the graph of the function is positive for $x < -1$ and $x > 3$, the intervals where $f(x)$ is increasing are $(-\infty, -1)$ and $(3, \infty)$.

   (b) Since the graph of the function is negative for $-1 < x < 3$, the interval where $f(x)$ is decreasing is $(-1, 3)$.

11. (a) Since the graph of the function is positive for $x < -8, -6 < x < -2.5$ and $x > -1.5$, the intervals where $f(x)$ is increasing are $(-\infty, -8), (-6, -2.5),$ and $(-1.5, \infty)$.

    (b) Since the graph of the function is negative for $-8 < x < -6$ and $-2.5 < x < -1.5$, the intervals where $f(x)$ is decreasing are $(-8, -6)$ and $(-2.5, -1.5)$.

13. $y = 2.3 + 3.4x - 1.2x^2$

    (a) $y' = 3.4 - 2.4x$

    y' is zero when

    $$3.4 - 2.4x = 0$$

    $$x = \frac{3.4}{2.4} = \frac{17}{12}$$

    and there are no values of x where y' does not exist, so the only critical number is $x = \frac{17}{12}$.

    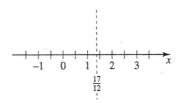

    Test a point in each interval.

    When $x = 0, y' = 3.4 - 2.4(0) = 3.4 > 0$.
    When $x = 2, y' = 3.4 - 2.4(2) = -1.4 < 0$.

    (b) The function is increasing on $\left(-\infty, \frac{17}{12}\right)$.

    (c) The function is decreasing on $\left(\frac{17}{12}, \infty\right)$.

15. $f(x) = \frac{2}{3}x^3 - x^2 - 24x - 4$

    (a) $f'(x) = 2x^2 - 2x - 24$
    $\quad = 2(x^2 - x - 12)$
    $\quad = 2(x + 3)(x - 4)$

    $f'(x)$ is zero when $x = -3$ or $x = 4$, so the critical numbers are -3 and 4.

    Test a point in each interval.

    $$f'(-4) = 16 > 0$$
    $$f'(0) = -24 < 0$$
    $$f'(5) = 16 > 0$$

    (b) f is increasing on $(-\infty, -3)$ and $(4, \infty)$.

    (c) f is decreasing on $(-3, 4)$.

17. $f(x) = 4x^3 - 15x^2 - 72x + 5$

    (a) $f'(x) = 12x^2 - 30x - 72$
$$= 6(2x^2 - 5x - 12)$$
$$= 6(2x + 3)(x - 4)$$

$f'(x)$ is zero when $x = -\frac{3}{2}$ or $x = 4$, so the critical numbers are $-\frac{3}{2}$ and 4.

$$f'(-2) = 36 > 0$$
$$f'(0) = -72 < 0$$
$$f'(5) = 78 > 0$$

    (b) f is increasing on $\left(-\infty, -\frac{3}{2}\right)$ and $(4, \infty)$.

    (c) f is decreasing on $\left(-\frac{3}{2}, 4\right)$.

19. $f(x) = x^4 + 4x^3 + 4x^2 + 1$

    (a) $f'(x) = 4x^3 + 12x^2 + 8x$
$$= 4x(x^2 + 3x + 2)$$
$$= 4x(x + 2)(x + 1)$$

$f'(x)$ is zero when $x = 0$, $x = -2$, or $x = -1$, so the critical numbers are 0, -2, and -1.

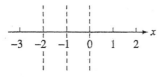

Test a point in each interval.

$$f'(-3) = -12(-1)(-2) = -24 < 0$$
$$f'(-1.5) = -6(.5)(-.5) = 1.5 > 0$$
$$f'(-.5) = -2(1.5)(.5) = -1.5 < 0$$
$$f'(1) = 4(3)(2) = 24 > 0$$

    (b) f is increasing on $(-2, -1)$ and $(0, \infty)$.

    (c) f is decreasing on $(-\infty, -2)$ and $(-1, 0)$.

21. $y = -3x + 6$

    (a) $y' = -3 < 0$

There are no critical numbers since y' is never 0 and always exists.

    (b) Since y' is always negative, the function is increasing on no interval.

    (c) y' is always negative, so the function is decreasing everywhere, or on the interval $(-\infty, \infty)$.

23. $f(x) = \dfrac{x + 2}{x + 1}$

    (a) $f'(x) = \dfrac{(x + 1)(1) - (x + 2)(1)}{(x + 1)^2}$
$$= \dfrac{-1}{(x + 1)^2}$$

The derivative is never 0, but it fails to exist at $x = -1$. Since -1 is not in the domain of f, however, -1 is not a critical number.

$$f'(-2) = -1 < 0$$
$$f'(0) = -1 < 0$$

    (b) f is increasing on no interval.

    (c) f is decreasing everywhere that it is defined, on $(-\infty, -1)$ and on $(-1, \infty)$.

25. $y = \sqrt{x^2 + 1}$
$$= (x^2 + 1)^{1/2}$$

    (a) $y' = \dfrac{1}{2}(x^2 + 1)^{-1/2}(2x)$
$$= x(x^2 + 1)^{-1/2}$$
$$= \dfrac{x}{\sqrt{x^2 + 1}}$$

$y' = 0$ when $x = 0$.
Since y does not fail to exist for any x, and since $y' = 0$ when $x = 0$, 0 is the only critical number.

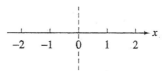

$$y'(1) = \dfrac{1}{\sqrt{2}} > 0$$
$$y'(-1) = \dfrac{-1}{\sqrt{2}} < 0$$

    (b) y is increasing on $(0, \infty)$.

    (c) y is decreasing on $(-\infty, 0)$.

27. $f(x) = x^{2/3}$

(a) $f'(x) = \dfrac{2}{3}x^{-1/3} = \dfrac{2}{3x^{1/3}}$

$f'(x)$ is never zero, but fails to exist when $x = 0$, so 0 is the only critical number.

$$f'(-1) = -\frac{2}{3} < 0$$

$$f'(1) = \frac{2}{3} > 0$$

(b) f is increasing on $(0, \infty)$.

(c) f is decreasing on $(-\infty, 0)$.

29. $y = x - 4 \ln(3x - 9)$

(a) $y' = 1 - \dfrac{12}{3x - 9} = 1 - \dfrac{4}{x - 3}$

$\quad = \dfrac{x - 7}{x - 3}$

y' is zero when $x = 7$. The derivative does not exist at $x = 3$, but note that the domain of f is $(3, \infty)$.

Thus, the only critical number is 7.

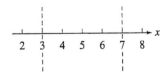

Choose values in the intervals $(3, 7)$ and $(7, \infty)$.

$$f'(4) = -3 < 0$$

$$f(8) = \frac{1}{5} > 0$$

(b) The function is increasing on $(7, \infty)$.

(c) The function is decreasing on $(3, 7)$.

31. $f(x) = xe^{-3x}$

(a) $f'(x) = e^{-3x} + x(-3e^{-3x})$

$\quad = (1 - 3x)e^{-3x}$

$\quad = \dfrac{1 - 3x}{e^{3x}}$

$f'(x)$ is zero when $x = \frac{1}{3}$ and there are no values of x where $f'(x)$ does not exist, so the critical number is $\frac{1}{3}$.

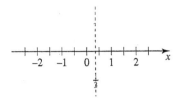

Test a point in each interval.

$$f'(0) = \frac{1 - 3(0)}{e^{3(0)}} = 1 > 0$$

$$f'(1) = \frac{1 - 3(1)}{e^{3(1)}} = -\frac{2}{e^3} < 0$$

(b) The function is increasing on $\left(-\infty, \frac{1}{3}\right)$.

(c) The function is decreasing on $\left(\frac{1}{3}, \infty\right)$.

33. $f(x) = x^2 2^{-x}$

(a) $f'(x) = x^2[\ln 2(2^{-x})(-1)] + (2^{-x})2x$

$\quad = 2^{-x}(-x^2 \ln 2 + 2x)$

$\quad = \dfrac{x(2 - x \ln 2)}{2^x}$

$f'(x)$ is zero when $x = 0$ or $x = \frac{2}{\ln 2}$ and there are no values of x where $f'(x)$ does not exist. The critical numbers are 0 and $\frac{2}{\ln 2}$.

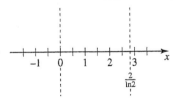

Test a point in each interval.

$$f'(-1) = \frac{(-1)(2 - (-1)\ln 2)}{2^{-1}} = -2(2 + \ln 2) < 0$$

$$f'(1) = \frac{(1)(2 - (1)\ln 2)}{2^1} = \frac{2 - \ln 2}{2} > 0$$

$$f'(3) = \frac{(3)(2 - (3)\ln 2)}{2^3} = \frac{3(2 - 3\ln 2)}{8} < 0$$

(b) The function is increasing on $\left(0, \frac{2}{\ln 2}\right)$.

(c) The function is decreasing on $(-\infty, 0)$ and $\left(\frac{2}{\ln 2}, \infty\right)$.

35. $y = x^{2/3} - x^{5/3}$

(a) $y' = \dfrac{2}{3}x^{-1/3} - \dfrac{5}{3}x^{2/3} = \dfrac{2 - 5x}{3x^{1/3}}$

$y' = 0$ when $x = \frac{2}{5}$. The derivative does not exist at $x = 0$. So the critical numbers are 0 and $\frac{2}{5}$.

Test a point in each interval.

$$y'(-1) = \frac{7}{-3} < 0$$

$$y'\left(\frac{1}{5}\right) = \frac{1}{3\left(\frac{1}{5}\right)^{1/3}} = \frac{5^{1/3}}{3} > 0$$

$$y'(1) = \frac{-3}{3} = -1 < 0$$

(b) y is increasing on $\left(0, \frac{2}{5}\right)$.

(c) y is decreasing on $(-\infty, 0)$ and $\left(\frac{2}{5}, \infty\right)$.

39. $f(x) = ax^2 + bx + c,\ a < 0$

$f'(x) = 2ax + b$

Let $f'(x) = 0$ to find the critical number.

$$2ax + b = 0$$
$$2ax = -b$$
$$x = \frac{-b}{2a}$$

Choose a value in the interval $\left(-\infty, -\frac{b}{2a}\right)$. Since $a < 0$,

$$\frac{-b}{2a} - \frac{-1}{2a} = \frac{-b+1}{2a} < \frac{-b}{2a}.$$
$$f'\left(\frac{-b+1}{2a}\right) = 2a\left(\frac{-b+1}{2a}\right) + b$$
$$= 1 > 0$$

Choose a value in the interval $\left(\frac{-b}{2a}, \infty\right)$. Since $a < 0$,

$$\frac{-b}{2a} - \frac{-1}{2a} = \frac{-b-1}{2a} < \frac{-b}{2a}.$$
$$f'\left(\frac{-b-1}{2a}\right) = 2a\left(\frac{-b-1}{2a}\right) + b$$
$$= -1 < 0$$

f' is increasing on $\left(-\infty, \frac{-b}{2a}\right)$ and decreasing on $\left(\frac{-b}{2a}, \infty\right)$.

This tells us that the curve opens downward and $x = \frac{-b}{2a}$ is the x-coordinate of the vertex.

$$f\left(\frac{-b}{2a}\right) = a\left(\frac{-b}{2a}\right)^2 + b\left(\frac{-b}{2a}\right) + c$$
$$= \frac{ab^2}{4a^2} - \frac{b^2}{2a} + c$$
$$= \frac{b^2}{4a} - \frac{2b^2}{4a} + \frac{4ac}{4a}$$
$$= \frac{4ac - b^2}{4a}$$

The vertex is $\left(\frac{-b}{2a}, \frac{4ac-b^2}{4a}\right)$ or $\left(-\frac{b}{2a}, \frac{4ac-b^2}{4a}\right)$.

41. $f(x) = \ln x$

$$f'(x) = \frac{1}{x}$$

$f'(x)$ is undefined at $x = 0$. $f'(x)$ never equals zero. Note that $f(x)$ has a domain of $(0, \infty)$. Pick a value in the interval $(0, \infty)$.

$$f'(2) = \frac{1}{2} > 0$$

$f(x)$ is increasing on $(0, \infty)$.
$f(x)$ is never decreasing.
Since $f(x)$ never equals zero, the tangent line is horizontal nowhere.

43. $f(x) = e^{0.001x} - \ln x$

$$f'(x) = 0.001e^{0.001x} - \frac{1}{x}$$

Note that $f(x)$ is only defined for $x > 0$. Use a graphing calculator to plot $f'(x)$ for $x > 0$.

(a) $f'(x) > 0$ about $(567, \infty)$, so $f(x)$ is increasing about $(567, \infty)$.

(b) $f'(x) < 0$ about $(0, 567)$, so $f(x)$ is decreasing about $(0, 567)$.

45. $H(r) = \dfrac{300}{1 + 0.03r^2} = 300(1 + 0.03r^2)^{-1}$

$$H'(r) = 300[-1(1 + 0.03r^2)^{-2}(0.06r)]$$
$$= \frac{-18r}{(1 + 0.03r^2)^2}$$

Since r is a mortgage rate (in percent), it is always positive. Thus, $H'(r)$ is always negative.

(a) H is increasing on nowhere.

(b) H is decreasing on $(0, \infty)$.

47. $C(x) = 0.32x^2 - 0.00004x^3$

$R(x) = 0.848x^2 - 0.0002x^3$

$P(x) = R(x) - C(x)$

$\quad = (0.848x^2 - 0.0002x^3) - (0.32x^2 - 0.00004x^3)$

$\quad = 0.528x^2 - 0.00016x^3$

$P'(x) = 1.056x - 0.00048x^2$

$1.056x - 0.00048x^2 = 0$

$x(1.056 - 0.00048x) = 0$

$\quad\quad\quad\quad x = 0 \text{ or } x = 2200$

Choose $x = 1000$ and $x = 3000$ as test points.

$P'(1000) = 1.056(1000) - 0.00048(1000)^2 = 576$

$P'(3000) = 1.056(3000) - 0.00048(3000)^2$

$\quad\quad\quad = -1152$

The function is increasing on $(0, 2200)$.

49. (a) These curves are graphs of functions since they all pass the vertical line test.

(b) The graph for particulates increases from April to July; it decreases from July to November; it is constant from January to April and November to December.

(c) All graphs are constant from January to April and November to December. When the temperature is low, as it is during these months, air pollution is greatly reduced.

51. $A(x) = 0.003631x^3 - 0.03746x^2$

$\quad\quad\quad + 0.1012x + 0.009$

$A'(x) = 0.010893x^2 - 0.07492x + 0.1012$

Solve for $A'(x) = 0$.

$x \approx 1.85 \text{ or } x \approx 5.03$

Choose $x = 1$ and $x = 4$ as test points.

$A'(1) = 0.010893(1)^2 - 0.07492(1)$

$\quad\quad\quad + 0.1012$

$\quad\quad = 0.037173$

$A'(4) = 0.010893(4)^2 - 0.07492(4)$

$\quad\quad\quad + 0.1012$

$\quad\quad = -0.024192$

(a) The function is increasing on $(0, 1.85)$.

(b) The function is decreasing on $(1.85, 5)$.

53. $K(t) = \dfrac{5t}{t^2 + 1}$

$K'(t) = \dfrac{5(t^2 + 1) - 2t(5t)}{(t^2 + 1)^2}$

$\quad\quad = \dfrac{5t^2 + 5 - 10t^2}{(t^2 + 1)^2}$

$\quad\quad = \dfrac{5 - 5t^2}{(t^2 + 1)^2}$

$K'(t) = 0$ when

$$\frac{5 - 5t^2}{(t^2 + 1)^2} = 0$$

$$5 - 5t^2 = 0$$

$$5t^2 = 5$$

$$t = \pm 1.$$

Since t is the time after a drug is administered, the function applies only for $[0, \infty)$, so we discard $t = -1$. Then 1 divides the interval into two intervals.

$K'(0.5) = 2.4 > 0$

$K'(2) = -0.6 < 0$

(a) K is increasing on $(0, 1)$.

(b) K is decreasing on $(1, \infty)$.

55. (a) $F(t) = -10.28 + 175.9te^{-t/1.3}$

$F'(t) = (175.9)(e^{-t/1.3})$

$\quad\quad\quad + (175.9 \cdot 9t)\left(-\dfrac{1}{1.3}e^{-t/1.3}\right)$

$\quad\quad = (175.9)(e^{-t/1.3})\left(1 - \dfrac{t}{1.3}\right)$

$\quad\quad \approx 175.9e^{-t/1.3}(1 - 0.769t)$

(b) $F'(t)$ is equal to 0 at $t = 1.3$. Therefore, 1.3 is a critical number. Since the domain is $(0, \infty)$, test values in the intervals from $(0, 1.3)$ and $(1.3, \infty)$.

$F'(1) \approx 18.83 > 0$ and $F'(2) \approx -20.32 < 0$

$F'(t)$ is increasing on $(0, 1.3)$ and decreasing on $(1.3, \infty)$.

57. $f(x) = \dfrac{1}{\sqrt{2\pi}} e^{-x^2/2}$

$f'(x) = \dfrac{1}{\sqrt{2\pi}} e^{-x^2/2}(-x)$

$\qquad = \dfrac{-x}{\sqrt{2\pi}} e^{-x^2/2}$

$f'(x) = 0$ when $x = 0$.

Choose a value from each of the intervals $(-\infty, 0)$ and $(0, \infty)$.

$f'(-1) = \dfrac{1}{\sqrt{2\pi}} e^{-1/2} > 0$

$f'(1) = \dfrac{-1}{\sqrt{2\pi}} e^{-1/2} < 0$

The function is increasing on $(-\infty, 0)$ and decreasing on $(0, \infty)$.

59. As shown on the graph,

(a) horsepower increases with engine speed on $(1500, 6250)$;

(b) horsepower decreases with engine speed on $(6250, 7200)$;

(c) torque increases with engine speed on $(1500, 2500)$ and $(3500, 4400)$;

(d) torque decreases with engine speed on $(3000, 3500)$ and $(6000, 7200)$.

13.2   Relative Extrema

1. As shown on the graph, the relative minimum of -4 occurs when $x = 1$.

3. As shown on the graph, the relative maximum of 3 occurs when $x = -2$.

5. As shown on the graph, the relative maximum of 3 occurs when $x = -4$ and the relative minimum of 1 occurs when $x = -2$.

7. As shown on the graph, the relative maximum of 3 occurs when $x = -4$; the relative minimum of -2 occurs when $x = -7$ and $x = -2$

9. Since the graph of the function is zero at $x = -1$ and $x = 3$, the critical numbers are -1 and 3.

Since the graph of the function is positive on $(-\infty, -1)$ and negative on $(-1, 3)$, there is a relative maximum at -1. Since the graph of the function is negative on $(-1, 3)$ and positive on $(3, \infty)$, there is a relative minimum at 3.

11. Since the graph of the function is zero at $x = -8, x = -6, x = -2.5$ and $x = -1.5$, the critical numbers are $-8, -6, -2.5$, and -1.5.

Since the graph of the function is positive on $(-\infty, -8)$ and negative on $(-8, -6)$, there is a relative maximum at -8. Since the graph of the function is on $(-8, -6)$ and positive on $(-6, -2.5)$, there is a relative minimum at -6. Since the graph of the function is positive on $(-6, -2.5)$ and negative on $(-2.5, -1.5)$, there is a relative maximum at -2.5. Since the graph of the function is negative on $(-2.5, -1.5)$ and positive on $(-1.5, \infty)$, there is a relative minimum at -1.5.

13. $f(x) = x^2 - 10x + 33$
$\qquad f'(x) = 2x - 10$

$f'(x)$ is zero when $x = 5$.

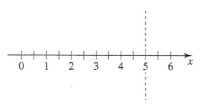

$f'(0) = -10 < 0$
$f'(6) = 2 > 0$

f is decreasing on $(-\infty, 5)$ and increasing on $(5, \infty)$. Thus, a relative minimum occurs at $x = 5$.

$f(5) = 8$

Relative minimum of 8 at 5

15. $f(x) = x^3 + 6x^2 + 9x - 8$
$\quad f'(x) = 3x^2 + 12x + 9 = 3(x^2 + 4x + 3)$
$\qquad\quad = 3(x + 3)(x + 1)$

$f'(x)$ is zero when $x = -1$ or $x = -3$.

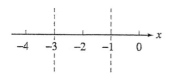

$f'(-4) = 9 > 0$
$f'(-2) = -3 < 0$
$f'(0) = 9 > 0$

Thus, f is increasing on $(-\infty, -3)$, decreasing on $(-3, -1)$, and increasing on $(-1, \infty)$. f has a relative maximum at -3 and a relative minimum at -1.

$f(-3) = -8$
$f(-1) = -12$

Relative maximum of -8 at -3; relative minimum of -12 at -1

17. $f(x) = -\dfrac{4}{3}x^3 - \dfrac{21}{2}x^2 - 5x + 8$

$f'(x) = -4x^2 - 21x - 5$
$\quad\;\; = (-4x - 1)(x + 5)$

$f'(x)$ is zero when $x = -5$, or $x = -\frac{1}{4}$.

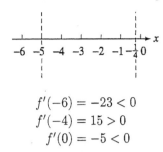

$f'(-6) = -23 < 0$
$f'(-4) = 15 > 0$
$f'(0) = -5 < 0$

f is decreasing on $(-\infty, -5)$, increasing on $\left(-5, -\frac{1}{4}\right)$, and decreasing on $\left(-\frac{1}{4}, \infty\right)$. f has a relative minimum at -5 and a relative maximum at $-\frac{1}{4}$.

$$f(-5) = -\frac{377}{6}$$

$$f\left(-\frac{1}{4}\right) = \frac{827}{96}$$

Relative maximum of $\frac{827}{96}$ at $-\frac{1}{4}$; relative minimum of $-\frac{377}{6}$ at -5

19. $f(x) = x^4 - 18x^2 - 4$
$f'(x) = 4x^3 - 36x$
$\quad\;\;\; = 4x(x^2 - 9)$
$\quad\;\;\; = 4x(x + 3)(x - 3)$

$f'(x)$ is zero when $x = 0$ or $x = -3$ or $x = 3$.

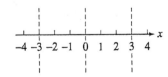

$f'(-4) = 4(-4)^3 - 36(-4) = -112 < 0$
$f'(-1) = -4 + 36 = 32 > 0$
$f'(1) = 4 - 36 = -32 < 0$
$f'(4) = 4(4)^3 - 36(4) = 112 > 0$

f is decreasing on $(-\infty, -3)$ and $(0, 3)$; f is increasing on $(-3, 0)$ and $(3, \infty)$.

$$f(-3) = -85$$
$$f(0) = -4$$
$$f(3) = -85$$

Relative maximum of -4 at 0; relative minimum of -85 at 3 and -3

21. $f(x) = 3 - (8 + 3x)^{2/3}$

$f'(x) = -\dfrac{2}{3}(8 + 3x)^{-1/3}(3)$

$\quad\;\; = -\dfrac{2}{(8 + 3x)^{1/3}}$

Critical number:

$$8 + 3x = 0$$

$$x = -\frac{8}{3}$$

$f'(-3) = 2 > 0$
$f'(0) = -1 < 0$

f is increasing on $\left(-\infty, -\frac{8}{3}\right)$ and decreasing on $\left(-\frac{8}{3}, \infty\right)$.

$$f\left(-\frac{8}{3}\right) = 3$$

Relative maximum of 3 at $-\frac{8}{3}$

23. $f(x) = 2x + 3x^{2/3}$
$f'(x) = 2 + 2x^{-1/3}$

$\quad\;\;\; = 2 + \dfrac{2}{\sqrt[3]{x}}$

Find the critical numbers.

$f'(x) = 0$ when

$$2 + \frac{2}{\sqrt[3]{x}} = 0$$

$$\frac{2}{\sqrt[3]{x}} = -2$$

$$\frac{1}{\sqrt[3]{x}} = -1$$

$$x = (-1)^3$$

$$x = -1.$$

$f'(x)$ does not exist when

$$\sqrt[3]{x} = 0$$

$$x = 0.$$

$$f'(-2) = 2 + \frac{2}{\sqrt[3]{-2}} \approx 0.41 > 0$$

$$f'\left(-\frac{1}{2}\right) = 2 + \frac{2}{\sqrt[3]{-\frac{1}{2}}}$$

$$= 2 + \frac{2\sqrt[3]{2}}{-1} \approx -0.52 < 0$$

$$f'(1) = 2 + \frac{2}{\sqrt[3]{1}} = 4 > 0$$

f is increasing on $(-\infty, -1)$ and $(0, \infty)$.
f is decreasing on $(-1, 0)$.

$$f(-1) = 2(-1) + 3(-1)^{2/3} = 1$$
$$f(0) = 0$$

Relative maximum of 1 at -1; relative minimum of 0 at 0

25. $f(x) = x - \dfrac{1}{x}$

$f'(x) = 1 + \frac{1}{x^2}$ is never zero, but fails to exist at $x = 0$.
Since $f(x)$ also fails to exist at $x = 0$, there are no critical numbers and no relative extrema.

27. $f(x) = \dfrac{x^2 - 2x + 1}{x - 3}$

$$f'(x) = \frac{(x-3)(2x-2) - (x^2 - 2x + 1)(1)}{(x-3)^2}$$

$$= \frac{x^2 - 6x + 5}{(x-3)^2}$$

Find the critical numbers:

$$x^2 - 6x + 5 = 0$$
$$(x-5)(x-1) = 0$$
$$x = 5 \quad \text{or} \quad x = 1$$

Note that $f(x)$ and $f'(x)$ do not exist at $x = 3$, so the only critical numbers are 1 and 5.

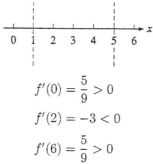

$$f'(0) = \frac{5}{9} > 0$$

$$f'(2) = -3 < 0$$

$$f'(6) = \frac{5}{9} > 0$$

$f(x)$ is increasing on $(-\infty, 1)$ and $(5, \infty)$.
$f(x)$ is decreasing on $(1, 5)$.

$$f(1) = 0$$
$$f(5) = 8$$

Relative maximum of 0 at 1; relative minimum of 8 at 5

29. $f(x) = x^2 e^x - 3$
$$f'(x) = x^2 e^x + 2x e^x$$
$$= xe^x(x + 2)$$

$f'(x)$ is zero at $x = 0$ and $x = -2$.

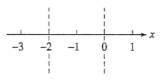

$$f'(-3) = 3e^{-3} = \frac{3}{e^3} > 0$$

$$f'(-1) = -e^{-1} = \frac{-1}{e} < 0$$

$$f'(1) = 3e^1 > 0$$

f is increasing on $(-\infty, -2)$ and $(0, \infty)$.
f is decreasing on $(-2, 0)$.

$$f(0) = 0 \cdot e^0 - 3 = -3$$
$$f(-2) = (-2)^2 e^{-2} - 3$$

$$= \frac{4}{e^2} - 3$$

$$\approx -2.46$$

Relative minimum of -3 at 0; relative maximum of -2.46 at -2

31. $f(x) = 2x + \ln x$

$$f'(x) = 2 + \frac{1}{x} = \frac{2x + 1}{x}$$

$f'(x)$ is zero at $x = -\frac{1}{2}$. The domain of $f(x)$ is $(0, \infty)$. Therefore $f'(x)$ is never zero in the domain of $f(x)$.
$f'(1) = 3 > 0$. Since $f(x)$ is always increasing, f has no relative extrema.

33. $f(x) = \dfrac{2^x}{x}$

$$f'(x) = \frac{(x)\ln 2(2^x) - 2^x(1)}{x^2}$$

$$= \frac{2^x(x \ln 2 - 1)}{x^2}$$

Find the critical numbers:

$$x \ln 2 - 1 = 0 \qquad \text{or} \quad x^2 = 0$$

$$x = \frac{1}{\ln 2} \qquad \quad x = 0$$

Since f is not defined for $x = 0$, 0 is not a critical number. $x = \frac{1}{\ln 2} \approx 1.44$ is the only critical number.

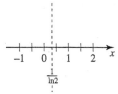

$$f'(1) \approx -0.6137 < 0$$
$$f'(2) \approx 0.3863 > 0$$

f is decreasing on $\left(0, \frac{1}{\ln 2}\right)$ and increasing on $\left(\frac{1}{\ln 2}, \infty\right)$.

$$f\left(\frac{1}{\ln 2}\right) = \frac{2^{1/\ln 2}}{\frac{1}{\ln 2}} = e \ln 2$$

Relative minimum of $e \ln 2$ at $\dfrac{1}{\ln 2}$.

35. $y = -2x^2 + 12x - 5$
$y' = -4x + 12$
$ = -4(x - 3)$

The vertex occurs when $y' = 0$ or when

$$x - 3 = 0$$
$$x = 3$$

When $x = 3$,

$$y = -2(3)^2 + 12(3) - 5 = 13$$

The vertex is $(3, 13)$.

37. $f(x) = x^5 - x^4 + 4x^3 - 30x^2 + 5x + 6$

$f'(x) = 5x^4 - 4x^3 + 12x^2 - 60x + 5$

Graph f' on a graphing calculator. A suitable choice for the viewing window is $[-4, 4]$ by $[-50, 50]$, Yscl $= 10$.

Use the calculator to estimate the x-intercepts of this graph. These numbers are the solutions of the equation $f'(x) = 0$ and thus the critical numbers for f. Rounded to three decimal places, these x-values are 0.085 and 2.161.

Examine the graph of f' near $x = 0.085$ and $x = 2.161$. Observe that $f'(x) > 0$ to the left of $x = 0.085$ and $f'(x) < 0$ to the right of $x = 0.085$. Also observe that $f'(x) < 0$ to the left of $x = 2.161$ and $f'(x) > 0$ to the right of $x = 2.161$. The first derivative test allows us to conclude that f has

a relative maximum at $x = 0.085$ and a relative minimum at $x = 2.161$.

$$f(0.085) \approx 6.211$$
$$f(2.161) \approx -57.607$$

Relative maximum of 6.211 at 0.085; relative minimum of -57.607 at 2.161.

39. $f(x) = 2|x + 1| + 4|x - 5| - 20$

Graph this function in the window $[-10, 10]$ by $[-15, 30]$, Yscl $= 5$.

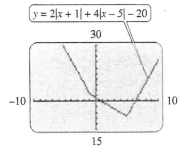

The graph shows that f has no relative maxima, but there is a relative minimum at $x = 5$.
(Note that the graph has a sharp point at $(5, -8)$, indicating that $f'(5)$ does not exist.)

41. $C(q) = 80 + 18q$; $p = 70 - 2q$

$P(q) = R(q) - C(q) = pq - C(q)$
$ = (70 - 2q)q - (80 + 18q)$
$ = -2q^2 + 52q - 80$

(a) Since the graph of P is a parabola that opens downward, we know that its vertex is a maximum point. To find the q-value of this point, we find the critical number.

$P'(q) = -4q + 52$
$P'(q) = 0$ when

$$-4q + 52 = 0$$
$$4q = 52$$
$$q = 13$$

The number of units that produce maximum profit is 13.

(b) If $q = 13$,

$$p = 70 - 2(13)$$
$$ = 44$$

The price that produces maximum profit is $44.

(c) $P(13) = -2(13)^2 + 52(13) - 80 = 258$

The maximum profit is $258.

43. $C(q) = 100 + 20qe^{-0.01q}$; $p = 40e^{-0.01q}$

$$\begin{aligned} P(q) &= R(q) - C(q) = pq - C(q) \\ &= (40e^{-0.01q})q - (100 + 20qe^{-0.01q}) \\ &= 20qe^{-0.01q} - 100 \end{aligned}$$

(a) $\begin{aligned} P'(q) &= 20e^{-0.01q} + 20qe^{-0.01q}(-0.01) \\ &= (20 - 0.2q)e^{-0.01q} \end{aligned}$

Solve $P'(q) = 0$.

$$\begin{aligned} (20 - 0.2q)e^{-0.01q} &= 0 \\ 20 - 0.2q &= 0 \\ q &= 100 \end{aligned}$$

Since $e^{-0.01q} > 0$ for all values of q, the sign of $P'(q)$ is the same as the sign of $20 - 0.2q$. For $q < 100, P'(q) > 0$; for $q > 100, P'(q) < 0$. Therefore, the number of units that produces maximum profit is 100.

(b) If $q = 100$,

$$\begin{aligned} p &= 40e^{-0.01(100)} \\ &= 40e^{-1} \\ &\approx 14.72 \end{aligned}$$

The price per unit that produces maximum profit is \$14.72.

(c) $\begin{aligned} P(100) &= 20(100)e^{-0.01(100)} - 100 \\ &= 2000e^{-1} - 100 \\ &\approx 635.76 \end{aligned}$

The maximum profit is \$635.76.

45. $P(t) = -0.01432t^3 + 0.3976t^2 - 2.257t + 23.41$
$P'(t) = -0.04296t^2 + 0.7952t - 2.257$

Solve for $P'(t) = 0$:

$t \approx 3.5001$ or $t \approx 15.0101$

Test points $t = 1, t = 4, t = 16$.

$$\begin{aligned} P'(1) &\approx -1.5048 \\ P'(4) &\approx 0.2364 \\ P'(16) &\approx -0.5316 \end{aligned}$$

P is decreasing on $(0, 3.5)$ and $(15, 18)$ and increasing on $(3.5, 15)$.

$$\begin{aligned} P(0) &= 23.41 \\ P(3.5) &\approx 19.767 \\ P(15) &\approx 30.685 \\ P(18) &\approx 28.092 \end{aligned}$$

Relative maximum of 23,410 megawatts at midnight; relative minimum of 19,767 megawatts at 3:30 a.m.; relative maximum of 30,685 at 3:00 p.m.; relative minimum of 28,092 at 6:00 p.m.

47. $p = D(q) = 200e^{-0.1q}$

$$\begin{aligned} R(q) &= pq \\ &= 200qe^{-0.1q} \\ R'(q) &= 200qe^{-0.1q}(-0.1) + 200e^{-0.1q} \\ &= 20e^{-0.1q}(10 - q) \end{aligned}$$

$R'(q) = 0$ when $q = 10$, the only critical number. Use the first derivative test to verify that $q = 10$ gives the maximum revenue.

$$\begin{aligned} R'(9) &= 20e^{-0.9} > 0 \\ R'(11) &= -20e^{-1.1} < 0 \end{aligned}$$

The maximum revenue results when $q = 10$

$p = D(10) = \frac{200}{e} \approx 73.58$, or when telephones are sold at \$73.58.

49. $C(x) = 0.002x^3 = 9x + 6912$

$$\overline{C}(x) = \frac{C(x)}{x} = 0.002x^2 + 9 + \frac{6912}{x}$$

$$\overline{C}'(x) = 0.004x - \frac{6912}{x^2}$$

$\overline{C}'(x) = 0$ when

$$\begin{aligned} 0.004x - \frac{6912}{x^2} &= 0 \\ 0.004x^3 &= 6912 \\ x^3 &= 1,728,000 \\ x &= 120 \end{aligned}$$

A product level of 120 units will produce the minimum average cost per unit.

51. (a) $M(t) = 6.281t^{0.242}e^{-0.025t}$
$$\begin{aligned} M'(t) &= (1.520002t^{-0.758})(e^{-0.025t}) \\ &\quad + (6.281t^{0.242})(-0.025e^{-0.025t}) \\ &= e^{-0.025t}(1.520002t^{-0.758} - 0.157025t^{0.242}) \end{aligned}$$

$M'(t) = 0$ when

$$\begin{aligned} 1.520002t^{-0.758} - 0.157025t^{0.242} &= 0 \\ t &= 9.68 \end{aligned}$$

Let $t = 9.68$ in $M(t)$.

$$\begin{aligned} M(9.68) &= 6.281(9.68)^{0.242}e^{-0.025(9.68)} \\ &\approx 8.54 \text{ kg} \end{aligned}$$

The maximum daily consumption is 8.54 kg and it occurs at 9.68 weeks.

(b) $M(t) = at^b e^{-ct}$

$M'(t) = (bat^{b-1})(e^{-ct}) + (at^b)(-ce^{-ct})$

$\qquad = ae^{-ct}(bt^{b-1} - ct^b)$

$M'(t) = 0$ when

$$bt^{b-1} - ct^b = 0$$
$$bt^{b-1} = ct^b$$
$$\frac{b}{c} = t$$

Let $t = \dfrac{b}{c}$ in $M(t)$.

$$M\left(\frac{b}{c}\right) = a\left(\frac{b}{c}\right)^b e^{-c\left(\frac{b}{c}\right)}$$

$$= a\left(\frac{b}{c}\right)^b e^{-b}$$

The maximum daily consumption is $a(b/c)^b e^{-b}$ kg and it occurs at b/c weeks.

53. $F(t) = -10.28 + 175.9te^{-t/1.3}$

$$F'(t) = (175.9t)\left(-\frac{1}{1.3}\right)e^{-t/1.3} + (175.9)(e^{-t/1.3})$$

$$= e^{-t/1.3}\left(-\frac{175.9t}{1.3} + 175.9\right)$$

$F'(t) = 0$ when

$$-\frac{175.9t}{1.3} + 175.9 = 0$$
$$t = 1.3$$

At 1.3 hours, the termal effect of the food is maximized.

55. $R(t) = \dfrac{20t}{t^2 + 100}$

$$R'(t) = \frac{20(t^2 + 100) - 20t(2t)}{(t^2 + 100)^2} = \frac{2000 - 20t^2}{(t^2 + 100)^2}$$

$R'(t) = 0$ when

$$2000 - 20t^2 = 0$$
$$-20t^2 = -2000$$
$$t^2 = 100$$
$$t = \pm 10.$$

Disregard the negative value.

Use the first derivative test to verify that $t = 10$ gives a maximum rating.

$$R'(9) = 0.0116 > 0$$
$$R'(11) = -0.0086 < 0$$

The film should be 10 minutes long.

13.3   Higher Derivatives, Concavity, and the Second Derivative Test

1. $f(x) = 5x^3 - 7x^2 + 4x + 3$

$f'(x) = 15x^2 - 14x + 4$

$f''(x) = 30x - 14$

$f''(0) = 30(0) - 14 = -14$

$f''(2) = 30(2) - 14 = 46$

3. $f(x) = 4x^4 - 3x^3 - 2x^2 + 6$

$f'(x) = 16x^3 - 9x^2 - 4x$

$f''(x) = 48x^2 - 18x - 4$

$f''(0) = 48(0)^2 - 18(0) - 4 = -4$

$f''(2) = 48(2)^2 - 18(2) - 4 = 152$

5. $f(x) = 3x^2 - 4x + 8$

$f'(x) = 6x - 4$

$f''(x) = 6$

$f''(0) = 6$

$f''(2) = 6$

7. $f(x) = \dfrac{x^2}{1+x}$

$$f'(x) = \frac{(1+x)(2x) - x^2(1)}{(1+x)^2}$$

$$= \frac{2x + x^2}{(1+x)^2}$$

$$f''(x) = \frac{(1+x)^2(2 + 2x) - (2x + x^2)(2)(1+x)}{(1+x)^4}$$

$$= \frac{(1+x)(2 + 2x) - (2x + x^2)(2)}{(1+x)^3}$$

$$= \frac{2}{(1+x)^3}$$

$f''(0) = 2$

$$f''(2) = \frac{2}{27}$$

9. $f(x) = \sqrt{x^2 + 4} = (x^2 + 4)^{1/2}$

$f'(x) = \dfrac{1}{2}(x^2 + 4)^{-1/2} \cdot 2x$

$\quad = \dfrac{x}{(x^2 + 4)^{1/2}}$

$f''(x) = \dfrac{(x^2 + 4)^{1/2}(1) - x\left[\frac{1}{2}(x^2 + 4)^{-1/2}\right]2x}{x^2 + 4}$

$\quad = \dfrac{(x^2 + 4)^{1/2} - \frac{x^2}{(x^2+4)^{1/2}}}{x^2 + 4}$

$\quad = \dfrac{(x^2 + 4) - x^2}{(x^2 + 4)^{3/2}}$

$\quad = \dfrac{4}{(x^2 + 4)^{3/2}}$

$f''(0) = \dfrac{4}{(0^2 + 4)^{3/2}}$

$\quad = \dfrac{4}{4^{3/2}} = \dfrac{4}{8} = \dfrac{1}{2}$

$f''(2) = \dfrac{4}{(2^2 + 4)^{3/2}}$

$\quad = \dfrac{4}{8^{3/2}} = \dfrac{4}{16\sqrt{2}} = \dfrac{1}{4\sqrt{2}}$

11. $f(x) = 32x^{3/4}$

$f'(x) = 24x^{-1/4}$

$f''(x) = -6x^{-5/4} = -\dfrac{6}{x^{5/4}}$

$f''(0)$ does not exist.

$f''(2) = -\dfrac{6}{2^{5/4}}$

$\quad = -\dfrac{3}{2^{1/4}}$

13. $f(x) = 5e^{-x^2}$

$f'(x) = 5e^{-x^2}(-2x) = -10xe^{-x^2}$

$f''(x) = -10xe^{-x^2}(-2x) + e^{-x^2}(-10)$

$\quad = 20x^2 e^{-x^2} - 10e^{-x^2}$

$f''(0) = 20(0^2)e^{-0^2} - 10e$

$\quad = 0 - 10 = -10$

$f''(2) = 20(2^2)e^{-(2^2)} - 10c^{-(2^2)}$

$\quad = 80e^{-4} - 10e^{-4} = 70e^{-4}$

$\quad \approx 1.282$

15. $f(x) = \dfrac{\ln x}{4x}$

$f'(x) = \dfrac{4x\left(\frac{1}{x}\right) - (\ln x)(4)}{(4x)^2}$

$\quad = \dfrac{4 - 4\ln x}{16x^2} = \dfrac{1 - \ln x}{4x^2}$

$f''(x) = \dfrac{4x^2\left(-\frac{1}{x}\right) - (1 - \ln x)8x}{16x^4}$

$\quad = \dfrac{-4x - 8x + 8x\ln x}{16x^4} = \dfrac{-12x + 8x\ln x}{16x^4}$

$\quad = \dfrac{4x(-3 + 2\ln x)}{16x^4} = \dfrac{-3 + 2\ln x}{4x^3}$

$f''(0)$ does not exist because $\ln 0$ is undefined.

$f''(2) = \dfrac{-3 + 2\ln 2}{4(2)^3} = \dfrac{-3 + 2\ln 2}{32} \approx 0.050$

17. $f(x) = 7x^4 + 6x^3 + 5x^2 + 4x + 3$

$f'(x) = 28x^3 + 18x^2 + 10x + 4$

$f''(x) = 84x^2 + 36x + 10$

$f'''(x) = 168x + 36$

$f^{(4)}(x) = 168$

19. $f(x) = 5x^5 - 3x^4 + 2x^3 + 7x^2 + 4$

$f'(x) = 25x^4 - 12x^3 + 6x^2 + 14x$

$f''(x) = 100x^3 - 36x^2 + 12x + 14$

$f'''(x) = 300x^2 - 72x + 12$

$f^{(4)}(x) = 600x - 72$

21. $f(x) = \dfrac{x - 1}{x + 2}$

$f'(x) = \dfrac{(x + 2) - (x - 1)}{(x + 2)^2} = \dfrac{3}{(x + 2)^2}$

$f''(x) = \dfrac{-3(2)(x + 2)}{(x + 2)^4} = \dfrac{-6}{(x + 2)^3}$

$f'''(x) = \dfrac{(-6)(-3)(x + 2)^2}{(x + 2)^6}$

$\quad = 18(x + 2)^{-4} \quad \text{or} \quad \dfrac{18}{(x + 2)^4}$

$f^{(4)}(x) = \dfrac{-18(4)(x + 2)^3}{(x + 2)^8}$

$\quad = -72(x + 2)^{-5} \quad \text{or} \quad \dfrac{-72}{(x + 2)^5}$

23. $f(x) = \dfrac{3x}{x-2}$

$f'(x) = \dfrac{(x-2)(3) - 3x(1)}{(x-2)^2} = \dfrac{-6}{(x-2)^2}$

$f''(x) = \dfrac{-6(-2)(x-2)}{(x-2)^4} = \dfrac{12}{(x-2)^3}$

$f'''(x) = \dfrac{-12(3)(x-2)^2}{(x-2)^6} = -36(x-2)^{-4}$

or $\dfrac{-36}{(x-2)^4}$

$f^{(4)}(x) = \dfrac{-36(-4)(x-2)^3}{(x-2)^8} = 144(x-2)^{-5}$

or $\dfrac{144}{(x-2)^5}$

25. $f(x) = \ln x$

(a) $f'(x) = \dfrac{1}{x} = x^{-1}$

$f''(x) = -x^{-2} = \dfrac{-1}{x^2}$

$f'''(x) = 2x^{-3} = \dfrac{2}{x^3}$

$f^{(4)}(x) = -6x^{-4} = \dfrac{-6}{x^4}$

$f^{(5)}(x) = 24x^{-5} = \dfrac{24}{x^5}$

(b) $f^{(n)}(x) = \dfrac{(-1)^{n-1}(n-1)!}{x^n}$

27. Concave upward on $(2, \infty)$
Concave downward on $(-\infty, 2)$
Inflection point at $(2, 3)$

29. Concave upward on $(-\infty, -1)$ and $(8, \infty)$
Concave downward on $(-1, 8)$
Inflection point at $(-1, 7)$ and $(8, 6)$

31. Concave upward on $(2, \infty)$
Concave downward on $(-\infty, 2)$
No points of inflection

33. $f(x) = x^2 + 10x - 9$
$f'(x) = 2x + 10$
$f''(x) = 2 > 0$ for all x.

Always concave upward
No inflection points

35. $f(x) = -2x^3 + 9x^2 + 168x - 3$
$f'(x) = -6x^2 + 18x + 168$
$f''(x) = -12x + 18$
$f''(x) = -12x + 18 > 0$ when
$-6(2x - 3) > 0$
$2x - 3 < 0$
$x < \dfrac{3}{2}.$

Concave upward on $\left(-\infty, \frac{3}{2}\right)$

$f''(x) = -12x + 18 < 0$ when
$-6(2x - 3) < 0$
$2x - 3 > 0$
$x > \dfrac{3}{2}.$

Concave downward on $\left(\frac{3}{2}, \infty\right)$

$f''(x) = -12x + 18 = 0$ when
$-6(2x + 3) = 0$
$2x + 3 = 0$
$x = \dfrac{3}{2}.$

$f\left(\dfrac{3}{2}\right) = \dfrac{525}{2}$

Inflection point at $\left(\frac{3}{2}, \frac{525}{2}\right)$

37. $f(x) = \dfrac{3}{x-5}$

$f'(x) = \dfrac{-3}{(x-5)^2}$

$f''(x) = \dfrac{-3(-2)(x-5)}{(x-5)^4} = \dfrac{6}{(x-5)^3}$

$f''(x) = \dfrac{6}{(x-5)^3} > 0$ when
$(x-5)^3 > 0$
$x - 5 > 0$
$x > 5.$

Concave upward on $(5, \infty)$

$f''(x) = \dfrac{6}{(x-5)^3} < 0$ when

$(x-5)^3 < 0$
$x - 5 < 0$
$x < 5.$

Concave downward on $(-\infty, 5)$

$f''(x) \neq 0$ for any value for x; it does not exist when $x = 5$. There is a change of concavity there, but no inflection point since $f(5)$ does not exist.

39. $f(x) = x(x+5)^2$

$f'(x) = x(2)(x+5) + (x+5)^2$

$\quad = (x+5)(2x+x+5)$

$\quad = (x+5)(3x+5)$

$f''(x) = (x+5)(3) + (3x+5)$

$\quad = 3x + 15 + 3x + 5 = 6x + 20$

$f''(x) = 6x + 20 > 0$ when

$\quad 2(3x+10) > 0$

$\quad\quad 3x > -10$

$\quad\quad\quad x > -\dfrac{10}{3}.$

Concave upward on $\left(-\dfrac{10}{3}, \infty\right)$

$f''(x) = 6x + 20 < 0$ when

$\quad 2(3x+10) < 0$

$\quad\quad 3x < -10$

$\quad\quad\quad x < -\dfrac{10}{3}.$

Concave downward on $\left(-\infty, -\dfrac{10}{3}\right)$

$$f\left(-\frac{10}{3}\right) = -\frac{10}{3}\left(-\frac{10}{3} + 5\right)^2$$

$$= \frac{-10}{3}\left(\frac{-10+15}{3}\right)^2$$

$$= -\frac{10}{3} \cdot \frac{25}{9} = -\frac{250}{27}$$

Inflection point at $\left(-\dfrac{10}{3}, -\dfrac{250}{27}\right)$

41. $f(x) = 18x - 18e^{-x}$

$f'(x) = 18 - 18e^{-x}(-1) = 18 + 18e^{-x}$

$f''(x) = 18e^{-x}(-1) = -18e^{-x}$

$f''(x) = -18e^{-x} < 0$ for all x

$f(x)$ is never concave upward and always concave downward. There are no points of inflection since $-18e^{-x}$ is never equal to 0.

43. $f(x) = x^{8/3} - 4x^{5/3}$

$f'(x) = \dfrac{8}{3}x^{5/3} - \dfrac{20}{3}x^{2/3}$

$f''(x) = \dfrac{40}{9}x^{2/3} - \dfrac{40}{9}x^{-1/3} = \dfrac{40(x-1)}{9x^{1/3}}$

$f''(x) = 0$ when $x = 1$

$f''(x)$ fails to exist when $x = 0$

Note that both $f(x)$ and $f'(x)$ exist at $x = 0$.

Check the sign of $f''(x)$ in the three intervals determined by $x = 0$ and $x = 1$ using test points.

$$f''(-1) = \frac{40(-2)}{9(-1)} = \frac{80}{9} > 0$$

$$f''\left(\frac{1}{8}\right) = \frac{40\left(-\frac{7}{8}\right)}{9\left(\frac{1}{2}\right)} = -\frac{70}{9} < 0$$

$$f''(8) = \frac{40(7)}{9(2)} = \frac{140}{9} > 0$$

Concave upward on $(-\infty, 0)$ and $(1, \infty)$; concave downward on $(0, 1)$

$$f(0) = (0)^{8/3} - 4(0)^{5/3} = 0$$
$$f(1) = (1)^{8/3} - 4(1)^{5/3} = -3$$

Inflection points at $(0, 0)$ and $(1, -3)$

45. $f(x) = \ln(x^2 + 1)$

$f'(x) = \dfrac{2x}{x^2 + 1}$

$f''(x) = \dfrac{(x^2+1)(2) - (2x)(2x)}{(x^2+1)^2}$

$\quad\quad = \dfrac{-2x^2 + 2}{(x^2+1)^2}$

$f''(x) = \dfrac{-2x^2 + 2}{(x^2+1)^2} > 0$ when

$\quad -2x^2 + 2 > 0$

$\quad\quad -2x^2 > -2$

$\quad\quad\quad x^2 < 1$

$\quad\quad -1 < x < 1$

Concave upward on $(-1, 1)$

$f''(x) = \dfrac{-2x^2 + 2}{(x^2+1)^2} < 0$ when

$\quad -2x^2 + 2 < 0$

$\quad\quad -2x^2 < -2$

$\quad\quad\quad x^2 > 1$

$\quad\quad x > 1$ or $x < -1$

Concave downward on $(-\infty, -1)$ and $(1, \infty)$

$$f(1) = \ln[(1)^2 + 1] = \ln 2$$
$$f(-1) = \ln[(-1)^2 + 1] = \ln 2$$

Inflection points at $(-1, \ln 2)$ and $(1, \ln 2)$

47. $f(x) = x^2 \log |x|$

$$f'(x) = 2x \log |x| + x^2 \left(\frac{1}{x \ln 10} \right)$$

$$= 2x \log |x| + \frac{x}{\ln 10}$$

$$f''(x) = 2 \log |x| + 2x \left(\frac{1}{x \ln 10} \right) + \frac{1}{\ln 10}$$

$$= 2 \log |x| + \frac{3}{\ln 10}$$

$f''(x) = 2 \log |x| + \frac{3}{\ln 10} > 0$ when

$$2 \log |x| + \frac{3}{\ln 10} > 0$$

$$2 \log |x| > -\frac{3}{\ln 10}$$

$$\log |x| > -\frac{3}{2 \ln 10}$$

$$\frac{\ln |x|}{\ln 10} > -\frac{3}{2 \ln 10}$$

$$\ln |x| > -\frac{3}{2}$$

$$|x| > e^{-3/2}$$

$$x > e^{-3/2} \text{ or } x < -e^{-3/2}$$

Concave upward on $(-\infty, -e^{-3/2})$ and $(e^{-3/2}, \infty)$

$f''(x) = 2 \log |x| + \frac{3}{\ln 10} < 0$ when

$$2 \log |x| + \frac{3}{\ln 10} < 0$$

$$2 \log |x| < -\frac{3}{\ln 10}$$

$$\log |x| < -\frac{3}{2 \ln 10}$$

$$\frac{\ln |x|}{\ln 10} < -\frac{3}{2 \ln 10}$$

$$\ln |x| < -\frac{3}{2}$$

$$|x| < e^{-3/2}$$

$$-e^{-3/2} < x < e^{-3/2}$$

Note that $f(x)$ is not defined at $x = 0$.

Concave downward on $(-e^{-3/2}, 0)$ and $(0, e^{-3/2})$.

$$f(-e^{-3/2}) = (-e^{-3/2})^2 \log \left| -e^{-3/2} \right|$$

$$= e^{-3} \log e^{-3/2} = -\frac{3e^{-3}}{2 \ln 10}$$

$$f(e^{-3/2}) = (e^{-3/2})^2 \log \left| e^{-3/2} \right|$$

$$= e^{-3} \log e^{-3/2} = -\frac{3e^{-3}}{2 \ln 10}$$

Inflection points at $\left(-e^{-3/2}, -\frac{3e^{-3}}{2 \ln 10} \right)$

and $\left(e^{-3/2}, -\frac{3e^{-3}}{2 \ln 10} \right)$

49. Since the graph of $f'(x)$ is increasing on $(-\infty, 0)$ and $(4, \infty)$, the function is concave upward on $(-\infty, 0)$ and $(4, \infty)$. Since the graph of $f'(x)$ is decreasing on $(0, 4)$, the function is concave downward on $(0, 4)$. The inflection points are at 0 and 4.

51. Since the graph of $f'(x)$ is increasing on $(-7, 3)$ and $(12, \infty)$, the function is concave upward on $(-7, 3)$ and $(12, \infty)$. Since the graph of $f'(x)$ is decreasing on $(-\infty, -7)$ and $(3, 12)$, the function is concave downward on $(-\infty, -7)$ and $(3, 12)$. The inflection points are at $-7, 3$, and 12.

53. Choose $f(x) = x^k$, where $1 < k < 2$.

If $k = \frac{4}{3}$, then

$$f'(x) = \frac{4}{3} x^{1/3} \qquad f''(x) = \frac{4}{9} x^{-2/3} = \frac{4}{9x^{2/3}}$$

Critical number: 0
Since $f'(x)$ is negative when $x < 0$ and positive when $x > 0$, $f(x) = x^{4/3}$ has a relative minimum at $x = 0$.

If $k = \frac{5}{3}$, then

$$f'(x) = \frac{5}{3} x^{2/3} \qquad f''(x) = \frac{10}{9} x^{-1/3} = \frac{10}{9x^{1/3}}$$

$f''(x)$ is never 0, and does not exist when $x = 0$; so, the only candidate for an inflection point is at $x = 0$.
Since $f''(x)$ is negative when $x < 0$ and positive when $x > 0$, $f(x) = x^{5/3}$ has an inflection point at $x = 0$.

55. (a) The slope of the tangent line to $f(x) = e^x$ as $x \to -\infty$ is close to 0 since the tangent line is almost horizontal, and a horizontal line has a slope of 0.

(b) The slope of the tangent line to $f(x) = e^x$ as $x \to 0$ is close to 1 since the first derivative represents the slope of the tangent line, $f'(x) = e^x$, and $e^0 = 1$.

57. $f(x) = -x^2 - 10x - 25$
$f'(x) = -2x - 10$
$\qquad = -2(x + 5) = 0$

Critical number: -5

$f''(x) = -2 < 0$ for all x.

The curve is concave downward, which means a relative maximum occurs at $x = -5$.

59. $f(x) = 3x^3 - 3x^2 + 1$
$f'(x) = 9x^2 - 6x$
$\quad\quad = 3x(3x - 2) = 0$

Critical numbers: 0 and $\frac{2}{3}$

$f''(x) = 18x - 6$
$f''(0) = -6 < 0$, which means that a relative maximum occurs at $x = 0$.

$f''\left(\frac{2}{3}\right) = 6 > 0$, which means that a relative minimum occurs at $x = \frac{2}{3}$.

61. $f(x) = (x + 3)^4$
$f'(x) = 4(x + 3)^3 = 0$

Critical number: $x = -3$

$f''(x) = 12(x + 3)^2$
$f''(-3) = 12(-3 + 3)^2 = 0$

The second derivative test fails.
Use the first derivative test.

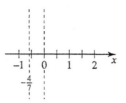

$f'(-4) = 4(-4 + 3)^2$
$\quad\quad = 4(-1)^3 = -4 < 0$

This indicates that f is decreasing on $(-\infty, -3)$.

$f'(0) = 4(0 + 3)^3$
$\quad\quad = 4(3)^3 = 108 > 0$

This indicates that f is increasing on $(-3, \infty)$.
A relative minimum occurs at -3.

63. $f(x) = x^{7/3} + x^{4/3}$

$f'(x) = \frac{7}{3}x^{4/3} + \frac{4}{3}x^{1/3}$

$f'(x) = 0$ when

$$\frac{7}{3}x^{4/3} + \frac{4}{3}x^{1/3} = 0$$

$$\frac{x^{1/3}}{3}(7x + 4) = 0$$

$$x = 0 \text{ or } x = -\frac{4}{7}.$$

Critical numbers: $-\frac{4}{7}, 0$

$f''(x) = \frac{28}{9}x^{1/3} + \frac{4}{9}x^{-2/3}$

$f''\left(-\frac{4}{7}\right) = \frac{28}{9}\left(-\frac{4}{7}\right)^{1/3} + \frac{4}{9}\left(-\frac{4}{7}\right)^{-2/3} \approx -1.9363$

Relative maximum occurs at $-\frac{4}{7}$.

$f''(0)$ does not exist, so the second derivative test fails.

Use the first derivative test.

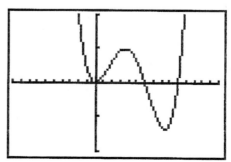

$f'\left(-\frac{1}{2}\right) = \frac{7}{3}\left(-\frac{1}{2}\right)^{4/3} + \frac{4}{3}\left(-\frac{1}{2}\right)^{1/3} \approx -0.1323$

This indicates that f is decreasing on $\left(-\frac{4}{7}, 0\right)$.

$f'(1) = \frac{7}{3}(1)^{4/3} + \frac{4}{3}(1)^{1/3} = \frac{11}{3}$

This indicates that f is increasing on $(0, \infty)$.
Relative minimum occurs at 0.

65. There are many examples. The easiest is $f(x) = \sqrt{x}$. This graph is increasing and concave downward.

$$f'(x) = \frac{1}{2}x^{-1/2} = \frac{1}{2\sqrt{x}}$$

$f'(0)$ does not exist, while $f'(x) > 0$ for all $x > 0$. (Note that the domain of f is $[0, \infty)$.)
As x increases, the value of $f'(x)$ decreases, but remains positive. It approaches zero, but never becomes zero or negative.

67. $f'(x) = 10x^2(x - 1)(5x - 3)$
$\quad\quad = 10x^2(5x^2 - 8x + 3)$
$\quad\quad = 50x^4 - 80x^3 + 30x^2$
$f''(x) = 200x^3 - 240x^2 + 60x$
$\quad\quad = 20x(10x^2 - 12x + 3)$

Graph f' in the window $[-1, 1.5]$ by $[-2, 2]$, Xscl = 0.1.

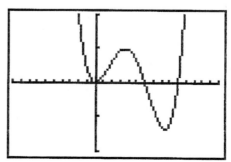

This window does not give a good view of the graph of f'', so we graph f'' in the window $[-1, 1.5]$ by $[-20, 20]$, Xscl $= 0.1$. Yscl $= 5$.

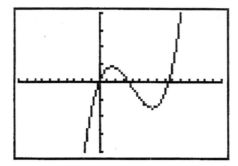

(a) The critical numbers of f are the x-intercepts of the graph of f'. (Note that there are no values where $f'(x)$ does not exist.) From the graph or by examining the factored expression for f', we see that the critical numbers of f are $0, 0.6$, and 1.

By either looking at the graph of f' and applying the first derivative test or by looking at the graph of f'' and applying the second derivative test, we see that f has a relative minimum at 1 and a relative maximum at 0.6.

(At $x = 0$, the second derivative test fails since $f''(0) = 0$, and the first derivative does not change sign, so there is no relative extremum at 0.)

(b) Examine the graph of f' to determine the intervals where the graph lies above and below the x-axis. We see that $f'(x) \geq 0$ on $(-\infty, 0.6)$, $f'(x) < 0$ on $(0.6, 1)$, and $f'(x) > 0$ on $(1, \infty)$. Therefore, f is increasing on $(-\infty, 0.6)$ and $(1, \infty)$ and decreasing on $(0.6, 1)$.

(c) Examine the graph of f''. We see that this graph has three x-intercepts, so there are three values where $f''(x) = 0$. These x-values are 0, about 0.36, and about 0.85. Because the sign of f'' and thus the concavity of f changes at these three values, we see that the x-values of the inflection points of the graph of f are 0, about 0.36, and about 0.85.

(d) We observe from the graph of f'' that $f''(x) > 0$ on $(0, 0.36)$ and $(0.85, \infty)$, so f is concave upward on the same intervals. Likewise, $f''(x) < 0$ on $(-\infty, 0)$ and $(0.36, 0.85)$, so f is concave downward on the same intervals.

69. $f'(x) = x^2 + x \ln x$

$$f''(x) = 2x + (x)\left(\frac{1}{x}\right) + (1)(\ln x)$$

$$= 2x + 1 + \ln x$$

Graph f' and f'' in the window $[0, 1]$ by $[-2, 3]$, Xscl $= 0.1$, Yscl $= 1$.

Graph of f':

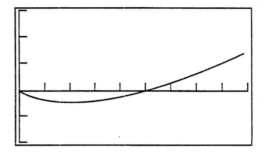

Graph of f'':

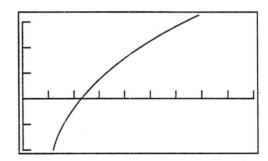

(a) The critical number of f is the x-intercept of the graph of f'. Using the graph, we find a critical number of f is about 0.5671. By looking at the graph of f'' and applying the second derivative test, we see f has a minimum at 0.5671.

(b) Examine the graph of f' to determine the intervals where the graph lies above and below the x-axis. We see that $f'(x) > 0$ on about $(0.5671, \infty)$, indicating that f is increasing on about $(0.5671, \infty)$. We also see that $f'(x) < 0$ on about $(0, 0.5671)$, indicating that f is decreasing on about $(0, 0.5671)$.

(c) Examine the graph of f''. We see that the graph has one x-intercept, so there is one x-value where $f''(x) = 0$. This value is about 0.2315. Because the sign of f'' changes at this value, we see that x-value of the inflection point of the graph of f is about 0.2315.

(d) We observe from the graph f'' that $f'' > 0$ on about $(0.2315, \infty)$, so f is concave upward on about $(0.2315, \infty)$. Likewise, we observe from the graph f'' that $f'' < 0$ on about $(0, 0.2315)$, so f is concave downward on about $(0, 0.2315)$.

71. $f(t) = 1.5207t^4 - 19.166t^3 + 62.91t^2$
$$+ 6.0726t + 1026$$
$$f'(t) = 6.0828t^3 - 57.498t^2 + 125.82t + 6.0726$$
$$f''(t) = 18.2484t^2 - 114.996t + 125.82$$

Solve $f''(x) = 0$ to find the inflection points.

$$18.2484t^2 - 114.996t + 125.82 = 0$$

$$t = \frac{114.996 \pm \sqrt{(-114.996)^2 - 4(18.2484)(125.82)}}{2(18.2484)}$$

$$t \approx 1.409 \text{ or } t \approx 4.892$$

$$f'(1.409) = 6.0828(1.409)^3 - 57.498(1.409)^2$$
$$+ 125.82(1.409) + 6.0726$$
$$\approx 86.218$$
$$f'(4.892) = 6.0828(4.892)^3 - 57.498(4.892)^2$$
$$+ 125.82(4.892) + 6.0726$$
$$\approx -42.303$$

Rents were increasing most rapidly when $t \approx 1.409$ or about mid 1999.

73. $R(x) = \frac{4}{27}(-x^3 + 66x^2 + 1050x - 400)$

$$0 \le x \le 25$$

$$R'(x) = \frac{4}{27}(-3x^2 + 132x + 1050)$$

$$R''(x) = \frac{4}{27}(-6x + 132)$$

A point of diminishing returns occurs at a point of inflection, or where $R''(x) = 0$.

$$\frac{4}{27}(-6x + 132) = 0$$

$$-6x + 132 = 0$$
$$6x = 132$$
$$x = 22$$

Test $R''(x)$ to determine whether concavity changes at $x = 22$.

$$R''(20) = \frac{4}{27}(-6 \cdot 20 + 132) = \frac{16}{9} > 0$$

$$R''(24) = \frac{4}{27}(-6 \cdot 24 + 132) = -\frac{16}{9} < 0$$

$R(x)$ is concave upward on $(0, 22)$ and concave downward on $(22, 25)$.

$$R(22) = \frac{4}{27}[-(22)^3 + 66(22)^2 + 1060(22) - 400]$$
$$\approx 6517.9$$

The point of diminishing returns is $(22, 6517.9)$.

75. $R(x) = -0.6x^3 + 3.7x^2 + 5x, \ 0 \le x \le 6$
$$R'(x) = -1.8x^2 + 7.4x + 5$$
$$R''(x) = -3.6x + 7.4$$

A point of diminishing returns occurs at a point of inflection or where $R''(x) = 0$.

$$-3.6x + 7.4 = 0$$
$$-3.6x = -7.4$$
$$x = \frac{-7.4}{-3.6} \approx 2.06$$

Test $R''(x)$ to determine whether concavity changes at $x = 2.05$.

$$R''(2) = -3.6(2) + 7.4$$
$$= -7.2 + 7.4 = 0.2 > 0$$
$$R''(3) = -3.6(3) + 7.4$$
$$= -10.8 + 7.4 = -3.4 < 0$$

$R(x)$ is concave upward on $(0, 2.06)$ and concave downward on $(2.06, 6)$.

$$R(2.06) = -0.6(2.06)^3 + 3.7(2.06)^2 + 5(2.06)$$
$$\approx 20.8$$

The point of diminishing returns is $(2.06, 20.8)$.

77. Let $D(q)$ represent the demand function. The revenue function, $R(q)$, is $R(q) = qD(q)$. The marginal revenue is given by

$$R'(q) = qD'(q) + D(q)(1)$$
$$= qD'(q) + D(q).$$
$$R''(q) = qD''(q) + D'(q)(1) + D'(q)$$
$$= qD''(q) + 2D'(q)$$

gives the rate of decline of marginal revenue.
$D'(q)$ gives the rate of decline of price.
If marginal revenue declines more quickly than price,

$$qD''(q) + 2D'(q) - D'(q) < 0$$
$$\text{or} \quad qD''(q) + D'(q) < 0.$$

79. (a) $R(t) = t^2(t - 18) + 96t + 1000;\ 0 < t < 8$
$$= t^3 - 18t^2 + 96t + 1000$$
$$R'(t) = 3t^2 - 36t + 96$$

Set $R'(t) = 0$.

$$3t^2 - 36t + 96 = 0$$
$$t^2 - 12t + 32 = 0$$
$$(t - 8)(t - 4) = 0$$
$$t = 8 \quad \text{or} \quad t = 4$$

8 is not in the domain of $R(t)$.
$R''(t) = 6t - 36$
$R''(4) = -12 < 0$ implies that $R(t)$ is maximized at $t = 4$, so the population is maximized at 4 hours.

(b) $R(4) = 16(-14) + 96(4) + 1000$
$$= -224 + 384 + 1000$$
$$= 1160$$

The maximum population is 1160 million.

81. $K(x) = \dfrac{3x}{x^2 + 4}$

(a) $K'(x) = \dfrac{3(x^2 + 4) - (2x)(3x)}{(x^2 + 4)^2}$
$$= \dfrac{-3x^2 + 12}{(x^2 + 4)^2} = 0$$
$$-3x^2 + 12 = 0$$
$$x^2 = 4$$
$$x = 2 \quad \text{or} \quad x = -2$$

For this application, the domain of K is $[0, \infty)$, so the only critical number is 2.

$K''(x) = \dfrac{(x^2+4)^2(-6x) - (-3x^2+12)(2)(x^2+4)(2x)}{(x^2 + 4)^4}$
$$= \dfrac{-6x(x^2 + 4) - 4x(-3x^2 + 12)}{(x^2 + 4)^3}$$
$$= \dfrac{6x^3 - 72x}{(x^2 + 4)^3}$$

$K''(2) = \frac{-96}{512} = -\frac{3}{16} < 0$ implies that $K(x)$ is maximized at $x = 2$.
Thus, the concentration is a maximum after 2 hours.

(b) $K(2) = \dfrac{3(2)}{(2)^2 + 4} = \dfrac{3}{4}$

The maximum concentration is $\frac{3}{4}\%$.

83. $G(t) = \dfrac{10{,}000}{1 + 49e^{-0.1t}}$

$G'(t) = \dfrac{(1 + 49e^{-0.1t})(0) - (10{,}000)(-4.9e^{-0.1t})}{(1 + 49e^{-0.1t})^2}$
$$= \dfrac{49{,}000e^{-0.1t}}{(1 + 49e^{-0.1t})^2}$$

To find $G''(t)$, apply the quotient rule to find the derivative of $G'(t)$.
The numerator of $G''(t)$ will be

$(1 + 49e^{-0.1t})^2(-4900e^{-0.1t})$
$\quad - (49{,}000e^{-0.1t})(2)(1 + 49e^{-0.1t})(-4.9e^{-0.1t})$
$\quad = (1 + 49e^{-0.1t})(-4900e^{-0.1t})$
$\qquad \cdot\ [(1 + 49e^{-0.1t}) - 20(4.9e^{-0.1t})]$
$\quad = (-4900e^{-0.1t})[1 + 49e^{-0.1t} - 98e^{-0.1t}]$
$\quad = (-4900e^{-0.1t})(1 - 49e^{-0.1t}).$

Thus,

$$G''(t) = \dfrac{(-4900e^{-0.1t})(1 - 49e^{-0.1t})}{(1 + 49e^{-0.1t})^4}.$$

$G''(t) = 0$ when $-4900e^{-0.1t} = 0$ or $1 - 49e^{-0.1t} = 0$.
$-4900e^{-0.1t} < 0$, and thus never equals zero.

$$1 - 49e^{-0.1t} = 0$$
$$1 = 49e^{-0.1t}$$
$$\frac{1}{49} = e^{-0.1t}$$
$$\ln\left(\frac{1}{49}\right) = -0.1t$$
$$\ln 1 - \ln 49 = -0.1t$$
$$-\ln 49 = -0.1t$$
$$\ln 49 = 0.1t$$
$$\ln 7^2 = 0.1t$$
$$2\ln 7 = 0.1t$$
$$20\ln 7 = t$$
$$38.9182 \approx t$$

The point of inflection is $(38.9182, 5000)$.

85. $L(t) = Be^{-ce^{-kt}}$
$\quad L'(t) = Be^{-ce^{-kt}}(-ce^{-kt})'$
$\qquad = Be^{-ce^{-kt}}[-ce^{-kt}(-kt)']$
$\qquad = Bcke^{-ce^{-kt}-kt}$
$\quad L''(t) = Bcke^{-ce^{-kt}-kt}(-ce^{-kt} - kt)'$
$\qquad = Bcke^{-ce^{-kt}-kt}[-ce^{-kt}(-kt)' - k]$
$\qquad = Bcke^{-ce^{-kt}-kt}(cke^{-kt} - k)$
$\qquad = Bck^2e^{-ce^{-kt}-kt}(ce^{-kt} - 1)$

$L''(t) = 0$ when $ce^{-kt} - 1 = 0$

$$ce^{-kt} - 1 = 0$$

$$\frac{c}{e^{kt}} = 1$$

$$e^{kt} = c$$

$$kt = \ln c$$

$$t = \frac{\ln c}{k}$$

Letting $c = 7.267963$ and $k = 0.670840$

$$t = \frac{\ln 7.267963}{0.670840} \approx 2.96 \text{ years}$$

Verify that there is a point of inflection at $t = \frac{\ln c}{k} \approx 2.96$. For

$$L''(t) = Bck^2 e^{-ce^{-kt} - kt}(ce^{-kt} - 1),$$

we only need to test the factor $ce^{-kt} - 1$ on the intervals determined by $t \approx 2.96$ since the other factors are always positive.
$L''(1)$ has the same sign as

$$7.267963e^{-0.670840(1)} - 1 \approx 2.72 > 0.$$

$L''(3)$ has the same sign as

$$7.267963e^{-0.670840(3)} - 1 \approx -0.029 < 0.$$

Therefore L, is concave up on $\left(0, \frac{\ln c}{k} \approx 2.96\right)$ and concave down on $\left(\frac{\ln c}{k}, \infty\right)$, so there is a point of inflection at $t = \frac{\ln c}{k} \approx 2.96$ years.
This signifies the time when the rate of growth begins to slow down since L changes from concave up to concave down at this inflection point.

87. $v(x) = -35.98 + 12.09x - 0.4450x^2$
$v'(x) = 12.09 - 0.89x$
$v''(x) = -0.89$

Since $-0.89 < 0$, the function is always concave down.

89. Since the rate of violent crimes is decreasing but at a slower rate than in previous years, we know that $f'(t) < 0$ but $f''(t) > 0$. Note that since $f'(t) < 0$, f is decreasing, and since $f''(t) > 0$, the graph of f is concave upward.

91. $s(t) = -16t^2$
$v(t) = s'(t) = -32t$

(a) $v(3) = -32(3) = -96$ ft/sec

(b) $v(5) = -32(5) = -160$ ft/sec

(c) $v(8) = -32(8) = -256$ ft/sec

(d) $a(t) = v'(t) = s''(t)$
$= -32$ ft/sec^2

93. $s(t) = 256t - 16t^2$
$v(t) = s'(t) = 256 - 32t$
$a(t) = v'(t) = s''(t) = -32$

To find when the maximum height occurs, set $s'(t) = 0$.

$$256 - 32t = 0$$
$$t = 8$$

Find the maximum height.

$$s(8) = 256(8) - 16(8^2)$$
$$= 1024$$

The maximum height of the ball is 1024 ft.
The ball hits the ground when $s = 0$.

$$256t - 16t^2 = 0$$
$$16t(16 - t) = 0$$
$$t = 0 \quad \text{(initial moment)}$$
$$t = 16 \text{ (final moment)}$$

The ball hits the ground 16 seconds after being thrown.

95. The car was moving most rapidly when $t \approx 6$, because acceleration was positive on $(0, 6)$ and negative after $t = 6$, so velocity was a maximum at $t = 6$.

13.4 Curve Sketching

1. Graph $y = x \ln |x|$ on a graphing calculator. A suitable choice for the viewing window is $[-1, 1]$ by $[-1, 1]$, Xscl $= 0.1$, Yscl $= 0.1$.

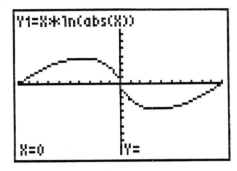

The calculator shows no y-value when $x = 0$ because 0 is not in the domain of this function. However, we see from the graph that

$$\lim_{x \to 0^-} x \ln |x| = 0$$

and

$$\lim_{x \to 0^+} x \ln |x| = 0.$$

Thus,

$$\lim_{x \to 0} x \ln |x| = 0.$$

3. $f(x) = -2x^3 - 9x^2 + 108x - 10$

Domain is $(-\infty, \infty)$.

$$f(-x) = -2(-x)^3 - 9(-x)^2 + 108(-x) - 10$$
$$= 2x^3 - 9x^2 - 108x - 10$$

No symmetry

$$f'(x) = -6x^2 - 18x + 108$$
$$= -6(x^2 + 3x - 18)$$
$$= -6(x + 6)(x - 3)$$

$f'(x) = 0$ when $x = -6$ or $x = 3$.
Critical numbers: -6 and 3
Critical points: $(-6, -550)$ and $(3, 179)$

$$f''(x) = -12x - 18$$
$$f''(-6) = 54 > 0$$
$$f''(3) = -54 < 0$$

Relative maximum at 3, relative minimum at -6
Increasing on $(-6, 3)$
Decreasing on $(-\infty, -6)$ and $(3, \infty)$

$$f''(x) = -12x - 18 = 0$$
$$-6(2x + 3) = 0$$
$$x = -\frac{3}{2}$$

Point of inflection at $(-1.5, -185.5)$
Concave upward on $(-\infty, -1.5)$
Concave downward on $(-1.5, \infty)$
y-intercept:
$$y = -2(0)^3 - 9(0)^2 + 108(0) - 10 = -10$$

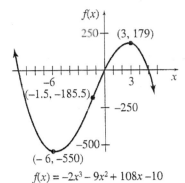

$$f(x) = -2x^3 - 9x^2 + 108x - 10$$

5. $f(x) = -3x^3 + 6x^2 - 4x - 1$

Domain is $(-\infty, \infty)$.

$$f(-x) = -3(-x)^3 + 6(-x)^2 - 4(-x) - 1$$
$$= 3x^3 + 6x^2 + 4x - 1$$

No symmetry

$$f'(x) = -9x^2 + 12x - 4$$
$$= -(3x - 2)^2$$
$$(3x - 2)^2 = 0$$
$$x = \frac{2}{3}$$

Critical number: $\frac{2}{3}$

$$f\left(\tfrac{2}{3}\right) = -3\left(\tfrac{2}{3}\right)^3 + 6\left(\tfrac{2}{3}\right)^2 - 4\left(\tfrac{2}{3}\right) - 1 = -\tfrac{17}{9}$$

Critical point: $\left(\tfrac{2}{3}, -\tfrac{17}{9}\right)$

$$f'(0) = -9(0)^2 + 12(0) - 4 = -4 < 0$$
$$f'(1) = -9(1)^2 + 12(1) - 4 = -1 < 0$$

No relative extremum at $\left(\tfrac{2}{3}, -\tfrac{17}{9}\right)$

Decreasing on $(-\infty, \infty)$
$$f''(x) = -18x + 12$$
$$= -6(3x - 2)$$
$$3x - 2 = 0$$
$$x = \frac{2}{3}$$

Point of inflection at $\left(\tfrac{2}{3}, -\tfrac{17}{9}\right)$

$$f''(0) = -18(0) + 12 = 12 > 0$$
$$f''(1) = -18(1) + 12 = -6 < 0$$

Concave upward on $\left(-\infty, \tfrac{2}{3}\right)$

Concave downward on $\left(\tfrac{2}{3}, \infty\right)$

Point of inflection at $\left(\frac{2}{3}, -\frac{17}{9}\right)$

y-intercept: $y = -3(0)^3 + 6(0)^2 - 4(0) - 1 = -1$

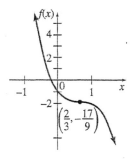

$$f(x) = -3x^3 + 6x^2 - 4x - 1$$

7. $f(x) = x^4 - 24x^2 + 80$

Domain is $(-\infty, \infty)$.

$$f(-x) = (-x)^4 - 24(-x)^2 + 80$$
$$= x^4 - 24x^2 + 80 = f(x)$$

The graph is symmetric about the y-axis.

$f'(x) = 4x^3 - 48x$

$$4x^3 - 48x = 0$$
$$4x(x^2 - 12) = 0$$
$$4x(x - 2\sqrt{3})(x + 2\sqrt{3}) = 0$$

Critical numbers: $-2\sqrt{3}, 0,$ and $2\sqrt{3}$
Critical points: $(-2\sqrt{3}, -64), (0, 80),$ and $(2\sqrt{3}, -64)$

$f''(x) = 12x^2 - 48$

$$f''(-2\sqrt{3}) = 12(-2\sqrt{3})^2 - 48 = 96 > 0$$
$$f''(0) = 12(0)^2 - 48 = -48 < 0$$
$$f''(2\sqrt{3}) = 12(2\sqrt{3})^2 - 48 = 96 > 0$$

Relative maximum at 0, relative minima at $-2\sqrt{3}$
and $2\sqrt{3}$
Increasing on $(-2\sqrt{3}, 0)$ and $(2\sqrt{3}, \infty)$
Decreasing on $(-\infty, -2\sqrt{3})$ and $(0, 2\sqrt{3})$

$$12x^2 - 48 = 0$$
$$12(x^2 - 4) = 0$$
$$x = \pm 2$$

Points of inflection at $(-2, 0)$ and $(2, 0)$
Concave upward on $(-\infty, -2)$ and $(2, \infty)$
Concave downward on $(-2, 2)$

x-intercepts: $0 = x^4 - 24x^2 + 80$

Let $u = x^2$.

$$u^2 - 24u + 80 = 0$$
$$(u - 4)(u - 20) = 0$$

$u = 4$ or $u = 20$
$x = \pm 2$ or $x = \pm 2\sqrt{5}$

y-intercept: $y = (0)^4 - 24(0)^2 + 80 = 80$

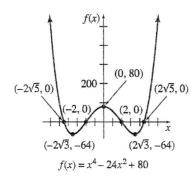

$$f(x) = x^4 - 24x^2 + 80$$

9. $f(x) = x^4 - 4x^3$

Domain is $(-\infty, \infty)$.

$$f(-x) = (-x)^4 - 4(-x)^3 = x^4 + 4x^3 \neq f(x) \text{ or } -f(x)$$

The graph is not symmetric about the y-axis or
the origin.

$f'(x) = 4x^3 - 12x^2$

$$4x^3 - 12x^2 = 0$$
$$4x^2(x - 3) = 0$$

Critical numbers: 0 and 3
Critical points: $(0, 0)$ and $(3, -27)$

$f''(x) = 12x^2 - 24x$

$$f''(0) = 12(0)^2 - 24(0) = 0$$
$$f''(3) = 12(3)^2 - 24(3) = 36 > 0$$

Second derivative test fails for 0. Use first derivative test.

$$f'(-1) = 4(-1)^3 - 12(-1)^2 = -16 < 0$$
$$f'(1) = 4(1)^3 - 12(1)^2 = -8 < 0$$

Neither a relative minimum nor maximum at 0
Relative minimum at 3
Increasing on $(3, \infty)$
Decreasing on $(-\infty, 3)$

$$12x^2 - 24x = 0$$
$$12x(x - 2) = 0$$
$$x = 0 \text{ or } x = 2$$

Points of inflection at $(0, 0)$ and $(2, -16)$
Concave upward on $(-\infty, 0)$ and $(2, \infty)$
Concave downward on $(0, 2)$

x-intercepts: $x^4 - 4x^3 = 0$
$$x^3(x - 4) = 0$$
$$x = 0 \text{ or } x = 4$$

y-intercept: $y = (0)^4 - 4(0)^3 = 0$

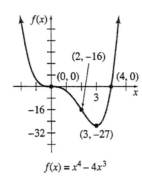

$$f(x) = x^4 - 4x^3$$

11. $f(x) = 2x + \dfrac{10}{x}$

$$= 2x + 10x^{-1}$$

Since $f(x)$ does not exist when $x = 0$, the domain is $(-\infty, 0) \cup (0, \infty)$.

$$f(-x) = 2(-x) + 10(-x)^{-1}$$
$$= -(2x + 10x^{-1})$$
$$= -f(x)$$

The graph is symmetric about the origin.

$$f'(x) = 2 - 10x^{-2}$$

$$2 - \frac{10}{x^2} = 0$$

$$\frac{2(x^2 - 5)}{x^2} = 0$$

$$x = \pm\sqrt{5}$$

Critical numbers: $-\sqrt{5}$ and $\sqrt{5}$
Critical points: $(-\sqrt{5}, -4\sqrt{5})$ and $(\sqrt{5}, 4\sqrt{5})$

Test a point in the intervals $(-\infty, -\sqrt{5})$, $(-\sqrt{5}, 0)$, $(0, \sqrt{5})$, and $(\sqrt{5}, \infty)$.

$$f'(-3) = 2 - 10(-3)^{-2} = \frac{8}{9} > 0$$

$$f'(-1) = 2 - 10(-1)^{-2} = -8 < 0$$
$$f'(1) = 2 - 10(1)^{-2} = -8 < 0$$

$$f'(3) = 2 - 10(3)^{-2} = \frac{8}{9} > 0$$

Relative maximum at $-\sqrt{5}$
Relative minimum at $\sqrt{5}$
Increasing on $(-\infty, -\sqrt{5})$ and $(\sqrt{5}, \infty)$
Decreasing on $(-\sqrt{5}, 0)$ and $(0, \sqrt{5})$
(Recall that $f(x)$ does not exist at $x = 0$.)

$$f''(x) = 20x^{-3} = \frac{20}{x^3}$$

$f''(x) = \dfrac{20}{x^3}$ is never equal to zero.

There are no inflection points.
Test a point in the intervals $(-\infty, 0)$ and $(0, \infty)$.

$$f''(-1) = \frac{20}{(-1)^3} = -20 < 0$$

$$f''(1) = \frac{20}{(1)^3} = 20 > 0$$

Concave upward on $(0, \infty)$
Concave downward on $(-\infty, 0)$
$f(x)$ is never zero, so there are no x-intercepts.
$f(x)$ does not exist for $x = 0$, so there is no y-intercept.

Vertical asymptote at $x = 0$
$y = 2x$ is an oblique asymptote.

$$f(-x) = 2(-x) + 10(-x)^{-1}$$
$$= -(2x + 10x^{-1})$$
$$= -f(x)$$

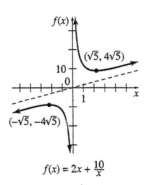

$$f(x) = 2x + \frac{10}{x}$$

13. $f(x) = \dfrac{-x + 4}{x + 2}$

Since $f(x)$ does not exist when $x = -2$, the domain is $(-\infty, -2) \cup (-2, \infty)$.

$$f(-x) = \frac{-(-x) + 4}{(-x) + 2} = \frac{x + 4}{-x + 2}$$

The graph is not symmetric about the y-axis or the origin.

$$f'(x) = \frac{(x + 2)(-1) - (-x + 4)(1)}{(x + 2)^2}$$

$$= \frac{-6}{(x + 2)^2}$$

$f'(x) < 0$ and is never zero. $f'(x)$ fails to exist for $x = -2$.

No critical numbers; no relative extrema

Decreasing on $(-\infty, -2)$ and $(-2, \infty)$

$$f''(x) = \frac{12}{(x+2)^3}$$

$f''(x)$ fails to exist for $x = -2$.

No points of inflection

Test a point in the intervals $(-\infty, -2)$ and $(-2, \infty)$.

$$f''(-3) = -12 < 0$$
$$f''(-1) = 12 > 0$$

Concave upward on $(-2, \infty)$

Concave downward on $(-\infty, -2)$

x-intercept: $\dfrac{-x+4}{x+2} = 0$

$$x = 4$$

y-intercept: $y = \dfrac{-0+4}{0+2} = 2$

Vertical asymptote at $x = -2$

Horizontal asymptote at $y = -1$

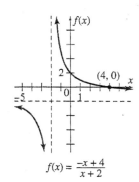

$$f(x) = \frac{-x+4}{x+2}$$

15. $f(x) = \dfrac{1}{x^2 + 4x + 3}$

$$= \frac{1}{(x+3)(x+1)}$$

Since $f(x)$ does not exist when $x = -3$ and $x = -1$, the domain is $(-\infty, -3) \cup (-3, -1) \cup (-1, \infty)$.

$$f(-x) = \frac{1}{(-x)^2 + 4(-x) + 3} = \frac{1}{x^2 - 4x + 3}$$

The graph is not symmetric about the y-axis or the origin.

$$f'(x) = \frac{0 - (2x+4)}{(x^2 + 4x + 3)^2} = \frac{-2(x+2)}{[(x+3)(x+1)]^2}$$

Critical number: -2

Test a point in the intervals $(-\infty, -3), (-3, -2), (-2, -1)$, and $(-1, \infty)$.

$$f'(-4) = \frac{-2(-4+2)}{[(-4+3)(-4+1)]^2} = \frac{4}{9} > 0$$

$$f'\left(-\frac{5}{2}\right) = \frac{-2\left(-\frac{5}{2}+2\right)}{\left[\left(-\frac{5}{2}+3\right)\left(-\frac{5}{2}+1\right)\right]^2} = \frac{16}{9} > 0$$

$$f'\left(-\frac{3}{2}\right) = \frac{-2\left(-\frac{3}{2}+2\right)}{\left[\left(-\frac{3}{2}+3\right)\left(-\frac{3}{2}+1\right)\right]^2} = -\frac{16}{9} < 0$$

$$f'(0) = \frac{-2(0+2)}{[(0+3)(0+1)]^2} = -\frac{4}{9} < 0$$

$$f(-2) = \frac{1}{(-2+3)(-2+1)} = -1$$

Relative maximum at $(-2, -1)$

Increasing on $(-\infty, -3)$ and $(-3, -2)$

Decreasing on $(-2, -1)$ and $(-1, \infty)$

$$f''(x) = \frac{(x^2 + 4x + 3)^2(-2) - (-2x - 4)(2)(x^2 + 4x + 3)(2x + 4)}{(x^2 + 4x + 3)^4}$$

$$= \frac{-2(x^2 + 4x + 3)[(x^2 + 4x + 3) + (-2x - 4)(2x + 4)]}{(x^2 + 4x + 3)^4}$$

$$= \frac{-2(x^2 + 4x + 3 - 4x^2 - 16x - 16)}{(x^2 + 4x + 3)^3}$$

$$= \frac{-2(-3x^2 - 12x - 13)}{(x^2 + 4x + 3)^3}$$

$$= \frac{2(3x^2 + 12x + 13)}{[(x+3)(x+1)]^3}$$

Since $3x^2 + 12x + 13 = 0$ has no real solutions, there are no x-values where $f''(x) = 0$. $f''(x)$ does not exist where $x = -3$ and $x = -1$. Since $f(x)$ does not exist at these x-values, there are no points of inflection.

Test a point in the intervals $(-\infty, -3), (-3, -1)$, and $(-1, \infty)$.

$$f''(-4) = \frac{2[3(-4)^2 + 12(-4) + 13]}{[(-4+3)(-4+1)]^3} = \frac{26}{27} > 0$$

$$f''(-2) = \frac{2[3(-2)^2 + 12(-2) + 13]}{[(-2+3)(-2+1)]^3} = -2 < 0$$

$$f''(0) = \frac{2[3(0)^2 + 12(0) + 13]}{[(0+3)(0+1)]^3} = \frac{26}{27} > 0$$

Concave upward on $(-\infty, -3)$ and $(-1, \infty)$

Concave downward on $(-3, -1)$

$f(x)$ is never zero, so there are no x-intercepts.

y-intercept: $y = \dfrac{1}{(0+3)(0+1)} = \dfrac{1}{3}$

Vertical asymptotes where $f(x)$ is undefined at $x = -3$ and $x = -1$.

Horizontal asymptote at $y = 0$

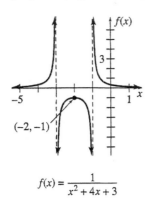

$$f(x) = \frac{1}{x^2 + 4x + 3}$$

17. $f(x) = \dfrac{x}{x^2 + 1}$

Domain is $(-\infty, \infty)$

$$f(-x) = \frac{-x}{(-x)^2 + 1} = -\frac{x}{x^2 + 1} = -f(x)$$

The graph is symmetric about the origin.

$$f'(x) = \frac{(x^2 + 1)(1) - x(2x)}{(x^2 + 1)^2}$$

$$= \frac{1 - x^2}{(x^2 + 1)^2}$$

$$1 - x^2 = 0$$

Critical numbers: 1 and -1

Critical points: $\left(1, \frac{1}{2}\right)$ and $\left(-1, -\frac{1}{2}\right)$

$$f''(x) = \frac{(x^2+1)^2(-2x) - (1-x^2)(2)(x^2+1)(2x)}{(x^2 + 1)^4}$$

$$= \frac{-2x^3 - 2x - 4x + 4x^3}{(x^2 + 1)^3}$$

$$= \frac{2x^3 - 6x}{(x^2 + 1)^3}$$

$$f''(1) = -\frac{1}{2} < 0$$

$$f''(-1) = \frac{1}{2} > 0$$

Relative maximum at 1
Relative minimum at -1
Increasing on $(-1, 1)$

Decreasing on $(-\infty, -1)$ and $(1, \infty)$

$$f''(x) = \frac{2x^3 - 6x}{(x^2 + 1)^3} = 0$$
$$2x^3 - 6x = 0$$
$$2x(x^2 - 3) = 0$$
$$x = 0, \ x = \pm\sqrt{3}$$

Inflection points at $(0, 0)$, $\left(\sqrt{3}, \frac{\sqrt{3}}{4}\right)$ and $\left(-\sqrt{3}, -\frac{\sqrt{3}}{4}\right)$

Concave upward on $(-\sqrt{3}, 0)$ and $(\sqrt{3}, \infty)$
Concave downward on $(-\infty, -\sqrt{3})$ and $(0, \sqrt{3})$

x-intercept: $0 = \dfrac{x}{x^2 + 1}$

$$0 = x$$

y-intercept: $y = \dfrac{0}{0^2 + 1} = 0$

Horizontal asymptote at $y = 0$

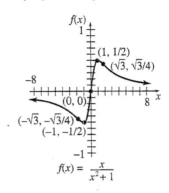

$$f(x) = \frac{x}{x^2 + 1}$$

19. $f(x) = \dfrac{1}{x^2 - 9}$

$$= \frac{1}{(x + 3)(x - 3)}$$

Since $f(x)$ does not exist when $x = -3$ and $x = 3$, the domain is $(-\infty, -3) \cup (-3, 3) \cup (3, \infty)$.

$$f(-x) = \frac{1}{(-x)^2 - 9} = \frac{1}{x^2 - 9} = f(x)$$

The graph is symmetric about the y-axis.

$$f'(x) = \frac{-2x}{(x^2 - 9)^2}$$

Critical number: 0

Critical point: $\left(0, -\frac{1}{9}\right)$

Test a point in the intervals $(-\infty, -3)$, $(-3, 0), (0, 3)$, and $(3, \infty)$.

$$f'(-4) = \frac{-2(-4)}{[(-4)^2 - 9]^2} = \frac{8}{49} > 0$$

$$f'(-1) = \frac{-2(-4)}{[(-1)^2 - 9]^2} = \frac{1}{32} > 0$$

$$f'(1) = \frac{-2(1)}{[(1)^2 - 9]^2} = -\frac{1}{32} < 0$$

$$f'(4) = \frac{-2(4)}{[(4)^2 - 9]^2} = -\frac{8}{49} < 0$$

Relative maximum at $\left(0, -\frac{1}{9}\right)$

Increasing on $(-\infty, -3)$ and $(-3, 0)$

Decreasing on $(0, 3)$ and $(3, \infty)$

$$f''(x) = \frac{(x^2 - 9)^2(-2) - (-2x)(2)(x^2 - 9)(2x)}{(x^2 - 9)^4}$$

$$= \frac{-2(x^2 - 9)[(x^2 - 9) + (-2x)(2x)]}{(x^2 + 4)^4}$$

$$= \frac{-2(x^2 - 9 - 4x^2)}{(x^2 - 9)^3}$$

$$= \frac{-2(-3x^2 - 9)}{(x^2 - 9)^3}$$

$$= \frac{6(x^2 + 3)}{[(x + 3)(x - 3)]^3}$$

Since $x^2 + 3 = 0$ has no solutions, there are no x-values where $f''(x) = 0$. $f''(x)$ does not exist where $x = -3$ and $x = 3$. Since $f(x)$ does not exist at these x-values, there are no points of inflection.

Test a point in the intervals $(-\infty, -3), (-3, 3)$, and $(3, \infty)$.

$$f''(-4) = \frac{6[(-4)^2 + 3]}{[(-4 + 3)(-4 - 3)]^3} = \frac{114}{343} > 0$$

$$f''(0) = \frac{6[(0)^2 + 3]}{[(0 + 3)(0 - 3)]^3} = -\frac{2}{81} < 0$$

$$f''(4) = \frac{6[(4)^2 + 3]}{[(4 + 3)(4 - 3)]^3} = \frac{114}{343} > 0$$

Concave upward on $(-\infty, -3)$ and $(3, \infty)$
Concave downward on $(-3, 3)$

$f(x)$ is never zero, so there are no x-intercepts.

y-intercept: $y = \dfrac{1}{0^2 - 9} = -\dfrac{1}{9}$

Vertical asymptotes where $f(x)$ is undefined at $x = -3$ and $x = 3$.
Horizontal asymptote at $y = 0$

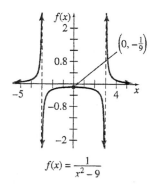

$$f(x) = \frac{1}{x^2 - 9}$$

21. $f(x) = x \ln|x|$

The domain of this function is $(-\infty, 0) \cup (0, \infty)$.

$$f(-x) = -x \ln|-x||-x|$$
$$= x \ln|-x| = -f(x)$$

The graph is symmetric about the origin.

$$f'(x) = x \cdot \frac{1}{x} + \ln|x|$$

$$= 1 + \ln|x|$$

$f'(x) = 0$ when

$$0 = 1 + \ln|x|$$
$$-1 = \ln|x|$$
$$e^{-1} = |x|$$

$$x = \pm\frac{1}{e} \approx \pm 0.37.$$

Critical numbers: $\pm\frac{1}{e} \approx \pm 0.37$.

$$f'(-1) = 1 + \ln|-1| = 1 > 0$$
$$f'(-0.1) = 1 + \ln|-0.1| \approx -1.3 < 0$$
$$f'(0.1) = 1 + \ln|0.1| \approx -1.3 < 0$$
$$f'(1) = 1 + \ln|1| = 1 > 0$$

$$f\left(\frac{1}{e}\right) = \frac{1}{e} \ln\left|\frac{1}{e}\right| = -\frac{1}{e}$$

$$f\left(-\frac{1}{e}\right) = -\frac{1}{e} \ln\left|-\frac{1}{e}\right| = \frac{1}{e}$$

Relative maximum of $\left(-\frac{1}{e}, \frac{1}{e}\right)$; relative minimum of $\left(\frac{1}{e}, -\frac{1}{e}\right)$.

Increasing on $\left(-\infty, -\frac{1}{e}\right)$ and $\left(\frac{1}{e}, \infty\right)$ and decreasing on $\left(-\frac{1}{e}, 0\right)$ and $\left(0, \frac{1}{e}\right)$.

$$f''(x) = \frac{1}{x}$$

$$f''(-1) = \frac{1}{-1} = -1 < 0$$

$$f''(1) = \frac{1}{1} = 1 > 0$$

Concave downward on $(-\infty, 0)$;
Concave upward on $(0, \infty)$.
There is no y-intercept.

x-intercept: $0 = x \ln|x|$

$$x = 0 \quad \text{or} \quad \ln|x| = 0$$
$$|x| = e^0 = 1$$
$$x = \pm 1$$

Since 0 is not in the domain, the only x-intercepts are -1 and 1.

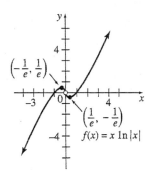

23. $f(x) = \dfrac{\ln x}{x}$

Note that the domain of this function is $(0, \infty)$.

$f(-x) = \dfrac{\ln(-x)}{-x}$ does not exist when $x \geq 0$, no symmetry.

$$f'(x) = \frac{x\left(\frac{1}{x}\right) - \ln x(1)}{x^2}$$
$$= \frac{1 - \ln x}{x^2}$$

Critical numbers:

$$1 - \ln x = 0$$
$$1 = \ln x$$
$$e^1 = x$$

$$f(e) = \frac{\ln e}{e} = \frac{1}{e}$$

Critical points: $\left(e, \dfrac{1}{e}\right)$

$$f'(1) = \frac{1 - \ln 1}{1^2} = \frac{1}{1} = 1 > 0$$

$$f'(3) = \frac{1 - \ln 3}{3^2} = -0.01 < 0$$

There is a relative maximum at $\left(e, \frac{1}{e}\right)$.
The function is increasing on $(0, e)$ and decreasing on (e, ∞).

$$f''(x) = \frac{x^2\left(-\frac{1}{x}\right) - (1 - \ln x)2x}{x^4}$$
$$= \frac{-x - 2x(1 - \ln x)}{x^4}$$
$$= \frac{-x[1 + 2(1 - \ln x)]}{x^4}$$
$$= \frac{-(1 + 2 - 2 \ln x)}{x^3}$$
$$= \frac{-3 + 2 \ln x}{x^3}$$

$f''(x) = 0$ when $-3 + 2 \ln x = 0$

$$2 \ln x = 3$$
$$\ln x = \frac{3}{2} = 1.5$$
$$x = e^{1.5} \approx 4.48.$$

$$f''(1) = \frac{-3 + 2 \ln 1}{1^3} = -3 < 0$$

$$f''(5) = \frac{-3 + 2 \ln 5}{5^3} \approx 0.0018 > 0$$

Inflection point at $\left(e^{1.5}, \frac{1.5}{e^{1.5}}\right) \approx (4.48, 0.33)$

Concave downward on $(0, e^{1.5})$; concave upward on $(e^{1.5}, \infty)$

$$f(e^{1.5}) = \frac{\ln e^{1.5}}{e^{1.5}} = \frac{1.5}{e^{1.5}}$$
$$= \frac{3}{2e^{1.5}} \approx 0.33$$

Since $x \neq 0$, there is no y-intercept.

x-intercept: $f(x) = 0$ when $\ln x = 0$
$$x = e^0 = 1$$

Vertical asymptote at $x = 0$
Horizontal asymptote at $y = 0$

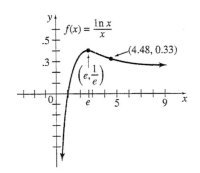

25. $f(x) = xe^{-x}$

Domain is $(-\infty, \infty)$.

$f(-x) = -xe^{x}$

The graph has no symmetry.

$$f'(x) = -xe^{-x} + e^{-x}$$
$$= e^{-x}(1 - x)$$

$f'(x) = 0$ when $e^{-x}(1 - x) = 0$
$$x = 1$$

Critical numbers: 1

Critical points: $\left(1, \frac{1}{e}\right)$

$f'(0) = e^{-0}(1-0) = 1 > 0$

$f'(2) = e^{-2}(1-2) = \dfrac{-1}{e^2} < 0$

Relative maximum at $\left(1, \frac{1}{e}\right)$

Increasing on $(-\infty, 1)$; decreasing on $(1, \infty)$

$f''(x) = e^{-x}(-1) + (1-x)(-e^{-x})$
$= -e^{-x}(1+1-x)$
$= -e^{-x}(2-x)$

$f'' = 0$ when $-e^{-x}(2-x) = 0$
$$x = 2.$$

$f''(0) = -e^{-0}(2-0) = -2 < 0$

$f''(3) = -e^{-3}(2-3) = \dfrac{1}{e^3} > 0$

Inflection point at $\left(2, \frac{2}{e^2}\right)$

Concave downward on $(-\infty, 2)$, concave upward on $(2, \infty)$

x-intercept: $0 = xe^{-x}$
$$x = 0$$

y-intercept: $y = 0 \cdot e^{-0} = 0$

Horizontal asymptote at $y = 0$

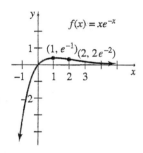

27. $f(x) = (x-1)e^{-x}$

Domain is $(-\infty, \infty)$

$f(-x) = (-x-1)e^{x}$

The graph has no symmetry.

$f'(x) = -(x-1)e^{-x} + e^{-x}(1)$
$= e^{-x}[-(x-1)+1]$
$= e^{-x}(2-x)$

$f'(x) = 0$ when $e^{-x}(2-x) = 0$
$$x = 2.$$

Critical number: 2

Critical point: $\left(2, \frac{1}{e^2}\right)$

$f''(x) = -e^{-x} + (2-x)(-e^{-x})$
$= -e^{-x}[1 + (2-x)]$
$= -e^{-x}(3-x)$

$f''(2) = -e^{-2}(3-2) = \dfrac{-1}{e^2} < 0$

Relative maximum at $\left(2, \frac{1}{e^2}\right)$

$f'(0) = e^{-0}(2-0) = 2 > 0$

$f'(3) = e^{-3}(2-3) = \dfrac{-1}{e^3} < 0$

Increasing on $(-\infty, 2)$; decreasing on $(2, \infty)$.

$f''(x) = 0$ when $-e^{-x}(3-x) = 0$
$$x = 3.$$

$f''(0) = -e^{-0}(3-0) = -3 < 0$

$f''(4) = -e^{-4}(3-4) = \dfrac{1}{e^4} > 0$

Inflection point at $\left(3, \frac{2}{e^3}\right)$

Concave downward on $(-\infty, 3)$; concave upward on $(3, \infty)$

$$f(3) = (3-1)e^{-3} = \frac{2}{e^3}$$

y-intercept: $y = (0-1)e^{-0}$
$= (-1)(1) = -1$

x-intercept: $0 = (x-1)e^{-x}$
$$x - 1 = 0$$
$$x = 1$$

Horizontal asymptote at $y = 0$

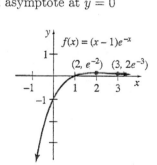

29. $f(x) = x^{2/3} - x^{5/3}$

Domain is $(-\infty, \infty)$.

$f(-x) = x^{2/3} + x^{5/3}$

The graph has no symmetry.

$f'(x) = \dfrac{2}{3}x^{-1/3} - \dfrac{5}{3}x^{2/3}$

$= \dfrac{2 - 5x}{3x^{1/3}}$

$f'(x) = 0$ when $2 - 5x = 0$

Critical number: $x = \dfrac{2}{5}$

$$f\left(\frac{2}{5}\right) = \left(\frac{2}{5}\right)^{2/3} - \left(\frac{2}{5}\right)^{5/3}$$

$$= \frac{3 \cdot 2^{2/3}}{5^{5/3}} \approx 0.326$$

Critical point: $(0.4, 0.326)$

$$f''(x) = \frac{3x^{1/3}(-5) - (2 - 5x)(3)\left(\frac{1}{3}\right)x^{-2/3}}{(3x^{1/3})^2}$$

$$= \frac{-15x^{1/3} - (2 - 5x)x^{-2/3}}{9x^{2/3}}$$

$$= \frac{-15x - (2 - 5x)}{9x^{4/3}}$$

$$= \frac{-10x - 2}{9x^{4/3}}$$

$$f''\left(\frac{2}{5}\right) = \frac{-10\left(\frac{2}{5}\right) - 2}{9\left(\frac{2}{5}\right)^{4/3}}$$

$$\approx -2.262 < 0$$

Relative maximum at $\left(\frac{2}{5}, \frac{3 \cdot 2^{2/3}}{5^{5/3}}\right) \approx (0.4, 0.326)$

$f'(x)$ does not exist when $x = 0$

Since $f''(0)$ is undefined, use the first derivative test.

$$f'(-1) = \frac{2 - 5(-1)}{3(-1)^{1/3}} = \frac{7}{-3} < 0$$

$$f'\left(\frac{1}{8}\right) = \frac{2 - 5\left(\frac{1}{8}\right)}{3\left(\frac{1}{8}\right)^{1/3}} = \frac{11}{12} > 0$$

$$f'(1) = \frac{2 - 5}{3 \cdot 1^{1/3}} = -1 < 0$$

Relative minimum at $(0, 0)$

f increases on $\left(0, \frac{2}{5}\right)$.
f decreases on $(-\infty, 0)$ and $\left(\frac{2}{5}, \infty\right)$.

$f''(x) = 0$ when $-10x - 2 = 0$

$$x = -\frac{1}{5}$$

$f''(x)$ undefined when $9x^{4/3} = 0$

$$x = 0$$

$$f''(-1) = \frac{-10(-1) - 2}{9(-1)^{4/3}} = \frac{8}{9} > 0$$

$$f''\left(-\frac{1}{8}\right) = \frac{-10\left(-\frac{1}{8}\right) - 2}{9\left(-\frac{1}{8}\right)^{4/3}} = -\frac{4}{3} < 0$$

$$f''(1) = \frac{-10(1) - 2}{9(1)^{4/3}} = -\frac{4}{3} < 0$$

Concave upward on $\left(-\infty, -\frac{1}{5}\right)$

Concave downward on $\left(-\frac{1}{5}, \infty\right)$

Inflection point at $\left(-\frac{1}{5}, \frac{6}{5^{5/3}}\right) \approx (-0.2, 0.410)$

y-intercept: $y = 0^{2/3} - 0^{5/3} = 0$
x-intercept: $0 = x^{2/3} - x^{5/3}$

$$= x^{2/3}(1 - x)$$

$$x = 0 \text{ or } x = 1$$

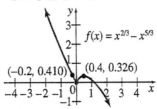

31. For Exercises 3, 7, and 9, the relative maxima or minima are outside the vertical window of $-10 \leq y \leq 10$.

For Exercise 11, the default window shows only a small portion of the graph.

For Exercise 15, the default window does not allow the graph to properly display the vertical asymptotes.

33. For Exercises 17, 19, 23, 25, and 27, the y-coordinate of the relative minimum, relative maximum, or inflection point is so small, it may be hard to distinguish.

For Exercises 35–39 other graphs are possible.

35. **(a)** indicates a smooth, continuous curve except where there is a vertical asymptote.

(b) indicates that the function decreases on both sides of the asymptote, so there are no relative extrema.

(c) gives the horizontal asymptote $y = 2$.

(d) and **(e)** indicate that concavity does not change left of the asymptote, but that the right portion of the graph changes concavity at $x = 2$ and $x = 4$. There are inflection points at 2 and 4.

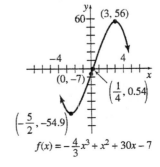

37. (a) indicates that there can be no asymptotes, sharp "corners", holes, or jumps. The graph must be one smooth curve.

(b) and (c) indicate relative maxima at -3 and 4 and a relative minimum at 1.

(d) and (e) are consistent with (g).

(f) indicates turning points at the critical numbers -3 and 4.

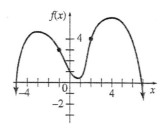

39. (a) indicates that the curve may not contain breaks.

(b) indicates that there is a sharp "corner" at 4.

(c) gives a point at $(1, 5)$.

(d) shows critical numbers.

(e) and (f) indicate (combined with (c) and (d)) a relative maximum at $(1, 5)$, and (combined with (b)) a relative minimum at 4.

(g) is consistent with (b).

(h) indicates the curve is concave upward on $(2, 3)$.

(i) indicates the curve is concave downward on $(-\infty, 2), (3, 4)$ and $(4, \infty)$.

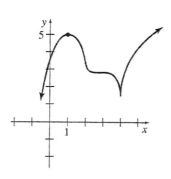

Chapter 13 Review Exercises

5. $f(x) = x^2 + 9x + 8$
$f'(x) = 2x + 9$

$f'(x) = 0$ when $x = -\frac{9}{2}$ and f' exists everywhere.

Critical number: $-\frac{9}{2}$

Test an x-value in the intervals $\left(-\infty, -\frac{9}{2}\right)$ and $\left(-\frac{9}{2}, \infty\right)$.

$f'(-5) = -1 < 0$
$f'(-4) = 1 > 0$

f is increasing on $\left(-\frac{9}{2}, \infty\right)$ and decreasing on $\left(-\infty, -\frac{9}{2}\right)$.

7. $f(x) = -x^3 + 2x^2 + 15x + 16$
$f'(x) = -3x^2 + 4x + 15$
$\quad = -(3x^2 - 4x - 15)$
$\quad = -(3x + 5)(x - 3)$

$f'(x) = 0$ when $x = -\frac{5}{3}$ or $x = 3$ and f' exists everywhere.

Critical numbers: $-\frac{5}{3}$ and 3

Test an x-value in the intervals $\left(-\infty, -\frac{5}{3}\right), \left(-\frac{5}{3}, 3\right)$, and $(3, \infty)$.

$f'(-2) = -5 < 0$
$f'(0) = 15 > 0$
$f'(4) = -17 < 0$

f is increasing on $\left(-\frac{5}{3}, 3\right)$ and decreasing on $\left(-\infty, -\frac{5}{3}\right)$ and $(3, \infty)$.

9. $f(x) = \dfrac{16}{9 - 3x}$

$f'(x) = \dfrac{16(-1)(-3)}{(9 - 3x)^2} = \dfrac{48}{(9 - 3x)^2}$

$f'(x) > 0$ for all x $(x \neq 3)$, and f is not defined for $x = 3$.

f is increasing on $(-\infty, 3)$ and $(3, \infty)$ and never decreasing.

11. $f(x) = \ln|x^2 - 1|$

$$f'(x) = \frac{2x}{x^2 - 1}$$

f is not defined for $x = -1$ and $x = 1$.

$f'(x) = 0$ when $x = 0$.

Test an x-value in the intervals $(-\infty, -1), (-1, 0)$, $(0, 1)$, and $(1, \infty)$.

$$f'(-2) = -\frac{4}{3} < 0$$

$$f'\left(-\frac{1}{2}\right) = \frac{4}{3} > 0$$

$$f'\left(\frac{1}{2}\right) = -\frac{4}{3} < 0$$

$$f'(2) = \frac{4}{3} > 0$$

f is increasing on $(-1, 0)$ and $(1, \infty)$ and decreasing on $(-\infty, -1)$ and $(0, 1)$.

13. $f(x) = -x^2 + 4x - 8$

$f'(x) = -2x + 4 = 0$

Critical number: $x = 2$

$f''(x) = -2 < 0$ for all x, so $f(2)$ is a relative maximum.

$$f(2) = -4$$

Relative maximum of -4 at 2

15. $f(x) = 2x^2 - 8x + 1$

$f'(x) = 4x - 8 = 0$

Critical number: $x = 2$

$f''(x) = 4 > 0$ for all x, so $f(2)$ is a relative minimum.

$$f(2) = -7$$

Relative minimum of -7 at 2

17. $f(x) = 2x^3 + 3x^2 - 36x + 20$

$f'(x) = 6x^2 + 6x - 36 = 0$

$6(x^2 + x - 6) = 0$

$(x + 3)(x - 2) = 0$

Critical numbers: -3 and 2

$f''(x) = 12x + 6$

$f''(-3) = -30 < 0$, so a maximum occurs at $x = -3$.

$f''(2) = 30 > 0$, so a minimum occurs at $x = 2$.

$f(-3) = 101$

$f(2) = -24$

Relative maximum of 101 at -3

Relative minimum of -24 at 2

19. $f(x) = \dfrac{xe^x}{x - 1}$

$$f'(x) = \frac{(x - 1)(xe^x + e^x) - xe^x(1)}{(x - 1)^2}$$

$$= \frac{x^2 e^x + xe^x - xe^x - e^x - xe^x}{(x - 1)^2}$$

$$= \frac{x^2 e^x - xe^x - e^x}{(x - 1)^2}$$

$$= \frac{e^x(x^2 - x - 1)}{(x - 1)^2}$$

$f'(x)$ is undefined at $x = 1$, but 1 is not in the domain of $f(x)$.

$f'(x) = 0$ when $x^2 - x - 1 = 0$

$$x = \frac{1 \pm \sqrt{1 - 4(1)(-1)}}{2}$$

$$= \frac{1 \pm \sqrt{5}}{2}$$

$$\frac{1 + \sqrt{5}}{2} \approx 1.618 \quad \text{or} \frac{1 - \sqrt{5}}{2} = -0.618$$

Critical numbers are -0.618 and 1.618.

$$f'(1.4) = \frac{e^{1.4}(1.4^2 - 1.4 - 1)}{(1.4 - 1)^2} \approx -11.15 < 0$$

$$f'(2) = \frac{e^2(2^2 - 2 - 1)}{(2 - 1)^2} = e^2 \approx 7.39 > 0$$

$$f'(-1) = \frac{e^{-1}[(-1)^2 - (-1) - 1]}{(-1 - 1)^2} \approx 0.09 > 0$$

$$f'(0) = \frac{e^0(0^2 - 0 - 1)}{(0 - 1)^2} = -1 < 0$$

There is a relative maximum at $(-0.618, 0.206)$ and a relative minimum at $(1.618, 13.203)$.

21. $f(x) = 3x^4 - 5x^2 - 11x$

$f'(x) = 12x^3 - 10x - 11$

$f''(x) = 36x^2 - 10$

$f''(1) = 36(1)^2 - 10 = 26$

$f''(-3) = 36(-3)^2 - 10 = 314$

23. $f(x) = \dfrac{4x + 2}{3x - 6}$

$$f'(x) = \frac{(3x - 6)(4) - (4x + 2)(3)}{(3x - 6)^2}$$

$$= \frac{12x - 24 - 12x - 6}{(3x - 6)^2}$$

$$= \frac{-30}{(3x - 6)^2}$$

$$= -30(3x - 6)^{-2}$$

$$f''(x) = -30(-2)(3x - 6)^{-3}(3)$$

$$= 180(3x - 6)^{-3} \quad \text{or} \quad \frac{180}{(3x - 6)^3}$$

$$f''(1) = 180[3(1) - 6]^{-3} = -\frac{20}{3}$$

$$f''(-3) = 180[3(-3) - 6]^{-3} = -\frac{4}{75}$$

25. $f(t) = \sqrt{t^2 + 1} = (t^2 + 1)^{1/2}$

$$f'(t) = \frac{1}{2}(t^2 + 1)^{-1/2}(2t) = t(t^2 + 1)^{-1/2}$$

$$f''(t) = (t^2 + 1)^{-1/2}(1)$$

$$+ t\left[\left(-\frac{1}{2}\right)(t^2 + 1)^{-3/2}(2t)\right]$$

$$= (t^2 + 1)^{-1/2} - t^2(t^2 + 1)^{-3/2}$$

$$= \frac{1}{(t^2 + 1)^{1/2}} - \frac{t^2}{(t^2 + 1)^{3/2}} = \frac{t^2 + 1 - t^2}{(t^2 + 1)^{3/2}}$$

$$= (t^2 + 1)^{-3/2} \quad \text{or} \quad \frac{1}{(t^2 + 1)^{3/2}}$$

$$f''(1) = \frac{1}{(1 + 1)^{3/2}} = \frac{1}{2^{3/2}} \approx 0.354$$

$$f''(-3) = \frac{1}{(9 + 1)^{3/2}} = \frac{1}{10^{3/2}} \approx 0.032$$

27. $f(x) = -2x^3 - \dfrac{1}{2}x^2 + x - 3$

Domain is $(-\infty, \infty)$
The graph has no symmetry.

$$f'(x) = -6x^2 - x + 1 = 0$$
$$(3x - 1)(2x + 1) = 0$$

Critical numbers: $\frac{1}{3}$ and $-\frac{1}{2}$

Critical points: $\left(\frac{1}{3}, -2.80\right)$ and $\left(-\frac{1}{2}, -3.375\right)$

$$f''(x) = -12x - 1$$

$$f''\left(\frac{1}{3}\right) = -5 < 0$$

$$f''\left(-\frac{1}{2}\right) = 5 > 0$$

Relative maximum at $\frac{1}{3}$

Relative minimum at $-\frac{1}{2}$

Increasing on $\left(-\frac{1}{2}, \frac{1}{3}\right)$

Decreasing on $\left(-\infty, -\frac{1}{2}\right)$ and $\left(\frac{1}{3}, \infty\right)$

$$f''(x) = -12x - 1 = 0$$

$$x = -\frac{1}{12}$$

Point of inflection at $\left(-\frac{1}{12}, -3.09\right)$

Concave upward on $\left(-\infty, -\frac{1}{12}\right)$

Concave downward on $\left(-\frac{1}{12}, \infty\right)$

y-intercept:

$$y = -2(0)^3 - \frac{1}{2}(0)^2 + (0) - 3 = -3$$

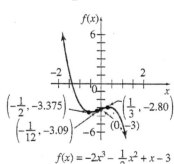

$$f(x) = -2x^3 - \frac{1}{2}x^2 + x - 3$$

29. $f(x) = x^4 - \dfrac{4}{3}x^3 - 4x^2 + 1$

Domain is $(-\infty, \infty)$
The graph has no symmetry.

$$f'(x) = 4x^3 - 4x^2 - 8x = 0$$
$$4x(x^2 - x - 2) = 0$$
$$4x(x - 2)(x + 1) = 0$$

Critical numbers: 0, 2, and -1
Critical points: $(0, 1)$, $\left(2, -\frac{29}{3}\right)$ and $\left(-1, -\frac{2}{3}\right)$

$$f''(x) = 12x^2 - 8x - 8$$
$$= 4(3x^2 - 2x - 2)$$
$$f''(-1) = 12 > 0$$
$$f''(0) = -8 < 0$$
$$f''(2) = 24 > 0$$

Relative maximum at 0
Relative minima at -1 and 2
Increasing on $(-1, 0)$ and $(2, \infty)$
Decreasing on $(-\infty, -1)$ and $(0, 2)$

$$f''(x) = 4(3x^2 - 2x - 2) = 0$$

$$x = \frac{2 \pm \sqrt{4 - (-24)}}{6}$$

$$= \frac{1 \pm \sqrt{7}}{3}$$

Points of inflection at $\left(\frac{1+\sqrt{7}}{3}, -5.12\right)$ and

$\left(\frac{1-\sqrt{7}}{3}, 0.11\right)$

Concave upward on $\left(-\infty, \frac{1-\sqrt{7}}{3}\right)$ and $\left(\frac{1+\sqrt{7}}{3}, \infty\right)$

Concave downward on $\left(\frac{1-\sqrt{7}}{3}, \frac{1+\sqrt{7}}{3}\right)$

y-intercept:

$$y = (0)^4 - \frac{4}{3}(0)^3 - 4(0)^2 + 1 = 1$$

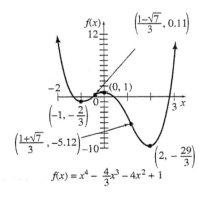

$$f(x) = x^4 - \frac{4}{3}x^3 - 4x^2 + 1$$

31. $f(x) = \dfrac{x - 1}{2x + 1}$

Domain is $\left(-\infty, -\frac{1}{2}\right) \cup \left(-\frac{1}{2}, \infty\right)$
The graph has no symmetry.

$$f'(x) = \frac{(2x + 1)(1) - (x - 1)(2)}{(2x + 1)^2}$$

$$= \frac{3}{(2x + 1)^2}$$

f' is never zero.
$f'(-\frac{1}{2})$ does not exist, but $-\frac{1}{2}$ is not a critical number because $-\frac{1}{2}$ is not in the domain of f. Thus, there are no critical numbers, so $f(x)$ has no relative extrema.

Increasing on $\left(-\infty, \frac{1}{2}\right)$ and $\left(\frac{1}{2}, \infty\right)$

$$f''(x) = \frac{-12}{(2x + 1)^3}$$

$$f''(0) = -12 < 0$$
$$f''(-1) = 12 > 0$$

No inflection points

Concave upward on $\left(-\infty, -\frac{1}{2}\right)$

Concave downward on $\left(-\frac{1}{2}, \infty\right)$

x-intercept: $\dfrac{x - 1}{2x + 1} = 0$

$$x = 1$$

y-intercept: $y = \dfrac{0 - 1}{2(0) + 1} = -1$

Vertical asymptote at $x = -\frac{1}{2}$

Horizontal asymptote at $y = \frac{1}{2}$

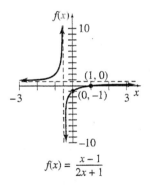

$$f(x) = \frac{x - 1}{2x + 1}$$

33. $f(x) = -4x^3 - x^2 + 4x + 5$

Domain is $(-\infty, \infty)$
The graph has no symmetry.

$$f'(x) = -12x^2 - 2x + 4$$
$$= -2(6x^2 + x - 2) = 0$$
$$(3x + 2)(2x - 1) = 0$$

Critical numbers: $-\frac{2}{3}$ and $\frac{1}{2}$

Critical points: $\left(-\frac{2}{3}, 3.07\right)$ and $\left(\frac{1}{2}, 6.25\right)$

$$f''(x) = -24x - 2$$
$$= -2(12x + 1)$$

$$f''\left(-\frac{2}{3}\right) = 14 > 0$$

$$f''\left(\frac{1}{2}\right) = -14 < 0$$

Relative maximum at $\frac{1}{2}$

Relative minimum at $-\frac{2}{3}$

Increasing on $\left(-\frac{2}{3}, \frac{1}{2}\right)$

Decreasing on $\left(-\infty, -\frac{2}{3}\right)$ and $\left(\frac{1}{2}, \infty\right)$

$$f''(x) = -2(12x + 1) = 0$$

$$x = -\frac{1}{12}$$

Point of inflection at $\left(-\frac{1}{12}, 4.66\right)$

Concave upward on $\left(-\infty, -\frac{1}{12}\right)$

Concave downward on $\left(-\frac{1}{12}, \infty\right)$

y-intercept:

$$y = -4(0)^3 - (0)^2 + 4(0) + 5 = 5$$

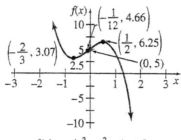

$$f(x) = -4x^3 - x^2 + 4x + 5$$

35. $f(x) = x^4 + 2x^2$

Domain is $(-\infty, \infty)$

$$f(-x) = (-x)^4 - 2(-x)^2$$
$$= x^4 + 2x^2 = f(x)$$

The graph is symmetric about the y-axis.

$$f'(x) = 4x^3 + 4x$$
$$= 4x(x^2 + 1) = 0$$

Critical number: 0
Critical point: $(0, 0)$

$$f''(x) = 12x^2 + 4 = 4(3x^2 + 1)$$
$$f''(0) = 4 > 0$$

Relative minimum at 0
Increasing on $(0, \infty)$
Decreasing on $(-\infty, 0)$

$$f''(x) = 4(3x^2 + 1) \neq 0 \text{ for any } x$$

No points of inflection

$$f''(-1) = 16 > 0$$
$$f''(1) = 16 > 0$$

Concave upward on $(-\infty, \infty)$
x-intercept: 0; y-intercept: 0

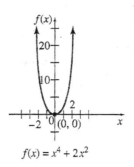

$$f(x) = x^4 + 2x^2$$

37. $f(x) = \dfrac{x^2 + 4}{x}$

Domain is $(-\infty, 0) \cup (0, \infty)$

$$f(-x) = \frac{(-x)^2 + 4}{-x}$$

$$= \frac{x^2 + 4}{-x} = -f(x)$$

The graph is symmetric about the origin.

$$f'(x) = \frac{x(2x) - (x^2 + 4)}{x^2}$$

$$= \frac{x^2 - 4}{x^2} = 0$$

Critical numbers: -2 and 2
Critical points: $(-2, -4)$ and $(2, 4)$

$$f''(x) = \frac{8}{x^3}$$
$$f''(-2) = -1 < 0$$
$$f''(2) = 1 > 0$$

Relative maximum at -2
Relative minimum at 2
Increasing on $(-\infty, -2)$ and $(2, \infty)$
Decreasing on $(-2, 0)$ and $(0, 2)$

$$f''(x) = \frac{8}{x^3} > 0 \text{ for all } x.$$

No inflection points
Concave upward on $(0, \infty)$
Concave downward on $(-\infty, 0)$
No x- or y-intercepts
Vertical asymptote at $x = 0$

Oblique asymptote at $y = x$

Horizontal asymptote at $y = -2$

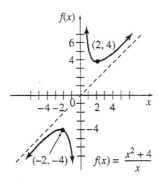

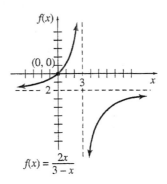

39. $f(x) = \dfrac{2x}{3 - x}$

Domain is $(-\infty, 3) \cup (3, \infty)$
The graph has no symmetry.

$$f'(x) = \frac{(3 - x)(2) - (2x)(-1)}{(3 - x)^2}$$

$$= \frac{6}{(3 - x)^2}$$

$f'(x)$ is never zero. $f'(3)$ does not exist, but since 3 is not in the domain of f, it is not a critical number.
No critical numbers, so no relative extrema

$$f'(0) = \frac{2}{3} > 0$$

$$f'(4) = 6 > 0$$

Increasing on $(-\infty, 3)$ and $(3, \infty)$

$$f''(x) = \frac{12}{(3 - x)^3}$$

$f''(x)$ is never zero. $f''(3)$ does not exist, but since 3 is not in the domain of f, there is no inflection point at $x = 3$.

$$f''(0) = \frac{12}{27} > 0$$

$$f''(4) = -12 < 0$$

Concave upward on $(-\infty, 3)$
Concave downward on $(3, \infty)$

x-intercept: 0; y-intercept: 0

Vertical asymptote at $x = 3$

41. $f(x) = xe^{2x}$

Domain is $(-\infty, \infty)$.

$f(-x) = -xe^{-2x}$

The graph has no symmety.

$$f'(x) = (1)(e^{2x}) + (x)(2e^{2x})$$
$$= e^{2x}(2x + 1)$$

$f'(x) = 0$ when $x = -\frac{1}{2}$.

Critical number: $-\frac{1}{2}$

Critical point: $\left(-\frac{1}{2}, -\frac{1}{2e}\right)$

$$f'(-1) = e^{2(-1)}[2(-1) + 1] = -e^{-2} < 0$$
$$f'(0) = e^{2(0)}[2(0) + 1] = 1 > 0$$

No relative maximum

Relative minimum at $\left(-\frac{1}{2}, -\frac{1}{2e}\right)$

Decreasing on $\left(-\infty, -\frac{1}{2}\right)$ and increasing on $\left(-\frac{1}{2}, \infty\right)$

$$f''(x) = 2e^{2x}(2x + 1) + e^{2x}(2)$$
$$= 4e^{2x}(x + 1)$$

$f''(x) = 0$ when $x = -1$.

$$f''(-2) = 4e^{2(-2)}[(-2) + 1] = -4e^{-4} < 0$$
$$f''(0) = 4e^{2(0)}[(0) + 1] = 4 > 0$$

Inflection point at $(-1, -e^{-2})$

Concave upward on $(-1, \infty)$
Concave downward on $(-\infty, -1)$

x-intercept: $xe^{2x} = 0$
$$x = 0$$

y-intercept: $y = (0)e^{2(0)} = 0$

Since $\lim\limits_{x \to -\infty} xe^{2x} = 0$, there is a horizontal asymptote at $y = 0$.

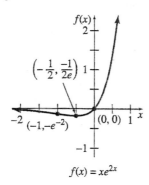

$$f(x) = xe^{2x}$$

43. $f(x) = \ln(x^2 + 4)$

Domain is $(-\infty, \infty)$.

$f(-x) = \ln[(-x)^2 + 4] = \ln(x^2 + 4) = f(x)$

The graph is symmetric about the y-axis.

$f'(x) = \dfrac{2x}{x^2 + 4}$

$f'(x) = 0$ when $x = 0$.

Critical number: 0

Critical point: $(0, \ln 4)$

$f'(-1) = \dfrac{2(-1)}{(-1)^2 + 4} = -\dfrac{2}{5} < 0$

$f'(1) = \dfrac{2(1)}{(1)^2 + 4} = \dfrac{2}{5} > 0$

No relative maximum

Relative minimum at $(0, \ln 4)$

Increasing on $(0, \infty)$

Decreasing on $(-\infty, 0)$

$f''(x) = \dfrac{(x^2 + 4)(2) - (2x)(2x)}{(x^2 + 4)^2}$

$\qquad = \dfrac{-2(x^2 - 4)}{(x^2 + 4)^2}$

$f''(x) = 0$ when

$x^2 - 4 = 0$

$\quad x = \pm 2$

$f''(-3) = \dfrac{-2[(-3)^2 - 4]}{[(-3)^2 + 4]^2} = -\dfrac{10}{169} < 0$

$f''(0) = \dfrac{-2[(0)^2 - 4]}{[(0)^2 + 4]^2} = \dfrac{1}{2} > 0$

$f''(3) = \dfrac{-2[(3)^2 - 4]}{[(3)^2 + 4]^2} = -\dfrac{10}{169} < 0$

Inflection points at $(-2, \ln 8)$ and $(2, \ln 8)$

Concave upward on $(-2, 2)$

Concave downward on $(-\infty, -2)$ and $(2, \infty)$

Since $f(x)$ never equals zero, there are no x-intercepts.

y-intercept: $y = \ln[(0)^2 + 4] = \ln 4$

No horizontal or vertical asymptotes.

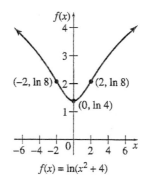

$$f(x) = \ln(x^2 + 4)$$

45. $f(x) = 4x^{1/3} + x^{4/3}$

Domain is $(-\infty, \infty)$.

$f(-x) = 4(-x)^{1/3} + (-x)^{4/3} = -4x^{1/3} + x^{4/3}$

The graph is not symmetric about the y-axis or origin.

$f'(x) = \dfrac{4}{3}x^{-2/3} + \dfrac{4}{3}x^{1/3}$

$f'(x) = 0$ when

$$\dfrac{4}{3}x^{-2/3} + \dfrac{4}{3}x^{1/3} = 0$$

$$\dfrac{4}{3}x^{-2/3}(1 + x) = 0$$

$$x = -1$$

$f'(x)$ is not defined when $x = 0$

Critical numbers: -1 and 0

Critical points: $(-1, -3)$ and $(0, 0)$

$f'(-8) = \dfrac{4}{3}(-8)^{-2/3} + \dfrac{4}{3}(-8)^{1/3} = -\dfrac{7}{3} < 0$

$f'\left(-\dfrac{1}{8}\right) = \dfrac{4}{3}\left(-\dfrac{1}{8}\right)^{-2/3} + \dfrac{4}{3}\left(-\dfrac{1}{8}\right)^{1/3} = \dfrac{14}{3} > 0$

$f'(1) = \dfrac{4}{3}(1)^{-2/3} + \dfrac{4}{3}(1)^{1/3} = \dfrac{8}{3} > 0$

No relative maximum

Relative minimum at $(-1, -3)$

Increasing on $(-1, \infty)$

Decreasing on $(-\infty, -1)$

$$f''(x) = -\frac{8}{9}x^{-5/3} + \frac{4}{9}x^{-2/3}$$

$f''(x) = 0$ when

$$-\frac{8}{9}x^{-5/3} + \frac{4}{9}x^{-2/3} = 0$$

$$\frac{4}{9}x^{-5/3}(-2 + x) = 0$$

$$x = 2$$

$f''(x)$ is not defined when $x = 0$

$$f''(-1) = -\frac{8}{9}(-1)^{-5/3} + \frac{4}{9}(-1)^{-2/3} = \frac{4}{3} > 0$$

$$f''(1) = -\frac{8}{9}(1)^{-5/3} + \frac{4}{9}(1)^{-2/3} = -\frac{4}{9} < 0$$

$$f''(8) = -\frac{8}{9}(8)^{-5/3} + \frac{4}{9}(8)^{-2/3} = \frac{1}{12} > 0$$

Inflection points at $(0,0)$ and $(2, 6 \cdot 2^{1/3})$

Concave upward on $(-\infty, 0)$ and $(2, \infty)$
Concave downward on $(0, 2)$

x-intercept: $4x^{1/3} + x^{4/3} = 0$

$$x^{1/3}(4 + x) = 0$$

$$x = 0 \text{ or } x = -4$$

y-intercept: $y = 4(0)^{1/3} + (0)^{4/3} = 0$

No horizontal or vertical asymptotes

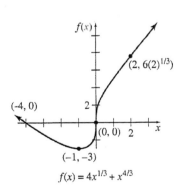

$$f(x) = 4x^{1/3} + x^{4/3}$$

47.

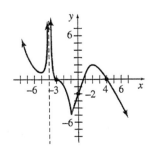

Other graphs are possible.

49. (a)-(b) If the price of the stock is falling faster and faster, $P(t)$ would be decreasing, so $P'(t)$ would be negative. $P(t)$ would be concave downward, so $P''(t)$ would also be negative.

51. (a) Profit = Income − Cost

$$\begin{aligned} P(q) &= qp - C(q) \\ &= q(-q^2 - 3q + 299) - (-10q^2 + 250q) \\ &= -q^3 - 3q^2 + 299q + 10q^2 - 250q \\ &= -q^3 + 7q^2 + 49q \end{aligned}$$

(b) $\begin{aligned} P'(q) &= -3q^2 + 14q + 49 \\ &= (-3q - 7)(q - 7) \\ &= -(3q + 7)(q - 7) \end{aligned}$

$$q = \frac{7}{3} \text{ (nonsensical) or } q = 7$$

$$P''(q) = -6q + 14$$

$$P''(7) = -6 \cdot 7 + 14 = -28 < 0$$

(indicates a maximum)

7 brushes would produce the maximum profit.

(c) $p = -7^2 - 3(7) + 299$
$= -49 - 21 + 299 = 229$

$229 is the price that produces the maximum profit.

(d) $P(7) = -7^3 + 7(7^2) + 49(7) = 343$

The maximum profit is $343.

(e) $P''(q) = 0$ when $-6q + 14 = 0$

$$q = \frac{7}{3}$$

$$P''(2) = -6(2) + 14 = 2 > 0$$

$$P''(3) = -6 \cdot 3 + 14 = -4 < 0$$

The point of diminishing returns is $q = \frac{7}{3}$ (between 2 and 3 brushes).

53. (a) $\begin{aligned} Y(M) &= Y_0 M^b \\ Y'(M) &= bY_0 M^{b-1} \\ Y''(M) &= b(b-1)Y_0 M^{b-2} \end{aligned}$

When $b > 0$, $Y' > 0$, so metabolic rate and life span are increasing function of mass.
When $b < 0$, $Y' < 0$, so heartbeat is a decreasing function of mass.
When $0 < b < 1$, $b(b-1) < 0$, so metabolic rate and life span have graphs that are concave downward.
When $b < 0$, $b(b-1) > 0$, so heartbeat has a graph that is concave upward.

(b) $\dfrac{dY}{dM} = bY_0 M^{b-1} = \dfrac{b}{M}Y_0 M^b$

$$= \frac{b}{M}Y$$

55. Let $a = 0.25$.

$$f(v) = v(0.25 - v)(v - 1)$$
$$f'(v) = -v^3 + 1.25v^2 - 0.25v$$
$$= -3v^2 + 2.5v - 0.25$$

By the quadratic formula, $f'(v) = 0$ when $v = 0.12$ and $v \approx 0.72$.
Critical numbers $v \approx 0.12$ and $v \approx 0.72$.

$$f'(0.1) = -0.03 < 0$$
$$f'(0.5) = 0.25 > 0$$
$$f'(1) = -0.75 < 0$$

The function is decreasing on $(0, 0.12)$ and $(0.72, 1)$ and increasing on $(0.12, 0.72)$.

$$f(0.12) \approx -0.01$$
$$\approx 0.09$$

Relative minimum at $(0.12, -0.01)$, relative maximum at $(0.72, 0.09)$.

$$f''(v) = -6v + 2.5$$

$f''(v) = 0$ when

$$-6v + 2.5 = 0$$
$$v \approx 0.42$$

$$f''(0.1) = 1.9 > 0$$
$$f''(0.5) = -0.5 < 0$$

Concave upward on $(0, 0.42)$
Concave downward on $(0.42, 1)$

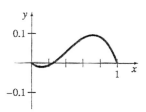

57. **(a)** Set the two formulas equal to each other.

$$1486S^2 - 4106S + 4514 = 1486S - 825$$
$$1486S^2 - 5592S + 5339 = 0$$

Take the derivative.

$$2972S - 5592 = 0$$

Solve for S.

$$S \approx 1.88$$

For males with 1.88 square meters of surface area, the red cell volume increases approximately 1486 ml for each additional square meter of surface area.

(b) Set the formulas equal to each other.

$$995e^{0.6085S} = 1578S$$
$$995e^{0.6085S} - 1578S = 0$$

Take the derivative.

$$605.4575e^{0.6085S} - 1578 = 0$$
$$e^{0.6085S} \approx 2.6063$$
$$0.6085S \approx \ln 2.6063$$
$$S \approx 1.57 \text{ square meters}$$

By plugging the exact value of S into the two formulas given for PV, we get about 2593 ml (Hurley) and 2484 for Pearson et al.

(c) For males with 1.57 square meters of surface area, the red cell volume increases approximately 1578 ml for each additional square meter of surface area.

(d) When f and g are closest together, their absolute difference is minimized.

$$\frac{d}{dx}|f(x_0) - g(x_0)| = 0$$
$$|f'(x_0) - g'(x_0)| = 0$$
$$f'(x_0) = g'(x_0)$$

59. **(a)** $P(t) = 325 + 7.475(t + 10)e^{-(t+10)/20}$

$$P'(t) = 7.475\left[1 \cdot e^{-(t+10)/20}\right.$$
$$\left. + (t + 10)e^{-(t+10)/20} \cdot \frac{-1}{20}\right]$$
$$= 7.475\left[1 - \frac{1}{20}(t + 10)\right]e^{-(t+10)/20}$$
$$= 7.475\left(\frac{1}{2} - \frac{t}{20}\right)e^{-(t+10)/20}$$

$P'(t)$ is zero when

$$\frac{1}{2} - \frac{t}{20} = 0$$
$$t = 10$$

$$P'(9) \approx 0.39$$
$$P'(11) \approx -0.36$$

$P(t)$ is increasing at 9 and decreasing at 11. So, a relative maximum occurs at $t = 10$. The population is largest in the year 2010.

(b) $P'(t) = 7.475 \left(\dfrac{1}{2} - \dfrac{t}{20} \right) e^{-(t+10)/20}$

$P''(t) = 7.475 \left[-\dfrac{1}{20} e^{-(t+10)/20} \right.$

$\left. + \left(\dfrac{1}{2} - \dfrac{t}{20} \right) e^{-(t+10)/20} \cdot \dfrac{-1}{20} \right]$

$= 7.475 \left(-\dfrac{1}{20} - \dfrac{1}{40} + \dfrac{t}{400} \right) e^{-(t+10)/20}$

$= 7.475 \left(\dfrac{t}{400} - \dfrac{3}{40} \right) e^{-(t+10)/20}$

$P'(t)$ is zero when

$$\dfrac{t}{400} - \dfrac{3}{40} = 0$$

$$t = 30$$
$$P'(29) = -0.0027$$
$$P'(31) = 0.0024$$

P' is decreasing at 29, and increasing at 31. So, a relative minimum occurs at $t = 30$. The population is declining most rapidly in the year 2010.

(c) As time t approaches infinity, the population P approaches

$\lim\limits_{t \to \infty} P(t) = \lim\limits_{t \to \infty} \left(325 + 7.475(t + 10)e^{-(t+10)/20} \right)$

$\qquad = 325 + 7.475 \lim\limits_{t \to \infty} \dfrac{t + 10}{e^{(t+10)/20}}$

$\qquad = 325 + 7.475(0)$

$\qquad = 325.$

The population is approaching 325 million.

61. (a) $s(t) = 512t - 16t^2$
$\qquad v(t) = s'(t) = 512 - 32t$
$\qquad a(t) = v'(t) = s''(t) = -32$

(b) The maximum height is attained when $v(t) = 0$.

$$512 - 32t = 0$$
$$t = 16$$

$\qquad v(0) = 512 > 0$
$\qquad v(20) = 512 - 640 = -128 < 0$

The height reaches a maximum when $t = 16$.

$$s(16) = 512 \cdot 16 - 16(16^2) = 4096$$

The maximum height is 4096 ft.

(c) The projectile hits the ground when $s(t) = 0$.

$$512t - 16t^2 = 0$$
$$16t(32 - t) = 0$$
$$t = 0 \text{ or } t = 32$$

$$v(32) = 512 - 32(32) = -512$$

The projectile hits the ground after 32 seconds with a velocity of -512 ft/sec.

Chapter 13 Test

[13.1]

Find the largest open intervals where each of the following functions is (a) increasing or (b) decreasing.

1.

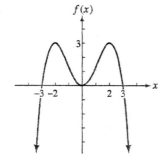

2. $f(x) = 3x^3 - 3x^2 - 3x + 5$

3. $f(x) = \dfrac{x - 2}{x + 3}$

4. For a certain product, the demand equation is $p = 10 - .05x$. The cost function for this product is $C = 5x - .75$. Over what interval(s) is the profit increasing?

5. What are the critical numbers? Why are they significant?

[13.2]

Find the x-values of all points where the following functions have any relative extrema. Find the value(s) of any relative extrema.

6. $f(x) = 4x^3 - \dfrac{9}{2}x^2 - 3x - 1$

7. $f(x) = 4x^{2/3}$

8. A manufacturer estimates that the cost in dollars per unit for a production run of x thousand units is given by $C = 3x^2 - 60x + 320$. How many thousand units should be produced during each run to minimize the cost per unit, and what is the minimum cost per unit?

9. Use the derivative to find the vertex of the parabola $y = -2x^2 + 6x + 9$.

[13.3]

Find $f'''(x)$, the third derivative of f, and $f^{(4)}(x)$, the fourth derivative of f, for each of the following functions.

10. $f(x) = 2e^{3x}$

11. $f(x) = \dfrac{x}{3x + 2}$

12. $f(x) = 4x^6 - 3x^4 + 2x^3 - 7x + 5$

Find any points of inflection for the following functions.

13. $f(x) = 6x^3 - 18x^2 + 12x - 15$ **14.** $f(x) = (x-1)^3 + 2$ **15.** $f(x) = \dfrac{2}{1+x^2}$

Find the second derivative of each function; then find $f''(-3)$.

16. $f(x) = -5x^3 + 3x^2 - x + 1$ **17.** $f(t) = \sqrt{t^2 - 5}$

18. Find the largest open intervals where $f(x) = x^3 - 3x^2 - 9x - 1$ is concave upward or concave downward. Find the location of any points of inflection. Graph the function.

19. The function $s(t) = t^3 - 9t^2 + 15t + 25$ gives the displacement in centimeters at time t (in seconds) of a particle moving along a line. Find the velocity and acceleration functions. Then find the velocity and acceleration at $t = 0$ and $t = 2$.

20. What is the difference between velocity and acceleration?

[13.4]

Find the horizontal asymptotes for the graphs of the following functions.

21. $f(x) = \dfrac{2x}{7x+1}$ **22.** $f(x) = \dfrac{3x^2 + 2x}{4x^3 - x}$

23. Graph $f(x) = \frac{2}{3}x^3 - \frac{5}{2}x^2 - 3x + 1$. Give critical points, intervals where the function is increasing or decreasing, points of inflection, and intervals where the function is concave upward or concave downward.

24. Graph $f(x) = x + \frac{32}{x^2}$. Give relative extrema, regions where the function is increasing or decreasing, points of inflection, regions where the function is concave upward or downward, intercepts where possible, and asymptotes where applicable.

25. Sketch a graph of a single function that has all the properties listed.

  (a) Continuous for all real numbers

  (b) Increasing on $(-3, 2)$ and $(4, \infty)$

  (c) Decreasing on $(-\infty, -3)$ and $(2, 4)$

  (d) Concave upward on $(-\infty, 0)$

  (e) Concave downward on $(0, 4)$ and $(4, \infty)$

  (f) Differentiable everywhere except $x = 4$

  (g) $f'(-3) = f'(2) = 0$

  (h) An inflection point at $(0, 0)$

Chapter 13 Test Answers

1. (a) $(-\infty, -2)$ and $(0, 2)$ (b) $(-2, 0)$ and $(2, \infty)$

2. (a) $(-\infty, -\frac{1}{3})$ and $(1, \infty)$ (b) $(-\frac{1}{3}, 1)$

3. (a) $(-\infty, -3)$ and $(-3, \infty)$ (b) Never decreasing

4. $(0, 50)$

5. Critical numbers are x–values for which $f'(x) = 0$ or $f'(x)$ does not exist. They tell where the derivative may change signs, and are used to tell where a function is increasing or decreasing.

6. Relative maximum of $-\frac{19}{32}$ at $-\frac{1}{4}$; relative minimum of $-\frac{9}{2}$ at 1

7. Relative minimum of 0 at 0

8. 10 thousand units; \$20 per unit

9. $\left(\frac{3}{2}, \frac{27}{2}\right)$

10. $f'''(x) = 54e^{3x}$, $f^{(4)}(x) = 162e^{3x}$

11. $f'''(x) = \frac{108}{(3x+2)^4}$, $f^{(4)}(x) = \frac{-1296}{(3x+2)^5}$

12. $f'''(x) = 480x^3 - 72x + 12$, $f^{(4)} = 1440x^2 - 72$

13. $(1, -15)$

14. $(1, 2)$

15. $\left(-\frac{\sqrt{3}}{3}, \frac{3}{2}\right), \left(\frac{\sqrt{3}}{3}, \frac{3}{2}\right)$

16. $f''(x) = -30x + 6$; $f''(-3) = 96$

17. $f''(t) = \frac{-5}{(t^2-5)^{3/2}}$; $f''(-3) = -\frac{5}{8}$

18. Concave upward on $(1, \infty)$; concave downward on $(-\infty, 1)$; point of inflection $(1, -12)$

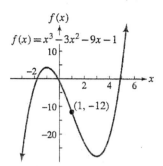

19. $v(t) = 3t^2 - 18t + 15$; $a(t) = 6t - 18$; $v(0) = 15$ cm/sec; $a(0) = -18$ cm/sec^2; $v(2) = -9$ cm/sec; $a(2) = -6$ cm/sec^2

20. Velocity gives the rate of change of position relative to time. Acceleration gives the rate of change of velocity.

21. $y = \frac{2}{7}$

22. $y = 0$

23. Critical points: $\left(-\frac{1}{2}, \frac{43}{24}\right)$ or $(-0.5, 1.79)$ (relative maximum) and $\left(3, -\frac{25}{2}\right)$ or $(3, -12.5)$ (relative minimum); increasing on $\left(-\infty, -\frac{1}{2}\right)$ or $(-\infty, -0.5)$ and $(3, \infty)$; decreasing on $\left(-\frac{1}{2}, 3\right)$ or $(-0.5, 3)$; point of inflection: $\left(\frac{5}{4}, -\frac{257}{48}\right)$ or $(1.25, -5.35)$; concave up on $\left(\frac{5}{4}, \infty\right)$ or $(1.25, \infty)$; concave down on $\left(-\infty, \frac{5}{4}\right)$ or $(-\infty, 1.25)$

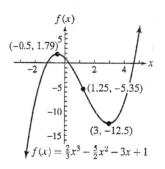

24. Relative extrema: Relative minimum at $(4,6)$, no relative maxima; increasing on $(-\infty,0)$ and $(4,\infty)$; decreasing on $(0,4)$; points of inflection: none; concave upward on $(-\infty,0)$ and $(0,\infty)$; concave downward nowhere; x–intercept: $-\sqrt[3]{32} \approx -3.17$; y–intercept: none; oblique asymptote: $y = x$

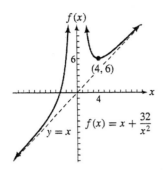

25. One such graph is shown. Other answers are possible.

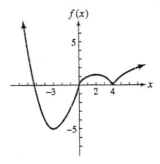

APPLICATIONS OF THE DERIVATIVE

14.1   Absolute Extrema

1. As shown on the graph, the absolute maximum occurs at x_3; there is no absolute minimum. (There is no functional value that is less than all others.)

3. As shown on the graph, there are no absolute extrema.

5. As shown on the graph, the absolute minimum occurs at x_1; there is no absolute maximum.

7. As shown on the graph, the absolute maximum occurs at x_1; the absolute minimum occurs at x_2.

11. $f(x) = x^3 - 6x^2 + 9x - 8;\ [0, 5]$

Find critical numbers:

$$f'(x) = 3x^2 - 12x + 9 = 0$$
$$x^2 - 4x + 3 = 0$$
$$(x - 3)(x - 1) = 0$$
$$x = 1 \quad \text{or} \quad x = 3$$

x	$f(x)$	
0	-8	Absolute minimum
1	-4	
3	-8	Absolute minimum
5	12	Absolute maximum

13. $f(x) = \dfrac{1}{3}x^3 + \dfrac{3}{2}x^2 - 4x + 1;\ [-5, 2]$

Find critical numbers:

$$f'(x) = x^2 + 3x - 4 = 0$$
$$(x + 4)(x - 1) = 0$$
$$x = -4 \quad \text{or} \quad x = 1$$

x	$f(x)$	
-4	$\dfrac{59}{3} \approx 19.67$	Absolute maximum
1	$-\dfrac{7}{6} \approx -1.17$	Absolute minimum
-5	$\dfrac{101}{6} \approx 16.83$	
2	$\dfrac{5}{3} \approx 1.67$	

15. $f(x) = x^4 - 18x^2 + 1;\ [-4, 4]$
$$f'(x) = 4x^3 - 36x = 0$$
$$4x(x^2 - 9) = 0$$
$$4x(x + 3)(x - 3) = 0$$

$$x = 0 \quad \text{or} \quad x = -3 \quad \text{or} \quad x = 3$$

x	$f(x)$	
-4	-31	
-3	-80	Absolute minimum
0	1	Absolute maximum
3	-80	Absolute minimum
4	-31	

17. $f(x) = \dfrac{1 - x}{3 + x};\ [0, 3]$

$$f'(x) = \dfrac{-4}{(3 + x)^2}$$

No critical numbers

x	$f(x)$	
0	$\dfrac{1}{3}$	Absolute maximum
3	$-\dfrac{1}{3}$	Absolute minimum

19. $f(x) = \dfrac{x - 1}{x^2 + 1};\ [1, 5]$

$$f'(x) = \dfrac{-x^2 + 2x + 1}{(x^2 + 1)^2}$$

$f'(x) = 0$ when

$$-x^2 + 2x + 1 = 0$$
$$x = 1 \pm \sqrt{2},$$

but $1 - \sqrt{2}$ is not in $[1, 5]$.

x	$f(x)$	
1	0	Absolute minimum
5	$\dfrac{2}{13} \approx 0.15$	
$1 + \sqrt{2}$	$\dfrac{\sqrt{2} - 1}{2} \approx 0.21$	Absolute maximum

21. $f(x) = (x^2 - 4)^{1/3};\ [-2, 3]$

$$f'(x) = \frac{1}{3}(x^2 - 4)^{-2/3}(2x)$$

$$= \frac{2x}{3(x^2 - 4)^{2/3}}$$

$f'(x) = 0$ when $2x = 0$

$$x = 0$$

$f'(x)$ is undefined at $x = -2$ and $x = 2$, but $f(x)$ is defined there, so -2 and 2 are also critical numbers.

x	$f(x)$	
-2	0	
0	$(-4)^{1/3} \approx -1.587$	Absolute minimum
2	0	
3	$5^{1/3} \approx 1.710$	Absolute maximum

23. $f(x) = 5x^{2/3} + 2x^{5/3};\ [-2, 1]$

$$f'(x) = \frac{10}{3}x^{-1/3} + \frac{10}{3}x^{2/3}$$

$$= \frac{10}{3x^{1/3}} + \frac{10x^{2/3}}{3}$$

$$= \frac{10x + 10}{3x^{1/3}}$$

$$= \frac{10(x + 1)}{3\sqrt[3]{x}}$$

$f'(x) = 0$ when $10(x + 1) = 0$

$$x + 1 = 0$$

$$x = -1.$$

$f'(x)$ is undefined at $x = 0$, but $f(x)$ is defined at $x = 0$, so 0 is also a critical number.

x	$f(x)$	
-2	1.587	
-1	3	
0	0	Absolute minimum
1	7	Absolute maximum

25. $f(x) = x^2 - 8 \ln x;\ [1, 4]$

$$f'(x) = 2x - \frac{8}{x}$$

$f'(x) = 0$ when $2x - \frac{8}{x} = 0$

$$2x = \frac{8}{x}$$

$$2x^2 = 8$$

$$x^2 = 4$$

$$x = -2 \text{ or } x = 2$$

but $x = -2$ is not in the given interval.

Although $f'(x)$ fails to exist at $x = 0$, 0 is not in the specified domain for $f(x)$, so 0 is not a critical number.

x	$f(x)$	
1	1	
2	-1.545	Absolute minimum
4	4.910	Absolute maximum

27. $f(x) = x + e^{-3x};\ [-1, 3]$

$$f'(x) = 1 - 3e^{-3x}$$

$f'(x) = 0$ when $1 - 3e^{-3x} = 0$

$$-3e^{-3x} = -1$$

$$e^{-3x} = \frac{1}{3}$$

$$-3x = \ln \frac{1}{3}$$

$$x = \frac{\ln 3}{3}$$

x	$f(x)$	
-1	19.09	Absolute maximum
$\dfrac{\ln 3}{3}$	0.6995	Absolute minimum
3	3.000	

29. $f(x) = \dfrac{-5x^4 + 2x^3 + 3x^2 + 9}{x^4 - x^3 + x^2 + 7};\ [-1, 1]$

The indicated domain tells us the x-values to use for the viewing window, but we must experiment to find a suitable range for the y-values. In order to show the absolute extrema on $[-1, 1]$, we find that a suitable window is $[-1, 1]$ by $[0, 1.5]$ with Xscl = 0.1, Yscl = 0.1.

From the graph, we se that on $[-1, 1]$, f has an absolute maximum of 1.356 at about 0.6085 and an absolute minimum of 0.5 at -1.

31. $f(x) = 2x + \dfrac{8}{x^2} + 1,\ x > 0$

$$f'(x) = 2 - \frac{16}{x^3}$$

$$= \frac{2x^3 - 16}{x^3}$$

$$= \frac{2(x - 2)(x^2 + 2x + 4)}{x^3}$$

Since the specified domain is $(0, \infty)$, a critical number is $x = 2$.

x	$f(x)$
2	7

There is an absolute minimum at $x = 2$; there is no absolute maximum, as can be seen by looking at the graph of f.

33. $f(x) = -3x^4 + 8x^3 + 18x^2 + 2$

$$\begin{aligned} f'(x) &= -12x^3 + 24x^2 + 36x \\ &= -12x(x^2 - 2x - 3) \\ &= -12x(x - 3)(x + 1) \end{aligned}$$

Critical numbers are 0, 3, and -1.

x	$f(x)$
-1	9
0	2
3	137

There is an absolute maximum at $x = 3$; there is no absolute minimum, as can be seen by looking at the graph of f.

35. $f(x) = \dfrac{x - 1}{x^2 + 2x + 6}$

$$\begin{aligned} f'(x) &= \frac{(x^2 + 2x + 6)(1) - (x - 1)(2x + 2)}{(x^2 + 2x + 6)^2} \\ &= \frac{x^2 + 2x + 6 - 2x^2 + 2}{(x^2 + 2x + 6)^2} \\ &= \frac{-x^2 + 2x + 8}{(x^2 + 2x + 6)^2} \\ &= \frac{-(x^2 - 2x - 8)}{(x^2 + 2x + 6)^2} \\ &= \frac{-(x - 4)(x + 2)}{(x^2 + 2x + 6)^2} \end{aligned}$$

Critical numbers are 4 and -2.

x	$f(x)$
-2	$-\dfrac{1}{2}$
4	0.1

There is an absolute maximum at $x = 4$ and an absolute minimum at $x = -2$. This can be verified by looking at the graph of f.

37. $f(x) = \dfrac{\ln x}{x^3}$

$$\begin{aligned} f'(x) &= \frac{x^3 \cdot \frac{1}{x} - 3x^2 \ln x}{x^6} \\ &= \frac{x^2 - 3x^2 \ln x}{x^6} \\ &= \frac{x^2(1 - 3\ln x)}{x^6} \\ &= \frac{1 - 3\ln x}{x^4} \end{aligned}$$

$f'(x) = 0$ when $x = e^{1/3}$, and $f'(x)$ does not exist when $x \le 0$. The only critical number is $e^{1/3}$.

x	$f(x)$
$e^{1/3}$	$\dfrac{1}{3}\, e^{-1} \approx 0.1226$

There is an absolute maximum of 0.1226 at $x = e^{1/3}$. There is no absolute minimum, as can be seen by looking at the graph of f.

39. Let $P(x)$ be the perimeter of the rectangle with vertices $(0, 0), (x, 0), (x, f(x))$, and $(0, f(x))$ for $x > 0$ when $f(x) = e^{-2x}$.

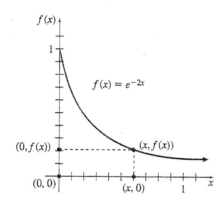

The length of the rectangle is x and the width is given by e^{-2x}. Therefore, an equation for the perimeter is

$$P(x) = x + e^{-2x} + x + e^{-2x} = 2(x + e^{-2x}).$$

$P'(x) = 2 - 4e^{-2x}$

$P'(x) = 0$ when $2 - 4e^{-2x} = 0$

$$\begin{aligned} -4e^{-2x} &= -2 \\ e^{-2x} &= \frac{1}{2} \\ e^{2x} &= 2 \\ 2x &= \ln 2 \\ x &= \frac{\ln 2}{2} \end{aligned}$$

x	$P(x)$
$\dfrac{\ln 2}{2}$	$1 + \ln 2 \approx 1.693$

There is an absolute minimum of 1.693 at $x = \frac{\ln 2}{2}$. There is no absolute maximum, as can be seen by looking at the graph of P. Therefore, the correct statement is **a**.

41. (a) By looking at the graph, there are relative maxima of 413 in 1997, 341 in 2000, and 134 in 2004. There are relative minima of 290 in 1996, 313 in 1998, and 131 in 2003.

(b) Annual bank burglaries reached an absolute maximum of 413 in 1997 and an absolute minimum of 131 in 2003.

43. $P(x) = -x^3 + 9x^2 + 120x - 400,\ x \geq 5$

$$P'(x) = -3x^2 + 18x + 120$$
$$= -3(x^2 - 6x - 40)$$
$$= -3(x - 10)(x + 4) = 0$$
$$x = 10 \quad \text{or} \quad x = -4$$

-4 is not relevant since $x \geq 5$, so the only critical number is 10.

The graph of $P'(x)$ is a parabola that opens downward, so $P'(x) > 0$ on the interval $[5, 10)$ and $P'(x) < 0$ on the interval $(10, \infty)$. Thus, $P(x)$ is a maximum at $x = 10$.
Since x is measured in hundred thousands, 10 hundred thousand or 1,000,000 tires must be sold to maximize profit.
Also,
$$P(10) = -(10)^3 + 9(10)^2 + 120(10) - 400$$
$$= 700.$$
The maximum profit is $700 thousand or $700,000.

45. $C(x) = x^3 + 37x + 250$

(a) $1 \leq x \leq 10$

$$\overline{C}(x) = \frac{C(x)}{x} = \frac{x^3 + 37x + 250}{x}$$
$$= x^2 + 37 + \frac{250}{x}$$

$$\overline{C}'(x) = 2x - \frac{250}{x^2}$$
$$= \frac{2x^3 - 250}{x^2} = 0 \text{ when}$$
$$2x^3 = 250$$
$$x^3 = 125$$
$$x = 5.$$

Test for relative minimum.

$$\overline{C}'(4) = -7.625 < 0$$
$$\overline{C}'(6) \approx 5.0556 > 0$$
$$\overline{C}(5) = 112$$
$$\overline{C}(1) = 1 + 37 + 250 = 288$$
$$\overline{C}(10) = 100 + 37 + 25 = 162$$

The minimum on the interval $1 \leq x \leq 10$ is 112.

(b) $10 \leq x \leq 20$

There are no critical values in this interval. Check the endpoints.

$$\overline{C}(10) = 162$$
$$\overline{C}(20) = 400 + 37 + 12.5 = 449.5$$

The minimum on the interval $10 \leq x \leq 20$ is 162.

47. The value $x = 11$ minimizes $\frac{f(x)}{x}$ because this is the point where the line from the origin to the curve is tangent to the curve.
A production level of 11 units results in the minimum cost per unit.

49. The value $x = 100$ maximizes $\frac{f(x)}{x}$ because this is the point where the line from the origin to the curve is tangent to the curve.
A production level of 100 units results in the maximum profit per item produced.

51. $S(x) = -x^3 + 3x^2 + 360x + 5000;\ 6 \leq x \leq 20$

$$S'(x) = -3x^2 + 6x + 360$$
$$= -3(x^2 - 2x - 120)$$
$$S'(x) = -3(x - 12)(x + 10) = 0$$
$$x = 12 \quad \text{or} \quad x = -10 \ \text{(not in the}$$
$$\text{interval)}$$

x	$f(x)$
6	7052
12	8024
10	7900

$12°$ is the temperature that produces the maximum number of salmon.

53. The function is defined on the interval $[15, 46]$. We look first for critical numbers in the interval. We find

$$R'(T) = -0.00021T^2 + 0.0802T - 1.6572$$

Using our graphing calculator, we find one critical number in the interval at about 21.92

T	$R(T)$
15	81.01
21.92	79.29
46	98.89

The relative humidity is minimized at about $21.92°C$.

55. $M(x) = -0.015x^2 + 1.31x - 7.3, 30 \le x \le 60$
$M'(x) = -0.03x + 1.31 = 0$
$$x \approx 43.7$$

x	$M(x)$
30	18.5
43.7	21.30
60	17.3

The absolute maximum of 21.30 mpg occurs at 43.7 mph. The absolute minimum of 17.3 mpg occurs at 60 mph.

57. Total area $= A(x)$

$$= \pi \left(\frac{x}{2\pi}\right)^2 + \left(\frac{12-x}{4}\right)^2$$

$$= \frac{x^2}{4\pi} + \frac{(12-x)^2}{16}$$

$$A'(x) = \frac{x}{2\pi} - \frac{12-x}{8} = 0$$

$$\frac{4x - \pi(12-x)}{8\pi} = 0$$

$$x = \frac{12\pi}{4+\pi} \approx 5.28$$

x	Area
0	9
5.28	5.04
12	11.46

The total area is maximized when all 12 feet of wire are used to form the circle.

59. (a) $I(p) = -p \ln p - (1-p) \ln (1-p)$

$$I'(p) = -p \left(\frac{1}{p}\right) + (\ln p)(-1)$$

$$- \left[(1-p)\frac{-1}{1-p} + [\ln (1-p)](-1)\right]$$

$$= -1 - \ln p + 1 + \ln (1-p)$$
$$= -\ln p + \ln (1-p)$$

(b) $-\ln p + \ln (1-p) = 0$
$$\ln (1-p) = \ln p$$
$$1 - p = p$$
$$1 = 2p$$
$$\frac{1}{2} = p$$

$$I'(0.25) = 1.0986$$
$$I'(0.75) = -1.099$$

There is a relative maximum of 0.693 at $p = \frac{1}{2}$.

14.2 Applications of Extrema

1. $x + y = 180, P = xy$

(a) $y = 180 - x$

(b) $P = xy = x(180 - x)$

(c) Since $y = 180 - x$ and x and y are nonnegative numbers, $x \ge 0$ and $180 - x \ge 0$ or $x \le 180$. The domain of P is $[0, 180]$.

(d) $P'(x) = 180 - 2x$
$$180 - 2x = 0$$
$$2(90 - x) = 0$$
$$x = 90$$

(e)

x	P
0	0
90	8100
180	0

(f) From the chart, the maximum value of P is 8100; this occurs when $x = 90$ and $y = 90$.

3. $x + y = 90$

Minimize $x^2 y$.

(a) $y = 90 - x$

(b) Let $P = x^2 y = x^2(90 - x)$
$$= 90x^2 - x^3.$$

(c) Since $y = 90 - x$ and x and y are nonnegative numbers, the domain of P is $[0, 90]$.

(d) $P' = 180x - 3x^2$

$$180x - 3x^2 = 0$$
$$3x(60 - x) = 0$$
$$x = 0 \text{ or } x = 60$$

(e)

x	P
0	0
60	108,000
90	0

(f) The maximum value of $x^2 y$ occurs when $x = 60$ and $y = 30$. The maximum value is 108,000.

5. $C(x) = \frac{1}{2}x^3 + 2x^2 - 3x + 35$

The average cost function is

$$A(x) = \overline{C}(x) = \frac{C(x)}{x}$$

$$= \frac{\frac{1}{2}x^3 + 2x^2 - 3x + 35}{x}$$

$$= \frac{1}{2}x^2 + 2x - 3 + \frac{35}{x}$$

$$\text{or} \quad \frac{1}{2}x^2 + 2x - 3 + 35x^{-1}.$$

Then

$$A'(x) = x + 2 - 35x^{-2}$$

$$\text{or} \quad x + 2 - \frac{35}{x^2}.$$

Graph $y = A'(x)$ on a graphing calculator. A suitable choice for the viewing window is $[0, 10]$ by $[-10, 10]$. (Negative values of x are not meaningful in this application.) Using the calculator, we see that the graph has an x-intercept or "zero" at $x \approx 2.722$. Thus, 2.722 is a critical number.
Now graph $y = A(x)$ and use this graph to confirm that a minimum occurs at $x \approx 2.722$.

Thus, the average cost is smallest at $x \approx 2.722$.

7. $p(x) = 160 - \frac{x}{10}$

(a) Revenue from sale of x thousand candy bars:

$$R(x) = 1000xp$$

$$= 1000x\left(160 - \frac{x}{10}\right)$$

$$= 160{,}000x - 100x^2$$

(b) $R'(x) = 160{,}000 - 200x$

$$160{,}000 - 200x = 0$$
$$160{,}000 = 200x$$
$$800 = x$$

The maximum revenue occurs when 800 thousand bars are sold.

(c) $R(800) = 160{,}000(800) - 100(800)^2$
$$= 64{,}000{,}000$$

The maximum revenue is 64,000,000 cents.

9. Let $x =$ the width
and $y =$ the length.

(a) The perimeter is

$$P = 2x + y$$
$$= 1400,$$

so

$$y = 1400 - 2x.$$

(b) Area $= xy = x(1400 - 2x)$
$$A(x) = 1400x - 2x^2$$

(c) $A' = 1400 - 4x$
$$1400 - 4x = 0$$
$$1400 = 4x$$
$$350 = x$$

$A'' = -4$, which implies that $x = 350$ m leads to the maximum area.

(d) If $x = 350$,

$$y = 1400 - 2(350) = 700.$$

The maximum area is $(350)(700) = 245{,}000$ m^2.

11. Let $x =$ the width of the rectangle
$y =$ the total length of the
rectangle.

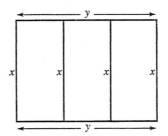

An equation for the fencing is

$$3600 = 4x + 2y$$
$$2y = 3600 - 4x$$
$$y = 1800 - 2x.$$

Area $= xy = x(1800 - 2x)$
$$A(x) = 1800x - 2x^2$$

$A' = 1800 - 4x$

$$1800 - 4x = 0$$
$$1800 = 4x$$
$$450 = x$$

$A'' = -4$, which implies that $x = 450$ is the location of a maximum.

If $x = 450$, $y = 1800 - 2(450) = 900$.
The maximum area is

$$(450)(900) = 405,000 \text{ m}^2.$$

13. Let $x =$ length at \$1.50 per meter
$y =$ width at \$3 per meter.

$$xy = 25,600$$

$$y = \frac{25,600}{x}$$

$$\text{Perimeter} = x + 2y = x + \frac{51,200}{x}$$

$$\text{Cost} = C(x) = x(1.5) + \frac{51,200}{x}(3)$$

$$= 1.5x + \frac{153,600}{x}$$

Minimize cost:

$$C'(x) = 1.5 - \frac{153,600}{x^2}$$

$$1.5 - \frac{153,600}{x^2} = 0$$

$$1.5 = \frac{153,600}{x^2}$$

$$1.5x^2 = 153,600$$

$$x^2 = 102,400$$

$$x = 320$$

$$y = \frac{25,600}{320} = 80$$

320 m at \$1.50 per meter will cost \$480. 160 m
at \$3 per meter will cost \$480. The total cost will
be \$960.

15. Let $x =$ the number of days to wait.

$$\frac{12,000}{100} = 120 = \text{the number of 100-lb groups}$$
$$\text{collected already.}$$

Then $7.5 - 0.15x =$ the price per 100 lb;
$4x =$ the number of 100-lb groups
collected per day;
$120 + 4x =$ total number of 100-lb groups
collected.

$$\text{Revenue} = R(x)$$
$$= (7.5 - 0.15x)(120 + 4x)$$
$$= 900 + 12x - 0.6x^2$$

$$R'(x) = 12 - 1.2x = 0$$
$$x = 10$$

$R''(x) = -1.2 < 0$ so $R(x)$ is maximized at
$x = 10$.

The scouts should wait 10 days at which time their
income will be maximized at

$$R(10) = 900 + 12(10) - 0.6(10)^2 = \$960.$$

17. Let $x =$ the number of refunds.
Then $535 - 5x =$ the cost per passenger
and $85 + x =$ the number of passengers.

(a) Revenue $= R(x) = (535 - 5x)(85 + x)$
$$= 45,475 + 110x - 5x^2$$
$$R'(x) = 110 - 10x = 0$$
$$x = 11$$

$R''(x) = -10 < 0$, so $R(x)$ is maximized when
$x = 11$.
Thus, the number of passengers that will maximize revenue is $85 + 11 = 96$.

(b) $R(11) = 45,475 + 110(11) - 5(11)^2$
$$= 46,080$$

The maximum revenue is \$46,080.

19. Let $x =$ the length of a side of
the top and bottom.
Then $x^2 =$ the area of the top and
bottom
and $(3)(2x^2) =$ the cost for the top and
bottom.

Let $y =$ depth of box.
Then $xy =$ the area of one side,
$4xy =$ the total area of the
sides,
and $(1.50)(4xy) =$ the cost of the sides.

The total cost is

$$C(x) = (3)(2x^2) + (1.50)(4xy) = 6x^2 + 6xy.$$

The volume is

$$V = 16,000 = x^2y.$$

$$y = \frac{16,000}{x^2}$$

$$C(x) = 6x^2 + 6x\left(\frac{16,000}{x^2}\right) = 6x^2 + \frac{96,000}{x}$$

$$C'(x) = 12x - \frac{96,000}{x^2} = 0$$

$$x^3 = 8000$$

$$x = 20$$

$C''(x) = 12 + \dfrac{192,000}{x^3} > 0$ at $x = 20$, which implies that $C(x)$ is minimized when $x = 20$.

$$y = \frac{16,000}{(20)^2} = 40$$

So the dimensions of the box are x by x by y, or 20 cm by 20 cm by 40 cm.

$$C(20) = 6(20)^2 + \frac{96,000}{20} = 7200$$

The minimum total cost is $7200.

21. (a) $S = 2\pi r^2 + 2\pi rh$, $V = \pi r^2 h$

$$S = 2\pi r^2 + \frac{2V}{r}$$

Treat V as a constant.

$$S' = 4\pi r - \frac{2V}{r^2}$$

$$4\pi r - \frac{2V}{r^2} = 0$$

$$\frac{4\pi r^3 - 2V}{r^2} = 0$$

$$4\pi r^3 - 2V = 0$$
$$2\pi r^3 - V = 0$$
$$2\pi r^3 = V$$
$$2\pi r^3 = \pi r^2 h$$
$$2r = h$$

23. Let $\quad x =$ the length of the side of the cutout square.

Then $3 - 2x =$ the width of the box and $\quad 8 - 2x =$ the length of the box.

$$V(x) = x(3 - 2x)(8 - 2x)$$
$$= 4x^3 - 22x^2 + 24x$$

The domain of V is $\left(0, \frac{3}{2}\right)$.

Maximize the volume.

$$V'(x) = 12x^2 - 44x + 24$$
$$12x^2 - 44x + 24 = 0$$
$$4(3x^2 - 11x + 6) = 0$$
$$4(3x - 2)(x - 3) = 0$$
$$x = \frac{2}{3} \quad \text{or} \quad x = 3$$

3 is not in the domain of V.

$$V''(x) = 24x - 44$$

$$V''\left(\frac{2}{3}\right) = -28 < 0$$

This implies that V is maximized when $x = \frac{2}{3}$. The box will have maximum volume when $x = \frac{2}{3}$ ft or 8 in.

25. Let $x =$ the width of printed material and $y =$ the length of printed material.

Then, the area of the printed material is

$$xy = 36,$$

$$\text{so} \quad y = \frac{36}{x}.$$

Also, $x + 2 =$ the width of a page and $y + 3 =$ the length of a page.

The area of a page is

$$A = (x + 2)(y + 3)$$
$$= xy + 2y + 3x + 6$$
$$= 36 + 2\left(\frac{36}{x}\right) + 3x + 6$$
$$= 42 + \frac{72}{x} + 3x.$$
$$A' = -\frac{72}{x^2} + 3 = 0$$
$$x^2 = 24$$
$$x = \sqrt{24}$$
$$= 2\sqrt{6}$$

(We discard $x = -2\sqrt{6}$ once we must have $x > 0$.) $A'' = \frac{216}{x^3} > 0$ when $x = 2\sqrt{6}$, which implies that A is minimized when $x = 2\sqrt{6}$.

$$y = \frac{36}{x} = \frac{36}{2\sqrt{6}} = \frac{18}{\sqrt{6}} = \frac{18\sqrt{6}}{6} = 3\sqrt{6}$$

The width of a page is

$$x + 2 = 2\sqrt{6} + 2$$
$$\approx 6.9 \text{ in.}$$

The length of a page is

$$y + 3 = 3\sqrt{6} + 3$$
$$\approx 10.3 \text{ in.}$$

27. Distance on shore: $7 - x$ miles
Cost on shore: $400 per mile
Distance underwater: $\sqrt{x^2 + 36}$
Cost underwater: $500 per mile
Find the distance from A, that is, $7 - x$, to minimize cost, $C(x)$.

$$C(x) = (7 - x)(400) + (\sqrt{x^2 + 36})(500)$$
$$= 2800 - 400x + 500(x^2 + 36)^{1/2}$$

$$C'(x) = -400 + 500\left(\frac{1}{2}\right)(x^2 + 36)^{-1/2}(2x)$$

$$= -400 + \frac{500x}{\sqrt{x^2 + 36}}$$

If $C'(x) = 0$,

$$\frac{500x}{\sqrt{x^2 + 36}} = 400$$

$$\frac{5x}{4} = \sqrt{x^2 + 36}$$

$$\frac{25}{16}x^2 = x^2 + 36$$

$$\frac{9}{16}x^2 = 36$$

$$x^2 = \frac{36 \cdot 16}{9}$$

$$x = \frac{6 \cdot 4}{3} = 8.$$

(Discard the negative solution.)
$x = 8$ is impossible since Point A is only 7 miles from point C.
Check the endpoints.

x	$C(x)$
0	5800
7	4610

The cost is minimized when $x = 7$.

$7 - x = 7 - 7 = 0$, so the company should angle the cable at Point A.

29. From Example 4, we know that the surface area of the can is given by

$$S = 2\pi r^2 + \frac{2000}{r}.$$

Aluminum costs 3¢/cm^2, so the cost of the aluminum to make the can is

$$0.03\left(2\pi r^2 + \frac{2000}{r}\right) = 0.06\pi r^2 + \frac{60}{r}.$$

The perimeter (or circumference) of the circular top is $2\pi r$. Since there is a 2¢/cm charge to seal the top and bottom, the sealing cost is

$$0.02(2)(2\pi r) = 0.08\pi r.$$

Thus, the total cost is given by the function

$$C(r) = 0.06\pi r^2 + \frac{60}{r} + 0.08\pi r$$

$$= 0.06\pi r^2 + 60r^{-1} + 0.08\pi r.$$

Then

$$C'(r) = 0.12\pi r - 60r^{-2} + 0.08\pi$$

$$= 0.12\pi r - \frac{60}{r^2} + 0.08\pi.$$

Graph

$$y = 0.12\pi x - \frac{60}{x^2} + 0.08\pi$$

on a graphing calculator. Since r must be positive in this application, our window should not include negative values of x. A suitable choice for the viewing window is $[0, 10]$ by $[-10, 10]$. From the graph, we find that $C'(x) = 0$ when $x \approx 5.206$.
Thus, the cost is minimized when the radius is about 5.206 cm.

We can find the corresponding height by using the equation

$$h = \frac{1000}{\pi r^2}$$

from Example 4.
If $r = 5.206$,

$$h = \frac{1000}{\pi(5.206)^2} \approx 11.75.$$

To minimize cost, the can should have radius 5.206 cm and height 11.75 cm.

31. In Exercises 29 and 30, we found that the cost of the aluminum to make the can is $0.06\pi r^2 + \frac{60}{r}$, the cost to seal the top and bottom is $0.08\pi r$, and the cost to seal the vertical seam is $\frac{10}{\pi r^2}$.
Thus, the total cost is now given by the function

$$C(r) = 0.06\pi r^2 + \frac{60}{r} + 0.08\pi r + \frac{10}{\pi r^2}$$

$$\text{or}\quad 0.06\pi r^2 + 60r^{-1} + 0.08\pi r + \frac{10}{\pi}r^{-2}.$$

Then

$$C'(r) = 0.12\pi r - 60r^{-2} + 0.08\pi - \frac{20}{\pi}r^{-3}$$

$$\text{or}\quad 0.12\pi r - \frac{60}{r^2} + 0.08\pi - \frac{20}{\pi r^3}.$$

Graph

$$y = 0.12\pi r - \frac{60}{r^2} + 0.08\pi - \frac{20}{\pi r^3}$$

on a graphing calculator. A suitable choice for the viewing window is $[0, 10]$ by $[-10, 10]$. From the graph, we find that $C'(x) = 0$ when $x \approx 5.242$.
Thus, the cost is minimized when the radius is about 5.242 cm.

To find the corresponding height, use the equation

$$h = \frac{1000}{\pi r^2}$$

from Example 4.
If $r = 5.242$,

$$h = \frac{1000}{\pi (5.242)^2} \approx 11.58.$$

To minimize cost, the can should have radius 5.242 cm and height 11.58 cm.

33. $N(t) = 20\left[\dfrac{t}{12} - \ln\left(\dfrac{t}{12}\right)\right] + 30;$
$1 \le t \le 15$

$$N'(t) = 20\left[\frac{1}{12} - \frac{12}{t}\left(\frac{1}{12}\right)\right]$$

$$= 20\left(\frac{1}{12} - \frac{1}{t}\right)$$

$$= \frac{20(t-12)}{12t}$$

$N'(t) = 0$ when

$$t - 12 = 0$$
$$t = 12.$$

$N''(t)$ does not exist at $t = 0$, but 0 is not in the domain of N.
Thus, 12 is the only critical number.

To find the absolute extrema on $[1, 15]$, evaluate N at the critical number and at the endpoints.

t	$N(t)$
1	81.365
12	50
15	50.537

Use this table to answer the questions in (a)-(d).

(a) The number of bacteria will be a minimum at $t = 12$, which represents 12 days.

(b) The minimum number of bacteria is given by $N(12) = 50$, which represents 50 bacteria per ml.

(c) The number of bacteria will be a maximum at $t = 1$, which represents 1 day.

(d) The maximum number of bacteria is given by $N(1) = 81.365$, which represents 81.365 bacteria per ml.

35. $H(S) = f(S) - S$
$f(S) = 12S^{0.25}$
$H(S) = 12S^{0.25} - S$
$H'(S) = 3S^{-0.75} - 1$

$H'(S) = 0$ when

$$3S^{-0.75} - 1 = 0$$

$$S^{-0.75} = \frac{1}{3}$$

$$\frac{1}{S^{0.75}} = \frac{1}{3}$$

$$S^{0.75} = 3$$

$$S^{3/4} = 3$$

$$S = 3^{4/3}$$

$$S = 4.327.$$

The number of creatures needed to sustain the population is $S_0 = 4.327$ thousand.

$H''(S) = \frac{-2.25}{S^{1.75}} < 0$ when $S = 4.327$, so $H(S)$ is maximized.

$$H(4.327) = 12(4.327)^{0.25} - 4.327$$
$$\approx 12.98$$

The maximum sustainable harvest is 12.98 thousand.

37. (a) $H(S) = f(S) - S$
$\qquad = Se^{r(1-S/P)} - S$

$$H'(S) = Se^{r(1-S/P)}\left(\frac{-r}{P}\right) + e^{r(1-S/P)} - 1$$

Note that

$$f(S) = Se^{r(1-S/P)}$$

$$f'(S) = Se^{r(1-S/P)}\left(-\frac{r}{P}\right) + e^{r(1-S/P)}.$$

$$H'(S) = Se^{r(1-S/P)}\left(-\frac{r}{P}\right) + e^{r(1-S/p)} - 1 = 0$$

$$Se^{r(1-S/P)}\left(-\frac{r}{P}\right) + e^{r(1-S/P)} = 1$$

$$f'(S) = 1$$

(b) $f(S) = Se^{r(1-S/P)}$

$$f'(S) = \left(-\frac{r}{P}\right)Se^{r(1-S/P)} + e^{r(1-S/P)}$$

Set $f'(S_0) = 1$

$$e^{r(1-S_0/P)}\left[\frac{-rS_0}{P} + 1\right] = 1$$

$$e^{r(1-S_0/P)} = \frac{1}{\frac{-rS_0}{P} + 1}$$

Using $H(S)$ from part (a), we get

$$H(S_0) = S_0 e^{r(1-S_0/P)} - S_0$$
$$= S_0(e^{r(1-S_0/P)} - 1)$$
$$= S_0\left(\frac{1}{1 - \frac{rS_0}{P}} - 1\right).$$

39. $r = 0.4$, $P = 500$

$$f(S) = Se^{r(1-S/P)}$$

$$f'(S) = -\frac{0.4}{500}Se^{0.4(1-S/500)} + e^{0.4(1-S/500)}$$

$$f'(S_0) = -0.0008S_0 e^{0.4(1-S_0/500)} + e^{0.4(1-S_0/500)}$$

Graph

$$Y_1 = -0.0008x_0 e^{0.4(1-x/500)} + e^{0.4(1-x/500)}$$

and

$$Y_2 = 1$$

on the same screen. A suitable choice for the viewing window is $[0, 300]$ by $[0.5, 1.5]$ with Xscl $= 50$, Yscl $= 0.5$. By zooming or using the "intersect" option, we find that the graphs intersect when $x \approx 237.10$.

The maximum sustainable harvest is 237.10.

41. Let $x =$ distance from P to A.

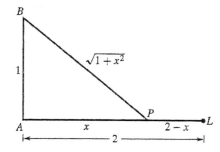

Energy used over land: 1 unit per mile
Energy used over water: $\frac{10}{9}$ units per mile
Distance over land: $(2 - x)$ mi
Distance over water: $\sqrt{1 + x^2}$ mi
Find the location of P to minimize energy used.

$$E(x) = 1(2 - x) + \frac{10}{9}\sqrt{1 + x^2}, \text{ where } 0 \leq x \leq 2.$$

$$E'(x) = -1 + \frac{10}{9}\left(\frac{1}{2}\right)(1 + x^2)^{-1/2}(2x)$$

If $E'(x) = 0$,

$$\frac{10}{9}x(1 + x^2)^{-1/2} = 1$$

$$\frac{10x}{9(1 + x^2)^{1/2}} = 1$$

$$\frac{10}{9}x = (1 + x^2)^{1/2}$$

$$\frac{100}{81}x^2 = 1 + x^2$$

$$\frac{19}{81}x^2 = 1$$

$$x^2 = \frac{81}{19}$$

$$x = \frac{9}{\sqrt{19}}$$

$$= \frac{9\sqrt{19}}{19}$$

$$\approx 2.06.$$

This value cannot give the absolute maximum since the total distance from A to L is just 2 miles. Test the endpoints of the domain.

x	$E(x)$
0	$3\frac{1}{9} \approx 3.1111$
2	2.4845

Point P must be at Point L.

43. (a) Solve the given equation for effective power for T, time.

$$\frac{kE}{T} = aSv^3 + I$$

$$\frac{kE}{aSv^3 + I} = T$$

Since distance is velocity, v, times time, T, we have

$$D(v) = v\frac{kE}{aSv^3 + I}$$

$$= \frac{kEv}{aSv^3 + I}.$$

(b) $D'(v) = \dfrac{(aSv^3 + I)kE - kEv(3aSv^2)}{(aSv^3 + I)^2}$

$$= \frac{kE(aSv^3 + I - 3aSv^3)}{(aSv^3 + I)^2}$$

$$= \frac{kE(I - 2aSv^3)}{(aSv^3 + I)^2}$$

Find the critical numbers by solving $D'(v) = 0$ for v.

$$I - 2aSv^3 = 0$$
$$2aSv^3 = I$$
$$v^3 = \frac{I}{2aS}$$
$$v = \left(\frac{I}{2aS}\right)^{1/3}$$

45. Let $8 - x =$ the distance the hunter will travel on the river.

Then $\sqrt{9 + x^2} =$ the distance he will travel on land.

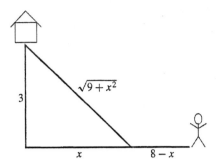

The rate on the river is 5 mph, the rate on land is 2 mph. Using $t = \frac{d}{r}$,

$\frac{8-x}{5} =$ the time on the river,

$\frac{\sqrt{9+x^2}}{2} =$ the time on land.

The total time is

$$T(x) = \frac{8 - x}{5} + \frac{\sqrt{9 + x^2}}{2}$$

$$= \frac{8}{5} - \frac{1}{5}x + \frac{1}{2}(9 + x^2)^{1/2}.$$

$$T' = -\frac{1}{5} + \frac{1}{4} \cdot 2x(9 + x^2)^{-1/2}$$

$$-\frac{1}{5} + \frac{x}{2(9 + x^2)^{1/2}} = 0$$

$$\frac{1}{5} = \frac{x}{2(9 + x^2)^{1/2}}$$

$$2(9 + x^2)^{1/2} = 5x$$

$$4(9 + x^2) = 25x^2$$

$$36 + 4x^2 = 25x^2$$

$$36 = 21x^2$$

$$\frac{6}{\sqrt{21}} = x$$

$$\frac{6\sqrt{21}}{21} = \frac{2\sqrt{21}}{7} = x$$

x	$T(x)$
0	3.1
$\frac{2\sqrt{21}}{7}$	2.98
8	4.27

Since the minimum time is 2.98 hr, the hunter should travel $8 - \frac{2\sqrt{21}}{7} = \frac{56 - 2\sqrt{21}}{7}$ or about 6.7 miles along the river.

14.3 Further Business Applications: Economic Lot Size; Economic Order Quantity; Elasticity of Demand

1. When $q < \sqrt{\frac{2fM}{k}}$, $T'(q) < -\frac{k}{2} + \frac{k}{2} = 0$; and when $q > \sqrt{\frac{2fM}{k}}$, $T'(q) > -\frac{k}{2} + \frac{k}{2} = 0$. Since the function $T(q)$ is decreasing before $q = \sqrt{\frac{2fM}{k}}$ and increasing after $q = \sqrt{\frac{2fM}{k}}$, there must be a relative minimum at $q = \sqrt{\frac{2fM}{k}}$. By the critical point theorem, there is an absolute minimum there.

3. The economic order quantity formula assumes that M, the total units needed per year, is known. Thus, c is the correct answer.

5. Use equation (3) with $k = 9, M = 13,950$, and $f = 31$.

$$q = \sqrt{\frac{2fM}{k}}$$

$$= \sqrt{\frac{2(31)(13,950)}{9}}$$

$$= \sqrt{96,100} = 310$$

310 cases should be made in each batch to minimize production costs.

7. From Exercise 5, $M = 13,950$ and $q = 310$. The number of batches per year is

$$\frac{M}{q} = \frac{13,950}{310} = 45$$

45 cases should be made in each batch to minimize production costs.

9. Here $k = 1, M = 900$, and $f = 5$. We have

$$q = \sqrt{\frac{2fM}{k}}$$

$$= \sqrt{\frac{2(5)(900)}{1}}$$

$$= \sqrt{9000} \approx 94.9$$

$T(94) \approx 94.872$ and $T(95) \approx 94.868$, so ordering 95 bottles per order minimizes the annual costs.

11. Use $q = \sqrt{\frac{fM}{k}}$ from Exercise 10 with $k = 6$, $M = 5000$, and $f = 1000$.

$$q = \sqrt{\frac{fM}{k}} = \sqrt{\frac{(1000)(5000)}{6}} \approx 912.9$$

with $T(q) = \frac{fM}{q} + kq$ (assume $g = 0$ since the subsequent cost per book is so low that it can be ignored), $T(912) \approx 10{,}954.456$ and $T(913) \approx 10{,}954.451$. So, 913 books should be printed in each print run.

13. Use $q = \sqrt{\frac{2fM}{k_1 + 2k_2}}$ from Exercise 12 with $k_1 = 1$, $k_2 = 2$, $M = 30{,}000$, and $f = 750$. Also, note that $g = 8$.

$$q = \sqrt{\frac{2fM}{k_1 + 2k_2}}$$

$$= \sqrt{\frac{2(750)(30{,}000)}{1 + 2(2)}}$$

$$= \sqrt{9{,}000{,}000} = 3000$$

The number of production runs each year to minimize her total costs is

$$\frac{M}{q} = \frac{30{,}000}{3000} = 10.$$

15. $q = 50 - \dfrac{p}{4}$

(a) $\dfrac{dq}{dp} = -\dfrac{1}{4}$

$$E = -\frac{p}{q} \cdot \frac{dq}{dp}$$

$$= -\frac{p}{50 - \frac{p}{4}}\left(-\frac{1}{4}\right)$$

$$= -\frac{p}{\frac{200-p}{4}}\left(-\frac{1}{4}\right)$$

$$= \frac{p}{200 - p}$$

(b) $R = pq$

$$\frac{dR}{dp} = q(1 - E)$$

When R is maximum, $q(1 - E) = 0$.
Since $q = 0$ means no revenue, set $1 - E = 0$.

$$E = 1$$

From (a),

$$\frac{p}{200 - p} = 1$$

$$p = 200 - p$$

$$p = 100.$$

$$q = 50 - \frac{p}{4}$$

$$= 50 - \frac{100}{4}$$

$$= 25$$

Total revenue is maximized if $q = 25$.

17. (a) $q = 37{,}500 - 5p^2$

$$\frac{dq}{dp} = -10p$$

$$E = \frac{-p}{q} \cdot \frac{dq}{dp}$$

$$= \frac{-p}{37{,}500 - 5p^2}(-10p)$$

$$= \frac{10p^2}{37{,}500 - 5p^2}$$

$$= \frac{2p^2}{7500 - p^2}$$

(b) $R = pq$

$$\frac{dR}{dp} = q(1 - E)$$

When R is maximum, $q(1 - E) = 0$. Since $q = 0$ means no revenue, set $1 - E = 0$.

$$E = 1$$

From (a),

$$\frac{2p^2}{7500 - p^2} = 1$$

$$2p^2 = 7500 - p^2$$

$$3p^2 = 7500$$

$$p^2 = 2500$$

$$p = \pm 50.$$

Since p must be positive, $p = 50$.

$$q = 37{,}500 - 5p^2$$

$$= 37{,}500 - 5(50)^2$$

$$= 37{,}500 - 5(2500)$$

$$= 37{,}500 - 12{,}500$$

$$= 25{,}000.$$

19. $p = 400e^{-0.2q}$

In order to find the derivative $\frac{dq}{dp}$, we first need to solve for q in the equation $p = 400e^{-0.2q}$.

(a) $\qquad \dfrac{p}{400} = e^{-0.2q}$

$$\ln\left(\frac{p}{400}\right) = \ln\left(e^{-0.2q}\right) = -0.2q$$

$$q = \frac{\ln\frac{p}{400}}{-0.2} = -5\ln\left(\frac{p}{400}\right)$$

Now

$$\frac{dq}{dp} = -5\frac{1}{\frac{p}{400}} \cdot \frac{1}{400} = \frac{-5}{p}, \text{ and}$$

$$E = -\frac{p}{q} \cdot \frac{dq}{dp} = -\frac{p}{q} \cdot \frac{-5}{p} = \frac{5}{q}.$$

(b) $\quad R = pq$

$$\frac{dR}{dp} = q(1 - E)$$

When R is maximum, $q(1 - E) = 0$. Since $q = 0$ means no revenue, set $1 - E = 0$.

$$E = 1$$

From part (a),

$$\frac{5}{q} = 1$$

$$5 = q$$

21. $\quad q = 400 - 0.2p^2$

$$\frac{dq}{dp} = 0 - 0.4p$$

$$E = -\frac{p}{q} \cdot \frac{dq}{dp}$$

$$E = -\frac{p}{400 - 0.2p^2}(-0.4p)$$

$$= \frac{0.4p^2}{400 - 0.2p^2}$$

(a) If $p = \$20$,

$$E = \frac{(0.4)(20)^2}{400 - 0.2(20)^2}$$

$$= 0.5.$$

Since $E < 1$, demand is inelastic. This indicates that total revenue increases as price increases.

(b) If $p = \$40$,

$$E = \frac{(0.4)(40)^2}{400 - 0.2(40)^2}$$

$$= 8.$$

Since $E > 1$, demand is elastic. This indicates that total revenue decreases as price increases.

23. (a) $q = 55.2 - 0.022p$

$$\frac{dq}{dp} = -0.022$$

$$E = -\frac{p}{q} \cdot \frac{dq}{dp}$$

$$= \frac{-p}{55.2 - 0.022p} \cdot (-0.022)$$

$$= \frac{0.022p}{55.2 - 0.022p}$$

When $p = \$166.10$,

$$E = \frac{3.6542}{55.2 - 3.6542}$$

$$\approx 0.071.$$

(b) Since $E < 1$, the demand for airfare is inelastic at this price.

(c) $\quad R = pq$

$$\frac{dR}{dp} = q(1 - E)$$

When R is a maximum, $q(1 - E) = 0$.
Since $q = 0$ means no revenue, set $1 - E = 0$.

$$E = 1$$

From (a),

$$\frac{0.022p}{55.2 - 0.022p} = 1$$

$$0.022p = 55.2 - 0.022p$$

$$0.044p = 55.2$$

$$p \approx 1255$$

Total revenue is maximized if $p \approx \$1255$.

25. $q = m - np$ for $0 \le p \le \dfrac{m}{n}$

$$\frac{dq}{dp} = -n$$

$$E = -\frac{p}{q} \cdot \frac{dq}{dp}$$

$$E = -\frac{p}{m - np}(-n)$$

$$E = \frac{pn}{m - np} = 1$$

$$pn = m - np$$

$$2np = m$$

$$p = \frac{m}{2n}$$

Thus, $E = 1$ when $p = \frac{m}{2n}$, or at the midpoint of the demand curve on the interval $0 \le p \le \frac{m}{n}$.

27. (a) $q = Cp^{-k}$

$$\frac{dq}{dp} = -Ckp^{-k-1}$$

$$E = \frac{-p}{q} \cdot \frac{dq}{dp}$$

$$= \frac{-p}{Cp^{-k}}(-Ckp^{-k-1})$$

$$= \frac{kp^{-k}}{p^{-k}} = k$$

29. The demand function $q(p)$ is positive and increasing, so $\frac{dq}{dp}$ is positive. Since p_0 and q_0 are also positive, the elasticity $E = -\frac{p_0}{q_0} \cdot \frac{dq}{dp}$ is negative.

14.4   Implicit Differentiation

1. $6x^2 + 5y^2 = 36$

$$\frac{d}{dx}(6x^2 + 5y^2) = \frac{d}{dx}(36)$$

$$\frac{d}{dx}(6x^2) + \frac{d}{dx}(5y^2) = \frac{d}{dx}(36)$$

$$12x + 5 \cdot 2y\frac{dy}{dx} = 0$$

$$10y\frac{dy}{dx} = -12x$$

$$\frac{dy}{dx} = -\frac{6x}{5y}$$

3. $8x^2 - 10xy + 3y^2 = 26$

$$\frac{d}{dx}(8x^2 - 10xy + 3y^2) = \frac{d}{dx}(26)$$

$$16x - \frac{d}{dx}(10xy) + \frac{d}{dx}(3y^2) = 0$$

$$16x - 10x\frac{dy}{dx} - y\frac{d}{dx}(10x) + 6y\frac{dy}{dx} = 0$$

$$16x - 10x\frac{dy}{dx} - 10y + 6y\frac{dy}{dx} = 0$$

$$(-10x + 6y)\frac{dy}{dx} = -16x + 10y$$

$$\frac{dy}{dx} = \frac{-16x + 10y}{-10x + 6y}$$

$$\frac{dy}{dx} = \frac{8x - 5y}{5x - 3y}$$

5. $5x^3 = 3y^2 + 4y$

$$\frac{d}{dx}(5x^3) = \frac{d}{dx}(3y^2 + 4y)$$

$$15x^2 = \frac{d}{dx}(3y^2) + \frac{d}{dx}(4y)$$

$$15x^2 = 6y\frac{dy}{dx} + 4\frac{dy}{dx}$$

$$\frac{15x^2}{6y + 4} = \frac{dy}{dx}$$

7. $3x^2 = \frac{2 - y}{2 + y}$

$$\frac{d}{dx}(3x^2) = \frac{d}{dx}\left(\frac{2 - y}{2 + y}\right)$$

$$6x = \frac{(2 + y)\frac{d}{dx}(2 - y) - (2 - y)\frac{d}{dx}(2 + y)}{(2 + y)^2}$$

$$6x = \frac{(2 + y)\left(-\frac{dy}{dx}\right) - (2 - y)\frac{dy}{dx}}{(2 + y)^2}$$

$$6x = \frac{-4\frac{dy}{dx}}{(2 + y)^2}$$

$$6x(2 + y)^2 = -4\frac{dy}{dx}$$

$$-\frac{3x(2 + y)^2}{2} = \frac{dy}{dx}$$

9. $2\sqrt{x} + 4\sqrt{y} = 5y$

$$\frac{d}{dx}(2x^{1/2} + 4y^{1/2}) = \frac{d}{dx}(5y)$$

$$x^{-1/2} + 2y^{-1/2}\frac{dy}{dx} = 5\frac{dy}{dx}$$

$$(2y^{-1/2} - 5)\frac{dy}{dx} = -x^{-1/2}$$

$$\frac{dy}{dx} = \frac{x^{-1/2}}{5 - 2y^{-1/2}}\left(\frac{x^{1/2}y^{1/2}}{x^{1/2}y^{1/2}}\right)$$

$$= \frac{y^{1/2}}{x^{1/2}(5y^{1/2} - 2)}$$

$$= \frac{\sqrt{y}}{\sqrt{x}(5\sqrt{y} - 2)}$$

11. $x^4y^3 + 4x^{3/2} = 6y^{3/2} + 5$

$$\frac{d}{dx}\left(x^4y^3 + 4x^{3/2}\right) = \frac{d}{dx}\left(6y^{3/2} + 5\right)$$

$$\frac{d}{dx}\left(x^4y^3\right) + \frac{d}{dx}\left(4x^{3/2}\right) = \frac{d}{dx}\left(6y^{3/2}\right) + \frac{d}{dx}(5)$$

$$4x^3y^3 + x^4\cdot 3y^2\frac{dy}{dx} + 6x^{1/2} = 9y^{1/2}\frac{dy}{dx} + 0$$

$$4x^3y^3 + 6x^{1/2} = 9y^{1/2}\frac{dy}{dx} - 3x^4y^2\frac{dy}{dx}$$

$$4x^3y^3 + 6x^{1/2} = \left(9y^{1/2} - 3x^4y^2\right)\frac{dy}{dx}$$

$$\frac{4x^3y^3 + 6x^{1/2}}{9y^{1/2} - 3x^4y^2} = \frac{dy}{dx}$$

13. $e^{x^2y} = 5x + 4y + 2$

$$\frac{d}{dx}\left(e^{x^2y}\right) = \frac{d}{dx}(5x + 4y + 2)$$

$$e^{x^2y}\frac{d}{dx}\left(x^2y\right) = \frac{d}{dx}(5x) + \frac{d}{dx}(4y) + \frac{d}{dx}(2)$$

$$e^{x^2y}\left(2xy + x^2\frac{dy}{dx}\right) = 5 + 4\frac{dy}{dx} + 0$$

$$2xye^{x^2y} + x^2e^{x^2y}\frac{dy}{dx} = 5 + 4\frac{dy}{dx}$$

$$x^2e^{x^2y}\frac{dy}{dx} - 4\frac{dy}{dx} = 5 - 2xye^{x^2y}$$

$$\left(x^2e^{x^2y} - 4\right)\frac{dy}{dx} = 5 - 2xye^{x^2y}$$

$$\frac{dy}{dx} = \frac{5 - 2xye^{x^2y}}{x^2e^{x^2y} - 4}$$

15. $x + \ln y = x^2y^3$

$$\frac{d}{dx}(x + \ln y) = \frac{d}{dx}(x^2y^3)$$

$$1 + \frac{1}{y}\frac{dy}{dx} = 2xy^3 + 3x^2y^2\frac{dy}{dx}$$

$$\frac{1}{y}\frac{dy}{dx} - 3x^2y^2\frac{dy}{dx} = 2xy^3 - 1$$

$$\left(\frac{1}{y} - 3x^2y^2\right)\frac{dy}{dx} = 2xy^3 - 1$$

$$\frac{dy}{dx} = \frac{2xy^3 - 1}{\frac{1}{y} - 3x^2y^2}$$

$$= \frac{y(2xy^3 - 1)}{1 - 3x^2y^3}$$

17. $x^2 + y^2 = 25$; tangent at $(-3, 4)$

$$\frac{d}{dx}\left(x^2 + y^2\right) = \frac{d}{dx}(25)$$

$$2x + 2y\frac{dy}{dx} = 0$$

$$2y\frac{dy}{dx} = -2x$$

$$\frac{dy}{dx} = -\frac{x}{y}$$

$$m = -\frac{x}{y} = -\frac{-3}{4} = \frac{3}{4}$$

$$y - y_1 = m(x - x_1)$$

$$y - 4 = \frac{3}{4}[x - (-3)]$$

$$4y - 16 = 3x + 9$$

$$4y = 3x + 25$$

$$y = \frac{3}{4}x + \frac{25}{4}$$

19. $x^2y^2 = 1$; tangent at $(-1, 1)$

$$\frac{d}{dx}\left(x^2y^2\right) = \frac{d}{dx}(1)$$

$$x^2\frac{d}{dx}\left(y^2\right) + y^2\frac{d}{dx}\left(x^2\right) = 0$$

$$x^2(2y)\frac{dy}{dx} + y^2(2x) = 0$$

$$2x^2y\frac{dy}{dx} = -2xy^2$$

$$\frac{dy}{dx} = \frac{-2xy^2}{2x^2y} = -\frac{y}{x}$$

$$m = -\frac{y}{x} = -\frac{1}{-1} = 1$$

$$y - 1 = 1[x - (-1)]$$

$$y = x + 1 + 1$$

$$y = x + 2$$

21. $2y^2 - \sqrt{x} = 4$; tangent at $(16, 2)$

$$\frac{d}{dx}\left(2y^2 - \sqrt{x}\right) = \frac{d}{dx}(4)$$

$$4y\frac{dy}{dx} - \frac{1}{2}x^{-1/2} = 0$$

$$4y\frac{dy}{dx} = \frac{1}{2x^{1/2}}$$

$$\frac{dy}{dx} = \frac{1}{8yx^{1/2}}$$

$$m = \frac{1}{8yx^{1/2}} = \frac{1}{8(2)(16)^{1/2}} = \frac{1}{8(2)(4)} = \frac{1}{64}$$

$$y - 2 = \frac{1}{64}(x - 16)$$

$$64y - 128 = x - 16$$

$$64y = x + 112$$

$$y = \frac{x}{64} + \frac{7}{4}$$

23. $e^{x^2+y^2} = xe^{5y} - y^2e^{5x/2}$; tangent at $(2, 1)$

$$\frac{d}{dx}(e^{x^2+y^2}) = \frac{d}{dx}(xe^{5y} - y^2e^{5x/2})$$

$$e^{x^2+y^2} \cdot \frac{d}{dx}(x^2 + y^2) = e^{5y} + x\frac{d}{dx}(e^{5y}) - \left[2y\frac{dy}{dx}e^{5x/2} + y^2e^{5x/2}\frac{d}{dx}\left(\frac{5x}{2}\right)\right]$$

$$e^{x^2+y^2}\left(2x + 2y\frac{dy}{dx}\right) = e^{5y} + x \cdot 5e^{5y}\frac{dy}{dx} - 2ye^{5x/2}\frac{dy}{dx} - \frac{5}{2}y^2e^{5x/2}$$

$$(2ye^{x^2+y^2} - 5xe^{5y} + 2ye^{5x/2})\frac{dy}{dx} = -2xe^{x^2+y^2} + e^{5y} - \frac{5}{2}y^2e^{5x/2}$$

$$\frac{dy}{dx} = \frac{-2xe^{x^2+y^2} + e^{5y} - \frac{5}{2}y^2e^{5x/2}}{2ye^{x^2+y^2} - 5xe^{5y} + 2ye^{5x/2}}$$

$$m = \frac{-4e^5 + e^5 - \frac{5}{2}e^5}{2e^5 - 10e^5 + 2e^5} = \frac{-\frac{11}{2}e^5}{-6e^5} = \frac{11}{12}$$

$$y - 1 = \frac{11}{12}(x - 2)$$

$$y = \frac{11}{12}x - \frac{5}{6}$$

25. $\ln(x + y) = x^3y^2 + \ln(x^2 + 2) - 4$; tangent at $(1, 2)$

$$\frac{d}{dx}[\ln(x + y)] = \frac{d}{dx}[x^3y^2 + \ln(x^2 + 2) - 4]$$

$$\frac{1}{x + y} \cdot \frac{d}{dx}(x + y) = 3x^2y^2 + x^3 \cdot 2y\frac{dy}{dx} + \frac{1}{x^2 + 2} \cdot \frac{d}{dx}(x^2 + 2) - \frac{d}{dx}(4)$$

$$\left(\frac{1}{x + y} - 2x^3y\right)\frac{dy}{dx} = 3x^2y^2 + \frac{2x}{x^2 + 2} - \frac{1}{x + y}$$

$$\frac{dy}{dx} = \frac{3x^2y^2 + \frac{2x}{x^2+2} - \frac{1}{x+y}}{\frac{1}{x+y} - 2x^3y}$$

$$m = \frac{3 \cdot 1 \cdot 4 + \frac{2 \cdot 1}{3} - \frac{1}{3}}{\frac{1}{3} - 2 \cdot 1 \cdot 2} = \frac{\frac{37}{3}}{\frac{-11}{3}} = -\frac{37}{11}$$

$$y - 2 = -\frac{37}{11}(x - 1)$$

$$y = -\frac{37}{11}x + \frac{59}{11}$$

27. $y^3 + xy - y = 8x^4$; $x = 1$

First, find the y-value of the point.

$$y^3 + (1)y - y = 8(1)^4$$
$$y^3 = 8$$
$$y = 2$$

The point is $(1, 2)$.

Find $\frac{dy}{dx}$.

$$3y^2 \frac{dy}{dx} + x\frac{dy}{dx} + y - \frac{dy}{dx} = 32x^3$$
$$(3y^2 + x - 1)\frac{dy}{dx} = 32x^3 - y$$
$$\frac{dy}{dx} = \frac{32x^3 - y}{3y^2 + x - 1}$$

At $(1, 2)$,

$$\frac{dy}{dx} = \frac{32(1)^3 - 2}{3(2)^2 + 1 - 1} = \frac{30}{12} = \frac{5}{2}.$$

$$y - 2 = \frac{5}{2}(x - 1)$$
$$y - 2 = \frac{5}{2}x - \frac{5}{2}$$
$$y = \frac{5}{2}x - \frac{1}{2}$$

29. $y^3 + xy^2 + 1 = x + 2y^2$; $x = 2$

Find the y-value of the point.

$$y^3 + 2y^2 + 1 = 2 + 2y^2$$
$$y^3 + 1 = 2$$
$$y^3 = 1$$
$$y = 1$$

The point is $(2, 1)$.

Find $\frac{dy}{dx}$.

$$3y^2\frac{dy}{dx} + x\,2y\frac{dy}{dx} + y^2 = 1 + 4y\frac{dy}{dx}$$
$$3y^2\frac{dy}{dx} + 2xy\frac{dy}{dx} - 4y\frac{dy}{dx} = 1 - y^2$$
$$(3y^2 + 2xy - 4y)\frac{dy}{dx} = 1 - y^2$$
$$\frac{dy}{dx} = \frac{1 - y^2}{3y^2 + 2xy - 4y}$$

At $(2, 1)$,

$$\frac{dy}{dx} = \frac{1 - 1^2}{3(1)^2 + 2(2)(1) - 4(1)} = 0.$$

$$y - 0 = 0(x - 2)$$
$$y = 1$$

31. $2y^3(x - 3) + x\sqrt{y} = 3$; $x = 3$

Find the y-value of the point.

$$2y^3(3 - 3) + 3\sqrt{y} = 3$$
$$3\sqrt{y} = 3$$
$$\sqrt{y} = 1$$
$$y = 1$$

The point is $(3, 1)$

Find $\frac{dy}{dx}$.

$$2y^3(1) + 6y^2(x - 3)\frac{dy}{dx}$$
$$+ x\left(\frac{1}{2}\right)y^{-1/2}\frac{dy}{dx} + \sqrt{y} = 0$$
$$6y^2(x - 3)\frac{dy}{dx} + \frac{x}{2\sqrt{y}}\frac{dy}{dx} = -2y^3 - \sqrt{y}$$
$$\left[6y^2(x - 3) + \frac{x}{2\sqrt{y}}\right]\frac{dy}{dx} = -2y^3 - \sqrt{y}$$
$$\frac{dy}{dx} = \frac{-2y^3 - \sqrt{y}}{6y^2(x - 3) + \frac{x}{2\sqrt{y}}}$$
$$= \frac{-4y^{7/2} - 2y}{12y^{5/2}(x - 3) + x}$$

At $(3, 1)$,

$$\frac{dy}{dx} = \frac{-4(1) - 2}{12(1)(3 - 3) + 3} = \frac{-6}{3} = -2.$$

$$y - 1 = -2(x - 3)$$
$$y - 1 = -2x + 6$$
$$y = -2x + 7$$

33. $x^2 + y^2 = 100$

(a) Lines are tangent at points where $x = 6$. By substituting $x = 6$ in the equation, we find that the points are $(6, 8)$ and $(6, -8)$.

$$\frac{d}{dx}(x^2 + y^2) = \frac{d}{dx}(100)$$
$$2x + 2y\frac{dy}{dx} = 0$$
$$2y\frac{dy}{dx} = -2x$$
$$dy = -\frac{x}{y}$$
$$m_1 = -\frac{x}{y} = -\frac{6}{8} = -\frac{3}{4}$$
$$m_2 = -\frac{x}{y} = -\frac{6}{-8} = \frac{3}{4}$$

First tangent:

$$y - 8 = -\frac{3}{4}(x - 6)$$

$$y = -\frac{3}{4}x + \frac{25}{2}$$

Second tangent:

$$y - (-8) = \frac{3}{4}(x - 6)$$

$$y + 8 = \frac{3}{4}x - \frac{18}{4}$$

$$y = \frac{3}{4}x - \frac{25}{2}$$

(b)

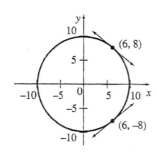

35. $3(x^2 + y^2)^2 = 25(x^2 - y^2);\ (2, 1)$

Find $\dfrac{dy}{dx}$.

$$6(x^2 + y^2)\frac{d}{dx}(x^2 + y^2) = 25\frac{d}{dx}(x^2 - y^2)$$

$$6(x^2 + y^2)\left(2x + 2y\frac{dy}{dx}\right) = 25\left(2x - 2y\frac{dy}{dx}\right)$$

$$12x^3 + 12x^2y\frac{dy}{dx} + 12xy^2 + 12y^3\frac{dy}{dx} = 50x - 50y\frac{dy}{dx}$$

$$12x^2y\frac{dy}{dx} + 12y^3\frac{dy}{dx} + 50y\frac{dy}{dx} = -12x^3 - 12xy^2 + 50x$$

$$(12x^2y + 12y^3 + 50y)\frac{dy}{dx} = -12x^3 - 12xy^2 + 50x$$

$$\frac{dy}{dx} = \frac{-12x^3 - 12xy^2 + 50x}{12x^2y + 12y^3 + 50y}$$

At $(2, 1)$,

$$\frac{dy}{dx} = \frac{-12(2)^3 - 12(2)(1)^2 + 50(2)}{12(2)^2 + 12(1)^3 + 50(1)}$$

$$= \frac{-20}{110}$$

$$= -\frac{2}{11}$$

$$y - 1 = -\frac{2}{11}(x - 2)$$

$$y - 1 = -\frac{2}{11}x + \frac{4}{11}$$

$$y = -\frac{2}{11}x + \frac{15}{11}$$

37. $2(x^2 + y^2)^2 = 25xy^2;\ (2, 1)$

Find $\dfrac{dy}{dx}$.

$$4(x^2 + y^2)\frac{d}{dx}(x^2 + y^2) = 25\frac{d}{dx}(xy^2)$$

$$4(x^2 + y^2)\left(2x + 2y\frac{dy}{dx}\right) = 25\left(y^2 + 2xy\frac{dy}{dx}\right)$$

$$8x^3 + 8x^2y\frac{dy}{dx} + 8xy^2 + 8y^3\frac{dy}{dx} = 25y^2 + 50xy\frac{dy}{dx}$$

$$8x^2y\frac{dy}{dx} + 8y^3\frac{dy}{dx} - 50xy\frac{dy}{dx} = -8x^3 - 8xy^2 + 25y^2$$

$$(8x^2y + 8y^3 - 50xy)\left(\frac{dy}{dx}\right) = -8x^3 - 8xy^2 + 25y^2$$

$$\frac{dy}{dx} = \frac{-8x^3 - 8xy^2 + 25y^2}{8x^2y + 8y^3 - 50xy}$$

At $(2, 1)$,

$$\frac{dy}{dx} = \frac{-8(2)^3 - 8(2)(1)^2 + 25(1)^2}{8(2)^2(1) + 8(1)^3 - 50(2)(1)}$$

$$= \frac{-55}{-60} = \frac{11}{12}$$

$$y - 1 = \frac{11}{12}(x - 2)$$

$$y - 1 = \frac{11}{12}x - \frac{11}{6}$$

$$y = \frac{11}{12}x - \frac{5}{6}$$

39. $y^2 = x^3 + ax + b$

$$\frac{d}{dx}(y^2) = \frac{d}{dx}(x^3 + ax + b)$$

$$2y\frac{dy}{dx} = 3x^2 + a$$

$$\frac{dy}{dx} = \frac{3x^2 + a}{2y}$$

41. $\sqrt{u} + \sqrt{2v+1} = 5$

$$\frac{dv}{du}\left(\sqrt{u} + \sqrt{2v+1}\right) = \frac{dv}{du}(5)$$

$$\frac{1}{2}u^{-1/2} + \frac{1}{2}(2v+1)^{-1/2}(2)\frac{dv}{du} = 0$$

$$(2v+1)^{-1/2}\frac{dv}{du} = -\frac{1}{2}u^{-1/2}$$

$$\frac{dv}{du} = -\frac{(2v+1)^{1/2}}{2u^{1/2}}$$

43. $C^2 = x^2 + 100\sqrt{x} + 50$

(a) $2C\dfrac{dC}{dx} = 2x + \dfrac{1}{2}(100)x^{-1/2}$

$$\frac{dC}{dx} = \frac{2x + 50x^{-1/2}}{2C}$$

$$\frac{dC}{dx} = \frac{x + 25x^{-1/2}}{C} \cdot \frac{x^{1/2}}{x^{1/2}}$$

$$\frac{dC}{dx} = \frac{x^{3/2} + 25}{Cx^{1/2}}$$

When $x = 5$, the approximate increase in cost of an additional unit is

$$\frac{(5)^{3/2} + 25}{(5^2 + 100\sqrt{5} + 50)^{1/2}(5)^{1/2}} = \frac{36.18}{(17.28)\sqrt{5}}$$

$$\approx 0.94.$$

(b) $900(x-5)^2 + 25R^2 = 22{,}500$

$$R^2 = 900 - 36(x-5)^2$$

$$2R\frac{dR}{dx} = -72(x-5)$$

$$\frac{dR}{dx} = \frac{-36(x-5)}{R} = \frac{180 - 36x}{R}$$

When $x = 5$, the approximate change in revenue for a unit increase in sales is

$$\frac{180 - 36(5)}{R} = \frac{0}{R} = 0.$$

45. $b - a = (b+a)^3$

$$\frac{d}{db}(b-a) = \frac{d}{db}[(b+a)^3]$$

$$1 - \frac{da}{db} = 3(b+a)^2\frac{d}{db}(b+a)$$

$$1 - \frac{da}{db} = 3(b+a)^2\left(1 + \frac{da}{db}\right)$$

$$1 - \frac{da}{db} = 3(b+a)^2 + 3(b+a)^2\frac{da}{db}$$

$$-\frac{da}{db} - 3(b+a)^2\frac{da}{db} = 3(b+a)^2 - 1$$

$$[-1 - 3(b+a)^2]\frac{da}{db} = 3(b+a)^2 - 1$$

$$\frac{da}{db} = \frac{3(b+a)^2 - 1}{-1 - 3(b+a)^2}$$

$$\frac{da}{db} = 0$$

$$3(b+a)^2 - 1 = 0$$

$$(b+a)^2 = \frac{1}{3}$$

$$b + a = \frac{1}{\sqrt{3}}$$

Since $b - a = (b+a)^3 = \left(\dfrac{1}{\sqrt{3}}\right)^3 = \dfrac{1}{3\sqrt{3}}$.

$$b + a = \frac{1}{\sqrt{3}}$$

$$-(b-a) = -\frac{1}{3\sqrt{3}}$$

$$\overline{}$$

$$2a = \frac{2}{3\sqrt{3}}$$

$$a = \frac{1}{3\sqrt{3}}$$

47. $s^3 - 4st + 2t^3 - 5t = 0$

$$3s^2\frac{ds}{dt} - \left(4t\frac{ds}{dt} + 4s\right) + 6t^2 - 5 = 0$$

$$3s^2\frac{ds}{dt} - 4t\frac{ds}{dt} - 4s + 6t^2 - 5 = 0$$

$$\frac{ds}{dt}(3s^2 - 4t) = 4s - 6t^2 + 5$$

$$\frac{ds}{dt} = \frac{4s - 6t^2 + 5}{3s^2 - 4t}$$

14.5   Related Rates

1. $y^2 - 8x^3 = -55$; $\dfrac{dx}{dt} = -4, x = 2, y = 3$

$$2y\frac{dy}{dt} - 24x^2\frac{dx}{dt} = 0$$

$$y\frac{dy}{dt} = 12x^2\frac{dx}{dt}$$

$$3\frac{dy}{dt} = 48(-4)$$

$$\frac{dy}{dt} = -64$$

3. $2xy - 5x + 3y^3 = -51$; $\dfrac{dx}{dt} = -6, x = 3, y = -2$

$$2x\frac{dy}{dt} + 2y\frac{dx}{dt} - 5\frac{dx}{dt} + 9y^2\frac{dy}{dt} = 0$$

$$(2x + 9y^2)\frac{dy}{dt} + (2y - 5)\frac{dx}{dt} = 0$$

$$(2x + 9y^2)\frac{dy}{dt} = (5 - 2y)\frac{dx}{dt}$$

$$\frac{dy}{dt} = \frac{5 - 2y}{2x + 9y^2} \cdot \frac{dx}{dt}$$

$$= \frac{5 - 2(-2)}{2(3) + 9(-2)^2} \cdot (-6)$$

$$= \frac{9}{42} \cdot (-6) = \frac{-54}{42} = -\frac{9}{7}$$

5. $\dfrac{x^2 + y}{x - y} = 9$; $\dfrac{dx}{dt} = 2, x = 4, y = 2$

$$\frac{(x-y)\left(2x\frac{dx}{dt} + \frac{dy}{dt}\right) - (x^2 + y)\left(\frac{dx}{dt} - \frac{dy}{dt}\right)}{(x - y)^2} = 0$$

$$\frac{2x(x-y)\frac{dx}{dt} + (x-y)\frac{dy}{dt} - (x^2+y)\frac{dx}{dt} + (x^2+y)\frac{dy}{dt}}{(x - y)^2} = 0$$

$$[2x(x-y) - (x^2+y)]\frac{dx}{dt} + [(x-y) + (x^2+y)]\frac{dy}{dt} = 0$$

$$\frac{dy}{dt} = \frac{[(x^2 + y) - 2x(x - y)]\frac{dx}{dt}}{(x - y) + (x^2 + y)}$$

$$\frac{dy}{dt} = \frac{(-x^2 + y + 2xy)\frac{dx}{dt}}{x + x^2}$$

$$= \frac{[-(4)^2 + 2 + 2(4)(2)](2)}{4 + 4^2}$$

$$= \frac{4}{20} = \frac{1}{5}$$

7. $xe^y = 3 + \ln x$; $\dfrac{dx}{dt} = 6, x = 2, y = 0$

$$e^y\frac{dx}{dt} + xe^y\frac{dy}{dt} = 0 + \frac{1}{x}\frac{dx}{dt}$$

$$xe^y\frac{dy}{dt} = \left(\frac{1}{x} - e^y\right)\frac{dx}{dt}$$

$$\frac{dy}{dt} = \frac{\left(\frac{1}{x} - e^y\right)\frac{dx}{dt}}{xe^y}$$

$$= \frac{(1 - xe^y)\frac{dx}{dt}}{x^2 e^y}$$

$$= \frac{[1 - (2)e^0](6)}{2^2 e^0}$$

$$= \frac{-6}{4} = -\frac{3}{2}$$

9. $C = 0.2x^2 + 10{,}000$; $x = 80, \dfrac{dx}{dt} = 12$

$$\frac{dC}{dt} = 0.2(2x)\frac{dx}{dt} = 0.2(160)(12) = 384$$

The cost is changing at a rate of \$384 per month.

11. $R = 50x - 0.4x^2$; $C = 5x + 15$; $x = 40$; $\frac{dx}{dt} = 10$

(a) $\dfrac{dR}{dt} = 50\dfrac{dx}{dt} - 0.8x\dfrac{dx}{dt}$

$$= 50(10) - 0.8(40)(10)$$
$$= 500 - 320$$
$$= 180$$

Revenue is increasing at a rate of \$180 per day.

(b) $\dfrac{dC}{dt} = 5\dfrac{dx}{dt} = 5(10) = 50$

Cost is increasing at a rate of \$50 per day.

(c) Profit = Revenue − Cost
$$P = R - C$$

$$\frac{dP}{dt} = \frac{dR}{dl} - \frac{dC}{dt} = 180 - 50 = 130$$

Profit is increasing at a rate of \$130 per day.

13. $pq = 8000$; $p = 3.50, \frac{dp}{dt} = 0.15$

$$pq = 8000$$

$$p\frac{dq}{dt} + q\frac{dp}{dt} = 0$$

$$\frac{dq}{dt} = \frac{-q\frac{dp}{dt}}{p}$$

$$= \frac{-\left(\frac{8000}{3.50}\right)(0.15)}{3.50}$$

$$\approx -98$$

Demand is decreasing at a rate of approximately 98 units per unit time.

15. $V = k(R^2 - r^2)$; $k = 555.6$, $R = 0.02$ mm, $\frac{dR}{dt} = 0.003$ mm per minute; r is constant.

$$V = k(R^2 - r^2)$$
$$V = 555.6(R^2 - r^2)$$
$$\frac{dV}{dt} = 555.6\left(2R\frac{dR}{dt} - 0\right)$$
$$= 555.6(2)(0.02)(0.003)$$
$$= 0.067 \text{ mm/min}$$

17. $b = 0.22m^{0.87}$

$$\frac{db}{dt} = 0.22(0.87)m^{-0.13}\frac{dm}{dt}$$
$$= 0.1914m^{-0.13}\frac{dm}{dt}$$
$$\frac{dm}{dt} = \frac{m^{0.13}}{0.1914}\frac{db}{dt}$$
$$= \frac{25^{0.13}}{0.1914}(0.25)$$
$$\approx 1.9849$$

The rate of change of the total weight is about 1.9849 g/day.

19. $r = 140.2m^{0.75}$

(a) $\frac{dr}{dt} = 140.2(0.75)m^{-0.25}\frac{dm}{dt}$
$$= 105.15m^{-0.25}\frac{dm}{dt}$$

(b) $\frac{dr}{dt} = 105.15(250)^{-0.25}(2)$
$$\approx 52.89$$

The rate of change of the average daily metabolic rate is about 52.89 kcal/day^2.

21. $C = \frac{1}{10}(T - 60)^2 + 100$

$$\frac{dC}{dt} = \frac{1}{5}(T - 60)\frac{dT}{dt}$$

If $T = 76°$ and $\frac{dT}{dt} = 8$,

$$\frac{dC}{dt} = \frac{1}{5}(76 - 60)(8) = \frac{1}{5}(16)(8)$$
$$= 25.6.$$

The crime rate is rising at the rate of 25.6 crimes/month.

23. Let $x =$ the distance of the base of the ladder from the base of the building;
$y =$ the distance up the side of the building to the top of the ladder.

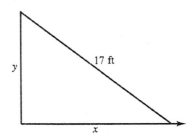

Find $\frac{dy}{dt}$ when $x = 8$ ft and $\frac{dx}{dt} = 9$ ft/min.

Since $y = \sqrt{17^2 - x^2}$, when $x = 8$,

$$y = 15.$$

By the Pythagorean theorem,

$$x^2 + y^2 = 17^2.$$

$$\frac{d}{dt}(x^2 + y^2) = \frac{d}{dt}(17^2)$$
$$2x\frac{dx}{dt} + 2y\frac{dy}{dt} = 0$$
$$2y\frac{dy}{dt} = -2x\frac{dx}{dt}$$
$$\frac{dy}{dt} = \frac{-2x}{2y}\cdot\frac{dx}{dt} = -\frac{x}{y}\cdot\frac{dx}{dt}$$
$$= -\frac{8}{15}(9)$$
$$= -\frac{24}{5}$$

The ladder is sliding down the building at the rate of $\frac{24}{5}$ ft/min.

25. Let $r =$ the radius of the circle formed by the ripple.

Find $\frac{dA}{dt}$ when $r = 4$ ft and $\frac{dr}{dt} = 2$ ft/min.

$$A = \pi r^2$$
$$\frac{dA}{dt} = 2\pi r\frac{dr}{dt}$$
$$= 2\pi(4)(2)$$
$$= 16\pi$$

The area is changing at the rate of 16π ft^2/min.

27. $V = x^3, x = 3$ cm, and $\frac{dV}{dt} = 2$ cm^3/min

$$\frac{dV}{dt} = 3x^2 \frac{dx}{dt}$$

$$\frac{dx}{dt} = \frac{1}{3x^2} \frac{dV}{dt}$$

$$= \frac{1}{3 \cdot 3^2}(2)$$

$$= \frac{2}{27} \text{ cm/min}$$

29. Let $y =$ the length of the man's shadow;
  $x =$ the distance of the man from the lamp post;
  $h =$ the height of the lamp post.

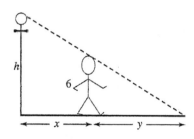

$$\frac{dx}{dt} = 50 \text{ ft/min}$$

Find $\frac{dy}{dt}$ when $x = 25$ ft.

Now $\frac{h}{x+y} = \frac{6}{y}$, by similar triangles.

When $x = 8$, $y = 10$,

$$\frac{h}{18} = \frac{6}{10}$$

$$h = 10.8.$$

$$\frac{10.8}{x+y} = \frac{6}{y},$$

$$10.8y = 6x + 6y$$
$$4.8y = 6x$$
$$y = 1.25x$$

$$\frac{dy}{dt} = 1.25 \frac{dx}{dt}$$

$$= 1.25(50)$$

$$\frac{dy}{dt} = 62.5$$

The length of the shadow is increasing at the rate of 62.5 ft/min.

31. Let $x =$ the distance from the docks
  $s =$ the length of the rope.

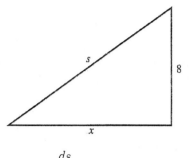

$$\frac{ds}{dt} = 1 \text{ ft/sec}$$

$$s^2 = x^2 + (8)^2$$

$$2s \frac{ds}{dt} = 2x \frac{dx}{dt} + 0$$

$$s \frac{ds}{dt} = x \frac{dx}{dt}$$

If $x = 8$,

$$s = \sqrt{(8)^2 + (8)^2} = \sqrt{128} = 8\sqrt{2}.$$

Then,

$$8\sqrt{2}(1) = 8 \frac{dx}{dt}$$

$$\frac{dx}{dt} = \sqrt{2} \approx 1.41$$

The boat is approaching the deck at $\sqrt{2} \approx 1.41$ ft/sec.

14.6   Differentials: Linear Approximation

1. $y = 2x^3 - 5x$; $x = -2$, $\Delta x = 0.1$
  $dy = (6x^2 - 5)\,dx$
  $\Delta y \approx (6x^2 - 5)\,\Delta x \approx [6(-2)^2 - 5](0.1) \approx 1.9$

3. $y = x^3 - 2x^2 + 3$, $x = 1$, $\Delta x = -0.1$
  $dy = (3x^2 - 4x)\,dx$
  $\quad \approx (3x^2 - 4x)\,\Delta x$
  $\quad = [3(1^2) - 4(1)](-0.1)$
  $\quad = 0.1$

5. $y = \sqrt{3x + 2}$, $x = 4$, $\Delta x = 0.15$

$$dy = 3\left(\frac{1}{2}(3x + 2)^{-1/2}\right) dx$$

$$\Delta y \approx \frac{3}{2\sqrt{3x + 2}} \Delta x \approx \frac{3}{2(3.74)}(0.15) \approx 0.060$$

7. $y = \dfrac{2x - 5}{x + 1}$; $x = 2$, $\Delta x = -0.03$

$$dy = \dfrac{(x+1)(2) - (2x-5)(1)}{(x+1)^2}\,dx$$

$$= \dfrac{7}{(x+1)^2}\,dx$$

$$= \dfrac{7}{(x+1)^2}\,\Delta x$$

$$= \dfrac{7}{(2+1)^2}(-0.03)$$

$$= -0.023$$

9. $\sqrt{145}$

We know $\sqrt{144} = 12$, so $f(x) = \sqrt{x}$, $x = 144$, $dx = 1$.

$$\dfrac{dy}{dx} = \dfrac{1}{2}x^{-1/2}$$

$$dy = \dfrac{1}{2\sqrt{x}}\,dx$$

$$dy = \dfrac{1}{2\sqrt{144}}(1) = \dfrac{1}{24}$$

$$\sqrt{145} \approx f(x) + dy = 12 + \dfrac{1}{24}$$

$$\approx 12.0417$$

By calculator, $\sqrt{145} \approx 12.0416$.
The difference is $|12.0417 - 12.0416| = 0.0001$.

11. $\sqrt{0.99}$

We know $\sqrt{1} = 1$, so $f(x) = \sqrt{x}$, $x = 1$, $dx = -0.01$.

$$\dfrac{dy}{dx} = \dfrac{1}{2}x^{-1/2}$$

$$dy = \dfrac{1}{2\sqrt{x}}\,dx$$

$$dy = \dfrac{1}{2\sqrt{1}}(-0.01) = -0.005$$

$$\sqrt{0.99} \approx f(x) + dy = 1 - 0.005$$

$$= 0.995$$

By calculator, $\sqrt{0.99} \approx 0.9950$.
The difference is $|0.995 - 0.9950| = 0$.

13. $e^{0.01}$

We know $e^0 = 1$, so $f(x) = e^x$, $x = 0$, $dx = 0.01$.

$$\dfrac{dy}{dx} = e^x$$
$$dy = e^x\,dx$$
$$dy = e^0(0.01) = 0.01$$
$$e^{0.01} \approx f(x) + dy = 1 + 0.01 = 1.01$$

By calculator, $e^{0.01} \approx 1.0101$.
The difference is $|1.01 - 1.0101| = 0.0001$.

15. $\ln 1.05$

We know $\ln 1 = 0$, so $f(x) = \ln x$, $x = 1$, $dx = 0.05$.

$$\dfrac{dy}{dx} = \dfrac{1}{x}$$

$$dy = \dfrac{1}{x}\,dx$$

$$dy = \dfrac{1}{1}(0.05) = 0.05$$

$$\ln 1.05 \approx f(x) + dy = 0 + 0.05 = 0.05$$

By calculator, $\ln 1.05 \approx 0.0488$.
The difference is $|0.05 - 0.0488| = 0.0012$.

17. Let $D =$ the demand in thousands of pounds;
$x =$ the price in dollars.
$$D(q) = -3q^3 - 2q^2 + 1500$$

(a) $q = 2$, $\Delta q = 0.10$

$$dD = (-9q^2 - 4q)\,dq$$
$$\Delta D \approx (-9q^2 - 4q)\,\Delta q$$
$$\approx [-9(4) - 4(2)](0.10)$$
$$\approx -4.4 \text{ thousand pounds}$$

(b) $q = 6$, $\Delta q = 0.15$

$$\Delta D \approx [-9(36) - 4(6)](0.15)$$
$$\approx -52.2 \text{ thousand pounds}$$

19. $R(x) = 12{,}000 \ln(0.01x + 1)$
$x = 100$, $\Delta x = 1$

$$dR = \dfrac{12{,}000}{0.01x + 1}(0.01)\,dx$$

$$\Delta R \approx \dfrac{120}{0.01x + 1}\,\Delta x$$

$$\approx \dfrac{120}{0.01(100) + 1}(1)$$

$$\approx \$60$$

21. If a cube is given a coating 0.1 in. thick, each edge increases in length by twice that amount, or 0.2 in. because there is a face at both ends of the edge.

$$V = x^3, \ x = 4, \ \Delta x = 0.2$$

$$dV = 3x^2\,dx$$
$$\Delta V \approx 3x^2\,\Delta x$$
$$= 3(4^2)(0.2)$$
$$= 9.6$$

For 1000 cubes $9.6(1000) = 9600$ in.^{3} of coating should be ordered.

23. **(a)** $A(x) = y = 0.003631x^3 - 0.03746x^2 + 0.1012x + 0.009$

Let $x = 1$, $dx = 0.2$.

$$\frac{dy}{dx} = 0.010893x^2 - 0.07492x + 0.1012$$

$$dy = (0.010893x^2 - 0.07492x + 0.1012)\,dx$$

$$\begin{aligned}
\Delta y &\approx (0.010893x^2 - 0.07492x + 0.1012)\,\Delta x \\
&\approx (0.010893 \cdot 1^2 - 0.07492 \cdot 1 + 0.1012) \cdot 0.2 \\
&\approx 0.007435
\end{aligned}$$

The alcohol concentration increases by about 0.74 percent.

(b) $\Delta y \approx (0.010893 \cdot 3^2 - 0.07492 \cdot 3 + 0.1012) \cdot 0.2 \approx -0.005105$

The alcohol concentration decreases by about 0.51 percent.

25. $P(x) = \dfrac{25x}{8 + x^2}$

$$dP = \frac{(8 + x^2)(25) - 25x(2x)}{(8 + x^2)^2}\,dx = \frac{(8 + x^2)(25) - 25x(2x)}{(8 + x^2)^2}\,\Delta x$$

(a) $x = 2$, $\Delta x = 0.5$

$$dP = \frac{[(8 + 4)(25) - (25)(2)(4)](0.5)}{(8 + 4)^2} = 0.347 \text{ million}$$

(b) $x = 3$, $\Delta x = 0.25$

$$dP = \frac{[(8 + 9)(25) - 25(3)(6)]0.25}{(8 + 9)^2} \approx -0.022 \text{ million}$$

27. r changes from 14 mm to 16 mm, so $\Delta r = 2$.

$$V = \frac{4}{3}\pi r^3$$

$$dV = \frac{4}{3}(3)\pi r^2\,dr$$

$$\Delta V \approx 4\pi r^2\,\Delta r = 4\pi(14)^2(2) = 1568\pi \text{ mm}^3$$

29. r increases from 20 mm to 22 mm, so $\Delta r = 2$.

$$A = \pi r^2$$

$$dA = 2\pi r\,dr$$

$$\Delta A \approx 2\pi r\,\Delta r = 2\pi(20)(2) = 80\pi \text{ mm}^2$$

31. $W(t) = -3.5 + 197.5e^{-e^{-0.01394(t - 108.4)}}$

(a) $dW = 197.5e^{-e^{-0.01394(t-108.4)}}(-1)e^{-0.01394(t-108.4)}(-0.01394)dt = 2.75315e^{-e^{-0.01394(t-108.4)}}e^{-0.01394(t-108.4)}dt$

We are given $t = 80$ and $dt = 90 - 80 = 10$.

$$dW \approx 9.258$$

The pig will gain about 9.3 kg.

(b) The actual weight gain is calculated as

$$W(90) - W(80) \approx 50.736 - 41.202 = 9.534$$

or about 9.5 kg.

33. $r = 3$ cm, $\Delta r = -0.2$ cm

$$V = \frac{4}{3}\pi r^3$$

$$dV = 4\pi r^2 \, dr$$

$$\Delta V \approx 4\pi r^2 \, \Delta r$$

$$= 4\pi(9)(-0.2)$$

$$= -7.2\pi \text{ cm}^3$$

35. $V = \frac{1}{3}\pi r^2 h$; $h = 13, dh = 0.2$

$$V = \frac{1}{3}\pi \left(\frac{h}{15}\right)^2 h$$

$$= \frac{\pi}{775}h^3$$

$$dV = \frac{\pi}{775} \cdot 3h^2 \, dh$$

$$= \frac{\pi}{225}h^2 \, dh$$

$$\Delta V \approx \frac{\pi}{225}h^2 \, \Delta h$$

$$\approx \frac{\pi}{225}(13^2)(0.2)$$

$$\approx 0.472 \text{ cm}^3$$

37. $A = x^2$; $x = 4, dA = 0.01$

$$dA = 2x \, dx$$

$$\Delta A \approx 2x \, \Delta x$$

$$\Delta x \approx \frac{\Delta A}{2x} \approx \frac{0.01}{2(4)} \approx 0.00125 \text{ cm}$$

39. $V = \frac{4}{3}\pi r^3$; $r = 5.81$, $\Delta r = \pm 0.003$

$$dV = \frac{4}{3}\pi(3r^2) \, dr$$

$$\Delta V \approx \frac{4}{3}\pi(3r^2) \, \Delta r$$

$$= 4\pi(5.81)^2(\pm 0.003)$$

$$= \pm 0.405\pi \approx \pm 1.273 \text{ in.}^3$$

41. $h = 7.284$ in., $r = 1.09 \pm 0.007$ in.

$$V = \frac{1}{3}\pi r^2 h$$

$$dV = \frac{2}{3}\pi r h \, dr$$

$$\Delta V \approx \frac{2}{3}\pi r h \, \Delta r$$

$$= \frac{2}{3}\pi(1.09)(7.284)(0.007)$$

$$= \pm 0.116 \text{ in.}^3$$

Chapter 14 Review Exercises

1. $f(x) = -x^3 + 6x^2 + 1$; $[-1, 6]$
$f'(x) = -3x^2 + 12x = 0$ when $x = 0, 4$.

$$f(-1) = 8$$
$$f(0) = 1$$
$$f(4) = 33$$
$$f(6) = 1$$

Absolute maximum of 33 at 4; absolute minimum of 1 at 0 and 6.

3. $f(x) = x^3 + 2x^2 - 15x + 3$; $[-4, 2]$
$f'(x) = 3x^2 + 4x - 15 = 0$ when
$$(3x - 5)(x + 3) = 0$$

$$x = \frac{5}{3} \quad \text{or} \quad x = -3.$$

$$f(-4) = 31$$
$$f(-3) = 39$$
$$f\left(\frac{5}{3}\right) = -\frac{319}{27}$$
$$f(2) = -11$$

Absolute maximum of 39 at -3; absolute minimum of $-\frac{319}{27}$ at $\frac{5}{3}$

7. (a) $f(x) = \frac{2\ln x}{x^2}$; $[1, 4]$

$$f'(x) = \frac{x^2\left(\frac{2}{x}\right) - (2\ln x)(2x)}{x^4}$$

$$= \frac{2x - 4x\ln x}{x^4}$$

$$= \frac{2 - 4\ln x}{x^3}$$

$f'(x) = 0$ when

$$2 - 4\ln x = 0$$
$$2 = 4\ln x$$
$$0.5 = \ln x$$
$$e^{0.5} = x$$
$$x \approx 1.6487.$$

x	$f(x)$
1	0
$e^{0.5}$	0.36788
4	0.17329

Maximum is 0.37; minimum is 0.

(b) $[2, 5]$

Note that the critical number of f is not in the domain, so we only test the endpoints.

x	$f(x)$
2	0.34657
5	0.12876

Maximum is 0.35, minimum is 0.13.

11. $x^2 - 4y^2 = 3x^3y^4$

$$\frac{d}{dx}(x^2 - 4y^2) = \frac{d}{dx}(3x^3y^4)$$

$$2x - 8y\frac{dy}{dx} = 9x^2y^4 + 3x^3 \cdot 4y^3\frac{dy}{dx}$$

$$(-8y - 3x^3 \cdot 4y^3)\frac{dy}{dx} = 9x^2y^4 - 2x$$

$$\frac{dy}{dx} = \frac{2x - 9x^2y^4}{8y + 12x^3y^3}$$

13. $2\sqrt{y - 1} = 9x^{2/3} + y$

$$\frac{d}{dx}[2(y - 1)^{1/2}] = \frac{d}{dx}(9x^{2/3} + y)$$

$$2 \cdot \frac{1}{2} \cdot (y - 1)^{-1/2}\frac{dy}{dx} = 6x^{-1/3} + \frac{dy}{dx}$$

$$[(y - 1)^{-1/2} - 1]\frac{dy}{dx} = 6x^{-1/3}$$

$$\frac{1 - \sqrt{y - 1}}{\sqrt{y - 1}} \cdot \frac{dy}{dx} = \frac{6}{x^{1/3}}$$

$$\frac{dy}{dx} = \frac{6\sqrt{y - 1}}{x^{1/3}(1 - \sqrt{y - 1})}$$

15. $\dfrac{6 + 5x}{2 - 3y} = \dfrac{1}{5x}$

$$5x(6 + 5x) = 2 - 3y$$
$$30x + 25x^2 = 2 - 3y$$

$$\frac{d}{dx}(30x + 25x^2) = \frac{d}{dx}(2 - 3y)$$

$$30 + 50x = -3\frac{dy}{dx}$$

$$-\frac{30 + 50x}{3} = \frac{dy}{dx}$$

17. $\ln(xy + 1) = 2xy^3 + 4$

$$\frac{d}{dx}[\ln(xy + 1)] = \frac{d}{dx}(2xy^3 + 4)$$

$$\frac{1}{xy + 1} \cdot \frac{d}{dx}(xy + 1) = 2y^3 + 2x \cdot 3y^2\frac{dy}{dx} + \frac{d}{dx}(4)$$

$$\frac{1}{xy + 1}\left(y + x\frac{dy}{dx} + \frac{d}{dx}(1)\right) = 2y^3 + 6xy^2\frac{dy}{dx}$$

$$\frac{y}{xy + 1} + \frac{x}{xy + 1} \cdot \frac{dy}{dx} = 2y^3 + 6xy^2\frac{dy}{dx}$$

$$\left(\frac{x}{xy + 1} - 6xy^2\right)\frac{dy}{dx} = 2y^3 - \frac{y}{xy + 1}$$

$$\frac{dy}{dx} = \frac{2y^3 - \frac{y}{xy+1}}{\frac{x}{xy+1} - 6xy^2}$$

$$= \frac{2y^3(xy + 1) - y}{x - 6xy^2(xy + 1)}$$

$$= \frac{2xy^4 + 2y^3 - y}{x - 6x^2y^3 - 6xy^2}$$

21. $y = 8x^3 - 7x^2$, $\frac{dx}{dt} = 4$, $x = 2$

$$\frac{dy}{dt} = \frac{d}{dt}(8x^3 - 7x^2)$$

$$= 24x^2\frac{dx}{dt} - 14x\frac{dx}{dt}$$

$$= 24(2)^2(4) - 14(2)(4)$$

$$= 272$$

23. $y = \dfrac{1 + \sqrt{x}}{1 - \sqrt{x}}$, $\dfrac{dx}{dt} = -4$, $x = 4$

$$\frac{dy}{dt} = \frac{d}{dt}\left[\frac{1 + \sqrt{x}}{1 - \sqrt{x}}\right]$$

$$= \frac{(1 - \sqrt{x})\left(\frac{1}{2}x^{-1/2}\frac{dx}{dt}\right) - (1 + \sqrt{x})\left(-\frac{1}{2}\right)\left(x^{-1/2}\frac{dx}{dt}\right)}{(1 - \sqrt{x})^2}$$

$$= \frac{(1 - 2)\left(\frac{1}{2 \cdot 2}\right)(-4) - (1 + 2)\left(\frac{1}{2 \cdot 2}\right)(-4)}{(1 - 2)^2}$$

$$= \frac{1 - 3}{1} = -2$$

25. $y = xe^{3x}$; $\dfrac{dx}{dt} = -2, x = 1$

$$\frac{dy}{dt} = \frac{d}{dt}(xe^{3x})$$

$$= \frac{dx}{dt} \cdot e^{3x} + x \cdot \frac{d}{dt}(e^{3x})$$

$$= \frac{dx}{dt} \cdot e^{3x} + xe^{3x} \cdot 3\frac{dx}{dt}$$

$$= (1 + 3x)e^{3x}\frac{dx}{dt}$$

$$= (1 + 3 \cdot 1)e^{3(1)}(-2) = -8e^3$$

29. $y = \dfrac{3x - 7}{2x + 1}$; $x = 2$, $\Delta x = 0.003$

$$dy = \frac{(3)(2x + 1) - (2)(3x - 7)}{(2x + 1)^2}\, dx$$

$$dy = \frac{17}{(2x + 1)^2}\, dx$$

$$\approx \frac{17}{(2x + 1)^2}\, \Delta x$$

$$= \frac{17}{(2[2] + 1)^2}(0.003)$$

$$= 0.00204$$

33. Let $x =$ the length and width of a side
of the base;
$h =$ the height.

The volume is 32 m^3; the base is square and there is no top. Find the height, length, and width for minimum surface area.

$$\text{Volume} = x^2 h$$
$$x^2 h = 32$$
$$h = \frac{32}{x^2}$$

$$\text{Surface area} = x^2 + 4xh$$
$$A = x^2 + 4x\left(\frac{32}{x^2}\right)$$
$$= x^2 + 128x^{-1}$$
$$A' = 2x - 128x^{-2}$$

If $A' = 0$,

$$\frac{2x^3 - 128}{x^2} = 0$$
$$x^3 = 64$$
$$x = 4.$$

$$A''(x) = 2 + 2(128)x^{-3}$$
$$A''(4) = 6 > 0$$

The minimum is at $x = 4$, where

$$h = \frac{32}{4^2} = 2.$$

The dimensions are 2 m by 4 m by 4 m.

35. Volume of cylinder $= \pi r^2 h$
Surface area of cylinder open
at one end $= 2\pi rh + \pi r^2$.

$$V = \pi r^2 h = 27\pi$$
$$h = \frac{27\pi}{\pi r^2} = \frac{27}{r^2}$$
$$A = 2\pi r\left(\frac{27}{r^2}\right) + \pi r^2$$
$$= 54\pi r^{-1} + \pi r^2$$
$$A' = -54\pi r^{-2} + 2\pi r$$

If $A' = 0$,

$$2\pi r = \frac{54\pi}{r^2}$$
$$r^3 = 27$$
$$r = 3.$$

If $r = 3$,

$$A'' = 108\pi r^{-3} + 2\pi > 0,$$

so the value at $r = 3$ is a minimum.

For the minimum cost, the radius of the bottom should be 3 inches.

37. Here $k = 0.15$, $M = 20{,}000$, and $f = 12$. We have

$$q = \sqrt{\frac{2fM}{k}} = \sqrt{\frac{2(12)20{,}000}{0.15}}$$
$$= \sqrt{3{,}200{,}000} \approx 1789$$

Ordering 1789 rolls each time minimizes annual cost.

39. Use equation (3) from Section 6.3 with $k = 1$, $M = 128{,}000$, and $f = 10$.

$$q = \sqrt{\frac{2fM}{k}} = \sqrt{\frac{2(10)(128{,}000)}{1}}$$
$$= \sqrt{2{,}560{,}000} = 1600$$

The number of lots that should be produced annually is

$$\frac{M}{q} = \frac{128{,}000}{1600} = 80.$$

41. $A = \pi r^2$; $\frac{dr}{dt} = 4$ ft/min, $r = 7$ ft

$$\frac{dA}{dt} = 2\pi r \frac{dr}{dt}$$

$$\frac{dA}{dt} = 2\pi(7)(4)$$

$$\frac{dA}{dt} = 56\pi$$

The rate of change of the area is 56π ft^2/min.

43. (a)

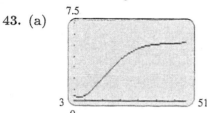

(b) We use a graphing calculator to graph

$$M'(t) = -0.4321173 + 0.1129024t - 0.0061518t^2$$
$$+ 0.0001260t^3 - 0.0000008925t^4$$

on $[3, 51]$ by $[0, 0.3]$. We find the maximum value of $M'(t)$ on this graph at about 15.41, or on about the 15th day.

45. Let $x =$ the distance from the base of the ladder to the building;
$y =$ the height on the building at the top of the ladder.

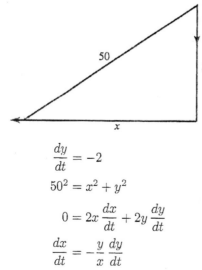

$$\frac{dy}{dt} = -2$$

$$50^2 = x^2 + y^2$$

$$0 = 2x\frac{dx}{dt} + 2y\frac{dy}{dt}$$

$$\frac{dx}{dt} = -\frac{y}{x}\frac{dy}{dt}$$

When $x = 30$, $y = \sqrt{2500 - (30)^2} = 40$.
So

$$\frac{dx}{dt} = \frac{-40}{30}(-2) = \frac{80}{30} = \frac{8}{3}$$

The base of the ladder is slipping away from the building at a rate of $\frac{8}{3}$ ft/min.

47. Let $x =$ one-half the width of the triangular cross section;
$h =$ the height of the water;
$V =$ the volume of the water.

$$\frac{dV}{dt} = 3.5 \ ft^3/\text{min}$$

Find $\frac{dV}{dt}$ when $h = \frac{1}{3}$.

$$V = \begin{pmatrix} \text{Area of} \\ \text{triangular} \\ \text{side} \end{pmatrix} (\text{length})$$

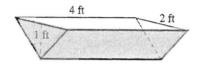

Area of triangular cross section

$$= \frac{1}{2}(\text{base})(\text{altitude})$$

$$= \frac{1}{2}(2x)(h) = xh$$

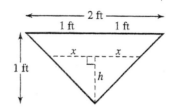

By similar triangles, $\frac{2x}{h} = \frac{2}{1}$, so $x = h$.

$$V = (xh)(4)$$
$$= h^2 \cdot 4$$
$$= 4h^2$$

$$\frac{dV}{dt} = 8h\frac{dh}{dt}$$

$$\frac{1}{8h} \cdot \frac{dV}{dt} = \frac{dh}{dt}$$

$$\frac{1}{8\left(\frac{1}{3}\right)}(3.5) = \frac{dh}{dt}$$

$$\frac{dh}{dt} = \frac{21}{16} = 1.3125$$

The depth of water is changing at the rate of 1.3125 ft/min.

49. $A = s^2$; $s = 9.2$, $\Delta s = \pm 0.04$

$$ds = 2s \, ds$$

$$\Delta A \approx 2s \, \Delta s$$

$$= 2(9.2)(\pm 0.04)$$

$$= \pm 0.736 \text{ in.}^2$$

51. We need to minimize y. Note that $x > 0$.

$$\frac{dy}{dx} = \frac{x}{8} - \frac{2}{x}$$

Set the derivative equal to 0.

$$\frac{x}{8} - \frac{2}{x} = 0$$

$$\frac{x}{8} = \frac{2}{x}$$

$$x^2 = 16$$

$$x = 4$$

Since $\lim\limits_{x \to 0} y = \infty$, $\lim\limits_{x \to \infty} y = \infty$, and $x = 4$ is the only critical value in $(0, \infty)$, $x = 4$ produces a minimum value.

$$y = \frac{4^2}{16} - 2\ln 4 + \frac{1}{4} + 2\ln 6$$

$$= 1.25 + 2(\ln 6 - \ln 4)$$

$$= 1.25 + 2\ln 1.5$$

The y coordinate of the Southern most point of the second boat's path is $1.25 + 2\ln 1.5$.

53. Distance on shore: $40 - x$ feet
Speed on shore: 5 feet per second
Distance in water: $\sqrt{x^2 + 40^2}$ feet
Speed in water: 3 feet per second

The total travel time t is $t = t_1 + t_2 = \dfrac{d_1}{v_1} + \dfrac{d_2}{v_2}$.

$$t(x) = \frac{40 - x}{5} + \frac{\sqrt{x^2 + 40^2}}{3}$$

$$= 8 - \frac{x}{5} + \frac{\sqrt{x^2 + 1600}}{3}$$

$$t'(x) = -\frac{1}{5} + \frac{1}{3} \cdot \frac{1}{2}(x^2 + 1600)^{-1/2}(2x)$$

$$= -\frac{1}{5} + \frac{x}{3\sqrt{x^2 + 1600}}$$

Minimize the travel time $t(x)$. If $t'(x) = 0$:

$$\frac{x}{3\sqrt{x^2 + 1600}} = \frac{1}{5}$$

$$5x = 3\sqrt{x^2 + 1600}$$

$$\frac{5x}{3} = \sqrt{x^2 + 1600}$$

$$\frac{25}{9}x^2 = x^2 + 1600$$

$$\frac{16}{9}x^2 = 1600$$

$$x^2 = \frac{1600 \cdot 9}{16}$$

$$x = \frac{40 \cdot 3}{4} = 30$$

(Discard the negative solution.)

To minimize the time, he should walk $40 - x = 40 - 30 = 10$ ft along the shore before paddling toward the desired destination. The minimum travel time is

$$\frac{40 - 30}{5} + \frac{\sqrt{30^2 + 40^2}}{3} \approx 18.67 \text{ seconds.}$$

Chapter 14 Test

[14.1]

Find the locations of any absolute extrema for the functions with graphs as follows.

1.

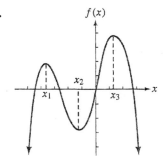

2.

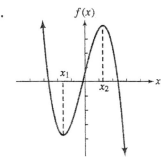

Find the locations of all absolute extrema for the functions defined as follows, with the specified domains.

3. $f(x) = x^3 - 12x$; $[0, 4]$

4. $f(x) = \dfrac{x^2 + 4}{x}$; $[-3, -1]$

5. Why is it important to check the endpoints of the domain when checking for absolute extrema?

[14.2]

6. Find two nonnegative numbers x and y such that $x + y = 50$ and $P = xy$ is maximized.

7. A travel agency offers a tour to the Bahamas for 12 people at \$800 each. For each person more than the original 12, up to a total of 30 people, who signs up for the cruise, the fare is reduced by \$20. What tour group size produces the greatest revenue for the travel agency?

8. \$320 is available for fencing for a rectangular garden. The fencing for the two sides parallel to the back of the house costs \$6 per linear foot, while the fencing for the other sides costs \$2 per linear foot. Find the dimensions that will maximize the area of the garden.

[14.3]

9. A small beauty supply store sells 600 hair dryers each year. It costs the store \$3 a year to store one hair dryer for one year. The fixed cost of placing the order is \$36. Find the optimum number of dryers per order.

10. A demand function is given by

$$q = 100 - 2p$$

where q is the number of units produced and p is the price in dollars.

(a) Find E, the elasticity of demand.   (b) Find the price that maximizes the total revenue.

11. What does elasticity of demand measure? What does it mean for demand to be elastic? What does it mean for demand to be inelastic?

[14.4]

12. Find $\frac{dy}{dx}$ for $x^3 - x^2y + y^2 = 0$.

13. Find an equation for the tangent line to the graph of $2x^2 - y^2 = 2$ at $(3, 4)$.

14. Suppose $3x^2 + 4y^2 + 6 = 0$. Use implicit differentiation to find $\frac{dy}{dx}$. Explain why your result is meaningless.

[14.5]

15. Find $\frac{dy}{dt}$ if $y = 3x^3 - 2x^2$, $\frac{dx}{dt} = -2$, and $x = 3$.

16. Find $\frac{dy}{dt}$ if $2xy - y^2 = x$, $x = -1$, $y = 3$, and $\frac{dx}{dt} = 2$.

17. When solving related rates problems, how should you interpret a negative derivative?

18. A 25-foot ladder is leaning against a vertical wall. If the bottom of the ladder is pulled horizontally away from the wall at 3 feet per second, how fast is the top of the ladder sliding down the wall when the bottom is 15 feet from the wall?

19. A real estate developer estimates that his monthly sales are given by

$$S = 30y - xy - \frac{y^2}{3000},$$

where y is the average cost of a new house in the development and x percent is the current interest rate for home mortgages. If the current rate is 11% and is rising at a rate of $\frac{1}{4}$% per month and the current average price of a new home is \$90,000 and is increasing at a rate of \$500 per month, how fast are his expected sales changing.

Given the revenue and cost functions $R = 30q - 0.2q^2$ and $C = 12q + 20$, where q is the daily production (and sales), answer Problems 20-22 when 50 units are produced and the rate of change of production is 8 units per day.

20. Find the rate of change of revenue with respect to time.

21. Find the rate of change of cost with respect to time.

22. Find the rate of change of profit with respect to time.

[14.6]

23. If $y = \frac{2x-3}{x^2-7}$, find dy.

24. Find the value of dy if $y = \sqrt{5 - x^2}$, $x = 1$, $\triangle x = 0.01$.

25. The radius of a sphere is claimed to be 5 cm with a possible error of 0.01 cm. Use differentials to estimate the possible error in the volume of the sphere.

Chapter 14 Test Answers

1. Absolute maximum at x_3

2. No absolute extrema

3. Absolute maximum of 16 at 4; absolute minimum of -16 at 2

4. Absolute maximum of -4 at -2; absolute minimum of -5 at -1

5. The smallest or largest value in a closed interval may occur at the endpoints.

6. 25 and 25

7. 26 people

8. $13\frac{1}{2}$ ft on each side parallel to the back, 40 ft on each of the other sides

9. 120

10. (a) $E = \frac{2p}{100-2p}$ (b) $25

11. The instantaneous responsiveness of demand to price; the relative change in demand is greater than the relative change in price; the relative change in demand is less than the relative change in price.

12. $\frac{dy}{dx} = \frac{2xy-3x^2}{2y-x^2}$

13. $3x - 2y = 1$

14. $\frac{-3x}{4y}$; no function exists such that $3x^2+4y^2+6 = 0$.

15. -138

16. 1.25

17. A decrease in rate

18. $-\frac{9}{4}$ ft/sec

19. Decreasing at $43,000/month

20. Increasing at a rate of $80/day

21. Increasing at a rate of $96/day

22. Decreasing at a rate of $16/day

23. $dy = \frac{-2x^2+6x-14}{(x^2-7)^2}dx$

24. -0.005

25. 3.14 cm^3

INTEGRATION

15.1 Antiderivatives

1. If $F(x)$ and $G(x)$ are both antiderivatives of $f(x)$, then there is a constant C such that

$$F(x) - G(x) = C.$$

The two functions can differ only by a constant.

5. $\displaystyle\int 6\,dk = 6\int 1\,dk$

$$= 6\int k^0\,dy$$

$$= 6 \cdot \frac{1}{1}k^{0+1} + C$$

$$= 6k + C$$

7. $\displaystyle\int (2z + 3)\,dz$

$$= 2\int z\,dz + 3\int z^0\,dz$$

$$= 2 \cdot \frac{1}{1+1}z^{1+1} + 3 \cdot \frac{1}{0+1}z^{0+1} + C$$

$$= z^2 + 3z + C$$

9. $\displaystyle\int (6t^2 - 8t + 7)\,dt$

$$= 6\int t^2\,dt - 8\int t\,dt + 7\int t^0\,dt$$

$$= \frac{6t^3}{3} - \frac{8t^2}{2} + 7t + C$$

$$= 2t^3 - 4t^2 + 7t + C$$

11. $\displaystyle\int (4z^3 + 3z^2 + 2z - 6)\,dz$

$$= 4\int z^3\,dz + 3\int z^2\,dz + 2\int z\,dz$$

$$- 6\int z^0\,dz$$

$$= \frac{4z^4}{4} + \frac{3z^3}{3} + \frac{2z^2}{2} - 6z + C$$

$$= z^4 + z^3 + z^2 - 6z + C$$

13. $\displaystyle\int (5\sqrt{z} + \sqrt{2})\,dz = 5\int z^{1/2}\,dz + \sqrt{2}\int dz$

$$= \frac{5z^{3/2}}{\frac{3}{2}} + \sqrt{2}z + C$$

$$= 5\left(\frac{2}{3}\right)z^{3/2} + \sqrt{2}z + C$$

$$= \frac{10z^{3/2}}{3} + \sqrt{2}z + C$$

15. $\displaystyle\int 5x(x^2 - 8)\,dx = \int (5x^3 - 40x)\,dx$

$$= \frac{5x^4}{4} - \frac{40x^2}{2} + C$$

$$= \frac{5x^4}{4} - 20x^2 + C$$

17. $\displaystyle\int (4\sqrt{v} - 3v^{3/2})\,dv$

$$= 4\int v^{1/2}\,dv - 3\int v^{3/2}\,dv$$

$$= \frac{4v^{3/2}}{\frac{3}{2}} - \frac{3v^{5/2}}{\frac{5}{2}} + C$$

$$= \frac{8v^{3/2}}{3} - \frac{6v^{5/2}}{5} + C$$

19. $\displaystyle\int (10u^{3/2} - 14u^{5/2})\,du$

$$= 10\int u^{3/2}\,du - 14\int u^{5/2}\,du$$

$$= \frac{10u^{5/2}}{\frac{5}{2}} - \frac{14u^{7/2}}{\frac{7}{2}} + C$$

$$= 10\left(\frac{2}{5}\right)u^{5/2} - 14\left(\frac{2}{7}\right)u^{7/2} + C$$

$$= 4u^{5/2} - 4u^{7/2} + C$$

21. $\displaystyle\int \left(\frac{7}{z^2}\right)dz = \int 7z^{-2}\,dz$

$$= 7\int z^{-2}\,dz$$

$$= 7\left(\frac{z^{-2+1}}{-2+1}\right) + C$$

$$= \frac{7z^{-1}}{-1} + C$$

$$= -\frac{7}{z} + C$$

23. $\displaystyle\int \left(\frac{\pi^3}{y^3} - \frac{\sqrt{\pi}}{\sqrt{y}} \right) dy = \int \pi^3 y^{-3} \, dy - \int \sqrt{\pi} y^{-1/2} \, dy$

$\displaystyle = \pi^3 \int y^{-3} \, dy - \sqrt{\pi} \int y^{-1/2} \, dy$

$\displaystyle = \pi^3 \left(\frac{y^{-2}}{-2} \right) - \sqrt{\pi} \left(\frac{y^{1/2}}{\frac{1}{2}} \right) + C$

$\displaystyle = -\frac{\pi^3}{2y^2} - 2\sqrt{\pi y} + C$

25. $\displaystyle\int \left(-9t^{-2.5} - 2t^{-1} \right) dt$

$\displaystyle = -9 \int t^{-2.5} \, dt - 2 \int t^{-1} \, dt$

$\displaystyle = \frac{-9t^{-1.5}}{-1.5} - 2 \int \frac{dt}{t}$

$\displaystyle = 6t^{-1.5} - 2 \ln |t| + C$

27. $\displaystyle\int \frac{1}{3x^2} dx = \int \frac{1}{3} x^{-2} dx$

$\displaystyle = \frac{1}{3} \int x^{-2} dx$

$\displaystyle = \frac{1}{3} \left(\frac{x^{-1}}{-1} \right) + C$

$\displaystyle = -\frac{1}{3} x^{-1} + C$

$\displaystyle = -\frac{1}{3x} + C$

29. $\displaystyle\int 3e^{-0.2x} \, dx = 3 \int e^{-0.2x} \, dx$

$\displaystyle = 3 \left(\frac{1}{-0.2} \right) e^{-0.2x} + C$

$\displaystyle = \frac{3(e^{-0.2x})}{-0.2} + C$

$\displaystyle = -15e^{-0.2x} + C$

31. $\displaystyle\int \left(-\frac{3}{x} + 4e^{-0.4x} + e^{0.1} \right) dx$

$\displaystyle = -3 \int \frac{dx}{x} + 4 \int e^{-0.4x} \, dx + e^{0.1} \int dx$

$\displaystyle = -3 \ln |x| + \frac{4e^{-0.4x}}{-0.4} + e^{0.1} x + C$

$\displaystyle = -3 \ln |x| - 10e^{-0.4x} + e^{0.1} x + C$

33. $\displaystyle\int \left(\frac{1 + 2t^3}{4t} \right) dt = \int \left(\frac{1}{4t} + \frac{t^2}{2} \right) dt$

$\displaystyle = \frac{1}{4} \int \frac{1}{t} \, dt + \frac{1}{2} \int t^2 \, dt$

$\displaystyle = \frac{1}{4} \ln |t| + \frac{1}{2} \left(\frac{t^3}{3} \right) + C$

$\displaystyle = \frac{1}{4} \ln |t| + \frac{t^3}{6} + C$

35. $\displaystyle\int (e^{2u} + 4u) \, du = \frac{e^{2u}}{2} + \frac{4u^2}{2} + C$

$\displaystyle = \frac{e^{2u}}{2} + 2u^2 + C$

37. $\displaystyle\int (x+1)^2 \, dx = \int (x^2 + 2x + 1) \, dx$

$\displaystyle = \frac{x^3}{3} + \frac{2x^2}{2} + x + C$

$\displaystyle = \frac{x^3}{3} + x^2 + x + C$

39. $\displaystyle\int \frac{\sqrt{x} + 1}{\sqrt[3]{x}} \, dx = \int \left(\frac{\sqrt{x}}{\sqrt[3]{x}} + \frac{1}{\sqrt[3]{x}} \right) dx$

$\displaystyle = \int \left(x^{(1/2 - 1/3)} + x^{-1/3} \right) dx$

$\displaystyle = \int x^{1/6} \, dx + \int x^{-1/3} \, dx$

$\displaystyle = \frac{x^{7/6}}{\frac{7}{6}} + \frac{x^{2/3}}{\frac{2}{3}} + C$

$\displaystyle = \frac{6x^{7/6}}{7} + \frac{3x^{2/3}}{2} + C$

41. $\displaystyle\int 10^x \, dx = \frac{10^x}{\ln 10} + C$

43. Find $f(x)$ such that $f'(x) = x^{2/3}$, and $\left(1, \frac{3}{5}\right)$ is on the curve.

$\displaystyle\int x^{2/3} dx = \frac{x^{5/3}}{\frac{5}{3}} + C$

$\displaystyle f(x) = \frac{3x^{5/3}}{5} + C$

Since $\left(1, \frac{3}{5}\right)$ is on the curve,

$\displaystyle f(1) = \frac{3}{5}.$

$\displaystyle f(1) = \frac{3(1)^{5/3}}{5} + C = \frac{3}{5}$

$\displaystyle \frac{3}{5} + C = \frac{3}{5}$

$\displaystyle C = 0.$

Thus,

$$f(x) = \frac{3x^{5/3}}{5}.$$

45. $C'(x) = 4x - 5$; fixed cost is $8.

$$C(x) = \int (4x - 5)\, dx$$

$$= \frac{4x^2}{2} - 5x + k$$

$$= 2x^2 - 5x + k$$

$$C(0) = 2(0)^2 - 5(0) + k = k$$

Since $C(0) = 8$, $k = 8$.

Thus,

$$C(x) = 2x^2 - 5x + 8.$$

47. $C'(x) = 0.03e^{0.01x}$; fixed cost is $8.

$$C(x) = \int 0.03e^{0.01x}\, dx$$

$$= 0.03 \int e^{0.01x}\, dx$$

$$= 0.03 \left(\frac{1}{0.01} e^{0.01x} \right) + k$$

$$= 3e^{0.01x} + k$$

$$C(0) = 3e^{0.01(0)} + k = 3(1) + k$$

$$= 3 + k$$

Since $C(0) = 8$, $3 + k = 8$, and $k = 5$.
Thus,

$$C(x) = 3e^{0.01x} + 5.$$

49. $C'(x) = x^{2/3} + 2$; 8 units cost $58.

$$C(x) = \int (x^{2/3} + 2)\, dx$$

$$= \frac{3x^{5/3}}{5} + 2x + k$$

$$C(8) = \frac{3(8)^{5/3}}{5} + 2(8) + k$$

$$= \frac{3(32)}{5} + 16 + k$$

Since $C(8) = 58$,

$$58 - 16 - \frac{96}{5} = k$$

$$\frac{114}{5} = k.$$

Thus,

$$C(x) = \frac{3x^{5/3}}{5} + 2x + \frac{114}{5}.$$

51. $C'(x) = 5x - \dfrac{1}{x}$; 10 units cost $94.20, so

$$C(10) = 94.20.$$

$$C(x) = \int \left(5x - \frac{1}{x} \right) dx = \frac{5x^2}{2} - \ln|x| + k$$

$$C(10) = \frac{5(10)^2}{2} - \ln(10) + k$$

$$= 250 - 2.30 + k.$$

Since $C(10) = 94.20$,

$$94.20 = 247.70 + k$$

$$-153.50 = k.$$

Thus, $C(x) = \dfrac{5x^2}{2} - \ln|x| - 153.50$.

53. $R'(x) = 175 - 0.02x - 0.03x^2$

$$R = \int (175 - 0.02x - 0.03x^2)\, dx$$

$$= 175x - 0.01x^2 - 0.01x^3 + C.$$

If $x = 0$, then $R = 0$ (no items sold means no revenue), and

$$0 = 175(0) - 0.01(0)^2 - 0.01(0)^3 + C$$

$$0 = C.$$

Thus, $R = 175x - 0.01x^2 - 0.01x^3$

gives the revenue function. Now, recall that $R = xp$, where p is the demand function. Then

$$175x - 0.01x^2 - 0.01x^3 = xp$$

$$175x - 0.01x - 0.01x^2 = p, \text{ the demand function.}$$

55. $R'(x) = 500 - 0.15\sqrt{x}$

$$R = \int (500 - 0.15\sqrt{x})\, dx$$

$$= 500x - 0.1x^{3/2} + C.$$

If $x = 0$, $R = 0$ (no items sold means no revenue), and

$$0 = 500(0) - 0.1(0)^{3/2} + C$$

$$0 = C.$$

Thus, $R = 500x - 0.1x^{3/2}$ gives the revenue function. Now, recall that $R = xp$, where p is the demand function. Then

$$500x - 0.1x^{3/2} = xp$$

$$500 - 0.1\sqrt{x} = p, \text{ the demand function.}$$

57. $f'(t) = 1.498t + 1.626$

(a) $f(t) = \int (1.498t + 1.626)dt$

$= 0.749t^2 + 1.626t + C$

In 1992 $(t = 2)$, $f(t) = 8.893$, and

$8.893 = 0.749(2)^2 + 1.626(2) + C$

$8.893 = 6.248 + C$

$2.645 = C$

Thus, $f(t) = 0.749t^2 + 1.626t + 2.645$.

(b) In 2006, $t = 16$, and

$f(16) = 0.749(16)^2 + 1.626(16) + 2.645$

$= 191.744 + 26.016 + 2.645$

$= 220.405$

The function predicted approximately 220 million subscribers in 2006.

59. (a) $P'(x) = 50x^3 + 30x^2$; profit is -40 when no cheese is sold.

$P(x) = \int (50x^3 + 30x^2)\, dx$

$= \dfrac{25x^4}{2} + 10x^3 + k$

$P(0) = \dfrac{25(0)^4}{2} + 10(0)^3 + k$

Since

$P(0) = -40,$

$-40 = k.$

Thus,

$P(x) = \dfrac{25x^4}{2} + 10x^3 - 40.$

(b) $P(2) = \dfrac{25(2)^4}{2} + 10(2)^3 - 40 = 240$

The profit from selling 200 lbs of Brie cheese is $240.

61. $\displaystyle\int \frac{g(x)}{x}\, dx = \int \frac{a - bx}{x}\, dx$

$= \displaystyle\int \left(\frac{a}{x} - b \right) dx$

$= a \displaystyle\int \frac{dx}{x} - b \int dx$

$= a \ln |x| - bx + C$

Since x represents a positive quantity, the absolute value sign can be dropped.

$\displaystyle\int \frac{g(x)}{x}\, dx = a \ln x - bx + C$

63. $N'(t) = Ae^{kt}$

(a) $N(t) = \dfrac{A}{k} e^{kt} + C$

$A = 50$, $N(t) = 300$ when $t = 0$.

$N(0) = \dfrac{50}{k} e^0 + C = 300$

$N'(5) = 250$

Therefore,

$N'(5) = 50e^{5k} = 250$

$e^{5k} = 5$

$5k = \ln 5$

$k = \dfrac{\ln 5}{5}.$

$N(0) = \dfrac{50}{\frac{\ln 5}{5}} + C = 300$

$\dfrac{250}{\ln 5} + C = 300$

$C = 300 - \dfrac{250}{\ln 5} \approx 144.67$

$N(t) = \dfrac{50}{\frac{\ln 5}{5}} e^{(\ln 5/5)t} + 144.67$

$= 155.3337 e^{0.321888t} + 144.67$

(b) $N(12) = 155.3337 e^{0.321888(12)} + 144.67$

≈ 7537

There are 7537 cells present after 12 days.

65. $a(t) = 5t^2 + 4$

$v(t) = \displaystyle\int (5t^2 + 4)\, dt$

$= \dfrac{5t^3}{3} + 4t + C$

$v(0) = \dfrac{5(0)^3}{3} + 4(0) + C$

Since $v(0) = 6$, $C = 6$.

$v(t) = \dfrac{5t^3}{3} + 4t + 6$

67. $a(t) = -32$

$$v(t) = \int -32 \, dt = -32t + C_1$$

$$v(0) = -32(0) + C_1$$

Since $v(0) = 0$, $C_1 = 0$.

$$v(t) = -32t$$

$$s(t) = \int -32t \, dt$$

$$= \frac{-32t^2}{2} + C_2$$

$$= -16t^2 + C_2$$

At $t = 0$, the plane is at 6400 ft.
That is, $s(0) = 6400$.

$$s(0) = -16(0)^2 + C_2$$
$$6400 = 0 + C_2$$
$$C_2 = 6400$$
$$s(t) = -16t^2 + 6400$$

When the object hits the ground, $s(t) = 0$.

$$-16t^2 + 6400 = 0$$
$$-16t^2 = -6400$$
$$t^2 = 400$$
$$t = \pm 20$$

Discard -20 since time must be positive.
The object hits the ground in 20 sec.

69. $a(t) = \dfrac{15}{2}\sqrt{t} = 3e^{-t}$

$$v(t) = \int \left(\frac{15}{2}\sqrt{t} + 3e^{-t} \right) dt$$

$$= \int \left(\frac{15}{2}t^{1/2} + 3e^{-t} \right) dt$$

$$= \frac{15}{2}\left(\frac{t^{3/2}}{\frac{3}{2}} \right) + 3\left(\frac{1}{-1}e^{-t} \right) + C_1$$

$$= 5t^{3/2} - 3e^{-t} + C_1$$

$$v(0) = 5(0)^{3/2} - 3e^{-0} + C_1 = -3 + C_1$$

Since $v(0) = -3$, $C_1 = 0$.

$$v(t) = 5t^{3/2} - 3e^{-t}$$

$$s(t) = \int (5t^{3/2} - 3e^{-t}) \, dt$$

$$= 5\left(\frac{t^{5/2}}{\frac{5}{2}} \right) - 3\left(-\frac{1}{1}e^{-t} \right) + C_2$$

$$= 2t^{5/2} + 3e^{-t} + C_2$$
$$s(0) = 2(0)^{5/2} + 3e^{-0} + C_2 = 3 + C_2$$

Since $s(0) = 4$, $C_2 = 1$.
Thus,

$$s(t) = 2t^{5/2} + 3e^{-t} + 1.$$

71. First find $v(t)$ by integrating $a(t)$:

$$v(t) = \int (-32) dt = -32t + k.$$

When $t = 5, v(t) = 0$:

$$0 = -32(5) + k$$
$$160 = k$$

and

$$v(t) = -32t + 160.$$

Now integrate $v(t)$ to find $h(t)$.

$$h(t) = \int (-32t + 160) dt = -16t^2 + 160t + C$$

Since $h(t) = 412$ when $t = 5$, we can substitute these values into the equation for $h(t)$ to get $C = 12$ and

$$h(t) = -16t^2 + 160t + 12.$$

Therefore, from the equation given in Exercise 70, the initial velocity v_0 is 160 ft/sec and the initial height of the rocket h_0 is 12 ft.

73. **(a)** First find $B(t)$ by integrating $B'(t)$:

$$B(t) = \int 9.2935e^{0.02955t} dt$$

$$\approx 314.5e^{0.02955t} + k$$

When $t = 0, B(t) = 792.3$:

$$792.3 = 314.5e^{0.02955(0)} + k$$
$$477.8 = k$$

and

$$B(t) = 314.5e^{0.02955t} + 477.8.$$

(b) In 2012, $t = 42$.

$$B(42) = 314.5e^{0.02955(42)} + 477.8$$
$$\approx 1565.8$$

About 1,566,000 bachelor's degrees will be conferred in 2012.

15.2   Substitution

3. $\int 4(2x+3)^4\,dx = 2\int 2(2x+3)^4\,dx$

Let $u = 2x + 3$, so that $du = 2\,dx$.

$$= 2\int u^4\,du$$

$$= \frac{2 \cdot u^5}{5} + C$$

$$= \frac{2(2x+3)^5}{5} + C$$

5. $\int \frac{2\,dm}{(2m+1)^3} = \int 2(2m+1)^{-3}\,dm$

Let $u = 2m + 1$, so that $du = 2\,dm$.

$$= \int u^{-3}\,du$$

$$= \frac{u^{-2}}{-2} + C$$

$$= \frac{-(2m+1)^{-2}}{2} + C$$

7. $\int \frac{2x+2}{(x^2+2x-4)^4}\,dx$

$$= \int (2x+2)(x^2+2x-4)^{-4}\,dx$$

Let $w = x^2 + 2x - 4$, so that $dw = (2x+2)\,dx$.

$$= \int w^{-4}\,dw$$

$$= \frac{w^{-3}}{-3} + C$$

$$= -\frac{(x^2+2x-4)^{-3}}{3} + C$$

$$= -\frac{1}{3(x^2+2x-4)^3} + C$$

9. $\int z\sqrt{4z^2-5}\,dz = \int z(4z^2-5)^{1/2}\,dz$

$$= \frac{1}{8}\int 8z(4z^2-5)^{1/2}\,dz$$

Let $u = 4z^2 - 5$, so that $du = 8z\,dz$.

$$= \frac{1}{8}\int u^{1/2}\,du$$

$$= \frac{1}{8} \cdot \frac{u^{3/2}}{\frac{3}{2}} + C$$

$$= \frac{1}{8} \cdot \left(\frac{2}{3}\right) u^{3/2} + C$$

$$= \frac{(4z^2-5)^{3/2}}{12} + C$$

11. $\int 3x^2\,e^{2x^3}\,dx = \frac{1}{2}\int 2 \cdot 3x^2\,e^{2x^3}\,dx$

Let $u = 2x^3$, so that $du = 6x^2\,dx$.

$$= \frac{1}{2}\int e^u\,du$$

$$= \frac{1}{2}e^u + C$$

$$= \frac{e^{2x^3}}{2} + C$$

13. $\int (1-t)e^{2t-t^2}\,dt$

$$= \frac{1}{2}\int 2(1-t)e^{2t-t^2}\,dt$$

Let $u = 2t - t^2$, so that $du = (2-2t)\,dt$.

$$= \frac{1}{2}\int e^u\,du$$

$$= \frac{e^u}{2} + C$$

$$= \frac{e^{2t-t^2}}{2} + C$$

15. $\int \frac{e^{1/z}}{z^2}\,dz = -\int e^{1/z} \cdot \frac{-1}{z^2}\,dz$

Let $u = \frac{1}{z}$, so that $du = \frac{-1}{z^2}\,dx$.

$$= -\int e^u\,du$$

$$= -e^u + C$$

$$= -e^{1/z} + C$$

17. $\int (x^3+2x)(x^4+4x^2+7)^8\,dx$

$$= \frac{1}{4}\int (x^4+4x^2+7)^8(4x^3+8x)\,dx$$

Let $u = x^4 + 4x^2 + 7$, so that $du = (4x^3+8x)\,dx$

$$= \frac{1}{4}\int u^8\,du = \frac{1}{4}\int \left(\frac{u^9}{9}\right) + C$$

$$= \frac{u^9}{36} + C = \frac{(x^4+4x^2+7)^9}{36} + C$$

19. $\int \frac{2x+1}{(x^2+x)^3}\,dx$

$$= \int (2x+1)(x^2+x)^{-3}\,dx$$

Let $u = x^2 + x$, so that $du = (2x+1)\,dx$.

$$= \int u^{-3}\,du = \frac{u^{-2}}{-2} + C$$

$$= \frac{-1}{2u^2} + C = \frac{-1}{2(x^2+x)^2} + C$$

21. $\int p(p+1)^5 \, dp$

Let $u = p + 1$, so that $du = dp$; also, $p = u - 1$.

$$= \int (u-1)u^5 \, du$$

$$= \int (u^6 - u^5) \, du$$

$$= \frac{u^7}{7} - \frac{u^6}{6} + C$$

$$= \frac{(p+1)^7}{7} - \frac{(p+1)^6}{6} + C$$

23. $\int \frac{u}{\sqrt{u-1}} \, du$

$$= \int u(u-1)^{-1/2} \, du$$

Let $w = u - 1$, so that $dw = du$ and $u = w + 1$.

$$= \int (w+1)w^{-1/2} \, dw$$

$$= \int (w^{1/2} + w^{-1/2}) \, dw$$

$$= \frac{w^{3/2}}{\frac{3}{2}} + \frac{w^{1/2}}{\frac{1}{2}} + C$$

$$= \frac{2(u-1)^{3/2}}{3} + 2(u-1)^{1/2} + C$$

25. $\int (\sqrt{x^2 + 12x})(x+6) \, dx$

$$= \int (x^2 + 12x)^{1/2}(x+6) \, dx$$

Let $x^2 + 12x = u$, so that

$$(2x + 12) \, dx = du$$
$$2(x + 6) \, dx = du.$$

$$= \frac{1}{2} \int u^{1/2} \, du = \frac{1}{2}\left(\frac{2}{3}\right) u^{3/2} + C$$

$$= \frac{(x^2 + 12x)^{3/2}}{3} + C$$

27. $\int \frac{t}{t^2 + 2} \, dt$

Let $t^2 + 2 = u$, so that $2t \, dt = du$.

$$= \frac{1}{2} \int \frac{du}{u}$$

$$= \frac{1}{2} \ln |u| + C$$

$$= \frac{\ln(t^2 + 2)}{2} + C$$

29. $\int \frac{(1 + 3\ln x)^2}{x} \, dx$

Let $u = 1 + 3\ln x$, so that $du = \frac{3}{x} \, dx$.

$$= \frac{1}{3} \int \frac{3(1 + 3\ln x)^2}{x} \, dx$$

$$= \frac{1}{3} \int u^2 \, du$$

$$= \frac{1}{3} \cdot \frac{u^3}{3} + C$$

$$= \frac{(1 + 3\ln x)^3}{9} + C$$

31. $\int \frac{e^{2x}}{e^{2x} + 5} \, dx$

Let $u = e^{2x} + 5$, so that $du = 2e^{2x} \, dx$.

$$= \frac{1}{2} \int \frac{du}{u}$$

$$= \frac{1}{2} \ln |u| + C$$

$$= \frac{1}{2} \ln \left| e^{2x} + 5 \right| + C$$

$$= \frac{1}{2} \ln (e^{2x} + 5) + C$$

33. $\int \frac{\log x}{x} \, dx$

Let $u = \log x$, so that $du = \frac{1}{(\ln 10)x} \, dx$.

$$= (\ln 10) \int \frac{\log x}{(\ln 10)x} \, dx$$

$$= (\ln 10) \int u \, du$$

$$= (\ln 10) \left(\frac{u^2}{2}\right) + C$$

$$= \frac{(\ln 10)(\log x)^2}{2} + C$$

35. $\int x 8^{3x^2 + 1} \, dx$

Let $u = 3x^2 + 1$, so that $du = 6x \, dx$.

$$= \frac{1}{6} \int 6x \cdot 8^{3x^2 + 1} \, dx$$

$$= \frac{1}{6} \int 8^u \, du$$

$$= \frac{1}{6} \left(\frac{8^u}{\ln 8}\right) + C$$

$$= \frac{8^{3x^2 + 1}}{6 \ln 8} + C$$

39. (a) $R'(x) = 4x(x^2 + 27{,}000)^{-2/3}$

$$R(x) = \int 4x(x^2 + 27{,}000)^{-2/3}\,dx = 2\int 2x(x^2 + 27{,}000)^{-2/3}\,dx$$

Let $u = x^2 + 27{,}000$, so that $du = 2x\,dx$.

$$R = 2\int u^{-2/3}\,du = 2\cdot 3u^{1/3} + C = 6(x^2 + 27{,}000)^{1/3} + C$$
$$R(125) = 6(125^2 + 27{,}000)^{1/3} + C$$

Since $R(125) = 29.591$,

$$6(125^2 + 27{,}000)^{1/3} + C = 29.591$$
$$C = -180$$

Thus,

$$R(x) = 6(x^2 + 27{,}000)^{1/3} - 180.$$

(b)

$$R(x) = 6(x^2 + 27{,}000)^{1/3} - 180 \geq 40$$
$$6(x^2 + 27{,}000)^{1/3} \geq 220$$
$$(x^2 + 27{,}000)^{1/3} \geq 36.6667$$
$$x^2 + 27{,}000 \geq 49{,}296.43$$
$$x^2 \geq 22{,}296.43$$
$$x \geq 149.4$$

For a revenue of at least \$40,000, 150 players must be sold.

41. $C'(x) = \dfrac{60x}{5x^2 + e}$

(a) Let $u = 5x^2 + e$, so that $du = 10x\,dx$.

$$C(x) = \int C'(x)\,dx = \int \frac{60x}{5x^2 + e}\,dx = 6\int \frac{du}{u} = 6\ln|u| + C = 6\ln|5x^2 + e| + C$$

Since $C(0) = 10, C = 4$.

Therefore,

$$C(x) = 6\ln|5x^2 + e| + 4 = 6\ln(5x^2 + e) + 4.$$

(b) $C(5) = 6\ln(5\cdot 5^2 + e) + 4 \approx 33.099$

Since this represents \$33,099 dollars which is greater than \$20,000, a new source of investment income should be sought.

43. $f'(t) = 4.0674 \cdot 10^{-4} t(t - 1970)^{0.4}$

(a) Let $u = t - 1970$. To get the t outside the parentheses in terms of u, solve $u = t - 1970$ for t to get $t = u + 1970$. Then $dt = du$ and we can substitute as follows.

$$f(t) = \int f'(t) dt = \int 4.0674 \cdot 10^{-4} t(t - 1970)^{0.4} dt$$

$$= \int 4.0674 \cdot 10^{-4} (u + 1970)(u)^{0.4} du$$

$$= 4.0674 \cdot 10^{-4} \int (u + 1970)(u)^{0.4} du$$

$$= 4.0674 \cdot 10^{-4} \int (u^{1.4} + 1970 u^{0.4}) du$$

$$= 4.0674 \cdot 10^{-4} \left(\frac{u^{2.4}}{2.4} + \frac{1970 u^{1.4}}{1.4} \right) + C$$

$$= 4.0674 \cdot 10^{-4} \left[\frac{(t - 1970)^{2.4}}{2.4} + \frac{1970(t - 1970)^{1.4}}{1.4} \right] + C$$

Since $f(1970) = 61.298, C = 61.298$.

Therefore,

$$f(t) = 4.0674 \cdot 10^{-4} \left[\frac{(t - 1970)^{2.4}}{2.4} + \frac{1970(t - 1970)^{1.4}}{1.4} \right] + 61.298.$$

(b) $f(2015) = 4.0674 \cdot 10^{-4} \left[\dfrac{(2015 - 1970)^{2.4}}{2.4} + \dfrac{1970(2015 - 1970)^{1.4}}{1.4} \right] + 61.298 \approx 180.9.$

In the year 2015, there will be about 181,000 local transit vehicles.

15.3   Area and the Definite Integral

3. $f(x) = 2x + 5$, $x_1 = 0$, $x_2 = 2$, $x_3 = 4$, $x_4 = 6$, and $\Delta x = 2$

(a) $\displaystyle\sum_{i=1}^{4} f(x_i)\Delta x$

$= f(x_1)\Delta x + f(x_2)\Delta x + f(x_3)\Delta x + f(x_4)\Delta x$
$= f(0)(2) + f(2)(2) + f(4)(2) + f(6)(2)$
$= [2(0) + 5](2) + [2(2) + 5](2) + [2(4) + 5](2) + [2(6) + 5](2)$
$= 10 + 9(2) + 13(2) + 17(2)$
$= 88$

(b)

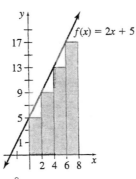

The sum of these rectangles approximates $\displaystyle\int_{0}^{8} (2x + 5)\, dx$.

7. $f(x) = 2x + 5$ from $x = 2$ to $x = 4$

For $n = 4$ rectangles:

$$\Delta x = \frac{4 - 2}{4} = 0.5$$

(a) Using the left endpoints:

i	x_i	$f(x_i)$
1	2	9
2	2.5	10
3	3	11
4	3.5	12

$$A = \sum_{1}^{4} f(x_i)\Delta x = 9\,(0.5) + 10\,(0.5) + 11\,(0.5) + 12\,(0.5) = 21$$

(b) Using the right endpoints:

i	x_i	$f(x_i)$
1	2.5	10
2	3	11
3	3.5	12
4	4	13

$$A = 10(0.5) + 11(0.5) + 12(0.5) + 13(0.5) = 23$$

(c) Average $= \dfrac{21 + 23}{2} = \dfrac{44}{2} = 22$

(d) Using the midpoints:

i	x_i	$f(x_i)$
1	2.25	9.5
2	2.75	10.5
3	3.25	11.5
4	3.75	12.5

$$A = \sum_{1}^{4} f(x_i)\Delta x = 9.5\,(0.5) + 10.5\,(0.5) + 11.5\,(0.5) + 12.5\,(0.5) = 22$$

9. $f(x) = -x^2 + 4$ from $x = -2$ to $x = 2$

For $n = 4$ rectangles:

$$\Delta x = \frac{2 - (-2)}{4} = 1$$

(a) Using the left endpoints:

i	x_i	$f(x_i)$
1	-2	$-(-2)^2 + 4 = 0$
2	-1	$-(-1)^2 + 4 = 3$
3	0	$-(0)^2 + 4 = 4$
4	1	$-(1)^2 + 4 = 3$

$$A = \sum_{i=1}^{4} f(x_i)\Delta x = (0)(1) + (3)(1) + (4)(1) + (3)(1) = 10$$

(b) Using the right endpoints:

i	x_i	$f(x_i)$
1	-1	3
2	0	4
3	1	3
4	2	0

$$\text{Area} = 1(3) + 1(4) + 1(3) + 1(0) = 10$$

(c) $\text{Average} = \dfrac{10 + 10}{2} = 10$

(d) Using the midpoints:

i	x_i	$f(x_i)$
1	$-\dfrac{3}{2}$	$\dfrac{7}{4}$
2	$-\dfrac{1}{2}$	$\dfrac{15}{4}$
3	$\dfrac{1}{2}$	$\dfrac{15}{4}$
4	$\dfrac{3}{2}$	$\dfrac{7}{4}$

$$A = \sum_{i=1}^{4} f(x_i)\Delta x = \frac{7}{4}(1) + \frac{15}{4}(1) + \frac{15}{4}(1) + \frac{7}{4}(1) = 11$$

11. $f(x) = e^x + 1$ from $x = -2$ to $x = 2$

For $n = 4$ rectangles:

$$\Delta x = \frac{2 - (-2)}{4} = 1$$

(a) Using the left endpoints:

i	x_i	$f(x_i)$
1	-2	$e^{-2} + 1$
2	-1	$e^{-1} + 1$
3	0	$e^0 + 1 = 2$
4	1	$e^1 + 1$

$$A = \sum_{i=1}^{4} f(x_i)\Delta x = \sum_{i=1}^{4} f(x_i)(1) = \sum_{i=1}^{4} f(x_i) = (e^{-2} + 1) + (e^{-1} + 1) + 2 + e^1 + 1 \approx 8.2215 \approx 8.22$$

(b) Using the right endpoints:

i	x_i	$f(x_i)$
1	-1	$e^{-1} + 1$
2	0	2
3	1	$e + 1$
4	2	$e^2 + 1$

$$\text{Area} = 1(e^{-1} + 1) + 1(2) + 1(e + 1) + 1(e^2 + 1) \approx 15.4752 \approx 15.48$$

(c) $\text{Average} = \dfrac{8.2215 + 15.4752}{2} = 11.84835 \approx 11.85$

(d) Using the midpoints:

i	x_i	$f(x_i)$
1	$-\dfrac{3}{2}$	$e^{-3/2}+1$
2	$-\dfrac{1}{2}$	$e^{-1/2}+1$
3	$\dfrac{1}{2}$	$e^{1/2}+1$
4	$\dfrac{3}{2}$	$e^{3/2}+1$

$$A = \sum_{i=1}^{4} f(x_i)\Delta x = (e^{-3/2}+1)(1) + (e^{-1/2}+1)(1) + (e^{1/2}+1)(1) + (e^{3/2}+1)(1) \approx 10.9601 \approx 10.96$$

13. $f(x) = \dfrac{2}{x}$ from $x = 1$ to $x = 9$

   For $n = 4$ rectangles:

$$\Delta x = \frac{9-1}{4} = 2$$

(a) Using the left endpoints:

i	x_i	$f(x_i)$
1	1	$\dfrac{2}{1} = 2$
2	3	$\dfrac{2}{3}$
3	5	$\dfrac{2}{5} = 0.4$
4	7	$\dfrac{2}{7}$

$$A = \sum_{i=1}^{4} f(x_i)\Delta x = (2)(2) + \frac{2}{3}(2) + (0.4)(2) + \left(\frac{2}{7}\right)(2) \approx 6.7048 \approx 6.70$$

(b) Using the right endpoints:

i	x_i	$f(x_i)$
1	3	$\dfrac{2}{3}$
2	5	$\dfrac{2}{5}$
3	7	$\dfrac{2}{7}$
4	9	$\dfrac{2}{9}$

$$\text{Area} = 2\left(\frac{2}{3}\right) + 2\left(\frac{2}{5}\right) + 2\left(\frac{2}{7}\right) + 2\left(\frac{2}{9}\right) = \frac{4}{3} + \frac{4}{5} + \frac{4}{7} + \frac{4}{9} \approx 3.1492 \approx 3.15$$

(c) Average $= \dfrac{6.7 + 3.15}{2} = 4.93$

(d) Using the midpoints:

i	x_i	$f(x_i)$
1	2	1
2	4	$\dfrac{1}{2}$
3	6	$\dfrac{1}{3}$
4	8	$\dfrac{1}{4}$

$$A = \sum_{i=1}^{4} f(x_i)\Delta x = 1(2) + \frac{1}{2}(2) + \frac{1}{3}(2) + \frac{1}{4}(2) \approx 4.1667 \approx 4.17$$

15. $\displaystyle\int_{0}^{5} (5 - x)\, dx$

Graph $y = 5 - x$.

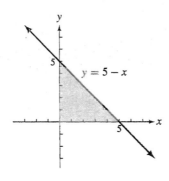

$\displaystyle\int_{0}^{5} (5 - x)\, dx$ is the area of a triangle with base $= 5 - 0 = 5$ and altitude $= 5$.

$$\text{Area} = \frac{1}{2}(\text{altitude})(\text{base}) = \frac{1}{2}(5)(5) = 12.5$$

17. (a) $\displaystyle\int_{0}^{2} f(x)dx$ is the area of a rectangle with width $x = 2$ and length $y = 4$. The rectangle has area $2 \cdot 4 = 8$.

$\displaystyle\int_{2}^{6} f(x)dx$ is the area of one-fourth of a circle that has radius 4. The area is $\frac{1}{4}\pi r^2 = \frac{1}{4}\pi(4)^2 = 4\pi$.

Therefore, $\displaystyle\int_{2}^{6} f(x)dx = 8 + 4\pi$.

(b) $\displaystyle\int_{0}^{2} f(x)dx$ is the area of one-fourth of a circle that has radius 2. The area is $\frac{1}{4}\pi r^2 = \frac{1}{4}\pi(2)^2 = \pi$.

$\displaystyle\int_{2}^{6} f(x)dx$ is the area of a triangle with base 4 and height 2. The triangle has area $\frac{1}{2} \cdot 4 \cdot 2 = 4$.

Therefore, $\displaystyle\int_{0}^{6} f(x)dx = 4 + \pi$.

19. $\displaystyle\int_{-4}^{0} \sqrt{16 - x^2}\, dx$

Graph $y = \sqrt{16 - x^2}$.

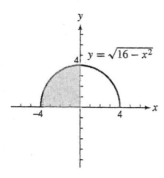

$\displaystyle\int_{-4}^{0} \sqrt{16 - x^2}\, dx$ is the area of the portion of the circle in the second quadrant, which is one-fourth of a circle. The circle has radius 4.

$$\text{Area} = \frac{1}{4}\pi r^2 = \frac{1}{4}\pi(4)^2 = 4\pi$$

21. $\displaystyle\int_{2}^{5} (1 + 2x)\, dx$

Graph $y = 1 + 2x$.

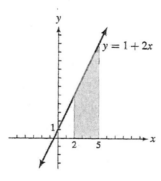

$\displaystyle\int_{2}^{5} (1 + 2x)\, dx$ is the area of the trapezoid with $B = 11$, $b = 5$, and $h = 3$. The formula for the area is

$$A = \frac{1}{2}(B + b)h,$$

so we have

$$A = \frac{1}{2}(11 + 5)(3) = 24.$$

23. (a) With $n = 10$, $\Delta x = \frac{1-0}{10} = 0.1$, and $x_1 = 0 + 0.1 = 0.1$, use the command seq(X^3,X, 0.1, 0.1, 0.1) →L1. The resulting screen is:

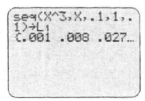

(b) Since $\sum\limits_{i=1}^{n} f(x_i)\Delta x = \Delta x \left(\sum\limits_{i=1}^{n} f(x_i) \right)$, use the command 0.1*sum(L1) to approximate $\int_0^1 x^3 dx$. The resulting screen is:

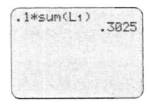

$$\int_0^1 x^3 dx \approx 0.3025$$

(c) With $n = 100$, $\Delta x = \frac{1-0}{100} = 0.01$, and $x_1 = 0 + 0.01 = 0.01$, use the command seq(X^3,X, 0.01, 0.1, 0.01) →L1. The resulting screen is:

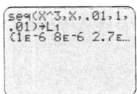

Use the command 0.01*sum(L1) to approximate $\int_0^1 x^3 dx$. The resulting screen is:

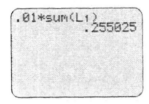

$$\int_0^1 x^3 dx \approx 0.255025$$

(d) With $n = 500$, $\Delta x = \frac{1-0}{500} = 0.002$, and $x_1 = 0 + 0.002 = 0.002$, use the command seq (X^3,X, 0.002, 1, 0.002) →L1. The resulting screen is:

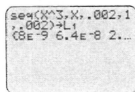

Use the command 0.002*sum(L1) to approximate $\int_0^1 x^3\,dx$. The resulting screen is:

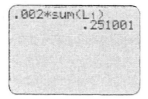

$$\int_0^1 x^3\,dx \approx 0.251001$$

(e) As n gets larger the approximation for $\int_0^1 x^3\,dx$ seems to be approaching 0.25 or $\frac{1}{4}$. We estimate $\int_0^1 x^3\,dx = \frac{1}{4}$.

For Exercises 25–33, readings on the graphs and answers may vary.

25. Left endpoints:

Read values of the function on the graph every 5 years from 1980 to 2000. These values give us the heights of 5 rectangles. The width of each rectangle is $\Delta x = 5$. We estimate the area under the curve as

$$A = \sum_{i=1}^{5} f(x_i)\Delta x = 702.7(5) + 818(5) + 902.9(5) + 962.1(5) + 1084.1(5) = 22{,}349$$

Right endpoints:

Read values of the function from the graph every 5 years from 1985 to 2005. These values give the heights of 5 rectangles. The width of each rectangle is $\Delta x = 5$. We estimate the area under the curve as

$$A = \sum_{i=1}^{5} f(x_i)\Delta x = 818(5) + 902.9(5) + 962.1(5) + 1084.1(5) + 1128.3(5) = 24{,}477$$

Average:

$$\frac{22{,}349 + 24{,}477}{2} = \frac{46{,}826}{2} = 23{,}413$$

The area under the curve represents the total U.S. coal consumption. We estimate this consumption as about 23,413 million short tons.

27. (a) Left endpoints:

Read values of the function from the graph for every 14 days from 18 Feb. through 30 Apr. The values give the heights of 6 rectangles. The width of each rectangle is $\Delta x = 14$. We estimate the area under the curve as

$$A = \sum_{i=1}^{6} f(x_i)\Delta x = 0(14) + 15(14) + 33(14) + 40(14) + 16(14) + 5(14) = 1526.$$

Right endpoints:

Read values of the function from the graph for every 14 days from 4 Mar. through 13 May. Now we estimate the area under the curve as

$$A = \sum_{i=1}^{6} f(x_i)\Delta x = 15(14) + 33(14) + 40(14) + 16(14) + 5(14) + 1(14) = 1540.$$

Average:

$$\frac{1526 + 1540}{2} = 1533$$

There were about 1533 cases of the disease.

(b) Left endpoints:

Read values of the function from the graph for every 14 days from 18 Feb. through 30 Apr. The values give the heights of 6 rectangles. The width of each rectangle is $\Delta x = 14$. We estimate the area under the curve as

$$A = \sum_{i=1}^{6} f(x_i)\Delta x = 0(14) + 10(14) + 15(14) + 10(14) + 3(14) + 1(14) = 546.$$

Right endpoints:

Read values of the function from the graph for every 14 days from 4 Mar. through 13 May. Now we estimate the area under the curve as

$$A = \sum_{i=1}^{6} f(x_i)\Delta x = 10(14) + 15(14) + 10(14) + 3(14) + 1(14) + 1(14) = 560.$$

Average:

$$\frac{546 + 560}{2} = 553$$

There would have been about 553 cases of the disease.

29. Read the value of the function for every 5 sec from $x = 2.5$ to $x = 12.5$. These are the midpoints of rectangles with width $\Delta x = 5$. Then read the function for $x = 17$, which is the midpoint of a rectangle with width $\Delta x = 4$.

$$\sum_{i=1}^{4} f(x_i)\Delta x \approx 36(5) + 63(5) + 84(5) + 95(4) \approx 1295$$

$$\frac{1295}{3600}(5280) \approx 1900$$

The Porsche 928 traveled about 1900 ft.

31. Left endpoints:

Read values of the function from the table for every number of seconds from 2.0 to 19.3. These values give the heights of 10 rectangles. The width of each rectangle varies. We estimate the area under the curve as

$$\sum_{i=1}^{10} f(x_i)\Delta x$$

$$= 30(2.9 - 2.0) + 40(4.1 - 2.9) + 50(5.3 - 4.1) + 60(6.9 - 5.3) + 70(8.7 - 6.9) + 80(10.7 - 8.7)$$
$$+ 90(13.2 - 10.7) + 100(16.1 - 13.2) + 110(19.3 - 16.1) + 120(23.4 - 19.3)$$
$$= 1876$$

$$\frac{5280}{3600}(1876) \approx 2751$$

Right endpoints:

Read values of the function from the table for every number of seconds from 2.0 to 23.4. These values give the heights of 11 rectangles. The width of each rectangle varies. We estimate the area under the curve as

$$\sum_{i=1}^{11} f(x_i)\Delta x$$

$$= 30(2.0 - 0) + 40(2.9 - 2.0) + 50(4.1 - 2.9) + 60(5.3 - 4.1) + 70(6.9 - 5.3) + 80(8.7 - 6.9)$$
$$+ 90(10.7 - 8.7) + 100(13.2 - 10.7) + 110(16.1 - 13.2) + 120(19.3 - 16.1) + 130(23.4 - 19.3)$$
$$= 2150$$

$$\frac{5280}{3600}(2150) \approx 3153$$

Average:

$$\frac{2751 + 3153}{2} = \frac{5904}{2}$$
$$= 2952$$

The distance traveled by the Mercedes-Benz S550 is about 2952 ft.

33. (a) Read values of the function on the plain glass graph every 2 hr from 6 to 6. These are at midpoints of the widths $\Delta x = 2$ and represent the heights of the rectangles.

$$f(x_i)\Delta x = 132(2) + 215(2) + 150(2) + 44(2) + 34(2) + 26(2) + 12(2) \approx 1226$$

The total heat gain was about 1230 BTUs per square foot.

(b) Read values on the ShadeScreen graph every 2 hr from 6 to 6.

$$\sum f(x_i)\Delta x = 38(2) + 25(2) + 16(2) + 12(2) + 10(2) + 10(2) + 5(2) \approx 232$$

The total heat gain was about 230 BTUs per square foot.

35. (a) The area of a trapezoid is

$$A = \frac{1}{2}h(b_1 + b_2) = \frac{1}{2}(6)(1 + 2) = 9.$$

Car A has traveled 9 ft.

(b) Car A is furthest ahead of car B at 2 sec. Notice that from $t = 0$ to $t = 2$, $v(t)$ is larger for car A than for car B. For $t > 2$, $v(t)$ is larger for car B than for car A.

(c) As seen in part (a), car A drove 9 ft after 2 sec. The distance of car B can be calculated as follows:

$$\frac{2 - 0}{4} = \frac{1}{2} = \text{width}$$

$$\text{Distance} = \frac{1}{2} \cdot v(0.25) + \frac{1}{2}v(0.75) + \frac{1}{2}v(1.25) + \frac{1}{2}v(1.75) = \frac{1}{2}(0.2) + \frac{1}{2}(1) + \frac{1}{2}(2.6) + \frac{1}{2}(5) = 4.4$$

$$9 - 4.4 = 4.6$$

The furthest car A can get ahead of car B is about 4.6 ft.

(d) At $t = 3$, car A travels $\frac{1}{2}(6)(2 + 3) = 15$ ft and car B travels approximately 13 ft. At $t = 3.5$, car A travels $\frac{1}{2}(6)(2.5 + 3.5) = 18$ ft and car B travels approximately 18.25 ft. Therefore, car B catches up with car A between 3 and 3.5 sec.

37. Using the left endpoints:

$$\text{Distance} = v_0(1) + v_1(1) + v_2(1) = 10 + 6.5 + 6 = 22.5 \text{ ft}$$

Using the right endpoints:

$$\text{Distance} = v_1(1) + v_2(1) + v_3(1) = 6.5 + 6 + 5.5 = 18 \text{ ft}$$

39. **(a)** Read values from the graph for every hour from 1 A.M. through 11 P.M. The values give the heights of 23 rectangles. The width of each rectangle is $\Delta x = 1$. We estimate the area under the curve as

$$A = \sum_{i=1}^{23} f(x_i)\Delta x$$

$$\begin{aligned}
= &\; 500(1) + 550(1) + 800(1) + 1600(1) + 4000(1) + 7000(1) + 7000(1) + 5900(1) + 4500(1) \\
&+ 3500(1) + 3100(1) + 3100(1) + 3500(1) + 3800(1) + 4100(1) + 4800(1) + 4750(1) \\
&+ 4000(1) + 2500(1) + 2250(1) + 1800(1) + 1500(1) + 1050(1) \\
= &\; 75,600
\end{aligned}$$

(b) Read values from the graph for every hour from 1 A.M. through 11 P.M. The values give the heights of 23 rectangles. The width of each rectangle is $\Delta x = 1$. We estimate the area under the curve as

$$A = \sum_{i=1}^{23} f(x_i)\Delta x$$

$$\begin{aligned}
= &\; 500(1) + 400(1) + 400(1) + 700(1) + 1500(1) + 3000(1) + 4100(1) + 3900(1) + 3200(1) \\
&+ 3600(1) + 4000(1) + 4000(1) + 4300(1) + 5200(1) + 6000(1) + 6500(1) + 6400(1) \\
&+ 6000(1) + 4700(1) + 3100(1) + 2600(1) + 1900(1) + 1300(1) \\
= &\; 77,300
\end{aligned}$$

15.4   The Fundamental Theorem of Calculus

1. $\displaystyle\int_{-2}^{4}(-3)\,dp = -3\int_{-2}^{4}dp = -3\cdot p\big|_{-2}^{4} = -3[4-(-2)] = -18$

3. $\displaystyle\int_{-1}^{2}(5t-3)\,dt = 5\int_{-1}^{2}t\,dt - 3\int_{-1}^{2}dt$

$$= \frac{5}{2}t^2\bigg|_{-1}^{2} - 3t\bigg|_{-1}^{2}$$

$$= \frac{5}{2}[2^2 - (-1)^2] - 3[2-(-1)]$$

$$= \frac{5}{2}(4-1) - 3(2+1)$$

$$= \frac{15}{2} - 9$$

$$= \frac{15}{2} - \frac{18}{2}$$

$$= -\frac{3}{2}$$

5. $\displaystyle\int_0^2 (5x^2 - 4x + 2)\,dx$

$$= 5\int_0^2 x^2\,dx - 4\int_0^2 x\,dx + 2\int_0^2 dx$$

$$= \frac{5x^3}{3}\Big|_0^2 - 2x^2\Big|_0^2 + 2x\Big|_0^2$$

$$= \frac{5}{3}(2^3 - 0^3) - 2(2^2 - 0^2) + 2(2 - 0)$$

$$= \frac{5}{3}(8) - 2(4) + 2(2)$$

$$= \frac{40 - 24 + 12}{3} = \frac{28}{3}$$

7. $\displaystyle\int_0^2 3\sqrt{4u + 1}\,du$

Let $4u + 1 = x$, so that $4\,du = dx$.

When $u = 0$, $x = 4(0) + 1 = 1$.
When $u = 2$, $x = 4(2) + 1 = 9$.

$$\int_0^2 3\sqrt{4u + 1}\,du$$

$$= \frac{3}{4}\int_0^2 \sqrt{4u + 1}\,(4\,du)$$

$$= \frac{3}{4}\int_1^9 x^{1/2}\,dx$$

$$= \frac{3}{4}\cdot\frac{x^{3/2}}{3/2}\Big|_1^9$$

$$= \frac{3}{4}\cdot\frac{2}{3}(9^{3/2} - 1^{3/2})$$

$$= \frac{1}{2}(27 - 1) = \frac{26}{2}$$

$$= 13$$

9. $\displaystyle\int_0^4 2(t^{1/2} - t)\,dt = 2\int_0^4 t^{1/2}\,dt - 2\int_0^4 t\,dt$

$$= 2\cdot\frac{t^{3/2}}{\frac{3}{2}}\Big|_0^4 - 2\cdot\frac{t^2}{2}\Big|_0^4$$

$$= \frac{4}{3}(4^{3/2} - 0^{3/2}) - (4^2 - 0^2)$$

$$= \frac{32}{3} - 16$$

$$= -\frac{16}{3}$$

11. $\displaystyle\int_1^4 (5y\sqrt{y} + 3\sqrt{y})\,dy$

$$= 5\int_1^4 y^{3/2}\,dy + 3\int_1^4 y^{1/2}\,dy$$

$$= 5\left(\frac{y^{5/2}}{\frac{5}{2}}\right)\Big|_1^4 + 3\left(\frac{y^{3/2}}{\frac{3}{2}}\right)\Big|_1^4$$

$$= 2y^{5/2}\Big|_1^4 + 2y^{3/2}\Big|_1^4$$

$$= 2(4^{5/2} - 1) + 2(4^{3/2} - 1)$$
$$= 2(32 - 1) + 2(8 - 1)$$
$$= 62 + 14$$
$$= 76$$

13. $\displaystyle\int_4^6 \frac{2}{(2x - 7)^2}\,dx$

Let $u = 2x - 7$, so that $du = 2\,dx$.

When $x = 6$, $u = 2\cdot 6 - 7 = 5$.
When $x = 4$, $u = 2\cdot 4 - 7 = 1$.

$$\int_4^6 \frac{2}{(2x - 7)^2}\,dx = \int_1^5 u^{-2}\,du$$

$$= \frac{u^{-1}}{-1}\Big|_1^5$$

$$= -u^{-1}\Big|_1^5$$

$$= -\left(\frac{1}{5} - 1\right)$$

$$= -\left(-\frac{4}{5}\right)$$

$$= \frac{4}{5}$$

15. $\displaystyle\int_1^5 (6n^{-2} - n^{-3})\,dn = 6\int_1^5 n^{-2}\,dn - \int_1^5 n^{-3}\,dn$

$$= 6\cdot\frac{n^{-1}}{-1}\Big|_1^5 - \frac{n^{-2}}{-2}\Big|_1^5$$

$$= \frac{-6}{n}\Big|_1^5 + \frac{1}{2n^2}\Big|_1^5$$

$$= \frac{-6}{5} - \left(\frac{-6}{1}\right) + \left[\frac{1}{2(25)} - \frac{1}{2(1)}\right]$$

$$= \frac{-6}{5} + \frac{6}{1} + \frac{1}{50} - \frac{1}{2}$$

$$= \frac{108}{25}$$

17. $\int_{-3}^{-2} \left(2e^{-0.1y} + \dfrac{3}{y}\right) dy$

$= 2\int_{-3}^{-2} e^{-0.1y}\,dy + \int_{-3}^{-2} \dfrac{3}{y}\,dy$

$= 2 \cdot \left.\dfrac{e^{-0.1y}}{-0.1}\right|_{-3}^{-2} + 3\ln|y|\,\bigg|_{-3}^{-2}$

$= -20e^{-0.1y}\,\big|_{-3}^{-2} + 3\ln|y|\,\bigg|_{-3}^{-2}$

$= 20e^{0.3} - 20e^{0.2} + 3\ln 2 - 3\ln 3$

≈ 1.353

19. $\int_{1}^{2} \left(e^{4u} - \dfrac{1}{(u+1)^2}\right) du$

$= \int_{1}^{2} e^{4u}\,du - \int_{1}^{2} \dfrac{1}{(u+1)^2}\,du$

$= \left.\dfrac{e^{4u}}{4}\right|_{1}^{2} - \left.\dfrac{-1}{u+1}\right|_{1}^{2}$

$= \dfrac{e^8}{4} - \dfrac{e^4}{4} + \dfrac{1}{2+1} - \dfrac{1}{1+1}$

$= \dfrac{e^8}{4} - \dfrac{e^4}{4} - \dfrac{1}{6}$

≈ 731.4

21. $\int_{-1}^{0} y(2y^2 - 3)^5\,dy$

Let $u = 2y^2 - 3$, so that

$$du = 4y\,dy \text{ and } \dfrac{1}{4}\,du = y\,dy.$$

When $y = -1$, $u = 2(-1)^2 - 3 = -1$.
When $y = 0$, $u = 2(0)^2 - 3 = -3$.

$\dfrac{1}{4}\int_{-1}^{-3} u^5\,du = \dfrac{1}{4} \cdot \left.\dfrac{u^6}{6}\right|_{-1}^{-3}$

$= \left.\dfrac{1}{24} u^6\right|_{-1}^{-3}$

$= \dfrac{1}{24}(-3)^6 - \dfrac{1}{24}(-1)^6$

$= \dfrac{729}{24} - \dfrac{1}{24}$

$= \dfrac{728}{24}$

$= \dfrac{91}{3}$

23. $\int_{1}^{64} \dfrac{\sqrt{z} - 2}{\sqrt[3]{z}}\,dz$

$= \int_{1}^{64} \left(\dfrac{z^{1/2}}{z^{1/3}} - 2z^{-1/3}\right) dz$

$= \int_{1}^{64} z^{1/6}\,dz - 2\int_{1}^{64} z^{-1/3}\,dz$

$= \left.\dfrac{z^{7/6}}{\tfrac{7}{6}}\right|_{1}^{64} - 2\left.\dfrac{z^{2/3}}{\tfrac{2}{3}}\right|_{1}^{64}$

$= \left.\dfrac{6z^{7/6}}{7}\right|_{1}^{64} - 3z^{2/3}\,\bigg|_{1}^{64}$

$= \dfrac{6(64)^{7/6}}{7} - \dfrac{6(1)^{7/6}}{7}$

$\qquad - 3(64^{2/3} - 1^{2/3})$

$= \dfrac{6(128)}{7} - \dfrac{6}{7} - 3(16 - 1)$

$= \dfrac{768 - 6 - 315}{7}$

$= \dfrac{447}{7} \approx 63.86$

25. $\int_{1}^{2} \dfrac{\ln x}{x}\,dx$

Let $u = \ln x$, so that

$$du = \dfrac{1}{x}\,dx.$$

When $x = 1$, $u = \ln 1 = 0$.
When $x = 2$, $u = \ln 2$.

$\int_{0}^{\ln 2} u\,du = \left.\dfrac{u^2}{2}\right|_{0}^{\ln 2}$

$= \dfrac{(\ln 2)^2}{2} - 0$

$= \dfrac{(\ln 2)^2}{2}$

≈ 0.2402

27. $\int_{0}^{8} x^{1/3}\sqrt{x^{4/3} + 9}\,dx$

Let $u = x^{4/3} + 9$, so that

$$du = \dfrac{4}{3}x^{1/3}\,dx \text{ and } \dfrac{3}{4}\,du = x^{1/3}\,dx.$$

When $x = 0$, $u = 0^{4/3} + 9 = 9$.
When $x = 8$, $u = 8^{4/3} + 9 = 25$.

$$\frac{3}{4}\int_{9}^{25}\sqrt{u}\,du = \frac{3}{4}\int_{9}^{25}u^{1/2}\,du$$

$$= \frac{3}{4}\cdot\frac{u^{3/2}}{\frac{3}{2}}\Big|_{9}^{25}$$

$$= \frac{1}{2}u^{3/2}\Big|_{9}^{25}$$

$$= \frac{1}{2}(25)^{3/2} - \frac{1}{2}(9)^{3/2}$$

$$= \frac{125}{2} - \frac{27}{2}$$

$$= 49$$

29. $\displaystyle\int_{0}^{1}\frac{e^{2t}}{(3+e^{2t})^2}\,dt$

Let $u = 3 + e^{2t}$, so that $du = 2e^{2t}\,dt$.

When $x = 1, u = 3 + e^{2\cdot 1} = 3 + e^2$.
When $x = 0, u = 3 + e^{2\cdot 0} = 4$.

$$\int_{0}^{1}\frac{e^{2t}}{(3+e^{2t})^2}\,dt = \frac{1}{2}\int_{4}^{3+e^2}u^{-2}\,du$$

$$= \frac{1}{2}\cdot\frac{u^{-1}}{-1}\Big|_{4}^{3+e^2}$$

$$= \frac{-1}{2u}\Big|_{4}^{3+e^2}$$

$$= \frac{1}{8} - \frac{1}{2(3+e^2)}$$

$$\approx 0.07687$$

31. $f(x) = 2x - 14;\ [6, 10]$

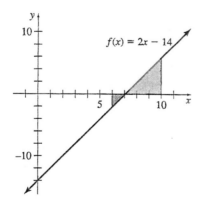

The graph crosses the x-axis at

$$0 = 2x - 14$$
$$2x = 14$$
$$x = 7.$$

This location is in the interval. The area of the region is

$$\left|\int_{6}^{7}(2x-14)dx\right| + \int_{7}^{10}(2x-14)dx$$

$$= \left|(x^2 - 14x)\big|_{6}^{7}\right| + (x^2 - 14x)\big|_{7}^{10}$$

$$= \left|(7^2 - 98) - (6^2 - 84)\right|$$
$$\quad + (10^2 - 140) - (7^2 - 98)$$

$$= |-1| + (-40) - (-49)$$

$$= 10.$$

33. $f(x) = 2 - 2x^2;\ [0, 5]$

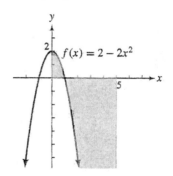

Find the points where the graph crosses the x-axis by solving $2 - 2x^2 = 0$.

$$2 - 2x^2 = 0$$
$$2x^2 = 2$$
$$x^2 = 1$$
$$x = \pm 1.$$

The only solution in the interval $[0, 5]$ is 1. The total area is

$$\int_{0}^{1}\left(2 - 2x^2\right)dx + \left|\int_{1}^{5}\left(2 - 2x^2\right)dx\right|$$

$$= \left(2x - \frac{2x^3}{3}\right)\Big|_{0}^{1} + \left|\left(2x - \frac{2x^3}{3}\right)\Big|_{1}^{5}\right|$$

$$= 2 - \frac{2}{3} + \left|10 - \frac{2(5^3)}{3} - 2 + \frac{2}{3}\right|$$

$$= \frac{4}{3} + \left|\frac{-224}{3}\right|$$

$$= \frac{228}{3}$$

$$= 76.$$

35. $f(x) = x^3;\ [-1, 3]$

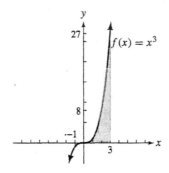

The solution

$$x^3 = 0$$
$$x = 0$$

indicates that the graph crosses the x-axis at 0 in the given interval $[-1, 3]$.

The total area is

$$\left| \int_{-1}^{0} x^3\, dx \right| + \int_{0}^{3} x^3\, dx$$

$$= \left| \frac{x^4}{4} \Big|_{-1}^{0} \right| + \left| \frac{x^4}{4} \Big|_{0}^{3} \right|$$

$$= \left| \left(0 - \frac{1}{4} \right) \right| + \left(\frac{3^4}{4} - 0 \right)$$

$$= \frac{1}{4} + \frac{81}{4} = \frac{82}{4}$$

$$= \frac{41}{2}.$$

37. $f(x) = e^x - 1;\ [-1, 2]$

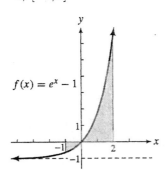

Solve

$$e^x - 1 = 0.$$
$$e^x = 1$$
$$x \ln e = \ln 1$$
$$x = 0$$

The graph crosses the x-axis at 0 in the given interval $[-1, 2]$.

The total area is

$$\left| \int_{-1}^{0} (e^x - 1)\, dx \right| + \int_{0}^{2} (e^x - 1)\, dx$$

$$= \left| (e^x - x) \Big|_{-1}^{0} \right| + (e^x - x) \Big|_{0}^{2}$$

$$= |(1 - 0) - (e^{-1} + 1)| \\ + (e^2 - 2) - (1 - 0)$$
$$= |1 - e^{-1} - 1| + e^2 - 2 - 1$$

$$= \frac{1}{e} + e^2 - 3$$

$$\approx 4.757.$$

39. $f(x) = \dfrac{1}{x} - \dfrac{1}{e};\ [1, e^2]$

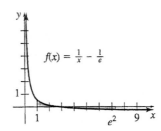

The graph crosses the x-axis at

$$0 = \frac{1}{x} - \frac{1}{e}$$
$$\frac{1}{x} = \frac{1}{e}$$
$$x = e.$$

This location is in the interval. The area of the region is

$$\int_{1}^{e} \left(\frac{1}{x} - \frac{1}{e} \right) dx + \left| \int_{e}^{e^2} \left(\frac{1}{x} - \frac{1}{e} \right) dx \right|$$

$$= \left| \ln |x| - \frac{x}{e} \Big|_{1}^{e} \right| + \left| \left(\ln |x| - \frac{x}{e} \right) \Big|_{e}^{e^2} \right|$$

$$= 0 - \left(-\frac{1}{e} \right) + |(2 - e) - 0|$$

$$= \frac{1}{e} + |2 - e|$$

$$= e - 2 + \frac{1}{e}.$$

41. $y = 4 - x^2$; $[0, 3]$

From the graph, we see that the total area is

$$\int_0^2 (4 - x^2)\, dx + \left| \int_2^3 (4 - x^2)\, dx \right|$$

$$= \left(4x - \frac{x^3}{3} \right) \Big|_0^2 + \left| \left(4x - \frac{x^3}{3} \right) \Big|_2^3 \right|$$

$$= \left[\left(8 - \frac{8}{3} \right) - 0 \right]$$

$$\quad + \left| \left[(12 - 9) - \left(8 - \frac{8}{3} \right) \right] \right|$$

$$= \frac{16}{3} + \left| 3 - \frac{16}{3} \right|$$

$$= \frac{16}{3} + \frac{7}{3}$$

$$= \frac{23}{3}$$

43. $y = e^x - e$; $[0, 2]$

From the graph, we see that the total area is

$$\left| \int_0^1 (e^x - e)\, dx \right| + \int_1^2 (e^x - e)\, dx$$

$$= \left| (e^x - xe) \Big|_0^1 \right| + (e^x - xe) \Big|_1^2$$

$$= \left| (e^1 - e) - (e^0 + 0) \right|$$
$$\quad + (e^2 - 2e) - (e^1 - e)$$
$$= |-1| + e^2 - 2e$$
$$= 1 + e^2 - 2e$$
$$\approx 2.952.$$

45. $\displaystyle\int_a^c f(x)\, dx = \int_a^b f(x)\, dx + \int_b^c f(x)\, dx$

47. $\displaystyle\int_0^{16} f(x)\, dx = \int_0^2 f(x)\, dx + \int_2^5 f(x)\, dx$

$$\quad + \int_5^8 f(x)\, dx + \int_8^{16} f(x)\, dx$$

$$= \frac{1}{2} \cdot 2(1 + 3) + \frac{\pi(3^2)}{4} - \frac{\pi(3^2)}{4} - \frac{1}{2}(3)(8)$$

$$= 4 + \frac{9}{4}\pi - \frac{9}{4}\pi - 12$$

$$= -8$$

49. Prove: $\displaystyle\int_a^b f(x)\, dx = \int_a^c f(x)\, dx + \int_c^b f(x)\, dx$.

Let $F(x)$ be an antiderivative of $f(x)$.

$$\int_a^c f(x)\, dx + \int_c^b f(x)\, dx$$

$$= F(x) \Big|_a^c + F(x) \Big|_c^b$$

$$= [F(c) - F(a)] + [F(b) - F(c)]$$
$$= F(c) - F(a) + F(b) - F(c)$$
$$= F(b) - F(a)$$

$$= \int_a^b f(x)\, dx$$

51. $\displaystyle\int_{-1}^4 f(x)\, dx$

$$= \int_{-1}^0 (2x + 3)\, dx \int_0^4 \left(-\frac{x}{4} - 3 \right) dx$$

$$= (x^2 + 3x) \Big|_{-1}^0 + \left(-\frac{x^2}{8} - 3x \right) \Big|_0^4$$

$$= -(1 - 3) + (-2 - 12)$$
$$= 2 - 14$$
$$= -12$$

53. (a) $g(t) = t^4$ and $c = 1$, use substitution.

$$f(x) = \int_c^x g(t)\, dt$$

$$= \int_1^x t^4\, dt$$

$$= \frac{t^5}{5} \Big|_1^x$$

$$= \frac{x^5}{5} - \frac{(1)^5}{5}$$

$$= \frac{x^5}{5} - \frac{1}{5}$$

(b) $f'(x) = \dfrac{d}{dx}(f(x))$

$$= \frac{d}{dx}\left(\frac{x^5}{5} - \frac{1}{5} \right)$$

$$= \frac{1}{5} \cdot \frac{d}{dx}(x^5) - \frac{d}{dx}\left(\frac{1}{5} \right)$$

$$= \frac{1}{5} \cdot 5x^4 - 0$$

$$= x^4$$

Since $g(t) = t^4$, then $g(x) = x^4$ and we see $f'(x) = g(x)$.

(c) Let $g(t) = e^{t^2}$ and $c = 0$, then $f(x) = \int_0^x e^{t^2} dt$.

$$f(1) = \int_0^1 e^{t^2} dt \text{ and } f(1.01) = \int_0^{1.01} e^{t^2} dt.$$

Use the fnInt command in the Math menu of your calculator to find $\int_0^1 e^{x^2} dx$ and $\int_0^{1.01} e^{x^2} dx$. The resulting screens are:

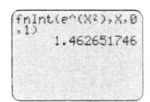

$f(1) \approx 1.46265$

$f(1.01) \approx 1.49011$

Use $\dfrac{f(1+h) - f(1)}{h}$ to approximate $f'(1)$ with $h = 0.01$

$$\frac{f(1+h) - f(1)}{h} = \frac{f(1.01) - f(1)}{0.01}$$
$$\approx \frac{1.49011 - 1.46265}{0.01}$$
$$= 2.746$$

So $f'(1) \approx 2.746$, and $g(1) = e^{1^2} = e \approx 2.718$.

55. $P'(t) = (3t + 3)(t^2 + 2t + 2)^{1/3}$

(a) $\displaystyle\int_0^3 3(t+1)(t^2 + 2t + 2)^{1/3} dt$

Let $u = t^2 + 2t + 2$, so that

$$du = (2t + 2) dt \text{ and } \frac{1}{2} du = (t+1) dt.$$

When $t = 0$, $u = 0^2 + 2 \cdot 0 + 2 = 2$.

When $t = 3$, $u = 3^2 + 2 \cdot 3 + 2 = 17$.

$$\frac{3}{2} \int_2^{17} u^{1/3} du = \frac{3}{2} \cdot \frac{u^{4/3}}{\frac{4}{3}} \Big|_2^{17}$$
$$= \frac{9}{8} u^{4/3} \Big|_2^{17}$$
$$= \frac{9}{8}(17)^{4/3} - \frac{9}{8}(2)^{4/3}$$
$$\approx 46.341$$

Total profits for the first 3 yr were

$$\frac{9000}{8}(17^{4/3} - 2^{4/3}) \approx \$46,341.$$

(b) $\displaystyle\int_3^4 3(t+1)(t^2 + 2t + 2)^{1/3} dt$

Let $u = t^2 + 2t + 2$, so that

$$du = (2t + 2) dt = 2(t + 1) dt \text{ and}$$
$$\frac{3}{2} du = 3(t + 1) dt.$$

When $t = 3$, $u = 3^2 + 2 \cdot 3 + 2 = 17$.

When $t = 4$, $u = 4^2 + 2 \cdot 4 + 2 = 26$.

$$\frac{3}{2} \int_{17}^{26} u^{1/3} du = \frac{9}{8} u^{4/3} \Big|_{17}^{26}$$
$$= \frac{9}{8}(26)^{4/3} - \frac{9}{8}(17)^{4/3}$$
$$\approx 37.477$$

Profit in the fourth year was

$$\frac{9000}{8}(26^{4/3} - 17^{4/3}) \approx \$37,477.$$

(c) $\displaystyle\lim_{t \to \infty} P'(t)$
$$= \lim_{t \to \infty} (3t + 3)(t^2 + 2t + 2)^{1/3}$$
$$= \infty$$

The annual profit is slowly increasing without bound.

57. $P'(t) = 140t^{5/2}$

$$\int_0^4 140t^{5/2} dt = 140 \cdot \frac{t^{7/2}}{\frac{7}{2}} \Big|_0^4$$
$$= 40t^{7/2} \Big|_0^4$$
$$= 5120$$

Since 5120 is above the total level of acceptable pollution (4850), the factory cannot operate for 4 years without killing all the fish in the lake.

59. Growth rate is $0.6 + \frac{4}{(t+1)^3}$ ft/yr.

(a) Total growth in the second year is

$$\int_1^2 \left[0.6 + \frac{4}{(t+1)^3}\right] dt$$

$$= \left[0.6t + \frac{4}{-2(t+1)^2}\right]\Big|_1^2$$

$$= \left[0.6(2) - \frac{2}{(2+1)^2}\right]$$

$$\quad - \left[0.6(1) - \frac{2}{(1+1)^2}\right]$$

$$= \frac{44}{45} - \frac{1}{10}$$

$$\approx 0.8778 \text{ ft.}$$

(b) Total growth in the third year is

$$\int_2^3 \left[0.6 + \frac{4}{(t+1)^3}\right] dt$$

$$= \left[0.6t + \frac{4}{-2(t+1)^2}\right]\Big|_2^3$$

$$= \left[0.6(3) - \frac{2}{(3+1)^2}\right]$$

$$\quad - \left[0.6(2) - \frac{2}{(2+1)^2}\right]$$

$$= \frac{67}{40} - \frac{44}{45}$$

$$\approx 0.6972 \text{ ft.}$$

61. $R'(t) = \frac{5}{t+1} + \frac{2}{\sqrt{t+1}}$

(a) Total reaction from $t = 1$ to $t = 12$ is

$$\int_1^{12} \left(\frac{5}{t+1} + \frac{2}{\sqrt{t+1}}\right) dt$$

$$= \left[5\ln(t+1) + 4\sqrt{t+1}\,\right]\Big|_1^{12}$$

$$= (5\ln 13 + 4\sqrt{13}) - (5\ln 2 + 4\sqrt{2})$$

$$\approx 18.12.$$

(b) Total reaction from $t = 12$ to $t = 24$ is

$$\int_{12}^{24} \left(\frac{5}{t+1} + \frac{2}{\sqrt{t+1}}\right) dt$$

$$= \left[5\ln(t+1) + 4\sqrt{t+1}\,\right]\Big|_{12}^{24}$$

$$= (5\ln 25 + 4\sqrt{25}) - (5\ln 13 + 4\sqrt{13})$$

$$\approx 8.847.$$

63. (b) $\int_0^{60} n(x)\,dx$

(c) $\int_5^{10} \sqrt{5x+1}\,dx$

Let $u = 5x + 1$. Then $du = 5\,dx$.
When $x = 5$, $u = 26$; when $x = 10$, $u = 51$.

$$\frac{1}{5}\int_{26}^{51} u^{1/2}\,du$$

$$= \frac{1}{5} \cdot \frac{u^{3/2}}{\frac{3}{2}}\Big|_{26}^{51}$$

$$= \frac{2}{15}u^{3/2}\Big|_{26}^{51}$$

$$= \frac{2}{15}\left(51^{3/2} - 26^{3/2}\right)$$

$$\approx 30.89 \text{ million}$$

65. $v = k(R^2 - r^2)$

(a) $Q(R) = \int_0^R 2\pi vr\,dr$

$$= \int_0^R 2\pi k(R^2 - r^2)r\,dr$$

$$= 2\pi k \int_0^R (R^2 r - r^2)\,dr$$

$$= 2\pi k \left(\frac{R^2 r^2}{2} - \frac{r^4}{4}\right)\Big|_0^R$$

$$= 2\pi k \left(\frac{R^4}{2} - \frac{R^4}{4}\right)$$

$$= 2\pi k \left(\frac{R^4}{4}\right)$$

$$= \frac{\pi k R^4}{2}$$

(b) $Q(0.4) = \frac{\pi k(0.4)^4}{2}$

$$= 0.04k \text{ mm/min}$$

67. $E(t) = 753t^{-0.1321}$

(a) Since t is the age of the beagle in years, to convert the formula to days, let $T = 365t$, or $t = \frac{T}{365}$.

$$E(T) = 753\left(\frac{T}{365}\right)^{-0.1321} \approx 1642T^{-0.1321}$$

Now, replace T with t.

$$E(t) = 1642t^{-0.1321}$$

(b) The beagle's age in days after one year is 365 days and after 3 years she is 1095 days old.

$$\int_{365}^{1095} 1642t^{-0.1321}\,dt$$

$$= 1642\,\frac{1}{0.8679}t^{0.8679}\Big|_{365}^{1095}$$

$$\approx 1892(1{,}095^{0.8679} - 365^{0.8679})$$

$$\approx 505{,}155$$

The beagle's total energy requirements are about 505,000 kJ/W$^{0.67}$.

69. (a) $f(x) = 40.1 + 2.03x - 0.741x^2$

$$\int_0^9 (40.1 + 2.03x - 0.741x^2)\,dx$$

$$= \left(40.1x + \frac{2.03}{2}x^2 - \frac{0.741}{3}x^3\right)\Big|_0^9$$

$$= (40.1x + 1.015x^2 - 0.247x^3)\Big|_0^9$$

$$\approx 263$$

This integral shows that the total population aged 0 to 90 was about 263 million.

(b) $\int_{3.5}^{5.5} (40.1 + 2.03x - 0.741x^2)\,dx$

$$= (40.1x + 1.015x^2 - 0.247x^3)\Big|_{3.5}^{5.5}$$

$$\approx 210.1591 - 142.1936$$

$$= 67.9655$$

The number of baby boomers is about 68 million.

71. $c'(t) = ke^{rt}$

(a) $c'(t) = 1.2e^{0.04t}$

(b) The amount of oil that the company will sell in the next ten years is given by the integral $\int_0^{10} 1.2e^{0.04t}\,dt$.

(c) $\int_0^{10} 1.2e^{0.04t}\,dx = \frac{1.2e^{0.04t}}{0.04}\Big|_0^{10}$

$$= 30e^{0.04t}\Big|_0^{10}$$

$$= 30e^{0.4} - 30$$

$$\approx 14.75$$

This represents about 14.75 billion barrels of oil.

(d) $\int_0^T 1.2e^{0.04t}\,dt = 30e^{0.04t}\Big|_0^T$

$$= 30e^{0.04T} - 30$$

Solve

$$20 = 30e^{0.04T} - 30.$$
$$50 = 30e^{0.04T}$$

$$\frac{5}{3} = e^{0.04T}$$

$$\ln\frac{5}{3} = 0.04T\,\ln e$$

$$T = \frac{\ln\frac{5}{3}}{0.04}$$

$$\approx 12.8$$

The oil will last about 12.8 years.

(e) $\int_0^T 1.2e^{0.02t}\,dt = 60e^{0.02t}\Big|_0^T$

$$= 60e^{0.02T} - 60$$

Solve

$$20 = 60e^{0.02T} - 60.$$
$$80 = 60e^{0.02T}$$

$$\frac{4}{3} = e^{0.02T}$$

$$\ln\frac{4}{3} = 0.02T\,\ln e$$

$$T = \frac{\ln\frac{4}{3}}{0.02} \approx 14.4$$

The oil will last about 14.4 years.

15.5   The Area Between Two Curves

1. $x = -2, x = 1, y = 2x^2 + 5, y = 0$

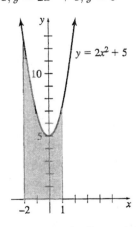

$$\int_{-2}^1 [(2x^2 + 5) - 0] = \left(\frac{2x^3}{3} + 5x\right)\Big|_{-2}^1$$

$$= \left(\frac{2}{3} + 5\right) - \left(-\frac{16}{3} - 10\right)$$

$$= 21$$

3. $x = -3, x = 1, y = x^3 + 1, y = 0$

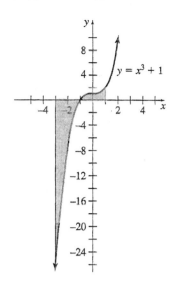

To find the points of intersection of the graphs, substitute for y.

$$x^3 + 1 = 0$$
$$x^3 = -1$$
$$x = -1$$

The region is composed of two separate regions because $y = x^3 + 1$ intersects $y = 0$ at $x = -1$.
Let $f(x) = x^3 + 1$, $g(x) = 0$.
In the interval $[-3, -1]$, $g(x) \geq f(x)$.
In the interval $[-1, 1]$, $f(x) \geq g(x)$.

$$\int_{-3}^{-1} [0 - (x^3 + 1)] dx + \int_{-1}^{1} [(x^3 + 1) - 0] dx$$

$$= \left(\frac{-x^4}{4} - x \right) \Big|_{-3}^{-1} + \left(\frac{x^4}{4} + x \right) \Big|_{-1}^{1}$$

$$= \left(-\frac{1}{4} + 1 \right) - \left(-\frac{81}{4} + 3 \right) + \left(\frac{1}{4} + 1 \right) - \left(\frac{1}{4} - 1 \right)$$

$$= 20$$

5. $x = -2, \ x = 1, \ y = 2x, \ y = x^2 - 3$

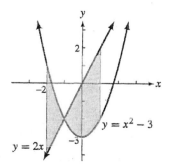

Find the points of intersection of the graphs of $y = 2x$ and $y = x^2 - 3$ by substituting for y.

$$2x = x^2 - 3$$
$$0 = x^2 - 2x - 3$$
$$0 = (x - 3)(x + 1)$$

The only intersection in $[-2, 1]$ is at $x = -1$.
In the interval $[-2, -1]$, $(x^2 - 3) \geq 2x$.
In the interval $[-1, 1]$, $2x \geq (x^2 - 3)$.

$$\int_{-2}^{-1} [(x^2 - 3) - (2x)] dx + \int_{-1}^{1} [(2x) - (x^2 - 3)] dx$$

$$= \int_{-2}^{-1} (x^2 - 3 - 2x) dx$$

$$+ \int_{-1}^{1} (2x - x^2 + 3) dx$$

$$= \left(\frac{x^3}{3} - 3x - x^2 \right) \Big|_{-2}^{-1}$$

$$+ \left(x^2 - \frac{x^3}{3} + 3x \right) \Big|_{-1}^{1}$$

$$= -\frac{1}{3} + 3 - 1 - \left(-\frac{8}{3} + 6 - 4 \right) + 1 - \frac{1}{3} + 3$$

$$- \left(1 + \frac{1}{3} - 3 \right)$$

$$= \frac{5}{3} + 6 = \frac{23}{3}$$

7. $y = x^2 - 30$
$y = 10 - 3x$

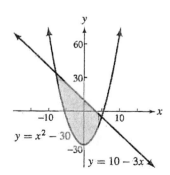

Find the points of intersection.

$$x^2 - 30 = 10 - 3x$$
$$x^2 + 3x - 40 = 0$$
$$(x + 8)(x - 5) = 0$$
$$x = -8 \quad \text{or} \quad x = 5$$

Let $f(x) = 10 - 3x$ and $g(x) = x^2 - 30$.

The area between the curves is given by

$$\int_{-8}^{5} [f(x) - g(x)] \, dx$$

$$= \int_{-8}^{5} [(10 - 3x) - (x^2 - 30)] \, dx$$

$$= \int_{-8}^{5} (-x^2 - 3x + 40) \, dx$$

$$= \left(\frac{-x^3}{3} - \frac{3x^2}{2} + 40x \right) \Big|_{-8}^{5}$$

$$= \frac{-5^3}{3} - \frac{3(5)^2}{2} + 40(5)$$

$$- \left[\frac{-(-8)^3}{3} - \frac{3(-8)^2}{2} + 40(-8) \right]$$

$$= \frac{-125}{3} - \frac{75}{2} + 200 - \frac{512}{3} + \frac{192}{2} + 320$$

$$\approx 366.1667.$$

9. $y = x^2$, $y = 2x$

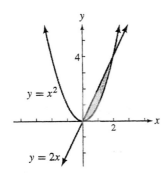

Find the points of intersection.

$$x^2 = 2x$$
$$x^2 - 2x = 0$$
$$x(x - 2) = 0$$
$$x = 0 \quad \text{or} \quad x = 2$$

Let $f(x) = 2x$ and $g(x) = x^2$.
The area between the curves is given by

$$\int_{0}^{2} [f(x) - g(x)] \, dx = \int_{0}^{2} (2x - x^2) \, dx$$

$$= \left(\frac{2x^2}{2} - \frac{x^3}{3} \right) \Big|_{0}^{2}$$

$$= 4 - \frac{8}{3} = \frac{4}{3}.$$

11. $x = 1, x = 6, y = \dfrac{1}{x}, y = \dfrac{1}{2}$

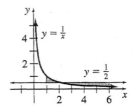

To find the points of intersection of the graphs, substitute for y.

$$\frac{1}{x} = \frac{1}{2}$$
$$x = 2$$

The region is composed of two separate regions because $y = \frac{1}{x}$ intersects $y = \frac{1}{2}$ at $x = 2$.

Let $f(x) = \frac{1}{x}, g(x) = \frac{1}{2}$.

In the interval $[1, 2]$, $f(x) \geq g(x)$.
In the interval $[2, 6]$, $g(x) \geq f(x)$.

$$\int_{1}^{2} \left(\frac{1}{x} - \frac{1}{2} \right) dx + \int_{2}^{6} \left(\frac{1}{2} - \frac{1}{x} \right) dx$$

$$= \left(\ln |x| - \frac{x}{2} \right) \Big|_{1}^{2} + \left(\frac{x}{2} - \ln |x| \right) \Big|_{2}^{6}$$

$$= (\ln 2 - 1) - \left(0 - \frac{1}{2} \right) + (3 - \ln 6) - (1 - \ln 2)$$

$$= 2 \ln 2 - \ln 6 + \frac{3}{2}$$

$$\approx 1.095$$

13. $x = -1, \; x = 1, \; y = e^x, \; y = 3 - e^x$

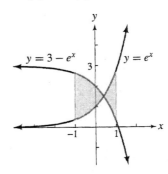

To find the point of intersection, set $e^x = 3 - e^x$ and solve for x.

$$e^x = 3 - e^x$$
$$2e^x = 3$$
$$e^x = \frac{3}{2}$$
$$\ln e^x = \ln \frac{3}{2}$$
$$x \ln e = \ln \frac{3}{2}$$
$$x = \ln \frac{3}{2}$$

The area of the region between the curves from $x = -1$ to $x = 1$ is

$$\int_{-1}^{\ln 3/2} [(3 - e^x) - e^x]\,dx + \int_{\ln 3/2}^{1} [e^x - (3 - e^x)]\,dx$$

$$= \int_{-1}^{\ln 3/2} (3 - 2e^x)\,dx + \int_{\ln 3/2}^{1} (2e^x - 3)\,dx$$

$$= (3x - 2e^x)\Big|_{-1}^{\ln 3/2} + (2e^x - 3x)\Big|_{\ln 3/2}^{1}$$

$$= \left[\left(3 \ln \frac{3}{2} - 2e^{\ln 3/2}\right) - [3(-1) - 2e^{-1}]\right]$$

$$+ \left[2e^1 - 3(1) - \left(2e^{\ln 3/2} - 3 \ln \frac{3}{2}\right)\right]$$

$$= \left[\left(3 \ln \frac{3}{2} - 3\right) - \left(-3 - \frac{2}{e}\right)\right]$$

$$+ \left[2e - 3 - \left(3 - 3 \ln \frac{3}{2}\right)\right]$$

$$= 6 \ln \frac{3}{2} + \frac{2}{e} + 2e - 6 \approx 2.605.$$

15. $x = -1, x = 2, y = 2e^{2x}, y = e^{2x} + 1$

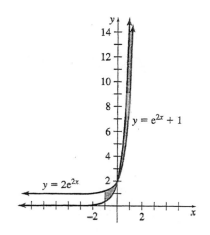

To find the points of intersection of the graphs, substitute for y.

$$2e^{2x} = e^{2x} + 1$$
$$e^{2x} = 1$$
$$2x = 0$$
$$x = 0$$

The region is composed of two separate regions because $y = 2e^{2x}$ intersects $y = e^{2x} + 1$ at $x = 0$. Let $f(x) = 2e^{2x}, g(x) = e^{2x} + 1$.
In the interval $[-1, 0]$, $g(x) \geq f(x)$.
In the interval $[0, 2]$, $f(x) \geq g(x)$.

$$\int_{-1}^{0} (e^{2x} + 1 - 2e^{2x})\,dx + \int_{0}^{2} [2e^{2x} - (e^{2x} + 1)]\,dx$$

$$= \left(-\frac{e^{2x}}{2} + x\right)\Big|_{-1}^{0} + \left(\frac{e^{2x}}{2} - x\right)\Big|_{0}^{2}$$

$$= \left(-\frac{1}{2} + 0\right) - \left(-\frac{e^{-2}}{2} - 1\right) + \left(\frac{e^4}{2} - 2\right) - \left(\frac{1}{2} - 0\right)$$

$$= \frac{e^{-2} + e^4}{2} - 2$$

$$\approx 25.37$$

17. $y = x^3 - x^2 + x + 1, \; y = 2x^2 - x + 1$

Find the points of intersection.

$$x^3 - x^2 + x + 1 = 2x^2 - x + 1$$
$$x^3 - 3x^2 + 2x = 0$$
$$x(x^2 - 3x + 2) = 0$$
$$x(x - 2)(x - 1) = 0$$

The points of intersection are at $x = 0$, $x = 1$, and $x = 2$.

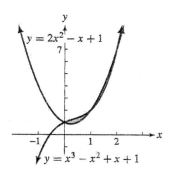

Area between the curves is

$$\int_0^1 [(x^3 - x^2 + x + 1) - (2x^2 - x + 1)]\,dx$$

$$+ \int_1^2 [(2x^2 - x + 1) - (x^3 - x^2 + x + 1)]\,dx$$

$$= \int_0^1 (x^3 - 3x^2 + 2x)\,dx + \int_0^1 (-x^3 + 3x^2 - 2x)\,dx$$

$$= \left(\frac{x^4}{4} - x^3 + x^2\right)\Big|_0^1 + \left(\frac{-x^4}{4} + x^3 - x^2\right)\Big|_1^2$$

$$= \left[\left(\frac{1}{4} - 1 + 1\right) - (0)\right]$$

$$+ \left[(-4 + 8 - 4) - \left(-\frac{1}{4} + 1 - 1\right)\right]$$

$$= \frac{1}{4} + \frac{1}{4}$$

$$= \frac{1}{2}.$$

19. $y = x^4 + \ln(x + 10),$
$y = x^3 + \ln(x + 10)$

Find the points of intersection.

$$x^4 + \ln(x + 10) = x^3 + \ln(x + 10)$$
$$x^4 - x^3 = 0$$
$$x^3(x - 1) = 0$$
$$x = 0 \quad \text{or} \quad x = 1$$

The points of intersection are at $x = 0$ and $x = 1$.
The area between the curves is

$$\int_0^1 [(x^3 + \ln(x + 10)) - (x^4 + \ln(x + 10))]\,dx$$

$$= \int_0^1 (x^3 - x^4)\,dx$$

$$= \left(\frac{x^4}{4} - \frac{x^5}{5}\right)\Big|_0^1$$

$$= \left(\frac{1}{4} - \frac{1}{5}\right) - (0) = \frac{1}{20}.$$

21. $y = x^{4/3}, \; y = 2x^{1/3}$

Find the points of intersection.

$$x^{4/3} = 2x^{1/3}$$
$$x^{4/3} - 2x^{1/3} = 0$$
$$x^{1/3}(x - 2) = 0$$
$$x = 0 \quad \text{or} \quad x = 2$$

The points of intersection are at $x = 0$ and $x = 2$.

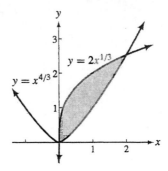

The area between the curves is

$$\int_0^2 (2x^{1/3} - x^{4/3})\,dx = 2\frac{x^{4/3}}{\frac{4}{3}} - \frac{x^{7/3}}{\frac{7}{3}}\Big|_0^2$$

$$= \frac{3}{2}x^{4/3} - \frac{3}{7}x^{7/3}\Big|_0^2$$

$$= \left[\frac{3}{2}(2)^{4/3} - \frac{3}{7}(2)^{7/3}\right] - 0$$

$$= \frac{3(2^{4/3})}{2} - \frac{3(2^{7/3})}{7}$$

$$\approx 1.62.$$

23. $x = 0, x = 3, y = 2e^{3x}, y = e^{3x} + e^6$

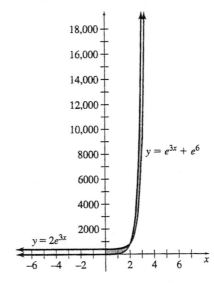

To find the points of intersection of the graphs, substitute for y.

$$2e^{3x} = e^{3x} + e^6$$
$$e^{3x} = e^6$$
$$3x = 6$$
$$x = 2$$

The region is composed of two separate regions because $y = 2e^{3x}$ intersects $y = e^{3x} + e^6$ at $x = 2$.

Let $f(x) = 2e^{3x}$, $g(x) = e^{3x} + e^6$.
In the interval $[0, 2]$, $g(x) \geq f(x)$.
In the interval $[2, 3]$, $f(x) \geq g(x)$.

$$\int_0^2 (e^{3x} + e^6 - 2e^{3x})dx + \int_2^3 [2e^{3x} - (e^{3x} + e^6)]dx$$

$$= \left(-\frac{e^{3x}}{3} + e^6 x \right)\Big|_0^2 + \left(\frac{e^{3x}}{3} - e^6 x \right)\Big|_2^3$$

$$= \left(-\frac{e^6}{3} + 2e^6 \right) - \left(-\frac{1}{3} + 0 \right) + \left(\frac{e^9}{3} - 3e^6 \right) - \left(\frac{e^6}{3} - 2e^6 \right)$$

$$= \frac{e^9 + e^6 + 1}{3}$$

$$\approx 2836$$

25. Graph $y_1 = e^x$ and $y_2 = -x^2 - 2x$ on your graphing calculator. Use the intersect command to find the two intersection points. The resulting screens are:

These screens show that $e^x = -x^2 - 2x$ when $x \approx -1.9241$ and $x \approx -0.4164$.
In the interval $[-1.9241, -0.4164]$,

$$e^x < -x^2 - 2x.$$

The area between the curves is given by

$$\int_{-1.9241}^{-0.4164} [(-x^2 - 2x) - e^x]dx.$$

Use the fnInt command to approximate this definite integral.
The resulting screen is:

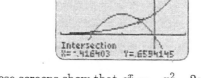

The last screen shows that the area is approximately 0.6650.

27. (a) It is profitable to use the machine until
$S'(x) = C'(x)$.

$$150 - x^2 = x^2 + \frac{11}{4}x$$

$$2x^2 + \frac{11}{4}x - 150 = 0$$

$$8x^2 + 11x - 600 = 0$$

$$x = \frac{-11 \pm \sqrt{121 - 4(8)(-600)}}{16}$$

$$= \frac{-11 \pm 139}{16}$$

$$x = 8 \quad \text{or} \quad x = -9.375$$

It will be profitable to use this machine for 8 years. Reject the negative solution.

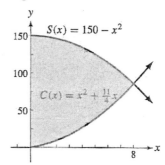

(b) Since $150 - x^2 > x^2 + \frac{11}{4}x$, in the interval $[0, 8]$, the net total savings in the first year are

$$\int_0^1 \left[(150 - x^2) - \left(x^2 + \frac{11}{4}x \right) \right] dx$$

$$= \int_0^1 \left(-2x^2 - \frac{11}{4}x + 150 \right) dx$$

$$= \left(\frac{-2x^3}{3} - \frac{11x^2}{8} + 150x \right)\Big|_0^1$$

$$= -\frac{2}{3} - \frac{11}{8} + 150$$

$$\approx \$148.$$

(c) The net total savings over the entire period of use are

$$\int_0^8 \left[(150 - x^2) - \left(x^2 + \frac{11}{4}x \right) \right] dx$$

$$= \left(\frac{-2x^3}{3} - \frac{11x^2}{8} + 150x \right)\Big|_0^8$$

$$= \frac{-2(8^3)}{3} - \frac{11(8^2)}{8} + 150(8)$$

$$= \frac{-1024}{3} - \frac{704}{8} + 1200$$

$$\approx \$771.$$

29. (a) $E'(x) = e^{0.1x}$ and $I'(x) = 98.8 - e^{0.1x}$

To find the point of intersection, where profit will be maximized, set the functions equal to each other and solve for x.

$$e^{0.1x} = 98.8 - e^{0.1x}$$
$$2e^{0.1x} = 98.8$$
$$e^{0.1x} = 49.4$$
$$0.1x = \ln 49.4$$
$$x = \frac{\ln 49.4}{0.1}$$
$$x \approx 39$$

The optimum number of days for the job to last is 39.

(b) The total income for 39 days is

$$\int_0^{39} (98.8 - e^{0.1x})\, dx$$

$$= \left(98.8x - \frac{e^{0.1x}}{0.1} \right) \Bigg|_0^{39}$$

$$= \left(98.8x - 10e^{0.1x} \right) \Bigg|_0^{39}$$

$$= [98.8(39) - 10e^{3.9}] - (0 - 10)$$
$$= \$3369.18.$$

(c) The total expenditure for 39 days is

$$\int_0^{39} e^{0.1x}\, dx = \frac{e^{0.1x}}{0.1} \Bigg|_0^{39}$$

$$= 10e^{0.1x} \Bigg|_0^{39}$$

$$= 10e^{3.9} - 10$$
$$= \$484.02.$$

(d) Profit = Income − Expense
$$= 3369.18 - 484.02$$
$$= \$2885.16$$

31. $S(q) = q^{5/2} + 2q^{3/2} + 50$; $q = 16$ is the equilibrium quantity.

Producers' surplus $= \displaystyle\int_0^{q_0} [p_0 - S(q)]\, dq$, where p_0 is the equilibrium price and q_0 is equilibrium supply.

$$p_0 = S(16) = (16)^{5/2} + 2(16)^{3/2} + 50$$
$$= 1202$$

Therefore, the producers' surplus is

$$\int_0^{16} [1202 - (q^{5/2} + 2q^{3/2} + 50)]\, dq$$

$$= \int_0^{16} (1152 - q^{5/2} - 2q^{3/2})\, dq$$

$$= \left(1152q - \frac{2}{7}q^{7/2} - \frac{4}{5}q^{5/2} \right) \Bigg|_0^{16}$$

$$= 1152(16) - \frac{2}{7}(16)^{7/2} - \frac{4}{5}(16)^{5/2}$$

$$= 18,432 - \frac{32,768}{7} - \frac{4096}{5}$$

$$= 12,931.66.$$

The producers' surplus is 12,931.66.

33. $D(q) = \dfrac{200}{(3q+1)^2}$; $q = 3$ is the equilibrium quantity.

Consumers' surplus $= \displaystyle\int_0^{q_0} |D(q) - p_0|\, dq$

$$p_0 = D(3) = 2$$

Therefore, the consumers' surplus is

$$\int_0^3 \left[\frac{200}{(3q+1)^2} - 2 \right] dq = \int_0^3 \frac{200}{(3q+1)^2}\, dq - \int_0^3 2\, dq.$$

Let $u = 3q + 1$, so that

$$du = 3\, dq \text{ and } \frac{1}{3}\, du = dq.$$

$$\int_0^3 \frac{200}{(3q+1)^2}\, dq - \int_0^3 2\, dq = \frac{1}{3} \int_1^{10} \frac{200}{u^2}\, du - \int_0^3 2\, dq$$

$$= \frac{200}{3} \int_1^{10} u^{-2} du - \int_0^3 2\, dq$$

$$= \frac{200}{3} \cdot \frac{u^{-1}}{-1} \Bigg|_1^{10} - 2q \Bigg|_0^3$$

$$= -\frac{200}{3u} \Bigg|_1^{10} - 6$$

$$= -\frac{200}{30} + \frac{200}{3} - 6$$

$$= 54$$

35. $S(q) = q^2 + 10q$
$D(q) = 900 - 20q - q^2$

(a) The graphs of the supply and demand functions are parabolas with vertices at $(-5, -25)$ and $(-10, 1900)$, respectively.

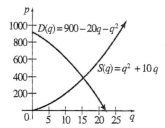

(b) The graphs intersect at the point where the y-coordinates are equal.

$$q^2 + 10q = 900 - 20q - q^2$$
$$2q^2 + 30q - 900 = 0$$
$$q^2 + 15q - 450 = 0$$
$$(q + 30)(q - 15) = 0$$
$$q = -30 \quad \text{or} \quad q = 15$$

Disregard the negative solution.
The supply and demand functions are in equilibrium when $q = 15$.

$$S(15) = 15^2 + 10(15) = 375$$

The point is $(15, 375)$.

(c) Find the consumers' surplus.

$$\int_0^{q_0} [D(q) - p_0)] \, dq$$

$$p_0 = D(15) = 375$$

$$\int_0^{15} [(900 - 20q - q^2) - 375] \, dq$$

$$= \int_0^{15} (525 - 20q - q^2) \, dq$$

$$= \left(525q - 10q^2 - \frac{1}{3}q^3 \right) \Big|_0^{15}$$

$$= \left[525(15) - 10(15)^2 - \frac{1}{3}(15)^3 \right] - 0$$

$$= 4500$$

The consumer's surplus is $4500.

(d) Find the producers' surplus.

$$\int_0^{q_0} [p_0 - S(q)] \, dq$$

$$p_0 = S(15) = 375$$

$$\int_0^{15} [375 - (q^2 + 10q)] \, dq$$

$$= \int_0^{15} (375 - q^2 - 10q) \, dq$$

$$= \left(375q - \frac{1}{3}q^3 - 5q^2 \right) \Big|_0^{15}$$

$$= \left[375(15) - \frac{1}{3}(15)^3 - 5(15)^2 \right] - 0$$

$$= 3375$$

The producer's surplus is $3375.

37. (a) $S(q) = q^2 + 10q$; $S(q) = 264$ is the price the government set.

$$264 = q^2 + 10q$$
$$0 = q^2 + 10q - 264$$
$$0 = (q - 12)(q + 22)$$
$$q = 12 \quad \text{or} \quad q = -22$$

Only 12 is a meaningful solution here. Thus, 12 units of oil will be produced.

(b) The consumers' surplus is given by

$$\int_0^{12} (900 - 20q - q^2 - 264) \, dq$$

$$= \int_0^{12} (636 - 20q - q^2) \, dq$$

$$= \left(636q - 10q^2 - \frac{1}{3}q^3 \right) \Big|_0^{12}$$

$$= 636(12) - 10(12)^2 - \frac{1}{3}(12)^3 - 0$$

$$= 5616$$

Here the consumer' surplus is $5616. In this case, the consumers' surplus is $5616 - 4500 = 1116$ larger.

(c) The producers' surplus is given by

$$\int_0^{12} [264 - (q^2 + 10q)] \, dq$$

$$= \int_0^{12} (264 - q^2 - 10q) \, dq$$

$$= \left(264q - \frac{1}{3}q^3 - 5q^2 \right) \Big|_0^{12}$$

$$= 264(12) - \frac{1}{3}(12)^3 - 5(12)^2 - 0$$

$$= 1872$$

Here the producers' surplus is \$1872. In this case, the producers' surplus is $3375 - 1872 = \$1503$ smaller.

(d) For the equilibrium price, the total consumers' and producers' surplus is

$$4500 + 3375 = \$7875$$

For the government price, the total consumers' and producers' surplus is

$$5616 + 1872 = \$7488.$$

The difference is

$$7875 - 7488 = \$387.$$

39. (a) The pollution level in the lake is changing at the rate $f(t) - g(t)$ at any time t. We find the amount of pollution by integrating.

$$\int_0^{12} [f(t) - g(t)]dt$$

$$= \int_0^{12} [10(1 - e^{-0.5t}) - 0.4t]dt$$

$$= \left(10t - 10 \cdot \frac{1}{-0.5}e^{-0.5t} - 0.4 \cdot \frac{1}{2}t^2\right)\Big|_0^{12}$$

$$= (20e^{-0.5t} + 10t - 0.2t^2)\Big|_0^{12}$$

$$= [20e^{-0.5(12)} + 10(12) - 0.2(12)^2]$$
$$\quad - [20e^{-0.5(0)} + 10(0) - 0.2(0)^2]$$
$$= (20e^{-6} + 91.2) - (20)$$
$$= 20e^{-6} + 71.2 \approx 71.25$$

After 12 hours, there are about 71.25 gallons.

(b) The graphs of the functions intersect at about 25.00. So the rate that pollution enters the lake equals the rate the pollution is removed at about 25 hours.

(c) $\int_0^{25} [f(t) - g(t)]dt$

$$= (20e^{-0.5t} + 10t - 0.2t^2)\Big|_0^{25}$$

$$= [20e^{-0.5(25)} + 10(25) - 0.2(25)^2] - 20$$

$$= 20e^{-12.5} + 105$$

$$\approx 105$$

After 25 hours, there are about 105 gallons.

(d) For $t > 25, g(t) > f(t)$, and pollution is being removed at the rate $g(t) - f(t)$. So, we want to solve for c, where

$$\int_0^c [f(t) - g(t)]dt = 0.$$

(Altternatively, we could solve for c in

$$\int_{25}^c [g(t) - f(t)dt = 105.$$

One way to do this with a graphing calculator is to graph the function

$$y = \int_0^x [f(t) - g(t)]dt$$

and determine the values of x for which $y = 0$. The first window shows how the function can be defined.

A suitable window for the graph is $[0, 50]$ by $[0, 110]$.

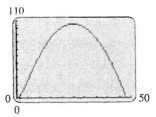

Use the calculator's features to approximate where the graph intersects the x-axis. These are at 0 and about 47.91. Therefore, the pollution will be removed from the lake after about 47.91 hours.

41. $I(x) = 0.9x^2 + 0.1x$

(a) $I(0.1) = 0.9(0.1)^2 + 0.1(0.1)$
$= 0.019$

The lower 10% of income producers earn 1.9% of total income of the population.

(b) $I(0.4) = 0.9(0.4)^2 + 0.1(0.4) = 0.184$

The lower 40% of income producers earn 18.4% of total income of the population.

(c) The graph of $I(x) = x$ is a straight line through the points $(0,0)$ and $(1,1)$. The graph of $I(x) = 0.9x^2 + 0.1x$ is a parabola with vertex $\left(-\frac{1}{18}, -\frac{1}{360}\right)$. Restrict the domain to $0 \le x \le 1$.

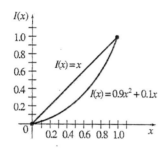

(d) To find the points of intersection, solve

$$x = 0.9x^2 + 0.1x.$$
$$0.9x^2 - 0.9x = 0$$
$$0.9x(x - 1) = 0$$
$$x = 0 \quad \text{or} \quad x = 1$$

The area between the curves is given by

$$\int_0^1 [x - (0.9x^2 + 0.1x)]\, dx$$

$$= \int_0^1 (0.9x - 0.9x^2)\, dx$$

$$= \left(\frac{0.9x^2}{2} - \frac{0.9x^3}{3}\right)\Big|_0^1$$

$$= \frac{0.9}{2} - \frac{0.9}{3} = 0.15.$$

15.6   Numerical Integration

1. $\displaystyle \int_0^2 (3x^2 + 2)\, dx$

$n = 4, b = 2, a = 0, f(x) = 3x^2 + 2$

i	x_i	$f(x_i)$
0	0	2
1	$\frac{1}{2}$	2.75
2	1	5
3	$\frac{3}{2}$	8.75
4	2	14

(a) Trapezoidal rule:

$$\int_0^2 (3x^2 + 2)\, dx$$

$$\approx \frac{2-0}{4}\left[\frac{1}{2}(2) + 2.75 + 5 + 8.75 + \frac{1}{2}(14)\right]$$

$$= 0.5(24.5)$$

$$= 12.25$$

(b) Simpson's rule:

$$\int_0^2 (3x^2 + 2)\, dx$$

$$\approx \frac{2-0}{3(4)}[2 + 4(2.75) + 2(5) + 4(8.75) + 14]$$

$$= \frac{2}{12}(72)$$

$$= 12$$

(c) Exact value:

$$\int_0^2 (3x^2 + 2)\, dx = (x^3 + 2x)\Big|_0^2$$

$$= (8 + 4) - 0$$

$$= 12$$

3. $\displaystyle \int_{-1}^3 \frac{3}{5 - x}\, dx$

$n = 4, b = 3, a = -1, f(x) = \dfrac{3}{5 - x}$

i	x_i	$f(x_i)$
0	-1	0.5
1	0	0.6
2	1	0.75
3	2	1
4	3	1.5

(a) Trapezoidal rule:

$$\int_{-1}^{3} \frac{3}{5-x}\, dx$$

$$\approx \frac{3-(-1)}{4}\left[\frac{1}{2}(0.5)+0.6+0.75+1+\frac{1}{2}(1.5)\right]$$

$$= 1(3.35)$$

$$= 3.35$$

(b) Simpson's rule:

$$\int_{-1}^{3} \frac{3}{5-x}\, dx$$

$$\approx \frac{3-(-1)}{3(4)}[0.5+4(0.6)+2(0.75)+4(1)+1.5]$$

$$= \frac{1}{3}\left(\frac{99}{10}\right)$$

$$= \frac{33}{10} \approx 3.3$$

(c) Exact value:

$$\int_{-1}^{3} \frac{3}{5-x}\, dx = -3\ln|5-x|\ \Big|_{-1}^{3}$$

$$= -3(\ln|2|-\ln|6|)$$

$$= 3\ln 3 \approx 3.296$$

5. $\displaystyle\int_{-1}^{2}(2x^3+1)\,dx$

$$n=4,\ b=2,\ a=-1,\ f(x)=2x^3+1$$

i	x_i	$f(x)$
0	-1	-1
1	$-\dfrac{1}{4}$	$\dfrac{31}{32}$
2	$\dfrac{1}{2}$	$\dfrac{5}{4}$
3	$\dfrac{5}{4}$	$\dfrac{157}{32}$
4	2	17

(a) Trapezoidal rule:

$$\int_{-1}^{2}(2x^3+1)\,dx$$

$$\approx \frac{2-(-1)}{4}\left[\frac{1}{2}(-1)+\frac{31}{32}+\frac{5}{4}+\frac{157}{32}+\frac{1}{2}(17)\right]$$

$$= 0.75(15.125)$$

$$\approx 11.34$$

(b) Simpson's rule:

$$\int_{-1}^{2}(2x^3+1)\,dx$$

$$\approx \frac{2-(-1)}{3(4)}\left[-1+4\left(\frac{31}{32}\right)+2\left(\frac{5}{4}\right)+4\left(\frac{157}{32}\right)+17\right]$$

$$= \frac{1}{4}(42)$$

$$= 10.5$$

(c) Exact value:

$$\int_{-1}^{2}(2x^3+1)\,dx$$

$$= \left(\frac{x^4}{2}+x\right)\Big|_{-1}^{2}$$

$$= (8+2)-\left(\frac{1}{2}-1\right)$$

$$= \frac{21}{2}$$

$$= 10.5$$

7. $\displaystyle\int_{1}^{5}\frac{1}{x^2}\,dx$

$$n=4,\ b=5,\ a=1,\ f(x)=\frac{1}{x^2}$$

i	x_i	$f(x_i)$
0	1	1
1	2	0.25
2	3	0.1111
3	4	0.0625
4	5	0.04

(a) Trapezoidal rule:

$$\int_{1}^{5}\frac{1}{x^2}\,dx$$

$$\approx \frac{5-1}{4}\left[\frac{1}{2}(1)+0.25+0.1111+0.0625+\frac{1}{2}(0.04)\right]$$

$$\approx 0.9436$$

(b) Simpson's rule:

$$\int_{1}^{5}\frac{1}{x^2}\,dx$$

$$\approx \frac{5-1}{12}[1+4(0.25)+2(0.1111)+4(0.0625)+0.04]$$

$$\approx 0.8374$$

(c) Exact value:

$$\int_1^5 x^{-2}\, dx = -x^{-1}\Big|_1^5$$

$$= -\frac{1}{5} + 1$$

$$= \frac{4}{5} = 0.8$$

9. $\int_0^1 4xe^{-x^2}\, dx$

$n = 4,\ b = 1,\ a = 0,\ f(x) = 4xe^{-x^2}$

i	x_i	$f(x_i)$
0	0	0
1	$\frac{1}{4}$	$e^{-1/16}$
2	$\frac{1}{2}$	$2e^{-1/4}$
3	$\frac{3}{4}$	$3e^{-9/16}$
4	1	$4e^{-1}$

(a) Trapezoidal rule:

$$\int_0^1 4xe^{-x^2}\, dx$$

$$\approx \frac{1-0}{4}\left[\frac{1}{2}(0) + e^{-1/16} + 2e^{-1/4} \right.$$

$$\left. + 3e^{-9/16} + \frac{1}{2}(4e^{-1})\right]$$

$$= \frac{1}{4}(e^{-1/16} + 2e^{-1/4} + 3e^{-9/16} + 2e^{-1})$$

$$\approx 1.236$$

(b) Simpson's rule:

$$\int_0^1 4xe^{-x^2}\, dx$$

$$\approx \frac{1-0}{3(4)}[0 + 4(e^{-1/16}) + 2(2e^{-1/4})$$

$$+ 4(3e^{-9/16}) + 4e^{-1}]$$

$$= \frac{1}{12}(4e^{-1/16} + 4e^{-1/4} + 12e^{-9/16} + 4e^{-1})$$

$$\approx 1.265$$

(c) Exact value:

$$\int_0^1 4xe^{-x^2}\, dx = -2e^{-x^2}\Big|_0^1$$

$$= (-2e^{-1}) - (-2)$$

$$= 2 - 2e^{-1} \approx 1.264$$

11. $y = \sqrt{4 - x^2}$

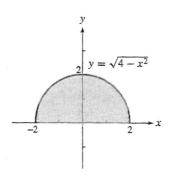

$n = 8,\ b = 2,\ a = -2,\ f(x) = \sqrt{4 - x^2}$

i	x_i	y
0	-2.0	0
1	-1.5	1.32289
2	-1.0	1.73205
3	-0.5	1.93649
4	0	2
5	0.5	1.93649
6	1.0	1.73205
7	1.5	1.32289
8	2.0	0

(a) Trapezoidal rule:

$$\int_{-2}^2 \sqrt{4 - x^2}\, dx$$

$$\approx \frac{2 - (-2)}{8}$$

$$\cdot \left[\frac{1}{2}(0) + 1.32289 + 1.73205 + \cdots + \frac{1}{2}(0)\right]$$

$$\approx 5.991$$

(b) Simpson's rule:

$$\int_{-2}^2 \sqrt{4 - x^2}\, dx$$

$$\approx \frac{2 - (-2)}{3(8)}$$

$$\cdot [0 + 4(1.32289) + 2(1.73205) + 4(1.93649) + 2(2)$$

$$+ 4(1.93649) + 2(1.73205) + 4(1.32289) + 0]$$

$$\approx 6.167$$

(c) Area of semicircle $= \frac{1}{2}\pi r^2$

$$= \frac{1}{2}\pi(2)^2$$

$$\approx 6.283$$

Simpson's rule is more accurate.

13. Since $f(x) > 0$ and $f''(x) > 0$ for all x between a and b, we know the graph of $f(x)$ on the interval from a to b is concave upward. Thus, the trapezoid that approximates the area will have an area greater than the actual area. Thus,

$$T > \int_a^b f(x)\, dx.$$

The correct choice is (b).

15. (a) $\int_0^1 x^4\, dx = \left(\dfrac{1}{5}\right) x^5 \Big|_0^1$

$$= \frac{1}{5}$$

$$= 0.2$$

(b) $n = 4$, $b = 1$, $a = 0$, $f(x) = x^4$

$$\int_0^1 x^4\, dx \approx \frac{1-0}{4}\left[\frac{1}{2}(0) + \frac{1}{256} + \frac{1}{16} + \frac{81}{256} + \frac{1}{2}(1)\right]$$

$$= \frac{1}{4}\left(\frac{226}{256}\right)$$

$$\approx 0.220703$$

$n = 8$, $b = 1$, $a = 0$, $f(x) = x^4$

$$\int_0^1 x^4\, dx \approx \frac{1-0}{8}\left[\frac{1}{2}(0) + \frac{1}{4096} + \frac{1}{256} + \frac{81}{4096}\right.$$

$$\left. + \frac{1}{16} + \frac{625}{4096} + \frac{81}{256} + \frac{2401}{4096} + \frac{1}{2}(1)\right]$$

$$= \frac{1}{8}\left(\frac{6724}{4096}\right)$$

$$\approx 0.20520$$

$n = 16$, $b = 1$, $a = 0$, $f(x) = x^4$

$$\int_0^1 x^4\, dx \approx \frac{1-0}{16}\left[\frac{1}{2}(0) + \frac{1}{65,536} + \frac{1}{4096}\right.$$

$$+ \frac{81}{65,536} + \frac{1}{256} + \frac{625}{65,536}$$

$$+ \frac{81}{4096} + \frac{2401}{65,536} + \frac{1}{16}$$

$$+ \frac{6561}{65,536} + \frac{625}{4096} + \frac{14,641}{65,536}$$

$$+ \frac{81}{256} + \frac{28,561}{65,536} + \frac{2401}{4096}$$

$$\left. + \frac{50,625}{65,536} + \frac{1}{2}(1)\right]$$

$$\approx \frac{1}{16}\left(\frac{211,080}{65,536}\right)$$

$$\approx 0.201302$$

$n = 32$, $b = 1$, $a = 0$, $f(x) = x^4$

$$\int_0^1 x^4\, dx$$

$$\approx \frac{1-0}{32}\left[\frac{1}{2}(0) + \frac{1}{1,048,576} + \frac{1}{65,536}\right.$$

$$+ \frac{81}{1,048,576} + \frac{1}{4096} + \frac{625}{1,048,576}$$

$$+ \frac{81}{65,536} + \frac{2401}{1,048,576} + \frac{1}{256} + \frac{6561}{1,048,576}$$

$$+ \frac{625}{65,536} + \frac{14,641}{1,048,576} + \frac{81}{4096} + \frac{28,561}{1,048,576}$$

$$+ \frac{2401}{65,536} + \frac{50,625}{1,048,576} + \frac{1}{16} + \frac{83,521}{1,048,576}$$

$$+ \frac{6561}{65,536} + \frac{130,321}{1,048,576} + \frac{625}{4096} + \frac{194,481}{1,048,576}$$

$$+ \frac{14,641}{65,536} + \frac{279,841}{1,048,576} + \frac{81}{256} + \frac{390,625}{1,048,576}$$

$$+ \frac{28,561}{65,536} + \frac{531,441}{1,048,576} + \frac{2401}{4096} + \frac{707,281}{1,048,576}$$

$$\left. + \frac{50,625}{65,536} + \frac{923,521}{1,048,576} + \frac{1}{2}(1)\right]$$

$$\approx \frac{1}{32}\left(\frac{6,721,808}{1,048,576}\right) \approx 0.200325$$

To find error for each value of n, subtract as indicated.

$n = 4$: $(0.220703 - 0.2) = 0.020703$

$n = 8$: $(0.205200 - 0.2) = 0.005200$

$n = 16$: $(0.201302 - 0.2) = 0.001302$

$n = 32$: $(0.200325 - 0.2) = 0.000325$

(c) $p = 1$

$$4^1(0.020703) = 4(0.020703)$$

$$= 0.082812$$

$$8^1(0.005200) = 8(0.005200)$$

$$= 0.0416$$

Since these are not the same, try $p = 2$.

$p = 2$:

$$4^2(0.020703) = 16(0.020703)$$

$$= 0.331248$$

$$8^2(0.005200) = 64(0.005200) = 0.3328$$

$$16^2(0.001302) = 256(0.001302)$$

$$= 0.333312$$

$$32^2(0.000325) = 1024(0.000325)$$

$$= 0.3328$$

Since these values are all approximately the same, the correct choice is $p = 2$.

17. (a) $\displaystyle\int_0^1 x^4\,dx = \left.\frac{1}{5}x^5\right|_0^1$

$$= \frac{1}{5}$$

$$= 0.2$$

(b) $n = 4,\ b = 1,\ a = 0,\ f(x) = x^4$

$$\int_0^1 x^4\,dx \approx \frac{1-0}{3(4)}\left[0 + 4\left(\frac{1}{256}\right) + 2\left(\frac{1}{16}\right)\right.$$
$$\left. + 4\left(\frac{81}{256}\right) + 1\right]$$

$$= \frac{1}{12}\left(\frac{77}{32}\right)$$

$$\approx 0.2005208$$

$n = 8,\ b = 1,\ a = 0,\ f(x) = x^4$

$$\int_0^1 x^4\,dx \approx \frac{1-0}{3(8)}\left[0 + 4\left(\frac{1}{4096}\right) + 2\left(\frac{1}{256}\right)\right.$$
$$+ 4\left(\frac{81}{4096}\right) + 2\left(\frac{1}{16}\right) + 4\left(\frac{625}{4096}\right)$$
$$\left. + 2\left(\frac{81}{256}\right) + 4\left(\frac{2401}{4096}\right) + 1\right]$$

$$= \frac{1}{24}\left(\frac{4916}{1024}\right)$$

$$\approx 0.2000326$$

$n = 16,\ b = 1,\ a = 0,\ f(x) = x^4$

$$\int_0^1 x^4\,dx$$

$$\approx \frac{1-0}{3(16)}\left[0 + 4\left(\frac{1}{65,536}\right) + 2\left(\frac{1}{4096}\right)\right.$$
$$+ 4\left(\frac{81}{65,536}\right) + 2\left(\frac{1}{256}\right) + 4\left(\frac{625}{65,536}\right)$$
$$+ 2\left(\frac{81}{4096}\right) + 4\left(\frac{2401}{65,536}\right) + 2\left(\frac{1}{16}\right)$$
$$+ 4\left(\frac{6561}{65,536}\right) + 2\left(\frac{625}{4096}\right) + 4\left(\frac{14,641}{65,536}\right)$$
$$+ 2\left(\frac{81}{256}\right) + 4\left(\frac{28,561}{65,536}\right) + 2\left(\frac{2401}{4096}\right)$$
$$\left. + 4\left(\frac{50,625}{65,536} + 1\right)\right]$$

$$= \frac{1}{48}\left(\frac{157,288}{16,384}\right) \approx 0.2000020$$

$n = 32,\ b = 1,\ a = 0,\ f(x) = x^4$

$$\int_0^1 x^4\,dx$$

$$\approx \frac{1-0}{3(32)}\left[0 + 4\left(\frac{1}{1,048,576}\right) + 2\left(\frac{1}{65,536}\right)\right.$$
$$+ 4\left(\frac{81}{1,048,576}\right) + 2\left(\frac{1}{4096}\right) + 4\left(\frac{625}{1,048,576}\right)$$
$$+ 2\left(\frac{625}{65,536}\right) + 4\left(\frac{14,641}{1,048,576}\right) + 2\left(\frac{81}{4096}\right)$$
$$+ 4\left(\frac{28,561}{1,048,576}\right) + 2\left(\frac{2401}{65,536}\right) + 4\left(\frac{50,625}{1,048,576}\right)$$
$$+ 2\left(\frac{1}{16}\right) + 4\left(\frac{83,521}{1,048,576}\right) + 2\left(\frac{6561}{65,536}\right)$$
$$+ 4\left(\frac{130,321}{1,048,576}\right) + 2\left(\frac{625}{4096}\right) + 4\left(\frac{194,481}{1,048,576}\right)$$
$$+ 2\left(\frac{14,641}{65,536}\right) + 4\left(\frac{279,841}{1,048,576}\right) + 2\left(\frac{81}{256}\right)$$
$$+ 4\left(\frac{390,625}{1,048,576}\right) + 2\left(\frac{28,561}{65,536}\right) + 4\left(\frac{531,441}{1,048,576}\right)$$
$$+ 2\left(\frac{2401}{4096}\right) + 4\left(\frac{707,281}{1,048,576}\right) + 2\left(\frac{50,625}{65,536}\right)$$
$$\left. + 4\left(\frac{923,521}{1,048,576}\right) + 1\right]$$

$$= \frac{1}{96}\left(\frac{50,033,168}{262,144}\right) \approx 0.2000001$$

To find error for each value of n, subtract as indicated.

$n = 4$: $(0.2005208 - 0.2) = 0.0005208$
$n = 8$: $(0.2000326 - 0.2) = 0.0000326$
$n = 16$: $(0.2000020 - 0.2) = 0.0000020$
$n = 32$: $(0.2000001 - 0.2) = 0.0000001$

(c) $p = 1$:

$\quad 4^1(0.0005208) = 4(0.0005208) = 0.0020832$
$\quad 8^1(0.0000326) = 8(0.0000326) = 0.0002608$

Try $p = 2$:

$\quad 4^2(0.0005208) = 16(0.0005208) = 0.0083328$
$\quad 8^2(0.0000326) = 64(0.0000326) = 0.0020864$

Try $p = 3$:

$\quad 4^3(0.0005208) = 64(0.0005208) = 0.0333312$
$\quad 8^3(0.0000326) = 512(0.0000326) = 0.0166912$

Try $p = 4$:

$\quad 4^4(0.0005208) = 256(0.0005208) = 0.1333248$
$\quad 8^4(0.0000326) = 4096(0.0000326) = 0.1335296$
$\quad 16^4(0.0000020) = 65536(0.0000020) = 0.131072$
$\quad 32^4(0.0000001) = 1048576(0.0000001) = 0.1048576$

These are the closest values we can get; thus, $p = 4$.

19. Midpoint rule:

$n = 4,\ b = 5,\ a = 1, f(x) = \dfrac{1}{x^2}, \Delta x = 1$

i	x_i	$f(x_i)$
1	$\dfrac{3}{2}$	$\dfrac{4}{9}$
2	$\dfrac{5}{2}$	$\dfrac{4}{25}$
3	$\dfrac{7}{2}$	$\dfrac{4}{49}$
4	$\dfrac{9}{2}$	$\dfrac{4}{81}$

$$\int_1^5 \frac{1}{x^2}\,dx \approx \sum_{i=1}^{4} f(x_i)\Delta x$$

$$= \frac{4}{9}(1) + \frac{4}{25}(1) + \frac{4}{49}(1) + \frac{4}{81}(1)$$

$$\approx 0.7355$$

Simpson's rule:

$m = 8, b = 5, a = 1, f(x) = \dfrac{1}{x^2}$

i	x_i	$f(x_i)$
0	1	1
1	$\dfrac{3}{2}$	$\dfrac{4}{9}$
2	2	$\dfrac{1}{4}$
3	$\dfrac{5}{2}$	$\dfrac{4}{25}$
4	3	$\dfrac{1}{9}$
5	$\dfrac{7}{2}$	$\dfrac{4}{49}$
6	4	$\dfrac{1}{16}$
7	$\dfrac{9}{2}$	$\dfrac{4}{81}$
8	5	$\dfrac{1}{25}$

$$\int_1^5 \frac{1}{x^2}\,dx$$

$$\approx \frac{5-1}{3(8)}\left[1 + 4\left(\frac{4}{9}\right) + 2\left(\frac{1}{4}\right) + 4\left(\frac{4}{25}\right)\right.$$

$$+ 2\left(\frac{1}{9}\right) + 4\left(\frac{4}{49}\right) + 2\left(\frac{1}{16}\right)$$

$$\left. + 4\left(\frac{4}{81}\right) + \frac{1}{25}\right]$$

$$\approx \frac{1}{6}(4.82906)$$

$$\approx 0.8048$$

From #7 part a, $T \approx 0.9436$, when $n = 4$. To verify the formula evaluate $\frac{2M+T}{3}$.

$$\frac{2M + T}{3} \approx \frac{2(0.7355) + 0.9436}{3}$$

$$\approx 0.8048$$

21. (a)

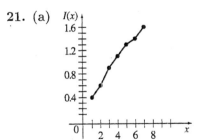

(b) $A = \dfrac{7-1}{6}\left[\dfrac{1}{2}(0.4) + 0.6 + 0.9 + 1.1\right.$

$$\left. + 1.3 + 1.4 + \frac{1}{2}(1.6)\right]$$

$$= 6.3$$

(c) $A = \dfrac{7-1}{3(6)}[0.4 + 4(0.6) + 2(0.9) + 4(1.1)$

$$+ 2(1.3) + 4(1.4) + 1.6]$$

$$\approx 6.27$$

23. $y = e^{-t^2} + \dfrac{1}{t+1}$

The total reaction is

$$\int_1^9 \left(e^{-t^2} + \frac{1}{t+1} \right) dt.$$

$n = 8,\ b = 9,\ a = 1,\ f(t) = e^{-t^2} + \frac{1}{t+1}$

i	x_i	$f(x_i)$
0	1	0.8679
1	2	0.3516
2	3	0.2501
3	4	0.2000
4	5	0.1667
5	6	0.1429
6	7	0.1250
7	8	0.1111
8	9	0.1000

(a) Trapezoidal rule:

$$\int_1^9 \left(e^{-t^2} + \frac{1}{t+1} \right) dt$$

$$\approx \frac{9-1}{8} \left[\frac{1}{2}(0.8679) + 0.3516 + 0.2501 \right.$$

$$\left. + \cdots + \frac{1}{2}(0.1000) \right]$$

$$\approx 1.831$$

(b) Simpson's rule:

$$\int_1^9 \left(e^{-t^2} + \frac{1}{t+1} \right) dt$$

$$\approx \frac{9-1}{3(8)} [0.8679 + 4(0.3516) + 2(0.2501)$$

$$+ 4(0.2000) + 2(0.1667) + 4(0.1429)$$
$$+ 2(0.1250) + 4(0.1111) + 0.1000]$$

$$= \frac{1}{3}(5.2739)$$

$$\approx 1.758$$

25. Note that heights may differ depending on the readings of the graph. Thus, answers may vary. $n = 10,\ b = 20,\ a = 0$

i	x_i	$f(x_i)$
0	0	0
1	2	5
2	4	3
3	6	2
4	8	1.5
5	10	1.2
6	12	1
7	14	0.5
8	16	0.3
9	18	0.2
10	20	0.2

Area under curve for Formulation A

$$= \frac{20-0}{10} \left[\frac{1}{2}(0) + 5 + 3 + 2 + 1.5 + 1.2 \right.$$

$$\left. + 1 + 0.5 + 0.3 + 0.2 + \frac{1}{2}(0.2) \right]$$

$$= 2(14.8)$$

$$\approx 30 \text{ mcgh/ml}$$

This represents the total amount of drug available to the patient for each ml of blood.

27. As in Exercise 25, readings on the graph may vary, so answers may vary. The area both under the curve for Formulation A and above the minimum effective concentration line in on the interval $\left[\frac{1}{2}, 6 \right]$.

Area under curve for Formulation A on $\left[\frac{1}{2}, 1 \right]$, with $n = 1$

$$= \frac{1 - \frac{1}{2}}{1} \left[\frac{1}{2}(2 + 6) \right]$$

$$= \frac{1}{2}(4) = 2$$

Area under curve for Formulation A on $[1, 6]$, with $n = 5$

$$= \frac{6-1}{5} \left[\frac{1}{2}(6) + 5 + 4 + 3 + 2.4 + \frac{1}{2}(2) \right]$$
$$= 18.4$$

Area under minimum effective concentration line
$\left[\frac{1}{2}, 6\right]$

$$= 5.5(2) = 11.0$$

Area under the curve for Formulation A and above minimum effective concentration line

$$= 2 + 18.4 - 11.0$$

$$\approx 9 \text{ mcgh/ml}$$

This represents the total effective amount of drug available to the patient for each ml of blood.

29. $y = b_0 w^{b_1} e^{-b_2 w}$

(a) If $t = 7w$ then $w = \dfrac{t}{7}$.

$$y = b_0 \left(\frac{t}{7}\right)^{b_1} e^{-b_2 t/7}$$

(b) Replacing the constants with the given values, we have

$$y = 5.955 \left(\frac{t}{7}\right)^{0.233} e^{-0.027t/7} dt$$

In 25 weeks, there are 175 days.

$$\int_0^{175} 5.955 \left(\frac{t}{7}\right)^{0.233} e^{-0.027t/7} dt$$

$n = 10, b = 175, a = 0,$

$$f(t) = 5.955 \left(\frac{t}{7}\right)^{0.233} e^{-0.027t/7}$$

i	t_i	$f(t_i)$
0	0	0
1	17.5	6.89
2	35	7.57
3	52.5	7.78
4	70	7.77
5	87.5	7.65
6	105	7.46
7	122.5	7.23
8	140	6.97
9	157.5	6.70
10	175	6.42

Trapezoidal rule:

$$\int_0^{175} 5.955 \left(\frac{t}{7}\right)^{0.233} e^{-0.027t/7} dt$$

$$\approx \frac{175 - 0}{10} \left[\frac{1}{2}(0) + 6.89 + 7.57 + 7.78 + 7.77 \right.$$

$$\left. + 7.65 + 7.46 + 7.23 + 6.97 + 6.70 + \frac{1}{2}(6.42) \right]$$

$$= 17.5(69.23)$$
$$= 1211.525$$

The total milk consumed is about 1212 kg.

Simpson's rule:

$$\int_0^{175} 5.955 \left(\frac{t}{7}\right)^{0.233} e^{-0.027t/7} dt$$

$$\approx \frac{175 - 0}{3(10)} [0 + 4(6.89) + 2(7.57) + 4(7.78)$$

$$+ 2(7.77) + 4(7.65) + 2(7.46) + 4(7.23)$$

$$+ 2(6.97) + 4(6.70) + 6.42]$$

The total milk consumed is about 1231 kg.

(c) Replacing the constants with the given values, we have

$$y = 8.409 \left(\frac{t}{7}\right)^{0.143} e^{-0.037t/7}.$$

In 25 weeks, there are 175 days.

$$\int_0^{175} 8.409 \left(\frac{t}{7}\right)^{0.143} e^{-0.037t/7} dt$$

$n = 10, \ b = 175, \ a = 0,$

$$f(t) = 8.409 \left(\frac{t}{7}\right)^{0.143} e^{-0.037t/7}$$

i	t_i	$f(t_i)$
0	0	0
1	17.5	8.74
2	35	8.80
3	52.5	8.50
4	70	8.07
5	87.5	7.60
6	105	7.11
7	122.5	6.63
8	140	6.16
9	157.5	5.71
10	175	5.28

Trapezoidal rule:

$$\int_0^{175} 8.409 \left(\frac{t}{7}\right)^{0.143} e^{-0.037t/7} dt$$

$$\approx \frac{175-0}{10}\left[\frac{1}{2}(0) + 8.74 + 8.80 + 8.50\right.$$

$$+ 8.07 + 7.60 + 7.11 + 6.63$$

$$\left. + 6.16 + 5.71 + \frac{1}{2}(5.28)\right]$$

$$= 17.5(69.96)$$

$$= 1224.30$$

The total milk consumed is about 1224 kg.

Simpson's rule:

$$\int_0^{175} 8.409 \left(\frac{t}{7}\right)^{0.143} e^{-0.037t/7} dt$$

$$\approx \frac{175-0}{3(10)}[0 + 4(8.74) + 2(8.80) + 4(8.50)$$

$$+ 2(8.07) + 4(7.60) + 2(7.11) + 4(6.63)$$

$$+ 2(6.16) + 4(5.71) + 5.28]$$

$$= \frac{35}{6}(214.28)$$

$$= 1249.97$$

The total milk consumed is about 1250 kg.

31. (a)

(b) $\dfrac{7-1}{6}\dfrac{1}{2}\left[(4) + 7 + 11 + 9 + 15 + 16 + \dfrac{1}{2}(23)\right]$

$$= 71.5$$

(c) $\dfrac{7-1}{3(6)}[4 + 4(7) + 2(11) + 4(9)$

$$+ 2(15) + 4(16) + 23]$$

$$= 69.0$$

33. We need to evaluate

$$\int_{12}^{36} (105e^{0.01\sqrt{x}} + 32)\, dx.$$

Using a calculator program for Simpson's rule with $n = 20$, we obtain 3413.18 as the value of this integral. This indicates that the total revenue between the twelfth and thirty-sixth months is about 3413.

35. Use a calculator program for Simpson's rule with $n = 20$ to evaluate each of the integrals in this exercise.

(a) $\displaystyle\int_{-1}^{1}\left(\frac{1}{\sqrt{2\pi}}e^{-x^2/2}\right)dx \approx 0.6827$

The probability that a normal random variable is within 1 standard deviation of the mean is about 0.6827.

(b) $\displaystyle\int_{-2}^{2}\left(\frac{1}{\sqrt{2\pi}}e^{-x^2/2}\right)dx \approx 0.9545$

The probability that a normal random variable is within 2 standard deviations of the mean is about 0.9545.

(c) $\displaystyle\int_{-3}^{3}\left(\frac{1}{\sqrt{2\pi}}e^{-x^2/2}\right)dx \approx 0.9973$

The probability that a normal random variable is within 3 standard deviations of the mean is about 0.9973.

Chapter 15 Review Exercises

5. $\displaystyle\int (2x + 3)\, dx = \frac{2x^2}{2} + 3x + C$

$$= x^2 + 3x + C$$

7. $\displaystyle\int (x^2 - 3x + 2)\, dx$

$$= \frac{x^3}{3} - \frac{3x^2}{2} + 2x + C$$

9. $\displaystyle\int 3\sqrt{x}\, dx = 3\int x^{1/2}\, dx$

$$= \frac{3x^{3/2}}{\frac{3}{2}} + C$$

$$= 2x^{3/2} + C$$

11. $\displaystyle\int (x^{1/2} + 3x^{-2/3})\, dx$

$$= \frac{x^{3/2}}{\frac{3}{2}} + \frac{3x^{1/3}}{\frac{1}{3}} + C$$

$$= \frac{2x^{3/2}}{3} + 9x^{1/3} + C$$

13. $\displaystyle\int \frac{-4}{x^3}\, dx = \int -4x^{-3}\, dx$

$$= \frac{-4x^{-2}}{-2} + C$$

$$= 2x^{-2} + C$$

15. $\int -3e^{2x}\,dx = \dfrac{-3e^{2x}}{2} + C$

17. $\int xe^{3x^2}\,dx = \dfrac{1}{6}\int 6xe^{3x^2}\,dx$

Let $u = 3x^2$, so that $du = 6x\,dx$.

$$= \frac{1}{6}\int e^u\,du$$

$$= \frac{1}{6}e^u + C$$

$$= \frac{e^{3x^2}}{6} + C$$

19. $\int \dfrac{3x}{x^2 - 1}\,dx = 3\left(\dfrac{1}{2}\right)\int \dfrac{2x\,dx}{x^2 - 1}$

Let $u = x^2 - 1$, so that $du = 2x\,dx$.

$$= \frac{3}{2}\int \frac{du}{u}$$

$$= \frac{3}{2}\ln |u| + C$$

$$= \frac{3\ln |x^2 - 1|}{2} + C$$

21. $\int \dfrac{x^2\,dx}{(x^3 + 5)^4} = \dfrac{1}{3}\int \dfrac{3x^2\,dx}{(x^3 + 5)^4}$

Let $u = x^3 + 5$, so that

$$du = 3x^2\,dx.$$

$$= \frac{1}{3}\int \frac{du}{u^4}$$

$$= \frac{1}{3}\int u^{-4}\,du$$

$$= \frac{1}{3}\left(\frac{u^{-3}}{-3}\right) + C$$

$$= \frac{-(x^3 + 5)^{-3}}{9} + C$$

23. $\int \dfrac{x^3}{e^{3x^4}}\,dx = \int x^3 e^{-3x^4}$

$$= -\frac{1}{12}\int -12x^3 e^{-3x^4}\,dx$$

Let $u = -3x^4$, so that $du = -12x^3\,dx$.

$$= -\frac{1}{12}\int e^u\,du$$

$$= -\frac{1}{12}e^u + C$$

$$= \frac{-e^{-3x^4}}{12} + C$$

25. $\int \dfrac{(3\ln x + 2)^4}{x}\,dx$

Let $u = 3\ln x + 2$ so that

$$du = \frac{3}{x}\,dx.$$

$$\int \frac{(3\ln x + 2)^4}{x}\,dx = \frac{1}{3}\int \frac{3(3\ln x + 2)^4}{x}\,dx$$

$$= \frac{1}{3}\int u^4\,du$$

$$= \frac{1}{3}\cdot\frac{u^5}{5} + C$$

$$= \frac{(3\ln x + 2)^5}{15} + C$$

27. $f(x) = 3x + 1$, $x_1 = -1$, $x_2 = 0$, $x_3 = 1$, $x_4 = 2$, $x_5 = 3$

$f(x_1) = -2$, $f(x_2) = 1$, $f(x_3) = 4$, $f(x_4) = 7$, $f(x_5) = 10$

$$\sum_{i=1}^{5} f(x_i)$$

$$= f(1) + f(2) + f(3) + f(4) + f(5)$$
$$= -2 + 1 + 4 + 7 + 10$$
$$= 20$$

29. $f(x) = 2x + 3$, from $x = 0$ to $x = 4$

$$\Delta x = \frac{4 - 0}{4} = 1$$

i	x_i	$f(x_i)$
1	0	3
2	1	5
3	2	7
4	3	9

$$A = \sum_{i=1}^{4} f(x_i)\Delta x$$

$$= 3(1) + 5(1) + 7(1) + 9(1)$$
$$= 24$$

31. (a) Since $s(t)$ represents the odometer reading, the distance traveled between $t = 0$ and $t = T$ will be $s(T) - s(0)$.

(b) $\int_0^T v(t)\,dt = s(T) - s(0)$ is equivalent to the Fundamental Theorem of Calculus with $a = 0$, and $b = T$ because $s(t)$ is an antiderivative of $v(t)$.

33. $\displaystyle\int_1^2 (3x^2 + 5)\,dx = \left(\frac{3x^3}{3} + 5x\right)\Big|_1^2$

$$= (2^3 + 10) - (1 + 5)$$
$$= 18 - 6$$
$$= 12$$

35. $\displaystyle\int_1^5 (3x^{-1} + x^{-3})\,dx = \left(3\ln|x| + \frac{x^{-2}}{-2}\right)\Big|_1^5$

$$= \left(3\ln 5 - \frac{1}{50}\right) - \left(3\ln 1 - \frac{1}{2}\right)$$

$$= 3\ln 5 + \frac{12}{25} \approx 5.308$$

37. $\displaystyle\int_0^1 x\sqrt{5x^2 + 4}\,dx$

Let $u = 5x^2 + 4$, so that

$$du = 10x\,dx \text{ and } \frac{1}{10}\,du = x\,dx.$$

When $x = 0$, $u = 5(0^2) + 4 = 4$.
When $x = 1$, $u = 5(1^2) + 4 = 9$.

$$= \frac{1}{10}\int_4^9 \sqrt{u}\,du = \frac{1}{10}\int_4^9 u^{1/2}\,du$$

$$= \frac{1}{10}\cdot\frac{u^{3/2}}{3/2}\Big|_4^9 = \frac{1}{15}u^{3/2}\Big|_4^9$$

$$= \frac{1}{15}(9)^{3/2} - \frac{1}{15}(4)^{3/2}$$

$$= \frac{27}{15} - \frac{8}{15}$$

$$= \frac{19}{15}$$

39. $\displaystyle\int_0^2 3e^{-2x}\,dx = \frac{-3e^{-2x}}{2}\Big|_0^2$

$$= \frac{-3e^{-4}}{2} + \frac{3}{2}$$

$$= \frac{3(1 - e^{-4})}{2} \approx 1.473$$

41. $\displaystyle\int_0^{1/2} x\sqrt{1 - 16x^4}\,dx$

Let $u = 4x^2$. Then $du = 8x\,dx$.
When $x = 0$, $u = 0$, and when $x = \frac{1}{2}$, $u = 1$.

Thus,

$$\int_0^{1/2} x\sqrt{1 - 16x^4}\,dx = \frac{1}{8}\int_0^1 \sqrt{1 - u^2}\,du.$$

Note that this integral represents the area of right upper quarter of a circle centered at the origin with a radius of 1.

Area of circle $= \pi r^2 = \pi(1^2) = \pi$

$$\int_0^1 \sqrt{1 - u^2}\,du = \frac{\pi}{4}$$

$$\frac{1}{8}\int_0^1 \sqrt{1 - u^2}\,du = \frac{1}{8}\cdot\frac{\pi}{4} = \frac{\pi}{32}$$

43. $\displaystyle\int_1^{\sqrt{7}} 2x\sqrt{36 - (x^2 - 1)^2}\,dx$

Let $u = x^2 - 1$. Then $du = 2x\,dx$.
When $x = \sqrt{7}, u = (\sqrt{7})^2 - 1 = 6$.
When $x = 1, u = (\sqrt{1})^2 - 1 = 0$.

Thus,

$$\int_1^{\sqrt{7}} 2x\sqrt{36 - (x^2 - 1)^2}\,dx = \int_0^6 \sqrt{36 - u^2}\,du.$$

Note that this integral represents the area of a right upper quarter of a circle centered at the origin with a radius of 6.

Area of circle $= \pi r^2 = \pi(6)^2 = 36\pi$

$$\int_0^6 \sqrt{36 - u^2}\,du = \frac{36\pi}{4} = 9\pi$$

45. $f(x) = (3x + 2)^6$; $[-2, 0]$

$$\text{Area} = \int_{-2}^0 (3x + 2)^6\,dx$$

$$= \frac{(3x + 2)^7}{21}\Big|_{-2}^0$$

$$= \frac{2^7}{21} - \frac{(-4)^7}{21}$$

$$= \frac{5504}{7}$$

47. $f(x) = 1 + e^{-x}$; $[0, 4]$

$$\int_0^4 (1 + e^{-x})\,dx = (x - e^{-x})\Big|_0^4$$

$$= (4 - e^{-4}) - (0 - e^0)$$
$$= 5 - e^{-4}$$
$$\approx 4.982$$

49. $f(x) = x^2 - 4x$; $g(x) = x - 6$

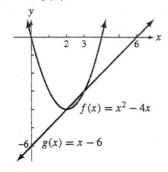

Find the points of intersection.

$$x^2 - 4x = x - 6$$
$$x^2 - 5x + 6 = 0$$
$$(x - 3)(x - 2) = 0$$
$$x = 2 \quad \text{or} \quad x = 3$$

Since $g(x) \geq f(x)$ in the interval $[2, 3]$, the area between the graphs is

$$\int_2^3 [g(x) - f(x)] \, dx$$

$$= \int_2^3 [(x - 6) - (x^2 - 4x)] \, dx$$

$$= \int_2^3 (-x^2 + 5x - 6) \, dx$$

$$= \left(\frac{-x^3}{3} + \frac{5x^2}{2} - 6x \right) \Big|_2^3$$

$$= \frac{-27}{3} + \frac{5(9)}{2} - 6(3) - \frac{-8}{3}$$

$$\quad - \frac{5(4)}{2} + 6(2)$$

$$= -\frac{19}{3} + \frac{25}{2} - 6 = \frac{1}{6}.$$

51. $f(x) = 5 - x^2$, $g(x) = x^2 - 3$, $x = 0$, $x = 4$

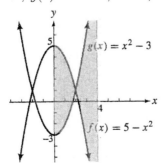

Find the points of intersection.

$$5 - x^2 = x^2 - 3$$
$$8 = 2x^2$$
$$4 = x^2$$
$$\pm 2 = x$$

The curves intersect at $x = 2$ and $x = -2$.

Thus, the area is

$$\int_0^2 [(5 - x^2) - (x^2 - 3)] \, dx$$

$$+ \int_2^4 [(x^2 - 3) - (5 - x^2)] \, dx$$

$$= \int_0^2 (-2x^2 + 8) \, dx + \int_2^4 (2x^2 - 8) \, dx$$

$$= \left(\frac{-2x^3}{3} + 8x \right) \Big|_0^2 + \left(\frac{2x^3}{3} - 8x \right) \Big|_2^4$$

$$= \frac{-16}{3} + 16 + \left(\frac{128}{3} - 32 \right) - \left(\frac{16}{3} - 16 \right)$$

$$= \frac{32}{3} + \frac{128}{3} - 32 - \frac{16}{3} + 16$$

$$= 32.$$

53. $\int_2^{10} \frac{x \, dx}{x - 1}$

Trapezoidal Rule:

$$n = 4, \ b = 10, \ a = 2, \ f(x) = \frac{x}{x-1}$$

i	x_i	$f(x_i)$
0	2	2
1	4	$\frac{4}{3}$
2	6	$\frac{6}{5}$
3	8	$\frac{8}{7}$
4	10	$\frac{10}{9}$

$$\int_2^{10} \frac{x}{x - 1} \, dx$$

$$\approx \frac{10 - 2}{4} \left[\frac{1}{2}(2) + \frac{4}{3} + \frac{6}{5} + \frac{8}{7} + \frac{1}{2} \left(\frac{10}{9} \right) \right]$$

$$\approx 10.46$$

Exact Value:

Let $u = x - 1$, so that $du = dx$ and $x = u + 1$.

Then

$$\int_2^{10} \frac{x}{x-1}\,dx = \int_1^9 \frac{u+1}{u}\,du$$

$$= \int_1^9 \left(1 + \frac{1}{u}\right) du$$

$$= \int_1^9 du + \int_1^9 \frac{1}{u}\,du$$

$$= u\Big|_1^9 + \ln|u|\Big|_1^9$$

$$= (9-1) + (\ln 9 - \ln 1)$$

$$= 8 + \ln 9 \approx 10.20.$$

55. $\int_1^3 \frac{\ln x}{x}\,dx$

Simpson's rule:

$n = 4,\ b = 3,\ a = 1,\ f(x) = \frac{\ln x}{x}$

i	x_i	$f(x_i)$
0	1	0
1	1.5	0.27031
2	2	0.34657
3	2.5	0.36652
4	3	0.3662

$$\int_1^3 \frac{\ln x}{x}\,dx$$

$$\approx \frac{3-1}{3(4)}[0 + 4(0.27031) + 2(0.34657)$$

$$+ 4(0.36652) + 0.3662]$$

$$\approx 0.6011$$

This answer is close to the value of 0.6035 obtained from the exact integral in Exercise 52.

57. $\int_0^1 e^x \sqrt{e^x + 4}\,dx$

Simpson's rule:

$n = 4,\ b = 1,\ a = 0,\ f(x) = e^x \sqrt{e^x + 4}$

i	x_i	$f(x_i)$
0	0	2.236
1	0.25	2.952
2	0.5	3.919
3	0.75	5.236
4	1	7.046

$$\int_0^1 e^x \sqrt{e^x + 4}\,dx$$

$$= \frac{1-0}{3(4)}[2.236 + 4(2.952) + 2(3.919)$$

$$+ 4(5.236) + 7.046$$

$$\approx 4.156$$

This answer is close to the answer of 4.155 obtained from the exact integral in Exercise 54.

59. $\int_{-2}^2 [x(x-1)(x+1)(x-2)(x+2)]^2\,dx$

(a) Trapezoidal Rule:

$n = 4,\ b = -2,\ a = 2,$

$$f(x) = [x(x-1)(x+1)(x-2)(x+2)]^2$$

i	x_i	$f(x_i)$
0	-2	0
1	-1	0
2	0	0
3	1	0
4	2	0

$$\int_{-2}^2 [x(x-1)(x+1)(x-2)(x+2)]^2\,dx$$

$$\approx \frac{2-(-2)}{4}\left[\frac{1}{2}(0) + 0 + 0 + 0 + \frac{1}{2}(0)\right]$$

$$= 0$$

(b) Simpson's Rule

$n = 4,\ b = 2,\ a = 2,$

$$f(x) = [x(x-1)(x+1)(x-2)(x+2)]^2$$

i	x_i	$f(x_i)$
0	-2	0
1	-1	0
2	0	0
3	1	0
4	2	0

$$\int_{-2}^2 [x(x-1)(x+1)(x-2)(x+2)]^2\,dx$$

$$\approx \frac{2-(-2)}{3(4)}[0 + 4(0) + 2(0) + 4(0) + 0]$$

$$= 0$$

61. $C'(x) = 3\sqrt{2x - 1}$; 13 units cost $270.

$$C(x) = \int 3(2x - 1)^{1/2}\, dx = \frac{3}{2} \int 2(2x - 1)^{1/2}\, dx$$

Let $u = 2x - 1$, so that

$$du = 2\, dx. = \frac{3}{2} \int u^{1/2}\, du = \frac{3}{2} \left(\frac{u^{3/2}}{3/2} \right) + C = (2x - 1)^{3/2} + C$$

$$C(13) = [2(13) - 1]^{3/2} + C$$

Since $C(13) = 270$,

$$270 = 25^{3/2} + C$$
$$270 = 125 + C$$
$$C = 145.$$

Thus,

$$C(x) = (2x - 1)^{3/2} + 145.$$

63. Read values for the rate of investment income accumulation for every 2 years from year 1 to year 9. These are the heights of rectangles with width $\Delta x = 2$.

Total accumulated income $= 11,000(2) + 9000(2) + 12,000(2) + 10,000(2) + 6000(2) \approx \$96,000$

65. $S'(x) = 3\sqrt{2x + 1} + 3$

$$S(x) = \int_0^4 (3\sqrt{2x + 1} + 3)\, dx = [(2x + 1)^{3/2} + 3x]\Big|_0^4 = (27 + 12) - (1 + 0) = 38$$

Total sales $= \$38,000$.

67. $S(q) = q^2 + 5q + 100$
$D(q) = 350 - q^2$
$S(q) = D(q)$ at the equilibrium point.

$$q^2 + 5q + 100 = 350 - q^2$$
$$2q^2 + 5q - 250 = 0$$
$$(-2q + 25)(q - 10) = 0$$
$$q = -\frac{25}{2} \quad \text{or} \quad q = 10$$

Since the number of units produced would not be negative, the equilibrium point occurs when $q = 10$.

Equilibrium supply
$= (10)^2 + 5(10) + 100 = 250$

Equilibrium demand
$= 350 - (10)^2 = 250$

(a) Producers' surplus $= \displaystyle\int_0^{10} [250 - (q^2 + 5q + 100)]\, dx = \int_0^{10} (-q^2 - 5q + 150)\, dx$

$$= \left(\frac{-q^3}{3} - \frac{5q^2}{2} + 150q \right)\Big|_0^{10} = \frac{-1000}{3} - \frac{500}{2} + 1500$$

$$= \cdot\frac{\$2750}{3} \approx \$916.67$$

(b) Consumers' surplus $= \displaystyle\int_0^{10} [(350 - q^2) - 250]\, dx = \int_0^{10} (100 - q^2)\, dx = \left(100q - \frac{q^3}{3} \right)\Big|_0^{10}$

$$= 1000 - \frac{1000}{3} = \frac{\$2000}{3} \approx \$666.67$$

69. (a) Total amount $= \frac{1}{2}(2.394) + 2.366 + 2.355 + 2.282 + 2.147 + 2.131 + 2.118 + 2.097 + 2.073 + 1.983 + \frac{1}{2}(1.869)$

$$\approx 21.684$$

This calculation yields a total of 21.684 billion barrels.

71. $f(t) = 100 - t\sqrt{0.4t^2 + 1}$

The total number of additional spiders in the first ten months is

$$\int_0^{10} (100 - t\sqrt{0.4t^2 + 1})dt,$$

where t is the time in months.

$$= \int_0^{10} 100 dt - \int_0^{10} t\sqrt{0.4t^2 + 1}\, dt.$$

Let $u = 0.4t^2 + 1$, so that

$$du = 0.8t\, dt \text{ and } \tfrac{1}{0.8}\, du = t\, dt.$$

When $t = 10, u = 41$.
When $t = 0, u = 1$.

$$= \int_0^{10} 100 dt - \frac{1}{0.8}\int_1^{41} u^{1/2}\, du$$

$$= 100t\Big|_0^{10} - \frac{5}{4}\cdot\frac{u^{3/2}}{\frac{3}{2}}\Big|_1^{41}$$

$$= 1000 - \frac{5}{6}u^{3/2}\Big|_1^{41}$$

$$\approx 782$$

The total number of additional spiders in the first 10 months is about 782.

73. (a) The total area is the area of the triangle on $[0, 12]$ with height 0.024 plus the area of the rectangle on $[12, 17.6]$ with height 0.024.

$$A = \frac{1}{2}(12 - 0)(0.024) + (17.6 - 12)(0.024) = 0.144 + 0.1344 = 0.2784$$

(b) On $[0, 12]$ we defined the function $f(x)$ with slope $\frac{0.024 - 0}{12 - 0} = 0.002$ and y-intercept 0.

$$f(x) = 0.002x$$

On $[12, 17.6]$, define $g(x)$ as the constant value.

$$g(x) = 0.024.$$

The area is the sum of the integrals of these two functions.

$$A = \int_0^{12} 0.002x\,dx + \int_{12}^{17.6} 0.024\,dx = 0.001x^2\Big|_0^{12} + 0.024x\Big|_{12}^{17.6}$$

$$= 0.001(12^2 - 0^2) + 0.024(17.6 - 12) = 0.144 + 0.1344 = 0.2784$$

75. (a) Total amount $= \frac{1}{2}(271{,}553) + 278{,}325 + 274{,}690 + 290{,}525 + 289{,}890$

$$+ \, 309{,}569 + 317{,}567 + 335{,}869 + 331{,}055 + \frac{1}{2}(331{,}208)$$

$$\approx 2{,}728{,}871$$

This calculation yields a total of about \$2,728,871.

77. For each month, subtract the average temperature from 65° (if it falls below 65°F), then multiply this number times the number of days in the month. The sum is the total number of heating degree days. Readings may vary, but the sum is approximately 4800 degree-days. (The actual value is 4868 degree-days.)

Chapter 15 Test

[15.1]

1. Explain what is meant by an antiderivative of a function $f(x)$.

Find each indefinite integral.

2. $\displaystyle\int \left(3x^3 - 5x^2 + x + 1\right)\, dx$

3. $\displaystyle\int \left(\frac{5}{x} + e^{0.5x}\right)\, dx$

4. $\displaystyle\int \frac{2x^3 - 3x^2}{\sqrt{x}}\, dx$

5. Find the cost function $C(x)$ if the marginal cost function is given by $C'(x) = 200 + 2x^{-1/4}$ and 16 units cost $4000.

6. A ball is thrown upward at time $t = 0$ with initial velocity of 64 feet per second from a height of 100 feet. Assume that $a(t) = -32$ feet per second per second. Find $v(t)$ and $s(t)$.

[15.2]

Use substitution to find each indefinite integral.

7. $\displaystyle\int 6x^2 \left(3x^3 - 5\right)^8\, dx$

8. $\displaystyle\int \frac{6x + 5}{3x^2 + 5x}\, dx$

9. $\displaystyle\int \sqrt[3]{2x^2 - 8x}\,(x - 2)\, dx$

10. $\displaystyle\int 4x^3 e^{-x^4}\, dx$

11. A city's population is predicted to grow at a rate of

$$P'(x) = \frac{400e^{10t}}{1 + e^{10t}}$$

people per year where t is the time in years from the present. Find the total population 3 years from now if $P(0) = 100,000$.

[15.3]

12. Evaluate $\displaystyle\sum_{i=1}^{4} \frac{2}{i^2}$.

13. Approximate the area under the graph of $f(x) = x^2 + x$ and above the x-axis from $x = 0$ to $x = 2$ using four rectangles of equal width. Let the height of each rectangle be the function value at the left endpoint.

14. Approximate the value of $\int_1^5 x^2\, dx$ by summing the areas of rectangles. Use four rectangles of equal width. Use the left endpoints, then the right endpoints; then give the average of these answers.

[15.4]

Evaluate the following definite integrals.

15. $\displaystyle\int_1^5 \left(4x^3 - 5x\right) dx$

16. $\displaystyle\int_1^5 \left(\frac{3}{x^2} + \frac{2}{x}\right) dx$

17. $\displaystyle\int_1^2 \sqrt{3r - 2}\, dr$

18. $\displaystyle\int_0^1 4xe^{x^2+1}\, dx$

19. $\displaystyle\int_{-2}^1 3x \left(x^2 - 4\right)^5 dx$

20. Find the area of the region between the x-axis and $f(x) = e^{x/2}$ on the interval $[0, 2]$.

21. The rate at which a substance grows is given by $R(x) = 500e^{0.5x}$, where x is the time in days. What is the total accumulated growth after 4 days?

[15.5]

22. Find the area of the region enclosed by $f(x) = -x + 4$, $g(x) = -x^2 + 6x - 6$, $x = 2$, and $x = 4$.

23. Find the area of the region enclosed by $f(x) = 5x$ and $g(x) = x^3 - 4x$.

24. A company has determined that the use of a new process would produce a savings rate (in thousands of dollars) of

$$S(x) = 2x + 7,$$

where x is the number of years the process is used. However, the use of this process also creates additional costs (in thousands of dollars) according to the rate-of-cost function

$$C(x) = x^2 + 2x + 3.$$

(a) For how many years does the new process save the company money?

(b) Find the net savings in thousands of dollars over this period.

25. Suppose that the supply function of a commodity is $p = 0.05q + 5$ and the demand function is $p = 12 - 0.02q$.

(a) Find the producers' surplus.   (b) Find the consumers' surplus.

[15.6]

26. Use $n = 4$ to approximate the value of the given integral by the following methods: **(a)** the trapezoidal rule and **(b)** Simpson's rule. **(c)** Find the exact value by integration.

$$\int_0^2 x\sqrt{x^2 + 1}\, dx$$

27. Use $n = 4$ to approximate the value of the given integral by the following methods: **(a)** the trapezoidal rule and **(b)** Simpson's rule.

$$\int_0^2 \frac{1}{\sqrt{1+x^3}}\, dx$$

28. Use Simpson's rule with $n = 6$ to approximate the value of $\int_0^3 \frac{1}{x^2+1}\, dx$.

29. Find the area between the curve $y = e^{-x^2}$ and the x-axis from $x = 0$ to $x = 3$, using the trapezoidal rule with $n = 6$.

30. Find the area between the curve $y = \frac{x}{x^2+1}$ and the x-axis from $x = 1$ to $x = 4$, using Simpson's rule with $n = 6$.

Chapter 15 Test Answers

1. An antiderivative of a function $f(x)$ is a function $F(x)$ such that $F'(x) = f(x)$.

2. $\frac{3}{4}x^4 - \frac{5}{3}x^3 + \frac{x^2}{2} + x + C$

3. $5\ln|x| + 2e^{0.5x} + C$

4. $\frac{4}{7}x^{7/2} - \frac{6}{5}x^{5/2} + C$

5. $C(x) = 200x + \frac{8}{3}x^{3/4} + 778.67$

6. $v(t) = -32t + 64$; $s(t) = -16t^2 + 64t + 100$

7. $\frac{2}{27}\left(3x^3 - 5\right)^9 + C$

8. $\ln\left|3x^2 + 5x\right| + C$

9. $\frac{3}{16}\left(2x^2 - 8x\right)^{4/3} + C$

10. $-e^{-x^4} + C$

11. 101,172

12. $2\frac{61}{72}$ or 2.85

13. 3.25

14. Left endpoints: 30; right endpoints: 54; average: 42

15. 564

16. 5.62

17. $\frac{14}{9}$ or 1.56

18. $2e^2 - 2e$ or 9.34

19. $\frac{729}{4}$ or 182.25

20. $2e - 2$ or 3.44

21. 6389

22. $\frac{10}{3}$ or 3.33

23. 40.5

24. (a) 2 years   (b) $5.33 thousand

25. (a) $250   (b) $100

26. (a) 3.457   (b) 3.392   (c) 3.393

27. (a) 1.397   (b) 1.405

28. 1.25

29. 0.89

30. 1.07

FURTHER TECHNIQUES AND APPLICATIONS OF INTEGRATION

16.1 Integration by Parts

1. $\int xe^x\,dx$

   Let $dv = e^x\,dx$ and $u = x$.

   Then $v = \int e^x\,dx$ and $du = dx$.

   $$v = e^x$$

   Use the formula

   $$\int u\,dv = uv - \int v\,du.$$

   $$\int xe^x\,dx = xe^x - \int e^x\,dx$$
   $$= xe^x - e^x + C$$

3. $\int (4x - 12)e^{-8x}\,dx$

   Let $dv = e^{-8x}\,dx$ and $u = 4x - 12$

   Then $v = \int e^{-8x}\,dx$ and $du = 4\,dx$.

   $$v = \frac{e^{-8x}}{-8}$$

   $$\int (4x - 12)e^{-8x}\,dx$$

   $$= (4x - 12)\left(\frac{e^{-8x}}{-8}\right) - \int \left(\frac{e^{-8x}}{-8}\right) \cdot 4\,dx$$

   $$= -\frac{4x}{8}e^{-8x} + \frac{12}{8}e^{-8x} - \left(-\frac{4}{8} \cdot \frac{e^{-8x}}{-8}\right) + C$$

   $$= -\frac{x}{2}e^{-8x} + \frac{3}{2}e^{-8x} - \frac{1}{16}e^{-8x} + C$$

   $$= \left(-\frac{x}{2} + \frac{23}{16}\right)e^{-8x} + C$$

5. $\int_0^1 \frac{2x + 1}{e^x}\,dx$

   $$= \int_0^1 (2x + 1)e^{-x}\,dx$$

   Let $dv = e^{-x}\,dx$ and $u = 2x + 1$.

   Then $v = \int e^{-x}\,dx$ and $du = 2\,dx$.

   $$v = -e^{-x}$$

   $$\int \frac{2x + 1}{e^x}\,dx$$

   $$= -(2x + 1)e^{-x} + \int 2e^{-x}\,dx$$

   $$= -(2x + 1)e^{-x} - 2e^{-x}$$

   $$\int_0^1 \frac{2x + 1}{e^x}\,dx$$

   $$= \left[-(2x + 1)e^{-x} - 2e^{-x}\right]\Big|_0^1$$

   $$= \left[-(3)e^{-1} - 2e^{-1}\right] - (-1 - 2)$$
   $$= -5e^{-1} + 3$$
   $$\approx 1.161$$

7. $\int \ln 3x\,dx$

   Let $dv = dx$ and $u = \ln 3x$.

   Then $v = x$ and $du = \frac{1}{x}\,dx$.

   $$\int \ln 3x\,dx = x\ln 3x - \int dx$$

   $$= x\ln 3x - x$$

   $$\int \ln 3x\,dx = (x\ln 3x - x)\Big|_1^9$$

   $$= (9\ln 27 - 9) - (\ln 3 - 1)$$
   $$= 9\ln 3^3 - 9 - \ln 3 + 1$$
   $$= 27\ln 3 - \ln 3 - 8$$
   $$= 26\ln 3 - 8 \approx 20.56$$

9. $\displaystyle\int x \ln dx$

Let $dv = x\, dx$ and $u = \ln x$.

Then $v = \frac{x^2}{2}$ and $du = \frac{1}{x}\, dx$.

$$\int x \ln dx = \frac{x^2}{2} \ln x - \int \frac{x}{2}\, dx$$

$$= \frac{x^2 \ln x}{2} - \frac{x^2}{4} + C$$

11. The area is $\displaystyle\int_2^4 (x-2)e^x\, dx$.

Let $dv = e^x\, dx$ and $u = x - 2$.
Then $v = e^x$ and $du = dx$.

$$\int (x-2)e^x\, dx = (x-2)e^x - \int e^x\, dx$$

$$\int_1^4 (x-2)e^x\, dx = [(x-2)e^x - e^x]\Big|_2^4$$

$$= (2e^4 - e^4) - (0 - e^2)$$

$$= e^4 + e^2 \approx 61.99$$

13. $\displaystyle\int x^2 e^{2x}\, dx$

Let $u = x^2$ and $dv = e^{2x}\, dx$.
Use column integration.

D		I
x^2	$+$	e^{2x}
$2x$	$-$	$\frac{e^{2x}}{2}$
2	$+$	$\frac{e^{2x}}{4}$
0		$\frac{e^{2x}}{8}$

$$\int x^2 e^{2x}\, dx = x^2\left(\frac{e^{2x}}{2}\right) - 2x\left(\frac{e^{2x}}{4}\right) + \frac{2e^{2x}}{8} + C$$

$$= \frac{x^2 e^{2x}}{2} - \frac{xe^{2x}}{2} + \frac{e^{2x}}{4} + C$$

15. $\displaystyle\int_0^5 x\sqrt[3]{x^2 + 2}\, dx$

$$= \int_0^5 x(x^2 + 2)^{1/2}\, dx$$

$$= \frac{1}{2}\int_0^5 2x(x^2 + 2)^{1/3}\, dx$$

Let $u = x^2 + 2$. Then $du = 2x\, dx$.
If $x = 5$, $u = 27$. If $x = 0$, $u = 2$.

$$= \frac{1}{2}\int_2^{27} u^{1/3}\, du$$

$$= \frac{1}{2}\left(\frac{u^{4/3}}{1}\right)\left(\frac{3}{4}\right)\Big|_2^{27}$$

$$= \frac{3}{8}(27)^{4/3} - \frac{3}{8}(2)^{4/3}$$

$$= \frac{243}{8} - \frac{3(2^{4/3})}{8}$$

$$= \frac{243}{8} - \frac{3\sqrt[3]{2}}{4} \approx 29.43$$

17. $\displaystyle\int (8x + 10)\ln(5x)dx$

Let $dv = (8x + 10)\, dx$ and $u = \ln(5x)$.

Then $v = 4x^2 + 10x$ and $du = \frac{1}{x}\, dx$.

$$\int (8x + 10)\ln(5x)dx$$

$$= (4x^2 + 10x)\ln(5x) - \int (4x^2 + 10x)\left(\frac{1}{x}\right)dx$$

$$= (4x^2 + 10x)\ln(5x) - \int (4x + 10)dx$$

$$= (4x^2 + 10x)\ln(5x) - 2x^2 - 10x + C$$

19. $\displaystyle\int x^2\sqrt{x + 4}\, dx$

Let $u = x^2$ and $dv = (x+4)^{1/2}$. Use column integration.

D		I
x^2	$+$	$(x+4)^{1/2}$
$2x$	$-$	$\frac{2}{3}(x+4)^{3/2}$
2	$+$	$(\frac{2}{3})(\frac{2}{5})(x+4)^{5/2}$
0		$(\frac{2}{3})(\frac{2}{5})(\frac{2}{7})(x+4)^{7/2}$

$$\int x^2 \sqrt{x+4}\,dx$$

$$= x^2(x+4)^{3/2}\left(\frac{2}{3}\right) - 2x(x+4)^{5/2}\left(\frac{2}{3}\right)\left(\frac{2}{5}\right)$$

$$+ 2(x+4)^{7/2}\left(\frac{2}{3}\right)\left(\frac{2}{5}\right)\left(\frac{2}{7}\right) + C$$

$$= \frac{2}{3}x^2(x+4)^{3/2} - \frac{8}{15}x(x+4)^{5/2}$$

$$+ \frac{16}{105}(x+4)^{7/2} + C$$

21. $\displaystyle\int_0^1 \frac{x^3\,dx}{\sqrt{3+x^2}} = \int_0^1 x^3(3x+x^2)^{-1/2}\,dx$

Let $dv = x(3+x^2)^{-1/2}\,dx$ and $u = x^2$.

Then $v = \frac{2(3+x^2)^{1/2}}{2}$

$\qquad v = (3+x^2)^{1/2}$ and $du = 2x\,dx$.

$$\int \frac{x^3\,dx}{\sqrt{3+x^2}}$$

$$= x^2(3+x^2)^{1/2} - \int 2x(3+x^2)^{1/2}\,dx$$

$$= x^2(3+x^2)^{1/2} - \frac{2}{3}(3+x^2)^{3/2}$$

$$\int_0^1 \frac{x^3\,dx}{\sqrt{3+x^2}}$$

$$= \left[x^2(3+x^2)^{1/2} - \frac{2}{3}(3+x^2)^{3/2}\right]\Big|_0^1$$

$$= 4^{1/2} - \frac{2}{3}(4^{3/2}) - 0 + \frac{2}{3}(3^{3/2})$$

$$= 2 - \frac{2}{3}(8) + \frac{2}{3}(3^{3/2})$$

$$= -\frac{10}{3} + 2\sqrt{3}$$

$$\approx 0.1308$$

23. $\displaystyle\int \frac{16}{\sqrt{x^2+16}}\,dx$

Use entry 5 from the table of integrals with $a = 4$.

$$\int \frac{16}{\sqrt{x^2+16}}\,dx = 16\int \frac{1}{\sqrt{x^2+4^2}}\,dx$$

$$= 16\ln\left| x + \sqrt{x^2+16}\right| + C$$

25. $\displaystyle\int \frac{3}{x\sqrt{121-x^2}}\,dx$

$$= 3\int \frac{dx}{x\sqrt{11^2-x^2}}$$

If $a = 11$, this integral matches entry 9 in the table.

$$= 3\left(-\frac{1}{11}\ln\left|\frac{11+\sqrt{121-x^2}}{x}\right|\right) + C$$

$$= -\frac{3}{11}\ln\left|\frac{11+\sqrt{121-x^2}}{x}\right| + C$$

27. $\displaystyle\int \frac{-6}{x(4x+6)^2}\,dx$

Use entry 14 from the table of integrals with $a = 4$ and $b = 6$.

$$\int \frac{-6}{x(4x+6)^2}\,dx$$

$$= -6\int \frac{1}{x(4x+6)^2}\,dx$$

$$= -6\left[\frac{1}{6(4x+6)} + \frac{1}{6^2}\ln\left|\frac{x}{4x+6}\right|\right] + C$$

$$= \frac{-1}{(4x+6)} - \frac{1}{6}\ln\left|\frac{x}{4x+6}\right| + C$$

31. First find the indefinite integral using integration by parts.

$$\int u\,dv = uv - \int v\,du$$

Now substitute the given values.

$$\int_0^1 u\,dv = uv\Big|_0^1 - \int_0^1 v\,du$$

$$= [u(1)v(1) - u(0)v(0)] - 4$$

$$= (3)(-4) - (2)(1) - 4 = -18$$

33. The area between the x-axis and the nonnegative function $h(x) = s(x)\frac{dr}{dx}$ on the interval $[0, 2]$ is

$$\int_0^2 s(x)\frac{dr}{dx}\,dx = \int_0^2 s(x)\,dr.$$

The area between the x-axis and the nonnegative function $d(x) = r(x)\frac{ds}{dx}$ on the interval $[0, 2]$ is

$$\int_0^2 r(x)\frac{ds}{dx}\,dx = \int_0^2 r(x)\,ds.$$

Rewrite the integration by parts formula in terms of r and s.

$$\int r\,ds = rs - \int s\,dr$$

Therefore, substituting the given values, we have

$$r(x)\,ds = \int_0^2 - \int_0^2 s(x)\,dr$$
$$10 = [r(2)s(2) - r(0)s(0)] - 5$$
$$15 = r(2)s(2) - 0 \cdot s(0)$$
$$15 = r(2)s(2).$$

35. $\displaystyle \int x^n \cdot \ln|x|\,dx, n \neq -1$

Let $u = \ln|x|$ and $dv = x^n\,dx$.

Use column integration.

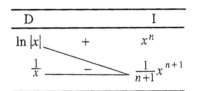

D		I
$\ln|x|$	$+$	x^n
$\dfrac{1}{x}$	$-$	$\dfrac{1}{n+1}x^{n+1}$

$$\int x^n \cdot \ln|x|\,dx$$

$$= \frac{1}{n+1}x^{n+1}\ln|x| - \int \left[\frac{1}{x} \cdot \frac{1}{n+1}x^{n+1}\right]dx$$

$$= \frac{1}{n+1}x^{n+1}\ln|x| - \int \frac{1}{n+1}x^n\,dx$$

$$= \frac{1}{n+1}x^{n+1}\ln|x| - \frac{1}{(n+1)^2}x^{n+1} + C$$

$$= x^{n+1}\left[\frac{\ln|x|}{n+1} - \frac{1}{(n+1)^2}\right] + C$$

37. $\displaystyle \int x\sqrt{x+1}\,dx$

(a) Let $u = x$ and $dv = \sqrt{x+1}\,dx$.
Use column integration.

D		I
x	$+$	$\sqrt{x+1}$
1	$-$	$\left(\frac{2}{3}\right)(x+1)^{3/2}$
0		$\left(\frac{4}{15}\right)(x+1)^{5/2}$

$$\int x\sqrt{x+1}\,dx$$

$$= \left(\frac{2}{3}\right)x(x+1)^{3/2} - \left(\frac{4}{15}\right)(x+1)^{5/2} + C$$

(b) Let $u = x+1$; then $u - 1 = x$ and $du = dx$.

$$\int x\sqrt{x+1}\,dx$$

$$= \int (u-1)u^{1/2}\,du = \int (u^{3/2} - u^{1/2})\,du$$

$$= \frac{2}{5}u^{5/2} - \frac{2}{3}u^{3/2} + C$$

$$= \frac{2}{5}(x+1)^{5/2} - \frac{2}{3}(x+1)^{3/2} + C$$

(c) Both results factor as $\frac{2}{15}(x+1)^{3/2}(3x-2)+C$, so they are equivalent.

39. $\displaystyle R = \int_0^{12} (x+1)\ln(x+1)\,dx$

Let $u = \ln(x+1)$ and $dv = (x+1)dx$.

Then $du = \dfrac{1}{x+1}\,dx$ and $v = \dfrac{1}{2}(x+1)^2$.

$$\int (x+1)\ln(x+1)\,dx$$

$$= \frac{1}{2}(x+1)^2\ln(x+1)$$
$$\quad - \int \left[\frac{1}{2}(x+1)^2 \cdot \frac{1}{x+1}\right]dx$$

$$= \frac{1}{2}(x+1)^2\ln(x+1) - \int \frac{1}{2}(x+1)dx$$

$$= \frac{1}{2}(x+1)^2\ln(x+1) - \frac{1}{4}(x+1)^2 + C$$

$$\int_0^{12} (x+1)\ln(x+1)\,dx$$

$$= \left[\frac{1}{2}(x+1)^2\ln(x+1) - \frac{1}{4}(x+1)^2\right]\Bigg|_0^{12}$$

$$= \frac{169}{2}\ln 13 - 42 \approx \$174.74$$

41. The total accumulated growth of the microbe population during the first 2 days is given by

$$\int_0^2 27xe^{3x}\,dx.$$

Let $\quad dv = e^{3x}\,dx \quad$ and $\quad u = 27x.$

Then $\quad v = \dfrac{e^{3x}}{3} \quad$ and $\quad du = 27\,dx.$

$$\int 27xe^{3x}\,dx = 27x \cdot \frac{e^{3x}}{3} - \int \frac{e^{3x}}{3} \cdot 27\,dx$$

$$= 9xe^{3x} - 3e^{3x}$$

$$\int_0^2 27xe^{3x}\,dx = \left(9xe^{3x} - 3e^{3x}\right)\Big|_0^2$$

$$= (18e^6 - 3e^6) - (0 - 3)$$

$$= 15e^6 + 3 \approx 6054$$

43. $\displaystyle\int_0^6 (-10.28 + 175.9te^{-t/1.3})\,dt$

$$= -10.28t + 175.9\int te^{-t/1.3}\,dt$$

Evaluate this integral using integration by parts.
Let $\quad u = t \quad$ and $\quad dv = e^{-t/1.3}dt.$
Then $\quad du = dt \quad$ and $\quad v = -1.3e^{-t/1.3}.$

$$\int te^{-t/1.3}\,dt$$

$$= (t)(-1.3e^{-t/1.3}) - \int(-1.3e^{-t/1.3})\,dt$$

$$= -1.3te^{-t/1.3} - 1.69e^{-t/1.3} + C$$

Substitute this expression in the earlier expression.

$$-10.28t + 175.9(-1.3te^{-t/1.3} - 1.69e^{-t/1.3})\Big|_0^6$$

$$= -10.28t - 228.67te^{-t/1.3} - 297.271e^{-t/1.3}\Big|_0^6$$

$$= (-61.68 - 1669.291e^{-6/1.3}) - (-297.271)$$

$$\approx 219.07$$

The total thermic energy is about 219 kJ.

16.2   Volume and Average Value

1. $f(x) = x,\ y = 0,\ x = 0,\ x = 3$

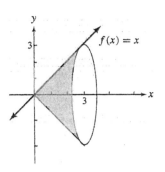

$$V = \pi \int_0^3 x^2\,dx = \frac{\pi x^3}{2}\bigg|_0^3 = \frac{\pi(27)}{3} - 0 = 9\pi$$

3. $f(x) = 2x + 1,\ y = 0,\ x = 0,\ x = 4$

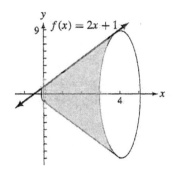

$$V = \pi \int_0^4 (2x + 1)^2\,dx$$

Let $u = 2x + 1$. Then $du = 2\,dx$.
If $x = 4$, $u = 9$. If $x = 0$, $u = 1$.

$$V = \frac{1}{2}\pi \int_0^4 2(2x + 1)^2\,dx$$

$$= \frac{1}{2}\pi \int_1^9 u^2\,du$$

$$= \frac{\pi}{2}\left(\frac{u^3}{3}\right)\bigg|_1^9$$

$$= \frac{\pi}{2}\left(\frac{729}{3} - \frac{1}{3}\right)$$

$$= \frac{728\pi}{6}$$

$$= \frac{364\pi}{3}$$

5. $f(x) = \frac{1}{3}x + 2$, $y = 0$, $x = 1$, $x = 3$

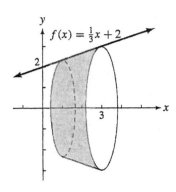

$$V = \pi \int_1^3 \left(\frac{1}{3}x + 2\right)^2 dx$$

$$= 3\pi \int_1^3 \frac{1}{3}\left(\frac{1}{3}x + 2\right)^2 dx$$

$$= 3\pi \frac{\left(\frac{1}{3}x + 2\right)^3}{3}\Big|_1^3$$

$$= \pi \left(\frac{1}{3}x + 2\right)^3\Big|_1^3$$

$$= 27\pi - \frac{343\pi}{27}$$

$$= \frac{386\pi}{27}$$

7. $f(x) = \sqrt{x}$, $y = 0$, $x = 1$, $x = 4$

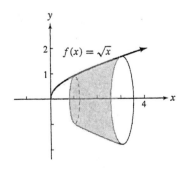

$$V = \pi \int_1^4 (\sqrt{x})^2 dx = \pi \int_1^4 x \, dx$$

$$= \frac{\pi x^2}{2}\Big|_1^4$$

$$= 8\pi - \frac{\pi}{2}$$

$$= \frac{15\pi}{2}$$

9. $f(x) = \sqrt{2x + 1}$, $y = 0$, $x = 1$, $x = 4$

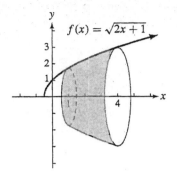

$$V = \pi \int_1^4 (\sqrt{2x + 1})^2 dx$$

$$= \pi \int_1^4 (2x + 1) dx$$

$$= \pi \left(\frac{2x^2}{2} + x\right)\Big|_1^4$$

$$= \pi[(16 + 4) - 2]$$

$$= 18\pi$$

11. $f(x) = e^x$; $y = 0$, $x = 0$, $x = 2$

$$V = \pi \int_0^2 e^{2x} dx = \frac{\pi e^{2x}}{2}\Big|_0^2$$

$$= \frac{\pi e^4}{2} - \frac{\pi}{2}$$

$$= \frac{\pi}{2}(e^4 - 1)$$

$$\approx 84.19$$

13. $f(x) = \frac{2}{\sqrt{x}}$, $y = 0$, $x = 1$, $x = 3$

$$V = \pi \int_1^3 \left(\frac{2}{\sqrt{x}}\right)^2 dx$$

$$= \pi \int_1^3 \frac{4}{x} dx$$

$$= 4\pi \ln|x|\Big|_1^3$$

$$= 4\pi(\ln 3 - \ln 1)$$

$$= 4\pi \ln 3 \approx 13.81$$

15. $f(x) = x^2$, $y = 0$, $x = 1$, $x = 5$

$$V = \pi \int_1^5 x^4 dx = \frac{\pi x^5}{5}\Big|_1^5 = 625\pi - \frac{\pi}{5} = \frac{3124\pi}{5}$$

17. $f(x) = 1 - x^2$, $y = 0$

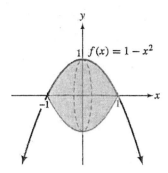

Since $f(x) = 1 - x^2$ intersects $y = 0$ where

$$1 - x^2 = 0$$
$$x = \pm 1,$$
$$a = -1 \quad \text{and} \quad b = 1.$$

$$V = \pi \int_{-1}^{1} (1 - x^2)^2 \, dx$$

$$= \pi \int_{-1}^{1} (1 - 2x^2 + x^4) \, dx$$

$$= \pi \left(x - \frac{2x^3}{3} + \frac{x^5}{5} \right) \Big|_{-1}^{1}$$

$$= \pi \left(1 - \frac{2}{3} + \frac{1}{5} \right) - \pi \left(-1 + \frac{2}{3} - \frac{1}{5} \right)$$

$$= 2\pi - \frac{4\pi}{3} + \frac{2\pi}{5}$$

$$= \frac{16\pi}{15}$$

19. $f(x) = \sqrt{1 - x^2}$
$r = \sqrt{1} = 1$

$$V = \pi \int_{-1}^{1} (\sqrt{1 - x^2})^2 \, dx$$

$$= \pi \int_{-1}^{1} (1 - x^2) \, dx$$

$$= \pi \left(x - \frac{x^3}{3} \right) \Big|_{-1}^{1}$$

$$= \pi \left(1 - \frac{1}{3} \right) - \pi \left(-1 + \frac{1}{3} \right)$$

$$= 2\pi - \frac{2}{3}\pi$$

$$= \frac{4\pi}{3}$$

21. $f(x) = \sqrt{r^2 - x^2}$

$$V = \pi \int_{-r}^{r} (\sqrt{r^2 - x^2})^2 \, dx$$

$$= \pi \int_{-r}^{r} (r^2 - x^2) \, dx$$

$$= \pi \left(r^2 x - \frac{x^3}{3} \right) \Big|_{-r}^{r}$$

$$= \pi \left(r^3 - \frac{r^3}{3} \right) - \pi \left(-r^3 + \frac{r^3}{3} \right)$$

$$= 2r^3 \pi - \left(\frac{2r^3 \pi}{3} \right)$$

$$= \frac{4r^3 \pi}{3}$$

23. $f(x) = r$, $x = 0$, $x = h$

Graph $f(x) = r$; then show the solid of revolution formed by rotating about the x-axis the region bounded by $f(x)$, $x = 0$, $x = h$.

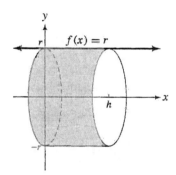

$$\int_{0}^{h} \pi r^2 \, dx = \pi r^2 x \Big|_{0}^{h}$$

$$= \pi r^2 h - 0$$

$$= \pi r^2 h$$

25. $f(x) = x^2 - 4$; $[0, 5]$

Average value

$$= \frac{1}{5 - 0} \int_{0}^{5} (x^2 - 4) \, dx$$

$$= \frac{1}{5} \left(\frac{x^3}{3} - 4x \right) \Big|_{0}^{5}$$

$$= \frac{1}{5} \left[\left(\frac{125}{3} - 20 \right) - 0 \right]$$

$$= \frac{13}{3} \approx 4.333$$

27. $f(x) = \sqrt{x+1};\ [3, 8]$

Average value

$$= \frac{1}{8-3} \int_3^8 \sqrt{x+1}\, dx$$

$$= \frac{1}{5} \int_3^8 (x+1)^{1/2}\, dx$$

$$= \frac{1}{5} \cdot \frac{2}{3} (x+1)^{3/2} \Big|_3^8$$

$$= \frac{2}{15} (9^{3/2} - 4^{3/2})$$

$$= \frac{2}{15} (27 - 8) = \frac{38}{15} \approx 2.533$$

29. $f(x) = e^{x/7};\ [0, 7]$

Average value

$$= \frac{1}{7-0} \int_0^7 e^{x/7}\, dx$$

$$= \frac{1}{7} \cdot 7 e^{x/7} \Big|_0^7$$

$$= e^{x/7} \Big|_0^7 = e^1 - e^0$$

$$e - 1 \approx 1.718$$

31. $f(x) = x^2 e^{2x};\ [0, 2]$

Average value $= \dfrac{1}{2-0} \displaystyle\int_0^2 x^2 e^{2x}\, dx$

Let $u = x^2$ and $dv = e^{2x}\, dx$.
Use column integration.

D	I
x^2 $\quad$ +	e^{2x}
$2x$ $\quad$ −	$\frac{1}{2} e^{2x}$
2 $\quad$ +	$\frac{1}{4} e^{2x}$
0	$\frac{1}{8} e^{2x}$

$$\frac{1}{2-0} \int_0^2 x^2 e^{2x}\, dx$$

$$= \frac{1}{2} \left[(x^2) \left(\frac{1}{2} \right) e^{2x} - (2x) \left(\frac{1}{4} \right) e^{2x} \right.$$

$$\left. + 2 \left(\frac{1}{8} \right) e^{2x} \right] \Bigg|_0^2$$

$$= \frac{1}{2} \left(2e^4 - e^4 + \frac{1}{4} e^4 - \frac{1}{4} \right)$$

$$= \frac{5e^4 - 1}{8} \approx 34.00$$

33. $f(x) = e^{-x^2},\ y = 0,\ x = -1,\ x = 1$

$$V = \pi \int_{-1}^1 (e^{-x^2})^2 dx$$

$$= \pi \int_{-2}^2 e^{-2x^2} dx$$

Using an integration feature on a graphing calculator to evaluate the integral, we obtain $3.758249634 \approx 3.758$.

35. Use the formula for average value with $a = 0$ and $b = 6$.

$$\frac{1}{6-0} \int_0^6 (37 + 6e^{-0.03t})\, dt$$

$$= \frac{1}{6} \left(37t + \frac{6}{-0.03} e^{-0.03t} \right) \Bigg|_0^6$$

$$= \frac{1}{6} (37t - 200e^{-0.03t}) \Bigg|_0^6$$

$$= \frac{1}{6} [(222 - 200e^{-0.18}) - (0 - 200)]$$

$$= \frac{1}{6} (422 - 200e^{-0.18})$$

$$\approx 42.49$$

The average price is $42.49.

37. Use the formula for average value with $a = 0$ and $b = 6$. The average price is

$$\frac{1}{30-0} \int_0^{30} (600 - 20\sqrt{30t})\, dt$$

$$= \frac{1}{30} \left(600t - 20\sqrt{30} \cdot \frac{2}{3} t^{3/2} \right) \Bigg|_0^{30}$$

$$= \frac{1}{30} \left(600t - \frac{40\sqrt{30}}{3} t^{3/2} \right) \Bigg|_0^{30}$$

$$= \frac{1}{30} (18{,}000 - 12{,}000)$$

$$= 200 \text{ cases}$$

39. $R(t) = te^{-0.1t}$

"During the nth hour" corresponds to the interval $(n-1, n)$.

The average intensity during nth hour is

$$\frac{1}{n-(n-1)} \int_{n-1}^{n} te^{-0.1t}\,dt = \int_{n-1}^{n} te^{-0.1t}\,dt$$

Let $u = t$ and $dv = e^{-0.1t}\,dt$.

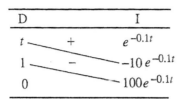

D	I
t $\quad+$	$e^{-0.1t}$
1 $\quad-$	$-10\,e^{-0.1t}$
0	$100e^{-0.1t}$

$$\int_{n-1}^{n} te^{-0.1t}\,dt$$

$$= (-10te^{-0.1t} - 100e^{-0.1t})\Big|_{n-1}^{n}$$

(a) Second hour, $n = 2$

Average intensity
$$= -10e^{-0.2}(12) + 10e^{-0.1}(11)$$
$$= 110e^{-0.1} - 120e^{-0.2}$$
$$\approx 1.284$$

(b) Twelfth hour, $n = 12$

Average intensity

$$= -10e^{-1.2}(12 + 10) + 10e^{-1.1}(11 + 10)$$
$$= 210e^{-1.1} - 220e^{-1.2}$$
$$\approx 3.640$$

(c) Twenty-fourth hour, $n = 24$

Average intensity
$$= -10e^{-2.4}(24 + 10) + 10e^{-2.3}(23 + 10)$$
$$= 330e^{-2.3} - 340e^{-2.4}$$
$$\approx 2.241$$

41. For each part below, use

Average value

$$= \frac{1}{b-a} \int_{a}^{b} 45\ln(t+1)\,dt$$

$$= \frac{45}{b-a} \int_{a}^{b} \ln(t+1)\,dt.$$

Solve the integral using integration by parts.

Let $\quad u = \ln(t+1) \quad$ and $\quad dv = dt$.

Then $\quad du = \dfrac{1}{t+1}\,dt \quad$ and $\quad v = t$.

$$\int \ln(t+1)\,dt$$

$$= t\ln(t+1) - \int \frac{t}{t+1}\,dt$$

$$= t\ln(t+1) - \int \left(1 - \frac{1}{t+1}\right)dt$$

$$= t\ln(t+1) - t + \ln(t+1) + C$$
$$= (t+1)\ln(t+1) - t + C$$

Therefore

Average value

$$= \frac{1}{b-a} \int_{a}^{b} 45\ln(t+1)\,dt$$

$$= \frac{45}{b-a}[(t+1)\ln(t+1) - t]\Big|_{a}^{b}.$$

(a) The average number of items produced daily after 5 days is

$$\frac{45}{5-0}[(t+1)\ln(t+1) - t]\Big|_{0}^{5}$$

$$= 9[(6\ln 6 - 5) - (\ln 1 - 0)]$$
$$= 9(6\ln 6 - 5)$$
$$\approx 51.76.$$

(b) The average number of items produced daily after 9 days is

$$\frac{45}{9-0}[(t+1)\ln(t+1) - t]\Big|_{0}^{9}$$

$$= 5(10\ln 10 - 9)$$
$$\approx 70.13.$$

(c) The average number of items produced daily after 30 days is

$$\frac{45}{30-0}[(t+1)\ln(t+1) - t]\Big|_{0}^{30}$$

$$= \frac{3}{2}(31\ln 31 - 30)$$

$$\approx 114.7.$$

43. From Exercise 22, the volume of an ellipsoid with horizontal axis of length 2a and vertical axis of length 2b is

$$V = \frac{4ab^2\pi}{3}.$$

For the Earth, $a = 6{,}356{,}752.3142$ and $b = 6{,}378{,}137$.

$$V = \frac{4(6{,}356{,}752.3142)(6{,}378{,}137)^2\pi}{3}$$

$$\approx 1.083 \times 10^{21}$$

The volume of the Earth is about 1.083×10^{21} cubic meters (m^3).

16.3 Continuous Money Flow

1. $f(x) = 1000$

(a) $P = \displaystyle\int_0^{10} 1000e^{-0.08x}\,dx$

$= \dfrac{1000}{-0.08}e^{-0.08x}\Big|_0^{10}$

$= -12{,}500(e^{-0.8} - e^0)$

$= -12{,}500(e^{-0.8} - 1)$

≈ 6883.387949

(We will use this value for P in part (b). Store it in your calculator without rounding.)

The present value is \$6883.39.

(b) $A = e^{0.08(10)}\displaystyle\int_0^{10} 1000e^{-0.08x}\,dx$

$= e^{0.8}P$

$\approx 15{,}319.26161$

The accumulated value is \$15,319.26.

3. $f(x) = 500$

(a) $P = \displaystyle\int_0^{10} 500e^{-0.08x}\,dx$

$= \dfrac{500}{-0.08}e^{-0.08x}\Big|_0^{10}$

$= -6250(e^{-0.8} - e^0)$

≈ 3441.693974

The present value is \$3441.69.

(b) $A = e^{0.08(10)}\displaystyle\int_0^{10} 500e^{-0.08x}\,dx$

$= e^{0.8}P$

≈ 7659.630803

The accumulated value is \$7659.63.

5. $f(x) = 400e^{0.03x}$

(a) $P = \displaystyle\int_0^{10} 400e^{0.03x}e^{-0.08x}\,dx$

$= 400\displaystyle\int_0^{10} e^{-0.05x}\,dx$

$= \dfrac{400}{-0.05}e^{-0.05x}\Big|_0^{10}$

$= -8000(e^{-0.5} - e^0)$

≈ 3147.754722

The present value is \$3147.75.

(b) $A = e^{0.08(10)}\displaystyle\int_0^{10} 400e^{0.03x}e^{-0.08x}\,dx$

$= e^{0.8}P$

≈ 7005.456967

The accumulated value is \$7005.46.

7. $f(x) = 5000e^{-0.01x}$

(a) $P = \displaystyle\int_0^{10} 5000e^{-0.01x}e^{-0.08x}\,dx$

$= 5000\displaystyle\int_0^{10} e^{-0.09x}\,dx$

$= \dfrac{5000}{-0.09}e^{-0.09x}\Big|_0^{10}$

$= -\dfrac{5000}{0.09}(e^{-0.9} - e^0)$

$\approx 32{,}968.35224$

The present value is \$32,968.35.

(b) $A = e^{0.08(10)}\displaystyle\int_0^{10} 5000e^{-0.01x}e^{-0.08x}\,dx$

$= e^{0.8}P$

$\approx 73{,}372.41725$

The accumulated value is \$73,372.42.

9. $f(x) = 25x$

(a) $P = \displaystyle\int_0^{10} 25xe^{-0.08x}\,dx$

$= 25\displaystyle\int_0^{10} xe^{-0.08x}\,dx$

Find the antiderivative using integration by parts.

Let $u = x$ and $dv = e^{-0.08x}\,dx$.

Then $du = dx$ and $v = \dfrac{1}{-0.08}e^{-0.08x}$

$= -12.5e^{-0.08x}.$

$$\int xe^{-0.08x}\,dx$$

$$= x(-12.5e^{-0.08x}) - \int(-12.5e^{-0.08x})\,dx$$

$$= -12.5xe^{-0.08x} + 12.5\int e^{-0.08x}\,dx$$

$$= -12.5xe^{-0.08x} + \frac{12.5}{-0.08}e^{-0.08x} + C$$

$$= -(12.5x + 156.25)e^{-0.08x} + C$$

Therefore;

$$P = 25[-(12.5x + 156.25)e^{-0.08x}]\Big|_0^{10}$$

$$= [-25(12.5x + 156.25)e^{-0.08x}]\Big|_0^{10}$$

$$= (-7031.25e^{-0.8}) - (-3906.25e^0)$$

$$\approx 746.9057211.$$

The present value is $746.91.

(b) $A = e^{0.08(10)} \displaystyle\int_0^{10} 25xe^{-0.08x}\,dx$

$$= e^{0.8}P$$

$$\approx 1662.269252$$

The accumulated value is $1662.27.

11. $f(x) = 0.01x + 100$

(a) $P = \displaystyle\int_0^{10} (0.01x + 100)e^{-0.08x}\,dx$

$$= \int_0^{10} 0.01xe^{-0.08x}\,dx + \int_0^{10} 100e^{-0.08x}\,dx$$

$$= 0.01\int_0^{10} xe^{-0.08x}\,dx + 100\int_0^{10} e^{-0.08x}\,dx$$

From Exercise 9, we know that

$$\int xe^{-0.08x}\,dx = -(12.5x + 156.25)e^{-0.08x} + C$$

From Exercise 1, we know that

$$\int e^{-0.08x}\,dx = -12.5e^{-0.08x} + C$$

Substitute the given expressions and simplify.

$$P = \{0.01[-(12.5x + 156.25)e^{-0.08x}]$$
$$\quad + 100(-12.5e^{-0.08x})\}\Big|_0^{10}$$
$$= [-(0.125x + 1251.5625)e^{-0.08x}]\Big|_0^{10}$$
$$= (-1252.8125e^{-0.8}) - (-1251.5625e^0)$$
$$\approx 688.6375571$$

The present value is $688.64.

(b) $A = e^{0.08(10)} \displaystyle\int_0^{10} (0.01x + 100)e^{-0.08x}\,dx$

$$= e^{0.8}P$$

$$\approx 1532.591068$$

The accumulated value is $1532.59.

13. $f(x) = 1000x - 100x^2$

(a) $P = \displaystyle\int_0^{10} (1000x - 100x^2)e^{-0.08x}\,dx$

$$= 1000\int_0^{10} xe^{-0.08x}\,dx - 100\int_0^{10} x^2e^{-0.08x}\,dx$$

From Exercise 9, we know that

$$\int xe^{-0.08x}\,dx = -(12.5x + 156.25)e^{-0.08x} + C$$

Evaluate the antiderivative $\displaystyle\int x^2e^{-0.08x}\,dx$ using column integration. (Note that $\frac{1}{-0.08} = -12.5$.)

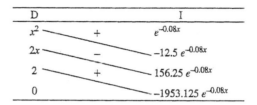

D	I
x^2 $\quad +$	$e^{-0.08x}$
$2x$ $\quad -$	$-12.5\,e^{-0.08x}$
2 $\quad +$	$156.25\,e^{-0.08x}$
0	$-1953.125\,e^{-0.08x}$

Thus,

$$\int_0^{10} x^2e^{-0.08x}\,dx$$

$$= (x^2)(-12.5e^{-0.08x}) - (2x)(156.25e^{-0.08x})$$
$$\quad + (2)(-1953.125e^{-0.08x}) + C$$
$$= -(12.5x^2 + 312.5x + 3906.25)e^{-0.08x} + C.$$

Therefore:

$$P = \{1000[-(12.5x + 156.25)e^{-0.08x}]$$
$$\quad - 100[-(12.5x^2 + 312.5x + 3906.25)e^{-0.08x}]\}\Big|_0^{10}$$

Collect like terms and simplify.

$$P = [(1250x^2 + 18{,}750x + 234{,}375)e^{-0.08x}]\Big|_0^{10}$$
$$= (546{,}875e^{-0.8}) - (234{,}375e^0)$$
$$\approx 11{,}351.77725$$

The principal value is $11,351.78.

(b) $A = e^{0.08(10)} \displaystyle\int_0^{10} (1000x - 100x^2)e^{-0.08x}\,dx$

$$= e^{0.8}P$$

$$\approx 25{,}263.84488$$

The accumulated value is $25,263.84.

15. $A = e^{0.14(3)} \int_0^3 20{,}000e^{-0.14x}\, dx$

$\quad = e^{0.42} \left(\dfrac{20{,}000}{-0.14} e^{-0.14x} \right) \Big|_0^3$

$\quad = e^{0.42} \left(\dfrac{20{,}000}{-0.14} e^{-0.42} + \dfrac{20{,}000}{0.14} \right)$

$\quad \approx \$74{,}565.94$

17. (a) Present value

$\quad = \displaystyle\int_0^8 5000e^{-0.01x} e^{-0.08x}\, dx$

$\quad = \displaystyle\int_0^8 5000e^{-0.09x}\, dx$

$\quad = \left(\dfrac{5000}{-0.09} e^{-0.09x} \right) \Big|_0^8$

$\quad = \dfrac{5000e^{-0.72}}{-0.09} + \dfrac{5000}{0.09}$

$\quad \approx \$28{,}513.76$

(b) Final amount

$\quad = e^{0.08(8)} \displaystyle\int_0^8 5000e^{-0.01x} e^{-0.08x}\, dx$

$\quad \approx e^{0.64}(28{,}513.76)$

$\quad \approx \$54{,}075.81$

19. $P = \displaystyle\int_0^5 (1500 - 60x^2)e^{-0.10x}\, dx$

$\quad = \displaystyle\int_0^5 1500e^{-0.10x}\, dx - \displaystyle\int_0^5 60x^2 e^{-0.10x}\, dx$

$\quad = 1500 \displaystyle\int_0^5 e^{-0.10x}\, dx$

$\qquad - 60 \displaystyle\int_0^5 x^2 e^{-0.10x}\, dx$

Find the second integral by column integration.

D		I
x^2	$+$	$e^{-0.10x}$
$2x$	$-$	$\dfrac{e^{-0.10x}}{-0.10}$
2	$+$	$\dfrac{e^{-0.10x}}{0.01}$
0		$\dfrac{e^{-0.10x}}{-0.001}$

$\displaystyle\int x^2 e^{-0.10x}\, dx$

$\quad = \dfrac{x^2 e^{-0.10x}}{-0.10} - \dfrac{2x e^{-0.10x}}{0.01} + \dfrac{2e^{-0.10x}}{-0.001} + C$

Now add the first integral to this result.

$1500 \displaystyle\int_0^5 e^{-0.10x}\, dx - 60 \displaystyle\int_0^5 x^2 e^{-0.10x}\, dx$

$\quad = \dfrac{1500}{-0.10} e^{-0.10x} \Big|_0^5$

$\qquad - 60 \left(-\dfrac{x^2 e^{-0.10x}}{0.10} - \dfrac{2x e^{-0.10x}}{0.01} \right.$

$\qquad\quad \left. - \dfrac{2e^{-0.10x}}{0.001} \right) \Big|_0^5$

$\quad = -15{,}000e^{-0.5} + 15{,}000$

$\qquad + 60 \left(\dfrac{25e^{-0.5}}{0.1} + \dfrac{10e^{-0.5}}{0.01} + \dfrac{2e^{-0.5}}{0.001} \right.$

$\qquad\quad \left. - 0 - 0 - \dfrac{2}{0.001} \right)$

$\quad = -15{,}000e^{-0.5} + 15{,}000 + 15{,}000e^{-0.5}$

$\qquad + 60{,}000e^{-0.5} + 120{,}000e^{-0.5} - 120{,}000$

$\quad = 180{,}000e^{-0.5} - 105{,}000$

$\quad \approx \$4175.52$

16.4 Improper Integrals

1. $\displaystyle\int_3^\infty \dfrac{1}{x^2}\, dx = \lim_{b\to\infty} \int_3^b x^{-2}\, dx = \lim_{b\to\infty} \int (-x^{-1}) \Big|_3^b$

$\quad = \displaystyle\lim_{b\to\infty} \left(-\dfrac{1}{b} + \dfrac{1}{3} \right)$

$\quad = \displaystyle\lim_{b\to\infty} \left(-\dfrac{1}{b} \right) + \lim_{b\to\infty} \dfrac{1}{3}$

As $b \to \infty$, $-\dfrac{1}{b} \to 0$. The integral is convergent.

$\displaystyle\int_3^\infty \dfrac{1}{x^2}\, dx = 0 + \dfrac{1}{3} = \dfrac{1}{3}$

3. $\displaystyle\int_4^\infty \dfrac{2}{\sqrt{x}}\, dx = \lim_{b\to\infty} \int_4^b 2x^{-1/2}\, dx$

$\quad = \displaystyle\lim_{b\to\infty} 4x^{1/2} \Big|_4^b$

$\quad = \displaystyle\lim_{b\to\infty} (4\sqrt{b} - 4\sqrt{4})$

$\quad = \displaystyle\lim_{b\to\infty} 4\sqrt{b} - 8$

As $b \to \infty$, $4\sqrt{b} \to \infty$. The integral diverges.

5. $\int_{-\infty}^{-1} \dfrac{2}{x^3}\,dx = \int_{-\infty}^{-1} 2x^{-3}\,dx = \lim_{a \to -\infty} \int_{a}^{-1} 2x^{-3}\,dx$

$$= \lim_{a \to -\infty} \left(\dfrac{2x^{-2}}{-2}\right)\bigg|_{a}^{-1} = \lim_{a \to -\infty} \left(-1 + \dfrac{1}{a^2}\right)$$

As $a \to -\infty$, $\dfrac{1}{a^2} \to 0$. The integral is convergent.

$$\int_{-\infty}^{-1} \dfrac{2}{x^3}\,dx = -1 + 0 = -1$$

7. $\int_{1}^{\infty} \dfrac{1}{x^{1.0001}}\,dx$

$$= \int_{1}^{\infty} x^{-1.0001}\,dx$$

$$= \lim_{b \to \infty} \int_{1}^{b} x^{-1.0001}\,dx$$

$$= \lim_{b \to \infty} \left(\dfrac{x^{-0.0001}}{-0.0001}\right)\bigg|_{1}^{b}$$

$$= \lim_{b \to \infty} \left(-\dfrac{1}{(0.0001)b^{0.0001}} + \dfrac{1}{0.0001}\right)$$

As $b \to \infty$, $-\dfrac{1}{0.0001 b^{0.0001}} \to 0$.

The integral is convergent.

$$\int_{1}^{\infty} \dfrac{1}{x^{1.0001}}\,dx = 0 + \dfrac{1}{0.0001} = 10{,}000$$

9. $\int_{-\infty}^{-10} x^{-2}\,dx = \lim_{a \to -\infty} \int_{a}^{-10} x^{-2}\,dx$

$$= \lim_{a \to -\infty} \left(-x^{-1}\right)\bigg|_{a}^{-10}$$

$$= \lim_{a \to -\infty} \left(\dfrac{1}{10} + \dfrac{1}{a}\right)$$

$$= \dfrac{1}{10} + 0$$

$$= \dfrac{1}{10}$$

The integral is convergent.

11. $\int_{-\infty}^{-1} x^{-8/3}\,dx = \lim_{a \to -\infty} \int_{a}^{-1} x^{-8/3}\,dx$

$$= \lim_{a \to -\infty} \left(-\dfrac{3}{5}x^{-5/3}\right)\bigg|_{a}^{-1}$$

$$= \lim_{a \to -\infty} \left(\dfrac{3}{5} + \dfrac{3}{5a^{5/3}}\right)$$

$$= \dfrac{3}{5} + 0$$

$$= \dfrac{3}{5}$$

The integral is convergent.

13. $\int_{0}^{\infty} 8e^{-8x}\,dx = \lim_{b \to \infty} \int_{0}^{b} 8e^{-8x}\,dx$

$$= \lim_{b \to \infty} \left(\dfrac{8e^{-8x}}{-8}\right)\bigg|_{0}^{b}$$

$$= \lim_{b \to \infty} \left(-e^{-8b} + 1\right)$$

$$= \lim_{b \to \infty} \left(-\dfrac{1}{e^{8b}} + 1\right)$$

$$= 0 + 1 = 1$$

The integral is convergent.

15. $\int_{-\infty}^{0} 1000e^{x}\,dx = \lim_{a \to -\infty} \int_{a}^{0} 1000e^{x}\,dx$

$$= \lim_{a \to -\infty} \left(1000e^{x}\right)\bigg|_{a}^{0}$$

$$= \lim_{a \to -\infty} \left(1000 - 1000e^{a}\right)$$

As a approaches $-\infty$, e^{a} is in the denominator of a fraction.

As $a \to \infty$, $-1000e^{a} \to 0$. The integral is convergent.

$$\int_{-\infty}^{0} 1000e^{x}\,dx = 1000 - 0 = 1000$$

17. $\int_{-\infty}^{-1} \ln|x|\,dx = \lim_{a \to -\infty} \int_{a}^{-1} \ln|x|\,dx$

Let $\quad u = \ln|x|$ and $dv = dx$.

Then $du = \dfrac{1}{x}\,dx$ and $v = x$.

$$\int \ln|x|\,dx = x\ln|x| - \int \dfrac{x}{x}\,dx$$

$$= x\ln|x| - x + C$$

$$\int_{-\infty}^{-1} \ln|x|\,dx = \lim_{a \to -\infty} \left(x\ln|x| - x\right)\bigg|_{a}^{-1}$$

$$= \lim_{a \to -\infty} \left(-\ln 1 + 1 - a\ln|a| + a\right)$$

$$= \lim_{a \to -\infty} \left(1 + a - a\ln|a|\right)$$

The integral is divergent, since as $a \to -\infty$.

$$(a - a\ln|a|) = -a(-1 + \ln|a|) \to \infty.$$

19. $\int_{0}^{\infty} \dfrac{dx}{(x+1)^2} = \lim_{b \to \infty} \int_{0}^{b} \dfrac{dx}{(x+1)^2}$ *Use substitution*

$$= \lim_{b \to \infty} -(x+1)^{-1}\bigg|_{0}^{b}$$

$$= \lim_{b \to \infty} \left(\dfrac{-1}{b+1} + 1\right)$$

As $b \to \infty$, $-\frac{1}{b+1} \to 0$. The integral is convergent.

$$\int_0^\infty \frac{dx}{(x+1)^2} = 0 + 1 = 1$$

21. $\int_{-\infty}^{-1} \frac{2x-1}{x^2-x} \, dx$

$$= \lim_{a \to -\infty} \int_a^{-1} \frac{2x-1}{x^2-x} \, dx \quad Use\ substitution$$

$$= \lim_{a \to -\infty} \ln \left| x^2 - x \right| \Big|_a^{-1}$$

$$= \lim_{a \to -\infty} \left(\ln 2 - \ln \left| a^2 - a \right| \right)$$

As $a \to -\infty$, $\ln \left| a^2 - a \right| \to \infty$. The integral is divergent.

23. $\int_2^\infty \frac{1}{x \ln x} \, dx$

$$= \lim_{b \to \infty} \int_2^b \frac{1}{x \ln x} \, dx \quad Use\ substitution$$

$$= \lim_{b \to \infty} \left[\ln (\ln x) \Big|_2^b \right]$$

$$= \lim_{b \to \infty} \left[\ln (\ln b) - \ln (\ln 2) \right]$$

As $b \to \infty$, $\ln (\ln b) \to \infty$. The integral is divergent.

25. $\int_0^\infty x e^{4x} \, dx = \lim_{b \to \infty} \int_0^b x e^{4x} \, dx$

Let $dv = e^{4x} \, dx$ and $u = x$.

Then $v = \frac{1}{4} e^{4x}$ and $du = dx$.

$$\int x e^{4x} \, dx = \frac{x}{4} e^{4x} - \int \frac{1}{4} e^{4x} \, dx$$

$$= \frac{x}{4} e^{4x} - \frac{1}{16} e^{4x} + C$$

$$= \frac{1}{16} (4x - 1) e^{4x} + C$$

$$\int_0^\infty x e^{4x} \, dx$$

$$= \lim_{b \to \infty} \left[\frac{1}{16} (4x - 1) e^{4x} \right] \Big|_0^b$$

$$= \lim_{b \to \infty} \left[\frac{1}{16} (4b - 1) e^{4b} - \frac{1}{16} (-1)(1) \right]$$

$$= \lim_{b \to \infty} \left[\frac{1}{16} (4b - 1) e^{4b} + \frac{1}{16} \right]$$

As $b \to \infty$, $\frac{1}{16} (4b - 1) e^{4b} \to \infty$. The integral is divergent.

27. $\int_{-\infty}^\infty x^3 e^{-x^4} \, dx = \int_{-\infty}^0 x^3 e^{-x^4} \, dx + \int_0^\infty x^3 e^{-x^4} \, dx$

We evaluate each of the two improper integrals on the right.

$$\int_{-\infty}^0 x^3 e^{-x^4} \, dx = \lim_{b \to -\infty} \int_b^0 x^3 e^{-x^4} \, dx \quad Use\ substitution$$

$$= \lim_{b \to -\infty} \left[-\frac{1}{4} e^{-x^4} \Big|_b^0 \right]$$

$$= \lim_{b \to -\infty} \left[-\frac{1}{4} + \frac{1}{4 e^{b^4}} \right]$$

As $b \to -\infty$, $\frac{1}{4 e^{b^4}} \to 0$. The integral is convergent.

$$\int_{-\infty}^0 x^3 e^{-x^4} \, dx = -\frac{1}{4} + 0 = -\frac{1}{4}$$

$$\int_0^\infty x^3 e^{-x^4} \, dx = \lim_{b \to \infty} \int_0^b x^3 e^{-x^4} \, dx \quad Use\ substitution$$

$$= \lim_{b \to \infty} \left[-\frac{1}{4} e^{-x^4} \Big|_0^b \right]$$

$$= \lim_{b \to \infty} \left[-\frac{1}{4 e^{b^4}} + \frac{1}{4} \right]$$

As $b \to \infty$, $-\frac{1}{4 e^{b^4}} \to 0$. The integral is convergent.

$$\int_0^\infty x^3 e^{-x^4} \, dx = 0 + \frac{1}{4} = \frac{1}{4}$$

Since each of the improper integrals converges, the original improper integral converges.

$$\int_{-\infty}^\infty x^3 e^{-x^4} \, dx = -\frac{1}{4} + \frac{1}{4} = 0$$

29. $\int_{-\infty}^\infty \frac{x}{x^2+1} \, dx = \int_{-\infty}^0 \frac{x}{x^2+1} \, dx + \int_0^\infty \frac{x}{x^2+1} \, dx$

We evaluate the first improper integral on the right.

$$\int_{-\infty}^0 \frac{x}{x^2+1} \, dx = \lim_{b \to -\infty} \int_b^0 \frac{x}{x^2+1} \, dx \quad Use\ substitution$$

$$= \lim_{b \to -\infty} \left[\frac{1}{2} \ln(x^2 + 1) \Big|_b^0 \right]$$

$$= \lim_{b \to -\infty} \left[0 - \frac{1}{2} \ln(b^2 + 1) \right]$$

As $b \to -\infty$, $\ln(b^2 + 1) \to \infty$. The integral is divergent. Since one of the two improper integrals on the right diverges, the original improper integral diverges.

31. $f(x) = \dfrac{1}{x-1}$ for $(-\infty, 0]$

$$\int_{-\infty}^{0} \frac{1}{x-1}\,dx = \lim_{a\to-\infty} \int_{a}^{0} \frac{dx}{x-1}$$

$$= \lim_{a\to-\infty} \left(\ln|x-1|\,\Big|_{a}^{0}\right)$$

$$= \lim_{a\to-\infty} \left(\ln|-1| - \ln|a-1|\right)$$

But $\displaystyle\lim_{a\to-\infty} (\ln|a-1|) = \infty.$

The integral is divergent, so the area cannot be found.

33. $f(x) = \dfrac{1}{(x-1)^2}$ for $(-\infty, 0]$

$$\int_{-\infty}^{0} \frac{1}{(x-1)^2}$$

$$= \lim_{a\to-\infty} \int_{a}^{0} \frac{1}{(x-1)^2} \quad Use\ substitution$$

$$= \lim_{a\to-\infty} -(x-1)^{-1}\,\Big|_{a}^{0}$$

$$= \lim_{a\to-\infty} \left(-\frac{1}{-1} + \frac{1}{a-1}\right)$$

As $a \to -\infty$, $\frac{1}{a-1} \to 0$. The integral is convergent.

$$= 1 + 0 = 1$$

Therefore, the area is 1.

35. $\displaystyle\int_{-\infty}^{\infty} xe^{-x^2}\,dx$

Let $u = -x^2$, so that $du = -2x\,dx$.

$$= \lim_{a\to-\infty} \left(-\frac{1}{2}\int_{a}^{0} -2xe^{-x^2}\,dx\right)$$

$$\quad + \lim_{b\to\infty} \left(-\frac{1}{2}\int_{0}^{b} -2xe^{-x^2}\,dx\right)$$

$$= \lim_{a\to-\infty} \left(-\frac{1}{2}e^{-x^2}\right)\Big|_{a}^{0}$$

$$\quad + \lim_{b\to\infty} \left(-\frac{1}{2}e^{-x^2}\right)\Big|_{0}^{b}$$

$$= \lim_{a\to-\infty} \left(-\frac{1}{2} + \frac{1}{2e^{-a^2}}\right)$$

$$\quad + \lim_{b\to\infty} \left(-\frac{1}{2e^{b^2}} + \frac{1}{2}\right)$$

$$= -\frac{1}{2} + \frac{1}{2} = 0$$

37. $\displaystyle\int_{1}^{\infty} \frac{1}{x^p}\,dx$

Case 1a $p < 1$:

$$\int_{1}^{\infty} \frac{1}{x^p}\,dx$$

$$= \int_{1}^{\infty} x^{-p}\,dx$$

$$= \lim_{a\to\infty} \int_{1}^{a} x^{-p}\,dx$$

$$= \lim_{a\to\infty} \left[\frac{x^{-p+1}}{(-p+1)}\Big|_{1}^{a}\right]$$

$$= \lim_{a\to\infty} \left[\frac{1}{(-p+1)}(a^{-p+1} - 1)\right]$$

$$= \lim_{a\to\infty} \left[\frac{1}{(-p+1)}a^{1-p} - \frac{1}{(-p+1)}\right]$$

Since $p < 1$, $1 - p$ is positive and, as $a \to \infty$, $a^{1-p} \to \infty$. The integral diverges.

Case 1b $p = 1$:

$$\int_{1}^{\infty} \frac{1}{x^p}\,dx = \int_{1}^{\infty} \frac{1}{x}\,dx$$

$$= \lim_{a\to\infty} \int_{1}^{a} \frac{1}{x}\,dx$$

$$= \lim_{a\to\infty} \left(\ln|x|\,\Big|_{1}^{a}\right)$$

$$= \lim_{a\to\infty} (\ln|a| - \ln 1)$$

$$= \lim_{a\to\infty} \ln|a|$$

As $a \to \infty$, $\ln|a| \to \infty$. The integral diverges.

Therefore, $\displaystyle\int_{1}^{\infty} \frac{1}{x^p}$ diverges when $p \le 1$.

Case 2 $p > 1$:

$$\int_{1}^{\infty} \frac{1}{x^p}\,dx = \lim_{a\to\infty} \int_{1}^{a} x^{-p}\,dx$$

$$= \lim_{a\to\infty} \left(\frac{x^{-p+1}}{-p+1}\Big|_{1}^{a}\right)$$

$$= \lim_{a\to\infty} \left[\frac{a^{-p+1}}{(-p+1)} - \frac{1}{(-p+1)}\right]$$

Since $p > 1$, $-p + 1 < 0$; thus as $a \to \infty$, $\dfrac{a^{-p+1}}{(-p+1)} \to 0.$

Hence,

$$\lim_{a \to \infty} \left[\frac{a^{-p+1}}{(-p+1)} - \frac{1}{(-p+1)} \right] = 0 - \frac{1}{(-p+1)}$$

$$= \frac{-1}{-p+1}$$

$$= \frac{1}{p-1}.$$

The integral converges.

39. (a) Use the *fnInt* feature on a graphing utility to obtain

$$\int_1^{20} \frac{1}{\sqrt{1+x^2}}\, dx \approx 2.808;$$

$$\int_1^{50} \frac{1}{\sqrt{1+x^2}}\, dx \approx 3.724;$$

$$\int_1^{100} \frac{1}{\sqrt{1+x^2}}\, dx \approx 4.417;$$

$$\int_1^{1000} \frac{1}{\sqrt{1+x^2}}\, dx \approx 6.720;$$

$$\int_1^{10,000} \frac{1}{\sqrt{1+x^2}}\, dx \approx 9.022.$$

(b) Since the values of the integrals in part a do not appear to be approaching some fixed finite number but get bigger, the integral $\int_1^\infty \frac{1}{\sqrt{1+x^2}}\, dx$ appears to be divergent.

(c) Use the *fnInt* feature on a graphing utility to obtain

$$\int_1^{20} \frac{1}{\sqrt{1+x^4}}\, dx \approx 0.8770;$$

$$\int_1^{50} \frac{1}{\sqrt{1+x^4}}\, dx \approx 0.9070;$$

$$\int_1^{100} \frac{1}{\sqrt{1+x^4}}\, dx \approx 0.9170;$$

$$\int_1^{1000} \frac{1}{\sqrt{1+x^4}}\, dx \approx 0.9260;$$

$$\int_1^{10,000} \frac{1}{\sqrt{1+x^4}}\, dx \approx 0.9269.$$

(d) Since the values of the integrals in part c appear to be approaching some fixed finite number, the integral $\int_1^\infty \frac{1}{\sqrt{1+x^4}}\, dx$ appears to be convergent.

(e) For large x, we may consider $1+x^2 \approx x^2$ and $1+x^4 \approx x^4$.
Thus,

$$\frac{1}{\sqrt{1+x^4}} \approx \frac{1}{\sqrt{x^2}} = \frac{1}{x} \quad \text{and}$$

$$\frac{1}{\sqrt{1+x^4}} \approx \frac{1}{\sqrt{x^4}} = \frac{1}{x^2}.$$

In Example 1(a) on page 455, we showed that $\int_1^\infty \frac{1}{x}\, dx$ diverges. Thus, we might guess that $\int_1^\infty \frac{1}{\sqrt{1+x^2}}\, dx$ diverges as well. In Exercise 1, we saw that $\int_2^\infty \frac{1}{x^2}\, dx$ converges. Thus, we might guess that $\int_1^\infty \frac{1}{\sqrt{1+x^4}}\, dx$ converges as well.

41. (a) Use the *fnInt* feature on a graphing utility to obtain

$$\int_0^{10} e^{-0.00001x}\, dx \approx 9.9995;$$

$$\int_0^{50} e^{-0.00001x}\, dx \approx 49.9875;$$

$$\int_0^{100} e^{-0.00001x}\, dx \approx 99.9500;$$

$$\int_0^{1000} e^{-0.00001x}\, dx \approx 995.0166.$$

(b) Since the values of the integrals in part a do not appear to be approaching some fixed finite number, the integral $\int_0^\infty e^{-0.00001x}\, dx$ appears to be divergent.

(c) $\int_0^\infty e^{-0.00001x}\, dx$

$$= \lim_{b \to \infty} \int_0^b e^{-0.00001x}\, dx$$

$$= \lim_{b \to \infty} \left[\frac{e^{-0.00001x}}{-0.00001} \Big|_0^b \right]$$

$$= \lim_{b \to \infty} \left[-\frac{1}{0.00001e^{0.00001b}} + \frac{1}{0.00001} \right]$$

$$= 0 + 100{,}000 = 100{,}000$$

43. $\displaystyle\int_0^\infty 1{,}000{,}000 e^{-0.05t}\, dt$

$$= \lim_{b \to \infty} \int_0^b 1{,}000{,}000 e^{-0.05t}\, dt$$

$$= \lim_{b \to \infty} \left(\frac{1{,}000{,}000}{-0.05} e^{-0.05t} \right) \Big|_0^b$$

$$= -20{,}000{,}000 \left[\lim_{b \to \infty} (e^{-0.05b}) - e^0 \right]$$

As $b \to \infty$, $e^{-0.05b} = \frac{1}{e^{0.05b}} \to 0$. The integral converges.

$$\int_0^\infty 1{,}000{,}000e^{-0.05t}\,dt$$

$$= -20{,}000{,}000(0-1)$$
$$= 20{,}000{,}000$$

The capital value is $20,000,000.

45. $\displaystyle\int_0^\infty 1200e^{0.03t}e^{-0.07t}\,dt$

$$= \lim_{b\to\infty} \int_0^b 1200e^{-0.04t}\,dt$$

$$= \lim_{b\to\infty} \left(\frac{1200}{-0.04}e^{-0.04t}\right)\bigg|_0^b$$

$$= -30{,}000\left[\lim_{b\to\infty}(e^{-0.04b}) - e^0\right]$$

As $b \to \infty$, $e^{-0.04b} = \frac{1}{e^{0.04b}} \to 0$. The integral converges.

$$\int_0^\infty 1200e^{0.03t}e^{-0.07t}\,dt = -30{,}000(0-1)$$
$$= 30{,}000$$

The capital value is $30,000.

47. $\displaystyle\int_0^\infty 3000e^{-0.1t}\,dt = \lim_{b\to\infty}\int_0^b 3000e^{-0.1t}\,dt$

$$= \lim_{b\to\infty} \frac{3000e^{-0.1b}}{-0.1}\bigg|_0^b$$

$$= \lim_{b\to\infty}\left(\frac{3000e^{-0.1b}}{-0.1} + \frac{3000}{0.1}\right)$$

$$= 0 + 30{,}000$$
$$= \$30{,}000$$

49. $\displaystyle S = N\int_0^\infty \frac{a(1-e^{-kt})}{k}e^{-bt}\,dt$

$$= \frac{Na}{k}\lim_{c\to\infty}\int_0^c (1-e^{-kt})(e^{-bt})\,dt$$

$$= \frac{Na}{k}\lim_{c\to\infty}\int_0^c (e^{-bt} - e^{-(b+k)t})\,dt$$

$$= \frac{Na}{k}\lim_{c\to\infty}\left[-\frac{1}{b}e^{-bt} + \frac{1}{b+k}e^{-(b+k)t}\right]\bigg|_0^c$$

$$= \frac{Na}{k}\lim_{c\to\infty}\left[\left(-\frac{1}{b}e^{-bc} + \frac{1}{b+k}e^{-(b+k)c}\right)\right.$$
$$\left. - \left(-\frac{1}{b}e^0 + \frac{1}{b+k}e^0\right)\right]$$

$$= \frac{Na}{k}\left(0 + 0 + \frac{1}{b} - \frac{1}{b+k}\right)$$

$$= \frac{Na}{k}\cdot\frac{(b+k)-b}{b(b+k)}$$

$$= \frac{Nak}{kb(b+k)}$$

$$= \frac{Na}{b(b+k)}$$

51. $\displaystyle\int_0^\infty 50e^{-0.06t}\,dt = 50\lim_{b\to\infty}\int_0^b e^{-0.06t}\,dt$

$$= 50\lim_{b\to\infty}\frac{e^{-0.06t}}{-0.06}\bigg|_0^b$$

$$= \frac{50}{-0.06}\lim_{b\to\infty}(e^{-0.06b} - e^0)$$

$$= -\frac{50}{0.06}(0-1)$$

$$= \frac{50}{0.06}$$

$$\approx 833.3$$

16.5   Solutions of Elementary and Separable Differential Equations

1. $\displaystyle\frac{dy}{dx} = -4x + 6x^2$

$$y = \int(-4x + 6x^2)\,dx$$

$$= -2x^2 + 2x^3 + C$$

3. $4x^3 - 2\dfrac{dy}{dx} = 0$

Solve for $\dfrac{dy}{dx}$.

$$\frac{dy}{dx} = 2x^3$$

$$y = 2\int x^3\, dx = 2\left(\frac{x^4}{4}\right) + C = \frac{x^4}{2} + C$$

5. $y\dfrac{dy}{dx} = x^2$

Separate the variables and take antiderivatives.

$$\int y\, dy = \int x^2\, dx$$

$$\frac{y^2}{2} = \frac{x^3}{3} + K$$

$$y^2 = \frac{2}{3}x^3 + 2K$$

$$y^2 = \frac{2}{3}x^3 + C$$

7. $\dfrac{dy}{dx} = 2xy$

$$\int \frac{dy}{y} = \int 2x\, dx$$

$$\ln|y| = \frac{2x^2}{2} + C$$

$$\ln|y| = x^2 + C$$

$$e^{\ln|y|} = e^{x^2+C}$$

$$y = \pm e^{x^2+C}$$

$$y = \pm e^{x^2} \cdot e^C$$

$$y = ke^{x^2}$$

9. $\dfrac{dy}{dx} = 3x^2y - 2xy$

$$\frac{dy}{dx} = y(3x^2 - 2x)$$

$$\int \frac{dy}{y} = \int (3x^2 - 2x)\, dx$$

$$\ln|y| = \frac{3x^3}{3} - \frac{2x^2}{2} + C$$

$$e^{\ln|y|} = e^{x^3 - x^2 + C}$$

$$y = \pm(e^{x^3 - x^2})e^C$$

$$y = ke^{x^3 - x^2}$$

11. $\dfrac{dy}{dx} = \dfrac{y}{x},\ x > 0$

$$\int \frac{dy}{y} = \int \frac{dx}{x}$$

$$\ln|y| = \ln x + C$$

$$e^{\ln|y|} = e^{\ln x + C}$$

$$y = \pm e^{\ln x} \cdot e^C$$

$$y = Me^{\ln x}$$

$$y = Mx$$

13. $\dfrac{dy}{dx} = y - 6$

$$\int \frac{dy}{y-6} = \int dx$$

$$\ln|y-6| = x + C$$

$$e^{\ln|y-6|} = e^{x+C}$$

$$y - 6 = \pm e^x \cdot e^C$$

$$y - 6 = Me^x$$

$$y = Me^x + 6$$

15. $\dfrac{dy}{dx} = y^2 e^{2x}$

$$\int y^{-2}\, dy = \int e^{2x}\, dx$$

$$-y^{-1} = \frac{1}{2}e^2 + C$$

$$-\frac{1}{y} = \frac{1}{2}e^2 + C$$

$$y = \frac{-1}{\frac{1}{2}e^2 + C}$$

17. $\dfrac{dy}{dx} + 3x^2 = 2x$

$$\frac{dy}{dx} = 2x - 3x^2$$

$$y = \frac{2x^2}{2} - \frac{3x^3}{3} + C$$

$$y = x^2 - x^3 + C$$

Since $y = 5$ when $x = 0$,

$$5 = 0 - 0 + C$$

$$C = 5.$$

Thus,

$$y = x^2 - x^3 + 5.$$

19. $2\dfrac{dy}{dx} = 4xe^{-x}$

$\dfrac{dy}{dx} = 2xe^{-x}$

Use the table of integrals or integrate by parts.

$y = 2(-x-1)e^{-x} + C$

Since $y = 42$ when $x = 0$,

$$42 = 2(0-1)(1) + C$$
$$42 = -2 + C$$
$$C = 44.$$

Thus,

$$y = -2xe^{-x} - 2e^{-x} + 44.$$

21. $\dfrac{dy}{dx} = \dfrac{x^3}{y}$; $y = 5$ when $x = 0$.

$$\int y\, dy = \int x^3\, dx$$

$$\frac{y^2}{2} = \frac{x^4}{4} + C$$

$$y^2 = \frac{1}{2}x^4 + 2C$$

$$y^2 = \frac{1}{2}x^4 + k$$

Since $y = 5$ when $x = 0$,

$$25 = 0 + k$$
$$k = 25.$$

So $y^2 = \dfrac{1}{2}x^4 + 25$.

23. $(2x+3)y = \dfrac{dy}{dx}$; $y = 1$ when $x = 0$.

$$\int (2x+3)\, dx = \int \frac{dy}{y}$$

$$\frac{2x^2}{2} + 3x + C = \ln|y|$$

$$e^{x^2+3x+C} = e^{\ln|y|}$$

$$y = (e^{x^2+3x})(\pm e^{C})$$
$$y = ke^{x^2+3x}$$

Since $y = 1$ when $x = 0$.

$$1 = ke^{0+0}$$
$$k = 1.$$

So $y = e^{x^2+3x}$.

25. $\dfrac{dy}{dx} = 4x^3 - 3x^2 + x$; $y = 0$ when $x = 1$.

$$y = \int (4x^3 - 3x^2 + x)\, dx$$

$$= x^4 - x^3 + \frac{x^2}{2} + C$$

Substitute.

$$0 = 1 - 1 + \frac{1}{2} + C$$

$$-\frac{1}{2} = C$$

$$y = x^4 - x^3 + \frac{x^2}{2} - \frac{1}{2}$$

27. $\dfrac{dy}{dx} = \dfrac{y^2}{x}$; $y = 3$ when $x = e$.

$$\int y^{-2}\, dy = \int \frac{dx}{x}$$

$$-y^{-1} = \ln|x| + C$$

$$-\frac{1}{y} = \ln|x| + C$$

$$y = \frac{-1}{\ln|x| + C}$$

Since $y = 3$ when $x = e$,

$$3 = \frac{-1}{\ln e + C}$$

$$3 = \frac{-1}{1 + C}$$

$$3 + 3C = -1$$
$$3C = -4$$

$$C = -\frac{4}{3}.$$

So $\qquad y = \dfrac{-1}{\ln|x| - \frac{4}{3}} = \dfrac{-3}{3\ln|x| - 4}$.

29. $\dfrac{dy}{dx} = (y-1)^2 e^{x-1}$; $y = 2$ when $x = 1$.

$$\frac{dy}{(y-1)^2} = e^{x-1}\, dx$$

$$\int (y-1)^{-2}\, dy = \int e^{x-1}\, dx$$

$$\frac{(y-1)^{-1}}{-1} = e^{x-1} + C$$

$$-\frac{1}{y-1} = e^{x-1} + C$$

$$-(y-1) = \frac{1}{e^{x-1} + C}$$

$$-y + 1 = \frac{1}{e^{x-1} + C}$$

$$1 - \frac{1}{e^{x-1} + C} = y$$

$$y = \frac{e^{x-1} + C}{e^{x-1} + C} - \frac{1}{e^{x-1} + C}$$

$$y = \frac{e^{x-1} + C - 1}{e^{x-1} + C}$$

$y = 2$, when $x = 1$.

$$2 = \frac{e^0 + C - 1}{e^0 + C}$$

$$2 = \frac{C}{1 + C}$$

$$2 + 2C = C$$
$$C = -2$$

$$y = \frac{e^{x-1} - 3}{e^{x-1} - 2}.$$

31. $\dfrac{dy}{dx} = \dfrac{k}{N}(N - y)y$

(a) $\dfrac{N\,dy}{(N - y)y} = k\,dx$

Since $\dfrac{1}{y} + \dfrac{1}{N - y} = \dfrac{N}{(N - y)y}$,

$$\int \frac{dy}{y} + \int \frac{dy}{N - y} = k\,dx$$

$$\ln\left|\frac{y}{N - y}\right| = kx + C$$

$$\frac{y}{N - y} = Ce^{kx}.$$

For $0 < y < N$, $Ce^{kx} > 0$.
For $0 < N < y$, $Ce^{kx} < 0$.
Solve for y.

$$y = \frac{Ce^{kx}N}{1 + Ce^{kx}} = \frac{N}{1 + C^{-1}e^{-kx}}$$

Let $b = C^{-1} > 0$ for $0 < y < N$.

$$y = \frac{N}{1 + be^{-kx}}$$

Let $-b = C^{-1} < 0$ for $0 < N < y$.

$$y = \frac{N}{1 - be^{-kx}}$$

(b) For $0 < y < N$; $t = 0$, $y = y_0$.

$$y_0 = \frac{N}{1 + be^0} = \frac{N}{1 + b}$$

Solve for b.
$$b = \frac{N - y_0}{y_0}$$

(c) For $0 < N < y$; $t = 0$, $y = y_0$.

$$y_0 = \frac{N}{1 - be^0} = \frac{N}{1 - b}$$

Solve for b.
$$b = \frac{y_0 - N}{y_0}$$

33. (a) $0 < y_0 < N$ implies that $y_0 > 0$, $N > 0$, and $N - y_0 > 0$.

Therefore,
$$b = \frac{N - y_0}{y_0} > 0.$$

Also, $e^{-kx} > 0$ for all x, which implies that $1 + be^{-kx} > 1$.

(1) $y(x) = \dfrac{N}{1 + be^{-kx}} < N$ since $1 + be^{-kx} > 1$.

(2) $y(x) = \dfrac{N}{1 + be^{-kx}} > 0$ since $N > 0$ and $1 + be^{-kx} > 0$.

Combining statements (1) and (2), we have

$$0 < \frac{N}{1 + be^{-kx}} = y(x)$$
$$= \frac{N}{1 + be^{-kx}} < N$$

or $0 < y(x) < N$ for all x.

(b) $\displaystyle\lim_{x\to\infty} \frac{N}{1 + be^{-kx}} = \frac{N}{1 + b(0)} = N$

$\displaystyle\lim_{x\to-\infty} \frac{N}{1 + be^{-kx}} = 0$

Note that as $x \to -\infty$, $1 + be^{-kx}$ becomes infinitely large.
Therefore, the horizontal asymptotes are $y = N$ and $y = 0$.

(c) $y'(x) = \dfrac{(1 + be^{-kx})(0) - N(-kbe^{-kx})}{(1 + be^{-kx})^2}$

$$= \frac{Nkbe^{-kx}}{(1 + be^{-kx})^2} > 0 \text{ for all } x.$$

Therefore, $y(x)$ is an increasing function.

(d) To find $y''(x)$, apply the quotient rule to find the derivative of $y'(x)$. The numerator of $y''(x)$ is

$$y''(x) = (1 + be^{-kx})^2(-Nk^2be^{-kx})$$
$$- Nkbe^{-kx}[-2kbe^{-kx}(1 + be^{-kx})]$$
$$= -Nkbe^{-kx}(1 + be^{-kx})$$
$$\cdot [k(1 + be^{-kx}) - 2kbe^{-kx}]$$
$$= -Nkbe^{-kx}(1 + be^{-kx})(k - kbe^{-kx}),$$

and the denominator is

$$[(1 + be^{-kt})^2]^2 = (1 + be^{-kx})^4.$$

Thus,

$$y''(x) = \frac{-Nkbe^{-kx}(k - kbe^{-kx})}{(1 + be^{-kx})^3}.$$

$y''(x) = 0$ when

$$k - kbe^{-kx} = 0$$
$$be^{-kx} = 1$$
$$e^{-kx} = \frac{1}{b}$$
$$-kx = \ln\left(\frac{1}{b}\right)$$
$$x = -\frac{\ln\left(\frac{1}{b}\right)}{k}$$
$$= \frac{\ln\left(\frac{1}{b}\right)^{-1}}{k} = \frac{\ln b}{k}.$$

When $x = \frac{\ln b}{k}$,

$$y = \frac{N}{1 + be^{-k\left(\frac{\ln b}{k}\right)}} = \frac{N}{1 + be^{(-\ln b)}}$$
$$= \frac{N}{1 + be^{\ln(1/b)}} = \frac{N}{1 + b\left(\frac{1}{b}\right)} = \frac{N}{2}.$$

Therefore, $\left(\frac{\ln b}{k}, \frac{N}{2}\right)$ is a point of inflection.

(e) To locate the maximum of $\frac{dy}{dx}$, we must consider, from part (d),

$$\frac{d}{dx}\left(\frac{dy}{dx}\right) = \frac{-Nkbe^{-kx}(k - kbe^{-kx})}{(1 + be^{-kx})^3}.$$

Since $y''(x) > 0$ for $x < \frac{\ln b}{k}$ and

$$y''(x) < 0 \text{ for } x > \frac{\ln b}{k},$$

we know that $x = \frac{\ln b}{k}$ locates a relative maximum of $\frac{dy}{dx}$.

35. $\dfrac{dy}{dx} = \dfrac{100}{32 - 4x}$

$$y = 100\left(-\frac{1}{4}\right)\ln|32 - 4x| + C$$
$$y = -25\ln|32 - 4x| + C$$

Now, $y = 1000$ when $x = 0$.

$$1000 = -25\ln|32| + C$$
$$C = 1000 + 25\ln 32$$
$$C \approx 1086.64$$

Thus,

$$y = -25\ln|32 - 4x| + 1086.64.$$

(a) Let $x = 3$.

$$y = -25\ln|32 - 12| + 1086.64$$
$$\approx \$1011.75$$

(b) Let $x = 5$.

$$y = -25\ln|32 - 20| + 1086.64$$
$$\approx \$1024.52$$

(c) Advertising expenditures can never reach \$8000. If $x = 8$, the denominator becomes zero.

37. $\dfrac{dy}{dt} = -0.05y$

See Example 4.

$$\int \frac{dy}{y} = \int -0.05\,dt$$
$$\ln|y| = -0.05t + C$$
$$e^{\ln|y|} = e^{-0.05t+C}$$
$$e^{\ln|y|} = e^{-0.05t} \cdot e^C$$
$$|y| = e^{-0.05t} \cdot e^C$$
$$y = Me^{-0.05t}$$

Let $y = 1$ when $t = 0$.

Solve for M:

$$1 = Me^0$$
$$M = 1.$$

So $y = e^{-0.05t}$.

If $y = 0.50$,

$$0.50 = e^{-0.05t}$$
$$\ln 0.5 = -0.05t \ln e$$
$$t = \frac{-\ln 0.5}{0.05}$$
$$\approx 13.9$$

It will take about 13.9 years for \$1 to lose half its value.

39. $E = -\dfrac{p}{q} \cdot \dfrac{dq}{dp}$ with $p > 0$ and $q > 0$

If $E = 2$,

$$2 = -\frac{p}{q} \cdot \frac{dq}{dp}$$
$$\frac{2}{p} dp = -\frac{1}{q} dq$$
$$\int \frac{2}{p} dp = -\int \frac{1}{q} dq$$
$$2 \ln p = -\ln q + K$$
$$\ln p^2 + \ln q = K$$
$$\ln (p^2 q) = K$$
$$p^2 q = e^K$$
$$p^2 q = C$$
$$q = \frac{C}{p^2}.$$

41.
$$\frac{dA}{dt} = Ai$$
$$\frac{dA}{A} = i\, dt$$
$$\int \frac{dA}{A} = \int i\, dt$$
$$\ln A = it + C$$
$$e^{\ln A} = e^{it+C}$$
$$A = Me^{it}$$

When $t = 0$, $A = 5000$. Therefore, $M = 5000$. Find i so that $A = 20{,}000$ when $t = 24$.

$$20{,}000 = 5000e^{24i}$$
$$4 = e^{24i}$$
$$\ln 4 = 24i$$
$$i = \frac{\ln 4}{24}$$
$$= \frac{2 \ln 2}{24}$$
$$= \frac{\ln 2}{12}$$

The answer is d.

43. (a) $\dfrac{dI}{dW} = 0.088(2.4 - I)$

Separate the variables and take antiderivatives.

$$\int \frac{dI}{2.4 - I} = \int 0.088\, dW$$
$$-\ln |2.4 - I| = 0.088W + k$$

Solve for I.

$$\ln |2.4 - I| = -0.088W - k$$
$$|2.4 - I| = e^{-0.088W - k} = e^{-k} e^{-0.088W}$$
$$I - 2.4 = Ce^{-0.088W}, \text{ where } C = \pm e^{-k}.$$
$$I = 2.4 + Ce^{-0.088W}$$

Since $I(0) = 1$, then

$$1 = 2.4 + Ce^0$$
$$C = 1 - 2.4 = -1.4.$$

Therefore, $I = 2.4 - 1.4e^{-0.088W}$.

(b) Note that as W gets larger and larger $e^{-0.088W}$ approaches 0, so

$$\lim_{W \to \infty} I = \lim_{W \to \infty} (2.4 - 1.4e^{-0.088W})$$
$$= 2.4 - 1.4(0) = 2.4,$$

so I approaches 2.4.

45. (a) $\dfrac{dw}{dt} = k(C - 17.5w)$

C being constant implies that the calorie intake per day is constant.

(b) pounds/day $= k$ (calories/day)

$$\frac{\text{pounds/day}}{\text{calories/day}} = k$$

The units of k are pounds/calorie.

(c) Since 3500 calories is equivalent to 1 pound, $k = \frac{1}{3500}$ and

$$\frac{dw}{dt} = \frac{1}{3500}(C - 17.5w).$$

(d) $\dfrac{dw}{dt} = \dfrac{1}{3500}(C - 17.5w)$; $w = w_0$ when $t = 0$.

$$\frac{3500}{C - 17.5w}\, dw = dt$$

$$\frac{3500}{-17.5} \int \frac{-17.5}{C - 17.5w}\, dw = \int dt$$

$$-200 \ln |C - 17.5w| = t + k$$

$$\ln |C - 17.5w| = -0.005t - 0.005k$$

$$|C - 17.5w| = e^{-0.005t - 0.005k}$$

$$|C - 17.5w| = e^{-0.005t} \cdot e^{-0.005k}$$

$$C - 17.5w = e^{-0.005M} e^{-0.005t}$$

$$-17.5w = -C + e^{-0.005M} e^{-0.005t}$$

$$w = \frac{C}{17.5} - \frac{e^{-0.005M}}{17.5} e^{-0.005t}$$

(e) Since $w = w_0$ when $t = 0$,

$$w_0 = \frac{C}{17.5} - \frac{e^{-0.005M}}{17.5} \quad (1)$$

$$w_0 - \frac{C}{17.5} = -\frac{e^{-0.005M}}{17.5}$$

$$\frac{e^{-0.005M}}{17.5} = \frac{C}{17.5} - w_0.$$

Therefore,

$$w = \frac{C}{17.5} - \left(\frac{C}{17.5} - w_0\right) e^{-0.005t}$$

$$w = \frac{C}{17.5} + \left(w_0 - \frac{C}{17.5}\right) e^{-0.005t}.$$

47. (a)

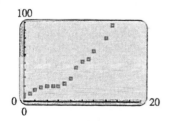

(b) The logistic regression equation is

$$y = \frac{258.70}{1 + 31.3998 e^{-0.1930x}}.$$

(c) The logistic equation fits the data well.

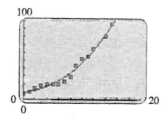

(d) As x gets very large, the value of the function approaches 259 (rounded to the nearest whole number).

49. $\dfrac{dy}{dt} = ky$

First separate the variables and integrate.

$$\frac{dy}{y} = k\, dt$$

$$\int \frac{dy}{y} = \int k\, dt$$

$$\ln |y| = kt + C.$$

Solve for y.

$$|y| = e^{kt + C_1} = e^{C_1} e^{kt}$$

$$y = Ce^{kt}, \text{ where } C = \pm e^{C_1}.$$

$$y(0) = 35.6, \text{ so } 35.6 = Ce^0 = C, \text{ and}$$

$$y = 35.6 e^{kt}.$$

Since $y(50) = 102.6$, then $102.6 = 35.6 e^{50k}$.

Solve for k.

$$e^{50k} = \frac{102.6}{35.6}$$

$$50k = \ln\left(\frac{102.6}{35.6}\right)$$

$$k = \frac{\ln\left(\frac{102.6}{35.6}\right)}{50} \approx 0.02117, \text{ so}$$

$$y = 35.6 e^{0.02117t}.$$

51. (a) $\dfrac{dy}{dt} = ky$

$$\int \frac{dy}{y} = \int k\, dt$$

$$\ln |y| = kt + C$$

$$e^{\ln |y|} = e^{kt + C}$$

$$y = \pm(e^{kt})(e^C)$$

$$y = Me^{kt}$$

If $y = 1$ when $t = 0$ and $y = 5$ when $t = 2$, we have the system of equations

$$1 = Me^{k(0)}$$

$$5 = Me^{2k}.$$

$$1 = M(1)$$

$$M = 1$$

Substitute.

$$5 = (1)e^{2k}$$
$$e^{2k} = 5$$
$$2k \ln e = \ln 5$$
$$k = \frac{\ln 5}{2}$$
$$\approx 0.8$$

(b) If $k = 0.8$ and $M = 1$,

$$y = e^{0.8t}.$$

When $t = 3$,

$$y = e^{0.8(3)}$$
$$= e^{2.4}$$
$$\approx 11.$$

(c) When $t = 5$,

$$y = e^{0.8(5)}$$
$$= e^4$$
$$\approx 55.$$

(d) When $t = 10$,

$$y = e^{0.8(10)}$$
$$= e^8$$
$$\approx 3000.$$

53. $\dfrac{dy}{dx} = 7.5e^{-0.3y}$, $y = 0$ when $x = 0$.

$$e^{0.3y}\, dy = 7.5\, dx$$
$$\int e^{0.3y}\, dy = \int 7.5\, dx$$
$$\frac{e^{0.3y}}{0.3} = 7.5x + C$$
$$e^{0.3y} = 2.25x + C$$
$$1 = 0 + C = C$$
$$e^{0.3y} = 2.25x + 1$$
$$0.3y = \ln(2.25x + 1)$$
$$y = \frac{\ln(2.25x + 1)}{0.3}$$

When $x = 8$,

$$y = \frac{\ln[2.25(8) + 1]}{0.3}$$
$$\approx 10 \text{ items.}$$

55. Let $t = 0$ be the time it started snowing. If h is the height of the snow and if the rate of snowfall is constant, $\frac{dh}{dt} = k_1$, where k_1 is a constant.

$$\frac{dh}{dt} = k_1 \text{ and } h = 0 \text{ when } t = 0.$$
$$dh = k_1\, dt$$
$$\int dh = \int k_1\, dt$$
$$h = k_1 t + C_1$$

Since $h = 0$ and $t = 0$, $0 = k_1(0) + C_1$. Thus, $C_1 = 0$ and $h = k_1 t$.

Since the snowplow removes a constant volume of snow per hour and the volume is proportional to the height of the snow, the rate of travel of the snowplow is inversely proportional to the height of the snow.

$$\frac{dx}{dt} = \frac{k^2}{h}, \text{ where } k_2 \text{ is a constant.}$$

When $t = T$, $x = 0$.
When $t = T + 1$, $x = 2$.
When $t = T + 2$, $x = 3$.

Since $\dfrac{dy}{dt} = \dfrac{k^2}{h}$ and $h = k_1 t$,

$$\frac{dy}{dt} = \frac{k_2}{k_1 t}$$
$$\frac{dx}{dt} = \frac{k_2}{k_1} \cdot \frac{1}{t}.$$

Let $k_3 = \dfrac{k_2}{k_1}$. Then

$$\frac{dx}{dt} = k_3 \frac{1}{t}$$
$$dx = k_3 \frac{1}{t}\, dt$$
$$\int dx = \int k_3 \frac{1}{t}\, dt$$
$$x = k_3 \ln t + C_2.$$

Since $x = 0$, when $t = T$,

$$0 = k_3 \ln T + C_2$$
$$C_2 = -k_3 \ln T.$$

Thus,

$$x = k_3 \ln t - k_3 \ln T$$
$$x = k_3(\ln t - \ln T)$$
$$x = k_3 \ln\left(\frac{t}{T}\right).$$

Since $x = 2$, when $t = T + 1$,

$$2 = k_3 \ln\left(\frac{T+1}{T}\right). \quad (1)$$

Since $x = 3$ when $t = T + 2$,

$$3 = k_3 \ln\left(\frac{T+2}{T}\right). \quad (2)$$

We want to solve for T, so we divide equation (1) by equation (2).

$$\frac{2}{3} = \frac{k_3 \ln\left(\frac{T+1}{T}\right)}{k_3 \ln\left(\frac{T+2}{T}\right)}$$

$$\frac{2}{3} = \frac{\ln(T+1) - \ln T}{\ln(T+2) - \ln T}$$

$$2\ln(T+2) - 2\ln T = 3\ln(T+1) - 3\ln T$$

$$\ln(T+2)^2 - \ln T^2 - \ln(T+1)^3 + \ln T^3 = 0$$

$$\ln\frac{(T+2)^2 T^3}{T^2(T+1)^3} = 0$$

$$\frac{T(T+2)^2}{(T+1)^3} = 1$$

$$T(T^2 + 4T + 4) = T^3 + 3T^2 + 3T + 1$$

$$T^3 + 4T^2 + 4T = T^3 + 3T^2 + 3T + 1$$

$$T^2 + T - 1 = 0$$

$$T = \frac{-1 \pm \sqrt{1+4}}{2}$$

$T = \frac{-1-\sqrt{5}}{2}$ is negative and is not a possible solution.

Thus, $T = \frac{-1+\sqrt{5}}{2} \approx 0.618$ hr.

0.618 hr $\approx$ 37 min and 5 sec
Now, 37 min and 5 sec before 8:00 A.M. is 7:22:55 A.M.
Thus, it started snowing at 7:22:55 A.M.

Chapter 16 Review Exercises

5. $\displaystyle\int \frac{3x}{\sqrt{x-2}}\,dx = \int 3x(x-2)^{-1/2}\,dx$

Let $\quad u = 3x$ and $dv = (x-2)^{-1/2}\,dx$.
Then $du = 3\,dx$ and $v = 2(x-2)^{1/2}$.

$$\int \frac{3x}{\sqrt{x-2}}\,dx$$

$$= 6x(x-2)^{1/2} - 6\int (x-2)^{1/2}\,dx$$

$$= 6x(x-2)^{1/2} - \frac{6(x-2)^{3/2}}{\frac{3}{2}} + C$$

$$= 6x(x-2)^{1/2} - 4(x-2)^{3/2} + C$$

7. $\displaystyle\int (3x+6)e^{-3x}\,dx$

Let $\quad u = 3x + 6 \quad$ and $\quad dv = e^{-3x}\,dx$.
Then $\quad du = 3\,dx \quad$ and $\quad v = \frac{1}{-3}e^{-3x}$.

$$\int (3x+6)e^{-3x}\,dx$$

$$= (3x+6)\left(-\frac{1}{3}e^{-3x}\right) - \int \left(-\frac{1}{3}e^{-3x}\right)3\,dx$$

$$= -(x+2)e^{-3x} + \int e^{-3x}\,dx$$

$$= -(x+2)e^{-3x} - \frac{1}{3}e^{-3x} + C$$

9. $\displaystyle\int (x-1)\ln|x|\,dx$

Let $\quad u = \ln|x|$ and $dv = (x-1)\,dx$.

Then $du = \frac{1}{x}\,dx$ and $\quad v = \frac{x^2}{2} - x$.

$$\int (x-1)\ln|x|\,dx$$

$$= \left(\frac{x^2}{2} - x\right)\ln|x| - \int \left(\frac{x}{2} - 1\right)dx$$

$$= \left(\frac{x^2}{2} - x\right)\ln|x| - \frac{x^2}{4} + x + C$$

11. $\displaystyle\int \frac{x}{\sqrt{16+8x^2}}\,dx$

Use substitution.
Let $u = 16 + 8x^2$. Then $du = 16x\,dx$.

$$\int \frac{x}{\sqrt{16+8x^2}}\,dx = \frac{1}{16}\int \frac{16x}{\sqrt{16+8x^2}}\,dx$$

$$= \frac{1}{16}\int \frac{1}{\sqrt{u}}\,du$$

$$= \frac{1}{16}\int u^{-1/2}\,du$$

$$= \frac{1}{16}(2)u^{1/2} + C$$

$$= \frac{1}{8}(16+8x^2)^{1/2} + C$$

$$= \frac{1}{8}\sqrt{16+8x^2} + C$$

13. $\displaystyle\int_0^1 x^2 e^{x/2}\, dx$

Let $u = x^2$ and $dv = e^{x/2}\, dx$.
Use column integration.

D	I
$x^2 \quad +$	$e^{x/2}$
$2x \quad -$	$2e^{x/2}$
$2 \quad +$	$4e^{x/2}$
0	$8e^{x/2}$

$$\int_0^1 x^2 e^{x/2}\, dx = (2x^2 e^{x/2} - 8xe^{x/2} + 16e^{x/2})\Big|_0^1$$

$$= 2e^{1/2} - 8e^{1/2} + 16e^{1/2} - 16$$
$$= 10e^{1/2} - 16 \approx 0.4872$$

15. $A = \displaystyle\int_1^3 x^3(x^2 - 1)^{1/3}\, dx$

Let $\quad u = x^2$ and $dv = x(x^2 - 1)^{1/3}\, dx$.
Then $du = 2x\, dx$ and $v = \frac{3}{8}(x^2 - 1)^{4/3}$.

$$\int x^3(x^2 - 1)^{1/3}\, dx$$

$$= \frac{3x^2}{8}(x^2 - 1)^{4/3} - \frac{3}{4}\int x(x^2 - 1)^{4/3}\, dx$$

$$= \frac{3x^2}{8}(x^2 - 1)^{4/3} - \frac{3}{4}\left[\frac{1}{2}\cdot\frac{3}{7}(x^2 - 1)^{7/3}\right]$$

$$= \frac{3x^2}{8}(x^2 - 1)^{4/3} - \frac{9}{56}(x^2 - 1)^{7/3} + C$$

$$A = \left[\frac{3x^2}{8}(x^2 - 1)^{4/3} - \frac{9}{56}(x^2 - 1)^{7/3}\right]\Big|_0^3$$

$$= \frac{3}{8}(144) - \frac{9}{56}(128) = 54 - \frac{144}{7} = \frac{234}{7} \approx 33.43$$

17. $f(x) = \sqrt{x - 4};\ y = 0;\ x = 13$

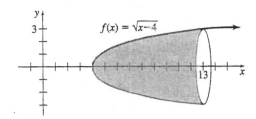

Since $f(x) = \sqrt{x - 4}$ intersects $y = 0$ at $x = 4$, the integral has lower bound $a = 4$.

$$V = \pi\int_4^{13}(\sqrt{x - 4})^2\, dx$$

$$= \pi\int_4^{13}(x - 4)\, dx$$

$$= \pi\left(\frac{x^2}{2} - 4x\right)\Big|_4^{13}$$

$$= \pi\left[\left(\frac{169}{2} - 52\right) - (8 - 16)\right]$$

$$= \pi\left(\frac{65}{2} + 8\right) = \frac{81}{2}\pi \approx 127.2$$

19. $f(x) = \dfrac{1}{\sqrt{x - 1}},\ y = 0,\ x = 2,\ x = 4$

$$V = \pi\int_2^4\left(\frac{1}{\sqrt{x - 1}}\right)^2\, dx$$

$$= \pi\int_2^4\frac{dx}{x - 1}$$

$$= \pi\left(\ln|x - 1|\right)\Big|_2^4$$

$$= \pi\ln 3 \approx 3.451$$

21. $f(x) = \dfrac{x^2}{4},\ y = 0,\ x = 4$

Since $f(x) = \frac{x^2}{4}$ intersects $y = 0$ at $x = 0$, the integral has a lower bound, $a = 0$.

$$V = \pi\int_0^4\left(\frac{x^2}{4}\right)^2\, dx = \pi\int_0^4\frac{x^4}{16}$$

$$= \frac{\pi}{16}\left(\frac{x^5}{5}\right)\Big|_0^4$$

$$= \frac{\pi}{16}\left(\frac{1024}{5}\right)$$

$$= \frac{64\pi}{5} \approx 40.21$$

25. Average value $= \dfrac{1}{2-0} \displaystyle\int_0^2 7x^2(x^3+1)^6 dx$

$$= \frac{7}{2}\int_0^2 x^2(x^3+1)^6 dx$$

Let $u = x^3 + 1$. Then $du = 3x^2 dx$.

$$\int x^2(x^3+1)^6 dx = \frac{1}{3}\int 3x^2(x^3+1)^6 dx$$

$$= \frac{1}{3}\int u^6 du$$

$$= \frac{1}{3}\cdot\frac{1}{7}u^7 + C$$

$$= \frac{1}{21}(x^3+1)^7 + C$$

$$\frac{7}{2}\int_0^2 x^2(x^3+1)^6 dx = \frac{7}{2}\cdot\frac{1}{21}(x^3+1)^7\Big|_0^2$$

$$= \frac{1}{6}(9^7 - 1^7)$$

$$= \frac{1}{6}(4{,}782{,}969 - 1)$$

$$= \frac{2{,}391{,}484}{3}$$

27. $\displaystyle\int_{-\infty}^{-5} x^{-2}dx = \lim_{a\to-\infty}\int_a^{-5} x^{-2}dx$

$$= \lim_{a\to-\infty}\left(\frac{x^{-1}}{-1}\right)\Big|_a^{-5}$$

$$= \lim_{a\to-\infty}\left(-\frac{1}{x}\right)\Big|_a^{-5}$$

$$= \frac{1}{5} + \lim_{a\to-\infty}\left(\frac{1}{a}\right)$$

As $a\to-\infty$, $\frac{1}{a}\to 0$. The integral converges.

$$\int_{-\infty}^{-5} x^{-2}dx = \frac{1}{5} + 0 = \frac{1}{5}$$

29. $\displaystyle\int_1^{\infty} 6e^{-x}\,dx = \lim_{b\to\infty}\int_1^b 6e^{-x}\,dx$

$$= \lim_{b\to\infty} -6e^{-x}\Big|_1^b$$

$$= \lim_{b\to\infty}(-6e^{-b} + 6e^{-1})$$

$$= \lim_{b\to\infty}\left(\frac{-6}{e^b} + \frac{6}{e}\right)$$

As $b\to\infty$, $e^b\to\infty$, so $\frac{-6}{e^b}\to 0$. The integral converges.

$$\int_1^{\infty} 6e^{-x}\,dx = 0 + \frac{6}{e} = \frac{6}{e} \approx 2.207$$

31. $\displaystyle\int_4^{\infty} \ln(5x)dx = \lim_{b\to\infty}\int_4^b \ln(5x)dx$

Let $\quad u = \ln(5x) \quad$ and $\quad dv = dx$.

Then $\quad du = \frac{1}{x}dx \quad$ and $\quad v = x$.

$$\int \ln(5x)dx = x\ln(5x) - \int x\cdot\frac{1}{x}\,dx$$

$$= x\ln(5x) - \int dx$$

$$= x\ln(5x) - x + C$$

$$\lim_{b\to\infty}\int_4^b \ln(5x)dx = \lim_{b\to\infty}\left[x\ln(5x) - x\right]\Big|_4^b$$

$$= \lim_{b\to\infty}\left[b\ln(5b) - b\right] - (4\ln 20 - 4)$$

As $b\to\infty$, $b\ln(5b) - b\to\infty$. The integral diverges.

33. $f(x) = 3e^{-x}$ for $[0,\infty)$

$$A = \int_0^{\infty} 3e^{-x}\,dx$$

$$= \lim_{b\to\infty}\int_0^b 3e^{-x}\,dx$$

$$= \lim_{b\to\infty}(-3e^{-x})\Big|_0^b$$

$$= \lim_{b\to\infty}\left(\frac{-3}{e^b} + 3\right)$$

As $b\to\infty$, $\frac{-3}{e^b}\to 0$.

$A = 0 + 3 = 3$

35. $R' = x(x-50)^{1/2}$

$$R = \int_{50}^{75} x(x-50)^{1/2}\,dx$$

Let $\quad u = x$ and $dv = (x-50)^{1/2}$.

Then $du = dx$ and $v = \frac{2}{3}(x-50)^{3/2}$.

$$\int x(x-50)^{1/2}\,dx$$

$$= \frac{2}{3}x(x-50)^{3/2} - \frac{2}{3}\int(x-50)^{3/2}\,dx$$

$$= \frac{2}{3}x(x-50)^{3/2} - \frac{2}{3}\cdot\frac{2}{5}(x-50)^{5/2}$$

$$R = \left[\frac{2}{3}x(x-50)^{3/2} - \frac{4}{15}(x-50)^{5/2}\right]\Big|_{50}^{75}$$

$$= \frac{2}{3}(75)(25^{3/2}) - \frac{4}{15}(25^{5/2})$$

$$= 6250 - \frac{2500}{3}$$

$$= \frac{16{,}250}{3} \approx 5416.67$$

37. $f(x) = 25,000$; 12 yr; 10%

$$P = \int_0^{12} 25,000e^{-0.10x}\, dx$$

$$= 25,000 \left. \left(\frac{e^{-0.10x}}{-0.10}\right) \right|_0^{12}$$

$$\approx 250,000(-0.3012 + 1)$$

$$\approx \$174,701.45$$

39. $f(x) = 15x$; 18 mo; 8%

$$P = \int_0^{1.5} 15xe^{-0.08x}\, dx$$

$$= 15 \int_0^{1.5} xe^{-0.08x}\, dx$$

Find the antiderivative using integration by parts.

Let $\quad u = x \quad$ and $\quad dv = e^{-0.08x}\, dx$.

Then $\quad du = dx \quad$ and $\quad v = \dfrac{1}{-0.08}e^{-0.08x}$

$$= -12.5e^{-0.08x}.$$

$$\int xe^{-0.08x}\, dx = -12.5xe^{-0.08x} - \int (-12.5e^{-0.08x})\, dx$$

$$= -12.5xe^{-0.08x} - 156.25e^{-0.08x} + C$$

$$P = 15 \int_0^{1.5} xe^{-0.08x}\, dx$$

$$= 15\left. \left(-12.5xe^{-0.08x} - 156.25e^{-0.08x}\right) \right|_0^{1.5}$$

$$= 15[(-18.75e^{-0.12} - 156.25e^{-0.12})$$

$$\quad - (0 - 156.25)]$$

$$= 15(-175e^{-0.12} + 156.25)$$

$$\approx 15.58385362$$

The present value is \$15.58.

41. $f(x) = 500e^{-0.04x}$; 8 yr; 10% per yr

$$A = e^{0.1(8)} \int_0^8 500e^{-0.04x} \cdot e^{-0.1x}\, dx$$

$$= e^{0.8} \int_0^8 500e^{-0.14x}\, dx$$

$$= e^{0.8} \left. \left(\frac{500}{-0.14}e^{-0.14x}\right) \right|_0^8$$

$$= e^{0.8} \left[\frac{500}{-0.14}(e^{-1.12} - 1)\right]$$

$$\approx 5354.971041$$

The accumulated value is \$5354.97.

43. $f(x) = 1000 + 200x$; 10 yr; 9% per yr

$$e^{(0.09)(10)} \int_0^{10} (1000 + 200x)e^{-0.09x}\, dx$$

$$= e^{0.9} \left. \left[\frac{1000}{-0.09}e^{0.09x} + \frac{200}{(0.09)^2}(-0.09x - 1)e^{-0.09x}\right] \right|_0^{10}$$

$$= e^{0.9} \left[\frac{1000}{-0.09}(e^{-0.9} - 1) + \frac{200}{(0.09)^2}(-1.9e^{-0.9} + 1)\right]$$

$$\approx \$30,035.17$$

45. $e^{0.105(10)} \int_0^{10} 10,000e^{-0.105x}\, dx$

$$= e^{1.05} \left. \left(\frac{10,000e^{-0.105x}}{-0.105}\right) \right|_0^{10}$$

$$= \frac{10,000e^{1.05}}{-0.105}(e^{-1.05} - 1)$$

$$\approx -272,157.25(-0.65006)$$

$$\approx \$176,919.15$$

47. $\int_0^5 0.5xe^{-x}\, dx$

$$= 0.5 \int_0^5 xe^{-x}\, dx$$

Let $\quad u = x$ and $dv = e^{-x}\, dx$.

Then $du = dx$ and $v = \dfrac{e^{-x}}{-1}$.

$$\int xe^{-x}\, dx = \frac{xe^{-x}}{-1} + \int e^{-x}\, dx$$

$$= -xe^{-x} + \frac{e^{-x}}{-1}$$

$$0.5 \int_0^5 xe^{-x}\, dx = 0.5\left. (-xe^{-x} - e^{-x}) \right|_0^5$$

$$= 0.5(-5e^{-5} - e^{-5} + e^0)$$

$$\approx 0.4798$$

The total reaction over the first 5 hr is 0.4798.

49. $\int_0^{320} 1.87t^{1.49}e^{-0.189(\ln t)^2}\, dt$

$n = 8, b = 320, a = 1, f(t) = 1.87t^{1.49}e^{-0.189(\ln t)^2}$

i	t_i	$f(t_i)$
0	1	1.8700
1	41	34.9086
2	81	33.9149
3	121	30.7147
4	161	27.5809
5	201	24.8344
6	241	22.4794
7	281	20.4622
8	321	18.7255

(a) Trapezoidal rule:

$$\int_1^{321} 1.87t^{1.49}e^{-0.189(\ln t)^2}\,dt$$

$$\approx \frac{321-1}{8}\left[\frac{1}{2}(1.87)+34.9086+33.9149+30.7147\right.$$

$$+\,27.5809+24.8344+22.4794+20.4622$$

$$\left.+\,\frac{1}{2}(18.7255)\right]$$

$$\approx 8208$$

The total amount of milk produced is about 8208 kg.

(b) Simpson's rule:

$$\int_1^{321} 1.87t^{1.49}e^{-0.189(\ln t)^2}\,dt$$

$$\approx \frac{321-1}{3(8)}[1.87+4(34.9086)+2(33.9149)$$
$$+\,4(30.7147)+2(27.5809)+4(24.8344)$$
$$+\,2(22.4794)+4(20.4622)+18.7255]$$
$$\approx 8430$$

The total amount of milk produced is about 8430 kg.

(c) Using a graphing calculator's fnInt feature:

$$\int_1^{321} 1.87t^{1.49}e^{-0.189(\ln t)^2}\,dt \approx 8558.$$

The total amount of milk produced is about 8558 kg.

Chapter 16 Test

[16.1]

Use integration by parts to find the following integrals.

1. $\displaystyle\int x \ln|5x|\, dx$
2. $\displaystyle\int x\sqrt{2x-1}\, dx$
3. $\displaystyle\int (2x-1)\, e^x\, dx$

4. $\displaystyle\int \frac{x+3}{(3x-1)^4}\, dx$
5. $\displaystyle\int_{4}^{11} x\sqrt[3]{x-3}\, dx$

6. Given that the marginal profit in dollars earned from the sale of x computers is

$$P'(x) = xe^{0.001x} - 100,$$

find the total profit from the sale of the first 500 computers.

[16.2]

Find the volume of the solid of revolution formed by rotating each of the following bounded regions about the x-axis.

7. $f(x) = 3x - 1,\ y = 0,\ x = 4$
8. $f(x) = \dfrac{1}{\sqrt{3x+1}},\ y = 0,\ x = 0,\ x = 5$

9. $f(x) = e^x,\ y = 0,\ x = -1,\ x = 2$
10. $f(x) = x^2,\ y = 0,\ x = 2$

11. Find the average value of $f(x) = x^3 - x^2$ on the interval $[-1, 1]$.

12. The rate of depreciation t years after purchase of a certain machine is

$$D = 10{,}000(t - 7),\ 0 \le t \le 6.$$

What is the average depreciation over the first 3 years?

[16.3]

13. The rate of flow of an investment is $2000 per year for 10 years. Find the present value if the annual interest rate is 11% compounded continuously.

14. The rate of flow of an investment is given by

$$f(x) = 200e^{-0.1x}$$

for 10 years. Find the present value if the annual interest rate is 8% compounded continuously.

15. The rate of flow of an investment is given by the function $f(x) = 1000$. Find the final amount at the end of 15 years at an interest rate of 11% compounded continuously.

16. An investment scheme is expected to produce a continuous flow of money, starting at $2000 and increasing exponentially at 4% per year for 8 years. Find the present value at an interest rate of 12% compounded continuously.

[16.4]

Find the value of each integral that converges.

17. $\displaystyle\int_2^\infty \frac{14}{(x+1)^2}\, dx$

18. $\displaystyle\int_{-\infty}^2 \frac{4}{x+1}\, dx$

19. $\displaystyle\int_1^\infty e^{-3x}\, dx$

20. $\displaystyle\int_1^\infty \frac{1}{x}\, dx$

21. $\displaystyle\int_1^\infty x^{-1/3}\, dx$

22. $\displaystyle\int_{100}^\infty x^{-3/2}\, dx$

Find the area between the graph of the function and the x-axis over the given interval, if possible.

23. $f(x) = 8e^{-x}$ on $[1, \infty)$

24. $f(x) = \dfrac{1}{(2x-1)^2}$ on $[2, \infty)$

25. Find the capital value of an asset that generates income at an annual rate of $4000 if the interest rate is 8% compounded continuously.

[16.5]

Find the general solutions for the following differential equations.

26. $\dfrac{dy}{dx} = 3x^2 + 4x - 5$

27. $\dfrac{dy}{dx} = 5e^{2x}$

28. $\dfrac{dy}{dx} = \dfrac{2x}{x^2 + 5}$

Find particular solutions for the following differential equations.

29. $\dfrac{dy}{dx} = 4x^3 + 2x^2 + 1$; $y = 4$ when $x = 1$

30. $\dfrac{dy}{dx} = \dfrac{1}{3x+1}$; $y = 4$ when $x = 3$

31. $\dfrac{dy}{dx} = 4\left(e^{-x} - 1\right)$; $y = 5$ when $x = 0$

32. After use of an insecticide, the rate of decline of an insect population is given by

$$\frac{dy}{dt} = \frac{-8}{1 + 4t},$$

where t is the number of hours after the insecticide is applied. If there were 40 insects initially, how many are left after 20 hours?

Find general solutions for the following differential equations. (Some solutions may give y implicitly.)

33. $\dfrac{dy}{dx} = \dfrac{2x + 1}{y - 1}$

34. $\dfrac{dy}{dx} = \dfrac{3y}{x + 1}$

35. $\dfrac{dy}{dx} = \dfrac{e^x - x}{y + 1}$

Find particular solutions of the following differential equations.

36. $\sqrt{y}\dfrac{dy}{dx} = xy$; $y = 4$ when $x = 2$

37. $\dfrac{dy}{dx} = e^y \cdot x^3$; $y = 0$ when $x = 0$

Chapter 16 Test Answers

1. $\frac{1}{2}x^2 \ln|5x| - \frac{1}{4}x^2 + C$

2. $\frac{x}{3}(2x-1)^{3/2} - \frac{1}{15}(2x-1)^{5/2} + C$

3. $(2x-3)e^x + C$

4. $-\frac{1}{9}\left[\frac{x+3}{(3x-1)^3}\right] - \frac{1}{54(3x-1)^2} + C$ or $-\frac{9x+17}{54(3x-1)^3} + C$

5. $\frac{2469}{28}$ or 88.18

6. $125,639.36

7. $\frac{1331}{9}\pi$ or 464.6

8. $\frac{4}{3}\pi \ln(2)$ or 2.903

9. $\frac{\pi}{2}(e^4 - e^{-2})$ or 85.55

10. $\frac{32\pi}{5}$ or 20.11

11. $-\frac{1}{3}$

12. 55,000 per year

13. $12,129.62

14. $927.45

15. $38,245.27

16. $11,817.69

17. $\frac{14}{3}$

18. Divergent

19. $\frac{1}{3e^3}$ or 0.017

20. Divergent

21. Divergent

22. $\frac{1}{5}$

23. $\frac{8}{e}$ or 2.943

24. $\frac{1}{6}$

25. $50,000

26. $y = x^3 + 2x^2 - 5x + C$

27. $y = \frac{5}{2}e^{2x} + C$

28. $y = \ln(x^2 + 5) + C$

29. $y = x^4 + \frac{2}{3}x^3 + x + \frac{4}{3}$

30. $y = \frac{1}{3}\ln|3x+1| + 4 - \frac{1}{3}\ln 10$

31. $y = -4e^{-x} - 4x + 9$

32. 31

33. $\frac{y^2}{2} - y = x^2 + x + C$

34. $y = C(x+1)^3$

35. $\frac{y^2}{2} + y = e^x - \frac{x^2}{2} + C$

36. $y^{1/2} = \frac{x^2}{4} + 1$

37. $e^{-y} = 1 - \frac{x^4}{4}$ or $y = -\ln\left|1 - \frac{x^4}{4}\right|$

MULTIVARIABLE CALCULUS

17.1 Functions of Several Variables

1. $f(x, y) = 2x - 3y + 5$

   (a) $f(2, -1) = 2(2) - 3(-1) + 5 = 12$

   (b) $f(-4, 1) = 2(-4) - 3(1) + 5 = -6$

   (c) $f(-2, -3) = 2(-2) - 3(-3) + 5 = 10$

   (d) $f(0, 8) = 2(0) - 3(8) + 5 = -19$

3. $h(x, y) = \sqrt{x^2 + 2y^2}$

   (a) $h(5, 3) = \sqrt{25 + 2(9)} = \sqrt{43}$

   (b) $h(2, 4) = \sqrt{4 + 32} = 6$

   (c) $h(-1, -3) = \sqrt{1 + 18} = \sqrt{19}$

   (d) $h(-3, -1) = \sqrt{9 + 2} = \sqrt{11}$

5. $x + y + z = 9$

   If $x = 0$ and $y = 0$, $z = 9$.
   If $x = 0$ and $z = 0$, $y = 9$.
   If $y = 0$ and $z = 0$, $x = 9$.

   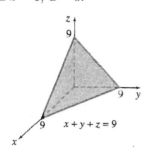

7. $2x + 3y + 4z = 12$

   If $x = 0$ and $y = 0$, $z = 3$.
   If $x = 0$ and $z = 0$, $y = 4$.
   If $y = 0$ and $z = 0$, $x = 6$.

   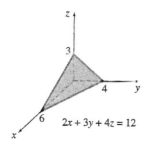

9. $x + y = 4$

   If $x = 0$, $y = 4$.
   If $y = 0$, $x = 4$.

   There is no z-intercept.

   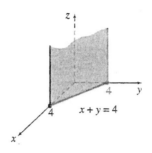

11. $x = 5$

    The point $(5, 0, 0)$ is on the graph.
    There are no $y-$ or z-intercepts.
    The plane is parallel to the yz-plane.

    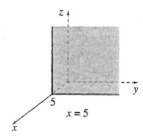

13. $3x + 2y + z = 24$

    For $z = 0$, $3x + 2y = 24$. Graph the line $3x + 2y = 24$ in the xy-plane.
    For $z = 2$, $3x + 2y = 22$. Graph the line $3x + 2y = 22$ in the plane $z = 2$.
    For $z = 4$, $3x + 2y = 20$. Graph the line $3x + 2y = 20$ in the plane $z = 4$.

    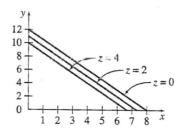

15. $y^2 - x = -z$

For $z = 0$, $x = y^2$. Graph $x = y^2$ in the xy-plane.
For $z = 2$, $x = y^2 + 2$. Graph $x = y^2 + 2$ in the plane $z = 2$.
For $z = 4$, $x = y^2 + 4$. Graph $x = y^2 + 4$ in the plane $z = 4$.

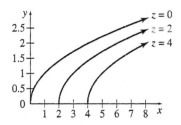

21. $z = x^2 + y^2$

The xz-trace is

$$z = x^2 + 0 = x^2.$$

The yz-trace is

$$z = 0 + y^2 = y^2.$$

Both are parabolas with vertices at the origin that open upward.
The xy-trace is

$$0 = x^2 + y^2.$$

This is a point, the origin.
The equation is represented by a paraboloid, as shown in (c).

23. $x^2 - y^2 = z$

The xz-trace is
$$x^2 = z,$$

which is a parabola with vertex at the origin that opens upward.
The yz-trace is
$$-y^2 = z,$$

which is a parabola with vertex at the origin that opens downward.

The xy-trace is
$$x^2 - y^2 = 0$$
$$x^2 = y^2$$
$$x = y \quad \text{or} \quad x = -y,$$

which are two lines that intersect at the origin.
The equation is represented by a hyperbolic paraboloid, as shown in (e).

25. $\dfrac{x^2}{16} + \dfrac{y^2}{25} + \dfrac{z^2}{4} = 1$

xz-trace:

$$\frac{x^2}{16} + \frac{z^2}{4} = 1, \text{ an ellipse}$$

yz-trace:

$$\frac{y^2}{25} + \frac{z^2}{4} = 1, \text{ an ellipse}$$

xy-trace:

$$\frac{x^2}{16} + \frac{y^2}{25} = 1, \text{ an ellipse}$$

The graph is an ellipsoid, as shown in (b).

27. $f(x, y) = 4x^2 - 2y^2$

(a) $\dfrac{f(x + h, y) - f(x, y)}{h}$

$$= \frac{[4(x + h)^2 - 2y^2] - [4x^2 - 2y^2]}{h}$$

$$= \frac{4x^2 + 8xh + 4h^2 - 2y^2 - 4x^2 + 2y^2}{h}$$

$$= \frac{h(8x + 4h)}{h} = 8x + 4h$$

(b) $\dfrac{f(x, y + h) - f(x, y)}{h}$

$$= \frac{[4x^2 - 2(y + h)^2] - [4x^2 - 2y^2]}{h}$$

$$= \frac{4x^2 - 2y^2 - 4yh - 2h^2 - 4x^2 + 2y^2}{h}$$

$$= \frac{h(-4y - 2h)}{h} = -4y - 2h$$

(c) $\displaystyle\lim_{h \to 0} \frac{f(x + h, y) - f(x, y)}{h} = \lim_{h \to 0} (8x + 4h)$
$$= 8x + 4(0) = 8x$$

(d) $\displaystyle\lim_{h \to 0} \frac{f(x, y + h) - f(x, y)}{h} = \lim_{h \to 0} (-4y - 2h)$
$$= -4y - 2(0) = -4y$$

29. $f(x,y) = xye^{x^2+y^2}$

(a) $\displaystyle\lim_{h\to 0}\frac{f(1+h,1) - f(1,1)}{h}$

$$= \lim_{h\to 0}\frac{(1+h)(1)e^{1+2h+h^2+1} - (1)(1)e^{1+1}}{h}$$

$$= \lim_{h\to 0}\frac{(1+h)e^{2+2h+h^2} - e^2}{h}$$

$$= e^2\lim_{h\to 0}\frac{(1+h)e^{2h+h^2} - 1}{h}$$

The graphing calculator indicates that
$\lim_{h\to 0}\frac{(1+h)e^{2h+h^2}-1}{h} = 3$, thus $\lim_{h\to 0}\frac{f(1+h,1)-f(1,1)}{h} = 3e^2$.
The slope of the tangent line in the direction of x
at $(1,1)$ is $3e^2$.

(b) $\displaystyle\lim_{h\to 0}\frac{f(1,1+h) - f(1,1)}{h}$

$$= \lim_{h\to 0}\frac{(1)(1+h)e^{1+1+2h+h^2} - (1)(1)e^{1+1}}{h}$$

$$= \lim_{h\to 0}\frac{(1+h)e^{2+2h+h^2} - e^2}{h}$$

$$= e^2\lim_{h\to 0}\frac{(1+h)e^{2h+h^2} - 1}{h}$$

So, this limit reduces to the exact same limit
as in part a. Therefore, since

$$\lim_{h\to 0}\frac{(1+h)e^{2h+h^2} - 1}{h} = 3,$$

then

$$\lim_{h\to 0}\frac{f(1,1+h) - f(1,1)}{h} = 3e^2.$$

The slope of the tangent line in the direction of y
at $(1,1)$ is $3e^2$.

31. $P(x,y) = 100\left[\frac{3}{5}x^{-2/5} + \frac{2}{5}y^{-2/5}\right]^{-5}$

(a) $P(32,1)$

$$= 100\left[\frac{3}{5}(32)^{-2/5} + \frac{2}{5}(1)^{-2/5}\right]^{-5}$$

$$= 100\left[\frac{3}{5}\left(\frac{1}{4}\right) + \frac{2}{5}(1)\right]^{-5}$$

$$= 100\left(\frac{11}{20}\right)^{-5}$$

$$= 100\left(\frac{20}{11}\right)^{5}$$

$$\approx 1986.95$$

The production is approximately 1987 cameras.

(b) $P(1,32)$

$$= 100\left[\frac{3}{5}(1)^{-2/5} + \frac{2}{5}(32)^{-2/5}\right]^{-5}$$

$$= 100\left[\frac{3}{5}(1) + \frac{2}{5}\left(\frac{1}{4}\right)\right]^{-5}$$

$$= 100\left(\frac{7}{10}\right)^{-5}$$

$$= 100\left(\frac{10}{7}\right)^{5}$$

$$\approx 595$$

The production is approximately 595 cameras.

(c) 32 work hours means that $x = 32$. 243 units
of capital means that $y = 243$.

$P(32,243)$

$$= 100\left[\frac{3}{5}(32)^{-2/5} + \frac{2}{5}(243)^{-2/5}\right]^{-5}$$

$$= 100\left[\frac{3}{5}\left(\frac{1}{4}\right) + \frac{2}{5}\left(\frac{1}{9}\right)\right]^{-5}$$

$$= 100\left(\frac{7}{36}\right)^{-5}$$

$$= 100\left(\frac{36}{7}\right)^{5}$$

$$\approx 359,767.81$$

The production is approximately 359,768 cameras.

33. $M = f(40, 0.06, 0.28)$

$$= \frac{(1 + 0.06)^{40}(1 - 0.28) + 0.28}{[1 + (1 - 0.28)(0.06)]^{40}}$$

$$= \frac{(1.06)^{40}(0.72) + 0.28}{[1 + (0.72)(0.06)]^{40}}$$

$$\approx 1.416$$

The multiplier is 1.42. Since $M > 1$, the IRA account grows faster.

35. $z = x^{0.6} y^{0.4}$ where $z = 500$

$$500 = x^{3/5} y^{2/5}$$

$$\frac{500}{x^{3/5}} = y^{2/5}$$

$$\left(\frac{500}{x^{3/5}}\right)^{5/2} = (y^{2/5})^{5/2}$$

$$y = \frac{(500)^{5/2}}{x^{3/2}}$$

$$y \approx \frac{5{,}590{,}170}{x^{3/2}}$$

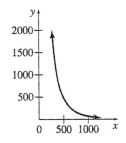

37. The cost function, C, is the sum of the products of the unit costs times the quantities x, y, and z. Therefore,

$$C(x, y, z) = 250x + 150y + 75z.$$

39. $F = \dfrac{v^2}{g\ell}$

(a) $2.56 = \dfrac{v^2}{(9.81)(0.09)}$

$$v^2 = (2.56)(9.81)(0.09)$$
$$v^2 \approx 2.260$$
$$v \approx 1.5$$

For a ferret, this change occurs at 1.5 m/sec.

$$2.56 = \frac{v^2}{(9.81)(1.2)}$$

$$v^2 = (2.56)(9.81)(1.2)$$
$$v^2 \approx 30.136$$
$$v \approx 5.5$$

For a rhinoceros, this change occurs at 5.5 m/sec.

(b) $0.025 = \dfrac{v^2}{(9.81)(4)}$

$$v^2 = (0.025)(9.81)(4)$$
$$v^2 = 0.981$$
$$v \approx 0.99$$

The sauropods were traveling at 1 m/sec.

41. $P(W, R, A) = 48 - 2.43W - 1.81R - 1.22A$

(a) $P(5, 15, 0) = 48 - 2.43(5) - 1.81(15) - 1.22(0)$
$$= 8.7$$

8.7% of fish will be intolerant to pollution.

(b) The maximum percentage will occur when the variable factors are a minimum, or when $W = 0$, $R = 0$, and $A = 0$.

$$P(0, 0, 0) = 48 - 2.43(0) - 1.81(0) - 1.22(0)$$
$$= 48$$

48% of fish will be intolerant to pollution.

(c) Any combination of values of W, R, and A that result in $P = 0$ is a scenario that will drive the percentage of fish intolerant to pollution to zero.

If $R = 0$ and $A = 0$:

$$P(W, 0, 0) = 48 - 2.43W - 1.81(0) - 1.22(0)$$
$$= 48 - 2.43W.$$

$$48 - 2.43W = 0$$

$$W = \frac{48}{2.43}$$

$$\approx 19.75$$

So $W = 19.75, R = 0, A = 0$ is one scenario.

If $W = 10$ and $R = 10$:

$$P(10, 10, A) = 48 - 2.43(10) - 1.81(10) - 1.22A$$
$$= 5.6 - 1.22A$$

$$5.6 - 1.22A = 0$$

$$A = \frac{5.6}{1.22}$$

$$\approx 4.59$$

So $W = 10, R = 10, A = 4.59$ is another scenario.

(d) Since the coefficient of W is greater than the coefficients of R and A, a change in W will affect the value of P more than an equal change in R or A. Thus, the percentage of wetland (W) has the greatest influence on P.

43. $A(L, T, U, C) = 53.02 + 0.383L + 0.0015T + 0.0028U - 0.0003C$

  (a) $A(266, 107{,}484, 31{,}697, 24{,}870) = 53.02 + 0.383(266) + 0.0015(107{,}484) + 0.0028(31{,}697) - 0.0003(24{,}870)$
$$\approx 397$$

The estimated number of accidents is 397.

45. $\ln(T) = 5.49 - 3.00\ln(F) + 0.18\ln(C)$

  (a) $e^{\ln(T)} = e^{5.49 - 3.00\ln(F) + 0.18\ln(C)}$

$$T = e^{5.49}e^{-3.00\ln(F)}e^{0.18\ln(C)} = \frac{e^{5.49}e^{\ln(C^{0.18})}}{e^{\ln(F^3)}}$$

$$T \approx \frac{242.257C^{0.18}}{F^3}$$

  (b) Replace F with 2 and C with 40 in the preceding formula.

$$T \approx \frac{242.257(40)^{0.18}}{(2)^3} \approx 58.82$$

T is about 58.8%. In other words, a tethered sow spends nearly 59% of the time doing repetitive behavior when she is fed 2 kg of food per day and a neighboring sow spends 40% of the time doing repetitive behavior.

47. Let the area be given by $g(L, W, H)$.
Then,
$$g(L, W, H) = 2LW + 2WH + 2LH \text{ ft}^2.$$

17.2   Partial Derivatives

1. $z = f(x, y) = 6x^2 - 4xy + 9y^2$

  (a) $\dfrac{\partial z}{\partial x} = 12x - 4y$

  (b) $\dfrac{\partial z}{\partial y} = -4x + 18y$

  (c) $\dfrac{\partial f}{\partial x}(2, 3) = 12(2) - 4(3) = 12$

  (d) $f_y(1, -2) = -4(1) + 18(-2) = -40$

3. $f(x, y) = -4xy + 6y^3 + 5$

$f_x(x, y) = -4y$
$f_y(x, y) = -4x + 18y^2$
$f_x(2, -1) = -4(-1) = 4$
$f_y(-4, 3) = -4(-4) + 18(3)^2 = 16 + 18(9) = 178$

5. $f(x, y) = 5x^2y^3$
$f_x(x, y) = 10xy^3$
$f_y(x, y) = 15x^2y^2$
$f_x(2, -1) = 10(2)(-1)^3 = -20$
$f_y(-4, 3) = 15(-4)^2(3)^2 = 2160$

7. $f(x,y) = e^{x+y}$

$f_x(x,y) = e^{x+y}$

$f_y(x,y) = e^{x+y}$

$f_x(2,-1) = e^{2-1}$

$\qquad = e^1 = e$

$f_y(-4,3) = e^{-4+3}$

$\qquad = e^{-1}$

$\qquad = \dfrac{1}{e}$

9. $f(x,y) = -6e^{4x-3y}$

$f_x(x,y) = -24e^{4x-3y}$

$f_y(x,y) = 18e^{4x-3y}$

$f_x(2,-1) = -24e^{4(2)-3(-1)} = -24e^{11}$

$f_y(-4,3) = 18e^{4(-4)-3(3)} = 18e^{-25}$

11. $f(x,y) = \dfrac{x^2 + y^3}{x^3 - y^2}$

$f_x(x,y) = \dfrac{2x(x^3 - y^2) - 3x^2(x^2 + y^3)}{(x^3 - y^2)^2}$

$\qquad = \dfrac{2x^4 - 2xy^2 - 3x^4 - 3x^2y^3}{(x^3 - y^2)^2}$

$\qquad = \dfrac{-x^4 - 2xy^2 - 3x^2y^3}{(x^3 - y^2)^2}$

$f_y(x,y) = \dfrac{3y^2(x^3 - y^2) - (-2y)(x^2 + y^3)}{(x^3 - y^2)^2}$

$\qquad = \dfrac{3x^3y^2 - 3y^4 + 2x^2y + 2y^4}{(x^3 - y^2)^2}$

$\qquad = \dfrac{3x^3y^2 - y^4 + 2x^2y}{(x^3 - y^2)^2}$

$f_x(2,-1) = \dfrac{-2^4 - 2(2)(-1)^2 - 3(2^2)(-1)^3}{[2^3 - (-1)^2]^2}$

$\qquad = -\dfrac{8}{49}$

$f_y(-4,3) = \dfrac{3(-4)^3(3)^2 - 3^4 + 2(-4)^2(3)}{[(-4)^3 - 3^2]^2}$

$\qquad = -\dfrac{1713}{5329}$

13. $f(x,y) = \ln\left|1 + 5x^3y^2\right|$

$f_x(x,y) = \dfrac{1}{1 + 5x^3y^2} \cdot 15x^2y^2 = \dfrac{15x^2y^2}{1 + 5x^3y^2}$

$f_y(x,y) = \dfrac{1}{1 + 5x^3y^2} \cdot 10x^3y = \dfrac{10x^3y}{1 + 5x^3y^2}$

$f_x(2,-1) = \dfrac{15(2)^2(-1)^2}{1 + 5(2)^3(-1)^2} = \dfrac{60}{41}$

$f_y(-4,3) = \dfrac{10(-4)^3(3)}{1 + 5(-4)^3(3)^2} = \dfrac{1920}{2879}$

15. $f(x,y) = xe^{x^2y}$

$f_x(x,y) = e^{x^2y} \cdot 1 + x(2xy)(e^{x^2y})$

$\qquad = e^{x^2y}(1 + 2x^2y)$

$f_y(x,y) = x^3 e^{x^2y}$

$f_x(2,-1) = e^{-4}(1 - 8) = -7e^{-4}$

$f_y(-4,3) = -64e^{48}$

17. $f(x,y) = \sqrt{x^4 + 3xy + y^4 + 10}$

$f_x(x,y) = \dfrac{4x^3 + 3y}{2\sqrt{x^4 + 3xy + y^4 + 10}}$

$f_y(x,y) = \dfrac{3x + 4y^3}{2\sqrt{x^4 + 3xy + y^4 + 10}}$

$f_x(2,-1) = \dfrac{4(2)^3 + 3(-1)}{2\sqrt{2^4 + 3(2)(-1) + (-1)^4 + 10}}$

$\qquad = \dfrac{29}{2\sqrt{21}}$

$f_y(-4,3) = \dfrac{3(-4) + 4(3)^3}{2\sqrt{(-4)^4 + 3(-4)(3) + 3^4 + 10}}$

$\qquad = \dfrac{48}{\sqrt{311}}$

19. $f(x,y) = \dfrac{3x^2y}{e^{xy} + 2}$

$f_x(x,y) = \dfrac{6xy(e^{xy} + 2) - ye^{xy}(3x^2y)}{(e^{xy} + 2)^2}$

$\qquad = \dfrac{6xy(e^{xy} + 2) - 3x^2y^2e^{xy}}{(e^{xy} + 2)^2}$

$f_y(x,y) = \dfrac{3x^2(e^{xy} + 2) - xe^{xy}(3x^2y)}{(e^{xy} + 2)^2}$

$\qquad = \dfrac{3x^2(e^{xy} + 2) - 3x^3ye^{xy}}{(e^{xy} + 2)^2}$

$f_x(2,-1) = \dfrac{6(2)(-1)(e^{2(-1)} + 2) - 3(2)^2(-1)^2e^{2(-1)}}{(e^{2(-1)} + 2)^2}$

$\qquad = \dfrac{-12e^{-2} - 24 - 12e^{-2}}{(e^{-2} + 2)^2}$

$\qquad = \dfrac{-24(e^{-2} + 1)}{(e^{-2} + 2)^2}$

$f_y(-4,3) = \dfrac{3(-4)^2(e^{(-4)(3)} + 2) - 3(-4)^3(3)e^{(-4)(3)}}{(e^{(-4)(3)} + 2)^2}$

$\qquad = \dfrac{48e^{-12} + 96 + 576e^{-12}}{(e^{-12} + 2)^2}$

$\qquad = \dfrac{624e^{-12} + 96}{(e^{-12} + 2)^2}$

21. $f(x, y) = 4x^2y^2 - 16x^2 + 4y$

$f_x(x, y) = 8xy^2 - 32x$
$f_y(x, y) = 8x^2y + 4$
$f_{xx}(x, y) = 8y^2 - 32$
$f_{yy}(x, y) = 8x^2$
$f_{xy}(x, y) = f_{yx}(x, y) = 16xy$

23. $R(x, y) = 4x^2 - 5xy^3 + 12y^2x^2$

$R_x(x, y) = 8x - 5y^3 + 24y^2x$
$R_y(x, y) = -15xy^2 + 24yx^2$
$R_{xx}(x, y) = 8 + 24y^2$
$R_{yy}(x, y) = -30xy + 24x^2$
$R_{xy}(x, y) = -15y^2 + 48xy$
$\qquad\quad\ = R_{yx}(x, y)$

25. $r(x, y) = \dfrac{6y}{x + y}$

$r_x(x, y) = \dfrac{(x + y)(0) - 6y(1)}{(x + y)^2} = -6y(x + y)^{-2}$

$r_y(x, y) = \dfrac{(x + y)(6) - 6y(1)}{(x + y)^2} = 6x(x + y)^{-2}$

$r_{xx}(x, y) = -6y(-2)(x + y)^{-3}(1) = \dfrac{12y}{(x + y)^3}$

$r_{yy}(x, y) = 6x(-2)(x + y)^{-3}(1) = -\dfrac{12x}{(x + y)^3}$

$r_{xy}(x, y) = r_{yx}(x, y)$
$\qquad\quad = -6y(-2)(x + y)^{-3}(1) + (x + y)^{-2}(-6)$
$\qquad\quad = \dfrac{12y - 6(x + y)}{(x + y)^3} = \dfrac{6y - 6x}{(x + y)^3}$

27. $\quad z = 9ye^x$

$z_x = 9ye^x$
$z_y = 9e^x$
$z_{xx} = 9ye^x$
$z_{yy} = 0$
$z_{xy} = z_{yx} = 9e^x$

29. $\quad r = \ln|x + y|$

$r_x = \dfrac{1}{x + y}$

$r_y = \dfrac{1}{x + y}$

$r_{xx} = \dfrac{-1}{(x + y)^2}$

$r_{yy} = \dfrac{-1}{(x + y)^2}$

$r_{xy} = r_{yx} = \dfrac{-1}{(x + y)^2}$

31. $\quad z = x \ln|xy|$

$z_x = \ln|xy| + 1$

$z_y = \dfrac{x}{y}$

$z_{xx} = \dfrac{1}{x}$

$z_{yy} = -xy^{-2} = \dfrac{-x}{y^2}$

$z_{xy} = z_{yz} = \dfrac{1}{y}$

33. $f(x, y) = 6x^2 + 6y^2 + 6xy + 36x - 5$

First, $f_x = 12x + 6y + 36$ and $f_y = 12y + 6x$.
We must solve the system

$$12x + 6y + 36 = 0$$
$$12y + 6x = 0.$$

Multiply both sides of the first equation by -2 and add.

$$
\begin{array}{r}
-24x - 12y - 72 = 0 \\
\underline{6x + 12y = 0} \\
-18x - 72 = 0 \\
x = -4
\end{array}
$$

Substitute into either equation to get $y = 2$.
The solution is $x = -4, y = 2$.

35. $f(x, y) = 9xy - x^3 - y^3 - 6$

First, $f_x = 9y - 3x^2$ and $f_y = 9x - 3y^2$.
We must solve the system

$$9y - 3x^2 = 0$$
$$9x - 3y^2 = 0.$$

From the first equation, $y = \frac{1}{3}x^2$.
Substitute into the second equation to get

$$9x - 3\left(\frac{1}{3}x^2\right)^2 = 0$$

$$9x - 3\left(\frac{1}{9}x^4\right) = 0$$

$$9x - \frac{1}{3}x^4 = 0.$$

Multiply by 3 to get

$$27x - x^4 = 0.$$

Now factor.

$$x(27 - x^3) = 0$$

Set each factor equal to 0.

$$x = 0 \quad \text{or} \quad 27 - x^3 = 0$$
$$x = 3$$

Substitute into $y = \frac{x^2}{3}$.

$$y = 0 \quad \text{or} \quad y = 3$$

The solutions are $x = 0$, $y = 0$ and $x = 3$, $y = 3$.

37. $f(x, y, z) = x^4 + 2yz^2 + z^4$
$$f_x(x, y, z) = 4x^3$$
$$f_y(x, y, z) = 2z^2$$
$$f_z(x, y, z) = 4yz + 4z^3$$
$$f_{yz}(x, y, z) = 4z$$

39. $f(x, y, z) = \dfrac{6x - 5y}{4z + 5}$

$$f_x(x, y, z) = \frac{6}{4z + 5}$$

$$f_y(x, y, z) = \frac{-5}{4z + 5}$$

$$f_z(x, y, z) = \frac{-4(6x - 5y)}{(4z + 5)^2}$$

$$f_{yz}(x, y, z) = \frac{20}{(4z + 5)^2}$$

41. $f(x, y, z) = \ln \left| x^2 - 5xz^2 + y^4 \right|$

$$f_x(x, y, z) = \frac{2x - 5z^2}{x^2 - 5xz^2 + y^4}$$

$$f_y(x, y, z) = \frac{4y^3}{x^2 - 5xz^2 + y^4}$$

$$f_z(x, y, z) = \frac{-10xz}{x^2 - 5xz^2 + y^4}$$

$$f_{yz}(x, y, z) = \frac{4y^3(10zx)}{(x^2 - 5xz^2 + y^4)^2}$$

$$= \frac{40xy^3 z}{(x^2 - 5xz^2 + y^4)^2}$$

43. $f(x, y) = \left(x + \dfrac{y}{2} \right)^{x + y/2}$

(a) $f_x(1, 2) = \lim\limits_{h \to 0} \dfrac{f(1 + h, 2) - f(1, 2)}{h}$

We will use a small value for h. Let $h = 0.00001$.

$$f_x(1, 2) \approx \frac{f(1.00001, 2) - f(1, 2)}{0.00001}$$

$$\approx \frac{\left(1.00001 + \frac{2}{2} \right)^{1.00001 + 2/2} - \left(1 + \frac{2}{2} \right)^{1 + 2/2}}{0.00001}$$

$$\approx \frac{2.00001^{2.00001} - 2^2}{0.00001}$$

$$\approx 6.773$$

(b) $f_y(1, 2) = \lim\limits_{h \to 0} \dfrac{f(1, 2 + h) - f(1, 2)}{h}$

Again, let $h = 0.00001$.

$$f_y(1, 2) \approx \frac{f(1, 200001) - f(1, 2)}{0.00001}$$

$$\approx \frac{\left(1 + \frac{2.00001}{2} \right)^{1 + 2.00001/2} - \left(1 + \frac{2}{2} \right)^{1 + 2/2}}{0.00001}$$

$$\approx \frac{2.000005^{2.000005} - 2^2}{0.00001}$$

$$\approx 3.386$$

45. $M(x, y) = 45x^2 + 40y^2 - 20xy + 50$

(a) $M_y(x, y) = 80y - 20x$
$M_y(4, 2) = 80(2) - 20(4) = 80$

(b) $M_x(x, y) = 90x - 20y$
$M_x(3, 6) = 90(3) - 20(6) = 150$

(c) $\dfrac{\partial M}{\partial x}(2, 5) = 90(2) - 20(5) = 80$

(d) $\dfrac{\partial M}{\partial y}(6, 7) = 80(7) - 20(6) = 440$

47. $f(p, i) = 99p - 0.5pi - 0.0025p^2$

(a) $f(19,400, 8)$
$$= 99(19,400) - 0.5(19,400)(8) - 0.0025(19,400)^2$$
$$= \$902,100$$

The weekly sales are $902,100.

(b) $f_p(p, i) = 99 - 0.5i - 0.005p$, which represents the rate of change in weekly sales revenue per unit change in price when the interest rate remains constant.

$f_i(p, i) = -0.5p$, which represents the rate of change in weekly sales revenue per unit change in interest rate when the list price remains constant.

(c) $p = 19,400$ remains constant and i changes by 1 unit from 8 to 9.

$$f_i(p, i) = f_i(19,400, 8)$$
$$= -0.5(19,400)$$
$$= -9700$$

Therefore, sales revenue declines by $9700.

49. $f(x,y) = \left(\frac{1}{4}x^{-1/4} + \frac{3}{4}y^{-1/4}\right)^{-4}$

(a) $f(16,81) = \left[\frac{1}{4}(16)^{-1/4} + \frac{3}{4}(81)^{-1/4}\right]^{-4}$

$$= \left(\frac{1}{4}\cdot\frac{1}{2} + \frac{3}{4}\cdot\frac{1}{3}\right)^{-4}$$

$$= \left(\frac{3}{8}\right)^{-4}$$

$$\approx 50.56790123$$

50.57 hundred units are produced.

(b) $f_x(x,y) = -4\left(\frac{1}{4}x^{-1/4} + \frac{3}{4}y^{-1/4}\right)^{-5}$

$$\left[\frac{1}{4}\left(-\frac{1}{4}\right)x^{-5/4}\right]$$

$$= \frac{1}{4}x^{-5/4}\left(\frac{1}{4}x^{-1/4} + \frac{3}{4}y^{-1/4}\right)^{-5}$$

$$f_x(16,81) = \frac{1}{4}(16)^{-5/4}\left[\frac{1}{4}(16)^{-1/4} + \frac{3}{4}(81)^{-1/4}\right]^{-5}$$

$$= \frac{1}{4}\left(\frac{1}{32}\right)\left(\frac{3}{8}\right)^{-5} = \frac{256}{243}$$

$$\approx 1.053497942$$

$f_x(16,81) = 1.053$ hundred units and is the rate at which production is changing when labor changes by one unit (from 16 to 17) and capital remains constant.

$$f_y(x,y) = -4\left(\frac{1}{4}x^{-1/4} + \frac{3}{4}y^{-1/4}\right)^{-5}\left[\frac{3}{4}\left(-\frac{1}{4}\right)y^{-5/4}\right]$$

$$= \frac{3}{4}y^{-5/4}\left(\frac{1}{4}x^{-1/4} + \frac{3}{4}y^{-1/4}\right)^{-5}$$

$$f_y(16,81) = \frac{3}{4}(81)^{-5/4}\left[\frac{1}{4}(16)^{-1/4} + \frac{3}{4}(81)^{-1/4}\right]^{-5}$$

$$= \frac{3}{4}\left(\frac{1}{243}\right)\left(\frac{3}{8}\right)^{-5} = \frac{8192}{19,683}$$

$$\approx 0.4161967180$$

$f_y(16,81) = 0.4162$ hundred units and is the rate at which production is changing when capital changes by one unit (from 81 to 82) and labor remains constant.

(c) If labor is increased by one unit, production would increase at the rate

$$f_x(x,y) = \frac{1}{4}x^{-5/4}\left(\frac{1}{4}x^{-1/4} + \frac{3}{4}y^{-1/4}\right)^{-5}$$

hundred units.

51. $z = x^{0.4}y^{0.6}$

The marginal productivity of labor is

$$\frac{\partial z}{\partial x} = 0.4x^{-0.6}y^{0.6} + x^{0.4}\cdot 0 = 0.4x^{-0.6}y^{0.6}.$$

The marginal productivity of capital is

$$\frac{\partial z}{\partial y} = x^{0.4}(0.6y^{-0.4}) + y^{0.6}\cdot 0 = 0.6x^{0.4}y^{-0.4}.$$

53. $f(w,v) = 25.92w^{0.68} + \dfrac{3.62w^{0.75}}{v}$

(a) $f(300,10) = 25.92(300)^{0.68} + \dfrac{3.62(300)^{0.75}}{10}$

$$\approx 1279.46$$

The value is about 1279 kcal/hr.

(b) $f_w(w,v) = 25.92(0.68)w^{-0.32} + \dfrac{3.62(0.75)w^{-0.25}}{v}$

$$= \frac{17.6256}{w^{0.32}} + \frac{2.715}{w^{0.25}v}$$

$$f_w(300,10) = \frac{17.6256}{(300)^{0.32}} + \frac{2.715}{(300)^{0.25}(10)}$$

$$\approx 2.906$$

The value is about 2.906 kcal/hr/g. This means the instantaneous rate of change of energy usage for a 300 kg animal traveling at 10 kilometers per hour to walk or run 1 kilometer is about 2.906 kcal/hr/g.

55. $A(W,H) = 0.202W^{0.425}H^{0.725}$

(a) $\dfrac{\partial A}{\partial W}(72,1.8) = 0.08585(72)^{-0.575}(1.8)^{0.725}$

$$= 0.01124$$

(b) $\dfrac{\partial A}{\partial H}(70,1.6) = 0.14645(70)^{0.425}(1.6)^{-0.275}$

$$\approx 0.7829$$

57. $f(n, c) = \frac{1}{8}n^2 - \frac{1}{5}c + \frac{1937}{8}$

(a) $f(4, 1200) = \frac{1}{8}(4)^2 - \frac{1}{5}(1200) + \frac{1937}{8}$

$$= 2 - 240 + \frac{1937}{8} = 4.125$$

The client could expect to lose 4.125 lb.

(b) $\dfrac{\partial f}{\partial n} = \frac{1}{8}(2n) - \frac{1}{5}(0) + 0 = \frac{1}{4}n,$

which represents the rate of change of weight loss per unit change in number of workouts.

(c) $f_n(3, 1100) = \frac{1}{4}(3) = \frac{3}{4}$ lb

represents an additional weight loss by adding the fourth workout.

59. $R(x, t) = x^2(a - x)t^2 e^{-t} = (ax^2 - x^3)t^2 e^{-t}$

(a) $\dfrac{\partial R}{\partial x} = (2ax - 3x^2)t^2 e^{-t}$

(b) $\dfrac{\partial R}{\partial t} = x^2(a - x) \cdot [t^2 \cdot (-e^{-t}) + e^{-t} \cdot 2t]$

$$= x^2(a - x)(-t^2 + 2t)e^{-t}$$

(c) $\dfrac{\partial^2 R}{\partial x^2} = (2a - 6x)t^2 e^{-t}$

(d) $\dfrac{\partial^2 R}{\partial x \partial t} = (2ax - 3x^2)(-t^2 + 2t)e^{-t}$

(e) $\frac{\partial R}{\partial x}$ gives the rate of change of the reaction per unit of change in the amount of drug administered.
$\frac{\partial R}{\partial t}$ gives the rate of change of the reaction for a 1-hour change in the time after the drug is administered.

61. $W(V, T)$

$$= 91.4 - \frac{(10.45 + 6.69\sqrt{V} - 0.447V)(91.4 - T)}{22}$$

(a) $W(20, 10)$

$$= 91.4 - \frac{(10.45 + 6.69\sqrt{20} - 0.447(20))(91.4 - 10)}{22}$$

$$\approx -24.9$$

The wind chill is $-24.9°F$ when the wind speed is 20 mph and the temperature is 10°F.

(b) Solve

$$-25 = 91.4 - \frac{(10.45 + 6.69\sqrt{V} - 0.447V)(91.4 - 5)}{22}$$

for V.

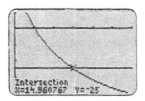

The wind speed is approximately 15 mph.

(c) $\qquad W_V = -\frac{1}{22}\left(\frac{6.69}{2\sqrt{V}} - 0.447\right)(91.4 - T)$

$$W_T = -\frac{1}{22}(10.45 + 6.69\sqrt{V} - 0.447V)(-1)$$

$$= \frac{1}{22}(10.45 + 6.69\sqrt{V} - 0.447V)$$

$$W_V(20, 10) = \frac{1}{22}\left(\frac{6.69}{2\sqrt{20}} - 0.447\right)(91.4 - 10)$$

$$\approx -1.114$$

When the temperature is held fixed at 10°F, the wind chill decreases approximately 1.1 degrees when the wind velocity increases by 1 mph.

$$W_T(20, 10) = \frac{1}{22}[10.45 + 6.69\sqrt{20} - 0.447(20)]$$

$$\approx 1.429$$

When the wind velocity is held fixed at 20 mph, the wind chill increases approximately 1.429°F when the temperature increases from 10°F to 11°F.

(d) A sample table is

T/V	5	10	15	20
30	27	16	9	4
20	16	3	−5	−11
10	6	−9	−18	−25
0	−5	−21	−32	−39

63. The rate of change in lung capacity with respect to age can be found by comparing the change in two lung capacity measurements to the difference in the respective ages when the height is held constant. So for a woman 58 inches tall, at age 20 the measured lung capacity is 1900 ml, and at age 25 the measured lung capacity is 1850 ml. So the rate of change in lung capacity with respect to age is

$$\frac{1900 - 1850}{20 - 25} = \frac{50}{-5} = -10 \text{ ml per year.}$$

The rate of change in lung capacity with respect to height can be found by comparing the change in two lung capacity measurements to the difference in the respective heights when the age is held constant. So for a 20-year old woman the measured lung capacity for a woman 58 inches tall is 1900 ml and the measured lung capacity for a woman 60 inches tall is 2100 ml. So the rate of change in lung capacity with respect to height is

$$\frac{1900 - 2100}{58 - 60} = \frac{-200}{-2} = 100 \text{ ml per in.}$$

The two rates of change remain constant throughout the table.

65. $F = \dfrac{mgR^2}{r^2} = mgR^2 r^{-2}$

(a) $F_m = \dfrac{gR^2}{r^2}$ is the approximate rate of change in gravitational force per unit change in mass while distance is held constant.

$F_r = \dfrac{-2mgR^2}{r^3}$ is the approximate rate of change in gravitational force per unit change in distance while mass is held constant.

(b) $F_m = \frac{gR^2}{r^2}$, where all quantities are positive. Therefore, $F_m > 0$.

$F_r = \frac{-2mgR^2}{r^3}$, where m, g, R^2, and r^3 are positive.

Therefore, $F_r < 0$.

These results are reasonable since gravitational force increases when mass increases (m is in the numerator) and gravitational force decreases when distance increases (r is in the denominator).

67. $T = (s, w) = 105 + 265 \log_2 \left(\dfrac{2s}{w}\right)$

(a) $T(3, 0.5) = 105 + 265 \log_2 \left[\dfrac{2(3)}{0.5}\right]$

$= 105 + 265 \log_2 12$

≈ 1055

(b) $T(s, w) = 105 + 265 \dfrac{\ln\left(\frac{2s}{w}\right)}{\ln 2}$

$= 105 + \dfrac{265}{\ln 2}[\ln(2s) - \ln(w)]$

$T_s(s, w) = \dfrac{265}{\ln 2}\left(\dfrac{1}{s}\right)$

$T_w(s, w) = -\dfrac{265}{\ln 2}\left(\dfrac{1}{w}\right)$

$T_s(3, 0.5) = \dfrac{265}{3 \ln 2} \approx 127.4 \text{ msec/ft}$

If the distance the object is being moved increases from 3 feet to 4 feet, while keeping w fixed at 0.5 foot, the time to move the object increases by approximately 127.4 msec.

$$T_w(3, 0.5) = -\frac{265}{0.5 \ln 2} \approx -764.5 \text{ msec/ft}$$

If the width of the target area is increased by 1 foot, while keeping the distance fixed at 3 feet, the movement time decreases by approximately 764.5 msec.

17.3   Maxima and Minima

1. $f(x, y) = xy + y - 2x$

$f_x(x, y) = y - 2$, $f_y(x, y) = x + 1$

If $f_x(x, y) = 0$, $y = 2$.

If $f_y(x, y) = 0$, $x = -1$.

Therefore, $(-1, 2)$ is the critical point.

$$f_{xx}(x, y) = 0$$
$$f_{yy}(x, y) = 0$$
$$f_{xy}(x, y) = 1$$

For $(-1, 2)$,

$$D = 0 \cdot 0 - 1^2 = -1 < 0.$$

A saddle point is at $(-1, 2)$.

3. $f(x, y) = 3x^2 - 4xy + 2y^2 + 6x - 10$

$f_x(x, y) = 6x - 4y + 6$

$f_y(x, y) = -4x + 4y$

Solve the system $f_x(x, y) = 0$, $f_y(x, y) = 0$.

$$
\begin{array}{rl}
6x - 4y + 6 = 0 & \\
-4x + 4y \quad\;\; = 0 & \\
\hline
2x \quad\quad + 6 = 0 & \\
x = -3 &
\end{array}
$$

$$
\begin{array}{rl}
-4(-3) + 4y = & 0 \\
y = & -3
\end{array}
$$

Therefore, $(-3, -3)$ is a critical point.

$$f_{xx}(x, y) = 6$$
$$f_{yy}(x, y) = 4$$
$$f_{xy}(x, y) = -4$$
$$D = 6 \cdot 4 - (-4)^2 = 8 > 0$$

Since $f_{xx}(x, y) = 6 > 0$, there is a relative minimum at $(-3, -3)$.

5. $f(x,y) = x^2 - xy + y^2 + 2x + 2y + 6$

$f_x(x,y) = 2x - y + 2, \ f_y(x,y) = -x + 2y + 2$

Solve the system $f_x(x,y) = 0, \ f_y(x,y) = 0.$

$$2x - \ y + 2 = 0$$
$$\underline{-x + 2y + 2 = 0}$$

$$2x - \ y + 2 = 0$$
$$\underline{-2x + 4y + 4 = 0}$$
$$3y + 6 = 0$$
$$y = -2$$

$$-x + 2(-2) + 2 = 0$$
$$x = -2$$

$(-2, -2)$ is the critical point.

$$f_{xx}(x,y) = 2$$
$$f_{yy}(x,y) = 2$$
$$f_{xy}(x,y) = -1$$

For $(-2, -2)$,

$$D = (2)(2) - (-1)^2 = 3 > 0.$$

Since $f_{xx}(x,y) > 0$, a relative minimum is at $(-2, -2)$.

7. $f(x,y) = x^2 + 3xy + 3y^2 - 6x + 3y$

$f_x(x,y) = 2x + 3y - 6, \ f_y(x,y) = 3x + 6y + 3$

Solve the system $f_x(x,y) = 0, \ f_y(x,y) = 0.$

$$2x + 3y - 6 = 0$$
$$3x + 6y + 3 = 0$$

$$-4x - 6y + 12 = 0$$
$$\underline{3x + 6y + \ 3 = 0}$$
$$-x + 15 = 0$$
$$x = 15$$

$$3(15) + 6y + 3 = 0$$
$$6y = -48$$
$$y = -8$$

$(15, -8)$ is the critical point.

$$f_{xx}(x,y) = 2$$
$$f_{yy}(x,y) = 6$$
$$f_{xy}(x,y) = 3$$

For $(15, -8)$,

$$D = 2 \cdot 6 - 9 = 3 > 0.$$

Since $f_{xx}(x,y) > 0$, a relative minimum is at $(15, -8)$.

9. $f(x,y) = 4xy - 10x^2 - 4y^2 + 8x + 8y + 9$

$f_x(x,y) = 4y - 20x + 8, \ f_y(x,y) = 4x - 8y + 8$

$$4y - 20x + 8 = 0$$
$$4x - \ 8y + 8 = 0$$

$$4y - 20x + \ 8 = 0$$
$$\underline{-4y + \ 2x + \ 4 = 0}$$
$$-18x + 12 = 0$$
$$x = \frac{2}{3}$$

$$4y - 20\left(\frac{2}{3}\right) + 8 = 0$$

The critical point is $\left(\frac{2}{3}, \frac{4}{3}\right)$.

$$f_{xx}(x,y) = -20$$
$$f_{yy}(x,y) = -8$$
$$f_{xy}(x,y) = 4$$

For $\left(\frac{2}{3}, \frac{4}{3}\right)$,

$$D = (-20)(-8) - 16 = 144 > 0.$$

Since $f_{xx}(x,y) < 0$, a relative maximum is at $\left(\frac{2}{3}, \frac{4}{3}\right)$.

11. $f(x,y) = x^2 + xy - 2x - 2y + 2$

$f_x(x,y) = 2x + y - 2, \ f_y(x,y) = x - 2$

$$2x + y - 2 = 0$$
$$x \qquad - 2 = 0$$
$$x = 2$$
$$2(2) + y - 2 = 0$$
$$y = -2$$

The critical point is $(2, -2)$.

$$f_{xx}(x,y) = 2$$
$$f_{yy}(x,y) = 0$$
$$f_{xy}(x,y) = 1$$

For $(2, -2)$,

$$D = 2 \cdot 0 - 1^2 = -1 < 0.$$

A saddle point is at $(2, -2)$.

13. $f(x, y) = 3x^2 + 2y^3 - 18xy + 42$
$f_x(x, y) = 6x - 18y$
$f_y(x, y) = 6y^2 - 18x$

If $f_x(x, y) = 0, 6x - 18y = 0$, or $x = 3y$. Substitute $3y$ for x in $f_y(x, y) = 0$ and solve for y.

$$6y^2 - 18(3y) = 0$$
$$6y(y - 9) = 0$$

$$y = 0 \quad \text{or} \quad y = 9$$
Then $\qquad x = 0 \quad \text{or} \quad x = 27.$

Therefore, $(0, 0)$ and $(27, 9)$ are critical points.

$$f_{xx}(x, y) = 6$$
$$f_{yy}(x, y) = 12y$$
$$f_{xy}(x, y) = -18$$

For $(0, 0)$,

$$D = 6 \cdot 12(0) - (-18)^2 = -324 < 0.$$

There is a saddle point at $(0, 0)$.

For $(27, 9)$,

$$D = 6 \cdot 12(9) - (-18)^2 = 324 > 0.$$

Since $f_{xx}(x, y) = 6 > 0$, there is a relative minimum at $(27, 9)$.

15. $f(x, y) = x^2 + 4y^3 - 6xy - 1$

$f_x(x, y) = 2x - 6y, \; f_y(x, y) = 12y^2 - 6x$

Solve $f_x(x, y) = 0$ for x.

$$2x + 6y = 0$$
$$x = 3y$$

Substitute for x in $12y^2 - 6x = 0$.

$$12y^2 - 6(3y) = 0$$
$$6y(2y - 3) = 0$$
$$y = 0 \quad \text{or} \quad y = \frac{3}{2}$$
Then $\qquad x = 0 \quad \text{or} \quad x = \frac{9}{2}.$

The critical points are $(0, 0)$ and $\left(\frac{9}{2}, \frac{3}{2}\right)$.

$$f_{xx}(x, y) = 2$$
$$f_{yy}(x, y) = 24y$$
$$f_{xy}(x, y) = -6$$

For $(0, 0)$,

$$D = 2 \cdot 24(0) - (-6)^2 = -36 < 0.$$

A saddle point is at $(0, 0)$.

For $\left(\frac{9}{2}, \frac{3}{2}\right)$,

$$D = 2 \cdot 24 \left(\frac{3}{2}\right) - (-6)^2$$

$$= 36 > 0.$$

Since $f_{xx}(x, y) > 0$, a relative minimum is at $\left(\frac{9}{2}, \frac{3}{2}\right)$.

17. $f(x, y) = e^{x(y+1)}$
$f_x(x, y) = (y + 1)e^{x(y+1)}$
$f_y(x, y) = xe^{x(y+1)}$

If $\qquad f_x(x, y) = 0$

$$(y + 1)e^{x(y+1)} = 0$$
$$y + 1 = 0$$
$$y = -1.$$

If $\quad f_y(x, y) = 0$

$$xe^{x(y+1)} = 0$$
$$x = 0.$$

Therefore, $(0, -1)$ is a critical point.

$$f_{xx}(x, y) = (y + 1)^2 e^{x(y+1)}$$
$$f_{yy}(x, y) = x^2 e^{x(y+1)}$$
$$f_{xy}(x, y) = (y + 1)e^{x(y+1)} \cdot x + e^{x(y+1)} \cdot 1$$
$$= (xy + x + 1)e^{x(y+1)}$$

For $(0, -1)$,

$$f_{xx}(0, -1) = (0)^2 e^0 = 0$$
$$f_{yy}(0, -1) = (0)^2 e^0 = 0$$
$$f_{xy}(0, -1) = (0 + 0 + 1)e^0 = 1$$
$$D = 0 \cdot 0 - 1^2 = -1 < 0$$

There is a saddle point at $(0, -1)$.

21. $z = -3xy + x^3 - y^3 + \dfrac{1}{8}$

$f_x(x, y) = -3y + 3x^2, \; f_y(x, y) = -3x - 3y^2$

Solve the system $f_x = 0, \; f_y = 0.$

$$-3y + 3x^2 = 0$$
$$-3x - 3y^2 = 0$$

$$-y + x^2 = 0$$
$$-x - y^2 = 0$$

Solve the first equation for y, substitute into the second, and solve for x.

$$y = x^2$$
$$-x - x^4 = 0$$
$$x(1 + x^3) = 0$$
$$x = 0 \quad \text{or} \quad x = -1$$
Then $\quad\quad y = 0 \quad \text{or} \quad y = 1.$

The critical points are $(0,0)$ and $(-1,1)$.

$$f_{xx}(x, y) = 6x$$
$$f_{yy}(x, y) = -6y$$
$$f_{xy}(x, y) = -3$$

For $(0,0)$,

$$D = 0 \cdot 0 - (-3)^2 = -9 < 0.$$

A saddle point is at $(0,0)$.
For $(-1,1)$,

$$D = -6(-6) - (-3)^2 = 27 > 0.$$

$$f_{xx}(x, y) = 6(-1) = -6 < 0.$$

$$f(-1, 1) = -3(-1)(1) + (-1)^3 - 1^3 + \frac{1}{8}$$
$$= \frac{9}{8}$$

A relative maximum of $\frac{9}{8}$ is at $(-1,1)$.
The equation matches graph (a).

23. $z = y^4 - 2y^2 + x^2 - \dfrac{17}{16}$

$$f_x(x, y) = 2x, \quad f_y(x, y) = 4y^3 - 4y$$

Solve the system $f_x = 0$, $f_y = 0$.

$$2x = 0 \quad (1)$$
$$4y^3 - 4y = 0 \quad (2)$$
$$4y(y^2 - 1) = 0$$
$$4y(y + 1)(y - 1) = 0$$

Equation (1) gives $x = 0$ and equation (2) gives $y = 0$, $y = -1$, or $y = 1$.
The critical points are $(0,0)$, $(0,-1)$, and $(0,1)$.

$$f_{xx}(x, y) = 2,$$
$$f_{yy}(x, y) = 12y^2 - 4,$$
$$f_{xy}(x, y) = 0$$

For $(0,0)$,

$$D = 2(12 \cdot 0^2 - 4) - 0 = -8 < 0.$$

A saddle point is at $(0,0)$.
For $(0,-1)$,

$$D = 2[12(-1)^2 - 4] - 0 = 16 > 0.$$

$$f_{xx}(x, y) = 2 > 0$$

$$f(0, -1) = (-1)^4 - 2(-1)^2 + 0^2 - \frac{17}{16}$$
$$= -2\frac{1}{16}$$

A relative minimum of $-2\frac{1}{16}$ is at $(0,-1)$.
For $(0,1)$,

$$D = 2(12 \cdot 1^2 - 4) - 0 = 16 > 0$$

$$f_{xx}(x, y) = 2 > 0$$

$$f(0, 1) = 1^4 - 2 \cdot 1^2 + 0^2 - \frac{17}{16}$$
$$= -\frac{33}{16}$$

A relative minimum of $-\frac{33}{16}$ is at $(0,-1)$.
The equation matches graph (b).

25. $z = -x^4 + y^4 + 2x^2 - 2y^2 + \dfrac{1}{16}$

$$f_x(x, y) = -4x^3 + 4x, \quad f_y(x, y) = 4y^3 - 4y$$

Solve $f_x(x, y) = 0$, $f_y(x, y) = 0$.

$$-4x^3 + 4x = 0 \quad (1)$$
$$4y^3 - 4y = 0 \quad (2)$$

$$-4x(x^2 - 1) = 0 \quad (1)$$
$$-4x(x + 1)(x - 1) = 0$$

$$4y(y^2 - 1) = 0 \quad (2)$$
$$4y(y + 1)(y - 1) = 0$$

Equation (1) gives $x = 0$, -1, or 1.
Equation (2) gives $y = 0$, -1, or 1.
Critical points are $(0,0)$, $(0,-1)$, $(0,1)$, $(-1,0)$, $(-1,-1)$, $(-1,1)$, $(1,0)$, $(1,-1)$, $(1,1)$.

$$f_{xx}(x, y) = -12x^2 + 4,$$
$$f_{yy}(x, y) = 12y^2 - 4$$
$$f_{xy}(x, y) = 0$$

For $(0,0)$,

$$D = 4(-4) - 0 = -16 < 0.$$

For $(0, -1)$,

$$D = 4(8) - 0 = 32 > 0,$$

and $f_{xx}(x, y) = 4 > 0$.

$$f(0, -1) = -\frac{15}{16}$$

For $(0, 1)$,

$$D = 4(8) - 0 = 32 > 0,$$

and $f_{xx}(x, y) = 4 > 0$.

$$f(0, 1) = -\frac{15}{16}$$

For $(-1, 0)$,

$$D = -8(-4) - 0 = 32 > 0,$$

and $f_{xx}(x, y) = -8 < 0$.

$$f(-1, 0) = \frac{17}{16}$$

For $(-1, -1)$,

$$D = -8(8) - 0 = -64 < 0.$$

For $(-1, 1)$,

$$D = -8(8) - 0 = -64 < 0.$$

For $(1, 0)$,

$$D = -8(-4) = 32 > 0,$$

and $f_{xx}(x, y) = -8 < 0$.

$$f(1, 0) = 1\frac{1}{16}$$

For $(1, -1)$,

$$D = -8(8) - 0 = -64 < 0.$$

For $(1, 1)$,

$$D = -8(8) - 0 = -64 < 0.$$

Saddle points are at $(0, 0)$, $(-1, -1)$, $(-1, 1)$, $(1, -1)$, and $(1, 1)$.
Relative maximum of $\frac{17}{16}$ is at $(-1, 0)$ and $(1, 0)$.
Relative minimum of $-\frac{15}{16}$ is at $(0, -1)$ and $(0, 1)$.
The equation matches graph (e).

27. $f(x, y) = 1 - x^4 - y^4$

$$f_x(x, y) = -4x^3, \ f_y(x, y) = -4y^3$$

The system

$$f_x(x, y) = -4x^3 = 0, f_y(x, y) = -4y^3 = 0$$

gives the critical point $(0, 0)$.

$$f_{xx}(x, y) = -12x^2$$
$$f_{yy}(x, y) = -12y^3$$
$$f_{xy}(x, y) = 0$$

For $(0, 0)$,

$$D = 0 \cdot 0 - 0^2 = 0.$$

Therefore, the test gives no information. Examine a graph of the function drawn by using level curves.
If $f(x, y) = 1$, then $x^4 + y^4 = 0$. The level curve is the point $(0, 0, 1)$.
If $f(x, y) = 0$, then $x^4 + y^4 = 1$. The level curve is the circle with center $(0, 0, 0)$ and radius 1.
If $f(x, y) = -15$, then $x^4 + y^4 = 16$. The level curve is the curve with center $(0, 0, -15)$ and radius 2.
The xz-trace is

$$z = 1 - x^4.$$

This curve has a maximum at $(0, 0, 1)$ and opens downward.
The yz-trace is

$$z = 1 - y^4.$$

This curve also has a maximum at $(0, 0, 1)$ and opens downward.

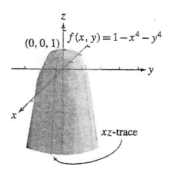

If $f(x, y) > 1$, then $x^4 + y^4 < 0$, which is impossible, so the function does not exist. Thus, the function has a relative maximum of 1 at $(0, 0)$.

31. $f(x, y) = x^2(y + 1)^2 + k(x + 1)^2 y^2$

(a) $f_x(x, y) = 2x + 2ky^2(x + 1)$
$f_y(x, y) = 2x^2(y + 1) + 2k(x + 1)^2 y$
$f_x(0, 0) = 2(0) + 2k(0)^2(0 + 1) = 0$
$f_y(0, 0) = 2(0)^2(0 + 1) + 2k(0 + 1)^2(0) = 0$

Thus, $(0, 0)$ is a critical point for all values of k.

(b) $f_{xx}(x, y) = 2 + 2ky^2$
$f_{yy}(x, y) = 2x^2 + 2k(x + 1)^2$
$f_{xy}(x, y) = 4ky(x + 1)$
$f_{xx}(0, 0) = 2 + 2k(0)^2 = 2$
$f_{yy}(0, 0) = 2(0)^2 = 2k(0 + 1)^2 = 2k$
$f_{xy}(0, 0) = 4k(0)(0 + 1) = 0$
$D = 2 \cdot 2k - 0^2 = 4k$

$(0, 0)$ is a relative minimum when $4k > 0$, hence when $k > 0$. When $k = 0$, $D = 0$ so the test for relative extrema gives no information. But if $k = 0$, $f(xy) = x^2(y+1)^2$, which is always greater than or equal to $f(0, 0) = 0$. So $(0, 0)$ is a relative minimum for $k \geq 0$.

33. $P(x, y) = 1500 + 36x - 1.5x^2 + 120y - 2y^2$
$P_x(x, y) = 36 - 3x$
$P_y(x, y) = 120 - 4y$

If $P_x = 0, x = 12$. If $P_y = 0, y = 30$. Therefore, $(12, 30)$ is a critical point.

$P_{xx}(x, y) = -3$
$P_{yy}(x, y) = -4$
$P_{xy}(x, y) = 0$
$\quad\quad D = (-3) \cdot (-4) - 0^2 = 12 > 0$

Since $P_{xx} = -3 < 0$, there is a relative maximum at $(12, 30)$.

$P(12, 30) = 1500 + 36(12) - 1.5(12)^2 + 120(30) - 2(30)^2$
$\quad\quad\quad = 3516 \text{ (hundred dollars)}$

The maximum profit is \$351,600 when the cost of a unit of labor is 12 and the cost of a unit of goods is 30.

35. $C(x, y) = 2x^2 + 2y^2 - 3xy$
$\quad\quad\quad\quad + 4x - 94y + 4200$

$C_x(x, y) = 4x - 3y + 4$
$C_y(x, y) = 4y - 3x - 94$

Solve the system $C_x(x, y) = 0, C_y(x, y) = 0$.
$$4x - 3y + 4 = 0$$
$$-3x + 4y - 94 = 0$$

$$12x - 9y + 12 = 0$$
$$\underline{-12x + 16y - 376 = 0}$$
$$7y - 364 = 0$$
$$y = 52$$

$$4x - 3(52) + 4 = 0$$
$$4x = 152$$
$$x = 38$$

Therefore, $(38, 52)$ is a critical point.

$C_{xx} = 4$
$C_{yy} = 4$
$C_{xy} = -3$
$\quad D = (4)(4) - (-3)^2 = 7 > 0$

Since $C_{xx} = 4 > 0$, there is a relative minimum at $(38, 52)$.

$C(38, 52)$
$\quad = 2(38)^2 + 2(52)^2 - 3(38)(52)$
$\quad\quad + 4(38) - 94(52) + 4200$
$\quad = 1832$

38 units of electrical tape and 52 units of packing tape should be produced to yield a minimum cost of \$1832.

37. $P(x, y) = 36xy - x^3 - 8y^3$
$P_x(x, y) = 36y - 3x^2$
$P_y(x, y) = 36x - 24y^2$
$P_x(x, y) = 0$
$36y - 3x^2 = 0$
$\quad 36y = 3x^2$

$$y = \frac{1}{12}x^2$$

$P_x(x, y) = 0$
$36x - 24y^2 = 0$
$\quad 36x = 24y^2$

$$x = \frac{2}{3}y^2$$

Use substitution to solve the system of equations

$$y = \frac{1}{12}x^2$$

$$x = \frac{2}{3}y^2.$$

$$y = \frac{1}{12}\left(\frac{2}{3}y^2\right)^2$$

$$y = \frac{1}{12}\left(\frac{4}{9}\right)y^4$$

$$y = \frac{1}{27}y^4$$

$$\frac{1}{27}y^4 - y = 0$$

$$\left(\frac{1}{27}y^3 - 1\right)y = 0$$

$$\frac{1}{27}y^3 - 1 = 0 \quad \text{or} \quad y = 0$$

$$\frac{1}{27}y^3 = 1 \quad \text{or} \quad y = 0$$

$$y^3 = 27 \quad \text{or} \quad y = 0$$

$$y = 3 \quad \text{or} \quad y = 0$$

If $y = 3, x = \frac{2}{3}(3)^2 = 6.$

If $y = 0, x = \frac{2}{3}(0)^2 = 0.$

The critical points are $(6, 3)$ and $(0, 0)$.

$$P_{xx}(x, y) = -6x$$
$$P_{yy}(x, y) = -48y$$
$$P_{xy}(x, y) = 36$$
$$P_{xx}(6, 3) = -36$$
$$P_{yy}(6, 3) = -144$$
$$P_{xy}(6, 3) = 36$$

$$D = (-36)(-144) - (36)^2 = 3888$$

Here $D > 0$ and $P_{xx} < 0$, so there is a relative maximum at $(6, 3)$.

$$P_{xx}(0, 0) = 0$$
$$P_{yy}(0, 0) = 0$$
$$P_{xy}(0, 0) = 36$$

$$D = 0 \cdot 0 - 36^2 = -1296$$

Since $D < 0$, there is a saddle point at $(0, 0)$.

$$P(6, 3) = 36(6)(3) - (6)^3 - 8(3)^3$$
$$= 648 - 216 - 216$$
$$= 216$$

So 6 tons of steel and 3 tons of aluminum produce a maximum profit of $216,000.

39. $P(\alpha, r, s) = \alpha(3r^2(1 - r) + r^3) + (1 - \alpha)$
$(3s^2(1 - s) + s^3)$

(a) $P(0.9, 0.5, 0.6) = 0.9[3(0.5)^2(1 - 0.5) + (0.5)^3]$
$+ (1 - 0.9)[3(0.6)^2(1 - 0.6)$
$+ (0.6)^3]$
$= 0.5148$

$P(0.1, 0.8, 0.4) = 0.1[3(0.8)^2(1 - 0.8) + (0.8)^3]$
$+ (1 - 0.1)[3(0.4)^2(1 - 0.4)$
$+ (0.4)^3]$
$= 0.4064$

The jury is less likely to make the correct decision in the second situation.

(b) If $r = s = 1$ then $P(\alpha, 1, 1) = 1$, so the jury always makes a correct decision. These values do not depend on α, but in a real-life situation α is likely to influence r and s.

(c) When P reaches a maximum, P_α, P_r, and P_s equal 0.

$$P_\alpha(\alpha, r, s) = 3r^2(1 - r) + r^3$$
$$- (3s^2(1 - s) + s^3)$$
$$= 3r^2 - 2r^3 - (3s^2 - 2s^3)$$

$$P_r(\alpha, r, s) = \alpha(6r(1 - r) - 3r^2 + 3r^2)$$
$$= 6\alpha r(1 - r)$$

$$P_s(\alpha, r, s) = (1 - \alpha)(6s(1 - s) - 3s^2 + 3s^2)$$
$$= 6s(1 - \alpha)(1 - s)$$

$$P_\alpha(\alpha, r, s) = 0 \quad \text{when} \quad r = s.$$

Since $P_r(\alpha, r, s) = 6\alpha r(1 - r)$, and $P_s(\alpha, r, s) = 6(1 - \alpha)s(1 - s)$, then P_α, P_r, and P_s are simultaneously 0 at the points $(\alpha, 1, 1)$ and $(\alpha, 0, 0)$. So $(\alpha, 1, 1)$ and $(\alpha, 0, 0)$ are critical points.
$P(\alpha, 0, 0) = 0$ while $P(\alpha, 1, 1) = 1$
Since $P(\alpha, r, s)$ represents a probability, $0 \leq P(\alpha, r, s) \leq 1$. Thus, $P(\alpha, 1, 1) = 1$ is a maximum value of the function.

41. $E(t, T) = 436.16 - 10.57t - 5.46T - 0.02t^2$
$+ 0.02T^2 + 0.08tT$

(a) $E(0, 0) = 436.16.$

The value of E before cooking is 436.16 kJ/mol.

(b) $E(10, 180) = 436.16 - 10.57(10) - 5.46(180)$
$- 0.02(10)^2 + 0.02(180)^2$
$+ 0.08(10)(180)$
$= 137.66$

After cooking for 10 minutes at 180°C, the total change in color is 137.66 kJ/mol.

(c) $E_t = -10.57 - 0.04t + 0.08T$
$E_T = -5.46 + 0.04T + 0.08t$

Solve the system $E_t = 0, E_T = 0$.

$$-0.04t + 0.08T - 10.57 = \quad 0$$
$$0.08t + 0.04T - \quad 5.46 = \quad 0$$

$$-0.04t + 0.08T - 10.57 = \quad 0$$
$$\underline{-0.16t - 0.08T + 10.92 = \quad 0}$$
$$-0.20t \qquad\qquad + 0.35 = \quad 0$$
$$t = 1.75$$

$$-0.04(1.75) + 0.08T - 10.57 = 0$$
$$0.08T - 10.64 = 0$$
$$T = 133$$

$(1.75, 133)$ is a critical point.

$$E_{tt} = -0.04$$
$$E_{TT} = 0.04$$
$$E_{tT} = 0.08$$
$$D = (-0.04)(0.04) - (0.08)^2 = -0.008 < 0$$

$(1.75, 133)$ is a saddle point.

17.4   Lagrange Multipliers

1. Maximize $f(x, y) = 4xy$,
subject to $x + y = 16$.

1. $g(x, y) = x + y - 16$

2. $F(x, y, \lambda) = 4xy - \lambda(x + y - 16)$.

3. $F_x(x, y, \lambda) = 4y - \lambda$
$F_y(x, y, \lambda) = 4x - \lambda$
$F_\lambda(x, y, \lambda) = -(x + y - 16)$

4. $\quad 4y - \lambda = 0 \quad (1)$
$\quad 4x - \lambda = 0 \quad (2)$
$x + y - 16 = 0 \quad (3)$

5. Equations (1) and (2) give $\lambda = 4y$ and $\lambda = 4x$.
Thus,
$$4y = 4x$$
$$y = x.$$

Substituting into equation (3),
$$x + (x) - 16 = 0$$
$$x = 8.$$
So $\qquad\qquad y = 8.$

Maximum is $f(8, 8) = 4(8)(8) = 256$.

3. Maximize $f(x, y) = xy^2$,
subject to $x + 2y = 15$.

1. $g(x, y) = x + 2y - 15$

2. $F(x, y, \lambda) = xy^2 - \lambda(x + 2y - 15)$

3. $F_x(x, y, \lambda) = y^2 - \lambda$
$F_y(x, y, \lambda) = 2xy - 2\lambda$
$F_\lambda(x, y, \lambda) = -(x + 2y - 15)$

4. $\qquad y^2 - \lambda = 0 \quad (1)$
$\quad 2xy - 2\lambda = 0 \quad (2)$
$x + 2y - 15 = 0 \quad (3)$

5. Equations (1) and (2) give $\lambda = y^2$ and $\lambda = xy$.
Thus,

$$y^2 = xy$$
$$y(y - x) = 0$$
$$y = 0 \quad \text{or} \quad y = x$$

Substituting $y = 0$ into equation (3),

$$x + 2(0) - 15 = 0$$
$$x = 15.$$

Substituting $y = x$ into equation (3)

$$x + 2(x) - 15 = 0$$
$$x = 5.$$
So $\qquad\qquad y = x = 5.$

Thus,

$$f(15, 0) = 15(0)^2 = 0, \text{ and}$$
$$f(5, 5) = 5(5)^2 = 125.$$

Since $f(5, 5) > f(15, 0), f(5, 5) = 125$ is a maximum.

5. Minimize $f(x, y) = x^2 + 2y^2 - xy$,
subject to $x + y = 8$.

1. $g(x, y) = x + y - 8$

2. $F(x, y, \lambda)$
$\quad = x^2 + 2y^2 - xy - \lambda(x + y - 8)$

3. $F_x(x, y, \lambda) = 2x - y - \lambda$
$F_y(x, y, \lambda) = 4y - x - \lambda$
$F_\lambda(x, y, \lambda) = -(x + y - 8)$

4. $2x - y - \lambda = 0$
$4y - x - \lambda = 0$
$x + y - 8 = 0$

5. Subtracting the second equation from the first equation to eliminate λ gives the new system of equations

$$x + y = 8$$
$$3x - 5y = 0.$$

Solve this system.

$$5x + 5y = 40$$
$$\underline{3x - 5y = 0}$$
$$8x = 40$$
$$x = 5$$

But $x + y = 8$, so $y = 3$.

Thus, $\qquad f(5, 3) = 25 + 18 - 15 = 28$

is a minimum.

7. Maximize $f(x, y) = x^2 - 10y^2$, subject to $x - y = 18$.

1. $g(x, y) = x - y - 18$

2. $F(x, y, \lambda)$
$$= x^2 - 10y^2 - \lambda(x - y - 18)$$

3. $F_x(x, y, \lambda) = 2x - \lambda$
$F_y(x, y, \lambda) = -20y - \lambda$
$F_\lambda(x, y, \lambda) = -(x - y - 18)$

4. $\qquad 2x - \lambda = 0$
$\quad -20y + \lambda = 0$
$\quad x - y - 18 = 0$

5. Adding the first two equations to eliminate λ gives

$$2x - 20y = 0$$
$$x = 10y.$$

Substituting $x = 10y$ in the third equation gives

$$10y - y = 18$$
$$y = 2$$

Thus, $\qquad x = 20.$

$$f(20, 2) = 20^2 - 10(2)^2$$
$$= 400 - 40 = 360.$$

$f(20, 2) = 360$ is a relative maximum.

9. Maximize $f(x, y, z) = xyz^2$, subject to $x + y + z = 6$.

1. $g(x, y, z) = x + y + z - 6$

2. $F(x, y, \lambda)$
$$= xyz^2 - \lambda(x + y + z - 6)$$

3. $F_x(x, y, z, \lambda) = yz^2 - \lambda$
$F_y(x, y, z, \lambda) = xz^2 - \lambda$
$F_z(x, y, z, \lambda) = 2zxy - \lambda$
$F_\lambda(x, y, z, \lambda) = -(x + y + z - 6)$

4. Setting F_x, F_y, F_z and F_λ equal to zero yields

$$yz^2 - \lambda = 0 \quad (1)$$
$$xz^2 - \lambda = 0 \quad (2)$$
$$2xyz - \lambda = 0 \quad (3)$$
$$x + y + z - 6 = 0. \quad (4)$$

5. $\lambda = yz^2$, $\lambda = xz^2$, and $\lambda = 2xyz$

$$yz^2 = xz^2$$
$$z^2(y - x) = 0$$
$$x = y \quad \text{or} \quad z = 0$$
$$yz^2 = 2xyz$$

$$2xyz - yz^2 = 0$$
$$yz(2x - z) = 0$$
$$y = 0 \quad \text{or} \quad z = 0 \quad \text{or} \quad z = 2x$$

In a similar way, the third equation

$$xz^2 = 2xyz$$

implies that $x = 0$ or $z = 0$ or $z = 2y$.

By the nature of the function to be maximized, $f(x, y, z) = xyz^2$, a nonzero maximum can come only from those points with nonzero coordinates. Therefore, assume $y = x$ and $z = 2y = 2x$. If $y = x$ and $z = 2x$ are substituted into equation (4), then

$$x + x + 2x - 6 = 0$$
$$x = \frac{3}{2}.$$

Thus, $y = \frac{3}{2}$ and $z = 3$, and

$$f\left(\frac{3}{2}, \frac{3}{2}, 3\right) = \frac{3}{2} \cdot \frac{3}{2} \cdot 9$$
$$= \frac{81}{4} > 0.$$

So, $f\left(\frac{3}{2}, \frac{3}{2}, 3\right) = \frac{81}{4} = 20.25$ is a maximum.

11. The problem can be restated as

Maximize $f(x,y) = 3xy^2$,
subject to $x + y = 24, x > 0, y > 0$.

1. $g(x,y) = x + y - 24$

2. $F(x,y,\lambda) = 3xy^2 - \lambda(x + y - 24)$

3. $F_x(x,y,\lambda) = 3y^2 - \lambda$
$F_y(x,y,\lambda) = 6xy - \lambda$
$F_\lambda(x,y,\lambda) = -(x + y - 24)$

4. $3y^2 - \lambda = 0$ (1)
   $6xy - \lambda = 0$ (2)
   $x + y - 24 = 0$ (3)

5. Equations (1) and (2) give $\lambda = 3y^2$ and $\lambda = 6xy$. Thus,

$$3y^2 = 6xy$$
$$3y^2 - 6xy = 0$$
$$3y(y - 2x) = 0$$
$$y = 0 \quad \text{or} \quad y = 2x.$$

Substituting $y = 0$ into equation (3),

$$x + (0) - 24 = 0$$
$$x = 24.$$

Substituting $y = 2x$ into equation (3),

$$x + (2x) - 24 = 0$$
$$3x - 24 = 0$$
$$x = 8.$$

So $y = 2x = 16.$

Thus,

$$f(24,0) = 3(24)(0)^2 = 0, \text{ and}$$
$$f(8,16) = 3(8)(16)^2 = 6144.$$

Since $f(8,16) > f(24,0), x = 8$ and $y = 16$ will maximize $f(x,y) = 3xy^2$.

13. Let x, y, and z be three numbers such that

$$x + y + z = 90$$
and $f(x,y,z) = xyz.$

1. $g(x,y,z) = x + y + z - 90$

2. $F(x,y,z)$
   $= xyz - \lambda(x + y + z - 90)$

3. $F_x(x,y,z,\lambda) = yz - \lambda$
$F_y(x,y,z,\lambda) = xz - \lambda$
$F_\lambda(x,y,z,\lambda) = xy - \lambda$
$F_\lambda(x,y,z,\lambda) = -(x + y + z - 90)$

4. $yz - \lambda = 0$ (1)
   $xz - \lambda = 0$ (2)
   $xy - \lambda = 0$ (3)
   $x + y + z - 90 = 0$ (4)

5. $\lambda = yz$, $\lambda = xz$, and $\lambda = xy$

$$yz = xz$$
$$yz - xz = 0$$
$$(y - x)z = 0$$
$$y - x = 0 \quad \text{or} \quad z = 0$$
$$xz - xy = 0$$
$$x(z - y) = 0$$
$$x = 0 \quad \text{or} \quad z - y = 0$$

Since $x = 0$ or $z = 0$ would not maximize
$f(x,y,z) = xyz$, then $y - x = 0$ and $z - y = 0$
imply that $y = x = z$.
Substituting into equation (4) gives

$$x + x + x - 90 = 0$$
$$x = 30.$$

$x = y = z = 30$ will maximize $f(x,y,z) = xyz$.
The numbers are 30, 30, and 30.

17. Consider the constraint and solve for x in terms of y.

$$x + 2y = 15$$
$$x = 15 - 2y$$

Then

$$f(x,y) = xy^2$$
$$= (15 - 2y)y^2$$
$$= -2y^3 + 15y^2$$

So, $f(x,y) = -2y^3 + 15y^2 = f(y)$. Notice that f is unbounded; more specifically,

$$\lim_{y \to \infty} f(y) = -\infty$$
$$\text{and} \quad \lim_{y \to -\infty} f(y) = \infty.$$

Therefore f, subject to the given constraint, has neither an absolute maximum nor an absolute minimum.

21. Let x be the length of the fence opposite the building and y be the length of each end. The area is then xy and the total cost is $25x + 15(2y)$. Restate the problem as follows.

Maximize xy,
subject to $25x + 30y = 2400$.

1. $g(x, y) = 25x + 30y - 2400$

2. $F(x, y, \lambda) = xy - \lambda(25x + 30y - 2400)$

3. $F_x(x, y, \lambda) = y - 25\lambda$
$F_y(x, y, \lambda) = x - 30\lambda$
$F_\lambda(x, y, \lambda) = -(25x + 30y - 2400)$

4.
$$y - 25\lambda = 0 \quad (1)$$
$$x - 30\lambda = 0 \quad (2)$$
$$25x + 30y - 2400 = 0 \quad (3)$$

5. Equations (1) and (2) give $\lambda = \frac{y}{25}$ and $\lambda = \frac{x}{30}$. Thus,

$$\frac{y}{25} = \frac{x}{30}$$
$$y = \frac{5}{6}x.$$

Substituting $y = \frac{5}{6}x$ into equation (3) gives

$$25x + 30\left(\frac{5}{6}x\right) - 2400 = 0$$
$$50x - 2400 = 0$$
$$x = 48.$$

So
$$y = \frac{5}{6}x = 40.$$

The dimensions are 48 ft (opposite the building) by 40 ft (the ends).

23. Maximize $P(x, y) = -x^2 - y^2 + 4x + 8y$, subject to $x + y = 6$.

1. $g(x, y) = x + y - 6$

2. $F(x, y, \lambda)$
$$= -x^2 - y^2 + 4x + 8y - \lambda(x + y - 6)$$

3. $F_x(x, y, \lambda) = -2x + 4 - \lambda$
$F_y(x, y, \lambda) = -2y + 8 - \lambda$
$F_\lambda(x, y, \lambda) = -(x + y - 6)$

4. $-2x + 4 - \lambda = 0$
$-2y + 8 - \lambda = 0$
$x + y - 6 = 0$

5. $\lambda = -(2x - 4)$ and $\lambda = -(2y - 8)$
$2x - 4 = 2y - 8$
$y = x + 2$

Substituting into $x + y - 6 = 0$, we have

$x + (x + 2) - 6 = 0$ so $x = 2$ and $y = 4$.

25. Maximize $f(x, y) = 12x^{3/4}y^{1/4}$,
subject to $100x + 180y = 25,200$.

1. $g(x, y) = 100x + 180y - 25,200$

2. $F(x, y, \lambda)$
$$= 12x^{3/4}y^{1/4}$$
$$- \lambda(100x + 180y - 25,200)$$

3. $F_x(x, y, \lambda) = \frac{3}{4}(12x^{-1/4}y^{1/4}) - 100\lambda$
$$= \frac{9y^{1/4}}{x^{1/4}} - 100\lambda$$
$F_y(x, y, \lambda) = \frac{1}{4}(12x^{3/4}y^{-3/4}) - 180\lambda$
$$= \frac{3x^{3/4}}{y^{3/4}} - 180\lambda$$
$F_\lambda(x, y, \lambda) = -(100x + 180y - 25,200)$

4. $\frac{9y^{1/4}}{x^{1/4}} - 100\lambda = 0$
$\frac{3x^{3/4}}{y^{3/4}} - 180\lambda = 0$
$100x + 180y - 25,200 = 0$

5. $\lambda = \frac{9y^{1/4}}{100x^{1/4}}$ and $\lambda = \frac{3x^{3/4}}{180y^{3/4}}$
$$= \frac{x^{3/4}}{60y^{3/4}}$$
$$\frac{9y^{1/4}}{100x^{1/4}} = \frac{x^{3/4}}{60y^{3/4}}$$
$$100x = 540y$$
$$x = \frac{27y}{5}$$

Substitute into

$$100x + 180y - 25,200 = 0.$$
$$100\left(\frac{27y}{5}\right) + 180y = 25,200$$
$$540y + 180y = 25,200$$
$$720y = 25,200$$
$$y = 35$$
$$x = \frac{27(35)}{5} = 189$$

Production will be maximized with 189 units of labor and 35 units of capital.

27. If x and y are the dimensions of the field, we must maximize $f(x,y) = xy$ subject to $x + 2y = 600$.

1. $g(x,y) = x + 2y - 600$

2. $F(x,y,\lambda)$
$$= xy - \lambda(x + 2y - 600)$$

3. $F_x(x,y,\lambda) = y - \lambda$
$F_y(x,y,\lambda) = x - 2\lambda$
$F_\lambda(x,y,\lambda) = -(x + 2y - 600)$

4. $x - \lambda = 0$
$x - 2\lambda = 0$
$x + 2y - 600 = 0$

5. $\lambda = y$ and $\lambda = \dfrac{x}{2}$

$$y = \frac{x}{2}$$

Substituting into $x + 2y - 600 = 0$, we have

$x + 2\left(\frac{x}{2}\right) - 600 = 0$, so $x = 300$ and $y = 150$.

The largest area is $(300)(150) = 45{,}000$ m^2.

29. Let x be the radius of the can and y be the height.

Minimize surface area $f(x,y) = 2\pi xy + 2\pi x^2$, subject to the constraint that $\pi x^2 y = 25$.

1. $g(x,y) = \pi x^2 y - 25$

2. $F(x,y,\lambda)$
$$= 2\pi xy + 2\pi x^2 - \lambda(\pi x^2 y - 25)$$

3. $F_x(x,y,\lambda) = 2\pi y + 4\pi x - 2\lambda\pi xy$
$F_y(x,y,\lambda) = 2\pi x - \lambda\pi x^2$
$F_\lambda(x,y,\lambda) = -(\pi x^2 y - 25)$

4. $2\pi y + 4\pi x - 2\lambda\pi xy = 0$
$2\pi x - \lambda\pi x^2 = 0$
$\pi x^2 y - 25 = 0$

5. $\lambda = \dfrac{2x + y}{xy}$ and $\lambda = \dfrac{2}{x}$

$$\frac{2x + y}{xy} = \frac{2}{x}$$

$$2x^2 + xy = 2xy$$
$$2x^2 - xy = 0$$
$$x = 0 \quad \text{or} \quad y = 2x$$

$x = 0$ is impossible.

Substituting $y = 2x$ into $\pi x^2 y - 25 = 0$, we have

$\pi x^2(2x) - 25 = 0$, so $x = \sqrt[3]{\frac{25}{2\pi}} \approx 1.585$ inches

and $y = 2\sqrt[3]{\frac{25}{2\pi}} \approx 3.169$ inches.

The can with minimum surface area will have a radius of approximately 1.585 inches and a height of approximately 3.169 inches.

31. If the box is x by x by y, we must minimize surface area $f(x,y) = 2x^2 + 4xy$, subject to $x^2 y = 185$.

1. $g(x,y) = x^2 y - 185$

2. $F(x,y,\lambda)$
$$= 2x^2 + 4xy - \lambda(x^2 y - 185)$$

3. $F_x(x,y,\lambda) = 4x + 4y - 2\lambda xy$
$F_y(x,y,\lambda) = 4x - \lambda x^2$
$F_\lambda(x,y,\lambda) = -(x^2 y - 185)$

4. $4x + 4y - 2\lambda xy = 0$
$4x - \lambda x^2 = 0$
$x^2 y - 185 = 0$

5. $\lambda = \dfrac{2x + 2y}{xy}$ and $\lambda = \dfrac{4}{x}$

$$\frac{2x + 2y}{xy} = \frac{4}{x}$$

$$2x^2 + 2xy = 4xy$$
$$2x^2 - 2xy = 0$$
$$2x(x - y) = 0$$
$$x = 0 \quad \text{or} \quad y = x$$

$x = 0$ is impossible.

Substituting $y = x$ into $x^2 y - 185 = 0$, we have

$$y = x = \sqrt[3]{185} \approx 5.698.$$

The dimensions are 5.698 inches by 5.698 inches by 5.698 inches.

33. Let the dimensions of the bottom be x by y, and let the height be z. We must minimize $f(x,y,z) = xy + 2xz + 2yz$ subject to $xyz = 32$.

1. $g(x,y,z) = xyz - 32$

2. $F(x,y,z,\lambda)$
$$= xy + 2xz + 2yz - \lambda(xyz - 32)$$

3. $F_x(x,y,z,\lambda) = y + 2z - \lambda yz$
$F_y(x,y,z,\lambda) = x + 2z - \lambda xz$
$F_z(x,y,z,\lambda) = 2x + 2y - \lambda xy$
$F_\lambda(x,y,z,\lambda) = -(xyz - 32)$

4. $y + 2z - \lambda yz = 0$
$x + 2z - \lambda xz = 0$
$2x + 2y - \lambda xy = 0$
$xyz - 32 = 0$

5.
$$\lambda = \frac{y+2z}{yz}$$
$$\lambda = \frac{x+2z}{xz}$$
$$\lambda = \frac{2x+2y}{xy}$$
$$xyz = 32$$
$$\frac{y+2z}{yz} = \frac{x+2z}{xz}$$
$$xyz + 2xz^2 = xyz + 2yz^2$$
$$2z^2(x-y) = 0$$
$$z^2 = 0 \quad \text{or} \quad x - y = 0$$
$$z = 0 \text{ (impossible) or } x = y$$
$$\frac{x+2z}{xz} = \frac{2x+2y}{xy}$$
$$x^2y + 2xyz = 2x^2z + 2xyz$$
$$x^2(y-2z) = 0$$
$$x^2 = 0 \quad \text{or} \quad y - 2z = 0$$
$$x = 0 \text{ (impossible)} \quad \text{or} \quad y = 2z$$

Since $x = y$ and $y = 2z$ and since $xyz = 32$, we have
$$(2z)(2z)z = 32$$
$$z^3 = 8$$
$$z = 2.$$

If $z = 2$, $y = 4$ and $x = 4$.
The dimensions are 4 feet by 4 feet for the base and 2 feet for the height.

35. (a)
$$P(r,s,t) = rs(1-t) + (1-r)st$$
$$\qquad + r(1-s)t + rst$$
$$g(r,s,t) = r + s + t - \alpha$$
$$F(r,s,t,\lambda) = rs(1-t) + (1-r)st$$
$$\qquad + r(1-s)t + rst$$
$$\qquad - \lambda(r+s+t-\alpha)$$

(b)

3. $F_r(r,s,t,\lambda) = s(1-t) - st + (1-s)t + st - \lambda$
$$= s + t - 2st - \lambda$$
$$F_s(r,s,t,\lambda) = r(1-t) + (1-r)t - rt + rt - \lambda$$
$$= r + t - 2rt - \lambda$$
$$F_t(r,s,t,\lambda) = -rs + (1-r)s + r(1-s) + rs - \lambda$$
$$= r + s - 2rs - \lambda$$
$$F_\lambda(r,s,t,\lambda) = -(r+s+t-\alpha)$$

4. $s + t - 2st - \lambda = 0$
$$r + t - 2rt - \lambda = 0$$
$$r + s - 2rs - \lambda = 0$$
$$r + s + t - \alpha = 0$$

5.
$$-\lambda = 2st - s - t$$
$$-\lambda = 2rt - r - t$$
$$-\lambda = 2rs - r - s$$
$$r + s + t = \alpha$$

$$2st - s - t = 2rt - r - t$$
$$s(2t-1) = r(2t-1)$$
$$(s-r)(2t-1) = 0$$

$$s = r \quad \text{or} \quad t = \frac{1}{2}$$

$$2rt - r - t = 2rs - r - s$$
$$t(2r-1) = s(2r-1)$$
$$(t-s)(2r-1) = 0$$

$$t = s \quad \text{or} \quad r = \frac{1}{2}$$

$$2st - s - t = 2rs - r - s$$
$$t(2s-1) = r(2s-1)$$
$$(t-r)(2s-1) = 0$$

$$t = r \quad \text{or} \quad s = \frac{1}{2}$$

Since r, s and t are probabilities, $0 \le r, s$, $t \le 1$. Also, $r + s + t = \alpha = 0.75$. If $t = \frac{1}{2}$, then either $t = s = \frac{1}{2}$ or $r = \frac{1}{2}$, both of which are impossible (the third value would have to be -0.25 to get a sum of 0.75). Thus, $r = s = t = 0.25$.

(c) Now we have $r + s + t = \alpha = 3$. If $t = \frac{1}{2}$, then either $t = s = \frac{1}{2}$ or $r = \frac{1}{2}$, both of which are impossible (the third value would have to be 2 to get a sum of 3). Thus, $r = s = t = 1.0$.

17.5   Total Differentials and Approximations

1. Let $z = f(x,y) = \sqrt{x^2 + y^2}$.

Then
$$dz = f_x(x,y)\,dx + f_y(x,y)\,dy$$
$$= \frac{1}{2}(x^2+y^2)^{-1/2}(2x)\,dx$$
$$\qquad + \frac{1}{2}(x^2+y^2)^{-1/2}(2y)\,dy$$
$$= \frac{x\,dx + y\,dy}{\sqrt{x^2+y^2}}.$$

To approximate $\sqrt{8.05^2 + 5.97^2}$, we let $x = 8$, $dx = 0.05$, $y = 6$ and $dy = -0.03$.

$$dz = \frac{8(0.05) + 6(-0.03)}{\sqrt{8^2 + 6^2}}$$

$$= \frac{4}{5}(0.05) + \frac{3}{5}(-0.03)$$

$$= 0.04 - 0.018 = 0.022$$

$$\begin{aligned} f(8.05, 5.97) &= f(8, 6) + \Delta z \\ &\approx f(8, 6) + dz \\ &= \sqrt{8^2 + 6^2} + 0.222 \\ &= 10.022 \end{aligned}$$

Thus, $\sqrt{8.05^2 + 5.97^2} \approx 10.022$.

Using a calculator, $\sqrt{8.05^2 + 5.97^2} \approx 10.0221$.

The absolute value of the difference of the two results is $|10.022 - 10.0221| = 0.0001$.

3. Let $z = f(x, y) = (x^2 + y^2)^{1/3}$.

Then

$$dz = f_x(x, y)\,dx + f_y(x, y)\,dy$$

$$dz = \frac{1}{3}(x^2 + y^2)^{-2/3}(2x)\,dx$$

$$+ \frac{1}{3}(x^2 + y^2)^{-2/3}(2y)\,dy$$

$$= \frac{2x}{3(x^2 + y^2)^{2/3}}\,dx + \frac{2y}{3(x^2 + y^2)^{2/3}}\,dy$$

To approximate $(1.92^2 + 2.1^2)^{1/3}$, we let $x = 2$, $dx = -0.08$, $y = 2$, and $dy = 0.1$.

$$dz = \frac{2(2)}{3[(2)^2 + (2)^2]^{2/3}}(-0.08)$$

$$+ \frac{2(2)}{3[(2)^2 + (2)^2]^{2/3}}(0.1)$$

$$= \frac{4}{12}(-0.08) + \frac{4}{12}(0.1)$$

$$= 0.00\overline{6}$$

$$\begin{aligned} f(1.92, 2.1) &= f(2, 2) + \Delta z \\ &\approx f(2, 2) + dz \\ &= 2 + 0.00\overline{6} \\ f(1.92, 2.1) &\approx 2.0067 \end{aligned}$$

Using a calculator, $(1.92^2 + 2.1^2)^{1/3} \approx 2.0080$. The absolute value of the difference of the two results is $|2.0067 - 2.0080| = 0.0013$.

5. Let $z = f(x, y) = xe^y$.

Then

$$\begin{aligned} dz &= f_x(x, y)\,dx + f_y(x, y)\,dy \\ &= e^y\,dx + xe^y\,dy. \end{aligned}$$

To approximate $1.03e^{0.04}$, we let $x = 1$, $dx = 0.03$, $y = 0$, and $dy = 0.04$.

$$dz = e^0(0.03) + 1 \cdot e^0(0.04)$$

$$= 0.07$$

$$\begin{aligned} f(1.03, 0.04) &= f(1, 0) + \Delta z \\ &\approx f(1, 0) + dz \\ &= 1 \cdot e^0 + 0.07 \\ &= 1.07 \end{aligned}$$

Thus, $1.03e^{0.04} \approx 1.07$.
Using a calculator, $1.03e^{0.04} \approx 1.0720$.
The absolute value of the difference of the two results is $|1.07 - 1.0720| = 0.0020$.

7. Let $z = f(x, y) = x \ln y$.

Then

$$\begin{aligned} dz &= f_x(x, y)\,dx + f_y(x, y)\,dy \\ &= \ln y\,dx + \frac{x}{y}\,dy \end{aligned}$$

To approximate $0.99 \ln 0.98$, we let $x = 1$, $dx = -0.01$, $y = 1$, and $dy = -0.02$.

$$dz = \ln(1) \cdot (-0.01) + \frac{1}{1}(-0.02)$$

$$= -0.02$$

$$\begin{aligned} f(0.99, 0.98) &= f(1, 1) + \Delta z \\ &\approx f(1, 1) + dz \\ &= 1 \cdot \ln(1) - 0.02 \\ &\approx -0.02 \end{aligned}$$

Thus, $0.99 \ln 0.98 \approx -0.02$.

Using a calculator, $0.99 \ln 0.98 \approx -0.0200$.
The absolute value of the difference of the two results is $|-0.02 - (-0.0200)| = 0$.

9. $z = f(x, y) = 2x^2 + 4xy + y^2$
$x = 5, dx = 0.03, y = -1, dy = -0.02$

$$f_x(x, y) = 4x + 4y$$
$$f_y(x, y) = 4x + 2y$$
$$\begin{aligned} dz &= (4x + 4y)dx + (4x + 2y)dy \\ &= [4(5) + 4(-1)](0.03) \\ &\quad + [4(5) + 2(-1)](-0.02) \\ &= 0.48 - 0.36 = 0.12 \end{aligned}$$

11. $z = \dfrac{y^2 + 3x}{y^2 - x}$, $x = 4$, $y = -4$,

$dx = 0.01$, $dy = 0.03$

$dz = \dfrac{(y^2 - x) \cdot 3 - (y^2 + 3x) \cdot (-1)}{(y^2 - x)^2}\, dx$

$\qquad + \dfrac{(y^2 - x) \cdot 2y - (y^2 + 3x) \cdot 2y}{(y^2 - x)^2}\, dx$

$\qquad = \dfrac{4y^2}{(y^2 - x)^2}\, dx - \dfrac{8xy}{(y^2 - x)^2}\, dy$

$\qquad = \dfrac{4(-4)^2}{[(-4)^2 - 4]^2}(0.01) - \dfrac{8(4)(-4)}{[(-4)^2 - 4]^2}(0.03)$

$\qquad \approx 0.0311$

13. $w = \dfrac{5x^2 + y^2}{z + 1}$

$x = -2$, $y = 1$, $z = 1$

$dx = 0.02$, $dy = -0.03$, $dz = 0.02$

$f_x(x, y) = \dfrac{(z + 1)10x - (5x^2 + y^2)(0)}{(z + 1)^2}$

$\qquad = \dfrac{10x}{z + 1}$

$f_y(x, y) = \dfrac{(z + 1)(2y) - (5x^2 + y^2)(0)}{(z + 1)^2}$

$\qquad = \dfrac{2y}{z + 1}$

$f_z(x, y) = \dfrac{(z + 1)(0) - (5x^2 + y^2)(1)}{(z + 1)^2}$

$\qquad = \dfrac{-5x^2 - y^2}{(z + 1)^2}$

$dw = \dfrac{10x}{z + 1}\, dx + \dfrac{2y}{z + 1}\, dy + \dfrac{-5x^2 - y^2}{(z + 1)^2}\, dz$

Substitute the given values.

$dw = \dfrac{-20}{2}(0.02) + \dfrac{2}{2}(-0.03) + \dfrac{[-5(4) - 1](0.02)}{(2)^2}$

$\qquad = -0.2 - 0.03 - \dfrac{21}{4}(0.02) = -0.335$

15. The volume of the can is

$$V = \pi r^2 h,$$

with $r = 2.5$ cm, $h = 14$ cm, $dr = 0.08$, $dh = 0.16$.

$dV = 2\pi r h\, dr + \pi r^2\, dh$

$\qquad = 2\pi(2.5)(14)(0.08) + \pi(2.5)^2(0.16)$

$\qquad \approx 20.73$

Approximately 20.73 cm³ of aluminum are needed.

17. The volume of the box is

$$V = LWH$$

with $L = 10$, $W = 9$, and $H = 18$.
Since 0.1 inch is applied to each side and each dimension has a side at each end,

$$dL = dW = dH = 2(0.1) = 0.2$$
$$dV = WH\, dL + LH\, dW + LW\, dH.$$

Substitute.

$dV = (9)(18)(0.2) + (10)(18)(0.2) + (10)(9)(0.2)$

$\qquad = 86.4$

Approximately 86.4 in.³ are needed.

19. $z = x^{0.65}y^{0.35}$

$x = 50$, $y = 29$,

$dx = 52 - 50 = 2$

$dy = 27 - 29 = -2$

$f_x(x, y) = y^{0.35}(0.65)(x^{-0.35})$

$\qquad = 0.65 \left(\dfrac{y}{x}\right)^{0.35}$

$f_y(x, y) = (x^{0.65})(0.35)(y^{-0.65})$

$\qquad = 0.35 \left(\dfrac{x}{y}\right)^{0.65}$

$dz = 0.65 \left(\dfrac{y}{x}\right)^{0.35} dx + 0.35 \left(\dfrac{x}{y}\right)^{0.65} dy$

Substitute.

$dz = 0.65 \left(\dfrac{29}{50}\right)^{0.35}(2) + 0.35 \left(\dfrac{50}{29}\right)^{0.65}(-2)$

$\qquad = 0.07694$ unit

21. The volume of the bone is

$$V = \pi r^2 h,$$

with $h = 7$, $r = 1.4$, $dr = 0.09$, $dh = 2(0.09) = 0.18$

$dV = 2\pi r h\, dr + \pi r^2\, dh$

$\qquad = 2\pi(1.4)(7)(0.09) + \pi(1.4)^2(0.18)$

$\qquad = 6.65$

6.65 cm³ of preservative are used.

23. $C = \dfrac{b}{a-v} = b(a-v)^{-1}$

$a = 160,$
$b = 200, \ v = 125$
$da = 145 - 160 = -15$
$db = 190 - 200 = -10$
$dv = 130 - 125 = 5$

$dC = -b(a-v)^{-2} \, da$
$\qquad + \dfrac{1}{a-v} \, db + b(a-v)^{-2} \, dv$

$\qquad = \dfrac{-b}{(a-v)^2} \, da + \dfrac{1}{a-v} \, db + \dfrac{b}{(a-v)^2} \, dv$

$\qquad = \dfrac{-200}{(160-125)^2}(-15) + \dfrac{1}{160-125}(-10)$

$\qquad + \dfrac{200}{(160-125)^2}(5)$

$\qquad \approx 2.98$ liters

25. $C(t,g) = 0.6(0.96)^{(210t/1500)-1}$

$\qquad + \dfrac{gt}{126t - 900}[1 - (0.96)^{(210t/1500)-1}]$

(a) $C(180, 8) = 0.6(0.96)^{(210(180)/1500)-1}$

$\qquad + \dfrac{(8)(180)}{126(180) - 900}$

$\qquad \cdot [1 - (0.96)^{(210(180)/1500)-1}]$

$\qquad \approx 0.2649$

(b) $C_t(t,g)$

$\qquad = 0.6(\ln 0.96)\left(\dfrac{210}{1500}\right)(0.96)^{(210t/1500)-1}$

$\qquad + \dfrac{g(126t - 900) - 126(gt)}{(126t - 900)^2}$

$\qquad \cdot [1 - (0.96)^{(210t/1500)-1}]$

$\qquad - \dfrac{gt}{126t - 900}(\ln 0.96)\left(\dfrac{210}{1500}\right)(0.96)^{(210t/1500)-1}$

$C_g(t,g)$

$\qquad = \dfrac{t}{126t - 900}[1 - (0.96)^{(210t/1500)-1}]$

$C(180 - 10, 8 + 1)$
$\qquad \approx C(180, 8) + C_t(180, 8) \cdot (-10)$
$\qquad + C_g(180, 8) \cdot (1)$
$\qquad \approx 0.2649 + (-0.00115)(-10) + 0.00519(1)$
$\qquad \approx 0.2816$

$C(170, 9) \approx 0.2817$

The approximation is very good.

27. $P(A, B, D) = \dfrac{1}{1 + e^{3.68 - 0.016A - 0.77B - 0.12D}}$

(a) Since bird pecking is present, $B = 1$.

$P(150, 1, 20) = \dfrac{1}{1 + e^{3.68 - 0.016(150) - 0.77(1) - 0.12(20)}}$

$\qquad = \dfrac{1}{1 + e^{-1.89}} \approx 0.8688$

The probability is about 87%.

(b) Since bird pecking is not present, $B = 0$.

$P(150, 0, 20) = \dfrac{1}{1 + e^{3.68 - 0.016(150) - 0.77(0) - 0.12(20)}}$

$\qquad = \dfrac{1}{1 + e^{-1.12}} \approx 0.7540$

The probability is about 75%.

(c) Let $B = 0$. To simplify the notation, let $X = 3.68 - 0.016A - 0.12D$. Then

$P(A, 0, D) = \dfrac{1}{1 + e^{3.68 - 0.016A - 0.12D}}$

$\qquad = \dfrac{1}{1 + e^X}.$

Some other values that we'll need are

$dA = 160 - 150 = 10$
$dD = 25 - 20 = 4$
$X(150, 20) = 3.68 - 0.016(150) - 0.12(20)$
$\qquad = -1.12$

$X_A = \dfrac{\partial X}{\partial A} = -0.016$

$X_D = \dfrac{\partial X}{\partial D} = -0.12.$

$P_A(A, 0, D) = \dfrac{X_A e^X}{(1 + e^X)^2} = \dfrac{0.016 e^X}{(1 + e^X)^2}$

$P_D(A, 0, D) = \dfrac{X_D e^X}{(1 + e^X)^2} = \dfrac{0.12 e^X}{(1 + e^X)^2}$

$dP = P_A(A, 0, D) \, dA + P_D(A, 0, D) \, dD$

$\qquad = \dfrac{0.016 e^X}{(1 + e^X)^2} \, dA + \dfrac{0.12 e^X}{(1 + e^X)^2} \, dD$

Substituting the given and calculated values,

$dP = \dfrac{0.016 e^{-1.12}}{(1 + e^{-1.12})^2}(10) + \dfrac{0.12 e^{-1.12}}{(1 + e^{-1.12})^2}(5)$

$\qquad = (0.016 \cdot 10 + 0.12 \cdot 5)\dfrac{e^{-1.12}}{(1 + e^{-1.12})^2}$

$\qquad \approx 0.76 \cdot 0.1855 \approx 0.14.$

Therefore,

$$P(160, 0, 25) = P(150, 0, 20) + \Delta P$$
$$\approx P(150, 0, 20) + dP$$
$$= 0.75 + 0.14 = 0.89.$$

The probability is about 89%.

Using a calculator, $P(160, 0, 25) \approx 0.8676$, or about 87%.

29. The area is $A = \frac{1}{2}bh$ with $b = 15.8$ cm, $h = 37.5$ cm, $db = 1.1$ cm, and $dh = 0.8$ cm.

$$dA = \frac{1}{2}b\,dh + \frac{1}{2}h\,db$$
$$= \frac{1}{2}(15.8)(0.8) + \frac{1}{2}(37.5)(1.1)$$
$$= 26.945$$

The maximum possible error is 26.945 cm^2.

31. Let $z = f(L, W, H) = LWH$

Then

$$dz = f_L(L, W, H)\,dL + f_W(L, W, H)\,dW$$
$$+ f_H(L, W, H)\,dH$$
$$= WH\,dL + LH\,dW + LW\,dH.$$

A maximum 1% error in each measurement means that the maximum values of dL, dW, and dH are given by $dL = 0.01L$, $dW = 0.01W$, and $dH = 0.01H$.
Therefore,

$$dz = WH(0.01L) + LH(0.01W) + LW(0.01H)$$
$$= 0.01LWH + 0.01LWH + 0.01LWH$$
$$= 0.03LWH.$$

Thus, an estimate of the maximum error in calculating the volume is 3%.

17.6   Double Integrals

1. $\displaystyle\int_0^5 (x^4y + y)\,dx = \left(\frac{x^5y}{5} + xy\right)\Big|_0^5$

$$= (625y + 5y) - 0 = 630y$$

3. $\displaystyle\int_1^7 \sqrt{x + 6y}\,dy = \frac{1}{6}\cdot\frac{2}{3}(x + 6y)^{3/2}\Big|_1^7$

$$= \frac{1}{9}[(x + 42)^{3/2} - (x + 6)^{3/2}]$$

5. $\displaystyle\int_4^5 x\sqrt{x^2 + 3y}\,dy$.

$$= \int_4^5 x(x^2 + 3y)^{1/2}\,dy$$

$$= \frac{2x}{9}(x^2 + 3y)^{3/2}\Big|_4^5$$

$$= \frac{2x}{9}[(x^2 + 15)^{3/2} - (x^2 + 12)^{3/2}]$$

7. $\displaystyle\int_4^9 \frac{3 + 5y}{\sqrt{x}}\,dx = (3 + 5y)\int_4^9 x^{-1/2}\,dx$

$$= (3 + 5y)2x^{1/2}\Big|_4^9$$

$$= (3 + 5y)2[\sqrt{9} - \sqrt{4}]$$

$$= 6 + 10y$$

9. $\displaystyle\int_2^6 e^{2x+3y}\,dx = \frac{1}{2}e^{2x+3y}\Big|_2^6$

$$= \frac{1}{2}(e^{12+3y} - e^{4+3y})$$

11. $\displaystyle\int_0^3 ye^{4x+y^2}\,dy$

Let $u = 4x + y^2$; then $du = 2y\,dy$.
If $y = 0$ then $u = 4x$.
If $y = 3$ then $u = 4x + 9$.

$$\int_{4x}^{4x+9} e^u \cdot \frac{1}{2}\,du = \frac{1}{2}e^u\Big|_{4x}^{4x+9}$$

$$= \frac{1}{2}(e^{4x+9} - e^{4x})$$

13. $\displaystyle\int_1^2 \int_0^5 (x^4y + y)\,dx\,dy$

From Exercise 1

$$\int_0^5 (x^4y + y)\,dx = 630y.$$

Therefore,

$$\int_1^2 \left[\int_0^5 (x^4y + y)\,dx\right]dy = \int_1^2 630y\,dy$$

$$= 315y^2\Big|_1^2$$

$$= 315(4 - 1) = 945.$$

15. $\displaystyle\int_0^1 \left[\int_3^6 x\sqrt{x^2+3y}\, dx \right] dy$

From Exercise 6,

$$\int_3^6 x\sqrt{x^2+3y}\, dx$$

$$= \frac{1}{3}[(36+3y)^{3/2} - (9+3y)^3].$$

$$\int_0^1 \left[\int_3^6 x\sqrt{x^2+3y}\, dx \right] dy$$

$$= \int_0^1 \frac{1}{3}[(36+3y)^{3/2} - (9+3y)^{3/2}]\, dy$$

Let $u = 36 + 3y$. Then $du = 3\, dy$.
When $y = 0$, $u = 36$.
When $y = 1$, $u = 39$.
Let $z = 9 + 3y$. Then $dz = 3\, dy$.
When $y = 0$, $z = 9$.
When $y = 1$, $z = 12$.

$$\frac{1}{9} \left[\int_{36}^{39} u^{3/2}\, du - \int_9^{12} z^{3/2}\, dz \right]$$

$$= \frac{1}{9} \cdot \frac{2}{5}[(39)^{5/2} - (36)^{5/2} - (12)^{5/2} + (9)^{5/2}]$$

$$= \frac{2}{45}[(39)^{5/2} - (12)^{5/2} - 6^5 + 3^5]$$

$$= \frac{2}{45}(39^{5/2} - 12^{5/2} - 7533)$$

17. $\displaystyle\int_1^2 \left[\int_4^9 \frac{3+5y}{\sqrt{x}}\, dx \right] dy$

From Exercise 7,

$$\int_4^9 \frac{3+5y}{\sqrt{x}}\, dx = 6 + 10y.$$

$$\int_1^2 \left[\int_4^9 \frac{3+5y}{\sqrt{x}}\, dx \right] dy$$

$$= \int_1^2 (6 + 10y)\, dy$$

$$= 6y \Big|_1^2 + 5y^2 \Big|_1^2$$

$$= 6(2-1) + 5(4-1)$$
$$= 6 + 15$$
$$= 21$$

19. $\displaystyle\int_1^3 \int_1^3 \frac{dy\, dx}{xy} = \int_1^3 \left[\int_1^3 \frac{1}{xy}\, dy \right] dx$

$$= \int_1^3 \left(\frac{1}{x} \ln|y| \right) \Big|_1^3 dx$$

$$= \int_1^3 \frac{\ln 3}{x}\, dx$$

$$= (\ln 3) \ln|x| \Big|_1^3$$
$$= (\ln 3)(\ln 3 - 0) = (\ln 3)^2$$

21. $\displaystyle\int_2^4 \int_3^5 \left(\frac{x}{y} + \frac{y}{3} \right) dx\, dy$

$$= \int_2^4 \left(\frac{x^2}{2y} + \frac{yx}{3} \right) \Big|_3^5 dy$$

$$= \int_2^4 \left[\frac{25}{2y} + \frac{5y}{3} - \left(\frac{9}{2y} + \frac{3y}{3} \right) \right] dy$$

$$= \int_2^4 \left(\frac{16}{2y} + \frac{2y}{3} \right) dy$$

$$= \left(8 \ln|y| + \frac{y^2}{3} \right) \Big|_2^4$$

$$= 8 (\ln 4 - \ln 2) + \frac{16}{3} - \frac{4}{3}$$

$$= 8 \ln \frac{4}{2} + \frac{12}{3}$$

$$= 8 \ln 2 + 4$$

23. $\displaystyle\iint_R (3x^2 + 4y)dx\, dy;\ 0 \le x \le 3, 1 \le y \le 4$

$$\iint_R (3x^2 + 4y)dx\, dy = \int_1^4 \int_0^3 (3x^2 + 4y)dx\, dy$$

$$= \int_1^4 (x^3 + 4xy) \Big|_0^3 dy$$

$$= \int_1^4 (27 + 12y)dy$$

$$= (27y + 6y^2) \Big|_1^4$$
$$= (108 + 96) - (27 + 6) = 171$$

25. $\iint\limits_{R} \sqrt{x+y}\, dy\, dx;\ 1 \leq x \leq 3,\ 0 \leq y \leq 1$

$$\iint\limits_{R} \sqrt{x+y}\, dy\, dx = \int_{1}^{3} \int_{0}^{1} (x+y)^{1/2}\, dy\, dx$$

$$= \int_{1}^{3} \left[\frac{2}{3}(x+y)^{3/2} \right]\Big|_{0}^{1} dx$$

$$= \int_{1}^{3} \frac{2}{3}[(x+1)^{3/2} - x^{3/2}]\, dx$$

$$= \frac{2}{3} \cdot \frac{2}{5}[(x+1)^{5/2} - x^{5/2}]\Big|_{1}^{3}$$

$$= \frac{4}{15}(4^{5/2} - 3^{5/2} - 2^{5/2} + 1^{5/2})$$

$$= \frac{4}{15}(32 - 3^{5/2} - 2^{5/2} + 1)$$

$$= \frac{4}{15}(33 - 3^{5/2} - 2^{5/2})$$

27. $\iint\limits_{R} \frac{3}{(x+y)^2}\, dy\, dx;\ 2 \leq x \leq 4,\ 1 \leq y \leq 6$

$$\iint\limits_{R} \frac{3}{(x+y)^2}\, dy\, dx = 3 \int_{2}^{4} \int_{1}^{6} (x+y)^{-2}\, dy\, dx$$

$$= -3 \int_{2}^{4} (x+y)^{-1}\Big|_{1}^{6} dx$$

$$= -3 \int_{2}^{4} \left(\frac{1}{x+6} - \frac{1}{x+1} \right) dx$$

$$= -3(\ln|x+6| - \ln|x+1|)\Big|_{2}^{4}$$

$$= -3 \left(\ln \left| \frac{x+6}{x+1} \right| \right)\Big|_{2}^{4}$$

$$= -3 \left(\ln 2 - \ln \frac{8}{3} \right) = -3 \ln \frac{2}{\frac{8}{3}}$$

$$= -3 \ln \frac{3}{4} \text{ or } 3 \ln \frac{4}{3}$$

29. $\iint\limits_{R} y e^{(x+y^2)}\, dx\, dy;\ 2 \leq x \leq 3,\ 0 \leq y \leq 2$

$$\iint\limits_{R} y e^{(x+y^2)}\, dx\, dy$$

$$= \int_{0}^{2} \int_{2}^{3} y e^{x+y^2}\, dx\, dy$$

$$= \int_{0}^{2} y e^{x+y^2}\Big|_{2}^{3} dy$$

$$= \int_{0}^{2} (y e^{3+y^2} - y e^{2+y^2})\, dy$$

$$= e^3 \int_{0}^{2} y e^{y^2}\, dy - e^2 \int_{0}^{2} y e^{y^2}\, dy$$

$$= \frac{e^3}{2} (e^{y^2})\Big|_{0}^{2} - \frac{e^2}{2} (e^{y^2})\Big|_{0}^{2}$$

$$= \frac{e^3}{2} (e^4 - e^0) - \frac{e^2}{2} (e^4 - e^0)$$

$$= \frac{1}{2} (e^7 - e^6 - e^3 + e^2)$$

31. $z = 8x + 4y + 3;\ -1 \leq x \leq 1,\ 0 \leq y \leq 3$

$$V = \int_{-1}^{1} \int_{0}^{3} (8x + 4y + 3)\, dy\, dx$$

$$= \int_{-1}^{1} (8xy + 2y^2 + 3y)\Big|_{0}^{3} dx$$

$$= \int_{-1}^{1} (24x + 18 + 9 - 0)\, dx$$

$$= \int_{-1}^{1} (24x + 27)\, dx$$

$$= (12x^2 + 27x)\Big|_{-1}^{1}$$

$$= (12 + 27) - (12 - 27) = 54$$

33. $z = x^2;\ 0 \leq x \leq 2,\ 0 \leq y \leq 5$

$$V = \int_{0}^{2} \int_{0}^{5} x^2\, dy\, dx$$

$$= \int_{0}^{2} x^2 y\Big|_{0}^{5} dx$$

$$= \int_{0}^{2} 5x^2\, dx$$

$$= \frac{5}{3} x^3\Big|_{0}^{2}$$

$$= \frac{40}{3}$$

35. $z = x\sqrt{x^2 + y};\ 0 \le x \le 1,\ 0 \le y \le 1$

$$V = \int_0^1 \int_0^1 x\sqrt{x^2 + y}\, dx\, dy$$

Let $u = x^2 + y$. Then $du = 2x\, dx$.
When $x = 0,\ u = y$.
When $x = 1,\ u = 1 + y$.

$$= \int_0^1 \left[\int_y^{1+y} u^{1/2}\, du \right] dy$$

$$= \int_0^1 \frac{1}{2}\left(\frac{2}{3}u^{3/2} \right)\Big|_y^{1+y} dy$$

$$= \int_0^1 \frac{1}{3}[(1 + y)^{3/2} - y^{3/2}]\, dy$$

$$= \frac{1}{3} \cdot \frac{2}{5}[(1 + y)^{5/2} - y^{5/2}]\Big|_0^1$$

$$= \frac{2}{15}(2^{5/2} - 1 - 1)$$

$$= \frac{2}{15}(2^{5/2} - 2)$$

37. $z = \dfrac{xy}{(x^2 + y^2)^2};\ 1 \le x \le 2,\ 1 \le y \le 4$

$$V = \int_1^2 \int_1^4 \frac{xy}{(x^2 + y^2)^2}\, dy\, dx$$

$$= \int_1^2 \left[\int_1^4 xy(x^2 + y^2)^{-2}\, dy \right] dx$$

$$= \int_1^2 \left[\int_1^4 \frac{1}{2}x(x^2 + y^2)^{-2}(2y)\, dy \right] dx$$

$$= \int_1^2 \left[-\frac{1}{2}x(x^2 + y^2)^{-1} \right]\Big|_1^4 dx$$

$$= \int_1^2 \left[-\frac{1}{2}x(x^2 + 16)^{-1} + \frac{1}{2}x(x^2 + 1)^{-1} \right] dx$$

$$= -\frac{1}{2}\int_1^2 \frac{1}{2}(x^2 + 16)^{-1}(2x)\, dx$$

$$\quad + \frac{1}{2}\int_1^2 \frac{1}{2}(x^2 + 1)^{-1}(2x)\, dx$$

$$= -\frac{1}{2} \cdot \frac{1}{2} \ln|x^2 + 16|\,\Big|_1^2 + \frac{1}{2} \cdot \frac{1}{2} \ln|x^2 + 1|\,\Big|_1^2$$

$$= -\frac{1}{4} \cdot \ln 20 + \frac{1}{4} \ln 17 + \frac{1}{4} \ln 5 - \frac{1}{4} \ln 2$$

$$= \frac{1}{4}(-\ln 20 + \ln 17 + \ln 5 - \ln 2)$$

$$= \frac{1}{4} \ln \frac{(17)(5)}{(20)(2)}$$

$$= \frac{1}{4} \ln \frac{17}{8}$$

39. $\displaystyle\iint_R xe^{xy}\, dx\, dy;\ 0 \le x \le 2;\ 0 \le y \le 1$

$$\iint_R xe^{xy}\, dx\, dy$$

$$= \int_0^2 \int_0^1 xe^{xy}\, dy\, dx$$

$$= \int_0^2 \frac{x}{x}e^{xy}\Big|_0^1 dx$$

$$= \int_0^2 (e^x - e^0)\, dx$$

$$= (e^x - x)\Big|_0^2$$

$$= e^2 - 2 - e^0 + 0$$
$$= e^2 - 3$$

41. $\displaystyle\int_2^4 \int_2^{x^2} (x^2 + y^2)\, dy\, dx$

$$= \int_2^4 \left(x^2 y + \frac{y^3}{3} \right)\Big|_2^{x^2} dx$$

$$= \int_2^4 \left(x^4 + \frac{x^6}{3} - 2x^2 - \frac{8}{3} \right) dx$$

$$= \left(\frac{x^5}{5} + \frac{x^7}{21} - \frac{2}{3}x^3 - \frac{8}{3}x \right)\Big|_2^4$$

$$= \frac{1024}{5} + \frac{16,384}{21} - \frac{2}{3}(64) - \frac{8}{3}(4)$$

$$\quad - \left(\frac{32}{5} + \frac{128}{21} - \frac{16}{3} - \frac{16}{3} \right)$$

$$= \frac{1024}{5} - \frac{32}{5} + \frac{16,384 - 128}{21}$$

$$\quad - \frac{128}{3} - \frac{32}{3} - \left(\frac{-32}{3} \right)$$

$$= \frac{992}{5} + \frac{16,256}{21} - \frac{128}{3}$$

$$= \frac{20,832}{105} + \frac{81,280}{105} - \frac{4480}{105}$$

$$= \frac{97,632}{105}$$

43. $\displaystyle\int_0^4 \int_0^x \sqrt{xy}\,dy\,dx$

$$= \int_0^4 \int_0^x (xy)^{1/2}\,dy\,dx$$

$$= \int_0^4 \left[\frac{2(xy)^{3/2}}{3x}\right]\Big|_0^x dx$$

$$= \frac{2}{3}\int_0^4 \left[\frac{(\sqrt{x^2})^3}{x} - \frac{0}{x}\right] dx$$

$$= \frac{2}{3}\int_0^4 x^2\,dx = \frac{2}{3}\cdot\frac{x^3}{3}\Big|_0^4 = \frac{2}{9}(64)$$

$$= \frac{128}{9}$$

45. $\displaystyle\int_2^6 \int_{2y}^{4y} \frac{1}{x}\,dx\,dy$

$$= \int_2^6 (\ln|x|)\Big|_{2y}^{4y} dy$$

$$= \int_2^6 (\ln|4y| - \ln|2y|)\,dy$$

$$= \int_2^6 \ln\left|\frac{4y}{2y}\right| dy$$

$$= \int_2^6 \ln 2\,dy$$

$$= (\ln 2)y\Big|_2^6$$

$$= (\ln 2)(6-2) = 4\ln 2$$

Note: We can write $4\ln 2$ as $\ln 2^4$, or $\ln 16$.

47. $\displaystyle\int_0^4 \int_1^{e^x} \frac{x}{y}\,dy\,dx$

$$= \int_0^4 (x\ln|y|)\Big|_1^{e^x} dx$$

$$= \int_0^4 (x\ln e^x - x\ln 1)\,dx$$

$$= \int_0^4 x^2\,dx = \frac{x^3}{3}\Big|_0^4 = \frac{64}{3}$$

49. $\displaystyle\int_0^{\ln 2} \int_{e^y}^2 \frac{1}{\ln x}\,dx\,dy$

Changing the order of integration,

$$\int_0^{\ln 2} \int_{e^y}^2 \frac{1}{\ln x}\,dx\,dy$$

$$= \int_1^2 \int_0^{\ln x} \frac{1}{\ln x}\,dy\,dx$$

$$= \int_1^2 \left[\frac{1}{\ln x}\,y\,\Big|_0^{\ln x}\right] dx$$

$$= \int_1^2 (1-0)\,dx$$

$$= x\,\Big|_1^2$$

$$= 2 - 1 = 1$$

51. $\displaystyle\iint_R (5x+8y)\,dy\,dx;\ 1\le x\le 3, 0\le y\le x-1$

$$\iint_R (5x+8y)\,dy\,dx$$

$$= \int_1^3 \int_0^{x-1} (5x+8y)\,dy\,dx$$

$$= \int_1^3 (5xy+4y^2)\Big|_0^{x-1} dx$$

$$= \int_1^3 [5x(x-1)+4(x-1)^2 - 0]\,dx$$

$$= \int_1^3 (9x^2 - 13x + 4)\,dx$$

$$= \left(3x^3 - \frac{13}{2}x^2 + 4x\right)\Big|_1^3$$

$$= \left(81 - \frac{117}{2} + 12\right) - \left(3 - \frac{13}{2} + 4\right)$$

$$= 34$$

53. $\displaystyle\iint_R (4 - 4x^2)\,dy\,dx;\ 0 \le x \le 1,$

$0 \le y \le 2 - 2x$

$\displaystyle\iint_R (4 - 4x^2)\,dy\,dx$

$\displaystyle = \int_0^2 \int_0^{2-2x} 4(1 - x^2)\,dy\,dx$

$\displaystyle = \int_0^1 [4(1 - x^2)y]\,\Big|_0^{2(1-x)}\,dx$

$\displaystyle = \int_0^1 4(1 - x^2)(2)(1 - x)\,dx$

$\displaystyle = 8\int_0^1 (1 - x - x^2 + x^3)\,dx$

$\displaystyle = 8\left(x - \frac{x^2}{2} - \frac{x^3}{3} + \frac{x^4}{4}\right)\Big|_0^1$

$\displaystyle = 8\left(1 - \frac{1}{2} - \frac{1}{3} + \frac{1}{4}\right)$

$\displaystyle = 8\left(\frac{1}{2} - \frac{1}{12}\right)$

$\displaystyle = 8 \cdot \frac{5}{12} = \frac{10}{3}$

55. $\displaystyle\iint_R e^{x/y^2}\,dx\,dy;\ 1 \le y \le 2,\ 0 \le x \le y^2$

$\displaystyle\iint_R e^{x/y^2}\,dx\,dy$

$\displaystyle = \int_1^2 \int_0^{y^2} e^{x/y^2}\,dx\,dy$

$\displaystyle = \int_1^2 [y^2 e^{x/y^2}]\,\Big|_0^{y^2}\,dy$

$\displaystyle = \int_1^2 (y^2 e^{y^2/y^2} - y^2 e^0)\,dy$

$\displaystyle = \int_1^2 (ey^2 - y^2)\,dy$

$\displaystyle = (e - 1)\frac{y^3}{3}\,\Big|_1^2$

$\displaystyle = (e - 1)\left(\frac{8}{3} - \frac{1}{3}\right)$

$\displaystyle = \frac{7(e - 1)}{3}$

57. $\displaystyle\iint_R x^3 y\,dy\,dx;\ R$ bounded by $y = x^2,\ y = 2x$

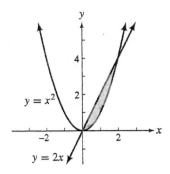

The points of intersection can be determined by solving the following system for x.

$$y = x^2$$
$$y = 2x$$

$$x^2 = 2x$$
$$x(x - 2) = 0$$
$$x = 0 \quad \text{or} \quad x = 2$$

Therefore,

$\displaystyle\iint_R x^3 y\,dx\,dy$

$\displaystyle = \int_0^2 \int_{x^2}^{2x} x^3 y\,dy\,dx$

$\displaystyle = \int_0^2 \left(x^3 \frac{y^2}{2}\right)\Big|_{x^2}^{2x}\,dx$

$\displaystyle = \int_0^2 \left[x^3 \frac{(4x^2)}{2} - x^3 \frac{(x^4)}{2}\right]\,dx$

$\displaystyle = \int_0^2 \left(2x^5 - \frac{x^7}{2}\right)\,dx$

$\displaystyle = \left(\frac{1}{3}x^6 - \frac{1}{16}x^8\right)\Big|_0^2$

$\displaystyle = \frac{1}{3} \cdot 2^6 - \frac{1}{16} \cdot 2^8$

$\displaystyle = \frac{64}{3} - 16$

$\displaystyle = \frac{16}{3}.$

59. $\displaystyle\iint\limits_{R} \frac{dy\,dx}{y}$; R bounded by $y = x$, $y = \dfrac{1}{x}$,

$x = 2$.

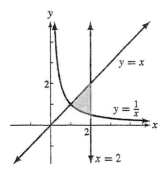

The graphs of $y = x$ and $y = \frac{1}{x}$ intersect at $(1, 1)$.

$$\int_1^2 \int_{1/x}^{x} \frac{dy}{y}\,dx = \int_1^2 \ln y \Big|_{1/x}^{x}\,dx = \int_1^2 \left(\ln x - \ln \frac{1}{x} \right) dx = \int_1^2 2\ln x\,dx$$

$$= 2(x \ln x - x) \Big|_1^2 = 2[(2\ln 2 - 2) - (\ln 1 - 1)] = 4\ln 2 - 2$$

63. $f(x, y) = x^2 + y^2$; $0 \le x \le 2$,
$0 \le y \le 3$

The area of region R is

$$A = (2 - 0)(3 - 0) = 6.$$

The average value of

$$f(x, y) = x^2 + y^2$$

over R is

$$\frac{1}{A} \iint\limits_{R} (x^2 + y^2)\,dy\,dx = \frac{1}{6} \int_0^2 \int_0^3 (x^2 + y^2)\,dy\,dx = \frac{1}{6} \int_0^2 \left(x^2 y + \frac{y^3}{3} \right) \Big|_0^3 dx = \frac{1}{6} \int_0^2 (3x^2 + 9)\,dx$$

$$= \frac{1}{6}(x^3 + 9x) \Big|_0^2 = \frac{1}{6}(8 + 18 - 0)_0^2 = \frac{1}{6} \cdot 26 = \frac{13}{3}.$$

65. $f(x, y) = e^{2x+y}$; $1 \le x \le 2$, $2 \le y \le 3$

The area of region R is

$$A = (2 - 1)(3 - 2) = 1.$$

The average value of f over R is

$$\frac{1}{A} \iint\limits_{R} e^{2x+y}\,dy\,dx = \iint\limits_{R} e^{2x+y}\,dy\,dx = \int_1^2 \int_2^3 e^{2x+y}\,dy\,dx = \int_1^2 \left(e^{2x+y} \right) \Big|_2^3 dx$$

$$= \int_1^2 \left(e^{2x+3} - e^{2x+2} \right) dx = \frac{1}{2} \left(e^{2x+3} - e^{2x+2} \right) \Big|_1^2$$

$$= \frac{1}{2}(e^{4+3} - e^{2+3} - e^{4+2} + e^4) = \frac{e^7 - e^6 - e^5 + e^4}{2}.$$

67. $C(x, y) = \dfrac{1}{9}x^2 + 2x + y^2 + 5y + 100$

The area of region R is

$A(80 - 40)(70 - 30) = 1600.$

The average cost is

$$\frac{1}{A} \iint\limits_{R} C(x, y)\, dy\, dx = \frac{1}{1600} \int_{30}^{70} \int_{40}^{80} \left(\frac{1}{9}x^2 + 2x + y^2 + 5y + 100 \right) dy\, dx$$

$$= \frac{1}{1600} \int_{30}^{70} \left(\frac{1}{27}x^3 + x^2 + xy^2 + 5xy + 100x \right) \Big|_{40}^{80} dy$$

$$= \int_{30}^{70} \left[\left(\frac{320}{27} + 4 + \frac{1}{20}y^2 + \frac{1}{4}y + 5 \right) - \left(\frac{40}{27} + 1 + \frac{1}{40}y^2 + \frac{1}{8}y + \frac{5}{2} \right) \right] dy$$

$$= \int_{30}^{70} \left(\frac{1}{40}y^2 + \frac{1}{8}y + \frac{857}{54} \right) dy = \left(\frac{1}{120}y^3 + \frac{1}{16}y^2 + \frac{857}{54}y \right) \Big|_{30}^{70}$$

$$= \left(\frac{8575}{3} + \frac{1225}{4} + \frac{29{,}995}{27} \right) - \left(225 + \frac{225}{4} + \frac{4285}{9} \right) = \frac{94{,}990}{27} \approx 3518.$$

The average cost is about \$3518.

69. $P(x, y) = -(x - 100)^2 - (y - 50)^2 + 2000$

$\text{Area} = (150 - 100)(80 - 40) = (50)(40) = 2000$

The average weekly profit is

$$\frac{1}{2000} \iint\limits_{R} [-(x - 100)^2 - (y - 50)^2 + 2000]\, dy\, dx$$

$$= \frac{1}{2000} \int_{100}^{150} \int_{40}^{80} [-(x - 100)^2 - (y - 50)^2 + 2000]\, dy\, dx$$

$$= \frac{1}{2000} \int_{100}^{150} \left[-(x - 100)^2 y - \frac{(y - 50)^3}{3} + 2000y \right] \Big|_{40}^{80} dx$$

$$= \frac{1}{2000} \int_{100}^{150} \left[-(x - 100)^2 (80 - 40) - \frac{(80 - 50)^3}{3} + \frac{(40 - 50)^3}{3} + 2000(80 - 40) \right] dx$$

$$= \frac{1}{2000} \int_{100}^{150} \left[-40(x - 100)^2 - \frac{30^3}{3} + \frac{(-10)^3}{3} + 2000(40) \right] dx$$

$$= \frac{1}{2000} \int_{100}^{150} \left[-40(x - 100)^2 - \frac{28{,}000}{3} + 80{,}000 \right] dx$$

$$= \frac{1}{2000} \cdot \left[\frac{-40(x - 100)^3}{3} - \frac{28{,}000}{3}x + 80{,}000x \right] \Big|_{100}^{150}$$

$$= \frac{1}{2000} \left[-\frac{40}{3}(150 - 100)^3 + \frac{40}{3}(100 - 100)^3 - \frac{28{,}000}{3}(150 - 100) + 80{,}000(150 - 100) \right]$$

$$= \frac{1}{2000} \left[-\frac{40}{3}(50)^3 + \frac{40}{3} \cdot 0 - \frac{28{,}000}{3}(50) + 80{,}000(50) \right]$$

$$= \frac{1}{2000} \left(\frac{-5{,}000{,}000 - 1{,}400{,}000 + 12{,}000{,}000}{3} \right)$$

$$= \frac{1}{2000} \left(\frac{5{,}600{,}000}{3} \right) = \$933.33.$$

71. $T(x,y) = x^4 + 16y^4 - 32xy + 40$

Area $= (4-0)(2-0) = 8$

The average time is

$$\frac{1}{8} \iint_R (x^4 + 16y^4 - 32xy + 40)\,dy\,dx$$

$$= \frac{1}{8} \int_0^4 \int_0^2 (x^4 + 16y^4 - 32xy + 40)\,dy\,dx$$

$$= \frac{1}{8} \int_0^4 \left(x^4 y + \frac{16y^5}{5} - \frac{32xy^2}{2} + 40y \right) \Big|_0^2 dx$$

$$= \frac{1}{8} \int_0^4 \left[x^4(2-0) + \frac{16(2-0)^5}{5} - \frac{32x(2-0)^2}{2} \right.$$
$$\left. + 40(2-0) \right] dx$$

$$= \frac{1}{8} \int_0^4 \left(2x^4 + \frac{512}{5} - 64x + 80 \right) dx$$

$$= \frac{1}{8} \left(\frac{2x^5}{5} + \frac{512}{5} - \frac{64x^2}{2} + 80 \right) \Big|_0^4$$

$$= \frac{1}{8} \left[\frac{2(4-0)^5}{5} + \frac{512}{5}(4-0) - \frac{64(4-0)^2}{2} \right.$$
$$\left. + 80(4-0) \right]$$

$$= \frac{1}{8} \left(\frac{2048}{5} + \frac{2048}{5} - 512 + 320 \right) = 78.4 \text{ hours}$$

Chapter 17 Review Exercises

5. $f(x,y) = 2x^2 y^2 - 7x + 4y$
$f(-1,2) = 2(1)(4) - 7(-1) + 4(2) = 23$
$f(6,-3) = 2(36)(9) - 7(6) + 4(-3) = 594$

7. $f(x,y) = \dfrac{\sqrt{x^2+y^2}}{x-y}$

$f(-1,2) = \dfrac{\sqrt{1+4}}{-1-2} = -\dfrac{\sqrt{5}}{3}$

$f(6,-3) = \dfrac{\sqrt{36+9}}{6+3} = \dfrac{\sqrt{45}}{9} = \dfrac{\sqrt{5}}{3}$

9. $x + 2y + 6z = 6$

x-intercept: $y = 0$, $z = 0$

$x = 6$

y-intercept: $x = 0$, $z = 0$

$2y = 6$
$y = 3$

z-intercept: $x = 0$, $y = 0$

$6z = 6$
$z = 1$

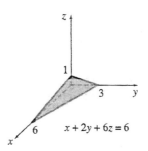

11. $4x + 3z = 12$

No y-intercept
x-intercept: $y = 0$, $z = 0$

$4x = 12$
$x = 3$

z-intercept: $x = 0$, $y = 0$

$3z = 12$
$z = 4$

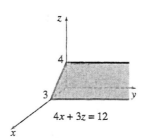

13. $y = 4$

No x-intercept, no z-intercept
The graph is a plane parallel to the xz-plane.

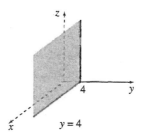

15. $z = f(x, y) = \dfrac{x + y^2}{x - y^2}$

(a) $\dfrac{\partial z}{\partial y} = \dfrac{(x - y^2) \cdot 2y - (x + y^2)(-2y)}{(x - y^2)^2}$

$= \dfrac{4xy}{(x - y^2)^2}$

(b) $\dfrac{\partial z}{\partial x} = \dfrac{(x - y^2) \cdot 1 - (x + y^2) \cdot 1}{(x - y^2)^2}$

$= \dfrac{-2y^2}{(x - y^2)^2}$

$= -2y^2(x - y^2)^{-2}$

$\left(\dfrac{\partial z}{\partial x}\right)(0, 2) = \dfrac{-8}{(-4)^2} = -\dfrac{1}{2}$

(c) $f_{xx}(x, y) = 4y^2(x - y^2)^{-3}$

$= \dfrac{4y^2}{(x - y^2)^3}$

$f_{xx}(-1, 0) = \dfrac{0}{1} = 0$

17. $f(x, y) = 5x^4y^3 - 6x^5y$

$f_x(x, y) = 20x^3y^3 - 30x^4y$

$f_y(x, y) = 15x^4y^2 - 6x^5$

19. $f(x, y) = \dfrac{2x + 5y^2}{3x^2 + y^2}$

$f_x(x, y) = \dfrac{(3x^2 + y^2) \cdot 2 - (2x + 5y^2) \cdot 6x}{(3x^2 + y^2)^2}$

$= \dfrac{2y^2 - 6x^2 - 30xy^2}{(3x^2 + y^2)^2}$

$f_y(x, y) = \dfrac{(3x^2 + y^2) \cdot 10y - (2x + 5y^2) \cdot 2y}{(3x^2 + y^2)^2}$

$= \dfrac{30x^2y - 4xy}{(3x^2 + y^2)^2}$

21. $f(x, y) = (y - 2)^2 e^{x + 2y}$

$f_x(x, y) = (y - 2)^2 e^{x + 2y}$

$f_y(x, y) = e^{x + 2y} \cdot 2(y - 2) + (y - 2)^2 \cdot 2e^{x + 2y}$

$= 2(y - 2)[1 + (y - 2)]e^{x + 2y}$

$= 2(y - 2)(y - 1)e^{x + 2y}$

23. $f(x, y) = \ln\left|2 - x^2y^3\right|$

$f_x(x, y) = \dfrac{1}{2 - x^2y^3} \cdot (-2xy^3)$

$= \dfrac{-2xy^3}{2 - x^2y^3}$

$f_y(x, y) = \dfrac{1}{2 - x^2y^3} \cdot (-3x^2y^2)$

$= \dfrac{-3x^2y^2}{2 - x^2y^3}$

25. $f(x, y) = -3x^2y^3 + x^3y$

$f_x(x, y) = -6xy^3 + 3x^2y$

$f_{xx}(x, y) = -6y^3 + 6xy$

$f_{xy}(x, y) = -18xy^2 + 3x^2$

27. $f(x, y) = \dfrac{3x + y}{x - 1}$

$f_x(x, y) = \dfrac{(x - 1) \cdot 3 - (3x + y) \cdot 1}{(x - 1)^2}$

$= \dfrac{-3 - y}{(x - 1)^2} = (-3 - y)(x - 1)^{-2}$

$f_{xx}(x, y) = -2(-3 - y)(x - 1)^{-3}$

$= \dfrac{2(3 + y)}{(x - 1)^3}$

$f_{xy}(x, y) = \dfrac{-1}{(x - 1)^2}$

29. $f(x, y) = ye^{x^2}$

$f_x(x, y) = 2xye^{x^2}$

$f_{xx}(x, y) = 2xy \cdot 2xe^{x^2} + e^{x^2} \cdot 2y$

$= 2ye^{x^2}(2x^2 + 1)$

$f_{xy}(x, y) = 2xe^{x^2}$

31. $f(x, y) = \ln\left|1 + 3xy^2\right|$

$f_x(x, y) = \dfrac{1}{1 + 3xy^2} \cdot 3y^2$

$= \dfrac{3y^2}{1 + 3xy^2}$

$= 3y^2(1 + 3xy^2)^{-1}$

$f_{xx}(x, y) = 3y^2 \cdot (-3y^2)(1 + 3xy^2)^{-2}$

$= \dfrac{-9y^4}{(1 + 3xy^2)^2}$

$f_{xy}(x, y) = \dfrac{(1 + 3xy^2) \cdot 6y - 3y^2(6xy)}{(1 + 3xy^2)^2}$

$= \dfrac{6y}{(1 + 3xy^2)^2}$

33. $z = x^2 + y^2 + 9x - 8y + 1$

$$z_x(x, y) = 2x + 9, \ z_y(x, y) = 2y - 8$$

$$2x + 9 = 0$$

$$x = -\frac{9}{2}$$

$$2y - 8 = 0$$

$$y = 4$$

$$z_{xx}(x, y) = 2, \ z_{yy}(x, y) = 2, \ z_{xy}(x, y) = 0$$

$$D = 2(2) - (0)^2 = 4 > 0 \text{ and } z_{xx}(x, y) > 0.$$

Relative minimum at $\left(-\frac{9}{2}, 4\right)$

35. $z = x^3 - 8y^2 + 6xy + 4$

$$z_x(x, y) = 3x^2 + 6y, \ z_y(x, y) = -16y + 6x$$

$$3x^2 + 6y = 0$$
$$x^2 + 2y = 0$$

$$y = -\frac{x^2}{2}$$

$$-16y + 6x = 0$$
$$-8y + 3x = 0$$

Substituting, we have

$$-8\left(-\frac{x^2}{2}\right) + 3x = 0$$

$$4x^2 + 3x = 0$$
$$x(4x + 3) = 0$$

$$x = 0 \quad \text{or} \quad x = -\frac{3}{4}$$

$$y = 0 \quad \text{or} \quad y = -\frac{9}{32}.$$

$$z_{xx}(x, y) = 6x, \ z_{yy}(x, y) = -16, \ z_{xy}(x, y) = 6$$

$$D = 6x(-16) - (6)^2 = -96x - 36$$

At $(0, 0)$, $D = -36 < 0$.
Saddle point at $(0, 0)$.

At $\left(-\frac{3}{4}, -\frac{9}{32}\right)$, $D = 36 > 0$ and $z_{xx}(x, y)$
$= -\frac{9}{2} < 0$.

Relative maximum at $\left(-\frac{3}{4}, -\frac{9}{32}\right)$

37. $f(x, y) = 2x^2 + 4xy + 4y^2 - 3x + 5y - 15$

$$f_x(x, y) = 4x + 4y - 3$$
$$f_y(x, y) = 4x + 8y + 5$$

Solve the system $f_x(x, y) = 0, f_y(x, y) = 0$.

$$4x + 4y - 3 = 0$$
$$4x + 8y + 5 = 0$$

$$-4x - 4y + 3 = 0$$
$$\underline{4x + 8y + 5 = 0}$$
$$4y + 8 = 0$$
$$y = -2$$

$$4x + 4(-2) - 3 = 0$$
$$4x = 11$$

$$x = \frac{11}{4}$$

Therefore, $\left(\frac{11}{4}, -2\right)$ is a critical point.

$$f_{xx}(x, y) = 4$$
$$f_{yy}(x, y) = 8$$
$$f_{xy}(x, y) = 4$$
$$D = 4 \cdot 8 - 4^2 = 16 > 0$$

Since $f_{xx} = 4 > 0$, there is a relative minimum at $\left(\frac{11}{4}, -2\right)$.

39. $f(x, y) = 7x^2 + y^2 - 3x + 6y - 5xy$

$$f_x(x, y) = 14x - 3 - 5y, \ f_y(x, y) = 2y + 6 - 5x$$

$$14x - 5y - 3 = 0$$
$$\underline{-5x + 2y + 6 = 0}$$
$$28x - 10y - 6 = 0$$
$$\underline{-25x + 10y + 30 = 0}$$
$$3x \qquad + 24 = 0$$
$$x = -8$$

$$-5(-8) + 2y + 6 = 0$$
$$2y = -46$$
$$y = -23$$

$$f_{xx}(x, y) = 14, \ f_{yy}(x, y) = 2, \ f_{xy}(x, y) = -5$$

$$D = 14(2) - (-5)^2 = 3 > 0 \text{ and } f_{xx}(x, y) > 0.$$

Relative minimum at $(-8, -23)$

41. $f(x, y) = x^2 y; \ x + y = 4$

1. $g(x) = x + y - 4$

2. $F(x, y, \lambda) = x^2 y - \lambda(x + y - 4)$

3. $F_x(x, y, \lambda) = 2xy - \lambda$
$F_y(x, y, \lambda) = x^2 - \lambda$
$F_\lambda(x, y, \lambda) = -(x + y - 4)$

4. $2xy - \lambda = 0 \quad (1)$
$x^2 - \lambda = 0 \quad (2)$
$x + y - 4 = 0 \quad (3)$

5. $\lambda = 2xy$
$\lambda = x^2$
$2xy = x^2$
$$2xy - x^2 = 0$$
$$x(2y - x) = 0$$
$$x = 0 \quad \text{or} \quad 2y = x$$

Substituting into equation (3) gives

$$y = 4 \quad \text{or} \quad y = \frac{4}{3}.$$

If $y = \frac{4}{3}, x = \frac{8}{3}$.

The critical points are $(0, 4)$ and $\left(\frac{8}{3}, \frac{4}{3}\right)$.

$$f(0, 4) = 0$$

$$f\left(\frac{8}{3}, \frac{4}{3}\right) = \frac{64}{9} \cdot \frac{4}{3} = \frac{256}{27}$$

Therefore, f has a minimum of 0 at $(0, 4)$ and a maximum of $\frac{256}{27}$ at $\left(\frac{8}{3}, \frac{4}{3}\right)$.

43. Let x and y be the numbers such that $x + y = 80$ and $f(x, y) = x^2 y$.

1. $g(x) = x + y - 80$

2. $F(x, y, \lambda) = x^2 y - \lambda(x + y - 80)$

3. $F_x(x, y, \lambda) = 2xy - \lambda$
$F_y(x, y, \lambda) = x^2 - \lambda$
$F_\lambda(x, y, \lambda) = -(x + y - 80)$

4. $2xy - \lambda = 0 \quad (1)$
$x^2 - \lambda = 0 \quad (2)$
$x + y - 80 = 0 \quad (3)$

5. $\lambda = 2xy$
$\lambda = x^2$
$2xy = x^2$
$$2xy - x^2 = 0$$
$$x(2y - x) = 0$$
$$x = 0 \quad \text{or} \quad x = 2y$$

Substituting into equation (3) gives

$$y = 80 \quad \text{or} \quad 2y + y - 80 = 0$$
$$3y = 80$$
$$y = \frac{80}{3}$$
$$\text{and} \quad x = \frac{160}{3}.$$

$$f(0, 80) = 0 \cdot 80^2 = 0$$

$$f\left(\frac{160}{3}, \frac{80}{3}\right) = \frac{(160)^2}{9} \frac{(80)}{3}$$

$$= \frac{2{,}048{,}000}{27} > f(0, 80)$$

f has a maximum at $\left(\frac{160}{3}, \frac{80}{3}\right)$.

Therefore, if $x = \frac{160}{3}$ and $y = \frac{80}{3}$, then $x^2 y$ is maximized.

45. No, a maximum does not exist without the requirement that x and y are positive. Consider maximizing $x^2 y$ with $x + y = 80$. If $x < 0$, then $y = 80 - x > 80$. The values of y and $x^2 y$ will increase as the value of $|x|$ increases, so there is no maximum. A similar situation occurs by xy^2 when $x + y = 50$, by taking $y < 0$ and considering what happens to the values of x and xy^2 as $|y|$ increases.

47. $z = (x, y) = \dfrac{x + 5y}{x - 2y}$

$x = 1, y = -2, dx = -0.04, dy = 0.02$

$$f_x(x, y) = \frac{(x - 2y)(1) - (x + 5y)(1)}{(x - 2y)^2}$$

$$= \frac{-7y}{(x - 2y)^2}$$

$$f_y(x, y) = \frac{(x - 2y)(5) - (x + 5y)(-2)}{(x - 2y)^2}$$

$$= \frac{7x}{(x - 2y)^2}$$

$$dz = \frac{-7y}{(x - 2y)^2} \, dx + \frac{7x}{(x - 2y)^2} \, dy$$

$$= \frac{-7(-2)}{[1 - 2(-2)]^2}(-0.04)$$

$$+ \frac{7(1)}{[1 - 2(-2)]^2}(0.02)$$

$$= -0.0224 + 0.0056 = -0.0168$$

49. Let $z = f(x, y) = \sqrt{x}\, e^y$.
Then

$$dz = f_x(x, y)\, dx + f_y(x, y)\, dy$$

$$= \frac{1}{2}x^{-1/2}e^y\, dx + x^{1/2}e^y\, dy$$

$$= \frac{e^y}{2\sqrt{x}}\, dx + \sqrt{x}e^y\, dy.$$

To approximate $\sqrt{4.06}\,e^{0.04}$, let $x = 4$, $dx = 0.06$, $y = 0$, and $dy = 0.04$.

Therefore,

$$dz = \frac{e^0}{2\sqrt{4}}(0.06) + \sqrt{4}e^0(0.04)$$

$$dz = \frac{1}{4}(0.06) + 2(0.04)$$

$$dz = 0.095.$$

$$f(4.06, 0.04) = f(4, 0) + \Delta z$$
$$\approx f(4, 0) + dz$$
$$= \sqrt{4}e^0 + 0.095$$
$$f(4.06, 0.04) \approx 2.095$$

Using a calculator, $\sqrt{4.06}\,e^{0.04} \approx 2.0972$. The absolute value of the difference of the two results is $|2.095 - 2.0972| = 0.0022$.

51. $\displaystyle\int_1^5 e^{3x+5y}\, dx = \frac{1}{3}e^{3x+5y}\Big|_1^5 = \frac{1}{3}(e^{15+5y} - e^{3+5y})$

53. $\displaystyle\int_1^3 y^2(7x + 11y^3)^{-1/2}\, dy$

$$= \frac{2}{33}(7x + 11y^3)^{1/2}\Big|_1^3$$

$$= \frac{2}{33}[(7x + 297)^{1/2} - (7x + 11)^{1/2}]$$

55. $\displaystyle\int_0^3 \int_0^5 (2x + 6y + y^2)\, dy\, dx$

$$= \int_0^3 \left(2xy + 3y^2 + \frac{1}{3}y^3\right)\Big|_0^5 dx$$

$$= \int_0^3 \left[\left(10x + 75 + \frac{125}{3}\right) - 0\right] dx$$

$$= \int_0^3 \left(10x + \frac{350}{3}\right) dx$$

$$= \left(5x^2 + \frac{350}{3}x\right)\Big|_0^3$$

$$= (45 + 350) - 0 = 395$$

57. (See Exercise 51.)

$$\int_1^2 \left[\int_3^5 (e^{2x-7y})\, dx\right] dy$$

$$= \int_1^2 \frac{1}{2}(e^{10-7y} - e^{6-7y})\, dy$$

$$= -\frac{1}{14}(e^{10-7y} - e^{6-7y})\Big|_1^2$$

$$= -\frac{1}{14}(e^{-4} - e^{-8} - e^3 + e^{-1})$$

$$= \frac{e^3 + e^{-8} - e^{-4} - e^{-1}}{14}$$

59. $\displaystyle\int_1^2 \int_1^2 \frac{dx\, dy}{x} = \int_1^2 \ln x\Big|_1^2 dy$

$$= \int_1^2 \ln 2\, dy$$

$$= y \ln 2\Big|_1^2$$

$$= 2\ln 2 - \ln 2$$

$$= \ln 2$$

61. $\displaystyle\iint_R \sqrt{2x + y}\, dx\, dy;\ 1 \le x \le 3, 2 \le y \le 5$

$$\iint_R \sqrt{2x + y}\, dx\, dy$$

$$= \int_1^3 \int_2^5 (2x + y)^{1/2}\, dy\, dx$$

$$= \int_1^3 \frac{2}{3}(2x + y)^{3/2}\Big|_2^5 dx$$

$$= \int_1^3 \frac{2}{3}[(2x + 5)^{3/2} - (2x + 2)^{3/2}]\, dx$$

$$= \frac{2}{15}[(2x + 5)^{5/2} - (2x + 2)^{5/2}]\Big|_1^3$$

$$= \frac{2}{15}(11^{5/2} - 8^{5/2} - 7^{5/2} + 4^{5/2})$$

$$= \frac{2}{15}(11^{5/2} - 8^{5/2} - 7^{5/2} + 32)$$

63. $\iint\limits_{R} ye^{y^2+x}\, dx\, dy; 0 \le x \le 1, 0 \le y \le 1$

$\iint\limits_{R} ye^{y^2+x}\, dx\, dy$

$= \int_{0}^{1} \int_{0}^{1} ye^{y^2+x}\, dy\, dx$

$= \int_{0}^{1} \frac{1}{2} e^{y^2+x} \bigg|_{0}^{1}\, dx$

$= \int_{0}^{1} \frac{1}{2}[e^{1+x} - e^{x}]\, dx$

$= \frac{1}{2}[e^{1+x} - e^{x}] \bigg|_{0}^{1}$

$= \frac{1}{2}[e^2 - e - e + 1]$

$= \frac{e^2 - 2e + 1}{2}$

65. $z = x^2 + y^2; 3 \le x \le 5, 2 \le y \le 4$

$V = \iint\limits_{R} (x^2 + y^2)\, dy\, dx$

$= \int_{3}^{5} \int_{2}^{4} (x^2 + y^2)\, dy\, dx$

$= \int_{3}^{5} \left[x^2 y + \frac{y^3}{3} \right] \bigg|_{2}^{4}\, dx$

$= \int_{3}^{5} \left[4x^2 + \frac{64}{3} - 2x^2 - \frac{8}{3} \right]\, dx$

$= \int_{3}^{5} \left(2x^2 + \frac{56}{3} \right)\, dx = \left(\frac{2x^3}{3} + \frac{56x}{3} \right) \bigg|_{3}^{5}$

$= \frac{250}{3} + \frac{280}{3} - 18 - 56$

$= \frac{308}{3}$

67. $\int_{1}^{2} \int_{2}^{2x^2} y\, dy\, dx$

$= \int_{1}^{2} \frac{1}{2} y^2 \bigg|_{2}^{2x^2}\, dx$

$= \int_{1}^{2} (2x^4 - 2)\, dx$

$= \left(\frac{2}{5} x^5 - 2x \right) \bigg|_{1}^{2}$

$= \left(\frac{64}{5} - 4 \right) - \left(\frac{2}{5} - 2 \right)$

$= \frac{52}{5}$

69. $\int_{0}^{1} \int_{y}^{\sqrt{y}} x\, dx\, dy = \int_{0}^{1} \frac{x^2}{2} \bigg|_{y}^{\sqrt{y}}\, dy$

$= \int_{0}^{1} \frac{1}{2} (y - y^2)\, dy$

$= \frac{1}{2} \left(\frac{y^2}{2} - \frac{y^3}{3} \right) \bigg|_{0}^{1}$

$= \frac{1}{2} \left(\frac{1}{2} - \frac{1}{3} \right) = \frac{1}{12}$

71. $\int_{0}^{8} \int_{x/2}^{4} \sqrt{y^2 + 4}\, dy\, dx$

Change the order of integration.

$\int_{0}^{8} \int_{x/2}^{4} \sqrt{y^2 + 4}\, dy\, dx$

$= \int_{0}^{4} \int_{0}^{2y} \sqrt{y^2 + 4}\, dx\, dy$

$= \int_{0}^{4} (y^2 + 4)^{1/2} x \bigg|_{0}^{2y}\, dy$

$= \int_{0}^{4} [(y^2 + 4)^{1/2}(2y)] - (y^2 + 4)(0)\, dy$

$= \int_{0}^{4} (y^2 + 4)^{1/2}(2y)\, dy$

$= \frac{(y^2 + 4)^{3/2}}{\frac{3}{2}} \bigg|_{0}^{4}$

$= \frac{2}{3}(4^2 + 4)^{3/2} - \frac{2}{3}(0^2 + 4)^{3/2}$

$= \frac{2}{3}(20^{3/2} - 4^{3/2})$

$= \frac{2}{3}(20\sqrt{20} - 8)$

$= \frac{2}{3}(40\sqrt{5} - 8)$

$= \frac{16}{3}(5\sqrt{5} - 1)$

73. $\displaystyle\iint\limits_{R} (2 - x^2 - y^2)\, dy\, dx;\ 0 \le x \le 1, x^2 \le y \le x$

$\displaystyle\int_0^1 \int_{x^2}^{x} (2 - x^2 - y^2)\, dy\, dx$

$\displaystyle = \int_0^1 \left(2y - x^2 y - \frac{y^3}{3}\right)\Big|_{x^2}^{x}\, dx$

$\displaystyle = \int_0^1 \left(2x - x^3 - \frac{x^3}{3} - 2x^2 + x^4 + \frac{x^6}{3}\right) dx$

$\displaystyle = \int_0^1 \left(2x - 2x^2 - \frac{4x^3}{3} + x^4 + \frac{x^6}{3}\right) dx$

$\displaystyle = \left(x^2 - \frac{2x^3}{3} - \frac{x^4}{3} + \frac{x^5}{5} + \frac{x^7}{21}\right)\Big|_0^1$

$\displaystyle = 1 - \frac{2}{3} - \frac{1}{3} + \frac{1}{5} + \frac{1}{21} = \frac{26}{105}$

75. $c(x, y) = 2x + y^2 + 4xy + 25$

(a) $c_x = 2 + 4y$

$\quad c_x(640, 6) = 2 + 4(6)$

$\qquad\qquad = 26$

For an additional 1 MB of memory, the approximate change in cost is $26.

(b) $c_y = 2y + 4x$

$\quad c_y(640, 6) = 2(6) + 4(640)$

$\qquad\qquad = 2572$

For an additional hour of labor, the approximate change in cost is $2572.

77. (a) Minimize $c(x, y)$

$\qquad = x^2 + 5y^2 + 4xy - 70x - 164y + 1800$

$c_x = 2x + 4y - 70$

$c_y = 10y + 4x - 164$

$$
\begin{aligned}
2x + 4y - 70 &= 0 \\
4x + 10y - 164 &= 0 \\
\hline
-4x - 8y + 140 &= 0 \\
4x + 10y - 164 &= 0 \\
\hline
2y - 24 &= 0 \\
y &= 12
\end{aligned}
$$

$\qquad 4x + 10(12) - 164 = 0$

$\qquad\qquad\qquad\quad 4x = 44$

$\qquad\qquad\qquad\quad x = 11$

Extremum at $(11, 12)$

$c_{xx} = 2,\ c_{yy} = 10,\ c_{xy} = 4$

For $(11, 12)$,

$D = (2)(10) - 16 = 4 > 0$ and

$c_{xx}(11, 12) = 2 > 0.$

There is a relative minimum at $(11, 12)$.

(b) $c(11, 12)$

$\quad = (11)^2 + 5(12)^2 + 4(11)(12)$

$\qquad - 70(11) - 164(12) + 1800$

$\quad = 121 + 720 + 528 - 770$

$\qquad - 1968 + 1800$

$\quad = \$431$

79. $V = \dfrac{1}{3}\pi r^2 h$

$r = 2\text{ cm},\ h = 8\text{ cm}$

$dr = 0.21\text{ cm},\ dh = 0.21\text{ cm}$

$dV = \dfrac{\pi}{3}(2rh\, dr + r^2\, dh)$

$\quad = \dfrac{\pi}{3}[2(2)(8)(0.21) + 4(0.21)]$

$\quad = \dfrac{\pi}{3}(6.72 + 0.84)$

$\quad = \dfrac{\pi}{3}(7.56)$

$\quad \approx 7.92\text{ cm}^3$

81. $V = \dfrac{1}{3}\pi r^2 h$

$r = 2.9\text{ cm},\ h = 11.4\text{ cm}$

$dr = dh = 0.2\text{ cm}$

$dV = \dfrac{\pi}{3}(2rh\, dr + r^2\, dh)$

$\quad = \dfrac{\pi}{3}[2(2.9)(11.4)(0.2) + (2.9)^2(0.2)]$

$\quad = \dfrac{\pi}{3}[13.224 + 1.682)$

$\quad = \dfrac{\pi}{3}(14.906)$

$\quad \approx 15.6\text{ cm}^3$

83. Assume that blood vessels are cylindrical.

$V = \pi r^2 h$

$r = 0.7,\ h = 2.7$

$dr = dh = \pm .1$

$dV = 2\pi rh\, dr + \pi r^2\, dh$

$\quad = 2\pi(0.7)(2.7)(\pm 0.1) + \pi(0.7)^2(\pm 0.1)$

$\quad \approx \pm 1.341$

The possible error is 1.341 cm^3.

85. $L(w, t) = (0.00082t + 0.0955)e^{(\ln w + 10.49)/2.842}$

    (a) $L(450, 4) = [0.00082(4) + 0.0955]e^{(\ln(450)+10.49)/2.842}$

           ≈ 33.982

The length is about 33.98 cm.

    (b) $\quad L_w(w, t) = (0.00082t + 0.0955)e^{(\ln w + 10.49)/2.842}$

$$\cdot \frac{1}{2.842w}$$

$$L_w(450, 7) \approx 0.02723$$

The approximate change in the length of a trout if its weight increases from 450 to 451 g while age is held constant at 7 yr is 0.027 cm.

$$L_t(w, t) = 0.00082e^{(\ln w + 10.49)/2.842}$$
$$L_t(450, 7) \approx 0.2821$$

The approximate change in the length of a trout if its age increases from 7 to 8 yr while weight is held constant at 450 g is 0.28 cm.

87. $f(a, b) = \frac{1}{4}b\sqrt{4a^2 - b^2}$

    (a) $f(3, 2) = \frac{1}{4}(2)\sqrt{4(3)^2 - 2^2}$

$$= \frac{1}{2}\sqrt{32}$$

$$= 2\sqrt{2}$$

$$\approx 2.828$$

The area of the bottom of the planter is approximately 2.828 ft².

    (b) $\quad A = \frac{1}{4}b\sqrt{4a^2 - b^2}$

$$dA = \frac{1}{4}b \cdot \frac{1}{2}(4a^2 - b^2)^{-1/2}(8a)da$$

$$+ \left[\frac{1}{4}b \cdot \frac{1}{2}(4a^2 - b^2)^{-1/2}(-2b)\right.$$

$$\left. + \frac{1}{4}(4a^2 - b^2)^{1/2}\right] db$$

$$dA = \frac{ab}{\sqrt{4a^2 - b^2}}da$$

$$+ \frac{1}{4}\left(\frac{-b^2}{\sqrt{4a^2 - b^2}} + \sqrt{4a^2 - b^2}\right) db$$

If $a = 3$, $b = 2$, $da = 0$, and $db = 0.5$,

$$dA = \frac{1}{4}\left(\frac{-2^2}{\sqrt{4(3)^2 - 2^2}} + \sqrt{4(3)^2 - 2^2}\right)(0.5)$$

$$dA \approx 0.6187.$$

The approximate effect on the area is an increase of 0.6187 ft².

89. Maximize $f(x, y) = xy$, subject to $2x + y = 400$.

1. $g(x, y) = 2x + y - 400$

2. $F(x, y, \lambda) = xy - \lambda(2x + y - 400)$

3. $F_x = y - 2\lambda$
    $F_y = x - \lambda$
    $F_\lambda = -(2x + y - 400)$

4. $y - 2\lambda = 0$
             $x - \lambda = 0$
     $2x + y - 400 = 0$

5. $\lambda = \frac{y}{2}$, $\lambda = x$

$$\frac{y}{2} = x$$

$$y = 2x$$

Substituting into $2x + y - 400$, we have

$$2x + 2x - 400 = 0,$$

so $x = 100, \ y = 200.$

Dimensions are 100 feet by 200 feet for maximum area of 20,000 ft².

Chapter 17 Test

[17.1]

1. Find $f(-2, 1)$ for $f(x, y) = 2x^2 - 4xy + 7y$.

2. Find $g(-1, 3)$ for $g(x, y) = \sqrt{2x^2 - xy^2}$.

3. Complete the ordered triples $(0, 0, \quad)$, $(0, \quad, 0)$ and $(\quad, 0, 0)$ for the plane $2x - 4y + 8z = 8$.

4. Graph the first octant portion of the plane $4x + 2y + 8z = 8$.

5. Let $f(x, y) = x^2 + 2y^2$. Find $\dfrac{f(x + h, y) - f(x, y)}{h}$.

[17.2]

6. Let $z = f(x, y) = 3x^3 - 5x^2y + 4y^2$. Find each of the following.

    (a) $\dfrac{\partial z}{\partial x}$ (b) $\dfrac{\partial z}{\partial y}(1, 1)$ (c) f_{xx}

7. Let $f(x, y) = \dfrac{x}{2x + y^2}$. Find each of the following.

    (a) $f_x(1, -1)$ (b) $f_y(2, 1)$

8. Let $f(x, y) = \sqrt{2x^2 - y^2}$. Find each of the following.

    (a) f_x (b) f_y (c) f_{xy}

9. Let $f(x, y) = xe^{y^2}$. Find each of the following.

    (a) f_x (b) f_y (c) f_{yx}

10. Let $f(x, y) = \ln(2x^2y^2 + 1)$. Find each of the following.

    (a) f_{xx} (b) f_{xy} (c) f_{yy}

11. The production function for a certain country is

$$z = 2x^5y^4,$$

where x represents the amount of labor and y the amount of capital. Find the marginal productivity of

    (a) labor;   (b) capital.

[17.3]

Find all points where the functions defined below have any relative extrema. Find any saddle points.

12. $z = 4x^2 + 2y^2 - 8x$ **13.** $z = 2x^2 - 4xy + y^3$ **14.** $z = 1 - 2y - x^2 - 2xy - 2y^2$

15. A company manufactures two calculator models. The total revenue from x thousand solar calculators and y thousand battery-operated calculators is

$$R(x, y) = 4000 - 5y^2 - 8x^2 - 2xy + 42y + 102x.$$

Find x and y so that revenue is maximized.

[17.4]

16. Use Lagrange multipliers to find extrema of $f(x, y) = x^2 - 6xy$, subject to the constraint $x + y = 7$.

17. Find two numbers whose sum is 30 such that x^2y is maximized.

18. A closed box with square ends must have a volume of 64 cubic inches. Find the dimensions of such a box that has minimum surface area.

[17.5]

19. Use the total differential to approximate the quantity $\sqrt{3.05^2 + 4.02^2}$. Then use a calculator to approximate the quantity and give the absolute value of the difference in the two results to four decimal places.

20. (a) Find dz for $z = e^{x+y} \ln xy$.

(b) Evaluate dz when $x = 1$, $y = 1$, $dx = 0.02$ and $dy = 0.01$.

21. A sphere of radius 3 feet is to receive an insulating coating $\frac{1}{2}$ inch thick. Use differentials to approximate the volume of the coating needed.

[17.6]

22. Evaluate $\int_0^4 \int_1^4 \sqrt{x + 3y}\, dx\, dy$.

23. Find the volume under the surface $z = 2xy$ and above the rectangle with boundaries $0 \le x \le 1$, $0 \le y \le 4$.

24. Evaluate $\int_0^2 \int_{y/2}^3 (x + y)\, dx\, dy$.

25. Use the region R with boundaries $0 \le y \le 2$ and $0 \le x \le y$ to evaluate

$$\iint_R (6 - x - y)\, dx\, dy.$$

Chapter 17 Test Answers

1. 23

2. $\sqrt{11}$

3. $(0,0,1)$, $(0,-2,0)$, $(4,0,0)$

4.

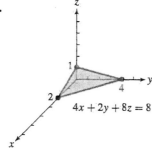

$4x + 2y + 8z = 8$

5. $2x + h$

6. (a) $9x^2 - 10xy$ (b) 3 (c) $18x - 10y$

7. (a) $\frac{1}{9}$ (b) $\frac{-4}{25}$

8. (a) $\frac{2x}{\sqrt{2x^2 - y^2}}$ (b) $\frac{-y}{\sqrt{2x^2 - y^2}}$ (c) $\frac{2xy}{(2x^2 - y^2)^{3/2}}$

9. (a) e^{y^2} (b) $2xye^{y^2}$ (c) $2ye^{y^2}$

10. (a) $\frac{-8x^2y^4 + 4y^2}{(2x^2y^2 + 1)^2}$ (b) $\frac{8xy}{(2x^2y^2 + 1)^2}$ (c) $\frac{4x^2 - 8x^4y^2}{(2x^2y^2 + 1)^2}$

11. (a) $10x^4y^4$ (b) $8x^5y^3$

12. Relative minimum of -4 at $(1,0)$

13. Relative minimum of $-\frac{32}{27}$ at $\left(\frac{4}{3}, \frac{4}{3}\right)$, saddle point at $(0,0)$

14. Relative maximum of 2 at $(1,-1)$

15. 6 thousand solar calculators and 3 thousand battery-operated calculators

16. Minimum of -63 at $(3,4)$

17. $x = 20$, $y = 10$

18. 4 inches by 4 inches by 4 inches

19. 5.046; 5.0461; 0.0001

20. (a) $\left(e^{x+y}\ln(xy) + \frac{e^{x+y}}{x}\right) dx + \left(e^{x+y}\ln xy + \frac{e^{x+y}}{y}\right) dy$ (b) $0.03e^2 \approx 0.222$

21. 4.71 cu ft

22. $\frac{4}{45}\left(993 - 13^{5/2}\right)$

23. 8

24. $\frac{40}{3}$

25. 8

PROBABILITY AND CALCULUS

18.1 Continuous Probability Models

1. $f(x) = \dfrac{1}{9}x - \dfrac{1}{18}$; $[2, 5]$

Show that condition 1 holds.

Since $2 \le x \le 5$,

$$\frac{2}{9} \le \frac{1}{9}x \le \frac{5}{9}$$

$$\frac{1}{6} \le \frac{1}{9}x - \frac{1}{18} \le \frac{1}{2}.$$

Hence, $f(x) \ge 0$ on $[2, 5]$.

Show that condition 2 holds.

$$\int_2^5 \left(\frac{1}{9}x - \frac{1}{18}\right) dx = \frac{1}{9}\int_2^5 \left(x - \frac{1}{2}\right) dx$$

$$= \frac{1}{9}\left(\frac{x^2}{2} - \frac{1}{2}x\right)\Big|_2^5$$

$$= \frac{1}{9}\left(\frac{25}{2} - \frac{5}{2} \cdot \frac{4}{2} + 1\right)$$

$$= \frac{1}{9}(8 + 1)$$

$$= 1$$

Yes, $f(x)$ is a probability density function.

3. $f(x) = \dfrac{1}{21}x^2$; $[1, 4]$

Since $x^2 \ge 0$, $f(x) \ge 0$ on $[1, 4]$.

$$\frac{1}{21}\int_1^4 x^2\, dx = \frac{1}{21}\left(\frac{x^3}{3}\right)\Big|_1^4$$

$$= \frac{1}{21}\left(\frac{64}{3} - \frac{1}{3}\right) = 1$$

Yes, $f(x)$ is a probability density function.

5. $f(x) = 4x^3$; $[0, 3]$

$$4\int_0^3 x^3\, dx = 4\left(\frac{x^4}{4}\right)\Big|_0^3$$

$$= 4\left(\frac{81}{4} - 0\right)$$

$$= 81 \neq 1$$

No, $f(x)$ is not a probability density function.

7. $f(x) = \dfrac{x^2}{16}$; $[-2, 2]$

$$\frac{1}{16}\int_{-2}^2 x^2\, dx = \frac{1}{16}\left(\frac{x^3}{3}\right)\Big|_{-2}^2$$

$$= \frac{1}{16}\left(\frac{8}{3} + \frac{8}{3}\right)$$

$$= \frac{1}{3} \neq 1$$

No, $f(x)$ is not a probability density function.

9. $f(x) = \dfrac{5}{3}x^2 - \dfrac{5}{90}$; $[-1, 1]$

Let $x = 0$. Then $f(x) = f(0) = -\dfrac{5}{90} < 0$.

So $f(x) < 0$ for at least one x-value in $[-1, 1]$.
No, $f(x)$ is not a probability density function.

11. $f(x) = kx^{1/2}$; $[1, 4]$

$$\int_1^4 kx^{1/2}\, dx = \frac{2}{3}kx^{3/2}\Big|_1^4$$

$$= \frac{2}{3}k(8 - 1)$$

$$= \frac{14}{3}k$$

If $\frac{14}{3}k = 1$,

$$k = \frac{3}{14}.$$

Notice that $f(x) = \frac{3}{4}x^{1/2} \ge 0$ for all x in $[1, 4]$.

13. $f(x) = kx^2$; $[0, 5]$

$$\int_0^5 kx^2\, dx = k\frac{x^3}{3}\Big|_0^5$$

$$= k\left(\frac{125}{3} - 0\right)$$

$$= k\left(\frac{125}{3}\right)$$

If $k\left(\frac{124}{3}\right) = 1$,

$$k = \frac{3}{125}.$$

Notice that $f(x) = \frac{3}{125}x^2 \ge 0$ for all x in $[0, 5]$.

15. $f(x) = kx;\ [0, 3]$

$$\int_0^3 kx\,dx = k\frac{x^2}{2}\Big|_0^3$$

$$= k\left(\frac{9}{2} - 0\right)$$

$$= \frac{9}{2}k$$

If $\frac{9}{2}k = 1$,

$$k = \frac{2}{9}.$$

Notice that $f(x) = \frac{2}{9}x \geq 0$ for all x in $[0, 3]$.

17. $f(x) = kx;\ [1, 5]$

$$\int_1^5 kx\,dx = k\frac{x^2}{2}\Big|_1^5$$

$$= k\left(\frac{25}{2} - \frac{1}{2}\right)$$

$$= 12k$$

If $12k = 1$,

$$k = \frac{1}{12}.$$

Notice that $f(x) = \frac{1}{12}x \geq 0$ for all x in $[1, 5]$.

19. For the probability density function $f(x) = \frac{1}{9}x - \frac{1}{18}$ on $[2, 5]$, the cumulative distribution function is

$$F(x) = \int_a^x f(t)\,dt$$

$$= \int_a^x \left(\frac{1}{9}t - \frac{1}{18}\right)dt$$

$$= \left(\frac{1}{18}t^2 - \frac{1}{18}t\right)\Big|_2^x$$

$$= \frac{1}{18}[(x^2 - x) - (4 - 2)]$$

$$= \frac{1}{18}(x^2 - x - 2),\ 2 \leq x \leq 5.$$

21. For the probability density function $f(x) = \frac{x^2}{21}$ on $[1, 4]$, the cumulative distribution function is

$$F(x) = \int_1^x \frac{t^2}{21}\,dt$$

$$= \frac{t^3}{63}\Big|_1^x$$

$$= \frac{1}{63}(x^3 - 1),\ 1 \leq x \leq 4.$$

23. The value of k was found to be $\frac{3}{14}$. For the probability density function $f(x) = \frac{3}{14}x^{1/2}$ on $[1, 4]$, the cumulative distribution function is

$$F(x) = \int_1^x \frac{3}{14}t^{1/2}\,dt$$

$$= \frac{3}{14}\cdot\frac{2}{3}t^{3/2}\Big|_1^x$$

$$= \frac{1}{7}(x^{3/2} - 1),\ 1 \leq x \leq 4.$$

25. The total area under the graph of a probability density function always equals 1.

29. $f(x) = \frac{1}{2}(1+x)^{-3/2};\ [0, \infty)$

$$\frac{1}{2}\int_0^\infty (1+x)^{-3/2}\,dx$$

$$= \lim_{a\to\infty}\frac{1}{2}\int_0^a (1+x)^{-3/2}\,dx$$

$$= \lim_{a\to\infty}\frac{1}{2}(1+x)^{-1/2}\left(\frac{-2}{1}\right)\Big|_0^a$$

$$= \lim_{a\to\infty}[-(1+a)^{-1/2} + 1]$$

$$= \lim_{a\to\infty}\left(\frac{-1}{\sqrt{1+a}} + 1\right)$$

$$= 0 + 1 = 1$$

Since $x \geq 0,\ f(x) \geq 0$.
$f(x)$ is a probability density function.

(a) $P(0 \leq X \leq 2)$

$$= \frac{1}{2}\int_0^2 (1+x)^{-3/2}\,dx$$

$$= -(1+x)^{-1/2}\Big|_0^2$$

$$= -3^{-1/2} + 1$$

$$\approx 0.4226$$

(b) $P(1 \leq X \leq 3)$

$$= \frac{1}{2}\int_1^3 (1+x)^{-3/2}\,dx$$

$$= -(1+x)^{-1/2}\Big|_1^3$$

$$= -4^{-1/2} + 2^{-1/2}$$

$$\approx 0.2071$$

(c) $P(X \geq 5)$

$$= \frac{1}{2} \int_5^\infty (1+x)^{-3/2} \, dx$$

$$= \lim_{a \to \infty} \frac{1}{2} \int_5^a (1+x)^{-3/2} \, dx$$

$$= \lim_{a \to \infty} \left[-(1+x)^{-1/2} \right]\Big|_5^a$$

$$= \lim_{a \to \infty} \left[-(1+a)^{-1/2} + 6^{-1/2} \right]$$

$$= \lim_{a \to \infty} \left(\frac{-1}{\sqrt{1+a}} + 6^{-1/2} \right)$$

$$\approx 0 + 0.4082$$

$$= 0.4082$$

31. $f(x) = \dfrac{1}{2} e^{-x/2}; \; [0, \infty)$

$$\frac{1}{2} \int_0^\infty e^{-x/2} \, dx$$

$$= \lim_{a \to \infty} \frac{1}{2} \int_0^a e^{-x/2} \, dx$$

$$= \lim_{a \to \infty} \frac{1}{2} \left(\frac{-2}{1} e^{-x/2} \right)\Big|_0^a$$

$$= \lim_{a \to \infty} -e^{-x/2} \Big|_0^a$$

$$= \lim_{a \to \infty} \left(\frac{-1}{e^{a/2}} + 1 \right)$$

$$= 0 + 1$$

$$= 1$$

$f(x) > 0$ for all x.
$f(x)$ is a probability density function.

(a) $P(0 \leq X \leq 1) = \dfrac{1}{2} \int_0^1 e^{-x/2} \, dx$

$$= -e^{-x/2} \Big|_0^1$$

$$= \frac{-1}{e^{1/2}} + 1$$

$$\approx 0.3935$$

(b) $P(1 \leq X \leq 3) = \dfrac{1}{2} \int_1^3 e^{-x/2} \, dx$

$$= -e^{-x/2} \Big|_1^3$$

$$= \frac{-1}{e^{3/2}} + \frac{1}{e^{1/2}}$$

$$\approx 0.3834$$

(c) $P(X \geq 2) = \dfrac{1}{2} \int_2^\infty e^{-x/2} \, dx$

$$= \lim_{a \to \infty} \frac{1}{2} \int_2^a e^{-x/2} \, dx$$

$$= \lim_{a \to \infty} (-e^{-x/2}) \Big|_2^a$$

$$= \lim_{a \to \infty} \left(\frac{-1}{e^{a/2}} + \frac{1}{e} \right)$$

$$\approx 0.3679$$

33. $f(x) = \begin{cases} \dfrac{x^3}{12} & \text{if } 0 \leq x \leq 2 \\[2mm] \dfrac{16}{3x^3} & \text{if } x > 2 \end{cases}$

First, note that $f(x) > 0$ for $x > 0$. Next,

$$\int_0^\infty f(x)dx = \int_0^2 \frac{x^3}{12}dx + \lim_{a \to \infty} \int_2^a \frac{16}{3x^3}dx$$

$$= \left(\frac{x^4}{48} \right)\Big|_0^2 + \lim_{a \to \infty} \left(-\frac{8}{3x^2} \right)\Big|_2^a$$

$$= \left(\frac{1}{3} - 0 \right) + \left[\lim_{a \to \infty} \left(-\frac{8}{3a^2} \right) - \left(-\frac{8}{12} \right) \right]$$

$$= \frac{1}{3} + \frac{2}{3}$$

$$= 1.$$

Therefore, $f(x)$ is a probability density function.

(a) $P(0 \leq X \leq 2) = \displaystyle\int_0^2 f(x)dx$

$$= \left(\frac{x^4}{48} \right)\Big|_0^2$$

$$= \frac{1}{3}$$

(b) $P(X \geq 2) = P(X > 2)$

$$= \int_2^\infty \frac{16}{3x^3} \, dx$$

$$= \lim_{a \to \infty} \int_2^a \frac{16}{3x^3} \, dx$$

$$= \lim_{a \to \infty} \left(-\frac{8}{3x^2} \right)\Big|_2^a$$

$$= \lim_{a \to \infty} \left(-\frac{8}{3a^2} \right) - \left(-\frac{8}{3 \cdot 2^2} \right)$$

$$= 0 - \left(-\frac{2}{3} \right)$$

$$= \frac{2}{3}$$

(c) $P(1 \le X \le 3) = \int_1^2 \frac{x^3}{12}\, dx + \int_2^3 \frac{16}{3x^3}\, dx$

$= \left(\frac{x^4}{48}\right)\Big|_1^2 + \left(-\frac{8}{3x^2}\right)\Big|_2^3$

$= \left(\frac{1}{3} - \frac{1}{48}\right) + \left(-\frac{8}{27} + \frac{2}{3}\right)$

$= \frac{295}{432}$

35. $f(x) = \frac{1}{2}e^{-x/2}$; $[0, \infty)$

(a) $P(0 \le X < 12) = \frac{1}{2}\int_0^{12} e^{-x/2}\, dx$

$= -e^{-x/2}\Big|_0^{12}$

$= \frac{-1}{e^6} + 1$

≈ 0.9975

(b) $P(12 \le X \le 20) = \frac{1}{2}\int_{12}^{20} e^{-x/2}\, dx$

$= -e^{-x/2}\Big|_{12}^{20}$

$= \frac{-1}{e^{10}} + \frac{1}{e^6}$

≈ 0.0024

(c) $F(x) = \int_0^x \frac{1}{2} e^{-1/2}\, dt$

$= \frac{1}{2}(-2)e^{-t/2}\Big|_0^x$

$= -(e^{-x/2} - 1)$

$= 1 - e^{-x/2}, x \ge 0.$

(d) $F(6) = 1 - e^{-6/2}$

$= 1 - e^{-3}$

≈ 0.9502

The probability is 0.9502.

37. If $f(x)$ is proportional to $(10+x)^{-2}$, then, for some value of k, $f(x) = k(10+x)^{-2}$ on $[0, 40]$. Find k. We know the total probability must equal 1.

$\int_0^{10} k(10+x)^{-2}\, dx = -k(10+x)^{-1}\Big|_0^{40}$

$= -k(50^{-1} - 10^{-1})^{-1}$

$= -k\left(\frac{1}{50} - \frac{1}{10}\right)$

$= \frac{2}{25}x$

If $\frac{2}{25}k = 1$, then $k = \frac{25}{2}$. Therefore

$$f(x) = \frac{25}{2}(10+x)^{-2}, 0 \le x \le 40$$

So the probability density function is

$F(x) = \int_0^x \frac{25}{2}(10+t)^{-2}\, dt$

$= -\frac{25}{2}(10+t)^{-1}\Big|_0^x$

$= -\frac{25}{2}[(10+x)^{-1} - 10^{-1}]$

$= -\frac{25}{2}\left(\frac{1}{10+x} - \frac{1}{10}\right)$

$= \frac{25}{2}\left(\frac{1}{10} - \frac{1}{10+x}\right)$

$F(6) = \frac{25}{2}\left(\frac{1}{10} - \frac{1}{106}\right)$

≈ 0.47

The correct answer choice is **c**.

39. $f(x) = \frac{1}{2\sqrt{x}}$; $[1, 4]$

(a) $P(3 \le X \le 4) = \int_3^4 \left(\frac{1}{2\sqrt{x}}\right) dx$

$= \frac{1}{2}\int_3^4 x^{-1/2}\, dx$

$= \frac{1}{2}(2)x^{1/2}\Big|_3^4$

$= 2 - 3^{1/2} \approx 0.2679$

(b) $P(1 \le X \le 2) = \int_1^2 \left(\frac{1}{2\sqrt{x}}\right) dx$

$= \frac{1}{2}(2)x^{1/2}\Big|_1^2$

$= 2^{1/2} - 1 = 0.4142$

(c) $P(2 \le X \le 3) = \int_2^3 \left(\frac{1}{2\sqrt{x}}\right) dx$

$= \frac{1}{2}(2)x^{1/2}\Big|_2^3$

$= 3^{1/2} - 2^{1/2} = 0.3178$

41. $f(x) = 1.185 \cdot 10^{-9}x^{4.5222} - 0.049846x$

   (a) $P(0 \le X \le 150) = \displaystyle\int_0^{150} 1.185 \cdot 10^{-9}x^{4.5222}e^{-0.049846x}\,dx \approx 0.8131$

   (b) $P(100 \le x \le 200) = \displaystyle\int_{100}^{200} 1.185 \cdot 10^{-9}x^{4.5222}e^{-0.049846x}\,dx \approx 0.4901$

43. (a)

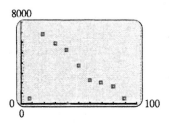

Of the types of functions available using the regression feature of a graphing utility, a polynomial function best matches the data.

(b) The function

$$N(x) = -0.00272454x^4 + 0.614038x^3 - 48.0160x^2 + 1418.53x - 7202.78$$

provided by the calculator models this data well, as illustrated by the following graph.

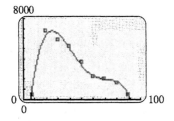

(c) Using the integration feature on our calculator, we find that

$$\int_{6.4}^{91.5} N(x)\,dx \approx 304{,}337.$$

So,

$$S(x) = \frac{1}{304{,}337}N(x) = \frac{1}{304{,}337}(-0.00272454x^4 + 0.614038x^3 - 48.0160x^2 + 1418.53x - 7202.78)$$

will be a probability density function for $[6.4, 91.5]$, because

$$\int_{6.4}^{91.5} S(x)\,dx = 1, \text{ and } S(x) \ge 0$$

for all x in $[6.4, 91.5]$.

(d) Again, using the integration feature on our calculator,

$$P(X < 25) = \int_{6.4}^{25} S(x)\,dx \approx 0.2917$$

$$P(45 < X < 65) = \int_{45}^{65} S(x)\,dx \approx 0.1919$$

$$P(X > 75) = \int_{75}^{91.5} S(x)\,dx \approx 0.0728.$$

From the table, the actual probabilities are

$$P(X < 25) = \frac{489 + 6795}{28,854} \approx 0.2524$$

$$P(45 < X < 65) = \frac{3722 + 2306}{28,854} \approx 0.2089$$

$$P(X > 75) = \frac{1738 + 561}{28,854} \approx 0.0797.$$

45. $f(x) = \dfrac{5.5 - x}{15};\ [0,\ 5]$

(a) $P(3 \le X \le 5) = \displaystyle\int_3^5 \frac{5.5 - x}{15}\, dx = 0.2 = \left(\frac{5.5}{15}x - \frac{1}{15} \cdot \frac{x^2}{2} \right)\Big|_3^5 = \left(\frac{5.5}{15} \cdot 5 - \frac{1}{15} \cdot \frac{5^2}{2} \right) - \left(\frac{5.5}{15} \cdot 3 - \frac{1}{15} \cdot \frac{3^2}{2} \right)$

(b) $P(0 \le X \le 2) = \displaystyle\int_0^2 \frac{5.5 - x}{15}\, dx = \left(\frac{5.5}{15}x - \frac{1}{15} \cdot \frac{x^2}{2} \right)\Big|_0^2 = -\left(\frac{5.5}{15} \cdot 2 - \frac{1}{15} \cdot \frac{2^2}{2} \right) - \left(\frac{5.5}{15} \cdot 0 - \frac{1}{15} \cdot \frac{0^2}{2} \right) = 0.6$

(c) $P(1 \le X \le 4) = \displaystyle\int_1^4 \frac{5.5 - x}{15}\, dx = \left(\frac{5.5}{15}x - \frac{1}{15} \cdot \frac{x^2}{2} \right)\Big|_1^4 = \left(\frac{5.5}{15} \cdot 4 - \frac{1}{15} \cdot \frac{4^2}{2} \right) - \left(\frac{5.5}{15} \cdot 1 - \frac{1}{15} \cdot \frac{1^2}{2} \right) = 0.6$

47. $f(t) = \dfrac{1}{3650.1} e^{-t/3650.1}$

(a) $P(365 < T < 1095) = \displaystyle\int_{365}^{1095} \frac{1}{3650.1} e^{-t/3650.1}\, dt = (-e^{-t/3650.1})\Big|_{365}^{1095} = -e^{-1095/3650.1} + e^{-365/3650.1} \approx 0.16$

(b) $P(T > 7300) = \displaystyle\int_{7300}^{\infty} \frac{1}{3650.1} e^{-t/3650.1}\, dT = \lim_{b \to \infty} \int_{7300}^{b} \frac{1}{3650.1} e^{-t/3650.1}\, dt = \lim_{b \to \infty} (-e^{-t/3650.1})\Big|_{7300}^{b}$

$$= \lim_{b \to \infty} (-e^{-b/3650.1} + e^{-7300/3650.1}) = 0 + e^{-7300/3650.1} \approx 0.14$$

49. $f(x) = 0.06049 e^{-0.03211};\ [16, 84]$

(a) $P(16 \le X \le 25) = \displaystyle\int_{16}^{25} f(x)\, dx = \int_{16}^{25} 0.06049 e^{-0.03211x}\, dx = \frac{0.06049}{-0.03211} (e^{-0.03211x})\Big|_{16}^{25}$

$$\approx -1.88384(e^{-0.03211x})\Big|_{16}^{25} = -1.88384(e^{-0.03211 \cdot 25} - e^{-0.03211 \cdot 16})$$

$$\approx 0.2829$$

(b) $P(35 \le X \le 84) = \displaystyle\int_{35}^{84} 0.06049 e^{-0.03211x}\, dx \approx -1.88384(e^{-0.03211x})\Big|_{35}^{84} \approx 0.4853$

(c) $P(21 \le X \le 30) = \displaystyle\int_{21}^{30} 0.06049 e^{-0.03211x}\, dx \approx -1.88384(e^{-0.03211x})\Big|_{21}^{30} \approx 0.2409$

(d) $F(x) = \displaystyle\int_{16}^{x} 0.06049 e^{-0.03211t}\, dt = 0.06049 \cdot \frac{1}{-0.03211t} e^{-0.03211t}\Big|_{16}^{x}$

$$= -1.8838(e^{-0.03211x} - e^{-0.03211 \cdot 16}) = 1.8838(0.5982 - e^{-0.03211x}), 16 \le x \le 84$$

(e) $F(21) = 1.8838(0.5982 - e^{-0.03211 \cdot 21}) = 1.8838(0.0887) = 0.1671$

The probability is 0.1671.

18.2 Expected Value and Variance of Continuous Random Variables

Use the alternative formula to find

$$\text{Var}(X) = \int_2^6 x^2 \left(\frac{x}{8} - \frac{1}{4} \right) dx - \left(\frac{14}{3} \right)^2$$

$$= \int_2^6 \left(\frac{x^3}{8} - \frac{x^2}{4} \right) dx - \frac{196}{9}$$

$$= \left(\frac{x^4}{32} - \frac{x^3}{12} \right) \Big|_2^6 - \frac{196}{9}$$

$$= \left(\frac{1296}{32} - \frac{216}{12} \right) - \left(\frac{16}{32} - \frac{8}{12} \right) - \frac{196}{9}$$

$$\approx 0.89.$$

$$\sigma = \sqrt{\text{Var}(X)} \approx \sqrt{0.89} \approx 0.94$$

1. $f(x) = \frac{1}{4}$; [3, 7]

$$E(X) = \mu = \int_3^7 \frac{1}{4} x \, dx = \frac{1}{4} \left(\frac{x^2}{2} \right) \Big|_3^7$$

$$= \frac{49}{8} - \frac{9}{8}$$

$$= 5$$

$$\text{Var}(X) = \int_3^7 (x-5)^2 \left(\frac{1}{4} \right) dx$$

$$= \frac{1}{4} \cdot \frac{(x-5)^3}{3} \Big|_3^7$$

$$= \frac{8}{12} + \frac{8}{12}$$

$$= \frac{4}{3} \approx 1.33$$

$$\sigma \approx \sqrt{\text{Var}(X)}$$

$$= \sqrt{\frac{4}{3}}$$

$$\approx 1.15$$

3. $f(x) = \frac{x}{8} - \frac{1}{4}$; [2, 6]

$$\mu = \int_2^6 x \left(\frac{x}{8} - \frac{1}{4} \right) dx$$

$$= \int_2^6 \left(\frac{x^2}{8} - \frac{x}{4} \right) dx$$

$$= \left(\frac{x^3}{24} - \frac{x^2}{8} \right) \Big|_2^6$$

$$= \left(\frac{216}{24} - \frac{36}{8} \right) - \left(\frac{8}{24} - \frac{4}{8} \right)$$

$$= \frac{208}{24} - 4$$

$$= \frac{26}{3} - 4$$

$$= \frac{14}{3} \approx 4.67$$

5. $f(x) = 1 - \frac{1}{\sqrt{x}}$; [1, 4]

$$\mu = \int_1^4 x(1 - x^{-1/2}) dx = \int_1^4 (x - x^{1/2}) dx$$

$$= \left(\frac{x^2}{2} - \frac{2x^{3/2}}{3} \right) \Big|_1^4 = \frac{16}{2} - \frac{16}{3} - \frac{1}{2} + \frac{2}{3}$$

$$= \frac{17}{6} \approx 2.83$$

$$\text{Var}(X) = \int_1^4 x^2(1 - x^{-1/2}) dx - \left(\frac{17}{6} \right)^2$$

$$= \int_1^4 (x^2 - x^{3/2}) dx - \frac{289}{36}$$

$$= \left(\frac{x^3}{3} - \frac{2x^{5/2}}{5} \right) \Big|_1^4 - \frac{289}{36}$$

$$= \frac{64}{3} - \frac{64}{5} - \frac{1}{3} + \frac{2}{5} - \frac{289}{36}$$

$$\approx 0.57$$

$$\sigma \approx \sqrt{\text{Var}(X)} \approx 0.76$$

7. $f(x) = 4x^{-5}$; [1, ∞)

$$\mu = \int_1^\infty x(4x^{-5}) dx$$

$$= \lim_{a \to \infty} \int_1^a 4x^{-4} \, dx$$

$$= \lim_{a \to \infty} \left(\frac{4x^{-3}}{-3} \right) \Big|_1^a$$

$$= \lim_{a \to \infty} \left(\frac{-4}{3a^3} + \frac{4}{3} \right)$$

$$= \frac{4}{3} \approx 1.33$$

$$\text{Var}(X) = \int_1^\infty x^2(4x^{-5})dx - \left(\frac{4}{3}\right)^2$$

$$= \lim_{a\to\infty} \int_1^a 4x^{-3}\,dx - \frac{16}{9}$$

$$= \lim_{a\to\infty} \left(\frac{4x^{-2}}{-2}\right)\bigg|_1^a - \frac{16}{9}$$

$$= \lim_{a\to\infty} \left(\frac{-2}{a^2} + 2\right) - \frac{16}{9}$$

$$= 2 - \frac{16}{9} = \frac{2}{9} \approx 0.22$$

$$\sigma = \sqrt{\text{Var}(X)} = \sqrt{\frac{2}{9}} \approx 0.47$$

11. $f(x) = \dfrac{\sqrt{x}}{18};\ [0,\ 9]$

(a) $E(X) = \mu = \displaystyle\int_0^9 \frac{x\sqrt{x}}{18}\,dx$

$$= \int_0^9 \frac{x^{3/2}}{18}\,dx$$

$$= \frac{2x^{5/2}}{90}\bigg|_0^9 = \frac{x^{5/2}}{45}\bigg|_0^9$$

$$= \frac{243}{45} = \frac{27}{5} = 5.40$$

(b) $\text{Var}(X) = \displaystyle\int_0^9 \frac{x^2\sqrt{x}}{18}\,dx - \left(\frac{27}{5}\right)^2$

$$= \int_0^9 \frac{x^{5/2}}{18}\,dx - \left(\frac{27}{5}\right)^2$$

$$= \frac{x^{7/2}}{63}\bigg|_0^9 - \left(\frac{27}{5}\right)^2$$

$$= \frac{2187}{63} - \left(\frac{27}{5}\right)^2 \approx 5.55$$

(c) $\sigma = \sqrt{\text{Var}(X)} \approx 2.36$

(d) $P(5.40 < X \le 9)$

$$= \int_{5.4}^9 \frac{x^{1/2}}{18}\,dx$$

$$= \frac{x^{3/2}}{27}\bigg|_{5.4}^9$$

$$= \frac{27}{27} - \frac{(5.4)^{1.5}}{27}$$

$$\approx 0.54$$

(e) $P(5.40 - 2.36 \le X \le 5.40 + 2.36)$

$$= \int_{3.04}^{7.76} \frac{x^{1/2}}{18}\,dx$$

$$= \frac{x^{3/2}}{27}\bigg|_{3.04}^{7.76}$$

$$= \frac{7.76^{3/2}}{27} - \frac{3.04^{3/2}}{27}$$

$$\approx 0.60$$

13. $f(x) = \dfrac{1}{2}x;\ [0,\ 2]$

(a) $E(X) = \mu = \displaystyle\int_0^2 \frac{1}{2}x^2\,dx = \frac{x^3}{6}\bigg|_0^2$

$$= \frac{8}{6} = \frac{4}{3} \approx 1.33$$

(b) $\text{Var}(X) = \displaystyle\int_0^1 \frac{1}{2}x^3\,dx - \frac{16}{9}$

$$= \frac{x^4}{8}\bigg|_0^2 - \frac{16}{9}$$

$$= 2 - \frac{16}{9} = \frac{2}{9} \approx 0.22$$

(c) $\sigma = \sqrt{\text{Var}(X)} = \sqrt{\dfrac{2}{9}} \approx 0.47$

(d) $P\left(\frac{4}{3} < X \le 2\right) = \displaystyle\int_{4/3}^2 \frac{x}{2}\,dx$

$$= \frac{x^2}{4}\bigg|_{4/3}^2$$

$$= 1 - \frac{16}{36} \approx 0.56$$

(e) $P\left(\frac{4}{3} - 0.47 \le X \le \frac{4}{3} + 0.47\right)$

$$= \int_{0.86}^{1.8} \frac{x}{2}\,dx = \frac{x^2}{4}\bigg|_{0.86}^{1.8}$$

$$= \frac{1.8^2}{4} - \frac{0.86^2}{4} \approx 0.63$$

15. $f(x) = \dfrac{1}{4};\ [3,\ 7]$

(a) $m = \text{median:}\ \displaystyle\int_3^m \frac{1}{4}\,dx = \frac{1}{2}$

$$\frac{1}{4}x\bigg|_3^m = \frac{1}{2}$$

$$\frac{m}{4} - \frac{3}{4} = \frac{1}{2}$$

$$m - 3 = 2$$

$$m = 5$$

(b) $E(X) = \mu = 5$ (from Exercise 1)

$$P(X = 5) = \int_5^5 \frac{1}{4} \, dx = 0$$

17. $f(x) = \frac{x}{8} - \frac{1}{4}$; $[2, 6]$

(a) m = median:

$$\int_2^m \left(\frac{x}{8} - \frac{1}{4}\right) dx = \frac{1}{2}$$

$$\left(\frac{x^2}{16} - \frac{x}{4}\right)\Big|_2^m = \frac{1}{2}$$

$$\frac{m^2}{16} - \frac{m}{4} - \frac{1}{4} + \frac{1}{2} = \frac{1}{2}$$

$$m^2 - 4m - 4 + 8 = 8$$

$$m^2 - 4m - 4 = 0$$

$$m = \frac{4 \pm \sqrt{16 + 16(1)}}{2}$$

Reject $\frac{4 - \sqrt{32}}{2}$ since it is not in $[2, 6]$.

$$m = \frac{4 + \sqrt{32}}{2} \approx 4.828$$

(b) $E(X) = \mu = \frac{14}{3}$ (from Exercise 3)

$$P\left(\frac{14}{3} \le X \le 4.828\right)$$

$$= \int_{4.667}^{4.828} \left(\frac{x}{8} - \frac{1}{4}\right) dx$$

$$= \left(\frac{x^2}{16} - \frac{x}{4}\right)\Big|_{4.667}^{4.828}$$

$$= \frac{4.828^2}{16} - \frac{4.828}{4} - \frac{4.667^2}{16} + \frac{4.667}{4}$$

$$\approx 0.0553$$

19. $f(x) = 4x^{-5}$; $[1, \infty)$

(a) m = median:

$$\int_1^m 4x^{-5} \, dx = \frac{1}{2}$$

$$\frac{4x^{-4}}{-4}\Big|_1^m = \frac{1}{2}$$

$$-m^{-4} + 1 = \frac{1}{2}$$

$$1 - \frac{1}{m^4} = \frac{1}{2}$$

$$2m^4 - 2 = m^4$$

$$m^4 = 2$$

$$m = \sqrt[4]{2} \approx 1.189$$

(b) $E(X) = \mu = \frac{4}{3}$ (from Exercise 7)

$$P(1.19 \le X \le \tfrac{4}{3}) \approx \int_{1.189}^{1.333} 4x^{-5} \, dx$$

$$\approx -x^{-4}\Big|_{1.189}^{1.333}$$

$$\approx -\frac{1}{1.333^4} + \frac{1}{1.189^4}$$

$$\approx 0.1836$$

21. $f(x) = \begin{cases} \dfrac{x^3}{12} & \text{if } 0 \le x \le 2 \\[2mm] \dfrac{16}{3x^3} & \text{if } x > 2 \end{cases}$

Expected value:

$$E(X) = \mu = \int_0^\infty x f(x) \, dx$$

$$= \int_0^2 x \left(\frac{x^3}{12}\right) dx + \lim_{a \to \infty} \int_2^a x \left(\frac{16}{3x^3}\right) dx$$

$$= \int_0^2 \frac{x^4}{12} \, dx + \lim_{a \to \infty} \int_2^a \frac{16}{3x^2} \, dx$$

$$= \left(\frac{x^5}{60}\right)\Big|_0^2 + \lim_{a \to \infty} \left(-\frac{16}{3x}\right)\Big|_2^a$$

$$= \left(\frac{8}{15} - 0\right) + \left[\lim_{a \to \infty} \left(-\frac{16}{3a}\right) - \left(-\frac{16}{6}\right)\right]$$

$$= \frac{16}{5}$$

Variance:

$$\text{Var}(X) = \int_0^\infty x^2 f(x) \, dx - \mu^2$$

$$= \int_0^2 x^2 \left(\frac{x^3}{12}\right) dx + \int_2^\infty \left(\frac{16}{3x^3}\right) dx - \left(\frac{16}{5}\right)^2$$

Examine the second integral.

$$\int_2^\infty x^2 \left(\frac{16}{3x^3}\right) dx = \lim_{a \to \infty} \int_2^a x^2 \left(\frac{16}{3x^3}\right) dx$$

$$= \lim_{a \to \infty} \int_2^a \frac{16}{3x} \, dx$$

$$= \lim_{a \to \infty} \frac{16}{3} \ln |a| - \frac{16}{3} \ln |2|$$

Since the limit diverges, neither the variance nor the standard deviation exists.

23. $f(x) = \begin{cases} \dfrac{|x|}{10} & \text{for } -2 \le x \le 4 \\ 0 & \text{otherwise} \end{cases}$

First, note that

$$|x| = \begin{cases} -x \text{ for } -2 \le x \le 0 \\ x \text{ for } 0 \le x \le 4 \end{cases}$$

The expected value is

$$E(X) = \mu = \int_{-2}^{4} x \cdot \frac{|x|}{10}\,dx = \int_{-2}^{0} x \cdot \frac{-x}{10}\,dx + \int_{0}^{4} x \cdot \frac{x}{10}\,dx = \int_{-2}^{0} -\frac{x^2}{10}\,dx + \int_{0}^{4} \frac{x^2}{10}\,dx$$

$$= -\frac{x^3}{30}\bigg|_{-2}^{0} + \frac{x^3}{30}\bigg|_{0}^{4} = -\left(0 - \frac{-8}{30}\right) + \left(\frac{64}{30} - 0\right) = \frac{56}{30} = \frac{28}{15}$$

The correct answer choice is **d**.

25. $f(x) = \dfrac{1}{11}\left(1 + \dfrac{3}{\sqrt{x}}\right);\ [4,\ 9]$

(a) From Exercise 6, $\mu \approx 6.41$ yr.

(b) $\sigma \approx 1.45$ yr.

(c) $P(X > 6.41) = \int_{6.41}^{9} \frac{1}{11}(1 + 3x^{-1/2})dx = \frac{1}{11}(x + 6x^{1/2})\bigg|_{6.41}^{9} = \frac{1}{11}[9 + 18 - 6.41 - 6(6.41)^{1/2}] \approx 0.49$

27. Using the hint, we have

$$\text{loss not paid} = \begin{cases} x \text{ for } 0.6 < x < 2 \\ 2 \text{ for } x > 2 \end{cases}$$

Therefore, the mean of the manufacturer's annual losses not paid will be

$$\mu = \int_{0.6}^{0} x \cdot f(x)dx + \int_{2}^{\infty} 2 \cdot f(x)\,dx = \int_{0.6}^{2} x\,\frac{2.5(0.6)^{2.5}}{x^{3.5}}\,dx + \int_{2}^{\infty} 2\,\frac{2.5(0.6)^{2.5}}{x^{3.5}}\,dx$$

$$= 2.5(0.6)^{2.5}\int_{0.6}^{2} \frac{1}{x^{2.5}}\,dx + 5(0.6)^{2.5}\int_{2}^{\infty} \frac{1}{x^{3.5}}\,dx = 2.5(0.6)^{2.5}\left(\frac{1}{-1.5}\right)\frac{1}{x^{1.5}}\bigg|_{0.6}^{2} + 5(0.6)^{2.5}\left(\frac{1}{-2.5}\right)\frac{1}{x^{2.5}}\bigg|_{2}^{\infty}$$

$$= -\frac{5}{3}(0.6)^{2.5}\left(\frac{1}{2^{1.5}} - \frac{1}{0.6^{1.5}}\right) - 2(0.6)^{2.5}\left(0 - \frac{1}{2^{2.5}}\right) \approx 0.8357 + 0.0986 \approx 0.93$$

The correct answer choice is **c**.

29. Since the probability density function is proportional to $(1+x)^{-4}$, we have $f(x) = k(1+x)^{-4}$, $0 < x < \infty$. To determine k, solve the equation $\int_{0}^{\infty} kf(x)dx = 1$.

$$\int_{0}^{\infty} k(1+x)^{-4}dx = 1$$

$$k\left(-\frac{1}{3}\right)(1+x)^{-3}\bigg|_{0}^{\infty} = 1$$

$$-\frac{k}{3}(0 - 1) = 1$$

$$\frac{k}{3} = 1$$

$$k = 3$$

Thus, $f(x) = 3(1+x)^{-4}, 0 < x < \infty$.

The expected monthly claims are

$$\int_0^\infty x \cdot 3(1+x)^{-4} dx = 3 \int_0^\infty \frac{x}{(1+x)^4} dx$$

The antiderivative can be found using the substitution $u = 1 - x$.

$$\int \frac{x}{(1+x)^4} dx = \int \frac{u-1}{u^4} du$$

$$= \int \left(\frac{1}{u^3} - \frac{1}{u^4} \right) du$$

$$= -\frac{1}{2u^2} + \frac{1}{3u^3}$$

Resubstitute $u = 1 + x$.

$$3 \int_0^\infty \frac{x}{(1+x)^4} dx = 3 \left(-\frac{1}{2(1+x)^2} + \frac{1}{3(1+x)^3} \right) \Big|_0^\infty$$

$$= 3 \left[0 - \left(-\frac{1}{2} + \frac{1}{3} \right) \right]$$

$$= 3 \left(\frac{1}{6} \right)$$

$$= \frac{1}{2}$$

The correct answer choice is **c**.

31. $f(x) = \dfrac{1}{(\ln 20)x}$; [1, 20]

(a) $\mu = \displaystyle\int_1^{20} x \cdot \frac{1}{(\ln 20)x} dx$

$$= \int_1^{20} \frac{1}{\ln 20} dx$$

$$= \frac{x}{\ln 20} \Big|_1^{20}$$

$$= \frac{19}{\ln 20} \approx 6.342 \text{ seconds}$$

(b) $\text{Var}(X) = \displaystyle\int_1^{20} x^2 \cdot \frac{1}{(\ln 20)x} dx - \mu^2$

$$= \int_1^{20} \frac{x}{\ln 20} dx - \mu^2$$

$$= \frac{x^2}{2 \ln 20} \Big|_1^{20} - (6.34)^2$$

$$= \frac{399}{2 \ln 20} - (6.34)^2$$

$$\approx 26.40$$

$$\sigma \approx \sqrt{26.40}$$

$$\approx 5.138 \text{ sec}$$

(c) $P(6.34 - 5.14 < X < 6.34 + 5.14)$
$= P(1.2 < X < 11.48)$

$$= \int_{1.2}^{11.48} \frac{1}{(\ln 20)x} dx$$

$$= \frac{\ln x}{\ln 20} \Big|_{1.2}^{11.48}$$

$$= \frac{1}{\ln 20} (\ln 11.48 - \ln 1.2)$$

$$\approx 0.7538$$

(d) The median clotting time is the value of m such that $\int_a^m f(x) dx = \frac{1}{2}$.

$$\int_1^m \frac{1}{(\ln 20)x} dx = \frac{1}{2}$$

$$\frac{1}{\ln 20} \ln x \Big|_1^m = \frac{1}{2}$$

$$\frac{1}{\ln 20} (\ln m - 0) = \frac{1}{2}$$

$$\ln m = \frac{\ln 20}{2}$$

$$m = e^{\ln 20/2} \approx 4.472$$

33. $f(x) = \dfrac{1}{2\sqrt{x}}$; [1, 4]

(a) $\mu = \displaystyle\int_1^4 x \cdot \frac{1}{2\sqrt{x}} dx$

$$= \int_1^4 \frac{x^{1/2}}{2} dx$$

$$= \frac{x^{3/2}}{3} \Big|_1^4$$

$$= \frac{1}{3}(8 - 1)$$

$$= \frac{7}{3} \approx 2.333 \text{ cm}$$

(b) $\text{Var}(X) = \displaystyle\int_1^4 x^2 \cdot \frac{1}{2\sqrt{x}} dx - \left(\frac{7}{3} \right)^2$

$$= \int_1^4 \frac{x^{3/2}}{2} dx - \frac{49}{9}$$

$$= \frac{x^{5/2}}{5} \Big|_1^4 - \frac{49}{9}$$

$$= \frac{1}{5}(32 - 1) - \frac{49}{9}$$

$$\approx 0.7556$$

$$\sigma = \sqrt{\text{Var}(X)}$$

$$\approx 0.8692 \text{ cm}$$

(c) $P(X > 2.33 + 2(0.87))$
     $= P(x > 4.07)$
     $= 0$

The probability is 0 since two standard deviations falls out of the given interval $[1, 4]$.

(d) The median petal length is the value of m such that $\int_a^m f(x)\,dx = \frac{1}{2}$.

$$\int_1^m \frac{1}{2\sqrt{x}}\,dx = \frac{1}{2}$$

$$\sqrt{x}\Big|_1^m = \frac{1}{2}$$

$$\sqrt{m} - 1 = \frac{1}{2}$$

$$\sqrt{m} = \frac{3}{2}$$

$$m = \frac{9}{4} = 2.25$$

The median petal length is 2.25 cm.

35. $f(x) = 1.185 \cdot 10^{-9} x^{4.5222} e^{-0.049846x}$

$$E(X) = \int_1^{1000} x\,f(x)\,dx$$

Using the integration function on our calculator.

$$E(X) \approx 110.80$$

The expected size is about 111.

37. $f(x) = \dfrac{5.5 - x}{15}$; $[0, 5]$

(a) $\mu = \displaystyle\int_0^5 x\left(\frac{5.5 - x}{15}\right)\,dx$

$$= \int_0^5 \left(\frac{5.5}{15}x - \frac{1}{15}x^2\right)\,dx$$

$$= \frac{5.5}{30}x^2 - \frac{1}{45}x^3 \Big|_0^5$$

$$= \left(\frac{5.5}{30} \cdot 25 - \frac{1}{45} \cdot 125\right) - 0$$

$$\approx 1.806$$

(b) $\mathrm{Var}(X) = \displaystyle\int_0^5 x^2 \left(\frac{5.5 - x}{15}\right)\,dx - \mu^2$

$$= \int_0^5 \left(\frac{5.5}{15}x^2 - \frac{1}{15}x^3\right)\,dx - \mu^2$$

$$= \left(\frac{5.5}{45}x^3 - \frac{1}{60}x^4\right)\Big|_0^5 - \mu^2$$

$$- \frac{5.5}{45} \cdot 125 - \frac{1}{60} \cdot 625 - 0 - \mu^2$$

$$\approx 1.60108$$
$$\sigma = \sqrt{\mathrm{Var}(X)}$$
$$\approx 1.265$$

(c) $P(X \le \mu - \sigma)$
     $= P(X \le 1.806 - 1.265)$
     $= P(X \le 0.541)$

$$= \int_0^{0.541} \frac{5.5 - x}{15}\,dx$$

$$= \left(\frac{5.5}{15}x - \frac{1}{30}x^2\right)\Big|_0^{0.541}$$

$$= \left(\frac{5.5}{15}(0.541) - \frac{1}{30}(0.541)^2 - 0\right)$$

$$\approx 0.1886$$

39. $f(x) = \dfrac{0.1906}{x^{0.5012}}$; or $[16, 44]$

(a) Expected value:

$$E(X) = \mu = \int_{16}^{44} x \frac{0.1906}{x^{0.5012}}\,dx$$

$$= \int_{16}^{44} 0.1906 x^{0.4988}\,dx$$

$$= \frac{0.1906}{1.4988} x^{1.4988}\Big|_{16}^{44}$$

$$= \frac{0.1906}{1.4988}(44^{1.4988} - 16^{1.4988})$$

$$\approx 28.8358$$
$$\approx 28.84 \text{ years}$$

(b) Standard deviation:

$$\text{Var}(X) = \int_{16}^{44} x^2 \frac{0.1906}{x^{0.5012}}\, dx - 28.8358^2 = \int_{16}^{44} 0.1906 x^{1.4988}\, dx - 28.8358^2$$

$$= \frac{0.1906}{2.4988} x^{2.4988} \Big|_{16}^{44} - 28.8358^2 = \frac{0.1906}{2.4988}(44^{2.4988} - 16^{2.4988}) - 28.8358^2$$

$$\approx 65.75$$

$$\sigma = \sqrt{\text{Var}(X)} = \sqrt{65.75} \approx 8.109$$

(c) $P(16 \le X \le 28.8 - 8.1) = \displaystyle\int_{16}^{20.7} \frac{0.1906}{x^{0.5012}}\, dx$

$$= \frac{0.1906}{0.4988} x^{0.4988} \Big|_{16}^{20.7}$$

$$= \frac{0.1906}{0.4988}(20.7^{0.4988} - 16^{0.4988})$$

$$\approx 0.2088$$

(d) To find the median age, find the value of m such that $\int_a^m f(x)dx = \frac{1}{2}$.

$$\int_{16}^{m} \frac{0.1906}{x^{0.5012}}\, dx = \frac{1}{2}$$

$$\frac{0.1906}{0.4988} x^{0.4988} \Big|_{16}^{m} = \frac{1}{2}$$

$$\frac{0.1906}{0.4988}(m^{0.4988} - 16^{0.4988}) = \frac{1}{2}$$

$$m^{0.4988} - 16^{0.4988} = \frac{0.4988}{2 \cdot 0.1906}$$

$$m^{0.4988} = \frac{0.4988}{0.3812} + 16^{0.4988}$$

$$0.4988 \ln m = \ln\left(\frac{0.4988}{0.3812} + 16^{0.4988}\right)$$

$$\ln m = \frac{1}{0.4988} \ln\left(\frac{0.4988}{0.3812} + 16^{0.4988}\right)$$

$$m = e^{\frac{1}{0.4988} \ln\left(\frac{0.4988}{0.3812} + 16^{0.4988}\right)} \approx 28.27$$

The median age is 28.27 years.

18.3   Special Probability Density Functions

1. $f(x) = \dfrac{5}{7}$ for x in $[3, \ 4.4]$

This is a uniform distribution:
$a = 3, \ b = 4.4$.

(a) $\mu = \dfrac{1}{2}(4.4 + 3) = \dfrac{1}{2}(7.4)$

$= 3.7$ cm

(b) $\sigma = \dfrac{1}{\sqrt{12}}(4.4 - 3)$

$= \dfrac{1}{\sqrt{12}}(1.4)$

≈ 0.4041 cm

(c) $P(3.7 < X < 3.7 + 0.4041)$
$= P(3.7 < X < 4.1041$

$= \displaystyle\int_{3.7}^{4.1041} \dfrac{5}{7} \, dx$

$= \dfrac{5}{7}x \Big|_{3.7}^{4.1041}$

≈ 0.2886

3. $f(t) = 4e^{-4t}$ for t in $[0, \ \infty)$

This is an exponential distribution: $a = 4$.

(a) $\mu = \dfrac{1}{4} = 0.25$ year

(b) $\sigma = \dfrac{1}{4} = 0.25$ year

(c) $P(0.25 < T < 0.25 + 0.25)$
$= P(0.25 < T < 0.5)$

$= \displaystyle\int_{0.25}^{0.5} 4e^{-4t} \, dt$

$= -e^{-4t} \Big|_{0.25}^{0.5}$

$= -\dfrac{1}{e^{-2}} + \dfrac{1}{e^{-1}}$

≈ 0.2325

5. $f(t) = \dfrac{e^{-t/3}}{3}$ for t in $[0, \ \infty)$

This is an exponential distribution: $a = \frac{1}{3}$.

(a) $\mu = \dfrac{1}{\frac{1}{3}} = 3$ days

(b) $\sigma = \dfrac{1}{\frac{1}{3}} = 3$ days

(c) $P(3 < T < 3 + 3) = P(3 < T < 6)$

$= \displaystyle\int_{3}^{6} \dfrac{e^{-t/3}}{3} \, dt$

$= -e^{-t/3} \Big|_{3}^{6}$

$= -\dfrac{1}{e^{-2}} + \dfrac{1}{e^{-1}}$

≈ 0.2325

In Exercises 7–13, use the table in the Appendix for areas under the normal curve.

7. $z = 3.50$

Area to the left of $z = 3.50$ is 0.9998. Given mean $\mu = z - 0$, so area to left of μ is 0.5.
Area between μ and z is
$$0.9998 - 0.5 = 0.4998.$$

Therefore, this area represents 49.98% of total area under normal curve.

9. Between $z = 1.28$ and $z = 2.05$

Area to left of $z = 2.05$ is 0.9798 and area to left of $z = 1.28$ is 0.8997.

$$0.9798 - 0.8997 = 0.0801$$

Percent of total area = 8.01%

11. Since $10\% = 0.10$, the z-score that corresponds to the area of 0.10 to the left of z is -1.28.

13. 18% of the total area to the right of z means $1 - 0.18$ of the total area is to the left of z.

$$1 - 0.18 = 0.82$$

The closest z-score that corresponds to the area of 0.82 is 0.92.

19. Let m be the median of the exponential distribution $f(x) = ae^{-ax}$ for $[0, \infty)$.

$$\int_0^m ae^{-ax}\, dx = 0.5$$

$$-e^{-ax}\Big|_0^m = 0.5$$

$$-e^{-am} + 1 = 0.5$$

$$0.5 = e^{-am}$$

$$-am = \ln\ 0.5$$

$$m = -\frac{\ln\ 0.5}{a}$$

or $-am = \ln\ \dfrac{1}{2}$

$$-am = -\ln\ 2$$

$$m = \frac{\ln\ 2}{a}$$

21. The area that is to the left of x is

$$A = \int_{-\infty}^x \frac{1}{\sigma\sqrt{2\pi}} e^{-\frac{(t-\mu)^2}{2\sigma^2}}\, dt.$$

Let $u = \frac{t-\mu}{\sigma}$. Then $du = \frac{1}{\sigma}\, dt$ and $dt = \sigma\, du$. If $t = x$,

$$u = \frac{x - \mu}{\sigma} = z.$$

As $t \to -\infty$, $\mu \to -\infty$.
Therefore,

$$A = \int_{-\infty}^z \frac{1}{\sigma\sqrt{2\pi}} e^{(-1/2)u^2}\sigma\, du$$

$$= \frac{\sigma}{\sigma}\int_{-\infty}^z \frac{1}{\sqrt{2\pi}} e^{-u^2/2}\, du$$

$$= \int_{-\infty}^z \frac{1}{\sqrt{2\pi}} e^{-u^2/2}\, du.$$

This is the area to the left of z for the standard normal curve.

In Exercises 23–25, use Simpson's rule with $n = 100$ or the integration feature on a graphing calculator to approximate the integrals. Answers may vary slightly from those given here depending on the method that is used.

23. (a) $\displaystyle\int_0^{50} 0.5e^{-0.5x}\, dx \approx 1.00000$

(b) $\displaystyle\int_0^{50} 0.5xe^{-0.5x}\, dx \approx 1.99999$

(c) $\displaystyle\int_0^{50} 0.5x^2 e^{-0.5x}\, dx = 8.00003$

25. $\displaystyle\int_{-\infty}^{\infty} \frac{1}{\sqrt{2\pi}} e^{-x^2/2}\, dx$

$$\approx \int_{-4}^4 \frac{1}{\sqrt{2\pi}} e^{-x^2/2}\, dx$$

(a) $\mu = 1.75 \times 10^{-14} \approx 0$

(b) $\sigma = 0.999433 \approx 1$

27. The probability density function for the uniform distribution is $f(x) = \frac{1}{b-a}$ for x in $[a, b]$.

The cumulative distribution function for f is

$$F(x) = P(X \le x)$$

$$= \int_a^x f(t)\, dt$$

$$= \int_a^x \frac{1}{b-a}\, dt$$

$$= \frac{1}{b-a}\, t\Big|_a^x$$

$$= \frac{1}{b-a}(x - a)$$

$$= \frac{x - a}{b - a},\ a \le x \le b.$$

29. For a uniform distribution,

$$f(x) = \frac{1}{b - a} \text{ for } [a,\ b].$$

Thus, we have

$$f(x) = \frac{1}{85 - 10} = \frac{1}{75}$$

for $[10, 85]$.

(a) $\mu = \dfrac{1}{2}(10 + 85) = \dfrac{1}{2}(95)$

$$= 47.5 \text{ thousands}$$

Therefore, the agent sells $47,500 in insurance.

(b) $P(50 < X < 85) = \displaystyle\int_{50}^{85} \frac{1}{75}\, dx$

$$= \frac{x}{75}\Big|_{50}^{85}$$

$$= \frac{85}{75} - \frac{50}{75}$$

$$= \frac{35}{75} = 0.4667$$

31. (a) Since we have an exponential distribution with $\mu = 4.25$,

$$\mu = \frac{1}{a} = 4.25$$

$$a = 0.235.$$

Therefore, $f(x) = 0.235e^{-0.235x}$ on $[0, \infty)$.

(b) $P(X > 10)$

$$= \int_{10}^{\infty} 0.235e^{-0.235x}dx$$

$$= \lim_{a \to \infty} \int_{10}^{a} 0.235e^{-0.235x}dx$$

$$= \lim_{a \to \infty} \left. (-e^{-0.235x}) \right|_{10}^{a}$$

$$= \lim_{a \to \infty} \left(-\frac{1}{e^{0.235a}} + \frac{1}{e^{2.35}} \right)$$

$$= \frac{1}{e^{2.35}} = 0.09537$$

33. (a) $\mu = 2.5$, $\sigma = 0.2$, $x = 2.7$

$$z = \frac{2.7 - 2.5}{0.2} = 1$$

Area to the right of $z = 1$ is

$$1 - 0.8413 = 0.1587.$$

Probability = 0.1587

(b) Within 1.2 standard deviations of the mean is the area between $z = -1.2$ and $z = 1.2$.
Area to left of $z = 1.2 = 0.8849$
Area to the left of $z = -1.2 = 0.1151$

$$0.8849 - 0.1151 = 0.7698$$

Probability = 0.7698

35. If X has a uniform distribution on $[0, 1000]$, then its density function is $f(x) = \frac{1}{1000}$ for x in $[0, 1000]$. The expected payment with no deductible is

$$E(X) = \int_0^{1000} x \cdot \frac{1}{1000}\, dx$$

$$= \frac{1}{1000} \cdot \frac{1}{2}x^2 \Big|_0^{1000}$$

$$= \frac{1}{2000} \cdot (1000^2 - 0)$$

$$= 500.$$

Now, let the deductible be D. According to the hint,

$$\text{payment} = \begin{cases} 0 & \text{for } x \le D \\ x - D & \text{for } x > D. \end{cases}$$

The expected payment with the deductible is therefore

$$E(X) = \int_0^{D} 0 \cdot \frac{1}{1000}\, dx + \int_D^{1000} (x - D) \cdot \frac{1}{1000}\, dx$$

$$= 0 + \frac{1}{1000} \cdot \left(\frac{1}{2}x^2 - Dx \right) \Big|_D^{1000}$$

$$= \frac{1}{1000} \left[\left(\frac{1}{2}1000^2 - 1000D \right) - \left(\frac{1}{2}D^2 - D^2 \right) \right]$$

$$= 500 - D + \frac{1}{2000}D^2.$$

For this amount to be 25% of the amount with no deductible, we must have

$$500 - D + \frac{1}{2000}D^2 = 0.25 \cdot 500$$

$$\frac{1}{2000}D^2 - D + 500 = 125$$

$$\frac{1}{2000}D^2 - D + 375 = 0$$

$$D^2 - 2000D + 750{,}000 = 0$$

$$(D - 1500)(D - 500) = 0$$

$$D = 1500 \quad \text{or} \quad D = 500$$

We reject $D = 1500$ since it is not in $[0, 1000]$. Therefore, $D = 500$. The correct answer choice is c.

37. Let the random variable X be the lifetime of the printer in years. Then it has exponential distribution $f(x) = ae^{-ax}$ for $x \ge 0$.

If the mean is 2 years, then $\frac{1}{a} = 2$, or $a = \frac{1}{2}$ and the function is $f(x) = \frac{1}{2}e^{-x/2}$ for $x \ge 0$.

We wish to find $P(0 \le X \le 1)$ and $P(1 \le X \le 2)$.

$$P(0 \le X \le 1) = \int_0^1 \frac{1}{2}e^{-x/2}dx$$

$$= -e^{-x/2}\Big|_0^1$$

$$= -e^{-1/2} + 1$$

$$P(1 \le X \le 2) = \int_1^2 \frac{1}{2}e^{-x/2}dx$$

$$= -e^{-x/2}\Big|_1^2$$

$$= -e^{-1} + e^{-1/2}$$

If 100 printers are sold, then

$(1 - e^{-1/2} + 1)(100)$ will fail in the first year,

and

$(e^{-1/2} - e^{-1/2})(100)$ will fail in the second year.

The manufacturer pays a full refund on those failing the first year and one-half refund on those failing during the second year.

Refunds $= (1 - e^{-1/2} + 1)(100)(\$200)$
$\quad + (e^{-1/2} - e^{-1})(100)(\$100)$
$\quad \approx \$10,255.90$

The correct answer choice is **d**.

39. For a uniform distribution,

$$f(x) = \frac{1}{b - a} \text{ for } x \text{ in } [a, \ b].$$

$$f(x) = \frac{1}{36 - 20} = \frac{1}{16} \text{ for } x \text{ in } [20, \ 36]$$

(a) $\mu = \frac{1}{2}(20 + 36) = \frac{1}{2}(56)$

$\quad = 28$ days

(b) $P(30 < X \leq 36)$

$$= \int_{30}^{36} \frac{1}{16} \, dx = \frac{1}{16} x \Big|_{30}^{36}$$

$$= \frac{1}{16}(36 - 30)$$

$$= 0.375$$

41. We have an exponential distribution, with a = 1.
$f(t) = e^{-t}, \ [0, \ \infty)$

(a) $\mu = \frac{1}{1} = 1$ hr

(b) $P(T < 30$ min$)$

$$= \int_0^{0.5} e^{-t} \, dt$$

$$= -e^{-t} \Big|_0^{0.5}$$

$$1 - e^{-0.5} \approx 0.3935$$

43. $f(x) = ae^{-ax}$ for $[0, \ \infty)$

Since $\mu = 25$ and $\mu = \frac{1}{a}$,

$$a = \frac{1}{25} = 0.04.$$

This, $f(x) = 0.04e^{-0.04x}$.

(a) We must find t such that $P(X \leq t) = 0.90$.

$$\int_0^t 0.04e^{-0.04x} \, dx = 0.90$$

$$-e^{-0.04x} \Big|_0^t = 0.90$$

$$-e^{-0.04t} + 1 = 0.90$$

$$0.10 = -e^{-0.04t}$$

$$-0.04t = \ln 0.10$$

$$t = \frac{\ln 0.10}{-0.04}$$

$$t \approx 57.56$$

The longest time within which the predator will be 90% certain of finding a prey is approximately 58 min.

(b) $P(X \geq 60)$

$$= \int_{60}^{\infty} 0.04e^{-0.04x} \, dx$$

$$= \lim_{b \to \infty} \int_{60}^{b} 0.04e^{-0.04x} \, dx$$

$$= \lim_{b \to \infty} (-e^{-0.04x}) \Big|_{60}^{b}$$

$$= \lim_{b \to \infty} [-e^{-0.04b} + e^{-0.04(60)}]$$

$$= 0 + e^{-2.4}$$

$$\approx 0.0907$$

The probability that the predator will have to spend more than one hour looking for a prey is approximately 0.09.

45. For an exponential distribution, $f(x) = ae^{-ax}$ for $[0, \infty)$.

Since $\mu = \frac{1}{a} = 12.1$, $a = \frac{1}{12.1}$

(a) $P(X \geq 20) = \int_{20}^{\infty} \frac{1}{12.1} e^{-x/12.1} dx$

$$= \lim_{b \to \infty} \int_{20}^{b} \frac{1}{12.1} e^{-x/12.1} dx$$

$$= \lim_{b \to \infty} \left(-e^{-x/12.1} \Big|_{20}^{b} \right)$$

$$= \lim_{b \to \infty} (-e^{-b/12.1} + e^{-20/12.1})$$

$$= e^{-20/12.1}$$

$$\approx 0.19$$

(b) $P(10 \leq X \leq 20) = \int_{10}^{20} \frac{1}{12.1} e^{-x/12.1} dx$

$$= -e^{-x/12.1} \Big|_{10}^{20}$$

$$= -e^{-20/12.1} + e^{-10/12.1}$$

$$\approx 0.25$$

47. We have an exponential distribution, with $a = 0.229$.
So $f(t) = 0.229 e^{-0.229t}$, for $[0, \infty)$.

(a) The life expectancy is

$$\mu = \frac{1}{a} = \frac{1}{0.229} \approx 4.36 \text{ millennia.}$$

The standard deviation is

$$\sigma = \frac{1}{a} = \frac{1}{0.229} \approx 4.36 \text{ millennia.}$$

(b) $P(T \geq 2) = \int_{2}^{\infty} 0.229 e^{-0.229t} dt$

$$= 1 - \int_{0}^{2} 0.229 e^{-0.229t} dt$$

$$= 1 + \left(e^{-0.229t} \Big|_{0}^{2} \right)$$

$$= 1 + [e^{-0.229(2)} - 1]$$

$$= e^{-0.458}$$

$$\approx 0.63$$

49. For an exponential distribution, $f(x) = ae^{-ax}$ for $[0, \infty)$.

Since $\mu = \frac{1}{a} = 8, a = \frac{1}{8}$.

(a) $P(X \geq 10) = \int_{10}^{\infty} \frac{1}{8} e^{-x/8} dx$

$$= 1 - \int_{0}^{10} \frac{1}{8} e^{-x/8} dx$$

$$= 1 + \left(e^{-x/8} \Big|_{0}^{10} \right)$$

$$= 1 + [e^{-10/8} - 1]$$

$$= e^{-10/8}$$

$$\approx 0.29$$

(b) $P(X < 2) = \int_{0}^{2} \frac{1}{8} e^{-x/8} dx$

$$= -e^{-x/8} \Big|_{0}^{2}$$

$$= -e^{-2/8} + 1$$

$$\approx 0.22$$

51. We have an exponential distribution $f(x) = ae^{-ax}$ for $x \geq 0$. Since $a = \frac{1}{90}$, $f(x) = \frac{1}{90} e^{-x/90}$ for $x \geq 0$.

(a) The probability that the time for a goal is no more than 71 minutes is

$$P(0 < X < 71) = \int_{0}^{71} \frac{1}{90} e^{-x/90} dx$$

$$= -e^{-x/90} \Big|_{0}^{71}$$

$$= -e^{-71/90} + 1$$

$$\approx 0.5457.$$

(b) The probability that the time for a goal is 499 minutes or more is

$$P(X \geq 499) = \int_{499}^{\infty} \frac{1}{90} e^{-x/90} dx$$

$$= -e^{-x/90} \Big|_{499}^{\infty}$$

$$= 0 + e^{-499/90}$$

$$\approx 0.003909.$$

Chapter 18 Review Exercises

1. In a probability function, the y-values (or function values) represent probabilities.

3. A probability density function f for $[a, b]$ must satisfy the following two conditions:

(1) $f(x) \geq 0$ for all x in the interval $[a, b]$;

(2) $\int_{a}^{b} f(x)\, dx = 1$.

5. $f(x) = \sqrt{x}$; $[4, 9]$

$$\int_{4}^{9} x^{1/2}\, dx = \frac{2}{3} x^{3/2} \Big|_{4}^{9}$$

$$= \frac{2}{3}(27 - 8)$$

$$= \frac{38}{3} \neq 1$$

$f(x)$ is not a probability density function.

7. $f(x) = 0.7 e^{-0.7x}$; $[0, \infty)$

$$\int_{0}^{\infty} 0.7 e^{-0.7x}\, dx = -e^{-0.7x} \Big|_{0}^{\infty}$$

$$= \lim_{b \to \infty} (-e^{-0.7b}) + e^{0}$$

$$= \lim_{b \to \infty} \left(-\frac{1}{e^{0.7b}} \right) + 1$$

$$= 0 + 1 = 1$$

$f(x) \geq 0$ for all x in $[0, \infty)$.

Therefore, $f(x)$ is a probability density function.

9. $f(x) = kx^2$; $[1, 4]$

$$\int_1^4 kx^2 \, dx = \left. \frac{kx^3}{3} \right|_1^4$$

$$= 21k$$

Since $f(x)$ is a probability density function,

$$21k = 1$$

$$k = \frac{1}{21}.$$

11. $f(x) = \dfrac{1}{10}$ for $[10, \ 20]$

(a) $P(10 \leq X \leq 12)$

$$= \int_{10}^{12} \frac{1}{10} \, dx$$

$$= \left. \frac{x}{10} \right|_{10}^{12}$$

$$= \frac{1}{5} = 0.2$$

(b) $P\left(\frac{31}{2} \leq X \leq 20\right)$

$$= \int_{31/2}^{20} \frac{1}{10} \, dx$$

$$= \left. \frac{x}{10} \right|_{31/2}^{20}$$

$$= 2 - \frac{31}{20}$$

$$= \frac{9}{20} = 0.45$$

(c) $P(10.8 \leq X \leq 16.2)$

$$= \int_{10.8}^{16.2} \frac{1}{10} \, dx$$

$$= \left. \frac{x}{10} \right|_{10.8}^{16.2} = 0.54$$

13. The distribution that is tallest or most peaked has the smallest standard deviation. This is the distribution pictured in graph (b).

15. $f(x) = \dfrac{2}{9}(x - 2)$; $[2, \ 5]$

(a) $\mu = \displaystyle\int_2^5 \frac{2x}{9}(x - 2)dx$

$$= \int_2^5 \frac{2}{9}(x^2 - 2x)dx$$

$$= \left. \frac{2}{9}\left(\frac{x^3}{3} - x^2\right) \right|_2^5$$

$$= \frac{2}{9}\left(\frac{125}{3} - 25 - \frac{8}{3} + 4\right) = 4$$

(b) $\text{Var}(X) = \displaystyle\int_2^5 \frac{2x^2}{9}(x - 2)dx - (4)^2$

$$= \int_2^5 \frac{2}{9}(x^3 - 2x^2)dx - 16$$

$$= \left. \frac{2}{9}\left(\frac{x^4}{4} - \frac{2x^3}{3}\right) \right|_2^5 - 16$$

$$= \frac{2}{9}\left(\frac{625}{4} - \frac{250}{3} - 4 + \frac{16}{3}\right) - 16$$

$$= 0.5$$

(c) $\sigma = \sqrt{0.5} \approx 0.7071$

(d) $\displaystyle\int_2^m \frac{2}{9}(x - 2) \, dx = \frac{1}{2}$

$$\left. \frac{1}{9}(m - 2)^2 \right|_2^m = \frac{1}{2}$$

$$\frac{1}{9}\left[(m - 2)^2 - 0\right] = \frac{1}{2}$$

$$m^2 - 4m + 4 = \frac{9}{2}$$

$$m^2 - 4m - \frac{1}{2} = 0$$

$$m = \frac{4 \pm 3\sqrt{2}}{2}$$

$$\approx -0.121, 4.121$$

We reject -0.121 since it is not in $[2, 5]$. So, $m = 4.121$

(e) $\displaystyle\int_2^x \frac{2}{9}(t - 2) \, dt = \left. \frac{1}{9}(t - 2)^2 \right|_2^x$

$$= \frac{1}{9}[(x - 2)^2 - 0]$$

$$= \frac{(x - 2)^2}{9}, 2 \leq x \leq 5$$

17. $f(x) = 5x^{-6}$; $[1, \infty)$

(a) $\mu = \int_1^\infty x \cdot 5x^{-6}\, dx = \int_1^\infty 5x^{-5}\, dx$

$= \lim_{b \to \infty} \int_1^b 5x^{-5}\, dx = \lim_{b \to \infty} \frac{5x^{-4}}{-4}\Big|_1^b$

$= \lim_{b \to \infty} \frac{5}{4}\left(1 - \frac{1}{b^4}\right) = \frac{5}{4}$

(b) $\text{Var}(X) = \int_1^\infty x^2 \cdot 5x^{-6}\, dx - \left(\frac{5}{4}\right)^2$

$= \lim_{b \to \infty} \int_1^b 5x^{-4}\, dx - \frac{25}{16}$

$= \lim_{b \to \infty} \frac{5x^{-3}}{-3}\Big|_1^b - \frac{25}{16}$

$= \lim_{b \to \infty} \frac{5}{3}\left(1 - \frac{1}{b^3}\right) - \frac{25}{16}$

$= \frac{5}{3} - \frac{25}{16} = \frac{5}{48} \approx 0.1042$

(c) $\sigma \approx \sqrt{\text{Var}(X)} \approx 0.32$

(d) $\int_1^m 5x^{-6}\, dx = \frac{1}{2}$

$-x^{-5}\Big|_1^m = \frac{1}{2}$

$-m^{-5} + 1 = \frac{1}{2}$

$m^{-5} = \frac{1}{2}$

$m^5 = 2$

$m = \sqrt[5]{2} \approx 1.149$

(e) $\int_1^x 5t^{-6}\, dt = -t^{-5}\Big|_1^x$

$= -x^{-5} + 1$

$= 1 - \frac{1}{x^5},\ x \geq 1$

19. $f(x) = 4x - 3x^2$; $[0, 1]$

$\mu = \int_0^1 x(4x - 3x^2)\, dx$

$= \int_0^1 (4x^2 - 3x^3)\, dx$

$= \left(\frac{4x^3}{3} - \frac{3x^4}{4}\right)\Big|_0^1$

$= \frac{4}{3} - \frac{3}{4} = \frac{7}{12} \approx 0.5833$

Find m such that

$\int_0^m (4x - 3x^2)\, dx = \frac{1}{2}.$

$\int_0^m (4x - 3x^2)\, dx = (2x^2 - x^3)\Big|_0^m$

$= 2m^2 - m^3 = \frac{1}{2}$

Therefore,

$$2m^3 - 4m^2 + 1 = 0.$$

This equation has no rational roots, but trial and error used with synthetic division reveals that $m \approx -0.4516$, 0.5970, and 1.855. The only one of these in $[0, 1]$ is 0.5970.

$P\left(\frac{7}{12} < X < 0.5970\right)$

$= \int_{7/12}^{0.5970} (4x - 3x^2)\, dx$

$= 2x^2 - x^3\Big|_{7/12}^{0.5970}$

$= 2(0.5970)^2 - (0.5970)^3 - 2\left(\frac{7}{12}\right)^2$

$+ \left(\frac{7}{12}\right)^3 \approx 0.0180$

21. $f(x) = 0.01e^{-0.01x}$ for $[0, \infty)$ is an exponential distribution.

(a) $\mu = \frac{1}{0.01} = 100$

(b) $\sigma = \frac{1}{0.01} = 100$

(c) $P(100 - 100 < X < 100 + 100)$
$= P(0 < X < 200)$

$= \int_0^{200} 0.01e^{-0.01x}\, dx$

$= -e^{-0.01x}\Big|_0^{200}$

$= 1 - e^{-2} \approx 0.8647$

For Exercises 23–29, use the table in the Appendix for the areas under the normal curve.

23. Area to the left of $z = -0.43$ is 0.3336.
Percent of area is 33.36%.

25. Area between $z = -1.17$ and $z = -0.09$ is

$$0.4641 - 0.1210 = 0.3431.$$

Percent of area is 34.31%.

27. The region up to 1.2 standard deviations below the mean is the region to the left of $z = -1.2$. The area is 0.1151, so the percent of area is 11.51%.

29. 52% of area is to the right implies that 48% is to the left.

$$P(z < a) = 0.48 \text{ for } a = -0.05$$

Thus, 52% of the area lies to the right of $z = -0.05$.

31. $f(x) = e^{-x}$ for $[0, \infty)$
$f(x) = 1e^{-1x}$

(a) This is an exponential distribution with $a = 1$.

(b) The domain of f is $[0, \infty)$.
The range of f is $(0, 1]$.

(c)

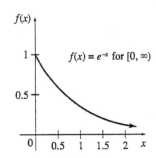

$f(x) = e^{-x}$ for $[0, \infty)$

(d) For an exponential distribution, $\mu = \frac{1}{a}$ and $\sigma = \frac{1}{a}$.
Thus

$$\mu = \frac{1}{1} = 1 \quad \text{and} \quad \sigma = \frac{1}{1} = 1.$$

(e) $P(\mu - \sigma \le X \le \mu + \sigma)$
$= P(1 - 1 \le X \le 1 + 1)$
$= P(0 \le X \le 2)$

$$= \int_0^2 e^{-x}\,dx$$

$$= -e^{-x}\Big|_0^2$$

$$= -e^{-2} + 1$$

$$\approx 0.8647$$

33. $f(x) = \dfrac{xe^{-x/2}}{4}$ for x in $[0, \infty)$

(a) $P(0 \le X \le \infty) = \displaystyle\int_0^\infty \dfrac{xe^{-x/2}}{4}\,dx$

For all $x \ge 0$, $e^{-x/2} > 0$ so that $f(x) \ge 0$ for x in $[0, \infty)$.

Evaluate $\displaystyle\int xe^{-x/2}\,dx$ using integration by parts.

Let $\quad u = x \quad$ and $\quad dv = e^{-x/2}\,dx$
Then $\quad du = dx \quad$ and $\quad v = -2e^{-x/2}$.

$$\frac{1}{4}\int xe^{-x/2}\,dx = \frac{1}{4}\left(-2xe^{-x/2} + \int 2e^{-x/2}\,dx\right)$$

$$= \frac{1}{4}\left(-2xe^{-x/2} - 4e^{-x/2}\right)$$

$$= -\frac{1}{2}xe^{-x/2} - e^{-x/2}$$

Therefore,

$$\int_0^\infty \frac{xe^{-x/2}}{4}\,dx = \left(-\frac{1}{2}xe^{-x/2} - e^{-x/2}\right)\Big|_0^\infty$$

$$= \lim_{b \to \infty}\left(-\frac{1}{2}xe^{-x/2} - e^{-x/2}\right)\Big|_0^b$$

$$= \lim_{b \to \infty}\left(-\frac{1}{2}be^{-b/2} - e^{-b/2}\right) - (0 - 1)$$

Recall from the section on l'Hospital's rule that $\lim_{b \to \infty} be^{-b/2} = 0$. Therefore,

$$\int_0^\infty \frac{xe^{-x/2}}{4}\,dx = 0 - (-1) = 1.$$

(b) $P(0 \le X \le 3) = \displaystyle\int_0^3 \dfrac{xe^{-x/2}}{4}\,dx$

$$= \left(-\frac{1}{2}xe^{-x/2} - e^{-x/2}\right)\Big|_0^3$$

$$= \left(-\frac{3}{2}e^{-3/2} - e^{-3/2}\right) - (0 - 1)$$

$$= 1 - \frac{5}{2}e^{-3/2}$$

$$\approx 0.4422$$

35. $f(x) = \dfrac{3}{4}(x^2 - 16x + 65)$ for $[8, 9]$

$P(8 \leq X < 8.50)$

$$= \int_8^{8.5} \frac{3}{4}(x^2 - 16x + 65)\,dx$$

$$= \frac{3}{4}\left(\frac{x^3}{3} - 8x^2 + 65x\right)\Big|_8^{8.5}$$

$$= \frac{3}{4}\Big[\frac{8.5^3}{3} - 8(8.5)^2 + 65(8.5) - \frac{8^3}{3}$$

$$+ 8(8)^2 - 65(8)\Big]$$

$$\approx 0.406$$

37. (a) $\mu = 8$

$$\frac{1}{a} = 8$$

$$a = \frac{1}{8}$$

$$f(x) = \frac{1}{8}e^{-x/8} \text{ for } [0, \infty)$$

(b) Expected number $= \mu = 8$

(c) $\sigma = \mu = 8$

(d) $P(5 < X < 10) = \displaystyle\int_5^{10} \frac{1}{8}e^{-x/8}\,dx$

$$= -e^{-x/8}\Big|_5^{10}$$

$$= -e^{-10/8} + e^{-5/8}$$

$$\approx 0.2488$$

39. $\mu = 46.2$, $\sigma = 15.8$, $x = 60$

$$z = \frac{x - \mu}{\sigma} = \frac{60 - 46.2}{15.8} \approx 0.8734$$

0.8734 is the z-score for the area of about 0.8078 (from the table).

$$P(X \geq 60) \approx P(z \geq 0.8734) \approx 1 - 0.8078 = 0.1922$$

41. $f(x) = 0.01e^{-0.01x}$ for $[0, \infty)$ is an exponential distribution.

$P(0 \leq X \leq 100)$

$$= \int_0^{100} 0.01e^{-0.01x}\,dx$$

$$= -e^{-0.01x}\Big|_0^{100}$$

$$= 1 - \frac{1}{e}$$

$$\approx 0.6321$$

43. $f(x) = \dfrac{6}{15{,}925}(x^2 + x)$ for $[20, 25]$

(a) $\mu = \displaystyle\int_{20}^{25} \frac{6}{15{,}925}(x^2 + x)\,dx$

$$= \frac{6}{15{,}925}\int_{20}^{25}(x^3 + x^2)\,dx$$

$$= \frac{6}{15{,}925}\left[\frac{x^4}{4} + \frac{x^3}{3}\right]\Big|_{20}^{25}$$

$$= \frac{6}{15{,}925}$$

$$\cdot \left[\frac{(25)^4}{4} + \frac{(25)^3}{3} - \frac{(20)^4}{4} - \frac{(20)^3}{3}\right]$$

$$\approx 22.68°\text{C}$$

(b) $P(X < \mu)$

$$= \int_{20}^{22.68} \frac{6}{15{,}925}(x^2 + x)\,dx$$

$$= \frac{6}{15{,}925}\left[\frac{x^3}{3} + \frac{x^2}{2}\right]\Big|_{20}^{22.68}$$

$$= \frac{6}{15{,}925}$$

$$\cdot \left[\frac{(22.68)^3}{3} + \frac{(22.68)^2}{2} - \frac{(20)^3}{3} - \frac{(20)^2}{2}\right]$$

$$\approx 0.4819$$

45. Normal distribution, $\mu = 2.2$ g, $\sigma = 0.4$ g, $X = $ tension

$$P(X < 1.9) = P\left(\frac{x - 2.2}{0.4} < \frac{1.9 - 2.2}{0.4}\right)$$

$$= P(z < -0.75)$$

$$\approx 0.2266.$$

47. (a)

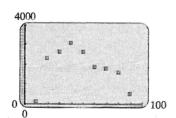

Of the types of functions available using the regression feature of a graphing utility, a polynomial function best matches the data.

(b) The function

$$N(x) = -0.000853613x^4 + 0.196608x^3$$
$$- 16.6309x^2 + 577.248x - 4040.47$$

provided by the calculator models this data well, as illustrated by the following graph.

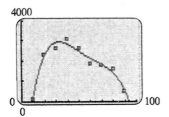

(c) Using the integration feature on our calculator, we find that

$$\int_{9.2}^{93.5} N(x)\,dx \approx 167{,}355.$$

So, $k = \frac{1}{167{,}355}$ and

$$S(x) = \frac{1}{167{,}355}N(x).$$

(d) Using the integration feature on our calculator,

$$P(25 \leq X \leq 35) = \int_{25}^{35} S(x)dx \approx 0.1731$$

$$P(45 \leq X \leq 65) = \int_{45}^{65} S(x)dx \approx 0.2758$$

$$P(X \geq 55) = \int_{55}^{93.5} S(x)dx \approx 0.348.$$

From the table, the actual probabilities are

$$P(25 \leq X \leq 35) = \frac{2521}{16{,}594} \approx 0.1579$$

$$P(45 \leq X \leq 65) = \frac{2677 + 1860}{16{,}594} \approx 0.2734$$

$$P(X \geq 55) = \frac{1860 + 1791 + 1606 + 524}{16{,}594}$$

$$\approx 0.3484.$$

(e) Using the integration feature on our calculator,

$$E(X) = \int_{9.2}^{93.5} x \cdot S(x)\,dx \approx 46.84 \text{ years.}$$

(f) Using the integration feature on our calculator,

$$\text{Var}(X) = \int_{9.2}^{93.5} x^2 \cdot S(x)\,dx - 46.84^2$$

$$\approx 409.5654$$

$$\sigma = \sqrt{\text{Var}(X)} = \sqrt{409.5654}$$
$$\approx 20.24 \text{ years.}$$

(g) $\displaystyle\int_{9.2}^{m} S(x)dx = \frac{1}{2}$

One method for approximating m using our calculator is to enter $\int_{9.2}^{m} S(x)\,dx$ into Y1, fnInt (Y1,X,9.2,X) (the left side of the equation) into Y2, and $\frac{1}{2}$ into Y3. Set the window to $0 \leq X \leq 100$ and $0 \leq Y \leq 1$. Turn off the graph of Y1 and graph Y2 and Y3 in the same window. (The process is slow.) Use the intersect feature to find where the graphs cross. This occurs at $m \approx 44.8$ years.

49. Normal distribution, $\mu = 40$, $\sigma = 13$, $x =$ "take"

$$P(X > 50) = P\left(\frac{X - 40}{13} > \frac{50 - 40}{13}\right)$$

$$= P(Z > 0.77)$$
$$= 1 - P(Z \leq 0.77)$$
$$= 1 - 0.7794$$
$$= 0.2206$$

Chapter 18 Test

[18.1]

Decide whether the functions defined as follows are probability density functions on the given intervals. If not, tell why.

1. $f(x) = 4$; $[3, 7]$

2. $f(x) = \dfrac{1}{5}$; $[5, 10]$

3. $f(x) = \dfrac{1}{2}(x - 4)$; $[4, 6]$

4. $f(x) = e^{-3x}$; $[0, \infty)$

5. Find the value of k that will make $f(x) = kx^{1/3}$ a probability density function on the interval $[1, 8]$.

6. The probability density function for a random variable x is defined by

$$f(x) = 0.2 \text{ for } x \text{ in } [12, 17].$$

Find the following probabilities.

(a) $P(X \le 14)$ (b) $P(13 < X < 16)$ (c) $P(X < 15)$

7. The probability density function of a random variable x is defined by

$$f(x) = 3x^2 \text{ for } x \text{ in } [0, 1].$$

Find the following probabilities.

(a) $P(X \le 1)$ (b) $P(0.2 \le X \le 0.7)$ (c) $P(X \ge 0.3)$

8. The time in years until a particular radioactive particle decays is a random variable with probability density function

$$f(t) = 0.04e^{-0.04t} \text{ for } t \text{ in } [0, \infty).$$

Find the probability that a certain such particle decays in less than 60 years.

[18.2]

Find the expected value and the standard deviation for each probability density function defined as follows.

9. $f(x) = \dfrac{1}{4}$; $[1, 5]$

10. $f(x) = 2x$; $[0, 1]$

11. $f(x) = \dfrac{3}{2}(x - 1)^2$; $[0, 2]$

12. $f(x) = \dfrac{1}{4\sqrt{x}}$; $[1, 9]$

13. Suppose that the time spent waiting in the waiting room of Dr. Jones' office is a random variable with probability density function defined by

$$f(x) = \frac{x(30-x)}{4500} \text{ for } x \text{ in } (0, 30).$$

Find the average waiting time. Find the standard deviation.

For the probability density functions defined as follows, find (a) the probability that the value of the random variable will be less than the mean and (b) the probability that the value of the random variable will be within one standard deviation of the mean.

14. $f(x) = 2(1-x)$ for x in $[0, 1]$

15. $f(x) = 1 - x^{-1/2}$ for x in $[1, 4]$

16. For the probability density function defined by

$$f(x) = \frac{1}{8}x \text{ for } x \text{ in } [0, 4],$$

find the probability that the value of the random variable will be between the median and the mean.

[18.3]

17. The maximum daily temperature in degrees Celsius in a certain city is uniformly distributed over the interval $[10, 40]$.

   (a) What is the expected maximum temperature?

   (b) What is the probability that the temperature will be less than $25°C$?

18. The number of repairs required by a new product each month is exponentially distributed with an average of 6.

   (a) What is the probability density function for this distribution?

   (b) What is the probability that the number of repairs is between 3 and 5?

   (c) What is the standard deviation?

Find the percent of area under a normal curve for each of the following.

19. The region to the left of $z = -0.57$

20. The region to the right of $z = 1.49$

21. The region between $z = 1.78$ and $z = -1.30$

Find a z-score satisfying the given condition.

22. 19% of the area to the left of z.

23. 91% of the area is to the right of z.

24. The heights of the students in a calculus class are normally distributed with mean 65 inches and standard deviation 10 inches.

   (a) Find the probability that a student's height is between 60 inches and 70 inches.

   (b) Find the probability that a student's height is more than 72 inches.

25. On a certain day, the amount spent for lunch in a particular restaurant was $3.75 with a standard deviation of $0.25. What percent of the customers spent between $3.60 and $4.10? Assume that the amount spent is normally distributed.

Chapter 18 Test Answers

1. No; $\int_3^7 4\,dx \neq 1$

2. Yes

3. Yes

4. No; $\int_0^\infty e^{-3x}\,dx \neq 1$

5. $\frac{4}{45}$

6. (a) 0.4   (b) 0.6   (c) 0.6

7. (a) 1   (b) 0.335   (c) 0.973

8. 0.9093

9. 03; 1.155

10. 0.6667; 0.2357

11. 1; 0.7746

12. 4.333; 2.329

13. 15; 6.708

14. (a) 0.5556   (b) 0.6285

15. (a) 0.4668   (b) 0.6059

16. 0.0556

17. (a) 25°   (b) 0.5

18. (a) $f(x) = 0.1667e^{-0.1667x}$ (b) 0.1719   (c) 6

19. 28.43%

20. 6.81%

21. 86.57%

22. −0.88

23. −1.34

24. (a) 0.383   (b) 0.242

25. 64.49%